TRANSPORTATION AND TRAFFIC ENGINEERING HANDBOOK

INSTITUTE OF TRAFFIC ENGINEERS

JOHN E. BAERWALD, *Editor*
Professor of Transportation and Traffic Engineering
Director, Highway Traffic Safety Center
University of Illinois at Urbana-Champaign

Associate Editors

Matthew J. Huber
Associate Professor, Department of Civil Engineering
University of Minnesota

Louis E. Keefer
Advanced Planning Director
Pennsylvania Department of Transportation
Harrisburg, Pennsylvania

PRENTICE-HALL, INC., *Englewood Cliffs. New Jersey*

Library of Congress Cataloging in Publication Data

Main entry under title:

Transportation and traffic engineering handbook.
    Published 1941–65 under title: Traffic engineering
handbook.
    Includes bibliographies.
    1. Traffic engineering. 2. Transportation.
I. Baerwald, John Edward   II. Institute of
Traffic Engineers.
HE333.T68 1975   387.3'1   75-6534
ISBN   0-13-930578-5

© 1976 by **THE INSTITUTE OF TRAFFIC ENGINEERS**

A complete revision of the Traffic Engineering Handbook, Third Edition, 1965

10 9 8 7 6 5 4 3 2 1

Printed in the United States of America

PRENTICE-HALL INTERNATIONAL, INC., *London*
PRENTICE-HALL OF AUSTRALIA, PTY. LTD., *Sydney*
PRENTICE-HALL OF CANADA, LTD., *Toronto*
PRENTICE-HALL OF INDIA PRIVATE LIMITED, *New Delhi*
PRENTICE-HALL OF JAPAN, INC., *Tokyo*
PRENTICE-HALL OF SOUTHEAST ASIA (PTE.) LTD., *Singapore*

CONTENTS

iii

PREFACE

A major broadening out of the role and responsibility of the Traffic Engineer has taken place since the publishing of the third edition of the *Traffic Engineering Handbook* in 1965. In this relatively short time, and because of the increasing complexity of transportation solutions primarily in urban areas, the traffic engineer has of necessity become a "transportation" engineer whose primary goal is achieving "balanced transportation". This edition of the Handbook thus reflects this change by the expansion of its scope to become a *Transportation and Traffic Engineering Handbook*. The Handbook emphasizes the major elements of total transportation planning, particularly as they relate to traffic engineering. First, the Handbook updates essential facts about the vehicle, the highway, and the driver, and all matters related to these three principal concerns of the traffic engineer. But, more important, it also discusses characteristics of the other modes of transportation in Chapters 5, 6, 12, 13, 21, 22 and 23. Other principal new chapters include: Computer Applications (Chapter 11); Statewide and Regional Transportation Planning (Chapter 13); Traffic Surveillance (Chapter 19); Environmental Considerations (Chapter 21); and Applications of Systems Concepts (Chapter 23).

Often in the past, the traffic engineer has been successful in improving the capacity and safety of highway facilities, only to have the improvements obliterated by the insatiable desire of motorists for individual freedom. It is important for traffic engineers, and everyone associated with transportation, to view each individual project in the light of the overall goals of society for an improved "quality of life." There is a movement of people to urban areas throughout the world. As an example, more than three quarters of the population of the United States is expected to be living in urban areas by 1980. The traffic engineer is now caught between the tremendous pressures to provide fast and convenient travel from suburbs to and from core areas, and the demands of the urban dwellers to stop "destroying" urban areas by constructing new facilities. For this reason, the traffic engineer must focus his attention on increasing street capacity to obtain the highest "level of service" out of the existing highway and street systems while simultaneously trying to decrease the auto traffic demands by actively aiding all other modes of transportation.

Similarly, the traffic engineer no longer can concern himself solely with moving people and goods from one place to another; he must keep in mind the relationship and effect that transportation facilities have on the social and physical environment.

Thus, he must give his attention to new standards for reducing air and noise pollution in the urban areas, and strive for compatibility of streets and highways with the environment. "Reserved lanes," "busways," "car pooling," "fringe parking," "ramp metering" are a few of the new, usually sophisticated techniques with which the urban traffic engineer is concerning himself today; all of these, in one form or another, will be useful in reducing the number of automobiles coming into core areas. This, in turn, will ameliorate the inevitable congestion, reduce the demand for additional downtown parking spaces and provide some, albeit small, relief in the current energy crisis.

Another major change that has taken place in the last few years relates to the decision making process itself; transportation and traffic engineers, both urban and rural, now find that they spend a large part of their time in the process of involving other agencies and the public in the early stages of a project so that adequate opportunity is provided for all to express their views and have an effect on the ultimate action taken. This participatory planning process is a direct result of public pressures to become involved at the earliest possible time in the decision making process. The resulting national guidelines for regional planning call for a comprehensive, cooperative, and continuous process to be established for all projects now under way.

These changes in transportation planning have taken place in recent years almost simultaneously throughout the world. As a result, the role of all traffic engineers has broadened to include thoughts of other modes of transportation and participation in joint development of regional and municipal public facilities with highway projects. In today's world, a good traffic engineer must be more than a good technician; he must be a first class diplomat, knowledgeable of the technical facts, and yet at the same time sensitive to people's concern for the total environment. While striving for technical perfection, the traffic engineer must be willing to accept the possible, recognizing that time does not stand still; he must be flexible, without giving up his principles.

This new Handbook recognizes all of these changes. The Institute of Traffic Engineers believes it goes a long way toward providing the technical background and base that will help make each traffic engineer successful in his endeavors to solve the transportation problems in a world of ever-increasing complexity.

JOHN E. BAERWALD

Urbana, Illinois

Chapter 1

VEHICLE, HIGHWAY, AND TRAVEL FACTS

Woodrow W. Rankin, Deputy Director, Transportation Development Division, Highway Users Federation, Washington, D.C.

This chapter summarizes the principal trends in vehicle and highway use and related factors in the United States. Selected data also are given for several representative countries that have a high level of motor vehicle use. A list of primary sources of this information is included to permit a periodic update.

POPULATION

Population growth and distribution are significant factors in the development of transportation demand and the systems that are provided to meet these demands. On July 1, 1970, the population of the United States, including the armed forces was 204,800,000, over twenty million more than in 1961 and almost 82 million more than in 1930. The growth since 1930 is shown in Fig. 1.1. Between 1946 and 1960 the growth rate was highest when it averaged 2.0 percent per year and was more than double the average rate before 1945. Since 1960 the growth rate has averaged 1.2 percent per year, approximately the same as the U.S. Bureau of Census Series D projection of population.

Increased urbanization is one of the most significant changes in population characteristics. In the United States between 1930 and 1970 the urban population increased from 56 percent of the total to 73 percent. The United States Department of Transportation has estimated that the urban population will reach 81 percent by 1990. Although there has been a net increase in population in all sections of urban areas, the greatest growth has been outside the central cities. In 1945 the population of urban areas was distributed 61 percent in the central cities and 39 percent outside. By 1970 the distribution was 54 percent central city and 46 percent outside. It is estimated that by 1985 the distribution will be 33 percent central city and 67 percent outside.

Growth and increased urbanization are characteristics of the population changes in most countries. Representative data on changes in urbanization for several countries that have a high level of motor vehicle use are given in Table 1.1.

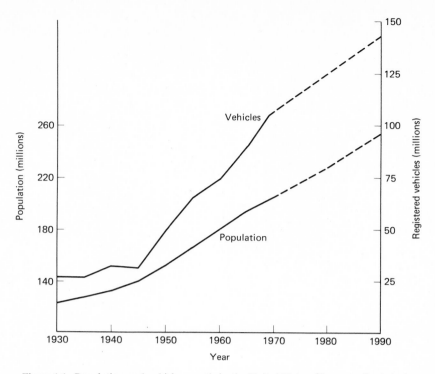

Figure 1.1. Population and vehicle growth in the United States. [Sources: *Statistical Abstract of the United States 1971* (Washington, D.C.: U.S. Government Printing Office). *Highway Statistics/1970 and Summary to 1965* (Washington, D.C.: U.S. Government Printing Office).]

TABLE 1.1
Population Growth and Urbanization, Selected Countries

Country	Year	Population		Population Growth		Urban Population Growth	
		Number (Millions)	Percent Urban*	Time Period	Annual Growth Percent	Time Period	Annual Growth Percent
1. Australia	1966	11.55	58	1961–66	1.84	1961–66	2.77
2. Canada	1966	20.01	74	1951–66	2.85	1951–66	4.95
3. Germany†	1969	60.84	80	1964–69	0.88	1964–69	1.36
4. Great Britain	1969	54.02	78	1951–69	0.58	1951–69	0.46
5. Japan	1970	103.72	72	1965–70	1.11	1965–70	2.84
6. Sweden	1968	7.95	77	1957–67	0.65	1960–65	1.10
7. United States	1970	204.80	73	1950–70	1.72	1950–70	2.71

*Usually residents of places over either 1,000 or 5,000.
†West Germany.

Sources:
  Canada 1970, and *Canada 1957,* Dominion Bureau of Statistics, Ottawa, Ontario. Reproduced by permission of Information Canada.
  Trends in Motorization and Highway Programs in 16 European Countries, International Road Federation, 1023 Washington Building, Washington, D.C. 20005, 1969.
  Statistical Abstracts of United States 1971, U.S. Department of Commerce, Bureau of Census. Published by U.S. Government Printing Office, Washington, D.C. 20402.
  Official sources various countries.

MOTOR VEHICLES IN USE

Worldwide, there were 246,400,000 motor vehicle registrations in 1970, a three-fold increase over the number of registrations in 1950. In 1970, 78 percent of these vehicles were passenger cars and 22 percent were trucks and buses. Fig. 1.2 illustrates the worldwide growth in registrations since 1946. The distribution of vehicle registrations in 1970 by areas of the world and by selected countries is given in Table 1.2. In the United States in 1970 there were 108,407,306 motor vehicles registered, an increase of over 300 percent since 1930. The pattern of this growth is shown in Fig. 1.1.

Persons per vehicle (population/registered vehicles) is a useful comparison measure of the potential level of motor vehicle use. In 1955, worldwide, there were 27 persons per vehicle. By 1970 this figure had dropped to 14. The persons per vehicle in 1970 and in 1955 for areas of the world and selected countries are given in Table 1.3.

In 1970, in the U.S. there was an average of 1.9 persons per registered vehicle and 2.3 persons per registered passenger car. On a state basis, per capita figures for all motor vehicles ranged from a high of 2.8 in New York to a low of 1.3 in Wyoming. For passenger cars, this range was from 3.1 in New York to 1.8 in Nevada. The population/vehicle ratio has been decreasing since 1946 as shown in Fig. 1.3.

Automobile ownership by households, although not a measure of potential motor vehicle use, is an indication of the significance of automobiles in personal transportation. In 1971, in the U.S. 50.2 percent of the households had one car, 25.0 percent had two, 4.8 percent had three or more, and 20.0 percent had no car. The percent of no car households, varying from a high of 28.8 percent in the Middle Atlantic Divi-

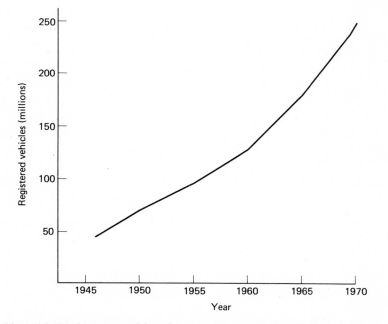

Figure 1.2. World motor vehicle registrations. [Source: *Automobile Facts and Figures 1946–1972* (Detroit: Motor Vehicle Manufacturers Association).]

TABLE 1.2
Motor Vehicle Registration, 1970

Location	Passenger Cars	Trucks and Buses	Total Vehicles	Percent of World Total
Africa	3,383,524	1,396,909	4,780,433	1.94
Asia	12,407,010	11,251,626	23,658,636	9.60
Europe	69,642,558	14,614,265	84,256,823	34.20
North and Central America	98,109,746	21,503,867	119,613,613	48.55
Oceania	4,732,841	1,162,121	5,894,962	2.39
South America	5,240,038	2,923,040	8,163,078	3.32
World Total	193,515,717	52,851,828	246,367,545	100.00
Australia	3,779,743	958,734	4,738,477	1.92
Brazil	2,234,500	1,305,200	3,539,700	1.44
Canada	6,602,176	1,481,197	8,083,373	3.28
France	12,290,000	2,114,750	14,404,750	5.85
Germany, West	14,376,484	1,228,406	15,604,890	6.33
Sweden	2,287,709	158,775	2,446,484	0.99
United Kingdom	11,792,500	1,910,000	13,702,500	5.56
United States	89,279,864	19,127,442	108,407,306	44.00

Source: *1972 Automobile Facts and Figures*, Motor Vehicle Manufacturers Association, 320 New Center Building, Detroit, Michigan 48202.

TABLE 1.3
Persons per Vehicle 1955 and 1970, World and Selected Countries

Location	Persons per Vehicle	
	1955	1970
Africa	114	73.4
Asia	644	83.7
Europe	32	8.3
North and Central America	3.5	2.6
Oceania	5.0	3.1
South America	66	23.6
World	27	14.4
Australia	4.3	2.6
Brazil	92	26.9
Canada	4.0	2.6
France	11	3.5
Germany, West	21	3.8
Great Britain	10	4.1
Sweden	10	3.3
United States	2.6	1.9

Source: *Automobile Facts and Figures*, 1956 and 1972, Motor Vehicle Manufacturers Association.

sion to a low of 12.3 percent in the Western Mountain Division, are given in Table 1.4 for the various U.S. Bureau of the Census Geographic Divisions. The comparable data for 1960 given in the table indicates the decrease in this percentage in the ten-year period. On a more localized basis within a metropolitan area, the central city

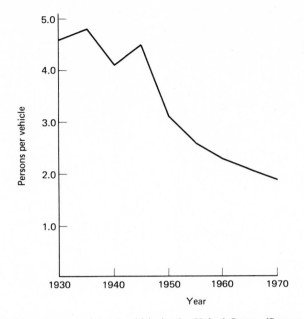

Figure 1.3. Persons per registered vehicle in the United States. (Source: *Highway Statistics/1970 and Summary to 1965.*)

TABLE 1.4
No Car Households, United States 1960, 1971

	Percent of Households with No Car	
Geographic Division	1960	1971
New England	20.7	19.6
Middle Atlantic	29.6	28.8
East North Central	18.0	16.6
West North Central	16.5	16.7
South Atlantic	23.4	22.0
East South Central	27.5	21.4
West South Central	21.2	18.4
Western Mountain	14.1	12.3
Pacific	15.9	14.7
All United States	21.5	20.0

Sources:
1960 Census of Housing, U.S. Department of Commerce, Bureau of the Census, U.S. Government Printing Office, Washington, D.C. 20402.

Current Population Reports, Series P-65 No. 40, "Household Ownership and Availability of Cars, Homes, and Selected Household Durables and Annual Expenditures on Cars and Other Durables: 1971," U.S. Bureau of the Census, U.S. Government Printing Office, Washington, D.C. 20402.

has the highest percent of no car households and the rural fringe the lowest. The 1971 distribution of no car households between the central cities and the suburban rings in Standard Metropolitan Areas is given in Table 1.5.

In the United States, since 1940 the average age of vehicles in use has fluctuated between 5.5 and 9.0 years for passenger cars and 5.6 and 8.6 for trucks. Average age figures for representative years since 1941 are given in Table 1.6. The distribution of passenger cars by age for representative years since 1950 is shown in Fig. 1.4.

In the United States the standard size car is the most predominant passenger car in use. Table 1.7 gives the distribution of passenger cars by size found in a 1969 study.

TABLE 1.5
Distribution of No Car Households in Standard
Metropolitan Areas, United States, 1971

Geographic Region*	Percent of No Car Households		
	Central City	Suburban Ring	Entire Area
Northeast	46.3	18.1	31.1
North Central	28.9	9.1	18.4
South	27.4	9.5	18.9
West	22.1	10.0	15.2

*U.S. Bureau of the Census geographic region in which the SMA's are located.

Source: *Current Population Reports*, Series P-65 No. 40, "Household Ownership and Availability of Cars, Homes, and Selected Household Durables and Annual Expenditures on Cars and Other Durables: 1971," U.S. Bureau of the Census, U.S. Government Printing Office, Washington, D.C. 20402.

TABLE 1.6
Average Age of Motor Vehicles in Use, United States, 1941–1970

Year	Average Age of Vehicles in Use*	
	Passenger Cars (Years)	Trucks (Years)
1941	5.5	5.6
1944	7.3	7.6
1946	9.0	8.6
1949	8.5	7.4
1950	7.8	7.0
1952	6.8	6.6
1954	6.4	6.6
1956	5.6	6.8
1958	5.6	7.2
1960	5.9	7.7
1962	6.0	8.0
1964	6.0	8.1
1966	5.7	7.8
1968	5.6	7.6
1970	5.6	7.4

*Estimated by Motor Vehicle Manufacturers Association.

Source: *1971 Automobile Facts and Figures*, Motor Vehicle Manufacturers Association of the U.S., Inc.

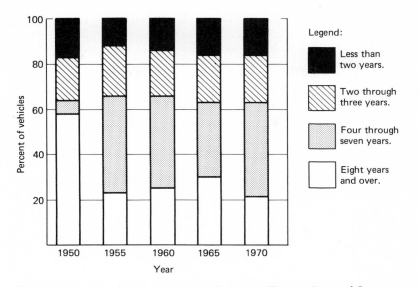

Figure 1.4. Age of vehicles in use in the United States. [Source: *Survey of Consumer Finances* (Ann Arbor: University of Michigan, Survey Research Center)].

TABLE 1.7
Size Distribution — Passenger Cars in Use, United States, 1969

Size	Percent of Total
Large Cars (Chrysler*)	20
Standard Cars (Plymouth*)	65
Compact Cars (Falcon*)	10
Small Cars (Volkswagen*)	5

*Typical car for size class.

Source: *Running Costs of Motor Vehicles as Affected by Road Design and Traffic*, NCHRP Report 111, Highway Research Board, National Academy of Sciences, 2101 Constitution Avenue, Washington, D.C. 20418.

MODE OF DOMESTIC PASSENGER TRAVEL—UNITED STATES

In the United States between 1940 and 1970 the amount of domestic intercity passenger travel increased from 304,000 million passenger miles to 1,185,000 million. Each year of that period, except during World War II, passenger cars accounted for over 85 percent of the travel, but there were significant shifts in the percent of the total travel carried by train, bus, and air. The share of the total by air rose from less than one to over ten percent, and the share by rail and bus fell off sharply. These patterns of growth and decline of intercity passenger travel by mode since 1940 are shown in Fig. 1.5.

It has been estimated that in 1970 only about 5 percent of all person trips in urban areas in the United States were by transit. The percentage for a specific area is dependent upon many factors and it varies widely between urban areas. The percent of

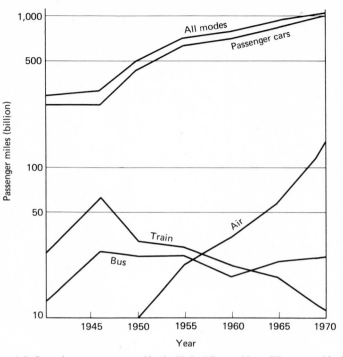

Figure 1.5. Intercity passenger travel in the United States. Note: Water travel included in total. (Source: *Statistical Abstract of the United States 1973.*)

TABLE 1.8
Person Trips by Urban Transit, Study Years 1955–1965

Urban Area	Percent of Person Trips by Transit
Chicago, Ill.	24.3
Detroit, Mich.	16.7
Pittsburgh, Pa.	21.7
Knoxville, Tenn.	4.3
Lexington, Ky.	5.4

Source: *Transportation and Parking for Tomorrow's Cities*, Wilbur Smith and Associates, New Haven, Connecticut 06504.

person trips by transit for five urban areas in the United States given in Table 1.8 illustrates this variation.

The trends of the annual total of transit riders and annual vehicle miles of travel by passenger cars in urban areas shown in Fig. 1.6 illustrate the relative importance of passenger cars and transit in intraurban passenger travel in the U.S. since 1940. During World War II, when the use of private vehicles was restricted, transit riding increased and travel by passenger car decreased. Conversely, since the war, passenger car travel in urban areas increased while transit use decreased until 1973. The annual totals of transit passengers and of passenger car miles of travel in urban areas for representative years since 1940 are given in Table 1.9.

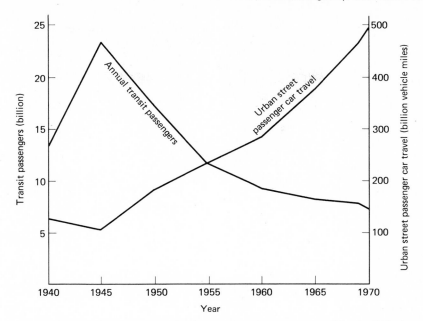

Figure 1.6. Transit passengers and urban street passenger car travel in the United States. [Sources: *'73–'74 Transit Fact Book* (Washington, D.C.: American Transit Association). *Highway Statistics/1970 and Summary to 1965.*]

TABLE 1.9
Transit Passengers and Urban Passenger Car Use,
United States, 1940–1973

Year	Transit Passengers (Millions)	Urban Passenger Car Travel (Million Vehicle Miles)
1940	13,098	129,060
1945	23,254	109,472
1950	17,246	182,518
1955	11,529	233,596
1960	9,395	284,800
1965	8,253	378,182
1970	7,332	494,543
1971	6,847	525,212
1972	6,567	567,541
1973	6,660	592,457

Sources:
1973–74 Transit Fact Book, American Transit Association, 465 L'Enfant Plaza West, SW, Washington, D. C. 20024.
Highway Statistics, Summary to 1965 and *Highway Statistics,* annual editions, Table VM-1, U. S. Department of Transportation, Federal Highway Administration. Published by U.S. Government Printing Office, Washington, D.C. 20402.

MODE OF DOMESTIC INTERCITY FREIGHT MOVEMENT—UNITED STATES

Between 1945 and 1970 the domestic intercity freight volume in the United States increased from 1,072,490 million ton miles to 1,936,000 million. During that period, although the movement of freight by each mode increased, there were major shifts in the distribution of shipments between modes. Movements by railroads dropped from 69 percent to 41 percent of the total, truck movements increased from 6 percent to 21 percent, and oil pipelines increased from 12 percent to 22 percent. Growth trends and shipment volumes of intercity freight movements for each mode from 1940 to 1970 are shown in Fig. 1.7 and Table 1.10.

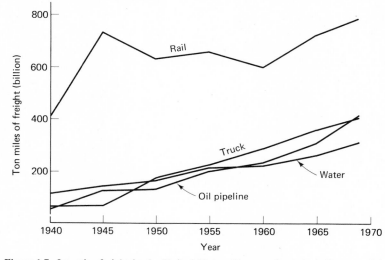

Figure 1.7. Intercity freight in the United States. (Source: *Statistical Abstract of the United States 1973.*)

TABLE 1.10
Domestic Intercity Freight Movement, United States, 1940–1970

| Year | Total | Ton-Mile Volume (Billions) | | | | |
		Railroads	Trucks	Inland Waterways	Oil Pipelines	Air
1940	651	412	62	118	59	*
1945	1,072	736	67	143	127	0.1
1950	1,094	628	173	163	129	0.3
1955	1,298	655	223	217	203	0.5
1960	1,330	595	285	220	229	0.8
1965	1,651	721	359	262	303	1.9
1970	1,936	771	412	319	431	3.3

*Less than 50 million ton-miles.

Source: *Statistical Abstract of United States 1973,* U.S. Department of Commerce, Bureau of the Census. Published by U.S. Government Printing Office, Washington, D.C. 20402.

ROAD AND STREET MILEAGE

The total mileage of roads and streets in the United States has increased very little since 1920. In 1971, there were 3,758,942 miles of roads and streets compared to 3,160,000 in 1921. In 1921 only seven percent of the total mileage was urban and by

1971 this had increased to 15 percent. In the same period the percent of surfaced mileage increased from 14 percent to 79 percent. Table 1.11 gives the distribution by surface type of the urban and rural mileage in the U.S. in 1941 and 1971.

Administrative responsibility for the roads and streets in the United States is divided between the various levels of government with the counties responsible for the most mileage and the federal government the least. This distribution of road and street administrative control in 1971 is given in Table 1.12. The number of miles on

TABLE 1.11
Road and Street Mileage by Surface Type, United States, 1941, 1971

Type of Surface		Rural		Urban	
		1941	1971	1941	1971
Nonsurfaced	Mi	1,620,598	750,017	81,081	25,853
	Km	2,607,542	1,206,777	130,459	41,597
Low Type	Mi	1,198,285	1,887,535	123,577	307,825
Surface	Km	1,928,040	3,037,044	198,835	495,290
High Type	Mi	186,904	528,343	99,233	259,369
Surface	Km	300,728	850,104	159,666	417,325
Total	Mi	1,385,189	2,415,878	282,810	567,194
Surfaced	Km	2,228,769	3,887,148	455,041	912,615
Total—All Roads	Mi	3,005,787	3,165,895	363,891	593,047
and Streets	Km	4,836,311	5,093,925	585,500	954,212

Source:
Highway Statistics/1971 and *Highway Statistics-Summary to 1965*, U.S. Department of Transportation, Federal Highway Administration. Published by U.S. Government Printing Office, Washington, D.C. 20402.

TABLE 1.12
Road and Street Mileage by Administrative Control, United States, 1971

Administrative Control		Total Mileage	Federal-aid Systems				Non-Federal Aid
			Interstate	Primary	Secondary	Urban*	
State {Rural	Mi	682,731	33,209	213,730	302,639	365	165,997
	Km	1,098,514	53,433	343,892	486,946	587	267,089
Urban	Mi	77,386	7,399	39,711	21,554	2,440	13,681
	Km	124,514	11,905	63,895	34,680	3,926	22,013
County	Mi	1,726,603	1	234	285,500	3,258	1,437,611
	Km	2,778,104	2	377	459,370	5,242	2,313,116
Towns, Other	Mi	529,668	5	138	5,938	96	523,496
Rural	Km	852,236	8	222	9,554	154	842,305
Cities	Mi	515,661	121	1,270	18,990	23,713	471,688
	Km	829,699	195	2,043	30,555	38,154	758,946
National Parks and	Mi	196,839	—	285	153	—	196,401
other Federal	Km	316,714	—	459	246	—	316,009
State Parks and	Mi	27,114	190	474	26	40	26,574
Forests	Km	43,626	306	763	42	64	42,758
Toll Agencies	Mi	2,940	2,078	2,119	6	1	814
	Km	4,730	3,344	3,409	10	2	1,310

*Federal-aid urban includes federal-aid urban highway system plus federal-aid primary urban Type II highways.

Source: *Highway Statistics/1971*, U.S. Department of Transportation.

the various federal-aid systems in each administrative classification are also given in Table 1.12.

In 1968, all roads and streets in the United States were classified in terms of their functional importance. The mileage and travel on each of these functional classes and the percent of each class on the present federal-aid systems are given in Table 1.13.

TABLE 1.13
Road and Street Mileage and Travel by Functional Classes, United States, 1968

Functional System	Length		Annual Travel		Percent on Federal-aid Systems		
	Miles	Kilometers	Million Vehicle Miles	Percent of Total	Pri-mary	Sec-ondary	Non Federal-aid
Rural							
Principal Arterial	112,918	181,685	191,162	19	92	5	3
Minor Arterial	169,061	272,019	108,780	11	58	39	3
Major Collector	265,321	426,901	72,637	7	7	78	15
Minor Collector	425,500	684,629	43,387	4	—	59	41
Local	2,106,814	3,389,863	69,121	7	—	3	97
Total Rural	3,079,614	4,955,097	485,087	48			
Urban							
Principal Arterial	47,698	76,746	288,714	29	61	19	20
Minor Arterial	46,991	75,608	106,931	10	6	29	65
Collector	43,973	70,752	42,740	4	3	8	89
Local	351,296	565,235	94,112	9	—	1	99
Total Urban	489,958	788,341	532,497	52			

Source: *1970 National Highway Needs Report with Supplement*, U.S. Department of Transportation, Federal Highway Administration. Published by U.S. Government Printing Office, Washington, D.C. 20402.

Roads and street mileage are related, in part at least, to the area served and its population density. The effect of these factors is illustrated in Table 1.14 which lists total highway system length, land area, and population density for five countries that have a relatively high level of motor vehicle use.

TABLE 1.14
Road and Street Mileage, Land Area, and Population Density, Selected Countries, 1967

Country	System Length		Land Area		Population Density	
	Miles (1,000)	Kilometers (1,000)	Mi² (1,000)	Km² (1,000)	Per Square Mile	Per Square Kilometer
France	487.2	784.5	212.7	551.2	234	90
Germany	251.5	405.0	95.9	248.5	623	241
Great Britain	203.8	328.1	94.2	244.0	585	226
Sweden	107.1	172.5	173.6	449.8	44	17
United States	3710.2	5969.7	3614.3	9363.4	54	21

Source: *Trends in Motorization and Highway Programs in 16 European Countries*, International Road Federation.

HIGHWAY FINANCE

In the United States between 1921 and 1971 the annual per capita expenditure for highway purposes increased from $13 to $109. Of a total expenditure of 22,504 million dollars in 1971, 68 percent was by state highway agencies. Capital outlays accounted for 55 percent of the total expenditures of all agencies. In Table 1.15, the total U.S. road and street expenditures have been broken down into their main component parts for each class of highway agency. With respect to the federal government, it should be noted that federal aid payments to the states are not considered an expenditure item.

Along with the increase in roads and street expenditures in the United States since 1921, there has been a major shift in the relative amount of spending by the various levels of government. In 1921, rural, local governments accounted for almost one half of all expenditures. In 1971, state governments accounted for almost 70 percent of all expenditures. Fig. 1.8 shows the trends in highway expenditures by state and local governments since 1921.

An element of highway finance in the United States not shown in Table 1.15 and Fig. 1.8 is the extent of federal government aid to state and local highway programs. This federal aid has risen from 6 percent of the total expenditures in 1921 to 23 percent in 1971.

In 1921, over 50 percent of highway revenues in the United States were derived from property taxes and less than nine percent from direct highway user imposts (taxes, tolls, etc.). Since 1921, along with the increase in highway revenues there has been a major shift in the sources of this revenue. In 1970, only 6.1 percent of the total was from property taxes and 70 percent was from highway user imposts. This shift in the source of highway funds is illustrated in Fig. 1.9.

An indication of the relative significance of highway expenditures in a country can be seen by a comparison of these expenditures on a per capita basis. The listing of these expenditures in Table 1.16 indicates the wide variations between countries.

TABLE 1.15
Road and Street Expenditures, United States, 1971

Type of Expenditure	(In Millions of Dollars*)				
	State Highway Agencies	Counties and Townships	Cities	Federal Agencies	Total
Capital Outlay	$ 9,926	$ 980	$1,145	$291	$12,342
Maintenance and Operation	2,140	1,555	1,350	56	5,101
Administration and Research	856	230	245	146	1,477
Enforcement and Safety	893	69	490	—	1,452
Interest on Debt	616	68	150	—	834
Debt Retirement	815	158	325	—	1,298
Total	15,246	3,060	3,705	493	22,504

*1971 dollars.

Source: *Highway Statistics/1971*, U.S. Department of Transportation.

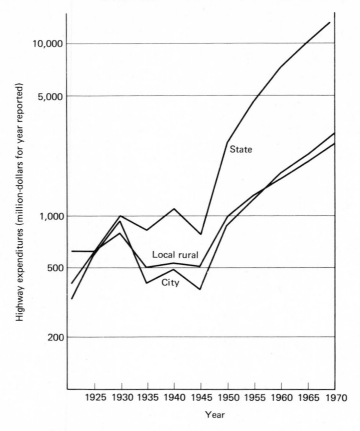

Figure 1.8. Highway expenditures by state and local governments in the United States. (Source: *Highway Statistics/1970 and Summary to 1965.*)

TABLE 1.16
Annual per Capita Highway Expenditures, Selected Countries

| Country | Year | Highway Expenditures | |
		Total (Million)	Per Capita
Canada	1967	$ 1,800	$ 86
France	1968	1,980	40
Germany	1968	2,910	50
Great Britain	1968	1,265	23
Sweden	1968	584	74
United States	1971	22,504	109

Sources:
Trends in Motorization and Highway Programs in 16 European Countries, International Road Federation.
Highway Statistics/1971, U.S. Department of Transportation.
Canada, 1970, Dominion Bureau of Statistics. Reproduced by permission of Information Canada.

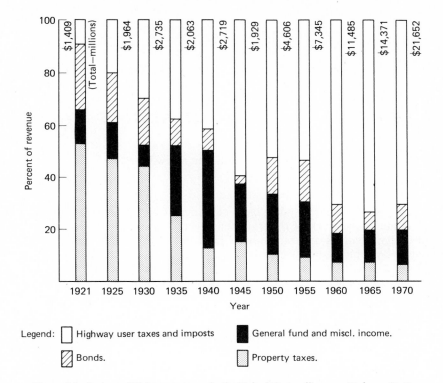

Figure 1.9. Source of highway revenue in the United States. (Source: *Highway Statistics/1970 and Summary to 1965*.)

REFERENCES FOR FURTHER READING

Automobile Facts and Figures and *Motor Truck Facts* (annual publications)
Motor Vehicle Manufacturers Association of the U.S., Inc.
320 New Center Building
Detroit, Michigan 48202

Highway Statistics—Summary to 1965
Highway Statistics (annual publication since 1945)
U.S. Department of Transportation
Federal Highway Administration
Published by U.S. Government Printing Office
Washington, D.C. 20402

Statistical Abstract of United States (annual publication)
U.S. Department of Commerce
Bureau of Census
Published by U.S. Government Printing Office
Washington, D.C. 20402

Transit Fact Book (annual publication)
American Transit Association
465 L'Enfant Plaza West, S.W.
Washington, D.C. 20024

Trends in Motorization and Highway Programs in 16 European Countries, 1969
International Road Federation
1023 Washington Building
Washington, D.C. 20005

Chapter 2

VEHICLE OPERATING CHARACTERISTICS

PAUL J. CLAFFEY, Consulting Engineer, Potsdam, New York.

This chapter deals with the physical characteristics of passenger cars and trucks and how they relate to the work of the traffic engineer. Because the traffic engineer must plan and design facilities and recommend laws that control and regulate the movement of highway vehicles, he must have knowledge of these vehicles, their performance limitations and capabilities, as well as their physical dimensions. Information is provided on road and vehicle resistances affecting motor vehicle operation, speed change considerations, vehicle measurements, and user operating costs.

RESISTANCE TO MOTION AND POWER REQUIREMENTS

The forces that must be overcome by motor vehicles if they are to proceed are rolling, air, grade, curve, and inertia resistance forces. Grade acts as a retarding force only when vehicles are on upgrades and inertia only when speed increases are involved. When vehicles are to be stopped or slowed, all of these resistances help braking action except downgrades and inertia. An additional resistance to motion during deceleration in gear is provided by engine compression forces.

ROLLING RESISTANCE

Rolling resistance results from the frictional slip between tire surfaces and the pavement, flexing of tire rubber at the surfaces of contact, rolling over rough particles (i.e., stones or broken asphalt particles), climbing out of road depressions, pushing wheels through sand, mud, or snow and internal friction at wheel, axle, and driveshaft bearings and in the transmission gears. For speeds up to 60 mph (96.5 kph) the rolling resistance of modern passenger cars on high-type pavement is constant at about 27 lb per ton of weight (13.5 kg/mton).[1] For higher speeds these values should be increased by 10 percent for each 10 mph (16 kph) increase in speed above 60 mph (96.5 kph). Typical passenger car rolling resistances at low speed on low types of surfaces are given in Table 2.1.

---

[1] J. C. KESSLER and S. B. WALLIS, "Aerodynamic Test Techniques," *S.A.E. Transactions*, Vol. 75, Sec. 3 (1967), p. 12.

TABLE 2.1
Rolling Resistances of Passenger Cars on Low-grade Road Surfaces*

Uniform Speed		Badly Broken and Patched Asphalt		Dry, Well-packed Gravel		Loose Sand	
mph	kph	lb/ton	kg/mt	lb/ton	kg/mt	lb/ton	kg/mt
20	32.1	29	14.5	31	15.5	35	17.5
30	48.3	34	17.0	35	17.5	40	20.0
40	64.4	40	20.0	50	25.0	57	28.5
50	80.5	51	25.5	62	31.0	76	38.0

*Computed using 27 lb per ton rolling resistance for high-type pavement and correcting for other pavement types using adjustment factors based on fuel consumption as given in Table 6B, p. 17, *Running Costs of Motor Vehicles as Affected by Road Design and Traffic* of Report 111, National Cooperative Highway Research Program (Washington, D.C.: Highway Research Board, 1971).

AIR RESISTANCE

Air resistance is composed of the direct effect of air in the pathway of vehicles, the frictional force of air passing over the surfaces of vehicles (including the under-surface), and the partial vacuum behind the vehicle. For the typical modern car having a projected frontal cross section of 30 sq ft (2.78 sq m) air resistance varies from zero at 10 mph (16 kph) to 55 lb (25 kg) at 55 mph (88.5 kph), approximately in proportion to the square of the velocity.[2] Eqs. (2.1) and (2.2) give the air resistance force acting on an automobile.

$$R_a = 0.0006 \, AV^2 \quad \text{(U.S. units)}. \tag{2.1}$$

R_a is air resistance in pounds.
A is frontal cross section area in square feet.
V is speed in miles per hour.

$$R_a = 0.0011 \, AV^2 \quad \text{(metric units)} \tag{2.2}$$

R_a is air resistance in kilograms.
A is frontal cross section in square meters.
V is speed in kilometers per hour.

GRADE RESISTANCE

Grade resistance, the force acting on a vehicle because it is on an incline, equals the component of the vehicle's weight acting down the grade. Eqs. (2.3) and (2.4) give the grade resistance force.

$$R_g = 20 \, Wg \quad \text{(U.S. units)}. \tag{2.3}$$

R_g is grade resistance in pounds.
W is gross vehicle weight in tons.
g is gradient in percent.

---

[2] *Ibid.*

$$R_g = 10 \ Wg \quad \text{(metric units).} \tag{2.4}$$

R_g is grade resistance in kilograms.
W is gross vehicle weight in metric tons.
g is gradient in percent.

CURVE RESISTANCE

Curve resistance is the force acting through the front wheel contact with the pavement needed to deflect a vehicle along a curvilinear path. This force is a function of speed because the faster an object is moving, the more difficult it is to change its direction. The results of recent fuel consumption studies show the curve resistance of standard or intermediate type passenger cars operating on high-type asphalt pavement to be as given in Table 2.2.[3]

TABLE 2.2
Curve Resistances of Passenger Cars on High-type Road Surfaces*

Curvature		U.S. Units		Metric Units	
Degree	Radius (ft)	Resistance (lb)	Speed (mph)	Resistance (kg)	Speed (kph)
5	1,146	40	50	18	80.5
5	1,146	80	60	36	96.5
10	573	40	30	18	48.3
10	573	120	40	54	64.4
10	573	240	50	108	80.5

*Determined from data on the effect of curvature on fuel consumption as reported in *Running Costs of Motor Vehicles as Affected by Road Design and Traffic*, NCHRP Report 111 (Washington, D.C.: Highway Research Board, 1971). The resistance forces were developed by evaluating the forces needed to produce the additional fuel consumption recorded for operation on curves.

INERTIA RESISTANCE

Inertia resistance is the force to be overcome in order to increase speed. It is a function of vehicle weight (regardless of type of vehicle) and the rate of acceleration and may be computed from the following equations:

$$R_i = 91.1 \ WA \quad \text{(U.S. units).} \tag{2.5}$$

R_i is inertia resistance in pounds.
W is gross vehicle weight in tons.
A is acceleration rate in miles per hour per second.

$$R_i = 28.0 \ WA \quad \text{(metric units).} \tag{2.6}$$

R_i is inertia resistance in kilograms.
W is gross vehicle weight in metric tons.
A is acceleration rate in kilometers per hour per second.

---

[3] PAUL J. CLAFFEY, *Running Costs of Motor Vehicles as Affected by Road Design and Traffic* (Washington D.C.: Highway Research Board, 1971), p. 17.

HORSEPOWER

Horsepower is the time rate of doing work, and the maximum an engine can deliver is a measure of its performance capability. The horsepower actually used by a motor vehicle for propulsion may be determined from Eqs. (2.7) and (2.8):

$$P = 0.0026 \ RV \quad \text{(U.S. units).} \tag{2.7}$$

P is horsepower actually used.
R is sum of resistances to motion in pounds.
V is speed in miles per hour.

$$P = 0.0036RV \quad \text{(metric units).} \tag{2.8}$$

P is horsepower actually used.
R is sum of resistances to motion in kilograms.
V is speed in kilometers per hour.

The maximum horsepower output available for propulsion at a given engine speed equals the maximum gross brake horsepower at the flywheel for that engine speed less the horsepower consumption of engine accessories including the alternator, automatic transmission, power steering, and air conditioner. For vehicles with typical accessories, maximum horsepower available for propulsion at 60 mph (96.5 kph) is about 50 percent of the manufacturer's nominal engine horsepower rating. This relationship may be used to estimate maximum acceleration rates and maximum speeds on grades given nominal engine horsepower in relation to engine speed and reliable values of resistances (particularly rolling and air resistance).

Empty weights and nominal horsepower ratings representative of major categories of motor vehicles are given in Table 2.3.

TABLE 2.3
Empty Weights and Nominal Horsepower Ratings Representative of Major Categories of Motor Vehicles

Motor Vehicle Category	Empty Weight with Driver Aboard		Nominal Horsepower	Engine Speed for Given Horsepower (rpm)
	(lb)	(kg)		
Intermediate type passenger car	4,000	1,814	195	4,800
Pickup truck	4,500	2,041	125	3,800
Two-axle, six-tire, single-unit truck	10,000	4,535	142	3,800
2S-2* tractor semi-trailer truck	20,000	9,070	175	3,200

*Two-axle tractor and two-axle semitrailer.

WEIGHT/HORSEPOWER RATIO

Weight/horsepower ratios are useful for indicating the overall performance characteristics of vehicles, particularly for making approximate performance comparisons among different vehicle types. The weight/horsepower ratio (the number of pounds of gross vehicle weight for each horsepower available for propulsion) is a direct measure of the sluggishness of vehicle operation. Because weight is a rough indicator of resistance to motion, the higher the weight/horsepower ratio, the more sluggish the action of the vehicle. A low weight/horsepower ratio means high performance because it reflects a high ratio of power capability to travel resistance. Weight/horsepower ratios may be expressed in metric units as kilograms per metric ton.

It would be inappropriate to present in a handbook specific values of truck weight/horsepower ratios by vehicle class. Vehicle weight depends on the weight of the carried load which, for the larger trucks and truck combinations, can vary from zero to an amount equal to twice the vehicle's weight. Furthermore, the horsepower available for propulsion depends on engine condition and size, transmission arrangement, and engine speed. Additional information on the weight/horsepower ratio as a factor in highway design and vehicle operation can be found in the report of the 1948 study of truck operation on grades,[4] in a 1955 report on climbing lane design,[5] and in the *Policy on Geometric Design on Rural Highways*.[6]

ACCELERATION PERFORMANCE

Information on vehicle acceleration capabilities is needed for evaluation of minimum sight distance requirements for passing and for determination of minimum lengths of acceleration lanes at stop and yield signs and in interchanges. Normal roadway acceleration rates are a factor in designing cycle lengths of traffic signals, in computing fuel economy and travel time values, and in estimating how normal traffic movement is resumed after a breakdown in traffic flow patterns.

MAXIMUM ACCELERATION RATES

Typical maximum level road acceleration rates for several groupings of passenger cars and for typical weight ranges of pickup trucks, of two-axle, single-unit trucks, and of tractor semitrailer combination trucks are shown in Table 2.4 for standing starts to 15 mph (24 kph) and 30 mph (48 kph) speeds. Maximum level road acceleration rates for representative small, compact, intermediate, and large passenger cars, for pickup and two-axle, single-unit trucks, and for tractor semitrailer combinations at normal weights for 10 mph (16 kph) increases in speed at running speeds of 30, 40, 50, and 60 mph (48, 64, 80, and 97 kph) are given in Table 2.5. The values in Tables 2.4 and 2.5 are for typical vehicles manufactured since 1965.

Maximum acceleration rates for operation on a series of plus gradients are presented in Table 2.6. These data were developed from the values of Tables 2.4 and 2.5 by computation as noted in the footnotes to Table 2.6.

The relationships between distance traveled and speed achieved for automobiles accelerating at their maximum rate from standing stop are given in Figure 2.1 for operation on level road and on 6 percent and 10 percent grades. Data are for the composite car described in Tables 2.4, 2.5, and 2.6.

---

[4] *Time and Gasoline Consumption in Motor Truck Operation*, Research Report 9-A (Washington, D.C.: Highway Research Board, 1950).

[5] T. S. HUFF and F. H. SCRIVER, "Simplified Climbing Lane Design Theory and Road-Test Results," *Vehicle Climbing Lanes*, Bulletin 104 (Washington, D.C.: Highway Research Board, 1955).

[6] *A Policy on Geometric Design of Rural Highways, 1965* (Washington, D.C.: American Association of State Highway Officials, 1965).

TABLE 2.4
Typical Maximum Motor Vehicle Acceleration Rates from Standing Starts for Various Vehicle Types*

Vehicle Type	Typical GVW		Net Engine Propulsion Capability				Typical Maximum Acceleration Rate on Level Roads†			
			Given by Manufacturer		At 15mph‡ (24 kph)		To 15 mph (24 kph)		To 30 mph (48 kph)	
	lb	kg	hp	rpm	hp	rpm	mphps	kphps	mphps	kphps
Large car	4,800	2,177	350	4,400	60	1,420	10.0	16.1	7.0	11.3
Intermediate car	4,000	1,814	195	4,800	40	1,180	8.0	12.9	5.0	8.0
Compact car	3,000	1,361	120	4,400	32	1,490	8.0	12.9	5.0	8.0
Small car	2,100	952	42	3,900	17	1,900	6.0	9.7(2)	4.0	6.4(3)
Composite car§	4,000	1,814	—	—	—	—	8.0	12.9	5.0	8.0
Pickup truck	5,000	2,268	125	3,800	30	1,300	8.0	12.9(3)	5.0	8.0
Two-axle, single-unit truck	12,000	5,443	142	3,800	43	1,500	2.0	3.2(3)	1.0	1.6
Tractor semi-trailer truck	45,000	20,411	175	3,200	140	2,660	2.0	3.2(3)	1.0	1.6(4)

*If transmission is other than highest gear (or is automatic), gear position is shown in parentheses for 0 to 15 mph in the To 15 mph column and for 15 to 30 mph in the To 30 mph column.
†These data were observed for vehicles used in the operating cost research study conducted for NCHRP Project 2-5A. They were not included in the report of that project (*Running Costs of Motor Vehicles as Affected by Road Design and Traffic*, NCHRP Report 111) since they were developed principally as part of the information needed for planning project activities.
‡Computed using typical graphs of engine speed vs. horsepower and known transmission and rear-axle ratios. The transmission and rear-axle ratios of the vehicles are given on pp. 7–8 of *Running Costs of Motor Vehicles as Affected by Road Design and Traffic*, NCHRP Report 111 (Washington, D.C.: Highway Research Board, 1971).
§The composite car represents the typical passenger car in traffic on American highways.

TABLE 2.5
Typical Maximum Motor Vehicle Acceleration Rates for 10 mph (16 kph) Speed Increases at Various Running Speeds on Level Roads*

Vehicles	Typical GVW		Running Speeds†							
			30 mph (48 kph)		40 mph (64 kph)		50 pmh (80 kph)		60 mph (97 kph)	
	lb	kg	mphps	kphps	mphps	kphps	mphps	kphps	mphps	kphps
Large car	4,800	2,177	5.0	8.0	4.0	6.4	3.0	4.8	2.5	4.0
Intermediate car	4,000	1,814	5.0	8.0	4.0	6.4	3.0	4.8	2.0	3.2
Compact car	3,000	1,361	4.0	6.4	3.0	4.8	2.2	3.5	1.1	1.8
Small car	2,100	952	2.0	3.2	1.2	1.9	0.7	1.1	—	—
Composite car‡	4,000	1,814	4.7	7.5	3.8	6.1	2.8	4.5	1.9	3.1
Pickup truck	5,000	2,268	2.0	3.2	1.8	2.9	1.5	2.4	0.7	1.1
Two-axle, single-unit truck	12,000	5,443	1.0	1.6	0.6	0.9	0.2	0.3	—	—
Tractor semi-trailer truck	45,000	20,411	0.8	1.3	0.4	0.6	—	—	—	—

*Determined, given the maximum running speeds on particular grades developed in connection with the research for NCHRP Project 2-5A and reported in *Running Costs of Motor Vehicles as Affected by Road Design and Traffic,* NCHRP Report 111 (Washington, D.C.: Highway Research Board, 1971). This was done by computing the accelerations that can be achieved on level roads if the forces needed to overcome the resistances of the grades of the NCHRP study are used to produce acceleration.
†Transmission is in highest gear (or automatic) except for the small car which is in second gear at 30 mph and in third gear at 40 and 50 mph, for the two axle, single-unit truck which is in third gear at 30 and 40 mph, and for the tractor semitrailer truck which is in third gear at 30 mph and in fourth gear at 40 mph.
‡The composite car represents the typical passenger car in traffic on American highways.

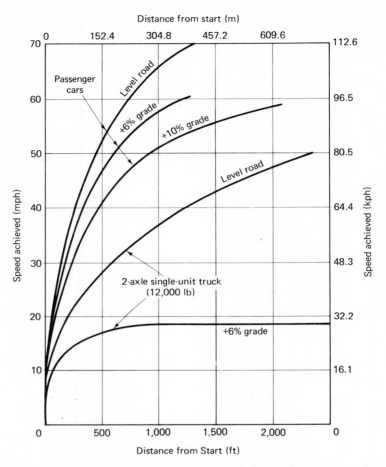

Figure 2.1. Speed-distance relationships observed during maximum rate accelerations. (Source: Tables 2.4, 2.5, and 2.6.)

Passing sight distances. Minimum passing sight distances on two-lane, two-way roadways are a function of maximum acceleration rates because the more quickly vehicles can accelerate while passing, the shorter the road length traversed during passing and the shorter the passing sight distance required. The minimum passing sight distances used for design are those recommended by the American Association of State Highway Officials shown in Table 14.5 (p. 612). The acceleration rates on which they are based are 1.40 mphps for an average passing speed of 34.9 mph, 1.43 mphps for 43.8 mph, 1.47 mphps for 52.6 mph, and 1.50 mphps for 62.0 mph.[7] At locations where maximum acceleration rates differ from those on which the AASHO policy passing sight distances are based (i.e., on parkways limited to passenger cars only) minimum passing distances may be computed by using the formulas from the policy manual and the maximum acceleration rates given in Table 2.5.

----

[7] *Ibid.*, p. 144.

TABLE 2.6
Typical Maximum Acceleration Rates of Representative Vehicles Operating
Upgrade on Various Grades*

	Vehicle Type†							
	Composite Passenger Car‡		Pickup Truck		Two-axle, Six-tire Truck		Tractor Semitrailer	
Gradient (%)	4,000 lb (mphps)	1,814 kg (kphps)	5,000 lb (mphps)	2,268 kg (kphps)	12,000 lb (mphps)	5,443 kg (kphps)	45,000 lb (mphps)	20,411 kg (kphps)
	Speed Change = 0–15 mph (0–24 kph)							
2	7.8	12.6	7.8	12.6	1.6	2.6	1.6	2.6
6	6.7	10.7	6.7	10.7	0.7	1.1	0.7	1.1
10	5.8	9.3	5.8	9.3	(14)	(23)	(4)	(6)
	Speed Change = 15–30 mph (24–48 kph)							
2	4.6	7.4	4.6	7.4	0.6	1.0	0.6	1.0
6	3.7	6.0	3.7	6.0	0.0	0.0	(23)	(37)
10	2.8	4.5	2.8	4.5	0.0	0.0	0.0	0.0
	Speed Change = 30–40 mph (48–64 kph)							
2	4.2	6.8	1.6	2.6	0.6	1.0	0.3	0.5
6	3.4	5.5	0.7	1.1	(30)	(48)	0.0	0.0
10	2.5	4.0	(30)	(48)	0.0	0.0	0.0	0.0
	Speed Change = 40–50 mph (64–80 kph)							
2	3.4	5.5	1.4	2.3	0.2	0.3	(45)	(72)
6	2.5	4.0	0.5	0.8	0.0	0.0	0.0	.0.0
10	1.6	2.6	0.0	0.0	0.0	0.0	0.0	0.0
	Speed Change = 50–60 mph (80–96 kph)							
2	2.4	3.8	1.0	1.6	(50)	(80)	0.0	0.0
6	1.5	2.4	0.2	3.2	0.0	0.0	0.0	0.0
10	0.6	1.0	0.0	0.0	0.0	0.0	0.0	0.0

*Computed, given the acceleration rates for level roads of Tables 2.4 and 2.5, by reducing the acceleration forces available on level roads by amounts equal to the corresponding grade resistances.
†Values given in parentheses in this table are typical maximum possible speeds in miles per hour (and kilometers per hour) for the given gradients.
‡The composite car represents the typical passenger car in traffic on American highways.

Normal acceleration rates. Observed normal roadway acceleration rates for passenger cars from standing stop to 15 mph (24 kph) and for 10 mph (16 kph) increases in speed at running speeds of 20, 30, 40, 50, and 60 mph (32, 48, 64, 80, and 96 kph) are given in Table 2.7. These acceleration rates were observed when drivers were not influenced to accelerate rapidly. They are typical of passenger cars starting up after a traffic signal turns green and those passing on four-lane divided highways. Observed normal deceleration rates of passenger cars are also given in Table 2.7.

TABLE 2.7
Observed Normal Acceleration and Deceleration Rates for Passenger Cars*

Speed Change		Accelerations		Decelerations	
mph	kph	mphps	kphps	mphps	kphps
0–15	0–24	3.3	5.3	5.3	8.5
0–30	0–48	3.3	5.3	4.6	7.3
30–40	48–64	3.3	5.3	3.3	5.3
40–50	64–80	2.6	4.2	3.3	5.3
50–60	80–97	2.0	3.2	3.3	5.3
60–70	97–113	1.3	2.1	3.3	5.3

*Determined by equipping a passenger car with an accelerometer, matching the speed change rates of this car with other cars in traffic, and observing acceleration and deceleration rates. These data were obtained in connection with the research conducted for NCHRP Project 2-5A. They were not included in the report of this project, *Running Costs of Motor Vehicles as Affected by Road Design and Traffic*, NCHRP Report 111 (Washington, D.C.: Highway Research Board, 1971) because they were developed principally as part of the information needed for planning project operations.

DECELERATION PERFORMANCE

Deceleration of motor vehicles occurs automatically when the accelerator pedal is released because of the retarding effect of the resistance to motion, including engine compression forces. For controlled deceleration and for maximum rates of deceleration, however, vehicle brakes are used to restrain vehicle motion.

DECELERATION WITHOUT BRAKES

Deceleration rates without brakes are much greater at the higher running speeds because the resistances to motion, particularly air resistance, are greater. This is important in planning for the control of high-speed traffic. For example, at speeds of 70 mph (113 kph) anything that causes a driver to take his foot off the accelerator will result in a rapid drop in speed of about 2.2 mphps (3.5 kphps) without a brake light warning to alert following motorists.[8] This effect may be a factor in many multi-car rear-end collisions on freeways.

An increase in one or more of the resistances to motion will cause a vehicle to decelerate automatically unless compensated for by an immediate increase in the throttle opening. For example, at points where a level or descending road changes to an upgrade, or where a straight road deflects onto a sharp curve, vehicles will decelerate appreciably unless the driver depresses the accelerator enough to offset the effect of the added resistance. Because sudden slowdowns of vehicles in fast high-volume traffic can disrupt traffic flow, drivers should be advised by means of a signal or a sign that they should maintain speeds at these critical points.

---

[8] E. E. WILSON, "Deceleration Distances for High-Speed Vehicles," *Proceedings of the 20th Annual Meeting of the Highway Research Board* (Washington, D.C.: Highway Research Board, 1940), pp. 393–97.

DECELERATION WITH BRAKES

Information on the deceleration rates of motor vehicles with braking (both maximum rates and the observed rates for normal slowdowns) is needed by traffic engineers. Maximum rates are used for estimating minimum stopping distances in emergencies. Normal slowdown rates provide the basis for estimating reasonable time and road lengths for stops at signs and signals where frequent normal stops are necessary.

MAXIMUM DECELERATION RATES

Retardation forces developed in brake drums or discs determine the braking deceleration rates of motor vehicles as long as slippage does not occur between pavement and tire surfaces. When the available braking force cannot be carried to the pavement without slippage, deceleration rates are determined by the effective coefficient of friction at the tire surface of contact. This coefficient is a function of pavement type, tire condition, and whether the pavement is wet or dry. Representative values are given in Table 2.8 along with values recommended for highway design purposes. Because braking systems in good order are usually able to provide more braking force than can be carried to the pavement, maximum deceleration depends primarily on this coefficient of friction between pavement and tire surfaces.

Equations (2.9) and (2.10) relate the coefficient of friction between pavement and tires and running speed to minimum stopping distance:

$$S = V^2/30\,(f \pm g) \quad \text{(U.S. units).} \tag{2.9}$$

S is the minimum stopping distance in feet.
V is running speed in miles per hour.
f is the coefficient of friction between pavement and tire surface.
g is the gradient.

$$S = V^2/255\,(f \pm g) \quad \text{(metric units).} \tag{2.10}$$

S is the minimum stopping distance in meters.
V is running speed in kilometers per hour.
f is the coefficient of friction between pavement and tire surface.
g is the gradient.

NORMAL DECELERATION RATES

Observed normal deceleration rates for passenger cars on dry pavements are presented in Table 2.7 for stops from running speeds of 15 and 30 mph (24 and 48 kph) and for 10 mph slowdowns (16 kph) from 40, 50, 60, and 70 mph running speeds (64, 80, 97, and 113 kph). Deceleration rates up to 5.5 mphps (8.8 kphps) are reasonably comfortable for car occupants.[9]

---

[9] *Ibid.*

TABLE 2.8

Average Passenger Car Skidding Friction Coefficients (a) on Various Dry Pavements and (b) as Recommended by AASHO for Design (Wet Surfaces)

	Dry Surface*		Wet Surface
Surface Description	New Standard Tires	Badly Worn Tires	Recommended by AASHO for Design†
Running Speed = 11 mph (17.7 kph)			
Dry bit. conc.	0.74	0.61	—
Sand asphalt	0.75	0.66	—
Rock asphalt	0.78	0.73	—
Port. cem. conc.	0.76	0.68	—
Running Speed = 20 mph (32.2 kph)			
Dry bit. conc.	0.76	0.60	0.40
Sand asphalt	0.75	0.57	0.40
Rock asphalt	0.76	0.65	0.40
Port. cem. conc.	0.73	0.50	0.40
Running Speed = 30 mph (48.2 kph)			
Dry bit. conc.	0.79	0.57	0.36
Sand asphalt	0.79	0.48	0.36
Rock asphalt	0.74	0.59	0.36
Port. cem. conc.	0.78	0.47	0.36
Running Speed = 40 mph (64.4 kph)			
Dry bit. conc.	0.75	0.48	0.33
Sand asphalt	0.75	0.39	0.33
Rock asphalt	0.74	0.50	0.33
Port. cem. conc.	0.76	0.33	0.33
Running Speed = 50 mph (80.5 kph)			
All pavements	—	—	0.31
Running Speed = 60 mph (96.5 kph)			
All pavements	—	—	0.30
Running Speed = 70 mph (112.6 kph)			
All pavements	—	—	0.29
Running Speed = 80 mph (128.7 kph)			
All pavements	—	—	0.27

*T. E. Shelbourne and R. L. Sheppe, "Skid Resistance Measurement of Virginia Highways," *Research Report 5–5*, *Highway Research Board* (Washington, D.C.: Highway Research Board, 1948), pp. 62–80.

†*A Policy on Geometric Design of Rural Highways* (Washington, D.C.: The American Association of State Highway Officials, 1965), p. 136.

VEHICLE OPERATING COSTS

The goal of the traffic engineer is to provide highway service that is rapid, safe, comfortable, convenient, and economical for motor vehicle users. In order for the engineer to plan and design traffic control facilities that are compatible with economical vehicle operation, he must know how vehicle operating costs are related to road geometry, suface conditions, traffic flows, and speed change requirements.

All of the information on vehicle operating costs presented in the following paragraphs and in Tables 2.9 to 2.16 was obtained from Report 111 of the National Cooperative Highway Research Program.[10]

FUEL CONSUMPTION

Vehicle fuel consumption is a major item of operating expense and one closely associated with road and traffic conditions. Table 2.9 presents the fuel consumption rates of the composite car for operation at various speeds on level roads and on gradients. The composite car reflects the vehicle distribution given in Table 1.7 (p. 7).

Table 2.10 gives the fuel consumed by the composite passenger car for stop-go and slowdown cycles in excess of that for continued operation at the given running speeds. Tabular values for stop-go cycles (upper values in each column of Table 2.10) do not include fuel consumption for stopped delays. Fuel consumption while stopped may be computed by using the composite passenger car idling fuel consumption rate of 0.58 gal per hr (2.20 liters per hr).

Table 2.11 is similar in form to Table 2.9, but it provides fuel consumption values for a two-axle, six-tire, single-unit truck at an average gross vehicle weight of 12,000 lb (5,443 kg).

Table 2.12 gives the excess fuel consumption for speed changes for the represen-

TABLE 2.9
Automobile Fuel Consumption as Affected by Speed and Gradient—
Straight, High-type Pavement and Free-flowing Traffic*

Uniform Speed		Gasoline Consumption on Plus Grades of†									
		Level		2%		4%		6%		8%		10%
mph	kph	gal/mi	lit./k	gal/mi	lit./k	gal/mi	lit./k	gal/mi	lit./k	gal/mi	lit./k	gal/mi	lit./k
10	16	0.072	0.169	0.087	0.204	0.103	0.242	0.121	0.285	0.143	0.336	0.179	0.421
20	32	0.050	0.117	0.070	0.164	0.086	0.202	0.104	0.245	0.128	0.301	0.160	0.376
30	48	0.044	0.103	0.060	0.141	0.078	0.183	0.096	0.226	0.124	0.291	0.154	0.362
40	64	0.046	0.108	0.062	0.145	0.078	0.183	0.096	0.226	0.124	0.291	0.156	0.367
50	80	0.052	0.122	0.070	0.164	0.083	0.195	0.104	0.244	0.130	0.306	0.162	0.381
60	97	0.058	0.136	0.076	0.179	0.093	0.219	0.112	0.263	0.138	0.325	0.170	0.400
70	113	0.067	0.158	0.084	0.198	0.102	0.240	0.122	0.287	0.148	0.348	0.180	0.423

*The composite passenger car represented here reflects the following vehicle distribution: large cars, 20%; standard cars, 65%; compact cars, 10%; small cars, 5%.
†The values of this table should be increased about 20% for speeds of 30 mph (48 kph) and 50% for speeds of 50 mph (80 kph) when operation is on badly broken and patched asphalt pavement.
They should also be increased for operation on curves:
On 5° curves, increase about 3% at 30 mph (48 kph) and 30% at 60 mph (97 kph).
On 10° curves, increase about 20% at 30 mph (48 kph) and 100% at 50 mph (80 kph).

---

[10] Claffey, *op. cit.*, pp. 16–39.

TABLE 2.10
Excess Gasoline Consumed per Stop or Slowdown Speed Change Cycle—Automobile*

Running Speed		Excess Gasoline Consumed by Amount of Speed Reduction Before Accelerating Back to Speed										
		10 mph	16 kph	20 mph	32 kph	30 mph	48 kph	40 mph	64 kph	50 mph	80 kph	60 mph	97 kph
mph	kph	gal	lit.	gal	lit.	gal	lit.	gal	lit.	gal	lit.	gal	lit.
10	16	0.0016†	0.0056†	—	—	—	—	—	—	—	—	—	—
20	32	0.0032	0.0121	0.0066†	0.0250†	—	—	—	—	—	—	—	—
30	48	0.0035	0.0132	0.0062	0.0235	0.0097†	0.0367†	—	—	—	—	—	—
40	64	0.0038	0.0144	0.0068	0.0257	0.0093	0.0352	0.0128†	0.0484†	—	—	—	—
50	80	0.0042	0.0158	0.0074	0.0280	0.0106	0.0401	0.0140	0.0530	0.0168†	0.0636†	—	—
60	97	0.0046	0.0174	0.0082	0.0310	0.0120	0.0454	0.0155	0.0587	0.0190	0.0719	0.0208†	0.0787†

*The composite passenger car represented here reflects the following vehicle distribution: large cars, 20%; standard cars, 65%; compact cars, 10%; small cars, 5%.

†Excess fuel consumed for stop-go cycles at given running speeds.

TABLE 2.11
Two-axle, Six-tire Truck Fuel Consumption as Affected by Speed and Gradient—Straight,
High-type Pavement and Free-flowing Traffic*

Uniform Speed†		Gasoline Consumption on Plus Grades of‡										
		Level		2%		4%		6%		8%		10%	
mph	kph	gal/mi	lit./k	gal/mi	lit./k	gal/mi	lit./k	gal/mi	lit./k	gal/mi	lit./k	gal/mi	lit./k
10	16	0.074	0.174	0.120	0.282	0.175	0.411	0.225	0.529	0.289	0.680	0.357	0.840
20	32	0.059	0.139	0.112	0.263	0.167	0.393	0.214	0.503	0.295	0.694	0.394	0.927
30	48	0.067	0.158	0.121	0.284	0.181	0.426	0.232	0.546	0.305	0.717	—	—
40	64	0.082	0.193	0.141	0.332	0.210	0.494	—	—	—	—	—	—
50	80	0.101	0.238	0.159	0.374	—	—	—	—	—	—	—	—
60	97	0.122	0.287	—	—	—	—	—	—	—	—	—	—

*The composite two-axle, six-tire truck represented here reflects the following vehicle distribution:
    Two-axle trucks at 8,000 lb GVW: 50%.
    Two-axle trucks at 16,000 lb GVW: 50%.
†Operation is in the highest gear possible for the grade and speed (fourth, third, or second). When vehicle approach speed exceeds the maximum sustainable speed on plus grades, speed is reduced to this maximum as soon as the vehicle gets on the grade.
‡The values of this table should be increased about 7% at 30 mph (48 kph) and 20% at 50 mph (80 kph) for operation on a badly broken and patched asphalt surface. They should also be increased for operation on curves:
    On 5° curves, increase about 3% at 30 mph (48 kph) and 23% at 50 mph (80 kph).
    On 10° curves, increase about 21% at 30 mph (48 kph) and 43% at 40 mph (64 kph).

TABLE 2.12
Excess Gasoline Consumed per Stop or Slowdown Speed Change Cycle—Two-axle,
Six-tire Truck*

Running Speed		Excess Gasoline Consumed by Amount of Speed Reduction Before Accelerating Back to Speed							
		10 mph	16 kph	20 mph	32 kph	30 mph	48 kph	40 mph	64 kph
mph	kph	gal	lit.	gal	lit.	gal	lit.	gal	lit.
10	16	0.0036†	0.0136†	—	—	—	—	—	—
20	32	0.0073	0.0276	0.0097†	0.0367†	—	—	—	—
30	48	0.0080	0.0303	0.0148	0.0560	0.0173†	0.0655†	—	—
40	64	0.0096	0.0363	0.0167	0.0632	0.0226	0.0855	0.0242†	0.0916†

*The composite two-axle, six-tire truck represented here reflects the following vehicle distribution:
    Two-axle trucks at 8,000 lb GVW: 50%.
    Two-axle trucks at 16,000 lb GVW: 50%.
†Excess fuel consumed for stop-go cycles at given running speeds.

tative two-axle, six-tire, single-unit truck and is similar in form to Table 2.10. Average idling fuel consumption while stopped is 0.65 gal per hr (2.46 liters per hr) for this vehicle.

TIRE WEAR

Tire wear cost values for the composite passenger car, based on a composite cost new (1970 prices) of $119 for a set of four tires, are presented in Table 2.13 by running speed, surface type, and curvature.

TABLE 2.13

Automobile Tire Cost (Four Tires) as Affected by Speed, Type of Road Surface, and Curvature*†

Uniform Speed		High-type Concrete						High-type Asphalt						Dry, Well-packed Gravel	
		Straight		5°		10°		Straight		5°		10°		Straight	
mph	kph	c/m	c/km	c/m	c/km	c/m	c/km	c/m	c/km	c/m	c/km	c/m	c/km	c/m	c/km
20	32	0.09	0.06	0.11	0.07	0.17	0.11	0.27	0.17	0.32	0.20	0.51	0.32	1.03	0.64
30	48	0.19	0.12	0.43	0.27	0.82	0.51	0.36	0.22	0.82	0.51	1.56	0.97	1.05	0.65
40	64	0.29	0.18	2.17	1.35	4.83	3.00	0.43	0.27	3.22	2.00	7.16	4.45	1.07	0.66
50	80	0.32	0.20	4.44	2.76	14.34	8.91	0.45	0.28	6.25	3.88	20.10	12.48	1.10	0.68
60	97	0.31	0.19	7.64	4.74	—	—	0.46	0.29	16.34	9.99	—	—	—	—
70	113	0.30	0.18	—	—	—	—	0.44	0.27	—	—	—	—	—	—
80	129	0.27	0.17	—	—	—	—	0.43	0.27	—	—	—	—	—	—

*The composite passenger car represented here reflects the following vehicle distribution: large cars, 20%; medium-quality tires based on the following unit tire costs by vehicle cars, 5%.

†Tire costs were computed by using a weighted average cost of $119 for a set of four new, standard-size cars, 65%; compact cars, 10%; small type (as noted in the northeastern states in 1969): large cars, $35 per tire; standard-size cars, $30 per tire; compact cars, $25 per tire; small cars, $15 per tire.

There are approximately 1,500 grams of usable tire tread in 80% of passenger car tires. This weight of usable tire tread was also recorded for the tires used in the tire wear test.

The excess tire cost for stop-go and 10 mph (16 kph) slowdown speed change cycles are shown in Table 2.14 for the composite passenger car for a series of running speeds on high-type road surfaces.

Tire wear cost values for the representative truck, a two-axle, six-tire, single-unit truck, for straight road operation at 45 mph (72 kph), for curve operation, and for operation on major urban arterials are given in Table 2.15.

MAINTENANCE COST

The total cost (1970 prices) of maintenance for vehicle components affected by operation (power train, exhaust system, and brakes, for example) is 1.15 cents per mile (0.71 cent per km) for passenger cars and 1.42 cents per mile (0.88 cents per km) for pickup trucks. The maintenance cost of intercity line-haul trucks is approximately 1.23 cents per mile (0.76 cent per km). The excess maintenance cost for passenger car stop-go cycles at 25 mph (40 kph) running speeds is 0.12 cent per stop.

OIL CONSUMPTION

Oil consumption results from oil contamination by use and oil loss through leakage and combustion. The combined oil consumption rates for contamination, leakage, and combustion for passenger cars and two-axle, six-tire, single-unit trucks are given in Table 2.16 for operation in free-flowing traffic on dust-free (high-type) roads.

TABLE 2.14
Excess Tire Cost per Speed Change Cycle—Automobile*†

| Running Speed | | Cost of Four Tires (Cents per Cycle) | | | |
| | | Stop-Go Speed Change Cycles | | 10 mph (16 kph) Slowdown Cycles | |
mph	kph	Concrete	Asphalt	Concrete	Asphalt
20	32	0.10	0.30	0.04	0.10
30	48	0.30	0.60	0.08	0.15
40	64	0.58	0.85	0.09	0.14
50	80	0.72	1.10	0.09	0.14
60	97	0.80	1.20	0.08	0.12
70	112	0.85	1.25	0.08	0.12

*The composite passenger car represented here reflects the following vehicle distribution: large cars, 20%; standard-size cars, 65%; compact cars, 10%; small cars, 5%.

†Tire costs were computed by using a weighted average cost of $119 for a set of four new, medium-quality tires based on the following unit tire costs by vehicle type (as noted in the northeastern states in 1969): large cars, $35 per tire; standard-size cars, $30 per tire; compact cars, $25 per tire; small cars, $15 per tire.

There are approximately 1,500 grams of usable tire tread in 80% of passenger car tires. This weight of usable tire tread was also recorded for the tires used in the tire wear test.

TABLE 2.15
Tire Wear Cost—Two-axle, Six-tire Truck*

Type of Operation and Road Surface	Tire Wear Costs (per Axle)			
	Rear Axle—Four Tires (12,000 lb) (5,443 kg)		Front Axle—Two Tires (4,000 lb) (1,814 kg)	
	c/mi	c/km	c/mi	c/km
Uniform speed of 45 mph (72 kph on high-type concrete	0.40	0.25	0.10	0.06
25–30 mph (40–48 kph) on four-lane major street urban arterial with high-type concrete surface (3 to 4 stops per mile)	1.96	1.22	0.28	1.74
25 mph (40 kph) on 30° curve with high-type surface	10.80	6.71	1.30	0.81
25 mph (40 kph) on 60° curve with high-type surface	108.00	67.10	9.20	5.72

*Tire wear costs based on a cost of $120 per tire for a medium-quality, 10-ply 8.25 × 20 transport tire (1970 price in the northeastern states). Each tire has approximately 4,500 grams of usable tread before recapping is necessary. Value of the tire carcass when recapped was assumed to be $20.

TABLE 2.16
Engine Oil Consumption Rates in Free-flowing Traffic on High-type Roads*

Speed†		Passenger Car‡		Two-axle, Six-tire Truck§	
mph	kph	qt/1,000 mi	lit./1,000 km	qt/1,000 mi	lit./1,000 km
30	48	0.97	0.57	2.77	1.63
35		0.97	0.57	2.77	1.63
40	64	1.12	0.66	2.84	1.67
45		1.28	0.75	3.00	1.76
50	80	1.45	0.85	3.16	1.86
55		1.64	0.96	3.33	1.96
60	97	1.78	1.05	3.50	2.06

*Oil consumption includes oil both for oil changes and for additions between oil changes.
†Minimum trip length = 10 mi (16 km).
‡An eight-cylinder Chevrolet sedan with an engine displacement of 283 cu in. represented the typical passenger car.
§A truck with a six-cylinder engine (engine displacement of 351 cu in.) represented the typical two-axle, six-tire truck.

DEPRECIATION

The magnitude of motor vehicle depreciation cost, when defined as the quotient resulting from dividing the difference between original cost and scrap value by life-time mileage, depends largely on nonhighway factors (new and used car market values and user travel desires). Because the work of the traffic engineer has little effect on these factors, vehicle depreciation cost factors are not included here.

CONGESTION COSTS

Highway congestion conditions affect vehicle operating costs for fuel and oil consumption, tire wear, and maintenance principally as a result of the speed changes and stopped delays associated with congestion. The frequency and severity of the speed changes and the duration of stopped delays caused by congestion vary widely from one congestion location to another. When it is desired to determine vehicle operating costs caused by congestion at a given location, the recommended procedure is to observe, at that location, both the frequency of speed changes by ranges of speed change and the frequency of vehicle stops by range of stop duration and to compute the resulting fuel and oil consumption, tire wear, and maintenance costs by using speed change and idling operating cost data given above.

VEHICLE DIMENSIONS

Information on the various dimensions of motor vehicles and current trends in these dimensions is needed by traffic engineers for planning road and street geometrics and parking lot layouts.

Figures 2.2 and 2.3 provide information on the overall lengths of passenger cars and their wheelbases. Figures 2.4 and 2.5 show the front and rear overhang distances of passenger cars. Overall passenger car width is shown in Figure 2.6 and height is shown in Figure 2.7. Figure 2.8 shows height of eye of driver of passenger cars. Design vehicle dimensions are presented in Table 14.3 (p. 609).

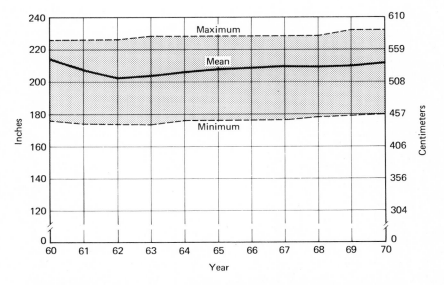

Figure 2.2. Trend in overall length of standard American passenger cars by years. [E. E. Seger and R. S. Brink, "Trends of Vehicle Dimensions and Performance Characteristics," *General Motors Proving Ground Engineering Publication* (Milford, Mich.: General Motors Proving Ground, 1971), p. 3.]

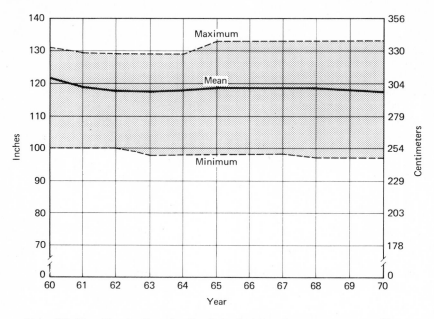

Figure 2.3. Trend in wheelbase length of standard American passenger cars. (Seger and Brink, p. 3.)

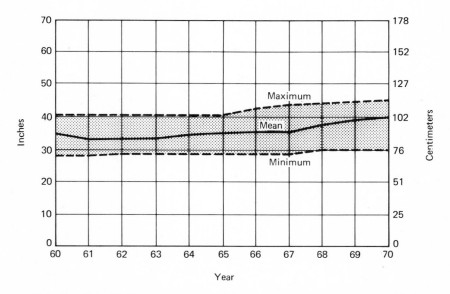

Figure 2.4. Trend in front overhang distance (measured from farthest point forward to center of front axle) of standard American passenger cars. (Seger and Brink, p. 3.)

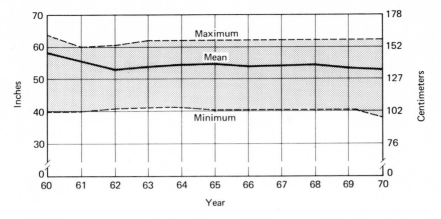

Figure 2.5. Trend in rear overhang dimension (measured from farthest point in rear to center of rear axle) of standard American passenger cars. (Seger and Brink, p. 4.)

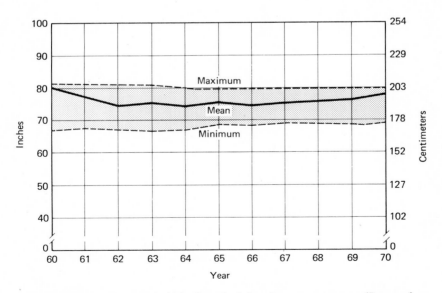

Figure 2.6. Trend in overall width of standard American passenger cars. (Seger and Brink, p. 6.)

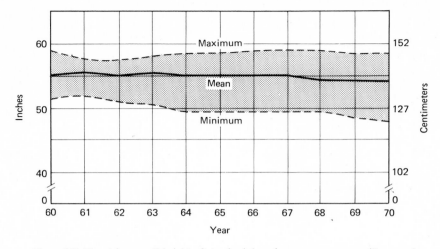

Figure 2.7. Trend in overall height of standard American passenger cars. (Seger and Brink, p. 7.)

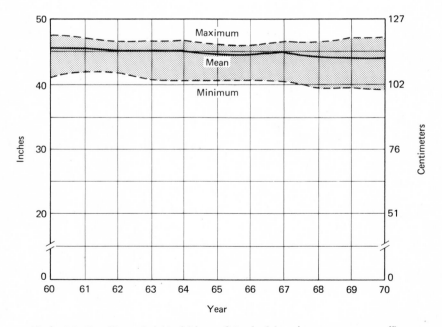

Figure 2.8. Trend in eye height of drivers of standard American passenger cars. (Seger and Brink, p. 8.)

REFERENCES FOR FURTHER READING

CLAFFEY, P. J., *Running Costs of Motor Vehicles as Affected by Road Design and Traffic*. Washington, D.C.: Highway Research Board (NCHRP Report 111), 1971.

MATSON, THEODORE M., WILBUR S. SMITH, and FREDERICK W. HURD, *Traffic Engineering*. New York: McGraw-Hill Book Company, 1955.

ODIER, LIONEL, *The Economic Benefits of Road Construction and Improvements*. Translated from the French document *Les Interets Economiques des Travaux Routiers* by Noel Lindsay. Paris: Bureau Central D'Etudes Pour les Equipments D'Outre-Mer, 1962.

Parking Dimensions, 1974 Model Cars, Engineering Notes. Detroit, Mich.: Motor Vehicle Manufacturers Association of the U.S., Inc. (Annual publication).

Policy on Geometric Design of Rural Highways. Washington, D.C.: American Association of State Highway Officials, 1965.

Policy on Design of Urban Highways and Arterial Streets, Washington, D.C.: American Association of State Highway Officials, 1973.

SEGER, E. E., and R. S. BRINK, *Trends of Vehicle Dimensions and Performance Characteristics 1960 Through 1970*. Milford, Mich.: General Motors Proving Ground, 1971.

WINFREY, ROBLEY, *Economic Analysis for Highways*. Scranton, Pa.: International Textbook Company, 1969.

Chapter 3

DRIVER AND PEDESTRIAN CHARACTERISTICS

SLADE HULBERT, Associate Research Psychologist, Institute of Transportation and Traffic Engineering, University of California, Los Angeles, California.

The human element of any traffic system, as represented by the driver and the pedestrian, is more variable and unpredictable than the vehicle and roadway elements. Thus it is essential that driver and pedestrian characteristics be identified and explained in order to better accommodate this element in traffic designs and programs.

This chapter describes various driver and pedestrian characteristics including behavior patterns and accident involvement factors. Bicyclists and bicycle usage as well as vehicle design are also discussed.

DRIVERS

Safety relationships in general as revealed in mass data analysis indicate that total traffic deaths have been rising steadily since 1925 (with a decrease during World War II). Nevertheless, registered vehicles and vehicle miles have been increasing in greater numbers than fatalities (Chapter 4); therefore, the *rate* of fatalities has been decreasing. Within those statistics, the raw numbers of pedestrian fatalities have been held nearly constant and are holding almost a constant rate per vehicle mile driven and per 100,000 population (Figure 3.1), while vehicle collisions and single vehicle fatal crash rates have steadily increased.

There were 118 million licensed drivers in 1973 in the U.S.[1], but because there are markedly different license standards among the 51 states, there is very little homogeneity in driving ability among these drivers. Most drivers received no formal training (Table 3.1), and even those who did, received so little that by far, the bulk of driving skills has been learned on a trial-and-error basis. This means that the street and highway facilities must accommodate all levels of competence ranging from the rankest beginners to the most experienced and skilled. National and local programs are underway to improve the training that drivers receive and accordingly increase the driver license requirements among the 50 states, but for some time to come, however, the range of driver capabilities must be recognized as being so great as to cause major concern and as much consideration by traffic engineers as feasible.

The wide range of driver skills and perceptual abilities is partially offset by the fact that driving is a self-paced task. The pace (or rate of driving speed) can of course greatly change the "task load." One aspect of task load is how frequently the roadway environment must be visually scanned or sampled. Beginning drivers have a low sampling rate because they must search for salient cues, whereas experienced drivers

---

[1] National Safety Council, *Accident Facts*, 1973, p. 40.

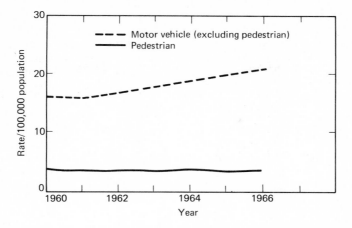

Figure 3.1. Pedestrian mortality from traffic accidents, United States, 1958–66 (Courtesy, Metropolitan Life Insurance Co., 1969).

TABLE 3.1
How Sample of California Drivers Learned to Drive

Age Group	N	Percent How Learned to Drive*						
		1	2	3	4	5	6	7
Under 20	1,921	47.1	0.9	11.9	0.7	9.9	29.1	0.4
20–24	2,168	60.3	2.0	9.5	1.1	6.0	20.2	0.9
25–29	1,831	74.7	3.2	6.9	1.7	4.2	8.7	0.6
30–34	1,755	85.5	4.3	3.4	1.4	1.3	3.8	0.3
35–39	1,820	90.5	4.1	1.8	1.3	0.5	1.6	0.2
40–44	1,848	89.8	4.0	1.8	2.1	0.3	1.5	0.6
45–49	1,587	93.1	2.7	1.1	1.3	0.1	1.4	0.3
50–54	1,443	92.6	3.7	0.6	1.1	0.2	1.5	0.3
55–59	1,022	93.8	2.7	0.8	1.3	0.1	1.1	0.2
60–64	774	92.0	4.3	1.2	1.2	0.5	0.8	0.1
65 and up	1,304	91.4	4.2	1.9	1.7	0.1	0.5	0.2
All Ages	17,473	80.5	3.2	4.3	1.4	2.5	7.7	0.4

*Legend:
1. Taught by family, friends, and/or "picked up"
2. Paid instruction
3. Driver education (classroom plus on-the-road)
4. 1 above plus paid instruction
5. 1 above plus classroom course only
6. 1 above plus driver education
7. Miscellaneous and not stated

Source: Albert Burg, "Characteristics of Drivers," *Human Factors in Highway Traffic Safety Research*, T. W. Forbes, ed. (New York: John Wiley and Sons, Inc., 1972), p. 79.

have gradually learned which cues are salient and can respond quickly to minimal cues in a more efficient search pattern. This is dramatically evident when comparing eye-movement patterns of novices with those of experienced drivers.[2] Similar results have

---

[2] R.R. Mourant, and T.H. Rockwell, "Strategies of Visual Search by Novice and Experienced Drivers," *Human Factors*, Vol. 14, No. 4 (1972), pp. 323–35.

been obtained from aircraft pilots[3] and it is a familiar experience for anyone who has ever attempted to teach another person to drive.

For the novice, the visual search pattern is more active and erratic and dwell-times tend to be longer. For the experienced driver, visual search patterns are generally less active and dwell-times much shorter because the central nervous system (CNS) processing takes place more rapidly and because the driver has learned what to expect.

With increasing age of the driver, the dwell-times increase because of a slowdown in CNS process time, but this slowdown occurs gradually over many years and is adjusted by changes in pace and offset by additional learning. Alcohol and other drugs also slow down the CNS;[4] this change, however, is rapid and may or may not be compensated for by a reduction in pace, depending on the degree of conscious awareness of the driver of the drug effects on his abilities. Alcohol is discussed in detail later in this chapter. Fatigue also slows CNS rate far more rapidly than age, but fatigue does have a large number of recognizable symptoms which drivers can use as a basis for deciding to reduce their pace. Kimball *et al.* clearly shows this trade off of pace for steering in a comparison of experienced drivers with novices.[5] The novices (with fewer than 10 hours of experience) also required more steering inputs to maintain "marginal control" of the vehicle steering task than did the experienced drivers.[6]

Unfortunately, reduction of the driving pace only reduces a portion of the driving task load because the remainder of the task load is determined by other drivers, pedestrians, other roadway users, and sudden or unexpected changes in the roadway environment such as illumination, alignment, traction, sight distance, and visibility.

AGE AND SEX OF DRIVERS

Age and sex of drivers are important variables in that they are easily determined and can therefore be used in driver licensing or other driver control efforts. Burg concludes the following from the data presented in Table 3.2:

> The trend in recent years shows that
>
> 1. Females are comprising an ever-increasing proportion of the driving population, with the male/female ratio approaching that which exists in the general population, [and that]
> 2. Both old and young drivers comprise an increasing proportion of the total driving population.[7]

It is clear that raw numbers of reported accidents and traffic law violation convictions decrease with age; when exposure (miles driven) is taken into account, however, both young and old drivers have poorer driving records than middle-aged drivers. Also, there are differences in violation patterns related to age.[8] Young drivers'

---

[3] Unpublished statement from Systems Technology Incorporated, Hawthorne, California, 1972.

[4] B.L. BELT, "Driver Eye Movement as a Function of Low Alcohol Concentrations," Driving Research Laboratory Technical Report, Department of Industrial Engineering, The Ohio State University, Columbus, Ohio, 1969.

[5] K. KIMBALL, V. ELLINGSTAD, and R. HAGEN, "Effects of Experience on Patterns of Driving Skill," *Journal of Safety Research*, Vol. 3, No. 3 (September 1971), pp. 129–35.

[6] *Ibid.*

[7] ALBERT BURG, "Characteristics of Drivers," *Human Factors in Highway Traffic Safety Research*, T.W. Forbes, ed. (New York: John Wiley and Sons, Inc., 1972), p. 76.

[8] DAVID M. HARRINGTON and ROBIN S. McBRIDE, "Traffic Violations by Type, Age, Sex and Marital Status," *Accident Analysis and Prevention*, Vol. 2, No. 1 (1960), pp. 67–69.

TABLE 3.2
Age and Sex Distribution of Licensed Drivers in the U.S. by Year*

Age Group	Percent of Total					
	1972	1971	1970	1969	1965	1960
Under 20	10.3	10.3	10.2	10.2	9.8	7.2
20–24	11.3	11.3	11.1	11.0	10.4	11.2
25–29	10.8	10.2	10.1	10.0	9.6	12.7
30–34	9.8	9.6	9.5	9.5	10.1	12.5
35–39	9.5	9.6	9.8	9.9	11.1	11.6
40–44	9.6	9.9	10.4	10.5	10.8	10.3
45–49	9.6	9.8	9.8	9.8	9.7	9.1
50–54	8.7	8.7	8.6	8.6	8.5	7.8
55–59	6.7	6.8	6.8	6.8	6.8	6.2
60–64	5.2	5.3	5.2	5.2	5.2	4.7
65–69	3.9	3.9	3.9	3.9	3.7	3.1
70–74	2.7	2.7	2.7	2.7	2.6	2.1
75 and up	1.9	1.9	1.9	1.9	1.7	1.5
Total Licensed Drivers (millions)	118.2	114.0	111.0	107.5	98.0	87.0
Percent Male	56	56	57	58	61	70
Percent Female	44	44	43	42	39	30

*From National Safety Council figures.

Sources: Burg, "Characteristics of Drivers," p. 76 and *Accident Facts* (Chicago: National Safety Council, 1973), p. 54.

violations, e.g., speeding, reflect a greater propensity for risk-taking behavior than older drivers' violations, e.g., signs, turning, passing, and right of way. This indicates decrements in physical and judgmental skills of both the younger and older driver. There is some evidence that accident type also varies with age, but this has not as yet been studied in detail.[9]

Miles driven (quantitative exposure to risk) is the single measured variable most highly correlated with accident and violation record. Accidents and convictions increase with increasing mileage, but not linearly.

Male drivers have more accidents and convictions than females; when miles driven is taken into account, however, the differences essentially disappear.

Married drivers have better driving records than single drivers for both males and females, at all age levels, and whether or not exposure (miles driven) is taken into account, although exposure does enter in and the relationship is less for older drivers.

Driver license laws and regulations in many states require older drivers to be re-examined more frequently, and some states encourage young drivers to complete driver education and driver training courses. No special requirements are set for single vs. married drivers although insurance rates do differ.

DRIVER BEHAVIOR AND ACCIDENTS

Some tabulations of types of improper driving as related to accidents have been made as shown in Table 4.8 (p. 111). These and other types of improper driving can be the results of either willful or inadvertent errors. Unfortunately, it is not easy to discover which type of behavior has caused an accident.

[9] ALBERT BURG, previously unreported findings from the UCLA Driver Vision Research Project, 1970.

An October 1970 report by the U.S. Department of Transportation deals with this difficult problem and concludes: "The negligence law usually treats 'driver error' as both avoidable and unreasonable, and imposes liability pursuant to an objective standard to which all drivers are held. But a review of the available research indicates that a significant gap exists between the standard of behavior required by the negligence law and the average behavior normally exhibited by most drivers."[10]

The report also says: "You will note that the standard of care required is that exercised by a person of reasonable and ordinary prudence, rather than that exercised by a person of extreme caution or exceptional skill. While exceptional skill is to be admired and encouraged, the law does not demand it as a general standard of conduct."[11]

Many programs of driver improvement seem to be based on an assumption of willful misbehavior and therefore a concentration on the multiple violator and accident repeater. However, recent studies by Campbell[12] show that there is little evidence to support this position. Most accidents involve drivers with good records who have not had any previous serious crashes. In other words, the old concept of the "accident-prone" driver is not supported by the facts.

Each year traffic accidents seem to be distributed among the states in about the same proportions per millions of vehicle miles driven. The national fatality rate seems to vary with the Gross National Product. Raw numbers of fatal crashes occur more frequently during the summer and fall when there is more travel. These facts all seem to point to the conclusion that the more driving we do, the more the chance of an accident occurring increases, and the more do occur. This would seem to argue for the major role of chance in the distribution of fatal crashes. Undoubtedly, chance factors are acting, but this does not at all mean that each accident was not caused and therefore could have been prevented.

Since the majority of motor vehicle accidents occur in daylight on dry roads with sound vehicles, the causes seem to be with the driver and the ways in which he interacts with the roadway. The more that is understood about drivers, the more likely are traffic control and remedial efforts to be successful. Burg deals with this question and presents the following conclusions:

1. *Biographical Descriptors:* A justification exists for differential licensing for both young and old drivers, and implementation of such a program is feasible. Not feasible, however, is differential licensing on the basis of such factors as marital status, education or annual mileage, although research results would suggest such a move.
2. *Chronic Medical Conditions:* There is sufficient evidence relating certain severe medical conditions to accidents to suggest that short-term licensing of such individuals might prove beneficial. However, final action of this sort should not be taken without confirmation of present findings through a carefully controlled study.
3. *Hearing:* Present evidence suggests that the deaf driver may be at a disadvantage, and that special training programs and/or special aids might be of benefit; however, additional research again is needed before action is warranted.

---

[10] U.S. Department of Transportation, *Driver Behavior and Accident Involvement: Implications for Tort Liability*, Insurance and Compensation Study, October 1970, p. 190.

[11] *Ibid.*, p. 191.

[12] B.J. CAMPBELL, "The Effects of Driver Improvement Actions on Driving Behavior," *Traffic Safety Research Review*, Vol. 3, No. 3 (1959), pp. 19–31.

4. *Loss of Limb*: There is no evidence to justify taking any action in this area.
5. *Vision*: Research results indicate that vision is indeed related to driving. However, the magnitude of the relationship appears to be small, and the question of practical significance arises. How much improvement in the traffic accident picture can be effected by more effective vision screening? By the same token, of what value are present licensing techniques such as written examinations and driver tests? These are questions that have no clear-cut answers, for definitive research has yet to be done, and other factors, such as "face validity" and "tradition" serve to confound the issue.[13]

Drivers can become involved in accidents even as innocent victims and yet be included in some records. Because of legal implications, many such records do not distinguish the "at fault" driver from those not at fault, which makes research in group behavior very difficult. Therefore, care must be exercised when examining studies of driving accident records to know what criteria were used.

THE DRIVING TASK

Man as a component in any transportation system has several important functions to perform in addition to actual control of the vehicle. For example, he must decide where and when he wants to go, and he must choose his route and type of vehicle. These various functions that man performs are described in the generalized framework shown in Figure 3.2 which can be used to analyze any transportation system (or subsystem). Using such a framework will help to keep in mind the many interactions that are involved in even the most simple motor trip. Pre-trip planning, for example, can be influenced by radio broadcast advisory information on traffic congestion. Many examples of the application of this framework are provided in Hulbert and Burg.[14]

Usually, however, the driving task is dealt with in a more narrow frame of reference that considers only the actual control of the vehicle along a roadway. From this viewpoint, research work has been done to describe man as a control element in a servo system. To date, several mathematical models have been developed to aid vehicle designers to produce automobiles with "good" handling characteristics so that drivers with less than average skill and strength can successfully accomplish the maneuvers they wish to perform.

These types of vehicle control maneuvers are influenced directly by the available information (cues) and the driver's ability to receive and process that information. A landmark study[15] reports on the information needs of drivers and classifies these needs into various categories that relate to pathway information and delineation which are primary needs and are related to route selection and tracking skills. The other major task is object avoidance about which much less has been published. Traffic control devices can help in both of these tasks in many ways; they are discussed by Hulbert,[16] for example, assigning right of way, warning of curves, and control of passing.

---

[13] BURG, "Characteristics of Drivers," pp. 91–92.

[14] SLADE F. HULBERT and ALBERT BURG, "Human Factors in Transportation Systems," *Systems Psychology*, Kenyon B. De Greene, ed. (New York: McGraw-Hill Book Company, 1970), pp. 471–509.

[15] *Development of Information Requirements and Transmission Techniques for Highway Users*, Volume I (Deer Park, New York: Airborne Instruments Laboratory, 1967).

[16] SLADE HULBERT, "Driver Information Systems," *Human Factors in Highway Traffic Safety Research*, T. W. FORBES, ed. (New York: John Wiley and Sons, Inc., 1972), pp. 110–132.

| | A. Trip planning | | B. Making vehicle control decisions | | | C. Executing vehicle control decisions | | D. Vehicle maintenance | |
	a. Schedule	b. Route	a. Path	b. Speed	c. System failure	a. Acceleration	b. Direction	a. At home	b. En route
1. Input (Sources of information)									
2. Output (Range of performance)									
3. Evaluation of performance									
4. Selection and placement									
5. Training of personnel									

Figure 3.2. A generalized framework for analysis of human functions and their characteristics in transportation systems. [Source: Slade F. Hulbert and Albert Burg, "Human Factors in Transportation Systems," *Systems Psychology*, Kenyon B. De Greene, ed. (New York: McGraw-Hill Book Company, 1970), p. 482.]

Tracking and object avoidance must be performed concurrently, but man has essentially a one-track (single-channel) mind and, therefore, he must *divide his attention* while driving. Most driving information comes to the driver visually in a stream of changing scenes that he must sample (because he cannot take it *all* in) and select from and use to make decisions on where he is going to be in the next instant and the next few seconds. This is a spatial commitment that Hulbert and Burg[17] describe as fan-shaped extending in front of the moving vehicle. Figure 3.3 describes this zone which, of course, varies in its exact shape but, nevertheless, is there and has been committed to by the driver.

Lyman Forbes[18] has extended the concept of the committed zone and has shown it in three-dimensional form. In this way, it is possible to portray the committed zone as it appears through the windshield to the driver. Forbes concludes:

> All portions of the driver's field of view are not of equal importance, nor are all portions equally easy for him to use, considering inherent constraints in the human visual system.
>
> However, committed zone analysis is a promising tool. . . . It should be remembered, however, that the size, shape and location of a committed zone will be deter-

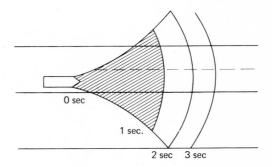

The exact configuration of this fan-shaped zone will depend on the vehicle speed, turning radius, and stopping distance as they interact with the driver's reaction time.

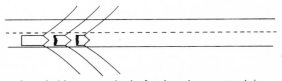

Figure 3.3 Spacial commitment of driver in moving vehicle. [Source: Slade Hulbert, "Driver Information Systems," *Human Factors in Highway Traffic Safety Research*, T. W. Forbes, ed. (New York: Wiley-Interscience, a division of John Wiley and Sons, Inc., 1972), p. 111.]

As each driver proceeds, the fan-shaped zone extends in front of his path and changes shape as velocity and pavement conditions vary.

[17] HULBERT and BURG, "Human Factors in Transportation Systems."

[18] HYMAN FORBES, "Geometric Vision Requirements in the Driving Task," *1970 International Automobile Safety Conference Compendium*, (New York: Society of Automotive Engineers, 1970), pp. 677–89.

mined by the statistical variables of vehicle velocities, accelerations, and road properties. Sizable samples of these statistics are available. But, more studies must be conducted before the descriptions have been thoroughly developed into normative samples. Then the relative importance of the zones will be known.[19]

As the time-frame of this commitment increases, the commitment becomes more and more provisional because the driver has more time to receive new (up dated) information and change his path or speed. Each driver knows that this is the case (although perhaps not consciously) and behaves accordingly, which explains why urban freeway drivers move with only 1 sec headway at speeds over 60 mph. Each driver knows that the driver ahead of him has already committed his vehicle to be somewhere far ahead in the next few seconds and therefore he is comfortable in making a similar spatial commitment allowing only 1 sec for reaction time and assuming that he can stop as quickly as the driver ahead of him can. If drivers did not behave this way, freeway volumes of as much as 1,900 veh/lane/hr could not be achieved. However, freeway accidents do occur and could be reduced by providing more information to the driver about headway.

EXPECTATION AND REACTION TIME

The role of expectancy is paramount in understanding driver behavior, and this understanding leads to an improved ability on the part of traffic engineers to shape and configure the environment in order to make traffic safe and efficient. Drivers' reactions to various roadway environments have been and are being studied in many research projects, and the results can be applied and useful only if they are understood in the context of the human perception process and the way it affects and changes driver behavior. Expectation is an explanatory concept that places all of the many things that traffic engineers do in a somewhat different frame of reference that can be extremely useful in applying results of research to everyday engineering practice.

For example, the addition of an amber phase to traffic signals clearly grew out of the fact that as vehicle speeds increased, the driver's limited range of abilities had to be assisted. At the same time there has been an increase in the number of signal heads because of the recognition of the driver's limited rate of sampling of his visual environment, so that if the driver's visual attention happened to be elsewhere (away from one signal head), there are other signal heads available for him to see within a normal cone (or field) of attention. This increases the likelihood that he will see at least one, and is sometimes called redundancy. This example of the evolution of traffic signals is so obvious that one is inclined to say, why go to all that length to explain a simple thing like that? However, if one considers recent trends to more complex, multi-phase, multi-indication signal displays, it soon becomes obvious that a conceptual and systematic understanding of driver behavior can be useful. For example, the addition of pedestrian heads to signals has become increasingly popular because they help to shorten the time necessary to allot to pedestrian movement. However, an understanding of driver behavior will lead to a recognition that drivers also can see the pedestrian heads and actually are using them as a pre-amber indication in much the same way as the traffic funnel concept. Pedestrian heads thus provide a possible increase in accuracy of driver expectation of a change in the signal indication. Evidence that pedestrian heads do serve this function is contained in a small study of the accident

---

[19] *Ibid.*

records at signalized intersections where there was found a reduction in car-to-car collisions after installation of pedestrian heads, and also a lower frequency of these crashes than on a comparable roadway where pedestrian heads were not installed.[20]

The extreme importance of the role of expectation in driver reaction time was documented in the 1971 report by Johansson and Rumar.[21] In a series of data collection experiments they gathered data from 321 drivers who expected to have to apply their brakes. These data are shown in Figure 3.4. The median value is 0.66 sec, the range being from 0.3 sec. to 2.0 sec.

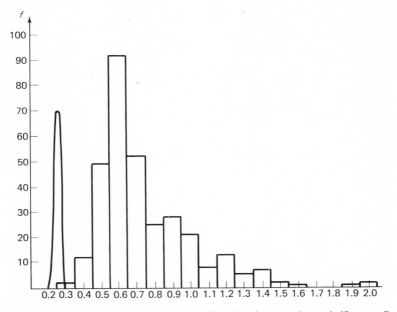

Figure 3.4. Brake reaction time distribution for 321 drivers on the road. [Source: G. Johansson and K. Rumar, "Drivers' Brake Reaction Times," *Human Factors*, Vol. 13, No. 1 (1971), p. 26.]

For a smaller group of drivers they gathered data on the stimulus coming as a surprise and also on the stimulus being anticipated. Table 3.3 shows the results. Every driver had longer reaction time when the signal occurred completely unexpectedly. The range of times also was smaller for the anticipated signal, although not as much smaller as might have been expected.

Using these results, Johannson and Rumar calculated a "correction factor" which they then applied to the data shown in Figure 3.4 in order to estimate how much longer these reaction times would have been if the need for braking had occurred unexpectedly. This resulted in an estimated brake reaction time of 0.9 sec or *longer* in in 50 percent of the drivers. For 10 percent of the drivers the time would have been 1.5 sec or longer and in a few cases it would have exceeded 2 sec. These authors state

[20] Los Angeles City Traffic Department, Florence Avenue Study, Informal Staff Report, March 1969, 5 pp.

[21] G. JOHANSSON and K. RUMAR, "Drivers' Brake Reaction Times," *Human Factors*, Vol. 13, No. 1 (1971), pp. 22–27.

TABLE 3.3
Brake Reaction Time for Surprise and Anticipated Situation

	Observations		Median		Range	
	s	a	s	a	s	a
Subject						
A	10	10	0.88	0.6	0.7–1.1	0.5–0.7
B	10	10	0.6	0.5	0.6–1.0	0.5–0.8
C	10	10	0.9	0.55	0.7–1.0	0.5–0.8
D	10	10	0.7	0.55	0.6–0.7	0.5–0.6
E	10	10	0.6	0.5	0.5–0.9	0.4–0.6
Total (mean)	50	50	0.73	0.54	0.5–1.1	0.4–0.8

s = surprise braking situation
a = anticipated braking situation

Source: G. Johansson and K. Rumar, "Driver's Brake Reaction Times," *Human Factors*, Vol. 12, No. 1 (1971), p. 99.

in conclusion:

> The distribution of brake reaction time obtained here is caused by interindividual differences (between drivers). It is not improbable to assume that the intraindividual (within drivers) distribution over various occasions could have the same form and range.[22]

Hulbert and Beers[23] in a study of driver responses to "wrong way" signs found that fully 5 percent of healthy, alert young drivers utterly and completely failed to see two sets of large red signs when they unexpectedly came upon them.

The role of expectation in human perception cannot be overemphasized. Indeed, one might even conclude that a major task of traffic engineering is to use all devices and techniques available in order to shape and manipulate driver expectation so that there will be as few surprises as possible. Design and construction errors create many unfortunate surprises that traffic engineers must try to render harmless by the use of warning devices, delineation, channelization, stop signs, and other techniques; for example, in some regions, signs have been placed warning "Dangerous Intersection." This it would seem is a very general alerting device and may be the only appropriate one in which a variety of "surprises" lie in wait.

SHARED TASKS

If man's single central processing channel is occupied with one kind of information, it cannot be utilized to process additional information. Therefore, time spent reading a highway sign or looking at a pedestrian is time taken away from the tracking task. For this reason designers of roadways and vehicles have striven to make the tracking task easier, and after many years of trial and error traffic control devices have been designed to be relatively unambiguous information generators, demanding a minimum of attention.

---

[22] *Ibid.*
[23] SLADE HULBERT and JINX BEERS, "Wrong-Way Driving: Off-Ramp Studies," *Highway Research Record No. 122*, 1966, pp. 35–49.

DRIVING TASK MODEL

Vanstrum and Caples have extended the concept of a spatial commitment projected fan-shaped ahead to describe their model of driver perception and how it relates to hazards on the road.[24] Figure 3.5 shows this zone of committed motion ahead divided into four segments or bands. Band 1 represents distance traveled during minimum perception time; band 2 distance traveled during minimum decision time; band 3 distance traveled during minimum reaction time; and band 4 the *minimum* committed motion area of the vehicle after activation has been made to turn or stop. Zone 4 out to arc *S* represents the minimum stopping distance for the vehicle based on vehicle speed and weight, brake efficiency, and coefficient of friction between tire and road should the driver choose to brake. On the right is a hazard of some sort as designated by the box marked *X*. This could be many things: a stalled vehicle, a pedestrian, debris on the road, an oncoming car. As a special case, it can also be considered a *potential hazard*, for example, an intersection, a curve, a car ahead just starting to slow down, a railroad crossing, or even the edge of the road.

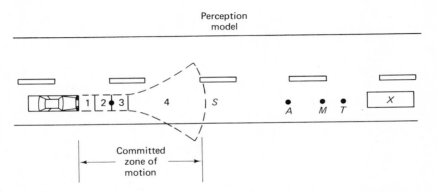

Figure 3.5. Conceptual model of driver perception. (Source: R. C. Vanstrum and B. G. Caples, "Perception Model for Describing and Dealing with Driver Involvement in Highway Accidents," *Improvement of Transportation Safety*, Highway Research Record No. 365 p. 19.)

T is designated as the *true point*, the last point at which action can be initiated in order to avoid the hazard. It is a point of no return and is determined by the zone of committed motion and the laws of physics. Action initiated after point *T* may help for injury reduction but will not be effective in avoiding the accident completely. Point *M* is termed the *mental point* which is the driver's perception of the true point *T*. It is where the driver believes the point of no return is. Point *A* is the *action point* or where the driver decides he actually will take action. The action involves slowing, stopping, steering, or accelerating. It must be noted that *M* and *A* are shown as points on the roadway for simplicity's sake, whereas, they are probably perceived by the driver more as areas. Also, the model represents just one moment in time and in the dynamic

24 R.C. VANSTRUM and B.G. CAPLES, "Perception Model for Describing and Dealing with Driver Involvement in Highway Accidents", *Improvement of Transportation Safety*, Highway Research Board, Record 365, Washington, D.C., 1971, pp. 17–24.

situation, the various points, the committed zone of motion, and the driver's perception of the relationships are changing from second to second.

Figure 3.6 is used by Vanstrum and Caples to discuss the concept of margin for error as follows:

> The distance *TM* is termed perceptual error. The driver's mental point *M* can be ahead or behind the true point *T* and if no perceptual error is involved, it coincides with *T* and *TM* is zero. But generally, *TM* is a plus or minus quantity; *M* can be either side of point *T*.[25]

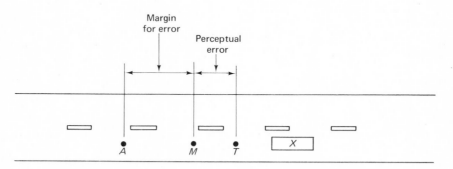

Figure 3.6. Perceptual error in driving task. (Source: Vanstrum and Caples, "Perception Model for Describing and Dealing with Driver Involvement in Highway Accidents," p. 20.)

In their model, the distance *AM*, or the difference between *M* the mental point and *A* the action point, is the driver's margin for error. This quantity is usually plus going from *M* to *A*, taking the direction from the hazard back to the driver as the positive direction. It is only minus when a driver deliberately tries to ram into something. A driver consciously trying to commit suicide would have a negative *AM*. For most drivers, however, *AM* is positive. In other words, the driver places *A*, his action point, ahead of *M*, his mental point of no return. He allows some margin for error.

The authors go on to show in Figure 3.7 that the interaction between *TM*, perceptual error, and *AM*, margin for error, determines whether or not an accident results. It determines whether point *A* is toward the driver from point *T* where no accident results or whether point *A* is on the other side of point *T* away from the driver in which case an accident does occur.

In the upper left, both *TM* and *AM* are positive, producing a safe situation. In the upper right, a larger *AM* compensates for a negative *TM*, producing a safe situation where point *A* comes before point *T*.

The unsafe situations are depicted in the middle portion of the figure. In the middle left, the *AM* does not compensate for a large *TM*, and an accident results.

There are many special cases for the unsafe condition. The middle right part of Figure 3.7 shows failure to set up points *M* and *A* entirely, or until after point *T* is reached, and it is a perceptual error. Failure to treat potential hazards as real hazards, resulting in failure to set up potential points *A* and *M*, is a perceptual error that results in a driving error that may be benign only if the driver is fortunate. Speeding through an intersection and passing on a curve of hill are indeed errors waiting to turn into collisions.

---

[25] *Ibid.*

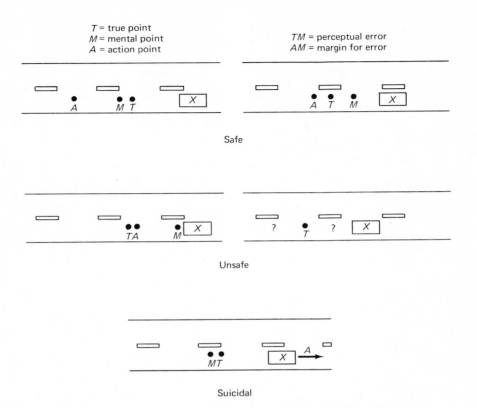

Figure 3.7. Perceptual model related to accidents. (Source: Vanstrum and Caples, "Perception Model for Describing and Dealing with Driver Involvement in Highway Accidents," p. 20.)

WILLFUL VS. INADVERTENT ERROR

The distinction made earlier in this chapter between willful vs. inadvertent errors is important here for the traffic engineer and highway designer to bear in mind. All too often, after the collision occurs it is impossible to determine which kind of error took place. Traffic control devices of all kinds can be considered as efforts to increase the AM and decrease the TM for all drivers. The importance of uniformity of traffic control devices and uniformity in their use and placement becomes clearly evident if AMs and TMs are to be influenced successfully.

This perceptual model will be referred to later in this chapter when drug effects and other unfortunate changes in perception take place. At this time, however, similar models will be used to describe some aspects of drivers' visual attention capability.

VISUAL ATTENTION

Hulbert and Burg discuss visibility of vehicles that are on a collision course with each other.[26] Figure 3.8 shows that as these vehicles approach the point of collision, they don't change their bearing to each other. Visually, this means that they remain

---

[26] HULBERT and BURG, "Human Factors in Transportation Systems."

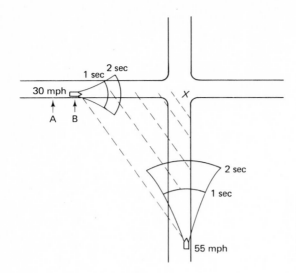

Figure 3.8. Committed zones for two vehicles approaching an intersection. [Source: Hulbert and Burg, "Human Factors in Transportation Systems," p. 485. *Systems Psychology*, Kenyon B. De Greene, ed. (N.Y: McGraw Hill Book Co., 1970) p. 485.]

stationary in each other's field of view. If one or both of these nonmoving images is hidden by the vehicle corner post (or other obstruction), it will remain there. Some experienced drivers take this into account in some situations and vigorously move their heads to see around the blind spot. But unsuspecting drivers who are on a collision course unfortunately, because of the geometry of the situation, have lost one of the major cues (namely, an object moving in their field of view) just at a time when they most need it.

The second visual attention factor in the driving task is the way in which the driver's eyes function. When the head and eyes are moved from one object to another (at different places in the visual field), an involuntary blink often occurs that blocks out what otherwise would be a streaming or blurred image (just as when a moving picture camera is panned too fast). This wonderfully timed blink is so natural that the driver is not aware that often as he transfers his gaze, the blink actually blanks out the visual scene that lies between the two points of visual attention. This is important for the placement of signs or signals particularly where some vehicle turning movements are being made. Figure 3.9 shows how a driver waiting to make a left turn against oncoming traffic will rapidly shift his gaze from the oncoming stream (when he finds an acceptable gap) over to his projected path as he executes his turn. In so doing, he will swing past a large segment of the visual field while his view is blocked by the involuntary blink. Assuming that the central field of attention is approximately 60°, the diagram shows how the driver can completely fail to see signs that are placed in the crosshatched area which is where many signs are placed or where a pedestrian may be crossing.

Robinson *et al.* studied drivers as they waited to cross a major highway and also as they made lane changes.[27] In the first case (stop and enter), visual search times ranged from 1.1 sec. to 2.6 sec. In the lane change situation, times ranged from 0.8 sec to 1.6 sec for minor use and 0.8 sec to 1.0 sec for "look back" time. The more

---

[27] G. H. ROBINSON *et al.*, "Visual Search by Automobile Drivers," *Human Factors*, Vol. 14, No. 4 (1972), pp. 315–323.

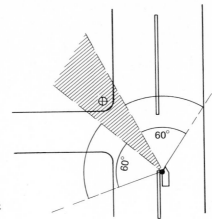

Figure 3.9. Change in visual field for left turn situation.

traffic, the greater number of "looks" and the "looks" tended to be longer (dwell-time increased).

ALCOHOL, DRUGS, AGE, AND FATIGUE

Buttiglieri, Brunse, and Case state: "Society today is a drug taking society."[28] The authors compared a 1959 survey with a 1968 survey and noted a fourfold increase in the percentage of people who reported using tranquilizers. Self-medication practices have increased[29] considerably according to a 1967 report by the California Medical Association, and many surveys describe the current popularity of stimulants, marijuana, and hallucinogens. Still the most widespread tranquilizer is alcohol, the use of which dates to prehistoric times on a nearly worldwide basis.

DRIVER CHARACTERISTICS, ALCOHOL, AND DRUGS

Since alcohol can be relatively easily detected and measured in the blood, breath, and urine, it has been most widely studied in relation to traffic accidents. Buttiglieri *et al.* point out that the effects of alcohol on the human body are much more complex than was first assumed.[30] They point out that although there is a good deal of variation among individuals, for most people a blood alcohol level (BAL) up to 0.05 percent induces some sedation or tranquility. From a 0.05 percent to a 0.15 percent BAL a lack of coordination may be apparent, as well as behavioral changes which seem to suggest stimulation of the brain (such as talkativeness, aggressiveness, and hyperactivity) but which actually result from the depression of brain centers that normally inhibit such behavior. As a driver's ability is impaired by higher BAL's, his self-judgment frequently worsens and he believes that he is performing normally or even better.[31]

---

[28] MATTHEW BUTTIGLIERI, ANTHONY J. BRUNSE, and HARRY W. CASE, "Effects of Alcohol and Drugs on Driving Behaviors," *Human Factors in Highway Traffic Safety Research*, T. W. Forbes, ed. (New York: John Wiley and Sons, Inc., 1972), p. 303.

[29] California Medical Association, "Self-Medication Practices," *California Medicine*, Vol. 107, No. 5 (November 1967), pp. 452–454.

[30] BUTTIGLIERI *et al.*, "Effects of Alcohol and Drugs on Driving Behaviors," p. 307.

[31] *Ibid.*

To the extent that traffic engineering techniques can provide an improved means for letting the drunk (or otherwise impaired) driver know that his performance is poor, this unfortunate effect of alcohol perhaps can be offset. In this respect, raised pavement markers, for example, not only provide a stronger visual cue that is more easily processed by the driver but also provide a rumble effect that adds another cue when tracking performance exceeds an error limit. This benefit of raised pavement markers has been praised by drivers but not documented by research.

In a traffic safety monograph by the National Safety Council it was stated that approximately 90 to 95 million persons in the United States drink alcoholic beverages at least occasionally.[32] It has been estimated that 80 percent of men and 67 percent of women over the age of 21 drink. How often they drive while drinking is not precisely known. However, it is estimated by the National Safety Council that alcohol plays a role in at least 1 million of the 16 million crashes annually,[33] and field studies have conclusively shown that the blood alcohol levels of drivers who were in serious or fatal crashes differ greatly from those drivers not involved in crashes.[34,35,36,37] In these studies only 1 percent to 4 percent of non-accident drivers were found with a BAL at or above 0.10 percent while from 48 percent to 57 percent of fatally injured drivers in one-car crashes were found to have BALs in excess of 0.10 percent. In multiple car crashes, similarly high BALs were found in 45 percent of fatally injured drivers.[38,39] Zylman has critically reviewed these studies and performed further analysis of the data which indicate the important role of traffic density, time of day, type of driver, and trip purpose.[40] For example, women were 26 percent of the control sample between noon and 3:00 p.m. but only 6 percent of the sample from 3:00 a.m. to 6:00 a.m. He speaks of the importance of drinking and driving experience such that those who are experienced will be safer than beginners.

Drivers with BALs of 0.10 percent will not usually show any marked outward evidence of impaired driving capability. This was clearly revealed in initial research at UCLA ITTE where drivers were intoxicated and then performance measured in the UCLA Driving Simulator.[41] It wasn't until a secondary visual task was added that the evidence of alcohol effects became clear. The underlying concept put forth by Moskowitz is that driving is a task that requires a division of attention.[42] In other

---

[32] National Safety Council, "On the Level," Traffic Safety Monograph No. 1, Chicago, Ill., 1969.

[33] National Safety Council, *Accident Facts*, 1972.

[34] R. L. HOLCOMB, "Alcohol in Relation to Traffic Accidents," *Journal of American Medical Association*, Vol. 111, No. 12 (Sept. 17, 1938), pp. 1076–1085.

[35] G. H. LUCAS, W. KALOW, J. D. McCOLL, B. A. GRIFFITH, and H. W. SMITH, "Quantitative Studies of the Relationship Between Alcohol Levels and Motor Vehicle Accidents," *Proceedings of the 2nd International Conference on Alcohol and Road Traffic*, Toronto, 1955, pp. 139–142.

[36] J. R. McCARROLL and W. HADDON, "A Controlled Study of Fatal Automobile Accidents in New York City," *Journal of Chronic Diseases*, Vol. 15, No. 8 (1962), pp. 811–826.

[37] R. F. BORKENSTEIN, R. F. CROWTHER, R. P. SHUMATE, B. W. ZIEL, and R. ZYLMAN, *The Role of the Drinking Driver in Traffic Accidents* (Indianapolis, Ind.: Indiana University Police Institute, February 1964).

[38] R. A. NIELSON, "*Alcohol Involvement in Fatal Motor Vehicle Accidents*," (San Francisco: California Traffic Safety Foundation, September 1965).

[39] R. A. NIELSON, "A Survey of Post-Mortem Blood-Alcohols from 41 California Counties in 1966," (San Francisco: California Traffic Safety Foundation, April 1967).

[40] RICHARD ZYLMAN, "Analysis of Studies Comparing Collision-Involved Drivers and Non-Involved Drivers," *Journal of Safety Research*, Vol. 3, No. 3 (September 1971), pp. 116–28.

[41] H. W. CASE, S. HULBERT, and H. A. MOSKOWITZ, *Alcohol Level and Driving Performance*, Institute of Transportation and Traffic Engineering Report No. 71–17, University of California, Los Angeles, April 1971.

[42] *Ibid.*

words, the single-track mental system described earlier in this chapter is used by the alert driver to sample the driving environment both outside and inside the vehicle and to look for cues that will enable him to correctly predict and anticipate what lies ahead.

The divided attention concept of why alcohol increases accident likelihood explains why simple reaction time may not be affected or may even be improved. Alcohol apparently narrows the field of attention which can actually improve the ability to respond to a simple and expected change in the environment. The UCLA work is showing that this holds true for auditory stimuli as well as for visual[43] which is evidence that the behavioral impairment takes place in the central nervous system and in particular reduces the driver's information handling capability. This concept helps to explain why visual acuity is not affected by BALs of 0.10 percent.

When the driving task description presented earlier in this chapter is considered in terms of the impairment in mental processing and environment sampling rate decrease caused by alcohol, it is readily understood how drunk drivers can fail to perform safely (see Figure 3.10). They can fail by completely "not seeing" obstacles or other vehicles because their visual scanning rate is simply too slow. They can, and do, fluctuate speed greatly and erratically because their rate of speed monitoring is too slow to detect speed changes as efficiently as normal. Their steering performance may not vary greatly but it can demand nearly all of their limited attention whereas normally (sober) they need devote only a fraction of attention to steering and have a great deal of attention available to devote to detection and processing of other cues from the environment.

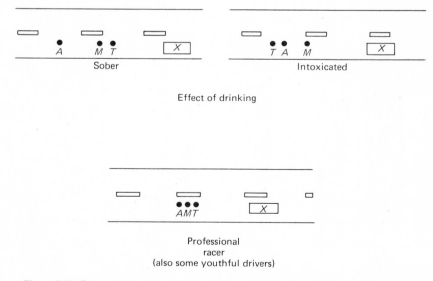

Effect of drinking

Professional
racer
(also some youthful drivers)

Figure 3.10. Perceptual model related to driving while intoxicated. (Source: Vanstrum and Caples, "Perception Model for Describing and Dealing with Driver Involvement in Highway Accidents," p. 21.)

---

[43] H. Moskowitz and D. DePry, "The Effect of Alcohol Upon Auditory Vigilance and Divided Attention Tasks," *Quarterly Journal of the Study of Alcohol*, Vol. 29, No. 1 (1968), pp. 54–63.

Similar effects on human performance are likely to occur as a result of fatigue, mental stress, age, and most certainly other drugs. It has been found that some tranquilizers interact with alcohol to produce a combined effect that is greater than the sum of their individual effects. Barbiturates in combination with alcohol can be lethal and even in small amounts can result in loss of consciousness. These and other unfortunate combination effects of medication are particularly dangerous and difficult to detect. Their frequency of use and role in traffic crashes is unknown but police officials do report many instances of drivers' unknowingly falling victim to these combination effects.

The widespread use of marijuana has focused attention on its possible effects on driving skills. Survey results have indicated that marijuana users receive more traffic tickets than do non-users.[44,45] Similar results have been derived from the traffic records of persons arrested for marijuana use, although the accident rate was not above average.[46] Of course, these findings are simply correlates of marijuana use and do not indicate a causal relationship. The user's own assessment of the effect of marijuana intoxication on driving performance is apparently related to age-related involvement in the current marijuana controversy—17 percent of a sample of student and other young marijuana users felt that their driving was impaired by the drug in comparison to 72 percent of a sample who began using marijuana some 20 years ago.[47]

One study compared the effects of alcohol (1.2 g/kg body weight) and smoked marijuana (22 mg THC) on driving simulator performance.[48] The alcohol dose significantly impaired simulator scores while the marijuana treatment produced minimal changes. Moskowitz *et al.* have examined the effect of marijuana on attentional aspects of driving, i.e., the ability to attend to peripheral cues while carrying out central tracking tasks.[49] Smoked marijuana containing 15 mg THC significantly impaired this function in laboratory tests of both the visual and auditory modalities. The extent of decrement was approximately equivalent to that produced by a blood-alcohol level of about 0.07 percent, i.e., the consumption of about 5 ounces of 80 proof liquor in less than one hour.

Recent unpublished results of Moskowitz's work at UCLA indicate impairment caused by marijuana is different in nature from that caused by alcohol. Peripheral attention and vision are affected differently and perhaps more seriously.[50]

AGE AND DRIVING

Age clearly changes driving behavior as well as other behavior. One study found that older drivers (51 years or more) had far more difficulty performing a simultane-

[44] J. L. HOCHMAN and N. Q. BRILL, *Marijuana Use and Psychosocial Adaption.* Unpublished paper delivered at American Psychiatric Association meeting in Washington, D.C., May, 1971.

[45] B. D. JOHNSON, *Social Determinants of the Use of "Dangerous Drugs" by College Students.* Doctoral Thesis, Columbia University, June, 1971.

[46] A. CRANCER and D. L. QUIRING, "Driving Records of Persons Arrested for Illegal Drug Use," *Police*, in press.

[47] W. H. McGLOTHLIN, *et al.*, "Marijuana Use Among Adults," *Psychiatry*, Vol. 33, No. 4 (1970), pp. 433–43.

[48] A. CRANCER, *et al.*, "Comparison of the Effects of Marijuana and Alcohol on Simulated Driving Performance," *Science*, Vol. 164, No. 3881 (1969), pp. 851–54.

[49] H. MOSKOWITZ, S. HULBERT, and W. H., McGLOTHLIN, *The Effects of Marijuana Upon Performance in a Driving Simulator*, unpublished paper, Institute of Transportation and Traffic Engineering, University of California, Los Angeles, 1973.

[50] *Ibid.*

ous sign reading and steering task on a simple driving simulator device.[51] They also drove more erratically in city traffic as well as in the UCLA driving simulator which is 31 miles of free moving through rural roads, arterials, and residential roadways. Although not enough data have yet been collected, there is strong evidence that age will prove to interact with alcohol effects on divided attention tasks. There also may be a simple age effect that is to be expected if CNS processing time and ability are affected in the way many researchers believe age related changes do come about.

SLEEP DEPRIVATION

Falling asleep at the wheel is a problem of major proportion in the cause of traffic accidents. The exact dimensions of this problem are not known; Hulbert, however, states that available data indicate from 35 percent to 50 percent of highway fatalities may well be caused by this factor.[52,53]

Alcohol, tranquilizers, antihistamines, barbiturates, and other substances promote drowsiness even in small doses for many persons. Driving in itself is soporific for some people. Trip habits and long stretches of superhighway keep drivers longer on the road.

All these factors are interacting to create a major source of driver impairment which is most clearly evident in single-vehicle crashes but is no doubt playing a part in many multiple vehicle crashes.

One research method is to use sleep loss to increase the likelihood of drowsiness at the wheel. Hulbert tested the effect of increasing the blood sugar level on sleep deprived (24 hours) drivers and found overwhelming evidence that increasing the sugar level prevented the drivers from falling asleep in using the UCLA driving simulator.[54]

Fifty percent of drivers in a study by Hulbert and Mellinger reported a tendency to fall asleep at the wheel even when they were not otherwise tired and had normal sleep.[55] Dr. Robert Yoss at the Mayo Clinic has studied drivers and aircraft pilots and concludes that many persons naturally are prone to such drowsiness.[56] He has been successful in correlating this unfortunate and dangerous tendency with a decrease in pupil size while looking into a pupilometer. Eventually, it may be feasible to screen out such drivers for medical treatment. At present, traffic engineers can only be aware that drugged and sleepy motorists need the most clear, unambiguous, and redundant traffic control devices that can feasibly be provided. Certain times and locations may be particularly likely to have these impaired drivers in the traffic stream. Accident record analyses and law enforcement efforts can help traffic engineers to locate such spots and perhaps can lead to effective countermeasures by the

[51] J. BEERS, H. W. CASE, and S. HULBERT, *Driving Ability as Affected by Age*, Institute of Transportation and Traffic Engineering Report No. 70–18, University of California, Los Angeles, March 1970.

[52] SLADE HULBERT, "Effects of Driver Fatigue," in *Human Factors in Highway Traffic Safety Research*, T. W. FORBES ed. (New York: John Wiley and Sons, Inc., 1972).

[53] S. HULBERT (with A. and H. Eisenberg), "Asleep at the Wheel," *Readers Digest*, Vol. 104, No. 619, November 1973, pp. 77–82.

[54] S. HULBERT, J. BEERS, J. HERZOG, and S. BLYDEN, *Blood Sugar Level and Fatigue Effects on a Simulated Driving Task*, Institute of Transportation and Traffic Engineering Report No. 63–53, University of California, Los Angeles, October 1963.

[55] S. HULBERT and R. L. MELLINGER, *Effects of Fatigue on Skills Related to Driving*, School of Engineering Report 70–60, University of California, Los Angeles, January 1970.

[56] D. YOSS, "A Test to Measure Ability to Maintain Alertness and Its Application in Driving," *Mayo Clinic Proceedings*, Vol. 44, No. 11 (1969), pp. 769–83.

community as a whole. Several federally financed trial programs are now underway that traffic engineers could assist both professionally and as private citizens.

PEDESTRIANS

Even though approximately 400,000 pedestrians annually are being struck by vehicles (resulting in about 10,000 fatalities),[57] they are being killed at consistently decreasing rates per vehicle miles. This is largely attributable to improved roadways, more sidewalks, and special law enforcement efforts.

Many pedestrians are killed in urban areas; in our largest cities they account for one-half of motor vehicle related deaths and one-third in middle-sized cities. For 18 cities the daily fatality rates reach as high as 11 per day per city.[58]

CHILDREN

One out of ten deaths of children between the ages of five and fourteen is a pedestrian traffic fatality. Child pedestrians receive most traffic engineering attention near school grounds, but the safety record at school crossings is excellent. Snyder found only 2 percent of 2,146 pedestrian accidents at school areas.[59] Small children are particularly vulnerable pedestrians because they are more easily hidden from the driver's view; and, conversely, from their lowly eye position there are more visual obstacles than oncoming vehicles. A 1968 statistic shows that primary-grade children have three times the death and injury rate of children twice that age.[60] In a recent book Sleight points out that only one of every four children killed and one in five injured were on their way to or from school.[61] Far more must be done away from school where children have all the problems of other pedestrians in addition to the use of streets as play areas. It is particularly important to educate children to understand that drivers have difficulty seeing them.

ELDERLY

It is clearly the elderly who are disproportionally the pedestrian victims. "About two out of three (total motor vehicle) deaths in 1971 occurred in places classified as rural. In urban areas, nearly two out of five of the victims were pedestrians; in rural areas, the victims were mostly occupants of motor vehicles."[62] The elderly are particularly vulnerable probably because of decreases in their abilities to perceive oncoming cars and because of their lessened agility and speed of movement to dodge or cross the roadway quickly. Snyder found higher fatality rates for older pedestrians.[63] Sleight cites a study in St. Petersburg, Florida, where reside three times the national average of persons over 65. The study shows that the elderly experienced nearly 70

---

[57] DIANE CHRAZANOWSKI ROBERTS, "Pedestrian Needs—Insights from a Pilot Survey of Blind and Deaf Individuals," *Public Roads*, Vol. 37, No. 1 (June 1972), pp. 29–31.

[58] ROBERT B. SLEIGHT, "The Pedestrian," *Human Factors in Highway Traffic Safety Research*, T. W. FORBES, ed. (New York: Wiley-Interscience, 1972), pp. 224–53.

[59] MONROE B. SNYDER and RICHARD L. KNOBLAUCH, *Pedestrian Safety: The Identification of Precipitating Factors and Possible Countermeasures* (Silver Spring, Maryland: Operations Research, Inc., January 1971).

[60] SLEIGHT, "The Pedestrian," p. 228.

[61] *Ibid.*, p. 229.

[62] *Accident Facts*, 1972, p. 41.

[63] SNYDER, "Pedestrian Safety."

percent of deaths and one-half the traffic injuries although they made up less than 30 percent of the population.[64]

PEDESTRIAN WALKING RATES

Walking rates are reported by Sleight from several studies where the average adult and elderly moved at approximately 4.5 ft per sec (1.4 m per sec). Children moved more rapidly at approximately 5.3 ft per sec (1.6 m per sec).[65] Some engineers use a 4.0 ft per sec rate (1.2 m per sec) but for the relatively slow walkers speeds of from 3.0 to 3.25 ft per sec (0.9 to 1.0 m per sec) would be more appropriate. Weiner found an average rate of 4.22 ft per sec (1.29 m per sec) however, and for women only 3.70 ft per sec (1.13 m per sec). When groups of pedestrians walked together, the male rate dropped to 3.83 (1.17 m per sec) and to 3.63 ft per sec (1.11 m per sec) for women.[66] Figure 3.11 shows these rates.

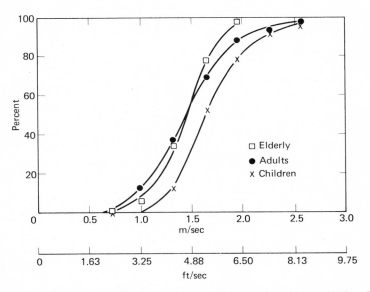

Figure 3.11. Typical speed of pedestrian movement at crossings. [Source: Robert B. Sleight, "The Pedestrian" *Human Factors in Highway Traffic Safety Research*, T. W. Forbes, ed. (New York: Wiley-Interscience, 1972); p. 235.]

PEDESTRIAN GAP ACCEPTANCE

Pedestrian habits while crossing roadways have been studied and reported by Jacobs in which the so-called threshold gap (defined as the gap accepted by 50 percent of pedestrians) for 20 mph traffic is 84 ft.[67] The distribution of gaps is shown in Figure 3.12.

---

[64] SLEIGHT, "The Pedestrian," pp. 229–30.

[65] *Ibid.*, p. 234.

[66] E. L. WEINER, "The Elderly Pedestrian: Response to an Enforcement Campaign," *Traffic Safety Research Review*, Vol. 12, No. 4 (1968).

[67] G. D. JACOBS, *The Effect of Vehicle Lighting on Pedestrian Movement in Well-Lighted Streets*, RRL Report LR214, Road Research Laboratory, Crowthorne, England, 1968.

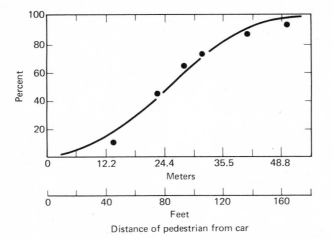

Figure 3.12. Percent of pedestrians accepting gaps of given size in crossing. (After Jacobs, 1968.) (Source: Robert B. Sleight, "The Pedestrian," p. 237.)

Snyder found a high incidence of pedestrian accidents in the center city outside the CBD.[68] "Such areas have been characterized as crowded, high-crime slums and ghettos."[69] It has been observed that large-city pedestrians at signalized intersections seem to respond more to gaps in traffic than to traffic signals. Police officers will even "turn away" when pedestrians cross narrow side streets against the light when no traffic is coming. This tacit recognition of human judgment is akin to using the 85 percentile value for setting roadway speed limits.

PEDESTRIAN VOLUME AND DENSITY

Pedestrian volume is defined as the number of persons passing a given point in a unit of time. Pedestrian density can either be expressed in tenths of a pedestrian per square foot or its reciprocal, the number of square feet of area per pedestrian.[70] Since it is easier to visualize the latter, it will be used here. Needless to say, pedestrian volume and density are interrelated.

Normal free walking speed increases as more area becomes available to pedestrians, i.e., as the density decreases. However, flow volume increases as the area per pedestrian decreases, until a critical point is reached at which movement is highly restricted because of lack of space.[71] Figure 3.13 shows this critical point for three categories of pedestrian traffic. Studies of pedestrian flows concluded that flows of 20 pedestrians per foot width per minute (PFM) were possible under a wide variety of conditions. Twenty-five PFM was attainable under favorable conditions and as high as 30 PFM could be attained under highly favorable conditions.[72]

Figure 3.14 is a graphic representation of longitudinal and lateral spacings of pedestrians. As density increases pedestrians maintain constrained spacing patterns

---

[68] SNYDER, "Pedestrian Safety." p. 4–2.

[69] *Ibid.*

[70] JOHN J. FRUIN, *Pedestrian Planning and Design,* (New York: Metropolitan Association of Urban Designers and Environmental Planners, Inc., 1971), p. 38.

[71] *Ibid.* p. 43.

[72] *Ibid.,* p. 45.

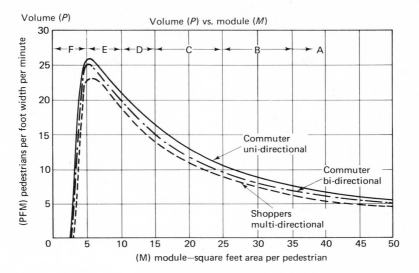

Figure 3.13. Level of service standards for walkways. (Source: John J. Fruin, *Pedestrian Planning and Design*, Metropolitan Association of Urban Designers and Environmental Planners, Inc., New York, 1971, p. 78.)

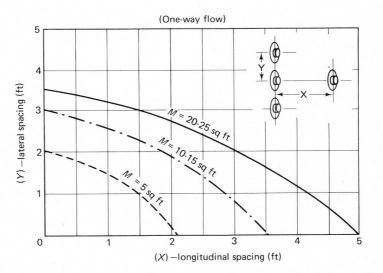

Figure 3.14. Average longitudinal and lateral spacing of pedestrians in a traffic stream. (Source: Fruin, *Pedestrian Planning and Design*, p. 48.)

in order to avoid brushing against others and to allow for pacing room. Spacing measurements suggest that average pedestrian areas greater than 25 sq ft per person are required before there is enough lateral space to freely bypass slower moving pedestrians.[73] Table 3.4 shows the distribution patterns for midtown Manhattan as an example of urban pedestrian traffic flow.

---

[73] *Ibid.*, p. 47.

OBSERVATION OF PEDESTRIAN REGULATIONS

Sleight cites a safety campaign to reduce jaywalking in a city known to have a large proportion of elderly pedestrians.[74] The before and after percentage of legal crossings improved from 61 percent to 73 percent when police were stationed there, but it did not increase when there were no police. At the height of the campaign, however, the legal crossings jumped to 93 percent when police were present and 75 percent when police were absent.[75] Apparently, the effects of campaigns and crackdowns are short-lived and directly related to the presence of police.

PEDESTRIAN ACCIDENT CHARACTERISTICS

Snyder found that an unusually high proportion of urban pedestrian accidents occurred in the afternoon.[76] About half of the 2,146 pedestrian accidents occurred in residential areas, 7 percent in mixed commercial-residential and 40 percent in primarily commercial areas.[77] Only 2 percent occurred in school areas. About half the accidents studied occurred at or near intersections. Traffic flow was generally normal (56%) or light (27%). Most accidents (78%) occurred on two-way streets. Crossing distances were generally not too great—57% were less than 40 ft. Average speed of vehicles during times that accidents occurred was not high (95% were under 35 mph and 58 percent were under 25 mph).[78] Hazlett and Allen point out that a three-year study of twelve U.S. cities having 500,000 or more population found that 50 percent of nighttime traffic fatalities were pedestrian deaths.[79] They report a "critical visibility distance" associated with vehicle speed.[80] This can be related to the "committed zone" described earlier in this chapter. Their data indicate that unless pedestrians wear light-colored clothing, they will not be revealed by automobile headlights in time for the motorist to avoid striking them. Hazlett and Allen also studied reflectorized material and found that this greatly increased the visibility distance of simulated pedestrians.[81]

PEDESTRIANS AND ALCOHOL

Snyder found alcohol present in only 4 percent of the 2,146 urban pedestrian accidents studied.[82] This does not necessarily conflict with Haddon's study of Westchester County pedestrian fatalities. In this study blood samples were used.[83] Haddon clearly described the role of alcohol when it is present, in greatly increasing pedestrian fatalities. He found that those pedestrians killed were several times more likely to have high (0.10 percent) blood alcohol than a random sample of pedestrians taken at

---

[74] SLEIGHT, "The Pedestrian," p. 233.
[75] WEINER, "The Elderly Pedestrian."
[76] SNYDER, "Pedestrian Safety," p. 4–2.
[77] Ibid., p. 4–4.
[78] Ibid., p. 4–4.
[79] R. D. HAZLETT and J. J. ALLEN, "The Ability to See a Pedestrian at Night: The Effects of Clothing, Reflectorization and Driver Intoxication," American Journal of Optometry and Archives of the American Academy of Optometry, Vol. 45, No. 4 (1968), pp. 246–258.
[80] Ibid., pp. 246–258.
[81] Ibid., pp. 246–258.
[82] SNYDER, "Pedestrian Safety." p. 4–4.
[83] W. HADDON and V. A. BRADESS, "Alcohol in the Single Vehicle Accident—Experience of Westchester County, New York," Journal of the American Medical Association, Vol. 14 (1959), pp. 127–133.

TABLE 3.4
Estimated Hourly Distribution of Person-Travel in
Midtown Manhattan by Surface Modes

Time	Percent of Daily Vehicular Flow*	Percent of Daily Pedestrian Flow†	Estimated Hourly PMT, Auto and Taxi Passengers‡	Estimated Hourly PMT, Bus Passengers§	Estimated Hourly PMT, Pedestrians	Pedestrian PMT as Percent of Total
4–5 a.m.	1.10	(0.03)	4,900	496	(288)	—‖
5–6	1.06	(0.03)	4,700	1,333	(288)	—‖
6–7	1.77	0.45	9,200	4,216	4,320	24
7–8	3.56	1.85	20,400	15,097	17,760	33
8–9	5.50	7.61	32,100	31,744	73,056	53
9–10	6.01	6.45	32,600	21,607	61,920	53
10–11	5.80	4.60	31,400	14,818	44,160	49
10–12	5.75	5.98	30,900	16,647	57,408	55
12–1 p.m.	5.52	11.76	29,700	17,763	112,896	70
1–2	5.50	11.90	29,900	17,980	114,240	70
2–3	5.75	8.38	31,300	19,809	80,448	61
3–4	5.86	6.60	32,400	22,010	63,360	54
4–5	6.17	8.05	36,400	26,691	77,280	55
5–6	6.39	11.36	39,600	31,372	109,056	61
6–7	6.03	5.32	38,300	22,258	51,072	45
7–8	5.12	3.09	33,500	12,927	29,664	39
8–9	4.54	2.27	29,700	9,362	21,792	36
9–10	3.86	(1.42)	25,100	6,851	(13,632)	—‖
10–11	3.81	(1.02)	25,000	6,541	(9,792)	—‖
11–12	3.92	(0.90)	25,400	5,673	(8,640)	—‖
12–1 a.m.	2.71	(0.50)	15,900	2,573	(4,800)	—‖
1–2	1.86	(0.30)	9,800	992	(2,880)	—‖
2–3	1.35	(0.10)	6,800	775	(960)	—‖
3–4	1.06	(0.03)	5,000	465	(288)	—‖
24 hr	100.00	100.00	580,000	310,000	960,000	1,850,000
Percent by each mode			31%	17%	52%	100%

*Based on NYC Dept. of Traffic counts.
†Based on averages from daily peaking, patterns extrapolated for night-time hours.
‡VMT (vehicle miles of travel) reduced by percent trucks and buses by hour, multiplied by occupancy counts by hour.
§PMT (person miles of travel) for period 1–3 PM expanded by hour in relation to peaking pattern of bus travel across 61 st street.
‖No percentages listed because pedestrian estimates before 6 AM and after 9 PM are not reliable.

Source: Boris S. Pushkarev and Jeffrey M. Zupan, "Walking Space for Urban Centers," unpublished paper, *Regional Plan Association,* (New York, Sept. 1971), p. 111.

the same location and time of day and day of week.[84] The same factors discussed earlier that caused alcohol and other drugs to reduce motorists' capabilities are active in pedestrian behavior.

PEDESTRIAN INJURIES

Injuries to pedestrians struck by vehicles relate directly to the profile of the vehicle and the size of the pedestrian. Some 89 percent of the pedestrians fatally struck by heavy trucks were killed by being run over by the wheels, while only 10 percent of

---

[84] *Ibid.*

those killed by passenger vehicles were run over by the wheels.[85] Modern styling of passenger vehicles results in adult pedestrians being tossed into the air when struck. Children, however, are more likely to be struck down and run over or propelled horizontally.[86] Studies by Robertson,[87] Severy,[88] and McKay[89] all find that many injuries are caused by pedestrians striking the pavement and the studies suggest that vehicles be constructed of more yielding materials that would absorb the striking energy and perhaps even carry the pedestrian along until the vehicle can be stopped.

SIDEWALKS AND UNDERPASSES

Traffic engineers are often pressured by the public for sidewalks and underpasses. Recently, overpasses have been provided largely for children en route to school.

Underpasses and overpasses should provide ramps for walking bicycles, but they should discourage bicycle riding because it is hazardous for pedestrians. Spiral ramps seem to do this effectively. Spiral ramps are also a help to the elderly who have difficulty climbing steps. If underpasses are used, they must be well lighted, cleaned, and even patrolled in areas where there are likely to be thieves and indolents. Overpasses may create nuisances and hazards because of pedestrians who throw objects at cars passing under. Fencing is sometimes required to combat this unfortunate aspect of pedestrian behavior.

PEDESTRIAN ACCIDENT COUNTERMEASURES

An important recent study by Snyder and Knoblauch on over 2,000 pedestrian accidents in 13 major cities led to the following list of the 5 most frequently noted types of accidents:[90]

Dart-Out (*First Half*) (24 percent). A pedestrian, not in an intersection crosswalk, appears suddenly from the roadside.

Dart-Out (*Second Half*) (9 percent). This is the same as the dart-out described for the first half above, except that the pedestrian covers half of a normal crossing before being struck.

Intersection Dash (8 percent). This category covers cases similar to dart-outs with regard to pedestrian exposure to view, but the incident occurs in or near a marked or unmarked crosswalk at an intersection.

Multiple Threat (3 percent). The pedestrian is struck by car *x* after other cars blocking the vision of car *x* stopped in other lanes going in the same direction, and avoided hitting the pedestrian.

Vehicle Turn Merge with Attention Conflict (7 percent). The driver is turning into or merging with traffic; the situation is such that he attends to auto traffic in one direction and hits the pedestrian who is in a different direction from his attention.

[85] J. S. ROBERTSON, A. J. McLEAN, and G. A. RYAN, *Traffic Accidents in Adelaide, South Australia, Summary 1963–64*. Australian Road Research Board Report No. 1, 1967.

[86] *Ibid.*

[87] *Ibid.*

[88] D. M. SEVERY, "Auto-Pedestrian Impact Experiments," *The Seventh Strapp Car Crash Conference*, 1963.

[89] G. M. McKAY, "Automobile Design and Pedestrian Safety," *International Road Safety and Traffic Review*, Summer 1965.

[90] SNYDER, "Pedestrian Safety," 1-3–1-4.

Countermeasures were also proposed to cope with these pedestrian accidents as follows:[91]

> *Street Parking Redeployment.* This countermeasure is aimed primarily at the dart-outs but would influence the other two types as well. The objective is to use parking control to remove some of the visual obstruction, provide a partial barrier to physically control the pedestrian course, and increase the likelihood of detection....
>
> Two steps would be taken. First, parking would be removed from one side of the street, probably the left. Second, head-in diagonal parking would replace parallel parking on the right.
>
> In appropriate locations this would accomplish the following. Visual obstructions would be removed from the left side of the road giving the driver increased view and more time to detect and react. The diagonal parking would provide a physical control that would tend to slow down the pedestrian as he ran across the street, but even more important, would angle him into traffic and direct his field of vision more in the direction of the threatening vehicles....
>
> *Meter Post Barrier.* In commercial areas with on-street parking meters, small fences or railings extending out a few feet from either side of the meter post could combine with parked cars to form a barrier to prevent dart-outs....
>
> *Signal Retiming or Modification.* One of the predisposing factors identified for the intersection dash was the inducement to risk-taking coming from the traffic signal. The pedestrian is wrong to cross against the light. He should wait until he has the proper signal, but it is apparent that some will become impatient when they must wait. Countermeasures include:
>
>> Resetting cycles to bring pedestrian waiting time in line with the norm, or lower if other considerations permit.
>> If rush hour volumes do not permit complete retiming, reduce pedestrian waiting periods during non-peak hours. (Two-thirds of intersection dashes occurred before or after the 4:00 p.m. to 6:00 p.m. rush period.)
>> Provide a signal indicating the waiting time remaining to green....

One additional recent possibility is to consider complex signal heads such as those that display "DONT WALK" and "WALK" simultaneously to different zones or portions of the crosswalk in order to reduce the total walk time for each walk period.

ACCIDENT TYPES INVOLVING SALIENT PREDISPOSING FACTORS

Four other accident types involved specific predisposing factors. They account for about 7 percent of the cases and offer possibilities for extreme reductions. The basic descriptions and countermeasure recommendations for each follow:[92]

> *Vendor—Ice Cream Truck* (2 percent). The pedestrian is struck going to or from a vendor in a vehicle on the street. This is usually similar to a dart-out with ice cream trucks being the most frequent attraction. This specific classification was given precedence over dart-out when assigning cases to types. The countermeasure is ice cream truck regulation and visual warning devices.
>
> *Pedestrian Exiting from Vehicle* (1 percent). The pedestrian had been a passenger or driver and is struck as he exits from a vehicle; all vehicles are included. The counter-measures are vehicle exit visual warning devices, regulation of licensed public vehicles, and exit platform design. Parking redeployment would also help.
>
> *Bus Stop Related* (3 percent). This type includes cases in which the location or

---

[91] *Ibid.*, 1-4-1-6.
[92] *Ibid.*, 1-8-1-9.

design of the stop appears to be a major factor in the causation, e.g., the pedestrian crosses in front of the bus standing at a stop on the corner, and the bus blocks the view of cars. It does not include those cases that may be considered as exiting from a vehicle, nor does it include cases in which the stop is only an attraction or distraction. The countermeasure is location of bus stops at the far side of the intersection.

Backing Up (2 percent). The pedestrian is struck by a vehicle which is backing up.

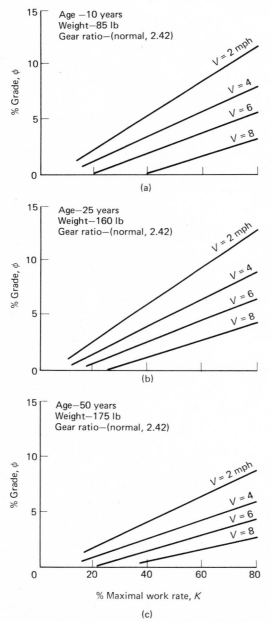

Figure 3.15. Example of grade capability for cyclists. (Source: "Bikeway Planning Criteria Guidelines," Institute of Transportation and Traffic Engineering, University of California, Los Angeles, April 1972, p. B-3.)

A case would not be so classified if the pedestrian were clearly aware of the movement of the vehicle; detection failure is important. This type was used even if the accident occurred off the street. The countermeasure is backup warning devices.

BICYCLISTS AND BICYCLE USE

Relatively little information on bicyclists and their characteristics is available in the literature on transportation engineering and highway operation. A recent study by the Institute of Transportation and Traffic Engineering, University of California, Los Angeles[93] contains data on cyclist dimensions, velocity, and grade capability.

The average dimensions of a bicycle and cyclist pertinent to minimum design standards are:[94]

Handlebar width:	1.96 ft (0.6 m)
Cycle length:	5.75 ft (1.75 m)
Pedal clearance:	0.5 ft (0.15 m)
Vertical space occupied by cycle/cyclist:	7.4 ft (2.5 m)

Mean measured separation between handlebars of two drivers riding abreast at a velocity of 10 mph is given as 2.5 ft.[95]

Measurements of bike speeds in Davis, California, have shown that speeds vary between 7 mph and 15 mph with the average between 10–11 mph.[96]

Figure 3.15 shows the variation of grade as a function of velocity and cycle characteristics. The curves, used for illustrative purposes only, relate the maximal work rate, K, to velocity and percent grade. A value of $K = 0.6$ would be equivalent to playing tennis for over 8 hours with frequent long rests, and a value of $K = 0.4$ would be equivalent to level bicycling for 8 hours with occasional 5–10 min rests.[97] The values are for a single gear ratio of 2.42 and for a male rider on a 35 lb bicycle that has 26 in. diameter wheels.

As an example of the use of such curves, for a design speed of 6 mph and a work level K of 50 percent, the 10-, 24- and 50-year old can tolerate grades of 2.9, 3.5, and 2.2 percent. Similarly, on a 5 percent grade with a 50 percent energy expenditure, speeds of 3.8, 4, and 2 mph can be negotiated by the three cyclists.

VEHICLE DESIGN

DRIVER DIMENSIONS

Versace in a Ford Motor Company report states in considering driver dimensions:

> Published anthropometric data are seldom applicable because they were obtained under standardized postural conditions which were not related to design needs. . . . As a result, a fair amount of on-site validation and checkout is required.[98]

---

[93] "Bikeway Planning Criteria and Guidelines," Institute of Transportation and Traffic Engineering, University of California, Los Angeles, April 1972.
[94] *Ibid.*, p. 22.
[95] *Ibid.*, p. 26.
[96] *Ibid.*, p. 21.
[97] *Ibid.*, p. B-7.
[98] JOHN VERSACE, University of Michigan Short Course in Engineering, unpublished lecture notes, 1967.

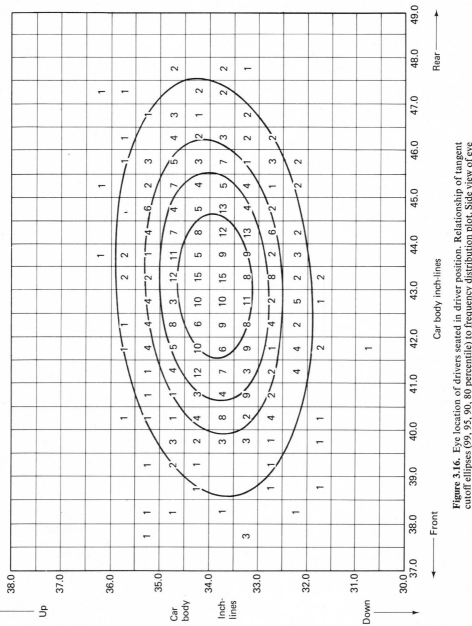

Figure 3.16. Eye location of drivers seated in driver position. Relationship of tangent cutoff ellipses (99, 95, 90, 80 percentile) to frequency distribution plot. Side view of eye position in 1963 Ford convertible referenced to car body inch-lines (population ratio, males only). (Source: John Versace, University of Michigan Short Course in Engineering, unpublished lecture notes, 1967.)

Figure 3.16 shows a scattergram of the locations of the eyes of 2,300 male drivers. Note that the coordinate system is related to standard car body layout lines. Each ellipse is a designated (by percentile) locus of instantaneous tangencies of any straight line drawn through the figure. In this way the vehicle designer is able to draw sighting rays from visual objects or obstacles in the car to the appropriate ellipse and know what proportion of drivers' eyes will be above or below that sighting ray. This is important for traffic engineers in their consideration of the placement of pavement markings and in dealing with vertical curves. Also of interest is the location of the driver's hip on the so-called "H-point" (Figure 3.17). When the automotive designer lays out the interior of a vehicle, he packages according to the location of the H-point.

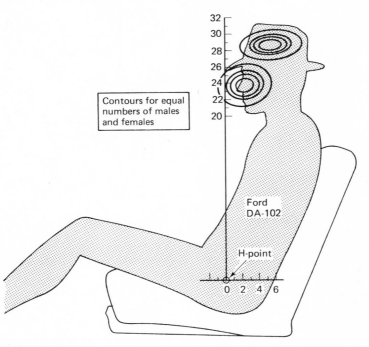

Figure 3.17. Distance from eye and top of head to hip-point. (Source: John Versace, University of Michigan Short Course in Engineering, 1967.)

For many years the U.S. Materiel Command has recognized that the distribution of human dimensions is not stationary in time, that people clearly are getting bigger. Figure 3.18 shows some projections of overall stature as a function of age, for males in 1961 as compared to 1980.

VISION

Discussing lights and vision, Versace points out that "the acceptable glare level of a light depends upon its brightness, size, and distance away."[99] Figure 3.19 shows that

---

[99] *Ibid.*

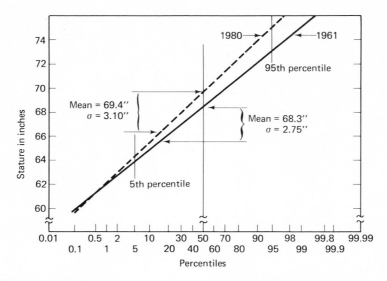

Figure 3.18. Civilian male stature in 1961 projected through 1980, age 15 years and above. (Source: J. Versace, University of Michigan Short Course in Engineering, 1967.)

any light which is bright enough to definitely be seen in the daytime will at some close distance be regarded as objectionable under very dark nighttime conditions. Because the human visual system responds to light approximately as the cube root of intensity, it takes about 10 times as much intensity for a light to appear twice as bright. And to assure positive absolute identification on a brightness scaling basis, intensity ratios greater than that are required. Brake lights are more intense than the taillight by ratios often exceeding 30: 1.

In vehicle lighting and signal light work, consideration has been directed to the effect of color on signal visibility. A signal light viewed at 400 ft, adjacent to an oncoming headlight, is less identifiable if it is amber than if it is red or green. Also to be considered in the color of signal lights are the penetrating quality in fog, the role of colorblindness, and the natural meaning of the color to the observer. Design changes, particularly those involving the signaling system, must be evaluated in terms of possible confusion in comprehension when there is a prolonged period in which both old and new systems are used concurrently on the road.

In a study of vehicle rear lighting Case, Hulbert, and Patterson conclude that rear-end accidents involve the following driver when he is either "coupled" or "uncoupled" to the car ahead.[100] In the first case he is aware of the lead vehicle and is performing what is called "car following." As discussed earlier, very short headways can safely be accomplished, particularly by commuter drivers on freeways, because each driver is aware of the committed zone in front of each moving vehicle. When uncoupled,

---

[100] HARRY W. CASE, SLADE F. HULBERT, and OSCAR E. PATTERSON, *Development and Evaluation of Vehicle Rear Lighting Systems*, Institute of Transportation and Traffic Engineering Report No. 68–24, University of California, Los Angeles, May 1968.

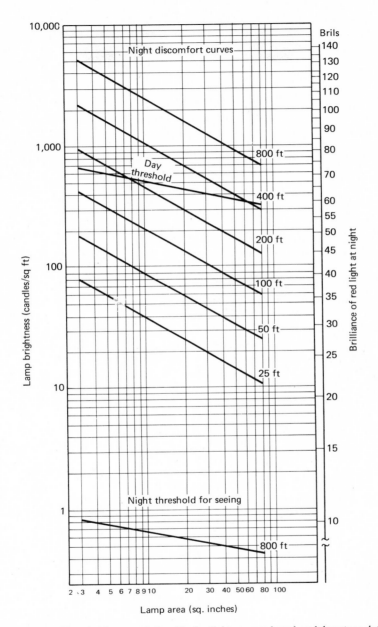

Figure 3.19. Glare-brightness relationship for light sources based on laboratory data with a high level of dark adaptation. (Source: J. Versace, University of Michigan Short Course in Engineering, 1967.)

however, the following (overtaking) driver has a complex series of visual detection and driving decision tasks to perform in a relatively short time.

Vehicle rear lighting is most important in helping drivers to move from the uncoupled to the coupled condition. Current systems of vehicle rear lighting do not take advantage of the many ways in which drivers can detect the presence and position of vehicles ahead of them. Color differences, position, brightness, and intermittence are also used to some extent.

Traffic engineers should follow with interest the development of vehicle rear lighting as it interacts with traffic signals and warning lights, particularly in rural areas where such devices are visible at great distances.

REFERENCES FOR FURTHER READING

Systems Psychology, KENYON B. DE GREENE, Ed., McGraw-Hill Book Company, New York, 1970.

ALLEN, MERRILL J., *Vision in Highway Safety*, Chilton Book Co., Philadelphia, 1970.

The Human Engineering Guide to Equipment Design, Rev. Edition, HAROLD P. VAN COTT AND ROBERT G. KINKADE, Eds., U.S. Superintendent of Documents, Washington, D. C., 1972.

Human Factors in Highway Traffic Safety Research, T. W. FORBES, Ed., John Wiley and Sons, Inc., New York, 1972.

HEIMSTRA, N. AND V. ELLINGSTAD, *Human Behavior, A Systems Approach*, Brooks/Cole Publishing Co., Monterey, California, 1972.

"Bikeway Planning Criteria and Guidelines," Institute of Transportation and Traffic Engineering, University of California, Los Angeles, April 1972.

Roadway Delineation Systems, National Cooperative Highway Research Program Report No. 130, Highway Research Board, Washington, D. C., 1972. (Especially note Appendix A and B.)

Factors Influencing Safety at Highway-Rail Grade Crossings, National Cooperative Highway Research Program Report No. 50, Highway Research Board, Washington, D. C., 1968. (Especially note Appendix B.)

Chapter 4

GENERAL TRAFFIC CHARACTERISTICS

JOSEPH C. OPPENLANDER, Professor and Chairman, Department of Civil Engineering, University of Vermont, Burlington, Vermont.

In the planning, design, and operation of a transportation system, the traffic characteristics describe the qualitative and quantitative nature of the vehicular and pedestrian flows that are being accommodated on that system of movement. The interactions of driver, vehicle, and facility culminate in various measurable parameters of traffic flow and safety. The general traffic characteristics described in this chapter are volume, speed, headway, parking, and accident patterns and trends.

Unless the general traffic characteristics are known or estimated in the planning, design, or operation of a highway or street, the traffic engineer has only a meager knowledge of the transportation system of interest. Therefore, an understanding of traffic characteristics is basic and fundamental to the development of any transportation system or traffic engineering activity. The purpose of this chapter is to describe the various measures of traffic flow and safety and to identify the various common patterns and trends that have been evaluated in engineering investigations. Studies for the collection of data as related to traffic characteristics are described in Chapter 10, Traffic Studies.

FUNDAMENTAL DEFINITIONS

The general characteristics of traffic movement are often quantitatively described by the flow rate or volume of vehicles, the speed or time rate of movement, and the density or concentration of vehicles. Volume (q) is defined as the time rate of traffic flow and is evaluated by counting the number of vehicles that pass a point in a unit of time. Although vehicular volume provides a direct "quantity" measure of traffic flow on a section of highway or street, the headway (h) or temporal spacing between vehicles is an indirect evaluation of the time rate of traffic movement. Average values of volume and headway are related by the following equation:

$$q = \frac{3,600}{h}$$

where q = volume (vehicles per hr),
h = headway (sec per vehicle).
Of course, the constant in this relationship depends on the various time units that are used in the measures of volume and headway.

Speed (u) is indicative of the "quality" of traffic movement and is described as the distance that a vehicle travels in a given interval of time. This time rate of vehicle movement over a distance is indirectly related to time (t) of travel by the following expression:

$$u = \frac{1.47d}{t}$$

where u = speed (mph),
$\quad\quad d$ = distance traveled (ft),
$\quad\quad t$ = travel time (sec).

The constant in this equation becomes 3.60 in the metric system in which speed is in kilometers per hour, distance is in meters, and time is in seconds. Thus, speed may be observed directly by an instantaneous measure of velocity at a point or indirectly by the measurement of time that elapses over a course of travel.

The concentration of vehicles on a highway or street is defined as density (k), which is the number of vehicles in a unit length of highway at any instant of time. Density numerically describes the relative occupancy of a specified roadway section and is indirectly related to the average spatial spacing between vehicles. The following equation defines the relationship between average density and average vehicular spacing at a particular point in time on a roadway section:

$$k = \frac{5,280}{d}$$

where k = density (vehicles per mi),
$\quad\quad d$ = spacing (ft per vehicle).

Of course, the constant in this equation becomes 1,000 when density is described as the number of vehicles per kilometer and spacing between vehicles is measured in meters.

The three fundamental measures of traffic flow are related in average terms by the following expression:

$$q = aku_s$$

where q = volume (vehicles per unit of time),
$\quad\quad k$ = density (vehicles per unit of distance),
$\quad\quad u_s$ = space-mean speed (distance per unit of time),
$\quad\quad a$ = constant.

The evaluation of the constant depends on the particular measures of distance and time that are used in quantifying volume, speed, and density.

In like manner, the relationship of headway, travel time, and spacing, which are indirect measures of volume, speed, and density, respectively, is described by the following function:

$$h = bdt$$

where h = headway (time per vehicle),
$\quad\quad d$ = spacing (distance per vehicle),
$\quad\quad t$ = travel time (time per unit of distance),
$\quad\quad b$ = constant.

Again, the constant in the equation reflects the choice of units that are selected to describe headway, spacing, and travel time. Detailed relationships between volume, density, speed, and headway are discussed in Chapter 7, Traffic Flow Theory.

TRAFFIC VOLUME CHARACTERISTICS

Like many dynamic systems, traffic facilities are subjected to volume loadings that possess spatial and temporal patterns. Because spatial distributions generally result from people's desires to make trips between selected origins and destinations, these patterns reflect community development and the fulfillment of those opportunities provided by the rural and urban environments in which the various origins and destinations are located. Temporal distributions are produced by the styles and standards of living that cause people to follow repetitive travel patterns on various time bases. In addition, the relative composition of the traffic stream more precisely defines the loading variability that results from a mixture of vehicle types with various operating characteristics.

SPATIAL DISTRIBUTIONS

Rural-urban distributions. Traffic distribution by routes is predicated on the trip-making desires of drivers between various origins and destinations, on the extent of the development of the various facilities in the highway system, and on the degree of urbanization in which the highway or street is located. The significant differences between rural and urban highways on the state primary system are shown in Figure 4.1, where the percentages of rural and urban mileages are plotted, respectively,

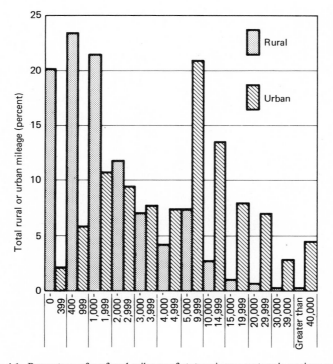

Figure 4.1. Percentage of surfaced mileage of state primary system in various volume groups, 1971. [Source: *Highway Capacity Manual,* Special Report 87 (Washington, D. C.: Highway Research Board, 1965), p. 29.]

against the average daily traffic volumes on these two area designations of highway location. Approximately 44 percent of the rural mileage accommodates traffic volumes of fewer than 1,000 vehicles per day, and 95 percent of these roads have volumes of fewer than 10,000 vehicles per day. Traffic volumes on more than 64 percent of the urban mileage exceed 4,000 vehicles per day, and the value of 10,000 vehicles per day is exceeded on approximately 36 percent of the urban streets. The concentration of vehicular travel on urban facilities is demonstrated by the fact that municipal extensions comprise only 13 percent of the total mileage in the state primary system but account for approximately 54 percent of the total urban and rural mileage where traffic volumes exceed 10,000 vehicles per day. These trends have remained fairly stable over the years.[1]

Distribution by direction. Although traffic flows by direction on most two-way facilities of reasonable length are approximately balanced over some specified time period, for example, a day or a week, differences in directional distribution occur during certain hours, days, and months of the year. These variations in route usage by direction are in large part explained by the temporal and cyclical patterns of trip making by the road users. A knowledge of the directional distributions during peak periods is essential for the proper planning, design, and operation of traffic facilities. These unbalanced flows generally represent the critical loads to which the highway is subjected, and the peak periods for each direction may occur at different hours of the day or even at different hours on different days of the week. An example of directional distribution in traffic movement is illustrated in Figure 4.2 for travel on I–91 in the vicinity of Springfield, Mass., during the 24 hours of three weekdays and the two days of the weekend. Each roadway location, however, has its own pattern of directional distributions, and these patterns may vary considerably along the route because of gains and losses in traffic volume at points of ingress and egress.

Lane distributions. The final spatial distribution is the variation of traffic flow by lane at the same point in time. When two or more lanes are provided for travel in the same direction, the distribution of vehicles among lanes for the same direction of movement depends on several factors, including traffic volume, medial and marginal friction, proportion of slow-moving vehicles, and the number and location of ingress and egress points. The lane distribution near points of entrance and exit is largely predicated on the origin and destination desires of the road users. In areas of uninterrupted flow, however, traffic distribution by lane is mainly influenced by the volume of vehicles and by the presence of large differentials in vehicular speed. The plot of variations in lane distribution as a function of traffic volume is shown in Figure 4.3 for travel in one direction on a six-lane freeway located in a metropolitan area.

TEMPORAL VARIATIONS

The variations in traffic volumes with time reflect the economic and social demands for transportation. These trip-making desires are cyclical in nature and reflect the hourly, daily, and seasonal patterns of living in the rural and urban communities. In addition, volumes vary over the years because of changes in population, number of

---

[1] *Highway Statistics*, U.S. Department of Transportation, Federal Highway Administration (Washington, D.C.: Government Printing Office, various years).

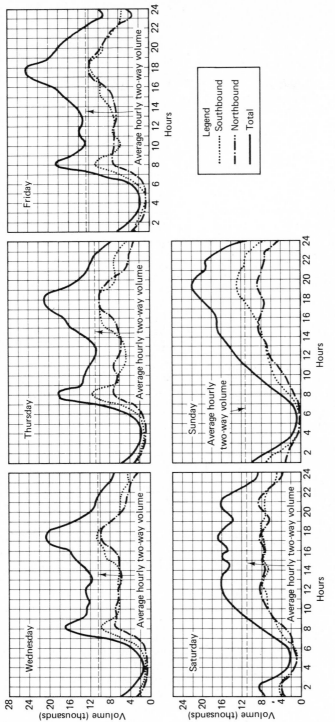

Figure 4.2. Hourly traffic volume variations on I-91 in Massachusetts, 1968. (Source: Wilbur Smith and Associates, "The Land Use, Transportation and Travel Inventories," *Springfield Urbanized Area Comprehensive Transportation Study*, Volume I, May 1969, p. 105.)

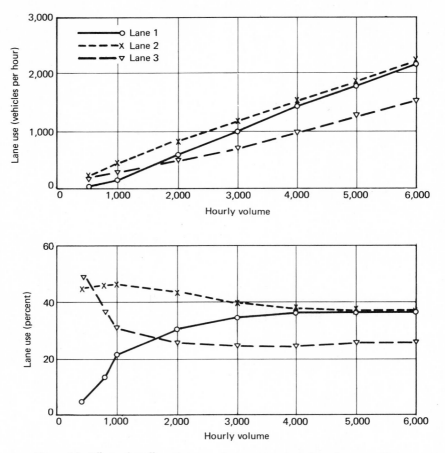

Figure 4.3. Effect of traffic volume on lane usage on six-lane freeways. [Source: A. D. May, Jr., "Traffic Characteristics and Phenomena on High Density Controlled Access Facilities," *Traffic Engineering*, XL, No. 6 (March, 1961), p. 13.]

vehicles per person, usage per vehicle, facilities available for travel, and land development. Except for the depression and war years, the general trend of these various factors has been reflected in annually increasing vehicular volumes on most highways and streets. The exact rate of growth on any facility, however, depends on many conditions including local economic conditions, status of community development, and the accompanying opportunities for satisfying trip desires, and extent of transportation improvements.

Season of year. If traffic volumes for a particular roadway are plotted according to the month of the year, the seasonal trend in traffic flow is developed for that facility. Typical traffic-volume variations by season of the year are illustrated in Figure 4.4 for three Connecticut highways having different functional characteristics. Greater seasonal variations are evident in the highway that serves recreational travel in a

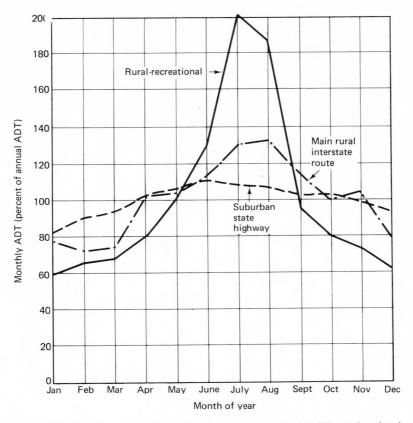

Figure 4.4. Monthly variation in traffic volume for routes having different functional characteristics. [Source: *An Introduction to Highway Transportation Engineering* (Washington, D.C.: Institute of Traffic Engineers, 1968), p. 22.]

rural area, and a more stable pattern of traffic movement is evident over the year on the suburban route. The rural interstate highway has seasonal traffic-volume variations that are intermediate with respect to the fluctuations of the recreational and urban routes. The seasonal difference between rural interstate and other main rural roads in Illinois may be observed in Figure 4.5.

Day of week. Another temporal distribution is the variation of traffic movement on a facility by day of the week. Typical weekly distributions are presented in Figure 4.6 for an urban route and a recreational type highway in a rural area. Although the patterns are reasonably consistent for each facility on weekdays, travel on the rural-recreational route is considerably greater on Saturday and Sunday than on the average day of the week. This trend is reversed on the urban street where traffic volumes are lower on the weekend in comparison with the average daily volume for the week. The peak daily flows on Friday for urban areas are indicative of current patterns of living, for which Friday is the last day of the work week and often pay day, is the time

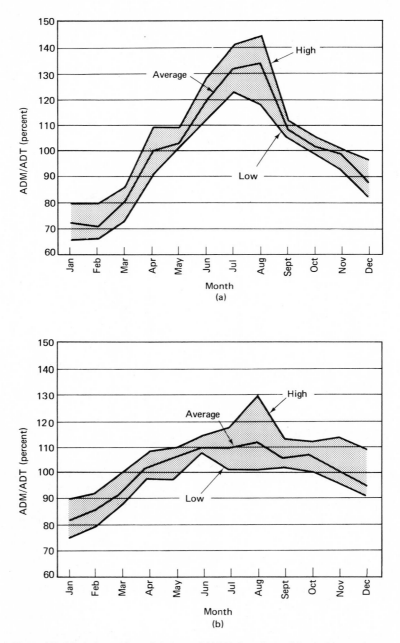

Figure 4.5. (a) Average day of the month/ADT for all rural interstate continuous count stations. (b) Average day of the month/ADT for all other main rural continuous count stations. [Source: *Traffic Characteristics on Illinois Highways/1971*, (Springfield: Illinois Department of Transportation), p. II-2.]

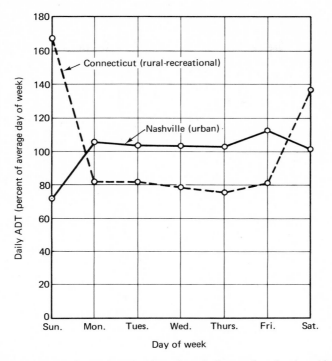

Figure 4.6. Volume fluctuations by day of week for rural and urban highways. [Source: *An Introduction to Highway Transportation Engineering* (Washington, D.C.: Institute of Traffic Engineers, 1968), p. 21.]

for nighttime retail shopping in the central business district (CBD) and in the regional shopping centers, and represents the period for exit trips from the urban area for weekend activities.

Hourly patterns. A further subdivision of time provides the distribution of traffic volumes according to the hour of the day. Volume distributions by hour of the day often describe the peak demands for service by our transportation facilities. Typical daily distributions by hour are provided in Figure 4.2 for Wednesday, Thursday, Friday, Saturday, and Sunday on an interstate highway passing through an urbanized area. The morning and evening peak-hour movements on weekdays represent the home-to-work and work-to-home trips, respectively. However, the volume variations throughout the day are less significant on the weekend except for late evening peak period on Sunday when weekend travelers are returning to metropolitan areas. In fact, this Sunday evening peak-hour volume is only exceeded by the work-to-home travel period on Friday afternoon. A graphical summary of time variations by month, day, and hour is shown in Figure 4.7 for the same street (Railroad St.) in a small New England city. A comparison with another street (Main St.) in the same town illustrates similar patterns on daily and hourly bases.[2]

---

[2] *St. Johnsbury TOPICS Improvement Program* (Newark, N.J.: Edwards and Kelcey, Inc., 1971).

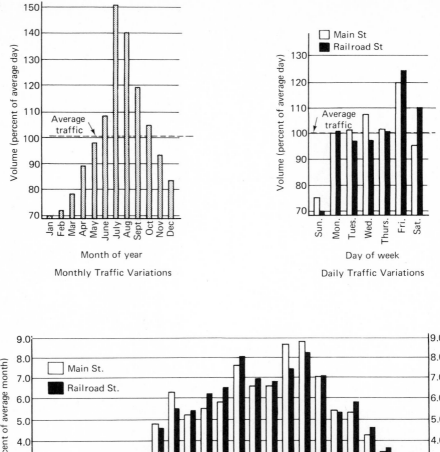

Monthly Traffic Variations

Daily Traffic Variations

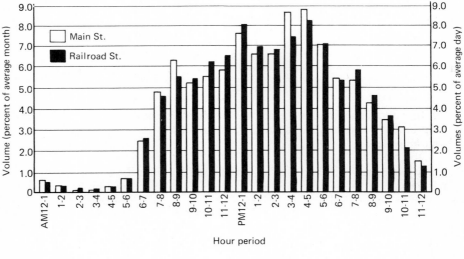

Hour period

Hourly Traffic Variations

Figure 4.7. Temporal variations in traffic volumes. [Source: *St. Johnsbury Topics Improvement Program* (Boston: Edwards and Kelcey, Inc., July, 1971), p. 11.]

Peak intervals within the hour. The variations of traffic flow within the peak hour are a function of the peaking characteristics of the highway facility. As the time period for the volume measure is reduced, the average number of vehicles is proportionally reduced for that period of time. However, the variability of smaller mean values is greater than that of larger mean values when the variability is expressed as a percentage of the mean. Therefore, the volume observed during a short-term period within a given peak hour has a greater probability of exceeding its mean by a specified percentage than does the volume observed during the whole hour. The evaluation of peaking characteristics is generally based on traffic volumes collected by 5-min periods, although some studies have recorded the peak-hour flows in time subsets of 6 min or 15 min. The determination of a peak-hour traffic volume based on the peak 5-min rate of flow and community population is illustrated by the nomograph in Figure 4.8. This procedure provides an estimate of an hourly volume that depends primarily on the magnitude of the short-term volume fluctuations. These variations are usually less than an hour in duration but represent the periods of greatest demand for use of the roadway. The application of peak-hour variations to problems of capacity analysis are presented in Chapter 8, Highway Capacity.

ADT and peak-hour volumes. In several studies of volume characteristics, relationships have been developed between various hourly volumes and the average daily traffic (ADT), which is the total volume during a given time period (in whole days greater than one day and less than one year) divided by the number of days in that time period. A typical series of these relationships is represented by the plots in Figure 4.9, where both percentage of annual average hourly volume and percentage of annual average daily traffic are plotted in decreasing order of magnitude against hours of the year. Although the various plots approximately represent a similar family of curves, the functional characteristics of a given route dictate the size of the ordinate differences in the range of the highest hourly volumes for the year. The very great demands for travel on recreational routes during only a few periods of the year account for a large proportion of the total annual traffic, but on urban routes the total annual volumes are more evenly distributed throughout the hours of the year.

Because of the economic considerations involved in the planning and design of highway facilities, design hourly volumes are often selected from a consideration of the relationships between the percentage of annual average daily traffic and the highest hours of the year for a given type of route. In general, a pronounced break in these curves, as illustrated in Figure 4.9, occurs in the range of the 20th to the 50th highest hour. If an hourly value is selected to the left of this range (that is, for some hour less than the 20th highest hour), the sizable increase in design requirements accommodates only several additional hours in the year. On the other hand, an hourly value to the right of the 50th highest hour produces only a very slight decrease in the percentage of the hourly value for design with respect to the existing or forecasted average daily traffic. Many highway agencies have selected the 30th highest hourly volume (30 HV) as this "point of diminishing returns" for the determination of a design hourly volume (DHV), although any highest hourly volume can be selected for design purposes.

The ratio of the design hourly volume to the average daily traffic for the design year is designated K. The standard practice is to determine the DHV by multiplying the estimated or forecasted ADT by K, as illustrated in Chapter 14, Geometric Design. The ratio of 30 HV to ADT on main rural highways, often used as an estimate of K, has an average value of approximately 15 percent with a range of from 12 percent to 18 percent. For urban facilities the average ratio is about 11 percent with a range of from 7 percent to 18 percent.[3] The ratio of any other highest hourly volume to the average daily traffic may also be selected as an estimate of K. The graph of hourly volume as a percentage of the average daily traffic for the highest hours of the year in Figure 4.10 is an illustration of the range within which the value of this ratio

---

[3] *A Policy on Geometric Design of Rural Highways* (Washington, D.C.: American Association of State Highway Officials, 1965), pp. 52–69, 87–98, 192–203.

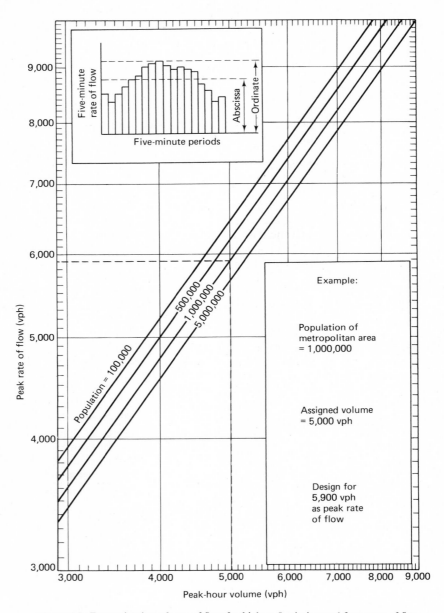

Figure 4.8. Determination of rate of flow for highest 5-min interval from rate of flow for the whole peak hour. [Source: D. R. Drew and C. J. Keese, "Freeway Level of Service as Influenced by Volume and Capacity Characteristics," *Freeway Characteristics, Operations and Accidents*, Record 99 (Washington, D.C.: Highway Research Board, 1965), p. 4.]

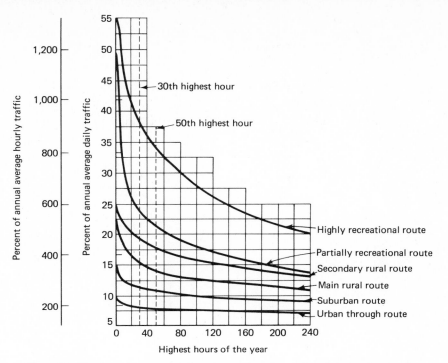

Figure 4.9. Hourly volumes on various types of routes as found during hours of highest volume on these routes. [Source: T. M. Matson, W. S. Smith, and F. W. Hurd, *Traffic Engineering* (New York: McGraw-Hill Book Company, 1955), p. 86.]

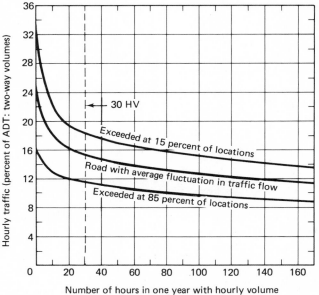

Figure 4.10. Relation between peak hour and average daily traffic volumes for main rural highways. [Source: *A Policy on Geometric Design of Rural Highways* (Washington, D.C.: American Association of State Highway Officials, 1965), p. 55.]

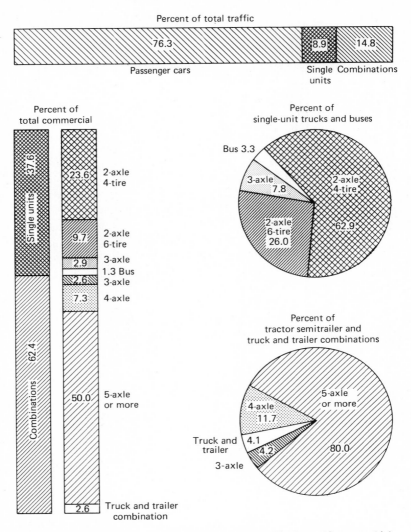

Figure 4.11(a). Proportions of traffic by vehicle type on Illinois rural interstate highways, 1971. (Source: *Traffic Characteristics on Illinois Highways/1971*, p. I-27.)

is expected to fall 70 percent of the time for typical traffic conditions encountered on main rural highways.

TRAFFIC COMPOSITION

The final factor relating to volume characteristics is the composition of the traffic stream. Traffic composition is generally measured by the percentage of trucks and buses at a given roadway location. Average percentage relationships of traffic composition on rural interstate and primary highways in Illinois are presented in Figure 4.11 for various vehicle classifications. Although passenger cars accounted for approxi-

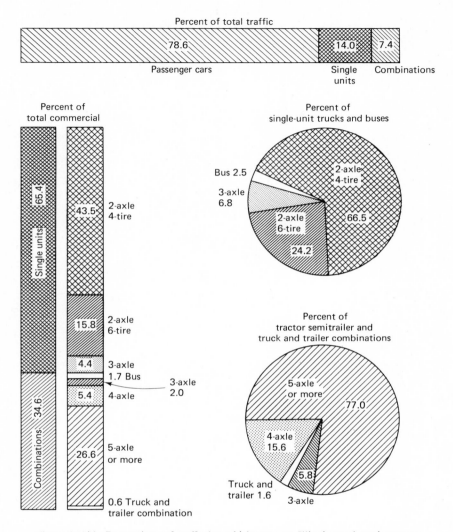

Figure 4.11(b). Proportions of traffic by vehicle type on Illinois rural noninterstate highways, 1971. (Source: *Traffic Characteristics on Illinois Highways/1971*, p. I-28.)

mately 79 percent of all vehicles on rural highways, approximately 56 percent of the commercial vehicles are single-unit trucks, and the remaining 44 percent are in some class of truck combination. The relative variations in size of truck combinations in Illinois over the years are illustrated in Figure 4.12 for semitrailers with three, four, and five axles. An increasing percentage of five-axle semitrailers is indicative of the trend toward larger commercial vehicles with greater load-carrying capacities.

The axles on a growing number of large trucks are being repositioned to increase the payload capabilities. For example, the tandem axles on vehicles used to carry heavy, high-density loads are being spaced farther apart than the current legal description. These "spread tandems" may then be legally considered single axles for deter-

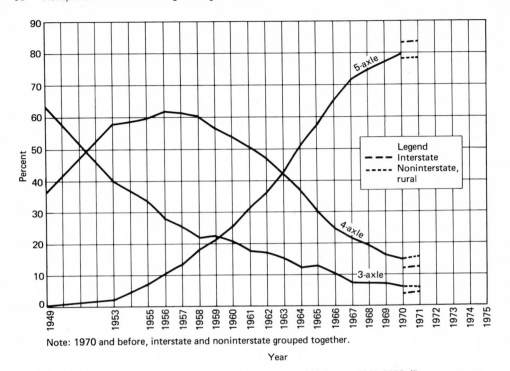

Figure 4.12. Proportion of semitrailers on Illinois rural highways, 1949–1975. (Source: *Traffic Characteristics on Illinois Highways/1971*, p. 30.)

mining maximum weight limits. Spread tandems are particularly popular on five-axle tractor-semitrailer combinations. A listing of axle spacings as observed in Illinois is shown in Table 4.1.

SPEED CHARACTERISTICS

The quality of travel is associated with speed or its reciprocal, travel time. In fulfilling their travel desires, drivers travel at those speeds deemed appropriate for the conditions under which the trips are made. Those variables that influence speed characteristics can be classified under the generic headings of driver, vehicle, roadway, traffic, and environment. Vehicular speed is an important consideration in highway transportation because the rate of vehicle movement has significant economic, safety, time, and service (comfort and convenience) implications to both the motoring and the general public.

SPEED FUNDAMENTALS

Vehicular speed is the rate of movement of traffic or of specified components of traffic and is usually expressed in miles per hour or kilometers per hour. In reality, the velocity (a vector) of a motor vehicle is the ratio of its displacement (a vector) to the time interval (a scalar) in which the displacement occurred. The traffic engineer,

TABLE 4.1
Percentage Distribution of Axle Spacings for Tractor-Semitrailer Combinations
Observed on Illinois Highways, 1969

Axle Spacings (ft)	3-Axle		4-Axle			5-Axle				5-Axle Spread Tandem			
	AB	BC	AB	BC	CD	AB	BC	CD	DE	AB	BC	CD	DE
2.0– 2.9													
3.0– 3.9					7.4		2.8		7.3		5.5		
4.0– 4.9					91.2		96.7		91.2		94.5		
5.0– 5.9					0.4		0.5		0.7				
6.0– 6.9					0.8				0.5			1.8	
7.0– 7.9					0.2				0.3			—	
8.0– 8.9	5.2		1.9			1.3		0.4				—	18.2
9.0– 9.9	8.9		11.3			23.0		0.2		23.6		—	72.7
10.0–10.9	10.4		19.6			43.1		0.7		36.4		—	1.8
11.0–11.9	40.7		33.6	0.4		20.7		0.4		32.7		—	—
12.0–12.9	25.9		23.1	0.4		7.4		0.5		3.7		—	—
13.0–13.9	4.4		8.2	0.6		2.5		0.1		1.8		—	3.7
14.0–14.9	4.1		1.3	0.2		1.0		0.3		1.8		1.8	1.8
15.0–15.9	0.4	1.1	0.4	1.3		0.7		0.3				1.8	—
16.0–16.9		3.0	0.4	0.8		0.3		0.2				5.5	—
17.0–17.9		3.3	0.2	2.3				0.5				9.1	—
18.0–18.9		3.7		2.3				0.7				18.2	—
19.0–19.9		2.2		3.2				0.9				7.3	—
20.0–20.9		2.6		4.2				1.5				9.1	—
21.0–21.9		3.0		5.0				4.0				10.9	—
22.0–22.9		2.6		13.0				4.2				23.6	—
23.0–23.9		4.8		18.3				5.1				9.1	—
24.0–24.9		7.8		16.4				7.4				1.8	1.8
25.0–25.9		4.4		11.6				14.9					
26.0–26.9		6.6		8.8				23.0					
27.0–27.9		5.9		5.1				23.3					
28.0–28.9		3.0		3.0				7.4					
29.0–29.9		9.3		2.3				2.2					
30.0–30.9		6.6		0.4				0.8					
31.0–31.9		11.1		0.2				0.4					
32.0–32.9		8.9		0.2				0.4					
33.0–33.9		3.0						—					
34.0–34.9		2.6						—					
35.0–35.9		1.5						—					
36.0–36.9		1.5						—					
37.0–37.9		1.5						0.1					
Over–38.0								0.1					
	270 Vehicles		524 Vehicles			2,272 Vehicles				55 Vehicles			

Note: A 5-axle spread-tandem tractor-semitrailer combination has the semitrailer tandem axles spaced greater than 8 ft apart. Axles are identified by lettering the axles in alphabetic order beginning with the front axle as Axle A.

Source: *Traffic Characteristics on Illinois Highways* (Springfield: Illinois Department of Transportation, 1971), p. 81.

however, is generally only interested in the magnitude and not the exact direction of the velocity vector. Therefore, the speed (a scalar) of a moving vehicle is defined as the ratio of the length of traveled path (a scalar) to the elapsed time (a scalar).

From the above definition of speed, two distinct types of average speed measures can be derived to express the rate of traffic movement. The first type of average speed is time-mean speed or spot speed, which is the mean value of a set of instantaneous vehicle speeds at some given location on a roadway. Time-mean speed is generally calculated as the average of several spot-speed observations at the particular highway location and is symbolically represented by the following equation:

$$\bar{u}_t = \frac{\sum_{i=1}^{n} u_i}{n}$$

where $\bar{u}_t$ = average time-mean speed,
  u_i = spot speed of the ith vehicle,
  n = number of vehicles that comprise the sample of speed observations.

A second expression of mean speed is space-mean speed or travel speed, which is computed as the specified travel distance divided by the mean travel time of several trips over this highway section and is expressed by the following relationship:

$$\bar{u}_s = \frac{dn}{\sum_{i=1}^{n} t_i}$$

where $\bar{u}_s$ = average space-mean speed,
  d = travel distance,
  n = number of trips that comprise the sample of time observations,
  t_i = travel time of the ith trip.

Time-mean speed is always greater than space-mean speed for a given sample of traffic flow except for the situation in which all vehicles are traveling at the same speed. The two speed measures are then equal in this special case. The results of speed measurements that are summarized as a time-mean speed and as a space-mean speed are illustrated by the computational example in Table 4.2. An approximate relationship between time-mean and space-mean speeds has been developed in accordance with the following expression:

$$\bar{u}_t = \frac{\bar{u}_s + \sigma_s^2}{\bar{u}_s}$$

where $\bar{u}_t$ = average time-mean speed,
  $\bar{u}_s$ = average space-mean speed,
  σ_s^2 = variance of the space-mean speeds.[4]
Space-mean speeds are a function of the density of vehicles on the highway; time-mean speeds are related only to the number of vehicles passing a given point on the roadway.

SPEED VARIATIONS

The influences of travel variables on speed characteristics are conveniently summarized under the categories of driver, vehicle, roadway, traffic, and environment.

Driver Aspects. Although driver variables have not been evaluated to great extents, trip distance has the most significant influence on spot-speed characteristics, whereas the presence of passengers in the car and the sex of the driver alter driving speeds to a lesser extent. In general, persons traveling long distances have newer cars and drive

---

[4] J. G. WARDROP, "Some Theoretical Aspects of Road Traffic Research" (London: *Proceedings Institute of Civil Engineers*, Road Paper No. 36, 1952.)

TABLE 4.2
Example for the Computation of Time-Mean and Space-Mean Speeds

Vehicle	Time to Travel 176 ft (sec)	Velocity (mph)
1	1.9	63
2	2.1	57
3	2.1	57
4	1.8	67
5	2.3	52
6	2.0	60
7	2.2	55
8	2.0	60
9	1.7	71
10	1.9	63

Mean time = 2.0 sec.
Space-mean speed = 60.0 mph.
Time-mean speed = 60.5 mph.

faster than local travelers, and their speeds increase with trip length. Lone drivers tend to travel at higher speeds than drivers with passengers. Finally, women drivers travel at about the same or at a slightly lower average speed than male operators. More men than women drive at dangerously high speeds, and divorced men and women and single women drive faster than married men and women. Of course, these driver variables influence vehicular speeds to different degrees in various parts of the country.[5]

Variations by vehicle types. Among the various vehicle variables, vehicle type (passenger car, single-unit truck, combination truck, or bus) and vehicle age appear to have predominant effects on spot speeds of highway motor vehicles. A further subdivision of single-unit trucks and combination trucks by gross weight is feasible in evaluating speed characteristics. The average speeds of buses are consistently higher than the average speeds of passenger cars, with trucks traveling noticeably slower. The widest speed ranges are associated with passenger cars, whereas a greater consistency of speed selection is evidenced by the drivers of trucks and buses. Average speeds of commercial vehicles decrease from light, single-unit trucks, to medium trucks, to heavy combination trucks, to heavy, single-unit trucks. Within both the single-unit truck and the combination truck classifications, average spot speeds generally decrease with an increase in the gross weight of the commercial vehicle. For a given travel distance, the average speed decreases as the vehicle age increases. Newer cars are generally driven faster than older vehicles because new cars go faster, ride more comfortably, travel more smoothly and quietly, handle better, and are generally in better mechanical condition.[6]

Variations by roadway elements. Actual speeds adopted by drivers are greatly affected by various aspects of the roadway. Functional classification of the roadway, curvature, gradient, length of grade, number of lanes, and surface type appear to have the most pronounced effects on speed characteristics. However, geographic

---

[5] J. C. OPPENLANDER, *Variables Influencing Spot-Speed Characteristics: Review of Literature*, Special Report 89 (Washington, D.C.: Highway Research Board, 1966).

[6] *Ibid.*

location, sight distance, lane position, lateral clearance, and frequency of intersections are also roadway elements that do influence speed patterns.

Highway types. On rural highways and urban expressways, drivers can operate their vehicles at safe speeds predicated on the geometric design elements of those roadways, whereas vehicular speeds on major streets are regulated by recurring peak traffic volumes, traffic-control devices, intersections, and other physical and psychological retarding forces peculiar to the urban environment. Thus, the functional classification of highway facilities with similar characteristics is a variable influencing speed characteristics. Speeds of all vehicles on rural highways progressively increase from primary feeder, to intercity, to interstate and interregional highway systems. These speed differences by highway type remain consistent throughout the range from low to high traffic volumes. As the transition from rural to urban travel is made, average spot speeds become lower. In extending the influence of highway functional classification on vehicular speeds to urban roadways, the following ranges of average speeds (also shown in Figure 4.13) were observed on various urban arterials in Detroit and Lansing, Michigan: freeways, 40 mph to 60 mph (64 kph to 97 kph); unsignalized

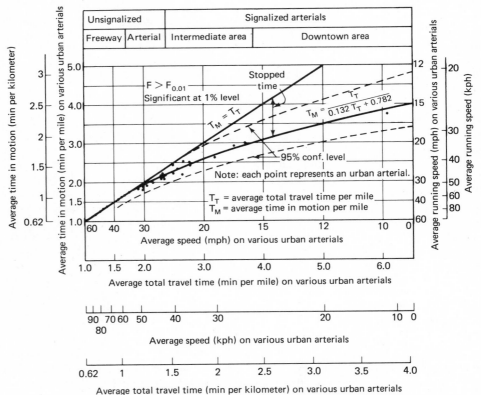

Figure 4.13. Running speed and stopped time related to average travel time for various urban arterials. [Source: A. D. May, Jr., "A Friction Concept of Traffic Flow," *Proceedings*, Vol. 38 (Washington, D.C.: Highway Research Board, 1959), p. 505.]

arterials, 32 mph to 40 mph (51 kph to 64 kph); signalized arterials in intermediate areas, 22 mph to 32 mph (35 kph to 51 kph); and signalized arterials in downtown areas, less than 22 mph (35 kph). Also, speeds are consistently higher on one-way streets than on two-way streets.

Horizontal alinement. Vehicular speeds are lower on horizontal curves than on tangent alinements, and the average spot speed approaches the calculated design speed as the degree of curvature increases. In addition, the average spot speed on a low-design speed curve is near the design speed, and the average spot speed on a high-design speed curve is substantially below the design speed and approximates the average speed observed on the tangent sections. A highly significant linear relationship has been observed between vehicular speed and degree of curve on rural highways and is represented by the following equation:

$$\bar{u} = 46.26 - 0.746D$$

where $\bar{u}$ = average spot speed (mph),
D = degree of curvature (degrees).[7]
A similar linearity on rural roadways in England was noted in studies of the influence of horizontal and vertical alinements on speeds. The following multiple linear expression was developed as a result of the study:

$$\Delta\bar{u} = 1.22D + 1.37G$$

where $\Delta\bar{u}$ = reduction in average speed (mph),
D = degree of curvature (degrees),
G = average gradient (percent).[8]

Vertical alinement. The vertical alinement of a roadway has a marked influence on vehicular speeds. However, the various effects are more pronounced on trucks than on passenger cars. Average spot speeds on downgrades, as compared to travel on level tangent sections, are increased on gradients up to 5 percent for trucks and 3 percent for buses and passenger cars. The speeds are reduced on downgrades in excess of these limits and on upgrades for all vehicle types. In studies of the gradeability of heavy commercial vehicles, the speed on a given grade was reduced almost linearly with an increase in the length of the grade until the crawl speed was reached. The truck then continued up the grade at this minimum speed. Typical values of these speed reductions, along with the respective crawl speeds, are summarized in Table 4.3 for various gradients. The speed-distance curves in Figure 4.14 provide heavy-truck gradeability characteristics that are conveniently presented for the design of vertical climbing lanes.

Number of lanes. Roadways with more than four lanes have operational characteristics similar to four-lane facilities. However, four-lane highways, on which passing is not restricted by opposing traffic, have higher average speeds than two- and three-lane highways. This discrepancy is even more pronounced for divided highways. In addition, observed speeds on three-lane facilities are only slightly higher than on

[7] A. TARAGIN, "Driver Performance on Horizontal Curves," *Proceedings*, Vol. 33 (Washington, D.C.: Highway Research Board, 1954), pp. 446–66.

[8] W. H. GLANVILLE, "Report of the Director of Road Research for the Year 1954," *Road Research 1954* (London: Department of Scientific and Industrial Research, 1955).

TABLE 4.3
Speed Reductions per 1,000-ft Length of Grade for Heavily Loaded Truck

% Grade	Speed Loss (mph)	Crawl Speed (mph)
2	2.0	23
3	5.0	17.5
4	9.5	12
5	15.5	9
6	23.0	7
7	33.5	6

Source: W. E. Willey, "Survey of Uphill Speeds of Trucks on Mountain Grades," *Proceedings*, Vol. 29 (Washington, D.C.: Highway Research Board, 1949), p. 304.

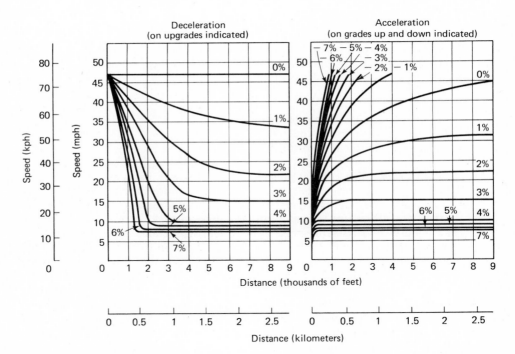

Figure 4.14. Speed-distance curves from road test of a typical heavy truck operating on various grades. [Source: T. S. Huff and F. H. Scrivner, "Simplified Climbing-Lane Design Theory and Road-Test Results," *Vehicle Climbing Lanes*, Bulletin 104 (Washington, D.C.: Highway Research Board, National Research Council, 1955), p. 3.]

similar two-lane roads. Spot speeds tend to increase as the road surface progresses from low to high types.

Geographic location. The average speeds on main rural highways in the central and western regions of the United States of America have been consistently from 4 mph to 7 mph (6 kph to 11 kph) higher than in the eastern region. These differences are probably explained by the varying topographic conditions and the different state restrictions on maximum speed limits.

Sight distance. Vehicular speeds decrease as the percentage of sight distance less than the passing sight distance increases for travel conditions on two-lane highways in rural areas. Because restricted sight distances limit the opportunities for passing maneuvers, the actual operating speed is determined by the combined influence of the traffic volume and the percentage of the total roadway length with sight distances insufficient to permit passing.

Lane position. The influence of lane position on speed characteristics has been evaluated in several investigations of traffic flow characteristics. Average speeds of inbound traffic are consistently from 2 mph to 4 mph (3 kph to 6 kph) faster than for outbound traffic on roadway approaches to urban areas. The distribution of spot speeds by lane position on three-lane, two-way highways has also been observed. The average speeds in the two outside lanes show the normal linear decrease with an increase in volume, but the average speed in the center lane is faster and does not change with variation in traffic volume. The average values of spot speeds on multilane freeways are reduced as the lane position progresses from the median, to the middle, to the shoulder lanes. The marked reduction in speeds for the vehicles in the curb lane is largely attributed to the presence of commercial vehicles in this lane, to the speed-change maneuvers performed by ingress and egress traffic in the outside lane, and to hazards of merging and diverging traffic anticipated by the through traffic in the right lane.

Lateral clearance. In general, restricted lateral clearances on two-lane highways cause reductions in average speed of from 1 mph to 3 mph (2 kph to 5 kph). Truck drivers are less influenced by lateral restrictions than drivers of passenger cars. Finally, spot speeds on urban roadways tend to decrease with an increase in the number of friction points passed per unit of distance. These points of friction include intersections, at-grade railroad crossings, hospital or school zones, and special pedestrian crossings.[9]

Speed-volume relation. The speed-volume relationship for a given type of roadway facility in a specific traffic area is often represented by a straight-line function with a negative slope when all other modifying variables are reasonably identical. As the volume on a given roadway increases, the average speed decreases approximately linearly until the traffic volume reaches the capacity of the particular facility under the prevailing roadway and traffic conditions. Although many mathematical expressions and graphical plots have been developed to describe the functional relation between average speed and average volume, the graph in Figure 4.15 is a typical representation of this relationship. This plot of operating speed vs. average lane volume represents travel in one direction under ideal uninterrupted flow conditions on multilane rural highways for various levels of average highway speed. At traffic

---

[9] J. C. Oppenlander, *Variables Influencing Spot-Speed Characteristics.*

densities greater than the critical density, both volume and speed are reduced, respectively, below the maximum volume and the optimum speed. This relationship is depicted in Figure 4.15 as a parabolic curve that begins at the point of capacity or maximum volume, decreases at a decreasing rate with a reduction in traffic volume, and ends at the origin representing no traffic flow.

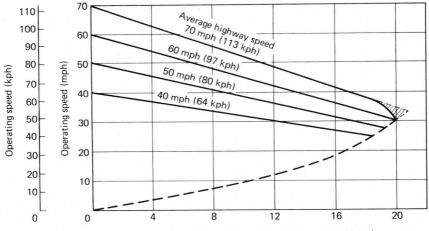

Figure 4.15. Typical relationships between volume per lane and operating speed in one direction of travel under ideal uninterrupted flow conditions on multilane rural highways. (Source: *Highway Capacity Manual*, p. 63.)

Speed-density relation. Speed is functionally related to the density of the traffic stream. This relationship, as shown in Figure 4.16, is often approximately linear with a negative slope between average space-mean speed and average density, and the maximum volume or capacity of a particular highway facility occurs at the mean or critical density which is located at the midpoint of the curve. The average speed at the critical density is defined as the optimum speed and is located halfway between the maximum average speed, which corresponds to minimum density, and the point of no traffic flow, which represents maximum density. Curvilinear relationships have recently been reported between average speed and average density, with speed decreasing at a decreasing rate for increasing values of density. These relationships are further explored in Chapter 7, Traffic Flow Theory.

Passing opportunities. As vehicular volume increases, the effect of increasing percentages of commercial vehicles on speed characteristics becomes more pronounced. When the increasing volume limits the opportunities for passing, average speeds decrease linearly with an increase in the percentage of commercial vehicles.

Drivers cannot maintain their desired speeds unless the faster moving vehicles can change lanes and pass the slower moving vehicles. Therefore, passing maneuvers necessitate speed changes and alter speed characteristics. In general, the average passing driver wants to travel approximately 10 mph (16 kph) faster than the vehicle

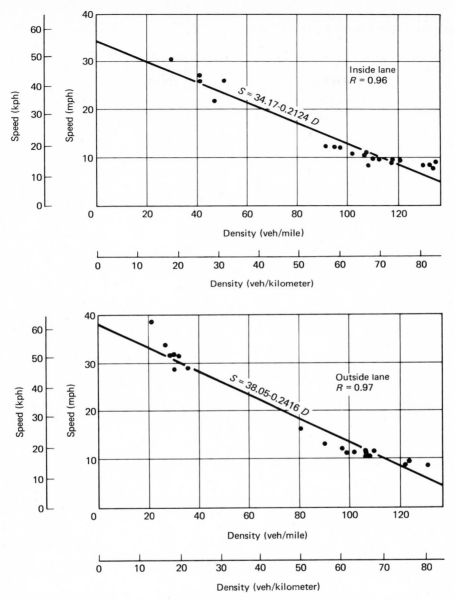

Figure 4.16. Speed vs. density relationships. [Source: M. J. Huber, "Effect of Temporary Bridge on Parkway Performance," *Highway Capacity Studies*, Bulletin 167 (Washington, D.C.: Highway Research Board, 1957), p. 67.]

that is passed and about 6 mph (10 kph) faster than the average speed of all traffic.[10] A comparison of passing practices shows little change over the years.

Access control. The control of access to roadways is often a prerequisite in the design of modern highways. In rural areas the degree of access control apparently

---

[10] C. W. PRISK, "Passing Practices on Rural Highways," *Proceedings*, Vol. 21 (Washington, D.C.: Highway Research Board, 1941), pp. 366–78.

has little influence on spot speeds, but in suburban and urban districts average speeds increase with greater control of access. Average speeds on full-controlled access highways in rural, suburban, and urban areas were reported, respectively, as 2.5 mph, 10.3 mph, and 20.9 mph (4.0 kph, 16.6 kph, and 33.6 kph) higher than corresponding speed values for travel on uncontrolled access roadways.[11]

Environmental variables. Environmental variables of time and weather present important considerations in the evaluation of speed characteristics. Since 1942 the average speeds on main rural highways have continued to increase at the approximate rate of 1.0 mph (1.6 kph) per year. Other measurements of vehicle speeds during different seasons of the year have indicated that average speeds are highest in the fall and winter, intermediate in the spring, and lowest in the summer. Conflicting statements on daily and hourly fluctuations in vehicular speeds on a given roadway appear in the literature. However, average spot speeds in the daytime are about 1 mph (2 kph) higher in urban areas and from 2 mph to 8 mph (3 kph to 13 kph) higher in rural areas than the corresponding speed values during the nighttime.

Unfavorable road surface conditions appear to produce greater speed reductions than does low visibility. Reductions in average spot speed attributed to weather conditions range from 7 percent to 23 percent for poor visibility, 4 percent to 38 percent for unfavorable road surface, and 10 percent to 24 percent for both impaired visibility and road surface.

SPEED EQUATIONS

In Illinois several studies have been conducted to evaluate the influence of travel variables on speed characteristics. A multivariate analysis of vehicular speeds permitted the statistical modeling of vehicular speeds in relation to those variables that influence traffic flow. The technique of factor analysis provided an exploratory appraisal of speed characteristics.

Two-lane rural highways. Variations in mean spot speeds on two-lane highways in rural areas were largely explained by the generated factors of horizontal resistance, long-distance travel, marginal friction, vertical resistance, and obsolete pavement. Because the results of the factor analysis permitted a knowledgeable selection of independent variables for a multiple linear regression with mean speed, the following expression was developed for the estimation of vehicular speeds on two-lane rural highways:

$$\bar{u} = 39.34 + 0.0267X_1 + 0.1396X_2$$
$$+ 0.8125X_3 + 0.1126X_4 + 0.0007X_5$$
$$+ 0.6444X_6 - 0.5451X_7 - 0.0082X_8$$

where $\bar{u}$ = mean spot speed (mph),
$\quad X_1$ = out-of-state passenger cars (percent),
$\quad X_2$ = truck combinations (tractor with one or more trailers) (percent),
$\quad X_3$ = degree of curvature (degree),

---

[11] A. D. MAY, JR., "Economics of Operation on Limited-Access Highways," *Vehicle Operation as Affected by Traffic Control and Highway Type*, Bulletin 107 (Washington, D.C.: Highway Research Board, 1955), pp. 49–62.

X_4 = gradient (percent),
X_5 = minimum sight distance (ft),
X_6 = lane width (ft),
X_7 = commercial roadside establishments (number per mile along both sides of the highway),
X_8 = total traffic volume (vehicles per hour).

The independent variables of horizontal curvature, gradient, sight distance, and lane width relate to geometric aspects of the highway, while the percentages of out-of-state passenger cars and truck combinations in the traffic stream, the number of commercial roadside establishments per mile, and the total traffic volume are measures of design and operating controls to which the highway is subjected.[12]

Four-lane rural highways. Another multiple linear regression analysis provided a speed estimation model for travel conditions on four-lane highways in rural areas. This functional relationship is expressed in the following equation:

$$\bar{u} = 20.51 + 0.1147X_1 + 0.0005X_2 + 0.4333X_3 - 0.4072X_4$$

where $\bar{u}$ = mean spot speed (mph),
X_1 = out-of-state passenger cars (percent),
X_2 = minimum sight distance (ft),
X_3 = posted speed limit (mph),
X_4 = roadside establishments (number per mile along both sides of the highway).

In this model for four-lane highways in rural areas, design and operation controls are quantified by the variables of the percentage of out-of-state passenger cars, the posted speed limit, and the number of roadside establishments per mile, while the only geometric variable is minimum sight distance.[13] The models for both the two-lane and the four-lane highways in rural areas permit the estimation of mean spot speeds with an acceptable degree of confidence.

TABLE 4.4
Variance in Open Highway Speeds on State Highways in Illinois, 1960

Type of Highway and Traffic Volume in ADT	Standard Deviation (mph)	Range of One Standard Error of Estimate (mph)
Two-lane, 2,000 ADT	9.1	±0.73
Two-lane, 5,000 ADT	8.3	±0.73
Two-lane, 8,600 ADT	7.3	±0.73
Four-lane highway	9.1	±0.84
Six-lane highway	6.2	±0.40

Source: J. C. Oppenlander, W. F. Bunte, and P. L. Kadakia, "Sample Size Requirements for Vehicular Speed Studies," *Traffic Volume and Speed Studies,* Highway Research Board Bulletin 281 (Washington, D.C.: Highway Research Board, 1961), p. 68.

---

[12] J. C. OPPENLANDER, "Multivariate Analysis of Vehicular Speeds," *Travel Time and Vehicle Speed,* Record 35 (Washington, D.C.: Highway Research Board, 1963), pp. 41–77.
[13] R. H. WORTMAN, "A Multivariate Analysis of Vehicular Speeds on Four-Lane Rural Highways," *Traffic Flow Characteristics, 1963 and 1964,* Record 72 (Washington, D.C.: Highway Research Board, 1965), pp. 1–18.

VARIABILITY OF SPOT SPEEDS

In addition to the variations in average measures of spot speed for various driver, vehicle, roadway, traffic, and environmental conditions, the variability of spot speeds is influenced by highway type and traffic volume conditions. Typical values of standard deviation and standard error of estimate are summarized in Table 4.4 for travel on rural highways in Illinois.

SPACING CHARACTERISTICS

The longitudinal distribution of vehicles in a traffic stream represents the pattern in spacing between common points of successive vehicles. This characteristic of traffic flow is measured either in some unit of time per vehicle or in some unit of distance per vehicle. A knowledge of spacing characteristics has application for estimating traffic delays and available gaps for vehicular or pedestrian crossings, for studying merging maneuvers between two streams of vehicles, for predicting vehicle arrivals at a point of interest, or for timing traffic signal systems.

SPACING FUNDAMENTALS

Headway, which is inversely proportional to traffic volume, is the time interval between the passage of successive vehicles going by a fixed point on the roadway. The unit of time selected for headway measurements is usually seconds.

The longitudinal arrangement of traffic may also be measured by the gap or distance interval between successive vehicles. Gap is recorded in some unit of length, usually feet in the English system and meters in the metric system, and is an inverse measure of traffic density. Average headway and average gap are related by the following equation:

$$\bar{d} = c\bar{h}\bar{u}$$

where $\bar{d}$ = average gap (distance per vehicle),
$\bar{h}$ = average headway (time per vehicle),
$\bar{u}$ = average speed (distance per unit of time),
c = constant.

The average relationship between gap and headway depends on the average speed of the traffic stream that is being observed. When distance, time, and speed are measured, respectively, in units of feet, seconds, and miles per hour, the value of the constant is 1.467. This constant has the value of 0.2778 in the metric system with distance in meters, time in seconds, and speed in kilometers per hour. In general, headway is the more frequently used measure of vehicular spacing and has the same units in the English and the metric systems of measurement.

OBSERVED VALUES

Typical summaries of spacing characteristics are presented in Figure 4.17 for travel in one direction on two-lane rural highways and in Figure 4.18 for travel in one direction on four-lane rural highways. The curves indicate the percentage of observed headways that are shorter than the noted headway values for a selected level of traffic volume. In addition, the average headway is plotted on each diagram for every hundred vehicles.

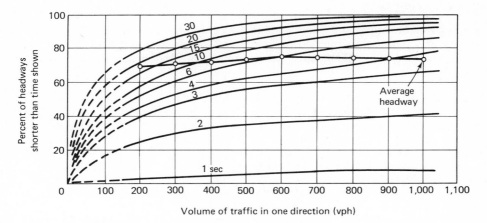

Figure 4.17. Headway distribution on two-lane, rural highways. (Source: *Highway Capacity Manual 1965*, p. 52.)

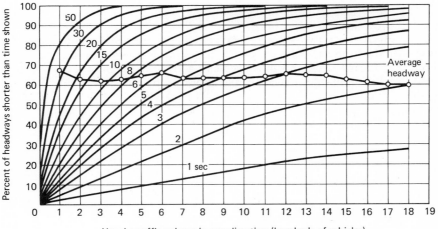

Figure 4.18. Headway distribution on four-lane, rural highways. (Source: *Highway Capacity Manual* 1965, p. 53.)

The relationship between speed and spacing is illustrated from traffic data collected at a temporary bridge on the Merritt Parkway in Connecticut where a headway of 4 sec per vehicle was determined to be the critical time-spacing value. Below a headway of 4 sec per vehicle the rear vehicle was usually traveling slower than the vehicle in front, but at headways above 4 sec per vehicle, the rear vehicle was often traveling at a faster speed than the front vehicle.[14]

---

[14] M. R. PALMER, "The Development of Traffic Congestion," *Quality and Theory of Traffic Flow* (New Haven: Bureau of Highway Traffic, Yale University, 1961), pp. 105–40.

POISSON DISTRIBUTION

Under certain frequently encountered traffic conditions, the number of vehicles arriving at a point in an interval of time follows a random or Poisson distribution. The following requirements define the conditions of randomness from the point of view of vehicle arrivals:

1. Each driver positions his vehicle independently of other vehicles.
2. The number of vehicles passing a point in a given length of time is independent of the number that passes this point in any other equal length of time.

A counting distribution that satisfies the above requirements is the Poisson distribution, which is described by the following expression:

$$P(n \mid \bar{q}t) = \frac{e^{-\bar{q}t}(\bar{q}t)^n}{n!}$$

where $P(n \mid \bar{q}t) =$ probability of the arrival of n vehicles at a point during the time interval t when the average volume is $\bar{q}$ vehicles per unit of time,

   $n =$ number of vehicle arrivals,
   $\bar{q} =$ average volume (vehicles per unit of time),
   $t =$ time interval (units of time),
   $e =$ base of natural logarithm $= 2.718$,
   $n! = n$ factorial $= (n)(n - 1)(n - 2) \cdots (2)(1)$.

If the spacing between two successive vehicles is considered as the occurrence of no arrival, then the headway between these vehicles must be equal to or greater than this time interval during which no other vehicles arrive. When the value of zero is substituted for the number of arrivals in the above probability statement, the following equation results:

$$P(h \geq t) = e^{-\bar{q}t}$$

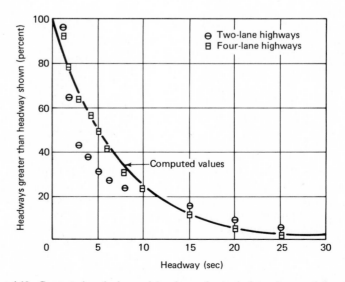

Figure 4.19. Computed and observed headways for typical two-lane and four-lane highways carrying 500 vph in one direction. (Source: *Highway Capacity Manual 1965*, p. 55.)

Comparisons of theoretical values computed for the Poisson distribution with observed headways evidence a close agreement for travel on four-lane rural highways. This goodness of fit is demonstrated in Figure 4.19 for a one-direction traffic volume of 500 vehicles per hour. However, the Poisson distribution is less effective in describing the actual headway patterns for two-lane highways in rural areas. This lack of agreement is explained by two factors that are characteristic of travel on two-lane roads: (1) Although the Poisson distribution is continuous to the time value of zero, this condition is impossible on a two-lane highway because each headway must contain the time for a vehicle to travel its own length and (2) the lack of passing opportunities, particularly as traffic volumes increase, causes the platooning or bunching of traffic behind the slower moving vehicles. These two situations explain the lack of fit in Figure 4.19 for computed and observed headways on two-lane rural highways.

Other probabilistic distributions have been proposed and validated for various traffic flow conditions. These relationships are fully described in several textbooks and are further developed in Chapter 7, Traffic Flow Theory.

ACCIDENT CHARACTERISTICS

Traffic accidents are indicative of failures in the travel interactions of driver, vehicle, roadway, traffic, and environmental conditions. Although the rate of traffic accidents is a common evaluation parameter in the description of accident characteristics, this index is often an insensitive measure because of the rare occurrence of an accident.

Accident indices for roadway sections may be expressed in terms of the numbers of accidents per person, per vehicle registration, or per vehicle mile of travel. The number of accidents per vehicle mile of travel is the most common index that is used to express the relative safety of highway or street sections. However, this measure does not represent a practical basis for quantifying the accident experience at a specific location without length, such as an intersection, a horizontal curve, or a merging area. When the site of risk is identified with a unit of risk, such as at an interchange, accident rates are expressed simply as volume based indices that relate the number of accidents occurring at a given location during some period of time to the volume of vehicles passing that location during the same period of time.

Although accident frequency describes the relative occurrence of collisions at a particular location or along a roadway section, a better measure of the accident pattern is realized when the accident frequency is combined with evaluations of the accident severity. Gross measures of severity are sometimes expressed in terms of the number of traffic fatalities per person, per vehicle registration, or per vehicle mile of travel or as the average total cost of the accidents per year. The product of the rate of involvement in traffic accidents (accidents per vehicle mile of travel) and the severity of the accidents (dollars per accident) provides an economic evaluation of the accident characteristics along a roadway section as well as a measure of the combined effect of accident frequency and accident severity. An example of the technique of combining accident involvement rates with accident severity data is given in Figures 4.20, 4.21, and 4.22. Rates of vehicular involvement and accident severity are shown in Figures 4.20 and 4.21, respectively, as a function of vehicular speed for travel on urban streets. The product of accident frequency and accident severity provides the accident cost curves of Figure 4.22, which more fully describes the accident characteristics as a function of travel speed on urban streets. In addition, these unit costs

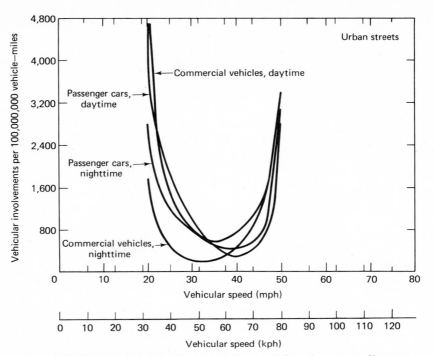

Figure 4.20. Vehicular involvements vs. vehicular speed for urban streets. [Source: J. C. Marcellis, "An Economic Evaluation of Traffic Movement at Various Speeds," *Travel Time and Vehicle Speed*, Record 35 (Washington, D.C.: Highway Research Board, 1963), p. 30.]

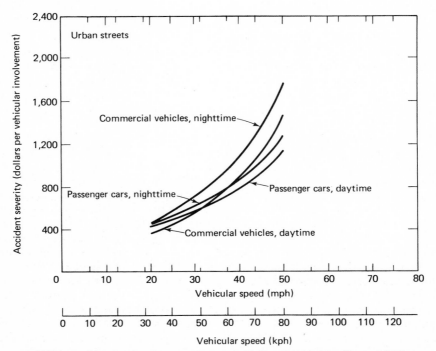

Figure 4.21. Accident severity vs. vehicular speed for urban streets. (Source: J. C. Marcellis, "An Economic Evaluation of Traffic Movement at Various Speeds," p. 30.)

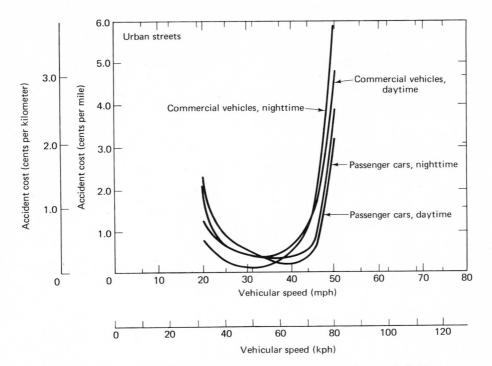

Figure 4.22. Accident costs vs. vehicular speed for urban streets. (Source: J. C. Marcellis, "An Economic Evaluation of Traffic Movement at Various Speeds," p. 31.)

permit the determination of the total accident cost per year for use in computing the road-user costs for a given street section. Further information related to the development and use of accident rates is found in Chapter 9, Accident Analysis.

ACCIDENT TRENDS

The validity of traffic accident statistics is predicated on the reliability of accident reporting. Because better uniformity exists in the reporting of fatal accidents, death or fatal accident rates are often used to compare traffic accident patterns among various governmental jurisdictions. If comparative analyses are based on total accidents, then personal injury and property damage only accidents must be reported on comparable bases. Considerable lack of uniformity exists in the reporting of property damage only accidents among various governmental agencies.

A general summary of fatal-accident statistics is presented in Table 4.5 for travel in the United States during the period 1913–1972. Although the numbers of vehicles and drivers and the amount of travel have been steadily increasing over the years, highway death rates on the basis of the number of motor vehicles and the vehicle miles of travel have been decreasing. Recently, the death rate on a population basis has begun to increase. The trends of these yearly accident data are graphically portrayed in Figures 4.23 and 4.24. A more detailed picture of the historical trend in traffic fatalities is summarized in Table 4.6 according to the nature of the collision.

TABLE 4.5

TABLE 4.5
Motor-Vehicle Death Rates and Costs, 1913 to 1972

Year	No. of Deaths	No. of Vehicles (millions)	Vehicle Miles (billions)	No. of Drivers (millions)	Death Rates Per 10,000 Motor Vehicles	Death Rates Per 100,000,000 Vehicle Miles	Death Rates Per 100,000 Population	Costs ($ million) *
1913–17 ave.	6,800	2.9	†	4.0	23.80	†	6.8	‡
1918–22 ave.	12,700	9.2	†	14.0	13.90	†	11.9	‡
1923–27 ave.	21,800	19.7	120	29.0	11.10	18.20	18.8	‡
1928–32 ave.	31,050	25.7	199	38.0	12.10	15.60	25.3	1,300
1933	31,363	24.2	201	35.0	12.98	15.60	25.0	1,350
1934	36,101	25.3	216	37.0	14.29	16.75	28.6	1,550
1935	36,369	26.5	229	39.0	13.70	15.91	28.6	1,550
1936	38,089	28.5	252	42.0	13.36	15.11	29.7	1,650
1937	39,643	30.1	270	44.0	13.19	14.68	30.8	1,800
1938	32,582	29.8	271	44.0	10.93	12.02	25.1	1,500
1939	32,386	31.0	285	46.0	10.44	11.35	24.7	1,500
1940	34,501	32.5	302	48.0	10.63	11.42	26.1	1,600
1941	39,969	34.9	334	52.0	11.45	11.98	30.0	1,900
1942	28,309	33.0	268	49.0	8.58	10.55	21.1	1,600
1943	23,823	30.9	208	46.0	7.71	11.44	17.8	1,250
1944	24,282	30.5	213	45.0	7.97	11.42	18.3	1,250
1945	28,076	31.0	250	46.0	9.05	11.22	21.2	1,450
1946	33,411	34.4	341	50.0	9.72	9.80	23.9	2,200
1947	32,697	37.8	371	53.0	8.64	8.82	22.8	2,650
1948	32,259	41.1	398	55.0	7.85	8.11	22.1	2,800
1949	31,701	44.7	424	59.3	7.09	7.47	21.3	3,050
1950	34,763	49.2	458	62.2	7.07	7.59	23.0	3,100
1951	36,996	51.9	491	64.4	7.13	7.53	24.1	3,400
1952	37,794	53.3	514	66.8	7.10	7.36	24.3	3,750
1953	37,955	56.3	544	69.9	6.74	6.97	24.0	4,300
1954	35,586	58.6	562	72.2	6.07	6.33	22.1	4,400
1955	38,426	62.8	606	74.7	6.12	6.34	23.4	4,500
1956	39,628	65.2	631	77.9	6.07	6.28	23.7	5,000
1957	38,702	67.6	647	79.6	5.73	5.98	22.7	5,300
1958	36,981	68.8	665	81.5	5.37	5.56	21.3	5,600
1959	37,910	72.1	700	84.5	5.26	5.41	21.5	6,200
1960	38,137	74.5	719	87.4	5.12	5.31	21.2	6,500
1961	38,091	76.4	738	88.9	4.98	5.16	20.8	6,900
1962	40,804	79.7	767	92.0	5.12	5.32	22.0	7,300
1963	43,564	83.5	805	93.7	5.22	5.41	23.1	7,700
1964	47,700	87.3	847	95.6	5.46	5.63	25.0	8,100
1965	49,163	91.8	888	99.0	5.36	5.54	25.4	8,900
1966	53,041	95.9	930	101.0	5.53	5.70	27.1	10,000
1967	52,924	98.9	962	103.2	5.35	5.50	26.8	10,700
1968	54,862	103.1	1,016	105.4	5.32	5.40	27.5	11,300
1969	55,791	107.4	1,071	108.3	5.19	5.21	27.7	12,200
1970	54,800	111.2	1,120	111.5	4.93	4.89	26.8	13,600
1971	54,700	116.3	1,186	114.4	4.70	4.61	26.5	15,800
1972	56,600	121.4	1,250	118.2	4.66	4.53	27.2	19,400

*Cost figures are not completely comparable through all years. As additional or more precise cost data have become available through the years, they have been used in developing cost estimates from that year forward, but previously estimated figures were not revised. Not included are costs of certain public agency activities such as police, fire, and courts; damages awarded in excess of direct cost; indirect costs to employers, etc.

†Mileage data inadequate prior to 1923.

‡Data insufficient for estimating costs.

Sources: *Accident Facts* (Chicago: National Safety Council, 1973), p. 59. Death totals from National Center for Health Statistics except 1964, 1970, 1971, and 1972 which are NSC estimates based on data from state traffic authorities. Motor vehicle registrations, mileage, and drivers from Federal Highway Administration. Costs are NSC estimates.

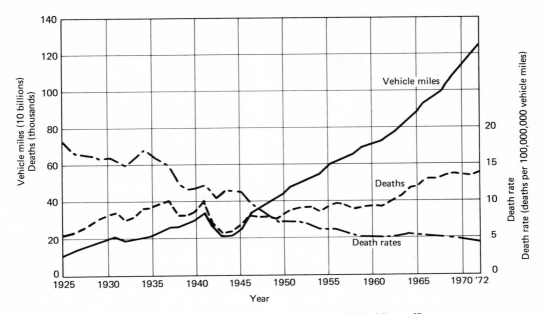

Figure 4.23. Trends in highway travel and traffic deaths in the United States. [Source: *Accident Facts* (Chicago: National Safety Council, 1973), p. 40.]

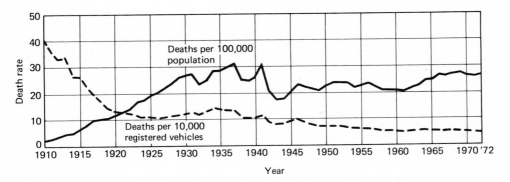

Figure 4.24. Trends in traffic death rates in the United States. (Source: *Accident Facts*, p. 41.)

ACCIDENT PATTERNS

For 1972, 17,000,000 traffic accidents were reported in the United States, and this total statistic represents 48,800 fatal, 1,400,000 nonfatal injury, and 15,600,000 property damage accidents on the various highways and streets. An analysis of the traffic deaths in 1972 is provided in Figure 4.25 for certain classifications of interest to the traffic engineer. Approximately 68 percent of the highway deaths occurred in rural areas, although less than half of all vehicular travel occurs in these locations. However, pedestrian fatalities are greater in urban than rural areas. Although traffic

TABLE 4.6
Motor-Vehicle Deaths by Type of Accident, 1913 to 1972

Year	Total Deaths*	Deaths from Collision with							Deaths from Non-collision Accidents
		Pedestrians	Other Motor Vehicles	Railroad Trains	Street Cars	Pedalcycles	Animal-drawn Veh. or Animal	Fixed Objects	
1913–17 ave.	6,800	†	†	†	†	†	†	†	†
1918–22 ave.	12,700	†	†	†	†	†	†	†	†
1923–27 ave.	21,800	†	†	1,200	480	†	†	†	†
1928–32 ave.	31,050	12,300	5,700	1,850	450	†	†	700	9,100
1933	31,363	12,840	6,470	1,437	318	400	310	900	8,680
1934	36,101	14,480	8,110	1,457	332	500	360	1,040	9,820
1935	36,369	14,350	8,750	1,587	253	450	250	1,010	9,720
1936	38,089	15,250	9,500	1,697	269	650	250	1,060	9,410
1937	39,643	15,500	10,320	1,810	264	700	200	1,160	9,690
1938	32,582	12,850	8,900	1,490	165	720	170	940	7,350
1939	32,386	12,400	8,700	1,330	150	710	200	1,000	7,900
1940	34,501	12,700	10,100	1,707	132	750	210	1,100	7,800
1941	39,969	13,550	12,500	1,840	118	910	250	1,350	9,450
1942	28,309	10,650	7,300	1,754	124	650	240	850	6,740
1943	23,823	9,900	5,300	1,448	171	450	160	700	5,690
1944	24,282	9,900	5,700	1,663	175	400	140	700	5,600
1945	28,076	11,000	7,150	1,703	163	500	130	800	6,600
1946	33,411	11,600	9,400	1,703	174	540	130	950	8,900
1947	32,697	10,450	9,900	1,736	102	550	150	1,000	8,800
1948	32,259	9,950	10,200	1,474	83	500	100	1,000	8,950
1949	31,701	8,800	10,500	1,452	56	550	140	1,100	9,100
1950	34,763	9,000	11,650	1,541	89	440	120	1,300	10,600
1951	36,996	9,150	13,100	1,573	46	390	100	1,400	11,200
1952	37,794	8,900	13,500	1,429	32	430	130	1,450	11,900
1953	37,955	8,750	13,400	1,506	26	420	120	1,500	12,200
1954	35,586	8,000	12,800	1,269	28	380	90	1,500	11,500
1955	38,426	8,200	14,500	1,490	15	410	90	1,600	12,100
1956	39,628	7,900	15,200	1,377	11	440	100	1,600	13,000
1957	38,702	7,850	15,400	1,376	13	460	80	1,700	11,800
1958	36,981	7,650	14,200	1,316	9	450	80	1,650	11,600
1959	37,910	7,850	14,900	1,202	6	480	70	1,600	11,800
1960	38,137	7,850	14,800	1,368	5	460	80	1,700	11,900
1961	38,091	7,650	14,700	1,267	5	490	80	1,700	12,200
1962	40,804	7,900	16,400	1,245	3	500	90	1,750	12,900
1963	43,564	8,200	17,600	1,385	10	580	80	1,900	13,800
1964	47,700	9,000	19,600	1,580	5	710	100	2,100	14,600
1965	49,163	8,900	20,800	1,556	5	680	120	2,200	14,900
1966	53,041	9,400	22,200	1,800	2	740	100	2,500	16,300
1967	52,924	9,400	22,000	1,620	3	750	100	2,350	16,700
1968	54,862	9,900	22,400	1,570	4	790	100	2,700	17,400
1969	55,791	10,100	23,700	1,495	2	800	100	3,900	15,700
1970	54,800	10,400	23,300	1,530	‡	820	100	4,450	14,200
1971	54,700	10,600	23,300	1,500	‡	850	100	4,650	13,700
1972	56,600	10,700	24,200	1,500	‡	1,100	100	4,600	14,400
Changes in Deaths									
1962–72	+39%	+35%	+48%	+20%	—	+120%	+11%	‡	‡
1971–72	+ 3%	+ 1%	+ 4%	0%	—	+ 29%	0%	−1%	+5%

*Yearly totals may not quite equal sums of the various types because totals for most types are estimated, and these have been rounded.

†Insufficient data for approximations.

‡Not available.

Sources: *Accident Facts*, p. 58. Deaths are based on data from National Center for Health Statistics, state traffic authorities, and Federal Railroad Administration.

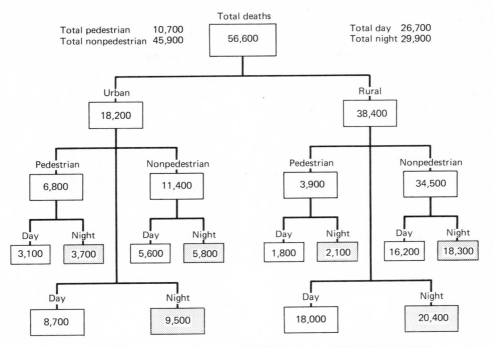

Figure 4.25. Principal classes of motor-vehicle deaths in the United States in 1972. (Source: *Accident Facts*, p. 41.)

volumes are generally heavier in the daytime as compared to the nighttime, about 53 percent of the traffic deaths occur at night. For all travel the fatality rate is 3.0 and 8.0 deaths per 100,000,000 vehicle miles of travel, respectively, for day and night conditions.

A directional analysis of motor-vehicle traffic accidents is summarized in Table 4.7. Data included are the location (urban or rural) and the movements of the vehicle(s) involved in the accident. The greatest number of pedestrian fatal accidents are in urban areas at nonintersection locations where little or no safety features are available for pedestrian traffic. Two motor-vehicle accidents represent the major category of traffic collisions. Accidents that result from vehicles running off the roadway are significant in terms of relative frequency and severity. Single-vehicle accidents are more than three times as likely to result in a fatality as other motor-vehicle accidents.[15]

Acts of improper driving that were related to the cause of accidents are presented in Table 4.8 for various accident classifications. Although speed too fast for conditions is an important factor in terms of accident severity, right-of-way violations are the most common driver error specified in the various categories of improper driving. Plots of persons injured and property damage as a function of travel speed are illustrated in Figure 4.26 for traffic accidents on main rural highways. A curvilinear

15 J. S. Baker, *Single-Vehicle Accidents—A Summary of Research Findings*, Automotive Safety Foundation, 1968.

TABLE 4.7
Directional Analysis of Motor-Vehicle Traffic Accidents, 1972

Location and Vehicle Movement	Fatal Accidents			All Accidents		
	Total	Urban	Rural	Total	Urban	Rural
	Percent Distribution of Accidents					
Total Accidents	100.0	100.0	100.0	100.0	100.0	100.0
PEDESTRIAN	21.6	39.9	11.8	2.6	3.0	1.1
Intersection	5.8	15.1	0.9	1.0	1.1	0.1
Car —going straight	5.1	13.2	0.9	0.8	0.8	0.1
—turning right	0.2	0.6	*	0.1	0.1	*
—turning left	0.4	0.9	*	0.1	0.2	*
—backing	0.1	0.4	*	*	*	*
All others	*	*	*	*	*	*
Nonintersection	15.8	24.8	10.9	1.6	1.9	1.0
Car —going straight	15.1	23.5	10.6	1.5	1.7	1.0
—turning right	0.1	0.2	*	*	*	*
—turning left	*	0.1	*	*	*	*
—backing	0.2	0.5	0.1	0.1	0.1	*
All others	0.4	0.5	0.2	*	0.1	*
TWO MOTOR VEHICLE	38.7	32.2	42.1	76.8	85.2	55.6
Intersection	14.9	18.1	13.2	31.7	36.5	19.8
Entering at angle	10.5	12.4	9.5	15.1	17.4	9.3
Entering same direction						
—both going straight	0.4	0.4	0.4	2.4	3.0	0.8
—one turn, one straight	0.5	0.4	0.5	3.4	3.5	3.3
—one stopped	0.4	0.3	0.5	4.4	5.1	2.4
—all others	0.1	*	0.1	0.8	0.9	0.7
Entering opposite direction						
—both going straight	0.8	1.2	0.6	0.9	1.1	0.4
—one left, one straight	2.1	3.3	1.5	4.3	5.1	2.5
—all others	0.1	0.1	0.1	0.4	0.4	0.4
Nonintersection	23.8	14.1	28.9	45.1	48.7	35.8
Opposite dir.—both moving	15.5	6.1	20.5	3.9	2.8	7.0
Same dir.—both moving	4.0	2.8	4.6	10.9	11.1	10.4
One car parked	1.2	2.4	0.6	11.3	14.5	3.0
One car stopped in traffic	1.2	1.3	1.1	10.4	12.1	6.3
One car ent. parked position	*	*	*	0.2	0.3	*
One car lv. parked position	0.1	0.1	0.1	1.6	1.7	1.3
One car ent. alley, driveway	0.8	0.5	1.0	2.4	1.8	3.4
One car lv. alley, driveway	0.5	0.5	0.5	2.9	3.0	2.5
All others	0.5	0.4	0.5	1.5	1.4	1.9
OTHER COLLISIONS	13.0	13.8	12.6	6.2	5.7	7.8
Intersection	1.6	3.6	0.7	1.5	1.7	1.2
Collision with						
—nonmotor-veh., train, bike	1.0	2.4	0.5	0.6	0.6	0.6
—fixed object in road	0.6	1.2	0.2	0.9	1.1	0.6
Nonintersection	11.4	10.2	11.9	4.7	4.0	6.6
Collision with						
—nonmotor-veh., train, bike	3.5	2.2	4.0	0.9	0.6	1.9
—fixed object in road	7.9	8.0	7.9	3.8	3.4	4.7
NONCOLLISION	26.7	14.1	33.5	14.4	6.1	35.5
Ran off road	24.3	12.1	31.0	11.9	4.9	29.4
At curve—nonintersection	9.7	2.8	13.5	3.3	0.6	10.0
On str. road—intersection	1.7	1.4	1.8	1.5	1.0	2.9
—nonintersection	12.9	7.9	15.7	7.1	3.3	16.5
Overturned in road	1.1	0.8	1.2	0.6	0.4	1.3
Fell from moving vehicle	0.7	0.8	0.6	0.1	0.1	0.1
All others	0.6	0.4	0.7	1.8	0.7	4.7

*Less than 0.05%.

Sources: Reports of city and state traffic authorities, as follows: Urban—100 cities over 50,000 population; Rural—14 states; Total—NSC estimates based on Urban and Rural reports. *Accident Facts*, p. 46.

TABLE 4.8
Improper Driving Reported in Accidents, 1972

Kind of Improper Driving	Fatal Accidents			Injury Accidents			All Accidents*		
	Total	Urban	Rural	Total	Urban	Rural	Total	Urban	Rural
Total	100.0%	100.0%	100.0%	100.0%	100.0%	100.0%	100.0%	100.0%	100.0%
Improper driving	78.5	76.4	79.2	87.8	83.5	91.0	88.4	85.8	92.8
Speed too fast†	26.9	20.1	28.4	19.3	11.5	25.1	14.6	8.2	24.9
Right of way	13.1	21.1	11.1	20.3	30.1	13.2	20.2	24.2	13.8
Failed to yield	9.1	14.7	7.7	14.5	21.0	9.8	14.9	17.7	10.3
Passed stop sign	2.8	2.9	2.8	2.9	3.7	2.3	2.6	2.8	2.4
Disregarded signal	1.2	3.5	0.6	2.9	5.4	1.1	2.7	3.7	1.1
Drove left of center	12.4	5.8	14.2	4.3	2.4	5.8	3.6	2.3	5.8
Improper overtaking	1.9	2.0	1.9	1.9	1.4	2.3	3.0	3.1	2.7
Made improper turn	0.6	1.0	0.6	1.3	2.1	0.7	3.0	4.1	1.1
Followed too closely	1.0	2.5	0.6	8.4	9.9	7.3	11.6	14.3	7.1
Other improper driving	22.6	23.9	22.4	32.3	26.1	36.6	32.4	29.6	37.4
No improper driving stated	21.5	23.6	20.8	12.2	16.5	9.0	11.6	14.2	7.2

*Principally property damage accidents, but also includes fatal and injury accidents.
†Includes "speed too fast for conditions."

Sources: Reports of state and city traffic authorities, as follows: Urban—43 cities; Rural—13 states; Total—NSC estimates based on Urban and Rural reports. *Accident Facts*, p. 48.

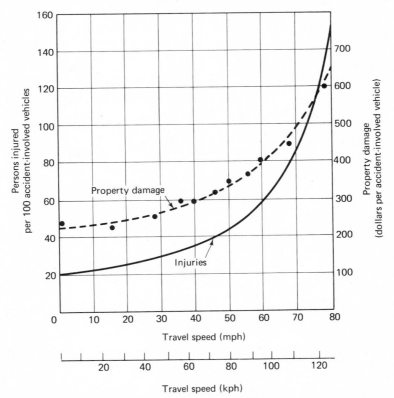

Figure 4.26. Speed of travel as related to persons injured and property damage for traffic accidents. [Source: D. Solomon, *Accidents on Main Rural Highways Related to Speed, Driver, and Vehicle*, (Washington, D.C.: Federal Highway Administration, July 1964), p. 11.]

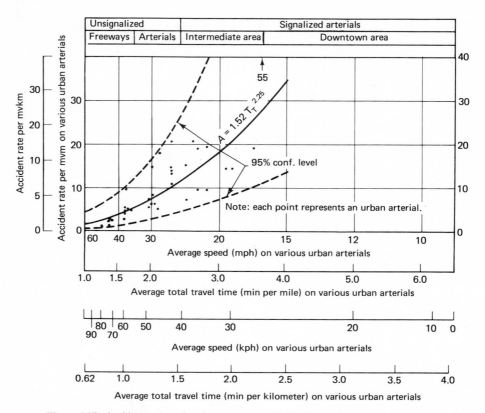

Figure 4.27. Accident rates related to average total time for various urban arterials. (Source: A. D. May, Jr., "A Friction Concept of Traffic Flow," p. 508.)

relationship between accident rate and speed or travel time has been developed for travel conditions on various classes of urban roadways and is presented in Figure 4.27. In several special studies of driver behavior, drinking drivers were involved in at least half of the fatal accidents. In addition, approximately 50 percent of the fatal accidents in which drinking was a factor were single-vehicle accidents. The corresponding statistic for nondrinking drivers was about 35 percent.

A summary of types of motor vehicles involved in traffic accidents is contained in Table 4.9. Although passenger cars represent about 79 percent of the registered vehicles, this type of vehicle is involved in 75 percent of the fatal accidents and 84 percent of all traffic accidents. In the truck category, combination trucks have greater accident frequency and accident severity than their proportionate share of the vehicle registrations. This discrepancy is probably because of the greater number of miles of travel for truck combinations as compared to the other types of motor vehicles.

Several correlations have been developed between motor-vehicle accidents and traffic volumes. Relationships of accident rates and average daily traffic volumes are shown in Figure 4.28 for travel on two- and four-lane highways in rural areas. Similar patterns exist for traffic accidents and vehicle miles of travel when these two

TABLE 4.9
Types of Motor Vehicles Involved in Accidents, 1972

Type of Vehicle	In Fatal Accidents		In All Accidents		Percent of Total Vehicle Registrations*	No. of Occupant Fatalities
	Number	%	Number	%		
All Types	70,900	100.0	29,100,000	100.0	100.0	§
Passenger cars	53,400	75.3	24,500,000	84.2	79.2	35,100
Trucks	12,700	17.9	3,500,000	12.0	17.1	5,500
Truck or truck tractor	8,000	11.3	2,830,000	9.7	16.3	‖
Truck tractor and semitrailer	3,700	5.2	440,000	1.5 ⎫	0.8	‖
Other truck combination	1,000	1.4	230,000	0.8 ⎭		‖
Farm tractors, equipment	200	0.3	22,000	0.1	†	180
Taxicabs	200	0.3	180,000	0.6	0.2	120
Buses, commercial	350	0.5	170,000	0.6	0.1	100
Buses, school	150	0.2	45,000	0.2	0.3	50
Motorcycles	2,600	3.7	320,000	1.1 ⎫	3.1	2,500
Motor scooters, motor bikes	200	0.3	23,000	0.1 ⎭		200
Other‡	1,100	1.5	340,000	1.1	‖	950

*Percentage figures are based on numbers of vehicles and do not reflect miles traveled or place of travel, both of which affect accident experience.
†These vehicles are not included in total vehicle registrations; estimated number—4,500,000.
‡Includes fire equipment, ambulances, special vehicles, other.
§In addition to these occupant fatalities, there were 10,700 pedestrian, 1,100 pedalcyclist, and 100 other deaths.
‖Data not available.

Sources: Based on reports from 16 state traffic authorities. Vehicle registrations based on data from Federal Highway Administration and International Taxicab Association. *Accident Facts*, p. 56.

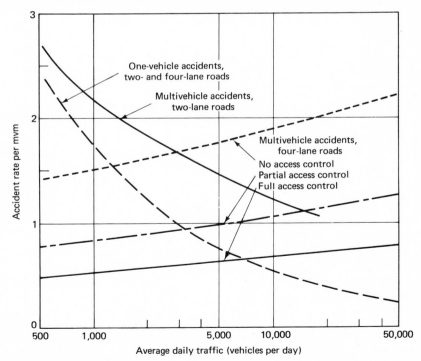

Figure 4.28. Predicted accident rates for two-mile segments of various roadway types. [Source: J. K. Kihlberg and K. J. Tharp, *Accident Rates as Related to Design Elements of Rural Highways*, Report 48 (Washington, D.C.: Highway Research Board, 1968), p. 23.]

statistics are plotted as a function of the time of day. In Figure 4.29, however, accidents respond with greater variability to time of day than to vehicle miles of travel.

An analysis of motor-vehicle death rates by states is summarized in Figure 4.30. Approximate comparisons of traffic safety in different countries around the world are provided by the summary data listed in Table 4.10.

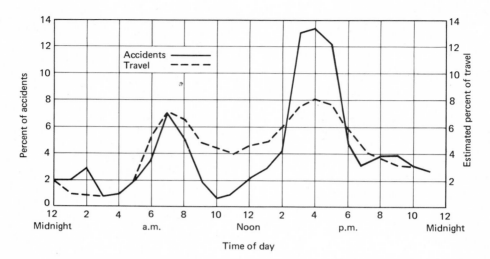

Figure 4.29. Distributions of travel accidents and travel by time of day. [Source: Gilbert T. Satterly, Jr., and Donald E. Cleveland, *Traffic Control and Roadway Elements —Their Relationship to Highway Safety* (Washington, D.C.: Highway Users Federation for Safety and Mobility, 1969), p. 6.]

TABLE 4.10
Motor-Vehicle Deaths by Nations

Nation	Year	Deaths	Rate*	Nation	Year	Deaths	Rate*
Mexico	1971	4,115†	8.1	Italy	1970	13,002‖	24.3
Belgium	1971	1,148‖	13.4	Netherlands	1971	3,171	24.4
Spain	1970	4,583§	13.4	Canada	1970	5,312†	24.8
United Kingdom	1971	8,182	14.7	Denmark	1970	1,221	24.8
Norway	1970	596	15.4	Switzerland	1970	1,617	26.1
Sweden	1970	1,376	17.1	Portugal	1971	2,359‖	26.6
Ireland	1971	533	17.9	United States	1969	55,791†	27.6
Japan	1971	21,020	20.1	Australia	1971	3,847	30.1
France	1970	11,852‡	23.3	Germany, Fed. Rep.	1970	19,143	31.1
Finland	1970	1,081	23.4	Austria	1971	2,731	36.6

*Deaths per 100,000 population. Death definition: In general, deaths are included if they occur within thirty days after the accident, but other time periods are used as follows:
†one year,
‡three days,
§twenty-four hours,
‖at accident scene only.

Source: *Accident Facts*, p. 71.

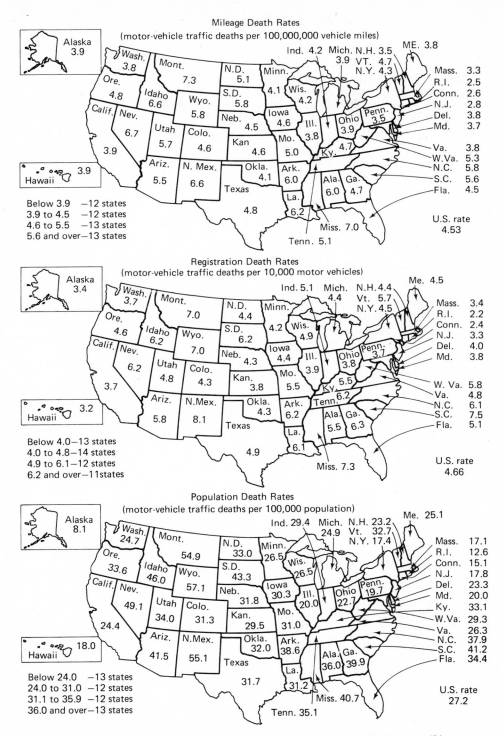

Figure 4.30. Motor-vehicle death rates by states, 1972. (Source: *Accident Facts*, p. 62.)

ACCIDENTS AND HIGHWAY DESIGN AND OPERATION

The design and operation of highways and streets interact with various driver, vehicle, traffic, environmental, and other roadway factors to affect accident frequency, severity, and type. In regard to the general design aspects of roadways, the safety advantages of access control have been demonstrated in many studies. Accidents and fatality rates on highways with full control of access are consistently one-third to one-half those of facilities with no access control. Although partial and full control of access are beneficial in rural areas, the greatest safety benefits are derived through the application of full access control in urban areas.[16]

Alinement. The horizontal and vertical alinement of roadways influence both the expediency and safety of highway travel. Horizontal curvature is strongly related to traffic accidents with significant increases in accidents for curvature in excess of 8°.[17] The combined effects of curvature and frequency of curvature on accident rates are summarized in Table 4.11. However, both curve warning signs with advisory speed signs and roadway delineators have been effective in reducing accidents on horizontal curves. Although the relationships between accident rates and upgrades and downgrades are not completely clear, the rate of traffic accidents appears to increase with increased gradient of the roadway. Increased accident rates have been recorded in both crest and sag vertical curves as compared to the tangent portion of the vertical alinement.[18]

Cross section.[19] Lane width is a significant factor in injury and fatal accidents, and accident rates increase significantly and uniformly for lane widths less than 11 ft.

TABLE 4.11
Accident Rates on Two-Lane Highways by Degree of Curvature
and Frequency of Curves

Number of Curves per Mile	Accident Rate* When Degree of Curvature Is			
	0–2.9	3–5.9	6–9.9	10 or More
0–0.9	3.0	5.4	4.2	8.9
1.0–2.9	2.3	3.7	4.5	4.2
3.0–4.9	2.1	2.9	3.3	4.3
5.0–6.9	3.3	3.2	2.8	4.6

*Accident rate per million vehicle-miles.

Source: M. S. Raff, "Interstate Highway Accident Study," *Traffic-Accident Studies*, Highway Research Board Bulletin 74 (Washington, D.C.: Highway Research Board, 1953), p. 35.

---

[16] ROBERT F. DAWSON and J. C. OPPENLANDER, "General Design," *Traffic Control and Roadway Elements—Their Relationship to Highway Safety/Revised* (Washington, D.C.: Highway Users Federation for Safety and Mobility, 1971), p. 8.

[17] JACK E. LEISCH AND ASSOCIATES, "Alinement," *Traffic Control and Roadway Elements—Their Relationship to Highway Safety/Revised* (Washington, D.C.: Highway Users Federation for Safety and Mobility, 1971), pp. 1–5.

[18] B. F. K. MULLINS and C. J. KEESE, "Freeway Traffic Accident Analysis and Safety Study," *Freeway Design and Operations*, Bulletin 291 (Washington, D.C.: Highway Research Board, 1961), pp. 26–78.

[19] JOHN A. DEARING and JOHN W. HUTCHINSON, "Cross Section and Pavement Surface," *Traffic Control and Roadway Elements—Their Relationship to Highway Safety/Revised* (Washington, D.C.: Highway Users Federation for Safety and Mobility, 1970).

A general reduction in accident rate based on vehicle miles of travel results as the median width of divided highways increases. In addition, this same measure of traffic accidents shows a significant increase with increasing frequency of the occurrence of median openings. Because approximately 80 percent of the vehicles leaving the roadway travel less than 30 ft from the pavement edge, at least a 30-ft zone should be clear of roadside obstacles along the highway. Unless practical or economic reasons do not allow for this refuge area, then properly designed guardrails may be warranted to reduce the severity of traffic accidents that result from vehicles running off the roadway.

Intersections.[20] Various features of the roadway, such as intersections, interchanges, driveways, and railroad grade crossings, provide discontinuities to the flow of traffic and often produce interrupted-flow conditions that have accident patterns different from those characteristics of the uninterrupted-flow sections of the roadway. The kind of traffic control selected for an intersection is predicated on the functional classifications of the intersecting roadways, topographic features, geometric arrangements, traffic volumes and speeds, accident patterns, and land use characteristics. Although yield signs are an effective control device under many low-volume situations, accident rates increase as cross-street volume increases and decrease as main-street volume increases for intersections controlled as two-way stops. A change from two-way to four-way stops provided a 56-percent reduction in total accidents for major intersecting streets with similar vehicular volumes. Various reports on the effect of traffic signals on traffic operations and safety have provided conflicting conclusions. Although total accidents have both increased and decreased at intersections after the installation of traffic signals, accident severity is often reduced because rear-end collisions are substituted for right-angle accidents.

Interchanges.[21] Although interchanges are designed to permit reasonably safe and efficient crossing, diverging, and merging of traffic flows, any discontinuity or change in driver, vehicle, roadway, traffic, or environmental conditions in the area of the interchange affords additional hazards to freeway travel. Accident rates in terms of vehicle miles of travel are shown in Figure 4.31 by type of interchange unit for interchanges located in both rural and urban areas. Relatively safer designs are produced when the main freeway passes over the minor facility and when the ramp terminals are at least 750 ft from the structure. On-ramps become high-accident locations in urban areas, but off-ramps represent the greatest accident problem in rural areas. The safety of entrance terminals is improved with geometric designs that provide auxiliary lanes or acceleration lanes of 800 ft or more in length. Deceleration lanes with a length of 800 ft or more eliminate traffic friction on the through lanes and account for reduced accident rates. In addition, adequate sight distances are essential at entrance and exit terminals. Geometric designs for weaving maneuvers should provide weaving sections that are at least 800 ft in length.

---

[20] PAUL C. BOX AND ASSOCIATES, "Intersections," *Traffic Control and Roadway Elements—Their Relationship to Highway Safety/Revised* (Washington, D.C.: Highway Users Federation for Safety and Mobility, 1970).

[21] J. C. OPPENLANDER and ROBERT F. DAWSON, "Interchanges," *Traffic Control and Roadway Elements—Their Relationship to Highway Safety/Revised* (Washington, D.C.: Highway Users Federation for Safety and Mobility, 1970).

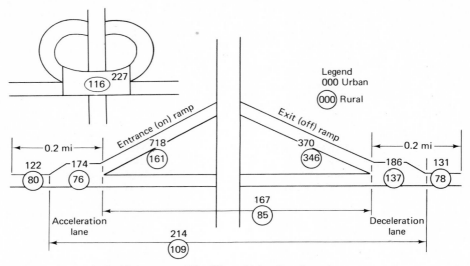

Accident rate per 100 million vehicle miles of travel

Figure 4.31. Accident rate by types of interchange unit. [Source: J. A. Cirillo, "Interstate System Accident Research Study II, Interim Report II," *Public Roads*, Vol. 35 (Washington, D.C.: Federal Highway Administration, 1968), p. 74.]

Driveways.[22] Although driveways are essential to provide access to various forms of land use, the resultant interruptions to traffic flow produce hazardous roadway situations. A positive relation exists between the number of commercial establishments along a roadway and the accident rate based on either vehicle miles of travel or route miles. Similar findings are evident in driveway accident characteristics on urban and rural routes for which the left-turn entry movement to the driveway is the most critical factor. However, left-turn lanes recessed in the median area are effective in reducing accidents.

Railroad crossings. Accidents that occur at railroad-highway grade crossings, although a numerically small part of the overall highway accident problem, are usually severe in terms of fatalities, personal injuries, and property damage. The relative hazard of these crossings in rural areas is influenced by the daily train and traffic volumes, the number of railroad tracks, the pavement width of the highways, and the number of roadside distractions on the approaches to the crossing. In an analysis of railroad-highway grade crossing accidents in urban areas, the daily train and traffic volumes, the corner sight triangle, the number of distractions along the roadside on the approaches to the crossing, and the type of crossing protection are important factors in the explanation of potential hazard.[23] The kinds and extent

---

[22] PAUL C. BOX AND ASSOCIATES, "Driveways," *Traffic Control and Roadway Elements—Their Relationship to Highway Safety/Revised* (Washington, D.C.: Highway Users Federation for Safety and Mobility, 1970).

[23] W. D. BERG, T. G. SCHULTZ and J. C. OPPENLANDER, "Proposed Warrants for Protective Devices at Railroad-Highway Grade Crossings," *Traffic Engineering*, XL, No. 8 (May 1970), pp. 38, 42–43 and No. 9 (June 1970), pp. 36–42.

TABLE 4.12
Unsafe Conditions Observed by Diagnostic Team
at Study Rail-Highway Grade Crossings

Conditions Observed	Percent of Crossings at Which Conditions Observed
1. Pavement markings missing, improperly located, or in need of maintenance.	72
2. Vehicles required by law to stop at all crossings would present a hazard to other vehicles by blocking traffic lanes and obstructing view of protective device.	60
3. Driver's visibility of railroad approach obstructed by growth of vegetation.	52
4. Under nighttime conditions lack of illumination presents additional hazards at grade crossing.	44
5. Conflicts for driver's attention created by traffic conditions and the location of traffic control devices on adjacent roadways.	40
6. Advanced warning signs missing, improperly located, or in need of maintenance.	40
7. Absence of area immediately adjacent to grade crossing in which driver may take evasive action.	36
8. Highway signs and fixed objects obstructing driver's view of protective and warning devices.	32
9. Fixed-mount protective devices or barriers presenting fixed-object hazard to vehicles.	32
10. Legally parked vehicle would block driver's view of protective and warning devices.	28
11. Geometrics of roadway design contribute to unsafe conditions at the crossing.	20
12. Railroad protective device not properly located or maintained.	12
13. Traffic conditions on adjacent roadway conducive to vehicles becoming stalled or stopped on railroad tracks.	8

Source: *Railroad-Highway Safety Part I: A Comprehensive Statement of the Problem*, Report to Congress, U.S. Department of Transportation, Nov. 1971 (Washington, D.C.: Government Printing Office, 1971), p. 74.

of unsafe conditions observed at 36 representative railroad-highway grade crossings by a diagnostic study team are listed in Table 4.12.

Parking.[24] Parking has been identified as primarily an urban problem with the highest incidence of accidents occurring on streets less than 40-ft wide, although parking related accidents are characterized by relatively low injury and property damage severity. Safety studies substantiate the finding that angle parking is more hazardous than parallel parking. As an example, the average number of parking accidents per block have been reduced from five with angle parking to one after a change to parallel parking.

One-way streets.[25] One-way street systems are often operated in urban areas to reduce traffic congestion and to improve vehicular movement. In addition, safety benefits are realized in reduced accident frequency and accident severity. In many cases, rear-end, sideswipe, head-on, turning, parking, and pedestrian accidents are reduced with the conversion to one-way movement. Mid-block accidents are generally

---

[24] Peter A. Mayer and Woodrow W. Rankin, "One-Way Streets and Parking," *Traffic Control and Roadway Elements—Their Relationship to Highway Safety/Revised* (Washington, D.C.: Highway Users Federation for Safety and Mobility, 1971), pp. 9–12.

[25] *Ibid.*, pp. 1–8.

reduced more than intersection accidents. In fact, the reduction in mid-block accidents may be as much as twice the reduction in intersectional accidents.

Illumination.[26] Highway illumination aids traffic safety as well as easing the task of driving at nighttime. Accident rates may be reduced after the installation of illumination at intersections and on roadways with high night-to-day accident ratios and low standards of geometric design. Illumination, over-the-roadway vehicle signals, and pedestrian signal indications have been effective safety measures at locations having significant pedestrian volumes.

ACCIDENT COSTS

A complete knowledge of the costs of traffic accidents is essential in the performance of economic analyses. Because the calculable costs of motor-vehicle accidents are subject to various uncertainties, the cost of accidents cannot be exactly determined. According to the U.S. National Safety Council,

the calculable costs of motor-vehicle accidents are wage loss, medical expense, insurance administrative costs, and property damage. The costs of all these items per case in 1972 were:

Death	$82,000
Nonfatal disabling injury	3,400
Property damage accident (including minor injuries)	480

The cost *per death* for ALL accidents—fatal, nonfatal, and property damage—differs for urban and rural accidents, due to differences in the ratios of nonfatal injuries and property damage accidents per death, as indicated below.

	All	Urban	Rural
Nonfatal injuries per death	35	70	20
Property damage accidents[27] per death	280	620	110
Cost, per death[28]	$330,000	$600,000	$210,000

These averages may be used to estimate the cost of motor-vehicle accidents in cities or states, but if the city is small they must be used with care. For example, if the year's total of deaths is five, of which four occurred in one accident, the average of $600,000 per death is not used. It includes the cost of 70 injuries and 620 property damage accidents, and in a "freak" experience these ratios will not hold. In such a case it would be better to use the unit costs for deaths, injuries, and property damage accidents separately.

If a city had only one or two deaths in the course of a year, it will be more satisfactory to use the following unit costs for each death:

Boy under 15 years	$ 70,800	Girl under 15 years	$44,800
Man 15 to 54 years old	118,000	Woman 15 to 54 years old	70,000
Man 55 years old and older	16,500	Woman 55 years and older	12,800

If a city has as many as 10 deaths, it will probably be satisfactory to use the all-ages average of $82,000 as given above.

Many cities and states do not have complete records of injuries and property damage accidents. If there is reason to believe the records of the city are incomplete, it will be better to use the unit cost of $600,000 per death. However, this should be used only if there were as many as 10 deaths and the year's experience was not "freak." If there were fewer deaths, the deaths, injuries and property damage cases are figured

---

[26] DONALD E. CLEVELAND, "Illumination," *Traffic Control and Roadway Elements — Their Relationship to Highway Safety/Revised* (Washington, D.C.: Highway Users Federation for Safety and Mobility, 1969).
[27] Excludes cases of minor damage which may not be repaired, such as hubcaps smashed against curbs in parking.
[28] Includes cost items listed above for one death and the number of injury and property damage accidents indicated by the appropriate ratios.

separately, and a reasonable amount is added to cover the estimated degree of incompleteness.[29]

A comprehensive analysis of accident severity and costs is summarized in Table 4.13 for accidents classified as fatal injury, nonfatal injury, property damage only, and total. These data are further subdivided by the vehicle categories of passenger car and truck.

TABLE 4.13
Ratios of Severity of Traffic Accidents and Cost per Accident
in Illinois for Illinois Registered Vehicles, 1958
(Cost Updated from 1958 to 1966 by a Factor of 1.25.)

Item and Class of Vehicle	Fatal Injury	Nonfatal Injury	Property Damage Only	Total
1. Number of accidents				
Passenger car	1,169	103,306	755,871	860,346
Truck	220	9,273	112,228	121,721
Total	1,389	112,579	868,099	982,067
2. Accidents per fatal injury				
Passenger car	1.00	88.37	646.60	735.97
Truck	1.00	42.15	510.13	553.28
Total	1.00	81.05	624.98	707.03
3. Percentage of all accidents				
Passenger car	0.12	10.52	76.97	87.61
Truck	0.02	0.94	11.43	12.39
Total	0.14	11.46	88.40	100.00
4. Cost of accidents				
Passenger car	$ 9,869,882	$168,894,225	$144,697,912	$323,462,019
Truck	1,519,212	8,406,034	12,676,309	22,601,555
Total	11,389,095	177,300,026	157,374,212	346,063,333
5. Ratio of cost to fatal injury cost				
Passenger car	1.00	17.11	14.66	32.77
Truck	1.00	5.53	8.34	14.87
Total	1.00	15.57	13.82	30.39
6. Percentage of total costs				
Passenger car	2.85	48.80	41.82	93.47
Truck	0.44	2.43	3.66	6.53
Total	3.29	51.23	45.48	100.00
7. Cost per accident				
Passenger car	$8,442	$1,635	$191	$376
Truck	6,905	906	112	186
Total	8,200	1,575	181	352

Source: R. Winfrey, *Economic Analysis for Highways* (Scranton, Pa.: International Textbook Company, 1969), p. 386.

PARKING CHARACTERISTICS

Information on parking characteristics is necessary in the formulation of parking programs and provides basic data for the planning and design of parking facilities. Applications of these parking characteristics are given in Chapter 15, Parking, Loading, and Terminal Facilities.

CBD PARKING DEMAND

The demand for parking space in the central business district is predicated on the local economy, the opportunities for employment and shopping, the availability

---

[29] "Estimating the Costs of Accidents," *Traffic Safety Memo 113* (Chicago: National Safety Council, July 1973).

of public transportation, and the supply of parking facilities. A graph of the number of daily parkers is presented in Figure 4.32 as a function of community population. This daily usage provides an approximation of the parking demand that is dependent on the available supply of parking facilities in the various communities. In Figure 4.33, the trip purposes of the parkers in the CBD are related to the population of the community. Shopping and business trips represent the major demands for parking in small urban areas, while the parking facilities in large communities accommodate the parkers whose trip purposes are predominantly work and business.

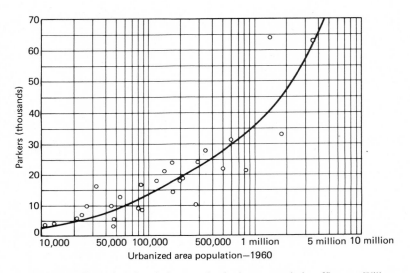

Figure 4.32. Daily parkers in relation to urbanized area population. [Source: Wilbur Smith and Associates, *Parking in the City Center* (Detroit: The Automobile Manufacturers Association, 1965), p. 11.]

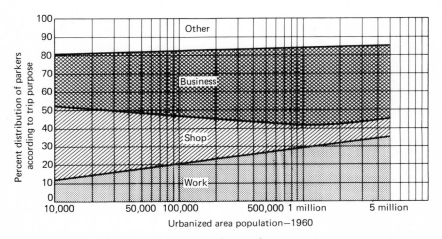

Figure 4.33. Trip purposes of parkers in relation to urbanized area population. (Source: Wilbur Smith and Associates, *Parking in the City Center*, p. 11.)

The estimation of parking demand is rather difficult in the planning of needed parking facilities because the actual usage of available parking spaces is based on the system of parking supply that exists. The real demand for parking facilities is determined only from an evaluation of the parking preferences of the drivers who desire to satisfy various trip purposes in the downtown areas.

CBD PARKING SUPPLY

The number of parking spaces provided in the CBD of cities located around the world can be related to the area of the CBD and to the population of the metropolitan area (see Table 4.14). Parking spaces in the CBD per thousand persons ranged from a low of 4.1 for Barcelona, Spain, to a high of 115.8 for Strasbourg, France. The 32 cities included in the study average nearly 33 spaces in the CBD per 1,000 population. Related to the area of the CBD, the number of parking spaces provided ranges from a low of about 7 spaces per acre (17.4 spaces per hectare) for Nashville and Rome to a high of 384 spaces per acre (950 spaces per hectare) for Toronto. It should be noted that the area defined as the CBD varies widely from city to city. Toronto's CBD was measured at 52.7 acres (21.3 hectares) but other cities with populations no greater than that of Toronto have CBD areas in excess of 2,000 acres (809 hectares).

The relationship between the number of CBD parking spaces provided in United States cities and city populations is shown in Figure 4.34. As the city size increases,

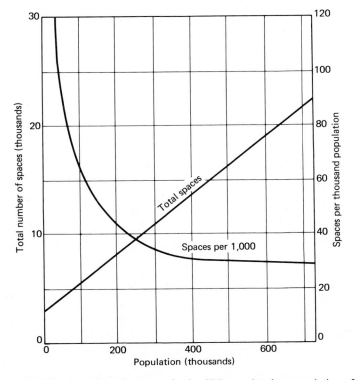

Figure 4.34. Number of parking spaces in the CBD as related to population of the urbanized area. [Source: *Parking Principles*, Special Report No. 125 (Washington, D.C.: Highway Research Board, 1971), p. 9.]

TABLE 4.14
Parking Spaces Existing in the CBD in Relation to CBD Area and Metropolitan Area Population

City	Metropolitan Area Population	CBD Area		CBD Parking Spaces Provided			Parking Space Ratios		
		Acres	Hectares	On-street	Off-street	Total	CBD Spaces per 1,000 pop.	Spaces per Acre	Spaces per Hectare
Adelaide (Aust.)	746,000	455	184	6,003	8,074	14,077	18.9	30.9	76.5
Perth (Aust.)	470,000	1,485	600	8,103	19,400	27,503	58.5	18.5	45.8
Toronto (Can.)	643,000	53	21	673	19,578	20,251	31.5	384.0	950.8
Vancouver (Can.)	782,000	545	220	3,502	22,430	25,932	33.1	47.5	117.6
Winnipeg (Can.)	496,000	1,089	440	5,305	24,204	29,509	59.5	27.1	67.1
Copenhagen (Den.)	1,400,000	1,455	588	17,141	2,025	19,166	13.7	13.2	32.6
Marseille (Fra.)	900,000	297	120	6,621	2,590	9,211	10.3	31.0	76.8
Nice (Fra.)	350,000	817	330	24,040	4,820	28,860	82.5	35.4	87.5
Strasbourg (Fra.)	250,000	2,426	980	26,550	2,386	28,936	115.8	11.9	29.5
Toulouse (Fra.)	375,000	1,435	580	8,435	8,175	16,610	44.3	11.6	28.7
Essen (Ger.)	727,000	89	36	2,122	4,941	7,063	9.7	79.4	196.4
Frankfurt (Ger.)	693,000	631	255	11,783	28,500	40,283	58.1	63.8	157.9
Hamburg (Ger.)	1,860,000	594	240	4,272	10,116	14,388	7.7	24.2	60.0
Coventry (Eng.)	330,000	275	111	621	3,067	3,688	11.2	13.4	33.2
Kingston Upon Hull, (Eng.)	310,000	280	113	2,400	2,300	4,700	15.2	16.8	41.6
Leeds (Eng.)	512,000	2,297	928	2,105	7,665	9,770	19.1	4.3	10.5
Tel Aviv–Yafo (Isr.)	400,000	359	145	6,000	5,200	11,200	28.0	31.2	77.3
Rome (It.)	2,489,000	4,950	2,000	25,252	9,582	34,834	14.0	7.0	17.4
Rotterdam (Neth.)	732,000	371	150	3,300	6,617	9,917	13.6	26.7	66.1
Auckland (N.Z.)	500,000	641	259	13,300	14,571	27,871	55.8	43.5	107.7
Cape Town (So. Af.)	508,000	394	159	7,718	4,174	11,892	23.4	30.2	74.8
Durban (So. Af.)	510,000	433	175	3,404	3,841	7,245	14.2	16.7	41.4
Barcelona (Sp.)	1,696,000	507	205	6,130	883	7,013	4.1	13.9	34.2
Madrid (Sp.)	2,700,000	547	221	8,250	12,500	20,750	7.7	38.0	94.0
Gothenburg (Swed.)	405,000	248	100	2,367	3,167	5,534	13.7	22.3	55.3
Malmo (Swed.)	245,000	604	244	5,300	3,400	8,700	35.5	14.4	35.7
Zurich (Swit.)	440,000	621	251	7,000	6,400	13,400	30.5	21.6	53.4
Detroit (U.S.)	1,620,000	678	274	4,000	36,000	40,000	24.7	59.0	146.0
Nashville (U.S.)	358,000	2,265	915	2,918	12,705	15,623	43.7	6.9	17.1
Norfolk (U.S.)	305,000	258	104	602	8,087	8,689	28.5	33.7	83.6
Pittsburgh (U.S.)	570,000	2,891	1,168	419	22,248	22,667	39.8	7.8	19.4
San Francisco (U.S.)	750,000	1,030	416	11,123	49,989	61,112	81.5	59.4	146.9

Source: Compiled from "Schemes for the Provision of Parking Spaces in Town Centres," Theme 6, Eighth International Study Week in Traffic Engineering, OTA, London, England, 1966.

the number of spaces provided in the CBD increases, but the spaces provided per thousand population decreases rapidly with city size until the city reaches a population of about 500,000. The index of spaces per thousand population then levels out and remains fairly constant at about 30 spaces per 1,000 for cities with population over 500,000.

The number of CBD parking spaces provided related to type of facility has been summarized for U.S. cities in Tables 4.15 and 4.16. The percentage of total spaces

TABLE 4.15
Number and Percent of Total Parking Spaces
Classified by Type of Facility

| Population Group of Urbanized Area | Type of Facility | | | Average Number of Total Spaces | Spaces per 1,000 Population |
| | Curb | Off-street | | | |
		Lot	Garage		
10,000–25,000	1,090 (43%)	1,530 (57%)	10 (0%)	2,630	150
25,000–50,000	1,430 (38%)	2,420 (59%)	140 (3%)	3,990	120
50,000–100,000	1,610 (35%)	2,790 (60%)	260 (5%)	4,660	70
100,000–250,000	2,130 (27%)	4,760 (62%)	820 (11%)	7,710	50
250,000–500,000	2,450 (20%)	7,910 (64%)	1,940 (16%)	12,300	30
500,000–1,000,000	3,200 (14%)	12,500 (56%)	6,900 (30%)	22,600	30
Over 1,000,000	8,000 (14%)	32,200 (55%)	18,600 (31%)	58,800	20

Source: *Parking Principles*, Highway Research Board Special Report No. 125 (Washington, D.C.: Highway Research Board, 1971), p. 9.

TABLE 4.16
Parking Spaces Classified by Type of Facility

| Population Group of Urbanized Area | Type of Facility | | | | | | |
| | Curb | | | Lot | | Garage | |
	Metered (%)	Non-metered (%)	Special (%)	Public (%)	Private (%)	Public (%)	Private (%)
10,000–25,000	47	51	2	18	82	93	7
25,000–50,000	55	40	5	27	73	50	50
50,000–100,000	55	41	4	42	58	56	44
100,000–250,000	47	46	7	52	48	89	11
250,000–500,000	49	40	11	66	34	95	5
500,000–1,000,000	54	38	8	68	32	87	13
Over 1,000,000	27	46	27	67	33	84	16

Source: *Parking Principles*, p. 10.

at the curb decreases as the city size increases. The corresponding increase in percentage of off-street spaces as the city size increases is attributable almost entirely to the increase in garage spaces. The percentage of total CBD spaces in off-street lots remains fairly constant, and peaks at a population of about 300,000 to 400,000 persons. The percentage distributions of parking spaces by curb, lot, and garage are summarized in Figure 4.35 as a function of community population.

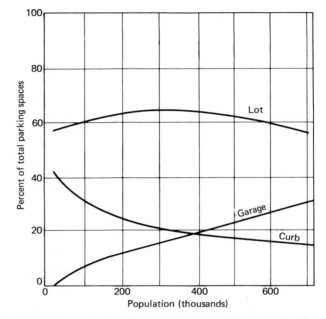

Figure 4.35. Percentage of all parking spaces in the CBD by type of facility. (Source: *Parking Principles*, p. 14.)

Local governments in the United States provide many parking spaces within their boundaries for use by residents and workers: in total, nearly 750,000 off-street spaces. Nearly 100,000 municipally owned spaces are provided in garages with the remainder in surface lots (see Table 4.17).

PARKING USAGE

The percentage of parkers at the curb in the CBD decreases greatly as the city size increases, while the amount of off-street parking increases. The percentage of parkers using garage facilities is very low in the smaller cities but increases in larger communities as shown in Table 4.18. As city size increases, the percentage of shopping trips decreases, but the percentage of work trips increases.

PARKING DURATION

The distribution of parking durations in the CBD is provided in Table 4.19 for cities of various sizes. Parking durations of less than one-half hour decrease rapidly

TABLE 4.17
Number and Type of Municipally Owned Off-street Parking Facilities

City Population	No. of Cities Surveyed	Parking in the CBD				Parking Outside the CBD			
		No. of Garages	No. of Spaces	No. of Lots	No. of Spaces	No. of Garages	No. of Spaces	No. of Lots	No. of Spaces
Under 2,500	42	0	0	61	3,968	0	0	6	595
2,500–5,000	122	1	29	282	13,341	1	24	41	2,487
5,000–10,000	251	5	780	680	41,640	3	70	131	13,447
10,000–25,000	302	22	2,231	1,141	76,361	0	0	218	22,001
25,000–50,000	166	18	3,558	929	72,894	0	0	210	25,040
50,000–100,000	106	34	13,227	761	65,187	3	368	205	37,934
100,000–250,000	63	40	15,504	454	50,844	2	1,100	203	32,582
250,000–500,000	21	23	12,589	124	19,038	1	389	67	22,785
500,000–1,000,000	18	35	24,603	41	13,488	4	2,960	129	47,217
Over 1,000,000	6	26	20,382	47	17,444	6	2,428	325	67,766
Total	1,097	202	92,063	4,466	382,674	19	6,950	1,527	251,890

Source: W. D. Heath, J. M. Hunnicutt, M. A. Neale, and L. A. Williams, *Parking in the United States— A Survey of Local Government Action*, National League of Cities, Department of Urban Studies, Washington, D.C., 1967, pp. 1–25.

TABLE 4.18
Percentage of Parkers Classified by Facility and Trip Purpose

Population Group of Urbanized Area	Curb					Off-street									
						Lot					Garage				
	Shopping	Personal Business	Work	Other	Total Curb	Shopping	Personal Business	Work	Other	Total Lot	Shopping	Personal Business	Work	Other	Total Garage
10,000–25,000	30	22	11	16	79	8	1	10	2	21	0	0	0	0	0
25,000–50,000	22	30	8	14	74	5	5	13	3	26	0	0	0	0	0
50,000–100,000	19	24	7	18	68	5	7	12	7	31	0	0	1	0	1
100,000–250,000	11	24	6	11	52	9	9	17	7	42	1	1	3	1	6
250,000–500,000	10	23	8	13	54	6	7	18	3	34	3	3	4	2	12
500,000–1,000,000	3	12	9	9	33	5	8	23	3	39	5	5	15	3	28
Over 1,000,000	3	15	4	8	30	4	13	29	8	54	3	2	8	3	16

Source: *Parking Principles*, pp. 12–13.

as city size increases, but long durations (over 5 hours) increase very substantially with community population.

Parking duration in the CBD is directly related to trip purpose in Table 4.20. Work trips have the longest parking duration and average 5 to 6 hours in the larger cities, while shopping and personal business trips produce parking durations that are much shorter in length. Regardless of trip purpose, parking durations increase greatly as city size increases.

TABLE 4.19
Parking Classified by Length of Time

Population Group of Urbanized Area	Length of Time Parked				
	0–0.5 hr (%)	0.5–1 hr (%)	1–2 hr (%)	2–5 hr (%)	Over 5 hr (%)
10,000–25,000	60	14	10	10	6
25,000–50,000	59	15	10	9	7
50,000–100,000	60	15	10	10	5
100,000–250,000	46	14	11	13	16
250,000–500,000	38	15	17	15	15
500,000–1,000,000	24	12	13	18	33
Over 1,000,000	16	12	20	12	40

Source: *Parking Principles*, p. 14.

TABLE 4.20
Length of Parked Time Classified by Trip Purpose

Population Group of Urbanized Area	Trip Purpose			Average All Trips (hr)
	Shopping (hr)	Personal Business (hr)	Work (hr)	
10,000–25,000	0.5	0.4	3.5	1.3
25,000–50,000	0.6	0.5	3.7	1.2
50,000–100,000	0.6	0.8	3.3	1.2
100,000–250,000	1.3	0.9	4.3	2.1
250,000–500,000	1.3	1.0	5.0	2.7
500,000–1,000,000	1.5	1.7	5.9	3.0
Over 1,000,000	1.1	1.1	5.6	3.0

Source: *Parking Principles*, p. 14.

PARKING TURNOVER

Parking turnover is a measure of the utilization of a parking space and indicates the number of different vehicles that use the space during a specified time period. A detailed tabulation of parking turnover is summarized in Table 4.21 for various classifications of curb and off-street parking. Because long-term parkers generally utilize off-street facilities, higher turnover rates are evidenced among the various classes of curb parking.

WALKING DISTANCE

Average walking distance of parkers in the CBD is related to such factors as city size, trip purpose, topography, type of parking facility used, fee charged, and the length of time parked. Walking distance is related to trip purpose and city size in Table 4.22. The work trip produces the longest walking distances. Average walking distance is related to parking facility type in Table 4.23. Curb parkers have shorter walking distances than off-street parkers, while off-street parkers using surface lots walk greater distances than those parked in garages.

TABLE 4.21
Parking Turnover Classified by Type of Facility*

Population Group of Urbanized Area	Type of Facility						
	Curb				Off-Street		
	Metered	Posted	Special	Average	Lot	Garage	Average
10,000–25,000	—	—	—	6.7	1.8	0.3	1.8
25,000–50,000	—	—	—	6.4	1.5	0.6	1.5
50,000–100,000	7.8	2.8	3.7	6.1	1.7	0.8	1.6
100,000–250,000	8.1	3.1	4.4	5.7	1.6	1.0	1.5
250,000–500,000	7.1	2.5	3.3	5.2	1.4	1.1	1.4
500,000–1,000,000	6.6	1.1	3.9	4.5	1.2	1.4	1.2
Over 1,000,000	5.5	3.6	2.9	3.8	1.1	1.0	1.1

*Parkers per 8-hr period between 10:00 a.m. and 6:00 p.m.

Source: *Parking Principles*, p. 16.

TABLE 4.22
Average Distance Walked from Parking Place to Destination
Classified by Trip Purpose

Population Group of Urbanized Area	Trip Purpose			
	Shopping (ft)	Personal Business (ft)	Work (ft)	Other (ft)
10,000–25,000	200	200	270	190
25,000–50,000	280	240	400	210
50,000–100,000	350	290	410	260
100,000–250,000	470	390	500	340
250,000–500,000	570	450	670	380
500,000–1,000,000	560	590	650	500

Source: *Parking Principles*, p. 15.

TABLE 4.23
Average Distance Walked from Parking Place to Destination
Classified According to Kind of Facility

Population Group of Urbanized Area	Kind of Facility			
	Curb (ft)	Off-street		Overall Average (ft)
		Lot (ft)	Garage (ft)	
10,000–25,000	210	210	—	210
25,000–50,000	250	350	100	280
50,000–100,000	280	380	240	280
100,000–250,000	370	540	330	420
250,000–500,000	390	760	700	550

Source: *Parking Principles*, p. 15.

TABLE 4.24
Average Distance Walked from Parking Place to Destination
Classified by Length of Time Parked

Population Group of Urbanized Area	Length of Time Parked			
	0.5–1 hr (ft)	1–2 hr (ft)	2–5 hr (ft)	Over 5 hr (ft)
10,000–25,000	220	250	280	330
25,000–50,000	270	290	370	500
50,000–100,000	310	350	370	430
100,000–250,000	420	380	500	440
250,000–500,000	440	510	590	740
500,000–1,000,000	480	480	560	910
Over 1,000,000	520	560	680	900*

*Estimated from limited sample.

Source: *Parking Principles*, p. 15.

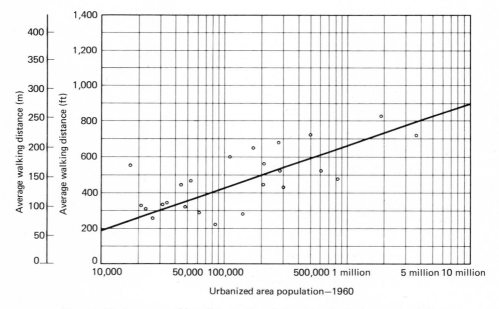

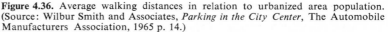

Urbanized area population—1960

Figure 4.36. Average walking distances in relation to urbanized area population. (Source: Wilbur Smith and Associates, *Parking in the City Center*, The Automobile Manufacturers Association, 1965 p. 14.)

Average walking distance is related to parking duration in Table 4.24. For each city size, walking distance increases as the parker's duration in the CBD increases. In all instances, the walking distance increases substantially with an increase in city size (see Figure 4.36).

ACCUMULATION

Typical parking accumulations for various city sizes show, in Figure 4.37, that accumulations remain relatively constant until about 3:00 p.m. when the exodus of

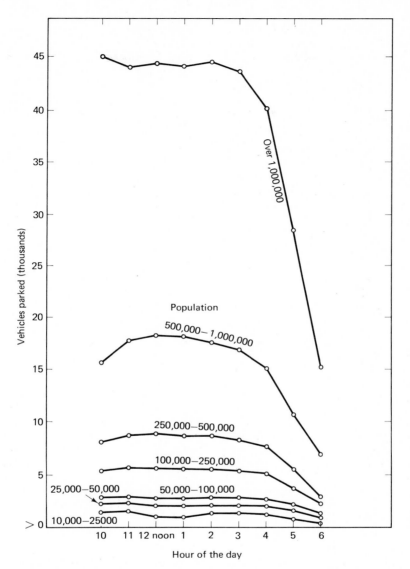

Figure 4.37. Hourly parking accumulation related to population group. (Source: *Parking Principles*, p. 12.)

parkers from the CBD begins. The typical accumulation curves in Figure 4.38 for several days of the week indicate that accumulations normally peak on Monday. The Wednesday accumulation curve is typical of days during midweek.

Other characteristics of typical accumulation curves include:[30]

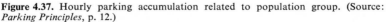

[30] *Parking Principles*, Special Report 125 (Washington, D.C.: Highway Research Board, 1971), pp. 12–13.

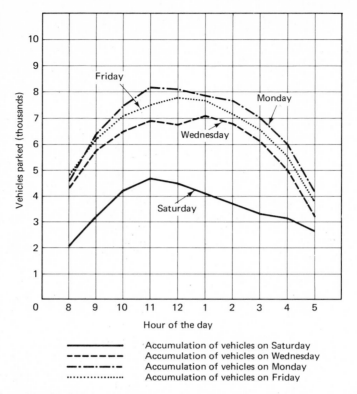

Figure 4.38. Typical CBD parking accumulation pattern on typical weekdays. (Source: *Nashville Metropolitan Area Transportation Study—Downtown Parking*, Nashville Parking Board, 1960, p. 57.)

1. Shopper and business trip accumulations peak in midafternoon (about 3:00 p.m.) for smaller cities and in early afternoon (about 1:00 p.m.) for larger cities.
2. Work trips peak in late morning (10:30 a.m. to 11:00 a.m.) for all size cities.
3. Curb accumulation is nearly constant from 11:00 a.m. to 3:00 p.m. in larger cities and until 4:00 p.m. in smaller cities.
4. In larger cities, curb accumulation at 9:00 a.m. and 5:00 p.m. is about 40 percent of maximum accumulation. This figure is 80 percent for smaller cities.
5. Off-street parking accumulation peaks between 10:00 a.m. and 11:00 a.m. for all city sizes.

INDUSTRIAL PLANTS

Vehicular traffic generated by industrial plants is indicative of the number of parking spaces that must be provided for employees. A reasonable relationship has been established in Figure 4.39 between the maximum parking demand and the maximum-shift employment (see Figure 4.39). Parking demand is based either on the

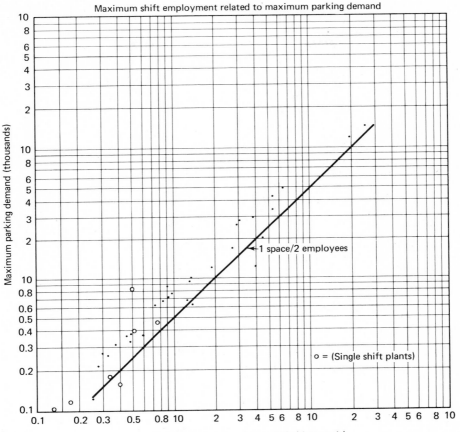

Figure 4.39. Parking demand as a function of employment at industrial plants.
[Source: *Parking Facilities for Industrial Plants* (Washington, D.C.: Institute of Traffic
Engineers, 1969), p. 15.]

total number of plant employees in single-shift operations or on the number employed
on the peak shift in multishift plants.

COMMERCIAL CENTERS

Several trip generation studies have been conducted to quantify parking space
requirements for commercial establishments and centers. Both average and range
values for the parking requirements in Table 4.25 are based on the gross floor area
for different kinds of commercial establishments located in the downtown area.
In Table 4.26, trip generation factors are listed for various commercial activities
found throughout the urban area. Rates of traffic generation are based on the peak
hour of traffic flow on the street.

TABLE 4.25
Average Parking Space Requirements for Selected
Downtown Establishments

Type of Establishment	Spaces per 1,000 sq ft	
	Average	Range
Banks	5.4	1.8–10.8
Bus depots	4.8	1.7– 7.9
Libraries	4.1	3.9– 4.3
Medical buildings	3.8	1.1– 8.6
Grocery stores	3.7	1.4– 7.5
City-County offices	3.6	1.2– 6.0
Post offices	3.4	2.0– 4.9
Utility company offices	2.9	0.4–10.7
Drug stores	2.9	1.4– 5.5
Department stores	2.8	1.4– 5.1
Clothing stores	2.5	1.1– 6.3
Restaurants	2.1	0.9– 3.3
YMCA-YWCA	1.6	1.2– 2.2
Offices	1.5	0.4– 2.9
Auto sales	1.2	0.9– 1.5
Variety stores	1.1	0.6– 1.9
Hotels	0.6	0.4– 1.0
Furniture stores	0.6	0.3– 1.2

Source: H. K. Evans, "Parking Study Applications," *Traffic Quarterly*, The Eno Foundation for Highway Traffic Control, April 1963, p. 277.

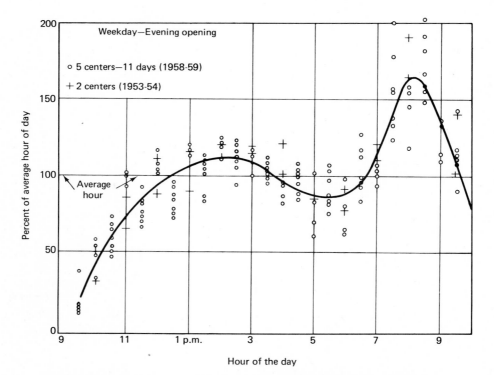

Figure 4.40. Daily parking accumulation at regional shopping centers with evening shopping hours. [Source: D. E. Cleveland and E. A. Mueller, *Traffic Characteristics at Regional Shopping Centers* (New Haven: Bureau of Highway Traffic, Yale University, 1961), p. 65.]

TABLE 4.26
Traffic Generation Rates for Various Urban Commercial Activities

Commercial Generator	Peak Hour of Operation	P.M. Peak Street-Hour
Drive-in restaurants	257 trips/1,000 sq ft GFA*	108 trips/1,000 sq ft GFA
Sit-down restaurants	35 trips/1,000 sq ft GFA	25 trips/1,000 sq ft GFA
Food stores	14 trips/1,000 sq ft GFA	12 trips/1,000 sq ft GFA
Neighborhood shopping centers	15 trips/1,000 sq ft GFA	14 trips/1,000 sq ft GFA
Automobile service stations	28 trips/hour	23 trips/hour
Motels	0.8 trips/unit	0.6 trips/unit
Office buildings	2.3 trips/1,000 sq ft GFA	2.3 trips/1,000 sq ft GFA
Hospitals	1.0 trips/bed	0.7 trips/bed

*GFA—Gross Floor Area of Building.

Source: Illinois Section, ITE, "Trip Generation Study of Selected Commercial and Residential Developments," *Traffic Engineering*, XL, No. 6 (1970), 40.

The parking accumulation curve in Figure 4.40 represents the daily parking demand typical for a regional shopping center with evening shopping hours. In Figure 4.41, the correlation of maximum accumulation of parked vehicles with the daily number of vehicles entering the shopping center provides an estimation of the parking

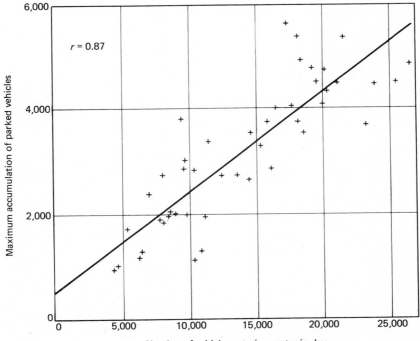

Figure 4.41. Relationship between accumulation of parked vehicles and daily traffic entering regional shopping centers. (Source: D. E. Cleveland and E. A. Mueller, *Traffic Characteristics at Regional Shopping Centers*, p. 70.)

requirements if the daily generated traffic volume has been developed from a marketing analysis of the commercial center. In another investigation, the total daily vehicle trips to a shopping center were related to the total floor space by the following equation:

$$Y = 1512 + 10.83\,X$$

where
Y = total daily trips (vehicles),
X = total floor area (sq ft per thousand).[31]

RESIDENTIAL AREAS

The accumulation of parked vehicles in residential areas is depicted throughout the day in Figure 4.42. The heavy demand for parking facilities is evident in high-density

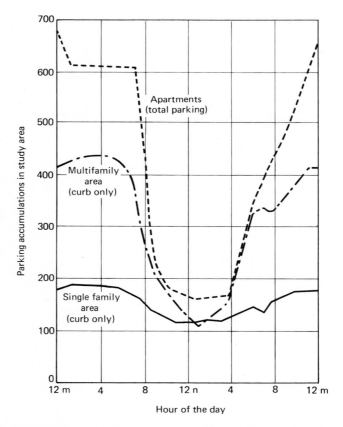

Figure 4.42. Variation of parking accumulation in residential areas by time of day. (Source: *Parking Principles*, p. 28.)

---

[31] L. E. KEEFER, *Urban Travel Patterns for Airports, Shopping Centers, and Industrial Plants*, National Cooperative Highway Research Program Report 24 (Washington, D.C.: Highway Research Board, 1966).

residential areas and represents the storage of vehicles over night. Trip generation rates are presented in Table 4.27 for single-family residences and multifamily apartments. These trip rates for residential areas account for both in and out movements of vehicles during morning and afternoon peak conditions.

TABLE 4.27
Traffic Generation Rates for Residential Areas

Residential Generator	A.M. Peak			P.M. Peak		
	In	Out	Total	In	Out	Total
Single-family residence subdivision	0.23	0.58	0.81 trips/unit	0.60	0.40	1.00 trips/unit
Multifamily apartments	0.08	0.49	0.57 trips/unit	0.46	0.23	0.69 trips/unit

Source: Illinois Section, ITE, "Trip Generation Study of Selected Commercial and Residential Developments," p. 40.

REFERENCES FOR FURTHER READING

"Accident Facts," National Safety Council, latest edition.

Highway Capacity Manual 1965, Highway Research Board, Special Report 87, 1965.

Highway Research Board, various *Records* devoted to traffic characteristics.

KENNEDY, N., J. H. KELL, and W. S. HOMBURGER, *Fundamentals of Traffic Engineering*, The Institute of Transportation and Traffic Engineering, 8th Edition, 1973.

OPPENLANDER, J. C., "Variables Influencing Spot-Speed Characteristics, Review of Literature," Highway Research Board, Special Report 89, 1966.

Parking Principles, Highway Research Board, Special Report 125, 1971.

SMITH, WILBUR, and Associates, *Parking in the City Center*, The Automobile Manufacturers Association, 1965.

"Traffic Control and Roadway Elements—Their Relationship to Highway Safety," Highway Users Federation for Safety and Mobility, Revised Edition, Chapters 1 through 12, 1968–1971.

WINFREY, R., *Economic Analysis for Highways*, Scranton, Pa., International Textbook Company, 1969.

Chapter 5

URBAN TRAVEL CHARACTERISTICS

Herbert S. Levinson, Senior Vice-President, Wilbur Smith & Associates, New Haven, Connecticut.

This chapter sets forth the broad characteristics of urban travel. It summarizes information on trip generation, trip purpose, trip lengths, and travel modes. It shows how population, density, income, car ownership, and age influence travel behavior. It summarizes travel characteristics of central business districts and other major generators.

The information can be used by traffic engineers and urban planners to analyze current conditions, to evaluate traffic impacts of proposed developments, and to derive broad-gauged estimates of future conditions. The various factors and relationships also provide important guidance in developing traffic modernization programs and in making key transport decisions.

An attempt has been made to introduce new, current material—where available—on urban travel, but of necessity much of the material is derived from comprehensive traffic studies conducted in the decade 1960 to 1970. These data may be used as a benchmark in evaluating changes in characteristics as they relate to revisions in transportation systems and/or availability of energy.

CHARACTERISTICS OF CITIES

The social, demographic, and land-use characteristics of the modern urban region influence the travel behavior of its residents and the demands for transportation facilities. The modern urban region is usually characterized by a continuing rapid rate of growth both in population and area of urbanization, a proliferation of commercial and industrial centers, an increase in car ownership which outpaces population growth, and a relative decline in many types of CBD attraction.

POPULATION AND DENSITY

Each urban area reflects historical, social, and economic antecedents. Likenesses and differences among cities mainly relate to economic base, topography, and age.

1. Population density patterns generally reflect city age and the modes of intra-urban transport that prevailed during formative years. Consequently, cities in Europe, Asia, and South America generally exhibit higher densities than American cities. Table 5.1 compares densities in large American, European, and Asian cities.
2. Within the United States, New York, Chicago, Philadelphia, Boston, and San Francisco are the only large central cities with densities exceeding 15,000 per-

TABLE 5.1
Population and Density of Selected Major Cities

United States and Canada

Central City	Year	Population	Area (sq mi)	Density (persons per sq mi)
New York (Manhattan)	1920	2,284,000	22	103,818
New York City	1923	5,927,625	299	19,825
Chicago	1923	2,886,971	195	14,805
Philadelphia	1923	1,922,788	127	15,141
Average*		3,579,130	207	17,300§
New York (Manhattan)	1970	1,539,233	22	69,965‡
New York City	1970	7,894,862	300	26,343‡
Chicago	1970	3,366,957	223	15,126‡
Los Angeles	1970	2,816,061	464	6,073‡
Philadelphia	1970	1,948,609	128	15,164‡
Detroit	1970	1,511,482	138	10,953‡
Montreal	1966	1,222,255	47	26,000‡
Average*		3,126,700	217	14,400§
Average excluding Los Angeles and Manhattan		3,188,833	167	19,100§

Rest of World

Central City	Year	Population	Area (sq mi)	Density (persons per sq mi)
London (excluding outer ring)	1921	4,484,523	117	38,329
Paris	1921	2,856,986	30	95,233
Berlin	1905	2,033,900	29	70,134
Tokyo	1905	1,969,833	30	65,661
Moscow	1902	1,092,360	32	34,136
Glasgow	1921	1,034,174	30	34,472
Average		2,245,300	45	50,000§
Tokyo	1960	9,124,217	207	44,078
Greater London	1960	8,210,000	722	11,377
Shanghai†	1953	6,204,000	345	17,982
Osaka	1960	5,158,010	123	41,935
Berlin	1960	4,244,600	344	12,339
Buenos Aires†	1955	3,575,000	74	48,310
London	1948	3,339,000	117	28,538
Bombay†	1960	3,000,000	30	100,000
Rio de Janeiro†	1955	2,900,000	60	48,333
Calcutta	1961	2,926,498	39	74,200
Paris	1960	2,800,000	34	83,580
Average		4,680,120	190	24,600§
Average excluding Greater London Conurbation		4,327,132	137	31,500§

*Excludes Manhattan.
†Denotes that some estimates were made to obtain consistent population and area values.
‡The original figures are densities as published by census *before* rounding from Table 20, pp. 1–76, U.S. Summary Census No. of Inhabitants.
§Average densities represent rounded values.

Sources: Compiled from U. S. and world census data from John C. Weaver and Fred E. Lukermann, *World Resource Statistics* (Minneapolis: Briggs Publishing Co., 1953); University of California, International Urban Research, *The World's Metropolitan Areas* (Berkeley: University of California Press, 1959); and 1920–1930 data from Herbert B. Dorau and Albert G. Hinman, *Urban Land Economics* (New York: Macmillan Co., 1928). Data shown only for cities where available. Adapted from Herbert S. Levinson and F. Houston Wynn "Effects of Density on Urban Transportation Requirements," *Community Values as Affected by Transportation*, Highway Research Record No. 2. (Washington, D.C.: Highway Research Board, 1963), p. 40.

sons per mile (1970). In Canada, Montreal and Toronto have central city densities of approximately 20,000 persons per mile. Central city population density generally relates to the year when the city population first reached 350,000.

3. Population density reflects the mix of dwelling units. The mix between single-family and multifamily units depends on the family life cycle, housing market conditions, land cost and/or availability, as well as city age, economy, and topography. New York City, for example, contains a high proportion of five-family or more apartments; Baltimore and Philadelphia contain high proportions of row-houses. Single-family houses predominate in Houston and Dallas despite the recent rise in apartment construction.

4. In most cities, net residential density declines exponentially with increasing distance/time from a principal city center,[1] i.e.,

$$d_x = d_0 e^{-bx} \qquad (5.1)$$

d_x = density at distance x
d_0 = density at distance 0 (central density)
b = density gradient

5. Car ownership increases with rising income and with decreasing density. The proportion of one-car households is approximately the same at all density levels; however, the proportion of multicar households is highest in low-density communities. Although car ownership is highest in suburban areas, the highest densities of car ownership (i.e., cars per square mile) are found in old central cities. Philadelphia and Boston, with approximately 15,000 people per square mile, average 4,000 cars per square mile, but Los Angeles, with 5,500 people per square mile, averages fewer than 2,500 cars per square mile.

LAND USE

Urban land uses in major metropolitan areas are summarized in Table 5.2 and Figure 5.1. Generally, approximately 33 to 50 percent of all developed land is devoted to residential purposes; approximately 10 percent for transportation, communication, manufacturing, and utilities; and up to 5 percent for commercial uses. The proportion of land devoted to public open space and institutional uses generally ranges from 10 to 25 percent.

Land-use studies indicate that street rights of way occupy one-fourth to one-third of all developed land regardless of city size and density (Table 5.2). This proportion has not changed appreciably even when freeways have been built. (In the Los Angeles and New York metropolitan areas, less than one-fourth of all land is devoted to streets, but within New York City's five boroughs streets occupy more than one-third of all developed land.)

L'Enfant's Washington, D.C. plan dedicated 49 percent of all land to arterial streets; Captain John Sutter's Sacramento plan reserved approximately 38 percent for street use.[2] In contrast, portions of Sacramento laid out between 1900 and 1930

---

[1] A high rate of decline shows that the city is compact, whereas a low-density gradient (increasingly typical of North American cities) reflects a dispersed development pattern. See, for example, T. F. C. Clark, "Urban Population Densities," *Journal of the Royal Statistical Society*, CXIV, Series A, pp. 490–96.

[2] K. MOSKOWITZ "Living and Travel Patterns in Automobile Oriented Cities," *The Dynamics of Urban Transportation* (Detroit: A National Symposium Sponsored by the Automobile Manufacturers, Inc., October 23–24, 1962).

TABLE 5.2
Land Use in Selected Urban Areas

Land Use	New York (Tri-state Region Intensity Developed Area)—1963 2,344,480 Acres		New York City—1960 204,681 Acres		Los Angeles—1960 5,776,000 Acres		Chicago—1956 791,343 Acres	
	Percent of Devel. Land	Percent of Total Land	Percent of Devel. Land	Percent of Total Land	Percent of Devel. Land	Percent of Total Land	Percent of Devel. Land	Percent of Total Land
Residential	53.8	29.0	26.4	23.2	38.3	6.4	32.1	14.6
Commercial	4.0	2.1	1.5	1.3	4.2	0.7	3.8	1.7
Manufacturing	2.8	1.5	5.2	4.6	9.6	1.6	4.4	2.0
Transp.-Commun.-Util.	3.6	2.0	5.4	4.7	2.4	0.4	9.0	4.1
Public and Semi-public Buildings	4.5	2.4	7.0	6.1			4.1	1.9
Open Space	13.7	7.4	19.7	17.3	24.0	4.0	20.4	9.3
Parking and Miscellaneous	—	—	0.2	0.2	—	NA	0.3	0.1
Streets and Alleys	17.6	9.4	34.6	30.1	21.5	3.6	25.9	11.8
Land in Urban Use	100.0	53.8	100.0	87.5	100.0	16.7	100.0	45.5
Vacant or Not in Urban Use		46.2		12.5		83.3		54.5
TOTAL		100.0		100.0		100.0		100.0

Source: Data on specific urban areas compiled from land-use and origin-destination studies in each urban area. Average data from Harland Bartholomew, *Land Uses in American Cities* (Cambridge, Mass.: Harvard University Press, 1955.)

TABLE 5.2 (*Continued*)

Land Use	Philadelphia—1960 752,010 Acres		Detroit—1953 436,121 Acres		Baltimore—1963 498,366 Acres		Pittsburgh—1958 268,570 Acres	
	Percent of Devel. Land	Percent of Total Land	Percent of Devel. Land	Percent of Total Land	Percent of Devel. Land	Percent of Total Land	Percent of Devel. Land	Percent of Total Land
Residential	47.6	22.4	45.6	22.0	39.3	14.2	44.5	19.3
Commercial	3.8	1.8	2.5	1.2	2.8	1.0	3.5	1.5
Manufacturing	6.9	3.2	9.1	4.4	16.5	6.0	4.5	2.0
Transp.-Commun.-Util.	3.8	1.8	NC	NC	—	—	5.8	2.5
Public and Semi-public Buildings	6.1	2.9	4.4	2.1	13.5	4.9	5.3	2.3
Open Space	16.7	7.9	5.4	2.6	11.3	4.1	11.4	4.9
Parking and Miscellaneous	—	—	2.2	1.0	—	—	—	—
Streets and Alleys	15.1	7.1	30.8	14.8	16.6	6.0	25.0	10.9
Land in Urban Use	100.0	47.1	100.0	48.1	100.0	36.2	100.0	43.4
Vacant or Not in Urban Use		52.9		51.9		63.8		56.6
TOTAL		100.0		100.0		100.0		100.0

* Recreational.

142

TABLE 5.2 (*Continued*)

Land Use	Minneapolis-St. Paul—1958 569,458 Acres		Buffalo—1962 996,941 Acres		Tucson Urban Area 58,912 Acres (1959)		Tucson Study Area 390,631 Acres (1959)	
	Percent of Devel. Land	Percent of Total Land	Percent of Devel. Land	Percent of Total Land	Percent of Devel. Land	Percent of Total Land	Percent of Devel. Land	Percent of Total Land
Residential	41.8	11.7	40.1	5.6	42.9	20.2	32.0	4.0
Commercial	3.0	.8	4.5	0.7	6.2	2.9	7.4	0.9
Manufacturing	3.0	.8	4.9	0.7	2.4	1.2	3.6	0.4
Transp.-Commun.-Util.	8.1	2.2	8.4	1.2	5.1	2.4	10.1	1.4
Public and Semi-public Buildings	3.7	1.1	3.3	0.4	10.7	5.2	8.2	1.0
Open Space	11.3	3.2	14.7	2.0	4.4	1.9	5.7	0.7
Parking and Miscellaneous	—	—	—	—	—	—	—	—
Streets and Alleys	29.1	8.1	24.1	3.3	28.3	13.3	33.0	4.2
Land in Urban Use	100.0	27.9	100.0	13.9	100.0	47.1	100.0	12.6
Vacant or Not in Urban Use		72.1		86.1		52.9		87.4
TOTAL		100.0		100.0		100.0		100.0

*Recreational.

143

TABLE 5.2 (Continued)

| Land Use | Nashville—1957 150,195.2 Acres | | Chattanooga—1960 90,085 Acres | | Average* | | |
	Percent of Devel. Land	Percent of Total Land	Percent of Devel. Land	Percent of Total Land	53 Central Cities Percent of Devel. Land	33 Satellite Cities Percent of Total Land	11 Urban Areas Percent of Devel. Land
Residential	53.2	26.2	51.3	20.2	39.6	42.2	28.0
Commercial	3.1	1.5	3.9	1.6	3.3	2.5	2.6
Manufacturing	3.5	1.7	13.5	5.4	6.4	7.8	5.7
Transp.-Commun.-Util.	6.9	3.4	—	—	4.8	4.6	6.2
Public and Semi-public Buildings	19.2	9.4	5.4	2.2	17.7†	15.3†	29.9†
Open Space	—	—	2.0*	0.8*	—	—	—
Parking and Miscellaneous	—	—	—	—	—	—	—
Streets and Alleys	14.1	6.9	23.9	9.5	28.2	27.6	27.6
Land in Urban Use	100.0	49.1	100.0	39.7	100.0	100.0	100.0
Vacant or Not in Urban Use		50.9		60.3			
TOTAL		100.0		100.0			

*Bartholomew, *Land Uses in American Cities.*
†Includes open spaces.

Source: Compiled from land-use and origin-destination studies in each urban area.

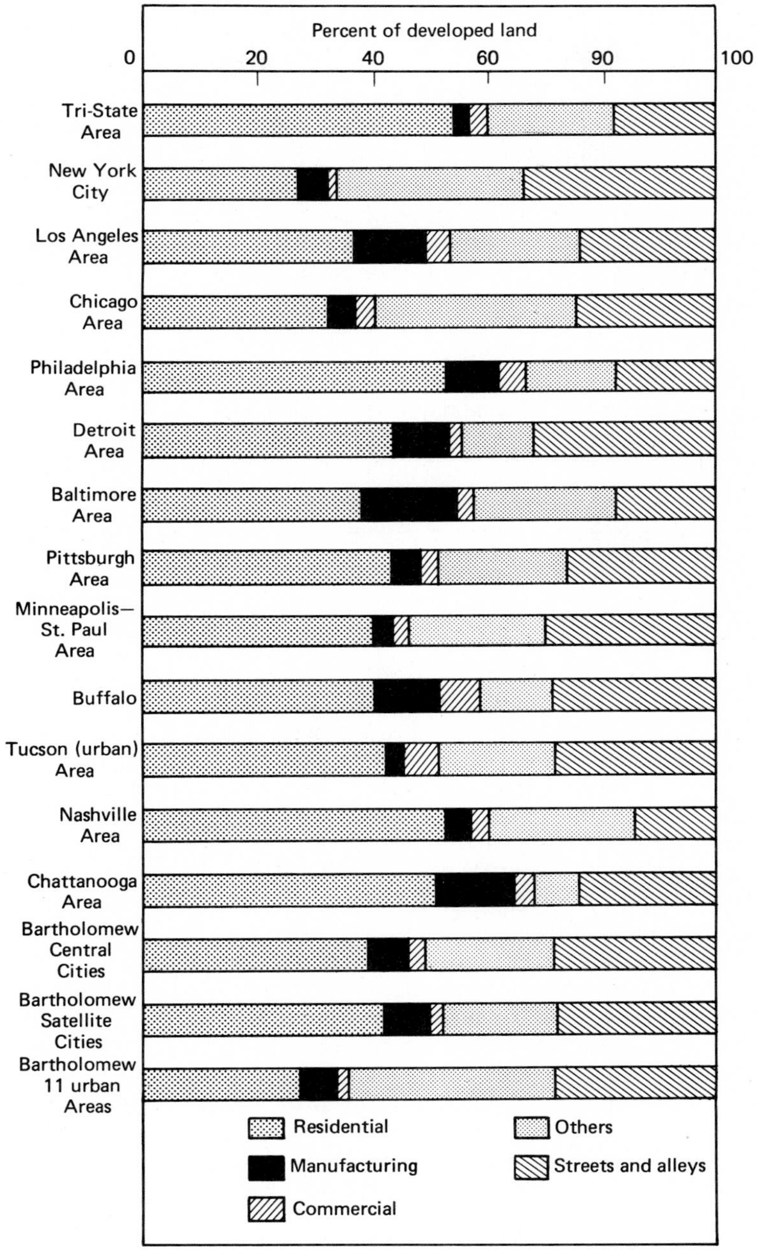

Figure 5.1. Land use in major urban areas. (Source: Data presented in Table 5.2.)

allocated only 21 percent of their area to streets, and some new areas developed since World War II have reserved only 15 percent of the subdivided land for transportation purposes.

CAR OWNERSHIP

There is approximately one passenger vehicle in the United States for every three persons. Expressed in slightly different terms, approximately 80 percent of all families in the United States own at least one automobile, and approximately 30 percent of the total family units own two or more automobiles.[3]

The incidence of multicar ownership varies with the age of household head. As shown in Table 5.3 the highest concentration of multicar ownership is found in the 45 to 54 age bracket, the lowest concentration in the over 65 bracket. These figures imply a correlation between family age, income, and multicar ownership.

TABLE 5.3
Percentage of Households Owning Two or More Cars By Age
of Household Head

Age (Years)	Multicar Households (Percent)
Under 25	20.3
25–29	26.6
30–34	31.7
35–44	42.2
45–54	43.2
55–64	28.6
65 and over	9.4

Source: *1972 Automobile Facts & Figures*, Motor Vehicle Manufacturers Association of the U.S., Inc., Detroit, Michigan, p. 38.)

Location of multicar households. The nationwide distribution of multicar ownership, as reported by the U.S. Census sample survey, denotes some regional variations in the frequency of multicar ownership in 1971. The highest overall incidence of multicar ownership, 34 and 33 percent of all households, is found in the Mountain and Pacific States respectively. This compares with 25 percent in the Middle Atlantic States, and 28 percent in the west south central states.[4] The high incidence of multicar ownership in the western states is particularly significant in view of the rapid growth there. The relatively low multicar ownership in the mid-Atlantic area is related to the low car ownership and continued reliance on public transport in the large, old, central cities of New York, Philadelphia, and Baltimore. In this context, the relatively high incidence of downtown residences in many of these metropolitan areas may also constrain multicar ownership.

Multicar ownership in a metropolitan area is, of course, highest in the suburban areas and lowest in the central city.[5] In the large central cities (3 million or more popu-

---

[3] The estimates of car ownership vary among different sample surveys and sources. The range, however, is not great, and all studies point toward the same general findings.

[4] U.S. Census data as reported in *1972 Automobile Facts and Figures*, p. 38., Motor Vehicle Manufacturers of the U.S., Inc., Detroit, Mich.

[5] Central city refers to the political or incorporated city as defined by the U.S. Dept. of Commerce, Bureau of the Census.

lation), according to 1970 Census data (weighted heavily by New York), only approximately 11 percent of all households have two or more cars; in urban areas with 1 million to 2,999,999 population approximately 17 percent of the central city residents own more than one car; in central cities with populations of 250,000 to 999,999, approximately 26 percent own more than one car; and in central cities within small urban areas (less than 250,000 population), this figure rises to approximately 30 percent.[6] In suburban areas, this figure approximates 40 percent. It is significant to note that continued expansion of suburban areas, coupled with relatively modest population increases in central cities, will, in all likelihood, further increase multicar ownership.

Population density is also related to multicar ownership. This relation, shown in Figure 5.2, shows how central city automobile availability varies with population density. In general, the availability of one car remains relatively constant at all population density levels, but multicar ownership tends to increase as population density declines. This is further verification of the earlier Census data. Lansing and Hendricks report similar findings on the effect of central city size, type, and configuration on the incidence of multicar ownership.[7] Their findings, summarized in Table 5.4, show the

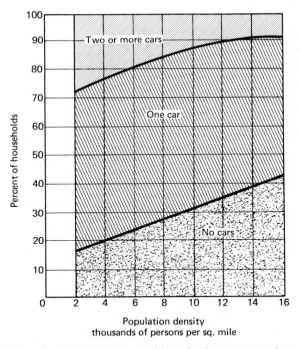

Figure 5.2. The effect of central city population density on automotive availability. [Source: Adapted from 1960 Census by H. S. Levinson and F. Houston Wynn, "The Effects of Density on Urban Transportation Requirements," *Community Values as Affected by Transportation*, Highway Research Record, No. 2 (Washington, D.C.: Highway Research Board, 1963) p. 52.]

---

[6] Only approximately 20 percent of nonmetropolitan area households of the United States have a second or third car.

[7] JOHN B. LANSING and GARY HENDRICKS, *Automobile Ownership and Residential Density* (Ann Arbor: Survey Research Center, University of Michigan, 1967).

TABLE 5.4
**Percentage of Families Owning 0, 1, and 2+ Cars, and Number of Families
in Sample Classified by Type of Central City and Family Income**

	Annual Family Income			
Area	All	Under $4,000	$4,000–7,499	$7,500 or more
*Old Central Cities of 11 Largest SMSAs**				
Family owns:				
No car	54%	85%	46%	14%
One car	38%	15%	46%	67%
Two or more cars	8%	†	8%	19%
Total	100%	100%	100%	100%
Number of families	117	53	28	36
*New Central Cities of 11 Largest SMSAs**				
Family owns:				
No car	16%	42%	19%	2%
One car	62%	58%	76%	53%
Two or more cars	22%	†	5%	45%
Total	100%	100%	100%	100%
Number of families	111	24	38	49
Smaller Central Cities				
Family owns:				
No car	20%	52%	12%	3%
One car	48%	41%	62%	38%
Two or more cars	32%	7%	26%	59%
Total	100%	100%	100%	100%
Number of families	474	130	177	167
All Suburban Areas				
Family owns:				
No car	4%	16%	3%	1%
One car	49%	64%	59%	38%
Two or more cars	47%	20%	38%	61%
Total	100%	100%	100%	100%
Number of families	772	101	287	384

*The 11 largest standard metropolitan statistical areas exclusive of the New York area have been divided into "old" and "new." The "old" cities are Baltimore, Boston, Chicago, Saint Louis, and Philadelphia. All had a population of over 500,000 as of 1900. The "new" cities are Cleveland, Detroit, Los Angeles, Pittsburgh, San Francisco, and Washington, D.C.
†Less than one-half of one percent.

Source: J. B. Lansing and G. Hendricks, *Automobile Ownership and Residential Density* (Ann Arbor: University of Michigan, 1967).

effects of both household income and density. In the old central cities of the eleven largest Standard Metropolitan Statistical Areas (SMSAs), 8 percent of all families own two or more cars, in contrast to 22 percent in new central cities of these large SMSAs and 32 percent in smaller central cities. In the suburban areas, 47 percent own two or more cars.

TRAFFIC APPROACHING CITIES

Travel between cities is related to city population, distance (or travel time) between cities, and the number of competing opportunities.

TABLE 5.5
Characteristics of Intercity Travel*

City	Population	Total Trips			Business Trips†			Non-business Trips‡		
		Trips/Capita	Avg. Trip Length (Hr)	Veh-hr/Capita	Trips/Capita	Avg. Trip Length (Hr)	Veh-hr/Capita	Trips/Capita	Avg. Trip Length (Hr)	Veh-hr/Capita
St. Louis, Mo.	1,456,673	0.0238	3.89	0.0925	0.0141	3.72	0.0524	0.0097	4.14	0.0401
Chattanooga, Tenn.	242,096	0.0792	2.19	0.1732	0.0612	2.02	0.1235	0.0180	2.76	0.0497
Madison, Wis.	169,236	0.1515	2.38	0.3600	0.0971	2.25	0.2180	0.0544	2.61	0.1420
Springfield, Mo.	109,768	0.1220	2.88	0.3520	0.0738	2.58	0.1910	0.0482	3.35	0.1610
Green Bay, Wis.	96,407	0.1271	1.96	0.2430	0.0725	1.76	0.1275	0.0546	2.12	0.1155
St. Joseph, Mo.	84,165	0.1633	2.21	0.3620	0.0744	2.37	0.1762	0.0850	2.09	0.1775
Sheboygan, Wis.	60,000	0.1160	2.06	0.2390	0.0765	1.61	0.1235	0.0395	2.93	0.1155
Joplin, Mo.	40,914	0.2750	2.02	0.5530	0.1596	1.94	0.3090	0.1154	2.10	0.2440
Morristown, Tenn.	27,000	0.2450	1.25	0.3061	0.1842	1.22	0.2253	0.0614	1.32	0.0812
Columbia, Tenn.	26,000	0.1459	1.47	0.2130	0.1062	1.25	0.1326	0.0396	2.04	0.0808
West Bend, Wis.	15,520	0.2650	1.08	0.2880	0.1545	1.06	0.1640	0.1105	1.12	0.1240
Athens, Tenn.	13,100	0.2161	1.32	0.2861	0.1698	1.31	0.2218	0.0468	1.37	0.0645
Dyersburg, Tenn.	12,499	0.2900	1.71	0.4950	0.2170	1.62	0.3514	0.0718	1.98	0.1420
Sturgeon Bay, Wis.	10,000	0.2395	2.40	0.5779	0.0609	2.44	0.1489	0.1786	2.40	0.4290
Burlington, Wis.	8,700	0.4080	1.08	0.4400	0.2360	1.12	0.2630	0.1720	1.03	0.1770
Humboldt, Tenn.	8,650	0.0669	4.03	0.2695	0.0626	2.64	0.1653	0.0043	24.30	0.1039
Monroe, Wis.	8,170	0.4810	1.29	0.6210	0.2488	1.37	0.3414	0.2319	1.20	0.2793
Oconomowoc, Wis.	8,000	0.2450	1.30	0.3190	0.2079	1.17	0.2430	0.0370	2.05	0.0759
Lake Geneva, Wis.	5,500	0.9340	2.93	2.7200	0.2691	2.36	0.6356	0.6664	3.16	2.1054
Waupaca, Wis.	4,500	0.5450	1.67	0.9060	0.2910	1.51	0.4420	0.2540	1.82	0.4640
Elkhorn, Wis.	3,600	—§	—§	—§	—§	—§	—§	—§	—§	—§
Rogersville, Tenn.	3,121	0.4480	3.00	1.3480	0.3762	1.99	0.7494	0.0727	8.67	0.6306

*Trips greater than 35 min.
†Includes work and business as defined in this study.
‡Includes all trips except work and business.
§Omitted because of inconsistent data.

(Source: Vogt, Ivers and Associates, *Social and Economic Factors Affecting Intercity Travel*, National Cooperative Highway Research Program Report 70 Washington, D.C.: Highway Research Board, 1969.)

TABLE 5.6
Typical Predictive Equations for Intercity Travel

Item	Total trips	Business Trips	Non-business Trips
Trips/Capita (over 35 min)	$\dfrac{11.0}{(\text{Cordon pop.})^{0.392}}$	$\dfrac{61}{(\text{Cordon pop.})^{0.599}}$	$\dfrac{435}{(\text{Cordon pop.})^{0.847}}$
Trip Length (Hr)	$\dfrac{(\text{Cordon pop.})^{0.278}}{11.05}$	$\dfrac{(\text{Cordon pop.})^{0.274}}{11.25}$	$\dfrac{(\text{Cordon pop.})^{0.315}}{15.4}$
Veh-hr/Capita	$\dfrac{1}{(\text{Cordon pop.})^{0.114}}$	$\dfrac{1}{(\text{Cordon pop.})^{0.325}}$	$\dfrac{28.25}{(\text{Cordon pop.})^{0.532}}$

Source: Vogt, Ivers and Associates, *Social and Economic Factors Affecting Intercity Travel.*

Illustrative characteristics of intercity travel based on a comparative analysis of 22 cities are shown in Tables 5.5 and 5.6.[8] These studies indicate the following:

1. The average trip time for some 664,000 cordon crossings approximated 50 min. Of this total, only 27.5 percent had trip lengths greater than 35 min, and approximately three-fourths had origins or destinations in the counties adjacent to the study area.
2. The number of counties linked to a given community increases with city size on an average day. Trips originating or terminating in the St. Louis study area, for example, had origins or destinations in 33 percent of the metropolitan areas in the United States.
3. The number of external trips per capita increases with decreasing city size. This inverse relationship emphasizes the role of the small city as a trip producer and the large city as a trip attractor. The longer average trip lengths (in hours) to larger cities further suggests their greater attractiveness and spheres of influence.

The destinations of external traffic *within* the urban area also vary with urban population (Figure 5.3 and Table 5.7.) In the smallest cities, more than half the approaching traffic goes beyond the city itself. As city size increases, there is an increase in the proportion of traffic bound for the city. Over 90 percent of the traffic approaching the largest cities has destinations within the urban areas, as compared with less than half in small cities.

Approximately 20 to 30 percent of all approaching traffic is bound for the central business district in small cities as compared with approximately 10 percent in large (over one million population) cities.

TRIP GENERATION

Comprehensive metropolitan area origin-destination land use and origin-destination studies have measured the nature of urban tripmaking and travel behavior. The number, length, purpose, mode, and orientation of urban trips relate closely to (1)

---

[8] Vogt, Ivers and Associates, *Social and Economic Factors Affecting Intercity Travel*, National Cooperative Highway Research Program Report No. 70, (Washington, D.C.: National Cooperative Highway Research Program, 1969).

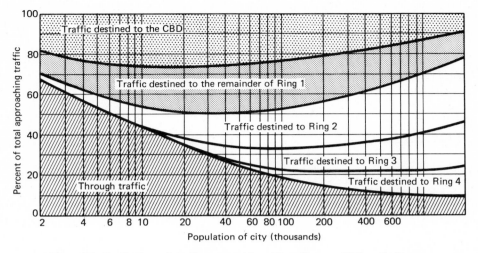

Figure 5.3. Destination of traffic approaching cities. [Source: Walter G. Hansen, "Traffic Approaching Cities," *Public Roads*, Vol. 31, No. 7 (April 1961), p. 156.]

TABLE 5.7
Destinations of External Traffic Entering Cities

Size of City (000's)	Percent of Through Traffic	Percent Bound for City	Percent Bound to CBD
0–5	57.7	42.3	24.3
10–25	32.7	67.3	31.1
25–50	19.9	80.1	33.2
100–250	15.4	84.6	22.1
250–500	10.4	89.6	24.2
500–1,000	5.9	94.1	18.2

Source: Walter G. Hansen, "Traffic Approaching Cities," *Public Roads*, Vol. 31, No. 7 (April 1961).

land-use patterns, (2) socioeconomic characteristics, and (3) available transportation services.

SUMMARY PATTERNS

Travel characteristics of residents in a cross section of American urban areas are summarized in Tables 5.8, 5.9, and 5.10.

1. In most American cities, more than three-quarters of all person-trips are made by car. The proportion of trips made by public transport tends to increase as the population and/or density increases. (The location of survey area boundaries influences the modal distribution—a "tight" cordon-line generally produces a higher proportion of trips by transit.)
2. Most trips in European, Asian, and South American cities with high population densities, low car ownership, and centrally oriented travel patterns are made by transit. This pattern is changing gradually as car ownership rises and

some deconcentration takes place. In Athens, for example, two-thirds of all person-trips were made by transit as compared with only one-third in the Tri-State New York Area (Table 5.8).

3. Average car occupancy was relatively constant at approximately 1.5 persons per car (Table 5.9).

4. Trucks constitute approximately 15 percent of all vehicle trips, ranging from 10 to 21 percent (Table 5.9).

TABLE 5.8
Travel Modes of Urban Residents

Urban Area	Year	Study Area Population	Thousands of Person Vehicle Trips			Percentage by	
			Auto	Transit	Total	Auto	Transit
New York (Tri-State/NY)	1963	16,302,000	19,840	9,730	29,570	67.1	32.9
Chicago, Ill.	1956	5,169,700	7,517	2,414	9,931	75.7	24.3
Philadelphia, Pa.	1960	4,007,000	6,477	1,283	7,760	83.5	16.5
Boston, Mass.	1963	3,584,400	6,351	1,500	7,851	80.2	19.8
Detroit, Mich.	1953	2,968,900	4,385	879	5,264	83.3	16.7
Cleveland, Ohio	1963	2,140,000	4,477	539	5,016	89.3	10.7
Dallas, Texas	1964	1,821,000	5,062	198	5,260	96.2	3.8
Toronto, Ont.	1964	1,800,000	3,124	753	3,877	80.6	19.4
Milwaukee, Wis.	1963	1,644,000	2,291	268	2,559	89.5	10.5
Baltimore, Md.	1962	1,600,800	2,150	455	2,605	82.5	17.5
Washington, D. C.	1955	1,568,500	1,987	639	2,626	75.7	24.3
Pittsburgh, Pa.	1958	1,472,100	1,895	482	2,377	79.7	20.3
Minneapolis-St. Paul, Minn.	1958	1,376,900	2,950	416	3,366	87.6	12.4
Buffalo, N.Y.	1962	1,350,000	2,476	284	2,760	90.3	9.7
Seattle, Wash.	1962	1,347,000	2,252	116	2,368	95.2	4.8
St. Louis, Mo.	1957	1,275,500	2,090	387	2,477	84.4	15.6
Houston, Texas	1953	878,600	1,701	252	1,953	87.1	12.9
Kansas City, Mo.	1957	857,600	1,685	185	1,870	90.1	9.9
New Orleans, La.	1960	855,500	1,115	402	1,517	73.6	26.4
Denver, Col.	1960	806,100	1,736	83	1,819	95.5	4.5
Louisville, Ky.	1964	768,900	1,279	78	1,357	94.3	5.7
San Juan P.R.	1964	758,800	737	439	1,196	62.7	37.3
Portland, Ore.	1960	715,100	1,624	144	1,768	91.9	8.1
Atlanta, Ga.	1961	700,000	1,084	162	1,246	87.0	13.0
Providence, R.I.	1961	685,600	1,319	85	1,404	94.0	6.0
Memphis, Tenn.	1964	647,700	1,345	85	1,430	94.1	5.9
Oklahoma City, Okla.	1965	574,013	1,636	87	1,723	95.0	5.0
Springfield, Mass.	1965	531,000	1,034	119	1,153	89.7	10.3
Miami, Fla.	1964	429,400	1,223	85	1,308	93.6	6.4
Honolulu, Haw.	1960	480,100	1,127	131	1,258	89.6	10.4
Richmond, Va.	1964	217,600	789	124	923	86.6	13.4
Salt Lake City, Utah	1960	394,300	924	54	978	94.5	5.5
Nashville, Tenn.	1959	357,600	756	63	819	92.3	7.7
Lehigh Valley, Pa.	1964	345,100	629	59	688	91.4	8.6
Tucson, Ariz.	1960	244,500	552	27	579	95.3	4.7
Chattanooga, Tenn.	1960	241,800	486	39	525	92.6	7.4
Fort Lauderdale, Fla.	1959	211,000	352	5	357	98.6	1.4
Charlotte, N.C.	1958	202,300	443	35	478	92.7	7.3
Columbia, S.C.	1964	196,000	496	36	532	93.3	6.7
Reno, Nev.	1955	55,000	134	2	136	98.5	1.5
London, England	1961	8,826,600	6,622	7,774	14,396	46.0	54.0
Bombay, India	1962	4,345,200	810	1,890	2,700	30.0	70.0
Athens, Greece	1962	1,900,000	1,120	2,080	3,200	35.0	65.0
Brisbane, Australia	1960	593,700	556	455	1,011	55.0	45.0

Source: Compiled from transportation planning studies in subject areas.

TABLE 5.9
Travel Modes, Car Occupancies, and Truck Trips—Selected Urban Areas

Urban Area	Year	Study Area Population	Thousands of Person Trips				Percentage by		Average Car Occupancy	Truck Trips	Total Vehicle Trips	Percent Trucks
			Auto Driver	Auto Truck Taxi Pass.	Transit Passenger	Total	Auto	Transit				
Chicago	1956	5,169,700	4,811	2,705	2,414	9,931	75.7	24.3	1.56	828	5,639	14.7
Philadelphia	1960	4,007,000	4,309	2,168	1,283	7,760	83.5	16.5	1.50	990	5,299	18.6
Boston	1963	3,584,400	4,444	1,907	1,500	7,851	80.2	19.8	1.43	878	5,322	16.4
Detroit	1953	2,968,900	2,991	1,394	879	5,264	83.3	16.7	1.46	495	3,486	14.2
Dallas	1964	1,821,000	3,337	1,724	198	5,260	96.2	3.8	1.52	418	3,755	11.1
Toronto	1964	1,800,000	2,244	880	753	3,877	80.6	19.4	1.44	NA	NA	NA
Baltimore	1962	1,600,800	1,426	724	455	2,605	82.5	17.5	1.50	378	1,804	20.9
Washington	1955	1,568,500	1,278	709	639	2,626	75.7	24.3	1.56	219	1,497	14.6
Pittsburgh	1958	1,472,100	1,292	603	482	2,377	79.7	20.3	1.47	229	1,521	15.1
Buffalo	1962	1,350,000	1,588	888	284	2,760	89.8	10.2	1.56	208	1,796	11.6
St. Louis	1957	1,275,500	1,349	731	387	2,477	84.4	15.6	1.54	280	1,639	17.1
Houston	1953	878,600	1,085	616	252	1,953	87.1	12.9	1.57	202	1,287	15.7
Kansas City	1957	857,600	1,108	577	185	1,870	90.1	9.9	1.52	181	1,289	14.0
Atlanta	1961	700,000	735	349	162	1,246	87.0	13.0	1.47	NA	NA	NA
Oklahoma City	1965	574,000	1,079	559	87	1,723	95.0	5.0	1.52	148	1,225	12.0
Springfield, Mass.	1965	531,000	725	309	119	1,153	89.7	10.3	1.43	81	806	10.0
Richmond	1964	417,600	542	247	124	923	86.6	13.4	1.47	94	636	14.7
Salt Lake City	1960	394,300	624	300	54	978	94.5	5.5	1.47	138	762	18.1
Nashville	1959	357,600	493	263	63	819	92.3	7.7	1.54	91	584	15.6
Lehigh Valley, Pa.	1964	345,100	443	186	59	688	91.4	8.6	1.42	82	525	15.6
Chattanooga	1960	241,800	312	174	39	525	92.6	7.4	1.46	64	376	17.0
Fort Lauderdale	1959	211,000	238	114	5	357	98.6	1.4	1.48	32	259	12.0
Charlotte	1958	202,300	303	140	35	478	92.7	7.3	1.46	52	355	14.6
Columbia, S.C.	1964	196,000	330	166	36	532	93.3	6.7	1.50	52	382	13.6
Reno	1955	55,000	81	53	2	136	98.5	1.5	1.68	22	103	21.4

Source: Compiled from transportation planning studies in subject areas.

The typical urban resident averages approximately two trips per day in vehicles. There is, however, a wide variation among cities, depending on city size, structure, economy, auto ownership levels, and year of study.[9]

Data in Table 5.10 indicate that

1. The number of daily trips ranges from 1.6 to 3.1 per person and from 5.3 to 9.9 per dwelling unit. (The extremes are Pittsburgh and Little Rock, respectively.)
2. The number of persons per car ranges from 2.4 to 3.8, and the number of persons per dwelling unit ranges from 2.2 to 3.5.

TABLE 5.10
Generation of Travel by Urban Residents in Selected American Cities

Urban Area	Year of Survey	Population	Trips per Person	Persons per Car	Trips per Dwelling	Persons per Dwelling	Cars per Dwelling
Chicago, Ill.	1956	5,170,000	1.92	3.85	5.96	3.10	0.80
Philadelphia, Pa.	1960	4,007,000	2.03	3.69	6.26	3.08	0.84
Boston, Mass.	1963	3,584,000	2.23	3.36	7.33	3.30	0.98
Detroit, Mich.	1953	2,969,000	1.77	3.51	5.88	3.31	0.94
S. E. Wis. (Milwaukee)	1963	1,644,000	2.07	3.10	7.05	3.41	1.12
Baltimore, Md.	1962	1,608,000	1.66	3.61	5.56	3.34	0.92
Washington, D.C.	1955	1,568,000	1.67	3.75	5.05	3.02	0.81
Pittsburgh, Pa.	1958	1,472,000	1.61	3.75	5.26	3.26	0.87
Minneapolis-St. Paul, Minn.	1958	1,377,000	2.45	3.15	8.25	3.37	1.07
Seattle, Wash.	1962	1,347,000	1.76	2.76	5.32	3.02	1.09
St. Louis, Mo.	1957	1,275,000	1.94	3.48	6.05	3.12	0.90
Houston, Texas	1953	879,000	2.22	3.43	7.16	3.22	0.94
Kansas City, Mo.	1957	858,000	2.18	3.26	6.69	3.07	0.95
Denver, Col.	1960	806,000	2.26	2.83	7.74	3.43	1.21
Atlanta, Ga.	1961	700,000	1.78	3.14	5.41	3.04	0.97
S. E. Va. (Norfolk)	1962	602,000	2.25	3.54	7.37	3.27	0.93
Oklahoma City	1965	574,000	3.00	2.49	9.51	3.17	1.27
Springfield, Mass.	1965	531,000	2.25	3.14	7.05	3.13	1.00
Richmond, Va.	1964	418,000	2.33	3.04	7.57	3.33	1.07
Phoenix, Ariz.	1957	397,000	2.29	2.87	6.88	3.01	1.05
Salt Lake City, Utah	1960	394,000	2.46	3.88	9.00	3.51	1.22
Nashville, Tenn.	1959	358,000	2.29	3.35	7.52	3.28	0.98
Lehigh Valley, Pa.	1964	345,000	2.00	2.84	6.40	3.21	1.13
Chattanooga, Tenn.	1960	242,000	2.03	3.31	7.58	3.33	1.01
Knoxville, Tenn.	1962	242,000	2.50	3.09	8.08	3.22	1.04
Little Rock, Ark.	1964	223,000	3.09	3.05	9.89	3.20	1.10
Fort Lauderdale, Fla.	1959	211,000	1.69	2.72	3.63	2.15	0.79
Charlotte, N.C.	1958	202,000	2.36	3.28	8.10	3.43	1.05
Columbia, S.C.	1964	196,000	2.96	2.83	9.42	3.17	1.12
Madison, Wis.	1962	169,000	2.25	3.00	7.10	3.14	1.05
Huntsville, Ala.	1964	123,000	2.87	2.52	9.20	3.21	1.27
Lexington, Ky.	1961	119,000	2.14	2.93	6.84	3.19	1.09
Rapid City, S.D.	1963	73,000	2.45	2.75	7.41	3.15	1.14
Reno, Nev.	1955	55,000	2.48	2.43	6.87	2.77	1.14

Source: Compiled fron transportation planning studies in subject areas.

[9] HERBERT S. LEVINSON and F. HOUSTON WYNN, "Some Aspects of Future Transportation in Urban Areas," *Urban Transportation Demand and Coordination, Bulletin 326*, (Washington, D.C.: Highway Research Board, 1962), pp. 1–31. Also see Wilbur Smith and Associates, *Future Highways and Urban Growth*, Detroit, 1961, pp. 61–156.

3. The number of cars per dwelling unit ranges from approximately 0.8 to approximately 1.3.

Generally, individuals in smaller communities report more trips than residents in larger ones. The number of daily trips in vehicles decreases gradually as city size and/or density rises. There is, however, considerable variation among cities in the same population range; the year of study and sampling variability also influence trip rates.

In large cities, such as Chicago, Philadelphia, and Detroit, the urban resident averages two trips daily in vehicles. In smaller cities, such as Oklahoma City and Little Rock, the urban resident averages three trips per day.

These differences are partially explained by the close relationship among car ownership, population density, and travel mode: The number of total person trips (in vehicles) decreases consistently with increasing population concentration and decreasing car ownership; however, the number of transit trips per capita increase. Many trips in high-density urban areas are pedestrian trips and are not reported in the basic trip data. When these pedestrian trips are superimposed on the trips in vehicles, it is likely that there is a rising rate of total trip generation, and hence interaction in high-density environments.

Results of origin-destination studies as summarized for the Bureau of Public Roads are shown in Tables 5.11 and 5.12. These data, based on studies conducted *prior* to 1958, provide a benchmark for comparative purposes.

TABLE 5.11
Average Travel and Ownership Ratios for 49 Cities (Prior to 1960)

Population Group	Number of Cities	Persons per Dwelling	Persons per Vehicle	Vehicles per Dwelling	Number of Trips		
					Per Dwelling	Per Vehicle	Per Person
Over 1,000,000	4	3.16	6.12	0.52	5.08	9.85	1.61
500,000–1,000,000	5	3.12	4.28	0.73	5.73	7.84	1.84
250,000–500,000	2	3.03	4.79	0.63	5.62	8.89	1.85
100,000–250,000	22	3.26	4.61	0.71	6.24	8.82	1.92
50,000–100,000	10	3.58	5.44	0.66	6.09	9.23	1.70
25,000–50,000	6	3.39	4.55	0.74	7.79	10.47	2.30
All cities	49	3.18	5.08	0.63	5.61	8.94	1.76

Source: Adapted from Frank B. Curran and Joseph T. Stegmaier, "Travel Patterns in 50 Cities," *Travel Characteristics in Urban Areas*, Bulletin 203 (Washington, D.C.: Highway Research Board, 1958).

TRIP PURPOSES: THE REASONS FOR TRAVEL

All urban trips are made for some gainful purpose, for example, a trip to work, to the doctor, or to visit a friend. Trip-purpose patterns, therefore, reflect the daily activities of urban residents.

1. The home or dwelling unit is the primary origin of most trips (Figure 5.4). Generally, more than three-fourths of all urban trips are to or from home.
2. Table 5.13 summarizes trip purposes according to travel mode and urban population. Data are shown for 50 American cities and are based on surveys conducted before 1961.

TABLE 5.12

Average Ratios per City Between Number of Trips by Each Mode of Travel and Selected Household Characteristics in Six Population Groups

Population Group (000 omitted)	Number of Cities	Trips per Dwelling Unit by Mode of Travel				Trips per Automobile Owned by Mode of Travel				Trips per Person by Mode of Travel			
		Automobile Driver	Automobile and Taxi Passenger	Mass Transit Passenger	Total	Automobile Driver	Automobile and Taxi Passenger	Mass Transit Passenger	Total	Automobile Driver	Automobile and Taxi Passenger	Mass Transit Passenger	Total
1,000 and over	4	1.62	0.91	2.50	5.03	3.15	1.76	4.85	9.76	0.51	0.29	0.79	1.59
500–1,000	6	2.35	1.16	1.99	5.50	3.45	1.70	2.91	8.06	0.74	0.37	0.63	1.74
250–500	2	2.54	1.49	1.41	5.44	4.02	2.35	2.22	8.59	0.84	0.49	0.46	1.79
100–250	20	3.15	1.79	1.35	6.29	4.38	2.48	1.88	8.74	0.97	0.55	0.42	1.94
50–100	12	3.09	1.75	1.41	6.25	4.50	2.54	2.05	9.09	0.88	0.50	0.40	1.78
Less than 50	5	3.73	2.10	0.85	6.60	4.61	2.60	1.06	8.27	1.08	0.61	0.25	1.94

Source: Curran and Stegmaier, "Travel Patterns in 50 Cities," p. 126.

TABLE 5.13
Percentage of Trips by Each Mode of Travel in Six Population Groups,
Classified According to Trip Purpose

Population Group (000 omitted)	Number of Cities	Mode of Travel	Work and Business	Social and Recreation	Shop	Miscellaneous	Home	Total
				Trip Purpose (percent)				
1,000 and over	4	Automobile drivers	35.4	9.3	6.2	12.9	36.2	100.0
		Automobile and taxi passengers	19.1	23.3	6.5	3.8	42.7	100.0
		Mass-transit passengers	28.2	7.0	6.2	17.8	40.8	100.0
		Total	28.9	10.7	6.3	14.5	39.6	100.0
500–1,000	6	Automobile drivers	31.5	8.8	8.4	12.4	38.9	100.0
		Automobile and taxi passengers	16.6	23.0	8.3	9.0	43.1	100.0
		Mass-transit passengers	33.0	6.2	6.8	6.4	47.6	100.0
		Total	28.9	10.9	7.8	9.5	42.9	100.0
250–500	3	Automobile drivers	33.9	10.4	8.4	12.2	35.1	100.0
		Automobile and taxi passengers	16.9	26.3	8.3	7.5	41.0	100.0
		Mass-transit passengers	30.2	9.0	7.2	10.4	43.2	100.0
		Total	28.4	13.8	8.0	10.4	30.4	100.0
100–250	20	Automobile drivers	30.4	9.8	9.3	14.8	35.7	100.0
		Automobile and taxi passengers	16.0	35.9	8.9	7.1	42.1	100.0
		Mass-transit passengers	26.3	9.2	8.3	8.6	47.6	100.0
		Total	25.4	14.3	8.9	11.3	40.1	100.0
50–100	12	Automobile drivers	30.0	10.9	8.0	14.6	36.5	100.0
		Automobile and taxi passengers	17.5	26.3	7.9	5.5	42.8	100.0
		Mass-transit passengers	29.8	9.1	7.8	6.6	46.7	100.0
		Total	26.5	14.8	7.9	10.2	40.6	100.0
Less than 50	5	Automobile drivers	29.0	10.8	9.0	16.9	34.3	100.0
		Automobile and taxi passengers	13.9	32.9	8.0	4.9	40.3	100.0
		Mass-transit passengers	26.6	9.2	9.9	7.6	46.7	100.0
		Total	24.0	17.5	8.8	11.9	37.8	100.0
All groups	50	Automobile drivers	32.3	9.5	8.0	13.4	36.8	100.0
		Automobile and taxi passengers	17.2	24.5	7.9	7.9	42.5	100.0
		Mass-transit passengers	29.5	7.2	6.8	12.5	44.0	100.0
		Total	27.9	12.0	7.5	11.8	40.8	100.0

Source: Curran and Stegmaier, "Travel Patterns in 50 Cities," p. 106.

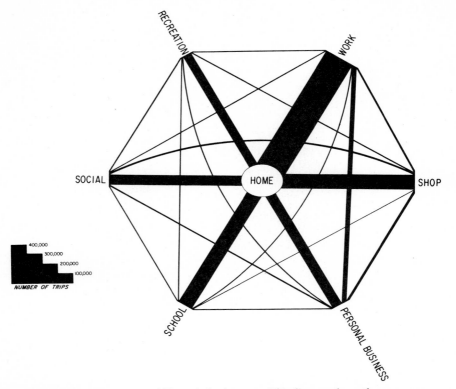

Figure 5.4. Trip purposes within typical urban area. This diagram shows the reasons for which trips are made in the Springfield, Mass. region on a typical 1965 day. More than three-fourths of the 1.15 million internal person trips were made either to or from home. (Source; Springfield Urbanized Area Comprehensive Transportation Study, Volume I, *The Land Use, Transportation, and Travel Inventories,* Wilbur Smith and Associates, 1969, p. 163.)

3. Table 5.14 summarizes the home-based trip purposes for urban areas in the United States and Canada. Approximately 32 percent of all trips are to or from work, 17 percent to or from shopping, 21 percent for social or recreational purposes, 11 percent for business purposes, 9 percent to and from school, and the remaining 10 percent are for other reasons.

4. Table 5.15 summarizes trip purposes "from the home" for a small British city according to household size. Approximately 37 percent of all trips were to and from work, 18 percent for shopping, 12 percent for social and recreational reasons, 5 percent for business purposes, 23 percent to or from schools, and 5 percent for other reasons.

5. Table 5.16 denotes the work travel proportions and modes in selected European, Asian, and Australian cities, as they relate to car ownership and total trip making. The data for Skopje suggest that work trips represent a larger proportion of total travel in cities with low overall trip making.

The number of daily work trips per person remains about the same from city to city in the United States. It averages 0.6 trips per person, despite the variations in

TABLE 5.14
Home-based Trips by Urban Residents According to Purpose

Urban Area	Year	Popu-lation (000)	Home-based Trips as Percent of All Linked Trips	Percent of Home-based Trip to and from:							Total Home-based Trips per Dwelling Unit
				Work	Busi-ness	Shop-ping	Social-Recrea-tional	School	Other	All Purpose	
Chicago, Ill.	1956	5,170	86.8	37.5	9.7	18.9	22.8	4.0	7.1	100.0	5.17
Philadelphia, Pa.	1960	4,007	85.4	34.8	9.8	12.7	17.1	6.6	19.0	100.0	3.90
Boston, Mass.	1963	3,584	81.0	28.3	9.9	17.5	16.0	11.5	15.8	100.0	5.30
Detroit, Mich.	1953	2,961	87.0	41.6	8.6	13.9	20.1	6.3	9.5	100.0	4.67
Toronto, Ont.	1964	1,800	88.7	48.5	8.9	17.3	16.2	4.6	4.5	100.0	5.64
Milwaukee, Wis.	1963	1,644	82.8	32.8	14.2	16.7	13.6	9.6	12.1	100.0	5.44
Baltimore, Md.	1962	1,608	88.0	39.7	8.1	15.1	15.6	12.0	9.5	100.0	4.58
Washington, D.C.	1955	1,568	91.6	43.1	9.6	14.2	12.5	9.4	11.2	100.0	4.23
Pittsburgh, Pa.	1948	1,472	87.0	37.7	21.6	14.9	13.8	12.0	—	100.0	4.21
Seattle, Wash.	1962	1,347	79.9	27.8	14.7	16.5	20.9	9.2	10.9	100.0	3.82
St. Louis, Mo.	1957	1,276	91.3	37.5	8.1	16.3	21.5	6.4	9.2	100.0	4.90
Houston, Texas	1953	879	91.0	33.1	8.9	17.3	18.6	10.8	11.3	100.0	5.51
Denver, Col.	1960	806	83.1	29.3	12.8	15.5	21.1	11.1	10.2	100.0	5.23
Oklahoma City, Okla.	1965	574	73.0	26.4	9.6	20.8	27.4	15.8	—	100.0	7.89
Springfield, Mass.	1964	531	81.1	30.1	9.6	16.5	18.3	13.6	11.9	100.0	5.37
Richmond, Va.	1964	418	80.3	34.2	13.0	16.6	18.1	11.0	7.1	100.0	5.51
Phoenix, Ariz.	1957	397	85.3	25.2	10.2	19.7	20.0	11.6	13.3	100.0	4.76
Salt Lake City	1960	394	85.0	22.9	4.8	18.4	28.7	8.1	17.1	100.0	6.99
Nashville, Tenn.	1959	358	85.5	30.3	8.5	16.9	23.9	7.4	13.0	100.0	5.48
Lehigh Valley, Pa.	1964	345	77.8	31.0	11.8	15.9	14.8	13.2	13.3	100.0	5.20
Knoxville, Tenn.	1962	242	79.5	25.9	11.1	23.8	30.9	9.3	—	100.0	6.42
Little Rock, Ark.	1964	223	83.2	25.8	8.9	20.1	31.2	14.0	—	100.0	8.23
Ft. Lauderdale, Fla.	1959	211	86.5	27.9	15.3	24.0	22.9	0.9	910	100.0	2.82
Charlotte, S.C.	1958	202	83.9	32.2	8.0	15.6	23.8	6.6	13.8	100.0	5.56
Columbia, S.C.	1964	196	77.5	25.3	7.9	16.9	20.2	14.9	14.8	100.0	6.60
Reno, Nev.	1958	55	86.5	29.2	12.7	18.1	26.3	0.5	13.2	100.0	4.88

Source: Compiled from transportation planning studies in subject areas.

TABLE 5.15

Persons per Household; Purposes and Numbers of Outward-bound Journeys (Trips from the Home in 24 Hours)

Purpose of Journey

Persons per Household*	To Work		On Business		Shopping		Education		Social–Recreation		Others Unclassified		Total	
	Trips per House-hold	Percent-age of All Trips	Trips per House-hold	Percent-age of All Trips	Trips per House-hold	Percent-age of All Trips	Trips per House-hold	Percent-age of All Trips	Trips per House-hold	Percent-age of All Trips	Trips per House-hold	Percent-age of All Trips	Trips per House-hold	Percent-age of All Trips
1 (79)*	0.27	36	0.04	5	0.27	36	0.01	1	0.09	12	0.07	10	0.75	100
2 (336)*	0.85	48	0.11	6	0.48	27	0.01	1	0.24	14	0.08	4	1.77	100
3 (247)*	1.28	43	0.17	6	0.53	18	0.50	17	0.39	13	0.09	3	2.96	100
4 (227)*	1.38	32	0.22	5	0.61	15	1.36	32	0.49	11	0.23	5	4.29	100
5 (80)*	1.24	27	0.14	3	0.51	11	1.86	41	0.48	10	0.35	8	4.58	100
6 (17)	0.71	13	0.24	4	0.88	17	2.29	44	0.53	10	0.64	12	5.29	100
7 (5)*	1.40	25	0.20	4	0.40	7	3.40	60	0.20	4	0.00	0	5.60	100
8 (1)*	2.00	50	0.00	0	1.00	25	1.00	25	0.00	0	0.00	0	4.00	100
Mean*	1.06	37	0.15	5	0.51	18	0.65	23	0.34	12	0.16	5	2.87	100

*Figures in parentheses give number of households in the group.

Source: T. E. H. Williams, G. G. Dobson, and H. W. T. White, "Traffic Generated by Households," *Traffic Engineering and Control*, Vol. 5, No. 3 (July 1963), p. 179.

TABLE 5.16

Trip Characteristics of Selected Cities (Europe-Asia-Australia)

City	Study Year	Population	Survey Area (sq mi)	Vehicle Registration	Persons per Vehicle	Total Person Trips (Internal)	Percent by Transit	Trips Per Person	Per-cent Work	Work Trips Per Person	% Work Trips by Transit	House-hold Inter-view Sample (%)
Hobart	1963	125,400	78	36,900	3.4	238,000	27	1.9	28	0.53	30	5.4
Skopje	1965	220,000	42	10,600	20.8	163,200	74	0.7	49	0.59	82	6.4
Brisbane	1960	593,668	375	151,560	3.9	1,011,200	45	1.7	35	0.34	67	5.0
Athens	1962	1,900,000	206	67,700	28	3,200,000	65	1.7	NA	NA	NA	0.33
Bombay	1962	4,345,202	328	62,200	70	2,700,000	70	0.6	NA	NA	NA	0.40
London	1961	8,826,620	941	1,454,000	6.1	14,396,000	54	1.6	52	0.83	63	1.7

Source: Wilbur S. Smith, "Research and Worldwide Urban Transportation", *Highway Transportation Research, Education and Technology Abroad*, Record 125 (Washington, D.C.: Highway Research Board, 1966), p. 32.

total trip making. This suggests that it is the nonwork trips that increase with rising car ownership and income.

Urban car occupancies by trip purpose related to car ownership is shown in Table 5.17. The occupancies of work and shopping trips increase as the number of persons per car increases (i.e., as car ownership in an urban area decreases). Social trip occupancies are the same for all levels of vehicle ownership, although fewer trips are made by families with low car ownership ratios.

TABLE 5.17
Urban Car Occupancies by Trip Purpose Related to Car Ownership*

Avg. No. Persons per Car	Average Auto Trip Occupancy for			
	Work	Shopping	Social	All Purposes
1.5–2.0	1.10	1.30	2.35	1.4–1.6
2.0–2.5	1.20	1.40	2.35	1.4–1.6
2.5–3.0	1.25	1.45	2.35	1.4–1.6
3.0–3.5	1.30	1.45	2.35	1.4–1.6
3.5–4.0	1.35	1.45	2.35	1.4–1.6
4.0–5.0	1.40	1.45	2.35	1.4–1.6
5.0–6.0	1.50	1.50	2.35	1.5–1.7
6.0+	1.60	1.55	2.35	1.6–1.8

*Compiled from origin-destination studies in various urban areas.

Source: Herbert S. Levinson and F. Houston Wynn, "Some Aspects of Future Transportation in Urban Areas," Bulletin 326 (Washington, D.C.: Highway Research Board, 1962), p. 8.

HOURLY VARIATIONS

The hourly variations of travel throughout the typical weekday reflect the basic purposes for which trips are made and the capabilities of the various travel modes.

Variations by mode. A composite of typical hourly variations by travel mode is shown in Figure 5.5 for Detroit, Washington, Chicago, Pittsburgh, and Toronto. Approximately 10 percent of all person travel takes place in the morning and evening peak periods. During these peaks, public transport carries a relatively large share of the total travel. Approximately from 8 to 10 percent of all person trips by car and from 12 to 16 percent by transit take place in the peak hour. During the evening shopping period (7:00–8:00 p.m.), more trips in autos are made than during the morning peak hour, largely because of higher car occupancies. Approximately 80 percent of all person trips in these areas are made by automobile.

Variations by city size. Generalized hourly variation patterns according to urban area size are shown in Figure 5.6. These patterns suggest a sharpening of peak period travel as urban population increases. In all cities, however, approximately 8 percent of the daily travel takes place in the evening peak hour.

Variations by vehicle class. Variations in motor truck travel are shown in Figure 5.7. Truck travel remains relatively constant throughout the midday periods, generally peaking at 10:00 a.m. This peak reflects increases in mid-morning deliveries.

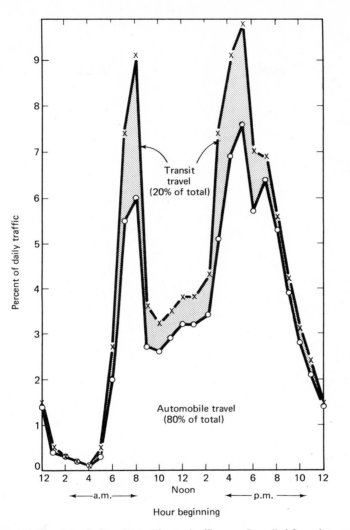

Figure 5.5. Hourly variation of travel by mode. (Source: Compiled from data on five North American cities.)

Cars in motion. Hourly variations in automobiles in motion and rest are shown in Figure 5.8. Two points are salient. (1) Comparatively few automobiles are in motion during any hour of the day. During the evening peak hour, 80 percent of all cars are at rest; (2) during the day, most cars are parked off-street in free facilities, but during the evening, as would be expected, most cars are parked on residential property.

Peak-hour vehicle trip factors. The peak-hour factors summarized in Table 5.18 provide a basis for estimating the factors to be applied to average daily vehicle travel to obtain peak-hour travel for specific trip purposes. During the evening peak hour, for example, approximately 14 percent of all work-to-home trips take place.

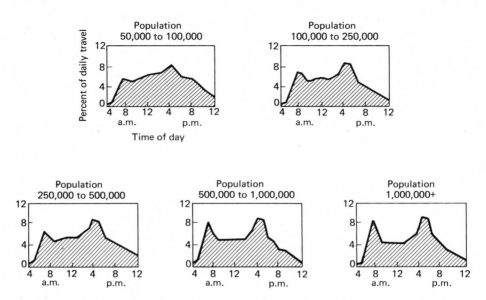

Figure 5.6. Travel by time of day related to urban area population. [Source: Harold Kassoff and David S. Gendell, "An Approach to Multi-Regional Urban Transport Policy Planning," (Washington, D.C.: Highway Research Record No. 348, 1971) p. 84.]

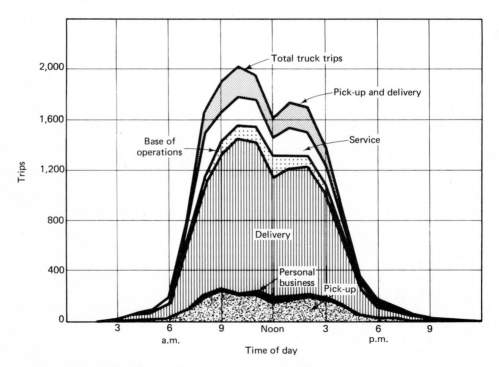

Figure 5.7. Truck trips by time of day. (Source: Pittsburgh Area Transportation Study, 1958, p. 36.)

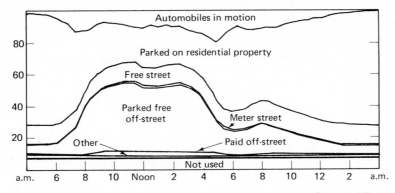

Figure 5.8. Automobiles in motion and at rest, by time of day. (Source: Niagara Frontier Transportation Study, Volume 2, *Travel 1962*, p. 19.)

TABLE 5.18
Peak-hour Vehicle Trip Factor Comparisons (Selected Urban Areas)

Trip Purpose	A.M. Peak-hour Factors				P.M. Peak-hour Factors			
	Swats*	Lehigh Valley†	Haiats‡	Reading§	Swats*	Lehigh Valley†	Haiats‡	Reading§
Home-Work	12.1	17.3	14.5	13.5	2.1	1.6	1.3	1.6
Home-School	18.3	19.4	20.5	NA	6.3	0.6	0.3	NA
Home-Other‖	1.6	2.1	1.5	1.5	4.1	5.0	3.5	2.9
Home-Shop	NA	NA	NA	.4	NA	NA	NA	2.1
Non-home Base	1.9	6.4	6.7	2.1	10.0	8.4	9.2	8.1
Work-Home	1.8	1.6	1.6	1.2	11.2	14.2	13.9	14.3
School-Home	2.9	3.7	7.6	NA	10.6	2.3	1.9	NA
Other-Home	0.6	0.5	0.3	0.2	9.2	8.1	5.2	3.9
Shop-Home	NA	NA	NA	0.3	NA	NA	NA	6.7
Internal-Commercial	7.1	NA	7.6	7.4	9.8	NA	6.5	7.5
External Vehicles	5.5	NA	4.8	—	7.6	NA	9.0	—
External-Local	NA	NA	NA	6.4	NA	NA	NA	7.7
External-Through	NA	NA	NA	5.6	NA	NA	NA	7.3

*Steubenville-Weirton, W. Va., 1966.
†Lehigh Valley, Pa., 1964.
‡Huntington-Ashland-Ironton, W. Va., 1965.
§Reading, Pa., 1965.
‖Home-based trips for purposes other than for which data are given in the table.

Source: Transportation planning studies in subject areas.

Peak-hour travel on specific facilities. Peak-hour travel on specific highway and transit facilities is summarized in Table 5.19. The peak-hour travel in the heavy direction exceeds 20 percent for commuter railroads, ranges from 12 to 16 percent for rapid transit lines, and ranges from 4 to 8 percent for urban express roads.

URBAN TRAVEL MAGNITUDES

Investment levels and capacity requirements of urban highway and transit facilities should relate to the total magnitudes of travel, as well as to the number of trips.

TABLE 5.19
Approximate Percentage of Transit Passengers or Vehicular Volumes
Traveling During Peak Hours in Selected Cities

	Percentage of Daily Volume		Percentage of Peak-hour Flow in Major Direc-tion (p.m.)	Peak Directional Volume as Percent of Daily Total
System and City*	During Four Peak Hours	During Maximum Peak Hour		
Bus Transit Systems				
Chicago	40	—	—	—
Washington, D.C. (3 Major Lines)	53	16	79.0	12.6
Rail Rapid Transit Systems				
Boston	44	—	—	—
New York City	49	14	86.5	12.1
Chicago	58	16	80.5	12.9
Toronto	51	18	80.5	14.5
Cleveland	58	19	84.0	16.0
Philadelphia	58	17	77.0	13.1
Railroad Commuter Systems				
Chicago	72	25	—	—
Washington, D.C. (Pennsylvania RR)	68	23	—	—
Philadelphia (Pennsylvania RR)	68	25	84.9	21.2
Highway Systems				
Detroit (Lodge-Ford Expressway)	28	7	57.4	4.0
Chicago (Congress Street Expressway)	30	8	62.6	5.0
Washington, D.C. (Memorial Bridge)	44	13	63.7	8.3
Boston (Route 128)	29	9	—	—

*Based on available data for 1959–1962 period.

Source: Adapted from J. R. Meyer, J. F. Kain, and M. Wohl, *The Urban Transportation Problem* (Cambridge, Mass.: Harvard University Press, 1965) pp. 95 and 98.

TABLE 5.20
Urban Person Travel in Selected Cities

Study Area	Year	Population (000)	Airline Person-Miles (000)	Person-Trips (000)	Miles per Capita	Average Trip Length (mi)
Chicago, Ill.	1956	5,170	49,164	10,525	9.5	4.7
Washington, D.C.	1955	1,569	11,536	2,589	7.4	4.5
Buffalo, N.Y.	1962	1,350	11,067	2,991	8.2	3.7
Philadelphia, Pa.	1960	4,007	27,700	7,760	6.9	3.6

Source: Comprehensive transportation studies in each urban area.

Thus, the person-miles and vehicle-miles of travel and urban trip lengths become important planning parameters.

Total person travel. Table 5.20 summarizes the daily airline (direct line as opposed to actual over-the-road miles) person-miles of travel in four major urban areas, the person-miles per capita, and the average airline trip lengths. In the four cities shown, the airline person-miles per capita ranged from approximately 7 to 10, and the average airline trip length from 3.6 to 4.7 miles.

Airline trip lengths. Airline trip lengths by urban travel mode are shown in Table 5.21 for five major metropolitan areas. The following ranges are indicated:

Auto (driver)	3.4–4.0 miles
Auto (passenger)	3.4–4.5 miles
Taxi (passenger)	1.9–2.1 miles
Bus, Streetcar	2.5–4.0 miles
Rapid transit	5.8–7.2 miles
Commuter railroad	10.2–17.6 miles

The increased length of rail transit trips results from (1) the greater speeds attainable, (2) the limited service coverage and greater distance between stations, and (3) the downtown focus of these trips. Trips to the city center are considerably longer than most other urban trips. This is apparent from Table 5.22 which sets forth comparative data for Philadelphia on downtown and overall trip length. It shows that all person trips to the CBD are approximately 60 percent longer. Auto trips to downtown Philadelphia are approximately 80 percent longer.

Table 5.23 indicates mean trip length for home-based trips in Chicago according to trip purpose. Table 5.24 indicates means and variances of trip lengths of home-

TABLE 5.21
Trip Length by Travel Mode (Miles)

	Urban Area				
Travel Mode	New York Tri-state (1963)	Chicago (1956)	Philadelphia (1960)	Buffalo (1962)	Detroit (1953)
Auto Driver	4.0	4.3	3.4	3.6	3.4
Auto Passenger	3.6	4.5	3.4	4.2	3.4
Taxi	2.1	—	—	1.9	—
Ferry	4.5	—	—	—	—
Bus, Street Car	2.5	3.6	3.0	3.2	4.0
Rapid Transit	5.8	7.2	5.9	—	—
Commuter Railroad	17.6	14.6	10.2	—	—

Source: Comprehensive transportation studies in each urban area.

TABLE 5.22
Airline Trip Lengths, Philadelphia, 1960

	Airline Trip Length (mi)		
Mode of Travel	All	CBD	RATIO
Auto Driver	3.4	5.9	1.7
Auto Passenger	3.4	6.0	1.8
All Auto Trips	3.4	6.0	1.8
Railroad	10.2	10.2	1.0
Subway-Elevated	5.9	6.0	1.0
Bus and Trolley	3.0	3.8	1.3
ALL TRANSIT*	4.4	5.7	1.3
ALL MODES	3.6	5.8	1.6

*"All transit trips" reflect the relative proportions of railroad, rapid transit, and surface transit trips in citywide and CBD travel.

Source: *Penn Jersey Transportation Study*, Vol. I: *The State of the Region*, a study conducted under the sponsorship of the Commonwealth of Pennsylvania and the State of New Jersey *et al.* (Philadelphia: Penn Jersey Transportation Study, 1964), p. 82.

TABLE 5.23
Characteristics of "Home-based" Trips, Chicago

Home to	Number of Trips	Avg. Trip Length (mi) (Airline Distance)	Percent of Total
Work	788,851	5.58	16.3
Shop	214,374	2.06	4.4
School	16,777	4.91	0.4
Social-Recreation	365,058	4.13	7.6
Eat Meal	39,091	3.21	.8
Personal Business	359,826	3.38	7.5
Serve Passenger	200,450	2.21	4.2
Ride	111	—	0.0
Home	1,106	—	0.0
Total	1,985,644	mean = 4.14	41.2%

Characteristics of "Home-Destination" Trips, Chicago

Home from	Number of Trips	Avg. Trip Length (mi) (Airline Distance)	Percent of Total
Work	756,520	5.57	15.7
Shop	235,688	1.99	4.9
School	15,523	5.34	.3
Social-Recreation	406,676	4.28	8.4
Eat Meal	46,134	3.60	1.0
Personal Business	332,460	3.51	6.9
Serve Passenger	204,446	2.22	4.2
Ride	—	—	0.0
Home	1,106	—	0.0
Total	1,998,553	mean = 4.15	41.4%

Source: Adapted from Chicago Area Transportation Study, 1956.

TABLE 5.24
Mean and Standard Deviation of Airline Trip Lengths, Columbia, S.C., 1964

Purpose (Home-based)	Mode							
	Auto Driver		Auto Passenger		Taxi Passenger		Public Bus Passenger	
	$\bar{X}$	(S_X)	$\bar{X}$	(S_X)	$\bar{X}$	(S_X)	$\bar{X}$	S_X
Work	3.78	(2.36)	3.36	(2.28)	2.14	(1.43)	2.76	(1.56)
Shopping-Personal Business	2.15	(1.88)	2.51	(1.97)	1.53	(1.14)	2.14	(1.25)
Recreation	3.30	(2.40)	2.63	(2.18)	3.14	(0.51)	2.57	(1.14)
School	2.78	(2.01)	1.61	(1.58)	1.96	(1.51)	2.10	(1.08)
Social	2.72	(2.17)	2.68	(2.30)	2.79	(1.49)	2.29	(1.78)
All Purposes	2.52	(2.14)	2.45	(2.10)	1.97	(1.37)	2.55	(1.48)

Source: Origin-destination data.

based trips according to purpose in Columbia, South Carolina. These parameters provide a basis for sample designs in other urban areas.

The relationship between airline and actual (over-the-road) trip lengths in selected urban areas is shown in Figure 5.9. This depends on the configuration of the specific

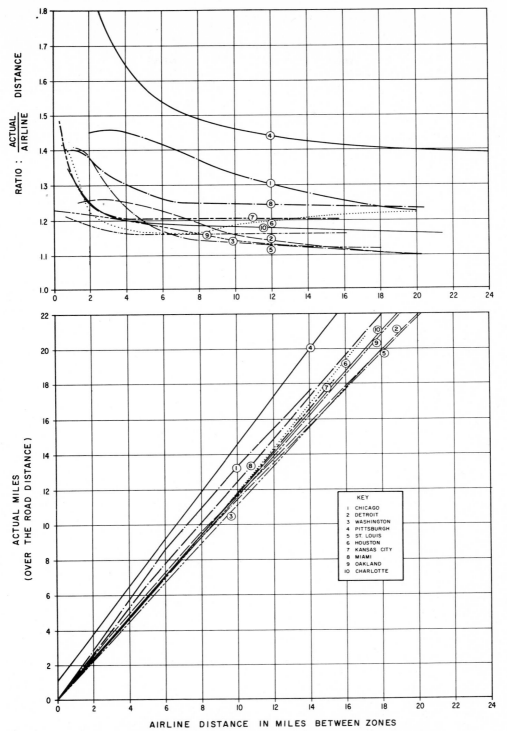

Figure 5.9. Trip length comparisons—airline mileage vs. over-the-road distance. (Source: H. S. Levinson and F. H. Wynn, "Some Aspects of Future Transportation in Urban Areas," Bulletin 326, Washington, D.C.: Highway Research Board, 1962, p. 8.)

street network. It ranges from approximately 1.2 to 1.4 miles; the higher figure reflects a complete grid, or irregular street system (Chicago, Pittsburgh), and the lower figure reflects radial-circumferential street systems (Detroit, Washington).

The effects of population on trip length are shown in Figure 5.10 and Table 5.25.

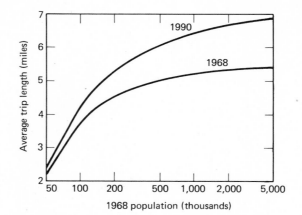

Figure 5.10. Average trip length in urban areas. (Source: Kassoff and Gendell, "An Approach to Multi-Regional Urban Transport Policy Planning," p. 83.)

TABLE 5.25
Urban Auto Driver Trip Length, Duration, and Speed

| | | Work Trip | | |
Urbanized Area	1960 Population (thousands)	Auto Driving Time (mins)	Length (mi)	Average Network Speed
1. Los Angeles	6,489	16.8	8.7	31.0
2. Philadelphia	3,635	20.1	7.2	21.5
3. Washington	1,808	14.3	5.9	24.7
4. Pittsburgh	1,804	12.6	4.2	20.0
5. Baltimore	1,419	16.7	7.0	24.6
6. Minneapolis-St. Paul	1,377	12.5	5.1	24.5
7. New Orleans	845	7.4	2.5	20.2
8. Fort Worth	503	15.7	8.1	30.9
9. Ottawa-Hull	406	12.6	5.3	25.2
10. Nashville	347	10.8	5.4	30.0
11. Edmonton	336	11.6	5.8	30.0
12. Davenport	227	7.7	3.2	24.9
13. Charlotte	210	11.0	5.5	30.0
14. Chattanooga	205	10.8	5.4	30.0
15. Erie	177	9.4	3.4	21.7
16. Waterbury	142	10.1	5.9	35.0
17. Pensacola	128	8.7	4.4	30.3
18. Greensboro	123	8.9	4.3	29.0
19. Lexington	112	9.1	5.7	37.6
20. Sioux Falls	67	7.0	2.9	24.8
21. Tallahassee	48	7.3	3.7	30.4
22. Hutchinson	38	6.1	2.0	19.2
23. Beloit	33	6.7	2.9	25.9

Source: Comprehensive urban transportation studies as reported in Alan M. Voorhees & Associates, *Factors and Trends in Trip Lengths*, National Cooperative Highway Research Program Report 48, (Washington, D.C.: Highway Research Board, 1968), p. 8.

Both trip lengths and trip times increase as urban populations rise. A regression analysis conducted by Voorhees, Bellomo, Schofer, and Cleveland[10] showed the following relationship between population and average work trip duration. The equations are based on data for 23 cities, ranging from Beloit, Wisconsin, to Los Angeles, California.

$$\log_e \bar{t} = -0.025 + 0.19 \log_e P \tag{5.2}$$

where
$\bar{t}$ = average trip duration in minutes;
P = urban area population.
This equation can be written as

$$\bar{t} = 0.98 P^{.019} \tag{5.3}$$

The standard error of the regression coefficient was 0.026, and the coefficient of determination (R^2) was 0.71.

The physical structure of an urban area seems to have the same general impact on trip lengths and durations. Fort Worth, for example, had an average work trip length of 8.1 miles and an average trip duration of 15.7 min. New Orleans, larger in population, but more compact, had an average work trip length of 2.5 miles and an average trip duration of 7.4 min.

URBAN TRUCK TRAVEL

Urban goods movement reflects the diverse needs of the many establishments in the metropolitan area. Although relatively little information on the characteristics of urban goods movement has been collected, considerable information has been obtained on urban truck travel.

Truck registrations. Truck ownership statistics for urban areas are shown in Figure 5.11. This chart reflects studies conducted between 1956 and 1965 for metropolitan areas ranging from approximately 50,000 to over 5,000,000 residents.

Urban motor truck registrations generally relate to the size of the urban area and its geographic setting. The higher per capita ownership of motor trucks in smaller cities largely reflects their use for personal transportation purposes. As cities grow, the need for a personal truck diminishes, truck services are more readily available from "for-hire" sources, and the affluence level of residents is somewhat higher.

Truck ownership rate is also influenced by geographic differences for cities under one million residents. For urban areas of approximately 100,000 population, in the East, South, and Midwest, the rate is approximately 60 per thousand population; for Western cities, the rate is approximately 100 per 1,000 residents. Above one million in population, the ownership rate is consistent at approximately 25 to 30 registered trucks per 1,000 persons.[11]

Summary characteristics. The following characteristics typify truck travel in American cities:

1. Motor truck travel represents approximately 15 percent of the total daily urban vehicle trips.

---

[10] A. M. Voorhees, S. J. Bellomo, J. L. Schofer, and D. E. Cleveland, "Factors in Work Trip Lengths," Highway Research Record No. 141 (Washington, D.C.: Highway Research Board, 1966) pp. 24–46.
[11] See *Motor Trucks in the Metropolis*, Wilbur Smith and Associates, 1969.

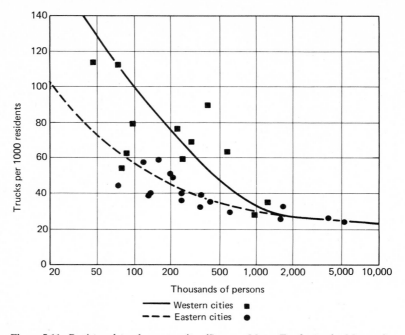

Figure 5.11. Registered trucks per capita. (Source: *Motor Trucks in the Metropolis*, prepared for Automobile Manufacturers Association by Wilbur Smith and Associates, 1969, p. 17.)

2. There are approximately 1.6 to 1.8 truck trips per acre of developed urban land.
3. Truck trips per capita decrease with increasing city size: for urban areas of 250,000 there are approximately 30 truck trips per 100 population; in urban areas of over 2,000,000, approximately 20.

Light trucks dominate the commercial traffic stream; they account for about two-thirds of the truck trips in Pittsburgh; medium trucks represent approximately 30 percent, and heavy trucks, the remainder. Data for eleven cities set forth in *Motor Trucks in the Metropolis* show that light trucks averaged 67 percent, medium trucks 28 percent, and heavy trucks 5 percent of the total. Light and medium trucks in Pittsburgh, for example, average six trips per vehicle per day, and heavy trucks average nearly ten.[12]

Motor truck movements are concentrated in the downtown area and its environs. Pittsburgh, for example, has nearly 20,000 truck trip ends in the Golden Triangle each day—60 trip ends per acre; they represent approximately 8 percent of the region's total truck trips.

Truck travel to, from, and within the central business district increases at a slower rate than overall urban population. There are approximately four CBD truck trips per 100 persons in urban areas of 250,000, as compared with fewer than two in areas

---

[12] *Pittsburgh Area Transportation Study*, 1958. Also see L. E. Keefer, "Trucks at Rest," *Origin and Destination Techniques and Evaluations*, Highway Research Record No. 41, (Washington, D. C.: Highway Research Board, 1963), pp. 29–38.

TABLE 5.26
Average Daily Truck Usage in 11 Urban Areas

Truck Class	Trucks Making Trips		Daily Trips		Daily Truck-Miles		Daily Mileage		Daily Trips Per Truck*
	Number	Percent	Number	Percent	Number	Percent	Per Truck	Per Trip	
Light	72,989	71.8	608,606	67.7	2,075,660	65.3	28.4	3.4	8.3
Medium-Heavy	28,691	28.2	289,810	32.3	1,104,742	34.7	36.5	3.8	10.1
TOTAL	101,680	100.0	898,416	100.0	3,180,402	100.0	31.3	3.5	8.8

*These values are for trucks making trips on a typical weekday. When related to all trucks registered in the urban area, the average is 5.9 trips per day, since a proportion of the registered trucks are idle on any given day.

Note: The values are summations of trip values for the 11 areas shown in source.

Source: Comprehensive transportation studies by Wilbur Smith and Associates in Albuquerque, New Mexico; Baltimore, Maryland; Baton Rouge, Louisiana; Columbia, South Carolina; Lewiston, Maine; Little Rock, Arkansas; Manchester, New Hampshire; Monroe, Louisiana; Richmond, Virginia; Sioux Falls, South Dakota; and Winston-Salem, North Carolina.

TABLE 5.27
Daily Truck Travel in 11 Urban Areas by Category of User and Truck Class

User Category	Percent of Trucks Making Trips			Percent of Daily Trips			Percent of Daily Vehicle-Miles		
	Light	Heavy	All	Light	Heavy	All	Light	Heavy	All
Industry:									
Agriculture	1.8	2.1	1.9	1.5	1.4	1.5	1.7	1.8	1.8
Construction	20.2	12.5	17.6	11.1	6.9	9.7	20.2	11.7	17.3
Manufacturing-Processing	4.4	11.8	6.5	9.4	13.5	10.8	5.8	11.2	7.7
Transportation-Public Utilities	6.2	23.3	11.0	12.0	16.5	13.4	7.4	25.2	13.5
Wholesale-Retail Trade	20.9	36.4	25.3	33.2	45.7	37.4	26.9	37.0	30.3
Service and Recreation	10.2	4.4	8.6	15.9	4.6	12.2	12.4	3.2	9.2
Government (Public Service)	2.9	6.5	3.9	3.8	9.1	5.5	3.7	7.4	5.0
Personal Use	33.4	3.0	25.2	13.1	2.4	9.6	21.9	2.5	15.2
ALL USERS	100.0	100.0	100.0	100.0	100.0	100.0	100.0	100.0	100.0

Source: See Source of Table 5.26.

of 2,000,000 or more. Trucks represent approximately 20 percent of all vehicles entering the CBD, 10–15 percent of the total of vehicles parking at any one time, and 3 percent of the maximum parking accumulation.

1. In large cities, some 60 percent of all deliveries are made at the front entrance along the curb, approximately 30 percent in the rear entrance or side alley, and approximately 10 percent from off-street service entrances. In Dallas, however, 80 percent of all truck deliveries were made from the curb. This higher use of curb loading results in part because there are no alleys.
2. Average durations of downtown truck parkers were reported as 22 min in Dallas, 33 min in Chattanooga, and 55 min in New Orleans. Trucks using loading zones generally stay less than one hour.
3. One-third of all trucks parking at curbs load or unload merchandise.
4. The percentage of trucks parking illegally varies widely among CBD's ranging up to 75 percent in large downtown areas.

Detailed characteristics. Detailed characteristics of urban truck travel in eleven urban areas are summarized in Tables 5.26 through 5.28. The eleven urban areas represented in these tabulations are Albuquerque, New Mexico, Baltimore, Maryland, Baton Rouge, Louisiana, Columbia, South Carolina, Lewiston, Maine, Little Rock, Arkansas, Manchester, New Hampshire, Monroe, Louisiana, Richmond, Virginia, Sioux Falls, South Dakota, and Winston-Salem, North Carolina.

1. Table 5.26 summarizes average daily truck usage. Light trucks represent two-thirds of the daily travel, heavy trucks one-third. On the average, each truck made nearly four trips per day.
2. Table 5.27 summarizes daily truck travel according to industry classifications. Approximately 30 percent of the total daily travel represents trips by trucks engaged in the wholesale and retail trades. Another 17 percent represents trucks used in construction, and 15 percent represents trucks in personal use.
3. Table 5.28 indicates the trip purposes of trucks. The greatest number of trucks, as would be expected, involve pickup and delivery operations.

TABLE 5.28
**Percentage Distribution of Purposes for Urban
Truck Trips in 11 Urban Areas**

Trip Purpose at Trip Destination		Percentage of Total Daily Trips
Home Base		19.3
Personal Use		9.1
All Pickup and Delivery:		41.1
Retail	17.3	
Wholesale	16.3	
Merchandise	7.5	
Mail and Express		6.1
Construction		4.9
Maintenance and Repair		8.0
Business Use		7.2
Other		4.3
All Purposes		100.0

Source: See data for Table 5.26.

CENTRAL BUSINESS DISTRICT TRAVEL CHARACTERISTICS

The central business district is the urban region's cultural, commercial, and financial center; the focus of its transportation system, and the area of greatest travel intensity. Downtown transportation planning should be based on existing and future land-uses, the travel demands and linkage requirements that these activities are likely to generate, adaptations of the downtown environment to optimize these activities and patterns, and the policies and technologies necessary to achieve these objectives.

A clear knowledge of downtown travel patterns is essential in assessing the capacity requirements of the urban transport system, in developing CBD circulation improvement plans, and in formulating regional transportation policy.

SUMMARY PATTERNS

Travel characteristics in a wide range of cities throughout the United States and Canada are summarized in Tables 5.29 and 5.30. The following characteristics typify

TABLE 5.29
Selected Characteristics of Major United States City Centers

Urbanized Area	Rank 1970 Census	Urbanized Area Population, 1970	Central City Population, 1970	Central City Density (Persons per sq mi)	Central Business District (sq mi)	Year of Survey	Central Business District Employment (000)
New York, N.Y.-Northeastern New Jersey	1	16,206,841	7,894,862	26,343	9.0	1963	1,777
Los Angeles-Long Beach	2	8,351,366	2,809,596	6,073	0.6	1960	130
Chicago, Ill.,-Northwestern Indiana	3	6,714,578	3,369,359	15,126	1.1	1956	300
Philadelphia, Pa.-N.J.	4	4,021,066	1,950,098	15,164	2.2	1960	225
Detroit	5	3,970,584	1,513,601	10,953	1.1	1953	114
SanFrancisco-Oakland	6	2,987,850	715,674	15,764	2.2	1965	282
Boston	7	2,652,575	641,071	13,936	1.4	1963	246
Washington, D.C.-Md.-Va.	8	2,481,489	756,510	12,321	1.7	1955	212
Cleveland	9	1,959,880	750,879	9,893	1.0	1963	117
St. Louis, Mo.-Ill.	10	1,887,944	622,236	10,167	0.8	1957	119
Pittsburgh	11	1,846,042	520,117	9,422	0.5	1958	84
Minneapolis-St. Paul*	12	1,704,423	434,400	8,135	0.9	1958	NA
Houston	13	1,677,863	1,232,802	3,102	0.9	1960	120
Baltimore	14	1,579,781	905,759	11,568	0.8	1962	78
Dallas	15	1,338,684	844,401	3,179	1.5	1964	164
Milwaukee	16	1,252,457	717,372	7,548	0.9	1963	NA
Seattle-Everett	17	1,238,107	530,831	6,350	0.6	1961	60
Miami	18	1,219,661	334,859	9,763	NA	1964	28
Atlanta	20	1,172,778	497,421	3,779	0.6	1961	75
Cincinnati	21	1,110,814	452,454	5,794	0.5	1965	NA
Kansas City, Mo.-Kans.	22	1,110,787	507,330	2,101	0.9	1957	65
Buffalo	23	1,086,594	462,768	11,205	0.9	1962	48
Denver	24	1,047,311	514,678	5,406	0.5	1959	50
Phoenix	27	863,357	581,562	2,346	0.7	1957	21
Nashville-Davidson	53	448,440	447,877	1,305	0.6	1959	34

TABLE 5.29 (*Continued*)

	CBD Floorspace sq ft (millions)	Person-Trips per Day to CBD (thousands)	Percent by		CBD Destinations per sq mi		Peak-hour One-Way Person Trips Across Cordon (000)
			Auto	Transit	All Modes	Auto Only	
New York, N.Y.-Northeastern New Jersey	800	1,300	5	95	1,300	66	827
Los Angeles-Long Beach	47	158	45	55	250	114	99
Chicago, Ill.,-Northwestern Indiana	92	466	29	71	439	128	210
Philadelphia, Pa.-N.J.	124	389	41	59	177	73	197
Detroit	50	253	56	44	234	132	68
SanFrancisco-Oakland	88	423	63	37	192	71	130
Boston	96	400	50	50	290	145	143
Washington, D.C.-Md.-Va.	NA	266	55	45	156	86	138
Cleveland	47	123	59	41	123	73	50
St. Louis, Mo.-Ill.	39	125	53	47	178	95	62
Pittsburgh	32	154	49	51	308	151	56
Minneapolis-St. Paul*	NA	188	73	27	209	153	NA
Houston	NA	113	80	20	125	100	57
Baltimore	33	130	54	46	162	88	68
Dallas	31	164	86	14	109	94	62
Milwaukee	31	104	66	34	115	76	NA
Seattle-Everett	27	126	71	29	210	167	NA
Miami	12	35	73	27	NA	NA	NA
Atlanta	30	94	72	28	156	42	31
Cincinnati	35	113	71	29	220	156	35
Kansas City, Mo.-Kans.	NA	107	70	30	118	82	NA
Buflalo	28	104	67	33	115	77	NA
Denver	24	105	80	20	210	167	31
Phoenix	NA	65	89	11	93	83	NA
Nashville-Davidson	NA	64	79	21	110	87	NA

*Central city and CBD characteristics relate to Minneapolis only.

Source: Compiled from Census data and comprehensive transportation studies in each urban area.

these American and Canadian cities:

1. Most central business districts are located near the population center of the urban region, except where inhibited by topography. They generally occupy from 1.0 to 1.5 square miles of area and contain less than 40 million square feet of floor space. Approximately half of all downtown land is devoted to streets and parking.
2. Peak-hour cordon volumes are generally less than 100,000 persons. This implies corridor movements by all modes which are usually less than 25,000 persons.
3. Per capita trip attractions to the center city generally decrease with increasing time and distance from the center city. Per capita attraction of workers is relatively constant in comparison.
4. In most cities, trips to the center city have grown at a slower rate than the overall growth in total metropolitan trips. This usually results in a decline of trips between established neighborhoods and the city center and an increase in

TABLE 5.30
Selected Characteristics of Major Canadian City Centers

Item	Vancouver	Toronto	Ottawa
1969 Population	1,000,000	1,916,100	445,800
CBD Employment	91,000*	118,000†	60,000
CBD Cordon Count of Two-Way Daily			
Person-Trips by All Modes	592,000	763,000	382,000
Percent by Car	64	48	78
Percent by Transit	36	52	22
CBD Cordon Count—Peak Hour, Peak			
Direction-Person Trips by All Modes	45,000	76,000	35,000
Percent by Car	60	32	60
Percent by Transit	40	68	40
CBD Peak Person Accumulation by All			
Modes	53,000‡	105,000	52,000‖
	62,000§		
Percent by Car	51	24	56‖
Percent by Transit	49	76	44‖

*Downtown Peninsula.
†0.85 sq mi CBD core.
‡Excluding pedestrians.
§Including pedestrians.
‖Estimated.

Source: Cordon count and origin-destination studies in each urban area.

trips from suburban areas. As a consequence, there is generally an increase in average trip lengths and a decrease in public transport use.

5. The importance of public transportation depends on: (1) historic antecedents, (2) barriers to and inhibitors of automobile travel, and (3) intensity of downtown land use and employment.
6. In cities with a high employment density, a major decline in public transport riding could seriously increase congestion and capital expenditure for transport.
7. In most American cities, half to two-thirds of all peak-hour traffic is passing through the center, with origins and destinations either in adjacent areas or elsewhere in the urban region.

Travel magnitudes and trip purposes. Generalized relationships between urban population and CBD trip destinations are shown in Figure 5.12. Figure 5.13 and Table 5.31 further illustrate the trip purpose characteristics of CBD travelers.

Travel to the central business district increases in magnitude as urban population rises, but at a decreasing rate. The proportion of trips for work purposes increases with size; more than half of all trips to the CBD's in urban areas of over a million persons are for work purposes.

1. The daily person-destinations in the central business district increase from approximately 40,000 in urban areas of up to 200,000 population to approximately 225,000 in urban areas with 2,000,000 inhabitants.
2. The proportion of work trips increases from approximately 33 percent in urban areas of 200,000 population to over 50 percent in the larger urban areas. This indicates that the center city increasingly becomes a work center as urban areas get larger.

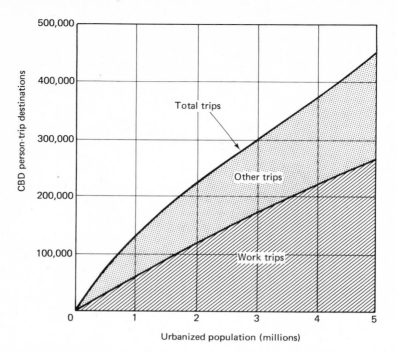

Figure 5.12. Generalized central business district trip generation. (Source: Urban Transportation Concepts. Wilbur Smith and Associates, Center City Transportation Project, Sept. 1970, p. 21.)

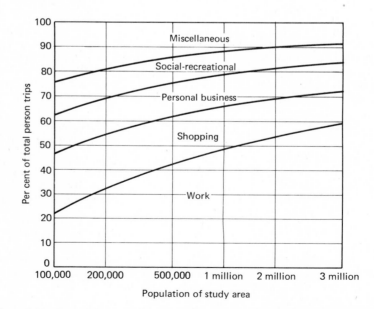

Figure 5.13. Central business district trip purposes. (Source: Transportation and Parking for Tomorrow's Cities, Prepared under Commission from the Automobile Manufacturers Association by Wilbur Smith & Associates, 1966, p. 55.)

TABLE 5.31
Purposes of Central Business District Trips

City	Year	Study Area Population (000)	Percentage Distribution					
			Work	Business	Shopping	Social-Recreational	Other	Total
Philadelphia	1960	4,007	59	13	11	10	7	100
Dallas	1964	1,821	55	15	5	5	20	100
Pittsburgh	1958	1,472	58	13	20	5	4	100
Minneapolis	1958	1,377	53	19	20	7	—	100
Denver	1962	806	47	12	12	12	17	100
Atlanta	1961	700	51	13	14	6	16	100
Tucson	1960	230	31	16	22	11	20	100
Chattanooga	1960	242	43	13	15	10	19	100

Source: Origin-destination studies in each urban area.

3. In urban areas in the 2 million population category, approximately 55 percent of all downtown person-destinations were for work purposes, 15 percent for shopping, 15 percent for personal business, and 15 percent for social, recreational, or miscellaneous reasons.

Walking distances. Pedestrian and parker walking distances vary according to city size and reflect the locations of transit stops and parking terminals in relation to major stores and offices. Walking distance patterns are generally consistent among cities of similar size (Figure 5.14). Within medium-sized cities like Dallas and Pittsburgh, the overall median walking distance approximated 500 ft, and 80 percent of the parkers walked less than 1,200 ft. In Boston, where pedestrian walking distances are influenced by locations of commuter railroad stations, 50 percent of all pedestrian trips were less than 1,000 ft and 80 percent were less than 2,000 ft. Even longer walking distances are found in New York because of the block spacing and location of subway stations. Most downtown walking trips are short, reflecting people's desires to minimize travel time and inconvenience.

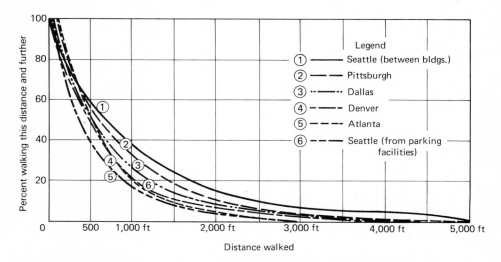

Figure 5.14. Walking distances in the center city. (Source: Available parking surveys in Atlanta and Pittsburgh. Special surveys in Dallas, Denver, and Seattle.)

Trip durations. The length of time spent in the center city varies with the basic trip purpose. Workers usually park from 6 to 8 hours, but shoppers and personal business trips usually average 2 hours. For example, overall parking durations averaged 3 hours, 9 min in Atlanta and 4 hours, 40 min in Denver. Workers stayed 7 hours, 33 min in Denver, as compared with 5 hours, 43 min in Atlanta. Shopping and business trips in both cities approximated 2 hours.

INTERNAL MOVEMENT PATTERNS

Most intracenter city person trips are pedestrian trips. Pedestrian trips are short in length and are highly concentrated in core areas. They reflect two major linkage patterns: (1) between principal transit stops or parking terminals and places of work and (2) between stores and offices within the retail core. Both linkages must be considered in center city circulation planning.

In center city Seattle, for example, 56 percent of all pedestrian trips were associated with line-haul transportation facilities; 44 percent were interbuilding trips (Table 5.32).

TABLE 5.32
Weekday Pedestrian Trips in Seattle Central Business District, 1970
(7: 00 a.m.–7: 00 p.m.)

Type of Trip	Pedestrian Trips	
	Number	Percent
To and From:		
Ferry	7,100	2.0
Parking	134,800*	39.0
Bus Stop	52,000†	15.1
Monorail Station	1,000‡	0.3
Subtotal	194,900	56.4
Buildings§	150,100	43.6
TOTAL	345,000	100.0

*Includes 17,200 trips to-from parking for use of the car within the CBD during the day; includes 1,000 trips to-from parking places for motorcycles, scooters, bicycles, etc.
†Low estimate reflecting only transit trips in and out of the CBD; transit trips within the CBD amount to an estimated additional 15 percent.
‡Based on a winter day operation; the annual daily average including travel to Seattle Center would approximate 3,000.
§Other than transportation facilities.

Source: Center City Transportation Project, Parking and Pedestrian Surveys, 1970; Ferry Walk-On Passenger Survey; Wilbur Smith and Associates, 1966.

Trip purpose. Pedestrian trip purposes in the CBD vary by location and time of day, reflecting the mix of land uses and the locations of major activity centers.

Retail stores generate more pedestrian trips per unit area than do offices (Table 5.33). Shoppers make multiple trips between stores; travel between office buildings is much less. Consequently, shoppers account for a higher proportion of total daily

TABLE 5.33
Pedestrian Trip Generation, 1970
(Seattle Central Business District)

Land Use	Pedestrian Trips per 1,000 sq ft of Floor Space			
	10 a.m.–4 p.m.		4 p.m.–6 p.m.	
	Total	Hourly Average	Total	Hourly Average
Office	5.1	0.85	1.8	0.90
Retail	15.6	2.60	4.8	2.40
Other	3.7	0.62	1.9	0.95

Source: Center City Transportation Project Pedestrian Survey, 1970.

pedestrian trips than they represent as a component of person-trips to CBD destinations. In Seattle, approximately 31 percent of all pedestrian trips were for shopping purposes as compared with an estimated 15 percent of all person destinations in the CBD (Table 5.34).

TABLE 5.34
Reported Purposes of Pedestrian Trips, 1970
(Seattle Central Business District)

Purpose	Percent
Work	24.1
Commercial Business	12.3
Personal Business	17.6
Sales and Service	2.2
Social-Recreational	2.4
Eat-Drink	5.8
Shopping	30.8
Other	4.8
TOTAL	100.0

Source: Center City Transportation Project, Parking and Pedestrian Surveys, 1970.

Pedestrian trip patterns observed in downtown Boston further illustrate these characteristics. During the noon hour, nearly half of all pedestrians reported "lunch" as their immediate trip purpose. During the evening peak hour, almost half reported "home" as their destination (Figure 5.15). The largest individual linkages were from work to lunch and from work to home.

Workers represented about two-thirds of all pedestrians at the noon hour and almost three-quarters of the total in the afternoon peak hour. During the lunch hours, over 60 percent of all pedestrian trips by workers represented lunch trips and 16 percent shopping trips. During the evening peak hour, 50 percent of all workers were on their way home and 16 percent were shopping.

Major internal linkages. Pedestrian destinations in the CBD are mainly concentrated in the retail and commercial cores. Comparatively little travel takes place between these core areas of intense demand and secondary activity concentrations on the edge of the CBD. Much intra CBD pedestrian travel is between parking space and building rather than between buildings. Thus, the location of major garages significantly influences the distribution of CBD walking trips. As a result of these factors, pedestrian flows are far more localized than either automobile or transit patterns.[13] For example,

1. In Seattle, daily crosswalk pedestrian volumes often exceed 20,000 persons in the core area but drop to 3,000 persons within two blocks of the core area.
2. In Dallas, daily crosswalk pedestrian volumes exceed 12,000 persons in the core area but drop to 3,000 persons within three blocks.

---

[13] HERBERT S. LEVINSON, "Pedestrian Way Concepts and Case Studies," *Pedestrians*, Highway Research Record, No. 355, (Washington, D.C.: Highway Research Board, 1971.)

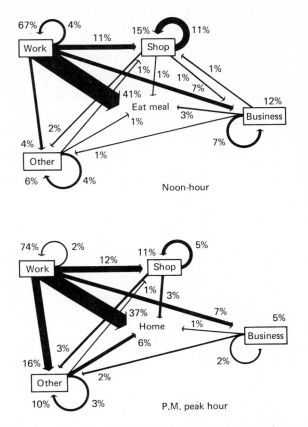

Figure 5.15. Pedestrian trip purposes, Boston Center City, 1963. (Source: Barton-Aschman Associates, *Traffic Circulation and Parking Plan, Central Business District Urban Renewal Area, Boston, Mass,* p. 23.)

3. In Philadelphia, daily crosswalk pedestrian volumes exceed 25,000 persons in the core area but drop to 1,000 persons within four blocks.

Reported CBD peak-hour pedestrian sidewalk volumes approximate 6,000 persons per hour in Boston, 5,000 in Philadelphia and Pittsburgh, and 3,000 in Atlanta, Dallas, and Seattle.

TRIP GENERATION FACTORS

OVERVIEW

The generation of urban travel reflects the activities of people in the urban setting. The number, type, and frequency of urban trips relate to the character and intensity of urban land use. Urban travel also depends on the socioeconomic characteristics and travel attitudes of urban residents.

The basic approaches to trip generation are widely documented in the many comprehensive urban area transportation studies. The role of trip generation in the comprehensive transportation planning process and approaches to its analyses, including statistical estimation techniques, are detailed in *Guidelines for Trip Generation Analysis* prepared by the Federal Highway Administration.[14]

Trip "production and attraction." Trip generation analysis generally includes two components:

1. *Trip production* relates to the home end of the trip and reflects the trips generated at home.
2. *Trip attraction* relates to the non-home end of the trips—to the commercial, industrial, school, or social-recreational activities that attract urban travelers.

For non-home based trips, the distinction between productions and attractions is not clear.

The following factors are related to trip production and attraction:

PRODUCTION OF TRIPS

1. Population
   a. total
   b. by age, sex, income, and family size
2. Number of dwelling units
3. Automobile ownership
4. Employed labor force
   a. white collar
   b. blue collar
5. Students

TRIP ATTRACTIONS

1. Employment
   a. total
   b. white collar
   c. blue collar
2. Floor space (sq ft)
3. Land use (acres)
4. School enrollment
5. Recreational attractiveness

Regression and category analysis. Trip generation analysis includes several basic approaches that may be used together to assess the relationship between the preceding parameters and urban trip making. (1) *Regression analysis* relates trip ends to land-use and socioeconomic characteristics; (2) *rate analysis* relates trip ends to land areas and/or socioeconomic characteristics of analysis units. (The latter is widely used by the traffic planner in estimating impacts of new developments.) In applying either approach, it is essential to evaluate results for reasonableness.

1. Multiple Regression—Multiple regression equations have been widely used to estimate trips. These equations take the form $Y = A_1x_1 + A_2x_2 + A_3x_3 \ldots A_nx_n + B$

---

[14] *Guidelines for Trip Generation Analysis*, Federal Highway Administration, Washington, D.C., 1967.

where, ideally, $B \rightarrow 0$; and x_1, x_2, x_3, x_n represent input variables that are ideally independent.

Collinearity among "input" variables is common when equations are developed on a zonal basis and the variables include population, school children, car ownership, and/or labor force. The more populous zone, for example, usually also has the greatest number of school children, workers, and cars. An analysis of intercorrelations among input variables is essential to identify and reduce the extent of collinearity.

2. Trip Rates—Category, trip rate, or cross classification analyses can be developed from specific household and/or land-use characteristics. In this procedure, households are grouped into specific categories and the rate of trip making in each category is derived. Once basic classifications are accomplished, regression equations also can be utilized. A typical cross-classification model might be as follows:

Trips per Dwelling Unit

Income range	Cars/Dwelling Unit			
	0	1	2	3+
Low	*	*	—	—
Low-Medium	*	*	*	—
Medium	*	*	*	*
Medium-High	—	*	*	*
High	—	*	*	*

*Denotes adequately sized sample.

Significant parameters. Typical variables found to be significant in developing trip production and attraction equations are summarized in Table 5.35 for eleven metropolitan areas. A further refinement in relation to *specific* types of urban trips suggests the following breakdown:

Trip Type	Production	Attraction
Work	Labor force	Employment
Shopping	Cars/DU	Sq ft of retail space, retail sales, and/or retail employment
School	Students/DU net residential density	School enrollment
Personal business	Cars/DU	Employment, retail sales
Social-recreational	Cars/DU	Population, and/or recreational attractiveness (i.e., sq ft/space)

TRIP GENERATION PARAMETERS

The forms of typical trip generation equations are summarized in Table 5.36. These equations reflect transportation studies done in the 1950's and 1960's and may involve some collinearity among input variables.

A review of these and similar trip generation analyses indicates the importance of population density, car ownership, household size, family income, and age of trip-

TABLE 5.35
Illustrative Variables Found to be Significant in Developing Trip
Production and Attraction Equations

Transportation Study	Aggregated Variables				Trip Production Variables Related to Dwelling Units				Trip Attraction Variables Related to Land Use				
									Sensitive to Intensity of Use				
	Gross Pop. by Zone	Gross Autos by Zone	Gross Employ. by Zone	Gross DU by Zone	Persons per DU	Autos per DU	DU per Acre	Income Index	Employ. by Type	Floor Area by Type	Retail Sales	Net Acres and Distance to CBD by Type	Insensitive to Intensity Net Acres by Type
1. Baltimore		X	X			X	X	X	X				
2. Chicago				X	X	X				X		X	
3. Erie	X	X	X						X				
4. Fort Wayne	X	X	X						X				
5. Los Angeles				X		X	X+		X				
6. Washington, D.C.			X	X			X		X	X			
7. Pittsburgh				X		X	X					X	
8. Philadelphia (Penn.-Jersey)				X	X	X	X	X	X				
9. Seattle (Puget Sound)				X	X	X	X+	X	X				
10. San Diego				X		X	X						X
11. Tucson				X	X	X	X					X	
12. Twin Cities				X	X	X	X		X°				

X Indicates variable was used.
+ Indicates a stratification of dwelling types; closely correlated to density.
° Incorporated the distance to the CBD.

Employ = Employment.
Pop. = Population.
DU = Dwelling unit (or family).
Acre = Net land-use acres

Source: Comprehensive origin-destination studies in each urban area as reported in *Traffic Planning for the North Central Freeway*, Wilbur Smith and Associates, Alan Voorhees and Associates, Washington, D.C., 1966, p. 22.

maker in estimating the number and mode of trips produced. Similarly, indices of land-use type and intensity govern trip attraction rates.

Dwelling unit type and population density. Studies have consistently shown the reductive effort of density on total trip generation, both between cities and within each city. Density in turn usually correlates with car ownership, city age, and travel mode. Pertinent findings from these studies are shown in Tables 5.37 and 5.38, and Figures 5.16 and 5.17.

TABLE 5.36
Typical Trip Attraction and Production Equations

City		Equation
Chicago	Person Trips/DU	$= A - B \text{ Log (DU/Net Acre)}$
	Auto Trips/DU	$= A - B \text{ Log (DU/Net Acre)}$
	Person Destinations/DU	$= A - B \text{ Log (DU/Net Acre)}$
	Auto Destinations/DU	$= A - B \text{ Log (DU/Net Acre)}$
	Person Trips/DU	$= A + B(\text{Autos/DU})$
	Auto Trips/DU	$= A + B(\text{Autos/DU})$
	Person Destinations/DU	$= A + B(\text{Autos/DU})$
	Auto Destinations/DU	$= A + B(\text{Autos/DU})$
Erie Production	Person Work Trips	$= A(\text{Labor Force})$
	Auto Driver Work Trips	$= A(\# \text{ of Autos}) - B(\text{Pop.})$
	Auto Driver Social-Recreational	$= A(\# \text{ of Autos}) + B(\text{Pop.})$
	Auto Driver Other Trips	$= A(\# \text{ of Autos}) - B(\text{Pop.})$
	Auto Driver N-H-B Trips	$= A(\# \text{ of Autos}) - B(\text{Pop.})$
Erie Attraction	Person Work Trips	$= A(\text{Total Employment})$
	Auto Shopping Trips	$= A(\text{Retail Employment})$
	Auto Social-Recreational Trips	$= A(\text{Pop.}) + B(\text{Retail} + \text{Other Employment})$
	Auto Other Trips	$= A(\text{Pop.}) + B(\text{Retail} + \text{Other Employment})$
	Auto N-H-B Trips	$= A(\text{Pop.}) + B(\text{Retail} + \text{Other Employment})$
	Commerical (Trucks)	$= A(\text{Pop.}) + B(\text{Retail} + \text{Other Employment})$ $+ C(\text{Manufacturing Employment})$
Washington Production	Total Person Trips/DU	$= A - B \text{ Log (DU/Net Acre)}$
	Home-Work Trips/DU	$= A - B \text{ Log (Total Person Trips/DU)}$
	Home-Social-Recreational Trips/DU	$= A - B \text{ Log (DU/Net Acre)}$
	Home-Shopping Trips/DU	$= A - B \text{ Log (DU/Net Acre)}$
	Home-School Trips/DU	$= A - B \text{ Log (DU/Net Acre)}$
	Home-Miscellaneous Trips/DU	$= A - B \text{ Log (DU/Net Acre)}$
	Non-Home-Based Trips	$= A - B \text{ Log (DU/Net Acre)}$

TABLE 5.36 (Continued)

City	Equation
Washington Attraction	
Home-Work	$= A + B$ (Employment)
Home-Shopping	$= A + B$(Retail Sales)
Home-Social-Recreational	$= A + B$(Retail Sales) $+ C$(DU)
Home-School	$= A + B$(School Acreage) $+ C$(DU)
Home-Misc.	$= A + B$(Employment) $+ C$(Retail Sales) $+ D$(DU)
Non-Home-Based	$= A + B$(Employment) $+ C$(Retail Sales) $+ D$(DU)
(This figure is artificially high)	
Pittsburgh	
Person Trips/DU	$= A - B$ Log (DU/Net Acre)
Auto Trips/DU	$= A - B$ Log (DU/Net Acre)
Person Destinations/DU	$= A - B$ Log (DU/Net Acre)
Auto Destinations/DU	$= A - B$ Log (DU/Net Acre)
Person Trips/DU	$= A - B$(Autos/DU)
Auto Trips/DU	$= A - B$(Autos/DU)
Person Destinations/DU	$= A - B$(Autos/DU)
Auto Destinations/DU	$= A - B$(Autos/DU)
Tucson	
Total Trips/DU	$= A + B$(Employed/DU)
	$= A + B$(Autos/DU)
	$= A + B$(Dist. CBD) $- C$(Dist. CBD)^2
Auto Trips/DU	$= A + B$(DU/Net Acre)
	$= A + B$(Employed/DU)
	$= A + B$(Autos/DU)
	$= A + B$(Dist. CBD) $- C$(Dist. CBD)^2
	$= A - B$(DU/Net Acre)
Hamilton, Ont.	
Total Trips/DU	$= A + B$(Autos/DU)
Work Trips/DU	$= A + B$(Persons/DU)
Other Home Trips/DU	$= A + B$(Autos/DU)
Non-Home Trips/DU	$= A + B$(Autos/DU)
Auto Driver Trips/DU	$= A + B$(Autos/DU)

Source: Comprehensive origin-destination studies in each area as reported in *Traffic Planning for the North Central Freeway*, p. 23.

TABLE 5.37
Person-Trip Generation Rates Related to Population Density, Puget Sound

Relative Density	Total Home-based	Home-based Work	Home-based Shop	Home-based Soc.-Rec.	Home-based Misc.	Home-based School	Non-Home-based	Total Person Trips
Low Relative Density	6.43	1.59	1.12	1.45	1.28	0.99	1.65	8.08
Medium Relative Density	6.03	1.69	1.11	1.41	1.23	0.59	1.55	7.58
High Relative Density	4.18	1.43	0.71	0.95	0.86	0.23	1.08	5.26

Source: Work papers, Puget Sound Regional Transportation Study, 1961.

TABLE 5.38
Structure Type Related to Total Person and Transit Trips
and Percent by Transit for Milwaukee and Kenosha, 1963

Structure Type	Milwaukee Home Interview Area			Kenosha Home Interview Area		
	Average Number of Trips per Household		Percent by Transit	Average Number of Trips per Household		Percent by Transit
	Total Person Trips	Transit Trips		Total Person Trips	Transit Trips	
1 Family	8.68	0.61	7.0	8.82	0.29	3.3
2 Family	5.77	0.88	15.3	5.59	0.21	3.5
3–4 Family	5.25	0.78	14.9	5.76	0.27	4.7
5–19 Family	4.47	0.84	18.8	5.33	0.30	5.6
20 or More Family	3.00	1.00	33.3	2.11	0.20	10.5
Trailer	5.13	0.13	2.5	5.64	0.05	0.9
Area Totals	7.05	0.72	10.2	7.72	0.27	3.5

Source: Southeastern Wisconsin Regional Planning Study as reported in *Proceedings of the Modal Choice and Transit Planning Conference*, Cleveland, 1966.

1. Table 5.37 shows the relationships found between "relative population density" and trip purpose in the Puget Sound (Seattle) area in 1961.
2. Table 5.38 shows the relationships found between type of dwelling unit, total person trips, and transit trips in the Milwaukee and Kenosha area.
3. Figure 5.16 summarizes various relationships between transit usage and population density. In almost every case, high density (in combination with low car ownership) tends to increase transit use, but "the relationships are also influenced by such factors as service and fare differentials, concentrations of activities within the central area, and alignment of population along specific corridors which all obviously influence patronage and affect the plotted values."[15]
4. Curves relating the percent of automobile driver trips to person trip end density are shown in Figure 5.17 for the Penn-Jersey, Pittsburgh, and Chicago areas. These curves indicate that below a density of 10,000 trips per square mile, approximately 57 percent of the total person trips are expected to be made as

[15] HERBERT S. LEVINSON AND F. HOUSTON WYNN, "Effects of Density on Urban Transportation Requirements," *Community Values as Affected by Transportation*, Highway Research Record, No. 2 (Washington, D.C.: Highway Research Board, 1962), p. 56.

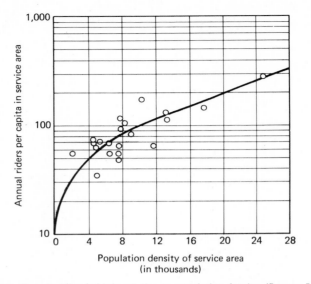

Figure 5.16. Transit riding habit in relation to population density. (Source: Levinson and Wynn, "Effects of Density on Urban Transportation Requirements," p. 58.)

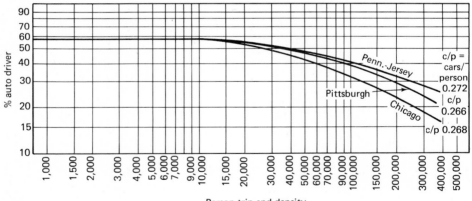
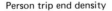
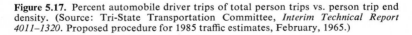

Figure 5.17. Percent automobile driver trips of total person trips vs. person trip end density. (Source: Tri-State Transportation Committee, *Interim Technical Report 4011–1320*. Proposed procedure for 1985 traffic estimates, February, 1965.)

auto driver trips. Above this density, there is a dropping off of auto driver trips at rates which appear to vary by area and by the cars per person ratio.

Car ownership. Almost every major metropolitan area transportation study has indicated the importance of car ownership, as well as population density on both the number of urban trips and the type of travel mode. See, for example, *Future Highways and Urban Growth*, Chapter 3, "Characteristics of Urban Travel."[16]

---

[16] Wilbur Smith and Associates, 1961.

Typical relationships among car ownership, trip generation, and travel mode for selected cities are summarized in Table 5.39. Members of zero-car households make slightly under one person trip per day, members of one-car households approximately two person trips per day, and members of multicar households approximately three to four person trips per day. In contrast, the number of person trips by public transit decreases from approximately 0.6 in large city zero-car households to 0.3 for one-car households, and to slightly under 0.3 for multicar households.

In smaller cities the overall rates of person tripmaking are approximately the same as for larger cities within a given level of car ownership. However, the proportion of trips by public transport is much less, even among zero-car households. It is important to note that a substantial proportion of travelers from zero-car households actually travel as automobile passengers. This implies a dependence on the neighbor, i.e., car sharing, as a means of traveling to work or to shop.

The data shown in Table 5.39 have been generalized by city size in Table 5.40. These figures provide a basis for broadly estimating urban travel behavior and modal

TABLE 5.39
**Effect of Car Ownership on Trip Generation and Travel Mode
in Selected Urban Areas**

| Item | Cars per Dwelling Unit | | | | |
	0	1	2	3+	All
Total Trips per Person					
London, England	0.93	2.03	2.56	3.06	NA
Chicago	0.78	1.76	2.92*		1.92
Detroit	0.70	2.08	3.20*		1.77
Pittsburgh	1.12	1.70	2.40*		1.61
Puget Sound	0.90	2.60	3.20	3.40	2.66
San Juan	0.99	2.00	3.28*		1.69
Knoxville	0.85	2.49	3.36	3.87	2.51
Lexington	0.76	2.06	2.91	3.26	2.13
Springfield, Mass.	0.71	2.27	3.13*		2.27
Transit Trips per Person					
London, England	0.78	0.60	0.55*		NA
Chicago	0.62	0.36	0.31*		0.47
Detroit	0.42	0.25	0.24*		0.30
Pittsburgh	0.74	0.30	0.26*		0.32
Puget Sound	0.37	0.12	0.06	0.06	NA
San Juan	0.73	0.38	0.24*		0.53
Knoxville	0.32	0.09	0.04	0.03	0.11
Lexington	0.28	0.10	0.07	0.05	0.12
Springfield, Mass.	0.20	0.07	0.04*		0.08
Percent of Person Trips by Transit					
London, England	83.9	29.5	20.7*		NA
Chicago	79.7	20.3	10.5*		24.3
Detroit	59.7	12.1	7.4*		16.7
Pittsburgh	64.3	17.6	10.8*		21.7
Puget Sound	41.3	4.6	2.0	1.8	NA
San Juan	73.3	18.9	7.3*		31.1
Knoxville	37.4	3.5	1.1	0.8	4.3
Lexington	36.9	4.7	2.3	1.6	5.4
Springfield, Mass.	28.7	3.1	1.2*		3.6

*Two or more cars per dwelling unit.

Source: Computed from origin-destination studies in each urban area. Data may vary slightly from that set forth in other tabulations. For Puget Sound, data are based on persons over five years old in households.

TABLE 5.40
Daily Trips per Person in Urban Areas—Generalized

	Cars per Household		
	0	1	2+
Large City			
Total Trips	1.0	2.0	3.0
Transit Trips	0.7	0.3	0.2
Medium-Sized City			
Total Trips	1.0	2.3	3.3
Transit Trips	0.4	0.2	0.1
Small City			
Total Trips	1.0	2.5	3.5
Transit Trips	0.3	0.1	0.1

Source: Comprehensive transportation studies, 1960–1970.

choice, either as macro estimates or as control-totals against which more refined estimates can be cross-checked.

The different levels of tripmaking in households owning zero, one, and two or more cars, according to trip purpose, are shown in Table 5.41. Here again, it is clear that a rise in multiple car ownership brings with it a concomitant rise in auto driver trips, particularly for non-work trips. The increase in tripmaking, however, also depends on family size.

TABLE 5.41
Effect of Car Ownership on Average Number of Trips per
Household by Trip Purpose, Cincinnati Urbanized Area

Trip Purpose	Noncar Households	One-car Households	Multicar Households	Ratio One/None	Ratio Multi/One
Home-based Work	.62	1.66	2.49	2.68	1.50
Home-based Shopping	.37	1.05	1.58	2.84	1.50
Home-based Social-Recreational	.30	1.11	2.10	3.70	1.89
Home-based School*	.17	0.44	1.04	2.59	2.36
Home-based other	.32	0.87	1.58	2.71	1.81
Nonhome-based	.19	1.37	2.86	7.20	2.09
All Purposes	1.97	6.50	11.65	3.30	1.79

*Based on trip and household data from households interviewed during school year.

Source: "Urban Transportation Models," Ohio-Kentucky-Indiana Regional Transportation and Development Plan, Wilbur Smith and Associates, 1972.

Household size and car ownership. The combined effects of automobile ownership and household size on the daily trips per household and daily trips per person are shown in Table 5.42. The larger the household size, the greater the total daily trip generation for any given level of car ownership. Also, for any given household size, the trips per household increase with increased car ownership. The trips per person generally increase markedly with each level of car ownership and more gradually with family size.

These interactions can be summarized as follows: There is an increase in per capita trip generation with increases in car ownership, workers, and income per family;

TABLE 5.42
Average Total Daily Person Trips per Household and per Person, Classified by Auto Ownership and Number of Persons in Household, Puget Sound, 1961

Number of Persons in Households*	Type Trip†	0 Car	1 Car	2 Cars	3 Cars	4+ Cars	Total
1	A	0.9	3.2	3.5	4.6‡	2.0‡	1.8
2	A	1.8	5.7	7.3	9.9	13.1‡	5.5
	B	0.9	2.8	3.6	5.0	6.5	2.8
3	A	3.5	7.8	10.2	11.1	14.1‡	8.5
	B	1.2	2.6	3.4	3.7	4.7	2.8
4	A	3.5	10.0	12.9	13.7	19.6	11.1
	B	0.9	2.5	3.2	3.4	4.9	2.8
5	A	4.3	11.5	15.0	17.2	16.3	13.1
	B	0.9	2.3	3.0	3.4	3.2	2.6
6–7 (6.5)	A	4.0	12.6	17.2	18.6	22.0	14.6
	B	0.6	1.9	2.6	2.9	3.4	2.2
8 and over	A	*3.9	15.1	21.9	23.6‡	29.3‡	18.1
	B	0.5	1.8	2.7	3.9	3.6	2.3
TOTAL	A	1.4	6.9	11.0	14.0	19.1	7.0

*Number of persons 5 years of age and over. Trip information was collected only for this group.
†A: Trips per household.
 B: Trips per person.
‡Average based on fewer than 25 samples.
  Source: Work Papers, Puget Sound Regional Transportation Study, 1961.

the per capita trip production declines slightly with each increase in family size. Although the aggregate number of trips generated by each household size increases as a household gets larger, the rate of increase generally slows down.

Trips per car. Many studies have documented the aggregate increase in the number of vehicle trips per car resulting from a comparable rise in the average number of cars per household, up to a maxima somewhere between one and two cars per household. Where car ownership is very low, most of the cars are used to travel to and from work and are consequently unavailable for other trips during most of the day. In multicar families, however, cars are available for short trips to non-work destinations. This, in turn, brings a rise in the average number of driver trips (vehicle trips) per car as car ownership rises. However, the law of diminishing returns appears to take effect at two or more cars per family. In Chicago, for example, three-car families averaged 2.86 driver trips per car as compared with 2.99 for one-car families. In Nashville, three-car families averaged 3.86 driver trips per car, one-car families 4.70.

Income. Household income affects trip generation directly and also influences car ownership. In all cities, car ownership increases as family incomes rise. Moreover, factors of net residential density and diffused working and living patterns, in combination with income, explain the high incidence of multicar ownership in suburban areas.

Lansing and Hendricks, for example, show that 20 percent of all families with incomes under $4,000, in suburban areas, owned two or more cars as compared with 7 percent in smaller central cities and none in the larger cities. At the other end of the scale, 19 percent of all families making $7,500 a year or more in large central cities owned two or more cars, 45 percent in the newer central cities, 59 percent in the smaller

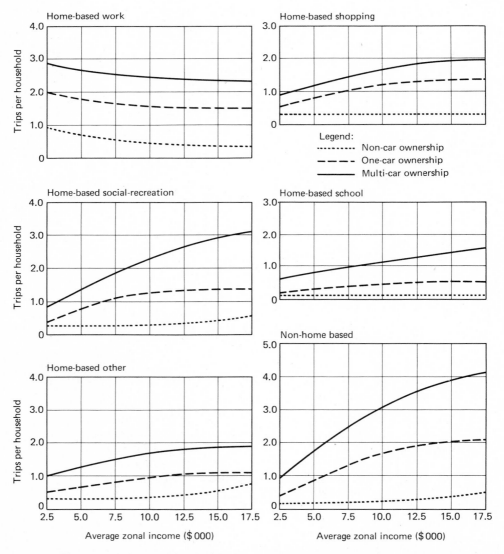

Figure 5.18. Trips per household by car ownership and income. (Source: *Ohio-Kentucky-Indiana Regional Transportation and Development Plan*, Wilbur Smith and Associates, p. 24.)

cities, and 61 percent in suburban areas. [These figures (1967) excluded New York City.][17] There has been a continued rise in the proportion of spending units owning two or more cars at each level of income. The underlying reasons are (1) increasing use of automobiles by women, particularly those going to work, (2) an increasing proportion of population living in suburban areas, and (3) continued dispersion of metropolitan employment opportunities. Wynn and Levinson indicate that an increasing number of workers in the family results in a growing number of automobiles at each income level.[18]

The combined effects of household income and car ownership on trip generation are shown in Figure 5.18. The curves suggest that rising income gradually increases household tripmaking at each level of car ownership for all types of non-work trips.

Age. Transportation planners are increasingly identifying the travel behavior patterns of various age groups. These patterns provide a basis for (1) identifying latent travel demands and (2) providing special mobility for the elderly.

Total tripmaking has been related to a combination of both age and income with

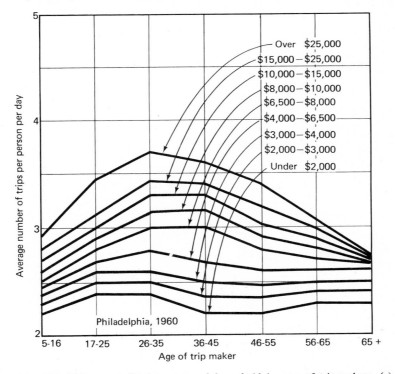

Figure 5.19. Trip rates related to age and household income of trip-makers. (a) Philadelphia (P-J) Urbanized area-1960. (Source: Wilbur Smith & Associates, "Patterns of Car Ownership, Trip Generation and Trip Sharing in Urbanized areas," pp. 119, 121, 125, 127.)

---

[17] J. B. LANSING AND G. HENDRICKS, *Automobile Ownership and Residential Density* (Ann Arbor: University of Michigan, 1967).

[18] F. H. WYNN AND H. S. LEVINSON, "Some Considerations in Appraising Bus Transit Potentials," *Passenger Transportation*, Highway Research Record 197, (Washington, D.C.: Highway Research Board, 1967.)

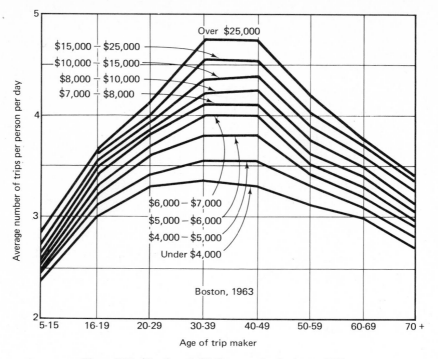

Figure 5.19. (Continued) (b) Boston urbanized area-1963

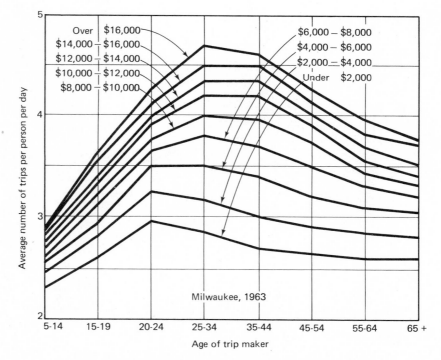

Figure 5.19. (Continued) (c) Milwaukee urbanized area-1963

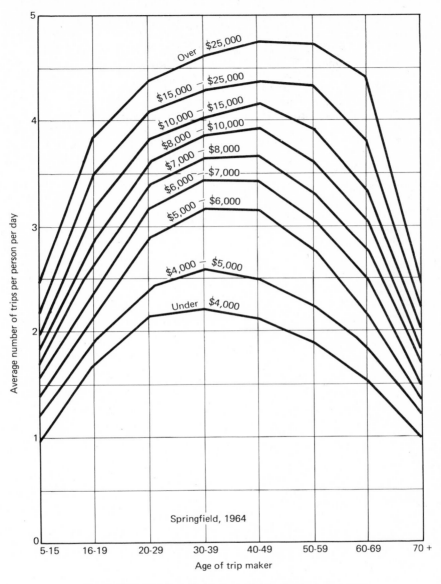

Figure 5.19. (Continued) (d) Springfield urbanized area-1964

maximum trip generation reported to occur during the middle years (between 30 and 49), depending on income. Typical relationships are shown in Figure 5.19a, b, c, d for Philadelphia, Boston, Milwaukee, and Springfield, Mass.[19]

A further analysis of age on urban travel behavior was conducted by Ashford and

[19] *Patterns of Car Ownership, Trip Generation and Trip Sharing in Urbanized Areas*, Wilbur Smith and Associates, New Haven, Connecticut, 1968.

Holloway, based on origin-destination data for Milwaukee, Wisconsin, and Albany, Augusta, Columbus, Macon and Savannah, Georgia. Principal findings are summarized in Tables 5.43 and 5.44, and in Figure 5.20.[20]

1. *CBD Orientation (Figure 5.20).* "Age had a direct influence on the percentage of a trip maker's CBD oriented journeys. As the trip maker's age increases, the CBD becomes a more dominant attractor."[21]

2. *Total Travel (Table 5.43).* "The work trip increases in importance up to the age of general retirement and then rapidly declines, while shopping trips increase in relative importance with age and increase sharply for the elderly. The relative importance of school trips declines rapidly, becoming negligible above the age of 25. Miscellaneous trips reach a minimum in middle age and rapidly increase in importance for elderly persons."[22]

TABLE 5.43
Unweighted Average Trips Generated per Tripmaker per Day for Six Selected Cities

| Age | Home-based Trip Purposes | | | | |
	Work	Shop	School	Misc.	All Purposes
5–14	.02	.24	.62	1.29	2.17
15–24	.51	.33	.34	1.00	2.18
25–34	.82	.48	.02	1.01	2.33
35–44	.89	.47	.01	1.04	2.41
45–54	.97	.44	0	.97	2.38
55–64	.92	.45	0	.95	2.32
65+	.38	.66	0	1.23	2.27
Unweighted Average	.65	.44	.14	1.06	2.32

Source: Norman Ashford and Frank M. Holloway, "The Effects of Age on Urban Travel Behavior," *Traffic Engineering*, Vol. XLI, No. 7 (April 1971), p. 49.

TABLE 5.44
Average Percentage of Trips Made by Transit for Six Selected Cities

	Work	Shop	Misc.	All Purposes
5–14	5.6	2.2	1.4	1.6
15–24	11.2	4.3	2.7	2.1
25–34	5.9	1.6	1.0	0.8
35–44	7.8	2.2	1.3	0.9
45–54	8.4	2.8	1.5	1.1
55–64	11.2	5.4	3.6	2.2
65 and over	11.3	8.2	6.9	2.7
Average all Ages	9.2	3.8	2.6	1.6

Source: Ashford and Holloway, "The Effect of Age on Travel Behavior," p. 47.

---

[20] NORMAN ASHFORD AND FRANK M. HOLLOWAY, "The Effect of Age on Travel Behavior," *Traffic Engineering*, Vol. XLI, No. 7 (April 1971), pp. 46–49, 67.
[21] *Ibid.*, p. 67.
[22] *Ibid.*

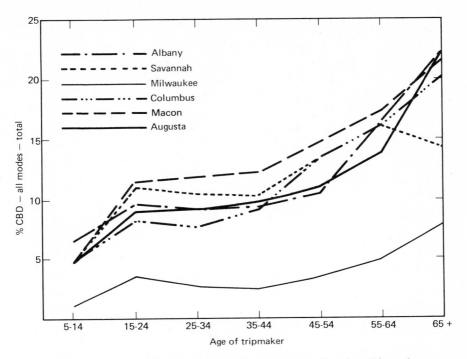

Figure 5.20. Percent of trips to CBD by various age groups, all modes, all home-base purposes. [Source: Norman Ashford and Frank M. Holloway, "The Effect of Age on Travel Behavior," *Traffic Engineering*, Vol. XLI, No. 7 (April 1971), p. 47.]

3. *Transit Trips* (*Table 5.44*). "In most cases the highest transit usage in all cities occurred within the 65-and-over age group."

Travel characteristics of elderly (i.e., over 65) in the New York Metropolitan Area are summarized in Table 5.45 and Figures 5.21 and 5.22. These findings are based on a sample of 57,000 households drawn from the 1963 home-interview data.[23]

Land-use factors. Each type of land use generates person and vehicular travel in accordance with its type and intensity of use. Commercial land may generate more than 150 person-destinations per acre; public open space, fewer than five.

Table 5.46 shows the daily person-destinations per 1,000 sq ft of floor space by type of use.

Table 5.47 presents the daily person-destinations per acre of land for each major class of use.

Traffic generation factors which can be applied in estimating the impacts of proposed developments are shown in Table 5.48. This tabulation expresses traffic generation in terms of vehicle trip ends per acre for various land-use types. They also relate

[23] JONI K. MARKOVITZ, "Transportation Needs of the Elderly," *Traffic Quarterly*, Vol. XXV, No. 2 (April 1971), pp. 237–254.

TABLE 5.45
Weekday Trips per Person by Purpose by Income, Tri-State (N.Y.) Area, 1963

Trip Purpose	Household Income*									
	$0-2,999		$3,000-5,999		$6,000-9,999		$10,000+		Total	
	Elderly	Total Population†	Elderly	Total Population†	Elderly	Total Population†	Elderly	Total Population†	Elderly	Total Population†
Home	0.22	0.35	0.36	0.59	0.41	0.84	0.55	1.00	0.35	0.77
Work	0.03	0.12	0.10	0.34	0.20	0.46	0.28	0.58	0.12	0.42
Shop	0.09	0.07	0.13	0.10	0.11	0.18	0.15	0.21	0.11	0.15
School	0.00	0.04	0.00	0.07	0.00	0.12	0.00	0.15	0.00	0.11
Social	0.05	0.06	0.07	0.07	0.07	0.08	0.06	0.10	0.06	0.08
Recreational	0.02	0.02	0.02	0.02	0.02	0.04	0.03	0.06	0.02	0.04
Personal Business	0.08	0.10	0.11	0.10	0.10	0.14	0.16	0.21	0.10	0.15
Other‡	0.00	0.01	0.02	0.02	0.02	0.06	0.04	0.10	0.03	0.06
TOTAL	0.49	0.77	0.81	1.31	0.93	1.92	1.27	2.41	0.79	1.78

*Income of household of which a person is a member.
†Population over 5 years old.
‡Ride, Serve Passenger trips without a primary trip purpose, and out-of-cordon change mode trips.

Source: Joni K. Markovitz, "Transportation Needs of the Elderly," *Traffic Quarterly*, Vol. XXV, No. 2 (April 1971), p. 240.

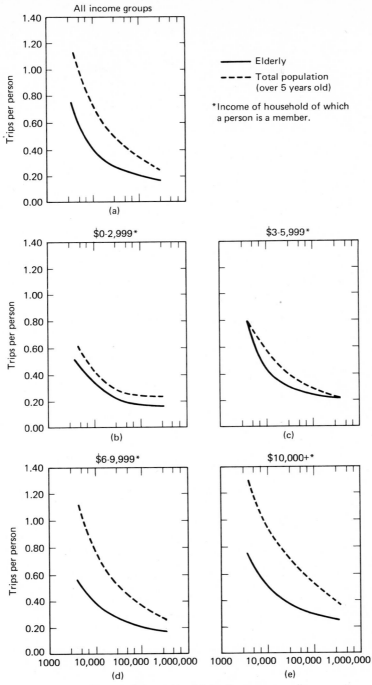

Figure 5.21. Trip characteristics of elderly and other persons as a function of net residential density and income—New York Metropolitan area. (Excluding home and work trip destinations.) [Source: Joni K. Markovitz, "Transportation Needs of the Elderly," *Traffic Quarterly*, Vol. XXV, No. 2 (April 1971) pp. 237–254.]

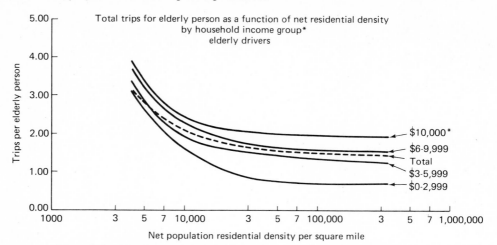

Figure 5.22. Trip characteristics of elderly persons as a function of net residential density and income*—New York Metropolitan area. [Source: Joni K. Markovitz, "Transportation Needs of the Elderly," *Traffic Quarterly*, Vol. XXV, No. 2 (April 1971), pp. 237–254.]

TABLE 5.46
Relation of Floor Space to Trip Generation

Type Land Use	Daily Person-Trips per 1,000 sq ft of Floor Space (Destinations)	
	Chicago Survey Area	Pittsburgh Central Business District
Retail	7.0	8.1
Services	5.4	5.2
Public Buildings	3.5	3.9
Residential	3.2	2.4
Manufacturing	2.1	1.0
Transportation	1.9	4.0
Wholesale	1.5	1.2
ALL TYPES	3.3	4.9

Source: Transportation and land-use studies in each urban area.

TABLE 5.47
Daily Person-Trip Destinations for Generalized Land-Use Categories, Selected Urban Areas

Land Use	Average Daily Destinations per Acre of Land in Each Class of Use			
	Chicago	Philadelphia (Penn.-Jersey)	Detroit	Pittsburgh
Residential	48	26	29	24
Commercial	181	238	269	158
Manufacturing	49	26	37	26
Transportation	9	20	NA	9
Public Buildings	53	35	33	46
Public Open Space	4	7*	3	2
ALL TYPES	38	30	36	27

*Trips to miscellaneous land, excluding streets and undeveloped land.

Source: Comprehensive urban transportation studies.

vehicular trip ends to pertinent parameters, such as dwelling units (residential), floor space (retail), or beds (hospitals).

Vehicular trip generation at regional shopping centers approximates 15 to 20 car destinations per day per 1,000 square foot of gross leasable floor space. This varies among shopping centers according to their location, productivity, store type, and parking supply. Approximately 15 to 20 percent of these movements take place

TABLE 5.48
Traffic Generation of Major Land Uses

Land Use	Density	Traffic Generation Rate Daily Trips	Range (Typical)
Residential	Dwelling Units/Acre	Vehicle Trip Ends/Acre	Vehicle Trip Ends/Dwelling Unit
Low Density (Single-Family Homes)	1–5	5–65 (40)	7–12 (9)
Medium Density (Patio Houses, Duplexes, Townhouses)	5–15	40–150 (75)	5–8 (7)
High Density (Apartments)	15–60	85–400 (180)	3–7 (5)
Commercial	Site	Vehicle Trip Ends/Acre	Vehicle Trip Ends/1,000 sq ft Floor Area
Retail Commercial Neighborhood Retail (Supermarket)	10 Acres	800–1,400 (1,000)	70–240 (130)
Community Retail (Junior Department Store)	10–30 Acres	700–1,000 (900)	60–140 (80)
Regional Retail (Regional Shopping Center)	30 Acres	400–700 (600)	30–50 (40)
Central Area Retail	High Density	600–1,300 (900)	10–50 (40)
Highway-Oriented Commercial (Motels, Service Stations)	Low Density	100–300 (240)	4–12 (10)
Commercial	Floor Area Ratio (FAR)	Vehicle Trip Ends/Acre	Vehicle Trip Ends/1,000 sq ft Floor Area
Service Commercial (Office Buildings) Single-Story Building with Surface Parking	0.5:1	120–1,200 (300)	6–60 (14)
Two-Story Building with Surface Parking	1:1	240–2,400 (600)	6–60 (14)
Three to Four-Story Building with Deck Parking	2:1	360–6,000 (1,200)	6–60 (14)
Three to Six-Story Building with Structure Parking	5:1	1,200–12,000 (2,600)	6–60 (14)
High-rise Office Building with Structure Parking, over 10 Stories	10:1	2,400–20,000	6–60 (14)

Source: Harold Marks, *Protection of Highway Utility*, National Cooperative Highway Research Program Report 121 (Washington, D.C.: Highway Research Board), 1971.

TABLE 5.48 (*Continued*)

Land Use	Density	Traffic Generation Rate Daily Trips	Range (Typical)
Industrial	Employees/ Acre	Vehicle Trip Ends/Acre	Vehicle Trip Ends/1,000 sq ft Floor Area
Highly Automated Industry; Low Employee Density (Refinery, Warehouse)	5	2–8 (4)	0.2–1.0 (0.6)
Light Service Industry; Single Lot Industry (Lumber Yard)	5–20	6–30 (16)	0.4–1.2 (0.8)
Industrial Tract (5 acres) (Machinery Factory)	20–100	30–160 (70)	0.6–4.0 (2.0)
Office Campus; Research and Development (Research Industry)	100	150–200 (170)	3–8 (4)
Mixed Central Industry; Small Industrial Plants	Varies	10–100	1–4
Public and Semi-public Uses	Varies	Vehicle Trips Ends/Acre	
Schools and Colleges	No. of students	Colleges: 7–600 (60)	0.4–1.0 (0.8) veh. trip ends/student
Places of Public Assembly (Theater, Stadium, Convention Center)	No. in attendance		Stadia: 2 veh. trip ends/4 seats
Administration Facilities (City Hall, State Offices, Post Offices)	Floor area ratio	70–600 (200)	10–60 (20) veh. trip ends/1,000 sq ft Floor Area
Recreation Facilities (Park, Zoo, Beach, Golf Course)		Parks: 1–10 (4)	Golf course: 2–10 (8) veh. trip ends/acre
Terminals (Bus Terminal, Airport)		3–30 (15)	Local airport: 6–12 (8) veh. trip ends/ based aircraft
Hospitals	No. of beds	16–70 (40)	6–16 (10) person trip ends/bed

inbound and also outbound during peak shopping hours.[24] A million sq ft regional shopping center would produce the following typical traffic values:

Daily vehicles entering	18,000
Vehicles Inbound, P.M. rush hour	1,600
Vehicles Outbound, P.M. rush hour	1,800
Vehicles Inbound Peak Shopping Hour	2,700
Vehicles Outbound Peak Exit Hour	3,200

---

[24] See, for example, Carle H. Burke, "An Approximation of Regional Shopping Center Traffic," *Traffic Engineering*, Vol. 42, No. 7 (April 1972).

REFERENCES FOR FURTHER READING

ASHFORD, NORMAN and FRANK M. HOLLOWAY, "The Effect of Age on Travel Behavior," *Traffic Engineering*, April, 1971.

BURKE, CARLE H, "An Approximation of Regional Shopping Center Traffic," *Traffic Engineering*, April, 1972.

CLARK, T.F.C., "Urban Population Densities," *Journal of the Royal Statistical Society*, CXIV, Series A, pp. 490–96.

Future Highways and Urban Growth, Prepared under commission from the Automobile Manufacturers' Association, Wilbur Smith and Associates, 1961.

Guidelines for Trip Generation Analysis, Federal Highway Administration, Washington, D.C., 1967.

LANSING, JOHN B. and GARY HENDRICKS, *Automobile Ownership and Residential Density*, Survey Research Center, University of Michigan, Ann Arbor, 1967.

LEVINSON, HERBERT S. "Pedestrian Way Concepts and Case Studies", Highway Research Record No. 355, *Pedestrians*, National Academy of Sciences, Highway Research Board, 1971.

LEVINSON, HERBERT S. and F. HOUSTON WYNN, "Effects of Density on Urban Transportation Requirements," *Community Values as Affected by Transportation*, Highway Research Board, Washington, D.C., Highway Research Record, No. 2, 1962, p. 56.

LEVINSON, HERBERT S. and F. HOUSTON WYNN, "Some Aspects of Future Transportation in Urban Areas," *Urban Transportation Demand and Coordination, Bulletin 326*, Highway Research Board, Washington, D.C., 1962, pp. 1–31. Also see Wilbur Smith and Associates, *Future Highways and Urban Growth*, Detroit, 1961, pp. 61–156.

MARKOVITZ, JONI K., "Transportation Needs of the Elderly," *Traffic Quarterly*, April, 1971.

MOSKOWITZ, K., "Living and Travel Patterns in Automobile Oriented Cities," *The Dynamics of Urban Transportation* (Detroit: A National Symposium Sponsored by the Automobile Manufacturers' Association, October 23–24, 1962).

Motor Trucks in the Metropolis, Prepared under commission from the Automobile Manufacturers' Association, Wilbur Smith and Associates, 1969.

National Cooperative Highway Research Program Report 70—*Social and Economic Factors Affecting Intercity Travel*, Vogt, Ivers and Associates, Cincinnati, Ohio, 1969.

Patterns of Car Ownership, Trip Generation and Trip Sharing in Urbanized Areas, Wilbur Smith and Associates, New Haven, Connecticut, 1968.

Pittsburgh Area Transportation Study, 1958. Also see L. E. Keefer, "Trucks at Rest," *Origin and Destination Techniques and Evaluations*, Highway Research Record No. 41, Highway Research Board, National Academy of Sciences, 1963, pp. 29–38.

U.S. Census as reported in *Automobile Facts*, 1967, p. 36, "Automobile Ownership by Selected Household Characteristics," 1964.

VOORHEES, A. M., S. J. BELLOMO, J. L. SCHOFER, and D. E. CLEVELAND, "Factors in Work Trip Lengths," Highway Research Board, No. 141, Washington, D.C.

WYNN, F. H. and H. S. LEVINSON, "Some Considerations in Appraising Bus Transit Potentials," Highway Research Record No. 197, *Passenger Transportation, 1967*.

ADDITIONAL REFERENCES FOR FURTHER READING

A Key to Change: Urban Transportation Research, Special Report 69 (1962).

Community Values as Affected by Transportation, Highway Research Record, Number 2, (1963).

CREIGHTON, ROGER, *Urban Transportation Planning*, University of Illinois Press, Urbana, Illinois, 1970.

Forecasting Highway Trips, Bulletin 297 (1961).

Freeway Operations, Bulletin 324 (1962).

Highway Traffic Estimation (1956) by *Robert E. Schmidt* and *M. Earl Campbell*, The Eno Foundation for Highway Traffic Control, Saugatuck, Connecticut.

MARKS, HAROLD, *Protection of Highway Utility*, National Cooperative Highway Research Program Report 121, Highway Research Board, Washington, 1971.

Traffic Characteristics and Traffic Assignment, Bulletin 347 (1962).

Traffic Characteristics at Regional Shopping Centers (1961) by Donald E. Cleveland and Edward A. Mueller.

Traffic Origin-and-Destination Studies: Appraisal of Methods, Bulletin 253 (1960).

Transportation and Parking for Tomorrow's Cities, Prepared under commission from the Automobile Manufacturers' Association, Wilbur Smith and Associates, December, 1966.

Travel Characteristics In Urban Areas, Bulletin 203 (1958).

Trip Generation and Urban Freeway Planning, Bulletin 230 (1959).

Urban Transportation Concepts, Center City Transportation Project, Wilbur Smith and Associates, 1970.

Urban Transportation: Demand and Coordination, Bulletin 326 (1962).

Urban Transportation Planning, Bulletin 293 (1961).

Chapter 6

MASS TRANSPORTATION CHARACTERISTICS

HENRY D. QUINBY, Deputy General Manager, Operations and Administration, San Francisco Minicipal Transit System, San Francisco, California.

This chapter is concerned with the physical dimensions, geometric characteristics, right-of-way requirements, capacities, accommodations, and performance of mass or public transportation (e.g., transit) vehicles. It is also concerned with transit stops, fares, system characteristics and performance, related economics, relevant environmental aspects, and novel and innovative transit methods and equipment. Other chapters in this handbook cover transit industry facts; travel characteristics; planning considerations, studies, techniques, and applications; and interfaces between transit and road facilities with respect to traffic engineering techniques and transit-highway joint use facilities. References for Further Reading at the end of this chapter and the footnotes cite sources of material developed for this chapter, and cite additional information and data on mass transportation characteristics.

Whenever data in this chapter are expressed in terms of typical ranges, the data involve ranges within which individual cases are usually or typically, but not necessarily always, encountered in current planning and operational practice. Sometimes, but not always, the most typical values encountered in current practice tend to fall toward the center of the range indicated. Other data are expressed in terms of single typical values. Some of the typical ranges shown involve wide, or relatively wide, spreads; this is indicative of wide ranges encountered in actual practice and should alert the reader to ranges of possibilities for alternative treatments of the variables involved.

Vehicle-types and system-types cited in tables and elsewhere in this chapter are generally self-explanatory. Some vehicle type nomenclature, however, may warrant amplification:

1. *Minibus.* A small self-propelled bus used in secondary distribution services such as the local CBD (central business district) shuttle service in Washington, D.C.
2. *Midibus.* A slightly larger version of the minibus.
3. *City-suburban transit bus.* The typical single-deck, self-propelled urban transit bus.
4. *Intercity bus.* A self-propelled bus used in longer-distance, over-the-road, intercity services.
5. *Electric buses* (*trolley-buses*). Generally similar to the corresponding self-propelled buses in most of the characteristics treated in this chapter.
6. *Tram.* A rail vehicle operated, singly or coupled into trains, on a tramway, at least part of the route of which usually, but not necessarily always, crosses or mixes with other vehicular traffic.

7. *Urban-regional rapid transit, rubber-tired.* Refers to the Paris, Montreal, and Mexico City type of local rubber-tired, side-and-rail guided rapid transit vehicle.
8. *Superregional rapid transit.* Includes existing and planned vehicles and facilities for the Washington–New York "Metroliner Line," the Tokyo–Osaka "New Tokaido Line," and other planned vehicles and facilities to serve superregional or megalopolitan conurbations.

VEHICLE DIMENSIONS

Typical ranges for key dimensional characteristics of transit vehicles currently in significant use are indicated in Table 6.1. Lengths and widths given are external body dimensions. Heights are from pavement or top of rail to top of roof. Gross floor areas include all area at the maximum horizontal area-plane. Weights are empty, dry vehicle weights. Weight per unit of length is a common measure of vehicle unit heaviness. Weight per unit of floor area is, however, a more accurate and indicative measure of vehicle unit heaviness. Floor area, but not usually cubic volume or cubage, units are often useful divisors in ratios when the dividends are units of weight or capacity.

GEOMETRIC AND RIGHT-OF-WAY CHARACTERISTICS

This section gives typical basic geometric characteristics and minimum right-of-way requirements for selected transit vehicles and train operation. Although the values indicated reflect current acceptable practice, it should be borne in mind that the individual circumstances of different applications will vary. The values shown for the selected cases, although typical, are subject to variation among different projects. Preparation of specific, detailed geometric criteria is desirable for each different kind of new transit service and facility application; these standards should reflect the specific conditions of that application.

Right-of-way requirements for different types of transit vehicle are subject to the factors discussed below and in Table 6.2. Particularly as to stations, these requirements will vary considerably in different applications and conditions. Typical minimum overall right-of-way requirements for tangent line sections are shown in item 3h of Table 6.2.

BUSES ON PUBLIC STREETS AND HIGHWAYS

Buses on public streets and highways can generally operate effectively within the ranges of geometric values set forth in Chapter 14, Geometric Design, for freeways, expressways, arterials, collectors, and local streets. Care should be taken on (1) the vertical clearances required for double-decker buses of different heights, (2) the paths and horizontal clearance requirements of various transit vehicles both while turning and in tangent operation, and (3) other characteristics of the transit vehicle involved, as such characteristics may affect geometric, right-of-way, and traffic operational considerations.

Generally, the operation of buses of 8 ft (2.4 m) or greater width is not desirable in traffic lanes of less than 10 ft (3.0 m) in width; wider lanes are preferred for these buses. However, practical compromises (involving reduced speeds and limited distances) sometimes have to be made in situations in which buses must use lanes of narrower width.

TABLE 6.1
Key Dimensions of Transit Vehicles, Expressed in Typical Ranges

Transit Vehicle Types	Length [ft (m)]	Width [ft (m)]	Height [ft (m)]	Gross Floor Area [sq ft (sq m)]	Weight [lb (kg)] (000)	Weight per Linear Foot (Meter) [lb per ft (kg per m)]	Weight per Square Foot (Meter) [lb per sq ft (kg per sq m)]
Bus, freewheeling							
1. Minibus-midibus	18–25 (5.5–7.6)	7.0–8.0 (2.1–2.4)	7–9 (2.1–2.8)	126–200 (11.7–18.6)	5–8 (2.3–3.6)	250–350 (372–521)	35–50 (171–244)
2. City-suburban transit bus: non-articulated	30–43 (9.1–13.1)	7.5–9.0 (2.3–2.8)	9–11 (2.8–3.4)	225–387 (20.9–35.9)	14–25 (6.3–11.3)	450–600 (670–893)	60–80 (293–391)
3. City-suburban transit bus: articulated	50–60 (15.2–18.3)	7.5–8.5 (2.3–2.6)	9–11 (2.8–3.4)	375–510 (34.8–47.8)	24–28 (10.9–12.7)	450–550 (670–818)	55–70 (269–342)
4. City-suburban transit bus: double-deck	25–35 (7.6–10.7)	7.5–8.5 (2.3–2.6)	13–15 (4.0–4.6)	375–595 (34.8–55.3)	15–22 (6.8–10.0)	580–680 (863–1,012)	35–45 (171–220)
5. Intercity	30–45 (9.1–13.7)	7.5–8.5 (2.3–2.6)	9–13 (2.8–4.0)	225–383 (20.9–35.6)	20–35 (9.1–15.9)	650–800 (967–1,190)	70–100 (342–488)
6. School	20–35 (6.1–10.7)	7.5–8.5 (2.3–2.6)	8–10 (2.4–3.1)	150–298 (13.9–27.7)	7–15 (3.2–6.8)	350–450 (521–670)	40–60 (195–293)
Rail, guided							
7. Tram: Non-articulated	40–50 (12.2–15.2)	7.5–9.5 (2.3–2.9)	10–11 (3.1–3.4)	300–415 (27.9–44.1)	33–38 (15.0–17.2)	700–900 (1,042–1,339)	80–110 (391–537)
8. Tram: articulated	60–130 (18.3–39.7)	7.5–9.5 (2.3–2.9)	10–11 (3.1–3.4)	450–1,235 (41.8–114.7)	43–85 (19.5–38.6)	550–750 (818–1,116)	65–100 (317–488)
9. Urban-regional rapid transit: rail	48–75 (14.6–22.9)	8.5–11.0 (2.6–3.4)	10–13 (3.1–4.0)	425–825 (39.5–76.6)	44–85 (20.0–38.6)	740–1,450 (1,101–2,158)	70–150 (342–732)
10. Urban-regional rapid transit: rubber-tired*	48–55 (15.2–16.8)	8.0–8.5 (2.4–2.6)	11–12 (3.4–3.7)	400–468 (37.2–43.5)	44–60 (20.0–27.2)	800–1,100 (1,190–1,637)	102–132 (498–644)
11. Commuter railroad: electric	85 (25.9)	10.0–10.5 (3.1–3.2)	12–13 (3.7–4.0)	850–893 (79.0–83.0)	85–115 (38.6–52.2)	1,000–1,400 (1,488–2,083)	100–135 (488–659)
12. Superregional rapid transit	80–85 (24.4–25.9)	10.0–12.0 (3.1–3.7)	12–13 (3.7–4.0)	800–1,020 (74.3–94.8)	100–160 (45.4–72.6)	1,200–1,900 (1,750–2,800)	120–180 (586–900)

*Lower weight values are for trailer cars; higher weight values are for motor cars. Usually, two motor and one trailer cars operate as a three-car unit.

Sources: Various databooks, inventories, reports, and specifications.

TABLE 6.2
Typical Basic Geometric and Right-of-way Characteristics for Selected Transit Facility Types

Characteristic [ft(m)]	Exclusive Busway	Express Tramway		Rapid Transit		
		In Street Center Reservation	In Exclusive Right-of-Way	Typical City Subway	Cleveland	Bay Area Rapid Transit (BART)
1. Access control	Full	Partial	Full	Full	Full	Full
2. Number of lanes or tracks	2	2	2	2	2	2
3. Widths						
a. Transit vehicle	8.5 (2.6)	9.5 (2.9)	9.5 (2.9)	10.3 (3.1)	10.3 (3.1)	10.5 (3.2)
b. Lane envelope clearance*	12.0 (3.7)	11.0 (3.4)	12.5 (3.8)	13.5 (4.1)	13.5 (4.1)	13.5 (4.1)
c. Track gauge	— —	4.71 (1.44)	4.71 (1.44)	4.71 (1.44)	4.71 (1.44)	5.5 (1.68)
d. Emergency walkway†	—	0.0 0.0	2.5 (0.8)	2.5 (0.8)	—	2.5 (0.8)
e. Median	None	None	None	None	None	None
f. Minimum shoulders (each)†	2.5 (0.8)	None	None	None	None	None
g. Border barriers or fencing (each)	2.0 (0.6)	0.0 0.0	1.0 (0.3)	1.0 (0.3)	1.0 (0.3)	1.0 (0.3)
h. Overall minimum right-of-way requirement‡						
i. Aerial	33 (9.5)	— —	24 (7.3)	26 (7.9)	26 (7.9)	26 (7.9)
ii. At-grade	33 (9.5)	22 (6.7)	30 (9.1)	32 (9.8)	32 (9.8)	40 (12.2)
iii. Subway	39 (11.9)	—	38 (11.6)	40 (12.2)	—	40 (12.2)
4. Stations§						
a. Side platform width	10 (3.0)	6 (1.8)	12 (3.7)	12 (3.7)	—	12 (3.7)
b. Center platform width	—	—	24 (7.3)	24 (7.3)	16 (4.9)	28 (8.5)
c. Platform length	200 (61)	400 (122)	400 (122)	500 (152)	400 (122)	700 (213)
5. Minimum vertical clearance	14 (4.3)	14 (4.3)	14 (4.3)	14 (4.3)	14.5 (4.4)	13 (4.0)
6. Minimum design speed [mph/(kph)]	60 (97)	40 (64)	60 (97)	60 (97)	60 (97)	80 (129)
7. Minimum horizontal curve radius (for new construction)	400 (122)	200 (61)	500 (152)	400 (122)	489 (149)	500 (152)
8. Maximum grade (%)	4.0	10.0	5.0	4.0	4.4	4.0

*Overall vehicle clearance requirements on tangent line.
†Emergency walkway for busways is incorporated in busway shoulders.
‡Minima based on normal structural requirements only, excluding unusual drainage provisions, sideslopes, or retaining walls on cuts and fills, and any subway lateral ventilation requirements. For tangent line sections without stations, and without station acceleration and deceleration lanes and tapers for busways.
§Typical line stations. In CBDs, busway station requirements may be much more severe and rail platform widths may be greater. Station acceleration and deceleration lanes and tapers are excluded from busway values.

Sources: Various databooks, inventories, reports, and criteria.

BUSWAYS FOR EXCLUSIVE BUS USE

Table 6.2 indicates typical basic geometric characteristics for exclusive busways. The technology of line, station, and terminal facilities for the exclusive use of buses is undergoing active development. Specific conditions of individual applications will require the development of detailed, specific geometric criteria in each case.

Busway stations normally require a loading-unloading lane separate from the through lane in each travel direction. Parallel, sawtooth, or diagonal loading berths are alternatives for design consideration and have different geometric and operational characteristics and right-of-way requirements. The number of loading berths to be provided depends on design patronage volumes, minimum bus headways, plans of operation, loading and fare collection characteristics, types and locations of stations, and other factors. Busway stations also frequently require acceleration and deceleration lanes and tapers for satisfactory transition between line and station operations.

Busway stations in CBDs usually require more elaborate facilities and space than stations located elsewhere. If CBD penetration involves subway structures, then ventilation, delivery, and CBD-station cross-sectional problems may be both severe and costly. Because of these factors, CBD delivery is often provided by means of the public street and highway system.

RAPID TRANSIT AND EXPRESS TRAMWAYS

Table 6.2 shows typical basic geometric characteristics for three examples of rapid transit: a typical city subway, Cleveland's city-suburban rapid transit system, and the regional San Francisco Bay Area Rapid Transit system. As with busways, express or limited tramways are receiving increasing planning and design attention in North America as well as in Europe and elsewhere in the world. Table 6.2 also shows typical basic geometric characteristics for two examples of express tramways: in the center-mall reservation of arterial streets and on exclusive rights-of-way.

VEHICLE CAPACITIES AND ACCOMMODATIONS

Table 6.3 shows typical ranges of capacity, in terms of numbers of seats, standees, and total passengers, for the transit vehicle-types and dimensional ranges shown in Table 6.1. In addition, Table 6.3 shows typical values for the maximum number of cars per train, and resulting total passengers per train, for those vehicle-types that lend themselves to train operation.

PASSENGER ACCOMMODATIONS ON VEHICLES: SEATS

Widths of seats per passenger typically vary from 16 in. (0.41 m) to 24 in. (0.61 m) with 17 in. to 20 in. (0.43 m to 0.51 m) typical of local urban transit vehicles. School bus seats per child typically vary from 13 in. to 17 in. (0.33 m to 0.43 m) in width.

The spacing distance between backs on transverse seats typically varies from 26 in. to 34 in. (0.66 m to 0.86 m), with from 26 in. to 30 in. (0.66 m to 0.76 m) typical of local urban transit vehicles. School bus transverse seats are typically spaced from 25 in. to 28 in. (0.64 m to 0.71 m) apart.

Accordingly, the area per seated passenger typically varies from 2.9 sq ft to 5.7 sq ft (0.27 sq m to 0.53 sq m), with from 3.1 sq ft to 4.2 sq ft (0.29 sq m to 0.39 sq m) typ-

TABLE 6.3
Passenger Capacities of Transit Vehicles, Expressed in Typical Ranges

Transit Vehicle Types	Per Single Unit			Per Maximum Train	
	Seats	Standees	Total Passengers*	Cars	Total Passengers*
Bus, freewheeling					
1. Minibus-midibus	15–25	10–20	25–45	—	—
2. City-suburban transit bus: non-articulated	30–55	10–50	45–100	—	—
3. City-suburban transit bus: articulated	35–75	30–120	100–170	—	—
4. City-suburban transit bus: double-deck	50–85	15–30	65–100	—	—
5. Intercity	30–50	10–25	40–70	—	—
6. School (children)	20–65	†	20–65	—	—
Rail, guided					
7. Tram: non-articulated	20–60	40–80	80–140	6	480–840
8. Tram: articulated	30–180	120–290	120–450	2–4	440–900
9. Urban-regional rapid transit: rail	40–85	50–250	100–330	10	1,000–3,300
10. Urban-regional rapid transit: rubber-tired	20–45	70–170	100–210	9	900–1,890
11. Commuter railroad: electric	100–130	50–125	100–240	15	1,500–3,600
12. Superregional rapid transit	30‡–110	0‡–50	30‡–160	18	1,300–2,900

*Higher portion of range shown represents crush capacity.
†No standees assumed.
‡Parlor (first class) car with no standees (minimum seats)
  Sources: Various databooks, inventories, reports, and specifications.

ical of local urban transit vehicles. School bus seated area per child is typically between 2.3 sq ft and 3.3 sq ft (0.21 sq m and 0.31 sq m).

Passenger seats are usually arranged transversely or longitudinally in relation to the transit vehicle body length. Transverse seating is usually preferred for passenger comfort, but vehicle narrowness, high planned proportions of standees, wheel housings, or other spatial factors often impose longitudinal seating in portions or the entirety of vehicles. Seats arranged at angles such as 45°, from the vehicle side are occasionally used because of spatial limitations and/or for passenger comfort. Pairs of transverse seats in transit vehicles occasionally are notched (longitudinally offset from each other by several inches) in order to save space or attempt to improve comfort. The dimensions of angled and notched seats generally fall within the ranges given above.

STANDEES, AISLES, DOORS

The area per standing passenger under near-crush or crush peak-period standing conditions typically falls within the range of from 1.3 sq ft to 1.6 sq ft (0.12 sq m to 0.15 sq m). The minimum area per standing passenger under easy standing conditions (such as may be used as a basis for peak-period service scheduling on some transit systems) varies typically between approximately 2.0 sq ft and 2.7 sq ft (0.19 sq m and 0.25 sq m).

Usually, minimum aisle widths on transit vehicles are between transversely positioned seats; here, aisle widths typically vary between 21 in. and 31 in. (0.53 m and

0.79 m). In order to discourage (but not prohibit, longitudinal passenger circulation between seat-door modules within one car, the aisle width may be between 17 in. and 21 in. (0.43 m and 0.53 m). Aisle widths between transversely-positioned seats in school buses vary between 11 in. and 16 in. (0.28 m and 0.41 m).

On most transit vehicles doorway clear widths per passenger lane (for boarding and alighting) typically vary between 22 in. and 30 in. (0.56 m and 0.76 m), with from 22 in. to 26 in. (0.56 m to 0.66 m) being typical of local urban transit vehicles. When there are two passenger lanes next to each other without a barrier or post between, the per-lane clear widths may sometimes be slightly less than for single-lane doors.

SEATED AND STANDING PASSENGER CAPACITIES

In order to compute the seated and standing passenger capacities of specific transit vehicles from the above data, the gross floor area, as defined for Table 6.1, may be taken as a point of departure. The thickness of the body shell and its related ducts, sash, and other appurtenances must be subtracted from the gross floor area; typically, from 5 percent to 10 percent of the gross floor area is subtracted. Floor areas occupied by motors and other related equipment must also be subtracted; modern transit vehicles, with the exception of some school buses, generally have such equipment installed below floor level. Any floor space occupied by a driver's and/or conductor's seat or cab and by related fare-collection equipment must be subtracted from the gross floor area; this space typically requires approximately from 2 percent to 6 percent of gross floor area per seat or cab, including compact fare-collection equipment. Special additional subtractive allowances have to be made for bulkier fare-collection equipment, such as turnstiles and automatic ticket-issuing and change-making machines, and for floor areas that must be made clear of standing and seated passengers, if such considerations apply. The resulting net floor area may then be apportioned by selected ratios between (and selected areal allowances for each of) standing and seated passengers.

TOTAL/SEATED PASSENGER RATIOS

Ratios of total (seated and standing) to seated passenger capacity on transit vehicles during peak periods vary widely because of both vehicle design and service scheduling policy. Few transit vehicles are incapable of accommodating standees if necessary. Because of space requirements for aisles, steps, and doors, usually at least 25 percent as many standees as seated passengers (total/seated ratio: 1.25) can be accommodated when seating space is maximized. At the other extreme, with minimal seating and crush standing, total/seated ratios of 7.00 or more may sometimes be obtained. Typically, however, city-suburban standard transit buses (single-deck, non-articulated) have crush total/seated capacity ratios in the range of from 1.70 to 2.00 and easy total/seated capacity ratios (frequently used for service scheduling purposes) approximating 1.50. Urban-regional rail rapid transit vehicles typically have crush total/seated capacity ratios in the range of from 2.00 to 6.00, with values in the lower half of this range typical of some recently constructed cars; easy total/seated capacity ratios (frequently used for service scheduling purposes) typically range from 1.50 to 3.50. Some types of transit service (particularly intercity, school bus, and superregional rapid transit) may frequently be scheduled for a 1.00 total/seated ratio during peak periods, with inevitable spatial and temporal maldistribution overloads accommodated either as standees or on other transit vehicles, depending on the policy in effect; a number of suburban services are similarly scheduled.

There are evident trade-offs between passenger comfort and ascending total/seated capacity ratios. As trip-lengths (especially in terms of travel time) increase with differ-

TABLE 6.4

Dimensional, Capacity, and Passenger Accommodation Characteristics Summarized for Three Specific Contemporary North American Urban Transit Vehicles of Different Types

Data Item [ft(m)] Unless Otherwise Specified	City-suburban transit bus: Non-articulated; Gen. Motors Corp; San Francisco Municipal System; 1970	Urban Rail Rapid Transit Car; Toronto Transit Commission; 1970	Regional Rail Rapid Transit Car; Bay Area Rapid Transit (BART) System, B-Type Car; 1969
Number of vehicles purchased	390	76	100
Length of body	40.0	74.5	70.0
	(12.2)	(22.7)	(21.3)
Width, maximum	8.5	10.4	10.5
	(2.6)	(3.2)	(3.2)
Height, wheels to roof, inclusive	10.0	11.3	10.5
	(3.0)	(3.4)	(3.2)
Floor area, gross [sq ft (sq m)]	340	775	750
	(31.6)	(72.0)	(69.7)
Weight, empty [lb (kg)] (000)	21.0	55.5	55.0
	(9.5)	(25.2)	(24.9)
Weight per linear foot (meter) [in lb per ft (kg per m)]	525 (781)	745 (1109)	791 (1177)
Weight per square foot (meter) [in lb per sq ft (kg per sq m)]	62 (303)	73 (356)	78 (381)
Seats	48*	83	72
Standees, easy capacity	29	139	60
Standees, crush capacity	47	217	156
Total passengers, easy capacity	77	222	132
Total passengers, crush capacity	95	300	228
Cars per train, maximum	—	6	10
Total passengers per maximum train			
Easy capacity	—	1,332	1,320
Crush capacity	—	1,800	2,280
Width of seats per passenger [in. (m)]	17 (0.43)	17.3 (0.44)	22 (0.56)
Spacing between backs of transverse seats [in. (m)]	28 (0.71)	26† (0.66)	34 (0.86)
Area per seated passenger [sq ft (sq m)]	3.3 (0.31)	3.1 (0.29)	5.2 (0.48)
Aisle width, minimum [in. (m)]	26.0 (0.66)	30.5 (0.77)	30.0 (0.76)
Doorway clear width, per passenger lane [in. (m)]	26.0 (0.66)	22.5 (0.57)	27.0 (0.69)
Doorways per side (times lanes per door)	2 × 1	4 × 2	2 × 2
Total/Seated passenger ratio			
Easy capacity	1.60*	2.67	1.83
Crush capacity	1.98*	3.61	3.17

*Maximum seating could be 53, but 5 seats were removed to permit more standees and hence higher peak total/seated capacity ratios.
†Estimated, based on seat plans; transverse seats not arranged in files.

Sources: Bus: General Motors Corporation. Rail cars: Institute for Rapid Transit, Rapid Transit Car Data Book Three, 1971.

ent types of transit service, vehicle accommodations and schedule policies tend generally to provide lower total/seated capacity ratios.

EXAMPLES OF SPECIFIC CONTEMPORARY VEHICLES

Table 6.4 summarizes dimensional, capacity, and passenger accommodation characteristics for three specific contemporary North American urban transit vehicles of different types. The 1970 San Francisco bus is typical of current urban transit buses in North America for the characteristics shown. The 1970 Toronto rapid transit car is typical of current city-suburban rail rapid transit practice on newer systems in North America. The San Francisco Bay Area Rapid Transit System (BART) rapid transit car, placed on order in 1969, is typical of current, modern, advanced regional rapid transit planning and design in North America.

VEHICLE PERFORMANCE

ACCELERATION, DECELERATION, AND JERK

Normal service maximum rates of acceleration, deceleration, and rate of change of acceleration and deceleration (jerk) in transit service must usually be related to the human tolerance of a standing passenger who is not able to hold onto any kind of hand grip. This condition frequently occurs on transit vehicles when passengers have both of their hands full with bundles, etc., and/or when such passengers cannot reach a hand grip or a hand grip is not available. The jerk rate (of change of acceleration or deceleration) is more critical to passenger comfort, but the actual rates of acceleration and deceleration themselves are also important.

Acceleration and deceleration rates of from 3.0 to 3.5 mph per sec (4.8 to 5.6 kph per sec) are usually considered appropriate upper limits for standing passengers not using stationary handholds. A preferred maximum jerk rate is 2.0 mph per sec per sec (3.2 kph per sec per sec) and an allowable maximum jerk rate is typically approximately 3.0 mph per sec per sec (4.8 kph per sec per sec).

Speed-time-distance curves indicating maximum acceleration, cruising, and deceleration rates for two different types of transit vehicle are shown in Figures 6.1 and 6.2. Figure 6.1 provides data for a typical standard city-suburban transit bus. Figure 6.2 provides the same data for San Francisco Bay Area Rapid Transit (BART) vehicles. Both figures represent operation with full seated loads on level traveled ways. Total loaded vehicle weight, grades, degree and type of moisture on the adhesion surfaces, wind speed and angle, air-conditioning power requirements, power transmission type, and other factors affect acceleration, cruising, and deceleration rates. Average daily maximum rates in normal service typically are from 5 percent to 25 percent less than maximum performance rates.

ADHESION

Typical values of adhesion for motor vehicles on highways are presented in Table 2.8. The range of adhesion values to be found in transit facilities is discussed in a report of the Venezuelan Ministry of Public Works:

> The value of rolling resistance for rubber tired wheels on dry, rough concrete is two
> to four times that of steel wheels on steel rails and has a value ranging from 0.35

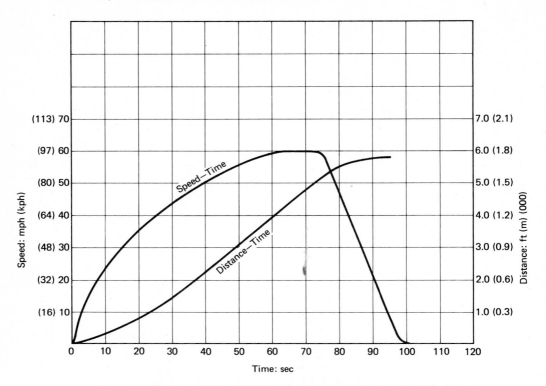

Figure 6.1. SPEED—TIME—DISTANCE PERFORMANCE CURVES. Typical Standard City-Suburban Transit Bus. Schematic representation for full seated load on level, tangent road; eight cylinder engine; VH-8V transmission. (Source: General Motors Corporation.)

to 0.65. . . . The adhesion coefficients . . . determined in connection with motor cars, trucks, and buses operating on highways . . . are not directly applicable to transit systems where service conditions are different. Transit service requires frequent use of high acceleration and deceleration rates in the same areas—such as leaving and entering stations where deposits of rubber or oil would adversely affect adhesion capability. It can be said, however, that the coefficients would be comparable as long as the surface remained clean and dry. In transit systems where these conditions prevail, such as the Paris and Montreal systems which are entirely underground, advantage can be taken of the superior dry adhesion characteristics of the rubber tire, within the limitations imposed by power supply and passenger comfort. . . . In many of the world's transit systems, however, a significant amount of the total system mileage is at grade or on aerial structure where the running surface is subject to wet and/or slippery conditions.[1]

The ordinarily accepted value of rolling friction for a steel wheel on steel rail is 0.18, but the available data on actual values for operating transit systems show a wide scatter, making it inappropriate to use a standard value for a particular system. Inasmuch as the determination of this value is of considerable importance to a transit system, the San Francisco Bay Area Rapid Transit District (BART) undertook a study of steel wheel adhesion under controlled conditions in order to assure a greater degree

---

[1] "A Comparative Study of Mass Rapid Transit Vehicles with Flanged Steel Wheels and Those with Pneumatic Rubber Tires," preliminary report, Venezuelan Ministry of Public Works, Ministerial Office of Transport, Caracas, January 1968, pp. 3–10 and 3–11.

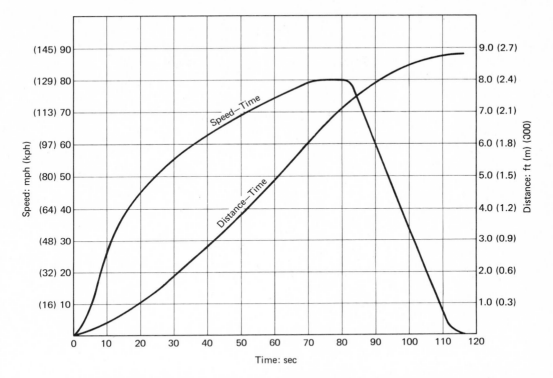

Figure 6.2. SPEED—TIME—DISTANCE PERFORMANCE CURVES. Four-Car San Francisco Bay Area Rapid Transit (BART) District Train. Schematic representation for full seated load on level, tangent track. (Source: Parsons, Brinckerhoff-Tudor Bechtel, General Engineering Consultants to BART.)

of accuracy than heretofore available. . . . Over the speed ranges contemplated for Caracas, average dry rail adhesion limits exceeded the rates required by the criteria traction systems studied. Wheel slippage did not occur at braking rates of 4 mph per sec (1.79 m per sec per sec) while spin-ups could not be produced at acceleration rates equivalent to 3.3 mph per sec (1.48 m per sec per sec). On both naturally and artificially wetted rail, accelerating adhesion values still exceeded the rates set forth in the criteria. In braking, however, minimum adhesion limits of 3.1 mph per sec (1.39 m per sec per sec) were obtained. On oily rail the adhesion limits dropped to slightly more than half the wet rail value.[2]

In addition to the types of materials used in the traveled way surface and wheels, the textures of the contact surfaces of these materials, the degree and condition of moisture (including possible slickness or slipperiness conditions when moist or wet), loaded vehicle weight, grades, and other factors influence adhesion characteristics.

VELOCITIES

Table 6.5 shows typical ranges of maximum and platform speeds for different types of transit vehicles and services. Maximum speed is typical maximum vehicle

---

[2] *Ibid.*, pp. 3–6 and 3–7.

TABLE 6.5
Vehicle Velocities

Transit Vehicle and Service Type	Typical Maximum Performance Speeds [mph (kph)]	Typical Platform Speeds [mph (kph)]
Local bus, urban	50–65 (80–105)	8–14 (13–23)
Limited-stop bus, urban	50–65 (80–105)	12–18 (19–29)
Express bus, urban	50–65 (80–105)	16–32 (26–51)
Intercity bus	60–75 (97–120)	25–55 (40–88)
School bus	50–70 (80–113)	10–20 (16–32)
Local tram, urban	40–60 (64–97)	8–15 (13–24)
Express tram, urban	50–65 (80–105)	15–35 (24–56)
Rapid transit, urban	50–70 (80–113)	15–35 (24–56)
Rapid transit, regional	70–85 (113–137)	35–55 (56–88)
Commuter railroad	70–100 (113–145)	25–65 (40–105)
Rapid transit, superregional	120–160 (193–257)	80–120 (129–193)

Sources: Various databooks, inventories, reports, and specifications.

performance capability with seated load on level roadways. Platform speed is typical normal service speed, including passenger stop and normal delay-recovery time but excluding terminal layovers, during specified periods. Schedule speed is platform speed plus the effect of terminal layovers and is typically from 5 percent to 15 percent slower than platform speed, unless terminal layovers are unusually long.

The following formula for approximating transit platform speed[3] may be used:

$$S = \frac{D}{T + \dfrac{D}{C} + C\left(\dfrac{1}{2a} + \dfrac{1}{2d}\right)}$$

where S = average transit vehicle speed,
T = stop time at stations or stops,
C = cruising speed,
a = rate of acceleration,
d = rate of deceleration,
D = average distance between stations.

"The preceding formula is only valid for cases where cruising speed is actually attained, and it assumes continuous acceleration and deceleration at the stated rates. If anything, this formula overstates average speeds."[4] Effective transit speeds are also subject

---

[3] R. L. CREIGHTON, *Urban Transportation Planning* (Urbana: University of Illinois Press, 1970), p. 111.
[4] *Ibid.*

to delays caused by congestion in mixed traffic, passenger volumes at stops and aboard transit vehicles, and other factors.

ENERGY-POWER REQUIREMENTS

In 1972, self-propelled buses in transit service in the United States consumed the following amounts of motor fuel, expressed as millions of gallons (liters): diesel oil, 247.3 (936.0); gasoline, 25.6 (96.9); propane, 24.4 (92.4); total, 297.3 (1,125.3). This consumption amounted to an average of 4.4 bus-miles per gallon (1.9 bus-kilometers per liter).[5]

In 1972, electrically-propelled vehicles in transit service in the United States consumed the following amounts of electrical energy (expressed as billions of kilowatt-hours): rapid transit, 2.149; tramways, 0.146; trolley coaches, 0.133; total 2.428. This consumption, in terms of kilowatt-hours per car-mile (car-kilometer), amounted to approximately 5.6 (3.5) for rapid transit, 4.6 (2.9) for tramways, and 4.5 (2.8) for trolley coaches.[6]

TRANSIT STOPS

The spacing, location, design, and operation of transit stops[7] (both intermediate line stops and terminal stops) have major effects on transit vehicle and system performance. Stops include stations.

Stop spacing is a primary determinant of transit schedule and platform speeds. Stop location and spacing also affect passenger door-to-door travel times and speeds.

SPACING

Typical ranges of stop spacings for different vehicle and service types are shown in Table 6.6. Because of its general tendency toward overall reduction in door-to-door transit travel times and toward specific increase in transit vehicle speeds and utilization, there is a trend in a number of metropolitan regions to increase the stop spacing distances on surface transit routes and in the planning of new rapid transit facilities.

Research has been conducted on optimum spacing of rapid transit stations. Figure 6.3 illustrates the type of result that will be sought with respect to passenger travel time.

Rapid transit systems sometimes employ skip-stop operation to accelerate transit service, especially when existing stops may be spaced relatively closely. Half the trains along a route section having skip-stop operation do not stop at one group of stations; the other half bypasses the other stations. All trains make important stops such as those in CBDs and major transfer or terminal points outside CBDs. Minimum possible train headways may be greater with skip-stop operation than without it. It is inconvenient for passengers who wish to travel between stations in different skip-stop groups, for they must transfer at a station where all trains stop.

---

[5] '72–'73 *Transit Fact Book* (Washington, D.C.: American Transit Association, 1971), pp. 15, 19.
[6] *Ibid.*
[7] Also see Chapter 15, Parking, Loading, and Terminal Facilities.

TABLE 6.6
Stop Spacings

Transit Vehicle and Service Type	CBDs	Non-CBD Typical Traditional Practice in North America and Elsewhere	Non-CBD Typical Contemporary Practice with Longer Stop Spacings
		Linear Spacing [ft (m)]	
Local bus, urban	400–800 (122–243)	500–800 (152–243)	1,000–1,500 (304–456)
Limited-stop bus, urban	400–800 (122–243)	1,200–3,000 (365–912)	2,000–5,000 (608–1,520)
Express bus, urban	500–1,000 (152–304)	4,000–30,000 (1,216–9,120)	1–30 mi (2–50 km)
Intercity bus	*	†	†
Local tram, urban	400–800 (122–243)	500–800 (152–243)	1,000–1,500 (304–456)
Express tram, urban	600–1,500 (182–457)	—	2,000–5,000 (608–1,520)
Rapid transit, urban	1,000–2,500 (304–608)	1,700–3,500 (517–1,064)	3,500–8,000 (1,064–2,432)
Rapid transit, regional	2,000–3,000 (608–912)	—	6,000–30,000 (1,824–9,120)
Commuter railroad	*	4,000–15,000 (1,216–4,560)	8,000–30,000 (2,432–9,120)
Rapid transit, superregional Super-express	*	—	50–150 mi (80–240 km)
Limited-express	*	—	10–50 mi (16–80 km)

*Usually stop at only one or two CBD terminal points.
†Widely variable, depending on route characteristics.

Sources: Various plans, schedules, reports, and other data.

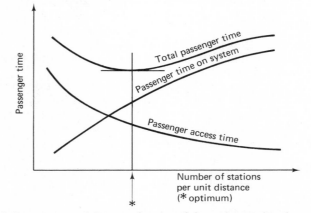

Figure 6.3. Passenger travel time as a function of the station spacing along a route. [Source: W. S. Homburger, ed., *Urban Mass Transit Planning* (Berkeley: Institute of Transportation and Traffic Engineering, University of California, 1967), p. 40, adapted from V. R. Vuchic, *Interstation Spacing for Line-Haul Passenger Transportation*, Graduate Report, University of California, 1966, Figure 11.]

LOCATION

Bus stops along streets are usually located along the street curb for direct, safe passenger access to and from the sidewalk waiting and walking area. Whether these stops should be located on the near or far side of intersections depends on a variety of factors.[8,9] Although bus stops are located at special turnouts along some freeway sections, their use may be limited because of the array of land uses within walking distance of these stops. Freeway bus stop turnouts may occur between or within interchange areas, the latter tending to be the case more often.

Tram stops along streets are usually located at special protected islands alongside the tracks if the track area is paved for motor vehicle use. If, much more preferably, the tram tracks along streets are located in a reserved center mall or median, then the stops are also located there. In order fully to utilize often limited available street cross-sectional width, the motor vehicle moving lanes alongside a tram center-street reservation may be eased over into the curb parking lane area in the vicinity of a tram stop, so that the stop platform alongside the track may occupy the space thus made available. Such stops should usually be located on either the near or far side of intersecting streets. If cross-street widths and peak volumes approach expressway proportions, consideration can be given to grade-separation of the tram tracks (and stop platforms), often by adjusting the profiles of both tram tracks and intersecting roadway and usually by taking advantage of the steep (up to 10 percent) grades possible with trams on short ramps.

Rapid transit, commuter railroad, and off-street express tramway stops (stations) should be located so that good access and circulation characteristics are provided for the area to be served. When stations serve major urban centers (such as CBDs) and subcenters, they should usually be located to maximize convenience of pedestrian delivery or access. Stations that depend on autos and buses for major portions of total passenger access should usually be located to optimize, insofar as possible, such vehicular access and general vehicular circulation in the immediate station vicinity and the service area of the station.[10]

BOARDING-ALIGHTING TIMES

Table 6.7 summarizes typical loading and unloading times per passenger for each separate door-lane for buses. These loading and unloading times would also apply to trams having steps.

For rapid transit trains and trams having level loading (i.e., not having steps between car floor and platform) an average of approximately 1.5 sec to 2.0 sec is required for each boarding and alighting passenger in each separate door-lane. If fare collection equipment is located directly at such point of entry, as on bus vehicles, then times will tend to approach the appropriate values shown in Table 6.7.

Pronounced or severe crowding inside or immediately outside the transit vehicle may markedly increase the boarding-alighting times cited in Table 6.7.

---

[8] W. S. Homberger, *Urban Mass Transit Planning* (Berkeley: Institute of Transportation and Traffic Engineering, University of California, 1967), pp. 34–35.

[9] *Highway Capacity Manual*, Special Report 87 (Washington, D.C.: Highway Research Board, 1965), p. 346.

[10] For further discussion of these aspects, see Henry D. Quinby, "Coordinated Highway-Transit Interchange Stations," *Origin and Destination: Methods and Evaluation*, Record No. 114 (Washington, D.C.: Highway Research Board, 1966), pp. 99–121.

TABLE 6.7
Bus Boarding and Alighting Intervals

Operation	Conditions	Time per Separate Door-lane per Passenger (Sec)
Loading	Single coin or token farebox	2.0–3.0
	Multiple-coin cash fares	3.0–4.0
	Multiple-zone fares; prepurchased tickets and registration on bus	4.0–6.0
	Multiple-zone fares; cash, including registration on bus	6.0–8.0
Unloading	Very little hand baggage and parcels; few transfers	1.5–2.5
	Moderate amount of hand baggage or many transfers	2.5–4.0
	Considerable baggage from racks (intercity runs)	4.0–6.0

Source: Adapted from *Highway Capacity Manual* (Washington, D.C.: Highway Research Board, 1965), p. 346.

VEHICLE DWELL TIMES

The standing time of a transit vehicle making a passenger service stop (or "dwell") comprises several different components. Time must be allowed for the opening and closing of doors (usually from 1 sec to 4 sec for each), for the alighting and boarding of passengers, and sometimes for additional components (such as waiting to start, maneuvering back into a moving lane for buses, or waiting for a bus stop area to be cleared for positioning the bus to stop).

Although passengers tend, especially in peak periods, to try to optimize their spatial distributions for boarding, alighting, and traveling on transit vehicles, imbalances or maldistributions in these functions, as between various available door-lanes and parts of vehicles (especially on trains) frequently occur. This imbalance factor has to be taken into consideration when computing dwell time requirements and/or threshold (boarding and alighting through doors) capacity potentials; an additional time allowance of about 5 percent to 15 percent during peak periods on trains having numerous doors is appropriate to account for this factor.

Human-mechanical interfacing on a transit system is most critical in the immediate vicinity of the vehicle doors at passenger stops. Human supervision of passenger boarding and alighting at all times, but especially during crowded peak conditions, is usually essential. Supervision may be in the form of an on-board crew member (a bus driver, train driver, or guard), assisted by mirrors, windows, and sometimes at some rapid transit stations by closed-circuit television. Inspectors or other platform or station personnel may also perform this function at heavy loading stops during peak periods and at other times and places.

BUS STOP CAPACITY

The number of buses that can be handled at curbside bus stops without unacceptably long queues (and associated waiting lines) being caused varies principally with the service [unloading and loading], time per bus and, to a lesser degree, with the number of loading positions. Additional loading spaces (or additional length of bus zones) increases the capacity, but at a decreasing rate as the number of spaces increases. . . . An acceptable rule of thumb might be to assume that the headways at a curbside bus stop (in minimum seconds of interval between vehicles) could be about twice the average service time per vehicle. Along any artery, the stop with the longest service time will be the bottleneck. The capacity of the artery itself could be increased by providing different bus stops for different routes, provided vehicles could overtake each other. . . . Headways can be approximately halved (frequency of service doubled) by providing alternate sets of bus stops far enough removed from each other so as not to cause interference in entering and leaving the loading zones . . . if exactly 50 percent of the buses are assigned to each set of stops, and if schedule reliability can be maintained. Of course, ample smoothly-operating stops help assure schedule reliability. However, it should be realized that in the case of the usual all-bus-lane operation, buses would be restricted to this lane, hence overtaking would be impossible and multiple stops would not be feasible.

Table 6.8 gives the minimum desirable lengths for bus curb-loading zones, for one and two-bus loading conditions.

TABLE 6.8
Minimum Desirable Lengths for Bus Curb Loading Zones*

Approx. Bus Seating Capacity	Approx. Bus Length (Ft)	Loading Zone Length† (Ft)					
		One-bus Stop			Two-bus Stop		
		Near Side‡	Far Side§	Mid-block	Near Side‡	Far Side§	Mid-block
30 and less	25	90	65	125	120	90	150
35	30	95	70	130	130	100	160
40–45	35	100	75	135	140	110	170
51–53	40	105	80	140	150	120	180

*Values shown as feet may be converted to meters by multiplying by 0.305.
†Measured from extension of building line or from an established stop line, whichever is appropriate. Based on side of bus positioned 1 ft from curb; if bus is as close as 0.5 ft from curb, 20 ft should be added to near-side stops, 15 ft to far-side stops, and 35 ft to mid-block stops.
‡Increase 15 ft where buses are required to make a right turn. If there is a heavy right-turn movement of other vehicles, near-side stop zone lengths should be increased 30 ft.
§Based on roadways 40 ft wide, which enable buses to leave the loading zone without passing over center line of street. Increase 15 ft if roadway is 36 ft wide and 30 ft if roadway is 32 ft wide.

Source: American Transit Association as reported in *Highway Capacity Manual*, p. 348.

Bus loading zones on an exclusive roadway within a freeway right-of-way have capacities similar to those of curbside loading zones. Here again, the length of the stop and the ability of buses to overtake others are important. Given similar loading facilities, any difference does not lie in the operation of the stop itself, but in the capacity of the roadway lane leading into and away from the stop.[11]

FARE SYSTEMS

A detailed treatment of transit fare structures, collection methods, and levels is beyond the scope of this discussion, but they are briefly treated below and the reader is referred to appropriate sources within the References for Further Reading at the end of this chapter.

---

[11] *Highway Capacity Manual*, pp. 347–48.

FARE STRUCTURES

In general, transit fares may be flat throughout a whole route network, zoned between different geographical portions of a route network, or graduated by distance traveled; or no fare at all may be charged, at least to the passenger. Negative fares, that is, paying people to ride transit, do not appear ever to have been tried, even on an experimental basis, but this concept at least deserves theoretical consideration.

A flat fare throughout a whole route network, or on a central major part of it, is self-descriptive. It tends to favor the longer passenger trip and penalize the shorter one, but it is simple to collect and to understand. It is nearly ubiquitous on North American city transit systems.

There are numerous kinds of zone fares. Zone fares fall between flat and graduated fares in their ability to relate length of trip with fare paid. A fine-grained zone-fare system approaches the equity of a distance-graduated fare, but it is usually cumbersome and expensive to administer with manual fare collection. Zones are often used for long route-sections extending beyond the usually large central area of a flat-fare system. Or an entire route network, as in Houston, Texas, can be divided by a series of concentric rings (representing fare-zone boundaries) centered on the CBD. Figure 6.4 illustrates the Houston zone-fare system. Such a zone-fare system may increase fare revenue, but may also slow operations and service because of additional fare or "hat-check" lifting at some zone boundaries, and it may also seem complex to the infrequent rider. In Europe, the purchase of season tickets good for one or more months and valid on all or only on a certain portion of a route network (as colored in on a diagrammatic network map on the ticket) permits the equity of zone (or stage) fares and the fast and convenient "flash-pass" of the ticket to the driver or conductor upon boarding or alighting. Stages (or route-segments) may be used as zones, as in London. A disadvantage of zone fares is that the fare is often relatively very high for short trips that happen to traverse a zone boundary.

Fares graduated by distance traveled provide the greatest equity to the passenger if distance traveled is the criterion. To date, there has not been a fully graduated fare system that is practical for continuous operations on one-man surface transit vehicles in which manual fare collection methods are used without the honor-inspector system described below. In order to measure distance traveled (and hence compute the fare), the passenger must be checked both at the beginning and end of his trip. This procedure is cumbersome, time-consuming, and expensive on surface transit vehicles with limited floor space, such as buses. If, however, either the season ticket (described above) and/or the honor-inspector system (described below) is used, a graduated fare can be applied to surface transit operations. Graduated fares find their most widespread current application, however, on some of the newer rapid transit systems, such as BART, and a few pioneering commuter railroads, such as the Illinois Central in Chicago. Passengers enter and leave "paid" areas in the stations of such systems by inserting magnetically encoded tickets in automatic turnstiles. The tickets are checked and the required fare is automatically subtracted.

Paper transfers may be issued between transit routes, with or without extra charge, usually at the time when the fare is paid. Transfers may also be dispensed before exiting at rapid transit stations for use on connecting routes. Some station layouts permit internal walking between different routes without additional payment or transfer issuance.

Various promotional and incentive schemes to increase transit patronage may be coordinated with the structure, collection, and level of transit fares. There is increas-

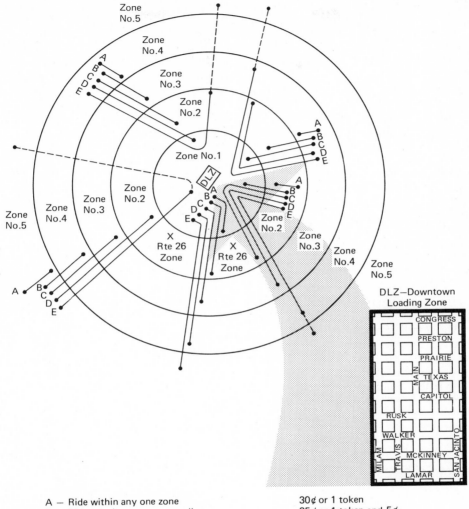

A — Ride within any one zone   30¢ or 1 token
B — Ride crossing any one zone line   35¢ or 1 token and 5¢
C — Ride crossing any two zone lines   40¢ or 1 token and 10¢
D — Ride crossing any three zone lines   45¢ or 1 token and 15¢
E — Ride crossing any four zone lines   50¢ or 1 token and 20¢
      — — — (No additional fare after crossing four zone lines)
An additional 5¢ deposit is required for transfers issued to adult passengers.
Students and children are not subject to zone fare charges or transfer deposits.

Figure 6.4. Zone Fare System Using Concentric Rings, Houston, Texas Rapid Transit Lines, Inc. Adult fares and zone fare system examples. (Source: Rapid Transit Lines, Inc., Houston.)

ing interest in common tariffs and transfer privileges, for example, between the routes of different transit systems within one metropolitan region. The Hamburger Verkehrsverbund (HVV), or Federation of Transit Systems, in the Hamburg, Germany urban region is an example. The HVV coordinates services, fares, and operating financial results for virtually all public transport within the Hamburg region.

Special fares may apply for regular commuters, children, students, senior citizens, public employees, and other special groups.

FARE COLLECTION

Fares may be paid in cash, tickets, or tokens, by flashing a prepaid pass or season ticket, by a ticket that is cancelled on board the vehicle, or by magnetically-encoded tickets as described above. Most North American surface transit systems now require the payment of exact fares, without change from the driver. The exact-fare system increases schedule and platform speed, reduces the driver's work, reduces travel time, and (the primary purpose of its original introduction) practically eliminates thefts of drivers' cash; however it represents some inconvenience to passengers. Bulk sale of tickets or tokens, often off-vehicle, partly alleviates this inconvenience.

Generally, fare collection systems may be divided into manual and automatic methods. Human sight-checking and handling are manual methods; at least substantial mechanical and/or electronic handling are automatic methods. There is a great variety of fare collection systems in each of these two methods.

There is a pronounced worldwide trend toward greater proportions of prepayment and automatic payment of fares, and away from on-board or other manual fare payment. In addition, there is also a trend, in Europe and other areas, toward greater use of crewless vehicles, for example, second and following cars on tram-trains, and toward transit stations without agents at least at offpeak times, coupled with the honor-inspector system.

The honor-inspector system permits the passenger to ride without having his fare payment checked except by roving inspectors at random times. Underpayment or nonpayment, when discovered by an inspector, subjects the offender to a substantial fine (and embarrassment) on the spot or, if there is failure to pay the on-the-spot fine, to a possibly much greater penalty in subsequent proceedings. The practicality of the honor-inspector system varies with different social environments.

FARE LEVELS

Fare levels refer to general or specific service pricing levels. They are usually based on policies and resulting calculations as determined in one or more of the following areas: (1) needs for patronage attraction, (2) competition from other modes, (3) notions of equity for the entire ridership or of relative equities for various subgroups of it, (4) needs for determined levels of operational and financial performance, (5) the different costs of providing service on different kinds of routes, and (6) constraints that may be imposed by the selected fare structure and methods of fare collection. The discussion in a subsequent section of the chapter, Characteristics of Transit Economy, includes consideration of fare levels in relation to other aspects of transit economy.

SYSTEM CHARACTERISTICS

Preceding sections of this chapter are concerned with the capabilities and characteristics of rolling (vehicle) and fixed facilities which are component parts of physical transit systems. This and following sections are concerned with transit systems rather than their component physical parts.

KINDS OF ROUTE

Transit systems involve several kinds of routes or route functions as described below. Tables 6.5 and 6.6 are relevant to this discussion of route types.

Local routes. Most frequently encountered are local routes operating along the surface of city streets with stops spaced approximately one or two blocks apart and serving entirely within one urban area. Such routes typically radiate from the CBD and may either terminate there or, usually (but not always) preferably, be routed through the CBD. Other local routes that do not reach CBDs may serve as crosstown or feeder-shuttle lines. Local routes usually provide the basic and preponderant transit service in an urban area.

Limited and express routes. When local routes exceed typically about 3 mi (5 km) in length from CBDs and when patronage volumes thereon are great enough (especially of longer-than-average trip-lengths), limited-stop and/or express service may complement local service along and/or beyond the extent of the local route. Often, such limited or express service warrants operation only during peak-volume periods. Sometimes route characteristics are such that only a limited or express service is operated and then often throughout the day, e.g., between a major airport and a CBD or between two major centers within one urban region. Limited-stop service usually involves operation along city streets with stops mainly at major transfer points beyond the CBD. Express service involves even faster operation with fewer stops than limited service and often on freeways or other major highways parallel to a complementary local route or routes in the same travel corridor.

Frequently, a radial limited-stop or express route will continue beyond a transitional point [which may typically be about 2 mi to 6 mi (3 km to 10 km) from the CBD] as a local route. In this pattern, a local transit route serves the intermediate area between the CBD and the transitional point, terminating at (or just beyond) the transitional point. Thus, in this urban transit network configuration, there are two (and sometimes more) "rings" of local service, the inner ring being served directly by local routes and the outer ring (or rings) being served locally by other routes which become limited or express inside the transitional point or area. These limited and express routes are termed "neighborhood" limited and express routes. Limited and express routes that do not provide local service outside CBDs on portions of their routes are termed "trunk" limited and express routes. This layering of local service in conjunction with limited and express service is highly developed in cities such as Cleveland, Ohio, and in the San Francisco Peninsula suburban service of Greyhound Lines.

Rapid transit lines. Grade-separated urban rapid transit lines generally serve the limited and/or express, rather than local, functions described above. Many urban areas having rapid transit systems arrange these systems to perform mainly limited and/or express functions, while surface transit facilities such as buses perform mainly local and feeder functions complementary to rapid transit. This arrangement usually involves a relatively marked degree of transferring, and hence time loss, between rapid and surface transit vehicles. In corridors where no grade-separated rapid transit exists, surface transit may perform limited and express functions as well, although often with reduced speed in congested mixed traffic during peak periods.

Tramways. There is increased interest in high-capacity tramways which, although not completely grade-separated, operate on reserved trackage such as in street and highway medians. With stop spacings beyond CBDs of from 1,000 ft to 1,500 ft (300 m to 450 m) and with complementary application of the range of traffic engineering techniques, tramways can provide both local and limited-stop or express functions in a single fast service. Platform speeds and track capacities can approach or equal those of fully grade-separated traditional urban rapid transit facilities. More direct service (with branch or parallel lines) can often be provided with generally less transferring; stops may be accessibly located on the street surface and still be spaced reasonably closely together. Thus, overall door-to-door travel times for urban trips up to approximately 8 mi to 15 mi (13 to 24 km) in length (depending on speed and other factors) may be lower than or in the same range as those fully grade-separated urban rapid transit facilities having relatively frequent stops and substantial amounts of transferring and station access by other modes.

Exclusive and priority treatment. The speed, capacity, and reliability of limited-stop and express buses can be enhanced by a variety of techniques such as (1) exclusive or preferential bus lanes on sections of streets and highways including freeways, (2) metered freeway entry with bus preference, (3) exclusive bus streets, malls, and ramps, (4) transit-actuated features in traffic signals, and (5) exclusive busways as previously discussed (pp. 210–211). Frequently the operation of exclusive transit lanes and other techniques just mentioned is limited to peak periods of relatively high transit patronage volumes.

If street and highway space is being considered for the exclusive use of transit vehicles, one test of reasonableness which may be employed is whether or not the combined auto and transit passenger volumes in the direction(s) and for the period(s) involved are at least equal to such volumes without the exclusive transit feature proposed. A similar test sometimes considered is whether or not total travel or delay times for the route section(s) and time period(s) involved are reduced.

Secondary distribution. Secondary distribution routes are increasing in importance; they include internal service within CBDs, airport, and university complexes, government centers, and other relatively large centers and subcenters; they may also provide access to these centers from nearby parking lots, transit stations, and other transportation terminals. Services may take the form of regular buses, minibuses, perhaps some form of elephant train, and the like. Special mechanical secondary distribution facilities, such as speed-walks, carveyors, monorails, and others, must be physically separated from street traffic, which may mean placing them underground if continuous overhead structures, especially in downtown areas, are unacceptable. Although the continuous operation of speed-walks and other people-movers is an attractive feature, it is usually offset by relatively slow speed and frequently by limited capacity and accessibility. Higher capacities usually involve relatively substantial facility widths and/or larger units with batch loading rather than continuous loading. Mechanical facilities prove most useful when their capabilities can be effectively realized and when they can be planned and designed into new developments.

Variable-route-and-schedule service. In addition to fixed routes and schedules, free-wheeling vehicles such as buses can be operated, within time and space limits, on routes and schedules partly or completely responsive to transitory changes in passenger demand. A bus operating on a fixed route and schedule may, for example, be permitted to deviate from that route up to certain distances (one-quarter mile or more) in order to pick up or deliver passengers and still operate within the general framework of the route and schedule involved on that bus trip. This is sometimes referred to as flexibly routed service.

Still more freedom from route and schedule constraints may be introduced by "dial-a-ride" service wherein routes and schedules are determined by actual demand patterns at one (usually the residential) end of passengers' trips or (with still greater freedom) at both ends of such trips. Dial-a-ride service approaches the characteristics of standard on-call taxi service and differs from it mainly in that several passengers (and their different origins and/or destinations) are served on the same general dial-a-ride cycle or round trip, and often with a vehicle larger than a taxi.

Either manual or computer dispatching, or both, are required to translate passengers' phone calls for service into efficient temporal and spatial distribution of that service. If fares are kept within or near the framework of customary transit fares, flexibly routed, and particularly dial-a-ride, services are likely to be expensive in relation to revenues generated.

Other transit routes. Other types of transit routes include special routes operated only at peak periods to major generators such as spectator facilities, large employment concentrations outside CBDs, universities, etc.; intercity longer-distance routes; special subscription services; and sightseeing and other special routes.

ROUTE SYSTEMS

Many factors influence the overall transit route network configuration in an urban area. The configuration is defined in terms of time (periods and schedules of service), space (routes, stops, and fixed-facilities geography), mode (the various kinds of transit facilities and services), and other characteristics. The scope of this discussion does not permit more than a listing of influencing factors, but a number of them are treated in other sections and subsections of this chapter. Such factors include:

1. The overall system service area
2. The geography of its land uses
3. Location of major generators by type and transit attractiveness
4. Streets and highways available for transit use
5. Available exclusive rights-of-way
6. Transit mode or modes available for use
7. Capacities of systems and vehicles
8. Interfacing with both macro-transportation and micro-transportation systems
9. Patronage volumes and patterns in time and space
10. Service areal coverage
11. Service time schedules
12. Operating and economic feasibility (including any fare subsidy policies contemplated or in effect)
13. The manner of utilization of various transit routing techniques, including those mentioned above
14. Stop spacing and location
15. Proper mixes of route types

16. Physical transfer policies
17. Through-routing
18. Terminal locations and functions
19. Short-turning of transit vehicle trips (e.g., turning back some vehicle or train trips short of the outermost terminal of a route)
20. The use made of branch lines and feeder lines

This listing is not exhaustive but does cover major considerations in the evolution of transit route systems in urban areas.

SERVICE AREA COVERAGE

. In order to reach an urban transit stop or station, patrons will typically travel up to the distance maxima shown in Table 6.9. A relatively few patrons may travel considerably farther than the distances shown, but they are not considered typical. The distance patrons will travel to and from transit stops or stations is affected by congestion, speed, weather, comfort, the location and type of stop or station involved, available alternatives, and other factors. The combined transit stop access time at both ends of a single trip may exceed the time spent on the transit vehicle itself, especially if the latter is relatively fast and involves no transfers.

TABLE 6.9
Typical Maximum Distance Traveled to Reach
Urban Transit Stops and Stations

Access Mode	Most Patrons [mi (km)]	Some Patrons [mi (km)]
Walk	0.4–0.6	0.6–1.0
	(0.6–1.0)	(1.0–1.6)
Bicycle	1.0–2.0	2.0–3.0
	(1.6–3.2)	(3.2–4.8)
Feeder transit;	2.0–4.0	4.0–8.0
motor-bicycle	(3.2–6.4)	(6.4–13.0)
Auto: kiss-ride;	3.0–4.0	4.0–6.0
taxi	(4.8–6.4)	(6.4–9.7)
Auto: parked	4.0–6.0	6.0–10.0
at station	(6.4–9.7)	(9.7–16.0)

Source: Henry D. Quinby, "Coordinated Highway-Transit Interchange Stations", *Origin and Destination: Methods and Evaluation*, Highway Research Record 114 (Washington, D. C.: Highway Research Board, 1966), pp. 99–121; and various other reference data.

The concept is sometimes advanced that different kinds of transit facilities and services require minimum areal population densities for their justification. Rather than average population densities in each urban area or corridor involved, a more significant test is the relation between estimated or measured volume-pattern travel demands and the transportation capacities planned or available to meet these demands. For this purpose, each of the most critical points or gateways in each of the identified travel corridors of the urban area, especially during peak periods and peak directions of travel, should be considered separately. Although there may or may not be significant correlations between transit usage and population density in a given area or corridor, the principal test of value is the ability of the transit facility and service in question to attract enough corridor traffic of sufficient length in order to minimize effectively the total transportation facilities and costs required to meet total peak corridor demands. An example is the volume of passenger traffic accumulated on Cleveland's West Side rapid transit route section across the critical Cuyahoga River gateway to downtown Cleveland. A large share of the volume is generated by outlying suburban and exurban areas with low population densities; this traffic reaches the rapid transit stations by various access modes, principally park-ride, kiss-ride, and feeder transit.

SERVICE SCHEDULING

Many factors enter into the preparation of schedules or timetables of transit service. Important in service scheduling are the characteristics described elsewhere in this chapter. In addition, the following factors are of special importance:

1. Loading criteria (including the ratio of total to seated passengers and the floor area allowance for each scheduled standee, if any)
2. Policy headways for periods (usually offpeak) when headways are not determined solely or at all by patronage volumes
3. The various characteristics of actual or estimated patronage (in time, space, and other stratifications)
4. The time required to complete a full cycle of one vehicle or train over a given route, including terminal layover allowances for delay-recovery time, crew rest and comfort time, and headway matchup time
5. The variations which occur in transit vehicle travel time during various times of day, especially for vehicles operating in or across other vehicular traffic
6. Policies of convenience and/or minimization in physical transferring between routes and vehicles
7. Schedule coordination of different routes (especially in common throat or trunk sections)
8. Schedule symmetry (including regular intervals) to enhance memorization
9. Reverse-peak service requirements
10. Opportunities for short-turning
11. The necessity and/or desirability of closed-door operations (i.e., route-sections where people are not permitted to board and/or alight)
12. Opportunities for successful electronic computerization of various portions or substantially all of the scheduling process
13. Other related factors

Transit scheduling involves separate but closely related preparation of schedules of the service itself (timetables), the rolling equipment to furnish that service, and the crews to operate the rolling equipment involved. Transit scheduling also has to consider the possibilities for loading imbalances, for example, between different vehicle doors and different parts of vehicles, especially in train operation. Furthermore, not only must the actual or estimated patronage demands (outside of policy headway periods) by day of week and direction of travel be determined past the maximum (and possibly other critical) load points on each route, but these volumes should also be available in 15-minute, 20-minute, or, at most, 30-minute increments at least during peak travel periods. Even when these data are available for periods not controlled by policy headways, consideration has to be given to the phenomena of surging or cresting which can occur within the time increments used. Offpeak transit scheduling, including scheduling based on policy headways, should take into consideration not only the possibilities for fluctuation and surging in patronage demands that can occur in these periods but also the practical means of coping with such demand variations.

SYSTEM PERFORMANCE

HEADWAYS

Transit vehicle or train headways, including minimum headways, are affected by (1) patronage, (2) vehicle, (3) geometric, (4) system, (5) performance, (6) capacity, (7) stops, (8) policy, and (9) other characteristics discussed elsewhere in this chapter. In addition and related thereto are (10) the dwell time required at a stop, (11) the number of independently accessible berths available at a stop, (12) the amounts and behavior of any other kinds of traffic that may be mixed in operation in and across the same lane or track, (13) the minimum safe stopping distance (including a specified safety margin) required, at the most critical point along the route-section in question, between the rear end of an instantaneously-stopped lead vehicle or train and the front end of its immediate follower operating in the same lane or track at the specified maximum speed and under the assumed minimum adhesion and climatic conditions, (14) stop spacing, (15) assumed acceleration and deceleration characteristics, (16) maximum vehicle or train length involved, and (17) other factors important in determining minimum headways.

Table 6.10 provides typical ranges of minimum practical average headways and resulting numbers of units or trains per hour encountered for various transit vehicles and facilities. Qualifications to the application of data in Table 6.10 are set forth therein. Because of the numerous factors affecting these headways, variations from and within the ranges shown are possible under specific individual conditions. Therefore, the values in Table 6.10 are not generally the absolute minimum headways that might be attainable under the most stringent conditions if other factors, such as stop dwell time and effects on other vehicles in mixed traffic, were severely constrained in favor of minimizing such transit headways. The ranges shown reflect varying platform and top operating speeds, varying requirements of other traffic when the transit operation involves mixture with such other traffic, and other factors mentioned above.

The conditions at stop areas, rather than line conditions, almost always determine the minimum headway along a given route-section of a single vehicle-facility-type. Factors that determine minimum headway under either stop or line conditions are set forth above. The highest minimum headway value for a given homogeneous route-section represents the most critical headway situation therein and determines the actual minimum headway value for the whole of such route-section. Headways in a given route-section may be affected by limiting conditions upstream or downstream from that route-section.

VELOCITIES

Table 6.5 shows typical maximum performance speeds and typical peak-period platform speeds for selected vehicle and service types. Platform speeds and schedule speeds are of greatest interest to transit system performance. Platform speeds are a

TABLE 6.10
Typical Minimum Average Headways, and Resulting Units or Trains per Hour,
per Moving Lane or Track in one Direction of Travel
(during the peak 15 min of the peak hour)[1]

Vehicle and Facility Type	Passenger Stops Along or Affecting Route Section Involved	Headways (in sec)	Resulting Nos. of Units or Trains per Hour
1. Buses in mixed street traffic	Yes	40–60	90–60
2. Buses in exclusive street lane	Yes	30–50	120–72
3. Buses in mixed freeway traffic	Yes	40–60	90–60
4. Buses in mixed freeway traffic	No[7,9]	10–30	360–120
5. Buses in exclusive busway	Yes[8]	25–35	144–103
6. Buses in exclusive busway	No[7]	10–30	360–120
7. Trams in mixed street traffic[2]	Yes	25–40	144–90
8. Trams in mixed street traffic[3,10]	Yes	50–60	72–60
9. Trams in reserved street medians[2]	Yes	25–40	144–90
10. Trams in reserved street medians[3]	Yes	50–60	72–60
11. Trams in exclusive right-of-way[2,4]	Yes	20–40	180–90
12. Trams in exclusive right-of-way[3,5,11]	Yes	40–120	90–30
13. Trams in exclusive right-of-way[2,4,12]	No[7]	10–30	360–120
14. Trams in exclusive right-of-way[3,5]	No[7]	60–80	60–45
15. Urban-regional rapid transit[6]	Yes	90–120	40–30
16. Urban-regional rapid transit[6]	No[7]	60–80	60–45
17. Commuter railroad[6]	Yes	120–240	30–15
18. Superregional rapid transit[6]	Yes	240–360	15–10

[1]For discussion of the content, application, and limitations of the values in this table, see the subsections on Headways and on Capacities in this chapter.

[2]Individual trams and tram-trains up to about 140 ft (43 m) in length of consist. In route-sections with stops, average headway values below about 40 sec (about 90 units or trains per hour) should not be used when individual consists are about 110 to 140 ft (34 to 43 m) in length.

[3]Tram-trains between about 140 and 280 ft (43 and 85 m) in length of consist. In route-sections with stops, average headway values below about 55 sec (about 65 trains per hour) should not be used when individual consists are about 230 to 280 ft (70 to 85 m) in length.

[4]Top speeds up to about 35 mph (56 kph).

[5]Top speeds over about 35 mph (56 kph).

[6]In exclusive right of way.

[7]The very extensive, expensive, and often cumbersome major-center terminal(s) or station(s) required, especially for the lower portions of the headway (and the higher portions of the "numbers of units or trains") value ranges shown for this condition, may make such portions of such ranges impractical or unattainable in specific circumstances.

[8]Average headway values below about 30 sec (about 120 units per hour) should not be used for articulated buses.

[9]Values in the lower portions of the headway (and the higher portions of the "numbers of units") value ranges shown for this condition involve very heavy preemption of available freeway lane capacity by bus vehicles and, therefore, may not be practical or attainable in specific circumstances.

[10]Operation of tram-trains of from 140 to 280 ft (43 to 85 m) in length or longer, in mixed street traffic, involves very heavy preemption of available street lane capacity and, therefore, may be possible only when such preemption is deemed acceptable and practical.

[11]Headway and resulting number of units or trains per hour are dependent principally on consist lengths, maximum line speeds, and performance in station areas and approaches, as well as on other factors.

[12]In route sections without stops, average headway values below about 20 sec (about 180 trains per hour) should not be used when individual consists are about 110 to 140 ft (34 to 43 m) in length.

Sources: Various manuals, reports, and observations.

function principally of stop dwell times, stop spacing, vehicular and pedestrian conflicts in mixed traffic situations, route geometry (including alignment and grades), rates of acceleration and deceleration, maximum allowable operating speeds, headways, and other factors. Transit schedule speeds are affected, in addition, by terminal layover durations required for crew rest and comfort, and for awaiting the scheduled moment of departure if this occurs after the normal crew rest and comfort interval.

The platform and schedule speeds of a transit system (and subdivisions thereof such as routes and route-sections) play a critical role in the operating economy of the system. Over the years there have been numerous cases in which reductions in the surface transit system schedule speed, caused by increased traffic congestion, have contributed significantly to worsening the economy of the system, sometimes to the point where financial performance has been jeopardized or severely impaired. A difference of 1 or 2 mph (1.6 or 3.2 kph) in system schedule speed of a labor-intensive surface transit system can make a significant difference in the system's financial and operating performance. A discussion of relevant characteristics of transit economy, including data on system speeds, is provided later in this chapter.

TRANSIT TRAVEL TIMES

The overall transit travel time envelope for a given passenger trip or trips may be thought of as the door-to-door travel time, i.e., all components of a transit passenger's travel from actual origin to actual destination on a single trip. Within door-to-door travel time there are: (1) access time from point of trip origin to the boarding transit stop, (2) waiting time for the transit vehicle, (3) travel time on the transit vehicle, (4) transfer time (waiting and possibly walking) required if more than one transit line is used for a single trip, and (3) access time from final alighting stop to point of destination. Time must also be allowed for walking through transit stations and terminals, and for using their escalators and stairs in order to reach desired boarding points along the loading platforms. Change-of-mode in these stations also involves access time from the point of leaving the arrival mode to the selected platform boarding point and vice versa.

Access travel to and from transit stops and stations may be by one or a combination of several access modes, each of which has different travel time characteristics: (1) walking; (2) driving or as a passenger in an automobile parked at or near the transit stop; (3) being driven to and dropped off at a transit stop with the vehicle involved not being parked there, and vice versa (frequently referred to as "kiss-ride"); (4) taxi; (5) feeder transit; (6) bicycle; (7) motorcycle; and (8) U-drive. Travel on the transit vehicle itself involves time increments for: (1) accelerations, (2) cruising, (3) decelerations, (4) dwell time for each intervening stop between the passenger's boarding and alighting stops, and (5) any delay times en route (which may depress normal running service performance and/or add special delay stops).

Transit operations in mixed vehicular traffic are particularly susceptible to delay time. Moreover, they are likely to have relatively high proportions of acceleration, deceleration, and stop or dwell time, particularly when stops are relatively close together. Conversely, longer-haul regional and superregional rapid transit services usually have higher proportions of cruising time and lower proportions of acceleration, deceleration, and dwell time; delays on these grade-separated services are usually minimal.

Table 6.11 classifies transit travel time spent en route, based on 600 surveyed trips in the Providence, Rhode Island area. Woonsocket and Pawtucket are outside the most central part of the Providence metropolitan area; traffic delays are smaller as a percentage of total trip time for transit lines with origins in these outer areas than in the case of lines with origins in the more congested central part of the urban area, that is, in Providence itself. Generally, loading times and traffic delays involved higher

TABLE 6.11
Classification of Transit Time Spent En Route, Providence, Rhode Island
(percent of total trip time, excluding layovers)

Lines with Origins in	Running		Loading		Traffic Delays	
	Out-bound	In-bound	Out-bound	In-bound	Out-bound	In-bound
Woonsocket						
Off-peak	68%	78%	20%	13%	12%	9%
Peak	70	71	19	22	11	7
Pawtucket						
Off-peak	67	65	19	23	14	12
Peak	61	59	24	26	15	15
Providence						
Off-peak	67	67	17	20	16	13
Peak	59	59	23	26	18	15

Source: *Trade, Transit and Traffic*, Providence, Rhode Island Study, Wilbur Smith & Associates, 1958.

percentages during peak periods than during off-peak times. The same study indicated that traffic signals, bus turning movements, and general traffic congestion were primary causes of traffic delays while intersectional conflicts and vehicular parking were secondary causes of delay.

Table 6.12 classifies transit delays in terms of traffic signals, stop signs, other traffic stops, and passenger stops, the latter being the largest component of delay, followed by traffic signals.

TABLE 6.12
Classification of Transit Delays, St. Louis Public Service Co.

Factor	Delay Time as Percent of Total Delay Time	Delay Time as Percent of Total Trip Time
Traffic delays		
Traffic signals	30.7%	9.1%
Stop signs	1.9	0.6
Other traffic stops	7.2	2.1
Total	39.8%	11.8%
Passenger stops	60.2	17.9
Total, all delays	100.0%	29.7%

Source: *St. Louis Metropolitan Area Transportation Survey Report*, W. C. Gilman & Company, Engineers, 1959.

Table 6.13 provides examples of the influence of access modes, times, and distances (to and from transit stops or stations) on the overall door-to-door travel speeds and times (from origin to destination) by transit. Various illustrative transit line-haul distances and platform speeds are considered in this table. It is of interest to note, for example, that if the transit line-haul distance is 8.0 miles (12.9 km) and if the platform speed is typically 20 mph (32 kph), then the actual door-to-door travel speed varies from 10.8 mph (17.4 kph) to 16.8 mph (27.0 kph), depending on the residential access mode and access distance chosen from the seven typical examples given. Toward the opposite end of the spectrum, if the transit line-haul distance is 40.0 miles (64.4 km)

TABLE 6.13
Influence of Access Modes, Times, and Distances on Overall Door-to-door Travel Speeds and Times (from Origin to Destination) by Transit[1]

Distance mi km Platform speed mph kph		Illustrative Transit Line-haul					
4.0							
6.4							
		4.0	8.0	12.0	16.0	20.0	40.0
		6.4	12.9	19.3	25.7	32.2	64.4
		10	20	30	40	50	60
		16	32	48	64	80	97
Typical Examples[2] of Residential Access	Value						
1. Walk 1,000 ft (305 m)[3]	SM	7.0	13.4	19.8	26.2	32.6	45.3
	SK	11.3	21.6	31.8	42.1	52.4	72.9
	TM	37.5	37.5	37.5	37.5	37.5	53.5
2. Walk 2,000 ft (610 m)[3]	SM	6.4	12.0	17.5	23.1	28.7	41.3
	SK	10.3	19.3	28.2	37.2	46.2	66.4
	TM	43.0	43.0	43.0	43.0	43.0	59.0
3. Walk 3,000 ft (914 m)[3]	SM	5.9	10.8	15.7	20.6	25.6	37.7
	SK	9.5	17.4	25.3	33.1	41.2	60.7
	TM	48.7	48.7	48.7	48.7	48.7	64.7
4. Bicycle 2.0 mi (3.2 km)[4]	SM	7.9	13.0	18.1	23.2	28.3	40.2
	SK	12.7	20.9	29.1	37.3	45.5	64.7
	TM	47.0	47.0	47.0	47.0	47.0	63.0
5. Feeder transit 2.0 mi (3.2 km)[5]	SM	7.7	12.7	17.7	22.7	27.7	39.6
	SK	12.4	20.4	28.5	36.5	44.6	63.7
	TM	48.0	48.0	48.0	48.0	48.0	64.0
6. Drive 2.0 mi (3.2 km)[6]	SM	9.1	14.9	20.8	26.6	32.5	44.4
	SK	14.6	24.0	33.5	42.8	52.3	71.4
	TM	41.0	41.0	41.0	41.0	41.0	57.0
7. Drive 5.0 mi (8.0 km)[7]	SM	11.7	16.8	21.9	27.0	32.2	43.0
	SK	18.8	27.0	35.2	43.4	51.8	69.2
	TM	47.0	47.0	47.0	47.0	47.0	63.0

[1]The overall door-to-door travel speeds and times tabulated opposite each of the seven typical examples of residential access are, from top to bottom: SM: speed in mph; SK: speed in kph; TM: time in min.
[2]Each example includes 5.5 min walking time [1,000 ft (305 m)] at the nonresidential trip-end plus 2.5 min wait for (i) line-haul transit vehicle; (ii) excludes transfers between lines (except for the feeder transit example).
[3]At 3 ft per sec (0.9 m per sec) average.
[4]At 10 mph (16 kph) average plus 3 min change-mode time.
[5]At 12 mph (19 kph) average plus 3 min feeder transit waiting time plus 3 min change-mode time.
[6]At 20 mph (32 kph) average plus 3 min change-mode time.
[7]At 25 mph (40 kph) average plus 3 min change-mode time.

Source: Data developed for this chapter.

and if the platform speed is 60 mph (97 kph), then the actual door-to-door travel speed varies from 37.7 mph (60.7 kph) to 45.3 mph (72.9 kph), depending on the same typical residential access examples. In a few examples (e.g., drive 5 miles, line-haul 4 miles) door-to-door travel speed is actually higher than line-haul transit speed; in other examples it is little more than half.

SERVICE INTERVALS

Typically, when transit patrons know that the service they wish to use is scheduled at headways of approximately 10 min or less, they will often go to the stop or station without referring to a timetable. Therefore, average waiting times of one-half the service headways may usually be assumed for headways of 10 min or less. When transit service headways lengthen to intervals over typically about 10 min, some patrons tend to consult timetables and arrive at the transit stop or platform 2 to 3 min before the vehicle is due to arrive. When headways exceed about 15 min, almost all patrons will usually follow this procedure.

Transit schedules that include headways of consistent intervals during major time periods of the day (preferably headways in multiples of 2, 5, or 10 min) are more easily remembered by patrons and minimize the needs for timetables and waiting time at stops. Good schedule adherence (particularly avoidance of running ahead of schedule but also minimizing running late) helps to minimize patron waiting time at stops. Such reliability can benefit patronage and is obtained by continuous careful supervision of line operations and scheduling.

CAPACITIES

Table 6.14 presents passenger capacities per lane or track, expressed in typical ranges, in terms of an hourly rate in one direction of travel during the peak 15 min of the peak hour. Several significant qualifications affecting the use of Table 6.14 are set forth therein and in the following text. Table 6.14 is derived directly from the values presented in Tables 6.3 and 6.10, and from the text relating to those tables. Table 6.14 provides ranges of seated and total capacity values for major transit vehicle-types operating on a variety of facilities.

TABLE 6.14
Passenger Capacities per Lane or Track, Expressed in Typical Ranges[1]
(hourly rate, in one direction of travel, during the peak 15 minutes of the peak hour)

Transit Vehicle and Facility Type	Seated Only	Total Seated Plus Standing[9]
1. Minibus-midibus		
a. in mixed street traffic	900– 2,300	1,500– 4,000
b. in exclusive street lane	1,100– 3,000	1,800– 5,400
2. City-suburban transit bus, non-articulated		
a. in mixed street traffic	1,800– 5,000	2,700– 9,000
b. in exclusive street lane	2,200– 6,600	3,200–12,000
c. in mixed freeway traffic[2]	1,800– 5,000	2,700– 9,000
d. in mixed freeway traffic[3,10]	3,600–19,800	5,400–36,000
e. in exclusive busway[2]	3,100– 7,900	4,600–14,400
f. in exclusive busway[3]	3,600–19,800	5,400–36,000
3. City-suburban transit bus, articulated		
a. in mixed street traffic	2,100– 6,700	6,000–15,300
b. in exclusive street lane	2,500– 9,000	7,200–20,400
c. in mixed freeway traffic[2]	2,100– 6,700	6,000–15,300
d. in mixed freeway traffic[3,10]	4,200–27,000	12,000–61,200
e. in exclusive busway[2]	3,600– 9,000	10,300–20,400
f. in exclusive busway[3]	4,200–27,000	12,000–61,200
4. City-suburban transit bus, double-deck		
a. in mixed street traffic	3,000– 6,700	3,900– 8,100
b. in exclusive street lane	3,600– 9,000	4,700–10,800
c. in mixed freeway traffic[2]	3,000– 6,700	3,900– 8,100
d. in exclusive busway[2]	5,200–10,800	6,700–13,000
5. Intercity bus		
a. in mixed street traffic	1,800– 4,500	2,400– 6,300
b. in mixed freeway traffic[2]	1,800– 4,500	2,400– 6,300
c. in exclusive busway[2]	3,100– 7,200	4,100–10,100
6. School bus		
a. in mixed street traffic	1,200– 5,800	1,200– 5,800
b. in mixed freeway traffic	1,200– 5,800	1,200– 5,800

TABLE 6.14 (Cont.)

Transit Vehicle and Facility Type	Seated Only	Total Seated Plus Standing[9]
7. Tram, non-articulated		
a. in mixed street traffic[4]	5,400–16,200	21,600–37,800
b. in mixed street traffic[5, 11]	7,800–23,400	31,200–54,600
c. in reserved street medians[4]	5,400–16,200	21,600–37,800
d. in reserved street medians[5]	7,800–23,400	31,200–54,600
e. in exclusive right-of-way[2, 4, 6]	5,400–16,200	21,600–37,800
f. in exclusive right-of-way[2, 5, 7]	7,800–23,400	31,200–54,600
g. in exclusive right-of-way[3, 4, 6]	10,800–32,400	43,200–75,600
h. in exclusive right-of-way[3, 5, 7]	5,400–21,600	21,600–50,400
8. Tram, articulated		
a. in mixed street traffic[4]	5,400–16,200	21,600–40,500
b. in mixed street traffic[5, 11]	7,800–23,400	31,200–58,500
c. in reserved street medians[4]	5,400–16,200	21,600–40,500
d. in reserved street medians[5]	7,800–23,400	31,200–58,500
e. in exclusive right-of-way[2, 4, 6]	5,400–16,200	21,600–40,500
f. in exclusive right-of-way[2, 5, 7]	7,800–23,400	31,200–58,500
g. in exclusive right-of-way[3, 4, 6]	10,800–32,400	43,200–81,000
h. in exclusive right-of-way[3, 5, 7]	5,400–21,600	21,600–54,000
9. Urban-regional rapid transit, rail		
a. [2, 8]	12,000–34,000	30,000–132,000
b. [3, 8]	18,000–51,000	45,000–198,000
10. Urban-regional rapid transit, rubber-tired		
a. [2, 8]	5,400–16,200	27,000–75,600
b. [3, 8]	8,100–24,300	40,500–113,400
11. Commuter railroad, electric[8]	22,500–58,500	22,500–108,000
12. Superregional rapid transit[8]	4,500–29,700	4,500–43,500

[1] Derived from Tables 6.3 and 6.10, and rounded to the nearest 100 passengers. The text accompanying this table and Tables 6.3 and 6.10 is to be carefully noted in applying these tables. See also the examples given of capacity calculations using these tables.

[2] Passenger stops along route section involved.

[3] No passenger stops along or affecting route section involved. The very extensive, expensive, and often cumbersome major-center terminal(s) or station(s) required, especially for the higher portions of the capacity value ranges shown for this condition, may make such higher values impractical or unattainable in specific circumstances.

[4] Individual trams and tram-trains up to about 140 ft (43 m) in length of consist.

[5] Tram-trains between about 140 ft and 280 ft (43 m and 85 m) in length of consist.

[6] Top speeds up to about 35 mph (56 kph).

[7] Top speeds over about 35 mph (56 kph).

[8] In exclusive right-of-way.

[9] Values in the higher portions of each range shown in this column represent crush capacity possible only with generally optimum passenger spatial distribution and generally optimum vehicle doorway area, platform, and platform access capabilities (including access walkways, escalators, stairs, fare collection equipment as necessary, station or stop access mode capabilities, etc.). If such optimum conditions cannot be assured but conditions approaching such optima can reasonably be postulated, then, for practical purposes, the highest value given for each range shown in this column should be multiplied by a factor varying between about 0.7 and about 0.9, depending on the circumstances and upon the vehicle-type and facility-type involved; in the absence of specific data or assumptions under such approaching conditions, a multiplying factor of about 0.75 would generally be appropriate.

[10] Values in the higher portions of the capacity ranges shown for this condition involve very heavy preemption of available freeway lane capacity by bus vehicles and, therefore, may not be practical or attainable in specific circumstances.

[11] Operation of tram-trains of from 140 ft to 280 ft (43 m to 85 m) in length or longer, in mixed street traffic, involves very heavy preemption of available street lane capacity and, therefore, may be possible only in circumstances in which such preemption is deemed acceptable and practical.

Sources: Derived from Tables 6.3 and 6.10.

Passenger capacities for specific combinations of vehicles and facilities can be computed by using values selected from Table 6.14, from its parent Tables 6.3 and 6.10, and by appropriate application of other factors and ratios discussed in this subsection and in those tables. The subsection, "Capacity Calculation Examples," on p. 240 illustrates these computations.

Total-to-seated passenger ratios. Ratios of total to seated passengers during critical peak periods in the peak direction can be selected using the following guides: (1) 1.00 if on the average there were no standees expected and just a "seated-load" (no standees) was scheduled; (2) somewhat less than, or equal to, 1.00 if no standees were permitted; (3) about 1.50 if half as many standees as seated passengers were scheduled (a ratio frequently used for the peak-period peak-directional scheduling of urban transit buses in North America); (4) between 1.50 and 6.50 for the peak-period peak-direction scheduling of urban rapid transit vehicles (values above 2.00 usually involve interior car designs having relatively fewer seats and with room for high proportions of standees; values above about 2.00 or 3.00 often involve crush standee conditions); or (5) other values as selected for specific applications.

Hourly rates for peak 60 minutes vs. peak 15 minutes. The hourly capacity rates during the peak 15 min of the peak hour (as shown in Table 6.14) may be converted to rates for the full peak hour by using factors that typically range from about 0.70 to about 0.95; values in the middle part of this range are frequently encountered. Such conversion factors reflect the facts that (1) it is often difficult for transportation systems to maintain peak 15-min capacity rates for much greater durations and (2) apart from pure capacity considerations, passenger traffic volumes usually do not sustain themselves at the peak 15-min rate for much greater durations unless the transportation facility and service involved are severely saturated with such demands, and often not even under those conditions.

Trams. Trams may be operated as single non-articulated or articulated units, or they may be coupled into trains. In Tables 6.3, 6.10, and 6.14, the longest tram-train consist that has been assumed is 280 ft (85 m), although longer individual consists are possible. This limitation in the tables reflects the typical operation of trams along portions of their routes where at-grade crossings with other vehicular traffic may occur. Operation of tram-trains of from 140 ft to 280 ft (43 m to 85 m) in length or longer in mixed street traffic involves very heavy preemption of available street lane capacity and therefore may be possible only when such preemption is deemed acceptable and practical.

Number of lanes or tracks. The passenger capacity of an entire transit facility in one direction of travel is, subject to the limitations discussed below, generally the product of the capacity of a single lane or track and the number of such similar lanes or tracks available in that same direction. Lanes or tracks involved in such a multiplication (and their access-egress areas) must not interfere with each other and must be similar to each other in capacity characteristics; such similarity is not always or necessarily present and must be verified in each case, with appropriate adjustments as found necessary.

Route-sections unaffected by stops. In Tables 6.10 and 6.14 values are shown for route-sections having no passenger stops along them or affecting them. These data are provided to illustrate headway and capacity potentials under such possible conditions, and are subject to the qualifications discussed in the next paragraph.

Most-limiting set of conditions. It should be borne in mind that the capacity of a given route-section, facility, or service usually can be no greater than the most limiting set of conditions directly or indirectly affecting such capacity. For example, if a route-section has no passenger stops, the operation of transit vehicles or trains along that section, particularly in terms of average headways, is often affected by passenger stops or other limiting conditions upstream or downstream from that route-section. Footnotes in Tables 6.10 and 6.14 provide further cautions in this respect and in the general treatment of transit capacities. (See particularly footnotes 7, 9, and 12 in Table 6.10, and footnotes 3 and 10 in Table 6.14.)

Capacity must not be confused with demand. The actual demand of transit service may not, and often will not, equal or even closely approach the potential capacity of that service. Maximum crush standee conditions, although frequently encountered at maximum load points or route-sections during crest peak periods in the peak direction or directions of travel, may not be tolerated by potential new transit users or permitted by some transit operating regulations. In any case, these conditions are definitely undesirable and are usually tolerated only because of stringent economic circumstances affecting the transit system and/or the community involved.

CAPACITY CALCULATION EXAMPLES

Examples of capacity calculations based on Table 6.14, its parent Tables 6.3 and 6.10, and other considerations raised in this chapter are given below. The factor, ratio, and Table 6.14 values selected for use in these calculations are only illustrative; ranges for such values are given in this chapter and appropriate values may be selected by the analyst to fit particular circumstances.

Example 1:

Find the approximate maximum crush capacity of large, standard non-articulated city-suburban transit buses operating on an exclusive busway having passenger stops, per lane in the peak direction of travel, in terms of an hourly rate for the full 60 min of the peak hour, and reflecting the typical case in which passenger spatial distribution and access characteristics only approach generally optimum conditions.

In Table 6.14, item 2 e covers such buses on exclusive busways having stops and shows a maximum capacity of 14,400. But in order to meet the specific typical conditions cited above, several adjustment factors cited in this chapter must be applied. Converting the hourly rate from the peak 15 min (as given in Table 6.14) to the full 60 min of the peak hour involves (in the absence of more specific input information) a multiplying factor in the middle of the 0.70–0.95 range cited in the above subsection on capacities; a value of 0.83 is selected. The 14,400-passenger maximum capacity value cited above is subject to footnote 9 of Table 6.14: passenger spatial distribution and access characteristics (as is often typical) only approach, and do not equal, generally optimum conditions; as indicated in footnote 9, in the absence of specific data or assumptions under such approaching conditions, a multiplying factor of approximately 0.75 would generally be appropriate.

When rounded, $14,400 \times 0.83 \times 0.75$ equals approximately 9,000. Since ranges are indicated above for each of the three component values, it could be appropriate to consider expressing the answer in terms of a reasonable range, say, from 8,000 to 10,000.

Example 2:

Find the same capacity as in Example 1, except assume easy loading conditions instead of crush loading conditions.

The same calculations are made as in Example 1, except that an additional multiplying factor is required to convert from crush to easy capacity. In Table 6.3, the maximum crush capacity of a large standard non-articulated city-suburban transit bus is about 100 passengers; 95, a value slightly below the maximum, is selected. Such a bus typically seats 50 passengers, which is a value near the high end of the range for seats given in Table 6.3. Standees may range from 10 at the low end of easy capacity to 50 at the high end of crush capacity. Twenty-five, a value in the lower half of this range, is selected, based on earlier discussions in this chapter in which a total-to-seated passenger ratio of 1.50 was suggested for easy capacity conditions. Fifty seated plus 25 standing is a total of 75 passengers per bus for easy capacity conditions, as opposed to approximately 95 under crush conditions. A crush-to-easy capacity conversion factor of 75/95, or 0.79, is thereby obtained.

The 9,000 average result of Example 1 is multiplied by 0.79 to provide an approximate rounded answer of 7,100. Because ranges are indicated for component values as discussed in Example 1, it could be appropriate to express the answer in terms of a reasonable range, say, from 6,400 to 7,800 or from 6,000 to 8,000.

Example 3:

Find the approximate maximum crush capacity of an urban-regional rail rapid transit line having passenger stops and one track in each direction of travel, per track in the peak direction of travel, in terms of an hourly rate for the full 60 min of the peak hour, and reflecting the typical case in which passenger spatial distribution and access characteristics only approach generally optimum conditions.

In Table 6.14, item 9a covers this condition, with stops, and shows a maximum crush capacity of 132,000, a value that is high in relation to observed capacities of existing systems with their various limitations. As in Example 1, conversion of the hourly rate from the peak 15 min (as given in Table 6.14) to the full 60 min of the peak hour involves (in the absence of more specific input information) a multiplying factor in the middle of the 0.70–0.95 range cited in the above subsection on capacities. A value of 0.83 is selected. The 132,000-passenger maximum crush capacity value is subject to footnote 9 of Table 6.14: passenger spatial distribution and access characteristics (as is often typical) only approach, and do not equal, generally optimum conditions; as indicated in footnote 9, in the absence of specific data or assumptions under such approaching conditions, a multiplying factor of about 0.75 would generally be appropriate.

When rounded, $132,000 \times 0.83 \times 0.75$ equals approximately 82,000. Since ranges are indicated above for each of the three component values above, it could be appropriate to consider expressing the answer in terms of a reasonable range, say, from 73,000 to 90,000 or from 70,000 to 90,000.

Example 4:

Find the same capacity as in Example 3, except assume easy loading conditions instead of crush loading conditions.

The same calculations are made as in Example 3, except that an additional multiplying factor is required to convert from crush to easy capacity. In Table 6.3, the maximum crush capacity of a 10-car urban-regional rail rapid transit train is about 3,300 passengers; 3,100, a value slightly below this maximum is selected. Individual cars in a 10-car train typically seat about 75 passengers, which is a value near the high end of the range for seats given in Table 6.3; when multiplied by 10 for a ten-car train, a total of 750 seats is obtained. A total-to-seated passenger ratio range of 1.50 to about 3.00 is suggested for easy capacity conditions with urban-regional rail rapid transit systems; a value of 2.00 is selected. This ratio times 750 seats per train equals 1,500 passengers per train, versus approximately 3,100 under crush conditions. A crush-to-easy capacity conversion factor of 1,500/3,100, or 0.48, is thereby obtained.

The 82,000 average result of Example 3 is multiplied by 0.48 to provide an approximate rounded answer of 39,000. Since ranges are indicated for component values as discussed in Example 1, it could be appropriate to express the answer in terms of a reasonable range, say, from 35,000 to 43,000.

OBSERVED VALUES FOR PEAK PASSENGER VOLUMES

Table 6.15, 6.16, and 6.17 provide observed values for peak-hour peak-direction passenger volumes for buses on city streets, buses on freeways, and rail rapid transit.[12]

TABLE 6.15
Observed Peak-hour Passenger Volumes on Urban Transit Routes
(buses on city streets in prevailing direction only)

Type	City	Facility	Buses per Hour	Actual Peak Hour	Hourly Rate for 15–20 Min
Local Buses	New York	Hillside Ave.	150	10,251	10,824
City streets,	San Francisco	Market St.	130	7,553	8,500
Parking	Cleveland	Euclid Ave.	90	4,316	5,600
prohibited	Chicago	Michigan Ave.	75	4,240	4,770
	Baltimore	Baltimore St.	76	4,387	4,758
Local Buses	Rochester	Main St.	93	4,982	—
City streets,	Chicago	Washington Blvd.	66	3,235	3,600
Reserved	Atlanta	Peachtree St.	67	2,807	3,504
transit	Dallas	Commerce St.	67	3,069	3,444
lane	Birmingham	2nd Ave., North	44	2,301	2,712
Express Buses	St. Louis	Gravois St.	66	2,918	4,185
City Streets,	Cleveland	Clifton Blvd.	32	1,872	2,700
Parking	Chicago	Archer Ave.	29	1,896	2,500
prohibited	San Francisco	Van Ness Ave.	17	1,234	1,784
	New Orleans	Earhart Blvd.	25	1,267	1,620

Source: *Capacities and Limitations of Urban Transportation Modes* (Washington, D. C.: Institute of Traffic Engineers, 1965), p. 23.

---

[12] Additional and more detailed data for buses are to be found in: *Highway Capacity Manual*, pp. 340–42. *The Potential for Bus Rapid Transit* (Detroit: Automobile Manufacturers Association, February 1970) Table 8, p. 43 lists statements from various sources as to the number of buses per hour (and the corresponding seat capacity) that have been or could be provided past a given point under widely varying roadway conditions, stop spacing, and average speeds. This tabulation includes both theoretical and actual values.

TABLE 6.16
Observed Peak-hour Passenger Volumes on Urban Transit Routes
(buses on freeways in prevailing direction only)

Type	City	Facility	Buses per Hour	Passenger Movement	
				Actual Peak Hour	Hourly Rate for 15–20 Min
Express Buses	Chicago	Lake Shore Dr.	99	5,595	6,350
On freeways	Cleveland	Shoreway West	32	1,872	2,700
	San Francisco	Bayshore Fwy.	35	2,270	2,700
	Los Angeles	Hollywood Fwy.	41	2,268	2,640
	St. Louis	Mark Twain Hwy.	52	1,767	2,295
	Atlanta	North Expressway	19	803	1,892
Express Buses	New York	Port Authority Bus Terminal	511	23,187	28,556
On terminal ramps,	Union City, N.J.	Route 3	397	17,800	23,000
tunnel approaches,	New York	Lincoln Tunnel	480	21,600	22,860
tunnels, and	San Francisco	Oakland Bay Bridge	216	7,812	10,945
bridges	New York	George Washington Bridge	136	6,939	9,468

Source: *Capacities and Limitations of Urban Transportation Modes*, p. 24.

TABLE 6.17
Observed Peak-hour Passenger Volumes on Urban Transit Routes
(rail rapid transit in prevailing direction only)

Type	City	Facility	Trains per Hour	Passenger Movement	
				Actual Peak Hour	Hourly Rate for 15–20 Min
Rail Rapid Transit	New York	IND 6th and 8th Ave. Express (10 car)	32	61,400	71,790
	New York	IND 8th Ave. Express (10 car)	30	62,030	69,570
	New York	IRT Lexington Ave. Express (9 car)	31	44,510	50,700
	Toronto	Yonge St. Subway (6 car-new)	28	35,166	39,850
	New York	IRT 7th Ave. Express (9 car)	24	36,770	38,520
	Chicago	Eisenhower Express-way* (6 car)	25	10,376	14,542
	Cleveland	Private R/W and Subway (6 car)	20	6,211	8,349

*Stretch represents densest "transit" mile of rail rapid transit operation on the freeway where two routes converge; consists of track and four auto lanes in each direction. (Transit passenger flow shown is for the prevailing direction only.)

Source: *Capacities and Limitations of Urban Transportation Modes*, p. 25.

SAFETY

In the three years 1970–1972, the United States transit industry had 25.58 disabling work injuries per million employee-hours of exposure (frequency rate) and 777 total chargeable days for work injuries per million employee-hours of exposure (severity rate). Deaths and permanent disabilities are included in these values at scheduled rates. On the basis of ascending frequency of injury rates, in 1972 the transit industry ranked 38th in a list of 41 major United States industries. Its 1972 severity rate was 784; the safest tabulated industry, Aero Space Communications, was 143, and the least-safe tabulated industry, Underground Coal Mining, had a severity rate of 4,272 (1971 rate).[13]

Table 6.18 shows accident data for 1970 for the United States transit industry, in terms of vehicle miles (km) of service and passengers carried, as well as estimated total fatalities.[14]

TABLE 6.18
U.S. Transit Industry Accidents, 1970

Accident Type	Rail Rapid Transit	All Surface Transit	Total
Collision-type accidents per million vehicle-mi (vehicle-km)	1.09 (0.68)	63.04 (39.40)	—
Passenger accidents per million passengers carried	7.52	7.79	—
Estimated fatalities	—	—	150

Source: American Transit Association.

CHARACTERISTICS OF TRANSIT ECONOMY

Figures 6.5 through 6.9 give trends in key aspects of the economics of mass transportation in the United States during the twenty-year period 1950–1970. Major characteristics and problems of the American transit industry are also illustrated in these figures.

Figure 6.5 shows that the full cost of operating a vehicle-mile of transit service rose over twice as rapidly as did both the full cost of driving an automobile and the consumer price index nationwide in urban areas during the two decades, 1950–1970. These contrasting trends and data below illustrate the most fundamental single problem of the transit industry in the current era. Unless the cost trend of transit service can be brought more nearly in line with the cost trend of auto travel and the overall consumer price index, mass transit in America will find itself increasingly priced out of the urban transportation market. This process is already well advanced, not only in the smaller American cities (where service is often already extinct), but also for many types of trips in larger urban areas. More recently, there have been signs of change. Many systems have increased their ridership, especially

---

[13] *Accident Facts*, 1973 Edition (Chicago: National Safety Council, 1973), pp. 35–37.
[14] American Transit Association.

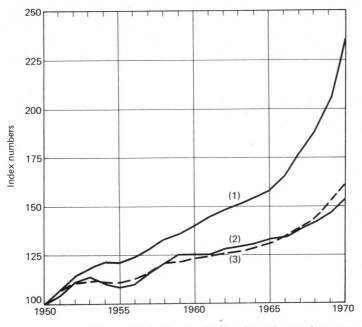

Figure 6.5. Operating Cost and Price Trends. (1) Transit total operating costs per revenue vehicle-mile. (2) Automobile total operating costs per urban vehicle-mile. (3) Consumer price index, U.S.A., 1950–1970 (Index: 1950 = 100). [Sources: (1) American Transit Association; (2) and (3) U.S. Bureau of Labor Statistics.]

with energy crisis impacts. Significant patronage increases can be one of the best ways of improving transit economy.

The average transit fare in the United States in 1972 was approximately 31.3 cents, with many systems charging a basic adult fare in the 35- to 50-cent range. The average transit trip-length in the United States is approximately 3 to 4 miles (4.8 to 6.4 km) and most American systems fall within the 2- to 5-mile (3.2 to 8.0 km) average trip-length range. Thus, the average fare per *passenger*-mile approximated 9 cents (5.6 cents per passenger-km) in 1972. Because total operating expenses exceeded operating revenues by more than 23 percent for United States transit systems in 1972, the actual operating cost per *passenger*-mile approximated 11 cents (6.8 cents per passenger-km). In contrast, the total operating cost per *vehicle*-mile for a suburban-based automobile in the United States in 1972 was 9.4 cents for subcompact cars, 10.8 cents for compact cars, and 13.6 cents for standard-size cars (respectively, 5.8, 6.7, and 8.5 cents per vehicle-km), including depreciation, maintenance, tires, fuel, oil, parking, tolls, insurance, and user taxes devoted largely to financing the capital costs of road improvements.

Under such conditions, two people traveling together in an automobile can often make an urban trip much less expensively than by transit, and without the travel time, schedule, and route-stop constraints imposed by transit service. When more than two people share the ride, transit is placed in an even more unfavorable competitive light. In fact, it is often less expensive for one person to travel alone on an urban trip in a modestly-priced automobile and to bear the full costs of that mode than to make the same trip by transit service when fares approach the levels indicated above, except when parking costs may be significant. If, as so often happens, people consider only the out-of-pocket costs of driving an auto (about one-half of the total costs), the choice of the auto becomes much more attractive. When transit fares are below the levels indicated above, the economic attractiveness of using transit tends to improve, but then outside subsidy to transit is often introduced or increased.

Figure 6.6 gives key trends of productivity in the American transit industry during the twenty-year period 1950-1970. Revenue passenger trips per transit employee and

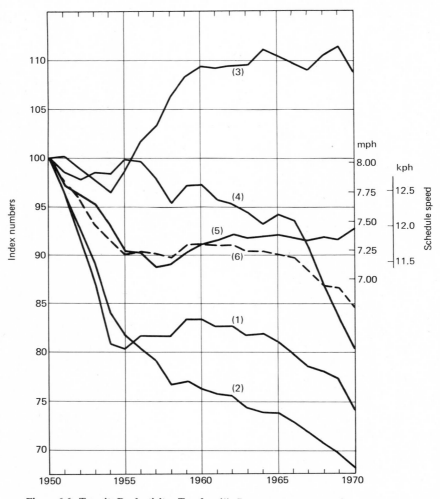

Figure 6.6. Transit Productivity Trends. (1) Revenue passenger trips per transit employee. (2) Revenue passenger trips per revenue transit vehicle-mile. (3) Revenue transit vehicle-miles per transit employee. (4) Total transit operating revenue (in 1950 constant dollars) per revenue transit vehicle-mile. (5) Total bus transit schedule speed (New York City Transit Authority only). (6) Composite transit productivity index [(1) through (5) above]. U.S.A. 1950–1970. (Except (5) above.) (Index: 1950 = 100.) [Sources: (1) through (5) American Transit Association data; (6) derived from (1) through (5) above.]

per vehicle-mile of transit service, and total transit operating revenue (in 1950 constant dollars) per vehicle-mile, all show pronounced long-term declines. These declines in productivity are not substantially offset economically by possible increases in the average transit trip-length (as fares have increased) or by any overall changes that may have occurred in average capacity per transit vehicle in these last two decades. Revenue vehicle-miles per transit employee have, however, shown modest improvement. Although data on system-wide speeds are unavailable for the industry as a whole, the experience of one of the largest American urban bus operators, the New York City Transit Authority, shown in Figure 6.6, was one of sharply declining system-wide bus schedule speed from 1950 to about 1957, followed by modest improvement from 1958 to 1962, and by a generally level trend from 1963 to 1970. Available data and experience suggest that many surface transit systems have had periods of substantial decline in schedule speeds because of increased traffic congestion, but that there have also been schedule speed increases from improved traffic engineering, more express operations, and greater use of improved and new freeways, expressways, and major arterials. A composite transit productivity index, giving equal weight to the five productivity trends discussed above, is shown as a dashed line in Figure 6.6. This index, although only an approximate measure, is generally indicative of the overall trend of productivity in the American transit industry.

Figure 6.7 compares the composite transit productivity index developed in Figure 6.6, with the output per man-hour in all manufacturing in the United States from 1950 to 1970. Manufacturing productivity rose constantly and substantially in the

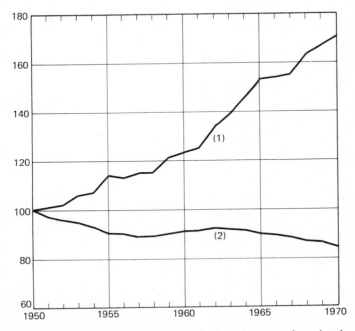

Figure 6.7. Productivity Trends Comparison. (1) Output per man-hour (total manufacturing); (2) Composite transit productivity index (from Figure 6.6). U.S.A., 1950–1970. (Index: 1950 = 100.) [Sources: (1) U.S. Bureau of Labor Statistics; (2) Figure 6.6.]

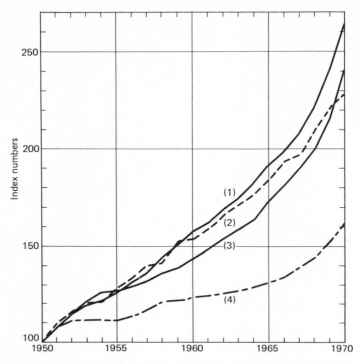

Figure 6.8. Labor Cost Trends: (1) Average annual earnings per transit employee. (2) Annual average weekly earnings of all production workers in manufacturing. (3) Total transit payroll per revenue vehicle-mile. (4) Consumer price index. U.S.A., 1950–1970. (Index: 1950 = 100). [Sources: (1) and (3) American Transit Association; (2) and (4) U.S. Bureau of Labor Statistics.]

past twenty years, but transit productivity generally declined. This contrast is most significant when compared with the trends shown in Figure 6.8. These contrasts and comparisons illustrate the most serious single problem of the American transit industry today.

Figure 6.8 illustrates the trend of transit labor costs in relation to both the labor costs of all U.S. production workers in manufacturing and the consumer price index. Since 1960, and especially since 1966, transit labor costs have risen faster than production manufacturing labor costs. By 1970, the total transit payroll per revenue transit vehicle-mile had risen nearly as fast as annual earnings per transit employee and had outstripped the trend of production manufacturing workers. The consumer price index is shown for comparative purposes for the same time span, 1950–1970. Remember that from 1950 to 1970 the full cost of driving an automobile in urban areas rose at only about the same rate as the consumer price index (Figure 6.5).

Figure 6.9 summarizes salient results of the economic conditions illustrated above. The American transit industry was transformed from a 1950 condition of marginal sufficiency (without, however, the means adequately to meet its continuing capital needs) to a massive and rapidly deepening overall deficit operation in 1970. These trends are illustrated in Figure 6.9 in terms of operating ratio (percent of transit operating expenses, including depreciation, to operating revenues) and operating

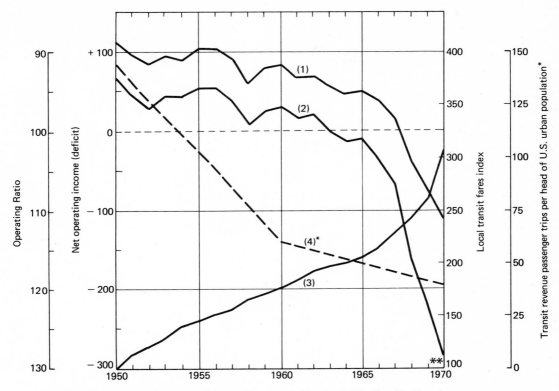

Figure 6.9. Transit Operating, Income, Fare, and Patronage Trends. (1) Transit operating ratio. (2) Transit Net Operating Income (Deficit). (3) Local Transit Fares Index. (4) Transit Revenue Passenger-Trips per Head of U.S. Urban Population (U.S.A., 1950–1970). [Sources: (1), (2), and (4) American Transit Association; (3) U.S. Bureau of Labor Statistics.]
*Trend line based on decennial data and revised definition of U.S. urban population.
‡$681,000,000 net operating deficit in 1973 (preliminary).

income-deficit. The rapid rise in fares paid for urban transit rides is also illustrated; from 1950 to 1970 these fares rose 3.9 times faster than did the full cost of driving an automobile in urban areas. The precipitous decline in transit rides per head of urban population (rides per capita) is also shown.

Approximately one-third of all U.S. transit rides in 1970 occurred in or to New York City which accounted for only 5.3 percent of United States urban population. Transit rides in American urban areas of up to 500,000 population declined by 75 percent between 1950 and 1970. For the nation as a whole, transit rides declined by 57 percent during the same period.

Generally, the above economic trends have not been nearly as severe in most other parts of the world, and they need not necessarily continue at the same pace and direction in the United States in the future. The economics and the patronage of some urban transit systems elsewhere in the world (and some, too, in the United States) have shown notable gains in the last twenty years. Major improvements in both facilities and services have been made and will continue to be made in response to demonstrated needs and increased utilization. If the United States can effectively come to terms with the problems discussed above, then it can participate to a greater extent in this kind of growth. Most recent developments offer some encouragement in this respect.

CAPITAL COSTS AND SERVICE LIVES

Table 6.19 presents typical ranges of order-of-magnitude total capital costs and useful service lives for major transit capital items in the United States in 1971. Many factors affect the actual capital costs and service lives of such items when related to specific, individual applications. The cost ranges shown in Table 6.19 could be exceeded or underrun in some circumstances. These variations in cost and service lives are particularly applicable in the case of construction of new fixed facilities such as rail rapid transit and busway lines and stations, where soil conditions, geometrics, utilities,

TABLE 6.19
Order of Magnitude Capital Costs and Service Lives of Selected
Transit Items in the United States, 1971
Expressed in Typical Ranges[1]

Capital Item	Total Capital Cost ($) (000,000)	Useful Service Life (Years)
1. Urban transit bus, non-articulated, 40 ft (13 m) long	0.034–0.043	13–20
2. Urban-regional rapid transit car, 60–75 ft (20–25 m) long	0.250–0.400	25–40
3. Urban-regional rapid transit, rail[2]		
Subway line[3]	15–35	Indefinite
Subway station[4, 5]	5–15	Indefinite
Aerial line[3]	4–10	Indefinite
Aerial station[4]	2–6	Indefinite
Surface line[3, 6]	2–9	Indefinite
Surface station[4, 6]	1–4	Indefinite
Main yard and shop[8]	5–10	Indefinite
4. Urban-regional busway[2]		
Subway line[3]	18–43	Indefinite
Subway station[4, 5]	6–25	Indefinite
Aerial line[3]	3–8	Indefinite
Aerial station[4, 7]	2–6	Indefinite
Surface line[3, 6]	2–7	Indefinite
Surface station[4, 6, 7]	1–4	Indefinite
Main yard and shop[8]	2–6	Indefinite

[1]Subject to the qualifications and cautions noted in text for Table 6.19.
[2]Includes construction, right-of-way, utilities, train or bus control, design, construction management and administration, and related items (including track and electrification for guided systems). Fully grade-separated.
[3]1.0 mi (1.6 km) in length with two tracks or lanes. Includes purely line items through station areas.
[4]450–700-ft (148–230-m)-long platforms. Excludes any non-transit functions and any station parking facilities.
[5]With mezzanine.
[6]Fully grade-separated.
[7]Exclusive of extensive terminal-type stations.
[8]Storage for 150–250 vehicles.

Source: Parsons, Brinckerhoff, Quade & Douglas (analysis of recent construction projects) and various data books.

right-of-way, the spacing, size, and elegance of stations, labor and materials costs and interrelationships, and other factors can substantially affect total capital costs. The relative quality and sophistication of any of the items listed in Table 6.19 also significantly affect their cost.

OPERATING EXPENSES

Transit operating expenses can vary significantly between different systems because of differences in schedule speed, condition of physical plant, maintenance policies, labor agreements, accounting practices, vehicle sizes, modal mixes, types of service provided, and other factors. Table 6.20 shows 1970 operating expense data per vehicle-mile (vehicle-km) for the large Chicago Transit Authority separately for its motor bus, electric trolley-bus, and rail rapid transit services. In comparing these figures, the different total capacities and platform speeds of these individual transit modes should be borne in mind.

Public transportation is a highly labor-intensive industry. Its expenses are also affected by varying but characteristically high ratios (as between peak and offpeak

TABLE 6.20
Operating Expenses, Chicago Transit Authority, per Vehicle-Mi (Km), 1970

Vehicle Type	Expense per Vehicle		Percent of Total
	Mi	Km	
Motor buses			
Maintenance of plant & equipment	$0.22	$0.14	16.1
Operating & garage expense, including fuel	0.06	0.04	4.4
Transportation, including drivers	0.69	0.42	50.3
Administrative, general, & miscellaneous	0.30	0.19	21.9
Depreciation	0.10	0.06	7.3
Total	$1.37	$0.85	100.0
Electric trolley coaches			
Maintenance of way & structures	$0.07	$0.04	4.9
Maintenance of equipment	0.19	0.12	13.3
Power	0.07	0.04	4.9
Transportation, including drivers	0.70	0.44	48.9
Administrative, general, & miscellaneous	0.30	0.19	21.0
Depreciation	0.10	0.06	7.0
Total	$1.43	$0.89	100.0
Rail rapid transit			
Maintenance of way & structures	$0.14	$0.09	11.5
Maintenance of equipment	0.12	0.07	9.8
Power	0.08	0.05	6.6
Transportation, including train crews	0.48	0.30	39.3
Administrative, general, & miscellaneous	0.30	0.19	24.6
Depreciation	0.10	0.06	8.2
Total	$1.22	$0.76	100.0

Source: American Transit Association.

periods) of traffic volume, vehicle-miles of service, and equipment utilization. In 1970, for the United States transit industry as a whole, the operating expense (including depreciation) per vehicle-mile was $1.004 ($0.627 per vehicle-km).

ENVIRONMENTAL ASPECTS OF TRANSIT

Mass transportation is often looked upon as a means for improving the urban environment. Frequently, the intent of this improvement is to reduce the spatial extent, visibility, and noise of urban transportation structures and vehicles, and to reduce the air pollution caused by these vehicles. A number of transit facilities, vehicles, and services are being considered, with emphasis frequently given to the use of electric vehicles, novel and innovative transit methods and equipment (see the following section of this chapter), and reduced reliance on private forms of vehicular transportation. The success of these improvements is usually dependent on the extent to which patronage can be attracted to transit and/or diverted from automobiles, and on the economics of these patronage shifts.

Pollution of the environment involves degradation of one or more of the elements that affect human and/or animal senses or intakes. Therefore, pollution may involve noise, air, vision, water, food, and touch. Mass transportation is trying to reduce noise, air, and visual pollution. Each of these kinds of pollution is considered below.

Environmental improvement involves not only the reduction or diminution of the negative factors in each element and situation, but also the creation and enhancement of positive factors whenever appropriate, for example, attractive and harmonious visual, aural, olfactory, and other surroundings.

Those interested further in this subject and in acceptable levels of environmental pollution are referred to Chapter 21, Environmental Considerations.

NOISE POLLUTION

Table 6.21 presents data on representative external and internal noise levels for typical types of transportation equipment and systems. The data shown are subject to substantial variation in different spatial, traffic, acoustical, and other situations, including the effects of acceleration, deceleration, and cruising at various speeds.

Figure 6.10 illustrates variations in sound pressure levels in trains arriving in selected Paris, Berlin, Hamburg, and Chicago rapid transit stations.

AIR POLLUTION

Table 6.22 presents data on the composition of dry burnt gases emitted by gasoline and diesel engines, expressed in terms of percent by volume. The data shown are subject to considerable variations with different operating, vehicle-type, vehicle load, maintenance, spatial, grade, traveled way, and other conditions. Electrically-driven vehicles and trains themselves generate no, or virtually no, air pollution.

Table 6.23 compares emissions produced by automobiles, diesel buses, and rapid transit, in terms of tons per million passenger-miles (metric tons per million passenger-km). Values in this table are based on 1.2 occupants per automobile, 20 passengers per bus or rapid transit car, and 5.5 kilowatt-hours per rapid transit car-mile.

TABLE 6.21
Transportation Noise Levels Representative of Typical
Equipment and Systems

Noise Levels, Outside	dB
Street (at curb), night, light traffic	30
Street (at curb), day, heavy traffic	75
Trolley bus at 20 ft (6 m)	79 $\pm$ 4
Street with rough pavement, heavy traffic	80
Automobiles, at 20 ft (6 m)	82 $\pm$ 4
Light trucks, at 20 ft (6 m)	82 $\pm$ 4
Electric train, at 20 ft (6 m)	83 $\pm$ 3
Diesel bus, accelerating, heavy traffic	88 $\pm$ 3
Old 'L' trains (at 20 ft) (6 m)	91 $\pm$ 4
Diesel and/or steam RR trains, at 100 ft (30 m)	91 $\pm$ 6
Heavy truck, accelerating, heavy traffic	93 $\pm$ 5
Subway trains (at 20 ft) (6 m)	94 $\pm$ 3

Noise Levels, Inside	dB
In coach, pullman, or lounge car of RR train at 70 mph (113 kph)	67–71
In club car or dining car of RR train at 70 mph (113 kph)	80
In vestibule of train at 70 mph (113 kph)	93–100
In subway at 15 mph (24 kph)	89
In subway at 25 mph (40 kph)	95
In subway at 35 mph (56 kph)	99
In subway at 45 mph (72 kph)	100

Source: George Bugliarello and Charles Wakstein, *Noise Pollution—A Review of its Techno-Sociological and Health Aspects*, Biotechnology Committee, Carnegie-Mellon University, unpublished study (1968).

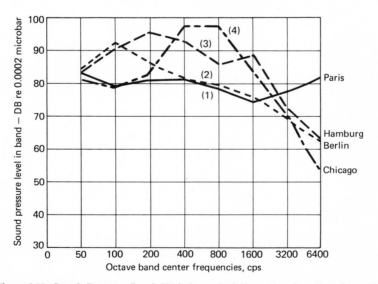

Figure 6.10. Sound Pressure Level Variations. Arriving trains in selected rapid transit stations in (1) Paris (2) Berlin (3) Hamburg (4) Chicago. [Source: Operations Research Incorporated, Silver Spring, Md., 1963.]

TABLE 6.22
Composition of Gasoline and Diesel Engine Emissions
Expressed in Terms of Percent by Volume

Component	Gasoline Engines				Diesel Engines	
	Full Load	Normal Use	Idling	Deceleration	Full Load	Normal Use
Nitrogen	84 to 86	85 to 86	≤86	≤85	≤86	≤86
Carbon dioxide	8 to 14	10 to 13.5	9 to 12	2 to 2.5	≤12	1 to 10
Carbon monoxide	0.2 to 5.5	0.2 to 4.5	3 to 9	4 to 5	≤1	<0.2
Hydrogen	0 to 8	0 to 3	1.5 to 4.5	?	≤0.5	<0.1
Methane	0 to 0.4	0 to 0.4	?	?	?	?
Hydrogen carbides or various carbohydrates	1 to 3	1 to 3	1 to 15	up to 60	≤0.5	?
SO_2	<0.004	<0.004	<0.004	<0.004	<0.06	<0.06
$PbBr_2$	0.0003	0.0003	0.0003	0.0003	nil	nil

Source: Prof. M. Serruys, as quoted in *International Commission on Standardization of Motorbuses,*
Report 5, 37th International Congress, International Union of Public Transport, Barcelona, 1967, p. 55.

TABLE 6.23
Emissions Produced by Automobiles, Diesel Buses, and Rapid Transit
In terms of tons per million passenger-mi
(metric tons per million passenger-km*)

Vehicle Type	Emissions				
	CO	HC	NO_x	Particulates	SO_2
Auto	80	15	3.9	0.36	0.22
Diesel bus	0.2	0.8	0.9	0.3	0.2
Rapid transit†	0.02	0.01	1.05	0.3–0.7	5.0
Rapid transit‡	0.001	0.01	1.1	0.2	3.5

*Multiply table-entry values by 0.564 to obtain rates in metric tons per million passenger-km.
†Utility coal fired thermal power plant (2.5% sulfur), assuming electrostatic precipitator in stack 90–95% efficient.
‡Residual oil fired thermal power plant (2.5% sulfur).

Source: S. Tilson, "Air Pollution," *Science and Technology,* No. 42 (1965), p. 22.

VISUAL POLLUTION

It is difficult to set forth, within the scope of this chapter, principles or criteria and other considerations of the visual aspects of mass transportation vehicles, structures, and systems. The exteriors and interiors of facilities and vehicles should be pleasing, harmonious, and tasteful because transit patrons, the general public, transit operators, planners, and designers are increasingly demanding such quality. Good balances should be struck between visual variety and continuity, and between restful and stimulative visual settings. The successful handling of the visual impact of bulk or mass is important in all structures, but probably more so for transit facilities that must attract patronage and that are usually constructed in or close to dense concentrations of human activity and settlement. In general, form must usually follow function in at least the basic parameters or envelopes and in much of the detail of mass transportation design. But there are important qualifications of visual design practice to this general observation, and successful visual results will always demand an intimate collaboration of both form and function in each of their varied aspects.

NOVEL AND INNOVATIVE TRANSIT METHODS AND EQUIPMENT

During and since the Industrial Revolution there has been a continuous public and professional interest in novel and innovative methods and equipment for mass transportation, as well as for other modes of transportation. This interest has heightened during the past two decades of transit development and since 1962 has been stimulated by federal financial assistance to transit demonstration and research programs in the United States.[15]

Any novel or innovative transit method or equipment should undergo tests before it is considered valid and practicable for a given proposed application. It is usually desirable to test such method or equipment against the best proven method or equipment (one or more) in its most modern form already available for practical use in the application under examination. Such testing involves the following sequence in which the candidate method or equipment is demonstrated to be equal or superior (or shows reasonable promise of being equal or superior) to one or more yardstick alternatives: (1) rigorous conceptual evaluation, (2) detailed analysis and costing, (3) simulation testing as necessary, (4) prototype development and testing, and (5) full operational demonstration development and extensive testing. Actual value or reasonable promise of value should be sufficiently demonstrated in each stage in order to provide a valid warrant for undertaking the next test stage.

Characteristics that should be tested include all relevant aspects of safety, reliability, capacity, speed, spatial requirements, adaptability, versatility, capital and operating costs, service life, noise, air pollution, visual factors, comfort, convenience, accessibility, community impacts, public acceptance, ability for staged development, and other factors.

Testing should include extensive use of economic and social benefit-cost, cost-effectiveness, and rate of return analysis embracing as wide a range of the above factors as possible.

Many novel and innovative concepts will fall out after just the first one or two testing stages listed above, when conscientiously and realistically applied. Others will warrant further stages of testing. A few will meet all reasonable tests and prove their

---

[15] Below are extensive recent inventories and evaluations of this subject:

Tomorrow's Transportation: New Systems for the Urban Future (Washington, D.C.: U.S. Department of Housing and Urban Development, 1968); summary report, 100 pp., plus 18 New Systems Study final project reports.

Urban Transportation: New Concepts (Washington, D.C.: Institute for Rapid Transit, 1970), 78 pp.

Study of New Systems of Public Transport, Report 7 (Brussels: International Union of Public Transport, 1969), 47 pp.

Urban Rapid Transit Concepts and Evaluation, Research Report 1 (Pittsburgh: Transportation Research Institute, Carnegie-Mellon University, 1968), 241 pp.

MICHAEL C. KLEIBER and LAWRENCE L. VANCE, JR., *New and Novel Passenger Transportation Systems: A List of Selected References* (Berkeley: Institute of Transportation and Traffic Engineering, University of California, 1971), 23 pp.

T. E. PARKINSON, "Passenger Transport in Canadian Urban Areas" (Canadian Transport Commission, Research Branch), December 1970, 119 pp.

practical applicability. Frequently many of the worthiest innovations pass through the necessary testing stages relatively quickly and then often receive wide utilization.

Prominent innovations in mass transportation in recent years are listed below. They are in varying stages of research, testing, development, and demonstration.

1. Demand-actuated, or variable-route-and-schedule, transportation systems, as discussed on pp. 228–229.
2. Personal rapid transit (PRT): "Small vehicles, traveling over exclusive rights-of-way, automatically routed from origin to destination over a network guideway system, primarily to serve low- and medium-population density areas of a metropolis."[16]
3. Dual-mode vehicle systems: Small or medium-sized vehicles that can be individually propelled and converted between driver-operated road travel and travel on automated guideway networks in individual units or trains.
4. Pallet or ferry systems: "An alternative to dual-mode vehicle systems ... to carry (or ferry) conventional automobiles, minibuses, or freight automatically on high-speed guideways."[17]
5. New conveyances for major activity centers: "Continuously moving belts; capsule transit systems, some on guideways, perhaps suspended above city streets."[18] See also the discussion on Secondary Distribution in this chapter.

This list is not exhaustive. Other novel and innovative methods and equipment, and combinations thereof, for mass transportation have been and will continue to be introduced and explored.[19]

REFERENCES FOR FURTHER READING

Listed below are selected references for further study of mass transportation characteristics. This bibliography is in addition to the overall basic traffic engineering references listed elsewhere in this Handbook.

"Transit Fact Book," American Transit Association, Washington, D. C. (Annual)

"Transit Operating Report(s)," American Transit Association, Washington, D. C. (Annual)

"Statistics of Urban Public Transport," 2nd Edition, International Union of Public Transport (UITP), Brussels, 1967

"International Commission for the Study of Motorbuses," *Report 4*, 39th International Congress, UITP, Rome, 1971

"Standardization of Motorbuses," *Report 5*, 38th International Congress, UITP, London, 1969

"International Commission for the Study of Motorbuses," *Report 5*, 37th International Congress, UITP, Barcelona, 1967

"A Consideration of Underground Urban Transport Systems," *Report 9*, 37th International Congress, UITP, Barcelona, 1967

PODOSKI, J., "Modern Tramways," *UITP Revue*, Brussels, Volume 19, No. 4, 1970, pages 263–282

"Rapid Transit Car Data Book," Institute for Rapid Transit (IRT), Washington, D.C. (5 volumes: April 1962 (2), May 1962, April 1965, 1971)

"Electric Commuter Car Data Book," IRT, Washington, D.C. 1970

"Rapid Transit Engineering Data Book," IRT, Washington, D.C. 1969

---

[16] "Tomorrow's Transportation: New Systems for the Urban Future" (Washington, D.C.: U.S. Department of Housing and Urban Development, 1968); summary report, p. 3.

[17] *Ibid.*

[18] *Ibid.*

[19] See footnote 15.

"Rapid Transit Fare Structure and Collection Methods," IRT, Washington, D.C., May 1970

HOMBURGER, W. S., "Urban Mass Transit Planning," Institute of Transportation and Traffic Engineering, University of California, Berkeley, 1967

"Highway Capacity Manual," *Special Report 87*, Highway Research Board, Washington, D.C., 1965, especially Chapter 11, pages 338–348

"Capacities and Limitations of Urban Transportation Modes," Institute of Traffic Engineers, Washington, D.C., 1965

RAINVILLE, W. S., and HOMBURGER, W. S., "Capacity of Urban Transportation Modes," *Journal of the Highway Division*, American Society of Civil Engineers, New York, N. Y., April 1963, pages 37–55

"Planning of Mass Transit Routes," Report of Technical Committee 3-D of the Institute of Traffic Engineers, *Traffic Engineering Magazine*, Washington, D.C. September 1957, pages 590–603

LEIBBRAND, K., "Stadtbahn Frankfurt am Main: Planerische Gesamtuebersicht," Volumes 1 and 2, June 1961

BERRY, D. S., BLOMME, G. W., and SHULDINER, P. W., "The Technology of Urban Transportation," Northwestern University Press, Evanston, Illinois, 1963

LANG, A. S., and SOBERMAN, R. M., "Urban Rail Transit," M.I.T. Press, Cambridge, Massachusetts, 1964

AYRES, R. U. and McKENNA, R. A., "Technology and Urban Transportation: Environmental Quality Considerations," 2 volumes, unpublished review draft, Hudson Institute, Inc., Croton-on-Hudson, N. Y., 1968

"The Potential for Bus Rapid Transit," Automobile Manufacturers Association, Detroit, Michigan, 1970

"Mass Transportation: 8 Reports," Highway Research Record No. 318, Highway Research Board, Washington, D.C., 1970

"Transportation Systems Planning: 12 Reports," Highway Research Record No. 293, Highway Research Board, Washington, D.C., 1969

"Origin and Destination: Methods and Evaluation: 10 Reports," Highway Research Record No. 114, Highway Research Board, Washington, D.C., 1966

SCHNEIDER, L. M., "Marketing Urban Mass Transit," Granduate School of Business Administration, Harvard University, Boston, Massachusetts, 1965

HAASE, R. H., and HOLDEN, W. H. T., "Performance of Land Transportation Vehicles," RAND Corporation, Santa Monica, California, 1964

Düwag Catalogue, Waggonfabrik Uerdingen A. G., Düsseldorf, 1968

GMC Catalogues: Transit, Light Transit, Suburban, Intercity; Truck and Coach Division, General Motors Corp., Pontiac, Michigan, 1970

"Basic Facts," London Transport Board, London, England, 1967, 20 pages mimeographed

"Statistics," London Transport Board, London, England, 1967, 17 pages mimeographed

"Bus Facts," 35th Edition, National Association of Motor Bus Operators, Washington, D. C., 1968

"The Measure of a Safer Bus," National Association of Motor Bus Operators, Washington, D. C., 1968

VIGRASS, J. W., "Articulated Bus Service in Germany," *Mass Transportation Magazine*, November 1959, pages 18–19

"The 'Londoner': London's New Double-Deck Bus," Metropolitan Magazine, July-August 1971, pages 10–11

"Minimum Standards for School Buses," 1964 Revised Edition, National Education Association, Washington, D.C., 1964

"Accident Facts," 1971 Edition, National Safety Council, Chicago, Illinois, 1971

HOMBURGER, W. S. and VUCHIC, V. R.: "Federation of Transit Agencies as a Solution for Service Integration," *Traffic Quarterly*, July 1970, pages 373–391.

"Proper Location of Bus Stops," ITE Recommended Practice, 1967

"Bus Stops for Freeway Operations," ITE Recommended Practice, 1971

"Demand-Actuated Transportation Systems," Highway Research Board (HRB), Special Report 124, 1971

"New Transportation Systems and Concepts," Highway Research Record 367, HRB, 1971

"New Transportation Systems and Technology," Highway Research Record 397, HRB, 1972

"Transit for the Poor, the Aged, and the Disadvantaged," Highway Research Record 403, HRB, 1972

"Mass Transportation: Application of Current Technology," Highway Research Record 415, HRB, 1972

"Public Transportation and Passenger Characteristics," Highway Research Record 417, HRB, 1972

Chapter 7

TRAFFIC FLOW THEORY

J. A. WATTLEWORTH, Professor of Civil Engineering, University of Florida, Gainesville, Florida.

INTRODUCTION

Traffic flow theory is one of the newest areas in the field of traffic engineering. In spite of its brief history there have been many significant developments in traffic flow theory. Some of these developments have led to very useful relationships while the applications of others have lagged.

Probably the most useful result of traffic flow theory is the development of the relationships among the macroscopic variables of traffic stream flow (flow rate, speed and density). The understanding of these relationships has carried over into most traffic engineering work and has led to the development of the level of service concept. It has led also to the development of an understanding of shock wave development and propagation and this has had important applications.

Workable models of delay have been developed and have had applications in freeway ramp merging situations, pedestrian crossing situations, stop sign applications and freeway bottleneck situations. In addition, car-following theories have provided insights into the behavior of individual vehicles in the traffic stream. Probability models have given traffic engineers tools to use in many situations. One other important application of traffic flow theory is in simulation. Flow theory has provided many useful inputs to the simulation process as well as some of the logical models which are internal to the overall simulation models.

There is some criticism of the traffic flow theory work regarding the lag between the theoretical developments and the applications of some portions of the flow theory work. The application lag is both regrettable and understandable. It is understandable because many of the persons who are working in flow theory are mathematicians, physicists or members of other disciplines than traffic engineering. Indeed, the traffic engineers owe a great deal to the contributions of these persons. The traffic flow theorist has largely looked into the basic relationships (*why* things happen) while the traffic engineer is concerned more with *what* happens. It, therefore, takes time for the applications to catch up to the theory. The applications of traffic flow theory which have most quickly followed the theoretical developments have come from traffic engineers who have developed a theory to help solve a particular problem.

Both the traffic engineers and flow theorists must accept some blame for the application lag. The flow theorist frequently investigates aspects of traffic flow which are not of immediate concern to the traffic engineers. Frequently, also, the results are written in mathematical terminology which is incomprehensible to most traffic engineers and/or are published in journals not generally read by traffic engineers. The traffic engineers have been slow to accept many of the flow theories and, in many instances, have not made an adequate effort to understand them.

Thus, the challenge facing the flow theorist is to address aspects of traffic flow that pose immediate problems to the traffic engineer and to output any practical results as soon as they are available and in an understandable form. The challenge to traffic engineers is to make greater attempts to understand traffic flow theory development and to communicate to the flow theorists the nature of their problem.

This chapter is intended to provide some of the basic developments in traffic flow theory and to present some applications. It is, of course, impossible to cover all of the subjects in the field or even to cover the selected subjects in great depth. It is hoped that the material in this chapter will be useful in providing not only an introduction to the field of traffic flow theory but also some insights into the applications of the theories.

FUNDAMENTAL DEFINITIONS

This section presents a description of some of the fundamental characteristics of traffic stream flow namely, *flow*, *speed*, *density*, and their related characteristics. In addition to the description of these basic characteristics, a discussion of their measurement at various types of locations is presented. A discussion of the relationships among these basic variables is presented later.

FLOW VARIABLES

1. *Flow rate*, q, is defined as the rate at which vehicles pass a point on a roadway. Flow rate is usually expressed as vehicles per hour (veh/hr), but it can be based on a shorter observation time.

   Example: In 15 min 900 vehicles pass a point on a roadway

$$q = \frac{900 \text{ veh}}{\frac{15}{60} \text{ hr}} = 3,600 \text{ vph}$$

2. The *volume*, Q, is the number of vehicles observed in a given time period, T. Volume is based on an actual count and expressed as vehicles.

   Example: For above data,

$$Q = 900 \text{ veh in 15 min}$$

   An hour volume is the actual count of vehicles for one hour. Unlike a flow rate, volume cannot be based on an expanded count.

3. *Time headway*, h_t, is the time between successive vehicles crossing a point on the roadway. It is usually expressed in seconds (sec).

4. *Average time headway*, $\bar{h}_t$, is the average of all headway times, h_t, on a roadway. Average time headway is expressed as seconds per vehicle (sec/veh) and can be determined using the volume, Q, or flow rate, q.

$$\bar{h}_t = \frac{3,600T}{Q} = \frac{3,600}{q} \tag{7.1}$$

SPEED VARIABLES

1. *Time mean speed*, u_t, is the speed on a roadway based on the average of the individual speeds of all vehicles on the roadway. Time mean speed is expressed as miles per hour (mph) or feet per second (fps).
2. *Travel time* is the time that it takes an individual vehicle to traverse a unit length of roadway.
3. *Total travel, TT*, is the sum of the individual travel distances of all vehicles in a given period of time or of all vehicles that have some other common characteristics in a given period of time.
4. *Total travel time, TTT*, is the sum of the individual travel times of all the vehicles crossing a length of roadway.
5. *Space mean speed*, u_s, is the speed on a roadway based on the average travel time across a length of roadway. Space mean speed is expressed as miles per hour (mph) or feet per second (fps). In any system, the space mean speed in a given time period is the total distance traveled by all vehicles while in the system divided by the total time that they were in the system, or

$$u_s = \frac{TT}{TTT} \tag{7.2}$$

DENSITY VARIABLES

1. *Density*, k, is the concentration of vehicles on a roadway. Density is expressed as vehicles per mile (veh/mi), but it can be based on a shorter length of roadway. Density can also be based on the total facility (all lanes) or on a per lane basis.

   Example: For a $\frac{1}{4}$ mi-length of a three-lane (one direction) roadway, 20 vehicles are observed in each lane at an instant of time.

$$k_{\text{lane}} = \frac{20 \text{ veh}}{\frac{1}{4} \text{ mi}} = 80 \text{ veh/mi/lane}$$

$$k_{\text{roadway}} = \frac{20 \text{ veh} \times 3 \text{ lanes}}{\frac{1}{4} \text{ mi lane}} = 240 \text{ veh/mi}$$

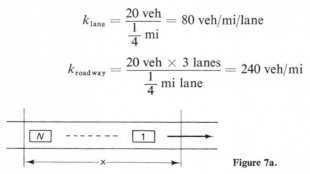

Figure 7a.

2. *Space headway*, h_d, is the distance between front bumpers of successive vehicles at a given instant of time. Space headway is expressed in feet.
3. *Average space headway*, $\overline{h}_d$, is the average of all space headways, h_d, on a roadway. Average space headway is expressed as feet per vehicle (ft/veh) and can be determined using density, k, or average time headway, $\overline{h}_t$, for a constant speed condition, u.

$$\overline{h}_d = \frac{5280}{k} = \overline{h}_t u \times \frac{5{,}280}{3{,}600} \tag{7.3}$$

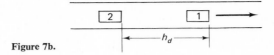

Figure 7b.

MEASUREMENTS AND INTERRELATIONSHIPS

General. As in any other fluid continuum, traffic stream flow is defined in terms of its three characteristic variables of flow, speed, and density. This definition is the interrelationship of these three variables and is applicable to any stream flow situation. In reference to roadway traffic this interrelationship is the basic equation of traffic flow, i.e.,

$$q = ku$$

where q = flow = veh/hr;
$\quad k$ = density = veh/mi;
$\quad u$ = speed = mi/hr.

The components of this equation are measured by employing one or more of the three basic types of measurements.

Point measurements are made at one location along the length of a roadway. This location or point is represented by a single line which is perpendicular to the roadway's length and extends across the roadway's width.

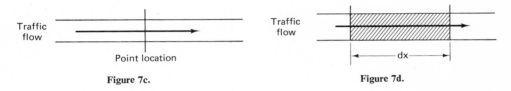

Figure 7c.   **Figure 7d.**

Section measurements are made between two locations on a roadway. The distance between the two locations is represented by length dx and defines a section of the roadway's length.

System measurements are of two types:

1. Physical or geometrical system measurements are made by treating the roadway as a system or unit. This involves establishing a boundary around the roadway and measuring the input and output of the system at this boundary.

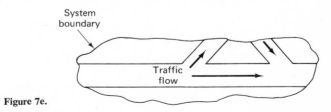

Figure 7e.

2. A certain group of vehicles can be considered a system. Measurements are made with respect to the number of vehicles, N, as they move along the roadway's length.

Upstream   Downstream   **Figure 7f.**

When measurements are in units of vehicles, hours, and miles, the following observations on the basic equation ($q = ku$) are made:

Quantity	Units	Measured at a
q	veh/hr	point
k	veh/mi	section
u	mi/hr	point or section

The basic equation ($q = ku$) is inconsistent because the components are measured in different ways. Therefore, an analysis of each of the three basic types of measurements will be made with respect to the characteristics of traffic stream flow. Ultimately, this analysis will remove the apparent inconsistency in the basic equation.

Measurements at a point.
Mean flow rate, $\bar{q}$, is defined as

$$\bar{q} = \frac{N}{T} \tag{7.4}$$

where N = the number of vehicles passing a point (veh);
$\quad T$ = elapsed time (*hr*).

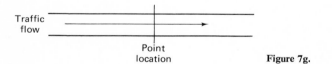

Traffic
flow

Point
location   **Figure 7g.**

Example:

100 vehicles are counted in 15 min.
Using Equation (7.4),

$$\bar{q} = \frac{1{,}000 \text{ veh}}{\frac{1}{4} \text{ hr}} = 4{,}000 \text{ vph}$$

The *volume*, Q, is defined for a time period, T, as the integral,

$$Q = \int_0^T q(t)\, dt \tag{7.5}$$

where t = time;
$\quad q$ = flow rate.
Integration gives

$$Q = \bar{q}T \tag{7.6}$$

Using equation (7.4), we get

$$Q = N \tag{7.7}$$

Example:

(For above data)
$Q = 1,000$ veh in 15 min
Mean time headway, $\overline{h}_t$, is defined as

$$\overline{h}_t = \frac{T}{N} \qquad \text{for } \overline{h}_t \text{ in hr/veh} \tag{7.8}$$

From Equation (7.4),

$$\overline{h}_t = \frac{1}{\overline{q}} \qquad \text{for } \overline{h}_t \text{ in hr/veh} \tag{7.9}$$

To calculate $\overline{h}_t$ in sec/veh, Equations (7.8) and (7.9) become

$$\overline{h}_t = \frac{3,600T}{N} \tag{7.10}$$

and

$$\overline{h}_t = \frac{3,600}{\overline{q}}, \text{ respectively} \tag{7.11}$$

Example:

(For above data),
Using Equation (7.8) or (7.9),

$$\overline{h}_t = \frac{.25 \text{ hr}}{1,000 \text{ veh}} = \frac{1}{4,000 \text{ veh/hr}} = 0.00025 \text{ hr/veh}$$

Using Equation (7.10) or (7.11),

$$\overline{h}_t = \frac{3600 \,(.25 \text{ hr})}{1000 \text{ veh}} = \frac{3600}{4000 \text{ veh/hr}} = 0.9 \text{ sec/veh}$$

Time mean speed, u_t, is defined as,

$$u_t = \frac{\sum\limits_{i=1}^{N} u_i}{N} \tag{7.12}$$

where $u_i = $ speed of vehicle i.
The *density*, k, cannot be defined at a point, i.e.,

$$\text{density} = \frac{\text{vehicles}}{\text{length}} \tag{7.13}$$

A characteristic similar to density can be made at a point by defining occupancy as the amount of time a point location has a vehicle present.

$$\text{Occupancy} = \frac{\sum\limits_{i=1}^{N} dt_i}{T} \tag{7.14}$$

where $dt_i = $ amount of time vehicle i is on the point location.
Occupancy cannot be used in the basic equation $q = ku$ for density, k, because it is dimensionless.

Mean distance headway, $\overline{h}_d$ is technically undefined since density cannot be defined at a point, i.e.,

$$\overline{h}_d = \frac{5{,}280}{k} \tag{7.15}$$

An approximation for $\overline{h}_d$ can be made at a point location using the time mean headway, $\overline{h}_t$, when the mean speed, u_t, is constant.

$$\overline{h}_{d\,\text{(approx.)}} = \frac{5{,}280}{3{,}600}\,\overline{h}_t u_t \tag{7.16}$$

for $\overline{h}_t$ in seconds, u_t in mi/hr, and $\overline{h}_{d\,\text{(approx.)}}$ in feet.

Measurements at a short section.

Mean flow rate, $\bar{q}$, is technically undefined in a section, i.e.,

$$\bar{q} = \frac{N}{T} \qquad \text{for } N \text{ at a point} \tag{7.17}$$

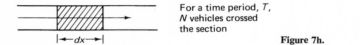

For a time period, T,
N vehicles crossed
the section

Figure 7h.

Volume, Q, in time T is technically undefined in a section, i.e.,

$$Q = N \qquad \text{which is the same as for } N \text{ at a point} \tag{7.18}$$

Mean time headway, $\overline{h}_t$, is technically undefined in a section, i.e.,

$$\overline{h}_t = \frac{T}{N} \tag{7.19}$$

and

$$\overline{h}_t = \frac{1}{\bar{q}} \qquad \bar{q} \text{ and } N \text{ are technically undefined in a section} \tag{7.20}$$

Time mean speed, u_t, is defined as

$$u_t = \frac{\sum\limits_{i=1}^{N} u_i}{N} \tag{7.21}$$

The speed, u_i, across a short section, dx, is

$$u_i = \frac{dx}{dt_i} \tag{7.22}$$

where dt_i is the time it takes vehicle, i, to travel the distance, dx.
Substituting Equation (7.22) into (7.21),

$$u_t = \frac{\sum\limits_{i=1}^{N} \dfrac{dx}{dt_i}}{N} = \frac{dx}{N}\left(\sum\limits_{i=1}^{N} \frac{1}{dt_i}\right) \tag{7.23}$$

Example:

Car 1 crosses a section $dx = 10$ ft in 0.5 sec.
Car 2 crosses in 1.0 sec.

Using Equation (7.23),

$$u_t = \frac{10 \text{ ft}}{2}\left(\frac{1}{0.5 \text{ sec}} + \frac{1}{1.0 \text{ sec}}\right) = 15 \text{ fps}$$

Space mean speed, u_s, is defined as

$$u_s = \frac{dx}{\overline{dt_i}} \qquad (7.24)$$

Since

$$\overline{dt_i} = \frac{\displaystyle\sum_{i=1}^{N} dt_i}{N} \qquad (7.25)$$

Equation (7.24) becomes

$$u_s = \frac{dx}{\displaystyle\sum_{i=1}^{N}\frac{dt_i}{N}} = \frac{N\, dx}{\displaystyle\sum_{i=1}^{N} dt_i} \qquad (7.26)$$

Example:

(For previous data)
Using equation (7.26), we get

$$u_s = \frac{2(10 \text{ ft})}{1.0 \text{ sec} + 0.5 \text{ sec}} = 13.3 \text{ fps}$$

Mean density, $\bar{k}$, over time period T is defined as

$$\bar{k} = \frac{\text{avg no. veh in } dx}{dx} \qquad (7.27)$$

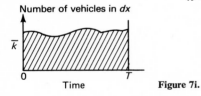

Figure 7i.

The probabilty that a vehicle, i, is in the section, dx, is

$$P(\text{veh } i \text{ is in } dx) = \frac{dt_i}{T} \qquad (7.28)$$

Using Equation (7.28), we get

$$\text{avg no. veh in } dx = \sum_{i=1}^{N} 1 \cdot P(\text{veh } i \text{ is in } dx) \qquad (7.29)$$

$$= \frac{1}{T} \sum_{i=1}^{N} dt_i \qquad (7.30)$$

Substituting Equation (7.30) into (7.27), we get

$$\bar{k} = \frac{\displaystyle\sum_{i=1}^{N} dt_i}{T\, dx} \qquad (7.31)$$

Example:

$$T = 10 \text{ min} = 600 \text{ sec}$$

$$dx = 10 \text{ ft}$$

$$N = 100 \text{ veh}$$

$$\overline{dt_i} = 0.1 \text{ sec}$$

Using equation (7.31), we get

$$\bar{k} = \frac{\sum\limits_{i=1}^{N} dt_i}{T \, dx} = \frac{N(\overline{dt_i})}{T \, dx} = \frac{100 \text{ veh}(0.1 \text{ sec})(5{,}280 \text{ ft/mi})}{600 \text{ sec}(10 \text{ ft})}$$

$$\bar{k} = 8.80 \text{ veh/mi}$$

Mean distance headway, $\overline{h_d}$, is technically undefined in a section, i.e.,

$$\overline{h_d} = \frac{5{,}280}{k} \qquad \text{at an instant of time} \tag{7.32}$$

where k is a function of time in a section,

$$k(t) = \frac{N(t)}{dx} \qquad N(t) \text{ is the number of vehicles in the system at time, } t \tag{7.33}$$

System measurements—physical system.

Flow rate, volume in time T and mean time headway are undefined except at points on the system boundary, i.e., they are not defined for the reasons given in Equations (7.17), (7.18), (7.19), and (7.20).

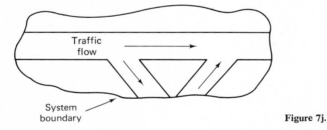

Traffic flow

System boundary

Figure 7j.

Time mean speed, u_t, is defined as

$$u_t = \frac{\sum\limits_{i=1}^{N} u_i}{N} \tag{7.34}$$

For a system, Equation (7.34) becomes

$$u_t(t) = \frac{\sum\limits_{i=1}^{N(t)} u_i(t)}{N(t)} \tag{7.35}$$

where $u_t(t)$ = time mean speed of the system at time t;

$N(t)$ = number of vehicles in the system at time t;

$\sum\limits_{i=1}^{N} u_i(t)$ = sum of the individual speeds, u_i, of the $N(t)$ vehicles at time t.

Space mean speed, u_s, over a short time interval, dt, is defined as

$$u_s = \frac{\sum_{i=1}^{N} dx_i}{N\,dt} \tag{7.36}$$

where dx_i = distance traveled by vehicle i in time dt.
Space mean speed over a section has units of vehicle miles per vehicle hours.
Space mean speed, u_s', over a time period, T, is defined as

$$u_s' = \frac{\text{vehicle miles of travel during } T}{\text{vehicle hours of total travel time during } T} \tag{7.37}$$

which becomes

$$u_s' = \frac{\text{total travel}}{\text{total travel time}} \tag{7.38}$$

Equation (7.38) is

$$u_s' = \frac{TT}{TTT} \tag{7.39}$$

Density, k, at an instant of time, t, is defined for a system as

$$k(t) = \frac{N(t)}{M} \tag{7.40}$$

where M = the length of the system in miles or lane miles.
Mean density, $\bar{k}$, over a period of time, T, is defined by using Equation (7.27) as

$$\bar{k} = \frac{1}{MT} \int_0^T N(t)\,dt \tag{7.41}$$

where $\frac{1}{T} \int_0^T N(t)\,dt$ is the average number of vehicles in the system during time period, T. M is the total distance in the system in miles.
Mean distance headway, $\bar{h}_d$, at an instant of time, t, is defined for a system as

$$\bar{h}_d = \frac{\text{total length of the system}}{\text{number of vehicles in the system at time, } t} = \frac{M}{N(t)} \tag{7.42}$$

Using Equation (7.40), we get

$$\bar{h}_d = \frac{1}{k(t)} \tag{7.43}$$

System measurement of a group of vehicles.

Flow rate, volume in time, T, and mean time headway are undefined for a group of N vehicles for the reasons given in Equations (7.17), (7.18), (7.19), and (7.20).

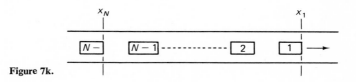

Figure 7k.

Mean speed, u, in a short interval of time, dt, is defined as

$$u(t) = \frac{\text{total distance traveled by the group during } dt}{\text{total travel time of group during } dt} \tag{7.44}$$

which is

$$u(t) = \frac{\sum\limits_{i=1}^{N} dx_i}{N\, dt} \tag{7.45}$$

Equations (7.36) and (7.45) are equal. Therefore,

$$u(t)_{\text{group veh}} = u_{s_{\text{system}}} = \frac{\sum\limits_{i=1}^{N} dx_i}{N\, dt} \tag{7.46}$$

Density, k, at an instant of time, t, is defined for a group of N vehicles as

$$k(t) = \frac{\text{total number of vehicles in group}}{\text{total length of group at time } t} \tag{7.47}$$

which is

$$k(t) = \frac{N-1}{x_1 - x_N} \tag{7.48}$$

where x_1 and x_N are shown in the above illustration.

The average space headway, $\overline{h}_d$, is defined for a group of N vehicles as

$$\overline{h}_d = \frac{\text{total length of group at time } t}{\text{total number of vehicles in group} - 1} \tag{7.49}$$

Equation (7.49) can be written as

$$\overline{h}_d = \frac{x_1 - x_N}{N-1} \tag{7.50}$$

and using Equation (7.42), we get

$$\overline{h}_d = \frac{1}{k(t)} \tag{7.51}$$

Returning to the basic equation of traffic ($q = ku$), we see from Equation (7.4) that

$$\bar{q} = \frac{N}{T} \tag{7.52}$$

and $\bar{k}$ from equation (7.31) is

$$\bar{k} = \frac{\sum\limits_{i=1}^{N} dt_i}{T\, dx} \tag{7.53}$$

Substituting Equations (7.52) and (7.53) into $q = ku$, we can determine which speed gives consistency to the $q = ku$. Then,

$$\frac{N}{T} = \frac{\sum\limits_{i=1}^{N} dt_i u}{T\, dx} \tag{7.54}$$

Equation (7.54) can be written as

$$u = \frac{N \, dx}{\sum\limits_{i=1}^{N} dt_i}$$

(7.55)

Equation (7.55) is identical to Equation (7.26) which is the equation for space mean speed. Therefore, the basic equation, $q = ku$, is consistent when space mean speed, u_s, is used for speed, u.

FUNDAMENTAL DEFINITIONS INDEPENDENT OF
METHOD OF MEASUREMENT

There are instances when the three previously discussed methods of measurement techniques are inappropriate, namely, platoons of vehicles traveling in space and time and for large values of section lengths, dx. Edie[1] developed procedures for extending the previous definitions to cover the additional cases, and the material in this section is extracted from Footnote 1. By combining the definitions derived for points, sections, and systems, fluid continuum equations that are independent of the method of measurement can be derived for platoons of vehicles and for long sections of roadway.

LARGE SECTION EQUATIONS

A space-time domain is defined as any enclosed portion of a space-time plane which is the area of the domain as shown in Figure 7.1.

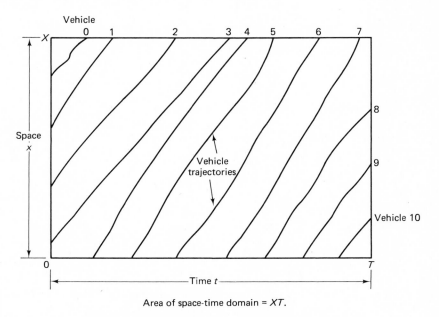

Area of space-time domain = XT.

Figure 7.1. Vehicle trajectories in space-time domain.

---

[1] L. C. EDIE, "Discussion of Traffic Stream Measurements and Definitions," *Proceedings, Second International Symposium on the Theory of Traffic Flow*, London, 1963.

The flow, q, of a traffic stream is defined as

$$q = \frac{\sum X_i}{A} \tag{7.56}$$

where $X_i =$ the distance traveled by the ith vehicle passing through the space-time domain;

$A =$ the area of the domain.

The density, k, of a traffic stream is defined as

$$k = \frac{\sum t_i}{A} \tag{7.57}$$

where $t_i =$ the time taken by the ith vehicle to pass through the space-time domain;

$A =$ the area of the domain.

The speed, u, of a traffic stream is

$$u = \frac{\sum X_i}{\sum t_i} \tag{7.58}$$

where X_i and t_i take on previously defined values.

These definitions are independent of measurement methods since t_i and X_i take on values for any configuration of the space-time domain whose area equals A. Also, these equations are independent of the statistical procedures used for analysis and therefore yield consistent results. These definitions have a reduced sensitivity to random errors which are inherent in point and short section measurements since they are caused by the measurement procedures themselves.

For application of these equations to combine N values of q_i, k_i, and u_i for domains with area a_i, the following identical rules apply;

$$q = \frac{\sum a_i q_i}{\sum a_i}, \qquad k = \frac{\sum a_i k_i}{\sum a_i}, \qquad u = \frac{\sum a_i q_i}{\sum a_i k_i} \tag{7.59}$$

EQUATIONS FOR PLATOONS OF VEHICLES

A space-time domain for a group of N vehicles traveling a distance X is shown in Figure 7.2. The bounds of the domain are the trajectory of vehicles 0 and N and the ends of the roadway. Neglecting the curvature of the trajectories for vehicles 0 and N, the area of the domain is defined as

$$A = TX - \frac{X^2}{2}\left(\frac{1}{u_0} + \frac{1}{u_N}\right) \tag{7.60}$$

where $T =$ total time and u_0 and u_N are the average speeds of vehicle 0 and N for traversing the domain.

Using Equations (7.56), (7.57), and (7.58), we see that the following definitions apply to the platoon of vehicles:

$$q = \frac{\sum x_i}{A} = \frac{NX}{TX - \dfrac{X^2}{2}\left(\dfrac{1}{u_0} + \dfrac{1}{u_N}\right)} \tag{7.61}$$

$$k = \frac{\sum t_i}{A} = \frac{X \sum \dfrac{1}{u_i}}{TX - \dfrac{X^2}{2}\left(\dfrac{1}{u_0} + \dfrac{1}{u_N}\right)} \tag{7.62}$$

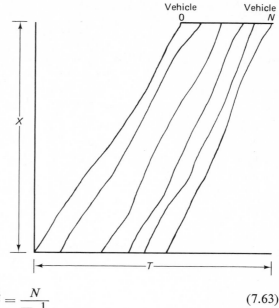

Figure 7.2. Trajectories for a group of N vehicles traveling distance X.

$$u = \frac{\sum X_i}{\sum t_i} = \frac{N}{\sum \frac{1}{u_i}} \qquad (7.63)$$

A space time domain for a group of N vehicles bound by a time period, T, and the trajectories of vehicles 0 and N is shown in Figure 7.3. The area and stream flow characteristics are defined as follows:

$$A = XT - \frac{T^2}{2}(u_0 + u_N) \qquad (7.64)$$

$$q = \frac{\sum X_i}{A} = \frac{\sum u_i T}{XT - \frac{T^2}{2}(u_0 + u_N)} = \frac{\sum u_i}{X - \frac{T}{2}(u_0 + u_n)} \qquad (7.65)$$

$$k = \frac{\sum t_i}{A} = \frac{N}{X - \frac{T}{2}(u_0 + u_N)} \qquad (7.66)$$

$$u = \frac{\sum X_i}{\sum t_i} = \frac{\sum u_i}{N} \qquad (7.67)$$

For Equations (7.61)–(7.63) and (7.65)–(7.67) the summations of distance, time, and velocity include data for only one (or the average) of the two vehicles (*zero*th and Nth) whose trajectories define the area of the domain.

TRAFFIC STREAM VARIABLE RELATIONSHIPS

GENERAL

The deterministic approach will be taken in this discussion of the relationships among the traffic-stream variables, flow, density, and speed. This implies that the functional relationships between the parameters measuring effectiveness and the input

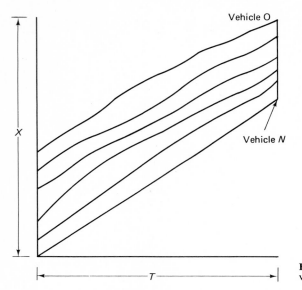

Figure 7.3. Trajectories for a group of N vehicles over time T.

variables are constant. This is to say that for any given set of values for the input variables, one and only one value of the measure of effectiveness function will occur. A dimensional analysis of the variables suggests the following relationship:

$$q \text{ (vehicles/hour)} = u \text{ (miles/hour)} \cdot k \text{ (veh/mile)} \tag{7.68}$$

where q = mean rate of flow;
u = space mean speed;
k = mean density.

If any two of these three variables are known, the third is uniquely determined. There is actually no single dependent variable; however, density is often considered the dependent variable because flow and speed are easier to measure and, therefore, serve as the independent variables. The relationship represented by Equation (7.68) can be visualized as a fundamental three-dimensional surface by plotting the equation on mutually perpendicular axes as shown in Figure 7.4.

Other terms requiring definition are as follows:

1. q_m is the maximum flow rate
2. u_f is the free speed or speed at free-flow conditions (level of service A).
3. u_m is the speed at which the flow rate is a maximum ($q = q_m$).
4. k_j is the jam density or density at which all vehicles are stopped ($u = 0$).
5. k_m is the density at which the flow rate is a maximum ($q = q_m$).

FLOW-DENSITY RELATIONSHIP

The fundamental traffic stream variable relationship is the flow-density relationship, which is depicted by the flow-density diagram, Figure 7.5(a). In general, density increases as flow increases until the capacity of the roadway is reached. Point C on the diagram represents the capacity or maximum flow rate, q_m. From this point on, flow decreases as density increases until jam density, k_j, is reached and flow equals

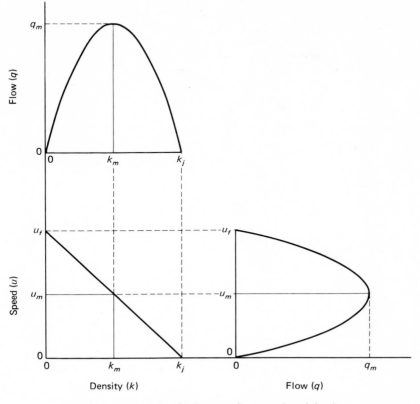

Figure 7.4. Relationship between flow, speed, and density.

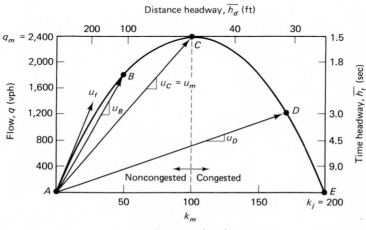

Figure 7.5(a). Flow-density relationship.

zero. The arrows from point A, the origin, to points B, C, and D on the curve are known as radius vectors. The slope of these radius vectors represents speed, u, in miles per hour of the operating conditions corresponding to the point at the end of the vector. The slope also represents travel time, which may be expressed in minutes per mile. The arrow that passes through point A is tangent to the curve and the slope of this arrow is the free speed, u_f. From fundamental definitions it can be shown that average time headways, $\overline{h}_t$, and average distance headways, $\overline{h}_a$, can be expressed as functions of flow and density, respectively, using Equations (7.1) and (7.3). Points on the flow-density curve for densities less than k_m represent noncongested conditions, and points with densities greater than k_m represent congested conditions.

A numerical example will illustrate the relationships portrayed by the flow-density diagram. Assume the average car length to be 20 ft and the average distance between cars at jam density to be 6.4 ft along a single lane of highway; therefore, $\overline{h}_a = 26.4$ ft. Since

$$\overline{h}_a = \frac{5{,}280}{k}$$

$k_j = 5{,}280/\overline{h}_a = 5{,}280/26.4 = 200$ vpm, the jam density value at point E on the curve. Next assume that $\overline{h}_t = 1.5$ sec. Since

$$\overline{h}_t = \frac{3{,}600}{q}$$

$q_m = 3{,}600/\overline{h}_t = 3{,}600/1.5 = 2{,}400$ vph, the maximum or capacity flow value at point C on the curve. The density, k_m, at point C may be read directly from the diagram and is equal to 100 vpm. Thus, the vertical and horizontal boundaries of the diagram are defined.

To determine the speed, u_m, at capacity flow, simply find the slope of the radius vector from the origin, A, to point C. Thus, $u_m = u_c = 2{,}400/100$ mph. Therefore at capacity, flow rate equals 2,400 vph, density equals 100 vpm and speed equals 24.0 mph in this example.

The values of other points on the flow density curve are found in a similar manner. Point B is a typical point representing noncongested conditions. From the diagram the flow at point B is 1,800 vph, the density is 50 vpm, and the speed (slope of radius vector AB) is 36 mph.

Point D is a typical point representing congested conditions. From the diagram the flow at point D is 1,224 vph, the density is 170 vpm, and the speed (slope of radius vector AD) is 7.2 mph. By definition, the values at point A equal zero for flow, density, and speed.

The specific values selected for this example are not intended to represent actual operating conditions on an actual highway facility but, instead, were chosen for convenience.

SPEED-DENSITY RELATIONSHIP

The speed-density relationship is depicted in Figure 7.5(b) where, for simplicity, a linear relationship is assumed. In general, speed decreases as density increases. Similarly, the average distance headway, $\overline{h}_a$, decreases as density increases. Free speed, u_f, occurs at point A. The speed-density curve does not intersect the vertical axis, but becomes asymptotic to it. This simply means that for there to be a free flowing speed,

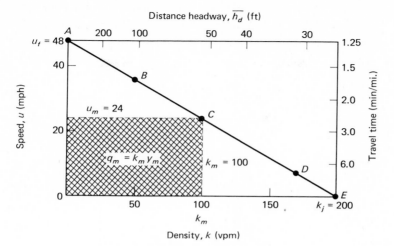

Figure 7.5(b). Speed-density relationship.

there must be at least one vehicle on the roadway traveling at that speed. This speed is generally governed by the geometric configuration and condition of the roadway.

Flow can also be interpreted in the speed-density diagram. For a given operating point in this diagram, the speed, u, and the density, k, are known. Since $q = ku$, the flow rate is the rectangular area which has the origin as one corner and the operating point as the opposite corner and which has the axes as two sides. This area is shown in in Figure 7.5(b) for point C.

The points A, B, C, D, and E correspond on both the flow-density and speed-density diagrams. Since these two diagrams share a common axis, density, it is easy to compare the respective positions of the typical points shown in Figures 7.5(a) and 5(b).

SPEED-FLOW RELATIONSHIP

The speed-flow relationship is depicted in Figure 7.5(c). This curve is similar in shape to the flow-density curve. In general, speed decreases as the flow increases until the capacity flow, q_m, is reached. In the congested portion of the curve, both flow and speed decrease. Points A, B, C, D, and E correspond to the same points on the flow-density and speed-density curves. The slope of the vectors from the origin, point E, to points on the curve represent the reciprocal of density, $1/k$, at that point. The portion of the speed-flow curve above point C represents noncongested conditions, while the portion of the curve below point C represents congested conditions. In summary, congested flow is represented by the right half of the flow-density curve and the lower half of the speed-flow curve in the region is defined by

$$q \leq q_m$$
$$k > k_m$$

and

$$u < u_m$$

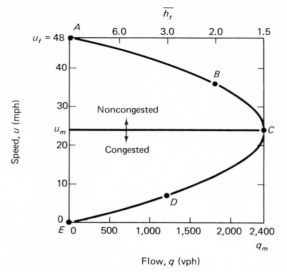

Figure 7.5.(c). Speed-flow relationship.

Noncongested flow is represented by the left half of the flow-density curve and the upper half of the speed-flow curve in the region defined by

$$q \leq q_m$$
$$k \leq k_m$$

and

$$u \geq u_m$$

MACROSCOPIC MODELS

Efforts to relate the various pairs of the three fundamental traffic variables have been based on: (1) measurement data and simple curve fitting, (2) deduction from the boundary conditions, and (3) physical analogies. These three approaches have resulted in the development of *macroscopic* models which assume a homogeneous movement of traffic or steady-state condition and describe the overall properties of the traffic stream.

MEASUREMENT AND CURVE FITTING

Greenshields,[2] in 1935, plotted speed, u, against density, k, for one lane of traffic, because this plane represented the only one of the three planes in the fundamental surface with a single valued relationship. He then fit a curve to the plotted data using the method of least squares which essentially estimates the true parameters of the curve, $u = f(k)$. Greenshields hypothesized that a linear relationship provided the best fit to the speed-density data. The linear model is expressed as

$$u = u_f - \frac{u_f}{k_j}(k) = u_f\left(1 - \frac{k}{k_j}\right) \tag{7.69}$$

---

[2] B. D. GREENSHIELDS, "A Study in Highway Capacity," Highway Research Board, *Proceedings*, Vol. 14, 1935, p. 468.

and a plot of this function is shown in Figure 7.6(a). Substituting this expression for u in Equation (7.68), we get

$$q = ku_f\left(1 - \frac{k}{k_j}\right) = u_f\left(k - \frac{k^2}{k_j}\right) \qquad (7.70)$$

which expresses q as a parabolic function of k. A plot of this function is shown in Figure 7.6(b). From this figure it is evident that q_m is a point on the curve where the slope of a line tangent to the curve is equal to zero and where $k = k_m$; therefore, differentiating Equation (7.70) with respect to k and setting it equal to zero, we get

$$\frac{dq}{dk} = u_f\left(1 - \frac{2k_m}{k_j}\right) = 0$$

Since u_f cannot equal zero,

$$1 - \frac{2k_m}{k_j} = 0$$

or

$$k_m = \frac{k_j}{2}$$

Next, we derive an expression for q as a function of u.
From Equation (7.69),

$$u - u_f = u_f\frac{-k}{k_j}$$

and

$$k = k_j\left(1 - \frac{u}{u_f}\right)$$

Substituting this expression for k in Equation (7.68), we get

$$q = uk_j\left(1 - \frac{u}{u_f}\right) = k_j\left(u - \frac{u^2}{u_f}\right) \qquad (7.71)$$

which expresses q as a parabolic function of u. A plot of this function is shown in Figure 7.6(c). Differentiating Equation (7.71) with respect to u and setting it equal to zero, we obtain

$$\frac{dq}{du} = k_j\left(1 - \frac{2u}{u_f}\right) = 0$$

Since k_j cannot equal zero at q_m,

$$1 - \frac{2u_m}{u_f} = 0$$

and

$$u_m = \frac{u_f}{2}$$

Therefore,

$$q_m = u_m k_m = \frac{u_f}{2}\frac{k_j}{2} = \frac{u_f k_j}{4} \qquad (7.72)$$

and is shown in Figure 7.6(a) as a rectangle of maximum area.

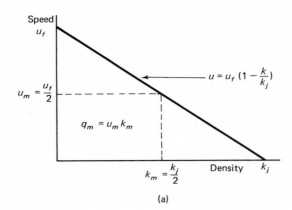

(a)

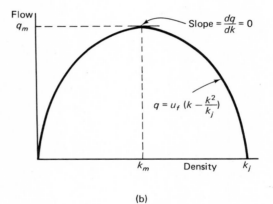

(b)

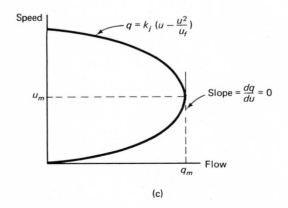

(c)

Figure 7.6. Relationships among the fundamental stream flow characteristics for the linear speed-density model.

Upon examination of field speed-density data, it is apparent that a strictly linear relationship does not always exist over the entire range of observation. Most data are better described by the curve shown in Figure 7.7. A linear model is appropriate for the center portion of the curve; however, the data suggests the possibility of other models that would take into account the apparent curvature that exists in the remaining portions. This led to the development of other models utilizing other approaches.

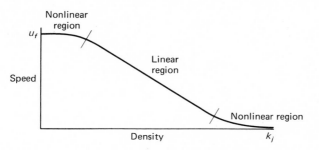

Figure 7.7. General type of speed-density curve obtained by field data.

THE DEDUCTIVE APPROACH

This approach simply employs the application of boundary conditions to points on the fundamental curves. Drew[3] demonstrates that if Equation (7.68) is differentiated with respect to k, the result is

$$\frac{dq}{dk} = k\frac{du}{dk} + u$$

At $k = k_m$,

$$\frac{dq}{dk} = 0$$

therefore,

$$k_m\frac{du}{dk} + u = 0$$

Separating the variables and integrating both sides results in

$$\ln u = \frac{k}{k_m} + c$$

where c is a constant of integration.

Applying the boundary condition at $k = 0$ where $u = u_f$, we get

$$\ln u_f = -\frac{0}{k_m} + c$$

Therefore,

$$c = \ln u_f$$

and

$$k = k_m \ln \frac{u_f}{u} \tag{7.73}$$

---

[3] D. R. DREW, *Traffic Flow Theory and Control* (New York: McGraw-Hill, 1968).

Substitution into Equation (7.68) gives

$$q = uk_m \ln \frac{u_f}{u} \qquad (7.74)$$

When $k = k_m$ and $u = u_m$, substitution into Equation (7.73) results in

$$k_m = k_m \ln \frac{u_f}{u_m}$$

Hence,

$$\ln \frac{u_f}{u_m} = 1$$

and

$$u_m = \frac{u_f}{e}$$

At $dq/dk = 0$,

$$u = u_m$$

Therefore,

$$k \frac{du}{dk} + u_m = 0$$

or

$$du + u_m \frac{dk}{k} = 0$$

Integrating, we get

$$u = -u_m \ln k + c$$

Applying the boundary condition at $k = k_j$ where $u = 0$ results in

$$0 = -u_m \ln k_j + c$$

and

$$c = u_m \ln k_j$$

Therefore,

$$u = u_m \ln \frac{k_j}{k} \qquad (7.75)$$

When $u = u_m$ and $k = k_m$, substitution into Equation (7.75) results in

$$u_m = u_m \ln \frac{k_j}{k_m}$$

$$\ln \frac{k_j}{k_m} = 1$$

and

$$k_m = \frac{k_j}{e}$$

Edie[4] obtained the same result by differentiating Equation (7.1) with respect to u. He hypothesized that there were two regimes of traffic flow: free flow and congested

[4] L. C. EDIE, "Car-following and Steady-state Theory for Non-congested Traffic," *Operations Research*, Vol. 9, No. 1 (January-February 1961), 66–75.

flow. He proposed that an exponential speed-density relation be used for the free-flow regime and that Equation (7.75) be used for the congested-flow regime giving rise to a "discontinuous" form of the steadystate surface.

In support of the discontinuous approach, May et al.,[5] define three zones that may be described as constant speed, constant volume, and constant rate of change of volume with density. In zone 1, the speed of the vehicle is determined by the facility itself and the volume matches the demand. Zone 2 represents impending poor operations; average speed drops but the flow rates may be sustained at a high level. In zone 3 both speed and volume rates decrease, which in itself may serve as a definition of congestion.

In 1960, Underwood[6] proposed an exponential speed-density relation

$$u = u_f e^{-k/k_m}$$

Many of these same results were obtained through the physical analogies approach.

THE PHYSICAL ANALOGY APPROACH

Analogies to traffic flow have been drawn from several other areas of engineering and science, particularly fluid flow and heat flow. The models that will be discussed here are primarily hydrodynamic models that are based on the flow of a one-dimensional, compressible fluid. Two conditions are assumed to be met in the development of the models. The first is that traffic behaves as a conserved system, i.e., if the flow rate decreases with distance, then density will increase with time. This principle is called the equation of continuity which stated mathematically is

$$\frac{\partial q}{\partial x} + \frac{\partial k}{\partial t} = 0 \qquad (7.76)$$

The second condition is that drivers adjust their speed in accordance with traffic conditions around them, i.e., if density increases with distance, speed decreases with time. This implies that speed could be negative; therefore, it is more logical to express an equation of motion in terms of acceleration rather than speed, so that the sign (positive or negative) would specify speeding up or slowing down and not forward or backward movement. The equation of motion is expressed mathematically as

$$\frac{du}{dt} = \frac{-c^2}{k} \frac{\partial k}{\partial x} \qquad (7.77)$$

In 1959, Greenberg[7] developed a relationship between speed and density by using the equation of continuity (Equation 7.70) and the equation of motion (Equation 7.77) which is expressed mathematically as

$$u = u_m \ln \frac{k_j}{k} \qquad (7.78)$$

This expression is commonly referred to as the Greenberg model, but note that it is identical to Equation (7.75) which was developed by using the deductive approach.

---

[5] A. D. May, P. Athol, W. Parker, and J. B. Rudden, "Development and Evaluation of Congress Street Expressway Pilot Detection System," Highway Research Record 21, 1963, pp. 48–68.

[6] R. T. Underwood, "Speed, Volume, and Density Relationships," *Quality and Theory of Traffic Flow* (New Haven: Yale Bureau of Highway Traffic, 1961), pp. 141–88.

[7] H. Greenberg, "An Analysis of Traffic Flow," *Operations Research*, Vol. 7, No. 1 (1959), 79–85.

Drew,[8] using the fluid-flow approach proposed by Greenberg, employs a more general derivation of this problem and proposes the following speed-density relation:

$$u = u_f\left(1 - \frac{k}{k_j}\right)^{\frac{n+1}{2}} \qquad \text{for } n > -1 \tag{7.79}$$

Drew[9] also discusses the case for $n = 0$ where the speed-density relation becomes

$$u = u_f\left[1 - \left(\frac{k_j}{k}\right)^{1/2}\right] \tag{7.80}$$

This expression is often referred to as the parabolic model. Substituting $n = 1$ and $n = -1$ into Drew's general model, Equation (7.79) results in Greenshields' linear model, Equation (7.69), and Greenberg's exponential model, Equation (7.78), respectively.

Drake, May, and Schofer[10] report on the application of a bell-shaped curve which gave satisfactory results when compared to speed-density measurements. This model is expressed mathematically as

$$u = u_f e^{-1/2(k/k_m)^2} \tag{7.81}$$

KINEMATIC WAVES

Using the method of kinematic waves, Lighthill and Whitham[11] suggest another relationship between flow and density. The hypothesis implies that slight changes in flow are propagated through the stream of vehicles along "kinematic waves," whose speed relative to the road is the slope of a line tangent to a point on the flow-density curve such as at points B and D shown in Figure 7.8. It is also pointed out that these waves can run together to form "shock waves" at which large reductions in speed occur very quickly. The speed of the shock wave is the slope of the line connecting two operating points on the flow-density curve. Shock waves often develop upstream of a bottleneck, which is defined as a section of roadway having a flow capacity less than the roadway sections to either side of it. This situation can be viewed graphically as shown in Figure 7.8 in which point B represents the conditions upstream of the shock wave and point D represents conditions downstream of the shock wave. The speed of the shock wave is interpreted as the slope of the line connecting these two points. Shock waves also form at the rear of traffic "humps" or sections of increased density. These humps will move slightly slower than the mean vehicle speed, and the vehicles passing through one of these sections will have to reduce speed quickly upon entering the section (at a shock wave), and they can increase speed only gradually as they leave it. A hump will gradually spread out along the road and with time will dissipate. Since a bottleneck can be thought of as a fixed hump, a period of low flow is required to dissipate the shock wave generated by it. The problem is further complicated when a traffic hump passes through a bottleneck. For a more detailed discussion of this problem and others related to it, the reader is referred to Lighthill and

---

[8] D. R. DREW, "Deterministic Aspects of Freeway Operations and Control," Texas Transportation Institute, Research Report 24–4, June 1965.

[9] D. R. DREW, "A Study of Freeway Traffic Congestion" (Doctoral dissertation, Texas A&M University, College Station, 1966).

[10] J. DRAKE, A. D. MAY, J. L. SCHOFER, "A Statistical Analysis of Speed Density Hypotheses," *Traffic Flow Characteristics*, Highway Research Record 154, 1967, pp. 53–87.

[11] M. T. LIGHTHILL and G. B. WHITHAM, "On Kinematic Waves. A Theory of Traffic Flow on Long Crowded Roads," *Proceedings of the Royal Society*, London, Series A, Vol. 229, 1955, pp. 317–345.

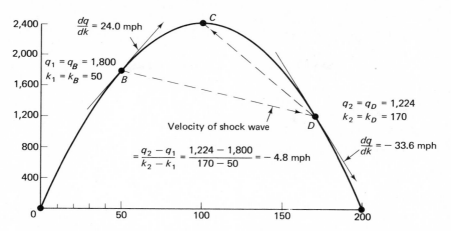

Figure 7.8. Kinematic and shock wave measurements related to flow-density curve. Point D represents the characteristics of a group of vehicles following a very slow (7.2 mph) truck where lane changing is not permitted. Upstream conditions at point B are for freely flowing traffic. The boundary line (shock wave) between free-flowing (B) and congested flow (D) will move backward down the roadway at 4.8 mph. If the truck were to exit, the flow characteristics at C will apply, and the shock wave between congested flow (D) and maximum flow (C) will move backward down the roadway at 16.8 mph.

Whitham, whose classic paper has been reprinted in Highway Research Board Special Report 79.[12]

ENERGY-MOMENTUM APPROACH

Drew[13] suggests yet another relationship between flow and density (which he calls the energy-momentum approach to level of service) in which he attempts to develop comparative methods to evaluate the effects of various designs and controls on the quality of flow. Again relying on the strong analogy between traffic flow and fluid flow, he employs the equation of continuity, motion, and momentum along with the law of conservation of energy to derive expressions of optimum flow, speed, and density that provide a rational basis for defining level of service and relating it to the fundamental traffic variables. Briefly, momentum in fluid flow is analogous to flow in traffic stream and is, therefore, defined as ku. Kinetic energy, E, is defined as αku^2 where α is a dimensionless constant. "Acceleration noise," σ, is considered as the disturbance of the vehicle's speed from a uniform speed and represents a measurement of the smoothness of traffic flow. Internal energy, I, or lost energy is represented by acceleration noise. The conservation of energy for the traffic stream over a section of road is a case of the total energy, T, being equal to the kinetic energy, plus the internal energy, or

$$T = E + I = \alpha ku^2 + \sigma \qquad (7.82)$$

From these principles and a consideration of geometrics and traffic interaction, an acceleration-noise-momentum-energy model was developed, resulting in the following

---

[12] *Introduction to Traffic Flow Theory*, Highway Research Board, Special Report 79, 1964.
[13] D. R. DREW, *Traffic Flow Theory and Control.*

parameters:

$$\text{Normalized kinetic energy } \frac{E}{T} = \frac{27}{4}\left[\left(\frac{u}{u_f}\right)^2 - \left(\frac{u}{u_f}\right)^3\right] \tag{7.83}$$

$$\text{Normalized internal energy } \frac{I}{T} = 1 - \frac{E}{T} \tag{7.84}$$

$$\text{Normalized flow } \frac{q}{q_m} = 4\left[\frac{u}{u_f} - \left(\frac{u}{u_f}\right)^2\right] \tag{7.85}$$

$$\text{Optimum speed } u_0 = \frac{2}{3}u_f = \frac{4}{3}u_m \quad \begin{array}{l}\text{based on maximizing } E \\ \text{and minimizing } I\end{array} \tag{7.86}$$

$$\text{Optimum service volume (flow) } q_0 = \frac{8}{9}q_m \tag{7.87}$$

$$\text{Optimum density } k_0 = \frac{2}{3}k_m \tag{7.88}$$

These parameters establish the four level-of-service zones defined by the 1965 *Highway Capacity Manual*—free, stable, unstable, and forced flow. Figure 7.9 illustrates the concept developed by Drew.

Numerous other models and approaches to macroscopic flow can be readily found in the literature. This discussion has been an attempt to highlight the work done in this area and to stimulate the reader to further reading and study.

MICROSCOPIC MODELS

This section describes the microscopic approach to traffic flow theory. The microscopic approach, sometimes referred to as the car-following theory, takes as its elements the spacing and speed of individual vehicles or vehicle pairs, while the macroscopic approach deals with traffic stream flows, densities, and average speeds.

APPLICABILITY OF CAR-FOLLOWING LAWS

Car-following theory applies to single-lane traffic with no overtaking. Pairs of vehicles are considered, one being the lead vehicle and the second being the trailing vehicle. It is important to point out that this theory can only be applied to two cars in actual traffic if the second driver is deemed to be following the first car. In practice, some drivers, even in reasonably dense traffic, will not attempt to follow the cars preceding them and hence their cars become leaders of separate groups of cars. It is to these groups of cars, moving under the close influence of each other, that the car-following theory applies.

The criterion for deciding whether or not a car is following another is difficult to formulate precisely. The *Highway Capacity Manual*[14] gives evidence of a "free zone" in which two cars are effectively independent and an "influenced zone" in which the cars' motions are closely dependent. The transition headway between the zones was measured to be 9 sec. From experiments carried out in Australia, George[15] has mea-

---

[14] *Highway Capacity Manual*, Highway Research Board, Special Report 87, 1965.

[15] H. P. GEORGE, "Methods of Measuring Traffic Volumes, Speed Delays, Traffic Capacities," paper presented at Summer School of Traffic Engineering, University of Melbourne, Australia, 1955.

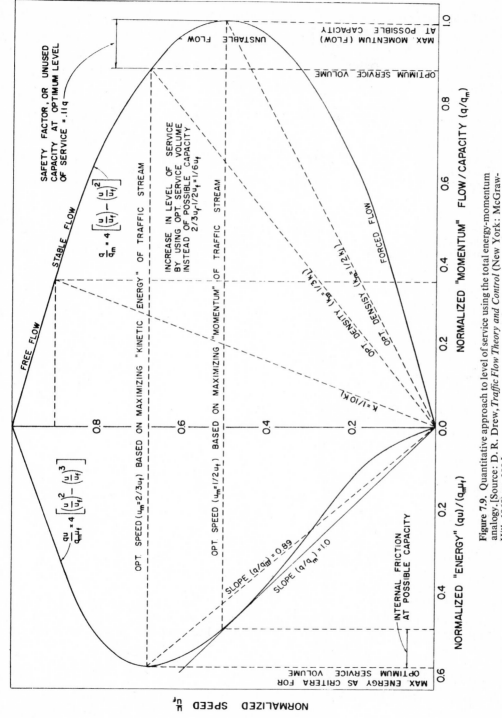

Figure 7.9. Quantitative approach to level of service using the total energy-momentum analogy. [Source: D. R. Drew, *Traffic Flow Theory and Control* (New York: McGraw-Hill, 1968), p. 380.]

285

sured this transition headway to be 6 sec. Also, the corresponding space headway has been statistically calculated as 200 ft.

The car-following theory attempts to describe mathematically the way vehicles proceed along a road and to determine qualitatively what happens to the dynamics of this chain when there is a fluctuation in the motion, i.e., in the speed. Does the disturbance increase or go away? If the former is the case, accidents may be caused or a major bottleneck may develop; this is important.

Car-following laws are derived from a stimulus-response approach, namely,

$$\text{response} = \text{sensitivity} \times \text{stimulus}$$

The response takes place in the trailing vehicle and is generally the acceleration of this vehicle. The stimulus is a function of the position, speed, or other attribute of the leading vehicle and the trailing vehicle. The sensitivity can take on several forms.

LINEAR CAR-FOLLOWING LAW

Reuschel[16] and Pipes[17] formulated the following equation for a pair of vehicles, one following the other:

$$x_n - x_{n+1} = L + D(\dot{x}_{n+1}) \tag{7.89}$$

where $x_n =$ the position of the lead vehicle;
$\quad x_{n+1} =$ the position of the trailing or following vehicle;
$\quad \dot{x}_{n+1} =$ the speed of the trailing vehicle;
$\quad L =$ the distance headway under jam conditions (when all vehicles are stopped);
$\quad D =$ a proportionality constant with the dimensions of time.

Equation (7.89) can be rewritten as follows:

$$\dot{x}_{n+1} = \frac{1}{D}(x_n - x_{n+1} - L) \tag{7.90}$$

This equation states that the speed of the trailing vehicle is proportional to the relative position of the leading and trailing vehicles.

The response in the car-following laws is generally the acceleration of the trailing vehicle and the differentiation of Equation (7.90) will produce the proper format. Equation (7.91) shows this result.

$$\ddot{x}_{n+1} = \frac{1}{D}(\dot{x}_n - \dot{x}_{n+1}) \tag{7.91}$$

This equation states that the acceleration of the following vehicle at any time is proportional to the relative speed of the leading and following vehicles at the same time. The following vehicle will accelerate if it is traveling more slowly than the lead vehicle and will decelerate if it is traveling faster than the lead vehicle.

This basic car-following model was extended by the research group at the General Motors Research Laboratory. Equation (7.91) states that there is an instantaneous response (acceleration) to the stimulus (relative speed). This group extended the model of Equation (7.91) to provide a time lag, T, between the stimulus and the response.

---

[16] A. REUSCHEL, "Fahrzeugbewegungen in der Kolonne," *Oesterreichisches Ingenieur-archiv,* Vol. 4, (1950), 193–215.

[17] L. A. PIPES, "An Operational Analysis of Traffic Dynamics," *Journal of Applied Physics,* Vol. 24, No. 3 (1953), 274–281.

The resulting model[18] is stated as follows:

$$\ddot{x}_{n+1}(t + T) = \lambda[\dot{x}_n(t) - \dot{x}_{n+1}(t)] \qquad (7.92)$$

where $\ddot{x}_{n+1}(t + T)$ = response (acceleration) of the $n + 1$th vehicle after time lag T
(this is of the order of 1.5 sec for 50 percent of the drivers, and in the range of 1.0 to 2.2 sec);

    $\dot{x}_n(t)$ = the speed of vehicle n (the lead vehicle) at time t;

    $\dot{x}_{n+1}(t)$ = the speed of vehicle $n + 1$ (the trailing vehicle) at time t;

    λ = sensitivity factor.

This model states that the acceleration of the following vehicle is proportional to the relative speed of the two vehicles at a time T earlier. Because the acceleration of the following vehicle is linearly related to the relative speed, this model is called the linear car-following model.

STABILITY OF THE LINEAR MODEL

The linear car-following model has two primary attributes that make it useful: its simplicity and its amenability to a stability analysis. In this section, results of stability analyses are presented.

Two types of stability will be described, local and asymptotic stability. Local stability is concerned with the response of one car to a change in motion of the car in front of it. Asymptotic stability is concerned with the manner in which a fluctuation of the motion of the lead car is propagated down a line of traffic, and stability limits in order for this fluctuation to be damped as it is propagated can be determined.

Local stability. Local stability limits have been determined theoretically from the car-following law with constant proportionality coefficient. Herman, Montroll, Potts, and Rothery[19] and Kometani and Sasaki[20] have shown that with this law the following situations, described in terms of the behavior of spacing between the two cars, can arise:

1. If $C = \lambda T > \pi/2 (\approx 1.57)$, the spacing is oscillatory with increasing amplitude.
2. If $C = \lambda T = \pi/2$, the spacing is oscillatory with undamped amplitude.
3. If $1/e (\approx .368) < C = \lambda T < \pi/2$, the spacing is oscillatory with damped amplitude.
4. If $C = \lambda T < 1/e$, the spacing is nonoscillatory and damped.

These local stability limits were checked by a numerical solution of the car-following law for two cars when the first decelerated and then accelerated back to its original speed. The responses for different values of C are illustrated in Figure 7.10.

Asymptotic stability. The limit for the asymptotic stability of a line of cars with respect to a fluctuation in the motion of the lead car has been investigated theoretically

[18] R. E. CHANDLER, R. HERMAN, and E. MONTROLL, "Traffic Dynamics-Studies in Car-Following," *Operations Research*, Vol. 6, No. 2 (1958), 165–84.

[19] R. HERMAN, E. W. MONTROLL, R. B. POTTS, and R. W. ROTHERY, "Traffic Dynamics: Analysis of Stability in Car-Following," *Operations Research*, Vol. 7, No. 1 (1959), 86–106.

[20] E. KOMETANI and T. SASAKI, "On the Stability of Traffic Flow," *Operations Research Society of Japan*, Vol. 2, No. 1 (1958), 11–16.

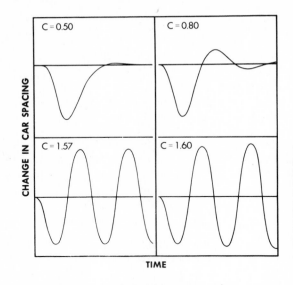

Figure 7.10. Change in car spacing of two cars for different values of $C = \lambda T$ when the first car maneuvers. For $C = 0.50$ and 0.80, the spacing is oscillatory and damped; for $C = 1.57$, oscillatory and undamped; for $C = 1.60$, oscillatory with increasing amplitude. (Source: R. Herman, E. W. Montroll, R. B. Potts, and R. W. Rothery, "Traffic Dynamics: Analysis of Stability in Car-Following," *Operations Research*, Vol. 7., No. 1, Jan-Feb, 1959, p. 94.)

for the linear car-following model by Chandler, Herman, and Montroll.[21] It was shown that if $C = \lambda T < \frac{1}{2}$, then the fluctuation is damped as it is propagated down the line, and that the rate with which the fluctuation is propagated is λ^{-1} sec per car.

Figure 7.11 illustrates the results of a numerical calculation for a line of eight cars subject to a change in motion of the first car for different values of C. It is interesting to note that when $C = 1/e$, the motion is both locally and asymptotically stable. It is important to point out, however, that in the theoretical and numerical calculations of

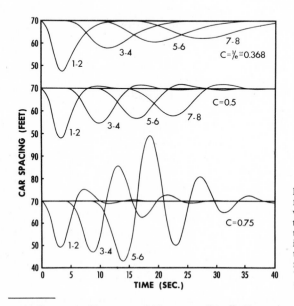

Figure 7.11. Car spacing of a line of cars for various values of $C = \lambda T$. The cars were originally spaced at 70 ft, the lead car then decelerated and then accelerated back to its original velocity, and the curves illustrate the propagation of the fluctuation down the line of cars. (Source; Herman, Montroll, Potts, and Rothery, "Traffic Dynamics: Analysis of Stability in Car-Following," p. 95.)

[21] CHANDLER, HERMAN, and MONTROLL, "Traffic Dynamics—Studies in Car-Following."

asymptotic stability, it has been tacitly assumed that the values of the proportionality coefficients for all cars in the line are the same. In practice this would not be true.

GENERAL FORMULATION OF THE CAR-FOLLOWING MODELS

The linear car-following model was useful in its simplicity and in its susceptibility to stability analyses. This model had one obvious shortcoming, namely, the reaction (acceleration) of the following vehicle was a function only of the relative speed of the two vehicles and was independent of the spacing of the vehicles.

The model formulation was refined in 1959 by Gazis, Herman, and Potts[22] by letting the sensitivity factor, λ, be inversely proportional to the distance of separation (distance headway).

$$\lambda = \frac{a_1}{x_n(t) - x_{n+1}(t)} \tag{7.93}$$

Then Equation (7.92) gives

$$\ddot{x}_{n+1}(t + T) = \frac{a_1}{x_n(t) - x_{n+1}(t)}[\dot{x}_n(t) - \dot{x}_{n+1}(t)] \tag{7.94}$$

In 1961, Gazis, Herman, and Rothery[23] proposed a more general expression for the sensitivity factor, λ,

$$\lambda = a \frac{\dot{x}_{n+1}^m(t + T)}{[x_n(t) - x_{n+1}(t)]^l} \tag{7.95}$$

The general expression for the microscopic theories thus becomes

$$\ddot{x}_{n+1}(t + T) = a \frac{\dot{x}_{n+1}^m(t + T)}{[x_n(t) - x_{n+1}(t)]^l} [\dot{x}_n(t) - \dot{x}_{n+1}(t)] \tag{7.96}$$

It can be seen that when $m = 0$ and $l = 0$, the general Equation (7.96) becomes Equation (7.92) (the linear model), while the condition $m = 0$ and $l = 1$ converts the general equation to Equation (7.94).

RELATIONSHIP BETWEEN MICROSCOPIC AND MACROSCOPIC MODELS

This section presents the generalized car-following model and its relationship to the various macroscopic models. It also evaluates different microscopic models by the use of traffic flow characteristics and statistical data as criteria.

MATRIX DEVELOPMENT AND RELATIONSHIP OF MICROSCOPIC AND MACROSCOPIC THEORIES

In order to derive the macroscopic models, steady-state flow is going to be considered; therefore, the lag-time, T, will be equal to zero.

The use of the general expression for the sensitivity factor in the stimulus-response equation (Equation 7.95) gives a very powerful tool for an evaluation of existing

---

[22] D. C. GAZIS, R. HERMAN, and R. POTTS, "Car-Following Theory of Steady-State Traffic Flow," *Operations Research*, Vol. 7, No. 4 (1959), 499–505.

[23] D. C. GAZIS, R. HERMAN, and R. W. ROTHERY, "Nonlinear Follow-the-Leader Models of Traffic Flow," *Operations Research*, Vol. 9, No. 4 (1961), 545–67.

models. All the models mentioned in the previous section can be described by the generalized equation (Equation 7.96) by using appropriate m and l values. Consider, for example, if $l = 0$, $m = 0$, and $\dot{x} = u$, then the above-mentioned generalized equation gives

$$\frac{du_{n+1}}{dt} = a(u_n - u_{n+1}) \qquad (7.97)$$

Since

$$x_n - x_{n+1} = S_{n+1} = \frac{1}{k_{n+1}}$$

then

$$u_n - u_{n+1} = \frac{dS_{n+1}}{dt} = \frac{1}{k_{n+1}^2} \frac{dk_{n+1}}{dt} \qquad (7.98)$$

where $S_{n+1} =$ spacing, between vehicles n and $n + 1$ in miles and

$\qquad k_{n+1} =$ density in vehicles per mile

If the subscripts are omitted, Equation (7.98) can be written

$$\frac{du}{dt} = -\frac{a}{k^2} \frac{dk}{dt} \qquad (7.99)$$

Integrating Equation (7.99) gives

$$u = \frac{a}{k} + C \qquad (7.100)$$

Boundary conditions are examined in order to evaluate C. Assume congested traffic flow, where $u = 0$ and $k = k_j$ (jam density); then from Equation (7.100) $C = -a/k_j$. Substituting C in Equation (7.100), we get

$$u = a\left(\frac{1}{k} - \frac{1}{k_j}\right) \qquad (7.101)$$

Now in the steady-state $q = uk$ and, hence,

$$q = a\left(1 - \frac{k}{k_j}\right) \qquad (7.102)$$

where a is a flow rate.

A matrix was developed to relate the various microscopic and macroscopic models. The matrix is presented in terms of the microscopic parameters (m and l), and the corresponding macroscopic model is the element of the matrix which is presented in Figure 7.12. An inspection of that matrix reveals that all of the earlier reported microscopic and macroscopic models and, in fact, several other possible models, can be located in terms of m and l combinations. For example, the models of Reuschel,[24] Pipes,[25] and Chandler, Herman, and Montroll[26] are obtained when $m = 0$ and $l = 0$. The Greenberg model[27] and the Gazis, Herman, Potts model[28] are obtained when $m = 0$ and $l = 1$. When $m = 0$ and $l = \frac{3}{2}$, the Drew model[29] is obtained. The Green-

[24] Reuschel, "Fahrzeugbewegungen in der Kilonne."
[25] Pipes, "An Operational Analysis of Traffic Dynamics."
[26] Chandler, Herman, and Montroll, "Traffic Dynamics—Studies in Car-Following."
[27] Greenberg, "An Analysis of Traffic Flow."
[28] Gazis, Herman, and Potts, "Car-Following Theory of Steady-State Traffic Flow."
[29] Drew, "Deterministic Aspects of Freeway Operations Control."

Microscopic parameters l \ m	$m = 0$	$m = 1$
$l = 0$	$u = a\left(\dfrac{1}{k} - \dfrac{1}{k_j}\right)$	—
$= 1$	$u = u_0 ln\left(\dfrac{k_j}{k}\right)$	—
$= 1.5$	$u = u_f\left[1 - \left(\dfrac{k}{k_j}\right)^{1/2}\right]$	—
$= 2$	$u = u_f\left[1 - \left(\dfrac{k}{k_j}\right)\right]$	$u = u_f e^{-(k/k_0)}$
$= 3$	—	$u = u_f e^{-1/2(k/k_0)^2}$

Figure 7.12. Matrix relationship between macroscopic and microscopic traffic flow models. (Source: A.D. May and H. E. M. Keller, "Non-Integer Car-Following Models," *Highway Research Record 199*, 1967, p. 23.)

shields model[30] results when $m = 0$ and $l = 2$, while the Edie[31] and Underwood[32] models result when $m = 1$ and $l = 2$. The bell-shaped curve proposed by Drake, May, and Schofer[33] is obtained when $m = 1$ and $l = 3$.

The matrix of m and l values not only shows that the existing traffic flow models can be reduced to the generalized car-following model, but also that by choosing particular m and l combinations, a wide variety of shaped curves for the speed-density relation can be selected. This can be seen in Figure 7.13. One also can recognize certain trends in the shape of the curves by keeping one of the exponents, m or l, constant. It should be noted that noninteger m and l values can be utilized and, consequently, an expression can be determined which more closely represents actual speed-density relations. This is shown in Figure 7.14 where for a constant exponent $m = 1$, the exponent l is changed in steps of $\frac{2}{10}$. One can see the gradual change from the exponential model ($m = 1$, $l = 2$) to bell-shaped model ($m = 1$, $l = 3$).

In examining the matrix, one should remember that the m value is the exponent of the following vehicle's speed $[x_{n+1}(t + T)]$ and the l value is the exponent of the spacing of the two vehicles $[x_n(t) - x_{n+1}(t)]$. Consequently, the fundamental difference between models is the weight given to the following vehicle's speed and the spacing between vehicles.

This matrix of m and l values has permitted the development of analytical techniques for evaluating deterministic traffic flow models using speed-density measurements. The results of the analytical techniques are described in the following section.

---

[30] GREENSHIELDS, "A Study in Highway Capacity."
[31] EDIE, "Car-following and Steady-state Theory for Non-congested Traffic."
[32] UNDERWOOD, "Speed, Volume, and Density Relationships."
[33] DRAKE, MAY, and SCHOFER, "A Statistical Analysis of Speed Density Hypothesis."

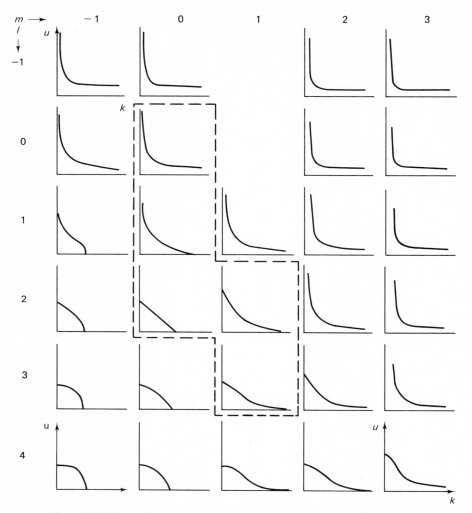

Figure 7.13. Matrix of speed-density relations for various *m*, *l* combinations of the general car-following equation. (Source: May and Keller, "Non-Integer Car-Following Models," p. 24.)

EVALUATION OF DETERMINISTIC INTEGER AND NONINTEGER TRAFFIC FLOW MODELS

The use of an *m*, *l* matrix gives the possibility of comparison of existing models due to given criteria. The *m*, *l* plane is the basis for the method of evaluation. Two different procedures have been used.[34] The statistical analysis is based on the minimizing of the mean deviations of the data points shown in Figure 7.15 from the determined regression curve. The preference to this method was given because of its clearness. Although

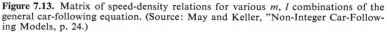

---

[34] A. D. MAY and H. E. M. KELLER, "Non-Integer Car-Following Models," *Mathematical and Statistical Aspects of Traffic*, Highway Research Record 199, 1967, pp. 19–32.

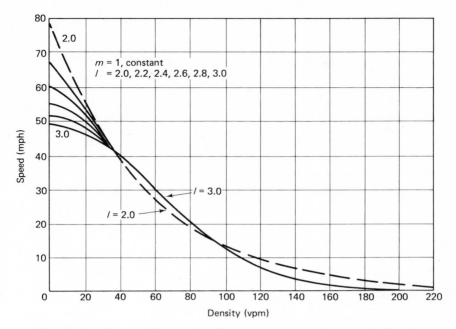

Figure 7.14. Influence of the use of non-integer exponents on the speed-density relation. (Source: May and Keller, "Non-Integer Car-Following Models," p. 24.)

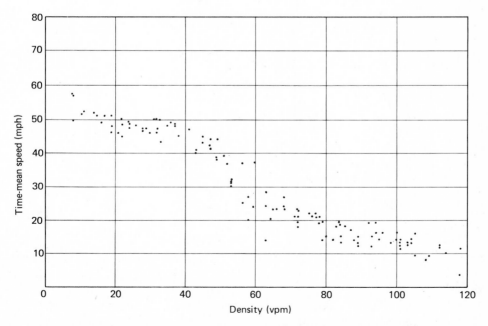

Figure 7.15. Speed-density data collected on the Eisenhower Expressway. (Source: May and Keller, "Non-Integer Car-Following Models," p. 25.)

there are very little differences in the mean deviations for particular m, l combinations, this method has been shown to be very effective.

The other procedure introduces the traffic flow characteristics as evaluation criteria. This resulted in a very helpful tool because it gives the control that certain physical restrictions or limitations of the flow characteristics are considered. For example, the free speed, jam density, and capacity flow values of the model can be examined to determine if the model is reasonable. The graphical superposition of results of both evaluation procedures allows a judgment about the goodness of fit of existing traffic flow models to given criteria and an estimate of those m, l combinations which best fit the investigated data. This is shown in Figure 7.16. The introduction of a continuum of noninteger exponents m and l implies a considerably greater variety of possible models and a more flexible adjustment to the assumed criteria of evaluation.

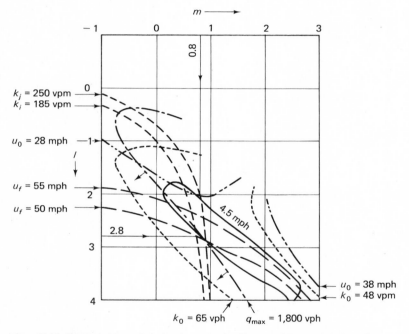

Figure 7.16. Superposition of evaluation criteria. (Source: May and Keller, "Non-Integer Car-Following Models," p. 28.)

The combination of $m = .8$ and $l = 2.8$ fulfills all requirements of the assumed evaluation criteria, mean deviation, free speed, jam density, optimum speed, optimum density, and maximum flow. The evaluation indicates that the area around the line between $m = .5$, $l = 2.5$, and $m = 2.5$, $l = 3.5$ covers models of very good fit, but with a maximum of flow rate of less than 1,800 vph. Models with a maximum flow rate greater than 1,800 vph appear in the area below this line. As can be seen from Figure 7.13, the speed-density relation tends to be bell-shaped in that area.

The investigation of the flow-density relation in these optimum combinations of m, l values indicates a unique shape of the flow-density curve. That is in the high-density regime, where the relationship exhibits a reversed curve.

STATISTICAL DISTRIBUTIONS AND TRAFFIC FLOW

Statistical distributions have been found to be quite useful in describing many of the processes with which traffic engineers must deal. These statistical distributions can be placed in two general categories:

1. Discrete or counting distributions.
2. Continuous or gap distribution.

Both categories of distributions are important and several types of each will be discussed.

DISCRETE DISTRIBUTIONS

In the discrete distributions the variable is the number of events, e.g., the number of vehicles in a given time period or length of highway, the number of accidents in a given time period, or the number of violations of a particular law. Four discrete distributions are considered:

1. Binomial distribution.
2. Negative binomial (or Pascal) distribution.
3. Geometric distribution.
4. Poisson distribution.

Binomial distribution. A binomial process is characterized by a series of n independent trials, each of which has only two possible outcomes and on each trial the probability of obtaining a particular outcome is constant. An example would be a simple coin-tossing experiment in which a coin is tossed ten times. There would be ten trials in this experiment and each trial has only two possible outcomes (heads or tails); the probability of obtaining a head remains constant from trial to trial. Thus, this is a binomial process.

Normally, one of the possible outcomes is defined as a "success" and p is the probability of a success on any trial. The binomial distribution gives the probability of the number of successes, x, in the n trials and can be stated mathematically as follows:

$$P(x) = \binom{n}{x} p^x q^{n-x} \tag{7.103}$$

where $P(x) =$ the probability of x successes in n trials;

$\quad\quad n =$ the number of trials;

$\quad\quad x =$ the number of successes;

$\quad\quad p =$ the probability of success on any given trial;

$\quad\quad q =$ the probability of failure on any given trial;

$\quad\quad\quad = 1 - p;$

$\quad\quad \binom{n}{x} = \dfrac{n!}{x!(n-x)!}$

The term $\binom{n}{x}$ is called the binomial coefficient and represents the number of ways n things can be placed in groups of x things.

The expected number of successes in n trials in which p is the probability of success on any trial is

$$E(x) = np \qquad (7.104)$$

and the variance of $x = npq$.

The binomial distribution is used to predict such events as the number of cars making turns at an intersection and the number of vehicles that will be traveling faster than the speed limit on a certain length of highway.

Negative binomial distribution (*Pascal distribution*). This distribution is similar in form to the binomial distribution, but it is nearly opposite in its application. In the binomial distribution the number of trials, "n," was fixed and the number of successes, "x," was distributed. In the negative binomial distribution the number of successes "x," is fixed and the number of trials, "n," required to obtain "x" successes is distributed. The form in which this distribution is most frequently used is to define $x = n - k$ so that the expression of the negative binomial distribution is the following:

$$P(x) = \binom{x + k - 1}{k - 1} p^k q^x \qquad (7.105)$$

where $P(x)$ = the probability that "x" failures occur in "n" trials before getting "k" successes;

p = the probability of success on a given trial;

$q = 1 - p$;

k = the number of successes in "n" trials where the last trial is a success.

The expected value of x is

$$E(x) = \frac{kq}{p}$$

and the variance of x is

$$\mathrm{var}(x) = \frac{kq}{p^2}$$

Geometric distribution. This distribution is a special case of the negative binomial distribution in which the process continues until there is one success ($k = 1$). The geometric distribution gives the probability that "n" trials are required to obtain one success and is expressed as

$$P(n) = pq^{n-1} \qquad (7.106)$$

The expected value of n is

$$E(n) = \frac{1}{p}$$

and the variance of n is

$$\mathrm{var}(n) = \frac{q}{p^2}$$

Poisson distribution. The Poisson distribution is the most frequently observed discrete distribution in traffic flow. This distribution gives the probability of "x" successes in terms of a single parameter, "m," where "m" is the expected number of "successes" in the process being analyzed.

In traffic studies this distribution gives the probability that a certain number of successes are observed in a period of time.

The equation of the probability distribution is

$$P(x) = \frac{m^x e^{-m}}{x!} \tag{7.107}$$

where x = the expected number of occurrences in a given time period;
m = the mean number of occurrences in a given time period.

The Poisson distribution is used to predict such events as the number of vehicles delayed at an intersection during the red phase, under different signal conditions, the delay caused by left-turning vehicles at an intersection, and the number of arrivals at an intersection.

A synopsis of discrete distributions is given in Table 7.1.

TABLE 7.1
Summary Table of Discrete Distributions

	Binomial Distribution	Negative Binomial Distribution	Poisson Distribution
Frequency	$\binom{n}{x} p^x q^{n-x}$	$\binom{x+k-1}{k-1} p^k q^x$	$\dfrac{m^x e^{-m}}{x!}$
Mean $E(x)$	np	$\dfrac{kq}{p}$	m
Variance var (x)	npq	$\dfrac{kq}{p^2}$	m
Mean/Variance	$1/q \ (> 1)$	$p(< 1)$	1

The method of fitting one of the discrete distributions to observed data is to first test which of the distributions is applicable and then to estimate the parameters of the selected distribution. The mean $\bar{x}$ and the variance s^2 of the observed data are calculated and the ratio mean/variance obtained. If this ratio is approximately one (1), the Poisson distribution will apply. If the ratio is appreciably greater than one (1), the binomial distribution may be tested, and if appreciably less than one (1), the negative binomial is appropriate.

For the Poisson distribution, the estimate of the single parameter "m" is given by

$$m = \bar{x} \tag{7.108}$$

For the binomial distribution, the parameter estimates are

$$n = \frac{\bar{x}^2}{\bar{x} - s^2} \tag{7.109}$$

$$p = \frac{\bar{x} - s^2}{\bar{x}} \tag{7.110}$$

For the negative binomial distribution, the parameter estimates are

$$k = \frac{\bar{x}^2}{s^2 - \bar{x}} \tag{7.111}$$

$$p = \frac{\bar{x}}{s^2} \tag{7.112}$$

CONTINUOUS DISTRIBUTIONS

Continuous distributions are used to describe variables that can assume any value over a range of values. Examples of continuous variables are gap size and time between accidents at a location.

Since the variable can take on any value over the specified range, it can essentially assume any of an infinite number of values. Consequently, the probability that a continuous variable takes on a particular value is zero. When a probability is calculated in a continuous distribution, it represents the probability that the value of the continuous variable falls within a range of values.

Some of the more useful continuous distributions are presented briefly in the following sections.

Negative exponential distribution. The negative exponential is used frequently in studying headways in a traffic stream or the interval between other events, such as accidents. It can also be used in cases in which the continuous variable is other than time, for example, distance.

The negative exponential distribution is used to describe the interval between events and, as such, can be derived from the Poisson distribution. The correspondence is that the interval between events is a period in which there are no events, e.g., a period in which $x = 0$ in the Poisson distribution.

$$P(x) = \frac{m^x e^{-m}}{x!} \tag{7.113}$$

The probability that an interval of time τ passes with no events occurring is the same as the probability that $x = 0$ in time τ.
Thus,

$$P(t \geq \tau) = P(x = 0 \text{ in time } \tau) = \frac{m^0 e^{-m}}{0!} = e^{-m}$$

Since m is the expected number of successes in time τ, $m = \lambda\tau$, where λ is the average rate of occurrence of events.
Thus,

$$P(t \geq \tau) = e^{-\lambda\tau} \tag{7.114}$$

and this is the probability that an interval between a pair of events is τ or longer. Correspondingly, the probability that an interval between a pair of events is less than τ is

$$P(t \leq \tau) = 1 - e^{-\lambda\tau} \tag{7.115}$$

The probability density function for the negative exponential distribution is

$$f(t) = \lambda e^{-\lambda\tau} \tag{7.116}$$

The negative exponential distribution is used to represent situations in which the events are distributed randomly. It is probably the most widely used distribution in traffic flow situations since the distribution of headways is of wide interest and traffic flow is frequently considered to be random. The latter assumption is generally considered to be valid for uninterrupted flow rates of 500 vehicles per lane per hour or less.

Shifted negative exponential distribution. For many applications, such as headway distributions in a traffic stream, a very low probability is associated with small values of the variable. The negative exponential has the highest probability density for these low values. In order to achieve a more realistic modeling of these cases, the shifted negative exponential distribution was developed. In the shifted negative exponential distribution, a minimum interval between events is specified and is termed c. This has the effect of simply shifting the negative exponential distribution to the right by a distance "c." Further, it will be observed that the flow rate $\lambda = 1/\bar{t}$ (the reciprocal of the mean headway).

The probability density function for the shifted negative exponential distribution can be stated

$$f(t) = \frac{1}{(\bar{t} - c)} e^{-[(t-c)/(\bar{t}-c)]} \qquad \text{for } t \geq c \tag{7.117}$$

The cumulative form of the shifted negative exponential distribution can be stated

$$P(t < c) = 1 - e^{-[(t-c)/(\bar{t}-c)]} \qquad \text{for } t \geq c \tag{7.118}$$

Erlang distribution. A more general distribution of headways is given by the Erlang distribution which can be varied by changing the parameter "a".

The cumulative form of the Erlang distribution can be stated

$$P(t < \tau) = 1 - e^{-\lambda \tau}\left(1 + \lambda\tau + \frac{(\lambda\tau)^2}{2!} + \frac{(\lambda\tau)^3}{3!} + \cdots \frac{(\lambda\tau)^{a-1}}{(a-1)!}\right) \tag{7.119}$$

The probability density function can be stated

$$f(t) = \lambda e^{-\lambda t} \frac{(\lambda t)^{a-1}}{(a-1)!} \qquad a = 1, 2, 3 \ldots \tag{7.120}$$

where t = the variable time;
τ = a specific value of time;
λ = rate of occurrence of events;
a = the coefficient of randomness.

The negative exponential distribution is a member of the Erlang family. It will be recalled that the negative exponential distribution represents a completely random distribution of events. In the Erlang family this corresponds to $a = 1$; if this value is substituted into Equations (7.119) and (7.120), the negative exponential distribution and probability density function are obtained.

A member of the Erlang family of distributions can be used to represent the distributions of intervals between events if the events are not randomly distributed. An approximate value of "a" can be determined for a process by using the following relationship:

$$a = \frac{\bar{t}^2}{s^2} \tag{7.121}$$

where for a set of observed headways or intervals,
$\bar{t}$ = the mean of the observed headways;
s^2 = the variance of the observed headways.

APPLICATIONS

This section presents some simple examples of the applications of some of the distributions that have been presented.

Binomial distribution. As an example, consider a police department which has found that on Saturday nights 25 percent of all drivers have been drinking. If they stop five drivers, what is the probability that exactly two will have been drinking? Let us define a "success" as finding a drinking driver. Thus, in this example

$$n = 5$$
$$x = 2$$
$$p = 0.25$$
$$q = 0.75$$

$$P(x) = \frac{n!}{x!(n-x)!} p^x q^{n-x}$$

$$= \frac{5!}{2!(5-2)!}(0.25)^2(0.75)^{5-2}$$

$$= \frac{5!}{2!3!}(0.25)^2(0.75)^3 = \frac{135}{512} = 0.264$$

For another example, consider an intersection approach with a left-turn bay and at which a separate left-turn phase is provided. An average of 20 vehicles arrive at the intersection per cycle and 25 percent turn left. What is the probability that the left-turn phase will not be used on a given cycle? In this case

$$n = 20 \text{ vehicles arriving}$$
$$x = 0 \text{ left-turning vehicles}$$
$$p = 0.25$$
$$q = 0.75$$

$$P(x) = \frac{n!}{x!(n-x)!} p^x q^{n-x}$$

$$P(o) = \frac{20!}{0!(20!)}(0.25)^0(0.75)^{20}$$

$$= (0.75)^{20}$$

$$= 0.0032$$

Negative binomial distribution. A traffic engineer has three new signal installations that require workable controllers. His experience indicates that when new controllers are taken from the warehouse, approximately 60 percent are immediately ready for use and about 40 percent require a week or so of adjustments. What is the probability that all five controllers from the warehouse will be required to produce the three workable controllers?

$$k = 3 \text{ workable controllers}$$
$$n = 5 \text{ controllers}$$

$$x = n - k = 2$$
$$p = 0.6$$
$$q = 0.4$$
$$P(x) = \frac{(x + k - 1)!}{(k - 1)!x!} p^k q^x$$
$$= \frac{4!}{2!2!}(0.6)^3(0.4)^2$$
$$= (6)(0.6)^3,(0.4)^2$$
$$= 0.207$$

Thus, the probability is 0.207 that exactly five controllers will have to be taken from the warehouse to obtain three workable controllers.

Poisson distribution. A given intersection has averaged two accidents per year for the past ten years. What is the probability that there will be five accidents at the intersection next year?

$$m = 2 \text{ accidents (the expected number of accidents in a year)}$$
$$x = 5 \text{ accidents}$$
$$P(x) = \frac{m^x e^{-m}}{x!}$$
$$P(5) = \frac{2^5 e^{-2}}{5!}$$
$$= 0.036$$

Thus, in approximately one year in 28 years the intersection will experience five accidents, even though the expected number of accidents there is only two.

Negative exponential distribution. At a stop sign location, vehicles require a 6 sec headway on the main stream flow in order to be able to cross the stream. If the flow rate on the main stream is 1,200 vph, what is the probability that any given headway will be 6 sec or greater?

$$\tau = 6 \text{ sec}$$
$$\lambda = 1,200 \text{ vph} = \frac{1}{3} \text{ veh/sec}$$
$$P(t \geq \tau) = e^{-\lambda\tau}$$
$$= e^{-(1/3)(6)} = e^{-2}$$
$$= 0.135$$

Thus, according to the negative exponential distribution approximately 13.5 percent of the headways are greater than 6 sec.

Shifted negative exponential distribution. In the previous example, if the minimum headway is 1.0 sec, what is the probability that a given headway will be greater than 6.0 sec?

$$c = 1.0 \text{ sec}$$

$$\tau = 6.0 \text{ sec}$$

$$\lambda = 1{,}200 \text{ vph} = \frac{1}{3} \text{ veh/sec}$$

$$\bar{t} = \frac{1}{\lambda} = 3 \text{ sec headway}$$

$$P(t \geq \tau) = e^{-[(\tau - c)/(\bar{t} - c)]}$$
$$= e^{-[(6 - 1)/(3 - 1)]}$$
$$= e^{-5/2}$$
$$= .082$$

Thus, with the shifted negative exponential distribution, approximately 8.2 percent of the headways are greater than 6.0 sec.

QUEUEING THEORY APPROACHES

GENERAL

Congestion, unfortunately, is an all too common attribute of the highway traffic system and the development of queues, or waiting lines, occurs in many portions of this system. Queues build up and dissipate normally during each phase of a traffic signal. Bunches of traffic, or moving queues, develop in moderately heavy to heavy traffic streams. And, of course, queues of pedestrians and vehicles develop while waiting to cross a major traffic stream. These are some of the more important and common examples of queueing in highway traffic situations.

In general, queueing situations arise either because of the variability (or stochastic nature) of the traffic flow or because the capacity varies with time. This section begins with a discussion of conventional queueing models and then presents some of the queueing models that are directly related to highway traffic.

CONVENTIONAL QUEUEING MODELS[35]

General. A queueing situation develops when items arrive at a serving channel for some type of service. The service of each arrival takes some length of time and can be provided from one or more than one serving channel. Figure 7.17 schematically illustrates this situation for a single channel queue. This figure shows a queueing system with a single serving channel. Units arrive at the system and enter the serving channel if it is idle or enter the queue to wait for service if the server is busy. The arrival rate is

Figure 7.17. Schematic of a single-channel queueing system.

[35] The material in this section can be found in most standard references on operations research, for example, M. Sasieni, A. Yaspan, and L. Friedman, *Operations Research—Methods and Problems* (New York: John Wiley and Sons, 1964).

usually termed λ and the average service rate (when service is being provided) is usually termed μ. Frequently, both the arrival rate and service rate vary and these variations cause the length of the queue to vary. The queue is defined as the number of units waiting for service and does not include any units actually being served (in Figure 7.17 the queue length is five units).

Classification of queueing systems. There are several means of classifying queueing systems. Several are discussed here.

1. *Single channel or multichannel queues.* Queues can be classed according to the number of serving channels—whether there is a single channel or several channels. Most queueing situations in highway traffic are single channel queues.
2. *Undersaturated or oversaturated queues.* If the arrival rate is less than the maximum serving rate of a queueing system, it is classed as an undersaturated queue. The queue length varies in length but reaches steady state. If the arrival rate is greater than the maximum service rate, the system never reaches steady state and the queue length is continually increasing. Most of the situations that will be dealt with in this chapter are undersaturated queues.
3. *Arrival and departure distributions.* The arrivals and service distributions can be one of several types ranging from uniform to random. For several reasons, most frequently both the arrival and service time distributions are treated as being random, i.e., Poisson distributed.
4. *Queue disciplines.* The queue can be operated in several ways, some of which are listed below.
   a. *First-in, first-out.* Under this system, the units are served in the order in which they arrive.
   b. *Last-in, first-out.* Under this system, the units are served in the reverse of their arrival order.
   c. *Queues with balking or reneging.* These are queues in which some arrivals will not wait if the queue is long and they will leave.
   d. *Truncated queues.* These are queueing situations in which there is a limited waiting area. When this maximum queue length is reached, all further arrivals leave.
   e. *Priority queue.* Under these priority queues, the arrivals are placed in various priority groups. The item to be selected for service is taken from the highest priority group in which there is a queue.

Conditions that determine queue length. The four conditions that affect the length of a queue are:

1. Probability distribution of arrivals.
2. Probability distribution of service times.
3. Number of serving channels.
4. Line discipline.

Figures of Merit—Single Channel Queue (Random Arrivals, Random Service Times)

1. *Probability distribution of the number in the system.* The number in the system refers to the number of items in the queue plus the number being served. The

probability of no units in the system (an idle system) is given as

$$P_0 = 1 - \frac{\lambda}{\mu} \tag{7.122}$$

and the probability of n units in the system is

$$P_n = \left(\frac{\lambda}{\mu}\right)^n P_0 \tag{7.123}$$

2. *Average queue length.* The average queue length, $E(m)$, is the average number of units waiting to be served.

$$E(m) = \frac{\lambda^2}{\mu(\mu - \lambda)} \tag{7.124}$$

3. *Average number in the system.* The average number in the system, $E(n)$, refers to the average number in the queue plus the average number being served.

$$E(n) = \frac{\lambda}{(\mu - \lambda)} \tag{7.125}$$

4. *Average waiting time of an arrival (time spent in the queue).* This refers to the average length of time that an arrival spends waiting before he begins to be served. It is termed $E(w)$.

$$E(w) = \frac{\lambda}{\mu(\mu - \lambda)} \tag{7.126}$$

5. *Average time a unit spends in the system.* The average time in the system is the total time that an arrival spends in the queue and being served. It is termed $E(v)$.

$$E(v) = \frac{1}{\mu - \lambda} \tag{7.127}$$

Figures of Merit—Multichannel Queueing System. Many queueing systems are characterized by more than one service channel. Although these situations are more frequently encountered in real-world applications, the mathematics of multichannel queueing systems are more complex than for single channel queues. Figure 7.18 shows a schematic of a multichannel queueing system. It should be mentioned at this time that all units waiting for service are shown waiting in a single queue and the unit at the head of the queue will go to the first available service channel. In actual practice, a separate queue usually forms at each service channel. It turns out, however, that these two situations are mathematically the same and it is easier to conceptualize the waiting units as being in a single queue.

In Figure 7.18, λ is again the arrival rate, μ_1 is the service rate of channel 1, μ_2 is the service rate of channel 2, and μ_k is the service rate of channel k. Thus, the queueing

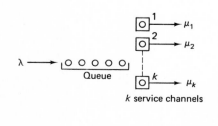

Figure 7.18. Schematic of a multichannel queueing system.

system would be undersaturated if

$$\lambda < \sum_{i=1}^{k} u_i$$

In many practical problems, the characteristics of all service channels are the same, i.e., $\mu_1 = \mu_2 = \ldots = \mu_k = \mu$. Thus, the system is undersaturated if $\lambda < k\mu$ where k is the number of channels and μ is the service rate of each channel.

Several figures of merit of a multichannel queueing system are presented below.

1. *Probability of no items in the system.* This is the probability that the queueing system is completely idle with no service being given and no units in the queue.

$$P_0 = \frac{1}{\left[\sum_{n=0}^{k-1} \frac{1}{n!} \left(\frac{\lambda}{\mu} \right)^n \right] + \frac{1}{k!} \left(\frac{\lambda}{\mu} \right)^k \left(\frac{k\mu}{k\mu - \lambda} \right)} \tag{7.128}$$

2. *Probability of* n *items in the system.* The probability that there are n items in the system is given by two formulas, one for n less than the number of channels and one for n greater than or equal to the number of channels.

$$P_n = \frac{1}{n!} \left(\frac{\lambda}{\mu} \right)^n P_0 \qquad \text{for } n \leq k - 1 \tag{7.129}$$

$$P_n = \frac{1}{k! k^{n-k}} \left(\frac{\lambda}{\mu} \right)^n P_0 \qquad \text{for } n \geq k \tag{7.130}$$

3. *Average queue length.* The average queue length, $E(m)$, is the average number of units waiting to be served.

$$E(m) = \frac{\lambda\mu \left(\frac{\lambda}{\mu} \right)^k}{(k-1)!(k\mu - \lambda)^2} \times P_0 \tag{7.131}$$

4. *Average number in the system.* The average number in the system, $E(n)$, is the average number in the queue plus the average number being served.

$$E(n) = E(m) + \frac{\lambda}{\mu} \tag{7.132}$$

5. *Average waiting time of an arrival* (*time spent in the queue*). The average waiting time, $E(w)$, is the average time an arrival spends before it begins to be served.

$$E(w) = \frac{\mu \left(\frac{\lambda}{\mu} \right)^k}{(k-1)!(k\mu - \lambda)^2} \times P_0 \tag{7.133}$$

6. *Average time a unit spends in the system.* The average time a unit spends in the system, $E(v)$, is the average waiting time plus the average service time.

$$E(v) = E(w) + \frac{1}{\mu} \tag{7.134}$$

INTERSECTION DELAY

General. Delay at intersections is a major problem to motorists and pedestrians alike. If no signal is present to control the assignment of right of way, the vehicles on the side street and pedestrians must wait for a gap of the proper size in the main street traffic stream before they can cross. If there is a signal controlling traffic at the intersection, vehicles and pedestrians must wait during their red phase(s) until the green phases begin and queues form and dissipate. The following sections present some theoretical approaches to the description of queueing at intersections. Pedestrian models generally differ from vehicular models in that pedestrians can accept a gap as a group (the entire queue is cleared as soon as an acceptable gap arrives), but vehicles must accept gaps singly or in small groups.

Nonsignalized intersection. Drew[36] developed a delay model for freeway entrance ramp merging situations, but the same model applies equally well to the use of pedestrians or vehicles crossing a stream of flow at an intersection. This is true since the model produces the delay to a vehicle that is first in line waiting for a gap. In other words, there is no accounting for the time the vehicle spends in the queue.

The Drew delay model is based on an Erlang headway distribution in the stream to be crossed and, thus, represents a more general approach than that taken by earlier investigators who generally assumed a negative exponential distribution. Also, it is assumed that the drivers have a constant critical headway, T, that is, all drivers will accept a headway larger than or equal to T and will reject a headway smaller than T. It is also assumed that the critical headway remains constant until the driver accepts a headway.

Drew found the average delay to a vehicle ready and waiting to merge or cross a stream to be

$$E(t) = \frac{e^{aqT} - \dfrac{(aqT)^i}{i}}{qT - \dfrac{(aqT)^i}{i}} \qquad (7.135)$$

where $T =$ the critical headway;
$\quad q =$ the flow rate in the stream to be crossed.
Thus, for the case in which the stream to be crossed has a negative exponential distribution ($a = 1$), the average delay becomes

$$E(t) = q^{-1}(e^{qT} - 1 - qT) \qquad (7.136)$$

It will be recalled that Equations (7.135) and (7.136) apply only to the delay after a vehicle (or pedestrian) is in a position to accept or reject a gap in the traffic stream. The above model can be extended to give an approximate value of the total waiting time in the queue for the case where only one vehicle at a time may cross or merge with a stream. The process may be treated as a single channel queue in which the vehicle that is ready to cross the stream is considered as being served. Thus, the service rate for this system, assuming a negative exponential distribution for the stream to be crossed, is

$$\mu = \frac{1}{E(t)} = [q^{-1}(e^{qT} - 1 - qT)]^{-1} = \frac{q}{e^{qT} - 1 - qT}$$

The arrival rate, λ, equals the flow rate on the side street. Since the average time in the system for a single channel queue (Equation 7.127) is

$$E(v) = \frac{1}{\mu - \lambda}$$

---

[36] DREW, *"Traffic Flow Theory and Control."*

(if we assume random arrivals and holding times), we find that

$$E(v) = \left(\frac{q}{e^{qT} - 1 - qT} - \lambda \right)^{-1} \tag{7.137}$$

This can be written as

$$E(v) = \frac{e^{qT} - 1 - qT}{q - \lambda(e^{qT} - 1 - qT)} \tag{7.138}$$

where q = the main stream flow rate;
$\quad \lambda$ = the side street flow rate;
$\quad T$ = the critical headway.

Technically, Equation (7.138) is correct only when a negative exponential waiting time distribution is found; therefore, the solution is considered to be approximate only.

Consider the following example: A stream of 720 vph is crossed by a minor stream of 180 vph. What is the delay to a vehicle (or pedestrian) waiting in position to cross? What is the total delay to the crossing vehicles, including time in queue?

$$q = \text{main stream flow rate} = 720 \text{ vph} = 0.2 \text{ veh/sec}$$

$$\lambda = \text{side street flow rate} = 180 \text{ vph} = 0.05 \text{ veh/sec}$$

$$T = \text{critical headway} = 5 \text{ sec}$$

From Equation (7.136)

$$E(t) = 5(e^1 - 1 - 1) = 5(.7183) = 3.59 \text{ sec}$$

The total delay per crossing vehicle by Equation (7.138) is

$$E(v) = \frac{(e^1 - 1 - 1)}{0.2 - .05(e^{-1} - 1 - 1)} = \frac{.7183}{0.2 - .05(.7183)} = 4.38 \text{ sec}$$

Signalized intersections. Queueing at signalized intersections occurs during the red phases on each approach and the queueing becomes more severe as the volume on an approach nears the capacity of the approach. Webster[37] developed an equation for the average delay per vehicle on an approach of an intersection controlled by a pre-timed signal controller. His formula is

$$d = \frac{c(1 - \lambda)^2}{2(1 - \lambda x)} + \frac{x^2}{2q(1 - x)} - 0.65\left(\frac{c}{q^2}\right)^{1/3} x^{(2+5\lambda)} \tag{7.139}$$

where d = average delay on the intersection approach; sec/veh;
$\quad c$ = cycle length, sec;
$\quad \lambda$ = proportion of the cycle which is green on the approach, i.e., the g/c ratio;
$\quad q$ = flow on the approach, vph;
$\quad s$ = saturation flow, vph, on the approach;
$\quad x$ = the degree of saturation on the approach = $g/\lambda s$.

It was found that a very good and simpler representation of the formula for average delay per vehicle is

$$d = 0.9\left[\frac{c(1 - \lambda)^2}{2(1 - \lambda x)} + \frac{x^2}{2q(1 - x)}\right] \tag{7.140}$$

[37] F. V. WEBSTER, *Traffic Signal Settings*, Road Research Laboratory, Technical Paper No. 39, 1958.

If d is the average delay per vehicle, the average queue length on the approach can can be found by

$$E(n) = \frac{qd}{3,600} \qquad (7.141)$$

where $E(n)$ = the expected queue length on the approach;

q = the flow rate on the approach, vph;

d = average delay per vehicle, sec/veh.

Of course, the queue length would vary considerably from the average length because of the periodic nature of the service rate.

SUGGESTED REFERENCES FOR FURTHER READING

DREW, D. R., *Traffic Flow Theory and Control*. New York: McGraw-Hill Book Company, 1968.

ASHTON, W. D., *The Theory of Road Traffic Flow*. New York: John Wiley & Sons, Inc., 1966.

HAIGHT, F. A., *Mathematical Theories of Traffic Flow*. New York: Academic Press, 1963.

Introduction to Traffic Flow Theory, Special Report 79. Washington, D.C.: Highway Research Board, 1964.

WOHL, M. and D. V. MARTIN, *Traffic System Analysis for Engineers and Planners*. New York: McGraw-Hill Book Company, 1967.

Chapter 8

HIGHWAY CAPACITY

CARLTON C. ROBINSON, Executive Vice President for Operations, Highway Users Federation for Safety and Mobility, Washington, D.C.

The capacity of a traffic facility is the measure of its ability to accommodate a stream of moving vehicles. It is a rate instead of a quantity and is not directly comparable to the capacity of a container or enclosed space. The maximum service rate, or "capacity," of a facility can be affected by a number of factors—the roadway, vehicle performance characteristics, operational controls, environmental elements. The basic determinant, however, is the driver and the summation of control decisions made by a group of drivers under the particular roadway, traffic stream, and environmental conditions present. This is inherently a variable because the control decisions made by members of any large group of drivers and even by the same driver in two successive time periods will not be identical.

CAPACITY FACTORS

Under the most ideal circumstances of roadway and environment and with the most homogeneous group of drivers and vehicles conceivable in current highway practice, the capacity of a single traffic lane would probably be in the range of 2,400 passenger vehicles per hour. This volume is produced by average headways of 1.5 sec. Such average headways have been observed for short time periods and, on rare occasions, for full hours in the center lanes of high design standard freeways. Because the combination of circumstances that permit these high flow rates is extremely rare and, if present, results in a highly unstable condition, this value is of little practical significance.

A more useful figure is the *maximum number of vehicles that have a reasonable expectation of passing over a given roadway in a given time period under the prevailing roadway and traffic conditions*. This is the definition of capacity adopted by the Highway Research Board[1] and generally followed in American practice. Although this value could change if there were a change in vehicle characteristics, driver competence, or other external factors, observations over 40 years in the United States as well as more recent studies in other parts of the world suggest that it is relatively stable. The capacities of various roadway configurations determined by these observations are given in Table 8.1.

---

[1] *Highway Capacity Manual*, Special Report 87, Highway Research Board (Washington, D.C.: Highway Research Board, 1965), p. 5.

TABLE 8.1
Capacity Under Ideal Conditions

Highway Type	Capacity (Passenger Vehicles per Hour)
Two or more lanes in one direction	2,000 average per lane
Two lanes in two directions	2,000 total both directions
Three lanes in two directions	4,000 total both directions
One lane, from a stopped condition	1,500

Source: *Highway Capacity Manual*, Special Report 87 (Washington, D.C.: Highway Research Board, 1965), pp. 76–77.

Even the values in Table 8.1 are of limited use to the engineer because:

1. All conditions necessary to sustain capacity flow are frequently not present and the more typical problem is to determine the capacity of a facility when some factors are less than ideal.
2. When a facility is serving capacity level volumes, the quality of service provided to the user is poor in terms of safety, freedom of maneuver, and speed. Normally, the engineer's problem is to determine the *service volume* that can be accommodated and to provide the user a selected *level of service*. This determination is discussed in a later section on level of service.

The use of computer programs to solve capacity problems is discussed in Chapter 11, Computer Applications.

TIME PERIOD

The number of vehicles that may pass a particular point on a highway facility is determined not only by the capacity of that section but also by the capacities of other portions of the system that meter traffic to and from the section and, importantly, by the demand for service that may or may not be present. As the time period under consideration is lengthened, this latter factor increasingly becomes the determinant.

It is known that traffic demand is cyclical on a daily, weekly, and seasonal basis and subject to other peaks caused by external influence. It is of little value to know the "annual capacity" of a facility based on 8,760 consecutive hours of maximum demand because such a pattern of demand will not occur (see Chapter 4, General Traffic Characteristics). At the other extreme, even with maximum demand present, there will be random variations in the rate of vehicles serviced during successive short time periods, for example, one minute. Therefore, most capacity computations are made on the basis of an intermediate time period which, in American practice, is usually one hour.

Adjustments to the pattern of demand over longer periods (one day, one week, or one year) are accomplished by applying factors based on known patterns of demand variation. This process is discussed in Chapter 13, Statewide and Regional Transportation Planning.

The traffic demand on a specific facility within a one-hour period can, frequently, be subject to fluctuations, depending on the relationship of that facility to traffic generators and other elements of the transportation system. A large number of observations of traffic flow on urban streets and on freeways have shown that only rarely will ten successive six-minute periods or four successive fifteen-minute periods be subject to equal and maximum demand. The practical effect of this demand variation is to reduce the number of vehicles that can be served during a one-hour period without those in the maximum demand period within that hour being subject to a lowered level of service.

The peak hour factor (phf) is a measure of these short-term variations in demand. It is defined as *the ratio of the volume occurring during an hour to the peak rate of flow during a given time period within that hour expressed as an hourly total.* Example 8.1 shows the computation of a six-minute peak hour factor (phf_6).

Example 8.1:

```
If during minute 0 to minute 6 flow equals 15
And  //      //   7 //    //  12 //    //    16
 //  //      //  13 //    //  18 //    //    18
 //  //      //  19 //    //  24 //    //    17
 //  //      //  25 //    //  30 //    //    21
 //  //      //  31 //    //  36 //    //    25
 //  //      //  37 //    //  42 //    //    23
 //  //      //  43 //    //  48 //    //    16
 //  //      //  49 //    //  54 //    //    13
 //  //      //  55 //    //  60 //    //    11
                                            ───
Total hour volume equals                    175
```

$$\text{and } phf_6 = \frac{\text{volume}}{10 \times \text{peak flow rate}} = \frac{175}{10 \times 25} = \frac{175}{250} = 0.70$$

The peak hour factor can be computed on different time bases. For example, the twelve-minute peak hour factor in Example 8.1 is

$$\frac{175}{5(25 + 23)} = 0.73$$

In American practice, a peak hour factor adjustment is made for freeways and expressways on a five- or six-minute base and for urban signalized intersections on a fifteen-minute base. Although there is some rationale for this practice, it is largely dictated by the format of available research and operational data.

All influences that affect the peak hour factor are not quantified; it is known, however, that the peak hour factor is a function of the type and diversity of traffic generators being served by the facility, the spatial relationship of the facility to those generators, and capacity restraints elsewhere in the system. Average peak hour factor has been shown to be closely correlated with metropolitan area population. Table 8.2 shows five-minute peak hour factors (phf_5) averaged from freeway facilities in 31 American cities.

Fifteen-minute peak hour factors (phf_{15}) have been compiled for a large number of urban signalized approaches as shown in Table 8.3. Although some intersections produced values near the theoretical maximum of 1.0, the observed values clustered between 0.80 and 0.95, with an average of 0.853 and a standard deviation of 0.06.

Consideration of variations in demand within the peak hour through use of the peak hour factor is important to the correct solution of many traffic design and operation problems. Whenever possible, a locally determined peak hour factor should

TABLE 8.2
Average Five-minute Peak Hour Factor on American Freeways

Metropolitan Area Population	phf$_5$*
100,000	0.77
500,000	0.80
1,000,000	0.85
5,000,000	0.88

*Standard deviation = 0.05.

Source: D. R. Drew and C. J. Keese, "Freeway Level of Service as Influenced by Volume and Capacity Characteristics," *Freeway Characteristics, Operations and Accidents*, Highway Research Record 99 (Washington, D.C.: Highway Research Board, 1965), p. 4.

TABLE 8.3
Distribution of Fifteen-minute Peak Hour Factors at Selected Urban Intersections

Peak Hour Factor	Frequency
.40 to .45	0
.45 to .50	1
.50 to .55	3
.55 to .60	4
.60 to .65	7
.65 to .70	10
.70 to .75	55
.75 to .80	58
.80 to .85	158
.85 to .90	224
.90 to .95	211
.95 to 1.00	42

Mean = 0.853.
Standard deviation = 0.06.

Source: O. K. Normann, "Variations of Flow at Intersections as Related to Size of City, Type of Facility and Capacity Utilization," *Traffic Characteristics and Intersection Capacities, II Intersection Capacity* (Washington, D.C.: Highway Research Board, 1962), p. 65.

be used because many characteristics of this measure are still unknown and the average values in Tables 8.2 and 8.3 can be inappropriate to specific situations.

LEVEL OF SERVICE

The objective of a transportation facility is to accommodate a *quantity* of traffic demand with an acceptable *quality* of service. This quality is apparent to the user in terms of his freedom to follow a path and speed of his choosing, the ease, and the physical and mental comfort of his operation. Although not directly conscious of these factors, the driver is affected also by the degree of hazard to which he is subject, the probability of failure to meet his transportation objectives, and the total cost of the service. All of these measures of quality vary as some function of the ratio of the rate of flow to the capacity of the facility. Some, such as driver comfort, are largely unmeasured—and perhaps unmeasurable. Others, such as speed capability, are better established.

Figure 8.1 shows the relationship of some of the better understood measures of quality of flow to the volume/capacity ratio and to the various levels of service for uninterrupted flow conditions. The relationship between speed, volume, and density implicit in Figure 8.1 is correct only if space mean speed is utilized. Because most capacity and service level descriptions are in terms of operating speed (which is always higher than mean speed), the relationship shown is conceptually but not mathematically correct. Operating speed as defined in the *Highway Capacity Manual* and used herein is *the highest over-all speed which a driver can travel on a given roadway under favorable weather conditions and prevailing traffic conditions without at any time exceeding the safe speed as determined by the design speed.* Its value will thus lie between average speed and design speed.

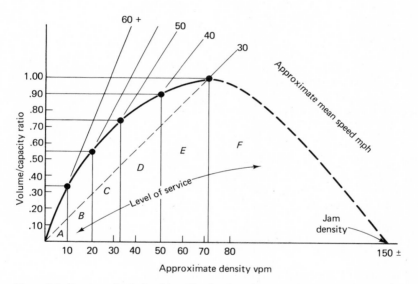

Figure 8.1. Conceptual relationship of levels of service to some measures of quality under uninterrupted flow conditions.

Level of service as used in this chapter is *a qualitative measure that represents the collective factors of speed, travel time, traffic interruptions, freedom to maneuver, safety, driving comfort and convenience, and operating costs provided by a highway facility under a particular volume condition.* The maximum volume associated with a particular level of service is termed the *service volume.*

It would be desirable to know and make use of all of the above factors in defining the various levels of service and associated service volumes. The present state of the art and the practicalities of traffic measurement have prevented this. In dealing with uninterrupted flow conditions, speed and the ratio of volume to capacity are the criteria most frequently cited because they are relatively simple to measure and understand. They are the best available but not entirely adequate surrogates for the remaining factors.

Similarly, because most of the factors cited presumably vary as a continuous, rather than step, function of volume or volume/capacity ratio, division of the continuum into steps or ranges is based on a combination of theoretical and practical considerations.

Figure 8.2 depicts a theoretical relationship between flow, speed, kinetic energy of the stream, and the "internal energy" of the stream which is analogous to the

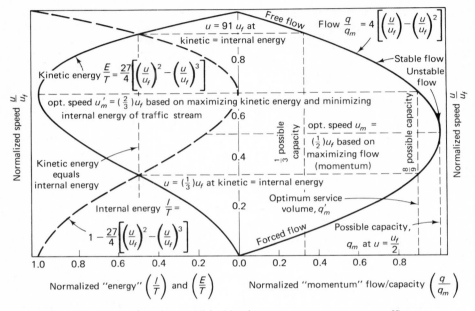

Figure 8.2. Levels of service established by the energy-momentum concept. [Source: Donald R. Drew, Conrad L. Dudek, and Charles J. Keese, "Freeway Level of Service as Described by an Energy-Acceleration Noise Model," *Geometric Aspects of Highways*, Highway Research Record 162 (Washington, D.C.: Highway Research Board, 1967), p. 67.]

acceleration noise (standard deviation of changes in speed). It is postulated that points at which the kinetic energy and internal energy curves cross and the point at which internal energy is minimized are significant parameters of traffic flow condition. These points coincide with the following speed and volume/capacity ratios:

	Normalized Speed	Normalized Volume
Upper limit of level of service A	0.91	0.35
B	0.83	0.55
C	0.75	0.75
D	0.67	0.90
E	0.33	1.0
F	<0.33	Not meaningful

Value for limits of service levels B and C are selected as logical subdivisions of the stable flow zone.

The practice of describing and labeling various service levels greatly facilitates communication both within and outside the profession and has been generally accepted. Its principal hazard lies in the attribution of a precision to the commonly cited boundary points between service levels which is unjustified by either the state of current knowledge or the basic characteristics of the highway traffic flow phenomenon.

The level of service descriptions that follow and the criteria given in later sections should be considered with the above caution in mind.

The definition and measurement of service levels under interrupted flow conditions pose different problems and are discussed in a later section on intersections.

Level of service A is the highest quality of service a particular class of highway can provide. It is a condition of free flow in which there is little or no restriction on speed or maneuverability caused by the presence of other vehicles. As shown in Figure 8.1, operating speed is in the highest range and density is low. On a freeway, lane density is approximately 10 vpm (6 vpk), and the volume/capacity ratio is typically about 1/3. Because speeds are high and volumes low, the occurrence rate of some kinds of accidents may be higher than at other service levels and total economic cost of providing the service may be excessive. Figure 8.3 shows a typical freeway operating at level of service A.

Level of service B is a zone of stable flow. However (as shown in Figure 8.1), operating speed is beginning to be restricted by other traffic. Under freeway conditions, density is under 20 vpm (12 vpk), restriction on maneuver is still negligible, and there is little probability of major reduction in speed or flow rate. This level of service approximates typical design volumes for high type rural highways, including freeways (see Figure 8.4).

Level of service C is still a zone of stable flow but at this volume and density level most drivers are becoming restricted in their freedom to select speed, change lanes, or pass. Operating speeds are still in the range of 2/3 to 3/4 of maximum; density is from 30 to 35 vehicles per lane mile on freeways (19 to 22 vehicles per lane kilometer). This service level is frequently selected as being an appropriate criterion for design purposes, particularly for urban freeways where the cost of providing the higher service levels during peak periods may be prohibitive (see Figure 8.5).

Level of service D approaches unstable flow. Tolerable average operating speeds are maintained but are subject to considerable and sudden variation. Freedom to maneuver and driving comfort are low because lane density has increased to between 45 and 50 vpm (28 and 31 vpk), and the probability of accidents has increased. Most drivers would probably consider this service level unsatisfactory (see Figure 8.6).

The upper limit of level of service E is the capacity of the facility. Operation in this zone is unstable, speeds and flow rates fluctuate, and there is little independence of speed selection or maneuver. Since headways are short and operating speeds subject to rapid fluctuation, driving comfort is low and accident potential high. Although circumstances may make operation of facilities under these conditions necessary, it is clearly undesirable and should be avoided whenever feasible (see Figure 8.7).

Level of service F describes forced flow operations after density has exceeded optimum which is normally in the range of 70 to 75 vpm (43 to 47 vpk) on free flowing facilities. Speed and rate of flow are below the levels attained in zone E and may, for short time periods, drop to zero. Figure 8.8 shows, pictorially, the operating conditions on a typical freeway under service volumes associated with the level of service F.

Figure 8.3. Level of service A as viewed looking up stream on a typical freeway: indicating no physical restrictions on operating speeds. (Source: Illinois Department of Transportation)

Figure 8.4. Level of service B as viewed looking up stream on a typical freeway: indicating stable flow with few restrictions on operating speed. (Source: Illinois Department of Transportation)

Figure 8.5. Level of service C as viewed looking up stream on a typical freeway: indicating stable flow, higher volume, and more restrictions on speed and lane changing. (Source: Illinois Department of Transportation)

Figure 8.6. Level of service D as viewed looking up stream on a typical freeway: indicating approaching unstable flow, little freedom to maneuver, and condition tolerable for short periods. (Source: Illinois Department of Transportation)

Figure 8.7. Level of service E as viewed looking up stream on a typical freeway: indicating unstable flow, lower operating speeds than level D and some momentary stoppages. (Source: Illinois Department of Transportation)

Figure 8.8. Level of service F as viewed looking up stream on a typical freeway: indicating forced flow operation at low speeds where the highway acts as a storage area and there are many stoppages. (Source Illinois Department of Transportation)

The values given above are typical for multilane facilities. Values for two-way, two-lane rural roads and urban arterials are given in later sections.

ROADWAY CAPACITY

FREEWAY SERVICE VOLUMES

The simplest and perhaps best understood traffic flow situation is the freeway or expressway roadway at locations well-removed from conflict points such as on- and off-ramps. There the quality of service is determined by vertical and horizontal alignment, cross-section dimensions, traffic volumes, peak hour factor (phf), composition of the traffic stream, and weather and lighting conditions.

Table 8.4 gives the operating speed and volume/capacity ratios that have been established in American practice as the criteria for each service level, as well as the maximum service volumes that can be accommodated under ideal conditions. Although operating speed and volume/capacity ratio are related, each measures a different aspect of the service provided to the user and each is considered a separate criterion that must be satisfied if that level of service is to be provided.

The service volumes given in Table 8.4 have a high probability of being accommodated at the designated service level when all factors conducive to high-volume production are present. The following section provides means to adjust service volume estimates for situations in which the more common of these factors are not present.

FACTORS AFFECTING FREEWAY SERVICE VOLUMES

Design speed—average highway speed. Horizontal alignment, superelevation, and sight distance are correlated in design through designation of a "design speed" (see Chapter 14, Geometric Design). *A weighted average of design speeds for all subsections of a section of highway* is termed the average highway speed (ahs). Table 8.4 gives service volumes and other characteristics for roadways with average highway speeds of 70 mph (113 kph) or greater. Freeways with 60 mph (97 kph) average highway speed cannot, by definition, provide level of service A and will provide levels of service B through D only at volume/capacity ratios considerably lower than those on a 70 mph (113 kph) design. Although detailed substantiating data are not available, current practice assumes a limiting volume/capacity ratio (and average lane volume) ranging from half that of a 70 mph (113 kph) roadway at level of service B to equal that of a 70 mph (113 kph) roadway at level of service E. Freeways or expressways of 50 mph (81 kph) or lower average highway speed are not common. It can be assumed that they would provide level of service D at about half the volume/capacity ratio (and average lane volume) of a 70 mph (113 kph) roadway. The service volume at level of service E is probably about five percent below that of a 70 mph (113 kph) design, although this probable reduction is frequently not considered in practice.

Figure 8.9 shows the relationship between operating speed and volume/capacity ratio on freeways of different average highway speed designs.

Grade. The effect of grade on level of service is caused by the differential speeding or slowing of components of the traffic stream. The extent of this effect is determined by the length and steepness of the grade and the hill climbing abilities of the vehicles which, in turn, are determined by their weight-horsepower ratios. Grades up to two

TABLE 8.4

Operating Criteria and Maximum Service Volumes Under Ideal Conditions on Freeways (One-way)

Level of Service	Description	Operating Speed (mph)	Volume/Capacity Ratio			Maximum Service Volume Passenger Cars per Lane	
			2-Lane Roadway	3-Lane Roadway	4-Lane Roadway	2-Lane Roadway	Each Added Lane
A	Free flow	$\geq$60	$\leq$0.35	$\leq$0.40	$\leq$0.43	700	1,000
B	Stable flow	$\geq$55	$\leq$0.50	$\leq$0.58	$\leq$0.63	1,000	1,500
C	Stable flow	$\geq$50	$\leq$0.75 × phf_s	$\leq$0.80 × phf_s	$\leq$0.83 × phf_s	1,500 × phf_s	1,800 × phf_s
D	Approaching unstable flow	$\geq$40		$\leq$0.90 × phf_s		1,800 × phf_s	
E	Unstable flow	<40		$\leq$1.00		2,000	
F	Forced flow	<30		Not meaningful		<2,000	

Source: Adapted from *Highway Capacity Manual 1965*, pp. 252–53.

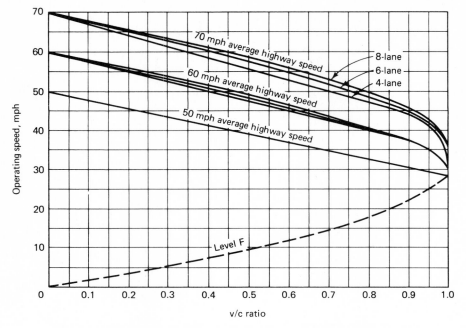

Figure 8.9. Relationship between operating speed and volume/capacity ratio on freeways. [Source: *Highway Capacity Manual*, Special Report 87 (Washington, D.C.: Highway Research Board, 1965), p. 264.]

percent have negligible effect on standard American passenger cars (weight-horsepower ratios of from 30 to 45 #/hp) or modern intercity buses. The effect of such grades is frequently ignored in capacity computations when truck percentages are low even though they have been found to have a small but discernible effect on capacity (level of service E) service volumes with a negative grade being noticeably beneficial.[2] Most American passenger cars and intercity buses can sustain speeds of from 30 to 40 mph on long grades up to seven percent (see Chapter 2, Vehicle Operating Characteristics) so that even such grades are considered to have negligible effect on capacity (level of service E) when only passenger cars or a low proportion of buses are involved.

The principal effect of grades is in conjunction with trucks and other vehicles, such as passenger vehicles pulling trailers, which have high weight-horsepower ratios. Hill climbing abilities of trucks in the 200, 300, and 400 #/hp classes are shown in Figure 8.10.

On a multilane highway the result of this slowing of one component of the traffic stream is complex and not completely understood. At the lower volume/capacity ratios and a low percentage of trucks, the effect on average operating speed is insignificant and only the increased accident hazard affects level of service. As the percentage of trucks and/or the length of the grade increases, however, faster vehicles tend to avoid the right lane completely and disproportionately increase the volume/capacity ratio of the remaining lane or lanes. This brings a lowering of operating speed and service level.

[2] K. Moskowitz and L. Newman, "Notes on Freeway Capacity," *Highway and Interchange Capacity*, Highway Research Record No. 27 (Washington, D.C.: Highway Research Board, 1963), pp. 44–68.

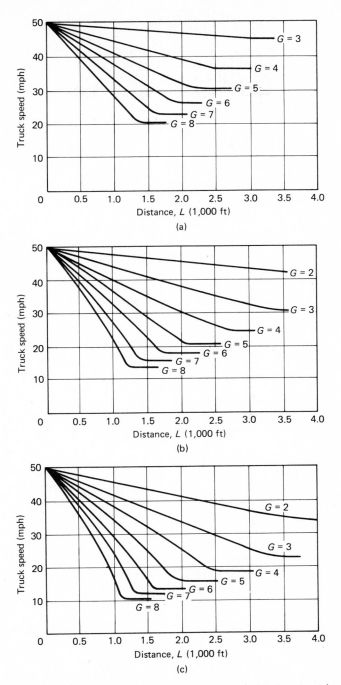

Figure 8.10. (a) Deceleration curves for trucks with weight/horsepower ratio = 200. (b) Deceleration curves for trucks with weight/horsepower ratio = 300. (c) Deceleration curves for trucks with weight/horsepower ratio = 400. [Source: John C. Glennon and Charles A. Joyner, Jr., "Re-evaluation of Truck Climbing Characteristics for Use in Geometric Design," *Highway Design Criteria—Research Study No. 2–8–68–134,* sponsored by The Texas Highway Department in cooperation with U.S. Department of Transportation (College Station, Texas: Texas A&M University, 1969), pp. 19, 20, 21.]

Adjusting for the combined effect of trucks and grades may be accomplished by multiplying the number of trucks by a factor known as the passenger car equivalent. This factor varies from a value of two for grades under two percent (including level and negative grades) to a value of 20 or more with high percentages of trucks on long, steep grades. A more convenient form of this adjustment for computational purposes is the truck factor:

$$T = \frac{100}{100 - P_t + E_t P_t} \tag{8.1}$$

where T = truck factor,
P_t = percentage of trucks,
E_t = passenger car equivalent.

Table 8.5 gives approximate truck factors for a number of grades, service levels, and percentages of trucks derived from limited observations. Further research may refine or alter these values and they should be used with discretion. The values at the bottom of the table may be used for approximate solutions for sections with numerous positive and negative grades.

Buses. Operating characteristics of intercity buses approximate those of passenger cars under most circumstances. Only their added length has an appreciable effect on capacity. On grades under four percent, an intercity bus is equivalent to about 1.6 passenger cars and this small effect is frequently disregarded when bus percentages are small. Table 8.5 provides bus adjustment factors that may be applied whenever warranted.

Lane width and lateral clearance. Narrow lanes and intermittent hazardous obstructions near the edge of the roadway increase driver tension and frequently cause drivers to select larger than normal headways and/or lower speeds. When this occurs, service volumes for all levels of service are lowered. Generally, low curbings or continuous regular features, such as guardrails, do not cause as severe a driver reaction as do intermittent obstacles. Accordingly, the factors given in Table 8.6 may be modified by assuming that continuous, low obstacles are from one to three feet farther from the lane edge than the actual distance. A paved shoulder adds to driver comfort and when a shoulder of at least 4 ft is available, it can be assumed to add 1 ft to the effective width of an adjacent lane.

Variations in flow rate. As previously discussed, the rate of flow during short time periods within the hour is seldom constant (see peak hour factor). At levels of service A and B, intervehicular influence is so small that these short fluctuations have little practical effect on overall operating speed or other measures of service. As volume and density increase, however, it becomes necessary to consider the effects of these short-term variations since a drastic reduction in speed or an instability in one time period can be expected to influence the service in one or more following periods and thus affect the overall level of service during the hour. As the flow increases to capacity levels, these short-term fluctuations are dampened by the capacity restraint and the facility increasingly operates under an artificial peak hour factor which approaches but rarely actually reaches 1.0.

It is frequent but not universal in current American practice to apply the service definitions of levels of service C and D on freeways to the five-minute period within

TABLE 8.5
Approximate Adjustment Factors (T) and (B) for Trucks and Buses
on Freeway and Multilane Highway Grades

Individual Grades		Levels of Service A, B, and C				Levels of Service D and E			
Grade (%)	Length (Miles)	Percentage of Trucks				Percentage of Trucks			
		3%	5%	10%	20%	3%	5%	10%	20%
		T				T			
Under 2	All	.97	.95	.91	.83	.97	.95	.91	.83
2	½	.89	.87	.77	.71	.89	.87	.77	.71
	1	.85	.83	.71	.63	.85	.83	.71	.63
	2	.85	.80	.67	.50	.85	.80	.67	.50
	4	.85	.77	.59	.42	.85	.77	.59	.42
3	½	.79	.74	.71	.63	.79	.74	.71	.63
	1	.79	.74	.67	.50	.79	.74	.67	.50
	2	.79	.71	.59	.42	.79	.71	.59	.42
	4	.79	.69	.50	.33	.79	.69	.50	.33
4	½	.75	.71	.71	.56	.74	.71	.71	.56
	1	.75	.69	.59	.42	.74	.69	.59	.42
	2	.75	.67	.50	.33	.74	.65	.50	.33
	4	.75	.63	.42	.28	.74	.61	.40	.26
5	½	.74	.67	.63	.45	.72	.67	.63	.45
	1	.74	.65	.53	.36	.72	.63	.53	.36
	2	.74	.61	.43	.28	.72	.59	.42	.26
	4	.70	.56	.36	.24	.69	.53	.32	.22
6	½	.72	.67	.59	.42	.70	.67	.59	.42
	1	.72	.63	.48	.33	.70	.61	.45	.33
	2	.72	.59	.40	.26	.70	.54	.37	.25
	4	.65	.53	.34	.21	.64	.48	.31	.19
Extended Sections									
Trucks in rolling terrain		.92	.87	.77	.63	.92	.87	.77	.63
Trucks in mountainous* terrain		.83	.74	.59	.42	.83	.74	.59	.42
		B				B			
Buses in rolling terrain		.94	.91	.83	.71	.94	.91	.83	.71
Buses in mountainous* terrain		.89	.83	.71	.56	.89	.83	.71	.56

*Combinations of length, steepness, and frequency of grades sufficient to reduce trucks to crawl speed (see Figure 8.10).

Source: *Highway Capacity Manual*, pp. 257–59, 260–61.

TABLE 8.6
**Adjustment Factors (W) for Narrow Lanes and/or Restricted
Lateral Clearances on Freeways**

Distance from Edge of Traffic Lane to Obstruction	Obstruction on One Side				Obstruction on Both Sides			
	12-ft Lanes	11-ft Lanes	10-ft Lanes	9-ft Lanes	12-ft Lanes	11-ft Lanes	10-ft Lanes	9-ft Lanes
Two-lane Roadway (one direction)								
				W				
6	1.00	.97	.91	.81	1.00	.97	.91	.81
4	.99	.96	.90	.80	.98	.95	.89	.79
2	.97	.94	.88	.79	.94	.91	.86	.76
0	.90	.87	.82	.73	.81	.79	.74	.66
Three or More Lane Roadway (one direction)								
				W				
6	1.00	.96	.89	.79	1.00	.96	.89	.78
4	.99	.95	.88	.77	.98	.94	.87	.77
2	.97	.93	.87	.76	.96	.92	.85	.75
0	.94	.91	.85	.74	.91	.87	.81	.70

Source: *Highway Capacity Manual*, p. 256.

the hour which is accommodating the maximum rate of flow. This is accomplished by multiplying the limiting volume/capacity ratio and service volume by an appropriate five-minute peak hour factor (phf_5). As previously indicated, peak hour factors between 0.75 and 0.95 have been found appropriate for urban freeways. One study of uncongested urban interstate freeways found an average value of 0.82.[3] Peak hour factors on nonmetropolitan freeways have not been extensively studied but probably lie in the lower portion of the range cited above. The peak hour factor adjustment for levels of service C and D is included in Table 8.4.

Weather and visibility. When a driver's visibility is restricted by darkness or inclement weather, his choice of speed and headway can be expected to be affected. The effect of this phenomenon on freeway capacity has not been extensively documented. A Connecticut[4] study of traffic performance under various levels of lighting suggests little significant difference in velocities with the average illumination range between 0.2 and 0.8 footcandles.

Rain, which decreases both visibility and surface friction available for stopping, can be expected to lower service volumes at all levels of service. A Texas study indicates a reduction of about 10 percent in capacity (level of service E) on freeways. British research[5] found a 14 percent reduction on very slippery pavement (wood block) but no corresponding reduction on rougher pavements. Snow and icy pavements can, in the extreme, reduce the capacity of a facility to zero. Any combination of visibility and surface conditions that reduces average freeway speeds to below 40 mph (64 kph) can be expected to have very significant reducing effects on capacity. When these conditions frequently occur during periods of maximum demand, special studies should be undertaken to develop appropriate adjustment factors.

---

[3] *An Investigation of Peak Hour Factors*, Missouri State Highway Department, Mimeo, 1970.
[4] *Effect of Illumination on Operating Characteristics of Freeways*, National Cooperative Highway Research Program Report No. 60 (Washington, D.C.: Highway Research Board, 1968), 150 pp.
[5] *Research on Road Traffic* (London: Her Majesty's Stationery Office, 1965), p. 213.

Example 8.2:

Problem: To determine service volume for levels of service C and E (capacity) for an urban freeway roadway of three lanes in one direction on a 3 percent, one-mile grade. The roadway has 11 ft lanes, a 10-ft right shoulder, and a continuous median barrier 4 ft from the left lane edge. The facility was designed for an average highway speed of 70 mph (113 kph) and carries 5 percent trucks and 1 percent intercity buses. The peak hour factor is 0.91.

Solution:
Maximum Service Volume$_E$ = 2,000 × 3 = 6000 (Table 8.4)

$$SV_E = (MSV_E)(W)(T)$$
$$SV_E = 6,000 \times 0.95 \times 0.74 = 4,218; \text{ use } 4,200 \text{ vph}$$

(where subscript E denotes level of service E)

Maximum Service Volume$_C$ = 1,500 + 1,500 + 1,800 (Table 8.4)

$$SV_C = (MSV_C)(\text{phf}_s)(W)(T)$$
$$= 4,800 \times 0.91 \times 0.95 \times 0.74 = 3,066; \text{ use } 3,100 \text{ vph}$$

(where subscript C denotes level of service C)

W for 3–11 ft lanes, 10-ft right clearance, and a 4-ft left clearance = 0.95 (Table 8.6). Because the median barrier is continuous, a 2-ft added distance could be assumed and the factor changed to 0.96. (Such refinement is seldom warranted in planning applications since volume estimates are seldom this precise.)

T for 5 percent trucks on a one-mile, 3 percent grade = 0.74 (Table 8.5). The small percentage of buses can be considered as passenger cars.

From Figure 8.9 it can be seen that with a v/c ratio of 0.74 and a 70 mph (113 kph) average highway speed design, the operating speed would be about 52 mph (84 kph), slightly above the controlling criterion of 50 mph (80 kph) for service level C. Had the average highway speed been 60 mph (97 kph), the operating speed at v/c = 0.74 would be well below the criterion speed. To prevent this, a v/c ratio of 0.43 would be required with a resulting reduction in service volume$_C$ to 1,800 vph (0.43 × 4,200).

SERVICE VOLUMES ON MULTILANE RURAL HIGHWAYS WITHOUT ACCESS CONTROL

A divided, multilane highway without access control can operate identically with a freeway in sections where crossing opportunities and marginal influences are negligible. At the other extreme, an undivided roadway with heavy crossing, turning, and entering movements operates in an entirely different manner and its capacity and service volumes are usually controlled by the characteristics of frequent major intersections.

Intermediate to these is the multilane roadway, usually undivided or with frequent turning opportunities, with driveways and intersections that offer the potential but infrequent occurrence of interference. The quality of service on this class of road is determined by the alignment, grades, cross section, volume, and composition of traffic just as in the freeway case. When operating at level of service E, capacity, these roads can carry traffic volumes comparable to a freeway; however, this condition would be extremely unstable and these maximum volumes may not be attainable in all cases.

At better service levels, the potential for interferences generally results in significantly lower maximum service volumes than on freeways. The criteria for each service level and the associated maximum service volumes are given in Table 8.7.

TABLE 8.7
Operating Criteria and Maximum Service Volumes Under Ideal Conditions on
Multilane Rural Highways

Level of Service	Description	Operating Speed (mph)	Average Highway Speed					
			70 mph		60 mph		50 mph	
			v/c Ratio	Maximum Service Volume*	v/c Ratio	Maximum Service Volume*	v/c Ratio	Maximum Service Volume*
A	Free flowing	≥60	≲.30	600				
B	Stable flow	≥55	≲.50	1000	≲.20	400		
C	Stable flow	≥45	≲.75	1500	≲.50	1000	≲.25	500
D	Approaching unstable flow	≥35	≲.90	1800	≲.85	1700	≲.70	1400
E	Unstable flow	≈30	1.00	2000	1.00	2000	1.00	2000
F	Forced flow	<30			Not meaningful			

*Passenger cars per lane.

Source: *Highway Capacity Manual*, p. 284.

FACTORS AFFECTING SERVICE VOLUMES ON MULTILANE RURAL HIGHWAYS

Design speed—average highway speed. Table 8.7 shows the marked reduction in maximum service volume at all levels below capacity that accompanies lower design standards. The better levels of service cannot be provided, by definition, and the maximum service volumes at lower levels are affected by driver reaction to restrictions in sight distance and to existence of curves and other impediments. The relationship of volume-capacity ratio, operating speed, and average highway speed on a typical rural multilane highway is shown in Figure 8.11.

Variations in flow. Few data are currently available on peaking characteristics on rural highways. Thus, a specific adjustment for peak hour factor is not usually made in level of service computations for rural highways. It is commonly assumed in American practice that flow variations within the hour will not adversely affect levels A, B, and C at the maximum service volumes listed. The maximum service volumes given for level of service D will, however, result in a strong likelihood of delays and breakdown in flow if any material peaking takes place. When this is known to be the case, the service volume D should be considered the maximum rate of flow that can

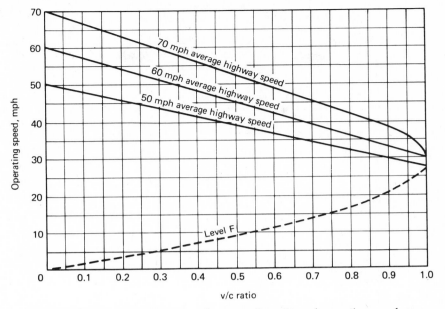

Figure 8.11. Relationship between volume-capacity ratio and operation speed on rural multilane highways. (Source: *Highway Capacity Manual*, p. 294.)

be accommodated during a *short period* of maximum demand. If a peak hour factor is known or can be reliably estimated, downward adjustment in maximum service volume should be made by following the procedures given in preceding sections.

Traffic composition. The effect of trucks and buses on capacity and service volumes of multilane highways is similar to that on freeways. Thus, the adjustment factors used for freeways are commonly used (Table 8.5). It should be noted that these factors are based on 200 #/hp trucks and on high-performance intercity buses. These are common on many multilane rural highways other than freeways, but they may not be appropriate in all cases. The adjustment factors for trucks on two-lane highways are based on the performance characteristics of a 325 #/hp truck typical of American farm vehicles. Should this be the predominant type vehicle expected on a particular multilane highway under study, appropriate adjustment factors should be utilized: on level to moderate rolling terrain, an equivalency of 3; in rolling terrain, 5; and mountainous terrain, from 10 to 12. For specific grades, reference to Figure 8.10 will assist in selecting appropriate factors. The factors given in a later section on two-lane roads should *not* be used since those factors also reflect the effect of passing restrictions on two-way roadways.

Lane widths and lateral clearance. On divided, multilane rural roads the effects of reduced lane widths and/or lateral clearances are the same as on freeways and the adjustment factors previously given may be used. Factors for undivided roads are given in Table 8.8. The factors at the bottom of the table should be used only when an

TABLE 8.8
Adjustment Factors (W) for Restricted Lane Widths and Lateral
Clearances on Undivided Multilane Highways

Distance (in ft) from Traffic Lane to Obstruction on Right Side Only	Two Lanes in Each Direction				Three or More Lanes in Each Direction			
	12-ft Lanes	11-ft Lanes	10-ft Lanes	9-ft Lanes	12-ft Lanes	11-ft Lanes	10-ft Lanes	9-ft Lanes
				W				
6	1.00	.95	.89	.77	1.00	.95	.89	.77
4	.98	.94	.88	.76	.99	.94	.88	.76
2	.95	.92	.86	.75	.97	.93	.86	.75
0	.88	.85	.80	.70	.94	.90	.83	.72
With Additional Obstruction on Left Side								
6								
4								
2	.94	.91	.75		.96	.92	.85	.75
0	.81	.79	.66		.91	.87	.81	.70

Source: *Highway Capacity Manual*, p. 286.

obstruction, such as a bridge pier or center line barrier, is introduced into the center of a normally undivided roadway.

Roadside development and intersections. The above procedures, based on uninterrupted traffic flow characteristics, can reasonably be applied where major intersections and/or roadside entrances are at least a mile apart and speed limits are 40 mph (64 kph) or greater. When this is not the case, procedures given later for urban arterials are more appropriate.

Example 8.3:

Problem: Determine the level of service provided on rural four-lane undivided highway with 11 ft lanes, no shoulders, and obstructions at the pavement edge. Average highway speed is 60 mph and there is a one-mile, 6 percent grade. The highway is serving 1,800 vph in the peak direction, with 7 percent trucks and 3 percent intercity buses. Determine the level of service with and without an added climbing lane.

Solution: Without climbing lane:

Base volume $= 2,000$ NWTB (Table 8.7)
Number of lanes (N) $= 2$
Width adjustment (W) $= 0.85$ (Table 8.8 for 2 11-ft lanes)
Truck adjustment (T) $= 0.52$ (Table 8.5 for level of service D–E)
Bus adjustment (B) $= 0.92$ (Table 8.5 for level of service D–E)

Base volume $= 1,627$
v/c ratio $= \dfrac{1,800}{1,627} = 1.10$

The grade could not accommodate this demand volume and level of service F would prevail.

With climbing lane:

Base volume	= 2,000 NWTB (Table 8.7)
Number of lanes (N)	= 3
Width adjustment (W)	= 0.90 (Table 8.8 for 3 11-ft lanes)
Truck adjustment (T)	= 0.55 (Table 8.5 for level of service C)
Bus adjustment (B)	= 0.85 (Table 8.5 for level of service C)
Base volume	= 2,525
v/c ratio	$= \dfrac{1,800}{2,525} = .71$

By reviewing Table 8.7 it is seen that this volume/capacity ratio exceeds level of service C on a 60 mph design; therefore, the section should be recomputed using truck and bus adjustments for level of service D, resulting in base volume = 2,713 and v/c ratio = .66. Level of service D would prevail.

An alternate computation would be to consider that all trucks and buses would be restricted to the climbing lane. The two remaining lanes would be serving 1,620 (1,800-126-54) passenger cars.

With all trucks and buses in climbing lane, for the remaining lanes:

Base volume	= 2,000 NWTB (Table 8.7)
Number of lanes (N)	= 2
Width adjustment (W)	= 0.85 (Table 8.8 for 2 11-ft lanes)
Truck and bus Adjustments (T) and (B)	= 1.00
Base volume	= 3,400
v/c ratio	$= \dfrac{1,620}{3,400} = .48$

Level of service C would prevail in these lanes. In the climbing lane, 126 trucks and 54 buses would be accommodated. Based on climbing performance shown in Figure 8.10 for 200 #/hp trucks, speeds would reduce to approximately 17 mph at about the half-way point on the grade and, since passing would not be permitted, a high proportion of buses and trucks would operate at or under this speed. Thus, level of service C would prevail in the outer two lanes but buses and trucks would be confined to low speeds and substantial delays, approximating level of service E.

SERVICE VOLUMES ON TWO-LANE RURAL HIGHWAYS

The overriding distinction between traffic operations on a two-lane, two-way roadway and on those roadway types previously discussed lies in the passing maneuver that must utilize a lane assigned to traffic traveling in the opposite direction. The resulting interaction requires that service volumes and capacities on two-way roadways be considered as total rather than one-way or one-lane values.

The need to pass in order to sustain a desired speed is a function of the amount of traffic in the lane and the distribution of speeds. The rate at which overtakings occurs is, per unit distance, per unit time approximately:[6]

$$r = 0.56\sigma_u k^2 \tag{8.2}$$

where r = rate of overtaking,
  σ_u = standard deviation of speed,
  k = density.
  In terms of flow rate and speed, this becomes:

$$r = c\frac{q^2}{\bar{u}} \tag{8.3}$$

where r = rate of overtaking,
  q = rate of flow,
  $\bar{u}$ = space mean speed,
  c = a constant dependent on variation of speed ($= 0.1$ when coefficient of variation $= 0.18$).
  The duration of the gap in opposing traffic required for the passing maneuver is a function of the speed and relative speed of passed and passing vehicles. The probability of such a gap occurring is a function of the opposing volume:

$$P_{(o)} = e^{-qt} \tag{8.4}$$

where $P_{(o)}$ = the probability of a gap of t duration,
  e = Napierian base,
  q = rate of flow,
  t = duration of gap.
  The ability to pass is further controlled by the existence or nonexistence of a sight distance sufficient to detect and utilize an available gap with safety. When passing sight distance is not available, all vehicles as they overtake will form into a queue traveling at the speed of the slowest vehicle in the stream.

The capacity and service volumes of a two-way, two-lane roadway are thus seen to be functions of the flow rates in each direction, the speed of travel, the variance in speed in the traffic stream, and the availability of passing sight distance. The variance of speed is, in turn, principally established by the presence of trucks or other low-performance vehicles and of grades.

The reduction of marginal frictions through control of access undoubtedly affects service volumes on this kind of roadway but, since data are not available to quantify this effect, it has not been possible to give separate treatment to controlled and non-controlled access facilities of this class. In general, the procedures given below are appropriate to roads having minor potential frictions, infrequently spaced. On sections having frequent turning and/or entering movements or having major crossing movements more frequent than one per mile, the procedures in a later section on urban streets should be used.

Table 8.9 gives the operating speed and volume/capacity ratios established in American practice as the criteria for each service level on a 70 mph (113 kph) average highway speed road with unlimited passing sight distance and all other factors ideal. It should be noted that volumes in both directions are combined in this analysis.

---

[6] J. G. WARDROP, "Some Theoretical Aspects of Road Traffic Research," *Proceedings of the Institution of Civil Engineers*, Vol. II (London: 1952), p. 325.

TABLE 8.9
**Operating Criteria and Maximum Service Volumes Under Ideal
Conditions on Two-lane Rural Highways**

Level of Service	Description	Operating Speed (mph)	Volume/ Capacity Ratio	Maximum Service Volume
		Passenger vehicles per hour in both directions		
A	Free flow	≥60	.20	400
B	Stable flow	≥50	.45	900
C	Stable flow	≥40	.70	1400
D	Approaching unstable flow	≥35	.85	1700
E	Unstable flow	≈30	1.00	2000
F	Forced flow	<30	Not meaningful	

Source: *Highway Capacity Manual*, pp. 302–3.

FACTORS AFFECTING SERVICE VOLUME

Passing sight distance. Since maintenance of the stipulated operating speed requires a high degree of freedom to pass slower vehicles with minimum delay, any restriction on passing sight distance will have a marked effect on potential service volume. This effect is highest at the better service levels and becomes less as the service level decreases since the differences between the stipulated operating speed and the average speed of trucks and other slower moving elements of the traffic stream becomes smaller.

Table 8.10 shows the approximate effect of restricted passing sight distance on maximum service volumes at each service level.

In this table, the limitation of passing sight distance is expressed as the percentage of the length of the section under study that provides a passing sight distance of 1,500 ft or more, averaged for both directions (see Chapter 14, Geometric Design, for definition and measurement of passing sight distance).

This method is also used in the *Highway Capacity Manual.* It should be noted that some publications in common use base sight distance restriction on the obverse definition: the percentage of roadway which does *not* provide the stipulated sight distance.

Average highway speed. The weighted average design speed of the highway is called the average highway speed. It affects the maximum service volumes in two ways: (1) it limits the operating speed of passenger vehicles and thus may make attainment of the better service levels impossible and (2) it reduces the average speed of trucks and slower moving vehicles and thereby increases the need for passing maneuvers. When combined with limitations on passing sight distance, which is often the case, a lower average highway speed design will have substantially lower maximum service volumes at all service levels below capacity level E. The approximate effect of these combined factors is shown in Table 8.10.

Lane width and lateral clearance. Because traffic in adjacent lanes of a two-lane, two-way roadway is moving in opposite directions, the effects of narrow lanes and/or

TABLE 8.10

Approximate Maximum Service Volumes on Two-lane, Two-way Rural Highways with Limited Passing Sight Distance and/or Average Highway Speeds Under 70 mph

Level of Service	Description	Operating Speed (mph)	Passing Sight Distance (%)*	Average Highway Speed — Maximum Service Volume Passenger (vehicles per hr. in both directions)			
				70 mph	60 mph	50 mph	40 mph
A	Free flow	≥60	100	400	—	—	—
			80	360	—	—	—
			60	300	—	—	—
			40	240	—	—	—
			20	160	—	—	—
			0	80	—	—	—
B	Stable flow	≥50	100	900	800	—	—
			80	840	700	—	—
			60	760	600	—	—
			40	680	480	—	—
			20	600	360	—	—
			0	480	240	—	—
C	Stable flow	≥40	100	1,400	1,320	1,120	—
			80	1,360	1,220	1,060	—
			60	1,300	1,120	940	—
			40	1,240	1,020	760	—
			20	1,180	900	560	—
			0	1,080	760	360	—
D	Approaching unstable flow	≥35	100	1,700	1,660	1,500	1,160
			80	1,680	1,620	1,440	1,100
			60	1,660	1,580	1,380	1,020
			40	1,640	1,520	1,320	900
			20	1,620	1,420	1,220	700
			0	1,600	1,320	1,020	380
E	Unstable flow	≈30	Any	2,000	2,000	2,000	2,000
F	Forced flow	<30	Any	Not meaningful			

*Percentage of roadway with passing sight distance exceeding 1,500 ft.

Source: *Highway Capacity Manual*, pp. 302–3.

restricted lateral clearances on vehicle-operator performance is more pronounced than on one-way or multilane roadways. The effect is also slightly more pronounced at the better service levels as indicated in Table 8.11.

Grades and trucks. The combination of trucks or other low performance vehicles and grades has a major effect on traffic operations on two-lane, two-way highways. Passenger car equivalents vary from a value of 2 on grades under 2 percent to a value of over 100 on long 7 percent or greater grades at or near capacity volumes. These equivalents are based on the hill climbing ability of loaded trucks in the 325 #/hp category. Higher performance trucks would have less effect on service volumes.

Table 8.12 gives truck factors for representative percentages of trucks on selected grades. The values at the bottom of the table may be used when studying the general effects of trucks on long sections with varying grades. It should be noted, however, that a critical grade within the section may control the capacity of the entire section to a level below that obtained by using these generalized values.

Normally, the computation for truck effect is based on the total percentage of trucks in the two-way traffic stream and with the assumption that the effect of plus and minus grades is the same. When the percentage of trucks in the two directions is appreciably different or when trucks travel loaded in one direction and return empty, this simplification cannot be used and the preceding tables are not valid.

Example 8.4:

Problem: Determine the maximum service volumes for levels of service C and E on a one-mile, 5 percent grade of a rural two-lane highway with 10-ft lanes, 4-ft paved shoulders, an average highway speed design of 50 mph, and 40 percent passing sight distance. The traffic composition includes 10 percent trucks of the 325 #/hp class.

Solution: For level of Service C,

$$SV_C = (MSV)(W)(T)$$

where $MSV = 760$ (Table 8.10)
$W = 0.86$ (Table 8.11) {considering the lanes 11 ft wide due to the presence of paved shoulders
$T = 0.18$ (Table 8.12)
$SV_C = (760)(.86)(.18) = 118$ vph in both directions.

For level of service E,

$$SV_E = (MSV)(W)(T)$$

where $MSV = 2,000$ (Table 8.10),
$W = 0.88$ (Table 8.11) (considering the lanes 11 ft
wide because of paved shoulders),

$T = 0.16$ (Table 8.12),
$SV_E = (2,000)(0.88)(0.16) = 282$ vph in both directions.

SERVICE VOLUMES ON URBAN ARTERIALS

The distinction between operation on a rural and an urban facility is due to the added frequency of marginal frictions, intersections, turning movements, and pedestrians on the typical urban facility. When these are not present in a situation under

TABLE 8.11
Adjustment Factors (W) for Narrow Lanes and/or Restricted Lateral Clearances on Two-lane Rural Highways

Distance from Traffic Lane Edge to Obstruction (ft)	Obstruction on One Side Only*								Obstructions on Both Sides*							
	12-ft Lanes		11-ft Lanes		10-ft Lanes		9-ft Lanes		12-ft Lanes		11-ft Lanes		10-ft Lanes		9-ft Lanes	
	Level B	Level E	Level B	Level E	Level B	Level E	Level B	Level E	Level B	Level E	Level B	Level E	Level B	Level E	Level B	Level E
6	1.00	1.00	0.86	0.88	0.77	0.81	0.70	0.76	1.00	1.00	0.86	0.88	0.77	0.81	0.70	0.76
4	0.96	0.97	0.83	0.85	0.74	0.79	0.68	0.74	0.92	0.94	0.79	0.83	0.71	0.76	0.65	0.71
2	0.91	0.93	0.78	0.81	0.70	0.75	0.64	0.70	0.81	0.85	0.70	0.75	0.63	0.69	0.57	0.65
0	0.85	0.88	0.73	0.77	0.66	0.71	0.60	0.66	0.70	0.76	0.60	0.67	0.54	0.62	0.49	0.58

*Includes allowance for opposing traffic.

Source: *Highway Capacity Manual*, p. 303.

TABLE 8.12

Approximate Adjustment Factors (T) and (B) for Trucks and Buses on Two-lane, Two-way Highways

Individual Grades		Levels of Service A and B				Level of Service C				Levels of Service D and E			
		Percentage of Trucks				Percentage of Trucks				Percentage of Trucks			
Grade (%)	Length (miles)	3%	5%	10%	20%	3%	5%	10%	20%	3%	5%	10%	20%
Under 2	All	.97	.95	.91	.83	.97	.95	.91	.83	.97	.95	.91	.83
3	½	.79	.69	.53	.36	.79	.69	.53	.36	.85	.77	.63	.45
	1	.68	.56	.38	.24	.63	.50	.33	.20	.64	.51	.34	.21
	2	.63	.50	.33	.20	.56	.44	.28	.17	.54	.42	.27	.16
	4	.60	.48	.31	.19	.52	.40	.25	.15	.51	.39	.25	.14
4	½	.69	.57	.40	.25	.64	.51	.34	.21	.64	.51	.34	.21
	1	.57	.44	.29	.17	.49	.37	.23	.13	.47	.35	.21	.11
	2	.53	.41	.26	.15	.45	.33	.20	.11	.42	.30	.18	.10
	4	.51	.39	.25	.14	.43	.31	.19	.10	.39	.28	.17	.09
5	½	.59	.47	.30	.18	.51	.39	.24	.14	.48	.36	.22	.12
	1	.51	.39	.25	.14	.42	.30	.18	.10	.38	.27	.16	.08
	2	.48	.36	.22	.12	.38	.27	.16	.08	.35	.24	.14	.07
	4	.46	.34	.20	.11	.37	.26	.15	.08	.33	.22	.13	.07
6	½	.51	.39	.25	.14	.42	.30	.18	.10	.38	.27	.16	.08
	1	.46	.34	.20	.11	.36	.25	.15	.08	.33	.22	.13	.07
	2	.43	.31	.19	.10	.34	.24	.14	.07	.30	.20	.11	.06
	4	.40	.29	.17	.09	.33	.22	.13	.07	.27	.18	.10	.05
Extended Sections													
Trucks in rolling terrain		.92	.87	.77	.63	.89	.83	.71	.56	.89	.83	.71	.56
Trucks in mountainous* terrain		.85	.77	.63	.45	.79	.69	.53	.36	.75	.65	.48	.31
Buses under 4% grade		.97	.95	.91	.83	.97	.95	.91	.83	.97	.95	.91	.83
Buses on 6% grade		.85	.77	.63	.45	.87	.80	.67	.50	.92	.87	.77	.63

*Combinations of length, steepness, and frequency of grades sufficient to reduce trucks to crawl speed.

Source: *Highway Capacity Manual*, pp. 305–7.

study, even though in an urban area, the procedures given in the preceding sections are appropriate and should be used.

Urban facilities are usually considered as those that have signalized intersections at less than one-mile intervals and sight distances and marginal frictions that hold the average (and usually legal maximum) speed to 35 mph (56 kph) or less. Under these conditions, marginal frictions and interruptions play a more significant role than stream frictions in determining capacity and service levels.

Traffic signal control at the major intersections is normally the capacity controlling factor. At volume levels below capacity, the interrelationship between adjacent signals frequently sets the maximum possible average speed irrespective of flow rate. Marginal frictions, bus operation, and midblock turning movements may further reduce these speeds, which are essentially independent of the rate of flow on the arterial.

Because signalization of major intersections is usually the practical determinant of capacity and service volumes, the most frequently used method of evaluating arterial service is to analyze the signalized approaches and assume that the results are representative of the entire section.

This is not entirely satisfactory at best and is totally unrealistic when number or width of lanes, parking control, or geometrics differ between intersection approaches and mid-section locations.

The use of operating, running, or average speed as the surrogate measure of service level also poses serious problems on arterials. Since top speeds are limited by environmental or legal factors, there is little apparent change in average overall speed through the range of densities up to 40 or 50 vpm which are associated with the gamut of service levels A to D on other facilities.

Other, more sensitive, measures of service level have been suggested but are not currently widely used or completely understood. One promising measure is the delay ratio, defined as the *ratio of amount of delay time to the total travel time on a section of roadway during a specified time period.*[7] In the cited study, delay ratios equal to or less than 0.05, 0.15, 0.25, 0.40, and 0.60 were proposed as upper limits of levels of service A through E. Even here, however, the problem of defining the "no-delay" condition still exists.

Several measures of smoothness of flow have been proposed. The first, acceleration noise, is the standard deviation of a vehicle's accelerations over time:

$$\sigma_a = \left[\frac{1}{T} \int_0^T a^2 \, dt \right]^{1/2} \tag{8.5}$$

where σ_a = acceleration noise,
T = running time (overall travel time minus delay time),
a = acceleration.

The mean velocity gradient (G) is the *ratio of acceleration noise to mean velocity*. The principal difficulty with these measures is the relatively high cost of data collection. Also, one study[8] found a highly correlated linear relationship between mean velocity gradient and travel time on urban arterials, indicating that the more easily measured travel time or its reciprocal, speed, was as accurate an index of quality of flow as mean velocity gradient.

---

[7] WALTER E. PONTIER ET AL., *Optimizing Flow on Existing Street Networks*, National Cooperative Highway Research Report No. 113 (Washington, D.C.: Highway Research Board, 1971).
   [8] *Ibid.*, p. 100.

Still another promising measure of quality of flow is the energy ratio,[9] defined as the ratio of *effective kinetic energy to measured (free flow) kinetic energy in the traffic stream:*

$$N_e = \frac{(Ku)^2}{(Ku_f)^2} \approx \left(\frac{u}{u_f}\right)^2 \tag{8.6}$$

where N_e = energy ratio,
$\quad u$ = effective speed (distance/time),
$\quad u_f$ = free-flowing (spot) speed,
$\quad K$ = density.

This more easily obtained measure of smoothness of flow is also highly correlated with acceleration noise.

The Highway Capacity Committee, although recognizing the disadvantages, has recommended average overall travel speed as the principal measure of service levels on urban arterials. These values, which are frequently followed in American practice, are given in Table 8.13.

The volume/capacity ratios listed in Table 8.13 were selected to be consistent with intersection load factor values discussed in a later section on signalized intersections.

TABLE 8.13
Operating Criteria for Urban Arterials

Level of Service	Description	Average Overall Travel Speed (mph)	Volume/ Capacity Ratio	Associated Load Factor
A	Free flow	$\gtrsim 30$	$\lesssim 0.60$	0.0
B	Stable flow	$\gtrsim 25$	$\lesssim 0.70$	$\lesssim 0.1$
C	Stable flow	$\gtrsim 20$	$\lesssim 0.80$	$\lesssim 0.3$
D	Approaching unstable flow	$\gtrsim 15$	$\lesssim 0.90$	$\lesssim 0.7$
E	Unstable flow	≈ 15	$\lesssim 1.00$	$\lesssim 1.0$
F	Forced flow	< 15	Not meaningful	

Source: *Highway Capacity Manual*, p. 323.

Another approach, suggested by the author, is described below. It is more satisfying in some respects than the above but has not been either tested or accepted generally in the field.

By definition, traffic on an arterial is regimented by traffic signals and proceeds through a section in a series of platoons separated by time gaps with few if any vehicles. If, for the base condition, it is assumed that the traffic signals have a 50 percent split and traffic flow is limited by the known performance of a line of cars starting from a stopped condition, then capacity, or the upper limit of level of service E, is 50 percent of 1,500, or 750 vph. At 10–11 mph (a criterion slightly below that established by the Highway Capacity Committee), the approximate density would be in the 70–75 vpm range. This is the range of densities commonly associated with level of service E on free-flowing facilities. Using this same approach with the other service levels results in the relationships given in Table 8.14.

[9] D. L. COOPER and R. J. WALINCHUS, "Measures of Effectiveness for Urban Traffic Control Systems," *Highway Capacity and Quality of Service*, Highway Research Record 321 (Washington, D.C.: Highway Research Board, 1970), pp. 46–59.

TABLE 8.14
Maximum Lane Service Volumes on Urban Arterials Based on
50 % Cycle Split and an Average Density and Speed Criteria

Level of Service	Overall Average Travel Speed (mph)	Density (vpm)	Approximate Volume* per Lane (vph)
A	$\geqslant 30$	10	< 300
B	$\geqslant 25$	20	500
C	$\geqslant 20$	30	600
D	≈ 15	45	675
E	≈ 10	75	750
F	< 10	> 75	Variable

*The speed-density-volume relationships implicit to this table are correct only if space mean speed is used. Overall average travel speed as usually measured only approximates this value.

FACTORS AFFECTING ARTERIAL SERVICE VOLUMES

Quantitative means of estimating the effects of various common factors on arterial service volumes are not readily available. Some of these factors are discussed below.

Signalized intersections. The operation of frequent signalized intersections and the extent of progressive timing will usually be the principal determinant of arterial capacities and service volumes. These may be estimated by methods given in the following section.

Unsignalized intersections. Turning movements and crossing volumes can reduce arterial service volumes. One analysis method is to assume a signalized intersection with green time proportionate to main and cross street volumes and geometrics.

Mid-block driveways. Both right and left turns into and from driveways reduce arterial service volumes. Residential driveways can usually be ignored, but heavily used commercial driveways become, for practical purposes, unsignalized intersections.

Curb parking or loading. The area occupied by parked vehicles is not available for traffic movement. The effect on capacity and service volumes is thus equivalent to a reduction in effective width of at least 8 ft. Additionally, in areas of heavy parking turnover, a sporadic interruption in the adjacent lane will result from vehicles entering or leaving parking spaces.

Even where parking or loading is legally prohibited, momentary stops to discharge passengers, transit movements, and the possibility of illegal parking decrease the desirability of a curb lane. The proximity to pedestrians and to fixed objects in the border area and, frequently, an irregular cross slope to accommodate drainage further decrease the attractiveness of a curb lane. In Australian practice, the curb lane of multilane approaches (over 2 lanes) is penalized up to 60 percent, depending on parking enforcement practices, prevalence of right turns, and downstream roadway conditions.

Double parking. Double parked trucks or passenger cars can drastically reduce the carrying capacity of an arterial because they occupy at least one lane and, on narrow streets, may require traffic to await gaps in opposing traffic before traffic can proceed.

Pedestrians. Light volumes of pedestrian traffic have marginal effect on service volumes since, by custom if not by law, they frequently adjust their crossing pattern to available gaps in the traffic stream. As pedestrian volumes increase, the effect is more pronounced because their presence interrupts both through movements and intersectional turning movements. Unregulated mid-block crossing will have further adverse effects.

Lane configuration and width. Major contributors to traffic delays are, as previously noted, left-turn and, to a lesser extent, right-turn movements. Higher service volumes can be accommodated if these movements are removed from the main through lanes by channelization or lane assignment.

The effect of lane markings and lane width on urban arterial flow is a subject of some uncertainty. American and British practice considers the carrying capacity of an arterial approaching a signalized intersection to be a function of approach width, irrespective of lane configuration. Australian practice considers the carrying capacity to be a function of number of lanes irrespective of width (within limits of 6.5 ft to 15 ft). If, however, all parameters of level of service are considered, including driving comfort, it seems appropriate to consider lane width and configuration as factors. At capacity flow levels, available data indicate about a 40 vph increment per foot of lane width.[10] This is substantially less than the effect of lane width on freeways and rural highways, which is not unreasonable in view of the lower speed on arteries.

One-way operation. One-way operation is generally more efficient than two-way operation for a given street width. Based primarily on data for signalized intersection approaches, it would appear that the advantage of one-way operation is of the magnitude of 10 percent to 20 percent. This advantage is caused primarily by the elimination of left-turn conflicts and, frequently, the simplification of intersectional controls. Other factors, such as signal progression, advantageous lane dimensions, multilane operation, changes in patterns of turning movements, etc., may or may not be present in a particular case and each situation should be studied individually.

Trucks and buses. Unless significant grades are present, trucks and express buses can operate at speeds comparable to passenger vehicles on arterials. Their effect on capacity and service volumes is caused primarily by their added size and somewhat lower acceleration capabilities. A passenger car equivalency of 2 is frequently used for trucks, although some evidence indicates[11] that an equivalency of between 1.6 and 2.0 would be more appropriate. Australian[12] practice uses an equivalency of 1.85. Research[13] indicates that the equivalency of a transit bus in express service lies between 1.4 and 1.7, with a value of 1.6 frequently used.

[10] Highway Research Record 289 (Washington, D.C.: Highway Research Board, 1969), pp. 5 and 15.

[11] *Highway Capacity Manual*, pp. 343–44.

[12] A. J. MILLER, "The Capacity of Signalized Intersections in Australia," *Australian Road Research Board*, Bulletin No. 3, Melbourne, 1968, p. 95.

[13] EUGENE F. REILLY and JOSEPH SEIFERT, "Truck Equivalency," *Highway Capacity*, Highway Research Record 289 (Washington, D.C.: Highway Research Board, 1969), pp. 25–37.

Transit buses in local service present a different case, depending on whether or not bus loading bays, bus zones in a parking lane, or curb stops in a normal driving lane are provided. The location of a bus stop in relation to a signalized intersection is also important and is treated in a later section.

In summary, methods of estimating capacity and service volumes on urban arterials are largely subjective because of the complex interrelationships involved and the paucity of scientific measurements in the past. Most capacity estimates are made in practice by relying on signalized intersection procedures discussed later.

INTERSECTIONS

The preceding sections have dealt with continuous movement of traffic, essentially in a single line. The situation at an intersection of two streams of traffic is a distinctly different phenomenon that requires drivers to perceive and enter a gap in an intersecting traffic stream. Geometric condition, including sight distance, angle of intersection, space available for the maneuver, and grades, have an important effect on the size of gap a driver will accept. The volume of traffic in the intersected stream will determine the frequency of gaps of the appropriate size (and the delay awaiting these gaps), and the volume of traffic in the intersecting stream will determine queueing delays.

MERGING

The simplest intersection situation is one in which a single traffic lane merges with another and leaves the merging area as a single lane. Under ideal geometric circumstances with ample sight distance, a small angle of convergence and space for both streams to adjust to a nearly equal speed prior to intersecting, a merge rate equal to or exceeding the free-flowing rate can be accommodated at each service level. Observations of single-lane merging areas on American freeways have revealed short-time merging rates up to 2,300 vph. These rates can seldom be sustained for a full hour nor reliably predicted; thus, the Highway Capacity Committee has recommended a value of 2,000 vph as the hourly volume that has a reasonable expectation of being served (capacity) on a properly designed freeway merging area. Recommended values for other service levels commonly used in American practice are given in Table 8.15.

It should be noted that the above values do not take into consideration the percentage split between the two traffic streams. On a theoretical basis, total delay to ramp vehicles will be greatest when ramp volume is near 40 percent of total volume.[14] At either higher or lower proportions of ramp traffic, total delay decreases. (Average delay for ramp vehicles only increases as freeway volume increases.) From simulation studies, it would appear that, at service level D (1,800 vph total on ramp and lane 1) average delay with 720 ramp vehicles (40 percent) would be 60 sec; with 200 or 1,150 ramp vehicles, average delay would be 20 sec.[15] These theoretical results seem to be consistent with observational studies.[16]

---

[14] ROBERT DAWSON and HAROLD MICHAEL, "Analysis of On-ramp Capacities by Monte Carlo Simulation," *Statistical and Mathematical Aspects of Traffic*, Highway Research Record 118 (Washington, D.C.: Highway Research Board, 1966), pp. 1–20.

[15] JOSEPH W. HESS, "Capacities and Characteristics of Ramp-Freeway Connections," Highway Research Record 27 (Washington, D.C.: Highway Research Board, 1963), pp. 86–89.

[16] See DAWSON and MICHAEL, op. cit.; K. A. BREWER, "Analysis of a Signal-Controlled Ramp's Characteristics," *Traffic Flow Characteristics*, Highway Research Record 308 (Washington, D.C.: Highway Research Board, 1970), pp. 48–61; and HESS, op. cit.

TABLE 8.15
Recommended Maximum Service Volumes for Single-lane Merging Areas on Freeways

Level of Service	Description	Approximate Freeway Speed (mph)	Maximum Service Volume*
A	Free flow	60	1,000
B	Some speed adjustment	55	1,200
C	Speed adjustment—limit of free flow	50	$1,700 \times$ phf†
D	Occasional queueing on ramp	40	$1,800 \times$ phf†
E	Unstable—frequent queueing	—	2,000
F	Stop and go alternate feed	—	$<2,000$ $\approx 1,800$

*Passenger car total on ramp plus first freeway lane. With favorable geometrics, flow can contain a small percentage of trucks.
†Peak hour factor.

Source: Adapted from *Highway Capacity Manual*, pp. 192–96.

The values in Table 8.15 are also predicated on the total downstream volume on the freeway (all lanes) not exceeding established freeway service volume levels. In practice, this will frequently be the limiting condition when ramp volumes are a high proportion of total volume.

Ramp geometrics affect capacity and other levels of service. The relationship of angle of entry, length and type of acceleration lane to the critical gap (t), and the critical gap to maximum service volumes for level of service D are given in Figure 8.12.

Estimation of ramp service volumes in a freeway merging situation requires that the freeway lane 1 volume and composition immediately upstream of the merging area be known or estimated. A variety of formulae and nomographs to assist the engineer in this determination are given in Chapter 8 of the *Highway Capacity Manual*. For the simple case of a single ramp entrance well-separated from other ramps operating at level of service A to C, the formulae given are:

$$\text{2-lane freeway roadway } V_1 = 136 + 0.345\, V_f - 0.115\, V_r \qquad (8.7)$$

$$\text{3-lane freeway roadway } V_1 = -120 + 0.244\, V_f \qquad (8.8)$$

$$\text{4-lane freeway roadway } V_1 = 312 + 0.201\, V_f + 0.127\, V_r \qquad (8.9)$$

where V_1 = lane 1 volume upstream of merge,
  V_f = total volume upstream of merge,
  V_r = on-ramp volume.

The different imputed effect of the on-ramp volumes under these three conditions is seemingly irrational, and statistical tests[17] of these prediction formulae indicate a fairly low reliability (average error of 19 percent). The tests show a marked tendency to underestimate lane 1 volumes in the higher volume range (over 1,500 vph) associated with levels of service D and E.

[17] Department of Transportation Planning, "Weaving Area Operations Study," Polytechnic Institute of Brooklyn, 1971.

Critical gap t based on entrance ramp geometrics

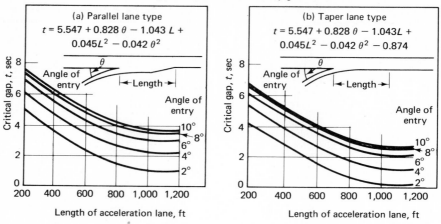

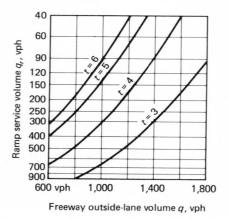

Relationship of service volumes to critical gap t

Note: Derivation by Drew was for a probability of 0.33 of an arriving ramp vehicle joining a queue. The relationship to level of service D is by the author.

Figure 8.12. Relationship of ramp geometrics to level D service volumes. [Source: Donald R. Drew, *Traffic Flow Theory and Control* (New York: McGraw-Hill Book Company, 1968), pp. 218–19.]

Research data from 49 study sites on American freeways[18] produced the lane distribution curves in Figure 8.13.

The report[19] provides a means to correct the values in Figure 8.13 for the influence of adjacent upstream downstream ramps.

[18] D. R. Drew and C. J. Keese, "Freeway Level of Service as Influenced by Volume and Capacity Characteristics," *Freeway Characteristics, Operations and Accidents*, Highway Research Record 99 (Washington, D.C.: Highway Research Board, 1965), pp. 1–39.

[19] *Ibid.*

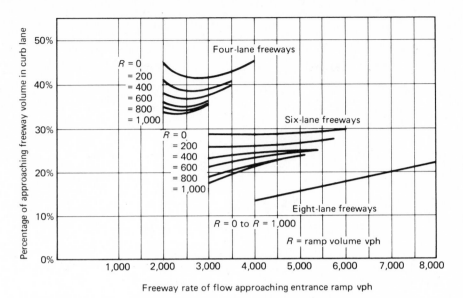

Figure 8.13. Lane distribution on freeways immediately upstream of an isolated on-ramp. (Source: Drew, *Traffic Flow Theory and Control*, p. 94.)

The highway designer who must estimate lane 1 volumes should use these formulae or graphs cautiously and should provide a generous factor of safety. In analyzing existing situations, lane 1 volumes should be obtained by actual count.

WEAVING: TRAFFIC CIRCLE

When two traffic streams that have merged divide again within a short distance, the relatively simple case previously described no longer prevails. The movement approaches a crossing as the distance between merge and diverge diminishes and the angle increases. In a relatively short weaving section, such as at a traffic circle, the maximum flow (capacity) has been found by British research[20] to be related to entrance width, weaving section width and length, and proportion of weaving traffic, as follows:

$$V_{max} = \frac{108w\left(\dfrac{1+m}{w}\right)\left(\dfrac{1-P}{3}\right)}{\dfrac{1+w}{l}} \tag{8.10}$$

where w = width of roadway in the weaving section (ft),
l = length of weaving section (ft),
m = total of widths of entrances to weaving section (ft) ($m_1 + m_2$),
P = proportion of weaving traffic to total traffic in the weaving section.
Measurements are taken as shown in Figure 8.14.

It has been found that this formula overestimates the capacity of small circles and the constant 108 should be reduced to 92 for circles having computed capacity

[20] *Research on Road Traffic*, Road Research Laboratory (London: Her Majesty's Stationery Office, 1965), pp. 220–28.

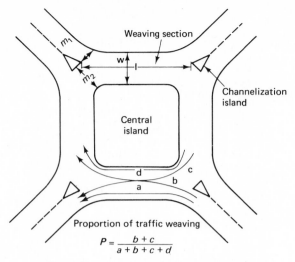

Note: The mirror image should be used
for right hand driving comparison.

Figure 8.14. Relevant dimensions of weaving section and proportion of weaving traffic for use in capacity formula at roundabout with square island. [Source: *Research on Road Traffic* (London: with permission of the Controller of Her Majesty's Stationery Office, 1965), p. 221.]

below 4,000 vph. The above values are maximum and are attained only with long queues and at slow speeds (about 10 mph or 16 kph). Values of approximately 80 percent of the above are considered satisfactory for design purposes (possibly level of service D). U.S. research[21] has resulted in the following formula for short-weaving section capacity:

$$V_{\max} = \frac{R+1}{R} \cdot \frac{3600}{t} \log (R+1) \tag{8.11}$$

where R = ratio of major weaving flow to minor weaving flow,
$\quad t$ = critical gap, sec.
Results of this formula are given in Figure 8.15. Although length and width of weaving section do not appear in this formula, they undoubtedly affect the value of t, which might increase from 2 to 5 or more with restricted designs. Empirical studies are required to determine the proper value for a specific design.

WEAVING: FREEWAYS

Traffic circle weaving takes place at relatively low speeds and involves a high proportion of all entering traffic. The weaving condition that frequently occurs between adjacent on- and off-ramps of a freeway presents a different case. Here, typically, multilanes are present and most traffic is not involved in the weaving movement. In an ideal, balanced design, sufficient length and width of weaving section are pro-

---

[21] E. B. SHARPE, "Testing a Traffic Circle for Possible Capacity," *Proceedings*, Highway Research Board, 1951.

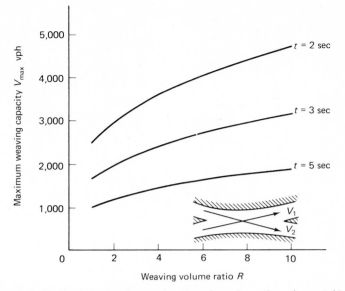

Figure 8.15. Traffic circle capacity as a function of weaving ratio and acceptable gap.
[Source: Martin Wohl and Brian V. Martin, *Traffic System Analysis for Engineers and Planners* (New York: McGraw-Hill Book Company, 1967), p. 417.]

vided so that speeds of the nonweaving traffic will not be adversely affected. When space is not available for such a design (or on an existing facility), the analysis method usually used in American practice is to introduce a factor k which is applied to the smaller weaving volume to account for the decremental effect of weaving friction. A factor of 1 denotes that the weaving volume has no adverse effect. A maximum factor of 3 is recommended in the *Highway Capacity Manual* although curves are provided for lower "quality of flow" conditions that invite use of higher factors.

The length, width, weaving volume, and k relationships are given in Figure 8.16. The basic relationship is:

$$\text{Equivalent volume} = V_t + (k - 1)V_{w2} \tag{8.12}$$

where V_t = total volume entering weaving section,
  k = weaving influence factor,
  V_{w2} = smaller weaving volume.

The number of lanes required to serve this equivalent volume at a particular level of service is, therefore:

$$N = \frac{V_t + (k - 1)V_{w2}}{SV} \tag{8.13}$$

where N = number of lanes,
  SV = service volume for a free-flowing lane on the type facility
    and selected service level,
  k = weaving influence factor,
  V_{w2} = smaller weaving volume.

The family of curves in Figure 8.16 are segregated by Roman numerals which denote, in a general sense, the quality of service to the weaving vehicles. The quality

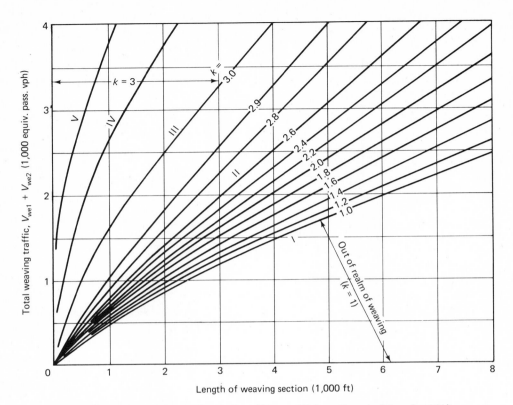

Figure 8.16. *k* values for weaving sections. (Source: *Highway Capacity Manual*, p. 166.)

of service provided may properly be lower than that provided to nonweaving vehicles but, in a balanced design, they should be related. The *Highway Capacity Manual* recommends the relationships given in Table 8.16 to equate quality of weaving service to level of service of the class of highway under consideration.

Quality of flow I is one in which operating conditions approach a normal operating section. On freeways, weaving will take place at 50 mph (80 kph) or over.

TABLE 8.16
Recommended Relationships Between Level of Service on Weaving
Sections and Overall Quality of Flow on the Facility

Level of Service	Freeway and Multilane Rural	Freeway Interchange Roadways	Two-lane Rural	Urban Arterial
A	I or II	II or III	II	III or IV
B	II	III	II or III	III or IV
C	II or III	III or IV	III	IV
D	III or IV	IV	IV	IV
E	IV or V	V	V	V
F		Unsatisfactory		

Source: *Highway Capacity Manual*, p. 173.

Quality of flow II involves minor restrictions and freeway weaving speeds between 45 and 50 mph (72 and 80 kph).

At quality of flow III, weaving speeds are between 40 and 45 mph (64 and 72 kph) and may vary considerably between individual vehicles and adjacent time periods.

Quality of flow IV provides weaving speeds of about 30 to 35 mph (48 and 56 kph). Occasional slowdowns and restrictions to maneuver can be expected.

Quality of flow V represents the weaving section capacity that is characterized by operations frequently averaging under 20 mph (32 kph), turbulence, and alternate feeding of the weaving lanes.

There are numerous difficulties, both theoretical and practical, in the procedures outlined above and the results can be considered only approximations. Research under way in 1974[22] suggests that a more satisfactory form of equation would be:

$$V_w = aL^b W^{(c+dR)} \qquad (8.14)$$

where V_w = weaving volume,

L = weaving section length,

W = weaving section width,

R = ratio of smaller to total weaving volume,

a, b, c, d = constants associated with service level.

Until this or other research provides improved methods, the engineer should utilize the method described above with some caution and not ascribe a high degree of accuracy to computational results.

UNCONTROLLED INTERSECTIONS

The intersection of two roadways at angles of approximately 90° without any control devices present is probably the most common of all intersections. Relatively little study, however, has been given to either capacity or service volume values under these conditions, undoubtedly because safety considerations usually require installation of control devices at volume levels well below those that would affect delay or other conventional measures of service.

The probability of an approaching vehicle being delayed and the amount of delay at an uncontrolled intersection depends on the traffic volumes on the various approaches, the sight distances available, and the gap size acceptable to approaching drivers. The latter, in turn, is undoubtedly influenced by grades, speed of traffic, and the geometry of the intersection. If random arrivals are assumed and volumes are sufficiently low so that queues usually will not occur, the probability that the vehicle on the lower volume street will be delayed is given by the formula:[23]

$$P = 1 - \frac{e^{-2.5q_s}e^{-2qt}}{1 - e^{-2.5q_s}(1 - e^{-qt})} \qquad (8.15)$$

where P = probability of delay,

e = Napierian base,

q_s = side street flow rate,

q = main street flow rate,

[22] Polytechnic Institute of New York, Department of Transportation Planning and Engineering, *Weaving Area Operations Study*, NCHRP Report (Washington D.C.: Transportation Research Board, 1975).

[23] MORTON S. RAFF, *A Volume Warrant for Urban Stop Signs* (Saugatuck, Conn.: The ENO Foundation for Highway Traffic Control, 1950), p. 48.

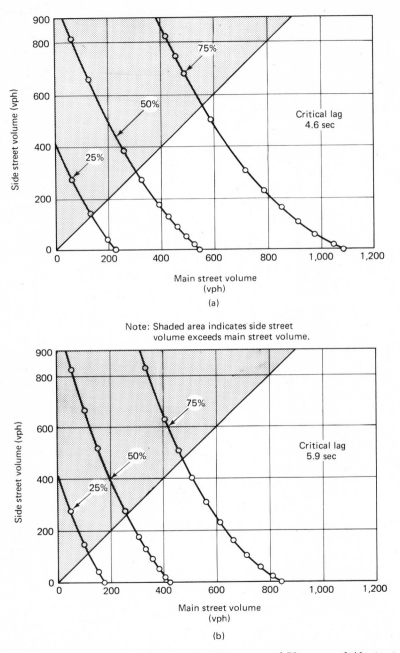

Figure 8.17. Volumes for which 25 percent, 50 percent, and 75 percent of side street vehicles are delayed. [Source: Morton S. Raff, *A Volume Warrant for Urban Stop Signs* (Naugatuck, Conn.: The ENO Foundation for Highway Traffic Control, 1950), pp. 94, 98.]

$t =$ the critical lag (the time spacing between arrival of side street and main street vehicles such that the number of rejected lags larger and the number of accepted lags smaller will be equal).

The percentage of side street vehicles delayed for various main and side street volumes is given in Figure 8.17(a, b) for critical lags of 4.6 and 5.9 sec which may be the typical range for urban intersections.

Another useful criterion is the mean delay to side street vehicles which is given by the formula:

$$\bar{d} = q^{-1}(e^{qt} - qt - 1) \qquad (8.16)$$

where $\bar{d} =$ mean delay caused by blockage(sec),

$e =$ Napierian base,

$q =$ main street flow rate,

$t =$ the critical gap (the time spacing between arrival of succeeding main street vehicles which half of the side street drivers will accept and half reject).

These values are plotted in Figure 8.18.

It should be noted that the above formula gives only the delay caused by blockage by main street cars. Time lost in deceleration and acceleration and any delay in queue awaiting a first-in-line position are not included. If queueing can be expected on the side street, the average delay, including the queueing delay, can be estimated from the

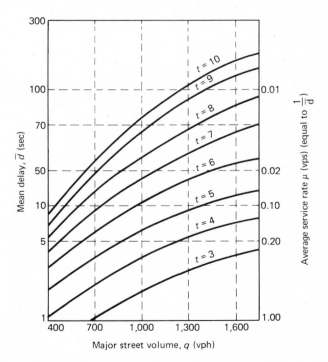

Figure 8.18. Mean delay to side street vehicles without queuing. [Source: Vasant H. Surti, "Operational Efficiency Evaluation of Selected At-Grade Intersections," *Highway Capacity and Quality of Service*, Highway Research Record 321 (Washington, D.C.: Highway Research Board, 1970), p. 68.]

formula:

$$\bar{d}_t = \frac{1}{\mu - \lambda} \qquad (8.17)$$

where $\bar{d}_t$ = average total delay (sec),

μ = average service rate, $1/\bar{d}$ (Figure 8.18) (vps)

λ = side street arrival rate (vps).

The delay value from this formula does not include deceleration-acceleration delays.

The expected number of vehicles in the queue can be estimated from the formula:

$$N = \frac{\lambda}{\mu - \lambda} \qquad (8.18)$$

where N = expected number in queue,

λ = side street arrival rate (vps),

μ = average service rate $1/\bar{d}$ (Figure 8.18) (vps)

None of the values from the above formula is an exact parallel of the speed or density criteria used in earlier sections to define various service levels. The engineer may, however, find them useful in evaluating service provided at an uncontrolled intersection. Because the flow rates on both main and side streets are significant to these formulae, flow variations within the peak hour should be considered, particularly since main and side street peak periods within the hour may not coincide.

YIELD AND STOP-SIGN CONTROLLED INTERSECTIONS

The principal effect of yield and stop-sign control is to assure that right of way will be assigned to the selected street and, in the case of stop signs, that all vehicles on the "minor" street will stop before entering the intersection. These rules affect the total intersection performance both in the initial delay imposed and in the increase in gap size required by vehicles entering from a stopped position.

Observations of queues of vehicles which each come to a stop at a stop sign have shown that the median minimum departure headway (with no cross-street interference) is 4 sec.[24] This provides a maximum stop sign capacity of 900 vehicles per lane per hour, providing that a continuous queue exists, no cross-street traffic is present, and all vehicles obey the stop regulation. Since such a combination of circumstances is rare, the above value is of more academic than real interest.

An intersection with two-way stop- or yield-sign control may be analyzed for average delay and queue length by using the formula given in the preceding section and appropriate critical gap values.

The above formulae are based on Poisson or random distribution of arrivals which is appropriate to low-volume conditions (under 400–600 vehicles per lane per hour). As density increases, other distributions are more descriptive (see Chapter 7, Traffic Flow Theory).

Because traffic volumes and densities at stop- or yield-sign controlled intersections are likely to be in the higher range or be affected by nearby traffic signals, the above formulae should be applied with discretion. Results of simulation studies given later

---

[24] JACQUES HEBERT, "A Study of Four-Way Stop Intersection Capacities," *Highway and Interchange Capacity*, Highway Research Record 27 (Washington, D.C.: Highway Research Board, 1963), pp. 130–47.

in this section may be more appropriate for higher volumes because the assumption of Poisson distribution was not followed.

The size of the critical gap value is important. Factors that affect this value are not well understood and, whenever possible, they should be determined from observations at the intersection under study (see Chapter 10, Traffic Studies).

Table 8.17 gives a range of values determined from past studies.

The average *waiting* delay to side street vehicles at a stop-sign controlled intersection (deceleration-acceleration delay not included) has been obtained from computer

TABLE 8.17
Observed Values of Critical Gaps at Urban Intersections

Situation-Location	Gap (sec)	Source
Minor street with major one-way (England)	8.0	1
Minor street with stop sign (New Haven)	6.1	2
Open intersection, no control (New Haven)	2.87	2
Blind intersection, no control (Hartford)	2.82	2
Stop sign, 39 ft one-way street	4.6	3
Stop sign, 34 ft one-way through street	4.7	3
Stop sign, 41 ft two-way street	5.9	3
Stop sign, 63 ft two-way street	6.0	3
Yield sign	6.2	4
Stop sign	6.5	4
Stop sign (through vehicles)	5.8	5
Stop sign (left turns)	6.2	5
Stop sign (right turns)	5.4	5
Left turn through opposing	4.25	6
Left turn through opposing (moving)	4.4	7
Left turn through opposing (from stop)	4.6	7
60° intersection—stop sign (right turns)	5.5	8
60° intersection—stop sign (left turns)	7.0	8
T intersection—stop sign (right turns)	5.7	8
T intersection—stop sign (left turns)	7.2	8
Stop sign—into one-way street (right turns)	4.0	8
Stop sign—into one-way street (left turns)	5.6	8

Sources:

1. *Research in Road Traffic* (London: Her Majesty's Stationery Office, 1965), p. 228.

2. Bruce D. Greenshields *et al.*, *Traffic Performance at Urban Street Intersections* (New Haven: Yale Bureau of Highway Traffic, 1944), pp. 67–70.

3. Morton Raff, *A Volume Warrant for Urban Stop Signs* (Saugatuck, Conn.: Eno Foundation, 1950), pp. 31–35. (The values given are for critical lag.)

4. De Leuw, Cather & Co., *Effect of Control Devices on Traffic Operations*, NCHRP Report 11 (Washington D.C.: Highway Research Board, 1964), p. 27.

5. H. H. Bissell, "Traffic Gap Acceptance for a Stop Sign" (Master's thesis, University of California, Berkeley, 1960).

6. F. J. Kaiser, Jr., "Left Turn Gap Acceptance" (Student thesis manuscript, Yale Bureau of Highway Traffic, 1951).

7. Olin K. Dart, "Left-turn Characteristics at Signalized Intersections on Four-lane Arterial Streets," *Characteristics of Traffic Flow*, Highway Research Record 230 (Washington, D.C.: Highway Research Board, 1968), p. 45.

8. Vasant H. Surti, "Operational Efficiency Evaluation of Selected At-grade Intersections," *Highway Capacity and Quality of Service*, Highway Research Record 321 (Washington, D.C.: Highway Research Board, 1970), p. 60.

simulation. In the model used, a two-way, two-lane street intersected a four-lane, two-way major street. Arrivals followed a modified binomial distribution, and critical lags of 5.8 sec and 4.8 sec were assumed. The volumes simulated were below approach capacity in all cases and a "backlog" limit of 20 vehicles was established. The resulting average waiting delays are shown in Figure 8.19. An acceleration-

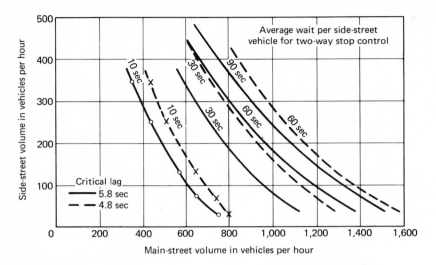

Figure 8.19. Waiting delay to side street vehicles at stop-sign controlled intersections. [Source: Russell M. Lewis and Harold L. Michael, "Simulation of Traffic Flow to Obtain Volume Warrants for Intersection Control," *Traffic Flow Theory*, Highway Research Record 15 (Washington, D.C.: Highway Research Board, 1963), p. 39.]

deceleration delay of 8.9 sec was found in addition to the waiting delay, based on approach speeds of 30 mph and 3 ft/sec^2 and 6 ft/sec^2 for acceleration and deceleration, respectively. No delay was imposed on main street vehicles by the side street traffic. When average total delay to all vehicles was considered, including the acceleration-deceleration delay and turning delays to main street vehicles (7 percent right and 7 percent left turns), results were as shown in Figure 8.20.

The values in Figures 8.19 and 8.20 may be used to estimate service volumes at levels of service which are better than level of service E (capacity).

FOUR-WAY STOP INTERSECTIONS

The interaction of vehicles at a four-way stop is complex. The criteria of gap acceptance are no longer appropriate since cross-street vehicles will be at a stop or expected to stop. With no cross-street traffic present, departure headways of 4 sec can be expected for single vehicles (900 vph) and approximately 4.5 sec for pairs of vehicles entering together (1,600 vph). When there was traffic present on all approaches, median headways of 7.32 sec. were observed at four-way stop intersections of two-way, two-lane suburban streets and median headways of 8.08 sec for vehicles crossing a four-lane intersecting street.[25] Based on these values and a tendency for the median

[25] *Ibid.*

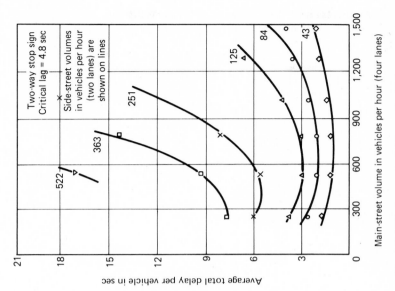

Figure 8.20(a). Average total delay to all vehicles at stop-sign controlled intersection (critical lag = 5.8 sec). (Source: Lewis and Michael, "Simulation of Traffic Flow to Obtain Volume Warrants for Intersection Control," p. 35.)

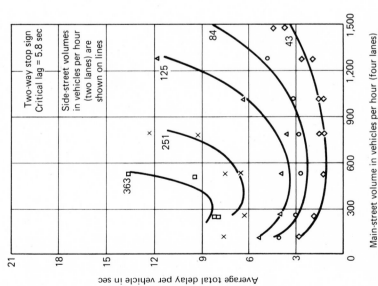

Figure 8.20(b). Average total delay to all vehicles at stop-sign controlled intersection (critical lag = 4.8 sec). (Source: Lewis and Michael, p. 36.)

353

headway to decrease on the major street when the split of traffic departs from 50/50, the capacity of an intersection (all approaches) has been computed as shown in Table 8.18. The capacities in Table 8.18 would be obtained only with near constant queues on all approaches. Left turns appear to not adversely affect these values and a high percentage of right turns probably acts to increase them.

TABLE 8.18
Estimated Capacity of Four-way Stop Intersection
of Two-way, Two-lane Streets

Split	Capacity (vph) (Total from All Approaches)
50/50	1,900
55/45	1,800
60/40	1,700
65/35	1,600
70/30	1,550

Source: *Highway Capacity Manual*, p. 158.

SIGNALIZED INTERSECTIONS

The capacity and service volumes that a signalized intersection can accommodate are dependent on the intersection geometrics, signal operation, and traffic factors. In the first category, the approach width and grades are most critical. The existence of parking, intersection width, exit width, turning radii, and lane configuration are also important. In the second category, the proportioning of green time is the single most important factor. Cycle length, phasing, and "lost time" features are somewhat less significant.

Traffic factors include the pattern and composition of arriving traffic, turning movements, presence of pedestrians, and general driver characteristics. The latter appear to be related to the size of and location within an urban area and, although not well quantified, can be estimated from those factors. The pattern of traffic arrivals is strongly influenced by nearby traffic signals and their coordination.

Of major concern in evaluating intersection capacity and service volume is the proper figure of merit to describe traffic performance at intersections. Speed and density (or v/c ratio), the criteria used in free-flowing traffic situations, are not directly applicable and surrogate measures must be employed. Although total delay time is probably the most satisfactory measure, its determination under field conditions is arduous.

LOAD FACTOR

In American practice, the measure most commonly used to describe intersection performance is load factor, which the *Highway Capacity Manual*[26] defines as *the ratio of the number of green phases that are loaded or fully utilized by traffic (usually during the peak hour) to the total number of green phases available for that approach during the same period.* A green phase is considered loaded when: (1) there are vehicles ready to enter the intersection in all lanes when the signal turns green and (2) they

[26] *Highway Capacity Manual*, p. 116.

continue to be available to enter in all lanes during the entire phase with no unused time or exceedingly long spacings between vehicles caused by lack of traffic. Load factor is relatively easy to determine in the field and is related in some manner to total delay, and/or average individual delay. The relationship between load factor and delay based on a simplified simulation model is given in Figure 8.21. The intersection

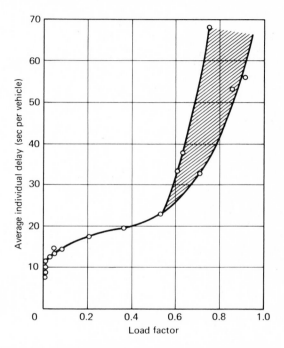

Figure 8.21. Average individual delay related to load factor. [Source: Adolf D. May, Jr., and David Pratt, "A Simulation Study of Load Factor at Signalized Intersections," *Traffic Engineering*, XXXVIII, No. 5 (1968), p. 44.]

modeled had a capacity of 600 vph, a 60-sec cycle divided into two equal phases, a Poisson arrival distribution, and a minimum departure headway of 3 sec. It can be seen that both the magnitude and variation of average individual delays increase rapidly at load factors above 0.5.

The Highway Capacity Committee has established the relationship between load factor and service levels (shown in Table 8.19). Because of the newness of load factor

TABLE 8.19
Relationship of Load Factor to Level of Service

Level of Service	Description	Load Factor
A	Free flow	0.0
B	Stable flow	$\lesssim 0.1$
C	Stable flow	$\lesssim 0.3$
D	Approaching unstable flow	$\lesssim 0.7$
E	Unstable flow	$\lesssim 1.0*$
F	Forced flow	—

*Usually $\lesssim 0.85$ in the absence of an exceptionally effective signal progression.

Source: *Highway Capacity Manual*, p. 131.

TABLE 8.20

Tabulation of $f_1 = \dfrac{\left(1 - \dfrac{G}{C}\right)^2}{2\left(1 - \dfrac{q}{s}\right)}$

$\dfrac{G}{C}$ \ x	0.1	0.2	0.30	0.35	0.40	0.45	0.50	0.55	0.60	0.65	0.70	0.80	0.90
0.1	0.409	0.327	0.253	0.219	0.188	0.158	0.132	0.107	0.085	0.066	0.048	0.022	0.005
0.2	0.413	0.333	0.261	0.227	0.196	0.166	0.139	0.114	0.091	0.070	0.052	0.024	0.006
0.3	0.418	0.340	0.269	0.236	0.205	0.175	0.147	0.121	0.098	0.076	0.057	0.026	0.007
0.4	0.422	0.348	0.278	0.246	0.214	0.184	0.156	0.130	0.105	0.083	0.063	0.029	0.008
0.5	0.426	0.356	0.288	0.256	0.225	0.195	0.167	0.140	0.114	0.091	0.069	0.033	0.009
0.55	0.429	0.360	0.293	0.262	0.231	0.201	0.172	0.145	0.119	0.095	0.073	0.036	0.010
0.60	0.431	0.364	0.299	0.267	0.237	0.207	0.179	0.151	0.125	0.100	0.078	0.038	0.011
0.65	0.433	0.368	0.304	0.273	0.243	0.214	0.185	0.158	0.131	0.106	0.083	0.042	0.012
0.70	0.435	0.372	0.310	0.280	0.250	0.221	0.192	0.165	0.138	0.112	0.088	0.045	0.014
0.75	0.438	0.376	0.316	0.286	0.257	0.228	0.200	0.172	0.145	0.120	0.095	0.050	0.015
0.80	0.440	0.381	0.322	0.293	0.265	0.236	0.208	0.181	0.154	0.128	0.102	0.056	0.018
0.85	0.443	0.386	0.329	0.301	0.273	0.245	0.217	0.190	0.163	0.137	0.111	0.063	0.021
0.90	0.445	0.390	0.336	0.308	0.281	0.254	0.227	0.200	0.174	0.148	0.122	0.071	0.026
0.92	0.446	0.392	0.338	0.312	0.285	0.258	0.231	0.205	0.179	0.152	0.126	0.076	0.029
0.94	0.447	0.394	0.341	0.315	0.288	0.262	0.236	0.210	0.183	0.157	0.132	0.081	0.032
0.96	0.448	0.396	0.344	0.318	0.292	0.266	0.240	0.215	0.189	0.163	0.137	0.086	0.037
0.98	0.449	0.398	0.347	0.322	0.296	0.271	0.245	0.220	0.194	0.169	0.143	0.093	0.042

Source: *Research on Road Traffic*, Department of Scientific and Industrial Research, Road Research Laboratory (London: Her Majesty's Stationery Office, 1965), p. 301.

as a measure of service and the limited research substantiation, these values should be treated as approximations.

DELAY

The average delay per vehicle on approaches to a fixed time signal may be computed from the following:

$$d = \left(Cf_1 + \frac{f_2}{q}\right)\frac{100 - f_3}{100} \qquad (8.19)$$

where d = average delay per vehicle on approach (sec),

C = cycle length (sec),

$f_1 = \dfrac{\left(1 - \dfrac{G}{C}\right)^2}{2\left(1 - \dfrac{q}{s}\right)}$ (see Table 8.20),

G = effective green time (sec),

q = approach flow (vehicles/sec),

s = saturation flow (vehicles/sec),

$f_2 = \dfrac{x^2}{2(1 - x)}$ (see Table 8.21),

$x = \dfrac{Cq}{Gs}$,

$f_3 = \dfrac{0.65\left(\dfrac{C}{q^2}\right)^{1/3} x^{2+5(G/C)}}{Cf_1 + \dfrac{f_2}{q}}$ (see Table 8.22).

TABLE 8.21
Tabulation of $f_2 = \dfrac{x^2}{2(1 - x)}$

x	0.00	0.01	0.02	0.03	0.04	0.05	0.06	0.07	0.08	0.09
0.1	0.006	0.007	0.008	0.010	0.011	0.013	0.015	0.017	0.020	0.022
0.2	0.025	0.028	0.031	0.034	0.038	0.042	0.046	0.050	0.054	0.059
0.3	0.064	0.070	0.075	0.081	0.088	0.094	0.101	0.109	0.116	0.125
0.4	0.133	0.142	0.152	0.162	0.173	0.184	0.196	0.208	0.222	0.235
0.5	0.250	0.265	0.282	0.299	0.317	0.336	0.356	0.378	0.400	0.425
0.6	0.450	0.477	0.506	0.536	0.569	0.604	0.641	0.680	0.723	0.768
0.7	0.817	0.869	0.926	0.987	1.05	1.13	1.20	1.29	1.38	1.49
0.8	1.60	1.73	1.87	2.03	2.21	2.41	2.64	2.91	3.23	3.60
0.9	4.05	4.60	5.28	6.18	7.36	9.03	11.5	15.7	24.0	49.0

Source: *Research on Road Traffic*, p. 302.

To use this formula, as demonstrated on page 372, demand flow rates for the approach, saturation flow rate, and effective green time must either be obtained from field measurements or be estimated.

The saturation flow rate, s, is the maximum rate at which vehicles enter the intersection in a single lane after the queue start-up delay has been eliminated and while a continuous demand exists. Studies of intersection performance in the U.S. indicate

TABLE 8.22

$$\text{Values of } f_3 = \frac{0.65\left(\frac{C}{q^2}\right)^{1/3} x^{2+5(G/C)}}{Cf_1 + \frac{f_2}{q}}$$

x	$\dfrac{G}{C}$ \\ qC	2.5	5	10	20	40
0.3	0.2	2	2	1	1	0
	0.4	2	1	1	0	0
	0.6	0	0	0	0	0
	0.8	0	0	0	0	0
0.4	0.2	6	4	3	2	1
	0.4	3	2	2	1	1
	0.6	2	2	1	1	0
	0.8	2	1	1	1	1
0.5	0.2	10	7	5	3	2
	0.4	6	5	4	2	1
	0.6	6	4	3	2	2
	0.8	3	4	3	3	2
0.6	0.2	14	11	8	5	3
	0.4	11	9	7	4	3
	0.6	9	8	6	5	3
	0.8	7	8	8	7	5
0.7	0.2	18	14	11	7	5
	0.4	15	13	10	7	5
	0.6	13	12	10	8	6
	0.8	11	12	13	12	10
0.8	0.2	18	17	13	10	7
	0.4	16	15	13	10	8
	0.6	15	15	14	12	9
	0.8	14	15	17	17	15
0.9	0.2	13	14	13	11	8
	0.4	12	13	13	11	9
	0.6	12	13	14	14	12
	0.8	13	13	16	17	17
0.95	0.2	8	9	9	9	8
	0.4	7	9	9	10	9
	0.6	7	9	10	11	10
	0.8	7	9	10	12	13
0.975	0.2	8	9	10	9	8
	0.4	8	9	10	10	9
	0.6	8	9	11	12	11
	0.8	8	10	12	13	14

Source: *Research on Road Traffic*, p. 303.

that under ideal circumstances headways between successive passenger vehicles stabilized at between 2.1 sec and 2.0 sec, yielding values of s of 0.48 to 0.50.

The effective green time, G, is the time available for flow at the rate s. It is thus equal to the green time plus amber time minus the time loss in queue start-up and the time taken for the last vehicle to cross the intersection. Studies of queues at urban intersections suggest that the first value averages about 3.7 sec and the latter about 2.0 sec (depending on speed and intersection geometrics).

Using the above values and a 120-sec cycle (30 cycles per hour) with 50 percent split, we see that the theoretical hourly capacity of a single lane signal approach under ideal conditions is approximately 800 [capacity $= 30\,(G - \text{lost time})/s$] or 1,600 passenger vehicles per hour of green. Such a flow rate would be accompanied by a high average delay (exceeding 4 min per vehicle). A value of 90 percent of the above, approximately 1,450 passenger vehicles per hour of green, has been suggested as a more realistic value.

The computed delay from the above formula at a representative intersection is shown in Figure 8.22. It can be seen that average delay increases very rapidly as volume exceeds about 80 percent of theoretical capacity and becomes nearly asymptotic above 90 percent of theoretical capacity.

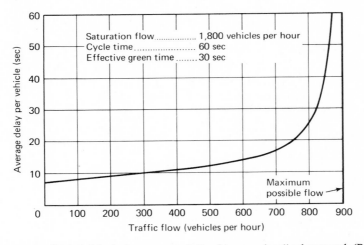

Figure 8.22. Typical delay/volume curve relationship on a signalized approach (British). (Source: *Research on Road Traffic*, p. 304.)

QUEUE LENGTH

Another possible measure of traffic performance at signalized intersections is the length of queues developed. Average queue length is approximated by the larger result of the two formulae:

$$n = qR \tag{8.20}$$

or

$$n = q\left(\frac{R}{2} + d\right) \tag{8.21}$$

where $n =$ average queue length,
  $q =$ approach flow (vehicles per sec),
  $R =$ red time (sec),
  $d =$ average individual delay from Equation (8.19).

A value of perhaps greater significance is the maximum probable queue length. This can be estimated from Table 8.23 in which the parameters C, G, q, and x are as defined for Equation (8.19).

TABLE 8.23
Estimate of Queue Length at Start of Green Which Will Not Be
Exceeded in More Than 1 Percent of Cycles

x	$\dfrac{G}{C}$ \\ qC	2.5	5.0	10.0	20.0	40.0
0.3	0.4	6	9	14	23	38
	0.6	5	6	11	17	28
	0.8	3	5	7	12	17
0.5	0.2	7	9	17	29	53
	0.4	6	9	14	23	38
	0.6	5	7	11	17	28
	0.8	4	5	7	12	18
0.7	0.2	9	12	17	28	50
	0.4	9	9	15	23	38
	0.6	8	9	12	18	28
	0.8	7	7	8	12	18
0.8	0.2	13	15	19	28	50
	0.4	12	13	17	24	39
	0.6	12	13	14	20	28
	0.8	11	12	12	15	18
0.9	0.2	29	25	29	38	55
	0.4	28	24	27	33	46
	0.6	27	24	26	28	42
	0.8	27	23	24	25	29
0.95	0.2	40	36	38	47	65
	0.4	40	34	37	44	55
	0.6	40	32	30	42	48
	0.8	39	32	34	36	40
0.975	0.2	82	70	79	69	93
	0.4	83	66	75	65	82
	0.6	82	70	69	59	79
	0.8	79	65	66	57	79

Source: *Research on Road Traffic*, p. 308.

CYCLE FAILURES

The percentage of cycles during which more vehicles arrive than depart can be estimated from the chart in Figure 8.23. Traffic arrivals are assumed to be Poisson distributed and departure rate 0.5 vehicles per sec with 4 sec lost time per phase. It should be noted that failures estimated by this method do not include the effect of queues carried over from previous cycle failures. It thus overestimates the possibility of a vehicle's being served during the cycle in which it arrives. Since overestimate becomes more serious as probability of failure increases, this estimate is reasonable only under the better level of service conditions.

The relationships between the failure rate determined by this method and levels of service computed by *Highway Capacity Manual* method for a representative condition are shown in Figure 8.24.

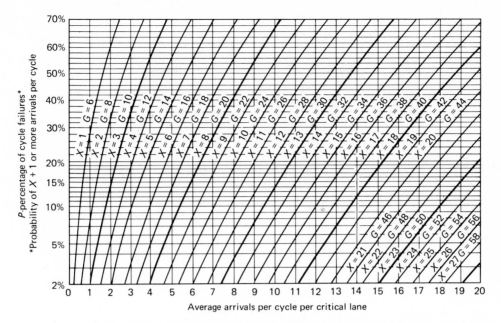

Figure 8.23. Probability of more vehicles arriving (X) than can be served during the green interval (G). (Source: Drew, *Traffic Flow Theory and Control*, p. 140.)

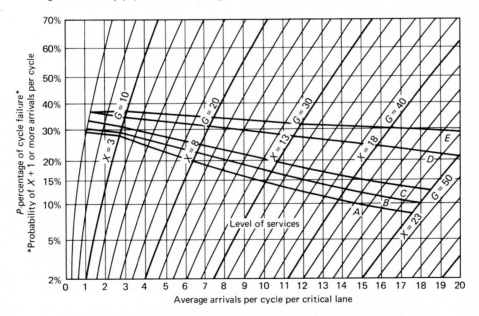

Figure 8.24. Typical relationship of computed cycle failure and level of service at a signalized intersection approach. [Source: John E. Tidwell, Jr., and Jack B. Humphreys, "Relation of Signalized Intersection Level of Service to Failure Rate and Average Individual Delay," *Highway Capacity and Quality of Service*, Highway Research Record 321 (Washington, D.C.: Highway Research Board, 1969), p. 23.]

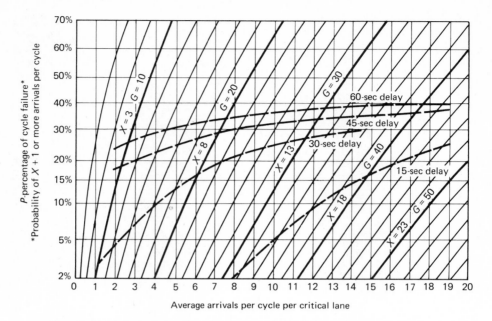

Figure 8.25. Typical relationship of a computed cycle failure and computed average individual delay at a signalized intersection approach. (Source: Tidwell and Humphreys, "Relation of Signalized Intersection Level of Service to Failure Rate and Average Individual Delay," p. 27.)

The relationship between failure rate and average individual delay computed by the method given previously in this chapter is shown in Figure 8.25.

At this time there is no general agreement as to which of the measures of intersection performance is most appropriate. In fact, different factors may be the most appropriate in specific cases. All of the methods described assume a distribution of vehicle arrivals which may not be descriptive of a specific case, particularly an urban situation with other nearby signals.

FACTORS AFFECTING SIGNALIZED INTERSECTION CAPACITY

Peaking characteristics. In American practice it is customary to consider the variation of flow within the hour in developing capacity and service volume values. A 15-min peak hour factor (phf_{15}) is usually used for this adjustment. The 15-min peak hour factor is the ratio of the volume in a full hour to four times the flow occurring in the highest 15 consecutive minutes. Use of this factor has the effect of developing capacity and service volumes on the basis of peak period rather than hourly flow rates. It has been found that traffic distributions in these peak periods are more likely to be random (and thus more consistent with theoretical models) than is the case with full peak hour flows.

Driver characteristics and city size. Some studies of traffic performance at intersections have found a consistent pattern of higher capacities for locations in large metropolitan areas than for similar intersections in rural areas or smaller communities. In metropolitan areas of one million population, capacities were found to be about 5 percent greater than in communities of 500,000; in areas of 250,000 population, capacities were found to average about 5 percent less. The reasons for this might be a more hurried pace of life, different speed characteristics, or greater driver experience. The relationship is not consistent, however, and other studies both in the United States and abroad have failed to detect a similar relationship. The *Highway Capacity Manual*, frequently followed in American practice, provides a correction factor that combines the effect of area population and peak hour factor. British and Australian practices do not utilize an adjustment for this characteristic.

Adjacent development. The culture and peripheral activities surrounding an intersection can also affect performance. In a busy central business district, the presence of many pedestrians, loading and unloading, and frequent parking maneuvers near the intersection have a deleterious effect on capacity. These factors are less frequent in fringe and outlying business districts and nearly absent in residential and rural areas. Thus, the functional capacity and service volume of two physically identical intersections can be markedly different depending on these influences which are difficult to predict or to quantify. Hopefully, future research will provide better insights and improved techniques for accounting for these influences. The *Highway Capacity Manual* provides gross adjustments in the 10 percent to 25 percent range which may be applied with considered engineering judgment. These adjustment factors are given in Table 8.24.

TABLE 8.24
Suggested Adjustment Factors to Account for Effect of Adjacent Development on Intersection Capacities and Service Volume

Type of Adjacent Development	One-way		Two-way	
	With Parking	No Parking	With Parking	No Parking
Central business district	1.00	1.10	1.25	1.25
Fringe of CBD	1.00	1.10	1.25	1.25
Outlying business district	1.15	1.10	1.25	1.25
Residential	1.25	1.20	1.25	1.25

Source: *Highway Capacity Manual*, pp. 134–36.

Approach width. The width available for approach traffic is critical to intersection capacity; it may be considered either by lane and lane width or as a total approach width. American and British practices favor the latter but Australian methods are developed around the former. Unquestionably, the lane configuration has a bearing on performance; only its relative importance is in question. Figure 8.26 shows the relative efficiency of different lane configurations from an American study.

A comparison of American, Australian, and British methods of considering width is shown in Figure 8.27. The British value, for widths above 17 ft, is $s = 160\,w$ where w is the width in feet measured from the curb to center line or edge of center island and s is the saturation flow in vehicles per hour. This value is reduced 6 percent for off-peak conditions. An approximate American figure (based on two-way street, no

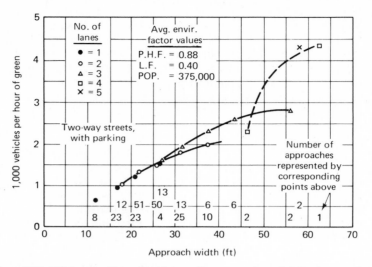

Figure 8.26. Efficiencies of signalized intersection approaches related to lane configuration. (Source: *Highway Capacity Manual*, p. 129.)

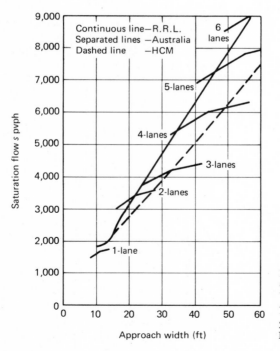

Figure 8.27. Saturation flow related to approach width at a signalized intersection. [Source: Alan J. Miller, "On the Australian Road Capacity Guide," *Highway Capacity*, Highway Research Record 289 (Washington, D.C.: Highway Research Board, 1969), p. 7.]

parking, no turns, no trucks, outlying location in metropolitan area of 250,000, phf 0.85) is $s = 125\ w + 185$.

PARKING

Parked vehicles in the vicinity of a signalized intersection reduce the space available for traffic movement and thus the carrying capacity of the intersection. Temporary blockage of an adjacent lane while vehicles enter or leave curb spaces will also reduce capacity.

In the British system,[27] the effect of a parked vehicle at z ft from the intersection stop line is considered as a width reduction equaling:

$$w^1 = 5.5 - \frac{0.9(z - 25)}{G} \tag{8.22}$$

where w^1 = width loss caused by parked vehicle,
z = clear distance between stop line and parked vehicle (ft),
G = green time (sec),
if $z < 25$ ft, the second term is disregarded and the width loss is 5.5 ft.

In the Australian[28] system, which considers capacity by lane, the capacity of the curb lane is reduced as follows:

1. No parking permitted, one and two lanes—no reduction
2. No parking permitted, three or more lanes with strict enforcement of stopping —0.6 vehicles per phase in curb lane
3. No parking permitted, three or more lanes—0.4 vehicles per phase
4. Parking on approach but not on exit—one vehicle per 30 ft of clear distance from stop line to parked vehicle per phase
5. Parking on exit (within 600 ft) but not on approach, one-and two-lane approach —no reduction
6. Parking on exit (within 600 ft) but not on approach, three or more lanes—$1\frac{1}{2}$ vehicles per phase in curb lane

The effect of parking in the American method is accounted for by a series of basic curves for different conditions (see Figures 8.28 through 8.32). An analysis of these curves shows the effect of parking to range from an equivalent width reduction of 6 ft to 8 ft on relatively narrow streets up to as much as 15 ft on wide approaches.

ONE-WAY OPERATION

The principal benefit of one-way operation is to permit left turns (right turns in England and Australia) to be made without interference from opposing traffic. Both right and left lanes, however, are curb lanes on a one-way approach and subject to adverse influence of parking and sidewalk activity.

In the British system, the effect of one-way street operation is accounted for by considering all turns as nonconflicting turns which, in that system, are equivalent to through vehicles.

---

[27] *Research on Road Traffic*, p. 219.
[28] ALAN J. MILLER, "On the Australian Road Capacity Guide," *Highway Capacity*, Highway Research Record 289 (Washington, D.C.: Highway Research Board, 1969), pp. 1–13.

In the Australian system, all turns from one-way streets are treated as nonconflicting and the curb lane reductions previously given are applied to both curb lanes.

In the American system, separate curves are given for one-way and two-way operation. The effect of left-turning vehicles under one-way operation is considered to be one-half as great as under two-way conditions (equivalent to right turns). The added capacity caused by removal of parking on a second (usually left) side of a one-way street is shown to be only about one-half as great as the increment gained from removal on one side.

NONCONFLICTING (RIGHT) TURNS

Right-turning vehicles have some effect on intersection capacity because, typically, a short-turn radius requires slowing considerably below through traffic speeds. Also, the potential or actual presence of pedestrians will affect headways and performance.

The British capacity computation procedure accounts for this effect only in the selection of a saturation flow value; nonconflicting-turn vehicles are considered through vehicles.

The Australian method considers a nonconflicting-turn passenger vehicle to be equivalent to 1.25 through passenger vehicles and a turning truck to be equivalent to 2.40 passenger vehicles.

The American system increases or reduces the capacity and service volumes around a central value of 10 percent turns by a factor which, for two-lane approaches (16 ft to 24 ft), is equal to 0.5 percent of right turns up to 30 percent. Above three lanes (34 ft or 39 ft) no adjustment is made. Adjustment factors for right turns are given in Table 8.25. Special procedures are used when a separate turning lane or separate turning phase is provided.

CONFLICTING (LEFT) TURNS

Left-turning vehicles not provided with a separate lane or phase must await gaps in the approaching traffic and thus create delays to themselves and to following vehicles in the same lane.

The British system, in a suburban condition in which saturation flow equals 160 passenger vehicles per hour of green per ft of width (for widths above 17 ft), considers one conflicting turn vehicle to be equivalent to 1.75 through vehicles. For downtown London conditions, a saturation flow of 115 passenger vehicles per ft of width per hour of green is assumed with 25 percent commercial vehicles and 10 percent conflicting turns. Each 1 percent of conflicting turns above or below 10 percent is considered to change this value by ± 0.6 percent up to a maximum reduction of 18 percent.

The Australian system provides a formula for computing the effect of conflicting turns as follows:

$$E = \frac{1.5}{f_G \dfrac{\varsigma G - qC}{(\varsigma - q)} + \dfrac{4.5}{G}} \tag{8.23}$$

where E = the through car equivalent of a conflicting turn (right in Australia) vehicle,

f = a function of opposing flow = 1 at low flows and 0.45 at $q = 800$,

ς = saturation flow of opposing traffic (vehicles/sec),

TABLE 8.25
Adjustment Factors for Right Turns (and Left Turns from One-way Streets)

| | Adjustment Factor | | | | | |
| | With No Parking* | | | With Parking† | | |
Turns (%)	Approach Width ≤15 ft	Approach Width 16 to 24 ft	Approach Width 25 to 34 ft	Approach Width ≤20 ft	Approach Width 21 to 29 ft	Approach Width 30 to 39 ft
0	1.20	1.050	1.025	1.20	1.050	1.025
1	1.18	1.045	1.020	1.18	1.045	1.020
2	1.16	1.040	1.020	1.16	1.040	1.020
3	1.14	1.035	1.015	1.14	1.035	1.015
4	1.12	1.030	1.015	1.12	1.030	1.015
5	1.10	1.025	1.010	1.10	1.025	1.010
6	1.08	1.020	1.010	1.08	1.020	1.010
7	1.06	1.015	1.005	1.06	1.015	1.005
8	1.04	1.010	1.005	1.04	1.010	1.005
9	1.02	1.005	1.000	1.02	1.005	1.000
10	1.00	1.000	1.000	1.00	1.000	1.000
11	0.99	0.995	1.000	0.99	0.995	1.000
12	0.98	0.990	0.995	0.98	0.990	0.995
13	0.97	0.985	0.995	0.97	0.985	0.995
14	0.96	0.980	0.990	0.96	0.980	0.990
15	0.95	0.975	0.990	0.95	0.975	0.990
16	0.94	0.970	0.985	0.94	0.970	0.985
17	0.93	0.965	0.985	0.93	0.965	0.985
18	0.92	0.960	0.980	0.92	0.960	0.980
19	0.91	0.955	0.980	0.91	0.955	0.980
20	0.90	0.950	0.975	0.90	0.950	0.975
22	0.89	0.940	0.980	0.89	0.940	0.980
24	0.88	0.930	0.985	0.88	0.930	0.985
26	0.87	0.920	0.990	0.87	0.920	0.990
28	0.86	0.910	0.995	0.86	0.910	0.995
30+	0.85	0.900	1.000	0.85	0.900	1.000

*No adjustment necessary for approach width of 35 ft or more; that is, use factor of 1.000.
†No adjustment necessary for approach width of 40 ft or more; that is, use factor of 1.000.

Source: *Highway Capacity Manual*, p. 140.

q = flow of opposing traffic (vehicles/sec),
C = cycle length (sec),
G = effective green (sec).

The average through car equivalent of a conflicting-turn passenger vehicle was found to be 2.9 and of a turning truck 3.9.

The American system applies a correction factor to capacity and service volumes for left-turning vehicles which, for average approach widths (from 16 ft to 34 ft of clear approach), equals 1 percent per percent of left turns up to 20 percent. A central value of 10 percent is used and adjustment factors vary according to approach width. Adjustment factors are given in Table 8.26.

It should be noted that the American system does not directly consider the delaying effect of increases in opposing traffic and is applied as a percentage correction to the entire approach capacity. The effect of a fixed *percentage* of left-turn vehicles thus decreases as volume increases. The opposite effect is found by the Australian method as shown in Table 8.27.

TABLE 8.26
Adjustment Factors for Left Turns on Two-way Streets

	Adjustment Factor					
	With No Parking			With Parking		
Turns (%)	Approach Width ≤15 ft	Approach Width 16 to 34 ft	Approach Width ≥35 ft	Approach Width ≤20 ft	Approach Width 21 to 39 ft	Approach Width ≥40 ft
0	1.30	1.10	1.050	1.30	1.10	1.050
1	1.27	1.09	1.045	1.27	1.09	1.045
2	1.24	1.08	1.040	1.24	1.08	1.040
3	1.21	1.07	1.035	1.21	1.07	1.035
4	1.18	1.06	1.030	1.18	1.06	1.030
5	1.15	1.05	1.025	1.15	1.05	1.025
6	1.12	1.04	1.020	1.12	1.04	1.020
7	1.09	1.03	1.015	1.09	1.03	1.015
8	1.06	1.02	1.010	1.06	1.02	1.010
9	1.03	1.01	1.005	1.03	1.01	1.005
10	1.00	1.00	1.000	1.00	1.00	1.000
11	0.98	0.99	0.995	0.98	0.99	0.995
12	0.96	0.98	0.990	0.96	0.98	0.990
13	0.94	0.97	0.985	0.94	0.97	0.985
14	0.92	0.96	0.980	0.92	0.96	0.980
15	0.90	0.95	0.975	0.90	0.95	0.975
16	0.89	0.94	0.970	0.89	0.94	0.970
17	0.88	0.93	0.965	0.88	0.93	0.965
18	0.87	0.92	0.960	0.87	0.92	0.960
19	0.86	0.91	0.955	0.86	0.91	0.955
20	0.85	0.90	0.950	0.85	0.90	0.950
22	0.84	0.89	0.940	0.84	0.89	0.940
24	0.83	0.88	0.930	0.83	0.88	0.930
26	0.82	0.87	0.920	0.82	0.87	0.920
28	0.81	0.86	0.910	0.81	0.86	0.910
30+	0.80	0.85	0.900	0.80	0.85	0.900

Source: *Highway Capacity Manual*, p. 141.

It seems highly probable that the results of the Australian methods are more reliable in this respect. Some users of the *Highway Capacity Manual* have adopted a value of 1,200 turning plus opposing vehicles as the capacity limit for ordinary intersections.

TRUCKS

The larger size and lower acceleration ability of trucks, through buses, and other commercial vehicles reduces the capacity of a signalized intersection carrying mixed traffic.

The British method adjusts the flow rate by converting all traffic to equivalent passenger car units (pcu) by the following factors:

$$\text{Heavy and medium commercial vehicle} = 1\tfrac{3}{4} \text{ pcu}$$

$$\text{Bus} = 2\tfrac{1}{4} \text{ pcu}$$

$$\text{Tram} = 2\tfrac{1}{2} \text{ pcu}$$

$$\text{Light commercial vehicle} = 1 \text{ pcu}$$

TABLE 8.27
Effect of Saturation Flow and Opposing Flow on Capacity Reduction Factors
for 10 Percent Left Turns on Two-way Streets with No Parking

No. of Lanes	Saturation Flow for Entering and Opposing Directions, vphg	Opposing Flow, q, vphg	E_1*	Approach Volume at Load Factor = 1, vphg		Approach Reduction Factors for 10 percent Left Turns	
				Total	Left Turns	Australia Data	United States Data
For Low Opposing Flow							
1	1,800	200	1.7	1,682	168	0.93	0.77
2	3,600	400	2.1	3,243	324	0.90	0.91
3	5,400	600	2.4	4,736	474	0.88	0.95
For Higher Opposing Flow							
1	1,800	400	2.3	1,593	159	0.89	0.77
2	3,600	800	3.1	2,970	297	0.83	0.91
3	5,400	1,200	4.1†	4,125	413	0.76	0.95‡

*Based on E_{rt} values from Appendix B, *Australian Road Capacity Guide*, for $C = 60$ and $g/c = 0.6$.
†From correspondence with Dr. Alan Miller.
‡Left-turn capacity is exceeded, because $q + LT > 1,200$.

Source: Y. B. Chang and Donald S. Berry, "Examination of Consistency in Signalized Intersection Capacity Charts of the Highway Capacity Manual," *Highway Capacity*, Highway Research Record No. 289 (Washington, D.C.: Highway Research Board, 1969), p. 19.

The Australian approach is the same. The factors applied are:

Through truck = 1.85 pcu

Nonconflict-turning truck = 2.40 pcu

Conflict-turning truck (average) = 3.90 pcu

The American method applies a factor that reduces or increases the capacity and service volumes 1 percent for each percentage of trucks and through buses in the approach stream above or below 5 percent. This value is based on observations of performance with truck percentages below 20 percent. There is some indication[29] that this overestimates the effect of trucks when they constitute over 20 percent of approach traffic. Table 8.28 gives truck adjustment factors.

BUS STOPS

The effect of a transit bus stop at a signalized intersection is complex, depending on the frequency of buses, frequency and duration of stops, existence of a curb zone that can be utilized by right-turn vehicles when a bus is not present, and size and location of the stop in relation to the intersection. This effect is not well-quantified, and the British and Australian methods are silent on the subject. The American

---

[29] D. W. GWYNN, "Truck Equivalency," *Mathematical and Statistical Aspects of Traffic*, Highway Research Record 199 (Washington, D.C.: Highway Research Board, 1967), p. 79.

TABLE 8.28
Truck and Through Bus Adjustment Factors

Trucks and Through Buses (%)	Correction Factor	Trucks and Through Buses (%)	Correction Factor	Trucks and Through Buses (%)	Correction Factor
0	1.05	7	0.98	14	0.91
1	1.04	8	0.97	15	0.90
2	1.03	9	0.96	16	0.89
3	1.02	10	0.95	17	0.88
4	1.01	11	0.94	18	0.87
5	1.00	12	0.93	19	0.86
6	0.99	13	0.92	20	0.85

Source: *Highway Capacity Manual*, p. 142.

Highway Capacity Manual presents a series of nomographs to develop correction factors. These are rationalizations based on limited data. For the simplest case, a near-side bus stop on a two-lane approach with no parking, the factor is approximately a 0.4 percent reduction per bus per hour.

SIGNAL TIMING

The effective green time available for movement is a critical determinant of the capacity of a signalized intersection approach. The total green time available in an hour is a function of both the cycle length and the green phase length. The British and Australian methods take both into account. The American method uses only the G/C ratio, assuming that the most effective cycle length will be employed. This is a questionable assumption since, frequently, cycle length is determined by considerations other than performance at the specific intersection under study.

In comparing and using the various methods presented here, it is important to note that effective green time is determined differently. In British and Australian methods, the effective green is considered to be the green plus clearance periods minus lost time at the beginning and end of each green phase. In the American system, the actual green time, excluding the clearance period, is used in determining the G/C ratio, although the values given assume some passage during the clearance interval. The values are similar but by no means identical. The capacities and service volumes determined from calculations are in terms of flow per hour of green and are reduced to hourly volumes by multiplying by the G/C ratio.

GRADES

A negative grade on a signalized intersection approach will increase and a positive grade decrease the approach capacity. British research has found a 3 percent increase or decrease for each 1 percent of negative or positive grade in the range from -5 percent to $+10$ percent.

SERVICE VOLUME CHARTS

Figures 8.28 through 8.32 give the basic service volumes and adjustment factors generally used in American practice. The chart values are modified by peak hour factor, metropolitan area population, a factor based on location within the metro-

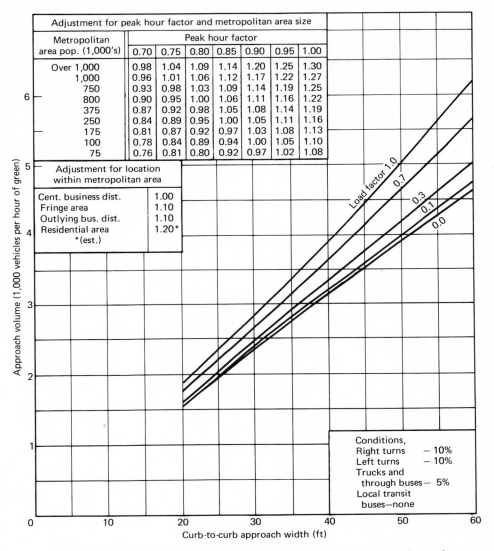

Figure 8.28. Urban intersection approach service volume, in vehicles per hour of green signal time, for one-way streets with no parking. (*Highway Capacity Manual*, p. 134.)

politan area, and percentages of turns and trucks. The load factor is selected to best represent the level of service desired (see Table 8.19, pg. 355).

Example

Given a 25 ft signalized approach with no parking, 15 percent left turns, 5 percent right turns, and 10 percent trucks on a level grade in an outlying business district of a 500,000-population city. The signal operates on a fixed time, 60-sec cycle with 26 sec

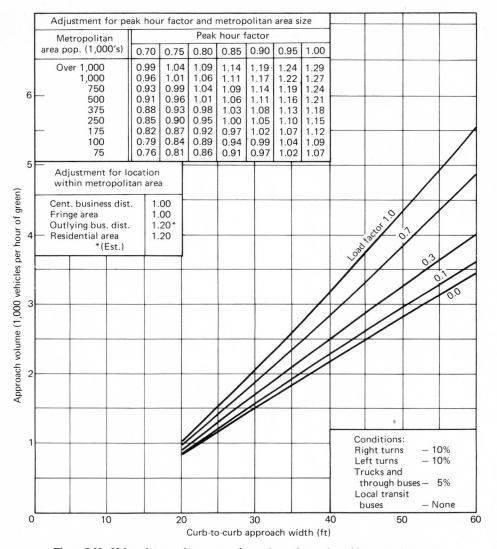

Figure 8.29. Urban intersection approach service volume, in vehicles per hour of green signal time, for one-way streets with parking one side. (*Highway Capacity Manual*, p. 134.)

green and 4 sec clearance periods. Determine the level of service or average delay that will be experienced with an approach volume of 800 vph.

British method

s = saturation flow = 160 pcu/ft/hour 160 × 25 =
4,000 pcu/hour = 1.11 veh/sec

q = approach flow in pcu =
800 (0.75 + .15 (1.75) + .10 (1.75)) = 950 pcu/hr = 0.264 veh/sec

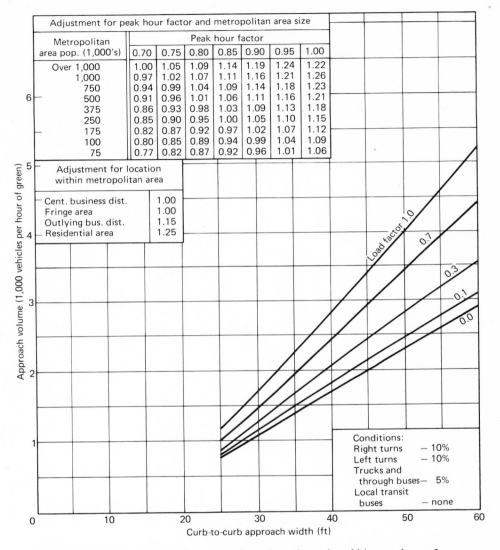

Adjustment for peak hour factor and metropolitan area size

Metropolitan area pop. (1,000's)	Peak hour factor						
	0.70	0.75	0.80	0.85	0.90	0.95	1.00
Over 1,000	1.00	1.05	1.09	1.14	1.19	1.24	1.22
1,000	0.97	1.02	1.07	1.11	1.16	1.21	1.26
750	0.94	0.99	1.04	1.09	1.14	1.18	1.23
500	0.91	0.96	1.01	1.06	1.11	1.16	1.21
375	0.86	0.93	0.98	1.03	1.09	1.13	1.18
250	0.85	0.90	0.95	1.00	1.05	1.10	1.15
175	0.82	0.87	0.92	0.97	1.02	1.07	1.12
100	0.80	0.85	0.89	0.94	0.99	1.04	1.09
75	0.77	0.82	0.87	0.92	0.96	1.01	1.06

Adjustment for location within metropolitan area

Cent. business dist.	1.00
Fringe area	1.00
Outlying bus. dist.	1.15
Residential area	1.25

Load factor 1.0   0.7   0.3   0.1   0.0

Conditions:
Right turns — 10%
Left turns — 10%
Trucks and
through buses — 5%
Local transit
buses — none

Approach volume (1,000 vehicles per hour of green)

Curb-to-curb approach width (ft)

Figure 8.30. Urban intersection approach service volume, in vehicles per hour of green signal time, for one-way streets with parking both sides. (*Highway Capacity Manual*, p. 135.)

C = cycle length = 60 sec

G = effective green = green + clearance − lost time. Assume 6 sec lost time = 24 sec

$$x = \frac{Cq}{Gs} = \frac{60 \times 0.264}{24 \times 1.11} = \frac{15.84}{26.64} = 0.59$$

$$G/C = \frac{24}{60} = .40$$

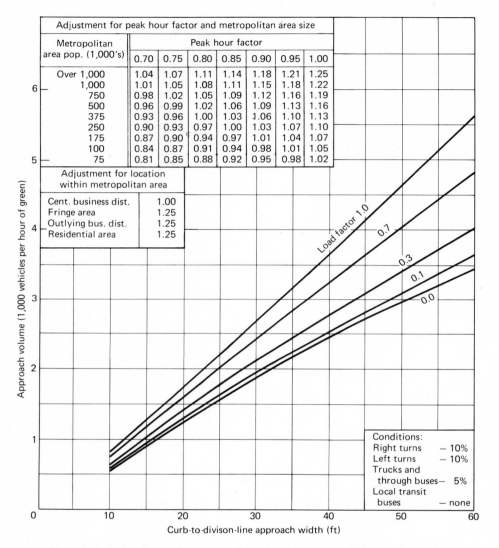

Figure 8.31. Urban intersection approach service volume, in vehicles per hour of green signal time, for two-way streets with no parking. (*Highway Capacity Manual*, p. 135.)

$f_1 = 0.236$ (Table 8.20, pg. 356)

$f_2 = 0.425$ (Table 8.21, pg. 357)

$f_3 = 5$ (Table 8.22, pg. 358)

$$d = \left(Cf_1 + \frac{f_2}{q} \right) \frac{100 - f_3}{100}$$

$$d = \left(60 \times .236 + \frac{0.425}{0.264} \right) \frac{95}{100}$$

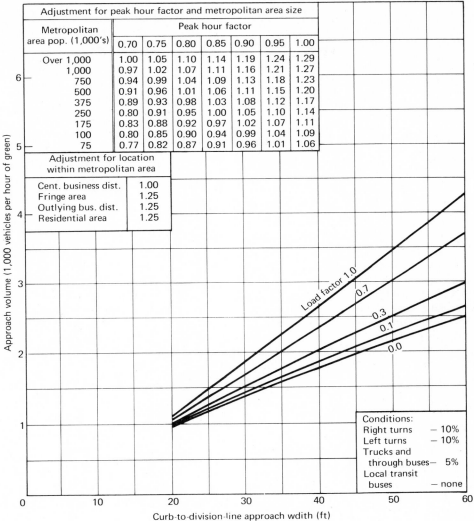

Adjustment for peak hour factor and metropolitan area size							
Metropolitan area pop. (1,000's)	Peak hour factor						
	0.70	0.75	0.80	0.85	0.90	0.95	1.00
Over 1,000	1.00	1.05	1.10	1.14	1.19	1.24	1.29
1,000	0.97	1.02	1.07	1.11	1.16	1.21	1.27
750	0.94	0.99	1.04	1.09	1.13	1.18	1.23
500	0.91	0.96	1.01	1.06	1.11	1.15	1.20
375	0.89	0.93	0.98	1.03	1.08	1.12	1.17
250	0.80	0.91	0.95	1.00	1.05	1.10	1.14
175	0.83	0.88	0.92	0.97	1.02	1.07	1.11
100	0.80	0.85	0.90	0.94	0.99	1.04	1.09
75	0.77	0.82	0.87	0.91	0.96	1.01	1.06

Adjustment for location within metropolitan area	
Cent. business dist.	1.00
Fringe area	1.25
Outlying bus. dist.	1.25
Residential area	1.25

Conditions:
Right turns — 10%
Left turns — 10%
Trucks and
  through buses— 5%
Local transit
  buses — none

Figure 8.32. Urban intersection approach service volume, in vehicles per hour of green signal time, for two-way streets with parking. (*Highway Capacity Manual*, p. 136.)

$$d = (15.77).95 = 14.98 \text{ sec}$$

American method

For level of service C, load factor = 0.3 (Table 8.19, pg. 355)

Basic service volume = 1,800 (Figure 8.31).

Assume $phf_{15} = 0.85$.

phf and metro population adjustment = 1.06 (Figure 8.31).

Location adjustment $= 1.25$ (Figure 8.31)

Right-turn adjustment $= 1.01$ (Table 8.25, pg. 367)

Left-turn adjustment $= 0.95$ (Table 8.26, pg. 368)

Truck adjustment $= 0.95$ (Table 8.28, pg. 370)

$G/C = \frac{26}{60} = 0.433$

Adjusted $SV_c = 1,800\,(1.06 \times 1.25 \times 1.01 \times 0.95 \times 0.95 \times 0.433)$

$\qquad\qquad = 940$

Approach is operating in the middle range of level C.

Note: The results of these two methods applied to this example are generally consistent with the simulation results shown in Figure 8.21 in which level of service C (load factor between 0.1 and 0.3) is associated with average individual delays of about 15 sec.

GENERAL COMMENTS

Although highway capacity and service levels have been studied for nearly 50 years, many questions remain unanswered. The analysis methods reported herein are adequate for gross examination but seldom provide the knowledgeable user with a complete sense of ease when a precise answer is required. In no area is this more true than with signalized intersections. The complexity of factors involved and the inherent variability of the traffic phenomenon have thus far limited both theoretical and empirical approaches to precise solution. In time this may be corrected. Until then, the engineer is well-advised to use the content of this chapter and the sources referred to as a guide to the application of his mature judgment in estimating the capacity and service that a highway facility can provide.

REFERENCES FOR FURTHER READING

Capacities and Limitations of Urban Transportation Modes. Washington, D.C.: Institute of Traffic Engineers, 1965.

"Signalized Intersection Capacity Parameters," *Traffic Engineering,* XLII, No. 9 (1972), pp. 50–55.

Highway Capacity Manual, Special Report 87. Washington, D.C.: Highway Research Board, 1965, p. 5.

K. MOSKOWITZ and L. NEWMAN, "Notes on Freeway Capacity," *Highway and Interchange Capacity,* Highway Research Record No. 27. Washington, D.C.: Highway Research Board, 1963, pp. 44–68.

Research on Road Traffic. London: Her Majesty's Stationery Office, 1965, p. 213.

A. J. MILLER, "The Capacity of Signalized Intersections in Australia," *Australian Road Research Board* Bulletin No. 3. Melbourne, 1968, p. 95.

Department of Transportation Planning, "Weaving Area Operations Study." Brooklyn: Polytechnic Institute of Brooklyn, 1971.

Chapter 9

TRAFFIC ACCIDENT ANALYSIS

J. Stannard Baker. Consultant, Traffic Institute, Northwestern University, Evanston, Illinois.

Accident analysis or accident studies are of two general kinds: (1) study in detail of individual accidents; (2) study of groups of accidents occurring at individual or similar locations. They serve different purposes.

STUDY OF INDIVIDUAL ACCIDENTS

Accident investgation—the study of an individual accident—as a source of data is discussed in Chapter 10. It is sufficient to point out here that five levels of investigation can be readily distinguished:[1]

1. *Accident reporting:* Identifying and briefly describing road location, vehicle, and people involved; describing damage and injury; identifying the kind of first harmful event; and specifying direction of travel and intended movement of each traffic unit involved. Factual data only. Reporting may be by police or drivers. Accident reporting is the source of data used for traffic engineering.
2. *Supplementary data collection* for selected accidents: Measurements, photography, informal statements. Factual data only.
3. *Technical data preparation*: Road and vehicle examination and tests; after-accident situation map. Factual data and opinions.
4. *Professional reconstruction*: Conclusions about how accident happened. Entirely opinion.
5. *Cause analysis.*

Engineers may participate in Levels 3, 4, and 5 but rarely in Levels 1 and 2, which are essentially police functions.

All except the first level of accident investigation may be directed toward or emphasize pre-crash circumstances (how accident happened), crash circumstances (how injury and damage were received), post-crash circumstances (what happened afterward to aggravate or ameliorate injury and damage), or any combination of these.

A traffic engineer may work on individual accidents in greater or lesser "depth" for any of a number of purposes: better to understand road-vehicle-driver relationships in accidents that occur for obscure reasons at a particular location; as a member of an interdisciplinary team studying a sample of accidents in great detail; or in connec-

---

[1] J. Stannard Baker, "Reconstruction of Accidents," *Traffic Digest and Review*, Vol. XVII, No. 3 (1969), pp. 9–10.

tion with a law suit arising from a motor-vehicle accident, in which case the engineer usually appears as an expert witness.[2]

ACCIDENT RECONSTRUCTION

Accident reconstruction (Level 4 of accident investigation) involves inferences about speeds, position on the road, observation or comprehension of traffic-control devices, and evasive tactics. It may result in velocity and acceleration diagrams, time-space diagrams, and other technical descriptions.

Often, data are insufficient for highly reliable conclusions. Then, whoever undertakes reconstruction may have to obtain additional information from personal observation or published sources. If measurements are lacking but photographs are available, reconstruction may involve elementary photogrammetry to obtain mapping data.

Essential to accident reconstruction is the ability to recognize the different kinds of tire marks on the road, for example, braking skidmarks, steering yaw marks, acceleration scuffs, and collision scrubs. Certain aspects of vehicle damage must also be recognized: contact damage areas, static and moving contact, and direction of thrust or force.

APPLICATIONS OF MECHANICS

Three concepts of dynamics are most frequently used: (1) moving bodies falling in air, (2) slowing of sliding bodies by energy dissipated as friction, and (3) centrifugal force on curves.

Bodies in motion through air provide formulas useful in estimating speeds of vehicles running off embankments, in determining where under-body debris will fall, and in estimating speeds from movement of granular cargo or even broken glass following collision.

When a vehicle falls or lofts and lands rightside up, the following formula gives its take-off speed:

$$v = \frac{k_1 s}{\sqrt{sn - h}} \tag{9.1}$$

where $k_1 = 2.73$ for v in mph; 4.01 for v in ft per sec,
  $= 7.97$ for v in kph; 2.21 for v in m per sec.

When a vehicle strikes a curb or furrows in and flips or vaults landing bottom up, the following formula gives *minimum* speed before take-off:

$$v = k_2\sqrt{s + h} \tag{9.2}$$

where $k_2 = 3.87$ for v in mph; 5.67 for v in ft per sec,
  $= 11.3$ for v in kph; 3.13 for v in m per sec.

Dissipation of energy by friction in skidding is the basis of operations relating speed, drag factor, and distance to *stop*:

$$v = k_3\sqrt{sf} \tag{9.3}$$

where $k_3 = 5.47$ for v in mph; 8.02 for v in ft per sec,
  $= 15.9$ for v in kph; 4.43 for v in m per sec.

---

[2] J. Stannard Baker, "The Expert Witness," *Traffic Digest and Review*, Vol. XVIII, No. 5 (1970), pp. 12–17.

Centrifugal force. The critical speed on a curve, that is, the speed at which *centrifugal force* equals tractive force and the vehicle begins to slip or yaw sidewise (this is not the design speed), can be found by Eq. (9.3) by substituting half the radius, *r*/2, of the curve for the distance slid (*s*). In this case, *f* represents the coefficient of friction plus the super-elevation. If the curve of the path of the center of mass of a car in a yaw is known from tire marks, the equation will give an approximation of its speed.

In Eqs. (9.1–9.3), the following symbols are used:

s = length or distance in feet (ft) or meters (m)

h = height or difference in elevation in ft or m, $+$ if change in elevation is upward, $-$ if downward

r = radius in ft or m

v = velocity or speed in feet or meters per second (ft per sec or m per sec) or miles or kilometers per hour (mph or kph)

n = grade or slope in feet per foot or meters per meter; tangent of angle of slope to level. (Not grade in percent or slope angle in degrees.) $+$ if change in elevation is upward, $-$ if downward

f = drag or acceleration factor is acceleration or deceleration divided by acceleration due to gravity (32.2 ft per sec² or 9.81 m per sec²), the number of *g*'s. It is equal to the coefficient of friction of a body sliding plus or minus the slope (*n*).

w = weight in pounds or kilograms

Equations for both critical or slideslip speed and braking to a stop speed require a drag factor based on coefficient of friction. It is best to determine coefficient of friction by test skids measuring stopping distance from known speed and solving Eq. (9.3) for *f*. For general estimates, however, a dry, smooth, clean, level, well-traveled pavement of Portland cement or bituminous concrete has a coefficient of friction of about 0.70 for speeds less than 30 mph and 0.60 for higher speeds. The same surface wet has a coefficient of friction of about 0.60 at speeds less than 30 mph and 0.55 at higher speeds.

COMBINING SPEED ESTIMATES

In most accidents, a collision occurs before skidding by braking stops the vehicle. Then, speed at the beginning of the skid will be greater than that estimated for a slide to an unrestricted stop in their measured length. If speed at the end of a skid is estimated from statements, falls, or collision results, that speed must *not* be added to the slide-to-stop speed determined from skidding. Speed at the beginning of the skid may be determined by this equation:

$$v = \sqrt{v_1^2 + v_2^2} \qquad (9.4)$$

where v = initial speed at beginning of skid,

v_1 = slide-to-stop speed from length of skid,

v_2 = speed at end of skid.

For example, suppose a vehicle slides 50 ft (15.2 m) with a drag factor of 0.7; its slide-to-stop speed for the skid would be Eq. (9.3) 32 mph (77 kph). Then, suppose that, instead of stopping at the end of the skid, the vehicle fell into a ditch with a "take-off" speed of 35 mph (56 kph) calculated from Eq. (9.1). Its speed at the beginning of the skid would be

$$v = \sqrt{32^2 + 35^2} = 47 \text{ mph } (= \sqrt{52^2 + 56^2} = 76 \text{ kph})$$

COLLISION SPEEDS

The concept of conservation of momentum is useful in estimating speed changes in collisions. Momentum is mass times velocity or speed times weight. Like velocity, it is a vector quantity which means its direction must be specified. Total momentum of two colliding vehicles is the same before and after collision. (But kinetic energy is lost doing damage in collision.) The general equation is:

$$w_1 v_1 + w_2 v_2 = w_1 v_1' + w_2 v_2' \tag{9.5}$$

where subscripts indicate vehicles 1 and 2 and primes indicate velocities after collision. This equation neglects coefficient of restitution which is very small in most vehicle collisions. For this and other reasons, the simple equation gives only a general, although often useful, approximation of velocity changes during collision. The equation is most easily solved graphically as illustrated in Figure 9.1. In this diagram, the length of each vector represents the amount of momentum (speed times weight) and its direction represents the direction of motion of the body. Then, if velocities and their directions after collision and directions of velocities before collision are known, momentums, and therefore velocities, before collision can be approximated.

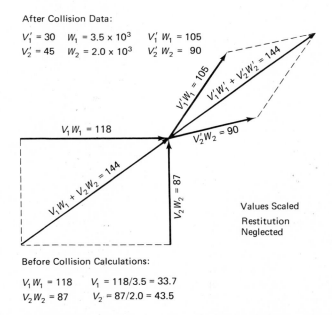

After Collision Data:

$V_1' = 30$ $W_1 = 3.5 \times 10^3$ $V_1' W_1 = 105$
$V_2' = 45$ $W_2 = 2.0 \times 10^3$ $V_2' W_2 = 90$

$V_1' W_1 = 105$
$V_1' W_1 + V_2' W_2 = 144$
$V_1 W_1 = 118$
$V_2' W_2 = 90$
$V_1 W_1 + V_2 W_2 = 144$
$V_2 W_2 = 87$

Values Scaled
Restitution
Neglected

Before Collision Calculations:

$V_1 W_1 = 118$ $V_1 = 118/3.5 = 33.7$
$V_2 W_2 = 87$ $V_2 = 87/2.0 = 43.5$

Figure 9.1. Graphical solution to momentum exchange between two vehicles in angle collision. Combined momentum of vehicles before collision is equal in amount and direction to combined momentum after collision.

ANALYZING ACCIDENTS AT SPECIFIC LOCATIONS

An engineering study of a group of accidents—as contrasted to the study of an individual accident—is usually directed toward accidents that occur at a particular location or a number of locations having similar characteristics. The purpose of these studies is to find out what can be done to prevent certain kinds of accidents at the locations studied. Studies of accidents at specific locations are part of what has come to be known as "spot improvement." Studies of locations with similar characteristics are to evaluate the effect of particular design or usage features of highways.

Relatively simple schemes for discovering and correcting hazardous locations have produced beneficial results, especially in small communities; but to be most effective, especially for extensive road networks, systematic search for, evaluation of, and attention to accident sore spots is necessary for best use of resources available for traffic accident prevention.

Efforts to reduce accidents at specific locations ordinarily involve the following five kinds of decisions:

1. Selecting locations to study.
2. Determining what can be done to improve each location studied.
3. Estimating cost benefits of the improvement, that is, comparing cost of the improvement with the value of harm that could be prevented by it.
4. Selecting locations to be improved.
5. Determining the success of the improvement after it has been made.

ACCIDENT BASIS FOR LOCATION IMPROVEMENT

The basis for any system of spot improvement is accident experience. Accident reporting and filing are described in Chapter 10, Traffic Studies. Chapter 11, Computer Applications, also discusses accident summaries.

Deficiencies in accident data may present three major obstacles in obtaining satisfactory information to use as a basis for traffic engineering studies. These are described below and may have to be overcome before accident information can be put to best use in improving the safety of roads.

First are deficiencies in identifying locations on accident reports. For machine processing, location of each accident must be unmistakable and expressed in standard terms; otherwise, the accident report is useless. There are several systems for identifying locations but, to be successful, all require special care in accident-report-form design and in training police and others to identify locations properly on reports. Some systems require maintaining special markings on the road network, such as mileposts; others require road log books or strip maps which identify locations by junctions, bridges, and other landmarks.

The second obstacle is incomplete or haphazard reporting. This can seriously bias the basis on which engineering judgments must be made, especially with machine processing of data received from numerous agencies. For example, if accidents on main routes are more fully reported than on minor routes, minor routes may be neglected. Also, incomplete reporting of lone-vehicle accidents may seriously affect treatment of nonjunction locations. Do not assume that whatever reports ordinarily come in are all that there are.

The third obstacle is the tendency of reporting agencies to omit reports of minor accidents. Since minor accidents are most numerous, omitting them may seriously reduce the data base for engineering judgments. Urge reporting of at least minimum data for all accidents coming to the attention of police, maintenance personnel, or other public employees.

Weighting accident data. Traditionally, the number of accidents reported for a location has been the basis for engineering studies. The accidents are unweighted, for a serious accident counts just as much as a minor one.

Giving each accident the same value in ranking locations for priority of study ignores the fact that some accidents are much more serious than others and that at some locations the proportion of serious accidents is greater than at others. Studying locations where accidents are severe offers greater possibilities for loss reduction than studying locations where accidents are less severe. Accident reporting, as commonly practiced, automatically gives some severity weighting because the less serious accidents are less fully reported; unreported accidents then have a severity weighting of zero; they just don't count.

Weighting by most serious injury. The standard way to describe accident severity is by the most serious injury to any person in the accident. This gives a severity scale as follows:[3]

1. Fatal accident
2. Incapacitating injury accident
3. Nonincapacitating injury accident   } Injury accident
4. Possible injury accident
5. Damage only accident

The three degrees of injury are often combined usually because they are not separated in accident reporting. Weighting by severity involves assignment of numerical weights to each degree of severity. No satisfactory statistical basis for these weights has been developed; therefore, they are arbitrary. A suggested weighting is 1 for damage only accidents, 3 for injury accidents, and 12 for fatal accidents. If fatal accidents have a high value, others do not count for much and the selection for study might almost as well be made on the basis of fatal accidents only; if fatal accidents have only slightly more value than injury accidents, they have so little additional effect that weighting is not worth the trouble. A more important consideration, however, is that "most serious injury" is only a makeshift, though traditional, evaluation of severity. For example, an accident in which an elderly pedestrian is nudged by a car and dies from hitting the ground is classified as more serious than an accident in which four persons are permanently disabled and three trucks are demolished.

Weighting by accident cost. Another method of weighting is to estimate the cost of accidents at the location. This is usually done by applying an estimated dollar value to casualties involved. The following values were estimated for 1972.[4]

---

[3] *Manual on Classification of Motor Vehicle Traffic Accidents* (Chicago: National Safety Council, 1970).
[4] *Estimating the Cost of Accidents*, Traffic Safety Memo 113 (Chicago: National Safety Council, 1973).

1. For each person killed $82,000
2. For each disabling injury $ 3,400
3. For each accident involving only damage $ 480

These values increase from year to year depending on changes in the value of the dollar and on other circumstances. They are estimated U.S.A. averages; local figures would be preferable if available. Note that damage value is for the entire accident regardless of number of vehicles involved, but death and injury values are for each person killed or injured in an accident.

A vehicle damage scale has been developed that classifies damage to passenger cars.[5] This scale is useful in describing vehicle damage but it has not yet been extended to trucks, motorcycles, or buses. Because of its limited use, it has not been adapted to weighting accident experience for traffic engineering purposes.

Until more satisfactory methods are developed, weighting by most severe damage or injury as a basis for selecting locations for study is a technical refinement of dubious value.

Weighting by number of traffic units involved. This system uses the number of involvements (motor vehicles, pedestrians, pedalcyclists) instead of the number of accidents. Thus, a two-vehicle collision counts twice as much as a lone-vehicle accident but the same as a pedestrian collision. No changes in accident reporting are required for this weighting. It gives rates in terms of involvements that more nearly represent true risk than rates in terms of accidents. Number of motor vehicles involved is a standard classification as a measure of accident severity,[3] but it is not exactly the same as number of involvements because it omits pedestrians and pedalcyclists.

Whether the simple number of reported accidents is used or the number of involvements or motor vehicles involved is used, the same procedures for selecting locations for study are applicable.

KINDS OF LOCATIONS

Road locations differ so much that intermingling them for purposes of accident study is both impractical and undesirable; nevertheless, methods of selecting locations for study apply generally to all kinds of locations. Roads may be classified for accident-study purposes in two ways: (1) as junctions or sections (spots or stretches) and (2) by character of service.

Junctions are spots on roads that naturally lend themselves to accident studies. Count as junction accidents:[3]

1. Collisions within the legally defined intersection[6]
2. Intersection-related accidents

Accidents on connector roads, in channelized junctions, and on merging and diverging roadways might also be included. Junctions of a street and alley or of two alleys are considered junctions, but a driveway access is not considered a junction.

---

[5] *Traffic Accident Data Project Vehicle Damage Rating Scale,* 2nd ed. (Chicago: National Safety Council, 1971).

[6] *Uniform Vehicle Code* (Washington, D.C.: National Committee on Uniform Traffic Laws and Ordinances, 1968).

Railroad grade crossings should be treated separately. List all having any collisions between trains and motor vehicles, pedestrians, or pedalcyclists.

Sections of road. In each road section count[3]

1. Nonjunction accidents
2. Driveway-access accidents

Because junctions are considered separately, do not count junction accidents in road sections. Junction accidents are so numerous that if they are counted in sections, they will predominate and obscure other kinds of accidents. Hence, junctions and accidents at them can be ignored in selecting road sections for study.

Because sections vary in length, the number of accidents used to rank sections for accident study must be the number of accidents per mile.

There are two methods of deciding how much road will be included in a section for accident-study purposes. (1) *homogeneity of characteristics* and (2) *standard section lengths.*

Homogeneity of characteristics. The entire road network is studied in order to divide it into sections each of which is similar or homogeneous throughout its length with respect to:

1. Cross-section characteristics such as number of lanes, width of lanes, median, and shoulders
2. Frequency of access driveways
3. Frequency and severity of curves and slopes
4. One-way or two-way traffic movement
5. Road surface characteristics
6. Adjacent land use

These sections should not be less than 0.1 mile (0.2 kilometer) long. Once established, the sections should not be changed unless there are changes in characteristics of the road.

Standard section lengths. A standard length of road section is chosen, for example, 0.1 mile (0.2 kilometer) in cities and 1.0 mile (2.0 kilometers) in rural areas.

With sections of standard lengths, the number of accidents does not need to be changed to accidents per mile for ranking purposes.

Character of location. If all spots on a road network are combined and all road sections are combined when ranking is done for accident study purposes, junctions and sections on major roads, where traffic is heavy, will be given priority. Locations on less-used roads, where inexpensive improvements can sometimes result in important accident reductions, will not be discovered. A functional road classification would seem to be most appropriate for the purpose, but functional classification in most places is incomplete and functional classification often does not correspond to design characteristics.

Hence, the following simplified eight categories of road sections are suggested:

Urban { two-lane / four or more lane undivided / four or more lane divided / freeway

Rural { two-lane / four or more lane undivided / four or more lane divided / freeway

Then urban junctions would be classified in ten categories and rural junctions in ten more, a category being a combination of any two kinds of roads. Thus, a junction of a rural two-lane road with a similar road would be one category; a junction of a rural two-lane road with a rural four-lane undivided road would be another category. For all but very large jurisdictions there would be few junctions in some categories. A simpler approximation of such a junction character classification can be achieved by classifying junctions according to the following control devices:

1. No control
2. Yield signs
3. Two-way stop signs
4. Three- or four-way stop signs
5. Traffic signals
6. Grade separation

RANKING LOCATIONS ACCORDING TO ACCIDENT EXPERIENCE

Except in very small places, resources will not permit study of all locations. The locations to be studied must, therefore, be selected. The first step in doing this is to list or rank the locations according to some measure of the risk or accident experience. The next step is to decide how far down the list to go in trying to study the locations. This cut off will be described later in this chapter.

The two principal methods of evaluating the risk at locations are (1) number of accidents (or involvements) and (2) accident (or involvement) rates.

Number of accidents is the simplest method. Locations are ranked according to the number of accidents experienced at each location for the same period, usually a year. The one having the most accidents is listed first, the one with next most second, and so on. Locations having three or fewer accidents are omitted because they have too little experience to be significant.

A list (Table 9.1, Line O) in which locations are ranked according to number of accidents can be prepared from a spot map, by going through a location file, or, if locations are numerous, by computer. No manipulation of data other than counting is required before beginning studies. This simple ranking by number of accidents is easy to explain. It is useful to point to such a list when somebody urges signs or signals "for safety" at an unimportant location. This method is a good one to begin with and will serve the purpose indefinitely in many jurisdictions.

The number of accidents in sections of road has little significance unless all sections are the same length. If sections are not the same length, accidents per mile instead of number of accidents must be used.

But ranking by number of accidents has disadvantages. For example, locations with few accidents at which inexpensive improvements could be very effective will be far down on the list and so may not be reached. Also, in extensive road networks, there may be many locations with nearly the same number of accidents. Then additional means are useful in choosing from among all those which might best be studied.

Rate or risk of accidents is a more useful method of ranking locations according to accident experience. A road location may have numerous accidents because it is much used rather than because it is especially hazardous. Thus, the location having the most accidents is not necessarily the most dangerous to use; conversely, lack of reported accidents for a specific period does not mean that there is no risk.

TABLE 9.1

Examples of Different Methods of Ranking Highway Sections According to Accidents Reported

Line	Section Number	1	2	3	4	5	6	7	8	9	10	Total	Average
	BASIC DATA												
A	Length in miles (km)	2.5	3.2	2.8	5.0	1.0	1.4	3.3	4.0	2.0	3.0	28.2	2.8
B	Average daily traffic × 10^{-2}	40	36	35	30	28	28	25	23	20	22	—	—
C	Accidents	23	12	10	7	2	5	7	9	15	8	98	9.8
D	Involvements	40	21	16	14	4	8	13	16	24	10	166	16.6
E	Killed	0	0	0	0	1	0	0	0	0	0	1	0.1
F	Injured	5	2	2	0	0	0	0	2	3	2	16	1.6
G	No-injury accidents	19	10	8	7	1	5	7	8	12	7	84	8.4
	RATES												
H	Accidents per mile (km)	9.20	3.75	3.57	1.40	2.00	3.57	2.12	2.25	7.50	2.67	—	3.5
I	Involvements per mile (km)	16.00	6.56	5.71	2.80	4.00	5.71	3.94	4.00	12.00	3.33	—	5.9
J	Vehicle-miles (km) × 10^{-6} = M	3.65	4.20	3.58	5.48	1.02	1.43	3.01	3.36	1.46	2.41	29.60	2.96
K	Estimated cost* × 10^{-3}	25.74	11.40	10.48	3.22	82.46	2.30	3.22	10.48	15.72	10.02	175.04	17.5
L	Accident rate, R_s	6.30	2.86	2.79	1.28	1.96	3.50	2.33	2.68	10.27	3.32	—	3.3
M	Involvement rate, R_s	10.96	5.00	4.47	2.55	3.92	5.59	4.32	4.76	16.44	4.15	—	5.6
N	Upper control limit, R_c	4.88	4.76	4.89	4.57	—5.25—		5.05	4.95	5.91	5.28	—	—
	RANKING BY												
O	Accidents	1	3	4	7	10	9	8	5	2	6		
P	Involvements	1	3	4	6	10	9	7	5	2	8		
Q	Costs (weighted)	2	4	5	8	1	10	9	6	3	7		
R	Accidents/mi (km)	1	3	4	10	9	5	8	7	2	6		
S	Involvements/mi (km)	1	3	4	6	10	9	7	5	2	8		
T	Accident rate	2	5	6	10	9	3	8	7	1	4		
U	Number-rate selection	1	0	0	0	0	0	0	0	2	0		
V	Out of control	2	0	0	0	0	0	0	0	1	0		

*Cost of accidents computed at $82,000 per death, $3,400 per injury, and $460 per no-injury accident.
Note: Accidents counted for 365 days.

Risk or hazard may be expressed as an accident rate: the number of those experiencing accidents (involvements) at a location in a specified time divided by the number using the location in the same period. Because accidents are rare events, the simple rate is a very small decimal fraction. Therefore, for ease in writing, the rate is multiplied by a million and quoted as accident involvements per million users.

Three kinds of rates are needed, one for junctions and two for sections of roads. The former (R_j) is the simple number of involvements per million users. But because road sections may vary in length and therefore give different exposure to accidents, rates for road sections must be in terms of accidents per mile or kilometer per year (R_m) or per million miles or kilometers traveled per year in the section (R_s). Equations for these three rates follow:

$$R_j = \frac{2A \times 10^6}{T(V_1 + V_2 \dots V_n)} \tag{9.6}$$

$$R_m = \frac{365A}{TL} \tag{9.7}$$

$$R_s = \frac{A \times 10^6}{TVL} \tag{9.8}$$

where R_j = junction rate in involvements (or accidents) per million vehicles entering,
R_m = section rate in accidents per mile or kilometer per year,
R_s = section rate in involvements (or accidents) per million vehicle miles or kilometers traveled,
A = involvements (or accidents) recorded in T days,
T = period (days) for which accidents are counted, usually exactly 365, a full year,
V = average annual daily traffic on a section (vehicles per day),
V_1 = average annual daily traffic on one junction leg (n = number of junction legs),
L = length of section in miles or kilometers.

If a section is exactly a mile long (standard section for rural area), the section rate (R_s) is the same as the junction rate (R_j); both are involvements per million vehicles entering.

The rate is untrustworthy if the number of accidents is small. In practice, any rate based on three or fewer accidents is likely to be untrustworthy. It is best to extend the period until five or more accidents have been accumulated before calculating the rate.

EFFECT OF TRAFFIC VOLUME ON ACCIDENT RISK

Many factors contribute to the risk of a location. The more important are:

1. Physical characteristics, such as geometric design, sight distances, control devices, and roadside obstacles
2. One-way or two-way traffic movement and irregular movements such as turning and backing
3. Speed
4. Quality of driving
5. Size and performance characteristics of vehicles
6. Volume of traffic

Risk is not directly proportional to volume. Because rates are intended to eliminate traffic volume as a variable in evaluating hazard, perhaps the last factor on the list needs explaining. To simplify, assume that the location is an ordinary, uncontrolled, four-leg junction without pedestrians, cyclists, or turning movements and that no driver pays any attention to vehicles on the cross street. Then the chance of a vehicle's entering one leg and colliding with another vehicle will be the proportion (percent) of the time that the vehicle's path is blocked by cross traffic. This time will increase almost in proportion to the volume of traffic on the cross road. If cross-road volume doubles, chance of collision doubles. The same is true for vehicles on every leg of the junction. Thus, the hazard or danger of collision to each unit of traffic entering a junction depends on the volume of traffic crossing. If cross traffic is eliminated (by grade separation, for example), this hazard is reduced to zero. If cross traffic is so great that it provides no gaps, the chance of collision (under these assumptions without evasive tactics) is infinite. Thus, in a road network, the *number* of angle and opposite-direction collisions tends to vary approximately as the *square* of the volume of traffic and so the accident *rate* varies directly as the volume of traffic. The *number* of lone-vehicle accidents varies as the volume of traffic and the accident *rate* is therefore independent of volume. The number and rate of same-direction collisions are less predictable because they depend more on the variance of speed among vehicles moving in the same direction.

Involvements compared to accidents. The numerator of an accident rate that truly represented risk of using a location would be the number of involvements, that is, the number of motor vehicles, pedalcycles, and pedestrians involved in accidents at that location during the period considered. The denominator would be the number of users (motor vehicles, pedalcyclists, and pedestrians) going through the location in the same time. In practice, for purposes of establishing priorities for study of accidents at locations, accident rates are simplified. In the numerator of the rate, number of accidents rather than the number of involvements is used. At ordinary junctions, because most accidents are two-car collisions, the number of accidents is about half of the number of involvements. But between junctions, where there are many lone-vehicle accidents, this is definitely not so. In the denominator of the simplified rate, only the number of motor vehicles is used, neglecting pedestrians and pedalcycle traffic. This makes little difference in rural nonjunction locations because nearly all of the traffic is motor vehicle; but in urban junctions, it can make an appreciable difference.

The volume of traffic is derived from the average annual daily traffic (AADT, see page 411). For junctions, the estimated number using the location in a day is half the sum of the AADT's of all the legs. Half the sum is used because the AADT for any leg is the total vehicles in both directions, that is, those entering and leaving on each leg. To use total AADT for all junction legs would be to count each vehicle twice.

From considerations of factors influencing risk or hazard, it is clear that risk or hazard may vary from season to season and even from moment to moment. Any rate representing such risk is, therefore, an average risk over a lengthy period.

The principal objection to using accident rates to establish priorities for studies of accidents at road locations is that the rates can give a low priority to locations where accidents are frequent and thus where even a small percent reduction will result in a substantial reduction in number of accidents. Another difficulty is that recent reliable traffic counts are often not available as a basis for rates and special counts for the purpose are tedious and costly.

SELECTION OF LOCATIONS TO BE STUDIED

Standards have not been developed that dichotomize road locations as hazardous or nonhazardous or that establish recognized levels of hazard for road locations. Yet, selecting accident locations for study implies a cutoff above which accidents are to be studied and below which they are not.

Generally, different kinds of locations are not considered together but are considered separately when establishing cutoffs to select locations for study. The most common groupings were discussed in the section on locations in this chapter.

Once locations have been evaluated or ranked according to hazard, some system or method is needed to choose the ones to study. Three methods will be described: (1) number of accidents, (2) number-rate combination, and (3) quality control.

Number of accidents is a natural cutoff, especially in small jurisdictions where the total number of locations to be studied is not large.

If the number of accidents at a location in the period is small, it probably represents chance occurrences that are unreliable for rates and are unlikely to indicate accident patterns. On this basis, locations with fewer than three accidents per year can be omitted (Table 9.1, Line C, Section 5, with only two accidents). If there is no possibility of studying all locations experiencing more than three accidents, the cutoff may be at whatever number of locations it seems feasible to study the accidents. If the number of locations having more than three accidents is in the hundreds or thousands, or if many locations have the same number of accidents or approximately the same accident rate, this method is not selective enough.

Number-rate combination is a useful method for state-wide and other large road networks where many locations have about the same number of accidents. It is especially useful for junctions. A first cutoff is made on the basis of the number of involvements of the locations thus selected; a second cutoff is made on the basis of involvement rates. Thus, of two locations having the same number of involvements, the one with the greatest risk (involvement rate) is given preference for study.

With large numbers of locations, locations of different kinds should be treated separately. Thus, sections of freeways will not be grouped with sections of minor roads; urban junctions will not be grouped with rural junctions.

The number of involvements and the rate per million vehicles entering are computed for all junctions; involvements per mile and the rate per million vehicle miles are computed for each section. Then, for each kind (or group) of location, the average number and rate are computed.

Cutoff criteria recommended[7] are as follows:

First cutoff
> Junctions: 2 times the average number of involvements
> Sections: 2 times the average number of involvements per mile

Second cutoff
> Junctions: average involvements per million vehicles entering
> Sections: average involvements per million vehicle miles

If the recommended cutoff points yield many more locations than can actually be

---

[7] *Highway Safety Improvement Projects*, Policy and Procedure Memorandum 21-16 (Washington, D.C.: Bureau of Public Roads, U.S. Department of Transportation, 1969).

studied, the cutoff point for number of accidents can be raised, for example, to three times average (Table 9.1, Line U).

Quality control is adapted from industry. It is mainly useful for sections of rural routes with fairly uniform traffic volumes, but it can also be used for junctions or any group of similar locations. A critical rate is calculated for each location based on the average for all locations in the group. If the actual accident rate is greater than the critical rate, the deviation is probably not due to chance but to an unfavorable characteristic of the location that warrants study. The location is then said to be "out of control."

The equation for the critical rate of a section is:

$$R_c = R_a + k\sqrt{\frac{R_a}{M}} + \frac{1}{2M} \tag{9.9}$$

where R_c = critical accident rate for the section,
$\quad\quad R_a$ = average accident rate for all sections in the group in accidents per million vehicle miles or kilometers,
$\quad\quad M$ = millions of vehicle miles or vehicle kilometers for the section,
$\quad\quad k$ = probability constant. A value of 1.5 is suggested.[7] A smaller value will tend to give more sections out of control and therefore a larger list of sections for study, but it will increase the probability that the rate is high by chance.

The same equation can be used for junctions. Then, R_a is the average accident rate in millions of vehicles entering for all junctions in the group, and M is the number of vehicles entering the particular junction.

For a series of sections on a route, the average rate for all sections and the critical rate for each section can be shown graphically, as in Figure 9.2. Points 1, 2, 3, etc., are the actual rates. Those above the critical value (1 and 9 in Figure 9.2) indicate sections that are out of control. The greater the difference between the actual and the critical rates, the less likely that the out-of-control situation is caused by chance and the more likely that it needs correction of some kind.

In the quality-control method, sections or junctions having very few or very many accidents are to be avoided. Hence, for this purpose, any section having fewer than seven accidents should be combined with the adjacent section having the fewest accidents. This has been done for sections 5 and 6 in Table 9.1, Line N. If the section

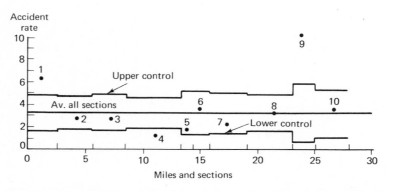

Figure 9.2. Example of control chart for ten sections of road. Data are from Table 9.1.

has more than 30 accidents, it can be divided into two sections for quality-control purposes.

The quality-control method offers an opportunity to discover locations at which accidents are much fewer than might be expected on the basis of chance. By examining these locations, it may be possible to discover road characteristics that are especially desirable. In Figure 9.2 a lower critical rate the same distance from but below the average rate for all sections is shown. Section 4 has an actual rate below the lower control limit and therefore would appear to have something about it that makes it unusually safe. The formula for the lower critical rate is:

$$R_c = R_a - k\sqrt{\frac{R_a}{M}} - \frac{1}{2M} \tag{9.10}$$

SPOTS OTHER THAN JUNCTIONS

Since road junctions are natural spots where accidents are likely to bunch, they are treated separately from road sections. But other kinds of spots where accidents cluster are initially included in section accident data and may well account for an unusual number of accidents in a section. If this special location can be identified, it can be separated from the section and treated individually, leaving the remainder of the section for more general consideration. Examples of nonjunction spots are driveway entrances, narrow bridges, and other points having limited clearance.

A "sliding length" of roadway may be used to reveal concentrations of accidents within sections. A length less than 0.3 mile (0.5 kilometer) is recommended so that a single feature of the section may be identified. This length is then moved or slid along the road and the number of accidents it brackets in any position is noted. If this number becomes twice the average for the length of sliding section used, the spot is marked for special separate study.

Searching for accident clusters can be done visually on a spot map or it can be done on a large scale by a computer.

OTHER CONSIDERATIONS

Realistically, cutoff is determined by the number of locations that can possibly be studied. Knowing approximately how many man days will be available for study and how many days on the average it takes a man to complete a study of a location, the number of locations that might be studied can be estimated. Then, all but this number can be omitted from the list of those to receive attention.

Cutoff may vary from year to year depending on resources available for study and for location improvement. If budgets for study and improvements change little from year to year, the number of locations studied in previous years is a good guide to cutoff in future years.

If time and money are unavailable for improvements of road locations, it might seem that there is no use in preparing a list of locations needing study, but this is not the case. A yearly list of at least the three percent of all locations having the most accidents will assist in getting money for study and improvement. The list will also help in resisting demands for changes in locations that are not ranked among the more dangerous.

Selection of locations for study must not be overemphasized. The several methods used do not produce greatly different results (see Table 9.1). Suppose that the 20 most "hazardous" locations in a road network were selected by each of several methods. Many locations would appear in all of the selections. Moreover, because accidents are few at most locations, ranking by any method involves a considerable element of chance.

A sensible approach, especially in road networks of moderate size, is to study the first location on the high-accident location list and see what can be done about it, how much it would cost, and what savings possible improvements might make. Then, study the next location on the list in the same way. The engineer chooses which among the hazardous locations will yield most readily to treatment and puts his money there. Often an important consideration is whether or not changes can be combined with improvements to increase capacity or reduce delays.

In the future, it is likely that states will collect accident reports from all cities and counties (except perhaps the larger ones) and prepare high-accident lists yearly for all jurisdictions as part of a nation-wide program.

DETERMINING POSSIBLE REMEDIES

Studying locations selected is essentially a matter of discovering accident patterns that suggest possible remedies.

Computer programs are of limited value for this purpose. Programs can, indeed, be written which classify accidents according to basic movements and conflicts involved and then print out (for each location to be studied) a list in which each movement-conflict combination is ranked in order of frequency. Sometimes this will adequately point up a difficulty, but often, especially with irregular locations, the computer printout has much less suggestive value than a graphic display does.

Computer programs require much additional data to be coded and entered into the system for all accidents. It is usually more economical to extract data manually for the relatively small number of accidents to be studied.

COLLISION DIAGRAMS

Collision diagrams are generally used to study accidents at a selected location. They permit accident patterns to be visualized quickly and easily. A collision diagram is a schematic drawing of the location showing, by conventional symbols, the traffic unit movements and conflicts involved in accidents at that place.

Reports of all accidents during a period of at least two years are used for collision diagrams. If this many accidents fail to suggest a pattern or predominance of some kind of accident, reports for additional years may be added if they are available.

On a schematic diagram of the roadways at the location, symbols are used to show the following for each accident:

1. Direction of travel of each traffic unit (motor vehicle, pedestrian, or pedal-cyclist) involved, including noncontact involved ("phantom") vehicles, so far as they are known. Direction of travel is that originally intended, not that resulting from evasive tactics.
2. Maneuvers (turns, backing, etc.) intended at the location but not those made as evasive tactics.
3. Day of week and date of each accident to relate it to traffic control, darkness, and other circumstances in effect at a particular time.

4. Severity of the damage or injury to each traffic unit at about two levels.

5. Special conditions, for example, slippery surfaces.

Symbols used (see Figure 10.30) are suggestive of their significance so that they are easily recognized. Location of the symbols does not represent the actual points on a map of first harmful event of the accidents represented. Instead, all symbols representing the same approach direction, and intended maneuver, are grouped so that patterns are more easily detected. Figure 11.15 shows an intersection collision diagram prepared by a computer-driven plotter.

A standard form can be used for ordinary right-angle junctions (see Figure 10.28). The same form can be used for T-junctions, but other configurations usually require special diagrams.

Collision diagrams of road junctions include all accidents in which the first hazardous event actually occurs "in the intersection" and also all which are "intersection related," that is, involve movement in or through the junction. Thus, collisions in a queue for a turn are included even though they may be a hundred feet or more from the actual turning area.

Collision diagrams of sections of roads between junctions show bridges, curves, railroad crossings, driveways, and other relevant features.

Study of collision diagrams consists mainly of looking for accident patterns, that is, a number of accidents having common circumstances. Sometimes these are conspicuous as, for example, a large number involving left turns; in other cases they are obscure, as when accidents occur most commonly under special light or road-sufrace conditions. Sometimes there are two or more patterns or different patterns at different times of day or week.

Remedies are sometimes suggested directly by accident patterns, but they are more often discovered by studying accident patterns in connection with study of the physical arrangement of the location. This can most easily be done by going to the location and watching traffic there, especially at times and under conditions that show up in the collision diagram. It is often useful to approach the location in the direction of vehicles in accident pattern movements to look for such things as view obstructions, obscured traffic control devices, confusing direction signs, illusions, and distractions.

CONDITION DIAGRAMS

If visits to locations are not practical, condition diagrams may give clues to possible remedies (see Figure 10.31). These are scale maps of the junction or other location showing roadway and shoulder widths, view obstructions, grades, traffic control, lighting, curbs, sidewalks, parking, access driveways, and other conditions. Road design drawings may be helpful for preparing condition diagrams, but, usually, special maps have to be made by traffic engineering technicians. These may also be useful as a basis for improved designs for the roadways or to show locations of traffic control devices.

TRAFFIC COUNTS

Detailed traffic counts show which movements are most common. Comparing movements that result in accidents with the movements at the location can indicate which movements are most dangerous and so suggest remedies.

SEVERITY REDUCTION

Possibilities of reducing severity of accidents as well as their frequency should be considered. Removing roadside obstacles, providing guardrails, and reducing speeds may lessen the harm done when actual accident prevention is impractical.

Discovering possible remedies is, perhaps, more of an art than a science. Certainly no step-by-step procedure is available for developing accident prevention measures at locations having a particular accident pattern. In general, numerous angle collisions suggest stop signs, yield signs, and traffic signals. If most of the collisions occur to vehicles approaching on two adjacent legs of a junction, removal of view obstructions in that angle should be considered. Numerous left-turn collisions suggest prohibition of left turns or special turn indications or signals. Same-direction collisions involving left-turning vehicles suggest the need for special left-turn lanes.

Reducing the amount of traffic at a given location will nearly always reduce accidents there, but it is rarely practical.

Diversion of traffic to alternate routes, prohibiting left turns, and similar remedies for a specific location may shift the problems to other locations.

ESTIMATING REDUCTIONS IN ACCIDENTS

Accident-reducing remedies at locations (spot improvement) are of two general kinds: (1) those to prevent specific kinds of accidents and (2) those to prevent accidents in general. Remedies for specific situations are usually more economical and often more effective.

Some remedies are obviously impractical because they are too costly. For example, a railroad grade separation will prevent all train-motor vehicle collisions, but unless traffic or train volumes are very great or accidents very numerous, the railroad grade separation may be much too expensive.

COUNTING METHODS

The customary way to assess the effect of improvements has been to count the number or estimate the percent of accidents likely to be avoided by the improvement in a specified period, regardless of severity of accidents. Often there is no alternative to this simple procedure because the only datum available is the total number of accidents.

Unfortunately, this estimate is insensitive to improvements in severity as well as frequency. Consider, for example, the effect of installing guardrails. It usually increases the number of collisions, but it will reduce the severity of accidents so much that there is a net gain.

Median guardrails, especially in narrow medians, produce many damage-only scrapes, but they prevent most cross-median, opposite-direction, full impacts which are likely to be fatal. Thus, with only total accidents to go by, median guardrails would appear to be more harmful than helpful; but if the number of accidents can be divided into two or more degrees of severity, the beneficial effect of the median guardrail becomes immediately apparent.

With respect to each involvement, two degrees of severity are practical and adequate. Customarily, total accidents are classified in three severity categories: (1) fatal, (2) injury, and (3) damage only. Because fatal accidents are so few, they can be combined with injury accidents to give two categories: (1) fatal-injury and (2) damage only. The newer standard classification[3] permits a better classification that has two categories: (1) disabling a vehicle or incapacitating a person (includes fatal) and (2) all others. This distinction makes a better approximation of dividing all accidents into high-cost and low-cost accidents.

TIME PERIODS

The experience accumulated for location studies may represent several years. If so, the estimated number of accidents prevented is divided by the number of years of experience studied to obtain the estimated saving in accidents per year. If full years are not used for the data base, the estimated number of accidents prevented for the whole period is divided by the number of days of experience and multiplied by 365 to obtain the estimated saving in accidents per year.

At *specific study locations*, reductions in involvements expected from proposed remedies can be estimated by counting the number of accidents in the past that might have been affected by the proposed remedies. For example, if 24 involvements were recorded for cars approaching on adjacent legs of a junction, removal of a view obstruction at that corner might be expected to remove that accident pattern and reduce the number of that kind of accident.

But it is unrealistic to believe that an improvement will entirely eliminate accidents of the kind it is intended to affect. Thus, in the example above, removal of the view obstruction would be unlikely to have prevented all 24 involvements because angle collisions also occur where there are no view obstructions. Hence, the apparent reduction must nearly always be discounted. Detailed rules by which to discount the effect of various improvements are unavailable, but satisfactory judgments can usually be based on the circumstances of the accident pattern and kind of improvement contemplated.

In some instances, improvements will reduce accidents in one pattern but increase those in another. For example, traffic lights installed to mitigate angle collisions often result in more same-direction collisions between vehicles approaching an intersection on the same junction leg. Hence, an estimate of the benefits of an improvement must take into consideration negative as well as positive effects and represent the net expected change.

Sometimes benefits achieved by improvements at one location may be at the cost of additional accidents elsewhere. Thus, prohibiting left turns can be expected to improve accident experience at that location, but turning movements then shift elsewhere and usually increase in number unless the other locations have much less traffic to interfere with added turning movements. Hence, accidents as a whole in an area may actually increase by prohibiting turning movements.

Estimate the effect of the improvement separately for involvements of each degree of severity.

General experience with various kinds of improvements has been studied and summarized. This summary is given in Table 9.2. Except where noted, the fractional reductions apply to all accidents, not just to those that might be affected by the improvement. Negative improvements are possible and are indicated by a minus sign in the table. These negative improvements may be expected to result in more accidents instead of fewer accidents and would, therefore, be undertaken only for a purpose other than accident reduction, for example, speed or capacity. Some of the fractional reductions in the table have question marks after them. These marks indicate that experience on which the fractional reductions were based is small or irregular and, therefore, that reductions estimated from them would be of questionable reliability. Refer to the original source of this table for detailed discussion of the reliability of these data.

TABLE 9.2
Forecast of Accident Reduction

Improvement Project	Urban or Rural	Number of Lanes	Fractional Reductions in		
			All Accidents	Fatal, Injury	Damage Only
Sections					
Eliminate parking	U	More than 2	.32	.03	
Install or improve edge markings	R	2	.14	.17?	
Install or improve warning signs	U	2	.14	.14?	
	U	More than 2	.20?	.26?	
	R	2	.36	.32?	
	R	More than 2	.18?	.03	
Install median barrier, cable, freeway	—	More than 2	−.33?	.04?	
Install median barrier, beam, freeway	—	More than 2	−.20?	−.22?	
Install center barrier	—	Median width 0 to 5 ft	−.53??	−.61?	
Median barrier, freeway	—	More than 2	−.44?	−.11?	
Painted or raised median	U	More than 2	.12		
Resurfacing	U	More than 2	.42	.46	
	R	2	.12	.21	
	R	More than 2	.44	.59	
Shoulder stabilization	R	2	.38	.46	
Widen shoulder, no dimensions	R	2	−.02	.07?	
Widen roadway, no dimensions	R	2	.38	.30	
Widen roadway from 9-ft lanes	R	2	.38	.16	
Widen roadway from 10-ft lanes	R	2	.05?	−.65??	−.37?
Livestock fencing	R	2 or more	.90	(Livestock accident only)	
Modernization to design standards	R	2	.10	−.06?	.40??
	R	More than 2	.15??	.22?	
Crests					
Centerline striping	R	2	.64		
Curves					
Install delineators	R	2	.02??	.16	
	R	More than 2	.46?	−.10??	.61
Install or improve warning signs	R	2	.57	.71	.23?
	R	More than 2	.52	.40	
Reconstruct curve	R	2	.88	.89	.96
Install warning signs and delineation	U	More than 2	.20	−.27?	
	R	2	.22?	.41?	

TABLE 9.2 (Cont.)

Improvement Project	Urban or Rural	Number of Lanes	Fractional Reduction in		
			All Accidents	Fatal, Injury	Damage Only
Junctions					
Install or improve signs, directional, warning	R	2	.37	.19	
	R	More than 2	.09	−.07	
	U	2	.29	.51?	
	U	More than 2	.41	.47?	.26?
Install or improve signs, T-junction	R	2	.61	.43??	
	R	More than 2	.65	.67	
Stop ahead sign	R	2	.47	.96	
Install yield sign	U	2	.59?	.80	
	U	More than 2	−.46		
Install minor leg stop control	U	2	.48	.71	
	U	More than 2	.38?	.18?	.22?
	R	2	.65	.89	
All-way stop signs	U	2	.68?	.67?	
Install warning signals	U	More than 2	−.27??	.73?	
	R	2	.56?	.29?	
Warning signals, T-junction	R	More than 2	.21??		
Add pedestrian signals	U	2	.13	.56?	
	U	More than 2	.02?	.42?	
Improve signals	U	2	.31	.35?	
	U	More than 2	−.02	.10??	
	R	More than 2	.42?	.45??	
Improve signals, T-junction	U	More than 2		.57	
Curtail turns	U	More than 2	.40	.39	
Left-turn lane without signal	U	2	.19?	.80?	
	U	More than 2	.06	.54?	.18?
	R	More than 2	−.06	−.01??	
Left-turn lane, T-junction without signal	U	2	.79	.79	
	U	More than 2	.51?	.62	
Left-turn lane, Y-junction without signal	R	2	.33	.05	−.15
Left-turn lane with signal	U	More than 2	.27	.01	.07??
	R	More than 2	.43?	.58?	
Left-turn lane, T-junction with signal	R	More than 2	−.42?	−.28??	
Add left-turn signal without lane	U	More than 2	.39	.57	
Add left-turn lane, signal, and illumination	U	More than 2	.46	.76	
New traffic signals		2 or more	.29	.50	
Deslicking	U	More than 2	.20	.15	
Rumble strips	R	2	.27??	.26??	.24??
Bridge or Underpass					
Install delineators	R	2	.47	−.08??	
	R	More than 2	.53?	.62	.89

Note: ? Moderate sample size limits confidence in this figure.
Note: ?? Small sample size limits confidence in this figure.

Source: Roy Jorgenson and Associates, *Evaluation of Criteria for Safety Improvements on the Highway* (Washington, D.C.: U.S. Bureau of Public Roads, Office of Highway Safety, 1966), p. 316.

Adjustments for traffic volume changes are needed to determine future saving in accidents due to any improvement. Average traffic volume for the next ten years will represent the volume during a reasonable life expectancy of the improvement well enough for estimating purposes.

$$S' = \frac{SV'}{V} \tag{9.11}$$

S is the estimated saving in involvements per year caused by the planned improvement; V is the average annual daily traffic for the period studied; V' is the average traffic volume anticipated over the next decade; and S' is the anticipated future yearly saving in involvements.

Elements of an estimate of savings in future accidents caused by an improvement in past experience are summarized for a hypothetical location in Table 9.3. If data on improvements and severity of accidents are not available, substitute accidents for involvement and omit the classification by severity.

TABLE 9.3
Anticipated Effect of Improvement at Specified Location

| | Involvements | | |
(Hypothetical)	No Disablements	Disablements	Total
Involvements reported 1971, 1972			
(31 Acc.) $A =$	41	15	56
Possibly affected by change	13	6	19
Probably prevented by change	10	5	15
Possibly induced by change	-3	-1	-4
Net improvement $2S =$	7	4	11
Per year $S =$	3.5	2.0	5.5
Percent improvement			
$2 \times 100S/A = P =$	17	27	20
AADT 1971–1972, $V = 5,200$			
AADT 1973–1983, $V' = 7,000$			
Anticipated yearly saving in			
involvements $S' = SV'/V$	4.7	2.7	7.4

ESTIMATES WITHOUT LOCATION STUDIES

Sometimes improvements expected to reduce accidents are incorporated in highway reconstruction projects without opportunity to make detailed studies of past experience. This may be for a road section of considerable length where the experience at most locations is so small or so poorly recorded that accident studies are impractical. Then, an estimate may be made based on general experience with such improvements. Such an estimate cannot be very accurate.

The number of accidents that might be expected to be prevented by improvements listed is the average number of accidents per year experienced in the past multiplied by the fractional reduction figure from the table.

$$S = \frac{PAT}{365} \tag{9.12}$$

A is the number of accidents in T days in the past; P is the fractional reduction (percent reduction/100); and S is the annual savings. To predict future accidents, corrections for changes in volume of traffic must be made as explained before. Then future annual savings are:

$$S = \frac{PATV'}{365V} \qquad (9.13)$$

If two or more improvements are applied without specific location studies, the total accident reduction would be much exaggerated if the estimated numerical reductions for each were simply added. There is a better way to estimate the combined effect of several improvements.

$$P_t = 1 - (1 - P_1)(1 - P_2)(1 - P_3), \text{ etc.} \qquad (9.14)$$

Here, P_t is the fractional reduction of the combined improvements; P_1 is the fractional reduction of improvement No. 1; P_2 is the fractional reduction of improvement No. 2; and so on.

For example, suppose that on a section of road, it is estimated that widening will give a fractional reduction of 0.38, resurfacing a fractional reduction of 0.12, and edge markings a reduction of 0.14. The sum of these reductions would seem to promise a total reduction of $0.38 + 0.12 + 0.14 = 0.64$; it would be more realistic, however, to compute the total expected reduction from Eq. (9.13), as follows:

$$P_t = 1 - (1 - 0.38)(1 - 0.12)(1 - 0.14)$$

$$= 1 - 0.62 \times 0.88 \times 0.86$$

$$= 0.53$$

DOLLAR VALUE OF ACCIDENTS PREVENTED

Good information for estimating dollar value of accidents is not readily available because data for this purpose are not regularly collected and tabulated.

National estimates for the cost of accidents are recommended[7] for general use. The values for 1972 were given in the section of this chapter on weighting accident experience. They are repeated here:

1. For each person killed — $82,000
2. For each person injured — 3,400
3. For each accident involving damage only — 480

These values increase from year to year with inflation and other influences. These values present problems in estimating costs at specific locations, especially in comparing costs before and after improvements. A single fatality, which is largely a matter of chance at a specific location, may inflate accident cost estimates. If a fatality occurs in the accidents at a location in the year before improvement and none in the year after, the improvement may show an excellent cost reduction; but if no fatality is among the before-improvement accidents and one occurs in the following year, the improvement is likely to show a dismaying increase in accident costs.

It is clear from the foregoing that estimating dollar costs of accidents involves at best loose estimates. Even with very loose estimates, however, useful judgments may be formed on likelihood of an improvement resulting in saving greater than its cost or on which of several possible improvements will be most beneficial.

COST-BENEFIT ESTIMATES

Essentially, cost benefits are a comparison of the dollar value of expected yearly reduction in accidents and the estimated yearly cost of the corresponding improvement.

The yearly cost of the improvement includes the cost of construction divided by the expected useful life. Life expectancy is a predominant element and the figure used for it has a large bearing on annual costs. Some capital improvements, such as realignment, have an indefinite life. Others, for example, new bridges, may have a life of three decades. Guardrails, signs, and signals might have to be replaced or renewed every 20 years.

Increased maintenance must be considered a cost. Some improvements may reduce maintenance, for example, replacing bridges that have to be painted with culverts that need no paint.

Annual costs are determined by methods of engineering economics. They represent initial cost times a capital recovery factor. Below are hypothetical examples of estimating costs of alternate improvements at the same location, a narrow bridge:

	New Bridge	Signs and Approaches
Construction cost	$45,000	$6,000
Life expectancy	30 years	15 years
Cost per year	$3,269	$618
Maintenance	$50	$30
Total yearly cost	$3,319	$648

The new bridge would more likely prevent accidents than would "narrow bridge" signs and guardrails at approaches to the bridge. The fractional reduction in accidents of each improvement may be estimated as explained earlier. The savings on each of the two improvements may be calculated. For example, in this hypothetical case, yearly average number of accidents is 3.70 and these represent an estimated loss of $4,100. All degrees of severity are assumed to be equally affected by the improvements. Then the cost benefits can be illustrated as follows:

	New Bridge	Signs and Approaches
Yearly accident cost	$4,100	$4,100
Fractional accident reduction	.90	.20
Safety benefit	$3,700	$820
Yearly cost	$3,319	$648
Yearly saving	$381	$172
Safety benefit-cost ratio	1.11	1.27

Safety benefit–cost ratio is the average yearly dollar savings in accidents divided by the average yearly cost. These are shown for the example used. Signs and approaches would be a better investment because they would save $1.27 for each dollar of cost, whereas a new bridge would save less, $1.11. With this small difference, other considerations might determine which improvement was made. The new bridge would save more than twice as much per year. Table 9.4 gives representative safety benefit-cost ratios compiled from a number of sources.

TABLE 9.4
Benefit-Cost Payoff for Various Kinds of Improvements

Project	Average Project Cost ($)	Annual Cost ($)	Annual Safety Benefits ($)*	Safety Benefit-Cost Ratio
Delineation	440	102	2,742	27.
Protective guardrail	2,230	289	3,751	13.
Flashing signals	2,740	355	1,250	3.5
Modified signals	7,600	1,176	3,069	2.9
Modified signals and channelization	20,330	2,633	6,722	2.6
Intersection channelization	13,240	1,275	3,120	2.4
New signals	14,530	1,165	2,248	1.9
New signals and channelization	34,470	2,764	2,648	.95
Reconstruction and miscellaneous	21,850	1,752	1,655	.95
Widening per mile per year	33,300	2,310	548	.24
Modernize to design standards per mile per year	126,300	8,800	177	.02

*Based on $662/accident saved (that is, all severity groups combined).

Source: Roy Jorgenson and Associates.

One consideration in deciding whether or not an improvement is warranted is possible savings other than reduction of accidents. For example, traffic signals might not be entirely warranted because of safety benefits alone, but additional benefits from reducing delays might well warrant signals at a particular location.

Future increases in traffic may also make a difference. The previous example supposes no increase in traffic. If it is estimated that during the next 15 years (life of the less expensive improvement) traffic will double and the average future traffic during that period will be 1.50 times what it is at present, then the example may be revised as follows:

Yearly	New Bridge	Signs and Approaches
Future accident costs (4,100 × 1.5)	$6,150	$6,150
Cost per year of improvement	$3,319	$648
Safety benefit—same traffic	$3,700	$820
Safety benefit—50% more traffic	$5,550	$1,230
Yearly saving	$2,231	$582
Safety benefit-cost ratio	1.67	1.90

DETERMINING RESULTS OF IMPROVEMENT

Results of every improvement made to reduce accidents should be evaluated. By doing so, experience is gained as to what kinds of improvement are successful and what kinds are failures in reducing accidents.

Results are judged by comparing accidents before and after the improvement. To avoid the effect of seasonal differences in accidents, full twelve-month periods before and after should be compared; this usually means a one-year period before and a one-year period afterward. If the improvement cannot be completed in a week or so, the months during which construction is in progress should not be counted but full twelve-month periods before and after the construction should be.

If there has been more than approximately 10 percent increase or decrease in traffic between the before and after periods, a correction in the number of accidents afterward should be made, as described for estimating possible saving in number of accidents or involvements. Then, the difference in accidents per twelve months before and after should be divided by the number of accidents before to get the fractional decrease (saving) or increase from before to after.

Confidence in this figure depends on how much experience it represents. A reduction from three involvements to two represents a 0.33 saving; so does a reduction from 30 to 20. But the former is much more likely to be a matter of chance than the latter.

There are statistical procedures for determining the significance of a difference. For practical purposes, the curve in Figure 9.3 will suffice.[8] It shows that a reduction from 3 to 2 is not significant but a reduction from 30 to 20 is. The same procedure can be used to determine whether or not an increase (corrected for traffic volume) really represents an unsuccessful improvement or may be caused by chance.

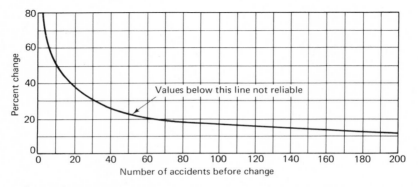

Figure 9.3. Chart for evaluating reliability of percent change in accidents at a location. [Source: Richard M. Michaels, "Two Simple Techniques for Determining the Significance of Accident-Reducing Measures," *Public Roads*, Vol. XXX, No. 10 (1959), p. 238.]

If the fractional (or percent) saving in accidents is not significant, greater experience is necessary. This may be obtained by increasing the periods before or after or both to give a greater number of events. If similar treatment is given to a number of locations, for example, yield signs, the before and after data for all locations so treated can be pooled in order to get an average improvement for the group. The larger number of cases will give a more reliable figure for the effect of the improvement.

Collision diagrams may be prepared for accidents occurring after the improvement to be compared with diagrams of experience before. These give a graphic and often striking illustration of what has been accomplished. Comparing collision diagrams has the advantage of indicating changes not only in the number of events but also in the severity and character of the accidents. They may indicate why an improvement failed to effect reductions in accidents and perhaps indicate what to do about it.

---

[8] RICHARD M. MICHAELS, "Two Simple Techniques for Determining the Significance of Accident-Reducing Measures," *Public Roads*, Vol. XXX, No. 10 (1959), p. 238.

REFERENCES FOR FURTHER READING

JAMES C. COLLINS and JOE L. MORRIS, *Highway Collision Analysis*. Charles C. Thomas, Springfield, Ill., 1967.

J. STANNARD BAKER, *Traffic Accident Investigation Manual*. Traffic Institute, Evanston, Ill., 1974.

JUD B. CARLISLE, *Speed and Acceleration Problems in Motor Vehicle Accidents*. Jud B. Carlisle, Austin, Tex., 1964.

Highway Safety Program Manual, Vol. 9, Identification and Surveillance of Accident Locations. National Highway Traffic Safety Administration. Washington, D.C., 1969.

DONALD A. MORIN, "Application of Statistical Concepts to Accident Data," *Highway Research Record, No. 188*, Highway Research Board, Washington, D.C., 1967.

Chapter 10

TRAFFIC STUDIES

Donald E. Cleveland, Professor of Civil Engineering, University of Michigan, Ann Arbor, Michigan.

Traffic studies are conducted to furnish the traffic engineer with the factual information he needs, both to identify the magnitude of traffic problems and to provide him with the data required for a quantitative approach to the solution of the problems. The studies must be so designed and carried out that the data produced are adequately accurate and unbiased and the cost of their collection and processing is within the limits of available manpower, funds, and time. This chapter emphasizes these aspects of the traffic study process; the application of the information obtained in traffic studies is discussed in the other chapters of this handbook. For more details on the conduct of studies, the reader is referred to the Institute's *Manual of Traffic Engineering Studies*[1] or the other references listed at the end of the chapter.

Studies made by the traffic engineer can be classified as *administrative studies*, those conducted which use data already available in office files, *inventories* of static features within the area of responsibility of the engineer, and *traffic studies* in which the dynamic and variable characteristics of traffic are studied.

ADMINISTRATIVE STUDIES

Records of inventories of physical items, operational parameters and legal controls, as well as other administrative records provide the foundation that is essential to the successful operation of any traffic engineering agency. Records of manpower and equipment requirements for task accomplishment are critically important in operational budgeting and planning.

INVENTORIES

Inventories of the physical plant and its fixed characteristics, when properly organized through good administrative practices, provide ready sources of data for the countless daily requirements of the traffic engineer. The development and maintenance of these data files require a special skill if the engineer is to make the best use of information on his physical resources.

---

[1] D. E. CLEVELAND, ed., *Manual of Traffic Engineering Studies* (Washington, D.C.: Institute of Traffic Engineers, 1964).

TRAFFIC STUDIES

These studies concentrate on the operational characteristics of the traffic stream and the individual drivers, pedestrians, and vehicles in it. For planning purposes, basic data on travel related activities such as purpose of trip, mode of travel, automobile ownership, and social-economic characteristics of these road users are also necessary. Traffic studies gather these data. In the operational and transportation planning process, however, the engineer does not make all of the studies providing useful data. Land-use inventories and plans, motor vehicle ownership trends and use, employment data and forecasts, and other physical, social, and economic data and forecasts are usually developed by other specialists for use by the traffic engineer.

CHAPTER ORGANIZATION

In this chapter attention is given to traffic inventory studies, a brief introduction to statistics (required for an understanding of the operational studies), the operational studies themselves, and finally to the more important planning studies. The equipment used, the methods of study, and techniques of data reduction and presentation are emphasized.

INVENTORIES

A readily available, accurate, and up-to-date inventory of all traffic facilities and controls provides essential information for both the planning and operating traffic engineer. This has been recognized recently by a detailed list of elements to be inventoried in a TOPICS study.[2] The most important elements for which information is necessary are:

1. An inventory of highways, streets, and alleys. Items included in this inventory are:
   a. location of boundary lines of all governmental units, state and federal parks, and reservations.
   b. recreational features adjacent to roadway.
   c. description of roadway, roadway right of way and surface width, surface type, section length, and information on pavement design.
   d. roadway condition: riding quality, condition of surface, and drainage facilities.
   e. roadway structures.
   f. railroad crossings.
   g. critical features of roadway geometry.
   h. services performed on the highway, such as mail, school bus, milk, and other special service routes.
   i. abutting land-use information.
2. An inventory of relevant laws, ordinances, and regulations as well as all legally authorized traffic control devices, showing the location and character of all

---

[2] U.S. BUREAU OF PUBLIC ROADS, *Urban Traffic Operations Program to Increase Capacity and Safety* (Policy and Procedure Memorandum 21–18), February, 1967.

traffic signs, signals, and pavement markings. The inventory should include such items as:

a. type of control device.

b. description of device—size, model number, special physical features, etc., as appropriate.

c. location information that can be referenced to the location identification systems used for other highway records.

d. authority for erection.

e. in-place history—initial installation date, dates and record of maintenance and repair work, dates of routine maintenance checks and trouble calls, and dates and record of any changes or modifications.

3. An inventory of parking facilities, summarizing the available curb and off-street parking spaces, mapping their location, indicating time limits, and parking or meter fees in effect; this inventory also includes the data on truck loading facilities at the curb and off-street loading docks (see also the Parking Studies section of this chapter, page 454).

4. An inventory of the physical dimensions of public transit routes, location and geometry of stops and off-street terminals, and special passenger facilities.

In preparing the initial inventory, records of the Department of Public Works, the Planning Department, the transit operator, utility companies, and other agencies are a source of much of the information required. For additional data, it is necessary to have field teams survey the control devices and transportation facilities, taking measurements, counts, and notes as needed. In recent years aerial photography at scales from 600 to 1,000 to 1 has been a very useful aid. Since much of the information collected will be of use to other governmental agencies, the collection of data may be done cooperatively by the staffs of the organizations concerned. The data are prepared for reference by placing relevant information in files, on maps, or in electronic data-processing systems. The U.S. Census Bureau has recently developed computerized systems of coding addresses and locating street intersections that should be explored as a possible aid in inventory studies.

It is essential that inventories be kept up-to-date. A program of continuing data gathering should be established. Changes affecting the physical plant originating outside the traffic engineer's office should be automatically reported to him. Alterations to traffic control devices, parking, loading provisions, and other traffic operations are posted in the inventory as a regular procedure. If the geographic limits of jurisdiction of the office are extended, for example, by annexation, an inventory of the added area is conducted and incorporated in the existing records.

Additional information on inventory procedures is contained in *Inventory of the Physical Street System.*[3]

STATISTICAL ANALYSIS

This section gives the engineer unfamiliar with the uses of probability and statistics a list of useful references. Many texts and self-study guides at all levels of mathematical difficulty exist in this field. The reader interested in an elementary treatment will find the text by Greenshields and Weida useful. The ITE *Manual of Traffic Engineering*

---

[3] NATIONAL COMMITTEE ON URBAN TRANSPORTATION (Procedure Manual 5A) (Chicago: Public Administration Service, 1958).

Studies contains a brief appendix with some of the highlights of elementary statistics. Wohl and Martin also present a good introduction. Moroney's interesting treatment is easily followed. Among the standard texts are Brownlee and Dixon and Massey.[4]

It is probably safe to say that the most useful mathematical tool of the traffic engineer is the knowledge of the methods of probability and statistics. The traffic engineer observes that the values with which he works are not exact or always the same. This is the problem that statistics deals with.

An important application is that of sampling from the traffic stream and using these data to infer the characteristics of the entire stream. In this chapter (particularly in speed studies) and elsewhere in this handbook, particular attention is given to the use of statistical methods in determining average values, differences, etc. There are many tools available, but only a few will be discussed. The traffic engineer is encouraged to master the elements of statistics and its important applications and to make use of the consulting services of skilled professionals.

VEHICLE VOLUME STUDIES

A volume or flow study is the measure of the time rate of vehicles passing a specific point on a roadway. It is one of the fundamental measurements of the importance of a road. This count may be stratified by time of day, direction of travel, and type of vehicle. TOPICS studies require that average daily traffic and peak hour traffic volumes be obtained.

SPOT COUNTS

Spot counts are conducted in order to obtain traffic volume data for a specific location on a road or street. Such information is required for highway design, capacity analysis, operations analysis, and signal timing.

Spot counts are conducted as required at the location for which the data are needed. Data are accumulated directionally or classified by vehicle type. Vehicle classes should be as few as possible. Values may be recorded either every 10 min or 15 min or for an entire hour. At temporary stations the duration of counting generally varies from one day to one week if automatic equipment is used. Manual counting usually extends only over several hours including the peak hours of traffic movement. In order to obtain daily variations, traffic flows during two to six periods are counted and the data averaged. A series of short counts within the period of interest can be made as follows.

Traffic at a location is counted for 5 min out of each hour of the day, the observers making a circuit of as many as six locations, utilizing 5 min to rest and travel between locations. The resulting total at each location is multiplied by 12 to give the estimated

[4] BRUCE D. GREENSHIELDS and FRANK WEIDA, *Statistics with Application to Highway Traffic Analysis* (Saugatuck, Conn.: ENO Foundation for Highway Traffic Control, 1952).

D. E. CLEVELAND, ed., *Manual of Traffic Engineering Studies*, Appendix Number 2.

MARTIN WOHL and BRIAN V. MARTIN, *Traffic System Analysis for Engineers and Planners* (New York: McGraw-Hill Book Company, 1967).

M. J. MORONEY, *Facts from Figures* (Baltimore: Penguin Books, 1957).

K. A. BROWNLEE, *Statistical Theory and Methodology in Science and Engineering* (New York: John Wiley and Sons, 1965).

W. J. DIXON and F. J. MASSEY, JR., *Introduction to Statistical Analysis* (New York: McGraw-Hill Book Company, 1957).

total hourly flow. This sampling procedure has been shown to yield estimates within several percent of the true totals for vehicular travel.

Intersection counts are usually conducted manually. Low-volume intersections may be counted by one person, but heavier volumes are usually counted by a team of two or more observers. Each observer counts vehicles entering the intersection from a maximum of two adjacent approaches, records both through and each turning movement, and classifies the vehicles by type. Counts are generally totaled every 15 min. Individual lane counts, desirable for certain purposes, require a relatively large crew of observers. A form often used in this type of study is shown in Figure 10.1.

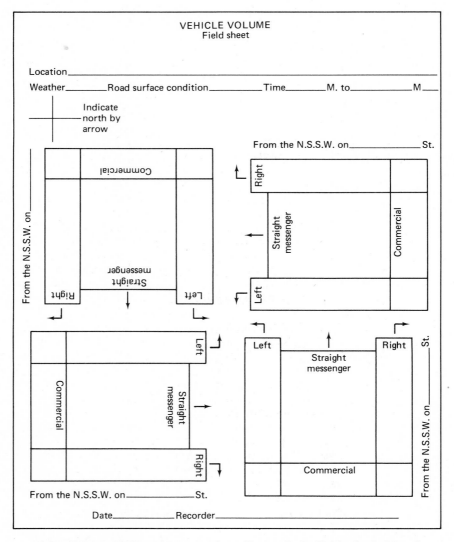

Figure 10.1. Intersection volume count form. [Source: D. E. Cleveland, ed., *Manual of Traffic Engineering Studies* (Washington, D.C.: Institute of Traffic Engineers, 1964), p. 11.]

Counts made between intersections are usually made automatically; one recorder is used if only total volumes are desired and two machines are used if direction of travel is also to be obtained. Short manual counts made at random in peak and off-peak periods can be used to estimate the proportions of vehicles of different types.

Spot count data processing depends on the information desired. Fluctuations in traffic during the day are often plotted to illustrate problems of peak flow as shown in Figure 10.2. Intersection counts may be graphically shown in an intersection volume flow diagram (see Figure 10.3). Separate diagrams or sets of figures on the same diagram can be used to show the volumes in the morning peak, evening peak, and during the entire period of the count.

An example of statistical analysis of spot volume studies is a comparison of the percent of trucks in two studies made at different times of the day at a single location. In one study of 400 vehicles, 40 trucks were observed. In another study of 600 vehicles, 100 trucks were seen. Details of statistical techniques used in a comparison of two percentages are described elsewhere.[5] The computations are made as follows:

1. Find the weighted average percent trucks

$$100(40 + 100)/(400 + 600) = 14\%$$

(We assume this percent of trucks is an accurate estimate for both studies.)

2. Find the expected number of trucks and passenger cars for each study based on this average:

Study 1   Trucks $(.14)(400) = 56$
             Cars $(.86)(400) = 344$

Study 2   Trucks $(.14)(600) = 84$
             Cars $(.86)(600) = 516$

3. Square each of the differences between the observed and expected values,

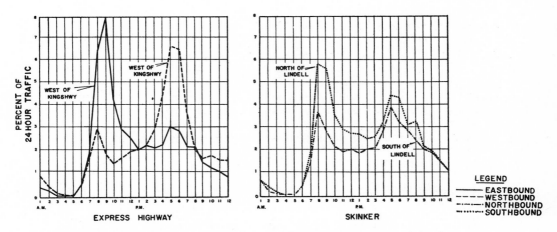

Figure 10.2. Typical example of hourly variations in traffic volume. (Source: *A Highway Planning Study for the St. Louis Metropolitan Area*, Columbia, South Carolina: Wilbur Smith and Associates, 1959, p. 36.)

---

[5] D. E. CLEVELAND, ed., *Manual of Traffic Engineering Studies*, Appendix Number 2.

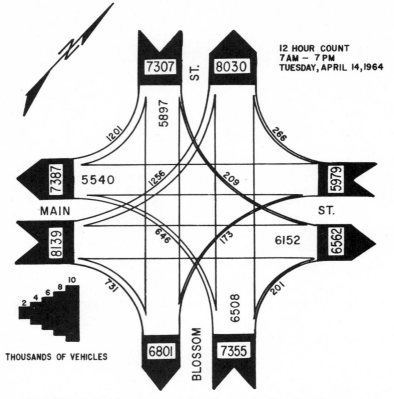

Figure 10.3. Typical intersection graphic directional volume flow diagram.

divide by the expected value, and total. The total is known as the chi-square statistic.

$$\frac{(40-56)^2}{56} + \frac{(360-344)^2}{344} + \frac{(100-84)^2}{84} + \frac{(500-516)^2}{516}$$

$$= 4.57 + 0.74 + 3.05 + 0.50 = 8.86$$

4. Use a table of the chi-square distribution, and note that this value exceeds 3.84 and hence the 10 percent of trucks measured in the first study and 16.7 percent measured in the second study are significantly different, that is, a value greater than 3.84 occurs infrequently if the two percentages are identical.

AREAWIDE STUDIES

Areawide vehicle volume counts are conducted in order to obtain spatial and time traffic patterns on the street and highway network in an area. By establishing a continuing program of sample counting at regular intervals, data are obtained to show the long-time trends in traffic volumes as well as short-term changes.

In many administrative areas of the United States and elsewhere, continuing traffic counting programs have been conducted for many years, and procedures are well established.

Traffic volume varies during the different hours of the day, days of the week, and weeks of the year. The basic theory of the method may be stated briefly as follows: The percent of daily traffic during any given period of the day is constant at all points along a section of a route (where traffic characteristics are homogeneous) or for routes of the same character in the same or similar districts. A short-time count can therefore be expanded by applying the appropriate factor to yield an estimate for any time interval other than that observed. This theory may be extended to include all fluctuations in movement including seasonal, daily, and hourly patterns of variation.

The conversion of short-time counts to average long-time counts may be made as follows:

Station A estimated long-time count = (Station A short-time count)

$$\times \frac{\text{Master Station B long-time count}}{\text{Master Station B short-time count}}$$

in which

$$\frac{\text{Master Station long-time count}}{\text{Master Station short-time count}} = \text{conversion factor}$$

Station A short-time count = known short-time count at location in question.

Station A long-time count = unknown long-time count at location in question, to be estimated.

Master Station B short-time count = known short-time count at a nearby location covering the same specific period as the short-time count at Station A. This count is usually a part of the Master Station long-time count.

Master Station B average long-time count = known average long-time count at the same nearby location, for the same long-time period as desired for station A.

In the same way that one- or two-hour counts may be converted to average all-day counts, a one-day count may be converted to an average weekly count or a monthly or annual count. Similarly, an average winter count may be converted to an average summer count. It is first necessary to classify the streets or highways by function as an aid to establishing routes of a similar character. The four basic urban street classifications often used are freeways or expressways, major arterials, collector streets, and local streets. Local streets may be further designated as residential, commercial, or industrial streets.[6] See Chapter 12 for more information on classes of streets.

Urban studies. The counting procedure used in urban areas is summarized below. Since it is impractical to count all locations simultaneously in the study area, continuing control counts are made to place the volume counts made at different times on a common basis. Typically, this involves the adjustment of sample counts to estimate the annual average daily traffic (ADT). Major control stations are selected to sample

---

[6] NATIONAL COMMITTEE ON URBAN TRANSPORTATION, *Determining Street Use* (Procedure Manual 1A) (Chicago: Public Administration Service, 1958).

the traffic movement on the major street system. At least one station should be located on each freeway, major arterial, and collector street. The minimum recommended duration and frequency of counting is a one-day weekday machine count every second year. Minor control stations are located in order to sample typical streets in the local street system. A minimum of three stations on each class of local street should be established in a small city. A one-day weekday nondirectional machine count is performed biennially at each of these stations.

Key count stations are selected control stations which are used to obtain hourly, daily, and seasonal variations in traffic volumes. At least one key station should be selected from each class of street. These stations are counted as follows: one nondirectional, seven-day machine count is performed annually, and one nondirectional one-day weekday machine count is made monthly or quarterly.

Coverage counts are used to estimate ADT volumes throughout the street system. On freeways and expressways, major arterial streets and collector streets, one nondirectional weekday count is taken within each control section. Since only the 24-hour total is needed, nonrecording counters may be satisfactory. The counts should be repeated every four years or less. Coverage counts can also be made by using the moving-vehicle method described on page 434. On local streets, a one-day nondirectional machine count is made for every mile of local street. Counts are repeated when local circumstances indicate a need.

Vehicle classification counts are made manually and are usually limited to a 14-hour weekday period from 6 a.m. to 8 p.m., with directional data being recorded for each hour.

As a part of the urban transportation planning process, it is often necessary to count the traffic on each of the links or sections of the highway network included in the traffic assignment network. In general, two one-day directional counts should be made on each link. Peak-period directional counts may be made at major control stations. The moving-vehicle method has been widely used in this situation.

Rural studies. Rural counting programs vary considerably, depending on the type and size of the area to be covered. Generally, the highways are classified into categories such as farm service routes, general purpose routes, recreational routes, winter resort routes, etc. The classification is based on land-service function, character of origins and destinations of traffic using the roads, continuity and administrative classification. One authority recommends use of a few control stations for each class of road. Traffic is counted continuously throughout the year at each of these stations.[7] Less important control stations are counted one week per month. Coverage stations are located on all highway sections and are counted two to four days per year. Standard errors less than 10 percent for roads with ADT's greater than 500 vehicles per day are recorded. At 100 vehicles per day, the standard error is approximately 20 percent (i.e., at about $\frac{2}{3}$ of the coverage stations, the true value of the ADT deviates from the estimate of the ADT by less than .20 × true ADT).

In order that all traffic counts made within the area be comparable, the ADT is usually computed for each highway section by using appropriate multiplicative conversion factors developed from control stations. The results are shown numerically on highway survey maps that use distinctive legends to indicate the volume ranges

---

[7] FEDERAL HIGHWAY ADMINISTRATION, *Volume Counting Manual* (Washington, D.C.: U.S. Government Printing Office).

into which the various highway sections fall or graphically on flow maps (see Figure 10.4 for an example) in which the width of each highway section is scaled to the traffic volume carried. Tabular presentation will generally show the percentage of change in traffic volumes over a period of a year or longer.

It should be noted that centralized statewide digital traffic reporting and processing systems have recently become operational. Detectors are connected to minicomputer peripheral components linked to a central digital computer by leased telephone lines. Data can be summarized in any desired manner.[8]

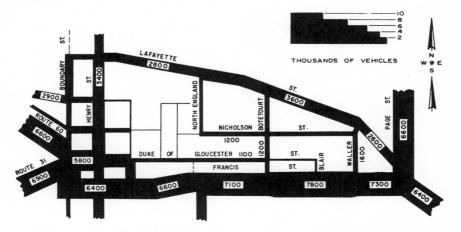

Figure 10.4. Typical traffic volume flow map. [Source: *Traffic—Parking—Transit, Colonial Williamsburg* (Columbia, South Carolina: Wilbur Smith and Associates, 1963)]

CORDON COUNTS

Cordon counts are conducted in order to obtain information on the total number of specific types of motor vehicles and/or passengers inside the cordon area and the number entering and leaving the area by time of day. Cordon counts are used to study traffic movements into and out of central business districts, other small areas of special interest, or entire cities or metropolitan areas. They may be a part of a comprehensive origin-destination survey, or they may be used at regular intervals to develop trend statistics.

The cordon line is defined as the boundary of the area being studied. Adjustments to the cordon line may be made to reduce the number of streets crossing it. Counting stations are established at the cordon line on all intercepted streets except that counts are seldom made on local streets known to be carrying negligible traffic volumes.

Appropriately classified directional counts are made at each station for 12- to 24-hour periods, depending on the information desired. If possible, all stations should be counted on the same day; however, if a set of control stations is being maintained, cordon counting may be spread over several weeks. All counts are then adjusted to a common day using the control counts.

---

[8] EMORY C. PARRISH, EDWYN D. PETERSON, and RAY THRELKELD, "Georgia's Program for Automated Acquisition and Analysis of Traffic Count Data," *Mathematical and Statistical Aspects of Traffic*, Highway Research Record Number 199 (Washington, D.C.: Highway Research Board, 1967), pp. 42–61.

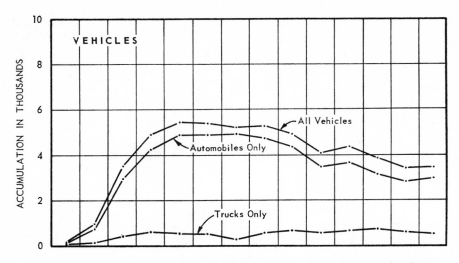

Figure 10.5. Typical cordon count analysis of vehicular accumulation by hourly periods. (Source: *San Diego Metropolitan Area Transportation Study*, 1958.)

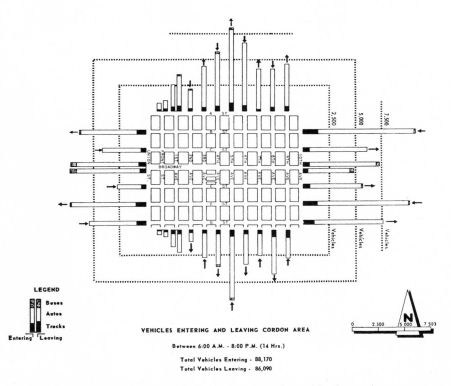

VEHICLES ENTERING AND LEAVING CORDON AREA

Between 6:00 A.M. - 8:00 P.M. (14 Hrs.)

Total Vehicles Entering - 88,170
Total Vehicles Leaving - 86,090

Figure 10.6. Typical graphical analysis of vehicles entering and leaving cordon area. (Source: *San Diego Metropolitan Area Transportation Study*, 1958.)

The analysis of cordon count data will depend on the information desired. For instance, the data may be broken down by hours and checked against the results of home interviews in an origin-destination survey. In studies of central business districts, the accumulation of vehicles within the cordon is of primary interest and is often shown graphically (see Figure 10.5). Information on the number of vehicles in the area at the beginning of the study is obtained by an occupancy study described under the parking studies section of this chapter. At equal intervals (usually every 15 minutes or every hour) the net number of vehicles entering or leaving the area during that interval is added to the number calculated at the end of the previous period. Information on hourly volumes crossing the cordon line in both directions is often used to present the results of this type of study as shown in Figure 10.6.

SCREEN LINE COUNTS

Screen line studies are made to determine the traffic crossing a major geographical barrier or moving between two areas. Some screen line studies are conducted at regular intervals so that long-range trends in traffic volumes at the screen line may be determined. Screen line studies are also used to check the accuracy of an origin-destination survey. A screen line drawn through the area will intercept trips moving from one side of the screen line to the other. Screen line counts are also used to aid in locating new crossings of barriers such as rivers.

Screen line studies are conducted in the same manner as cordon counts. The screen line may be naturally well defined by a river, ridge, railroad, or other geographical barrier. If established artificially, it is drawn to keep the number of highways crossing it to a minimum and to avoid multiple crossings of the line by heavy traffic flows.

Analysis and presentation of the data depend on the purpose of the study. Trend charts and modified flow charts may be drawn and the traffic crossing the screen line may be distributed among the various gateways crossing it. For an origin-destination survey check, the hourly measured volumes are compared to the traffic volumes estimated to cross the screen line, as indicated from the home interview data and the external survey. This comparison will enable the development of appropriate adjustment factors for the home interview data.

VOLUME STUDY EQUIPMENT

Equipment used in automatic vehicular volume studies normally consists of two elements: (1) a sensing device or detector to detect the presence or passage of vehicles or axles and (2) a counter to accumulate the number of vehicles or axles detected in a fixed time interval. A summary of volume study equipment can be found in an Institute of Traffic Engineers Committee information report.[9] Errors with automatic counters are usually less than 2 percent.

Detectors. There are many devices to detect vehicles or axles available commercially. Some are portable and therefore adaptable to temporary or short-time counting studies, and others are designed for permanent installations. In the latter category,

---

[9] INSTITUTE OF TRAFFIC ENGINEERS COMMITTEE 7-G, Information Report, "Volume Survey Devices," *Traffic Engineering*, Vol. 31 (March 1961), 44–51. No. 6; "Special Purpose Traffic Survey Devices," *Traffic Engineering*, Vol. 36, No. 5 (February 1966), pp. 29–41.

the detectors developed primarily for traffic signal control may be adaptable for accurate vehicle counting. The detectors may be sensitive to heat, light, pressure, sound or interference with electrical or magnetic forces as shown in Table 10.1. Several manufacturers have detectors that classify vehicles by height, length, or number of axles. Some detectors are inherently sensitive to the direction of movement of the vehicle. Errors of less than 2 percent are usually attainable.

The device most commonly used at temporary counting sites is the "road-tube." Flexible tubing is stretched tightly across the roadway. As the wheels of an axle cross the tube, an impulse of air is generated which closes the pneumatic switch, completing an electrical circuit and generating a signal for the counter.

Electrical tape detectors consist of two metallic contacts separated by spacers and molded into a flexible covering. The wheels of vehicles passing over the detector force the metallic contacts together. These devices can be permanently installed in shallow grooves cut in the pavement.

In both sensors, an error is introduced by vehicles having more than two axles. For example, a five-axle tractor, semitrailer combination is counted as $2\frac{1}{2}$ vehicles. The error increases as the proportion of heavy commercial traffic increases. The error can be compensated for by conducting short classification counts and computing correction factors. Some engineers prefer to use the original count in certain studies because the multi-axle vehicles occupy more road space than two-axle vehicles. The unadjusted count is sometimes referred to as the "equivalent passenger car count."

TABLE 10.1
Types of Detectors

Type and Name	Detection Agent
Permanent	
Tape	Electrical contact
Photoelectric	Light
Magnetic	Electrical field
Induction loop	Electrical field
Radar	Radio signal
Ultrasonic	Sound
Infrared	Heat
Temporary	
Road-tube	Pressure
Tape	Electrical contact

Some permanent count stations on rural two-lane highways use photoelectric detectors. One or two beams of light are directed across the roadway to photocells. Whenever a beam is interrupted, an electrical impulse is generated for the counter. These detectors have been used vertically in tunnel applications.

The magnetic detector and the induction loop detector work on the principle that the passage of a motor vehicle causes a disturbance in an electrical field or a change in the induction of the loop with the small change in potential being magnified by an amplifier, and an impulse sent to the counter.

Electronic detectors mounted above the pavement have also been developed for vehicular counting. These are of three general types: radar, ultrasonic, and infrared. The radar detector uses a high-frequency radio beam and the Doppler shift to detect the motion of a vehicle. Ultra-high frequency sound is the detection medium used in the ultrasonic detector. Infrared detectors use reflections of a transmitted beam of infrared light to detect vehicles. Counting models may be sidemounted or installed overhead.

Other models of electronic detectors designed primarily for traffic signal applications cover more than one lane. When these detectors are used for counting, undercounting results when two or more adjacent vehicles are simultaneously within the zone of detection.

Pressure-sensitive, magnetic, and induction loop detectors may be used at permanent counting stations. It is usually a simple matter to connect a counter either permanently or on a short-term basis to detectors used for traffic signal control.

Counters. In the simplest traffic counter, the nonrecording model, an accumulating register is manually read at desired intervals, perhaps every 24 hours, and the only data obtained are total counts over the length of time between readings. These counters may be equipped with time clocks that will start and end a count at preset times. Thus, if it is impractical to read the counter at certain hours, the clock will turn the register on and off at the beginning and end of the count period and keep the desired total on the register.

In one type of recording counter, a printing mechanism actuated by a clock transfers the count from the register to a paper tape at predetermined intervals. Commonly, subtotals are printed each 15 min, and the register is automatically reset to zero every hour.

A circular graphic chart recording counter is also commercially available. Traffic volumes from 0 to 1,000 vehicles for intervals of 5-, 10-, 15-, 20-, 30-, and 60-min counting periods may be recorded on a circular chart for periods of 24 hours or 7 days. The distance the recording pen moves out from the center of the chart is proportional to the volume, and the rotation is a function of time.

The punched-tape counter records the volume in binary-coded decimal configurations on a special tape. This tape can be analyzed manually or processed by a translating device to generate punched cards or tapes for computer analysis. The counting interval may be set for 5, 15, 30, or 60 min.

Special devices. Multiple-pen recorders may be used for special volume studies, particularly when it is of interest to obtain a record of fluctuations in traffic flow and other traffic stream information for small increments of time. The pens may be actuated by vehicle detectors and/or observers. The transcription of data into tabular form is time consuming.

MANUAL VOLUME COUNTS

Manual counts are made when the desired data cannot be obtained by mechanical or automatic counting equipment or when the costs of installing such equipment is greater than gathering the data manually. The primary groups of data in this category are turning movements at intersections, classification of vehicles by type, and relating vehicle counts to axle counts. Manual counting may also have to be used when more counts are scheduled at one time than there is equipment available and when vandalism may be a problem. Short-term counts on multilane facilities where permanent detectors are unavailable usually have to be conducted manually because of the difficulty of placing and maintaining portable counting equipment.

The counts are made by field checkers who record the data on appropriate forms.

For low volumes, tally marks on a form are adequate. Manually operated tally counters, which eliminate the need for checkers to take their eyes off the roadway to make tally marks, are especially useful for higher volumes. These counters are usually mounted in banks on a clipboard. For turning movement counts, four banks of three counters each are frequently mounted on a clipboard so that each bank is oriented on the board to represent an intersection approach. Counter totals are transferred to data forms at desired intervals, usually 15 min. Field checkers can probably count 1,000 to 1,500 vehicles per hour with an error less than 1 percent in simple counting situations.

VEHICLE SPEED STUDIES

Motor-vehicle speed studies are made in order to provide information on the speeds of traffic streams. The type of study undertaken depends on the results desired. TOPICS studies require both peak and off-peak hour travel speeds. In this section, speed studies made at one point and along a route are described and information on study techniques and equipment is presented.

SPOT SPEED STUDIES

Measuring vehicular speeds at a specific location is called a spot speed study. The number of vehicles measured, the sample size, must be adequate in order to estimate the desired characteristics accurately. Speed characteristics obtained from spot speed studies have many applications. By periodic studies at the same point, speed trends may be established. Spot speed studies are valuable in establishing (1) speed zones, (2) answering complaints on speeding, (3) determining a basis for requesting enforcement, (4) determining the need for posting advisory safe speed indications at curves, (5) providing useful information on the proper size and location of regulatory, warning, and guide signs, (6) establishing lengths of no-passing zones, and (7) analyzing accident experience. Speed studies made before and after a change in conditions provide a means of evaluating the effect of the change.

In measuring spot speeds, it is important to remember the difference between the time and space distributions, discussed in Chapter 4. If speed data are collected by measuring the instantaneous speeds of all vehicles on a section of highway (for example, by studying two closely timed aerial photographs), a distribution "in space" results, and successive measurement of the speeds of vehicles passing a point gives an "in time" distribution. These distributions and the values of their respective means are not the same because the time-speed distribution has a higher average than the space-speed distribution. This can be understood if a short section of highway is visualized. A spot speed sample will include some fast vehicles that have not yet entered the section and will exclude some of the slow vehicles within the highway section at the start of the study. An instantaneous study would include all vehicles within the highway section at the moment of exposure.

The relationship between the mean-time speed and the mean-space speed is expressed by:

$$\bar{u}_t = \bar{u}_s + \frac{\sigma_s^2}{\bar{u}_s}$$

where σ_s^2 = the variance of the space distribution,

$\bar{u}_t$ or $\bar{u}_s$ = the mean-time speed and the mean-space speed, respectively.

ROUTE SPEED STUDIES

Information desired on speeds over a route with length L (miles) may be simply the overall speed, $\bar{u}$; this is obtained from

$$\bar{u} = \frac{60L}{t}$$

where t is the travel time (minutes).

The description of travel-time studies will be found in the next section of this chapter. The moving-vehicle method described on page 434 may also be used to obtain overall route speeds. When the average speed while moving (the running speed) is desired, time spent at rest is deducted from the denominator of the formula.

When more detailed information is required on the speed profile over L, then it is convenient to describe speed by both the average speed and the measure of its dispersion called "acceleration noise" (see Chapter 7). Practically, acceleration noise is obtained by measuring the times between successive speed changes of 2 mph, Δt_i. When the duration of the study is T, the acceleration noise A is found from:[10]

$$A = 1.47 \frac{1}{T} \sum_i \frac{1}{\Delta t_i}$$

METHODS AND EQUIPMENT[11]

Two major techniques are used to obtain spot speed data. One method consists of measuring the time required by a vehicle to traverse a known short distance. The other method uses electronic means of detecting the speed of the moving vehicles.

Stopwatch timing is a widely used manual means of measuring spot speeds. A measured distance or "trap" is established on the roadway. The stopwatch is started when a vehicle enters the trap and is stopped as it leaves the trap. The length of the trap is determined by visibility, speed of the vehicles, and limitations of the observer in accurately measuring the time. For arithmetical convenience in reducing data (dividing a constant by the time spent in the trap to obtain the speed in miles per hour directly) the distance is usually a multiple of 88 ft (the distance covered in 1 sec by a vehicle traveling 60 mph; divide 60 by the trap time in seconds) when using ordinary stopwatches (0.1 sec reading) for timing. To ensure accuracy in the speed measurements, the following trap lengths are recommended:

88 ft for velocities below 25 mph (mph = $60/t$)

176 ft for velocities between 25 and 40 mph (mph = $120/t$)

264 ft for velocities over 40 mph (mph = $180/t$)

For a trap length of 147 ft the velocity in mph equals $100/t$ and in metric units, for a trap length of 27.8 m, the velocity in kph also equals $100/t$.

---

[10] T. R. JONES and R. B. POTTS, "The Measurement of Acceleration Noise: A Traffic Parameter," *Operations Research*, Vol. 10, No. 6 (1962), pp. 745–63.

[11] For additional information on traffic speed measuring equipment, see "Spot Speed Survey Devices," *Traffic Engineering*, Vol. 32, No. 8 (May 1962), pp. 45–53.

Trap limits may be delineated by means of painted markings on the pavement, a practice which may introduce an error resulting from parallax. This error may be overcome by using a simple device called an enoscope, flash box, or mirror, an L-shaped box, open at both ends, with a mirror set at a 45° angle. The enoscope bends the line of sight of the observer so that it is perpendicular to the path of the vehicles. The observer looks into the box and sees the vehicle at the exact time of passage as a flash of color. It is most desirable to use an enoscope at each end of the course with the observer inconspicuously stationed between them (see Figure 10.7). This procedure eliminates parallax error. Night studies may be made by placing small light sources directly opposite the enoscopes. Passing vehicles interrupt the light beam, thereby indicating the beginning or ending of a speed measurement. Accuracies of 2 mph can be attained.

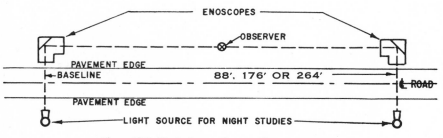

Figure 10.7. Typical setup for use of two enoscopes.

Mechanical devices that may be used to automatically start and stop the timing watch are commercially available. These devices use road-tubes for detection and solenoids for actuating the watch. They eliminate human error in detecting vehicle passages across the speed trap. The major disadvantage of this meter is the physical presence of road-tubes. Driver behavior, especially for vehicles being operated at higher speeds, is often influenced by two road-tubes spaced a distance apart on the roadway.

Radar meters are the most commonly used speed measuring equipment. The radar meter operates on the principle that a radio wave, reflected from a moving target, has its frequency changed in proportion to the speed of the moving target. Various commercially available models have some variation in their operating characteristics. Graphic recorders are available to provide a permanent record of the speeds; however, these recorders are more applicable to enforcement than engineering usage because lower speeds may not be recorded or may be difficult to distinguish. The major advantage of the radar meter is that it is inconspicuous and can be set up quickly. A disadvantage, in addition to relatively high cost and Federal Communications Commission license requirements, is the difficulty in distinguishing the speed of individual vehicles under high-volume conditions. Accuracies of 2 mph are often guaranteed.

Other equipment used to measure speeds include motion pictures (or pictures taken at predetermined short intervals), 20-pen graphic recorders, and photoelectric cell detectors. Photographic techniques are most often used when more than one stream characteristic is being measured. They are briefly discussed in the section on traffic stream studies.

Size and selection of sample. Normally, the speeds of at least 50 (preferably 100 or more) vehicles should be measured in any one sample. Techniques are available for estimating the size sample required at a specific location based upon the average daily traffic.[12] Figure 10.8 shows the relationship between sample size and variability for the median and 85 percentile speeds. Values of the standard deviation of the speed can be estimated from data presented in Chapter 4. When daily values are needed, studies during the 9–12 a.m., 3–6 p.m., and 8–10 p.m. periods are often made. Studies are usually made during good weather conditions.

When volumes are low (fewer than 200 vehicles per hour), the observer can probably measure the speed of 90 percent or more of the vehicles and can try to study every vehicle. At higher flows, it is necessary to select vehicles for the sample. To avoid bias in the results, the observer should select vehicles for the sample on a random basis from the traffic stream. Some common errors that tend to introduce bias and procedures to reduce these errors are:

1. Always selecting the first vehicle in a platoon. Since following cars desire to move at least as fast as the lead car, results are biased toward lower speeds. When traffic is platooned, one should select vehicles in varying positions in the platoon.
2. Selecting too large a proportion of trucks. These speeds may not be representative of the rest of the sample. Attempts should be made to obtain approximately the same proportion of trucks in the sample as exists in the traffic stream.
3. Tendency to obtain too large a proportion of higher-speed vehicles. Untrained observers have been known to ignore a normal-speed vehicle in order to "catch" a high-speed approaching vehicle or to measure all higher-speed vehicles to find the "fastest." Results are thus biased toward the upper speeds. Isolated free-moving vehicles should not be over-represented in the sample.

Analysis and presentation of data. In analyzing spot speed data, several characteristics can be developed. Some values are computed directly from the data and other values are easily determined from a graphic representation.

An example of the statistical treatment of an urban spot speed study is given below. Table 10.2 shows a set of 130 observations of spot speeds made by timing vehicles through a "trap" of 88 ft. In this example, the stopwatch data are grouped into 0.2 sec classes (a $\frac{1}{5}$ sec stopwatch was used to obtain the data). The first column of the table shows the midpoint of each group. The second column shows the frequency of the observations. The cumulative percent shown in the third column is obtained by dividing each cumulative frequency (for example, 8 vehicles took 5 sec or longer to pass through the trap) by the total number of observations and multiplying by 100 ($8 \times 100/130 = 6.2$ percent). The next two columns are computed to develop the average and variance of the space distribution. The sixth column indicates the speed of each time class and the last two columns are used to obtain the average speed and variance of the time distribution.

---

[12] J. C. OPPENLANDER, J. F. BUNTE, and P. L. KADAKIA, "Sample Size Requirements for Vehicular Speed Studies," Highway Research Board Bulletin 281, 1961, pp. 68–86. See also J. C. OPPENLANDER, "Sample Size Determination for Spot-Speed Studies at Rural, Intermediate, and Urban Locations," *Travel Time and Vehicle Speed*, Highway Research Record No. 35, 1963, pp. 78–80. A nomograph for relating allowable error, confidence level, and sample size for varying standard deviations of speed will be found in L. PIGNATARO ET AL., *Traffic Engineering Theory and Practice* (Englewood Cliffs, N.J.: Prentice-Hall, Inc., 1973), p. 127.

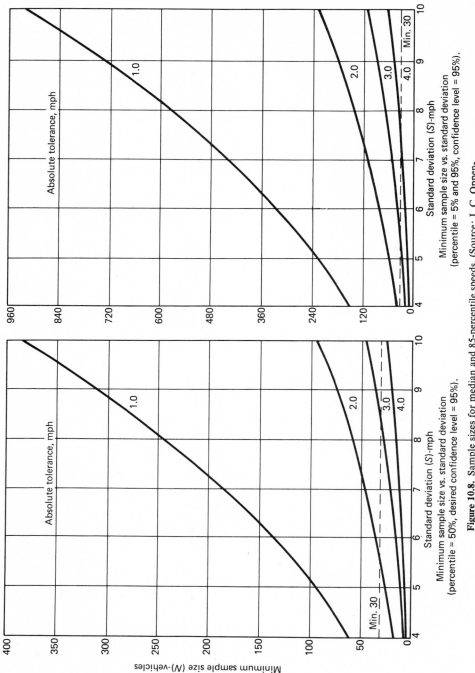

Figure 10.8. Sample sizes for median and 85-percentile speeds. (Source: J. C. Oppenlander et al., *Highway Research Board Bulletin 281*, 1961, pp. 68–86.)

TABLE 10.2
Spot Speed Study Calculations

n	(1) Time* (sec) t_n	(2) Number of Observations f_n	(3) Cumulative† Percent P_n	(4) $f_n t_n$	(5) $f_n t_n^2$	(6) Speed (mph) u_n	(7) $f_n u_n$	(8) $f_n u_n^2$
1.	2.0	1	100.0	2.0	4.0	30.0	30.0	900.
2.	2.2	2	99.2	4.4	9.7	27.3	54.6	1491.
3.	2.4	2	97.7	4.8	11.5	25.0	50.0	1250.
4.	2.6	4	96.2	10.4	27.0	23.1	92.4	2134.
5.	2.8	7	93.1	19.6	54.9	21.4	149.8	3206.
6.	3.0	15	87.7	45.0	135.0	20.0	300.0	6000.
7.	3.2	22	76.2	70.4	225.3	18.8	413.6	7776.
8.	3.4	16	59.2	54.4	185.0	17.6	281.6	4956.
9.	3.6	12	46.9	43.2	155.5	16.7	200.4	3347.
10.	3.8	11	37.7	41.8	158.8	15.8	173.8	2746.
11.	4.0	9	29.2	36.0	144.0	15.0	135.0	2025.
12.	4.2	7	22.3	29.4	123.5	14.3	100.1	1431.
13.	4.4	5	16.9	22.0	96.8	13.6	68.0	925.
14.	4.6	6	13.1	27.6	127.0	13.0	78.0	1014.
15.	4.8	3	8.5	14.4	69.1	12.5	37.5	469.
16.	5.0	0	6.2	0.0	0.0	12.0	0.0	0.
17.	5.2	2	6.2	10.4	54.1	11.5	23.0	264.
18.	5.4	3	4.6	16.2	87.5	11.1	33.3	370.
19.	5.6	1	2.3	5.6	31.4	10.7	10.7	114.
20.	5.8	0	1.5	0.0	0.0	10.3	0.0	0.
21.	6.0	1	1.5	6.0	36.0	10.0	10.0	100.
22.	6.2	1	0.8	6.2	38.4	9.7	9.7	94.
	Sum	130		469.8	1774.5		2251.5	40612.

*Range of times measured is value shown ± 0.1 sec.
†Accumulated from slow speed end of distribution.

Sample Averages. The arithmetic mean-time speed $\bar{u}$ is the most frequently used speed statistic. It is a measure of the central tendency of the time distribution, and it is computed from the formula

$$\bar{u} = \frac{\sum f_n \cdot u_n}{N}$$

where $\bar{u}$ = mean or "average" speed,
$\quad \Sigma$ = sigma, a statistical symbol meaning "the sum of,"
$\sum f_n \cdot u_n$ = sum of the speeds of all vehicles (total of Column 7),
$\quad N$ = total number of vehicles observed (total of Column 2).
This calculation is shown in Table 10.3 where a value of 17.3 mph has been obtained. The mean-space speed is calculated using the average time spent passing through the trap (3.61 sec) as shown in Table 10.3, giving a value of 16.6 mph.

Standard Deviation. The standard deviation of the population from which the sample was drawn is estimated by

$$\sigma = \sqrt{\frac{\sum f_n \cdot (u_n)^2}{N-1} - \left(\frac{\sum f_n u_n}{N}\right)^2}$$

where σ = standard deviation,
$\sum f_n \cdot (u_n)^2$ = sum of the squared frequencies (total of Column 8).
The variance, σ^2, is the square of the standard deviation, σ.

TABLE 10.3
Statistical Calculations Spot Speed Study

Average time $= \dfrac{469.8}{130} = 3.61$ sec

Space-mean speed $= \dfrac{60}{3.61} = 16.6$ mph

Time-mean speed $= \dfrac{2251.5}{130} = 17.3$ mph

Standard deviation	$= \sqrt{\dfrac{40{,}612}{129} - \left(\dfrac{2251.5}{130}\right)^2} = 3.86$ mph
Variance	$= (3.86)^2 = 14.9$ (mph)^2
7% speed	$= 12.3$ mph
15% speed	$= 13.7$ mph
Median (50% speed)	$= 17.5$ mph
85% speed	$= 20.3$ mph
93% speed	$= 22.3$ mph
Skewness index	$= 2\left(\dfrac{22.3 - 17.5}{22.3 - 12.3}\right) = 0.96$
Standard error of mean	$= \sqrt{\dfrac{14.9}{130}} = 0.34$ mph

Standard Error of the Mean. This is a statistic that indicates the likely range of the actual mean speed of all traffic at the same place and time as when the sample was taken.

$$\sigma_{\bar{u}} = \sqrt{\frac{\sigma^2}{N}}$$

where $\sigma_{\bar{u}} =$ standard error of the sample mean,
$N =$ number of observed vehicles.

It can then be inferred that with 95 percent confidence of being correct, the actual mean speed of all traffic at the same time and place is within the range defined by the observed sample mean, $\bar{u}$ plus and minus almost twice its standard error $\sigma_{\bar{u}}$. For the example data in Table 10.2 as summarized in Table 10.3, the actual mean lies in the range of 17.3 plus and minus approximately two times $\sigma_{\bar{u}}$ (0.34) or between 16.6 and 18.0 mph at the 95 percent confidence level.

Comparing averages for two speed studies. In order to determine whether or not the difference between the mean speed of two spot speed studies is significant, it is necessary to estimate the standard deviation of the difference in means by using the equation

$$\sigma_d = \sqrt{\sigma_{\bar{u}_1}^2 + \sigma_{\bar{u}_2}^2}$$

where $\sigma_d =$ standard deviation of the difference in means,
$\sigma_{\bar{u}_1}^2 =$ variance of the mean for Study 1,
$\sigma_{\bar{u}_2}^2 =$ variance of the mean for Study 2.

If the absolute value of the difference in sample mean speeds is greater than almost twice the standard deviation of the difference in means, i.e., if

$$|d = \bar{u}_1 - \bar{u}_2| > 2\sigma_d$$

where $\bar{u}_1 =$ mean speed of Study 1,
$\bar{u}_2 =$ mean speed of Study 2,
$>$ is symbol meaning greater than,
$|\ |$ is the absolute value,

it can be said with 95 percent confidence that the observed difference in mean speeds is statistically significant and not caused by chance. An example computation is presented in Table 10.4, where data not shown are used to estimate the parameters of the distribution.

TABLE 10.4
Significant Differences in Means

Parameter	Study 1	Study 2
Mean speed, $\bar{u}$	17.3	19.0
Standard error of mean	0.34	0.50

$$\sigma_d = \sqrt{(0.34)^2 + (0.50)^2} = 0.605$$
$$|d| = |17.3 - 19.0| = 1.7$$
Since $1.7 > 2 \cdot (0.604)$, the difference is significant

Graphic analysis. The cumulative percentage, calculated in Table 10.2, is plotted against the upper limit of each speed group. In this example the limits of each group are determined by the time intervals used to determine the speed. Starting at the slowest speed (longest travel time), we observe that no vehicles require more than 6.3 sec to traverse the "trap", or 0.0 percent travel less than 9.5 mph. Similarly, one vehicle (0.8 percent) requires at least 6.1 sec, corresponding to an upper limit of 9.8 mph, and a total of two vehicles (1.5 percent) require at least 5.9 sec, an upper limit of 10.2 mph for this class. At the upper end of the curve 100 percent of the vehicles require 1.9 sec or more, so that 100 percent of the vehicles travel at a speed slower than 31.5 mph. (A smooth S-shaped curve drawn through the points is called the cumulative speed curve.) Since the sample is used to estimate the total distribution, it is important that it be a smooth curve instead of a joining of points. The example in Table 10.2 is plotted, and the corresponding cumulative speed curve is drawn as shown in Figure 10.9. Significant values obtained from the curve shown on the figure are described below.

Median. One measure of the central tendency of the speed distribution is the median or middle speed. The median is that speed at which there are as many vehicles going faster as there are going slower. The median is obtained from the curve by reading the speed that corresponds to the 50 percent value, or 17.5 mph. It can be calculated by linear interpolation within the class in which the median falls (see Column 3 in Table 10.2).

Percentile Speeds. The ordinate of the graph is the percentage of vehicles traveling at or below the indicated speed. The speed corresponding to any percentage on this scale, say i percent, is called the ith percentile speed. The uses of some of the percentiles are as follows:

1. The 85th percentile is sometimes referred to as the critical speed. Drivers exceeding the 85th percentile speed are usually considered to be driving faster than is safe under existing conditions. In the example the 85th percentile speed is 20.3 mph (see also column 3 of Table 10.2).
2. The 15th percentile is assuming more importance as the use of minimum speed

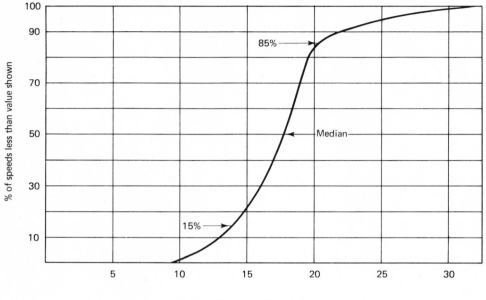

Figure 10.9. Cumulative speed distribution.

limits increases. Vehicles traveling below this value tend to obstruct the flow of traffic, thereby increasing the accident hazard. In the curve, a value of 13.7 mph is obtained (see Table 10.2).

3. The 50th percentile is the median speed, 17.5 mph.
4. The 7th, 50th, and 93rd percentile ranges are useful in determining the skewness of the distribution or lack of symmetry.[13] The skewness index is computed from the cumulative percentile values as follows:

$$\text{skewness index} = \frac{2 \times (P_{93} - P_{50})}{P_{93} - P_7}$$

where P_i is the ith percentile speed.

A skewness index of 1.0 indicates symmetry about the median, a value below 1.0 indicates that the distribution curve is skewed toward lower speeds, and one above 1.0 is found when the tail is at the higher speed values. On uncongested roads the distributions of speeds have very little skewness, but travel times are skewed toward longer travel times. On congested roads speed distri-

[13] D. S. Berry and D. M. Belmont, "Distribution of Vehicle Speeds and Travel Times," *Proceedings*, 2nd Symposium on Mathematical Statistics and Probability, University of California, Berkeley, 1951, pp. 589–602.

butions are skewed toward the higher speeds. In Table 10.2 the skewness index is found to be 0.96 (see Table 10.3).

A more detailed discussion of these statistical measures can be found in the references listed in footnote 4.

TRAVEL TIME AND DELAY STUDIES

Measurement of the time required to traverse a route is called a travel time study. If additional information is obtained on the location, duration, and cause of delays, it is called a speed and delay study. Studies ignoring time spent not moving are called running time studies.

Travel time studies are made when the total amount of time (or its inverse, average speed) required to traverse a particular route or section is desired. TOPICS studies require both travel time data and information on points of delay along major routes. Speed and delay studies are undertaken when precise information is needed on impediments to traffic stream flow within the section.

Travel time data are needed to evaluate existing levels of service in transportation planning studies and in economic studies. When they are made on a continuing basis, travel time studies provide trend data on changes in the level of service with the passage of time.

On congested streets, speed and delay data provide facts on the amount, location, and cause of delays for use in traffic control selection. This information may also indicate locations where other traffic studies are needed to determine the proper remedy. Before and after studies may use data in order to determine the effectiveness of a change in conditions, for example, parking prohibitions, signal timing revisions, new one-way streets, or turn prohibitions. Delay may be considered to be made up of fixed delay caused by traffic design and control and operational delay caused by traffic interference.

METHODS AND EQUIPMENT

Two major methods are used to obtain travel time information. In one method the test vehicle is in the stream of traffic. In the second method, individual vehicles are observed through the section of interest.

The test vehicle used in the first method is driven over the route in a series of "runs" in an attempt to obtain representative travel times. Two driving strategies are used. In the "floating car" technique, the driver attempts to estimate the median speed by passing as many vehicles as pass him. Some inaccuracies arise in using this technique, especially on multilane highways during periods of congested flow and on roads with very low volumes. The second driving strategy is the "average speed" technique in which the driver travels at a speed that, in his opinion, is representative of the speed of all traffic at a point and time. Tests of this method have shown excellent correlation with actual average travel times.[14]

[14] WILLIAM P. WALKER, "Speed and Travel Time Measurement in Urban Areas," *Traffic Speed and Volume Measurements*, Highway Research Board Bulletin 156, 1957, pp. 27–44. See also FELIX J. RIMBERG, "Urban Travel Time Measurement by Taxicab Speed Studies," *Traffic Engineering*, Vol. 31, No. 9 (June 1961), pp. 43–44.

MAJOR ROUTE CONGESTION
Field sheet

Trip no._____

Route _____ Direction _____

Trip started at _____ a.m. at _____ _____
                          p.m.   (Location)   (Mileage)

Trip ended at _____ a.m. at _____ _____
                         p.m.   (Location)   (Mileage)

STOPS			SLOWS	
Location	Stopped	Cause	Location	Cause

Total trip length_____ Total trip time_____ Average speed_____

Running time_____ Stopped time_____

Symbols of delay causes   TS—traffic signals   SS—stop sign   LT—left turns
              PK—parked cars   DP—double parking   T—general congestion
              PED—pedestrians   BP—bus passengers loading or unloading

Comments _____

Date_____ Recorder_____

Figure 10.10. Major route congestion field form. (Source: *Manual of Traffic Engineering Studies*, p. 43.)

Data obtained from both techniques are recorded either by an observer in the vehicle or by a mechanical recorder. The use of an observer with two stopwatches is the most common method. The observer starts the first stopwatch at the beginning of the test run and records the time at various control points along the route. In a speed and delay study the second stopwatch is used to determine the length of individual stopped-time delays (see Figure 10.10). In its simplest form, the second stopwatch is used to record the total stopped time so that running times can be obtained. The time, location, and cause of these delays is recorded either on forms developed for this purpose or by voice recording equipment. If only time data are desired, it is possible for the driver to obtain all the necessary information by using voice recording equipment and a stopwatch mounted on the dashboard of the vehicle, thus eliminating the need for an observer. Studies have shown that from 6 to 12 runs in each direction must be made for accuracy of the order of 10 percent in estimating the average travel time.

The moving-vehicle method described on page 434 can also be used to obtain average travel times.

Although individual vehicles can be watched by a single observer from a vantage point to obtain speed and delay information, the typical use of the observer method is to obtain travel time information. This technique requires one or more observers at each entrance and exit of a section for which information concerning travel time is desired. Each observer records the time and some identifying characteristic of the vehicle (such as the last three or four digits of the license number of each vehicle; hence the technique is often called the license plate method) as it passes his observation point. The identification numbers are matched later and the travel time for each vehicle, the difference between the two recorded times, is determined.

The equipment used in this study consists of either synchronized stopwatches and recording forms or voice recorders with or without audible time signals. Hand reduction of the field data is quite tedious. Large amounts of data can economically be analyzed by punched card techniques. A sample size of 50 matches usually provides good accuracy for most purposes.

The interview technique may be useful where a large amount of information is needed quickly with little expense for field observations. Usually, employees of establishments or municipal agencies are asked to record their travel time to and from work on a particular day. If they cooperate, the results obtained may be very satisfactory for the particular set of conditions under which the trips are made.

ANALYSIS AND PRESENTATION OF DATA

Travel time information may be presented as the average time or the overall travel speed maintained on the study route. In a study of travel from one point in all directions, isochronal charts are often drawn. Isochronal charts show the distance from a common origin (often the central business district) that can be reached in a given time. A typical isochronal chart is shown in Figure 10.11. Congested areas are apparent whenever the isochrons are close together. Free-flowing freeways and major arterials become evident when the lines become long "fingers" leading away from the common origin.

Another method of presenting travel time data is to compute the vehicle delay rate. The method consists of computing the observed travel time in minutes per mile and comparing this value with the level of service recommended for this type of street. The difference between the two values (minutes/mile) is the delay rate. The delay rate multiplied by the volume gives a vehicle delay rate in vehicle-minutes per mile. These vehicle delay rates may be plotted on a map by using either the color code or the flow-map principle; the width of line represents the magnitude of the vehicle delay rate.[15]

---

[15] NATIONAL COMMITTEE ON URBAN TRANSPORTATION, *Determining Travel Time* (Procedure Manual 3B) (Chicago: Public Administration Service, 1958).

TIME IN MINUTES

Figure 10.11. Typical isochronal chart showing travel times from central business district. [Source: *Baltimore Metropolitan Area Transportation Study* (Columbia, South Carolina: Wilbur Smith and Associates, 1963)]

Data gathered by speed and delay studies are subject to a wide variety of analyses. Graphical presentation of some of the more typical analyses are set forth in Figure 10.12. Overall speeds, running speeds, average delay, intersectional delay, mid-block delay, and duration and frequency of each type (cause) of delay by time and location are considered as significant.

OTHER TRAFFIC STREAM STUDIES

Although volume and speed studies are the basic tools of the traffic engineer, there are several other characteristics of traffic streams that are measurable and may be used directly for some purposes or that can be used to estimate either volume or speed or both volume and speed. In this section an introduction to the measurement of density, gaps, and vehicle spacing is given. Following these, studies that measure more than

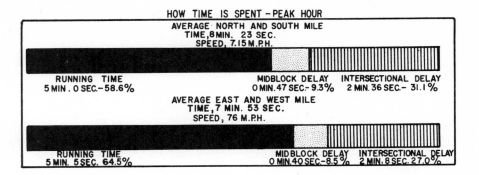

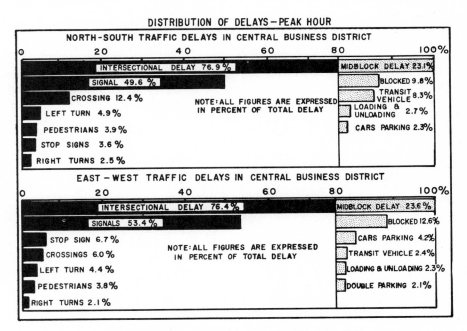

Figure 10.12. Typical graphical speed and delay analysis.

one characteristic and that can be used as a substitute for conventional measures are presented.

DENSITY, GAPS, AND SPACING

Density, gaps, and spacing are fundamental in traffic stream flow. Density, the number of vehicles per mile, k, is directly related to volume, q, and the space average speed, $\bar{u}$, by

$$k = \frac{q}{\bar{u}}$$

A gap or headway in seconds, h, is the time spacing between successive vehicles.

It is related to volume, q, (vehicle/hour), by

$$h = \frac{3,600}{q}$$

Spacing, in feet, s, is the reciprocal of density, k (vehicles/mile)

$$s = \frac{5,280}{k}$$

Density and spacing studies. Density at any instant of time may be obtained by observation or by photography. Density can also be computed from occupancy input-output studies or from the many studies providing joint information on volumes and speed. In observational studies, frequent samples of density counts on a road of known length are made. The average is taken as an unbiased estimate of the mean density during the sampling period. In photographic studies, spacing can be obtained by direct measurement of the distance between vehicles.

Gap studies. Time headways or gaps between vehicles (measured from the arrival of the front ends of successive vehicles) are important to traffic engineers. The frequent occurrence of gaps of adequate duration is necessary in many conflict situations. The frequency of adequate gaps is also used in warrants for traffic officer protection at school crossings and for traffic controls at other pedestrian crossings. Time headways are also important in capacity computations, in determining weaving from lane to lane, merging, safety from rear-end collisions, and in warrants for stop signs.

Gaps are measured with the aid of timing devices. The most satisfactory device is a digital or strip-chart recorder that is actuated by a detector or human observer. Complete records of the time of arrival of vehicles may be differenced in order to obtain gaps. Stopwatches and metronomes may also be used to measure individual gaps. Of course, time-lapse photography and closed-circuit television systems can provide records from which gaps may be measured. Figure 10.13 shows the results of a gap study plotted in an inverse cumulative form and fitted to a theoretical probability distribution.

For some advanced studies it is necessary to have the distribution of lags. A lag is the time from a random instant until the arrival of the next vehicle.

OCCUPANCY STUDIES

The availability of detectors that can signal the presence of a vehicle at the detector location has led to the widespread use of the occupancy measure. The occupancy, O, is defined as the percent of time that a vehicle is over the detector. After accounting for effective detector width, it can be easily seen that the occupancy is related to the average density and the average vehicle length. For example, a single lane with a flow of 1,000 vehicles per hour with average lengths of 19 ft operating at 30 mph over a detector with a 6 ft effective detection distance would register an occupancy of

$$\frac{(19 + 6)(1,000)(100)}{(30)(1.47)(3,600)} = 15.8\%$$

INPUT-OUTPUT STUDIES

Another study that can provide information on traffic flow characteristics is the input-output study, a study easily described graphically. In this study, continuous

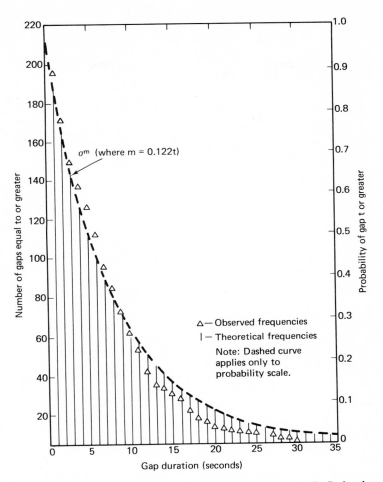

Figure 10.13. Observed and theoretical gap distributions. [Source: D. L. Gerlough and F. C. Barnes, *Poisson and Other Distributions in Traffic* (Saugatuck, Conn.: ENO Foundation for Transportation, 1971), p. 37.]

cumulative volume counts are recorded for short intervals (1, 2, 5 min) at each end of a road section. The results are cumulated and the study results plotted as shown in Figure 10.14. In this case the "normal" time spent in the section has been subtracted from the output time. It can be seen that the travel time for individual vehicles can be obtained by scaling horizontal distances (times); the number of vehicles in the section (its occupancy) can be scaled vertically, and the average delay time in the section determined by dividing the area between the two curves by the number of vehicles passing through the section during that time.

In the example, the 15,000th vehicle was in the system from 8 : 00 a.m. to 8 : 15 a.m., a delay of 15 min. At 8 : 30 a.m., the "occupancy" of the system was approximately 1,600 vehicles (the difference between the number having entered and left by 8 : 30 a.m., 18,300 and 16,700). The total delay was approximately 150,000 vehicle

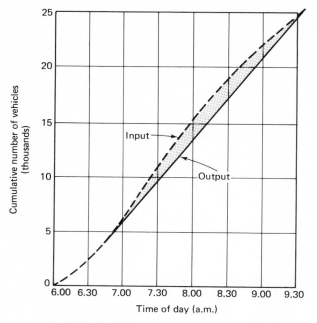

Figure 10.14. Example of input-output study.

minutes, an average of more than 8 min for each of the 18,000 vehicles entering the section during the period of congestion.

MULTIPLE CHARACTERISTIC STUDIES

Some traffic studies provide estimates of two traffic flow characteristics on a highway or street network. Since average volume, speed, and density, and their inverses, headways, travel time, and spacing are simply related, the measurement of any two independent characteristics can be used to derive the others. It is also possible to use other measures correlated with these characteristics. The number of possibilities is very great, and in this section only those that have shown practical cost effectiveness are described.

MOVING-VEHICLE METHOD

In the moving-vehicle method, the travel time of the test vehicle and the volumes relative to the moving vehicle are used to estimate speed and actual volume on a road section. This method yields reasonably accurate values for traffic volume. A 30-min sample on a street carrying 300 vehicles per hour has a standard error of 10 percent of the actual volume. Some difficulty has been reported when this method is used on low-volume streets.[16]

The highway or street network to be studied is divided into sections of length L

---

[16] R. C. BLENSLY, "Moving Vehicle Method of Estimating Traffic Volumes," *Traffic Engineering*, Vol. 27, No. 3 (December 1956), pp. 127–29, 147.

(miles or kilometers), so that each section will have approximately uniform traffic and physical characteristics. One or more test cars are driven through these sections in both directions, usually at the estimated average speed of traffic.

Routing of these vehicles may include several sections in one trip. Each car is manned by two observers, one of whom can be the driver if a travel-time meter or voice recorder is used. The following data are collected:

1. T_i: travel time (in hours) of the test vehicle through the section when moving in direction i; opposite direction is called j.
2. N_i: number of vehicles moving in direction i met by test vehicle when moving in the opposite direction j.
3. F_i: number of vehicles moving in direction i which overtake the test car when it is moving in direction i.
4. S_i: number of vehicles moving in direction i which are passed by the test car when it is moving in direction i.

At least six runs are required for each section for each period of the day being studied. If the sum of the differences between the individual travel times and the average value is greater than the average value, four or six additional runs are made and their data combined with the first set of runs. Averages for T, N, F, and S are taken and used in the calculations below.

The traffic volume for the ith direction of traffic, q_i, in a section of highway is computed from the formula[17]

$$q_i = \frac{N_i + F_i - S_i}{T_i + T_j}$$

and then the space mean speed for that same stream, $\bar{u}_i$, from the formula

$$u_i = \frac{L}{\dfrac{T_i - (F_i - S_i)}{q_i}}$$

On a one-way street, q_i is obtained by a spot count and only the second equation is used to obtain the average speed.

TOTAL VEHICLE-MILES STUDIES

The total vehicle-mile study, most appropriate in networks, can be accomplished indirectly by aerial photographic samples of density and test vehicle estimation of average speed. These values are multiplied to give an estimate of volume, which in turn is multiplied by the length of the route.

TOTAL TRAVEL-TIME STUDIES

In addition to the obvious technique of measuring the average travel time and volume and multiplying them together, the density can be sampled several times over an interval T, usually by photographic techniques. The average density is computed and multiplied by the study period T in order to obtain an estimate of average travel

---

[17] D. E. CLEVELAND, ed., *Manual of Traffic Engineering Studies.*

time. A variation of this technique is used to estimate travel time through an intersection (see page 438). This study can be very effective in networks.[18]

INTERSECTION STUDIES

There are several important special studies made by the traffic engineer at both signalized and unsignalized intersections. Pedestrian and vehicle volume studies at intersections have been described in their respective sections of this chapter. The special studies described in this section include studies of signalized intersections which are applicable to techniques described in the *Highway Capacity Manual.*[19] Related studies include intersection discharge rate studies. Studies made at both signalized and unsignalized intersections include queue length studies, delay studies, and conflict studies. Gap acceptance studies are usually made at unsignalized intersections.

CAPACITY STUDIES AT SIGNALIZED INTERSECTIONS

In addition to physical measurements of the intersectional characteristics, two operational studies, the peak-hour factor and load factor studies, are made to use the O. K. Normann method of measuring signalized intersections described in the *Highway Capacity Manual.*

Load factor. The load factor is a measurable characteristic of the degree of saturation or use of a signalized intersection approach and, indirectly, of the delay encountered by vehicles using that approach. It is defined as the percentage of the peak-hour green intervals that are fully utilized by traffic and can be measured by observing whether or not demand is sufficient to use all of the green time available and recording the result for each phase. It can be accomplished as a special part of the queue length study described later in this chapter.

Peak-hour factor. Since traffic volumes fluctuate substantially within an hour and highway capacity analyses often are for periods of less than one hour, it is necessary to develop a correction factor for this variability, called the peak-hour factor. The peak-hour factor is obtained at intersections by counting for each 15 min of the peak hour and developing the factor as follows:

$$\text{peak-hour factor}_{15} = \frac{\text{peak-hour volume}}{4 \times (\text{peak 15-min volume})}$$

Other counting periods are used. If t-min volumes are used, the multiplier in the denominator is $60/t \times (\text{peak } t\text{-min volume})$.

QUEUE STUDIES

There are two queue studies made at signalized intersections, the queue discharge and queue length studies.

---

[18] For recent developments using modern detectors, see A. CHRISTENSEN, "Use of Computer and Vehicle Loop Detectors to Measure Queues and Delay at Signalized Intersections," *Aspects of Traffic Control Devices*, Highway Research Record 211, 1967, pp. 34–53.

GEORGE SAGI, "Multi-Parameter Vehicle Flow Detection," *Traffic Engineering*, Vol. 38, No. 4 (January 1968), pp. 52–53.

[19] *Highway Capacity Manual, 1965*, Special Report 87 (Washington, D.C.: Highway Research Board, 1965).

Queue discharge. In this study, several green phases are studied in order to obtain the parameters needed to estimate capacity in special cases. In this study, a stopwatch is started when the green phase begins and is stopped when the last vehicle in the delayed queue is discharged or the phase ends. The number of vehicles entering the intersection and the time of the last vehicle's entrance are used to estimate the discharge capacity.

Queue length. Queue length studies provide an important source of data for the calculation of load factors required for highway capacity studies and to estimate delay as described in the intersectional delay studies. In such studies the number of vehicles in a standing or slowly moving queue is counted (or photographed) a number of times. When a count is made at a signalized intersection, it should be made at the start of the green phase for each approach being studied. It should also be made at the end of the yellow interval. It should also be noted if the queue is exhausted during the phase. At unsignalized intersections the samples are usually taken at equal intervals of time. An interval of 30 sec or 1 min is often used.

A queue length study can be used to measure the effectiveness of a change in traffic signal control as shown in Figure 10.15.

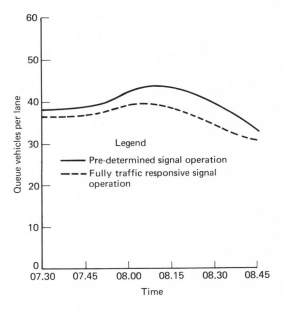

Figure 10.15. Comparison of effect of pre-timed and traffic responsive traffic signal operation. [Source: *Improved Street Utilization Through Traffic Engineering*, Special Report No. 93 (Washington, D.C.: Highway Research Board, 1967), p. 133.]

INTERSECTION DELAY STUDIES

A travel time or speed and delay study may indicate that certain intersections are the location of undue amounts of delay, and that these locations may need more intensive study. An intersection delay study provides factual information on the amount of delay at one or more approaches to an intersection.

This study is useful in design and in determining needed traffic control and signal timing changes such as prohibition of parking or turning movements, modified right-of-way controls, and additional lane lines and channelization.

There are certain basic values that are normally computed from an intersection delay study. These include total delay for each approach, average delay per vehicle, average delay per stopped vehicle, percentage of vehicles delayed, and approach volumes. These quantitities provide the factual data necessary for evaluating the operation of the intersection.

Methods and Equipment. Methods for field measurement of intersection travel time and delay are varied. The travel time study, when adapted to intersection delay studies, measures the travel time from a point in advance of the intersection to a point in or beyond the intersection on one or more approaches to the intersection. The direct methods used to obtain the travel time data are as follows:

1. Test cars operated between key points.
2. License plate numbers and times recorded at key points.
3. Time-lapse photographs taken from a vantage point to permit the timing of each vehicle shown on the film.
4. Strip-chart records actuated by road-tubes or observers operating switches.
5. Observers stationed at a vantage point tracing individual vehicles through an intersection and recording times at critical points.

All of these methods may require extensive personnel or time for the collection or analysis of the data.

A method that gives an estimate of the average travel time for all vehicles on the approach is based on the flow-density-speed relationship.[20] This method requires one observer with a stopwatch for each approach studied. The observer counts and records all vehicles in his approach at periodic intervals, for example, every 15 sec. The length of the approach is from a point farther back than any expected queue to the point at which no additional delay is incurred, possibly the far side of the intersection. This sampling of density, accompanied by volume counts, permits estimating the total and average vehicle-seconds of travel time or delay with considerable accuracy if the sampling interval is properly selected and is not an even subdivision of the length of the signal cycle. The total vehicle seconds of travel time are obtained by multiplying the total number of vehicles recorded (sum of periodic observations) by the sampling interval (seconds). The average travel time is obtained by dividing by the number of vehicles entering the approach. This can be expressed by

$$T = \frac{Nt}{V}$$

where T = average travel time through the study area,

N = total density count; the sum of vehicles observed during the periodic density counts each t sec,

t = time interval between density observation (seconds),

V = total volume entering trap during total elapsed study period.

Five minutes of data for such a travel time study are shown in Table 10.5. In this example the total volume in the 5-min period was 44 vehicles (V). There were 61 (N)

---

[20] DAVID SOLOMON, "Accuracy of the Volume-Density Method of Measuring Travel Time," *Traffic Engineering*, Vol. 27, No. 6 (March 1957), pp. 261–62, 288.

GEORGE SAGI and L. R. CAMPBELL, "Vehicle Delay at Signalized Intersections," *Traffic Engineering*, Vol. 39, No. 5 (February 1969), pp. 32–40.

vehicles recorded during observations made at 15-sec intervals (t). The average travel time was 20.8 sec as shown in the table.

TABLE 10.5
Field Data from Intersection Delay Study

Time (Minute Beginning)	Number of Vehicles Seen at				Volume
	+00 sec	+15 sec	+30 sec	+45 sec	
5:00 p.m.	0	0	2	4	11
5:01 p.m.	7	2	0	1	7
5:02 p.m.	3	6	3	0	5
5:03 p.m.	0	4	9	9	11
5:04 p.m.	3	0	2	6	10
Total for Period	61				44

$N = 61$

$t = 15$ $T = \dfrac{61 \times 15}{44} = 20.8 \text{ sec}$

$V = 44$

GAP ACCEPTANCE STUDIES

When two traffic streams conflict, it is often necessary to observe the gaps in the stream with the right of way accepted and rejected by the minor stream. It is not difficult to measure the gap duration (see section on traffic stream studies) and to determine the driver or pedestrian acceptance or rejection of the gap. There are several possible measures used to describe this behavior. The gap accepted by half of the drivers has been used. A critical gap (or lag) has also been used. The critical gap is defined as the size of gap for which the number of accepted gaps shorter is equal to the number of rejected lags longer (see Figure 10.16 for one way of computing this measure). The fact that a studied driver accepts only one gap and may reject any number of gaps can also be taken into account. One way to achieve this is to measure only the lag for a given driver. Another way is to average the gap accepted with the next shortest gap rejected by a given driver.

SCHOOL CROSSING STUDIES

Criteria for school crossing protection have evolved from surveys of hazards based upon an appraisal of street widths, vehicular speeds and volumes, pedestrian volumes, and gaps in the traffic stream. The recommended practice of the Institute of Traffic Engineers[21] suggests a study at school crossings where apparent hazards exist. Field studies are used to determine:

1. N, the number of rows of pedestrians walking five abreast when a group crosses.
2. W, the width of the pavement to be crossed by the group of pedestrians.
3. D, the actual pedestrian delay time created by the traffic flow at the study location expressed as a percent of the total survey time.

---

[21] "A Program for School Crossing Protection: A Recommended Practice of the Institute of Traffic Engineers," *Traffic Engineering*, Vol. 33, No. 1 (October 1962), pp. 45–52, 56, 58, and 60 (revised 1971 and reprinted as *A Program for School Crossing Protection*).

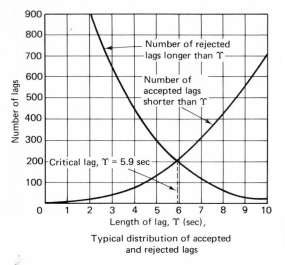

Typical distribution of accepted
and rejected lags

Figure 10.16. The critical lag. [Source: *An Introduction to Traffic Flow Theory*, Special Report 79 (Washington, D.C.: Highway Research Board, 1964), p. 57.]

It is assumed that five pedestrians will walk abreast when a group crosses a roadway. Therefore, if the number crossing in a group is measured and divided by five, the required number of rows, N, will be obtained. The recommended practice requires the development of the 85th percentage group size. A form similar to that shown in Figure 10.17 is used to record the crossing data. N is taken as an integer because even one pedestrian in excess of an even five will make an additional row and will require extra clearance time. These pedestrian counts are made on a normal school day during the heaviest hour of crossing activity in the morning or afternoon, but preferably both.

The pavement width, W, is the curb-to-curb width as measured at the crossing. If the roadway is divided and the center island is wide enough for the maximum sized group of pedestrians to stand on it in safety, the curb-to-curb width of only one roadway is used for W. This information is obtained at the same time the pedestrian group size study is made, and the pertinent data are recorded at the top of the form shown in Figure 10.17.

Before the field survey is made to determine pedestrian delay time at the study location, it is necessary to find the minimum length (in seconds) of a gap in traffic which will permit an 85th-percentile size group of pedestrians to cross the street of width W. This minimum gap in traffic, known as the adequate gap time, G, includes both perception and reaction time and the time needed to cross the roadway without coming in conflict with passing vehicles.

The adequate gap time is computed by using the following equation:

$$G = \frac{W}{3.5} + 3 + (N - 1)2$$

where G = adequate gap time in seconds,

$\frac{W}{3.5}$ = crossing time in seconds (critical width in feet of the pavement to be crossed, W, divided by the assumed juvenile pedestrian walking speed of 3.5 ft per sec),

| STUDY DATE ___3/25/74___ TIME: From _8 a.m._ To _9 a.m._ LOCATION _4th and L_ |
| CROSSWALK ACROSS ___4th Street___   CURB-TO-CURB DISTANCE ___40'___ |
| DIVIDED ROADWAY   Yes   <u>No</u>   WIDTH OF ISLAND ___NA___ |

| Group size | Number of groups | | Cumulative | Computations |
	Tally	Total		
5 or fewer	*I*	1	1	60 x .85 = 51
6-10	*THL*	5	6	
11-15	*THL THL II*	12	18	
16-20	*THL THL THL III*	18	36	
21-25	*THL THL III*	13	49	85% = 26-30
26-30	*THL II*	7	56	pedestrians
31-33	*III*	3	59	
36-40	*I*	1	60	
41-45				
46-50				
	TOTAL NUMBER OF GROUPS	60		

Figure 10.17. Sample pedestrian group size study field sheet. [Source: *A Program for School Crossing Protection* (Washington, D.C.: Institute of Traffic Engineers, 1971), p. 17.]

3 = pedestrian perception and reaction time (the number of seconds required for a child to look both ways, make a decision, and start to walk across the street),

$2N - 2$ = pedestrian clearance time (additional seconds of time required to clear large groups of children from the roadway). Children are assumed to cross the roadway in rows of five with two-second time intervals between each row. The clearance time interval is equal to $(N - 1)2$ where N is the number of rows, 1 represents the first row, and 2 represents the time interval between rows.

After the adequate gap time has been selected, the field study to determine the actual delay time to pedestrians by passing traffic can be undertaken. This study measures the gaps between the passing vehicles. Those intervals or traffic gaps that are equal to or greater than the adequate gap time are the periods when children may cross the roadway safely. The intervals between these gaps are the delay gaps, the sum of which is the actual pedestrian delay. If only part of the roadway must be crossed once the pedestrian leaves the curb, traffic flow in only those lanes crossed until a refuge is reached must be considered.

When the field survey is completed, the total time of all gaps when pedestrians could cross is found by adding together the length in seconds of each gap which was equal to or greater than the adequate gap time, G. This figure is known as t and is subtracted from the total survey time in seconds, T. The following equation, which is also indicated on the sample form shown in Figure 10.18, is then used to determine the percentage of actual pedestrian delay, D:

$$D \text{ (in percent)} = \left(\frac{T-t}{T}\right)100$$

The three parameters width (W), number of groups (N), and percent delay (D) are then used in the engineering analysis described in Chapter 18.

TRAFFIC CONFLICTS STUDIES

A recently developed technique[22] relates accident experience to the frequency of occurrence of observed intersectional vehicular conflicts of various types. The existence of a conflict is inferred from an easily identifiable driver response to the conflict. These responses include the application of the brake pedal that illuminates the brake light and lane changes. The causes of the action are categorized as left-turn conflicts, cross-traffic conflicts, or rear-end conflicts. The number of conflicts and the traffic volume are counted. Relationships developed from experience are used to evaluate the accident potential of the intersection.

INTERSECTION SPEED STUDIES

Special spot speed studies are sometimes made at intersections. An accurate mechanical method requires the use of a graphic pen recorder with a moving chart. The observer marks out the approach lanes at 20-ft intervals starting approximately 150 ft ahead of the intersection and continuing approximately 20 ft into the intersection. The marks should be inconspicuous to the drivers of approaching cars. As a vehicle crosses each line, the observer presses the corresponding button to actuate a pen. The moving chart provides a time scale from which the speed trajectory pattern of the car can be computed. A sample of 50 to 100 vehicles should be taken for each approach. A full stop should be indicated by actuating a special pen, since it will otherwise not be recorded.

If the approach being studied is signalized, counts are made to show the number of vehicles entering during green, yellow, and red signal intervals. A special pen actuated by the signal circuits can be used to indicate the signal phase at any time.

DRIVER, VEHICLE, AND ROAD STUDIES

The traffic engineer must often base his designs of the geometry of routes or traffic control devices on a "design" vehicle with a "design" driver. In some cases, knowledge

---

[22] STUART R. PERKINS, "Traffic Conflicts Technique Procedures Manual," Research Publication GMR-895 (Warren, Michigan: General Motors Research Laboratories, August, 1969).

STUART R. PERKINS and J. J. HARRIS, "Traffic Conflict Characteristics—Accident Potential at Intersections," *Traffic Safety and Accident Research*, Highway Research Record 225, 1968, pp. 35–47.

WILLIAM T. BAKER, "An Evaluation of the Traffic Conflicts Technique, *Traffic Records*, Highway Research Record 384, 1972, pp. 1–8.

PEDESTRIAN DELAY TIME STUDY

Study date _5/11/73_   Location _A ᵀᴴ and D_ Crosswalk across _D street_

End of survey (to nearest minute) _8:57 AM_   Number of rows. "N" _6_

Start of survey (to nearest minute) _8:02 AM_   Roadway width. "W" _40_ ft.

Total survey time (minutes) _55_   Adequate gap time. "G" _24_ secs.

Gap size (seconds)	Number of gaps		Multiply by gap size	Computations
	Tally	Total		
8 9 10 11 12 13 14 15 16 17 18 19 20 21 22 23	Discard gaps of less than 24 seconds from study.			
24	I	I	24	
25	IIII	4	100	
26	III	3	78	
27	II	2	54	T = total survey
28	I	I	28	time x 60
29	III	3	87	
30	HHT	5	150	T = _55_ x 60
31	II	II	62	
32	IIII	IIII	128	T = 3300 secs.
33		0	—	
34	III	3	102	
35	IIII	4	140	
36		0	—	
37	I	I	37	D = $\frac{T-t}{T}$ 100
38 39 40 41 42 43				D = $\left(\frac{3300-900}{3300}\right)100$
				D = 70
"t" (total time of all gaps equal or greater than "G")		_990_ secs.		D = _70_ %

Figure 10.18. Sample pedestrian delay time study field sheet. (Source: *A Program for School Crossing Protection*, p. 20.)

of human or vehicular performance is not available, and special studies must be made at the sites being studied. In this section, several studies of this type are described.

ACCELERATION AND DECELERATION STUDIES

Often, the engineer is interested in typical accelerations or decelerations at a specific location. When the acceleration is nearly constant, a study of this type can be made by timing the arrival of the vehicle at four or five marked locations, preferably located so that the elapsed time between locations is approximately 2 sec (for stopwatch observations). Three or four stopwatches are used, all being started at the first location and one stopped at the vehicle's arrival at each succeeding checkpoint. The trajectory of the vehicle is then plotted on a time-space (x, t) diagram by drawing a smooth curve between the points selected. The velocity, the time derivative (dx/dt) is obtained graphically by drawing a tangent to the curve at several locations. This, in turn, is plotted in a dx/dt vs. t diagram, a curve is drawn, and the acceleration is the slope of the velocity-time curve.

Skid mark studies. Skid marks can be used to estimate either the tire-pavement coefficient of sliding friction, f, or the initial speed of a vehicle at the beginning of the skid when its skidding distance is known. This method is based on the use of Newton's laws relating work done to energy lost (see Chapters 2 and 9). A recent summary of the technique is available.[23]

Field studies to measure f made with a test vehicle are conducted in such a way that the speed at both the beginning and at the end of the skid can be measured. If the vehicle skids to a complete stop, only the initial speed must be measured. The gradient of the pavement must also be known. It is recommended that the study be repeated three times, particularly if the results are to be used to estimate the speed of a vehicle involved in an accident. In this latter type of study, it is important to give consideration to the following:

1. Tests should be made with the road surface, vehicle, and tire conditions similar to the original vehicle.
2. Tests to determine the coefficient of friction should be made at a speed that will lay down skid marks close to the same length as the investigated skid marks.
3. The braking distance instead of the skidding distance should be used in the test run. Skid marks are not visible at the point that the brakes are applied, but large braking forces may be acting before the skidding begins. Ignoring this fact would give larger friction coefficients that would not be correct. Ignoring this fact in computing speed from skid marks (as must be done because initial deceleration nonskid distance is not known) gives a lower initial speed than actually existed and therefore is termed minimum initial speed.

As an example, a test vehicle similar to the one for which the speed was unknown was brought to a stop from a speed of 30 mph (48 kph) with an average skid mark length of 54 ft (21.3 m). Using

$$D = \frac{V^2}{30f}$$

where D = length of skid mark,
$\quad V$ = speed of vehicle (mph),
$\quad f$ = the average coefficient of friction,
and solving for f, we obtain

$$f = \frac{(30)^2}{54 \times 30} = 0.56$$

If the skid marks of the vehicle with the unknown speed averaged 90 ft, then the calculated speed is

$$= \sqrt{30 \cdot f \cdot D} = \sqrt{30 \cdot 0.56 \cdot 90} = 38.9 \text{ mph} \quad (62.6 \text{ kph})$$

[23] ROLANDS L. RIZENBERGS and HUGH A. WARD, "Skid Testing with an Automobile," *Design Performance and Surface Properties of Pavement*, Highway Research Record 189, 1967, pp. 115–36.

TRAVEL PATHS

In some cases it is desirable to measure the turning path characteristics of a vehicle in a special situation. This study is conducted at a location where a grid can be laid out on the pavement. Paint can be applied to certain tires, and the resulting marks on the pavement measured and later plotted. Water or paint drops from a container affixed to the vehicle can also be used in this study. Photographic techniques have also been utilized.

SIGHT DISTANCE STUDIES

Engineers must occasionally make special studies of sight distance. The eye height and obstacle height for the case being studied are established, and two test vehicles equipped with communication equipment and odometers or electronic distance measuring equipment are used.

SAFE SPEED STUDIES

Advisory speed indications are widely used on horizontal curves. The safe speed to be posted may be determined in the field by making several trial runs through the curves in a test vehicle equipped with a ball-bank indicator. The ball-bank reading is a measure of the amount of overturning force on the vehicle. Readings of 14° for speeds below 20 mph, of 12° for speeds between 20 and 35 mph, and of 10° for speeds above 35 mph are the usually accepted limits at which riding discomfort and loss of vehicle control begin.

PEDESTRIAN STUDIES

Pedestrian traffic streams are very similar to vehicular streams insofar as basic study techniques are concerned. Since reliable portable mechanical counting devices are unavailable, studies must be made using observers, stopwatches, chart or digital recorders, or photographic techniques. Studies that are particularly appropriate are listed below, along with the chapter section describing a similar vehicular study or the pedestrian study itself.

Study	Chapter Section
Pedestrian Volume	Vehicle Volume
Pedestrian Speed	Vehicle Speed
Travel Time	Travel Time and Delay
Density	Traffic Stream Studies
Gap Acceptance	Intersection Studies
School Crossings	Intersection Studies
Walking Distance	Parking

PUBLIC TRANSIT STUDIES

The general purpose of public transit studies is to evaluate the service provided by mass transit carriers within a given area and to estimate the extent to which mass transit service is utilized. Major regional transportation studies produce information pertaining to transit service and needs generally. Special studies may focus on local problems, such as the location of bus stops or the effect of one-way streets on transit routing. A number of transit studies are described in detail in *Measuring Transit Service*[24] and *Urban Mass Transportation Travel Surveys*[25] as well as in the *Manual of Traffic Engineering Studies.*

TRANSIT OPERATIONS INVENTORY

An inventory of public transit service supplies essential background information for other transit studies and for the evaluation of the service provided by transit carriers. In addition to the physical data map described in the section on inventories (page 406), schedules indicating the frequency and hours of service on each route and the trip times between various points on the system, a summary of the rolling stock used in providing the service (showing its capacity, age, and condition), and the schedule of fares charged are obtained.

TRANSIT OPERATIONS STUDIES

Public transit operations studies provide data on passenger volumes, transit vehicle trip characteristics, vehicle occupancy, travel times, and adherence to schedules. These studies are of concern since the data developed lead to improvement in the routing and scheduling of transit lines. Traffic engineers need some of these data in studying the location of bus stops, turn prohibitions, and one-way street systems. The data are useful in the investigation of street operating plans that favor transit vehicles such as signal timing and designation of exclusive bus lanes.

Three types of studies are used to determine transit operation characteristics: transit load checks, boarding and alighting checks, and speed and delay studies.

Transit load checks are made by observers stationed at one or more points along the transit routes being studied. Generally, one point is at that location along the transit route called the maximum loadpoint, where the number of passengers carried is known to be the greatest. If two or more points along one route are studied simultaneously, the travel time of each transit vehicle between these points can be measured. Each observer records vehicle identification, time of arrival and/or departure, number of persons on board the vehicle when arriving, number of persons alighting, and number of persons boarding.

Boarding and alighting checks are conducted by observers traveling on transit vehicles. Each observer records the number of persons boarding and alighting at each stop, the number of persons on board between stops, the time the vehicle passes certain time check points on route, and sometimes the types of fare paid (cash, school tickets, transfers). On lightly traveled transit runs, the observer may be able to keep track of each passenger in order to relate the passenger's boarding stop to his alighting stop to provide information on the transit portion of that person's trip.

Transit speed and delay studies parallel similar studies for the entire vehicle stream described earlier in the chapter. Data are obtained by observers riding on the transit vehicles at various hours of the day. The time each vehicle passes a check

---

[24] NATIONAL COMMITTEE ON URBAN TRANSPORTATION, *Measuring Transit Service* (Procedure Manual 4A) (Chicago: Public Administration Service, 1958).

[25] URBAN TRANSPORTATION SYSTEMS ASSOCIATES, "Urban Mass Transportation Travel Surveys" (Washington, D.C.: Federal Highway Administration, August 1972).

point and the cause and duration of delays are recorded. In addition to the types of delays found in the vehicle speed and delay studies, the transit study also shows delays caused at bus stops by passengers alighting and boarding, delays at time check points if the vehicle has arrived ahead of schedule, and other necessary operations requiring time.

Recently, there has been prototype development of on-board electronic sensing and communication equipment that makes it possible to conduct these studies continuously as a part of transit operations management.

TRANSIT PLANNING STUDIES

Transit planning studies are conducted in order to determine the origins, destinations, and characteristics of transit patrons. The home interview study described in the section of this chapter dealing with transportation planning studies is generally inadequate for a detailed description of transit user characteristics and tripmaking. The transit operations studies discussed above do not give any information about the portion of a passenger's trip between his origin and boarding place, his alighting stop, and final destination. In this study, a questionnaire is handed to each boarding passenger on one or more routes, with the request that he complete and return it to the driver or survey supervisor when alighting. Because this procedure may cause delays near the vehicle doors and inconvenience to passengers, return postcards are sometimes used. Passengers can then complete the postcards at a time convenient to them. A return of 30 to 50 percent of the postcards issued may be expected in these studies.

LAW OBSERVANCE STUDIES

Law observance studies are conducted in order to determine the degree of driver and pedestrian obedience to traffic regulations and control. The level of obedience is a useful indicator of the effectiveness of the regulations and devices employed, the adequacy of publicity and education dealing with these traffic controls, and the level of enforcement by local police officers. Results of these studies may lead to recommendations for changes in control devices or for increases in education and enforcement efforts. Samples of 100 are often adequate to indicate observance, except when violations are rare.

SPEED LIMIT OBSERVANCE

Observance of speed regulations is measured by spot speed studies (see page 418). The percent of vehicles traveling above the legal speed limit is obtained from that point (see Figure 10.9) corresponding to the speed limit. Before-and-after spot speed studies can indicate the effectiveness of a change in speed regulations or education and enforcement practices.

TRAFFIC SIGNAL OBSERVANCE

Such a study is particularly concerned with the response of drivers and pedestrians approaching the signal during the clearance interval and red phase. Care must be taken to consider perception-reaction times in classifying violations of traffic signals.

TRAFFIC REGULATION OBSERVANCE

Many traffic regulations require simple responses by motorists and pedestrians. These include turn prohibitions, pedestrian crosswalks, and rules requiring the yielding of the right of way by drivers to pedestrians or to other drivers at uncontrolled approaches. Obedience to these regulations is measured by making counts of the numbers of persons observing and disobeying the given regulation. Samples are taken to cover all applicable periods of the day. The percentage of drivers or pedestrians disobeying the regulations is computed and used for analysis and comparison purposes. In some cases, at stop and yield signs, for example, there may be several responses of interest (stop, crawling stop, slowing, no response). Each of these categories must be counted.

Since the objective of these studies is to count persons violating certain traffic regulations, it is important that observers conduct studies in an inconspicuous manner. Use of clearly visible equipment should be avoided, and consideration should be given to such apparently minor matters as parking official cars well away from the point of study.

PARKING REGULATIONS OBSERVANCE

Observance of parking regulations is measured by parking studies described on page 456. Overparking in time-restricted zones is obtained by a parking duration study. Double parking and parking in illegal zones are determined from a parking usage study.

TRANSPORTATION PLANNING STUDIES

Transportation planning studies include all of the major operational studies described in this chapter: volume, speed, travel time, transit, etc. Inventories of the transportation network are also obtained. Other data gathered for use in transportation planning are concerned with current population and its characteristics, present economic determinants, land-use types and intensities, and detailed surveys of existing travel demand and its characteristics. This includes information on person and truck movement by location, mode, purpose, land use, and time of day. It is not unusual for these studies to cost $1 per resident of the study area. Land-use, population, and economic studies are outside the scope of this handbook. For further information, see a recent text.[26] Also see Chapter 12, Urban Transportation Planning, and Chapter 13, Statewide and Regional Transportation Planning.

ORIGIN-DESTINATION STUDIES

In this section we deal with the origin-destination study, the long-time name for the detailed study of existing travel demand and characteristics. Figure 10.19 shows the general procedure used in these studies.

The comprehensive origin-destination study is the most complex traffic study. Limited studies of this type are occasionally conducted utilizing certain parts of the comprehensive study. This section will deal primarily with the larger study. For a summary and discussion of how the information derived from these studies is used, see

---

[26] ROGER L. CREIGHTON, *Urban Transportation Planning* (Urbana: University of Illinois Press, 1970).

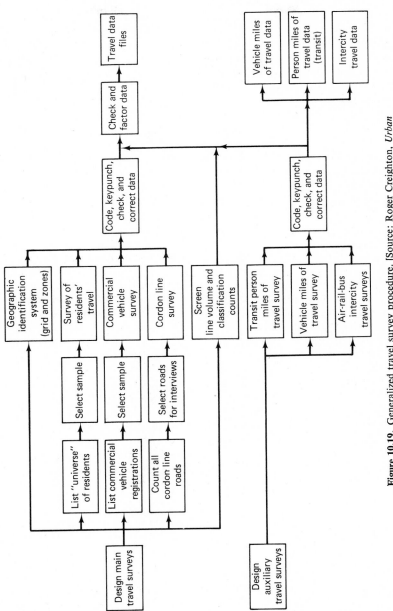

Figure 10.19. Generalized travel survey procedure. [Source: Roger Creighton, *Urban Transportation Planning* (Urbana: University of Illinois Press, 1970), p. 155.]

Chapters 5, 6, and 12. Detailed information on the conduct of this survey is contained in *Conducting a Home Interview Origin-Destination Survey.*[27]

The major purpose of an origin-destination survey is to obtain information on existing travel practices so that efficient transportation of people and goods can be planned and provided. The facts obtained from such a study include:

1. Where people begin and end their trips, their origins and destinations.
2. How they travel; via private automobile, public transit, truck (minor modes, for example, bicycles, walking, water craft, have usually not been studied intensively).
3. When they travel, by time of day.
4. Why they travel, to work, shop, eat, etc.
5. Where they park.

Planning agencies need origin-destination data in order to accomplish their functions. These data aid in the planning of major street systems, freeway location and design (number of lanes, location of interchanges, etc.), major bridge locations, public transit improvements, and terminal facilities (off-street parking, bus, and truck terminals, etc.).

Facilities are, of course, designed to meet both future needs and those that exist at the time of data collection. Origin-destination data are projected to the planning or design year in conjunction with projections of future economic and population growth, vehicle ownership and usage, land use, and similar factors as described in Chapter 12.

At this time, the use of census data gathered in the 1970 Decennial Census is beginning to be used for origin-destination data on trips made to work by mode of travel. In recent years there has been some use of origin-destination surveys conducted at the place of employment. This technique has the advantage of simplicity in data collection, but it is limited because it samples only workers and usually obtains only work trip data. Postcard surveys with appropriate follow-up techniques are beginning to be more widely used.

The first step in conducting an origin-destination survey is to define the survey area by means of a cordon line. There are three general types of person trips studied in different surveys: (1) the external-external trips, where both the origin and destination are outside the survey area (through trips); (2) external-internal trips with one end of the trip outside the area and the other within; (3) internal-internal trips with both origin and destination within the area. For convenience of study, trips made by taxicab and trucks housed in the survey area are inventoried in separate studies.

EXTERNAL STUDY

The main external study determines the origins and destinations of groups (1) and (2) above, those persons traveling through, entering, or leaving the study area by motor vehicle. Special studies are sometimes made at air, bus, and rail terminals in order to develop information on this aspect of urban travel demand.

Methods of conducting an external highway survey. The most widely used method of conducting an external highway study is the roadside interview. In the roadside interview study, stations are established at the cordon line on all major roads entering

---

[27] NATIONAL COMMITTEE ON URBAN TRANSPORTATION, *Conducting a Home Interview Origin-Destination Survey* (Procedure Manual 2B) (Chicago: Public Administration Service, 1958).

the survey area. A large sample of vehicles is stopped and the drivers are asked the origin and destination of their trips. In some studies, they are asked the purpose of their trips, the location of internal stops made, and what route they are taking. Figure 10.20 shows a widely used form.

In the return postcard method, the drivers are handed postcards and are requested to list the origin and destination of their trip and return the postcards by mail.

In limited studies of small areas, other methods can be used. License plate studies may be made. Observers are stationed at each entrance and exit of the area and they record the license number of each vehicle passing the location. In a recent development, a questionnaire is mailed to the owner of the vehicle within a few days after the license number of the vehicle is recorded.

The vehicle-intercept method also requires stations at all entrances and exits to the area. In this case, each entering vehicle is stopped and a precoded or colored card is handed to the driver (or affixed to the car, thus eliminating the need for stopping the vehicle a second time). In a variation of this method, when one entering station is studied, drivers are asked to turn on their headlights until they pass an exit station.

INTERNAL STUDY

The internal study provides information on trips made by residents of the area. These trips usually comprise the bulk of the travel in the area.

The home interview survey, developed by the Federal Highway Administration, is the most commonly used form of internal survey. Dwelling unit samples distributed over the entire area in proportion to housing density are carefully selected, preliminary contacts are made with the households, and interviewers are assigned the task of personally contacting the people living in these dwellings. Information obtained includes social and economic data on the household residents in addition to data on all trips made by all residents over 5 years of age on the previous weekday (see Figure 10.21).

There are several other forms of internal surveys that have been used. One is the controlled postcard survey, in which the vehicle owners of the area are sent postcards and are requested to list their trips for one day and to return the cards by mail.[28] The use of television as a replacement for the interview has been attempted.[29]

In the multiple cordon survey,[30] concentric cordon circles are established within the urban area. Drivers are stopped to be interviewed or given premarked cards which are surrendered at the next interception point.

SPECIAL STUDIES

Additional information is obtained on truck, taxi, and transit trips. The truck and taxi data are obtained by interviewing commercial organizations because much of the data are available from office records. See the section of this chapter on public transit studies for a description of origin-destination studies for transit users.

---

[28] R. C. BARKLEY, "Origin-Destination Surveys and Traffic Volume Studies," *Bibliography 11* (Washington, D.C.: Highway Research Board, 1951). See also HOWARD MCCANN and G. MARING, "Evaluation of Bias in License Plate Traffic Survey Response," *Travel Analysis*, Highway Research Record 322, 1970, pp. 77–83.

[29] WILLIAM R. MCGRATH and CHARLES GUINN, "Simulated Home Interview by Television," *Origin and Destination Techniques and Evaluations*, Highway Research Record 41, 1963, pp. 1–6.

[30] R. C. BARKLEY, "Origin-Destination Surveys and Traffic Volume Studies."

METROPOLITAN AREA TRAFFIC SURVEY
EXTERNAL TRIP REPORT

Card number [3] Station ---------- Date ----- Hour period beginning ----- a.m. p.m. Inbound 1 Outbound 2

1	2	3	4	5	6	7	8	9	10		11	
Serial number	Vehicle type	No. in vehicle	Where did this trip begin?	Where will this trip end?	Trip purpose	Where is car owned or garaged	Control points passed	Route exit or entrance	Intermediate stop		Exp. factor	
									Location	Purpose	Hour	24 hour
						In area 1 Out area 2						
						In area 1 Out area 3						
						In area 1 Out area 3						
						In area 1 Out area 3						
						In area 1 Out area 3						
						In area 1 Out area 3						

Figure 10.20. External trip interview form. (Source: Federal Highway Administration.)

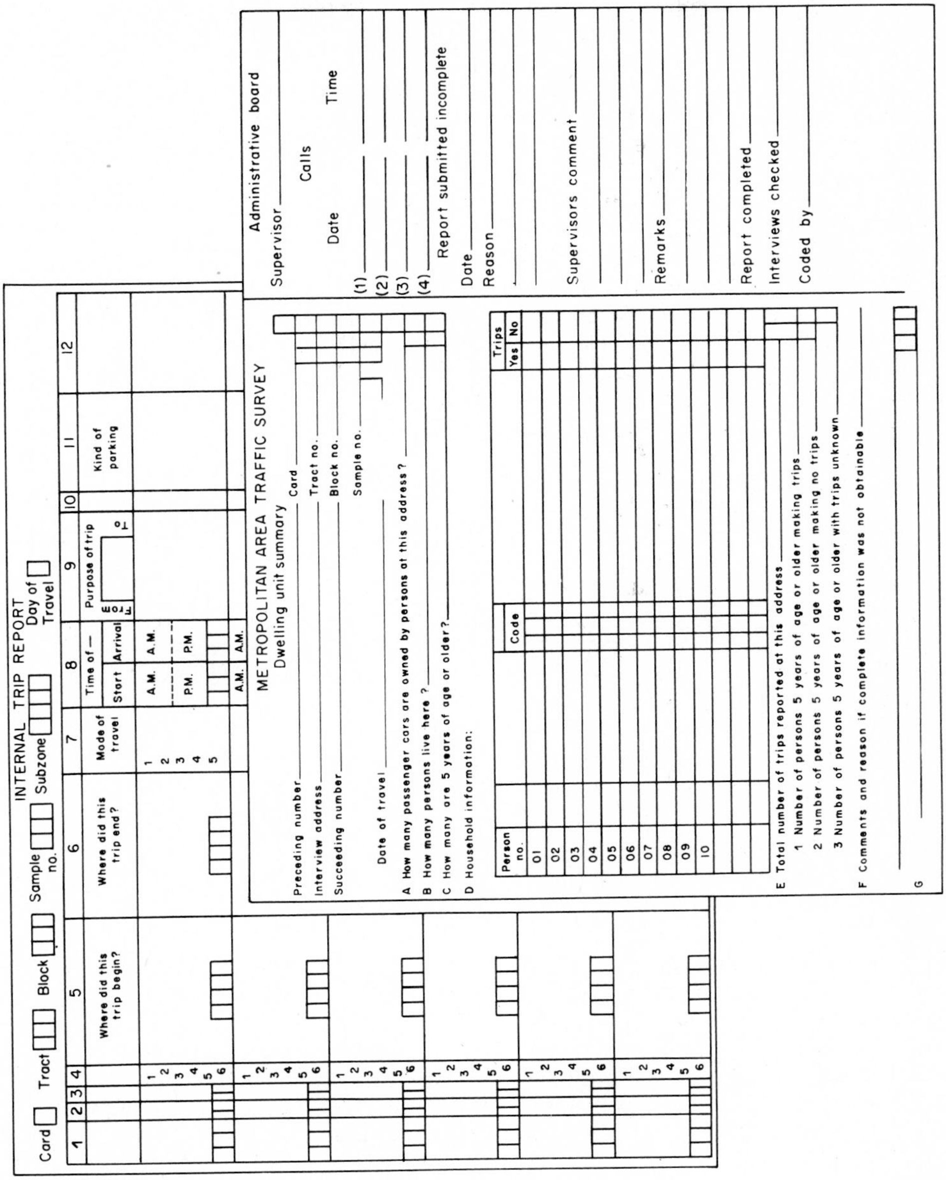
INTERNAL TRIP REPORT
Card | Tract | Block | Sample no. | Subzone | Day of Travel
1 | 2 | 3 | 4 | 5 Where did this trip begin? | 6 Where did this trip end? | 7 Mode of travel | 8 Time of — Start / Arrival | 9 Purpose of trip | 10 | 11 Kind of parking | 12
A.M. P.M.

METROPOLITAN AREA TRAFFIC SURVEY
Dwelling unit summary
Preceding number ____ Card
Interview address ____ Tract no.
Succeeding number ____ Block no.
Sample no.
Date of travel ____
A How many passenger cars are owned by persons at this address?
B How many persons live here?
C How many are 5 years of age or older?
D Household information:
Person no. | Code | Trips Yes No
01 02 03 04 05 06 07 08 09 10
E Total number of trips reported at this address
1 Number of persons 5 years of age or older making trips
2 Number of persons 5 years of age or older making no trips
3 Number of persons 5 years of age or older with trips unknown
F Comments and reason if complete information was not obtainable
G

Administrative board
Supervisor ____
Calls
Date Time
(1) (2) (3) (4)
Report submitted incomplete
Date ____
Reason ____
Supervisors comment ____
Remarks ____
Report completed ____
Interviews checked ____
Coded by ____

Figure 10.21. Internal trip household and trip forms. (Source: Federal Highway Administration.)

ANALYSIS AND PRESENTATION OF DATA

Origin-destination surveys produce large amounts of data that are analyzed by using electronic computers. Each trip is first coded by assigning numerical codes indicating the specific zones of origin and destination for the trip. Appropriate codes are used for the other information obtained in the study.

The data are expanded to a 100-percent sample for an average weekday in both the external and internal phases. The expanded data are then compared to the actual screen line counts (see earlier section in this chapter on volume studies) and possibly adjusted to match these counts. The data are then summarized and origin-destination tables are prepared showing the number of trips between each pair of zones. These tables indicate the total current person travel on an average weekday.

Extensive use is made of graphic presentation in summarizing origin-destination data. In a desire line chart, straight lines are drawn between zones of origin and destination with the width of line proportional to the number of trips made between those points on an average day. Desire line charts usually separate the type of trip into through trips, local trips, and internal trips. Figure 10.22 shows a typical external-internal desire line chart. Desire line charts may also be prepared for special zones that contain high-volume generators, such as the central business district or large industrial tracts.

Other data obtained in the survey are also presented graphically on maps. Examples of these data include population distribution, land use, and trip density.

PARKING STUDIES

Parking studies are conducted in order to accumulate the information essential in developing a picture of the parking problem in a problem area. The scale of this study may vary from the preparation of a citywide parking program to the determination of the loading zone requirements for a single store.[31]

The following types of information are almost always required:

1. Parking supply inventory.
2. Parking demand.
3. Characteristics of current parking usage.
4. Legal, financial, and administrative factors.

As a first step, the area to be studied is delineated. The limits include not only the source of the parking problem itself (the business district, the industrial park, etc.) but also the surrounding area within a reasonable walking distance from the parking generator. The reasonable walking distance may vary from 300 to 1,500 ft.

Detailed descriptions of the conduct of all parking studies, including studies dealing with specific problems (e.g., employee parking, parking lot analysis, etc.), are described in *Conducting a Limited Parking Study*, *A Comprehensive Parking Study*[32] and the *Manual of Traffic Engineering Studies*.[33]

---

[31] *Parking Principles*, Special Report 125 (Washington, D.C.: Highway Research Board, 1971), p. 205.
[32] NATIONAL COMMITTEE ON URBAN TRANSPORTATION, "Conducting a Limited Parking Study" and "Conducting a Comprehensive Parking Study" (Chicago: Public Administration Service, 1958).
[33] D. E. CLEVELAND, *Manual of Traffic Engineering Studies*.

Figure 10.22. Typical travel desire line chart (external-internal trips). [Source: *Major Route Plan. Montgomery Metropolitan Area* (Columbia, South Carolina: Wilbur Smith and Associates, 1960). Courtesy of the City of Montgomery and Alabama Highway Department.]

Each small section or block in the study area is identified by code number for locating and tabulating subsequent data. Curb spaces are inventoried by observers who traverse all streets (and often alleys) in order to estimate or measure the number of linear feet of curb cuts, prohibited parking zones, truck and passenger loading zones, parking spaces subject to time limits, and unrestricted parking spaces. Each off-street facility is classified to indicate whether or not it is private (e.g., employee parking only, not open to the public), commercial, public with restriction and/or charges, and public unrestricted. The space capacity and scale of parking fees are also noted because this information indicates whether or not the facility serves primarily short-time or all-day parking.

The inventory is tabulated and graphical representations are often superimposed on maps of the area (see Figure 10.23).

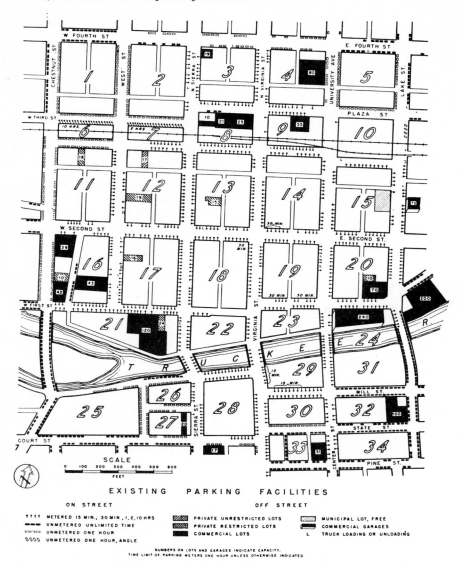

Figure 10.23. Typical curb and off-street parking inventory analysis. [Source: *Conducting a Comprehensive Parking Study*, Procedure Manual 3D (Chicago: National Committee on Urban Transportation, Public Administration Service, 1958).]

PARKING SPACE OCCUPANCY

Parking space occupancy studies are made in order to measure the number of parking spaces used at various times as a guide to the location and duration of high demand usage and surplus parking space supply.

Curb space occupancy can be measured by observing each block face at regular intervals, counting the number of parking spaces occupied, commercial vehicles in loading zones, vehicles parked in prohibited spaces or loading zones, vehicles double parked, and spaces made unavailable by improperly parked vehicles or for any other reason. Observers will generally travel in vehicles if the study area covers more than a few blocks. The interval between successive observations of each parking space depends on the needs of the study. Generally, one count every hour is sufficient, but a higher frequency may be necessary if sharp fluctuations in parking demand exist. In small communities, from two to five counts may adequately span the business day.

Off-street space occupancy can also be obtained by counting the number of vehicles parked at regular intervals. Another method is to use observers or recording traffic counters to conduct a cordon count of the single facility. If the parking facility is very large, such as a shopping center, and if it is of interest to compare the occupancy of different sections, visual counts may still be required. Parking tickets used in public facilities are a useful source of occupancy data if the times of arrival and departure of each vehicle are indicated on the tickets.

Parking space occupancy is summarized in tabular form and can also be shown graphically on maps. The percentage of the available parking space-hours used by parked vehicles is the most useful unit in presenting these data. Comparisons are also usually made between legal parking, loading, and illegal parking, using space-hours as the common unit.

PARKING DURATION AND TURNOVER STUDIES

Parking duration and turnover studies are conducted in order to measure the length of time vehicles are parked and the rate of usage of spaces in a facility (called the turnover). Time limits for metered curb spaces and the geometric and operational features of off-street parking places are influenced by parking duration and turnover information.

These data can be obtained by observing spaces at frequent and regular intervals and noting the license plate number or other identifying data for each vehicle parked at the time of observation (see Figure 10.24). Because most curb parking has a relatively short duration, some vehicles will park and unpark in between check times and will not be seen at all. Their contribution to the parking duration distribution will hence be lost. To minimize the number missed, the check interval between successive observation must be short, usually not more than 20 min.

In outlying shopping areas, even a 10-min interval may be too great. It has been found that about 20 percent of all vehicles parking at the curb in a neighborhood business district would not be seen if observations were made every 10 min. For a 15-min interval, the error was found to be approximately 30 percent.[34] There is, however, a technique based on the exponential distribution of the length of time parked that makes it possible to estimate the percent of parkers not seen in such a study. The correction technique makes use of the graph shown in Figure 10.25. It is based upon the relationship between the check interval, i, and the average parking duration, $\bar{t}$.[35]

In some high turnover areas it is necessary to observe the curb spaces continuously. The observer identifies the cars by the spaces they occupy and some easily recognized

[34] ERIC MOHR, "Euclid Avenue Parking Study" (Student Research Paper), Institute of Transportation and Traffic Engineering, University of California, Berkeley, 1949.

[35] D. E. CLEVELAND, "Accuracy of the Periodic Check Parking Study," *Traffic Engineering*, Vol. 33, No. 12 (September 1963), pp. 14–17.

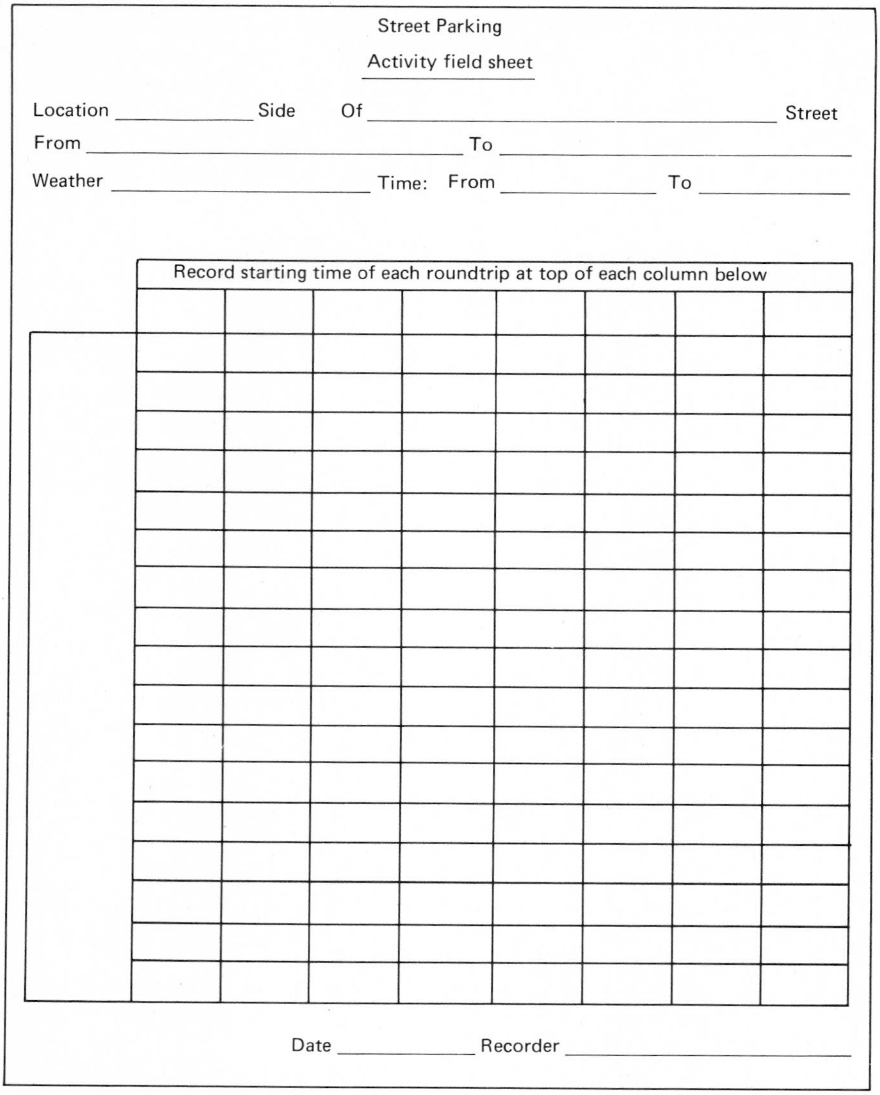

Street Parking

Activity field sheet

Location _____ Side   Of _____ Street

From _____ To _____

Weather _____ Time: From _____ To _____

Record starting time of each roundtrip at top of each column below							

Date _____ Recorder _____

Figure 10.24. Periodic check field form. (Source; *Manual of Traffic Engineering Studies*, p. 85.)

characteristic of the vehicle (make, color, etc.) Ten to twenty busy spaces are the practical limit for this study.

Commercial off-street parking durations are most easily obtained by analyzing parking tickets showing times of arrival and departure. For facilities not using tickets, the durations can be measured by continuous observation of a part or of all of the facility or by recording license plate numbers at periodic intervals. Turnover can be obtained by counting entering or leaving traffic at exits. Since parking durations in

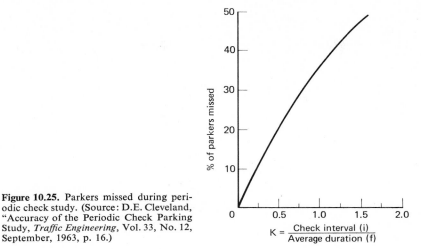

Figure 10.25. Parkers missed during periodic check study. (Source: D.E. Cleveland, "Accuracy of the Periodic Check Parking Study, *Traffic Engineering*, Vol. 33, No. 12, September, 1963, p. 16.)

off-street facilities are generally longer than the durations found at curb spaces, the interval between successive trips can be 20 or 30 min in facilities catering primarily to short-time parkers, and 1 or 2 hours in facilities having predominantly all-day parking. It is also possible to conduct the equivalent of a license plate travel time study by cordoning the facility and recording the license plate and time of entry and exit for each vehicle and matching the license plate numbers in the office.

In periodic check parking studies, the time parked is estimated by multiplying the number of times a vehicle is seen by the time interval between checks. Because of the missed parkers, the average duration calculated from these data is higher than is actually the case.

Parking durations are generally summarized in tables, and turnovers are often presented in map form. The cumulative distribution of parking duration is often plotted.

Success in the use of colored aerial photographs in estimating CBD durations and turnovers has been reported.[36] In this case, occupancy and duration were satisfactorily obtained at cost savings exceeding 70 percent. Also, the photographs of the central business district have been useful for other planning studies.

TRUCK-LOADING ZONE STUDIES

Before curb truck spaces can be allocated, a considerable amount of information must be collected. These data may include special cordon counts, speed and delay checks, origin and destination studies of truck movements, information on motor carrier operational practices (such as volume and type of truck deliveries), time of day, direction and points of origin, the needs of individual merchants (such as the number and size of deliveries), and an inventory of available loading facilities. Studies of conflicts and delays encountered are made, and spot checks of loading-space usage are conducted.

---

[36] THOMAS A. SYRAKIS AND JOHN R. PLATT, "Aerial Photographic Parking Study Techniques," *Parking*, Highway Research Record 267, 1969, pp. 15–28.

PARKING PLANNING STUDIES

A parking interview survey is made in order to determine the distances that vehicle occupants walk from their parking spaces to their ultimate destinations, the purposes for which the trips are made, and the origins or next destinations of the vehicular trips. These studies are used in the planning of new facilities. Parking occupancies, durations, and turnovers are often obtained at the same time.

In the standard study, interviews are conducted at all curb and off-street parking spaces in the study area. Because of manpower limitations, the area may be subdivided and one portion studied at a time. Each interviewer is assigned from 12 to 15 curb spaces and attempts to interview each driver parking in these spaces. At off-street facilities, interviewers are often placed at entrances and exits. Origin-destination study data have been used for this type study.

The interviewer obtains the following data:

1. Location to which vehicle occupants plan to go while vehicle is parked (or where they have been).
2. Origin of trip which has brought the vehicle to the parking space.
3. Purpose of the trip.[37]

The interviewer notes on his interview form the location of the parking space used and the times of arrival and departure of the vehicle. Occupants are sometimes counted and classified by sex and general age group.

If the purpose of the study is to ascertain the effect of an important traffic generator on parking, the interviews may be conducted at the traffic generator. The interviewers are stationed at entrances to the generator. The interview takes the same general form as in the previous study except that it is first determined whether the person interviewed used a private automobile or other means of transportation. The location where they parked their vehicle is determined. The same interview elicits information on use of transit and taxi service to the study area.

Destinations of parkers within the study area are often shown graphically on maps (see Figure 10.26). The trip purposes and distances walked are generally shown in tabular forms. Origins or destinations of vehicle trips to or from parking spaces in the study area are plotted on maps or on desire line charts.

Parking postcard study. A parking postcard survey obtains the same information secured through a parking interview study. Postcards are placed under the windshields of all vehicles parked in the study area, both at the curb and off-street. In the case of the off-street facilities, attendants may distribute the postcards to their customers.

The postcards include the questions asked in the personal interview. Each postcard is given an identifying number to indicate the location of the parking space where it was issued. Postcards may also be handed out at traffic generators in order to determine their effect on the parking characteristics of their area of influence. Experience shows that about one-third of the postcards issued are returned.

The value of this survey is limited by the fact that response biases cause expansion of the data on the postcards to be unreliable.

---

[37] *Parking Principles*, Special Report 125. See also LAWRENCE L. SCHULMAN and ROBERT W. STOUT, "A Parking Study Through the Use of Origin-Destination Data," *Parking Analyses*, Highway Research Record 317, 1970, pp. 14–29.

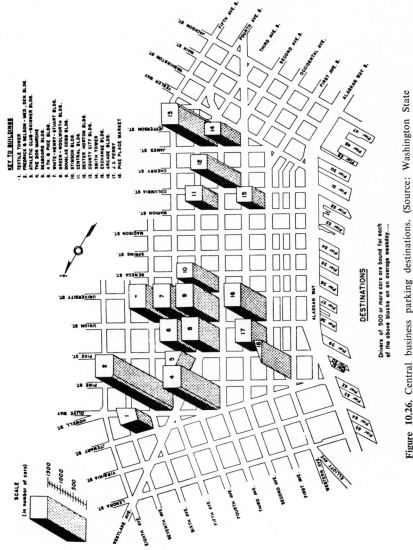

KEY TO BUILDINGS

1. TEXTILE TOWER
2. FREDERICK & NELSON - MED. DEN. BLDG.
3. ATHLETIC CLUB - SKINNER BLDG.
4. THE BON MARCHE
5. SEABOARD BLDG.
6. 4TH. & PIKE BLDG.
7. WHITE - HENRY - STUART BLDG.
8. GREEN - WOOLWORTH BLDG.
9. DOUGLAS COBB BLDG.
10. STIMSON BLDG.
11. CENTRAL BLDG.
12. DEXTER HORTON BLDG.
13. COUNTY CITY BLDG.
14. SMITH TOWER
15. EXCHANGE BLDG.
16. ARCADE BLDG.
17. J. C. PENNY
18. PIKE PLACE MARKET

DESTINATIONS

Drivers of 500 or more cars are bound for each
of the above blocks on an average weekday....

Figure 10.26. Central business parking destinations. (Source: Washington State Department of Highways.)

461

LEGAL, FINANCIAL, AND ADMINISTRATIVE PARKING STUDIES

The analysis of various solutions to a parking problem cannot proceed unless data on the legal, financial, and administrative factors bearing on the problem are available. This problem is unique to parking because of the mixed private and public sector responsibilities for parking. These data are obtained by a study of existing state laws and local ordinances dealing with curb and off-street parking and current administrative policies. Financial statistics, such as the revenue produced in existing off-street facilities, curb parking meter revenues, operating and maintenance costs of off-street facilities, and curb meters, are collected and tabulated. If curb parking meter revenues can be subtotaled by blocks or other small sections of the study area, they are often a useful indicator of parking space usage.

ACCIDENT STUDIES

This section covers methods of accident record analysis. Traffic engineering uses of accident records, accident trends, causes and costs, and the relationship of accident frequency and severity to highway design are described in Chapter 9.

COLLECTION OF DATA

Uniform definitions. To secure comparability of traffic accident data, the National Safety Council's Committee on Uniform Traffic Accident Statistics has developed the *Manual on Classification of Motor Vehicle Traffic Accidents.*[38] This manual has been approved as an American standard and is recommended for use in all jurisdictions.

Accident reporting procedures. Most state motor vehicle laws require drivers to submit a report for any motor vehicle traffic accident that results in death, personal injury, or property damage of more than a specified dollar value. In addition, many states require enforcement agencies to submit copies of reports of police investigation of accidents to a designated state agency. Laws in many states also require drivers involved in motor vehicle traffic accidents to notify the nearest enforcement agency by the quickest means possible. Some communities, by local ordinance, also require drivers to submit a written report of their accident to the local police department.

Other sources of accident information made available to the records agency in some states and cities are coroners' reports, reports from garages, and data on fatal accidents from the local vital statistics agency.

Standard accident report forms. The National Safety Council has developed standard specifications for police and driver traffic accident report forms. These are specifications of content rather than format and arrangement. They provide for all factual data needed by public officials to administer basic traffic safety programs. Model forms meeting these specifications, available from the National Safety Council, are shown in Figures 10.27 and 10.28.

A modification of this form has been developed for use by drivers. This driver form has two separable parts: one for instructions, which is detached by the person making

---

[38] COMMITTEE ON UNIFORM TRAFFIC ACCIDENT STATISTICS, *Manual on Classification of Motor Vehicle Traffic Accidents* (Chicago: National Safety Council, 1962).

POLICE REPORT OF MOTOR VEHICLE TRAFFIC ACCIDENT

National Safety Council
Chicago

TIME	DATE OF ACCIDENT ...., 19 .... **Day of Week** .... Hour .... A.M. .... P.M.	

LOCATION

PLACE WHERE
ACCIDENT OCCURRED: County .... **City, town or township** .... State ....

If accident was outside city limits,
indicate distance from nearest town .... miles ☐ ☐ ☐ ☐ of ....
North   S   E   W   City or Town

ROAD ON WHICH
ACCIDENT OCCURRED ....
Give name of street or highway number (U.S. or State). If no highway number, identify by name.

AT ITS INTERSECTION WITH ....
Name of intersecting street or highway number

IF NOT AT INTERSECTION .... feet ☐ ☐ ☐ ☐ of ....
North   S   E   W   Show nearest intersecting street or highway, house no., bridge, RR crossing, alley, driveway, culvert, milepost, underpass, or other landmark.

DO NOT WRITE
IN THIS SPACE

No. ....

VEHICLE NO. 1

VEHICLE .... **License Plate** ....
Year   Make   Type (sedan, truck, taxi, bus, etc.)   Year   State   Number

Parts of vehicle damaged .... Vehicle removed to: .... By: ....

OWNER .... Address ....
Print or type FULL name   Street or R.F.D.   City and State

DRIVER .... Address .... | AGE | SEX | INJURY |
Print or type FULL name   Street or R.F.D.   City and State
Driver's .... Regular Operator's License ☐   Date of
License .... Other Type License ☐   Birth ....
State   Number   Specify Type and/or Restrictions   Month, Day, Year

Total number vehicles involved ....

OCCUPANTS
Front Center .... Address ....
Name   Street or R.F.D.   City and State

Front Right .... Address ....

Rear Left .... Address ....

Rear Center .... Address ....

Rear Right .... Address ....

VEHICLE NO. 2 or PEDESTRIAN

VEHICLE .... **License Plate** ....
Year   Make   Type (sedan, truck, taxi, bus, etc.)   Year   State   Number

Parts of vehicle damaged .... Vehicle removed to: .... By: ....

OWNER .... Address ....
Print or type FULL name   Street or R.F.D.   City and State

DRIVER .... Address .... | AGE | SEX | INJURY |
(or Pedestrian)   Print or type FULL name   Street or R.F.D.   City and State
Driver's .... Regular Operator's License ☐   Date of
License .... Other Type License ☐   Birth ....
State   Number   Specify Type and/or Restrictions   Month, Day, Year

OCCUPANTS
Front Center .... Address ....
Name   Street or R.F.D.   City and State

Front Right .... Address ....

Rear Left .... Address ....

Rear Center .... Address ....

Rear Right .... Address ....

FIRST AID
GIVEN BY: ....   Injured
Taken to: ....

DAMAGE TO PROPERTY
OTHER THAN VEHICLES ....
Name object and state nature of damage

Name and address of
owner of object struck ....

WITNESSES
Name .... Address ....

Name .... Address ....

CODE FOR INJURY
(Use only the most serious one in each space for injury.)

K–Dead before report made.
A–Visible signs of injury, as bleeding wound or distorted member; or had to be carried from scene.
B–Other visible injury, as bruises, abrasions, swelling, limping, etc.
C–No visible injury but complaint of pain or momentary unconsciousness.
O–No indication of injury.

AGE	SEX

Figure 10.27. Standard police accident report form (front side). (Courtesy National Safety Council.)

the report, and one for motor vehicle liability insurance information, which is used by the state for financial responsibility purposes.

Standard state and city summary forms have also been developed by the Committee on Uniform Traffic Accident Statistics in order to provide for uniform periodic state-wide and citywide summaries.

KIND OF LOCALITY (Check one)	WEATHER (Check one)	WHAT DRIVERS WERE GOING TO DO BEFORE ACCIDENT

Driver No. 1 was headed ☐ ☐ ☐ ☐ on .... (Street or highway)
North S E W

Driver No. 2 was headed ☐ ☐ ☐ ☐ on .... (Street or highway)

KIND OF LOCALITY (Check one)
☐ Apartments, Stores, Factories
☐ One-family homes
☐ Farms, Fields
☐ No marginal development

ROAD SURFACE (Check one)
☐ Dry
☐ Wet
☐ Snowy or Icy
☐ .... Specify other

LIGHT CONDITIONS (Check one)
☐ Daylight
☐ Dawn or dusk
☐ Darkness

ROAD TYPE (Check one or more)
Driver 1 2
☐☐ 1 driving lane
☐☐ 2 driving lanes
☐☐ 3 driving lanes
☐☐ 4 or more lanes
☐☐ Divided roadway
☐☐ Expressway, parkway, toll road

WEATHER (Check one)
☐ Clear
☐ Raining
☐ Snowing
☐ Fog
☐ .... Specify other

TRAFFIC CONTROL (Check one or more)
☐ Stop sign
☐ Stop-and-go signal
☐ Officer or watchman
☐ R.R. gates or signals
☐ .... Specify other
☐ No traffic control

ROAD CHARACTER (Check two)
☐ Straight road
☐ Curve
☐ Level
☐ On grade
☐ Hillcrest

Driver 1 2 (Check one for each driver)
☐☐ Go straight ahead
☐☐ Overtake
☐☐ Make right turn
☐☐ Make left turn
☐☐ Make U turn
☐☐ Slow or stop
☐☐ Start in traffic lane
☐☐ Start from parked position
☐☐ Back
☐☐ Remain stopped in traffic lane
☐☐ Remain parked

WHAT PEDESTRIAN WAS DOING
Pedestrian was going ☐☐☐☐ (Check one) N S E W
☐ Along
☐ Across or into .... (Street name, highway No.)
From .... To .... (N.E. corner to S.E. corner, or west to east side, etc.)
☐ Crossing or entering at intersection
☐ Crossing or entering not at intersection
☐ Getting on or off vehicle
☐ Walking in roadway—with traffic
☐ Walking in roadway—against traffic
☐ Standing in roadway
☐ Pushing or working on vehicle
☐ Other working in roadway
☐ Playing in roadway
☐ Other in roadway
☐ Not in roadway

CONTRIBUTING CIRCUMSTANCES
Driver 1 2 (Check one or more for each driver)
☐☐ Speed too fast
☐☐ Failed to yield right of way
☐☐ Drove left of center
☐☐ Improper overtaking
☐☐ Passed stop sign
☐☐ Disregarded traffic signal
☐☐ Followed too closely
☐☐ Made improper turn
☐☐ Other improper driving
☐☐ Inadequate brakes
☐☐ Improper lights
☐☐ Had been drinking

SHOW NORTH BY ARROW ◯

INDICATE ON THIS DIAGRAM WHAT HAPPENED

Street or highway

Street or highway

Street or highway

DESCRIBE WHAT HAPPENED:
(Refer to vehicles by number) ....

POLICE ACTIVITY
Time notified of accident .... Date Hour
☐ A.M. ☐ P.M.
What was the source of accident information? .... (Officer at scene, No. 1 driver contacted station, both drivers contacted station, etc.)

Arrests
Name .... Charge ....
Name .... Charge ....

Other action taken: ....

SIGN HERE ....
Officers rank and name   Badge No.   Department   Date of report

Figure 10.28. Standard police accident report form (back side). (Courtesy National Safety Council.)

FILING PROCEDURES

To facilitate engineering use of accident data, the original reports of accidents must be available by location of accident occurrence.[39] Some authorities prefer to file original reports chronologically and provide a location cross-reference index file.

---

[39] *Highway Safety Program Manual,* Volume 10, "Traffic Records" (Washington, D.C.: U.S. National Highway Traffic Safety Administration, 1969).

In cities, the manual location file is generally an alphabetical file of street names, with intersection accidents filed behind a primary guide bearing the name of the street that comes first alphabetically, and a secondary guide bearing the name of the intersecting street or streets. Accidents between intersections are filed immediately behind a primary guide for the street on which the accident occurred, with subdivisions by block numbers as needed. Electronic files are maintained as shown in Figure 10.29.

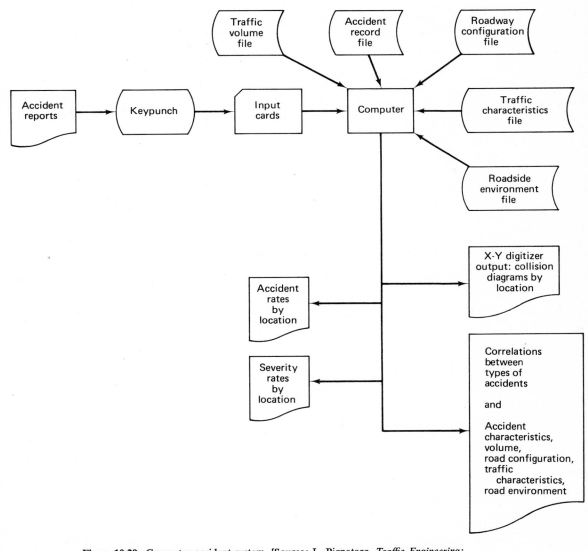

Figure 10.29. Computer accident system. [Source: L. Pignataro, *Traffic Engineering: Theory and Practice* (Englewood Cliffs, N.J.: Prentice-Hall, Inc., 1973), p. 279.]

The location of rural accidents is usually indexed first by county of occurrence, then by township, then by route number, and finally by route section. Mileage indicators are coming into increasing use and are used to locate accidents to 0.1 mile or less. These may be supplemented by logs of the highway.

In states, if original accident reports are filed by location for engineering purposes, a brief abstract is furnished the driver licensing agency for inclusion in the permanent driver record files. Another record of the accident, usually a detachable part of the driver's accident report goes to the financial responsibility division. If the original accident report is not readily available to the traffic and highway engineers, a copy of the pertinent data is obtained for the engineering location file.

In many states and large cities, electronic data processing provides flexibility and economy in making available the data necessary to develop accident prevention programs and to measure the effectiveness of such programs.

Coordination with accident records agency. Costly duplication of efforts can be avoided by designating a single agency within a governmental jurisdiction to be responsible for the collection and primary analysis of traffic accident data. A central accident records agency regularly prepares routine as well as specially requested summaries and reports for other official agencies having a responsibility in traffic control and accident prevention, including motor vehicle administration, engineering, enforcement, and educational activities. In some jurisdictions, the using agencies assign representatives to the central records agency who work on analyses for their respective departments under the supervision of the central records agency.

TRAFFIC ENGINEERING STUDIES OF ACCIDENT RECORDS

Accident analysis. Accident reports from drivers and police supply the necessary basic data for analysis. Periodic routine summaries and special analyses are essential to guide the safety activities of all agencies having a responsibility and interest in highway transportation and traffic safety.[40]

Accident spot maps. An accident spot map furnishes a quick visual index of the location of accident concentrations. The most common spot map is the one showing the location of accidents by pin, pasted spots, or symbols on the map. This map is posted currently as reports are received; different shapes, sizes, or colors are used to indicate different types and severities of accidents. To avoid distortion by unusually severe accidents involving multiple casualties (deaths and injuries), one spot should represent one accident instead of one casualty. The legend should be as simple as possible; no more than four or five types, sizes, or colors of spots should be used.

A simple street map at a scale of 1 : 5,000, showing streets and a few topographical features, is most satisfactory. In rural areas, a map at a scale of 1 : 50,000 (1 : 25,000 in congested areas near cities) is generally used. The State Highway Department "planning survey" county maps are usually the most desirable maps for this purpose.

Accident spot maps are ordinarily maintained for the calendar year. The previous year's map is held over during the current year in order to compare the experience with that of the current year. The map is then photographed and the picture is filed. Other methods of maintaining the spot map have been used successfully, including

---

[40] *Highway Safety Program Manual*, Volume 9, "Identification and Surveillance of Accident Locations" (Washington, D.C.: U.S. National Highway Traffic Safety Administration, 1969).

moving 12-month maps (on which January of the previous year is removed from the map before January of the current year is spotted, etc.), seasonal maps covering three months, and perpetual maps from which spots are removed only when a study or a physical change at a location is completed.

Types of special spot maps sometimes used are as follows:

1. Pedestrian-accident spot map.
2. Night-accident spot map.
3. School-child accident spot map.
4. Spot map of accidents involving nonresident drivers.
5. Spot map of accidents involving drinking drivers.
6. Residence spot map of drivers involved in accidents.
7. Residence spot map of pedestrians involved in accidents.
8. Spot map of employment of persons involved in accidents.
9. Spot map of origin of trip on which accident occurred.
10. Spot maps showing accidents involving specified vehicle types.

High-accident frequency location lists. With the aid of accident spot maps and the location file or electronic data-processing techniques, high-accident frequency locations are readily detected for detailed analysis and study. An annual or more frequent periodic listing of such locations, along with the location's traffic volume (to indicate exposure), serves as a guide for the detailed study of engineering alternatives. Details of the techniques are presented in Chapter 9.

Collision diagram and strip diagram. A collision diagram illustrates graphically, by means of directional arrows and symbols, the paths, types, and severities of collision of vehicles and pedestrians involved in accidents. Diagrams may be prepared for intersections or locations between intersections, as illustrated in Figure 10.30.

Collision diagrams are used to study accident patterns in order to determine the kinds of remedial measures required and the results that follow their application. When collision diagrams for equal periods before and after corrective treatment are compared, they show the types of accidents that have been eliminated, those that have continued to occur, and any new types that have developed. Strip diagrams are similar to collision diagrams but are prepared for homogeneous sections of highways several miles long.

Collision diagrams are seldom drawn to scale, and in the case of strip diagrams, distances between high-accident frequency locations may be shortened. The diagrams are schematic and arrows do not show exact paths because they would overlap or be otherwise confusing. It is essential, however, to show driveways and other physical features that can cause conflicts in the traffic stream. All roadway design features (channelization, markings, and other traffic control devices) are shown. On one of the arrows representing each accident the date and the time of day to the nearest hour is shown. Unusual conditions, such as "intoxicated driver," "ice," etc., should be noted. Fixed objects that are struck should be indicated. The generally approved and recommended symbols of representation are shown in Figure 10.30.

Condition diagrams. A condition diagram is a scaled drawing of the important physical conditions at a high-accident frequency location. It is used in conjunction with the collision diagram as an aid to interpreting accident patterns. Commonly used

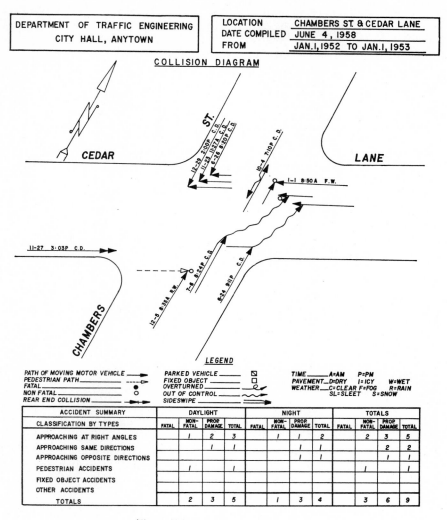

Figure 10.30. Collision diagram.

scales range from 1 to 100 to 1 to 250. All important features that can affect traffic movement are shown (see Figure 10.31). The following are usually shown:

1. Curb lines and roadway limits.
2. Property lines.
3. Sidewalks and driveways.
4. View obstructions.
5. Physical obstructions in and near the roadway.

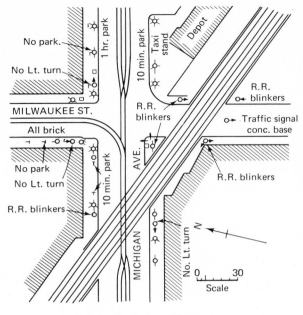

Existing physical conditions

Figure 10.31. Intersection condition diagram.

ENVIRONMENTAL STUDIES

In response to the National Environmental Policy Act of 1969, the traffic engineer may find it necessary to have access to or to obtain data on existing noise, air, and water quality and on ecological conditions in the region or at specific sites. Specialists should conduct all of these studies, although the engineer can learn to use a sound meter for gross measurements of noise. Techniques and details of analysis are available in recent publications.[41]

GENERAL COMMENTS

Proper application of the traffic studies described in this chapter will provide a factual basis for traffic engineering decisions, thus facilitating official and public acceptance. It is important that accurate and complete information be collected and

[41] WILLIAM J. GALLOWAY, WELDEN E. CLARK, and JEAN S. KERRICK, *Highway Noise: Measurement, Simulation, and Mixed Reactions*, National Cooperative Highway Research Program, Report 78 (Washington, D.C.: Highway Research Board, 1969).

COLIN, G. GORDON, WILLIAM J. GALLOWAY, B. ANDREW KUGLER and DANIEL L. NELSON, *Highway Noise: A Design Guide for Highway Engineers*, National Cooperative Highway Research Program, Report 117 (Washington, D.C.: Highway Research Board, 1971).

CYRIL M. HARRIS, ed., *Handbook of Noise Control* (New York: McGraw-Hill Book Company, Inc., 1957).

THEODORE J. SCHULTZ, "Noise Assessment Guidelines: Technical Guidelines" (prepared for the U.S. Department of Housing and Urban Development). (Washington, D.C.: U. S. Government Printing Office, 1972).

analyzed, that gathering of superfluous information merely because it might prove useful eventually be avoided, and that procedures should be initiated that will provide for a sound traffic engineering planning and operations program on a continuing basis.

REFERENCES FOR FURTHER READING

Manual of Traffic Engineering Studies (1964) Third Edition, edited by Donald E. Cleveland, Institute of Traffic Engineers.

Better Transportation for Your City (1958)

Supplemented by Series of Procedure Manuals (1958)
  1A. *Determining Street Use*
  2A. *Origin-Destination and Land Use*
  2B. *Conducting A Home Interview Origin-Destination Survey*
  3A. *Measuring Traffic Volumes*
  3B. *Determining Travel Time*
  3C. *Conducting a Limited Parking Study*
  3D. *Conducting a Comprehensive Parking Study*
  4A. *Measuring Transit Service*
  5A. *Inventory of the Physical Street System.*

National Committee on Urban Transportation, Public Administration Service, 1313 East 60th Street, Chicago, Illinois.

Parking Principles (1971) Special Report Number 125, Highway Research Board, Washington, D.C.

Research on Road Traffic (1965), Road Research Laboratory, Her Majesty's Stationery Office, London.

Chapter 11

COMPUTER APPLICATIONS

Robert L. Bleyl. Associate Professor, Department of Civil Engineering, The University of New Mexico, Albuquerque, New Mexico.

The rapid and widespread growth of computer utilization, especially in the field of traffic and transportation engineering, has made it necessary for traffic engineers to have a basic understanding of computers. This chapter is concerned with the application of electronic digital computers to the solution of traffic and transportation engineering problems. Two objectives will be served in this chapter:

1. To provide a brief sketch of the fundamentals that lie at the root of every computer application problem.
2. To show ways in which the computer is currently being used in the traffic and transportation engineering profession.

The material presented here does not attempt to serve as a primer for those who wish to learn how to program a computer and it does not delve into the interesting and intricate mathematics behind some of the more sophisticated applications. Instead, the attempt has been to persuade the traffic and transportation engineer that the computer is simply a tool that he may advantageously use in pursuit of his profession and to show him interesting ways in which the computer is currently being used in his field.

THE COMPUTER

There is nothing magical about an electronic digital computer. Like using the slide rule and desk calculator, solving problems using a computer consists basically of providing the numbers (or data) to be used in the computations, specifying the sequence of the operations to be performed, and reading the answers. The computer must be given precise instructions for every operation it performs.

The computer is capable of providing an answer to a problem in various ways. It can print the result on an electric typewriter or a high-speed printer, punch the solution into a data processing card, write the output on magnetic or paper tape, display the results on a cathode-ray tube or a plotter. By using an appropriate interface, the answer can control electrical circuits, for example, controlling traffic signals and ramp metering devices.

ADVANTAGES

Electronic digital computers have certain characteristics that make them particularly useful:

1. The speed at which they perform calculations is extremely fast and is measured in microseconds, or millionths of seconds.
2. They are extremely accurate.
3. Since the instructions to be followed, called the *program*, are read into and stored in the memory of the computer, repetitious problems are easily solved by simply instructing the computer to go back and follow the same instructions again and again.

DISADVANTAGES

The costs involved in using the computer relate to time and money:

1. The process of developing the programs is comparatively slow and time consuming. It may be quicker and less expensive in the long run to use a desk calculator to solve problems that may be run only once or twice.
2. Computers are expensive to buy or lease. The rental cost for typical machines used by traffic engineers ranges from a few hundred dollars per month upward to several hundred dollars per hour. Many organizations rent computer time from an organization that already has a computer.

COMPUTER OPERATION

The memory of an electronic digital computer might be compared to a series of pigeonholes. Each pigeonhole represents a storage location, called a *word* or *address*. Each storage location can contain one piece of information—a number to be used as data or an instruction.

In order to solve a problem, the *program*, or list of instructions to be followed in solving the problem, is first stored in the computer memory in sequential addresses. The program is then executed by following the stored instructions one after the other. Some instructions read data and store it elsewhere in the computer memory. Other instructions transmit data already stored in the computer memory to the printer or other output device. Arithmetic instructions perform computations using the stored data. Control instructions compare data and alter the sequence in which the instructions are executed. Eventually, the sequence of instructions should lead to an instruction to stop.

Figure 11.1 is an example representing the memory of a computer that contains a simple program to calculate average speed. The execution of this program begins by obeying the first instruction stored in the computer memory, i.e., place a zero in cell 8. Since an address can contain only one piece of information at a time, that which was previously stored in cell 8 will now be replaced by a zero. The content of a storage location is never erased but is changed only when it is replaced by another bit of information. The computer then moves to the next sequential instruction, cell 2, and then on to cell 3. In cell 3 the computer has to make a decision. If there are no more observations to be read, the computer goes to cell 10 for its next instruction. If, however, there are additional observations waiting to be read, the computer continues to the next sequential instruction.

When the computer reaches cell 7, the order of execution is altered by a control instruction to go back to cell 3 for the next instruction, thus forming a *loop* which, in this example, would be followed once for each speed observation. After the last

1	2	3	4
Place a zero in Cell 8.	Place a zero in Cell 9.	If no more speed observations to be read, jump to Cell 10 to continue.	Read a speed observation and place it in Cell 16.
5	6	7	8
Add the number in Cell 16 to the number in Cell 8.	Add the number one (1) to the number in Cell 9.	Jump to Cell 3 for the next instruction.	
9	10	11	12
	Divide the number in Cell 8 by the number in Cell 9. Place answer in Cell 15.	Return carriage on typewriter and print "THE AVERAGE SPEED IS."	Print on type writer the number in Cell 15.
13	14	15	16
Print on type writer "MILES PER HOUR."	Stop.		

Figure 11.1 Example of pigeonhole program to calculate average speed.

speed observation has been read and processed, the instruction in cell 3 will cause the computer to jump out of this loop by going to cell 10 for the next instruction. From that point the instructions will be executed sequentially until cell 14 is reached at which point the computer is instructed to stop.

COMPUTER ACCESSIBILITY

At this point in time computer facilities are widely and readily available to those who do not have their own equipment. These computer facilities can be classified into three groups:

1. Governmental agencies, business corporations, educational institutions, and others who have purchased or leased computer equipment. As a sideline to their incorporated purpose, these organizations often make their equipment available for use by others during idle periods, for example, during the second or third shift.
2. Data processing organizations that have been established in numerous cities throughout the country solely for the purpose of providing computer services. Usually they specialize either in engineering and scientific applications or in commercial (business) applications. Listings of local data processing services can be obtained by consulting the yellow pages of the local telephone directory.
3. Computing centers established by computer manufacturers. Their services are generally of a sales-promotion nature.

BASIC COMPUTER ELEMENTS

A complete computer system consists of three primary elements:

1. There must be some way of getting the program and data into the computer. This is accomplished by various kinds of input devices.

2. There must be some way of getting the results or answers out of the computer. This is accomplished by various kinds of output devices.
3. The third element of the system is the computer itself.

Figure 11.2 is a photograph of a basic computer system. The unit on the right is a card reader, the input device. The unit in the center is the computer. The unit on the left is a printer, the output device.

Figure 11.2. Basic computer system consisting of high-speed printer, main computer, and card reader. (Source: IBM Corporation, White Plains, New York.)

INPUT DEVICES

There are many ways of providing information to the computer depending on the information to be supplied:

1. Perhaps the most common input device is a card reader, which senses the location of coded holes punched in data processing cards.
2. Paper tape readers also sense the location of coded holes punched in paper tape. In both cases, the position of the holes is related to specific characters. Each column on a data processing card and each row across a paper tape corresponds to one character.
3. Special electric typewriters can be used to feed information to the computer. They are often used at locations remote from the computer; the link between the typewriter terminal and the computer could be a regular telephone line.
4. Other input media include magnetic tape, magnetic disc, and magnetic drum. A light pen and a cathode-ray tube can be used to supply input data. At times, a small computer is coupled to a much larger computer. In this case, the small computer feeds information from its memory to the larger computer. Information from traffic detectors can be transmitted to the computer by using an appropriate interface.

OUTPUT DEVICES

There are also numerous ways of obtaining results from the computer:

1. The high-speed printer is probably the most common method.

2. The remote typewriter terminal used for input can also serve as the output device.
3. Other output media include punched data processing cards, magnetic tape, magnetic disc, and magnetic drum. The results can be transmitted to another computer, displayed on a cathode-ray tube, or drafted on a plotter. The output can be channeled to an interface causing traffic signals and ramp metering devices to respond.

An important aspect of input and output devices is their response time. As indicated earlier, the speed at which computers execute instructions is measured in millionths of a second.

1. Information transmitted to and from the computer via a typewriter is limited to the speed of the typist or the speed of the typewriter, possibly five characters per second.
2. Data processing cards can be used to transmit information at the rate of roughly 1,000 characters per second.
3. Regular high-speed printers can print up to 2,500 characters per second.
4. Information transfer using magnetic tape occurs at the rate of approximately 100,000 characters per second.

Because of the slow speed at which most input and output devices operate, when compared to the speed of the computer in performing operations, input/output buffering is often employed. Buffering might be considered to be an intermediate storage area between the input or output device and the computer. On output, for example, information to be printed would be placed by the computer in the output buffer area. While the relatively slow typewriter or high-speed printer printed the information from the buffer, the computer could be performing other calculations, thus reducing the time required to execute the program. Buffers commonly used include special buffer storage, magnetic tape, and one or more smaller computers connected directly with the main computer.

MAIN COMPUTER

The main computer consists of three basic elements:

1. The *memory unit* stores the program and the other information in which the user is interested.
2. The *arithmetic unit* performs the adding, subtracting, multiplying, dividing, and logical operations, such as determining if one number is larger than, smaller than, or equal to a second number.
3. The *control unit* monitors the flow of information to and from the memory and arithmetic units, from the input units, and to the output units.

As far as the traffic engineer is concerned, these three units may be considered to comprise one big black box, the main computer. Figure 11.3 summarizes the basic elements of a computer system.

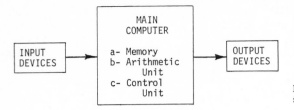

Figure 11.3. Basic elements of a computer system.

PROGRAMS AND PROGRAMMING

In order to have the computer accomplish what the traffic engineer desires, the computer must be instructed as to the specific steps it must take. The list of sequential instructions to be followed by the computer is called the *program*. The process of preparing the program is called *programming*.

Following are the four specific steps that are used in programming:

1. *Problem formulation and analysis.* This step involves a precise formulation of the problem, usually in mathematical terms. The kind of information to be supplied as input to the program and the information to be output by the program must be identified. The formulas or algorithms necessary to calculate the output information from the input information must also be specified.
2. *Outlining the sequence of operations.* The sequence of computer operations to be performed on the input data in order to achieve the desired results must be outlined. As an aid in visualizing the steps involved in the solution of the problem, a detailed flow chart is prepared for complicated programs; a list of steps outlining the logical flow of processes is prepared for simple programs.
3. *Program writing and coding.* If the computer is to solve the problem, it must be given precise instructions for every operation. Computer instructions following the flow chart or list must be written in a language that can be understood by the computer.
4. *Debugging.* Once the program has been written and put in a form that can be read by the computer hardware (usually keypunched into data processing cards), the program must be tested to insure that it operates properly and arrives at the correct solution to the problem. Keypunching errors, coding errors, and errors in logic must all be located and corrected. As a final test, the accuracy of the program is checked by comparing the computer results with hand calculated results. When all errors have been corrected and the program has been verified to yield accurate results, the program can then be placed into production.

PROGRAMMING LANGUAGES

There are three levels of programming languages used in writing computer programs:

1. *Machine languages.* Each computer has a set of codes that is used to instruct the computer. The set of symbols which the hardware can interpret for execution of the program is called the *machine language*. Machine languages

not only vary between computer manufacturers, but they also vary between models of a given computer manufacturer. To write a program in machine language would be a very difficult task because machine languages are extremely complicated and difficult to learn.

2. *Symbolic assembly languages.* In order to simplify the process of coding the specific instructions, an innovation was made enabling the computer to write its own machine language instructions based on key instructions it had been given. This development was the *symbolic assembly language* which employs mnemonic symbols for instructions and addresses. Although these languages greatly simplified the task of coding the specific instructions, they still required a knowledge of the number representation in the computer and the use of registers; they were also different for different computers.

3. *Compiler languages.* Higher-level languages, known as *compiler languages*, were developed in order to overcome the programming disadvantages of the symbolic assembly languages. Compiler languages are somewhat standardized so that a program could be compiled for different machines without having to rewrite or alter the program. No knowledge of machine language, number representation, or computer registers is necessary to write a program using a compiler language. Although the machine language instructions generated from a compiler language program are not as efficient as those generated from a symbolic assembly language program, the savings in programming time using a compiler programming language more than offsets the resulting inefficiency, except for programs that will be run hundreds of times.

   a. FORTRAN. Of the more than 200 different programming languages that have now been developed, perhaps the most well-known programming language is FORTRAN, one of the compiler languages. It so closely resembles the language of mathematics that traffic engineers have little difficulty coding problems for the computer after they have had a brief period of instruction.

   b. ALGOL. ALGOL is another widely used compiler language. It is the result of an attempt to develop a more powerful and more flexible language than FORTRAN. It has gained widespread acceptance in European countries and other parts of the world, but it has not been widely accepted in the United States probably because FORTRAN is so extensively used and many organizations have a heavy investment in FORTRAN programs.

COMPUTER CENTER MANAGEMENT

There are different modes of operation for a computer staff:

1. *Closed shop.* In the closed-shop arrangement, the computer center staff does all the problem analysis, programming, and operation of the computer equipment.

2. *Open shop.* In the open-shop arrangement, the computer user does his own analysis, programming, and operation of the equipment.

3. *Mixed shop.* Many organizations have an operating philosophy that falls between the closed-shop and open-shop arrangements. Because of differences in their backgrounds and education, it is sometimes difficult for a traffic engineer to communicate to a programmer his ideas about a program to be

developed. Similarly, it is difficult for the programmer to grasp the magnitude and scope of the traffic engineering problem without a background in traffic engineering. The general feeling is that it is more efficient to teach the traffic engineer to do his own programming than it is to teach the programmer to be proficient in traffic engineering and the many other related fields. Consequently, in many organizations programming is done by the traffic engineer, or user, and operation of the equipment is done by the computer section staff. Consultation and guidance in developing efficient programs may be provided by the computer section staff, when requested by the user. Unless the user has had many years of programming experience, he would be well-advised to seek the advice of the computer center staff during the development of the computer programs he writes.

DOCUMENTATION

A most important part of developing a computer program is preparing proper documentation for the program developed. Documentation is a brief report describing the program and its function, application, scope, computational approach, limitations, and requirements. The dissemination of the computer program to other users requires such documentation. Furthermore, unless a program is properly documented, it may easily become worthless if the programmer leaves for other employment. A program is considered to be only as good as its documentation.

The documentation report should contain essential information about the program. A suggested outline for a typical traffic engineering program is given below:

1. Title, author, date.
2. Statement of the problem solved by the program and the scope of the problem.
3. Computational approach used, including equations, assumptions, and simplifications made to facilitate programming. A flow chart may be included when appropriate.
4. Program limitations and machine requirements.
5. Data preparation requirements. Coding forms may be illustrated and coding instructions outlined.
6. Special instructions for the computer operator when necessary.
7. Interpretation of computer output, including error messages.
8. Illustrations showing the various stages of solution for a sample problem.
9. Listing of the program statements (optional).

PROGRAM EXCHANGE

The two biggest expenses in computer applications are the cost of the computer equipment and the cost of preparing programs. In some cases the programming costs are the greater of the two costs. Because of the high programming costs, many organizations have sought to exchange or purchase programs from other users. In some cases this approach to program acquisition has been successful. There may, however, be certain disadvantages:

1. At times a program written by another group may not do exactly what the new user desires. Program modifications are then necessary in order to adapt

the program to the peculiarities and desires of the new user. If too many alterations have to be made, it may be quicker and less expensive to develop the program from scratch instead of trying to follow the intricacies of someone else's work.

2. There may be differences in computer equipment. A computer program that requires three tape drives, for example, may be useless on a system having only two tape drives. A program requiring a certain amount of computer memory cannot be run on another computer if that computer has less than that amount.

3. Program documentation is often lacking or is inadequate. As a result, the new user is unable to use the program to its fullest extent.

There are several program sources that may have programs of interest to the traffic and transportation engineer:

1. The Federal Highway Administration.
2. The traffic engineering and the transportation planning units of state highway departments and some regional, city, and county traffic engineering or planning units.
3. In England there is the National Computing Center library of computer programs.
4. HEEP (Highway Engineering Exchange Program).
5. Numerous articles published in *Traffic Engineering*, *Transportation Research Record*, and other periodicals describing programs that have been developed.

EXAMPLE APPLICATION AND PROGRAM

This section will present a simplified example of a computer application showing the various steps in the process of developing the computer program. This example is included to illustrate the ease and simplicity with which this task can be accomplished.

An efficient computer application does not begin with the analysis of the data that has been collected; it begins before the collection of the data. Processing data can be greatly simplified if the data are originally recorded on forms ready for data processing. For some applications, mark-sense forms, which can be read directly by machine, are appropriate. For most traffic engineering applications, however, it is sufficient to record the data on a form ready for keypunching. In any case, thought must be given to the way in which the data can be most easily and most efficiently processed. The home-interview form used for origin-destination studies (Chapter 10) is an example of a form easily used for keypunching.

The application selected for this illustration is a radar speed check analysis. As mentioned earlier in this chapter, programming this and every application consists of four steps: (1) problem formulation and analysis, (2) outlining the sequence of operations, (3) program writing and coding, (4) debugging.

PROBLEM FORMULATION AND ANALYSIS

As a part of the problem analysis, the following questions should be considered:
1. What will be the input information?

2. In what form will this information be most efficiently presented considering the entire task of gathering and analyzing the data?
3. What information do we want to obtain by using this program?
4. What formulas can be used to compute these values?
5. What form of the equation should we use?

For this simplified example let us assume that radar speeds will be observed and recorded manually to the nearest even mile per hour (kilometer per hour in the metric system). Instead of recording each individual speed and later having to keypunch each speed observed, it would be more efficient, and nothing would be lost, to tally the number of observations at each speed. At the completion of the study, it would take less time for the field study personnel to manually count and code the number of tallies at each speed than it would take to keypunch each individual observation. In keeping with these suggestions, a field study form similar to that illustrated in Figure 11.4 might be developed. The left side of this form would be used in the field study. The right side of the form contains space for the summarized number of observations, which is the information that would be keypunched for computer

RADAR SPEED CHECK

LOCATION _Example   Test   Site_____

SPEED	TALLY OF OBSERVATIONS	SPEED (1-5)	NO.OBS. (6-10)
80		80	
78		78	
76	/	76	1
74		74	
72	//	72	2
70		70	
68	//	68	2
66		66	
64	//	64	2
62	⊦⊦⊦ ⊦⊦⊦	62	10
60	⊦⊦⊦ ⊦⊦⊦ ⊦⊦⊦	60	15
58	⊦⊦⊦ ⊦⊦⊦ ⊦⊦⊦ ⊦⊦⊦ ⊦⊦⊦ /	58	26
56	⊦⊦⊦ ⊦⊦⊦ ⊦⊦⊦ ⊦⊦⊦ ⊦⊦⊦ ///	56	28
54	⊦⊦⊦ ⊦⊦⊦ ⊦⊦⊦ ⊦⊦⊦ ⊦⊦⊦	54	25
52	⊦⊦⊦ ⊦⊦⊦ ⊦⊦⊦ ⊦⊦⊦ ////	52	24
50	⊦⊦⊦ ⊦⊦⊦ ⊦⊦⊦ //	50	17
48	⊦⊦⊦ ⊦⊦⊦ ⊦⊦⊦ /	48	16
46	⊦⊦⊦ ⊦⊦⊦	46	10
44	⊦⊦⊦ //	44	7
42	⊦⊦⊦	42	5
40	///	40	3
38	///	38	3
36	/	36	1
34	/	34	1
32	/	32	1
30		30	
28		28	
26		26	
24		24	
22		22	
20		20	

Figure 11.4. Field form and sample data for example program.

processing. The card columns into which each line containing one or more observations would be keypunched are shown on the field form. Sample data are shown in Figure 11.4 and illustrate how a typical speed study might be recorded and summarized for data processing. Figure 11.5 illustrates these data after keypunching.

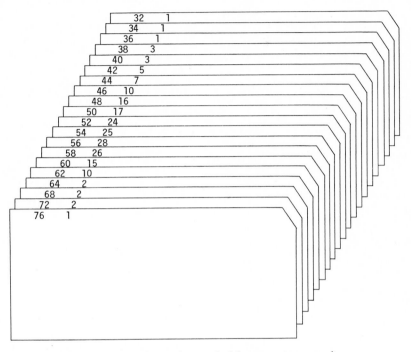

Figure 11.5. Example data keypunched for computer processing.

Let us assume for this example that we want to determine the average speed, standard deviation of speeds, standard error, and total number of observations. Of the various formulas that can be used to yield these results, the following seem to be most efficient for computer processing:

$$\text{Average speed} = \frac{\Sigma S}{N}$$

$$\text{Standard deviation} = \sqrt{\frac{N \Sigma S^2 - (\Sigma S)^2}{N(N-1)}}$$

$$\text{Standard error} = \frac{\text{standard deviation}}{\sqrt{N}}$$

where
  S = speed of each observation
  N = total number of observations

OUTLINING THE SEQUENCE OF OPERATIONS

The sequence of computer operations to be performed in achieving the desired results must be outlined. The steps for this simple program might be listed as follows:

1. Set accumulators to zero.
2. Read data using a loop and add values to accumulators.
3. Solve formulas.
4. Write results.
5. Stop.

Although a flow chart probably would not be constructed for a program as simple as this example, Figure 11.6 does illustrate the individual steps for this example in flow chart form.

PROGRAM WRITING AND CODING

A computer program written in FORTRAN programming language to solve this problem is shown in Figure 11.7. Below is an explanation of each step in the FORTRAN program. The numbers refer to the line numbers in Figure 11.7.

1. A zero is placed in the storage location to be identified throughout this program by the mnemonic FREQ. This location will be used to count the number of speed observations.

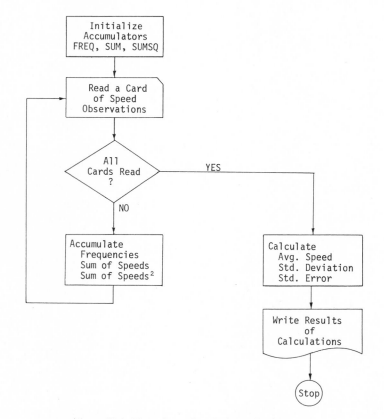

Figure 11.6. Flow chart of example application.

```
C        RADAR SPEED CHECK PROGRAM
C
1            FREQ = 0.
2            SUM = 0.
3            SUMSQ = 0.
4         10 READ(5,20,END=30) SPEED, OBS
5         20 FORMAT(2F5.0)
6            FREQ = FREQ + OBS
7            SUM = SUM + SPEED * OBS
8            SUMSQ = SUMSQ + SPEED * SPEED * OBS
9            GO TO 10
10        30 AVG = SUM / FREQ
11           SDEV = SQRT((FREQ*SUM-SUMSQ)/(FREQ*(FREQ-1.)))
12           SERROR = SDEV / SQRT(FREQ)
13           WRITE(6,40) FREQ, AVG, SDEV, SERROR
14        40 FORMAT('1SUMMARY OF SPEED CHECK DATA' /
             $       'ONUMBER OF OBSERVATIONS =', F6.0 /
             $       ' AVERAGE SPEED =', F6.2, ' MPH' /
             $       ' STANDARD DEVIATION =', F6.2, ' MPH' /
             $       ' STANDARD ERROR =', F5.2, ' MPH' )
15           STOP
16           END
```

Figure 11.7. FORTRAN program of example problem.

2. A storage location identified by SUM is also initialized at zero. This location will be used to sum the individual speed observations.

3. A zero is also placed in a third storage location identified as SUMSQ. This location will be used to sum the square of the individual speed observations.

4. This instruction is identified as statement number 10. It tells the computer to read two items from a data processing card. The first number is to be stored at a location identified henceforth as SPEED, and the second number is to be stored at a location referred to as OBS. The 5 in parentheses indicates the input device to be read. Unit 5 is the standard code for the card reader. The 20 in parentheses refers to a format statement, which indicates the way in which the two items to be read are positioned on the card. The END = 30 will send the computer to statement 30 when there are no more data cards left to read.

5. This statement is identified as statement 20. It identifies the way in which the data are positioned on the input card. The specification is 2F5.0 meaning that the data are located at the beginning of the card in two fields of 5 characters each. None of the digits is considered to be to the right of the decimal point.

6. The number stored at the location identified as FREQ (this storage location was set to zero in the first statement) is added to the number presently stored at the location identified as OBS and the result is stored back at the location identified as FREQ. Here we are adding up the number of speed observations.

7. The number presently stored at the location identified as SPEED (this value was just read from a data card) is multiplied by the number stored at the location identified as OBS and the resulting value is added to the number presently stored at SUM. The result is stored at SUM. Here we are summing the speeds.

8. The number stored at the location identified as SPEED is multiplied by itself, multiplied by the number stored at the location identified as OBS, and added to the number stored at SUMSQ. Here we are summing the square of the speeds.

9. This statement instructs the computer to return to statement number 10 for its next instruction. When it does so, the next card will be read from unit 5, the card reader, and the two values read will then replace the values presently stored at the two locations identified as SPEED and OBS. The same steps will be followed for this and the remaining cards as just described for the first data processing card. When there are no cards left to be read on unit 5, the END = 30 in statement number 10 will cause the computer to jump down to statement number 30 for the next instruction.

10. Statement number 30 divides the number stored at the location identified as SUM (which now contains the sum of all the observed speeds) by the number stored at the location identified as FREQ (which now contains the total number of observations) and stores the result in a location to be identified as AVG. This instruction calculates the average speed.

11. The standard deviation in this statement is calculated according to the formula determined in the analysis of the problem. SQRT takes the square root of the expression located within the outside set of parentheses. The result of all this arithmetic manipulation is stored at a location identified as SDEV.

12. The standard error is calculated by dividing the number stored at SDEV by the square root of the number stored at FREQ. The result is stored at a location identified as SERROR.

13. This statement instructs the computer to write the contents of four storage locations, FREQ, AVG, SDEV, and SERROR. The 6 in parentheses refers to the unit on which the results are to be written. Unit 6 is the standard code for the high-speed printer. The 40 in parentheses refers to a format statement, which will indicate the way in which the four values are to be positioned on the printed record.

14. This long, multiline statement is identified as statement number 40. It specifies the way in which the output is to be positioned on the printer. Characters that fall between apostrophes will be included on the printed record exactly as indicated in this format statement except for the first character on each line. The first character on each line is a carriage control character. The number 1 at the beginning of the format specification is the carriage control character to print that line at the top of a new page. The number 0 just below the 1 is the carriage control character to double space that line from the line above it. The blank character below the 0 is the carriage control character to single space that line from the line above it. The slash at the end of each line indicates the end of each printed line. The F6.0, F6.2, F6.2, and F5.2 indicate the location and size of the field where the four values are to be printed. The values are printed in the exact sequence listed in the previous write statement. In this case, there will be zero, two, two, and two digits following the decimal points for the four values respectively. The dollar signs in this statement indicate that the line on which they occur are continuations of the preceding line. The printer output obtained by processing the example data is illustrated in Figure 11.8. The printing of this entire output was produced by the write and format statements just described.

15. This statement terminates the execution of the computer program.

16. This statement is a signal to the compiler program that the end of the program statements has been reached. This statement is necessary since the STOP statement does not necessarily always fall as the last statement of the program.

SUMMARY OF SPEED CHECK DATA

NUMBER OF OBSERVATIONS =   199.
AVERAGE SPEED = 53.32 MPH
STANDARD DEVIATION =   6.25 MPH
STANDARD ERROR = 0.44 MPH

Figure 11.8. Printer output for sample data.

DEBUGGING

The fourth step in the programming process is the debugging stage. It is common to make errors in coding and keypunching the program. The commas, parentheses, and consistent spelling of the various mnemonics are critical. As a final test, since this is a new program just written, the accuracy of the four values just printed should be checked by manually calculating what the answers should have been.

Last, and perhaps most important, if this program were to be a production program, a documentation report would be prepared as outlined in an earlier section of this chapter.

SPECIFIC APPLICATIONS

There are hundreds of ways in which digital computers have helped in solving traffic engineering problems. The purpose of this section is to give the traffic engineer a glimpse of a few such applications. The examples that follow were not intended to be an exhaustive coverage of each subject; they were selected as illustrations of some of the more imaginative and interesting applications. The ideas and examples were obtained from program descriptions and documentation reports submitted by state highway departments, traffic consultants, universities, and other traffic and transportation planning agencies throughout the world.

Every successful computer application must satisfy one or more of the following functions:

1. It can provide *economy*. Economy is gained when less manpower is required to solve the problem, when the same program can be used over and over again, or when speed and accuracy of computation have an economic value.
2. It can provide *insight*. Insight is gained when the computer can be used to simulate situations in order to analyze the impact of various alternatives.
3. It can provide *feasibility*. Feasibility is gained when the problem could not be solved without using the computer because the results must be available a very short time after the input data are available or when millions of pieces of data must be manipulated with precision and reliability.

The computer applications presented in this section have been arranged into five functional categories.

COMPUTATIONAL APPLICATIONS

Computer applications that are computational in nature are differentiated from data processing applications in the extent of computations and the amount of data involved. Computational applications are those uses which usually have small amounts of data but require much calculation. They are primarily used in design and statistical analyses. Below are specific traffic engineering applications that are classed as computational programs:

1. *Designing guide signs.* The task of determining the legend positioning on highway guide signs is a difficult and tedious job because each guide sign is unique and there are numerous detailed rules and spacing criteria to be followed. The computer has been used to aid the traffic engineer in designing highway guide signs.

   At least one program[1] has been developed which not only determines the layout of all sign legend in accordance with commonly used spacing rules and criteria, but also causes a computer-driven plotter to draft the signs. The plots are made to any scale specified and the legend is shown in its true shape and proper position, as illustrated by the example plot in Figure 11.9.

Figure 11.9. Traffic guide sign designed and drafted by computer.

   One of the major benefits of these plots is the opportunity it affords the traffic engineer to visually check the balance and the positioning of the sign copy before the sign is constructed. These sign plots can be used in preparing the plans for signing contracts; their use aids the sign fabricators by showing them exactly what the fabricated sign should look like.

2. *Signal progression.* Chapter 17, Traffic Signals, describes trial-and-error and graphical methods for determining progressive timing of traffic signals. These methods can become involved under the following conditions:

   a. Signals are not uniformly spaced.
   b. Phase splits are not the same at all signals.
   c. Traffic speed patterns vary from one section of the system to another section.
   d. Traffic speeds differ by direction of travel.
   e. Travel distances between stop lines differ by direction of travel.
   f. Progression is unbalanced or proportional to directional traffic volumes.

   Computer programs have been designed to handle all of these variable elements in optimizing the widths of the progressive bands or other figures of merit. Output from these programs consists of printed tables and an intermediate file containing all the parameters necessary for a supplemental com-

---

[1] R. L. BLEYL and H. B. BOUTWELL, "A Computer Program for Guide Sign Designing and Drafting," *Traffic Engineering*, Vol. 38, No. 6 (1968), pp. 22–26, 57.

puter program to prepare a time-space diagram of the optimum solution. Figure 11.10 illustrates such a diagram prepared on the printer. In this instance, the progressive bands are later added by manually connecting the dots.

3. *Capacity computations.* The task of using the charts and tables available for making highway capacity calculations has been considered cumbersome and involved by some. One proposed solution to this problem has been the development of a series of five computer programs[2] to calculate the capacity at the following locations:

   a. Intersections
   b. Exclusive turning lanes
   c. Freeways and highways
   d. Ramps
   e. Weaving sections

Input for each program is coded on a single data processing card. Each program prints a summary of all input data, modification statements, if any, and the results.

DATA PROCESSING APPLICATIONS

Data processing applications serve primarily to reduce large amounts of raw data to a more convenient form. They are used in summarizing field observations, in preparing summary statistics, and in making graphs and plots of the data. Below are several specific applications that are classed as data processing programs:

1. *Speed check analysis.* One of the most common data processing applications is a program to summarize radar speed check observations. This application provides economy by being used repeatedly for each speed check undertaken.

    The field form used by one highway department, illustrated in Figure 11.11, is indicative of the efficiencies that can be incorporated into a computer application beginning at the data collection stage. Noteworthy characteristics of this field form include:

   a. It doubles as a field and computer coding form.

   b. Identifying conditions at the study location are coded at the top of the form. The codes are mnemonic representations that are interpreted by the program and printed as text (e.g., Weather code "C" would be printed "WEATHER—CLEAR TO CLOUDY").

   c. Speed observations are recorded by placing the appropriate vehicle type code, shown in the legend in the corner of the form, in the left-most blank box at the observed speed. This procedure produces a visual distribution of the observed speeds.

---

[2] ADOLF D. MAY, JR., GALE AHLBORN, and FREDERICK L. COLLINS, "A Computer Program for Intersection Capacity," *Traffic Engineering*, Vol. 38, No. 4 (1968), pp. 42–48.

    ADOLF D. MAY, JR., and ROBERT E. HOM, "Intersection Capacity for Exclusive Turning Lanes," *Traffic Engineering*, Vol. 38, No. 6 (1968), pp. 48–52.

    FREDERICK L. COLLINS and ADOLF D. MAY, JR., "A Computer Program for Freeway and Highway Capacity," *Traffic Engineering*, Vol. 38, No. 7 (1968), pp. 44–49.

    GALE AHLBORN, WILLIAM L. WOODIE, and ADOLF D. MAY, JR., "A Computer Program for Ramp Capacity," *Traffic Engineering*, Vol. 39, No. 3 (1968), pp. 38–44.

    WILLIAM L. WOODIE, GALE AHLBORN, and ADOLF D. MAY, JR., "A Computer Program for Weaving Capacity," *Traffic Engineering*, Vol. 39, No. 4 (1969), pp. 12–17.

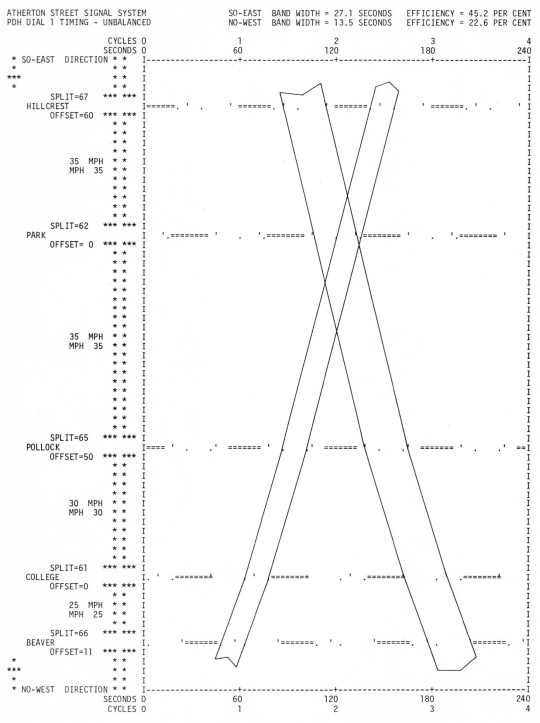

Figure 11.10. Time-space diagram prepared on a high-speed printer.

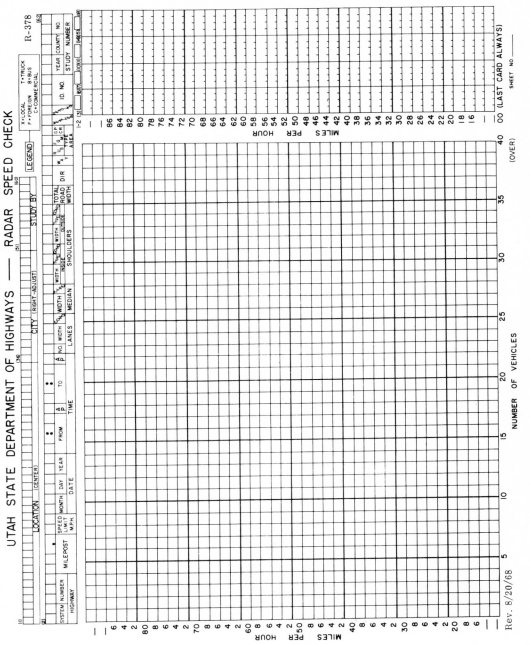

Figure 11.11. Efficient field form doubling as a keypunch coding sheet. (Source: Utah State Department of Highways, Salt Lake City, Utah.)

d. Space is provided at the top and bottom of the speed scale to include speeds falling beyond the limits on the form.

e. Heavy vertical rulings facilitate the process of tallying the observations at each speed for recording at the right side of the form. This manual summary process was considered more efficient than keypunching each observed speed individually.

f. Flexibility in summary categories is provided by including four summary columns to facilitate the summary of speeds by travel direction and/or by vehicle type, as appropriate.

Output from speed check programs usually include identifying conditions at the study site and the following statistics:

a. 85th (90th) percentile speed (maximum speed limit indication)

b. 15th (10th) percentile speed (minimum speed limit indication)

c. Average speed

d. Median speed

e. Standard deviation of speeds

f. Maximum speed

g. Percentages traveling faster than specified speeds

h. Sample size

2. *Intersection directional volume count.* Manual observation using a multi-counter board is the most efficient method yet devised to obtain a turning movement count at an intersection. Efficient computer methods, however, are being used in processing the count once it has been observed and recorded.

In order to eliminate the chances for errors in reading, transcribing, or keypunching the observed traffic volumes, a method of recording the count in the field that can be read directly by data processing equipment would be desirable. One of several possible approaches is to use "porta-punch" cards, which are the size of data processing cards, but which have prescored numbers that can be pushed out with a sharp pencil or stylus.

In preparing tabulations and tables for this and other applications, some thought should be given to sizing the resulting print-outs. Eventually the print-outs will be filed. If the tabulations have been formatted to fit on an $8\frac{1}{2} \times 11$ in. sheet, they will not only fit in standard size files or looseleaf binders, but they can also be reproduced on almost any copy machine and can be included full size in reports. The format illustrated in Figure 11.12 was designed to summarize a sixteen-hour intersection directional count on a single $8\frac{1}{2} \times 11$ in. sheet.

For quick, visual interpretation of intersection directional volume counts, an intersection flow chart is helpful. The computer has been used not only to tabulate the count, but also to prepare flow charts, as illustrated in Figure 11.13. Other programs compare the summarized volumes with the warrants for traffic signals and prepare traffic signal warrant graphs and traffic volume histograms.

3. *Accident summaries.* Because of the great number of accident reports to be processed, states and large cities have sought the aid of the computer in performing some of their accident recording and processing functions. Information pertaining to each accident is coded on keypunch forms or mark-sense sheets from which it is converted to a computer oriented file, such as data processing cards or magnetic tape records. This file is then sorted and interrogated to produce the desired accident summaries.

INTERSECTION DIRECTIONAL VOLUME COUNT SUMMARY

REEDVILLE   US-44   12/16/71   THURSDAY
CENTER STREET -- 2 LANES NORTH, 2 LANES SOUTH
FOSTER STREET -- 1 LANE EAST, 1 LANE WEST   STOP SIGN CONTROLLED
WEATHER CLEAR   COUNT BY R.L.B.

HOUR BEG.	...NORTH LEG... LEFT	THRU	RIGHT	...SOUTH LEG... LEFT	THRU	RIGHT	...EAST LEG... LEFT	THRU	RIGHT	...WEST LEG... LEFT	THRU	RIGHT
6AM	7	195	5	4	201	3	4	4	5	3	7	1
7AM	22	304	22	14	758	11	10	27	10	13	21	12
8AM	13	279	4	4	718	7	9	15	19	12	12	17
9AM	5	321	13	8	420	1	7	18	4	12	7	8
10AM	7	307	7	8	352	7	10	20	15	13	8	3
11AM	15	254	4	11	286	6	7	14	15	13	14	5
12-N	22	309	10	11	222	4	6	16	14	18	13	8
1PM	17	275	13	14	346	7	6	13	20	15	9	6
2PM	23	308	18	19	291	9	12	14	8	6	20	7
3PM	19	355	15	19	304	6	5	17	12	11	23	3
4PM	22	630	26	25	342	9	8	22	8	11	24	10
5PM	29	1041	41	20	404	3	9	13	9	21	19	13
6PM	21	561	24	20	251	5	4	14	4	18	22	8
7PM	14	288	15	7	229	6	7	14	20	17	17	10
8PM	16	253	12	10	169	8	15	17	13	13	15	3
9PM	6	219	17	16	153	10	15	10	7	12	19	3
TOT.	258	5899	246	210	5446	102	134	248	183	208	250	117
AVG.	17	369	16	14	341	7	9	16	12	14	16	8

HOUR BEG.	..........MOTOR VEHICLE TOTALS.......... N-LEG	S-LEG	E-LEG	W-LEG	N&S	E&W	ALL	....PEDESTRIAN TOTALS.... N-X	S-X	E-X	W-X	TOTAL
6AM	207	208	13	11	415	24	439	0	0	0	0	0
7AM	348	783	47	46	1131	93	1224	1	0	0	0	1
8AM	296	729	43	41	1025	84	1109	0	0	0	0	0
9AM	339	429	29	27	768	56	824	0	0	1	0	1
10AM	321	367	45	24	688	69	757	1	0	0	0	1
11AM	273	303	36	32	576	68	644	0	0	0	0	0
12-N	341	237	36	39	578	75	653	0	0	0	0	0
1PM	305	367	39	30	672	69	741	0	0	0	0	0
2PM	349	319	34	33	668	67	735	3	0	0	1	4
3PM	389	329	34	37	718	71	789	4	1	0	0	5
4PM	678	376	38	45	1054	83	1137	2	0	0	0	2
5PM	1111	427	31	53	1538	84	1622	0	0	1	0	1
6PM	606	276	22	48	882	70	952	0	0	1	0	1
7PM	317	242	41	44	559	85	644	0	0	0	0	0
8PM	281	187	45	31	468	76	544	0	0	0	0	0
9PM	242	179	32	34	421	66	487	0	0	0	0	0
TOT.	6403	5758	565	575	12161	1140	13301	11	1	3	1	16
AVG.	401	360	36	36	761	72	832	1	0	0	0	1

Figure 11.12. Intersection turning-movement summary formatted to fit on an $8\frac{1}{2} \times 11$ in. page.

   Traffic engineering uses of this accident data generally fall into the following six categories:
a. *Statistical accident summaries.* A statewide, countywide, or citywide statistical breakdown of the number of accidents occurring by time of day, day of week, month of year, severity, kind of collision, kind of vehicle, weather condition, surface condition, light condition, causal factors, vehicular movements, driver age, driver sex, and other factors.

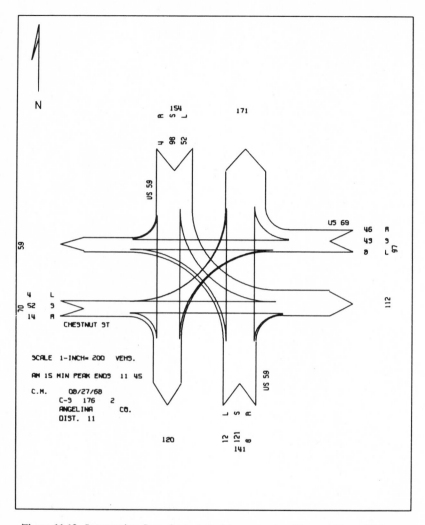

Figure 11.13. Intersection flow chart plotted by computer. (Source: Texas Highway Department, Austin, Texas.)

b. *Route summaries.* A one-line summary of each route showing route length, average daily traffic, kind and number of accidents, and accident rate. The listing may be arranged by consecutive route numbers or by descending accident rates.

c. *Sequential listing of individual accidents.* A one-line capsule summary of each accident. These listings are most helpful if the accident condition codes are interpreted, as illustrated in Figure 11.14. With such a decoded format, high-accident locations and accident patterns are more readily identified.

d. *Listing of high-accident locations.* A one-line summary of each intersection or roadway section having a cluster of *x* or more accidents during a given

1963 ACCIDENT SUMMARY TO APRIL 19, 1963

HIGHWAY	MILEPOST		DATE	DAY	TIME	LOCATION	CARS INV.	ACCIDENT TYPE	INJ.	KILL	SURF.	LIGHT
SR-68	55.1		2- 5-63	TUE	635 PM	2910 SOUTH	2	TURNING MOVEMENT			DRY	NITE
SR-68	55.3		1-20-63	SUN	430 PM	2770 SOUTH	2	REAR END	4		DRY	DAY
SR-68	55.4		2-14-63	THU	745 AM	2700 SOUTH	2	REAR END	1		DRY	DAY
SR-68	55.4		3-11-63	MON	555 AM	2700 SOUTH	2	REAR END			SNO	DAWN
SR-68	55.4		2-16-63	SAT	500 PM	2700 SOUTH	1	PEDESTRIAN	1		DRY	DAY
SR-68	55.5		2-14-63	THU	815 PM	2640 SOUTH	2	SIDESWIPE-SAME DIR			DRY	NITE
SR-68	55.5		1-25-63	FRI	610 PM	2500 SOUTH	2	REAR END			WET	DAY
SR-68	55.6		2-14-63	THU	400 PM	2320 SOUTH	2	REAR END			DRY	DAY
SR-68	56.0		4-17-63	WED	410 PM	2320 SOUTH	2	REAR END			WET	DAY
SR-68	56.0		3-14-63	THU	710 AM	S. OF 2100 SOUTH	2	TURNING MOVEMENT			DRY	DAY
SR-68	56.2		2-26-63	TUE	810 PM	S. OF 2100 SOUTH	2	PARKED CAR			WET	NITE
SR-68	56.2		4- 6-63	SAT	410 PM	2100 SOUTH	2	BACKING			DRY	DAY
SR-68	56.3		2-26-63	TUE	516 PM	N. OF 2100 SOUTH	2	REAR END			WET	DAY
SR-68	56.5		2-26-63	TUE	517 PM	N. OF 2100 SOUTH	2	REAR END			WET	DAY
SR-68	56.7		2-25-63	MON	450 PM	S. OF 1700 SOUTH	3	REAR END	1		DRY	DAY
SR-68	56.8		4-20-63	SAT	514 PM	1700 SOUTH	2	REAR END	1		WET	DAY
SR-68	56.8		4- 8-63	MON	712 AM	1700 SOUTH	2	REAR END			WET	DAY
SR-68	56.8		3-12-63	TUE	128 PM	1700 SOUTH	1	ROLL OVER	1		WET	DAY
SR-68	56.8	*	2-26-63	TUE	504 PM	1700 SOUTH	3	REAR END			ICE	DAY
SR-68	56.8	*	2-18-63	MON	734 AM	1700 SOUTH	2	REAR END			ICE	DAY
SR-68	56.8	*	2-26-63	TUE	548 PM	1700 SOUTH	2	REAR END			WET	DUSK
SR-68	56.8	#	1- 6-63	SUN	851 AM	1700 SOUTH	2	REAR END	3		DRY	FOG
SR-68	56.8	#	1-18-63	FRI	600 AM	1700 SOUTH	2	TURNING MOVEMENT			SNO	DAY
SR-68	56.8		3- 9-63	SAT	959 PM	1700 SOUTH	2	SIDESWIPE-SAME DIR			DRY	NITE
SR-68	57.0		4-19-63	FRI	1103 AM	1400 SOUTH	2	REAR END	1		DRY	DAY
SR-68	57.3		4-19-63	FRI	452 PM	CALIFORNIA AVE	3	REAR END			WET	DAY
SR-68	57.3		4-12-63	FRI	445 PM	CALIFORNIA AVE	2	REAR END	1		DRY	DAY
SR-68	57.3		4- 2-63	TUE	827 AM	CALIFORNIA AVE	2	REAR END			WET	DAY
SR-68	57.3	*	3- 6-63	WED	715 PM	CALIFORNIA AVE	3	REAR END			DRY	DUSK
SR-68	57.3	*	3-31-63	SUN	128 PM	CALIFORNIA AVE	2	TURNING MOVEMENT	6		DRY	DAY
SR-68	57.3	*	3-16-63	SAT	955 AM	CALIFORNIA AVE	3	REAR END	2		WET	DAY
SR-68	57.3	*	3- 1-63	FRI	731 AM	CALIFORNIA AVE	2	REAR END	2		WET	DAWN
SR-68	57.3	#	1-14-63	MON	654 PM	CALIFORNIA AVE	2	REAR END	3		ICE	NITE
SR-68	57.4		3-15-63	FRI	1124 PM	N. OF CALIFORNIA AVE	2	REAR END			ICE	NITE
SR-68	57.9		2-13-63	WED	520 PM	1050 SOUTH	2	PEDESTRIAN	1		DRY	DAY
SR-68	58.0		3-15-63	FRI	455 PM	INDIANA AVE	4	REAR END	1		WET	DAY

Figure 11.14. Capsule summary of individual accidents by succeeding locations.

time period. The listing may be arranged by sequential locations, by descending accident frequency, or by descending accident rate.

e. *Preparation of accident collision diagrams.* Several attempts have been made to let the computer prepare accident collision diagrams for high-accident locations. One such diagram of the collision pattern prepared on a computer-

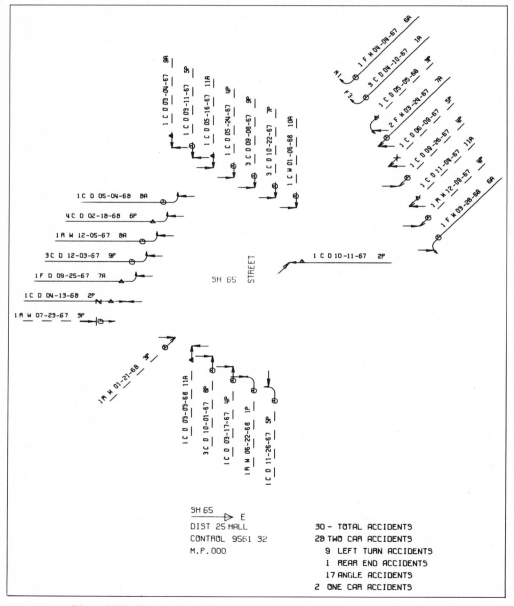

Figure 11.15. Intersection collision diagram prepared on a computer-driven plotter. (Source: Texas Highway Department, Austin, Texas.)

driven plotter is illustrated in Figure 11.15. This plot indicates the following accident characteristics:

(1) Direction of travel
(2) Vehicle maneuvers
(3) Kind of accident
(4) Light condition
(5) Road surface condition
(6) Weather condition
(7) Time of accident (date and time of day)
(8) Accident severity

f. *Special reports.* Special tabulations to summarize all accidents involving only railroad crossings, deer, construction zones, bicycles, pedestrians, or other special situations, as desired.

SIMULATION APPLICATIONS

Simulation applications attempt to model physical systems that are too complex for direct analytical evaluation. These applications permit a wide range of conditions to be analyzed at less cost and in less time than would be possible under actual conditions. Following are specific simulation applications:

1. *Traffic assignment.* One of the important tasks in transportation planning is to determine future traffic loads on the street and highway network. The method used in accomplishing this task is a computer simulation of future traffic flow. This simulation process involves the following basic steps:

   a. Establish computer files representing the street network.
   b. Establish computer files indicating future trips between origins and destinations.
   c. Simulate the drivers' decision processes in selecting routes between origins and destinations. Factors that might be considered in selecting routes include travel times, travel distances, travel speeds, and volume-capacity relationships.
   d. Assign the trips to the selected routes and count the number of trips passing over each link of the network.
   e. Prepare maps and tabulations for use in evaluating the network. An example of such a computer-driven plot is illustrated in Figure 11.16.

2. *Intersection flow simulation.* Numerous attempts have been made to predict traffic stops, delays, queue buildups, signal operation with bus-preemption, traffic conflicts, and accident potential at intersections by simulating the flow of vehicles on the computer. By using these simulation models, the traffic engineer is able to evaluate the effect of various traffic stream conditions, geometric configurations, and traffic control techniques without the time, expense, and dangers involved in undertaking similar field studies.

   An important step in using the computer for traffic simulation is model validation. The results of studies at field sites having characteristics similar to the intersection being simulated should compare favorably with the results of studies using the simulation model. A range of conditions should be compared in order to establish the validity of the model under the range of conditions to be simulated.

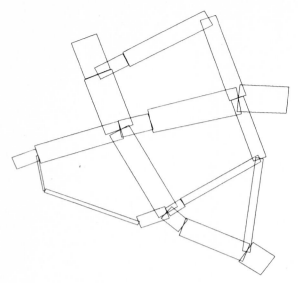

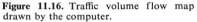

Figure 11.16. Traffic volume flow map drawn by the computer.

REAL-TIME CONTROL

In real-time control applications, the computer is an integral part of the system. Input data are fed to the computer from the system; the computer evaluates the data and returns output signals to control the system. Following are traffic engineering examples of real-time control applications:

1. *Merging control systems.* It is sometimes difficult for motorists entering a freeway to merge safely with freeway traffic. Various merging control systems have been designed to aid the driver in smoothly and safely joining the traffic stream. These systems detect the position and velocity of vehicles on both the entrance ramp and the freeway, determine when an acceptable gap for a ramp vehicle to merge into the freeway traffic stream will occur, and guide the ramp vehicle into the acceptable gap by the use of rampside driver displays.

2. *Traffic surveillance and control.* The efficient operation of an urban freeway and expressway system depends in part on the early detection of undesirable traffic conditions so that corrective measures can be initiated without unnecessary delay. Traffic volumes, speeds, or occupancies are continuously monitored and compared. When traffic speeds or densities reach specified values, or when the occupancy at a downstream detector falls off as the occupancy at an upstream detector builds up, signals are generated by the computer calling for corrective action. These signals may activate variable message diversion signs, ramp metering devices, or lights on a map board of the freeway system pinpointing the location of the trouble. These surveillance and control systems can also provide a wealth of information on traffic volumes, speeds, and occupancies throughout the roadway network.

3. *Traffic signal control.* A number of cities are now using digital computers for traffic signal control. The movement of vehicles throughout the street network under computer control is sensed by using hundreds of detectors, evaluated by the computer, and controlled by computer-driven traffic signals. These

systems are not only traffic responsive, but they also provide great flexibility in control strategies without requiring hardware changes in the field. They can have an unlimited number of timing plans and they can be expanded to include other locations for a minimum incremental cost.

OTHER APPLICATIONS RELEVANT TO TRAFFIC AND
TRANSPORTATION ENGINEERING

The following specific applications are not unique to this profession alone, for they have widespread application in many fields. Below are ways in which the traffic engineering profession is benefitting by these specific applications:

1. *Data storage and retrieval systems.* Situations occur in which it is desirable to store large amounts of information, or data, thereby making it available for people to retrieve the information when they need it. The public library is an example of such an information storage and retrieval system. The basic elements of the system include:
   a. Acquisition and selection of information to be stored
   b. Assignment of a storage location
   c. Establishment of indexes for providing later access to the materials stored
   d. Retrieval of specific information when needed
      Computer based data banks can be established to accomplish this same objective; they have the same four basic elements. The advantages of using the computer for this kind of application include:
   a. The computer can organize, store, and examine vast amounts of information at high speeds.
   b. It can retrieve information and print it out very quickly.
   c. Information in storage is always available; it is not "checked out," but is duplicated for each user.
   d. The computer can disseminate information to remote locations.
   e. It can undertake tasks too large to consider doing manually.
      The most widely known data bank application in the fields of traffic or transportation engineering is the Highway Research Information Service (HRIS) developed as NCHRP project 20-1 by the Transportation Research Board. The objectives of this specific application are:
   a. To select and store input information from current and past highway research that will be of value to users of highway information
   b. To disseminate current information to users
   c. To retrieve relevant information on request
   The HRIS system uses magnetic tape for information storage. Each entry on the tape includes a brief abstract and basic reference data about the report or document. A sample HRIS document record is shown in Figure 11.17. Retrieval of this record by the computer could be accomplished by referring to:
   a. Various index terms taken from a thesaurus list
   b. Specific subject area
   c. Publication source
   d. Research agency
   e. Author or principal investigator
   f. Other identifier terms

```
01/04/71        HRIS SELECTIONS      55 HRBA3A 02
```

```
1P53 089463

A PRACTICAL COMPUTER PROGRAM FOR DESIGNING TRAFFIC-SIGNAL-
SYSTEM TIMING PLANS

BLEYL RL

HIGHWAY RESEARCH RECORD, HWY RES BOARD       67
NO 211, PP 19-33, 8 FIG, 3 TAB, 5 REF

   THIS PAPER DISCUSSES THE ELEMENTS, TECHNIQUES, AND
CHARACTERISTICS OF A PRACTICAL COMPUTER PROGRAM DEVELOPED
FOR DESIGNING PROGRESSIVE TRAFFIC-SIGNAL-SYSTEM TIMING
PLANS. THE ELEMENTS DISCUSSED INCLUDE DIRECTIONAL TRAVEL
DISTANCE VARIATIONS, DIRECTIONAL AND SECTIONAL SPEED
DIFFERENCES, BAND WIDTHS, OFFSETS, CYCLE LENGTHS,
PROGRESSIVE SPEEDS, AND SPLITS.  THE COMPUTER PROGRAM,
WRITTEN IN FORTRAN IV PROGRAMMING LANGUAGE, CONVERTS ALL
SPEED AND DISTANCE UNITS TO TRAVEL TIME UNITS. THE TIMING
PLAN RESULTING IN THE GREATEST EFFICIENCIES IS THEN
DETERMINED FROM A TIME-TRAVEL TIME DIAGRAM. THE PROGRAM
FAVORS THE DIRECTIONAL BAND WIDTHS IN PROPORTION TO THE
DESIRED RELATIVE BAND WIDTHS AND PRINTS A SERIES OF TABLES
WHICH INDICATE, FROM THE RANGES SPECIFIED FOR THE NUMEROUS
VARIABLE ELEMENTS, THE OPTIMUM TIMING PLAN. /AUTHOR/
```

Figure 11.17. Sample document record from HRIS information storage and retrieval system. (Source: Highway Research Information Service, Highway Research Board, Washington, D.C.)

2. *Document writer.* The administrator in traffic or transportation engineering may encounter situations in which typewritten documents must be reprinted from time to time with only minor changes. Each revision generally requires retyping on new masters much or all of the material, incorporating the modifications. The addition of a short paragraph, for example, may require retyping all pages from that point until the end of that unit.

Since the computer output is usually printed by a high-speed printer, the administrator could take advantage of the computer's speed and accuracy characteristics. By typing the initial draft on a keypunch machine or a typewriter computer terminal instead of on a standard typewriter, a file (deck of data processing cards) could be prepared containing the information to be printed. Changes could easily be made by inserting, deleting, or replacing records (cards) in the file. The file could then be read by the computer and printed on the high-speed printer. Subsequent revisions and reprints would only involve making the necessary changes in the file and processing it as before.

At least three specific applications of this nature are presently in use within the traffic engineering profession:

a. *Speed zone descriptions.* A description of all current speed zones is needed by those involved in the establishment, posting, maintenance, and enforcement of speed regulations. An example of such a computer printed document is illustrated in Figure 11.18.

```
                    SPEED ZONE DESCRIPTIONS
                 AS OF        JUNE 1, 1965


      DISTRICT 1              U.S. ROUTE 91       STATE ROUTE NO 085

   MILEPOST                                    SPEED   DATE     DATE
 BEGIN    END          ZONE DESCRIPTION        D. N.  POSTED  APPROVED

 0.00-  0.50  FROM MAIN STREET AT 2ND SOUTH    40-40  2-25-64  2-10-64
              EASTERLY TO 6TH EAST STREET IN
              BRIGHAM CITY.

 0.50-  0.65  FROM 6TH EAST STREET EASTERLY TO  50-50  2-25-64  2-10-64
              A POINT 800 FT.* EAST OF 6TH EAST
              STREET IN BRIGHAM CITY.

 0.65- 17.10  FROM A POINT 800 FT.* EAST OF 6TH 60-50  11-05-64  9-21-64
              EAST STREET IN BRIGHAM CITY EAST-
              ERLY TO THE JCT. OF SR-23 NEAR
              WELLSVILLE.

17.10- 25.70  FROM THE JUNCTION OF SR-23 NORTH- 65-65  11-05-64  9-21-64
              EASTERLY TO A POINT 0.5 MILES *
              SOUTH OF THE LOGAN RIVER BRIDGE.

25.70- 26.20  FROM A POINT 0.5 MILES * SOUTH OF 50-50  11-05-64  9-21-64
              THE LOGAN RIVER BRIDGE NORTHEAST-
              ERLY TO THE LOGAN RIVER BRIDGE.

26.20- 26.40  FROM THE LOGAN RIVER BRIDGE       40-40  12-07-64  11-18-64
              NORTHERLY TO A POINT NEAR 400
              SOUTH STREET IN LOGAN.

26.40- 27.80  FROM A POINT NEAR 400 SOUTH IN    30-30  12-07-57  11-08-57
              LOGAN NORTHERLY TO A POINT 500
              FT.* NORTH OF 6TH NORTH STREET IN
              LOGAN.

27.80- 28.46  FROM A POINT 500 FT.* NORTH OF    40-40  1-02-63  11-06-62
              6TH NORTH STREET NORTHERLY TO A
              POINT 0.16 MILES * NORTH OF 10TH
              NORTH STREET IN LOGAN.

28.46- 28.81  FROM A POINT 0.16 MILES * NORTH   50-50  1-02-63  11-06-62
              OF 10TH NORTH STREET NORTHERLY TO
              A POINT 0.51 MILES * NORTH OF
              10TH NORTH STREET IN LOGAN.
```

Figure 11.18. Example of an easily revised and reprinted computer-prepared document. (Source: Utah State Department of Highways, Salt Lake City, Utah.)

b. *Specifications for contracts.* Installing traffic signals, highway lighting, or other items by contract requires the preparation of special provisions. In most cases, the bulk of the special provisions may be unchanged from contract to contract.

c. *Manuscripts.* There are two ways in which a computer application of this nature has facilitated preparing reports and other manuscripts:

   (1) Reprinting the same report material with only minor changes, as in the above applications.

   (2) Speeding up the editorial, review, and other preparation processes by reducing the time required to retype the interim and final drafts. Almost

all manuscripts undergo a number of editorial changes and reviews between the first typewritten draft and the final copy.

Manuscript typing programs usually include the following features:

(1) *Free formatted input.* Words are separated by one or more spaces, the exact number being unimportant. This greatly facilitates the process of revising or making editorial changes.

(2) *Margins.* The number of characters per line and the number of lines per page are variable. Except for the last line in each paragraph, the lines are justified on both the left and right margins.

(3) *Headings.* Several levels of headings, with or without underlining, are possible.

(4) *Upper and lower case letters.* A special printing chain containing both upper and lower case letters is available for many high-speed printers.

(5) *Typing guidelines.* The output should conform to standard format rules for typewritten material.

An example of material produced using such a program is illustrated in Figure 11.19.

3. *Inventories.* The purpose of an inventory is to account for the property of an organization; the condition and worth of the items inventoried are usually included in this accounting. One appropriate inventory application deals with traffic signs. Traffic sign inventories are necessary in order to:

a. Satisfy legal requirements to maintain records of traffic control devices
b. Provide a basis for proper maintenance of traffic signs
c. Provide a basis for periodic inspection, especially of regulatory and warning signs
d. Provide a basis for traffic sign upgrading and replacement programs

Information about each sign is punched in a data processing card. Information included on the card usually includes sign location, kind of sign, sign legend, mounting method, and installation date. Information on mounting height, sign size, sign materials, physical and reflective conditions of the sign, and other sign characteristics may also be included on the card.

The inventory cards could be processed by computer or by card sorting and tabulating equipment to provide the following summaries or listings:

a. Sequence of signs along specified streets
b. Signs of a specified kind
c. Signs mounted on a particular kind of pole
d. Age of sign
e. Various combinations of various categories, for example, listing of all school crossing signs over five years of age

REFERENCES FOR FURTHER READING

Programming language manuals of the various computer manufacturing companies.

BARRODALE, I., ROBERTS, F.D.K., and EHLE, B.L., *Elementary Computer Applications in Science, Engineering, and Business.* New York: John Wiley & Sons, Inc., 1971.

BATES, F., and M. L. DOUGLAS, *Programming Language One* (2nd ed.). Englewood Cliffs, N.J.: Prentice-Hall, Inc., 1970.

MINIMUM INFORMATION REQUIRED TO USE <u>TYPESCRIPT</u> PROGRAM
BY ROBERT L BLEYL

DESCRIPTION

      This program produces typewritten pages on the high speed printer. The text to be printed is submitted on free-formatted data processing cards. The printed output is right and left justified with appropriate margins for 8-1/2 by 11 inch pages. The material you are now reading was produced using this program.

TEXT

      Text is prepared for the program in paragraph size units. Up to 60 data processing cards can be used to write a paragraph of text. Words must be separated by one or more spaces. The number of spaces used is unimportant. Column 1 of the first card for each paragraph of text must contain the character "T." The next four columns of this first card must be left blank. Column 80 of the last card for each paragraph of text must contain the character "/" to designate the end of the paragraph. All other columns may be used for the text. Hyphenation of text between successive cards is not allowed.

HEADINGS

      Headings may either be centered or printed at the left side. For a centered heading, column 1 must contain the character "C." For a heading at the side, column 1 must contain the character "S." Column 4 must contain a code number as follows:

      1 = Heading will be placed at the top of a new page.
      2 = Heading will be placed next in sequence on the same page.
      3 = Heading will be placed next in sequence on the same page and underlined.

The heading to be printed is punched in columns 6-80 with at least one space between words.

OTHER FEATURES

      Many options and features are available with this program. The above material contains all that is necessary in order to properly prepare most of the material that would normally be produced using this program. The other features and options are described in a more comprehensive write-up.

Figure 11.19. Example of material produced by a manuscript-typing computer program.

BLATT, J. M., *Introduction to FORTRAN IV Programming*. Pacific Palisades, Calif.: Goodyear Publishing Co., 1971.

EARNSHAW, J. D., and W. BLACKFORD, *Computer Programming in ALGOL*. New York: Pitman Publishing Corp., 1970.

FARINA, M. V., *Programming in BASIC*. Englewood Cliffs, N.J.: Prentice-Hall, Inc., 1968.

FENVES, S. J., *Computer Methods in Civil Engineering*. Englewood Cliffs, N.J.: Prentice-Hall, Inc., 1967.

FOXLEY, E., *et al.*, *A First Course in ALGOL 60*. Reading, Mass.: Addison-Wesley, 1968.

PARSLOW, R. D., *et al.*, eds., *Computer Graphics*. New York: Plenum Press, 1969.

SILVER, G. A., *Simplified FORTRAN IV Programming*. New York: Harcourt, Brace, Jovanovich, 1971.

Chapter 12

URBAN TRANSPORTATION PLANNING

DAVID K. WITHEFORD. Assistant Program Director, National Cooperative Highway Research Program, Transportation Research Board, Washington, D.C.

Providing for the safe and efficient movement of people and goods is the aim of all traffic engineering, and planning is as essential as proper design, operation, maintenance, and administration in achieving this mission. It means anticipating needs, developing economical and acceptable methods of meeting them, recommending programs of investment to provide necessary facilities, and lastly, monitoring developments either to confirm the adequacy of past planning or to modify plans as conditions change.

Traffic engineers in planning frequently work with other professionals because transportation planning often calls for a multidisciplinary approach. When traffic planning is subsidiary to planning for new urban development proposals, they will work with architects and planners. Even when transportation plans are the end product, they can expect to work with planners, demographers, economists, and other specialists. Traffic engineers in other capacities, particularly those charged with operating facilities, also need to be aware of the planning process and to be capable of evaluating its effectiveness. They have an obvious stake in assuring that the plans provided for them are workable.

Like many other facets of traffic engineering, urban transportation planning changes constantly. New considerations arise as aspects of urban transportation planning translate from the realm of qualitative judgment to forms that can be treated by computers and quantitative analysis. Even the language changes. Environmental investigations were not part of the evaluation process in the United States until they were required by federal legislation in 1968 and 1969. The concepts of "joint development" and "multiple use" were not widely implemented until the late 1960's. Thus, this chapter will mainly identify those elements that describe urban transportation planning, outlining the activities involved and the role of the traffic engineer in them. In so doing, it will note other sources of information providing more detail for those who wish it.

The chapter begins by listing various studies classifiable as urban transportation planning. These include traffic and parking projects as well as long-range planning programs. The chapter then outlines the procedures of the long-range studies, without detailing survey specifics, because these are described in other chapters.

TRANSPORTATION PLANNING STUDIES

For convenience, the activities in transportation planning can be classed as short range and long range. Short-range projects tend to be specialized and identified with a single site, a single mode, or a single route. They include investigations of traffic needs related to:

1. *Individual traffic generators:* industrial plants, shopping centers, universities, hospitals, and so on.
2. *Major land use projects:* subdivisions, planned unit developments, joint development and multiple-use situations, urban renewal projects, transportation centers.
3. *Area studies:* TOPICS (Traffic Operations Program to Increase Capacity and Safety)-type studies, some forms of transit planning, goods movement studies, or small community planning analyses.

Long-range studies include those charged with developing transportation facilities for the long-term needs of urban areas. In small urban areas, long-range studies may not be very demanding of personnel and resources. In major metropolitan areas, however, the long-range studies call for large staffs, substantial budgets, and considerable time for completion. They may also require the establishment of a continuing transportation planning function where none existed before.

SHORT-RANGE STUDIES

Individual traffic generators. Traffic studies for specific sites are made for site developers to determine the parking needs and peak period movements that must be designed for within the site. This may be necessary in order to demonstrate adequate traffic design before gaining the approval of planning and zoning bodies for development. Similar studies might be made by public agencies for two reasons: (1) to ascertain the ability of the street system to absorb the traffic generated by new activities, and (2) to verify that adequate parking is being provided on site.

This kind of study,[1] whether made by public agencies or by consultants for a developer, usually requires:

1. Obtaining measures of site activity.
2. Determining the patterns of automobile and truck traffic resulting.
3. Selecting the peak design period to be accommodated.
4. Ascertaining maximum accumulations and peak in-and-out movements.

Activity may be measured by floor space or land area, if reasonable trip generation rates from experience elsewhere can be applied, but it is desirably measured in direct terms such as employees, visitors, customers, and so on. Their travel patterns are dependent on trip purposes and travel modes available, which in turn can be related to the activity and its location.

For example, automobile trip generation created by office buildings differs between downtown and suburban locations and between one-tenant and multitenant buildings.[2] More transit trips to work are likely at downtown locations than at suburban office parks. Fewer trips to work and more trips for personal business are likely at multitenant than at one-tenant buildings. Multitenant buildings are likely to have visitors as well as occupants who may be in and out during the day. As a result, they will have proportionally lower peaks for arrival and departure, greater turnover, and shorter parking durations than the typical corporate headquarters office building.

---

[1] Study techniques and typical data requirements are described in Chapter 10.

[2] L. E. Keefer and D. K. Witheford, *Urban Travel Patterns for Hospitals, Universities, Office Buildings, and Capitols,* National Cooperative Highway Research Program Report 62 (Washington, D.C.: Highway Research Board, 1969), p. 106.

Weekly or seasonal travel patterns require consideration for certain land uses like shopping centers. Their design needs may be based on tenth highest hour volumes that typically occur on pre-Christmas Saturdays.[3] Industrial plant traffic design, however, may be based on the average weekday (unless a cyclical industry), allowing for normal conditions of absenteeism and vacation schedules. Factory ingress-egress design requirements may be determined by the time relationship of peak employee movements to the normal access highway peak periods.[4]

How the typical major generator study may proceed is revealed by a study of the Memphis Medical Center.[5] First, the goals of the traffic and parking systems were defined. The key elements were efficiency, accessibility, safety, and feasibility. Next, urban transportation study reports provided information on the adequacy of present area parking and travel facilities as well as expected future travel demands. Next, future parking demand was projected from ratios relating parking to forecasted numbers of beds and employees. Site planning for structures to provide more than 4,000 additional spaces followed. Pedestrian distances, interior vehicle circulation, and access to freeways were important planning criteria considered in achieving a solution. (For travel characteristics associated with the major travel generators mentioned here, see Chapter 5.)

Major land use projects. Traffic planning for larger projects follows similar patterns. Here it is even more important that the causes of traffic be analyzed instead of merely applying standardized trip and parking rates to units of floor space or other physical measures. The latter may give guidance for a rough check of needs—order of magnitude estimates—but they are not adequate for careful planning.

Land use developments. New residential subdivisions, planned unit developments involving a variety of land uses, or urban renewal schemes all require selection of design standards and analysis of street layout, parking needs, and pedestrian movements.

Figure 12.1 illustrates some of the principles applicable to residential subdivisions. The layout of blocks, lots, and other areas for public use or open space in conformance with topography and other esthetic values is foremost. Traffic considerations include sight distance, intersection controls, layouts to minimize arterial connections and through traffic, adequate widths, parking provisions, and separation of pedestrian and vehicle movements. A recommended ITE (Institute of Traffic Engineers) practice contains specifications for design elements and other material relevant to the design of subdivision streets for capacity and safety.[6]

Planned unit developments frequently contain a wide mixture of land uses within a single project. Because standardized zoning and traffic controls are frequently inadequate for testing design solutions, there is an increasing need for traffic planning and analysis with such proposals. Traffic circulation, parking, and pedestrian

---

[3] ALAN M. VOORHEES and CAROLYN E. CROW, "Shopping Center Parking Requirements," *Shopping Centers and Parking*, Highway Research Record 130 (Washington, D.C.: Highway Research Board, 1966), pp. 20–38; and "Transportation Considerations of Regional Shopping Centers," *Traffic Engineering*, XLII, No. 11 (1972), pp. 14–21, 63.

[4] *Parking Facilities for Industrial Plants* (Washington, D.C.: Institute of Traffic Engineers, 1969), p. 18.

[5] WILLIAM H. CLAIRE and RANDALL A. P. JOHNSON, "Planning for Memphis Medical Center Parking," *Transportation Engineering Journal of ASCE,* XCVII, No. TE1 (1971), p. 51.

[6] *Recommended Practices for Subdivision Streets* (Washington, D.C.: Institute of Traffic Engineers, 1965), p. 15.

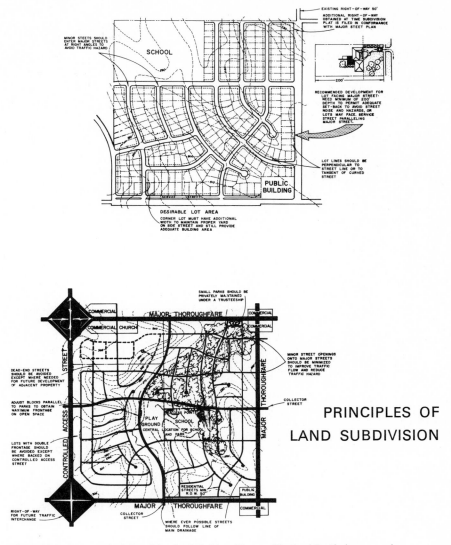

Figure 12.1. Some principles of land subdivision. [Source: Harland Bartholomew and Associates (Memphis, Tenn., 1961)].

impacts require investigation not only to insure the success of the development but also to prevent any later operational problems affecting neighboring activities and traffic facilities.

Questions that should be asked and answered are:

1. Can approach streets accommodate new traffic demands posed by the development?

2. Are modifications to existing operations necessary?
3. Do internal circulation patterns prevent internal congestion from feeding back into surrounding streets?
4. Are parking provisions adequate or can spillover be accommodated within reasonable distances?
5. Can goods movement requirements be accommodated without congestion?
6. Are pedestrian-vehicle conflict situations properly anticipated and treated?

Federally aided projects may routinely receive such examination or review. In other instances, the city traffic engineering staff should be called in when major proposals are submitted to planning and zoning agencies for approvals. Evidence from a survey of zoning practices suggests this is not often done.[7]

Transportation terminals. Intermodal transfer points, for both people and goods, are becoming increasingly important in the nation's cities. They may be relatively small in scale, for example, suburban parking lots coordinated with rail or bus transit service to CBD's (central business district), or they may be complex multimodal facilities in the CBD. Interfaces between intra and intercity travel, for example, airports and trucking terminals, are still other forms.

In all of them, solving the problems of person movement is essential to the design process. The centers may include the following services: commuter rail and other transit terminal facilities; bus terminals with access ramps to freeways; heliports; automobile parking; recreational and pedestrian facilities; airline offices and transfer points for airport service; and commercial or other land uses within the same structure.[8] Understanding and predicting the nature of demand by travel mode—the arrivals, the transfers, and the departures—is vital to later successful operation.

Suburban transit transfer points for downtown commuters require careful analysis. Measures of existing transit ridership and auto travel in the corridor to be served are prerequisite to estimating transit transfers from future all-day parkers, "kiss and ride," and previous transit users. On-site planning must insure adequate space for (1) automobile loading and unloading, (2) transfers from local buses, and (3) all-day parking demand. Furthermore, studies should be made of the impact of this new traffic generator on surrounding activities. Related experience and practice in representative cities are summed up in an ITE committee report.[9]

The complexities of downtown transportation center planning are suggested by Figure 12.2, which depicts a combined transportation-visitor center proposed for Washington D.C.[10] Although this preliminary proposal has been subsequently modified, the basic concept continues to incorporate a heliport, railroad terminal, rail transit and bus terminals, and a parking garage as the basic transportation facilities. Grouped around them are a visitors' center, department store, arena, hotel, offices, and educational activities. The plan would provide access and parking for 75,000 daily visitors, in addition to access, parking, and transfer facilities for 75,000

---

[7] DAVID K. WITHEFORD and GEORGE E. KANAAN, *Zoning, Parking and Traffic* (Saugatuck, Conn.: Eno Foundation for Transportation, 1972).

[8] WILBUR SMITH and ASSOCIATES, *Transportation and Parking for Tomorrow's Cities* (New Haven, Conn.: 1966).

[9] *Change of Mode Parking, A State of the Art* (Washington, D.C.: Institute of Traffic Engineers, 1973).

[10] ROBERT W. HARRIS, "Proposed Transportation and Visitor Center," *Transportation Engineering Journal of ASCE*, XCVI, No. 3 (1970), pp. 312–15.

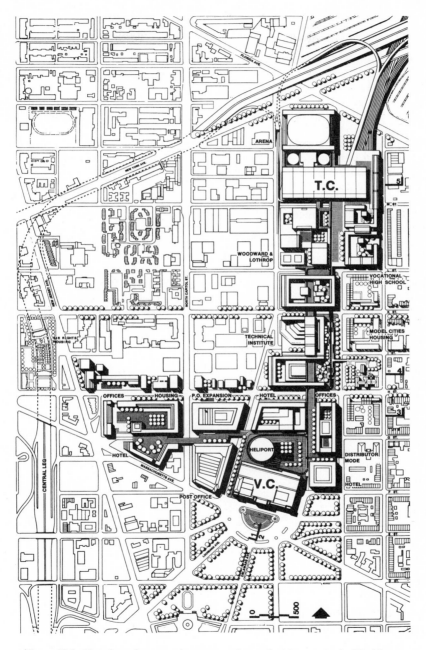

Figure 12.2. Site plan of proposed transportation and visitor center in Washington, D.C. [Source: Robert W. Harris, "Proposed Transportation and Visitor Center," *Transportation Engineering Journal of ASCE*, XCVI, No. TE3 (1970), p. 315.]

intercity passengers daily. Figure 12.3 shows the distribution between modes of the forecasted 1985 daily movements through the transportation center.

Airport ground traffic problems are equally complex. Planning for air passengers includes estimation of incoming traffic volumes (by auto, limousine, taxi, bus, or other transit), consideration of parking and loading/unloading areas, baggage handling, and minimizing pedestrian travel distances between unloading, ticketing, and boarding areas. Provisions for passenger and baggage interline transfer also may be a factor in traffic planning.

The air passenger ground transportation problem is only part of the traffic picture at airports. Employee travel can have greater impact. A study at John F. Kennedy International Airport in New York showed that "during the 6 a.m. to 9 a.m. peak period employee traffic outnumbered air passenger traffic nearly two to one and in the 3 p.m. to 6 p.m. peak, employee traffic predominated 54 percent to 46 percent. The inbound peak hour from 7 a.m. to 8 a.m. totalled 3,829 vehicles, of which 2,390 were employee vehicles."[11] Visitor travel related to sightseeing and other recreational activities must also be anticipated in planning traffic and parking facilities. Most important, it should be recognized that the origins of trips to airports are commonly dispersed rather than concentrated. Perhaps the paramount concern in traffic planning for airports is to see that sufficient capacity is provided for the highway approach volumes generated by growing airport development.

Trucking terminals present an entirely different set of concerns. Traffic planning investigations may be necessary for those transfer facilities that break down intercity shipments for local delivery or assemble loads for intercity movement. Even isolated

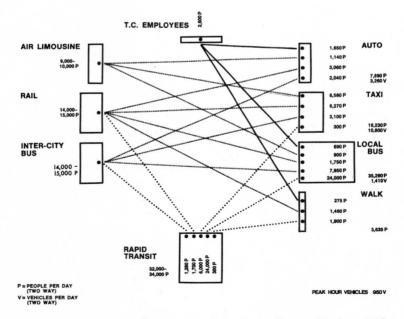

Figure 12.3. Transportation center travel demand 1985. [Source: Harris, p. 312.]

11 LOUIS E. BENDER, "Airport Surface Traffic Demands," *Traffic Quarterly*, XXIV, No. 3 (1970), p. 356.

facilities may affect local traffic patterns. One terminal near Interstate 294 outside Chicago can handle 1,000 units a day, or more than 12,000,000 pounds of freight daily. It covers 55 acres, has 258 loading doors and nearly 7 acres of covered storage.[12] When activities of this kind cluster together, as they often do, their impact on local traffic should be anticipated by appropriate planning studies. (For their internal design considerations, see Chapter 15.)

Joint development and multiple use. The concept of joint development has been defined as the "process of conceiving, designing, and carrying out a combination of urban development activities in a unified way, to the end that benefits are greater than if each individual activity were separately planned and executed."[13] Customarily used in conjunction with urban freeway planning, it often embraces the concept of "multiple use," meaning the employment of the airspace or other portions of the transportation rights of way for nontransportation uses such as buildings (see Figure 14-8), recreational facilities, or parking areas. Multiple use and joint development of transportation rights of way can be effective in bringing about a greater degree of compatibility between the facility and its environment.

Freeways or other major transportation corridor facilities are most likely to offer joint development opportunities. Favorable conditions have been listed by one study[14] as follows:

1. Scarcity of land in the immediate environment
2. Adjacent properties needing land for expansion
3. Ease of development for ancillary uses like parking or storage
4. Availability of land at low cost or with minimum difficulty
5. Willingness of transportation agency to improve roadway appearance.

There are four ways of undertaking joint development on urban freeways. Projects may make use of:

1. Areas under an elevated structure
2. Air rights above a freeway
3. Lands abutting a freeway
4. Combinations of the above in suburban or even rural areas.[15]

The kinds of multiple uses developable with an elevated freeway situation have been suggested in a Philadelphia study.[16] Of about 180 acres of air rights area around or beneath several miles of the Delaware Expressway, 72 acres were recommended for industrial-related usage, 38 acres were found suitable for park and recreational use, 12 acres for other public uses, and 4 acres for miscellaneous other uses. Only

---

[12] WILBUR SMITH and ASSOCIATES, *Motor Trucks in the Metropolis* (Detroit: Automobile Manufacturers' Association, 1969), p. 100.

[13] FREDERICK T. ASCHMAN, "Background and Objectives," *Joint Development and Multiple Use of Transportation Rights-of-Way*, Special Report 104 (Washington, D.C.: Highway Research Board, 1969), p. 23.

[14] *I-84 Environmental and Joint Use Study* (Hartford, Conn.: Connecticut Department of Transportation, 1970), p. 15.

[15] ROGER F. NUSBAUM, "Engineering Considerations," *Joint Development and Multiple Use of Transportation Rights-of-Way*, Special Report 104 (Washington, D.C.: Highway Research Board, 1969), p. 119.

[16] UELAND and JUNKER, ARCHITECTS, *Joint Development: Philadelphia* (Philadelphia: Pennsylvania Department of Transportation, 1971), p. 5.

2 acres were considered unusable because of inaccessibility. The remaining 51 acres were ineligible for development by the Pennsylvania Department of Transportation because, as "aerial easements," they were not under departmental control. In other circumstances, residential projects may be appropriate, depending on development costs, the costs of design for adequate light and air circulation, and necessary protection from pollution, dust, noise, and fire hazards.

Joint development projects have proved difficult to achieve. The projects are not easy to coordinate and finance; existing laws in many instances do not lend themselves to these types of development, and financing them may require resources greater than those available to the involved agencies at the time needed. Joint development, however, offers in many cases the best means of integrating transportation facilities into the urban environment. Instances in which joint development would be particularly advantageous are: (1) When entire blocks can be acquired at little additional cost over that for minimum highway right-of-way taking, the local agency can use the excess land for coordinated replacement housing or other purposes. (2) When redevelopment of a deteriorated area of a city is planned in the path of a major highway location, the public works projects can be coordinated in order to achieve a maximum degree of compatibility. (3) Planned educational facilities such as schools, colleges, and universities can be developed jointly with highways. (4) Park development, for example, a linear park between a highway and a riverfront, in conjunction with a freeway would benefit the entire community as well as the highway user. (5) Rapid rail transit or exclusive bus lanes can be developed in the medians of freeways. (6) Terminals for other transportation modes can be developed jointly with highways.

Design concept teams. Joint development opportunities have led to a new form of organization for planning many urban freeways. The "design team" provides a multidisciplinary approach including representatives from the fields of architecture, urban planning and design, highway and traffic engineering, economists, and sociologists. Representatives from other disciplines, for example, systems analysts, real estate specialists, lawyers, economists, educators, health planners, and others may be included as required. These teams are, therefore, capable of applying the expertise of numerous professions to help solve the problem of integrating the transportation facility into the urban community.

As well as greater capabilities, design concept teams have expanded tasks. They are responsible for acquiring greater community participation in the design process and for giving greater attention to social problems. The team is responsible for evaluating the total environment of the area in which the transportation facility is to be located. They are responsible for indicating changes in surrounding land uses when the existing uses might conflict with the transportation facility and to plan possible community improvements in conjunction with the transportation development.

Three kinds of teams may be needed at different stages of a project.[17] The *advance planning team* studies a project on a regional or corridor basis, examining the broad context of its possibilities at an early stage of planning. Next, the *project impact team* investigates more carefully the community impacts associated with more specific location proposals. Evaluating the social, economic, and design potentials at this stage would require maximum participation from all disciplines. The *implementation*

[17] Douglas C. Smith, *Urban Highway Design Teams* (Washington, D.C.: Highway Users Federation for Safety and Mobility, 1970), p. 46.

team concerns itself with the practical aspects of design and plan preparations for construction.

According to the findings of a special ITE committee, the traffic engineer's function in the interdisciplinary approach to urban highway design can be significant. The committee concluded that there are several roles for the traffic engineer in environmental design.[18] In order of increasing comprehensiveness they are:

1. Operational considerations only within the physical limits of the proposed facility
2. Operational considerations only for adjacent facilities as well as for the proposed facility
3. Location of interchanges plus all operational considerations
4. Estimates of traffic, concept design, operational considerations
5. Traffic impact, traffic estimate, concept design, operational considerations
6. Broad corridor planning, traffic impact, traffic estimate, concept design, operational considerations.

Certainly not least among the traffic engineer's functions in such large undertakings should be the examination of existing traffic requirements during the construction period.

Area studies. At a broader scale than the urban redevelopment projects or highway joint development proposals is another class of short-range traffic planning studies. TOPICS studies, parking studies, immediate action programs for improving transit services, and urban freight distribution studies are all logical interests of the traffic engineer.

The Traffic Operations Program to Increase Capacity and Safety. TOPICS was established by the United States Congress in 1968 as a means of reducing urban traffic congestion and facilitating traffic flow. Projects were "based on a continuing comprehensive transportation planning process . . . coordinated with other projects developed for public transportation fringe parking . . . ."[19]

The initial stage called for the designation of a street network (Primary Type II) supplementary to the Primary Type I and Secondary Federal Aid Systems. This network included arterial highways and major streets connecting at each end to other links in the Primary (Types I and II) or Secondary networks, or linking major traffic generators or rural arterials at city limits to other parts of the networks. As part of the Type II system selection (made by state highway officials in cooperation with appropriate local officials and subject to the approval of the Federal Highway Administration) the already established networks were also reevaluated.

After designating the system, areawide plans were developed for traffic operations improvements. Planning included:

1. Inventory of network characteristics and evaluation
2. Review of traffic laws and ordinances

---

[18] "The Traffic Engineer's Role in Environmental Design," *Traffic Engineering*, XIX, No. 7 (1969), pp. 26–29.

[19] "Urban Traffic Operations Program to Increase Capacity and Safety," Policy and Procedure Memorandum 21–18 (Washington, D.C.: Bureau of Public Roads, May 28, 1970).

3. Parking studies
4. Terminal facility studies
5. Airport access evaluation.

A wide range of recommended improvements was eligible for federal funding: signal and surveillance systems, improvements to other control devices and highway lighting; widening, channelization, grade separations for rail-highway or major pedestrian-vehicle crossing points and reconstruction to aid route continuity; separate traffic lanes to facilitate transit and modifications associated with restrictions on curb parking.

TOPICS funds were used for planning studies as well as for providing improvements, and they were also used for the required program evaluation. The latter measured the effect of improvements by comparing accidents, capacity, and travel speed and delay characteristics on a before and after basis. The evaluation additionally aided in developing choices for future improvements.

Parking studies. Various purposes and methods of parking studies were described in Chapter 10. Of particular interest in transportation planning is relating parking policy to other solutions to central business district congestion and downtown street capacity problems. Coordinated planning of highway, transit, and parking programs is part of the urban transportation planning process; fringe parking combined with transit service for CBD commuters, for example, is one alternative to providing downtown highway capacity and CBD all-day parking facilities.

Transit planning studies. Short-range transit studies may have one or both of the following purposes: (1) to develop immediate assistance to generally declining transit systems, (2) to develop new bus facilities to relieve local traffic congestion problems. Inventories of the following kinds may be necessary:

1. Operational data. Trends in annual revenues, passengers served, route miles, etc., compiled from transit operator records
2. Fare structures. Regular fares, transfer arrangements, special classes (commuter, school, elderly), and any unusual off-peak or low-cost inducements
3. Physical property. Equipment capacity, age, and adequacy; maintenance facilities and other structures
4. Management structure. Organization of personnel, labor relations, public relations
5. Routes and services. Headways by time of day, schedule adherence, operator and passenger safety, bus stop location, and transfer point services
6. Rider characteristics. Attitude surveys of current riders and potential riders[20]

Other data sources on transit tripmaking are home interview surveys and census data that provide further information on population and work trip characteristics. Analysis of present ridership, transit company practices, and population characteristics may lead to recommendations for changes in current service. New routes, revised schedules, possibly new equipment, or changes in fare structures and management procedures may result.

---

[20] *Transit Planning Guidelines* (Harrisburg: Pennsylvania Department of Transportation, 1974), p. 23.

Bikeway planning studies. Long an important mode of transportation in many European and Asian countries, the bicycle is rapidly becoming a viable transport mode in the United States. This movement, accelerated by the development of multispeed cycles that can easily climb hills while attaining 30 mph (48 kph) on level ground and average 10 mph (16 kph) or better for long periods of time, has resulted in growing public pressure for the provision of special bicycle facilities.

The term "bikeway" has been used to describe all facilities that explicitly provide for nonmotorized bicycle travel ranging from fully grade-separated rights of way designated for the exclusive use of bicycles to a shared right of way designated as such by traffic signs and/or markings. Three main classes of bikeways have been identified:

1. Class I (Exclusive Bikeway). A completely separated right of way designated for the exclusive use of bicycles. Cross flows by pedestrians and motorists are minimized. More commonly located in parks, recreation areas, rural areas, and new developments where the routes are designed to be completely separated from both roadways and pedestrian paths.
2. Class II (Restricted Bikeway). A restricted right of way designated for the exclusive or semiexclusive use of bicycles. Through travel by motor vehicles or pedestrians is not allowed; vehicle parking, however, may be permitted. Crossflows by motorists, for example, to gain access to driveways or parking facilities, are allowed; pedestrian crossings, for example, to gain access to parking facilities or associated land use, are permitted.
3. Class III (Shared Bikeway). A shared right of way designated as such by signs placed on vertical posts or stencilled on the pavement. Any bikeway that shares its through traffic right of way with either or both moving (not parking) motor vehicles and pedestrians is considered a Class III bikeway.[21]

Bikeway planning considerations have been described as follows:

> The planners of bikeways for existing and especially future communities must consider that bicycle movement is part of the larger transportation system, and, therefore, should be planned and designed to reflect this relationship. Realization that the bikeway is a related element in a many faceted transportation system relates the bikeway to such important factors as mixed-mode travel which implies providing planned options for combination with auto, train, walking, and bus facilities. These interfaces as well as the present and future relation of the bikeway to other systems of travel should be considered not only in long-range plans, but also in short-term planning of bikeways for areas where immediate action must be taken.

In planning bikeways, attention should be directed toward the following considerations:

1. Inventories should be made of existing bicycle facilities and their use in order to identify travel and user characteristics, land-use relationships, and physical conditions.

[21] *Bikeway Planning Criteria and Guidelines* (Los Angeles: Institute of Transportation and Traffic Engineering, University of California, 1972), pp. 19–20, 141–65.
VINCENT R. DESIMONE, "Planning Criteria for Bikeways," *Transportation Engineering Journal of ASCE*, XCIX, No. TE3, August 1973 (New York: American Society of Civil Engineers, 1973), pp. 609–25. These excellent reports are the major references for this section.

2. Forecasts of bicycle travel and demand should be made. Attention should also be directed toward interrelationships with other transport modes and seasonal influences on bike utilization.
3. Planning goals and objectives should be established to include safety, mobility, efficiency, route flexibility, adaptability, and imageability (passing through environments that will remain as vivid memories even after short exposure).
4. Providing bicycle storage areas at appropriate locations and locking devices in order to prevent casual and professional theft (see Chapter 15).

Goods movement studies. The transportation planner has become increasingly concerned with freight movement in cities because virtually all final distribution of goods to homes and businesses is by truck. Figure 12.4 shows how bypass routes may be set up to keep through truck movements out of the center of the city and how designated business routes may serve the central business district and other areas attracting numerous truck trips. Posted truck routes should be located so that they avoid residential areas, points of low overhead clearances, and structures with low load limits. At the same time, certain streets may be prohibited to truck travel because of the prevailing land uses or inadequacy of pavements (see, also, Chapter 18).

Other planning studies related to trucking may include surveys of downtown truck loading and curb usage patterns. With an objective of maximizing the separation of goods loading and unloading from goods and person movement, such studies would examine the benefits of:

1. Regulation of delivery hours
2. Relocation of loading zones
3. Designation of one-way alley systems
4. Limits on truck size in certain areas
5. Provision of off-street loading areas.

Another kind of short-range freight analysis may be performed by transportation planning organizations. Figure 12.5 illustrates a process for examining the short-range

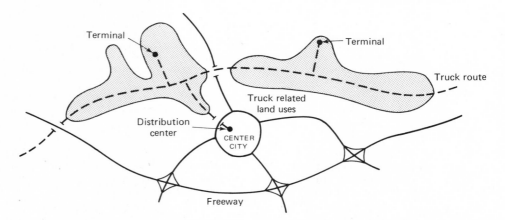

Figure 12.4. Truck routes may be designated to serve truck-related commercial areas. [Source: *The Freeway in the City, A Report to the Secretary, Department of Transportation* (Washington, D.C.: U.S. Government Printing Office, 1968), p. 31.]

REGIONAL SHORT-RANGE FREIGHT FACILITY STUDY

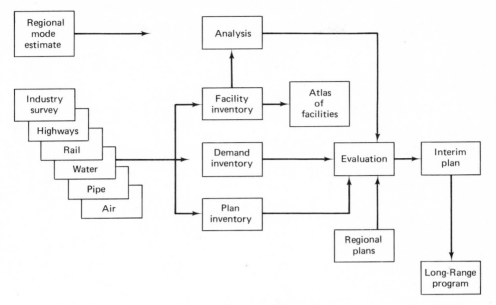

Figure 12.5. An organizational structure for goods movement studies. [Source: James R. Blaze and Nancy Raasch, *Planning for Freight Facilities* (Chicago: Chicago Area Transportation Study: June, 1970), p. 18.]

supply and demand for freight facilities in the Chicago metropolitan area. This investigation used questionnaires and interviews to obtain data from companies engaged in each mode of freight shipment, and then a regional Atlas of Freight Facilities was compiled. The 39-page Atlas, covers an area of more than 4,000 sq mi, at a scale of 1 : 24,000, and designates the following unique and highly-useful comprehensive inventory information:

1. Expressways, primary arterials, motor carrier terminals, and warehouses
2. Rail lines, classification yards, and truck-on-flatcar facilities
3. Navigational waterway channels, principal waterfront industry, and cargo-handling areas
4. Crude petroleum and petroleum product pipelines and natural gas pipelines
5. Airports, air cargo terminals, and consolidation points[22]

LONG-RANGE STUDIES

So far, only those studies necessary for short-range transportation planning have been discussed. Long-range studies are designed to develop programs for urban system development that may take decades to complete. Like the short-term studies, the long-range studies may cover a single mode or a single subject, for example, freight facility

[22] JAMES R. BLAZE and NANCY RAASCH, *Planning for Freight Facilities* (Chicago: Chicago Area Transportation Study, June 1970), pp. 18–20.

planning. However, they are more likely to be multimodal, and most important, to be integrated with other elements of urban planning.

The integrated planning projects may cover several problems, the range of which is suggested by the four practical studies described in "Traffic in Towns."[23] The first study dealt with the traffic problems of a community of about 30,000 population, the second with those of an area with 500,000 people. A third study dealt with an intermediate size city with an environment of historic and architectural features such that major traffic facilities could not be added. The last study investigated a densely developed mixed-use area of only 184 acres in central London.

Many of the problems encountered in these studies, and many of the procedures directed toward their solutions, are mirrored by the large metropolitan area studies conducted in urban communities of the United States.

The remainder of this chapter deals largely with such metropolitan area studies. A description of how they evolved, their objectives, and the order of magnitude of investment (in both the studies and their proposals for construction) will help to show their importance.

History of development. Urban transportation planning has been an evolutionary process. Its beginnings may be traced to the home-interview studies conducted in more than 100 cities in the United States during the decade following the end of World War II. The concept of small sample interviews was then combined with cordon line surveys in order to derive patterns of urban travel. Future traffic usage of urban highway projects was predicted by manually assigning selected origin-destination (O-D) movements to the routes being planned.

In the early 1950's there were studies investigating land use and traffic relationships because better estimating methods were needed in order to project design year travel. Methods of forecasting future population and its distribution, trip generation analysis relating travel to underlying household characteristics (car ownership, etc.), and planning for networks instead of single routes were introduced at this time. Improved procedures were facilitated by the growing use of punch card data processing systems and later by the increasing capabilities of electronic computers. The latter permitted greater sophistication in transportation planning because they permitted the examination of more alternatives. "Modelling" of future land-use plans and future highway and transit systems was combined with more elegant methods of evaluation. Criteria for determining if plans met community objectives (a concept itself not generally introduced until the mid-1960's) could be increasingly quantified.

Direction and funding for the development of urban transportation planning in the United States has been supplied principally through federal legislation. The Federal-Aid Highway Act of 1944 first permitted the use of federal funds for urban highway facilities. The Federal-Aid Highway Act of 1962 required that federally assisted highway projects must be "... based on a continuing comprehensive transportation planning process carried on cooperatively by states and local communities. ..." The so-called "3C" philosophy of *continuing*, *comprehensive*, and *cooperative* urban transportation planning best sums up the status of the process in the early 1970's.

Objectives and approach. The fundamental objective of the long-range urban transportation study is to develop a transportation plan serving the community of the

---

[23] *Traffic in Towns—A Study for the Long-Term Problems of Traffic in Urban Areas* (London: Her Majesty's Stationery Office, 1963).

future. The definition of a plan may vary from study to study. Some plans may include parking and terminal recommendations with proposals for new transit and highway facilities. Some may include "immediate action" programs for arterial improvements together with twenty-year or even longer-range freeway development plans. Some may detail precise highway locations with ramp connections pinpointed. Others may deal only in "corridors" and undetermined alignments.

Generally, the long-range studies are concerned not so much with precise locations or precise answers for design hour volumes as with the density and configuration of future freeway, arterial, and transit systems. Such plans are indications of the scale—in mileage, construction cost, and environmental impacts—of the facilities required to meet the travel expectations of a future population.

The approach is through systems analysis: studying the behavior of separate systems and the interactions between them. Thus, the transportation planner is concerned equally with the influence of land-use alternatives on transportation solutions and with the effect of different transportation alternatives on urban development patterns.

Within the framework of the total transportation system the planner is concerned with two subsystems, person transport and goods transport, although the latter has seldom received much attention. Within each of these systems he is concerned with the problems of passenger vehicle movement and storage (parking) and goods movement and storage (truck transfer and terminal facilities). Within the person movement system, particularly, he is concerned with analysis of the services, influences, and costs of different transport modes.

Planning scale. As understanding of the system interactions has grown and as computational capabilities have advanced, the planning process has become more complex. It has also become more costly. "Typically, transportation studies have cost, over the three-year (or longer) periods necessary to obtain data, prepare and test plans, and produce a final report, an amount equal to $1.00—$1.50 per capita. The large studies generally cost less than this (the Chicago study cost less than $1.00 per capita for a six-year initial effort) and the smaller studies sometimes more."[24] Looked at another way, the annual costs for seven very large studies (Baltimore, Boston, Chicago, Milwaukee, Minneapolis-St. Paul, New York metropolitan region, and Philadelphia) totalled about 5 million or about $750,000 each *per year.*[25]

These costs are best viewed in relation to the scale of the plans that they produce. Highways and transit lines can cost from $5 million per mile to many times that in highly developed districts, while combined mileages of recommended freeways and transit systems may total 100 or 200 miles or more. Thus, the cost for a major metropolitan area may easily exceed a figure of $500 million. Against such figures, planning costs (typically less than one percent of the total) are small if they lead to the desired direction for transportation and urban development.

ORGANIZING FOR LONG-RANGE STUDIES

Personnel organization. Several organizational forms have been used to carry out the planning process. An ITE report lists five types:

---

[24] ROGER L. CREIGHTON, *Urban Transportation Planning* (Urbana, Ill.: University of Illinois Press, 1970), p. 134.
[25] DAVID E. BOYCE, NORMAN D. DAY, and CHRIS McDONALD, *Metropolitan Plan Making* (Philadelphia: Regional Science Research Institute, 1970), p. 137.

1. *Centralized state staff.* This may be an existing agency or a new department incorporating the necessary multidisciplinary talents, as in New York State's Department of Transportation.
2. *Semi-independent organization.* The Chicago Area Transportation Study, established as an *ad hoc* joint effort and responsive to a multiagency board, illustrates this approach.
3. *Council of governments.* A study organization can be created under a council made up of elected representatives of communities within the region. The transportation planning programs in Washington, D.C., St. Louis, and Albuquerque are examples of this arrangement.
4. *Regional Planning Commission.* Established planning bodies for metropolitan regions are sometimes the organizations housing the transportation planning staff. Programs for Southeast Wisconsin, Southwest Pennsylvania, and the New York City metropolitan region are so organized.
5. *Contract Study Organization.* In this form, consultants under the supervision and monitoring of either a state representative or local study director perform all or some of the stages in the planning process. The procedure has been used extensively by most states.[26]

Direction and guidance. Regardless of organizational structure, there are comparable procedures for insuring that study activities are consistent with planning objectives. A common practice is to appoint a study director responsible to a committee structure made up of a policy committee, a technical committee, and a citizens advisory committee. Figure 12.6 illustrates the lines of communication possible with different organizational forms.

Policy Committee. The policy committee should include representatives of agencies participating financially in a study. In most cases, federal, state, and local officials will be members. In addition, executives of local transportation agencies and regional planning organizations should also be appointed. The viewpoint of small communities within the study area should be represented collectively by a designated elected or appointed official from one such jurisdiction. Total membership may be as few as four or five, but from ten to twelve is more common.

The function of the policy committee is to provide direction through:[27]

1. Budget control and decisions on major expenditures
2. Establishing rules and regulations for study personnel
3. Supervising key technical matters and providing liaison with other agencies
4. Aiding in plan development and judging alternative proposals by establishing objectives
5. Recommending a plan and a continuing planning program.

Technical Committee. Agencies represented on the policy committee normally appoint technical personnel to the technical committee. Other local agencies not directly represented on the policy committee, for example, airport management, traffic

---

[26] CLYDE E. SWEET, JR., *Guidelines for the Administration of Urban Transportation Planning* (Washington, D.C.: Institute of Traffic Engineers, 1969), p. 6.

[27] RALPH A. MOYER, "Comprehensive Urban Transportation Study Methods," *Journal of the Highway Division*, XCI, No. HW2 (1965), p. 62.

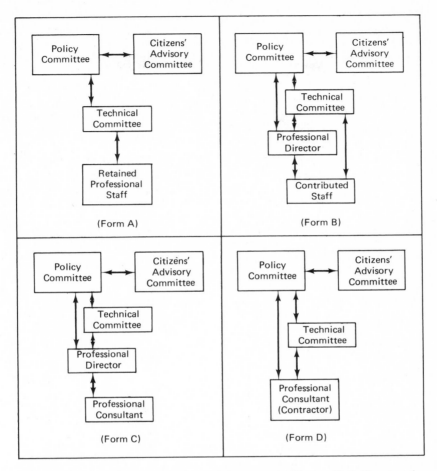

Figure 12.6. Alternate structures for study management. [Source: R. J. Hensen and W. L. Grecco, "Evaluation of the Effectiveness of Transportation Planning in the Smaller Urban Area," *Traffic Quarterly*, XXIV No. 3 (1970), 398. By permission of ENO Foundation for Transportation, Inc.]

engineering, urban renewal, and public utilities, might also be invited to participate.

The role of the technical committee is to guide the study organization in technical procedures and standards. Its principal functions are:

1. Reviewing and evaluating study methods and procedures
2. Assisting in developing alternative plans; performing technical evaluations and making recommendations to the policy committee
3. Coordinating technical service contributions of participating agencies
4. Enlisting and sustaining the interest of local agencies in the transportation planning process.

Citizens Advisory Committee. The composition and function of this committee are subject to wide variation. Its values depend on the size of the study area, the effectiveness of public relations in local government, and the interest of communications media in study objectives.

The communications media should certainly be represented on the committee, as should business leaders and leaders of civic groups interested in community betterment.

The committee serves an important role by providing information to the policy committee and study staff on public thinking. In this way it can assist in defining plan goals and objectives. It serves further to improve public understanding of the planning process, to encourage acceptance of study recommendations, and to build public support for plan implementation.

Study Staff. Staffing arrangements are clearly related to the organizational structure described earlier. Regardless of organizational form, however, the overall personnel requirements will be approximately the same. Figure 12.7 portrays personnel needs for a large metropolitan area study for a three-year period. The peak personnel demands occur during the first year when survey crews carry out roadside and home interviews and clerical staff code the resulting data for machine processing. The number of personnel and office space requirements then drop to a level of one-third to one-quarter of the peak. The maximum payroll for a study of the scope illustrated might be approximately 200 people. By the plan analysis stage, the staff might drop to 40 or fewer.

A variety of staff disciplines is needed throughout or at selected stages of the planning process. "This staff should be a team composed of urban planners, urban transportation analysts, traffic engineers, highway engineers, transit engineers, computer programmers, mathematicians, economists, and other specialists. Without an interdisciplinary staff it is nearly impossible to perform the broad range of work

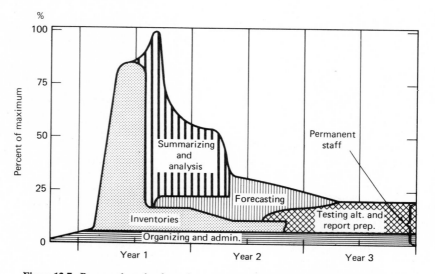

Figure 12.7. Personnel needs of a major transportation study expressed as a percent of maximum. [Source: Clyde E. Sweet, Jr., *Guidelines for the Administration of Urban Transportation Planning* (Washington, D.C.: Institute of Traffic Engineers, 1969), p. 26.]

involved in a transportation study."[28] Supporting staff will include secretarial and clerical help, data processing specialists, field crews and their supervisors, draftsmen, and administrative assistants. Additional staff services are commonly provided by participating agencies.

Funding and study costs. The Department of Transportation and the Department of Housing and Urban Development have been the principal sources of funds for urban transportation studies in the United States. Both departments have programs that require matching by state and/or local contributions. Contributions from the Department of Transportation normally come through the annually available Highway Planning and Research funds of the Federal Highway Administration, but special Urban Mass Transportation Administration grants may also be sought for special studies related to mass transit. In the first case, the federal-state matching requirements will be determined on a uniform statewide basis, and state departments of transportation or highways may then work out individual arrangements with participating local communities. In Pennsylvania, for example, local communities are required to provide 15 percent of data collection costs.

Section 701 funds from the Department of Housing and Urban Development are the other major source of federal contributions. These planning funds may be similarly matched by state or local agencies to carry on urban planning. Their use in transportation planning is generally related to land-use inventories or forecasts of population distribution and land use. In many cases they directly support a service contribution of a planning agency instead of being channeled through the transportation study organization.

Study costs, like personnel requirements, vary throughout the course of a typical study. Table 12.1 summarizes costs for each technical program phase, based on data from eight separate urban studies. Data collection and preparation expenditures far outweigh costs for other activities. In the future, however, as knowledge of urban travel characteristics improves and the availability of data from external sources like the census increases, inventory costs seem likely to drop. At the same time, greater emphasis will be placed on model testing and development.

TABLE 12.1
Study Costs by Phase

Phase	Average Cost	Percent of Total
Data collection and preparation	$250,000	49
Developing and testing model	31,000	6
Forecasting	31,000	6
Assignments	20,000	4
Analysis of alternative systems	80,000	15
Report preparation	37,000	7
Other	68,000	13
	517,000	100

Source: Thomas J. Hillegass, "Urban Transportation Planning—A Question of Emphasis," *Traffic Engineering*, XIX, No. 7 (1969), p. 47.

---

[28] *Principles and Practices of Urban Planning* (Washington, D.C.: International City Managers Association, 1968), p. 154.

THE TRANSPORTATION PLANNING PROCESS

The principal stages in the transportation planning process have already been introduced in the illustrations of personnel needs and study costs. They are summarized in Figure 12.8. The first item includes obtaining agreement on the funding, participation, and organizational form, setting up the committee structure, and arranging for staffing the study. The process itself then consists of six steps that lead to plan recommendation and implementation: setting goals and objectives, data collection, analysis, forecasting, alternate plan development, plan testing and evaluation.

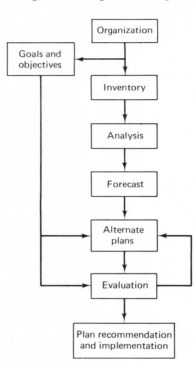

Figure 12.8. Major steps in the urban transportation planning process.

SETTING GOALS AND OBJECTIVES

Studies are established to carry out the continuing, comprehensive, and cooperative process for planning urban transportation facilities. Their success largely depends on agreement on the goals and objectives of the plan to be selected.

Goals and objectives are the essential bridge between technical evaluation and the community. The evaluation process combines quantitative standards or relatively precise criteria with judgments. Setting goals and objectives early in the planning process directs the quantitative rankings and judgments toward conformity with underlying values of the community. One study[29] has depicted the relationship as a hierarchy of values, goals, objectives, criteria, and standards as follows:

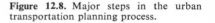

[29] EDWIN N. THOMAS and JOSEPH L. SCHOFER, *Strategies for the Evaluation of Alternative Transportation Plans*, National Cooperative Highway Research Program Report 96 (Washington, D.C.: Highway Research Board, 1970), p. 40.

1. *Values* are basic social drives that govern human behavior. They include the desire to survive, the need to belong, the need for order, and the need for security.
2. *Goals* define conditions to be achieved, as environments favorable to maximizing values. They can be stated, but the degree of their achievement may not be definable. "Equal opportunity," for example, is a goal based on the values of security and belonging.
3. *Objectives* are specific, attainable, and measurable. In relation to the goal of equal opportunity, a transportation objective might be equal public transportation costs for all citizens regardless of location within the city.
4. *Criteria* are the measures or tests to show whether or not objectives are attained. For example, the ratio of transit fares to personal income may be the criterion for determining whether or not the foregoing equal transportation cost objective has been met.
5. *Standards*, as with highway design standards, establish a performance level that must be equalled or surpassed. To continue with the previous example, transit service within one-quarter mile of every residence would be a standard.

It can be seen that each value may lead to several goal statements. In turn, each goal may lead to several objectives. One objective may serve more than one goal, and each objective may call for setting up one criterion or more than one. Collectively, there may be a dozen or more criteria by which to measure a plan's effectiveness in serving several goals.

The study's guiding committees should play the largest role in framing goals and objectives. Citizen advisory committees aid in reminding the administratively or technically oriented committees of community values and should help policy committees postulate the transportation planning goals. Policy committees and technical committees have a joint interest in defining objectives. Criteria and standards are a particular concern of technical committees and study staff. Study personnel should, desirably, also be called on to assist in the formulation of goals and objectives.

Each study should make its own unique assessment of goals, but three fundamentals may be suggested: (1) increased and more equitable accessibility, (2) greater urban opportunity, and (3) efficiency in the use of resources. More precise goals might be defined as follows:

Accessibility

1. Provide equitable access to employment and education
2. Enable a range of choice and diversity of urban experience
3. Enhance the opportunity for individuals to develop a stronger sense of community

Opportunity

1. Plan and locate activities in a rational arrangement that citizens can understand
2. Provide the amenities of convenience, safety, beauty, and health
3. Facilitate variety and ease of contact

Efficiency

1. Conserve human and natural resources
2. Practice efficiency and economy in the use of public funds[30]

Several objectives can be listed in relation to the foregoing goals.

1. Reduce transportation costs by reducing accident rates and severity, unit costs of vehicle operation, and time spent in travel
2. Provide transportation facilities for nonauto users
3. Foster compatibility of transportation and adjacent land uses
4. Minimize disruption of existing communities in new facility construction
5. Reduce air pollution, noise, and other nuisances from transportation activities
6. Obtain maximum transportation benefits for level of investment feasible

Criteria by which to determine the success of alternative plans will be outlined later under plan evaluation.

DATA COLLECTION

Understanding the relationships between land use, travel facility, and tripmaking characteristics is essential to good planning. Analyses will be simplified if geographic coding of study area zones and network link coding can be interrelated and if land-use categories are consistent between land-use surveys and travel survey coding. Data should be interrelated not only for internal use but also for the benefit of service agencies which may be able to draw on the transportation study files as a data bank. Zone boundary compatibility with census boundaries has particular value for the latter purposes.

Inventories of economic activity and population. Typical transportation study data include:

1. *Historic population patterns.* Past distribution, migrations, density, and trends in growth
2. *Present population.* Distribution by area and makeup by race, density, average income, car ownership, and dwelling unit type
3. *Employment trends.* Historic patterns of employment by occupation and location
4. *Present employment.* Employment by industry or activity and total labor force
5. *Economic activity.* Patterns of investment in manufacturing, services, redevelopment, and other real estate
6. *Transportation resources.* A review of past and present outlays for regional transportation facilities

Land-use inventories. Typical transportation study data include:

1. Historic development trends such as patterns of urbanization by decade

---

[30] RODNEY E. ENGELEN and DARWIN G. STUART, "Development Objectives for Urban Transportation Systems," *Traffic Quarterly,* XXIV, No. 2 (1970), pp. 250–53.

2. Topography and physical constraints on development
3. Acres of land in urban use, by detailed type of use
4. Acreage of vacant land, classified by unusable and usable and by public and private ownership
5. Location of major travel generators
6. Identification of neighborhood and community boundaries (social rather than political or physical)
7. Nature of existing land-use controls: zoning, official maps, subdivision regulations
8. Identification of redevelopment areas

Transportation facilities. Assessing the physical and performance characteristics of the existing transportation network is essential for making use of and augmenting the present systems in the planning process. Several studies are usually needed.

Classification of highways and streets by jurisdiction and function identifies primary and secondary networks (see Chapter 13). Physical inventories provide inputs for analysis of system spacing, coherence and unity, route capacity, and potentials for improvement. Speed and delay studies indicate quality of service, and accident compilations identify safety problems.

Transit service quality can be gauged by analysis of route maps and service schedules, but equipment inventories and analysis of management policies also may be necessary. In addition, vehicle occupancy and schedule adherence checks can be made in order to determine performance.

Parking and terminal facility surveys. Comparison of parking demand and supply in CBD's and other centers, for example, makes it possible to assess the potential for gaining street capacity through curb parking prohibition. Problems associated with mode transfer terminals (as at airports, railroad stations, and freight yards) require definition before solutions can be advanced.

Travel characteristics. Traffic volumes on existing streets establish a base condition against which trip survey and travel model estimates can be checked. Machine coverage counts, manual classification counts, and hourly screenline volume counts are needed.

The costliest surveys, however, are those that provide comprehensive data on person travel—by mode, by time, by purpose, by land use, and so on. Surveys of external trips which have at least one trip-end outside of the study area boundaries, home interview surveys of household tripmaking, and truck-taxi surveys call for the largest part of the study's manpower and budget. Once these surveys are reduced to cards and tape records, they are the basis for developing and testing trip generation and distribution models.

Table 12.2 summarizes the travel survey data collected by the Lehigh Valley (Pennsylvania) Transportation Study. The planning program covered a 173 sq mi area containing the three major cities of Allentown, Bethlehem, and Easton and a study area population of nearly 350,000. The home interview study included a 10 percent sample of households; the truck survey, a 20 percent sample of the trucks registered in the study area; and the taxi survey, a 100 percent sample of the vehicles operated on the survey day. Figure 12.9 suggests what is involved in setting up such a survey operation and in processing the raw data.

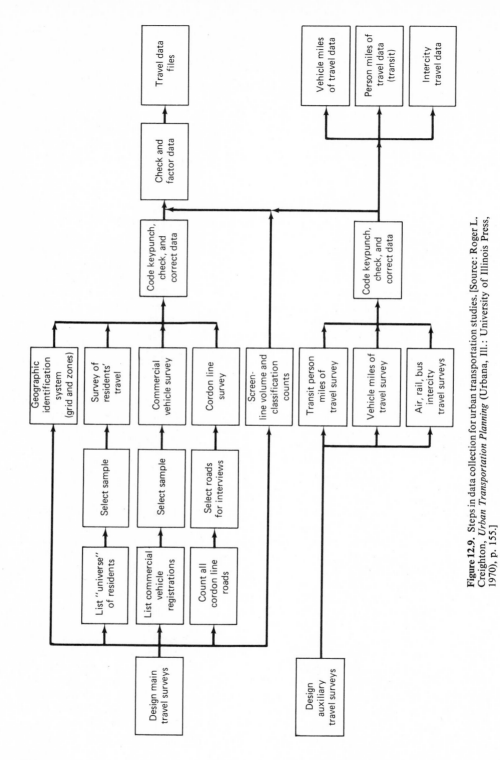

Figure 12.9. Steps in data collection for urban transportation studies. [Source: Roger L. Creighton, *Urban Transportation Planning* (Urbana, Ill.: University of Illinois Press, 1970), p. 155.]

TABLE 12.2
Travel Survey Characteristics, Lehigh Valley Study Area

Survey	Number of Completed Samples	Daily Trips*
Home interview	10,446	745,971
Truck	1,798	82,172
Taxi	111	8,877
Transit Questionnaire	2,532	—
External roadside	93,000	167,000

*Trips are vehicle trips except for home interview figure, which is person trips.

Source: Lehigh Valley Transportation Study, *Phase I Data Collection and Inventories*, Pennsylvavia Department of Highways, May 1966.

ANALYSIS

The first phase of a long-range transportation study is to assemble data on current conditions for the following purposes:

1. To identify the bases from which to forecast future land use and travel
2. To identify the scale of current inadequacies in transportation systems
3. To provide inputs from which to derive land-use–travel relationships
4. To provide inputs from which to calibrate trip distribution and network assignment models

Figure 12.10 shows schematically how the study progresses from inventories into analysis, and from these to forecasts.

Economic and population forecasting techniques. Specialists in economics, geography, demography, and other social sciences use various techniques for developing forecasts of population and economic activity. The simplest involve either extrapolations of current trends or projections of current relationships between the study area and a larger region whose future activities already have been forecast. More complex are methods that aggregate separate forecasts for manufacturing, service, and government sectors of the economy. The most advanced techniques examine the impacts of change in one sector on other sectors by using input-output analyses or "behavioral models" to generate future patterns of urban development.[31]

Land-use forecasting techniques. The future patterns of urban land use must be derived before future travel patterns can be estimated. Some models of economic and population forecasting are designed to project land use as the end result. When they are not used it is necessary to forecast the amounts and location of vacant land that will be converted to residential and other uses to support the activities of growing urban populations.

The procedure, as indicated in Figure 12.11, starts with population and economic growth, existing land uses, travel and network characteristics. Then it projects their influences to the forecast year and allocates the results to study area zones. Future land use may be projected by extension of trends, or alternatively, planned patterns of

---

[31] *Urban Development Models*, Special Report No. 97 (Washington, D.C.: Highway Research Board, 1968).

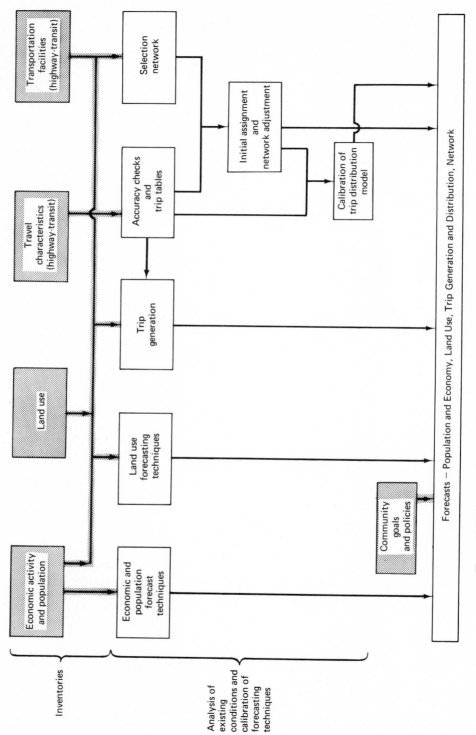

Figure 12.10. Steps in the analysis stage of urban transportation studies. [Source: Derived from *Guidelines for Trip Generation Analysis* (Washington, D.C.: U.S. Bureau of Public Roads: June, 1967).]

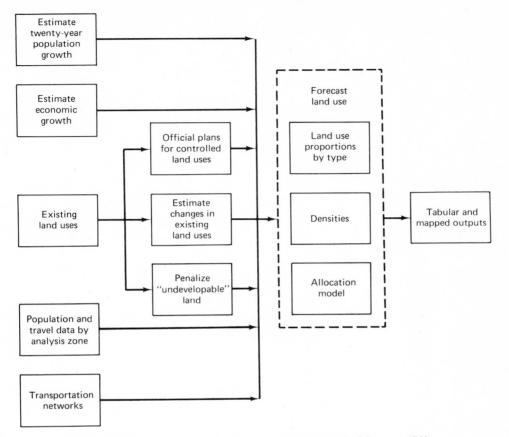

Figure 12.11. Steps in forecasting future land use. (Source: Creighton, p. 176.)

urban development may be projected. For example, new growth may be accommodated by development of high-density radial corridors serving an expanded CBD, or all new growth may be projected to take place in suburban satellite centers around a central city whose further growth would be restricted. How the accessibility offered by different transportation plans may affect locational choices can be examined by simulation models.[32]

Accuracy checks and trip tables. It is essential to check the validity of travel surveys and their expansions from sample survey data. The first checks are made by experienced personnel before interview forms are keypunched so that missing or inconsistent survey information can be quickly accounted for in the field. Machine contingency checks then verify the accuracy of coding and the consistency of travel and household data.

Next, survey data are factored to a base of average weekday travel according to the adjusted sampling rate, which accounts for households in which interviews could not

[32] GEORGE T. LATHROP and JOHN R. HAMBURG, "An Opportunity Accessibility Model for Allocating Regional Growth," *Journal of the American Institute of Planners,* XXXI, No. 2 (1965), p. 95.

be completed. The expanded trips can be checked against screenline volume counts. Trips that must have crossed screenlines between origin and destination can be compared to the classification counts of vehicles crossing in equivalent time periods. If discrepancies exceed 5–10 percent, trip survey data may be factored again.

Other checks will also be made. For example, household survey data on population and automobile ownership, when expanded and totalled by area, should be comparable to census information. Estimates of total auto travel, after factoring survey reported airline trip lengths to over-the-road distances, can also be compared to vehicle miles of travel estimates derived from ground counts.

Standard tables describing average daily travel and household characteristics are then prepared. The tables provide the measure of present conditions: the household characteristics and trip ends by district, the number of trips by mode, by trip purpose and by time of day, the zone-to-zone movements by vehicle type, and trip destinations by type of land use.

Selection of networks. The designation of the arterial highway network, which serves as a "given" in all subsequent traffic assignments, is critical to the planning process. Traffic volume counts identify routes that have major volumes, and network planners consider this information, together with other highway classification results and other factors, in establishing the basic network. For initial testing, all streets in major use should be included, even though some may be excluded later to suit desirable planning objectives. Additionally, because all zones must be interconnected adequately, a number of minor or local streets may be selected in outlying low-density districts.

Once it is designated, the network must be coded for traffic assignment purposes. Each link is identified by a pair of code numbers representing its terminal intersections. These may be geographically or otherwise systematically coded to facilitate later computer mapping or summary of outputs by geographic areas. They may be further identified by type of route (freeway, ramp, arterial, dummy, or zone centroid link). Additional information typically needed for each link is its length, average speed, travel time, average daily traffic, and capacity. A schematic diagram of a network section is given in Figure 12.12.

Testing the network. Initial traffic assignments by computer will verify the accuracy of network coding and the reasonableness of minimum paths between zones. The computer determines the quickest route between origin and destination by using a *minimum path algorithm* and assigns all trips between a given origin and destination to this route.[33] For example, trips will not follow a devious arterial path when a nearby freeway may be more logically used; if assignment results show that they do, the network should be checked for missing ramps or erroneous link travel times. Generally, so-called "capacity restraint" techniques are combined with assignment programs to limit maximum link loads in some relation to calculated link capacities.

The coded network provides an opportunity to depict the quality of existing traffic service. Ratios of volumes-to-capacity by link or by area show the location and degree of network capacity deficiencies or surplus. Traffic flow maps, sometimes constructed directly by data plotters, show congestion points and the relative importance of differ-

---

[33] For details on the minimum path algorithm, see Creighton, *op. cit.*, p. 248.

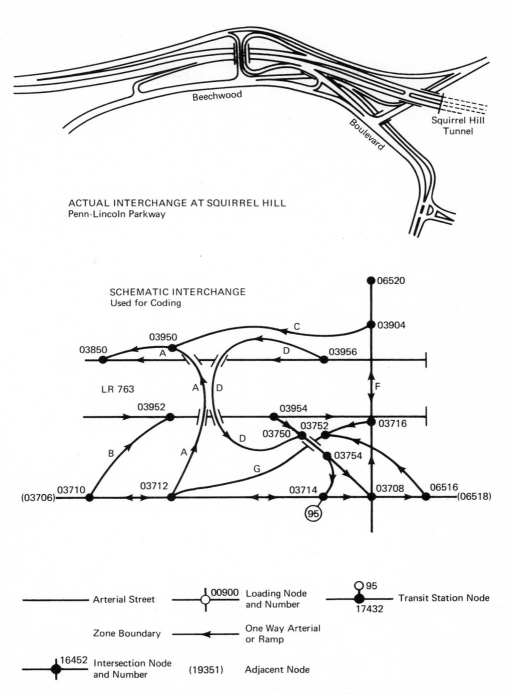

Figure 12.12. An example of network coding procedures. (Source: Southwestern Pennsylvania Regional Planning Commission.)

ent routes. Or, tabulations that compare travel and capacity by rings centered on the CBD can be made. (See Table 12.3.)

At this point the network is ready to serve as the base to which future transportation plan proposals will be added. Just as important, it serves as the framework upon which trip distribution models are calibrated and tested. (Essentially parallel procedures may be followed for developing a transit network.)

TABLE 12.3
Comparison of Vehicle-miles of Travel
and Design Capacity by Ring

Ring	Capacity (000)	Travel (000)	Capacity Minus Travel (000)	Travel to Capacity Ratio
0	164	209	− 45	1.27
1	328	454	−126	1.38
2	624	707	− 83	1.13
3	1,006	1,144	−138	1.14
4	1,055	1,133	− 78	1.07
5	1,833	1,778	+ 55	0.97
6	2,362	1,815	+547	0.77
7	1,202	790	+412	0.66
Total	8,574	8,030	+544	0.94

Source: *Study Findings*, Pittsburgh Area Transportation Study, Vol. 1, Pittsburgh, Pa., November 1961, p. 72.

Trip distribution. Current travel can be computer simulated by assigning survey-based trip movements to a transportation network. But what about the forecasted travel for the design year? Estimating the future travel pattern is necessary, and several methods are available.[34]

The Fratar Method. Current tripmaking between zonal pairs is expanded to future levels by a growth-factor method, which requires several iterations or adjusting steps so that interchanges ultimately balance by direction. Mathematically, the method may be described as:

$$T_{ij}(k + 1) = (T_{ijk}F_{jk})F_{ik} \qquad (12.1)$$

where

$$F_{jk} = \frac{T_j}{\sum\limits_{i=1}^{n} T_{ijk}}$$

and

$$F_{ik} = \frac{T_i}{\sum\limits_{i=1}^{n} T_{ijk}F_{jk}}$$

---

[34] For more detailed discussions, see *Calibrating and Testing a Gravity Model for Any Size Urban Area* (Washington, D.C.: U.S. Bureau of Public Roads, 1968) or *Traffic Assignment Manual* (Washington, D.C.: U.S. Bureau of Public Roads, June 1964) or *Urban Transportation Planning—General Information and Introduction to System/360* (Washington, D.C.: U.S. Department of Transportation, Federal Highway Administration, Bureau of Public Roads, June 1970).

where T_{ijk} = trips between i and j for iteration k,

$\quad F_{jk}$ = destination factor j,

$\quad F_{ik}$ = origin factor i,

$\quad T_j$ = final desired total for destination j,

$\quad T_i$ = final desired total for origin i,

$\quad i$ = origin zone number,

$\quad j$ = destination zone number.

Gravity Model. In this technique, the trip volume from Zone A to Zone B is a direct function of the product of trip volumes in both zones and an inverse function of the time or distance separating them. Mathematically, the standard model may be expressed as:

$$T_i = \frac{P_i A_j F_{ij} K_{ij}}{\sum\limits_{j=1}^{n} (A_j F_{ij} K_{ij})}, \quad i = 1, 2, \ldots, n \tag{12.2}$$

where T_{ij} = trip interchange between i and j,

$\quad A_j$ = attraction factor for zone j,

$\quad T_{ij}$ = trips produced at i and attracted at j,

$\quad P_i$ = total trips produced at i,

$\quad A_j$ = total trips produced at j,

$\quad F_{ij}$ = calibration term for interchange ij,

$\quad K_{ij}$ = socioeconomic adjustment factor for interchange ij,

$\quad i$ = an origin zone number,

$\quad n$ = number of zones.

The calibrating term, F_{ij}, is generally found to be an inverse exponential function of impedance. It is not obligated, however, to take that particular form. This elasticity is, perhaps, one of the major strengths of the model.

When all trip interchanges have been computed according to Eq. (12.2), production (row) totals will be correct because of the structure of Eq. (12.2), the gravity model formula. Attraction (column) totals, however, will not necessarily match their desired values. An iterative procedure is employed to refine calculated interchanges until actual attraction totals closely match the desired results.

After each iteration, adjusted attraction factors are calculated according to the following:

$$A_{jk} = \frac{A_j}{C_{j(k-1)}} A_{j(k-1)} \tag{12.3}$$

where A_{jk} = adjusted attraction factor for attraction zone (column) j, iteration k

$\quad A_{jk} = A_j$ when $k = 1$,

$\quad C_{jk}$ = actual attraction (column) total for zone j, iteration k,

$\quad A_j$ = desired attraction total for attraction zone (column) j,

$\quad j$ = attraction zone number, $j = 1, 2, \ldots, n$,

$\quad n$ = number of zones,

$\quad k$ = iteration number, $k = 1, 2, \ldots, m$,

$\quad m$ = number of iterations.

In each iteration, the gravity model equation is applied to calculate zonal trip interchanges using the adjusted attraction factors obtained from the preceding iteration. In practice, the gravity model equation thus becomes:

$$\left[T_{ijk} = \frac{P_i A_{jk} F_{ij} K_{ij}}{\sum\limits_{j=1}^{n} (A_{jk} F_{ij} K_{ij})} \right]_p \tag{12.4}$$

where T_{ijk} is the trip interchange between i and j for iteration k and $A_{jk} = A_j$ when $k = 1$. Subscript j goes through one complete cycle every time k changes, and i goes through one complete cycle every time j changes. Eq. (12.4) is enclosed in brackets which are subscripted p to indicate that the complete process is completed for each trip purpose. It is equivalent to placing a subscript p on every variable in Eq. (12.4). Trip volumes may be stratified by trip purpose, and the trip distribution process is completed separately for each purpose. Values for K_{ij} are determined in the calibration process so that simulated trips reasonably duplicate survey-recorded travel patterns. Several iterations may be required to balance the attractions and, ultimately, trip distributions.

Intervening Opportunities Model. This model describes the probability that a given destination zone will be accepted by a trip in relation to the potential trip ends available there and those that it may have encountered within a shorter travel time. The model is described as:

$$T_{ij} = O_i(e^{-LB} - e^{-LA}) \tag{12.5}$$

where T_{ij} = trips between origin zone i and destination zone j,
O_i = total trip origins produced at zone i,
e = constant base of natural logarithms = $2.71828\ldots$,
A = sum of all destinations for zones between, in terms of closeness, i and j and including j,
B = the sum of all destinations for zones between i and j but excluding j,
L = the probability density (probability per destination) of destination acceptability at the point of consideration.
The value of L may be derived for one or more trip categories so that the trip length distribution in the model compares to that of the survey data.

Direct Assignment. The newest (but least used) method does not calculate zone-to-zone movements prior to traffic assignment. Instead, this model assigns trips directly to a link based on its position with respect to surrounding trip density patterns. Descriptions of the model may be found in reports of the Tri-State Transportation Commission.[35]

Whichever method is used, the essential step is to calibrate the model so that it successfully reproduces the network loading patterns created by previous assignment of the trip survey data. When this is achieved, the model may then be applied to forecast year inputs.

Trip generation and other analysis of current data. Trip generation analysis is the key to obtaining future trip ends by zones. The basic procedure is first to relate survey-reported tripmaking to household characteristics and the land-use types by zone through regression or factor analysis, using either single variable or multivariate approaches.[36] The equations or rates thus derived may then be applied to forecasted

---

[35] "Direct Estimation of Traffic Volume at a Point," "Direct Traffic Estimation—Method of Computation," "Direct Traffic Estimation—Test Case Results," *Tri-State Transportation Commission Interim Technical Reports* (New York, Tri-State Transportation Commission) or Morton Schneider, "Direct Estimation of Traffic at a Point," *Origin and Destination—Advances in Transportation Planning*, Record No. 165 (Washington, D.C.: Highway Research Board, 1967).

[36] *Guidelines for Trip Generation Analysis* (Washington, D.C.: U.S. Bureau of Public Roads, June 1967) or A. A. Douglas and R. J. Lewis, "Trip Generation Techniques," *Traffic Engineering and Control*, XII, No. 7 (1970), 362–65 and XII, No. 10 (1971), pp. 532–35.

land-use data. The trip generation analysis requirements for any study are a function of the kind of trip distribution model to be used, the sequence of modal split analysis employed in the forecast process, and the data available in forecasting. Because gravity models may stratify trips by as many as eight purposes, for example, trip generation analysis may require the following to be forecasted by zone:

1. Home-based trips to work, shopping, social-recreation, schools, and all other purposes
2. Nonhome-based trips (those with neither origin nor destination at home)
3. Truck trips
4. Taxi trips

Almost all trip generation studies predict person trips. These may then be separated into categories of vehicle and transit trips before trip distribution, or this step may occur after trip distribution, as Figure 12.13 shows. Both techniques are commonly used. A description of the methodology followed by different urban studies can be found in *Modal Split—Documentation of Nine Methods for Estimating Transit Usage*.[37]

The foregoing analysis procedures are directed toward the long-range planning objectives of the transportation study. Certain analyses may be required for short-range, immediate-action programs or for other benefits to participating agencies. These could include:

1. Measurements of parking adequacy in central areas
2. Identification of accident and congestion locations (using speed and delay data, for example)
3. Reports on traffic flow and other volume count data
4. Reports on land-use and population data by small areas

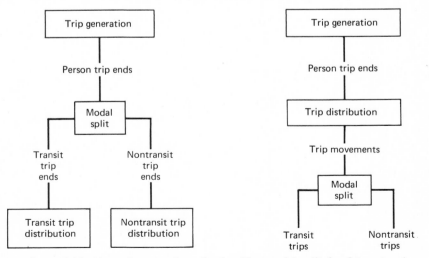

Figure 12.13. Alternative procedures for handling modal split for future travel. (Source: *Guidelines for Trip Generation Analysis*), p. 47.

[37] *Modal Split—Documentation of Nine Methods for Estimating Transit Usage* (Washington, D.C.: U.S. Bureau of Public Roads, December 1966).

FORECASTING

The interactions of land use and transportation are explicitly recognized at this point in the long-range study as forecasts of land use (which in themselves may be shaped by concepts of transportation improvements) and become the basis for travel estimates against which plans are tested for adequacy.

Target dates. Selecting a design year (or "horizon" year for planning) is usually a compromise. It must be far enough in the future so that major programs can be initiated and construction staged according to funding availability. Yet the design year cannot be so far in the future that forecasts of future development and traffic will have doubtful reliability. Thus, forecasts in the 15 to 25 year range are most common. In certain situations it will be easier for laymen to understand projected levels of anticipated transportation demand which may actually occur prior to, during, or after the design year.

Land-use projections. An example from the Lancaster Area (Pennsylvania) Transportation Study illustrates the scale of growth that may be predicted for a middle-sized metropolitan area. In Table 12.4 a mid-range forecast for 1975 accompanies the

TABLE 12.4
Projections of Social and Economic Statistics, 1963, 1975, 1990

Category	Year			Ratio	
	1963	1975	1990	1975/1963	1990/1963
Population	107,610	150,949	188,765	1.40	1.75
Employment	43,277	60,235	74,804	1.39	1.73
Labor force	44,733	59,364	72,940	1.33	1.63
Median family income	6,190	9,040	13,942	1.46	2.25
Dwelling units	34,925	46,874	58,891	1.34	1.69
Population density (per acre)	2.9	4.1	5.1	1.41	1.76
Persons per dwelling unit	3.1	3.2	3.2	1.03	1.03
Cars owned	34,946	60,502	89,933	1.73	2.57
Persons per car	3.1	2.5	2.1	0.81	0.68
Cars per dwelling unit	1.0	1.3	1.5	1.30	1.50

Source: Lancaster Area Transportation Study, Volume II, Analyses and Forecasts (Harrisburg, Pa.: Pennsylvania Department of Transportation, November 1970), p. 38.

long-range predictions of socioeconomic data for 1990. (Ratios serve to emphasize those elements that change the most; they serve further to check internal consistency among separate forecast elements.) The impact of the growth statistics may best be judged by their effect on land-use patterns. Residential land use was predicted to increase from 5,800 to 13,800 acres, industrial and commercial uses from 1,600 to 4,200 acres, and all other uses from 5,800 to 10,700 acres. The supply of agricultural and vacant land was expected to shrink from 23,700 acres to 8,200 acres.[38]

The forecasted land uses must be distributed next by zone. The nature of proposed transportation investments may affect this distribution, but the principal controls are the density of development by land-use type and associative links, for example, the relationship of street, public, and commercial space with residential patterns.

[38] *Lancaster Area Transportation Study, Volume II, Analyses and Forecasts* (Harrisburg, Pa.: Pennsylvania Department of Transportation, November 1970), p. 38.

ALTERNATE PLAN DEVELOPMENT

Preparing alternative land-use plans. Many of the major studies have developed several land-use alternatives instead of preparing a single forecast or plan of land use. The alternatives may be developed to:

1. Challenge or confirm existing recommended plans
2. Discover whether or not one land-use form offers particular advantages over another
3. Probe community values and provoke public discussion on key issues
4. Educate the public in the values of planning
5. Identify the need for change in financing or government organization to facilitate plan implementation[39]

Alternate concepts may be shaped by several factors. Different "plan-forms" of radial corridors, linear cities, dispersed or "spread" cities, or satellite nucleations around a central city may be the alternatives designed for testing. Or different patterns of development densities and dispersion may be tested. The forms of transportation development themselves may be determinants; possible alternatives are networks that emphasize:

1. Highways vs. transit
2. Arterials vs. freeways
3. Grid vs. radial layouts
4. Minimum vs. maximum construction

Table 12.5 lists alternatives developed by six studies in the late 1960's. Four of the six employed computer growth models to generate detailed alternatives for evaluation.

TABLE 12.5
General Type and Identifying Names of Alternatives in Programs Reviewed

Program	General Type of Alternative						
	Current Trends	Modified Trends	Composite Plan	Corridor Development	Dispersed Development	Nucleated Development	Satellite Towns
Baltimore	Unplanned development	Trend				Metrotowns	
Bay Area			Composite of local plans	Urban corridor	Suburban dispersal	City centered	
Boston			Composite of others	Corridor	Spread	Nucleated	
Chicago	Trends						
Milwaukee	Uncontrolled growth	Controlled existing trends		Corridor			Satellite city
Twin Cities	Present trends			Radial corridor	Spread city	Multiple centers	

Source: David E. Boyce, Norman D. Day, Chris McDonald, *Metropolitan Plan Making* (Philadelphia: Regional Science Research Institute, 1970), p. 40.

[39] BOYCE, DAY, McDONALD, *op. cit.*, p. 30.

Preparing highway plan alternatives. How many miles of new or improved facilities are needed to meet future travel needs? A rough yardstick would indicate one mile of freeway per 10,000 of forecast urban population. By adopting planning principles and design standards and recognizing the constraints on network development, however, much better answers can be found.

Freeway systems. Fundamentally, freeway spacing should relate to the characteristics of population density and freeway capacity and to the balance between construction costs and costs of travel. Table 12.6 illustrates the first principle and Figure 12.14 the second. Table 12.6 gives the planner an opportunity to determine not only the scale of the planning network, but also a spacing arrangement that can be related to population density.[40] It does not introduce cost criteria, however. Figure 12.14

TABLE 12.6
Freeway Spacing vs. Population Density

| | Grid Spacing in Miles | | |
Population Density	4-lane	6-lane	8-lane
4,000 psns/sq mi	5.0	7.5	10.0
8,000 psns/sq mi	2.5	3.8	5.0
12,000 psns/sq mi	1.7	2.5	3.3

Source: *System Considerations for Urban Freeways* (Washington, D.C.: Institute of Traffic Engineers, 1969), p. 4.

MINIMUM COST SOLUTION

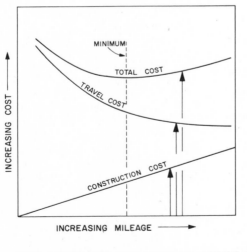

BEYOND A CERTAIN POINT, ADDED CONSTRUCTION INVESTMENT BRINGS A LESS THAN EQUAL SAVING IN TRAVEL COSTS.

Figure 12.14. The minimum total cost concept. [Source: *System Considerations for Urban Freeways* (Washington, D.C.: Institute of Traffic Engineers, 1967), p. 5.]

[40] *System Considerations for Urban Freeways* (Washington, D.C.: Institute of Traffic Engineers, 1967), pp. 4–5.

does, and the principle can be applied to derive desirable freeway spacing as a function of construction costs, travel costs (reflecting network speeds), and average trip lengths.[41] Minimum community costs for construction and travel combined (which could be a primary objective) may be achieved by interrelating the three underlying system mileage determinants of trip density, travel characteristics, and design criteria with such local factors as spacing of other arterials and physical characteristics. Because of the variation in these factors, each urban area will have a different optimum spacing. However, the optimum freeway spacing will normally fall between two and seven miles.

Freeway network design should recognize some basic principles, whether plans follow a grid, ring and radial, or irregular pattern for the system:

1. There should be continuity of both route and capacity. Stub ends and offset T intersections of freeways should be avoided. When freeways must end, direct connections with arterial routes of sufficient capacity to match that of the freeway should be provided.
2. Except under unusual conditions, interchanges should provide for movements in all directions.
3. Abrupt changes in capacity should be avoided. Where multimodal systems are involved, care must be directed toward providing continuity of capacity in terms of people rather than vehicles.
4. Staging of construction should consider continuity. Because by necessity completion of a system takes a number of years, care should be exercised to ensure continuity as each freeway portion is completed and added to the system.
5. As new freeways are built there should be a concurrent program of connecting arterial street improvements in order to avoid overloading arterials in the vicinity of the freeway ramps.
6. With the construction of freeway systems the primary functions of the individual arterial streets may change. Some streets will be most concerned with serving traffic; other will provide land service. Future arterial improvements should be such that they encourage use of the street consistent with its changed primary function.
7. Special attention should be given to providing freeway service to existing and planned major generators such as regional shopping centers, industrial parks, and major recreational facilities.
8. Care must be exercised to ensure that there is sufficient capacity at interchanges. In some cases collector-distributor facilities will be required to distribute the traffic from the freeway onto several arterial streets.
9. In undeveloped areas where there is potential for future developments, there should be enough flexibility to allow for addition of future interchanges should the need develop. The ultimate freeway plan should be developed initially, identifying the location of future interchanges.
10. The potential use of freeway surveillance and control to achieve maximum utility of the freeway network should be considered prior to the construction stage of the network development.

---

[41] ROGER L. CREIGHTON, IRVING HOCH, MORTON SCHNEIDER, and HYMAN JOSEPH, "Estimating Efficient Spacing for Arterials and Expressways," *Traffic Origin and Destination Studies*, Highway Research Board Bulletin 253 (Washington, D.C.: Highway Research Board, 1960).

11. Mass transit should be considered during the planning stage of freeway networks in order to ensure that the transportation modes complement one another.
12. The environmental effect of the total freeway system design on the land uses and urban development should be given careful consideration.

Levels of service—probably varying by regions within the study area—must be set for capacity standards. Similarly, speeds for network assignment purposes must be set by area and by route type.

Alternative network plans are influenced just as much by planning constraints as by planning principles. The constraints include:

1. Patterns of existing arterial networks, inside and outside the region
2. Commitments already made for future construction
3. Neighborhood development patterns and regional facilities, for example, reservoirs, parks, and cemeteries
4. Topographical and manmade impediments, for example, slopes, rivers, railroads, and high-density developments
5. Requirements for compatibility with mass transit development
6. Availability of funds for construction

With due recognition to these problems, sketch maps of designs can be worked up for subsequent coding and testing by traffic assignment.

Arterial streets. During this planning stage, consideration should be given to improving both arterial spacing and operational standards. While the upgrading (or downgrading) of existing streets will be constrained by considerations of coordination with freeways and transit and by abutting land-use relationships, arterial improvements are almost always necessary. Even with optimum freeway development,[42] Figure 12.15 shows that, depending on city size, arterials can be expected to bear from 40 to 60 percent of the total vehicle miles of travel. Any new arterial links should be added and old ones deleted before the networks are tested by traffic assignment.

The following principles are applicable in the system consideration of urban arterial streets:

1. In undeveloped areas arterial street needs should be anticipated and right of way should be reserved for their future construction.
2. Future arterial improvements should be made according to the primary function of the arterial street. Where through traffic demands are greatest, service drives or frontage roads can be developed, coordinated signalization can be installed, and special median treatment with left-turn bays can be provided. Conversely, where local service is an important street function, the roadway design can be adapted to provide for frequent turns and pedestrian activity.
3. Effort should be made to achieve optimum spacing of arterials. Desirable spacing is a function of capacity of the system, transit facilities, and the effect on the freeway system. Based on typical urban traffic demands, consideration of traffic

---

[42] *System Considerations for Urban Arterial Streets* (Washington, D.C.: Institute of Traffic Engineers, 1969).

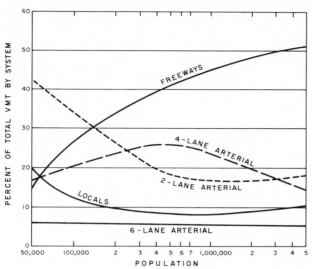

DISTRIBUTION OF VEHICLE-MILES OF TRAVEL IN RELATION
TO THE URBAN AREA POPULATION

Figure 12.15. The distribution of vehicle miles of travel varies by system according to area population. [Source: *System Considerations for Arterial Streets* (Washington, D.C.: Institute of Traffic Engineers, 1969), p. 23.]

signal timing, and freeway distribution requirements, recommended spacings are as follows:

Central business district	$\frac{1}{8}$–$\frac{1}{4}$ mi
Urban (central city)	$\frac{1}{4}$–$\frac{1}{2}$ mi
Suburban	$\frac{1}{2}$–1 mi
Rural	1–2 mi

4. Arterial street design should not adversely affect the integrity of neighborhoods, school areas, industrial sites, hospitals, and other areas where internal intrarelationships are important. Instead, arterial streets should be used as an effective buffer in desirably separating (both physically and psychologically) industrial, commercial, and residential areas, which may tend to infringe upon and conflict with each other.

5. The arterial street system should be fully integrated with the existing and planned mass transit systems. Normally, an arterial street system that is well-spaced and otherwise sufficient for auto trips will serve the needs of bus mass transit trips equally well. If, however, there are anomalies between auto and bus transit uses, the arterial street system should be reviewed with respect to bus transit needs. Integrating the arterial street system with rail rapid transit requires a different set of considerations than other forms of mass transit. With rail rapid transit each station is a major generator of auto and bus trips. Because a high proportion of rapid transit users rely on streets to reach the stations, the ability of the streets to accommodate these additional traffic volumes becomes a major consideration.

Local roads and streets (subdivision streets). Attention should be given to the local streets as the largest component of the urban transportation system. Consider the following principles:[43]

1. The primary function of local streets is to provide land access. Adequate vehicular and pedestrian access should, therefore, be provided to all parcels.
2. Because their main function is to provide land access, local streets within subdivisions should be designed to minimize through traffic movements.
3. The pattern of local streets should satisfy the needs of visitors, delivery trucks, and emergency vehicles, as well as local residents.
4. Local streets should be designed for low volumes and in a manner to encourage low speeds without reducing the overall safety.
5. If necessary, provisions for bus transit service should be established.

Preparing transit networks. Comparable principles, design standards, and constraints apply to the task of transit network planning. Transit routes must be particularly responsive to population density, to equipment performance characteristics (grade and alignment, for example, strongly affect rail system locations), and to station spacing and terminal transfer facilities (see also Chapter 6).

PLAN TESTING AND EVALUATION

The plan evaluation process should incorporate:

1. Assignment tests of network adequacy in terms of capacity, service quality, and accessibility
2. Economic tests of travel costs, construction costs, and feasibility of financing
3. Tests of compatibility with
   a. land-use forecasts or planned regional development
   b. community aesthetic values
   c. local neighborhood patterns
   d. economic development factors such as tax bases and availability of public services

Traffic assignment tests. The forecasts of future trip ends derived from the land-use forecasts, the calibrated trip distribution models, and coded network alternatives come together at this point. Before the performance of proposed new networks is studied, however, it is customary to assign future travel to the existing network augmented by those facilities whose construction is committed (see Figure 11.16). This base condition, showing the traffic impact if no further improvements are made, is useful for demonstrating to advisory groups and to the public that planning—and subsequent programming—is essential. (The computer simulation of future traffic flow was discussed in Chapter 11, Computer Applications.)

The "do-nothing" alternative may be compared also against the computer output from later traffic assignments to alternative plans. Usually the following will be checked:

---

[43] *Recommended Practices for Subdivision Streets* (Washington, D.C.: Institute of Traffic Engineers, 1967).

1. System efficiency in overall measures of capacity vs. volume, average travel speeds, travel costs, percent of trips using freeways, and accessibility measures
2. Traffic loadings of freeways vs. arterials, amounts of each network overloaded and underloaded, volumes at critical screenline areas and around the CBD
3. Network speeds by subarea, balance of capacity and travel by subarea and route type, etc.

These analyses identify plan alternative deficiencies. Systemwide excesses or deficiencies in capacity will become apparent. Minor deficiencies may be remedied by modifying travel corridor locations, spacing, and interchange connections. When arterial volumes are heavy and freeway volumes light, for example, new interchanges may bring about a better balance. When the opposite condition occurs, interchanges may be removed from the network.

Economic tests. Both travel and construction costs must be summed to find the minimum cost solution. Travel costs are derived from assignment results. Construction costs are separately determined on an average cost per mile basis, considering development densities, complexities of terrain, etc., or by engineering prefeasibility studies. Other benefit-cost analysis methods measure performance by incremental rates of return (savings in travel costs over other alternatives) on investment (construction cost differentials between the plans).[44]

One important economic test is to relate the estimated system construction costs to anticipated resources for construction programs. Program costs should be within the realm of financial feasibility before plans are submitted for agency approvals and public acceptance.

Other tests. The preceding tests can quantitatively compare the merits of one alternative with those of another. A number of necessary tests are more qualitative in nature. "It is particularly important for the decision-maker to recognize that analysis techniques now in use do not take into consideration a broad range of important environmental impacts that fall on nonusers or on the community as a whole. Therefore, the rate of return or benefit-cost ratio associated with each proposal must be only one of several factors weighted by the decision-maker in the process of selection. Other factors to be considered include noise, pollution, aesthetic intrusion, land consumption, economic impact, social disruption, and relocation."[45]

Environmental values have gained particular importance in recent years and a number of publications are useful references or starting points for further study.[46]

Values may be incorporated explicitly in planning goals and objectives or they may be only implicitly recognized within objectives related to land use, economic development, and aesthetics. Evaluation of alternatives in these terms can be difficult.

[44] *Costs and Benefits of Transportation Planning*, Highway Research Record 314 (Washington, D.C.: Highway Research Board, 1970), p. 182.

[45] *The Urban Transportation Planning Process* (Paris: Organisation for Economic Cooperation and Development, 1971), p. 29.

[46] *Transportation and Community Values*, Special Report 105 (Washington, D.C.: Highway Research Board, 1969). *Transportation Decision Making: A Guide to Social and Environmental Considerations*, National Cooperative Highway Research Program No. 156, (Washington, D.C.: Transportation Research Board, 1975). *Socio-Economic Considerations in Transportation Planning*, Highway Research Record No. 305 (Washington, D.C.: Highway Research Board, 1970), p. 178. THOMAS and SCHOFER, *op. cit.*

A ranking process (i.e., which plan best satisfies a given objective, which least) conducted by professionals familiar with each subject area is one approach. Because attaching relative weights to each measure is judgmental, the value of well-informed technical and advisory committees becomes readily apparent.

Imaginative use of transportation planning data is always helpful. Figure 12.16 shows how two alternatives might be compared in terms of job accessibility. The graph of average journey time to work by income groups shows that although Plan A and Plan B produce similar effects for high-income families, Plan A provides significantly shorter travel times for low-income families. Accessibility to recreational or medical facilities might be measured by similar applications of study data, if these were other social criteria.

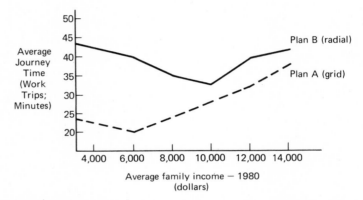

Figure 12.16. Possible relationships between work trip travel time and family income for alternate planning solutions. [Source: Edwin N. Thomas and Joseph L. Schafer, *Strategies for the Evaluation of Alternative Transportation Plans*, NCHRP Report 96 (Washington, D.C.: Highway Research Board, 1970), p. 93.]

Additional quantitative, although not monetary, measures of social impacts can be constructed. They include:

1. Estimates of land area taken up by proposed freeways or transit systems
2. Property values removed from tax rolls
3. Number of families to be relocated
4. School districts or neighborhood community services divided or disrupted

The importance of these latter evaluations must be recognized by transportation planners. A technically acceptable plan by the standards of traffic assignment or benefit-cost ratios is of little value if it is not acceptable to the community that eventually pays for its construction.

The results of planning. Figure 12.17 and its accompanying tabulation show the plans that evolved from the planning process in the major metropolitan areas of Detroit, Pittsburgh, and Houston. The three areas, different in population, topographical constraints, car ownership, and population densities, produced different forms of plans. Houston, with its high car ownership, low population density, high propor-

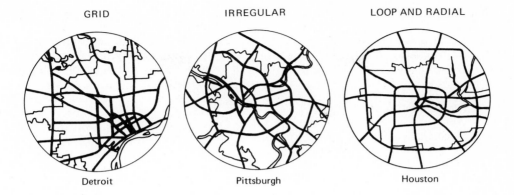

GRID   IRREGULAR   LOOP AND RADIAL

Detroit   Pittsburgh   Houston

	Detroit	Pittsburgh	Houston
1960 urbanized area population	3,538,000	1,804,000	1,140,000
Urbanized area (sq mi)	732	525	430
Population density	4800	3400	2600
Autos owned	1,116,000	485,000	390,000
Persons/auto	3.2	3.7	2.9
Annual transit revenue passengers 1966 (millions)	131	90	32
Annual transit passengers per capita	37	50	28
Freeway mileage existing plus proposed systems	329	269	281
Percent of freeways on interstate system	59	34	54
Freeway miles/10,000 population	0.93	1.49	2.46
Freeway miles/sq mi	0.45	0.51	0.65

Figure 12.17. Urban freeway plan forms may vary according to the characteristic of the community. [Source: From the front cover of *Urban Freeway Development in Twenty Major Cities* (Washington, D.C.: Highway Users Federation)]

tion of interstate highway mileage, and radial and loop system, has the highest plan mileage per capita. Detroit, with its greater transit usage and higher population density, has the lowest plan mileage per capita. With its highest per capita transit usage, lowest car ownership, but intermediate population density, Pittsburgh's plan falls between, but is closer to Detroit's in both freeway miles per capita and per square mile.

PLAN IMPLEMENTATION

Three elements make up the implementation process: approvals, programming, and coordination. To some extent these are interrelated; approval and participation by Federal funding agencies in transportation projects will be withheld unless they are demonstrably products of the planning process. Transportation facilities programming should clearly be coordinated among all implementing agencies. It should also be related to planning for other public services.

Plan approvals. The recommended plan must first win the approvals of the policy, technical, and citizen advisory committees. Then it can be presented to implementing agencies and to the public or representative organizations for acceptance and support. Citizen advisory groups may be invaluable in this latter respect. Policy committee members, usually leaders in the implementing agencies, would typically be expected to help secure the necessary approvals of the organizations they represent.

In the United States the early stages of proposed construction could further require the formality of "corridor public hearings" in compliance with procedures of the Federal Highway Administration (see Chapter 14). Corridor hearings, designed to afford full opportunity for effective public participation in the consideration of highway location proposals, are held before route locations are approved and before state highway organizations are committed to a specific proposal. They are intended to provide a public forum for presenting views on the social, economic, and environmental effects of alternate locations.

Programming. Several steps precede those of design and construction. They include (1) developing the construction stages for the major new facilities, (2) staging the necessary supporting facility improvements, and (3) integrating both processes with operating agency capital programs.

Long-term planning programs of up to two decades need to be broken down into shorter stages for financial planning and construction budgeting. The following principles guide in selecting the facilities to be scheduled:

1. Obtaining maximum traffic benefits in each stage
2. Coordinating with land use and other transportation programs (transit and terminals, for instance)
3. Scheduling construction work according to available funds

Obtaining the maximum traffic benefits often means scheduling improvements in the areas of worst congestion first. Usually, freeway systems should be built from the center of a city outward. Alternatively, new freeway sections could be scheduled to tie together parts of the existing and committed systems into a cohesively functioning whole. Programming usable segments that can be constructed and opened to traffic sequentially is fundamental.

Coordinating with land use and transit or parking programs is obviously important. Plans for areas of new development or redevelopment may have construction schedules that can be related to transportation facility needs for right of way. Or, new development and parking facilities may generate added traffic volumes that require timely construction of new street facilities.

Scheduling consistent with fund availability is vital because construction resources may include federal, state, and local contributions from different sources. Federal funds, allocated by system and apportioned some years ahead, require state or local matching funds and can affect the scale of local costs depending on the allocations to different cities within a state. As a result, project scheduling may be contingent on the system within which a route is classified and the ability of local governments to produce their share of construction costs.

Several criteria may be used to further establish the priorities for improvement scheduling. Sufficiency ratings, used by some state highway departments for many years, offer such opportunities based on assessments of roadway condition, safety, and service. The Arizona Highway Department has devised a plan for augmenting the rating scheme by assessing environmental and socioeconomic considerations.[47] The new ratings, based on a 100 point maximum like the existing roadway ratings, cover the subjects of environment (pollution, resource conservation, aesthetics, and recreation), economic development (stimulus to new development, relation to land-use planning, travel savings, and relocation costs), and traffic safety (with particular emphasis on hazardous locations). Combined with the other rating scores, the ratings can be used in ranking projects for program priorities.

Other means of deriving priorities for urban arterial improvements are suggested in *System Consideration for Urban Arterial Streets*[48] in which the following criteria are listed:

1. Congestion, as measured by present volume-to-capacity ratios
2. Structural adequacy, as reported by sufficiency ratings
3. Accident costs—dollars per vehicle mile of travel
4. Spacing and development needs, particularly in suburban growth areas

After priorities are set, projects must be scheduled with sufficient lead time to solve both engineering and social problems associated with new locations or major reconstruction. The end result will be a project listing and budget coordinated for the various participating governmental jurisdictions. One more consideration is essential. The programming process should include a procedure for annual reviews. Construction timetables may be upset by delays, unforeseen needs may arise, or shifts of emphasis may change priorities. An annual review permits flexibility and the opportunity to maintain maximum coordination with other elements of urban development.

An example of programming. Phoenix, Arizona, offers an example of arterial programming carried out in coordination with state highway programs. The state plan for development of freeways within the city is consistent with a major street and highway plan, for the Phoenix urban area and Maricopa County, first agreed upon in

---

[47] WILLIAM E. WILLEY, "Priority Programming for Arizona Highways," *Traffic Quarterly*, XXVI, No. 3 (1972), pp. 425–34.
[48] *System Considerations for Urban Arterial Streets*, p. 36.

1960. A state highway department map, periodically reissued, indicates freeways in use, those with confirmed location, those with temporary corridor locations, and temporary corridor locations for other freeways not in the state system. In addition, other current state or federal and proposed state highways are shown.

Meanwhile, Phoenix has an ongoing major street program.[49] Figure 12.18 illustrates program proposals for the years 1971–1977. Three groups of projects are identified, with their mileage and estimated costs. Table 12.7 summarizes the program scheduled for the 1971–1972 fiscal year. Fund sources for the 1971–1972 program were the city's major street fund, 63 percent federal aid (including TOPICS, secondary and urban system funds—31 percent), and other local sources 6 percent.

TABLE 12.7
Six-year Major Street Program, Programmed Projects
1971–1972, Phoenix, Arizona

Streets	Miles	Cost
Plans	3.50	$ 148,000
Right of way	5.50	246,000
Construction	5.00	2,640,000
TOPICS		
Plans	—	110,000
Right of way	—	—
Construction	—	300,000
Urban system*		
Plans	1.0	40,000
Right of way	1.0	51,000
Construction	1.0	590,000
Program contingency and advance construction		300,000
Total:		$4,425,000

*New system funded by 1970 Federal-Aid Highway Act.

Source: *Six-year Major Street Program* (Phoenix: City of Phoenix, 1971).

The annual report's foreword states: "This construction program implements the responsibility of the City of Phoenix as part of the Continuing Comprehensive Cooperative Urban Transportation Planning Program of the Maricopa Association of Governments (MAG) and is included in the Five-Year Major Street and Highway Improvement Program approved by MAG."

Achieving program coordination. Each level of government in metropolitan areas has certain powers that affect plan implementation. At the local level, these include licenses, covenants and agreements affecting private development plans, an official map defining rights of way for future streets, and zoning ordinances controlling land-use changes. At the state level, there is the power of eminent domain and sometimes the power to acquire advance rights of way. At the federal level, authority is provided to make or withhold grants according to the quality of planning and program coordination. Because urban transportation plans typically cover a region having many local jurisdictions, means must be found to coordinate their programming.

The Maricopa Association of Governments exemplifies one solution. This organizational form can also be found in other regions. In southern California 6 counties

---

[49] *Six Year Major Street Program* (City of Phoenix, Arizona, June 1971).

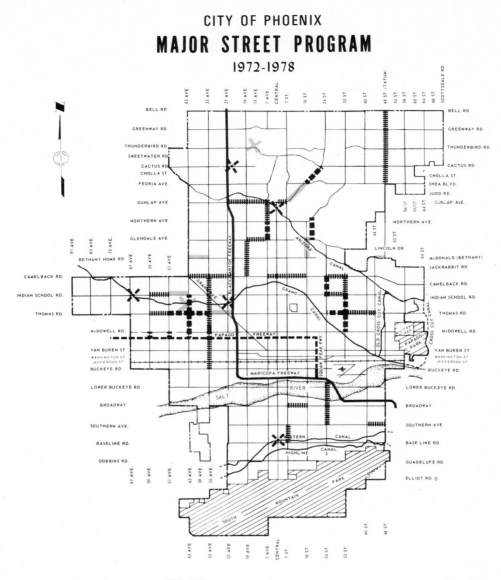

CITY OF PHOENIX
MAJOR STREET PROGRAM
1972-1978

LEGEND	JULY 1972	Miles	Cost*
▨ 1972-73 CONSTRUCTION PROJECTS		5.50	$ 5.4 Million
▰▰▰ 1973-75 CONSTRUCTION PROJECTS		12.50	10.4 Million
▥ 1975-78 HIGH PRIORITY PROJECTS		25.50	17.4 Million
✕ BRIDGES			
– – – CITY LIMITS	TOTAL	43.50	$ 33.2 Million

* Including Engineering, R W, Construction

JULY 1972

Figure 12.18. The major street program of Phoenix, Arizona, 1972–1978. [Source: *Six-year Major Street Program* (City of Phoenix, 1971), Figure 5.]

and 105 of the 147 cities form the Southern California Association of Governments (SCAG) which was established in 1965 to take a cooperative and coordinated approach to areawide problems. "Concurrently with its initial efforts in land use, population, housing, employment and economic trend studies, the Association has given first priority to coordinating transportation planning and the development of advisory transportation plans and programs for the region."[50] An advisory body to SCAG, the Comprehensive Transportation Planning Committee (CTPC) has absorbed the technical staff of the Los Angeles Regional Transportation Study, which provided the first regionwide data base for transportation planning.

A study of seven areas that have successful urban highway programs revealed a number of common factors in their coordination procedures.

1. Official civic advisory groups were important.
2. County governments exerted a strong coordinating role.
3. State highway organizations provided initial guidance.
4. Cooperatively developed policies, processes, and standards become part of official resolutions and formal documents.
5. Participants developed mutual trust and exhibited patience and a willingness to innovate.
6. Program and financing requirements were flexible.[51]

The report emphasized the importance of a regional view and good working relationships among the representatives of participating agencies.

CONTINUING TRANSPORTATION STUDIES

The expression "continuing comprehensive transportation planning process carried on cooperatively" states the need to carry on planning after a major initial study has developed an acceptable plan. Many *ad hoc* study groups as a result find themselves becoming part of a permanent planning structure. The objectives, organization, and activities of these agencies are briefly described here. For further detail, two ITE informational reports may be consulted.[52]

OBJECTIVES

The continuing study's purpose is to insure that the planning process is sufficiently flexible and dynamic. Its objectives could be listed as follows:

1. To keep plans moving toward implementation
2. To maintain current data bases
3. To provide services to local agencies
4. To update plans as needed

---

[50] WILBUR E. SMITH, "Southern California: Regional Transportation Coordination," *Traffic Engineering*, XXXIX, No. 11 (1969), p. 27.

[51] MARIAN T. HANKERD, "Coordinating Metropolitan Roadwork" (Washington, D.C.: Highway Users Federation for Safety and Mobility, 1971), p. 8.

[52] *The Continuing Cooperative Planning Process—Coordination with Day-to-day Traffic Operations*, and *Updating Transportation and Land Use Planning Programs* (Washington, D.C.: Institute of Traffic Engineers, 1969).

The mere existence of the study reminds operating agencies that a planning objective exists. Its functions logically include developing program priorities, but the study may exert more indirect influence toward plan implementation by its responsiveness to changing circumstances than it can by direct measures.

ORGANIZATION

Continuing studies may be carried out with the same organizational structures as the original studies. The Chicago Area Transportation Study, begun in 1956 as an *ad hoc* agency, continues as an agency of the Illinois Department of Transportation. Or, the continuing transportation study process may become a function of other planning agencies. The Pittsburgh Area Transportation Study organization was merged, along with other planning groups, into the Southwestern Pennsylvania Regional Planning Commission. The Los Angeles study group (as noted earlier) has become a staff arm of the Southern California Association of Governments. Other continuing studies may be carried out in special planning sections of state highway or transportation departments. Metropolitan councils of government also provide suitable administrative structures for the continuing transportation planning process.

ACTIVITIES

The continuing study has a multiple role: monitoring, servicing, and reporting. The initial surveys of land use, travel, and travel facilities should be kept up-to-date; for instance, an ITE report[53] recommends that the following be kept current:

1. Traffic counting programs
2. Land use, population, employment data files
3. Truck and taxi travel patterns
4. Transportation facility characteristics

The study should also keep abreast of changes in population characteristics (distribution, incomes, housing types, etc.) and changes in financial resources and administrative planning policies, for example, zoning.

By maintaining a data bank of current statistics, the continuing study can provide services to other organizations. A traffic agency may test the impact of a proposed detour by traffic assignment, or an implementing agency may test the impacts of alternative staging sequences. Social planning groups may be interested in job accessibility from different districts by travel mode, for instance. Civil defense agencies may benefit from population data by hour of day and district from travel survey and population files.

Continuing study reports can disseminate information as well as create greater awareness of the planning process. With regular reports, the study can periodically show progress toward plan completion and call attention to development conditions that may be contrary to plan objectives.

Nevertheless, plan updating is the principal function. Plans should be reviewed for their adequacy in meeting current needs at least every five years. Study groups must be prepared to develop new land-use forecasts and travel estimates for an adjusted target

---

[53] *Updating Transportation and Land Use Planning Programs*, pp. 69–70.

year, usually five years later than that first chosen. Updated plans must then be drawn up and evaluated if previous plans are no longer valid. The testing and evaluation process itself may require revision in the light of changing emphases on economic, social, and environmental considerations.

A general summary of continuing study activities is shown in Figure 12.19.[54] Surveillance, reappraisal, service, research and development, and reporting are shown as the prime concerns. The same elements can be found in a proposed one-year work program for a small continuing study, shown in Table 12.8. The costs are only those incurred by the continuing study and do not include those of participating agencies making service contributions.

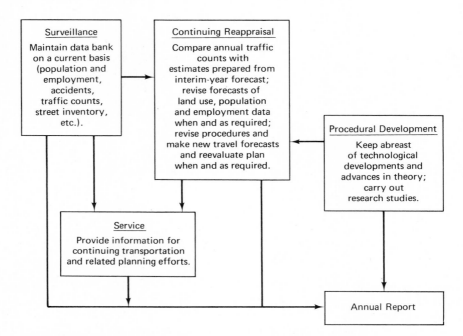

Figure 12.19. Activities of continuing transportation studies. [Source: Based on an illustration by Austin E. Brant, Jr., and Dana L. Low, "Continuing Urban Transportation Studies, *Traffic Quarterly*, XXIII No. 2 (1969), p. 214. By permission of Eno Foundation for Transportation, Inc.]

SMALL AREA TRANSPORTATION STUDIES

Transportation planning for areas of less than 50,000 population requires much the same elements as planning for larger areas, but with a simplified and scaled-down process. The scale and cost of improvements are less, the small area is less likely to have local traffic and planning expertise, and local contributions to transportation solutions may be minor or nonexistent. The important facilities are likely to be constructed, operated, and maintained by state level organizations.

---

[54] AUSTIN E. BRANT, JR., and DANA E. LOW, "Continuing Urban Transportation Studies," *Traffic Quarterly*, XXIII, No. 2 (1969), p. 214.

TABLE 12.8
**Estimated Costs by Major Project in the Recommended
One-year Work Program**

Major Projects Plus Recommended Tasks	Salary Cost	Other Cost	Total Cost
Transit analysis and planning Rail-commuter service extension study Joint bus-rail terminal study Post-strike bus passenger survey Review of bus route changes	$10,500	$ 3,700	$14,200
Continuous district planning Develop 1975 zone level forecasts Detail 1985 forecast by district Local area traffic assignment needs study Relate actual development trends to plan	22,500	7,000	29,500
Data bank Update and detail existing files— feasibility study Update land-use file Block-level data collection units study Street address coding guide Data file reorganization Collector and local streets inventory Detail data files for special districts Refine machine data-mapping capability	29,000	19,500	48,500
Special work Monthly newsletter and public presentations Machine-processed mailing address file Continuing information and data requests	4,000	2,000	6,000
Total:	$66,000	$32,200	$98,200

Source: "A Detailed One-year Work Program for 1967–1968," New Castle County Program (Wilmington, Del., 1967), preliminary and unpublished.

As a first step, integrating local and state objectives in transportation planning can be achieved through the guiding committee structures described earlier. Travel inventories may be limited to external surveys and estimates of household tripmaking. Other inventories may include surveys of population, land use, and employment patterns needed to develop trip generation data for use in trip distribution models. An inventory of transportation facilities with classification of streets into local, collector, and arterial categories is required, as are seasonally adjusted volume counts. When the community indicates a need for improvement, surveys of parking, transit, and accident conditions also may be warranted.

A number of forecasts are necessary. External traffic may be forecasted by applying growth factors based on larger regional expectations. Internal traffic estimates necessitate forecasts of future population and future commercial, industrial, and recreational activities. Future travel volumes may be derived from models of trip generation, mode split, zone to zone trip distribution, and network assignments, depending on the extent of technical resources and funds available. Major facility plans can then be tested for their adequacy in meeting travel demands.

The small area study output may include not only a major thoroughfare plan compatible with community development objectives but also a series of recommendations for immediate traffic engineering improvements. The latter may be the most important product of the study. In smaller communities, significant gains in traffic service may be achieved through modest outlays for channelization, signal improvements, and off-street parking. Because the last, at least, is more a matter for private or local public participation, the citizen advisory committee can be important in stimulating plan implementation.

THE ROLE OF THE TRAFFIC ENGINEER

The traffic engineer may and should play an important part in every one of the urban transportation planning studies previously described. But even if he is not directly so employed, he has responsibilities in obtaining realistic transportation planning programs.

TRAFFIC ENGINEERING WITHIN THE STUDIES

The traffic engineer may contribute to several phases of the major long-range study and in several ways to other planning studies as well.

Long-range study. The selection and description of an arterial network is basic to the planning process. Volumes, speed characteristics, capacity limitations, and system classifications are needed, and traffic engineering personnel are essential for this inventory function. Traffic engineers may appropriately be responsible also for parking and accident studies as well as for taking part in travel survey design and operation. Estimating capacity or other deficiencies in existing networks and setting performance standards for network elements are clearly traffic engineering assignments. Screenline accuracy checks, traffic assignment, network construction, and even trip model development and calibration may also be done by traffic engineers. Analyzing spacing, network configurations, interchange location, capacity designation, speed and other design standards and coordinating with transit network schemes are additional logical functions of the traffic engineer.

In the evaluation stage, reviewing link volume/capacity ratios, calculating benefit-cost ratios, and evaluating quality of service are tasks that can be accomplished by traffic engineering personnel. Their qualifications should particularly enable them to translate deficiencies in one plan to improvements in a subsequent one. Finally, in the long-range study, traffic engineering skills can be applied to jurisdictional assignment of projects, coordination of arterial and freeway improvements, and designation of priorities.

Role in the continuing study. Many of the foregoing tasks of the traffic engineer will continue during the continuing study. Here, a staff traffic engineer should be attuned to the needs of local traffic engineers. He can provide data collection, research studies, and forecasts of the traffic impacts of construction programs.

Role in short-range studies. Much of the short-range urban traffic planning work is entirely traffic engineering, for example, as in TOPICS studies and studies of major generators and passenger terminal or transfer facilities. Obviously, the traffic engineer should have a part in other planning too, as for example, in analyzing traffic generation for joint development or multiple use projects. Traffic engineering is also needed by design teams engaged in freeway corridor planning.

Traffic engineering has a part in transit planning. Testing the adequacy of local access to "park and ride" transit terminals and insuring their adequate internal traffic circulation are examples.

THE CITY TRAFFIC ENGINEER AND THE TRANSPORTATION STUDY

Cooperation between local traffic officials and transportation planning staffs results in mutual benefits. The traffic engineer can provide useful study inputs and he in turn can benefit from planning study output.[55]

Local traffic personnel assist the study during the inventory stage by providing data on volumes, parking, and control devices, and they can be particularly helpful in selecting the functional arterial network. The study's technical committee should include a representative from the traffic engineer's office, preferably the city traffic engineer, because he may have a better feel for the network behavior and operational feasibility of plan alternatives than any other participant in the study.

In return, the city staff can benefit from study information and capabilities. Study data sources often provide detailed data that are unavailable elsewhere. Network capacity inputs, O-D patterns, and even land-use information may be of value. Traffic assignments may test the effects of necessary street closings, etc., anticipated by the city. Moreover, the study may lend emphasis to the city traffic engineer's needs in providing immediate-action answers to congestion problems.

NEW DIRECTIONS IN URBAN TRANSPORTATION

There have been dramatic changes in urban transportation during the first two-thirds of the twentieth century, and they seem likely to continue. Future transportation plans will be influenced by at least the following considerations:

1. Public transportation is necessary for those who do not have access to automobiles.
2. Transportation facilities must be designed to avoid negative community impacts.
3. The environmental impacts of all transportation modes must be considered.
4. Transportation systems have a part to play in improving the quality of urban life.

The summary recommendations of a report entitled *Future Directions for Research in Urban Transportation*[56] bear out the foregoing by citing the principal needs as follows:

1. In all fields of urban transportation research, more attention should be paid to human factors, particularly to the needs and requirements of present and potential travellers . . . .
2. More attention should be paid in transportation research to the analysis of total effects of transportation . . . .

---

[55] Louis E. Keefer, "City Traffic Engineer and Urban Transportation Study," *Traffic Engineering*, XXXV, No. 7 (1965), p. 10.

[56] *Future Directions for Research in Urban Transportation* (Paris: Organisation for Economic Cooperation and Development, 1969), pp. 14–15.

3. Greater emphasis should be placed on the development of analytical techniques to predict social and economic consequences of transport proposals and decisions, thereby reducing the uncertainty inherent in such decisions . . . .

CHANGES IN URBAN ADMINISTRATION

It seems likely that there will be changes in urban government or else a shifting of planning powers. Metropolitan government with regional responsibilities may emerge, or alternatively, planning powers will be transferred from local, uncoordinated and spatially fragmented jurisdictions to regional, county, or state levels.

At the same time, current methods of transportation funding are likely to disappear and new ones emerge. Trust funds for highways alone may be replaced by total transportation trust funds which may be expended on modes from bicycles to tracked-aircushion vehicles. In the process, traditional federal, state, and local funding relationships may disappear. Disbursements may be made directly to the local level where development mode choice decisions may be made. The allocation of transportation construction or sustaining funds by mode may be decentralized.

Finally, realignment of national goals may change the total of funding resources available for transportation or may induce changes within the field of transportation itself. Cutbacks on research and development for supersonic air travel and on operations of atomic-powered vessels have been indicative of one trend. Government operation of passenger railroads and growing public ownership of urban transit systems suggest another trend. If increased attention is given to other urban problems, for example, health, education, welfare, and housing, lessened attention may be given to transportation concerns.

NEW TRANSPORTATION TECHNOLOGY

A 1968 publication listed the more promising of the new urban transit systems. Among them were the following:

1. *Dial-a-bus.* A bus system activated on demand of the potential passengers, perhaps by telephone, after which a computer logs the calls, origins, destinations, location of vehicles, and number of passengers and then selects the vehicle and dispatches it.
2. *Personal rapid transit.* Small vehicles, traveling over exclusive rights of way, automatically routed from origin to destination over a network guideway system, primarily to serve low-to-medium-population density areas of a metropolis.
3. *Dual mode vehicle system.* Small vehicles that can be individually driven and converted from street travel to travel on automatic guideway networks.
4. *Automated dual mode bus.* A large-vehicle system that would combine the high-speed capacity of a rail system operating on its private right of way with the flexibility and adaptability of a city bus.
5. *Pallet or ferry systems.* An alternative to dual mode vehicle systems is the use of pallets to carry (or ferry) conventional automobiles, minibuses, or freight automatically on high-speed guideways.
6. *Fast intraurban transit links.* Automatically controlled vehicles capable of operating either independently or coupling into trains to serve metropolitan area travel needs between major urban nodes.

7. *New systems for major activity centers.* Continuously moving belts, capsule transit systems, some on guideways, perhaps suspended above city streets.[57]

More exotic systems include tracked systems supported by an air cushion and propelled by linear induction motors and gravity-vacuum tube systems operating in tunnels under urban centers. "People movers" (continuous conveyors) of various kinds are being designed to facilitate pedestrian movement in central business districts.

NEW TECHNOLOGY AND TRANSPORTATION PLANNING

It is possible that the coming changes in urban development and transportation may be so significant that the currently accepted long-range transportation planning process may become invalid. Views are increasingly being expressed that planning should abandon its twenty-year forecasts, look only from five to ten years ahead, and make incremental plans.

There are a number of reasons for this. First, there is growing impetus from the public to change the existing pattern of urban transportation. Second, financial resources have not generally been sufficient to meet the costs of urban transportation proposals. Third, new funding arrangements reflecting greater local influence may result in choice of other mode alternatives. Fourth, choices in urban development forms (new towns, satellite centers, etc.) may become realistically possible with more powerful regional planning controls, so that urban planning could induce some change in urban travel patterns.

The last point has significance for transportation planners. They are likely to be called on to make more detailed tests of urban development choices and the transportation alternatives that will shape them and serve them. The art of modelling by computer, incorporating more variables and simulating growth more responsively, generates a continued challenge for their skills. Therefore, the urban transportation planner will continue to require versatility, insight, and an ability to communicate effectively with the other disciplines engaged in the planning process.

REFERENCES FOR FURTHER READING

The list of available reading in urban transportation planning is too long to present here. Instead, some of the most useful references have been footnoted throughout the chapter. Instead of listing references, it is more appropriate to list sources of current and recent information on the subject of urban transportation planning.

PUBLIC OR QUASI-PUBLIC AGENCIES

U.S. Department of Transportation, notably the Federal Highway Administration and Urban Mass Transportation Administration
   Transportation Research Board
   National Research Council of Canada
   Road Research Laboratories of Great Britain and Australia
   Organisation for Economic Cooperation and Development
   State level and regional transportation study groups such as Southeastern Wisconsin and Southwestern

---

[57] *Tomorrow's Transportation—New Systems for the Urban Future* (Washington, D.C.: U.S. Department of Housing and Urban Development, 1968), p. 3.

Pennsylvania Regional Planning Commissions, Chicago Area Transportation Study, and Tri-State Regional Planning Commission

Research organizations like the Centre for Environmental Studies.

Universities with graduate programs in transportation.

TECHNICAL SOCIETIES AND ASSOCIATIONS

American Institute of Planners
American Public Transportation Association
American Society of Civil Engineers
American Society of Planning Officials
Motor Vehicle Manufacturers Association
Highway Users Federation for Safety and Mobility
Institute of Traffic Engineers
Operations Research Society of America
Regional Planning Association
Counterpart technical and professional societies in other nations

PERIODICALS

Journal of the Town Planning Institute
Traffic Engineering (ITE)
Traffic Engineering and Control
Traffic Quarterly
Transportation Engineering Journal (ASCE)
Transportation Science (ORSA)

Chapter 13

STATEWIDE AND REGIONAL
TRANSPORTATION PLANNING

Roger L. Creighton. Creighton, Hamburg, Inc., Delmar, New York.

Statewide and regional transportation planning is relatively new in the United States, although its roots go back to the Federal Aid Highway Act of 1921 and the statewide highway planning surveys of the 1930's. Three forces have given it greater impetus. First, the urban transportation studies, with their technical advances of the 1950's and 1960's, have increased the confidence of professionals in using rational processes to attack problems of complex transportation systems. Second, the example of the Federal Department of Transportation and the increasingly obvious need to coordinate government policies across all modes of transportation have started a strong trend toward the creation of state departments of transportation. Third, the biennial National Transportation Needs Studies require *all* states to estimate their long-term needs for all forms of transportation.

Although statewide and regional transportation planning does not yet have the standardized procedures of more established processes such as urban transportation planning, sufficient work has been done, and the lessons of urban transportation planning are sufficiently applicable so that a basic framework can be described. This is the purpose of this chapter.

The chapter is organized into five parts. The first part contains a definition of the content of comprehensive statewide transportation planning together with statements that suggest practical limits for the process. The *scale* of transportation at the regional level is indicated through the use of statistics. The second part contains information on organization: powers of transportation departments, internal organization for statewide planning, and external organization. The third part deals with preparatory steps for statewide transportation planning—the process itself, stating goals, and obtaining forecasts. The fourth part discusses studies that should be undertaken for several of the major modes: key problems to be faced, data collection, planning methods, research, and means for effectuation. Coordination between the separate modal studies is taken up in the fifth part, which also presents coordination with land-use plans and implementation through capital programming.

DEFINITION

Statewide or regional transportation planning may be defined as an activity, or series of activities, that lead to recommendations for undertaking action

1. To attain a series of goals, or to improve performance as measured by a series of criteria

2. Of different groups, for example, people who use transportation facilities, shippers and receivers of goods, suppliers of transportation services, and particularly the public at large
3. Through coordinated changes in the construction, investment in, management, technology, pricing, subsidy, and regulation
4. Of transportation facilities and services of all types
5. For the movement of people and goods
6. Planned by means of an orderly, objective process that is based on measurement
7. Closely integrated with land use, economic, environmental, and other types of planning
8. For the geographic area of a state, several states, or a region of some other type
9. For a long-range period that may vary from 10 to 50 years or more in length

Clearly, statewide and regional transportation planning is not limited to one or at the most two modes of transportation, with a single criterion of performance, using the construction of new facilities as the sole means of effecting improvements and relying on a single agency to carry out recommendations. Instead, statewide and regional transportation planning is concerned with a variety of modes to achieve improvements toward several objectives, dealing with both public and private organizations, and using not only new construction but also the tools of pricing, regulation, and management in order to make improvements, and being seriously concerned with policies of population location and land development over regions with tens of thousands of square miles of area.

In dealing with larger and larger regions, with more modes of transportation and with more complex sets of goals, two things must be recognized. First, the process is moving toward and even beyond the limits of current planning skills. Although traffic can be simulated over entire state highway networks with varying degrees of success, it is not yet possible to simulate the operations of aviation, railroad, bus, and truck transportation systems with sufficient reliability to be useful in planning. A number of years and much research are needed to build up data, computer models, and other techniques for planning these other modes of transportation.

Second, real advances in transportation planning for states can only come when goals are stated clearly, when measurements define the current levels of performance, and when forecasts and simulation are used to estimate future performance. These are technical processes. They cannot eliminate all uncertainty and they cannot deal with all goals. But they do offer the opportunity of making better informed judgments.

QUALIFICATION OF DEFINITION

For various practical reasons, the above definition must be qualified in a number of ways.

Exclusion of urban areas. Although statewide or regional transportation planning must be concerned with *all* transportation throughout its defined area, urban areas for which separate transportation studies are in process should, for most purposes, be excluded from the concern of the statewide planning group. The reason for exclusion is the practical one of avoiding duplication of effort. (Some states include the review of urban plans as a component of their statewide work, however.) Urban area studies are primarily concerned with plans to serve internal person and vehicle travel and have advanced processes for preparing such plans. However,

1. Statewide planning should be concerned with the *level* of transportation investment in each urban area because this affects other areas as well as the state as a whole.
2. Statewide planning should be concerned with the *pattern* of freeway, rail and other model routes as these leave each urban area because these routes become part of the statewide system.
3. Statewide planning may be concerned, in some cases, with the *general location of airports, ports, and rail terminals* within or near urban areas (particularly COFC/TOFC[1] terminals) because these may in some cases affect service to other cities and to the state as a whole.
4. In some states very large planning regions with extensive rural areas having their own transportation planning or comprehensive planning organizations may have been designated. In such cases, careful allocations of responsibility must be made because a duplication of effort or a gap in planning could be encountered.

This leaves statewide transportation planning concerned chiefly with the nonurban transportation of people and goods.

Concerns of statewide transportation planning. Statewide transportation planning, like urban transportation planning, deals with transportation on a fairly generalized scale. For example, urban transportation planning is often considered to be the planning of *systems of arterials and freeway corridors* but not to be the planning of route locations within corridors. Similarly, statewide transportation planning is concerned with systems of *corridors* of rail and highway[2] facilities (not with their exact centerlines) and with the generalized locations of ports and airports (but not with their exact placement within a single urban area).

This general scale of statewide transportation planning should not be interpreted as a lack of concern for detail. Quite the opposite should be the case. The ability to plan the generalized locations of transportation facilities, the best levels of investment in new facilities, the mix and coordination of facilities of different types, and the kinds of regulations that should be imposed requires a solid mastery of many details. The urban transportation planner knows the facts on trip generation, trip length, the theories of trip propagation through space, and so forth. Similarly, the statewide transportation planner must understand the economics of the private carrier, the technical inventions on the horizon, the effective journey speeds of different modes, and many other facts and relationships.

Table 13.1 attempts to define the scale, focus, and concerns of statewide transportation planning—what it is, as well as what it is *not* concerned with. This table should be modified by each state to meet its particular needs and situations.

Generally, Table 13.1 indicates that statewide transportation should be concerned with matters of (1) systems design, (2) levels of investment, (3) service to users, and (4) relationships to land use, economic development, and the environment. Statewide

---

[1] Container on flatcar (COFC) and trailer on flatcar—"piggyback" (TOFC).

[2] In some states, however, such as Connecticut, the highway route location activity is located within the statewide planning organization. The advantages of having route location within statewide planning include the prestige and authority gained by having an action component within a planning organization; the greatest disadvantage probably is the tendency of immediate pressures to draw manpower away from long-range issues.

TABLE 13.1
Subject Matter of Statewide Transportation Planning

Subject Matter	Areas of Concern	Areas of Limited Concern
Highway	System design in principle for all systems (basically spacing and configuration); in corridor location for primary and interstate routes, investment levels by type, location and timing (both intraurban and statewide).	Route location; engineering design, corridors of secondary highways in counties (unless owned by state); traffic engineering and control.
Bus	Systems of routes (design and interline coordination); level of service (headways); generalized terminal location; pricing; bus size.	Detailed terminal location; scheduling; internal management and operations; safety.
Air passenger	Systems of air routes and airports; generalized airport location, size, and investment; airspace use; pricing; utilization of airports by type of airplane.	Detailed airport location; scheduling; internal operations; air traffic control, safety.
General aviation	Systems of airports; generalized airport location, size and investment; airspace use; pricing; utilization of airport by type of airplane.	Detailed airport location; scheduling, internal operations; safety; air traffic control.
Rail passenger	Rail passenger systems, generalized station locations; pricing; service levels (headways); public investment; grade crossing protection.	Scheduling and operations, safety.
Rail freight	Extent and design of system; investment; terminals (especially TOFC/-COFC); system speed and pickup frequency; rail-truck coordination; pricing; grade crossing protection.	Scheduling and operations, safety.
Truck	TOFC/COFC terminal locations; expressway location; truck size and pricing.	Operations; details of TOFC/COFC location; safety.
Canals	Investment and maintenance costs; systems as related to rail and highways; recreational use.	Operations
Ports	Investment; coordination with rail, highway; interport coordination and general location.	Design; management; operations.
Pipelines	Impact on rail, canals.	Safety; management and operations.
Land use	Relationship between accessibility (by mode) and the distribution and level of economic activity; population distribution.	
Environment	Preservation of natural resources; of historical and aesthetic resources.	

transportation should *not* be concerned directly with matters of engineering design, precise location of new facilities, scheduling of common carrier vehicles (as opposed to general frequency or level of service), or management and operations. This is not to say that the transportation planner need not understand matters of management and operations; quite the contrary. But the reason for this division of responsibility is simply that if the statewide transportation planner does not set bounds to his work, he will not be able to deal effectively with the significant issues.

Applicable planning areas; substate regions. State transportation departments are recognizing that the state may not be the best area for planning certain systems of transportation. Table 13.2 gives suggested optimum planning areas for each of the major modes of transportation. If the optimum planning area for any given mode is larger than a state, this does not mean that a statewide transportation planning organization can ignore that mode. The planning team must be competent in all modes so that it is able to represent the state's interests in that field. In many cases states should band together to undertake planning for multistate regions. Air passenger systems and rail freight systems can probably be planned more efficiently at the level of a multistate region than at the national level because the nation is too big a unit.

TABLE 13.2
Suggested Optimum Planning Areas for Different Modes of Transportation

Mode	Suggested Optimum Planning Area	Reason for Suggesting the Optimum Planning Area
Highway	State	State ownership
Bus	State	State franchising powers: use of state highways
Air passenger	Multistate region or nation	The air passenger system is bigger than a single state
General aviation	State	State general aviation improvement programs
Rail passenger	Nation	National operation
Rail freight	Multistate region	The system is bigger than a single state
Truck	State	State regulation; use of state highways
Canal	Nation	National ownership
Port	State	Ownership by state and local authorities
Pipeline	State	Franchise powers; safety regulation

In some states it has been proposed that statewide transportation planning should be undertaken on a substate regional level in order to obtain (1) greater understanding of and responsiveness to local needs, (2) greater local participation, and (3) better coordination with local land use or comprehensive planning. Some arguments against this approach are (1) that the sum of small area plans cannot be as good or creative as a single plan for a larger area (an urban area transit or expressway plan could not be developed by adding up the local plans of 50–100 small political jurisdictions), (2) organizational inefficiency in the face of the need for a whole new technical effort in new modes, and (3) inability to deal with systems that cover states or groups of states. This chapter concentrates on planning for regions that are at least as large as a single state.

SCALE OF PERSON TRAVEL AND FREIGHT MOVEMENT

At the beginning of a new program of statewide transportation planning, several states have found it desirable to make a reconnaissance of the *scale* of person travel and freight movement. There are three main purposes for this: First, the chief transportation variables (person trips, tonnages, miles of travel, length and capacity of the system, etc.) must be measured simply because planning must deal with facts. Second, the facts permit relationships to be developed, for example, the size of each mode of transportation related to all the other modes. Third, past trends are needed as a partial basis for making estimates of future trends.

Because person trips vary so greatly in length, the best unit for studying intercity and rural travel is the person-mile of travel. The person-mile of travel (PMT) is the equivalent of one person travelling one mile.

The movement of freight and goods is best studied by comparing such factors as ton-miles carried by each mode and/or routing. A ton-mile of travel is the equivalent of one ton of cargo being transported one mile.

Tables and graphs giving examples of the basic measurements of intercity transportation are found in Chapter 1. These proportions will vary substantially in other countries and in individual states.

Although precise data are difficult to obtain and even more difficult to put in comparable terms, it is valuable to know the relative dollar amounts being spent for different forms of transportation in a nation, state, or region.

For the planner, the relative size of capital expenditures for transportation of different modes is of considerable interest, as is the relationship between capital and operating expenditures. The rails, the air carriers, and the motor carriers are not as big as highways in the scale of their capital expenditures, but they are not dwarfed. In all of these modes even small percentage increases in efficiency resulting from better allocations of capital or operating expenses can produce major dollar savings.

A reconnaissance of the principal modes of transportation in each state is a logical first order of business. Sometimes this will be done prior to establishing a department of transportation because the structuring of a transportation department depends on the emphases to be given each mode. Once the reconaissance is complete, decisions on staff organization and the direction of technical studies can be made.

ORGANIZATION FOR STATEWIDE TRANSPORTATION PLANNING

Although comprehensive statewide transportation planning might conceivably be undertaken by a state without specific legislative mandate, effective planning should have an adequate basis in legal authority. The concern here is for two kinds of authority: (1) authority to plan for all modes of transportation and (2) authority to carry plans into being. Without some kind of power for effectuation, planning is only an exercise.

Many states in the United States have established departments of transportation, and legislation and/or studies to create new departments of transportation are pending in a number of other states. With the example of the U.S. Federal Department of Transportation, it is probable that more states will move in this direction.

ENABLING LEGISLATION FOR PLANNING

Of twenty two state departments of transportation (1974), twenty one have specific authorization to prepare comprehensive transportation plans. Six have already published plans, and the remainder are more or less actively engaged in preparing or revising comprehensive plans[3]. (See Table 13.3)

AUTHORITY TO CARRY OUT PLANS

Various kinds of authority have been given to the state transportation departments by which they can influence actions taken in the various modes. For convenience, four kinds of authority have been identified:

---

[3] ROGER L. CREIGHTON, "State of the Art in Statewide Transportation Planning," paper prepared for the Highway Research Board Conference on Statewide Transportation Planning, Williamsburg, Virginia, February 24, 1974.

1. *The power to construct, operate, and maintain.* This is the strongest power, for it springs from ownership. All states have this power with respect to state highway systems, but few have it with respect to ports, airports, and canals.
2. *The power to give financial aid.* This power is important because it gives states the potential to exercise considerable influence on the design and operation of transportation facilities, while leaving day-to-day management to others. Tax relief is a form of financial grant.
3. *The power to license and regulate.* States may delegate to their departments of transportation the power to license and regulate private carriers in various fashions, including regulating:
   a. a license to operate
   b. safety
   c. passenger fares and freight rates
   d. adequacy of service
   e. route(s) or territory served
   Typically, the power to license and regulate carriers has been held by state public service commissions or public utilities commissions. Only in the case of New York has this group of powers been transferred to the department of transportation.
4. *Power to license vehicles.* The power to license vehicles permits states to regulate size of vehicle and various safety features.

In addition to the above, state departments of transportation may exercise miscellaneous other powers or undertake other functions, for example, control or review over toll authorities, promotion of transportation, construction and operation of interstate bridges, conduct of research and demonstration studies, state highway patrol, and other nontransportation functions.

Twenty two states were polled individually in order to determine the extent of their powers in relation to the various modes of transportation. The data obtained in this survey is displayed in Table 13.3. Although care has been taken in the preparation of this table and many exceptions have been footnoted, the table should be treated as an approximation because it is almost impossible to capsule the meaning of complex laws and administrative practices into simple yes-no responses on a checklist. Nevertheless, this form of compilation does permit a number of observations.

First, state departments of transportation can have, and many do have, a large array of powers by which their policies and plans can be carried out. These powers include, in addition to outright ownership, the power to give or withhold funds and the power to establish various regulations. The means of effectuating plans, therefore, can be made available.

Second, the powers of departments of transportation are not uniform with respect to all modes. States generally have fewer powers governing private modes. Actually, uniformity should not be expected because there are so many historic, geographic, economic, and governmental factors at work.

In time, additional powers will probably be obtained by state transportation departments. If private carriers experience financial failure, then the government is generally required to assume greater financial responsibility and with it, greater control. This has important implications for statewide transportation planning. If government gains additional responsibility, it will be looking to planners for direction and coherence in its policies on the new modes. This will challenge planners to develop their objectives and to demonstrate that their plans will produce gains for the public.

TABLE 13.3

State	Authorization for Comprehensive Transportation Planning	Has a Comprehensive Transportation Plan Been Prepared? (Year)	Is Comprehensive Transportation Planning in Process?	Authority to Build, Operate, and Maintain: Highways (State)	Airports	Ports	Canals and Waterways	Urban Transit	Authority to Give Financial Aid to: County or Municipal Roads	General Aviation Airports	Commercial Airports	Ports	Urban Transit	Authority to Regulate, License, and Set Rates for: General Aviation Airports	Commercial Airlines	Bus Passenger Service	Rail Passenger Service	Truck Freight	Rail Freight	Urban Mass Transit	Authority to License Vehicles: Motor Vehicles	General Aviation Airplanes	Year DOT Established
Hawaii	X	1961i	X	X	X	X																	1959
New Jersey	Xm	1972	X	X			X		X	eb			X	X		n	n			n		X	1966
New York	X	1973	X	X					X	X			X			X	o	X	X	p			1967
Wisconsin	X	s	X	X	t				X	Xe	X		X								X		1967
Connecticut	X	1971d	X	X	X	X	X	X	X	X	X	X	X									g	1969
Florida	X	1973	X	X	X	X	X	X	X	X	X	X	X			f	f			f		g	1969
Oregon	X		X	X	X				X	X	X		X	X	X						X	X	1969
Delaware	X		X	X	X			Xh	X	X	X		X	X	X	Xh	Xh			Xh		X	1970
Pennsylvania	X	1970	X	X	X				X	X	X	q	X	r	r						X		1970
Rhode Island	X	hh	X	X	X	X	X		X	X	X	X	X	X							X	X	1970
Maryland	X	j	X	X	X	X	X	X	X	X	X	X	X	X	f	k	l			k	X	X	1971
Massachusetts	X		X	X		X	X	X	X	X			X										1971
North Carolina			cc	X	cc				Xgg	X	X	X									X		1971
Georgia	X	hh	X	X					X	X	X		X										1972

TABLE 13.3 (*Continued*)

State	Authorization for Comprehensive Transportation Planning	Has a Comprehensive Transportation Plan Been Prepared? (Year)	Is Comprehensive Transportation Planning in Process?	Authority to Build, Operate and Maintain					Authority to Give Financial Aid to:					Authority to Regulate, License, and Set Rates for:							Authority to License Vehicles		Year DOT Established
				Highways (State)	Airports	Ports	Canals and Waterways	Urban Transit	County or Municipal Roads	General Aviation Airports	Commercial Airports	Ports	Urban Transit	General Aviation Airports	Commercial Airlines	Bus Passenger Service	Rail Passenger Service	Truck Freight	Rail Freight	Urban Mass Transit	Motor Vehicles	General Aviation Airplanes	
Maine	X		X	X	X	X	X		X	X	X	X	X	X									1972
Ohio	X		X	X	X				X	X	X	X	X	X								X	1972
Illinois	X		X	X	X	y	y		X	e	e	e	z	aa	bb							X	1972
Tennessee	X		X	X	X				X	X	X		Xb	Xff									1972
California	X		X	X	a				X	X	X	X	X	X	X	X		X			c		1973
Kentucky	X	hh	X	X	X				X	X	X	X	X	X			Xcc		Xcc	Xdd	X	X	1973
Michigan	X	u	X	X	Xv				X	X	X	X	X	Xw								X	1973
South Dakota	Xee		X	X	X				X	X	X	X	X	X								X	1973

a. State law provides authority, but by policy it is not exercised.
b. Planning only.
c. By Department of Motor Vehicles.
d. Updated annually.
e. Publicly owned only.
f. Only for service under contract to state.
g. Authority to register, not license.
h. Power to create authorities, which in turn have administrative power.
i. Updated for major modes, but not republished in single document.
j. Presently in preparation.
k. Baltimore area only.
l. Commuter rail only.
m. Except water transportation.
n. Only when carrier is receiving public aid.
o. Authority to give financial aid to Amtrak for subsidy of additions to the "basic service".
p. Provided by public authorities exempt from NYSDOT regulations.
q. Authorized by implication only.
r. For airports, not airlines.
s. For highways and airports only.
t. Authority in statutes, but unclear.
u. 1968 for highways; airport plan now being prepared.
v. Authority only exercised on one airport.
w. No authority to set rates.
y. Specific case by case legislative authority needed.
z. Only capital improvements unless specific legislative authority given.
aa. Safety regulation only.
bb. Only intra-state third level carriers.
cc. Pending new legislation.
dd. At discretion of local operating agencies.
ee. Authorization but no funding.
ff. Airport licensing only.
gg. There are no county roads; only state and municipal.
hh. To be published in 1974.

INTERNAL ORGANIZATION FOR STATEWIDE PLANNING

A single staff organization for statewide planning is recommended for a department of transportation. Such an organization is shown in a recommended organization plan for the California Department of Transportation (see Figure 13.1). Such an organization is needed to develop and coordinate plans for all modes of transportation and to advise the chief executive of the transportation department on plans and programs.

Within a staff planning organization there can be a variety of arrangements. The exact organization will depend on the size of the state and the level of detail expected of the plans to be produced by the planning staff. In a large state the statewide staff may only deal with general corridor locations and other functions recommended in

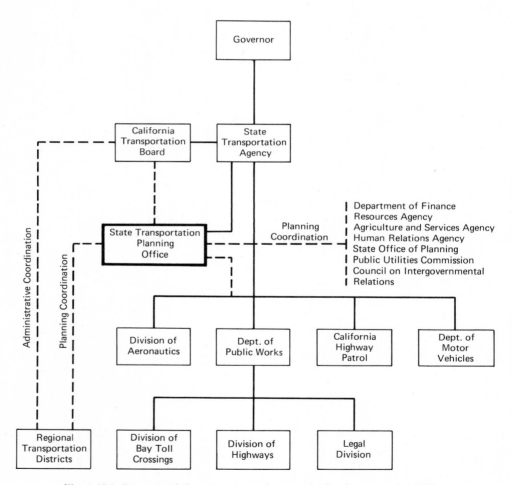

Figure 13.1. Recommended state transportation organizational structure for California. (Source: "Report of Governor's Task Force on Transportation," State of California, 1968.)

Table 13.1. In a small state or in a region the staff planning organization may carry plans down to the route location or preliminary engineering plan level of detail. Other goals influencing organization may be the desire to

1. Set up a single group, without specializing in modes, to develop plans with the accent on intermodal coordination
2. Employ existing modal specialists because of their experience, thus organizing on modal lines
3. Organize on functional lines: data collection, goal setting, systems analysts, systems designers, etc.
4. Organize on geographic lines

Generally, organizations reflect compromises between these different viewpoints. In New York's organization (see Figure 13.2.) the statewide planning section is a unit of 19 positions that was originally conceived as being organized on nonmodal lines. It has, however, since begun to specialize on modal lines. This planning section can avail itself of services of many specialists in other sections—people who are strong in research, computer programming, and in the modally organized project development bureau.

In Connecticut all statewide planning activities are undertaken by its Division of Transportation Planning, which operates within a larger Bureau of Planning and Research. (See Figure 13.3.) The Division of Transportation Planning can obtain the services of other divisions within the parent bureau. Its main functions are given below:

> . . . conducts comprehensive land use and transportation studies; establishes standards and procedures for the development of mathematical models, land use distribution and traffic distribution techniques for land use and transportation studies; develops research in mathematical model techniques and traffic operations; acts as liaison and coordinates activities with other elements of the department, and with public, local, state and federal agencies, consultants, and regional and local planning agencies on matters related to land use and transportation studies; prepares technical transportation planning reports and reviews those prepared by other sources; prepares and reviews transportation planning reports intended for public consumption.[4]

EXTERNAL ORGANIZATION FOR STATEWIDE PLANNING

State departments of transportation generally have regular channels for the coordination of their projects and programs with other state agencies. This is often managed through correspondence and informal meetings or through the capital budget review process. Sometimes the state planning organization has served in a coordinative capacity.

Direct cooperation and coordination with private carriers' planning operations is relatively unknown, however, even though it is particularly important that there be direct cooperation in technical planning with private carriers. In the future, the financial and operational realities of private carriers will have to become a factor in statewide system planning as well as state financial deliberations.

---

[4] *Organizational Manual* (Hartford, Conn.: Connecticut Department of Transportation, 1970), p. 20.

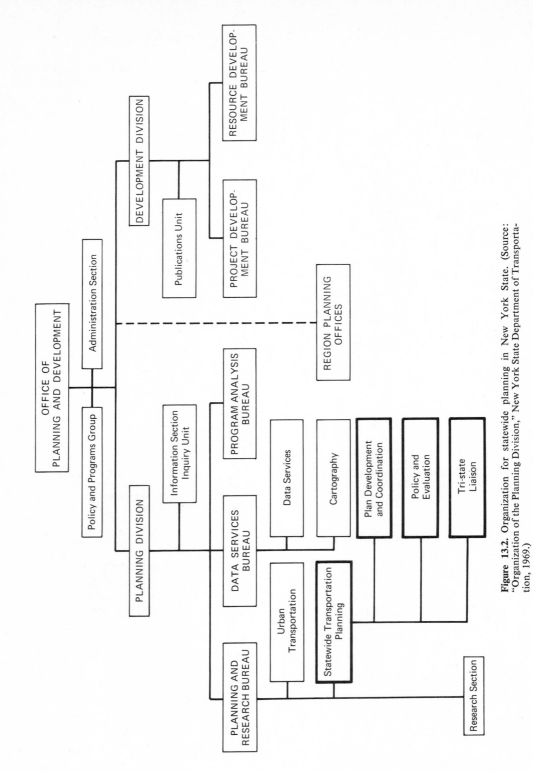

Figure 13.2. Organization for statewide planning in New York State. (Source: "Organization of the Planning Division," New York State Department of Transportation, 1969.)

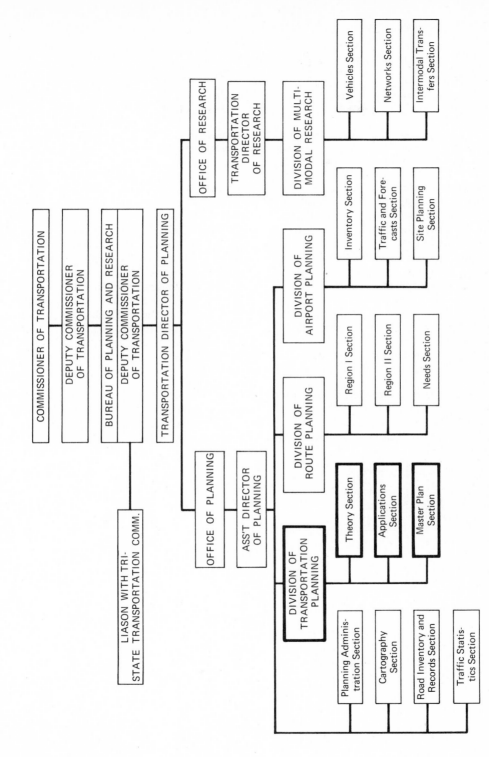

Figure 13.3. Organization for statewide planning in Connecticut. (Source: State of Connecticut Department of Transportation "Organization Manual", 1970.)

PREPARATORY STEPS FOR STATEWIDE
TRANSPORTATION PLANNING

ALTERNATIVE APPROACHES TO STATEWIDE TRANSPORTATION
PLANNING

The three basic approaches to statewide transportation planning are (1) the *needs-standards* approach, (2) the *single-mode simulation-evaluation* approach, and (3) the *multimode simulation-evaluation* approach. These are described briefly below.

The needs-standards approach. In this approach standards are set for each of the separate modes of transportation. These may include standards of physical design (as of roadway geometrics), standards of service levels (capacity in relationship to demand or frequency of mass transportation service), and safety standards. Surveys then measure existing conditions; forecasts estimate future demands. The difference between the standards and existing (or future) conditions is the *need*. Generally, needs exceed financial resources, and, therefore, priority projects are identified, which become the program for construction. This approach is diagrammed in Figure 13.4.

The advantages of the needs-standards approach, which is basically the approach used in the National Transportation Planning Manuals of the U.S. Department of Transportation (July, 1970), are its simplicity, directness, credibility, and the fact that it *can be done*. The disadvantages are that the standards can be debated, that the benefits to users and nonusers are not directly measured, and that comparisons of intermodal investment productivity cannot be made directly.

The single-mode simulation-evaluation approach. The single-mode simulation-evaluation approach is derived from the urban transportation planning process. It typically contains four major elements plus the elements of data collection and programming-implementation. The major elements are (1) *the statement of goals* or criteria, (2) *the*

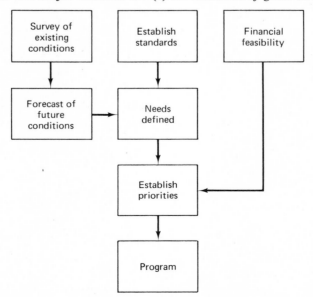

Figure 13.4. The needs-standards approach for statewide transportation planning.

preparation of plans to improve performance in relationship to those goals or criteria, (3) *the simulation* of present and/or future performance of the planned system, and (4) *the evaluation* of the results.

The basic distinction between the needs-standards approach and the simulation-evaluation approach is in the nature of the goals. *Standards* tend to be physically dimensioned, relating to the facility itself; the goals used in simulation-evaluation stem from performance *as observed by users or nonusers.* The sequence of steps used in single-mode *simulation-evaluation* can be diagrammed in many ways, but a simple diagram containing the essential feedback steps is shown in Figure 13.5.

Essential to the preceding process are data that represent the behavior of the system being planned and permit future behavior to be represented. Quantitative data, instead of word-data or descriptions, are needed.

The advantages of the single-mode simulation-evaluation approach are (1) that it evaluates plans directly in terms of user and nonuser goals, for example, minimizing construction and operating costs, cost of travel, time, and safety, (2) that it deals with and represents systems directly, thus leading to greater understanding, and (3) that it offers the ability to add up costs (e.g., time) on the same basis for several modes, thus permitting intermodal comparisons. The disadvantages of this approach are (1) its complexity and difficulty and (2) the fact that it is not operational for most statewide transportation systems except for highway traffic simulation.

The multimode simulation-evaluation approach. In this approach, which might be called a third-generation approach to statewide transportation planning, the demands for transportation, both of people and goods, are estimated for all parts of a state.[5] The demands are then allocated between modes and simulation would be undertaken for all modes, much as in the single-mode process described above, except that allowance would be made for feedbacks, as planned changes in service levels affect the choices of mode. This process is diagrammed in Figure 13.6.

The advantages of this approach are that it deals with all modes of transportation simultaneously and presumably would permit more effective planning and coordination across all modes. The disadvantages of this method are (1) its extreme complexity, (2) the inadequacy of necessary data, and (3) that there is so little experience, except for a few studies undertaken for foreign countries.[6]

The needs-standards approach is, in effect, a requirement of the U.S. Department of Transportation, and, as such, is being used by all states whether or not they have departments of transportation. A number of states are already involved in the second-generation approach through the simulation of traffic on statewide highway networks, and some preliminary work is being done in simulation of other modes at the national level. Simulation of travel over other modes is relatively unknown. The ultimate target for statewide planning is probably the third-generation, or multimode approach.

Case study: The 1972 National Transportation Needs Study. The process employed by the 1972 National Transportation Needs Study, shown in Figure 13.7, is based on

---

[5] Commonwealth of Pennsylvania by Transportation Research Institute of Carnegie-Mellon University and Pennsylvania Transportation and Traffic Center, *Methodological Framework for Comprehensive Transportation Planning*, Final report prepared for Governor's Committee on Transportation (Pennsylvania State University, 1968).

[6] John R. Meyer, ed., *Techniques of Transportation Planning, Volume Two, Systems Analysis and Simulation Models*, by David T. Kresge and Paul O. Roberts (Washington, D.C.: The Brookings Institution, 1971).

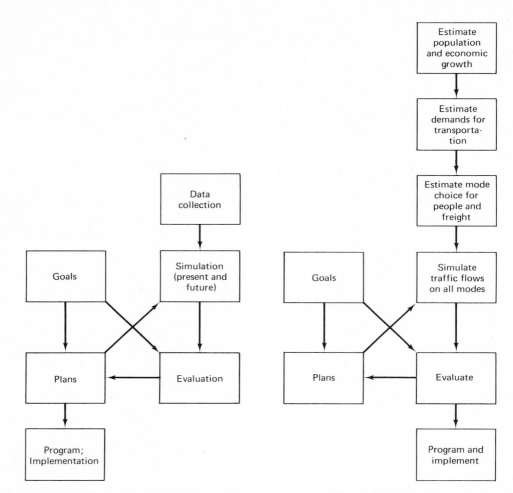

Figure 13.5. The single-mode simulation-evaluation approach to statewide transportation planning.

Figure 13.6. The multimode simulation-evaluation approach to statewide transportation planning.

a series of manuals[7] prepared by the U.S. Department of Transportation whose purpose was to insure uniformity of estimates for all states. The point of contact in each state is the governor's office, which is the logical coordinating point for the majority of states that do not have multimode transportation departments.

In the words of Manual A,

> 5. Each relevant state agency or other agency should do the functions relating to estimates of transportation needs described in each of the other manuals in the set and designed to meet the overall goals. In general, the steps are to:
> a. apply given standards of facility development, or develop and apply standards

[7] "National Transportation Planning Manual (1970–1990)," including Manual A (General Instructions); Manual B (National Highway Functional Classification and Needs Study Manual); Manual C (Urban Public Transportation); and Manual D (Airports and Other Intercity Terminals) (Washington, D.C.: U.S. Department of Transportation, 1970, 1971).

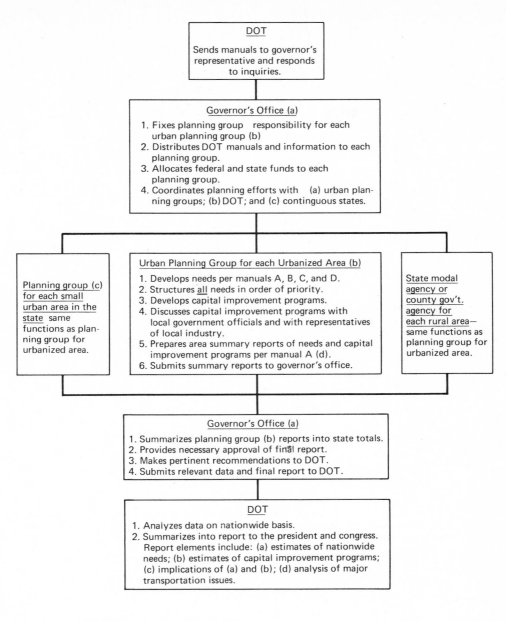

DOT

Sends manuals to governor's representative and responds to inquiries.

Governor's Office (a)

1. Fixes planning group responsibility for each urban planning group (b)
2. Distributes DOT manuals and information to each planning group.
3. Allocates federal and state funds to each planning group.
4. Coordinates planning efforts with (a) urban planning groups; (b) DOT; and (c) contiguous states.

Planning group (c) for each small urban area in the state same functions as planning group for urbanized area.

Urban Planning Group for each Urbanized Area (b)

1. Develops needs per manuals A, B, C, and D.
2. Structures all needs in order of priority.
3. Develops capital improvement programs.
4. Discusses capital improvement programs with local government officials and with representatives of local industry.
5. Prepares area summary reports of needs and capital improvement programs per manual A (d).
6. Submits summary reports to governor's office.

State modal agency or county gov't. agency for each rural area— same functions as planning group for urbanized area.

Governor's Office (a)

1. Summarizes planning group (b) reports into state totals.
2. Provides necessary approval of final report.
3. Makes pertinent recommendations to DOT.
4. Submits relevant data and final report to DOT.

DOT

1. Analyzes data on nationwide basis.
2. Summarizes into report to the president and congress. Report elements include: (a) estimates of nationwide needs; (b) estimates of capital improvement programs; (c) implications of (a) and (b); (d) analysis of major transportation issues.

Notes: (a) Or governor's representative.
     (b) An urban planning group for each urbanized area. For nonurban and small urban areas, the governor's office should designate a state or county agency, such as the state planning office to assume the responsibilities of the urban planning group.
     (c) State or county agencies will normally assume planning responsibilities for small urban areas.
     (d) Including supporting data when necessary.

Figure 13.7. Federal, state, and local responsibilities for developing needs and capital improvement programs. [Source: "National Transportation Planning Manual (1970–1990) Manual D: Airports and other Inter-City Terminals," U.S. Department of Transportation, Washington, D.C., 1971.]

> of facility development or public transportation service to describe the level of service which would be available on each mode.
>
> b. estimate the amounts of future travel by each mode assuming provision of facilities and services corresponding to the levels of service described above.
>
> c. calculate the amount of improvements required to evolve from the 1970 situation to one which meets the standards with the expected 1990 travel and also perform similar calculations for intermediate years.
>
> d. estimate the total costs in 1969 dollars . . . of achieving such a system by 1990 and also the cost of achieving intermediate year systems. . . .
>
> 6. The state should make a preliminary allocation of federal funds and state funds in each major program category *to each urbanized area* so that the urban planning group in that area can proceed to develop local urban area capital improvement programs in parallel with those being developed by the state for intercity and rural transportation and for transportation within smaller urban areas. There would be three separate allocations, one each for Alternatives I, II and III, the federal aid alternatives which are described in the next chapter.
>
> 7. The state should make a preliminary allocation of federal funds and state funds to the relevant branches of the state government so that they may develop capital improvement programs as required by this manual and manuals B, C, & D for intercity and rural transportation and for transportation within smaller urban areas. There would be three allocations, one each for Alternatives I, II and III.

Two federal funding alternatives are provided at the outset—a low-level and a high-level—with amounts prescribed for different programs for each state. A third alternative is allowed in which the funds available under Alternative II (high-level) can be redistributed among programs in accordance with a state's own policies.

The programs to which funds may be allocated under Alternatives 1 and 2 are:

Interstate highways
Primary and secondary highways
Urban extensions
TOPICS
Urban public transit
General aviation airports
Air carrier airports

Thus, at the end of preparing its portion of the 1972 National Transportation Needs Study, each state will have a *measurement* of its needs for most modes of *person transportation*. States will not necessarily have done any work in goods movement and some elements of long-distance person transportation (i.e., bus and high-speed rail) may not have been studied. But because the needs studies are expected to be repeated every two years, a strong and continuing first-generation planning approach is assured.

Case study: New York. Figure 13.8 illustrates the process used by New York State in 1968.[8] A legislative mandate was the initiating action, followed by the setting of more specific goals than those stated in legislation. Note the input of "state development goals," which define land use, economic, environmental, and other nontransportation criteria.

This is an example of the separate modal planning approach, with each mode

---

[8] *Policies and Plans for Transportation in New York State* (Albany, N.Y.: New York State Department of Transportation, 1968).

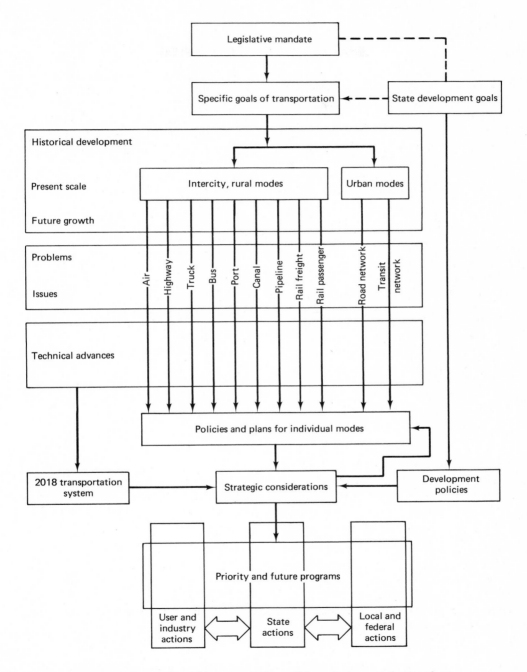

Figure 13.8. Process by which transportation policies and plans are developed, coordinated, and implemented.

viewed against a background of data on scale, historical development, future growth estimates, problems, issues, and potential technological advances.

At the end of each modal planning operation, the results are coordinated (between modes, and with other state development policies) and then programmed for accomplishment. Broad actions, private, and public, are suggested.

This process was highly generalized; it is not in the form of a detailed step-by-step procedure or even the macrodiagrams conventionally used to describe the urban transportation planning process. This level of generalization was necessary in 1968 because the procedures within many of the elements shown in Figure 13.8 were themselves not yet developed. This description of the process, however, displays all the important elements, in their proper relationships, in one comprehensible figure. This sets the stage for more detailed work.

In a more recent staff paper[9] a new statewide planning work program was proposed. This is diagrammed in Figure 13.9. Here, passenger and freight transportation

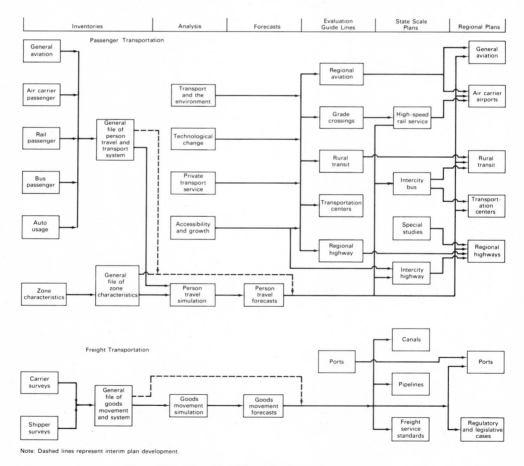

Note: Dashed lines represent interim plan development.

Figure 13.9. Elements of the statewide work program.

[9] ROBERT BREUER, *Prospectus for a Statewide Transportation Planning Program: Non-Urban Elements* (Albany, N.Y.: New York State Department of Transportation, 1970).

are dealt with separately. The work program is broken into manageable units. The two following items should be noted:

1. State-scale plans are to be developed *before* regional plans (in New York, regions are groupings of counties).
2. Simulation and forecasting activities are by-passed in the initial stage of preparing an interim plan. This reflects a realization of the present state-of-the-art and the necessity of getting a product at an early date.

STANDARDS AND GOALS

Standards and goals have two distinct meanings. A *standard* is a specific characteristic of a transportation facility, for example, a pavement width, a volume to capacity ratio, or a transit or airline service frequency. A *goal* is a performance rating of a transportation facility *seen from the viewpoint* of a person or group of people *in terms of its impact upon them.*

Both standards and goals must be established by transportation planners in each state.

Examples of standards are those set up in the manuals of the 1972 National Transportation Needs Study. Table 13.4 is an example of standards for rural arterials and collectors.

A basic difficulty with standards is in determining how much it is worth to attain standards, or even make improvements toward them. Here is where an approach that rates performance of transportation system in relationship to the things (goals) people want for themselves or for their organizations can provide an additional strength. Costs of making improvements can be related to reductions in costs to users or to gains in other recognized values.

The following goal statements may be used as a starting point for the development of goals for states or regions. For convenience, these goals have been divided into three groups, according to the viewpoint of the persons or groups most concerned with each goal. The three groups of viewers are: (1) the *users* of transportation facilities, including individuals who themselves use transportation facilities or the shippers of goods; (2) the *providers* of transportation facilities, both public and private; and (3) the *nonuser* public, which is the total population playing the roles of residents, conservationists, and so forth. As a matter of procedure, goal statements should be developed at the staff level and then reviewed at the policy level of government. Acceptance at the policy level can be very helpful later when plans have been drawn and are being reviewed because there will then be an accepted set of criteria.

User goals. Common user goals include some or all of the following:

1. *Increased mobility.* Greater mobility is desired for all the population. In addition, special consideration must be given to increasing the mobility of older citizens, the handicapped, and those who cannot afford to drive or who are incapable of driving to work, to obtain services and recreational facilities, or for other purposes.
2. *Increased dependability.* A dependable transportation system is desired which will be able to cope with bad weather and other adverse conditions and which will have a low incidence of breakdowns and late departures.

TABLE 13.4
Minimum Tolerable Conditions for Rural Arterials and Collectors

Functional Systems	Rural Principal Arterials			Rural Minor Arterials									Rural Collectors												
ADT for Analysis Year	All			Over 6,000			2,000–6,000			Under 2,000			1,000–6,000*			400–1,000			100–400			Below 100		
Terrain	F	R	M	F	R	M	F	R	M	F	R	M	F	R	M	F	R	M	F	R	M	F	R	M
Operating speed (peak hour)	55	50	45	50	45	40	50	45	40	40	40	35	—	—	—	—	—	—	—	—	—	—	—	—
Surface type	High			High			Inter.			Inter.			Inter.			Low			Low			Gravel		
Lane width	11			11			11			11			11			10			9			22′ roadway		
Shoulder type	Stab.			Stab.			Stab.			Earth			Earth			Earth			Earth			—		
Graded right shoulder width (ft)	8	8	6	8	8	6	6	6	4	6	6	4	4	4	4	4	4	4	2	2	2	—		
Safe speed†	65	55	45	60	50	40	60	50	40	60	50	40	50	45	35	50	40	30	50	40	30	40	35	25
Stopping sight distance	550	415	315	475	350	275	475	350	275	475	350	275	350	315	250	350	275	200	350	275	200	—		
Maximum curvature	5	6	10	5	8	13	5	8	13	5	8	13	8	10	18	8	13	23	8	13	23	—		
Maximum gradient‡	3	4	8	3	5	9	3	5	9	3	5	9	4	5.5	10	4	6	11	4	6	11	—		
Number of lanes	§			§			2			2			2‡			2			2			—		
Pavt. cond. rating (PSR or equivalent)	2.6			2.6			2.1			2.1			2.1			2.1			2.1			—		
Railroad crossing protection	See Table III-15																							
Structures Width (ft)			Traveled way width + 6 ft.			Traveled way width + 4 ft.									Traveled way width + 2 ft.									18	18	18
Vertical clearance (ft)	14			14			14			14			14			14			14			14		
Loading	H-20			H-20			H-15			H-15			H-15			H-15			H-15			H-15		

*Rural collectors with present ADT above 6,000 should be multilane where necessary to maintain peak hour operating speeds of 40, 35, and 30 in flat, rolling, and mountainous terrain respectively.

†Approximate speed on which minimum tolerable stopping sight distance curvature, and gradients are based.

‡Steeper grades may be considered tolerable if lengths are relatively short or climbing lanes are provided.

§As necessary to maintain the operating speed specified.

||For bridges over 250 ft. in length, widths 4 ft. less than shown, but in no case less than the width of the approach traveled way will be considered adequate.

Source: "National Highway Functional Classification and Needs Study Manual (1970–1990)" (Manual B of National Transportation Planning Study) (Washington, D.C.: U.S. Department of Transportation, 1970).

3. *Reduced time spent in travel.* Faster door-to-door transportation is desired for both people and goods. This includes time spent in transferring people and goods between transportation modes.
4. *Increased comfort.* The comfort of transportation can be increased by providing clean, air-conditioned, and well-maintained vehicles and terminals and by reducing congestion.
5. *Reduced accidents.* A major goal is to reduce fatalities, injuries, and property damage accidents for all modes of transportation.
6. *Reduce user costs.* User costs are monies spent for operating costs for private vehicles, fares on and operating subsidies for mass transportation systems, and prices for the shipment of goods. Clearly, it is an important goal to keep these costs as low as possible.
7. *Enhance aesthetics.* People spend a significant portion of their time travelling to and between cities. The travel experience should be kept as pleasant as possible for all modes of transportation by promoting beautification and the development of scenic facilities along rights of way and at terminals.

Goals of transportation agencies and companies. Public and private transportation groups may have the following goals:

1. *Reduce capital costs.* The construction of transportation facilities and the purchase of common carrier vehicles for the movement of people and goods are major investments. These should be kept as low as possible, consistent with the achievement of other desired goals.
2. *Reduce maintenance costs.* The costs of maintaining existing facilities and equipment should be kept as low as possible, consistent with the attainment of other goals, including the goal of reducing capital costs.
3. *Reduce operating costs.* Labor and energy costs are major components of the total cost of providing for the movement of goods and for the common carrier transportation people. These costs should be kept as low as possible, consistent with the attainment of other goals.
4. *Profit.* For the private carriers, making a profit is an important goal; in fact, profit is an indication of organizational vitality and efficiency.
5. *Increase coordination.* The total transportation system of a state should be planned to work as a system with the various modes interconnected for flexibility yet providing that efficiency which only specialization can produce.
6. *Preserve resources.* The transportation system should be planned so that natural resources such as prime agricultural lands, forests, shorelands, lakes, wildlife preserves are conserved. In addition, new transportation facilities should avoid sites that have historic and aesthetic value.
7. *Economic growth.* Transportation is a vital ingredient in promoting economic development.

Community goals. Depending on local circumstances, community goals may be to:

1. *Increase accessibility.* A major statewide goal is to increase accessibility to all parts of the state and especially to those parts not adequately served by transportation.
2. *Reduce pollution.* The emission of noise, gaseous and particulate pollutants,

and solid wastes from transportation sources should be kept below the levels set by state and federal law.

3. *Encourage desirable settlement patterns.* Because transportation affects the patterns of development within a state, it should be used to the extent feasible to promote desired forms of population settlement patterns and the desired location of economic activities.

4. *Reduce harmful environmental impact.* The construction of transportation facilities should be planned so that harmful impact on the natural environment, including marine life and wildlife, is minimized.

FORECASTS AND LAND-USE PLANS

Long-range transportation planning requires an input of forecasts, both of population and economic growth. Forecasts of the total population and economic activity of a state are derived from a number of subsidiary forecasts—vehicle registrations, vehicle miles of travel, expenditures for person travel, estimates of future goods movement, and the like. In addition, there must be forecasts, or plans, of the *location* of population and economic activity for fairly fine geographic subareas of the state. These should be at least down to the level of the minor civil division (city, town, or village); the county is at too gross a level of detail for usefulness in planning.

All forecasts on a state must be done on a uniform basis, and the various derived or subordinate forecasts should be consistent with the overall state forecasts. It is generally most efficient for a single state agency, for example, a state planning board, to prepare and *maintain up-to-date* population forecasts. These should be checked carefully against national forecasts such as the series used by the Federal Highway Agency. There is nothing more awkward from a staff point of view than having to work with two forecasts of population for the same area. Economic forecasts are also required and will play an increasingly important role as states begin to concern themselves more with the transportation of goods.

Given these two major inputs, forecasts of total person travel and of the demand for the movement of goods should be made. Generally, these are made from two directions: first, from the point of view of aggregate demand for transportation of people and goods and, second, from the historical records of the individual modes. These two independent forecasts must be reconciled.

Land-use information, both for present and future plans, should be compiled by local and/or regional agencies and made available to the department of transportation. Population location will be reflected in the forecasts of population of minor civil division. Plans for publicly owned parks, forest preserves, water conservation areas, institutions, and similar uses should be compiled. Mappings of prime agricultural lands should be obtained. These and other land-use plans have direct influence on transportation planning.

MODAL STUDIES

In this section each of the principal modes of nonurban transportation will be dealt with separately. In each case the discussion centers around (1) key facts and trends concerning the mode, (2) problems and issues, and (3) comments on planning for that mode. Because there are presently few advanced planning techniques for many modes, the comments on modal planning attempt to identify where advances in techniques are most needed.

The problems and issues described herein are issues as seen by the transportation planner or transportation executive working in government. This viewpoint tends to be technical, concerned with investment and allocation of resources, and concerned with systems in the long run. The issues as seen by the carriers tend to be much more concerned with individual, short-range problems (such as costs of labor, franchises, and other management and operational problems) and rarely concerned with systems. It will be important for the future of the transportation industry that a shared view of issues and problems be taken, acknowledging the existence of both short- and long-range issues.

RAILROAD TRANSPORTATION

Railroads in the United States are an enormous industry. In 1971 they had a net investment (Class I railroads alone) of $28.2 billion and their operating revenues were $12.7 billion. This compares with an estimated $21.9 billion available to all U.S. governmental units for highway purposes during the same year. In terms of system length, there were about 205,000 mi of railroad lines in the U.S. in 1971 compared with 32,988 mi of interstate highways and 480,332 mi of rural primary state highways.

Railroads are still the major transporter of intercity freight. In 1971 the number of tons of revenue freight originated on Class I railroads was estimated at 1.392 billion tons. Revenue ton-miles exceeded 739 billion. Tons originated have not changed significantly over the years, but ton-miles rose from the 447 billion ton-miles of 1929 to the level of 768 billion ton-miles in 1969 and dropped to 739 ton-miles in 1971. This implies that rails are hauling freight over longer distances, leaving more short-haul transportation to trucks.[10]

Problems and issues. Problems and issues as seen from the public viewpoint are:

1. *Railroad grade crossing elimination.*
2. *Branch line continuation.* How extensive a system of branch lines is ultimately needed? The aggregate length of roadway of all line-haul railroads in the United States has declined from 249,433 mi in 1929 to about 205,000 mi in 1971.[11] Are more reductions proper? Can branch line service be replaced by container on flatcar or trailer on flatcar (COFC/TOFC) service?
3. *Interline coordination.* Railroads grew up in an era of extreme competition, and in many areas (both urban and rural) the system appears to be poorly organized and wasteful. Can a more practical network be devised by using systems planning, without regard to past ownership? In some areas this is being attained through mergers; the ultimate may only be attained by nationalization.
4. *Railroad coordination.* "Piggy-back" (TOFC/COFC) locations need to be coordinated with arterial and interstate highway systems.
5. *Passenger service.* Should long-distance rail passenger service be maintained by the federal government over an entire continent?

---

[10] Railroad data from "Yearbook of Railroad Facts" (Washington, D.C.: Association of American Railroads, 1972 Edition); highway data from "1972 Automobile Facts & Figures" (Detroit, Mich.: Motor Vehicle Manufacturers Association of the United States, Inc.).

[11] *Ibid.*

6. *Relationship to state development.* Does the rail system, particularly the freight system, adequately serve the planned development of the state?

Comments on planning. Statewide rail planning should be concerned primarily with the extent and design of rail systems, investment, terminal locations (including TOFC/COFC) and frequency of service, together with concern for the relationship between rail systems and land development. Statewide transportation planning ought to develop the technical capacity to plan rail systems in the same way that urban freeway systems have been planned. The level of detail does not include management of operations over the rail lines, but it does include generalized corridor location and the extent of feeder lines.

In order to obtain this kind of capacity for planning, substantial additional data must be obtained. This data collection, along with rail system planning, could probably best be done on a multistate region basis. Data collection can be a very difficult operation. Aside from the ICC carload waybill statistics (published annually), it is extraordinarily difficult to obtain accurate and current information on the tonnage and kind of freight shipments from origin to destination and the number of trains, freight cars, and/or tons of goods moving over main lines and branch lines in a system. Extensive studies should be made to obtain these data. It may be that such data can be more readily obtained from the shippers and receivers of goods than from the carriers.

Passenger origin-destination data can be readily obtained by personal interview at the terminals or by means of postcards either mailed back or picked up at destination stations.

Extensive work must be done before origin-destination data over rail systems can be obtained or before freight movements can be simulated. Extensive research should be undertaken to develop this capability.

BUS TRANSPORTATION

With rails steadily declining as a means of moving people between cities, and with air transportation being fairly inefficient in terms of cost and speed for trips less than 150 mi, buses remain an important and economical common carrier for passengers making trips of medium length.

Bus transportation is a major industry. In the United States all Class I bus companies carried 14.1 billion passenger miles in 1971 with 856 million bus miles of service. Average vehicle loading was thus about 19.4 persons per vehicle.[12] Revenue per passenger mile was 3.83¢.

Approximately 70 percent of the revenues of intercity carriers comes from the passengers. This is supplemented by revenues from charter passengers (13 percent of total revenues) and by a brisk business in package express, which accounts for nearly 12 percent of operating revenues.

Public problems and issues. From the viewpoint of the public agency, the following are among the current principal problems and issues in the field of bus transportation:

1. *Terminal location.* Terminals must meet the needs of proximity to the state expressway system in order to decrease journey time, proximity to local transit

---

[12] "Bus Facts, 1970," National Association of Motor Bus Owners.

systems, adequate parking, and centrality with respect to the location of potential passengers.

2. *Vehicle design.* Bus owners and operators favor legislation authorizing bus widths of 102 in. instead of the 96 in. for buses that operate on the interstate system. The objective is to allow slightly wider seats. The impact of such change on truck widths and indirectly on the railroad industry, as well as impact on passenger car drivers using the interstate and noninterstate systems, are factors that will have to be determined.

3. *Increasing journey speeds.* The completion of the interstate system will help improve the speed of bus transportation. One study measured 20 percent increases in average scheduled speeds for buses using interstate roads as compared with other rural highways.[13]

4. *Coordination of schedules and routes.* Bus companies operate in a highly competitive world and with very little method in granting of franchises. As a result, service is often a hodgepodge of local and intercity routes that do not operate well as a system. Gains could be made by getting better coordination of routings and schedules.

Comments on planning. Factual data on origin and destination travel patterns, trip purpose, passenger characteristics, and volumes of intercity bus travel are generally inadequate. Yet means for collecting these data are simple and relatively inexpensive. Interviews can be taken at bus terminals or forms can be distributed to passengers. These forms may either be mailed back or picked up by the driver or agent at destination terminals. These O-D (origin-destination) data should always obtain information on the exact location of the passengers' origin and destination so that the data can be used not only for intercity system planning but also for the improvement of intraurban service to bus terminals.

Substantial data collection, research, and analysis must be undertaken before formal planning methods can be established. This work is vitally needed because bus transportation is the most important common carrier of persons in ground transportation.

AIR TRANSPORTATION

The growth of air travel in the United States has been phenomenal. Between 1950 and 1960 the number of public carrier air passenger miles of intercity travel in the United States more than tripled, from 9.3 billion PMT to 31.7 billion PMT. Between 1960 and 1971 air travel more than tripled, from 31.7 billion PMT to an estimated 110.6 billion PMT. By comparison, between 1950 and 1971, intercity rail passenger volumes declined from 32.5 billion PMT to an estimated 10.0 billion PMT.

Almost all forecasts indicate continued rapid growth in all components of air transportation. The *ATA Airline Airport Demand Forecasts*[14] foresees domestic airline passenger miles in the United States rising to 503 billion by 1985, or approximately five times the 1969 level. International air travel, according to the same report, will

---

[13] "Transportation in the Appalachian Region of New York State: Phase II," report prepared for New York State Office of Planning and Coordination and New York State Department of Transportation by Creighton, Hamburg, Inc., 1970.

[14] *Industry Report ATA Airline Airport Demand Forecasts* (Washington, D.C.: Air Transport Association of America, 1969).

rise from the 11 million enplaned and deplaned passenger level of 1965 to 115 million in 1985. This is a tenfold increase. General aviation aircraft are expected to increase in number from 92,000 in 1965 to 261,000 in 1985. And domestic air freight ton-miles are expected to grow from the 1.02 billion ton-mile level in 1965 to 30.3 billion ton-miles in 1985.

These startling changes call for very active state concern for air transportation because it is clear that such growth in demand will press strongly against the supply of facilities and of airspace.

Problems and issues. Problems and issues as seen from the public viewpoint are:

1. *Airspace and airport congestion in the large metropolitan centers.* "Experience has shown that planning for airports in the larger metropolitan areas is a much more complex task than planning for airports in the smaller communities. This is due, logically, to the greater demand for aeronautical services in the larger cities and the requirement to provide a sufficient number of adequate airports in densely populated areas. In addition, the aeronautical problems related to the use of the airspace, the problems of insufficient land, congestion of surface transportation, objections to aircraft noise, etc., highlight the complexities of metropolitan area airport planning."[15]
2. *Provision of commercial passenger service to smaller urban centers.* Lower levels of demand and rising costs make it economically difficult to provide air passenger service to many communities with less than 100,000 population.
3. *Preservation of general aviation airports.* This need is greatest in and near urban areas where competition for land becomes intense.
4. *Coordination of airport investments between airports and between airports and expressways.* Almost all airport planning is on a one-airport basis. In more densely populated states there needs to be more airport *system* planning taking into account the geographic positioning of airports and the changed accessibility provided by rural expressways.
5. *Relationship to state development.* Does the planned airport system work with or against state land development policies?
6. *Relationship to the environment.* How can airports be located in reasonably close proximity to major urban areas and still not be an environmental nuisance?

Comments on planning. A substantial amount of skillful work has been done in air traffic forecasting and in outlining methods for the preparation of statewide airport plans.[16,17,18] Inventories of physical facilities, of navigational aids, and of passenger movements can be routinely made.

However, too often planning is restricted to an airport-by-airport approach. How can *systems* of airports be planned—and planned in full knowledge of the changing ground accessibility caused by construction of interstate highways? How can *systems*

---

[15] "Planning and State Airport System" (Washington, D.C.: U.S. Department of Transportation, Federal Aviation Administration, 1968).

[16] "National Transportation Planning Manual—Manual D: Airports and Other Intercity Terminals" (Washington, D.C.: U.S. Department of Transportation, 1971).

[17] *ATA Airline Airport Demand Forecast* (Washington, D.C.: Air Transportation Association of America, 1969).

[18] "Airport Site Selection" (Washington, D.C.: U.S. Department of Transportation, 1967).

of air services be planned when different airlines are competing for identical routes? Can changing sizes and speeds of aircraft be taken into account? Solving these problems will take substantial effort and will require close cooperation between industry representatives and governmental officials.

HIGHWAYS: TRANSPORTATION FOR PEOPLE AND GOODS

The dominant transportation mode in every state of the United States is the highway system, which moves nationally over 95 percent of all person-miles of travel and about 50 percent of all ton-miles of freight. For 28 out of 50 states (as of 1974), highways were the function of a single department; in the remaining states highways were by far the largest single concern of state departments of transportation.

The growth of population and the national economy continue to create new pressures on highway systems, both urban and rural. Settlement patterns have been assuming new forms: the expansion and linkage of existing metropolitan centers into continuous clusters of cities; the spreading of rural, nonfarm settlement at very low densities around major centers; and the lack of growth of very small, independent towns and villages.

What form the future highway system should take in response to these challenges is one of the most important questions of our time. And this is one reason why statewide highway planning is so important. Many of the basic problems and issues listed below relate closely to how well the highway system of the future will fit together with, and serve, the land development patterns of each state.

Problems and issues. Major problems and issues as seen from the public viewpoint are:

1. *Urban-rural split.* How much should be spent for rural and intercity highways as compared with investment in urban transportation systems?
2. *Road-type split.* How much money should go to building (and/or maintaining) nonurban expressways as opposed to other kinds of rural roads?
3. *Economic impact.* How much effect do highways have on the location and growth of economic activities in a state?
4. *Land development.* Can highway planning promote better land development patterns? Conversely, what land development controls are needed to protect a highway system? What are state land development policies?
5. *Environmental impact.* How can the impact of highways on the total environment be reduced?
6. *Safety.* How can nonurban highways be made safer?

Comments on statewide highway planning. Planning for statewide highway systems has a longer history of public acceptance than planning for other nonurban modes in the United States. The first federal-aid systems were designated in 1921. In 1934 the Hayden-Cartwright Act provided a steady source of funds for planning and research. With these resources many statewide inventories of physical facilities and traffic were undertaken in the 1930's.

In 1956 the interstate system was established, and this became a dominant factor in the construction program of all the states for the next fifteen years. In the 1970's the states will look beyond the completion of the interstate system in order to determine what their next major targets will be. Quite likely, these targets will differ substantially from region to region. The problems of the Western and Mountain states are quite different from those of the East or South.

A variety of basic planning techniques are available to help states establish their highway construction policies. Several of these are described below, beginning with highway classification studies and needs studies.

Highway Classification. Highway classification is an important first step in nonurban statewide highway planning. There are several purposes for which classification is undertaken. One purpose is to define the systems to be examined in highway needs studies. A second purpose is to provide a basis for dividing the total road system among political jurisdictions (town, county, and state) and for use in establishing the various federal-aid systems. Third, classification may be used as the means for eliminating local roads from a network so that planning attention and computer capacity can be devoted to more important highways. Finally, classification studies provide the basis for determining the level of future highway expenditures and for the allocation of funds between states and among federal-aid highway networks. For a more detailed discussion of highway classification, see Chapter 14, Geometric Design.

The Needs Study: A Method of Highway Planning. After a classification study has been completed, a sensible method of statewide highway planning can be carried out in conjunction with a long-range evaluation of highway needs. This is the approach used in Manual B of the *U.S. National Transportation Planning Study.* According to this manual,

> the functional classification process, the process by which streets and highways are grouped into classes or systems according to the character of service they are intended to provide, has been widely used in conjunction with needs estimates to outline long-range highway plans. In this study, focus is being aimed at the development of nationwide 1990 functional plans in concert with 1990 nationwide needs evaluation.[19]

The first step in the needs study approach to highway planning is to complete a present-day highway classification study, as outlined in the previous section or as more fully described in a 1968 manual and in Chapter 14 of this handbook.[20]

Once the rural highway system has been classified according to type and into prescribed systems, a forecasting operation must be conducted. Working from consistent national, or state, population forecasts, planners should estimate the growth of cities of various sizes throughout a state, together with the populations of their rural areas. The boundaries of urban areas as they are likely to be at some future point in time must be estimated, say, for a planning period of twenty years.

Given the new sizes of cities and the present highway network, the highway planner must reexamine the network in order to determine where changes should be planned so that future travel between enlarged cities and towns can be accommodated. In many ways this is simply a reclassification based on changed population and land use. Manual B states:

---

[19] "National Highway Functional Classification and Needs Study Manual" (Washington, D.C.: U.S. Department of Transportation, 1970).

[20] "1968 National Highway Functional Classification Study Manual" (Washington, D.C.: U.S. Department of Transportation, 1969).

The (future) functional classification will differ from the 1968 classification in two basic respects: (1) it will be based on projected 1990 population, land use, and travel; and (2) it will include, in addition to existing facilities, such projected totally *new* facilities as will be needed to serve 1990 land use and travel.[21]

According to Manual B, the additional new miles of rural highways are expected to be (1) unbuilt interstate mileage, (2) belts and bypasses for smaller cities, (3) some new routes needed to serve new recreational areas or new towns, and (4) replacement or expanded facilities on new rights of way.

It can be seen that this is a fairly conservative approach to rural highway planning. Manual B states:

> Studies conducted over the years have indicated a large degree of stability in the routes and corridor locations of arterial systems. To a considerable extent, centers of the lower size range of places served by these systems . . . are not undergoing great or rapid change. Furthermore . . . if all centers were growing in [the same] proportion, . . . such growth would not affect the functional relationships in the road network.[22]

The same argument applies to rural collector routes.

Manual B is careful to point out that new routes, introduced into the existing network, will be likely to affect the travel choices of people and thus change traffic volumes and functional uses of other routes. Such changes in traffic volumes will have to be estimated carefully, and estimates of new needs must be adjusted accordingly.

Once a future classification plan has been prepared, the needs study proper can be completed for both rural and urban areas. For each link, or for a sample of links, studies are made to determine the costs of bringing these links up to a predetermined standard. The costs may either be to overcome an existing deficiency, called a "backlog" need, or to overcome a future deficiency, called an "accruing" need. Costs of maintenance and future repairs are also estimated. The product of these studies can be used directly in capital programming.

The simulation approach to rural highway system planning. Simulation is being used increasingly by state transportation and highway departments in the United States as a means for preparing and testing rural highway plans. As of mid-1971, 14 states have used simulation to assign traffic to statewide highway networks and 8 are in the process of developing or applying assignment techniques.[23] Simulation offers a way of estimating traffic on *all* links in a state's highway network, and it is one of the few methods of estimating the consequences of inserting new links within an existing system. Simulation also offers the potential for reading out certain consequences of vehicular travel (accident, time, operating, air pollution) that are obtainable only with difficulty by other techniques. At present it must be recognized that simulation of traffic over entire state networks is in a rudimentary stage (except in the smaller, more densely settled states like Connecticut and Rhode Island) and that substantial improvements will need to be recorded in the next few years.

---

[21] "National Highway Functional Classification and Needs Study Manual" (Washington, D.C.: U.S. Department of Transportation, 1970).

[22] *Ibid.*

[23] From conversation with Philip Hazen of the FHWA. See also Philip I. Hazen, "A Comparative Analysis of Statewide Transportation Studies," *Intermodal Transportation Planning at the State, Multistate, and National Scale*, Highway Research Record 401 (Washington, D.C.: Highway Research Board, 1972).

An early user of simulation in statewide highway planning was Connecticut.[24,25] The basic process used by Connecticut is illustrated in Figure 13.10. The Connecticut

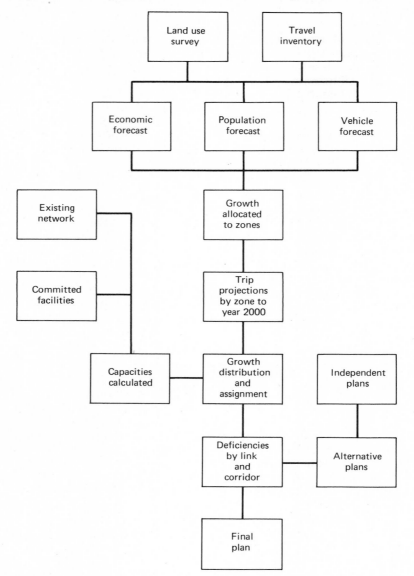

Figure 13.10. Connecticut highway planning process. (Source: Adapted from "Planning for the Future—Connecticut's Major Highway Needs, Present to Year 2000", by Connecticut Highway Department, September 1968.)

---

[24] "Planning for the Future—Connecticut's Major Corridor Needs, Present to Year 2000," Connecticut Highway Department, September 1968.

[25] "Planning for the Future—Connecticut's Major Arterial Needs, Present to Year 2000: Part II," Connecticut Highway Department, September 1969.

process is similar to the planning processes used in many urban areas. It involves data collecting, forecasting of trip ends (taking into account population, employment, land use, and vehicle registrations) and the assignment of trips by using a mathematical model (in this case the gravity model) over a network. The basic network consisted of the existing network plus the committed network, that is, those facilities that were programmed plus those facilities which, for various reasons, were considered to be definitely "committed."

Traffic was then assigned to the existing and committed highway network for every decade from 1970 through 2000.

Capacities of the existing and committed highway networks were calculated, and the differences were calculated between demand (as determined by the assignments for each decade) and the capacities. Thus, deficiencies were known for each decade. These deficiencies were obtained for highway links *grouped* by a series of screenlines created by a variable-width grid superimposed on a map of the state. Finer grid intervals were used in higher-density corridors.

Deficiencies by decade were used to develop two different sketch plans, one based on needs for additional lanes in each corridor, the other based on an analysis of link deficiencies on the network, with emphasis on the expressway network. In addition, a composite of regional plan highway proposals was examined and also the Connecticut Highway Department's then existing long-range proposals. Traffic volumes for the year 2000 were assigned to all four of these plans. As noted in the report, "Through the continuous process of testing and analyzing various combinations of proposed facilities, a plan evolved which best appeared to satisfy traffic demand of the 2000 decade."[26]

Truck weight studies. Truck weight studies are generally made each year by the states in order to provide data for use in the design of highway pavements, for allocations of highway costs in relation to usage and revenue, and for a variety of other studies, including goods movement. In the words of the *Guide for Truck Weight Study Manual,*[27]

> Truck weight data are obtained either by using permanent platform scales, most of which are capable of weighing an entire truck, or portable scales, generally used in pairs to weigh one axle of a truck at a time. The permanent scales are located on main roads entering a state so that a significant percentage of the truck traffic entering that state is intercepted. Portable truck weight stations can be set up in order to obtain a wider sample of data at different locations throughout a state and on different kinds of roads having different percentages of truck traffic.

Procedures for locating weight stations, sampling and classifying the stream of trucks, and conducting both weighing and driver interviewing are contained in the report, *Guide for Truck Weight Study Manual.*

PIPELINE TRANSPORTATION

Pipelines are the fourth largest mode carrying goods in the United States. Oil pipelines carried an estimated 444 billion ton-miles of oil and petroleum products in 1971, or 23 percent of the total ton-miles of goods moved in the country. The pipeline

---

[26] "Planning for the Future—Connecticut's Major Corridor Needs, Present to Year 2000," Connecticut Highway Department, September 1968, p. 16.

[27] "Guide for Truck Weight Study Manual" (Washington, D.C.: Federal Highway Administration, U.S. Department of Transportation, April 1971).

industry has been growing rapidly, having more than quadrupled its ton-miles since 1946.

Despite these impressive statistics, pipeline transportation will probably remain a low-priority subject with most state departments of transportation because (1) the pipes are buried and apparently do not interfere with other forms of transportation very much, or, with the use of land (perhaps being out of sight really does put pipelines out of mind) and (2) the pipelines are owned and used mainly by large corporations, which traditionally try to keep a low public visibility except when they are advertising their wares. Rate fights are carried out with the only press coverage being in the business section.

Right now the main public concern with pipelines is safety. Will a gas line explode? Will an oil pipeline spill large quantities of oil into a watercourse? This concern can be met, to a great extent, by careful supervision of construction and by periodic inspection. This is an engineering activity that can be carried out equally well in a state public service commission or in a department of transportation.

As departments of transportation increase their technical competence in dealing with rail, truck, and other freight-carrying modes of transportation, they should also keep aware of developments in pipeline transportation. A Pennsylvania report states,

> In some parts of the United States and Canada, grain, sulphur, sand, gravel, coal, and wood chips are being transported by this method. Moving liquified natural gas, phosphates, fertilizer, and cement materials by pipeline is under study. Consideration is being given also to the feasibility of using pipelines to collect and carry solid industrial and untreated household wastes in slurry form to storage and landfill areas. Experiments are also being conducted on the use of capsules, tightly sealed, to carry materials along with the liquids that are being transported. . . .[28]

Such potential developments, while still serving highly specialized groups of users, will add to a network of lines that is already impressively big in many states.

COORDINATION AND PROGRAMMING

As plans for individual modes of transportation are being prepared, there must be continuing review on the part of statewide and regional planners to see that these plans are coordinated (1) with all the other modes and (2) with land use, environmental, and other plans of appropriate governmental bodies. Each of these topics is taken up in this concluding section and is followed by a statement on implementation through capital programming.

INTERMODAL COORDINATION

Coordination is the art of making separate forces or actions work together harmoniously as, for example, in the coordinated use of several muscles of the body. There is a clear inference that coordinated actions are taken toward a common goal. Applied to the field of intermodal transportation, coordination means taking actions that will make the several modes work together in harmony for the achievement of one (or more) common goals.

The kinds of actions or efforts that can be coordinated in the statewide transporta-

---

[28] "Transportation Policies for Pennsylvania" (Harrisburg: Pennsylvania Department of Transportation, 1970).

tion planning field are all the kinds of actions that a state department of transportation derives from legislation. These include (1) the power to build, operate, and maintain, (2) the authority to give financial aid to the various modes of transportation, (3) the authority to regulate, license, and set rates for certain kinds of transportation facilities and services, and (4) the authority to license vehicles.

Examples of coordinated actions in statewide transportation include providing good highways to rail, air, bus, or marine terminals and helping truck lines and rail lines work together.

Examples of *un*coordinated actions in the transportation field might include failure to provide high-speed access roads to serve an airport and building an expensive bridge over a railroad line that is soon to be abandoned. Also, there is reason to suggest that granting state aid to two competing forms of transportation, one of which is operating below capacity, is a failure of coordination.

Coordination is effective when two actions work together to achieve the goals of transportation, and particularly the goals of economic efficiency. If actions are taken for two modes that combine to produce faster travel for people or cheaper transportation of goods, then coordination has been achieved. If facilities are built, regulations are enforced, or financial aid is given which do not combine to produce better performance toward goals, or which produce small gains at great cost, then there is a lack of coordination.

How can coordination be achieved? Basically, there are three means of coordination: (1) physical coordination, (2) financial coordination, and (3) coordination of regulations. Each is related to the kinds of powers that a department of transportation has.

Physical coordination. Physical coordination is related to the power of a state department of transportation to build new transportation facilities. This kind of coordination can be achieved by carefully reviewing all plans for new facilities, of whatever mode of transportation, during the time when these plans are being prepared and when the projects are being programmed for implementation. Provisions for interconnections, parking, loading and unloading, joint use of right of way, and mechanical transfers of goods between modes are examples of physical coordination.

Financial coordination. All grants of direct funds for construction, operation, or maintenance of transportation facilities, and for reductions of taxes, should be carefully examined to determine whether or not the effects of these aids are producing improvements at a fast enough rate in the overall quality of transportation in a state or region. This is, of course, difficult to determine in many cases. Yet, ultimately this is one of the most important aspects of coordination, since we want to make sure that every financial investment has an effective role in achieving the goals of transportation. The way in which this kind of coordination can probably best be achieved is through a process of testing and retesting, a process in which the question is asked in each state, "Will this financial aid produce a better result in terms of speed, safety, economy, etc., than if the money were given to this other project or mode of transportation?" Testing of this kind will probably have to wait until better means are available for simulating the performance of the various modes within computers.

Coordination of regulations. This also is a very difficult area to consider at present, although the end result is the same kind of test: will a regulation or other governmental administrative practice produce a better result if applied or discontinued?

In many ways states do not now have measures by which they can judge how well their transportation systems are coordinated. Instead, a few obvious cases of poor coordination stand out, for example, inadequate public transportation for the air or bus passenger at the end of his intercity journey. Better measures of performance for all modes used in a single journey, or in the shipment of a single product, are needed to enable state departments of transportation to better coordinate all modes of transportation.

COORDINATION WITH NONTRANSPORTATION PLANS

Increasing attention is being paid to the need for coordinating statewide transportation plans with official plans for land use, economic growth, the environment, social welfare (in the broadest sense of those words), and other functional programs of state government. This interest has developed for a number of reasons, among which the more important are (1) the various crises of urban areas, particularly central cities, (2) growing air and water pollution over large areas, (3) uncontrolled rural development, (4) the adverse impact of land development on the service characteristics of rural arterials and collectors, (5) concern over the preservation of forests, shores, and other diminishing natural resources, and (6) an awareness that piecemeal planning by local and even metropolitan planning agencies is not effective in shaping future growth or adequately reducing known problems.

All planners are aware that employment patterns, economic growth patterns, transportation patterns, and lifestyles are changing at unprecedented rates. Technological developments and increasing productivity, with resultant greater real wealth, are basic causes of change. People may live in the country 50 mi from work and yet have communications, public water, school transportation, sewage disposal facilities, and all the other attributes of an urban way of life. Computer technology has altered conventional forms of storing and delivering goods, with consequent organizational shifts and changes in land-use patterns. Transportation systems, with their significant increases in speed and availability, have played and are playing a major role in these social changes.

It is, therefore, generally accepted that transportation plans and plans for statewide development should be coordinated so that the goals of society can be achieved more rapidly and efficiently.

There are two kinds of coordination between transportation and nontransport elements: "hard coordination" and "strategic coordination." It is hard coordination when the physical presence of transportation facilities—rails, highways, airports, etc., —is coordinated with given land uses. Such coordination may be negative (as in the avoidance of parks, historic sites and buildings, forest preserves, Indian reservations, prime agricultural lands, lakes, shorelands, and wildlife sanctuaries) or it may be positive (as in the provision of service to major generators, mines, cities, new towns, and so forth). Hard coordination is most easily achieved when a transportation department is supplied with accurate documents (maps, historic atlases, plans, etc.) that give the precise locations of land uses that either must be avoided or served.

Strategic coordination is more difficult to attain. Here the concern is to relate transportation and such things as economic growth (differentially within a state as well as aggregate), the renewal of cities, social welfare, and the total environment. Major problems, both conceptual and practical, exist. Too little is known about the impact of transportation on land development. The relationship of economic growth and transportation is equally cloudy. And in many other areas there are too few measures of social welfare (especially historical series) to allow evaluation of the benefits of one or the other land use-transportation mixes at the state scale. Therefore, many intuitive judgments that reflect political views of what is best must be made.

State departments of transportation are generally aware of the need to coordinate their programs with those of other agencies. The mechanisms for such coordination, however, are often difficult to establish, especially since early warning is often critically important. Informal mechanisms, the luncheon meeting of agency heads, for example, may be an effective adjunct to formal paper work.

PROGRAMMING IMPROVEMENTS

Preparing long-range improvement programs should be a function of a statewide transportation planning unit. This function should be operational whether or not a statewide transportation plan has been prepared, although it gains extra importance as the means for scheduling the effectuation of a plan. Programming is the assignment of priorities to improvements and, as such, it requires using a number of the same processes (such as cost-benefit analyses) that go into planning itself.

Normally, programming is thought of only in terms of major capital investments. Minor construction items (costing less than some fixed limit) and maintenance are usually excluded from consideration at the state programming level because they are delegated to districts or regions for local programming. With multimode transportation planning, however, actions other than major capital construction should be considered for inclusion within a program, because actions such as fare changes, demonstration projects, tax abatements, and regulatory changes are means of carrying out plans, and some of these actions have higher priority than others.

Contacts with several states indicate no standard techniques for programming. The following several basic elements are commonly used, however, and should be considered by states establishing units for statewide planning and programming.

Programming cycle. There should be an orderly cycle for programming that is published and recognized not only by the state department of transportation but also by the state executive and the legislature. It should be an annual cycle if possible, but biannual cycles may be satisfactory if this fits better with the state budgeting process. The cycle should indicate cut-off dates for receiving suggestions for projects and dates for staff, departmental, and executive review.

Definition and grouping of projects. Projects submitted to the programming unit should meet the specifications of an acceptable project. This will usually include a lower cost limit. Projects should be categorized by type, i.e., highway construction, airport construction, demonstration project, urban projects, rural projects, etc.

Rating of projects. Projects within individual categories should then be priority rated. Rating within groups instead of across groups is probably the most practical approach, pending development of more satisfactory intermodal comparison methods. User benefits vs. costs will tend to predominate as a rating method.

Application of constraints. Rating methods alone cannot be used satisfactorily for the development of programs. For example, the employment of user benefits vs. cost in rating highway improvements would force almost all highway investment into urban areas because it is there that traffic congestion is greatest and great improvements in service can be provided. Therefore, constraints that express very practical and necessary factors are needed. The following constraints may be employed:

1. *Geographic distribution.* The idea of a fair-share allocation of construction (or other) monies to different parts of a state is an important consideration. California follows this practice for highway construction.
2. *Funding distribution.* Either legal or administrative factors may limit the amount of money that can be spent for each project. An example is the funding limits on secondary roads set by the U.S. Federal Highway Administration; although this does not fix state spending, it tends to force a particular policy. Rigid funding limits may give way in the future to more flexible policies based more on investment productivity in terms of various goals.
3. *System constraints.* There are economies in making new high-type facilities connect with existing high-type facilities because continuous systems serve the public better than do a series of discontinuous facilities or services.
4. *Intermodal constraints.* The construction of certain new kinds of transportation facilities may demand that other facilities be built to serve them. For example, the construction of a new airport will generally require the commitment of funds for new roads and possibly mass transit facilities to serve it.
5. *Related-development constraints.* Certain new nontransportation developments may require the budgeting of new transportation facilities. For example, a new reservoir may require that roads be relocated or a new city may require substantial investments in air, highway, and/or transit services to provide the needed new access.
6. *Past commitments.* Frequently, past commitments by top state officials serve as critical constraints, forcing the construction of facilities which may in fact be less important than others.
7. *Total fund limitations.* These may be suggested at the outset of the process, or they may be imposed at the completion of the process after negotiation with executive and/or legislative branches.
8. *Design time.* Obviously, scheduling the start of construction has to allow for a reasonable period for engineering design, project approvals and reviews, and taking bids.

Improvement programs are prepared for four-, five-, or six-year periods, with a list of projects by type expected to be funded or started in each of these years. Capital improvement programs must allow for adequate time for projects to be designed, approved by reviewing agencies, and for bids to be taken. This involves close integration with the program of internal design or consultant design.

Basically, improvement programming is the application of common sense to scheduling. But for success there must be complete dedication to this idea throughout a major department. Otherwise, programs will be prepared, but they will be wrecked by actions contrary to the program.

CONCLUSIONS AND RECOMMENDATIONS

This chapter has outlined the contents of statewide and regional transportation planning, providing a definition of the subject, looking at organization, examining alternative technical approaches, examining the problems of single-mode planning, and finally dealing with coordination and implementation. There have been significant achievements in such planning, but it is also clear that a substantial effort is needed to bring about improvements in that art. Society, the economy, transportation, communications, and land-use patterns are in a state of rapid, perhaps accelerating, change. Little is being done to mold that change toward higher levels of performance as suggested by the goals of society. Too much is still happenstance. But at the same time our ability to cope with larger systems is increasing, aided in part by much greater experience in obtaining and manipulating data via computers.

The trend toward creation of state departments of transportation is indicative of the gathering of forces to attack these larger problems. The probability is that extensive work will be undertaken in the following directions:[29]

1. Development of techniques for measuring performance of all the major modes of transportation in terms of basic user, community, environmental cost, and operator costs.
2. Development of data collection techniques for (a) systems represented and (b) travel of persons and transport of goods.
3. Improvement of simulation models for (a) auto and motor carrier systems, (b) rail freight and passenger systems, (c) air carrier systems, and (d) waterway systems, leading ultimately to development of a methodology that will deal simultaneously with all modes as required.
4. Development of better methods for simulating the mutual impact of transportation facilities and regional development.
5. Assistance in development of better means of estimating dollar costs and social benefits of alternative land development patterns, and the transportation systems to serve them. Also, development of better methods of implementation of transportation plans.

REFERENCES FOR FURTHER READING

GENERAL

"Statewide Transportation Planning—Needs and Requirements," *Synthesis of Highway Practice 15*, Report on NCHRP Project 20-5. Washington, D.C.: National Cooperative Highway Research Program, Highway Research Board, 1972.

Intermodal Transportation Planning at the State, Multistate, and National Scale. Washington, D.C.: Highway Research Record 401. Highway Research Board, 1972.

"A Statement on National Transportation Policy." Washington, D.C.: U.S. Department of Transportation, 1970.

"Policies and Plans for Transportation in New York." Albany, N.Y.: New York State Department of Transportation, 1968.

ORGANIZATION

"Organization of the Planning Division." Albany, N.Y.: New York State Department of Transportation, 1969.

"Department of Transportation Research Study," A Report to the State of Minnesota Transportation Task Force. St. Paul, Minnesota, 1970.

[29] "Statewide Transportation Planning—Needs and Requirements, National Cooperative Highway Research Program Report 15 (Washington, D.C.: Highway Research Board, 1972), pp. 38–39.

BREUER, ROBERT, "Prospectus for a Statewide Transportation Planning Program: Non-Urban Elements." Albany, N.Y.: New York State Department of Transportation (preliminary report), 1970.

"A Status Report of State Departments of Transportation." Washington, D.C.: Highway Users Conference for Safety and Mobility, 1970.

MODAL STUDIES

Yearbook of Railroad Facts, Annual Publication. Washington, D.C.: American Association of Railroads.

"The American Railroad Industry: A Prospectus." Washington, D.C.: America's Sound Transportation Review Organization, 1970.

Bus Facts, Annual Publication. Washington, D.C.: National Association of Motor Bus Owners.

"Manual D: Airports and Other Intercity Terminals," National Transportation Planning Manual (1970–1990). Washington, D.C.: U.S. Department of Transportation, 1971.

"1990 Statewide Airport System Plan Work Program." Harrisburg: Pennsylvania Department of Transportation, 1971.

"Industry Report—A.T.A. Airline Airport Demand Forecast." Washington, D.C.: Air Transportation Association of America, 1969.

"1968 National Highway Functional Classification Study Manual." Washington, D.C.: U.S. Department of Transportation, 1969.

"National Functional Classification and Needs Study Manual" (Manual B). Washington, D.C.: U.S. Department of Transportation, 1969.

"Guide for Truck Weight Study Manual." Washington, D.C.: U.S. Department of Transportation, 1971.

"Planning for the Future (Part II) Connecticut's Major Arterial Needs, Present to Year 2000," Hartford: Connecticut Highway Department, 1969.

"State Highway Plan—Scope and General Development Procedure" (together with nine additional technical reports). Madison: Wisconsin State Highway Commission (undated).

"1972 National Highway Needs Report," Parts I, II, and III. U.S. Department of Transportation (Washington, D.C.: U.S. Government Printing Office, 1972).

Chapter 14

GEOMETRIC DESIGN

D. W. LOUTZENHEISER, Formerly, Director, Office of Engineering, and C. L. KING, Highway Engineer, Geometric Standards Branch, Federal Highway Administration, U.S. Department of Transportation, Washington, D.C.

The term *geometric design* pertains to the dimensions and arrangements of the visible features of the highway. This includes pavement widths, horizontal and vertical alinement, slopes, channelization, interchanges, and other features the design of which significantly affect highway traffic operation, safety, and capacity. Structural design related to vehicle loads rather than to traffic operations is not included.

The guiding precepts for geometric design have changed significantly over the years. Once, the objective was simply to provide a traversible way between two points. Now many factors must be considered: safety, economy, environmental concerns, and social effects must all be fully considered in designing a highway.

Although emphases shift from time to time, the goal of good geometric design remains the same: to provide a safe, efficient, and economical system of highways, consistent with the volumes, speeds, and characteristics of the vehicles and drivers who use them. Highways constructed today will serve well into the future; consequently, designs must anticipate future vehicular characteristics and operational patterns.

Upgrading design standards and criteria to meet changing conditions has been made possible largely through a feedback and evaluation of operational information. No design can be completely evaluated until it has been subjected to traffic. If a facility does not operate as expected, it is imperative to find out why. Not only should the immediate problem be solved, but the answer should also be brought to the attention of the designer for consideration in future designs. This feedback should not be left to chance. There should be a systematic operational evaluation of newly constructed facilities within any highway organization. Subsequently, many operational problems can be eliminated at the design stage, thus reducing the accident potential and providing cost savings for future remedial construction.

A complete international treatise on geometric design reflecting the practice of all countries would take several volumes. Thus, this chapter covers only the important details relating to standards and guides used in the United States. A survey of other representative countries indicates that design values for the most part are similar.

HIGHWAY CLASSIFICATION

Classification is the tool by which a complex network of highways can be allocated into groups or systems of routes having similar characteristics. A single classification system satisfactory for all purposes would be advantageous but has not been found practical. Moreover, in any classification system the division between classes must be arbitrary and, consequently, opinions differ on the best definition of each class. Schemes for classifying highways are numerous, and the class definitions vary depending on the purpose of classification.

The principal purposes of road classification are to:

1. Establish logical, integrated systems which, because of their particular service, should be administered by the same jurisdiction
2. Relate geometric traffic control and other design standards to the roads in each class
3. Establish a basis for developing long-range programs, improvement priorities, and fiscal plans

TYPES OF CLASSIFICATION

Various classification systems must be recognized. In the past misunderstandings in terminology of reference to particular elements and classification systems have resulted in confusion in design standards.[1] Several of the more common classification systems are outlined under the following headings:

Administrative system. Based on that governmental responsibility for maintenance and construction, highways are classified as follows:

1. Township roads
2. County roads
3. City streets
4. State roads
5. Roads in federal lands
6. Toll roads and bridges

Federal-aid system. For the distribution and administration of federal-aid highway funds, federal-aid highways are classified as follows:

1. Secondary roads
2. Urban system roads
3. Primary roads
4. Interstate highways
5. Highways not on federal-aid system

Under this system, the various state highway departments determine such classifications subject to the approval of the Federal Highway Administration.

Numbered highway system. Main highways in the United States are numbered within given systems (interstate, U.S., state, county, or other local jurisdictions) for administrative reference and for the convenience of highway travelers, map makers, and information services. The designation and continued correlation of the interstate

---

[1] *Protection of Highway Utility*, NCHRP Report 121 (Washington, D.C.: Highway Research Board, 1971), p. 21.

and state route numbers is maintained by the American Association of State Highway and Transportation Officials (AASHTO).

FUNCTIONAL SYSTEM OF CLASSIFICATION

For transportation planning purposes, as well as for design purposes, highways are most effectively classified by function. Highways have two functions: to provide mobility and to provide land access. From a design standpoint, these functions are incompatible. For mobility, high or continued speeds are desirable and variable or low speeds undesirable; for land access, low speeds are desirable and high speeds undesirable. For example, freeways provide a high degree of mobility, with access provided only at spaced interchanges to preserve the high-speed, high-volume characteristics of the facility. If low-speed, land access traffic were permitted on these roads, extremely hazardous conditions would be created. The opposite is true for local, low-speed roads that must provide access to the adjacent land areas. Between these extremes are highways that comprise the bulk of existing highway mileage and are the most difficult to classify: those that must provide both mobility and land access. The general relationship of functionally classified systems in serving mobility and land access is illustrated in Figure 14.1.

Given a functional classification, design criteria can be applied to encourage the use of the road as intended. Design features that can convey the level of functional classification to the driver include width of roadway, continuity of alinement, spacing of intersections, frequency of driveways, building setbacks, alinement and grade standards, and traffic controls.[2]

Although there are many ways to classify highways functionally, the most widely

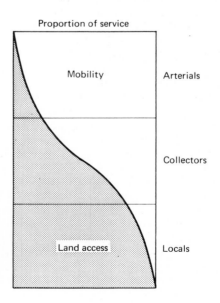

Figure 14.1 Relationship of functionally classified systems in serving traffic mobility and land access. [Source: U.S. Department of Transportation, Federal Highway Administration, *1968 National Highway Functional Classification Study Manual* (Washington, D.C.: U.S. Government Printing Office, 1969), p. II-6.]

[2] *Ibid.*

used method is that specified for the National Highway Functional Classification[3] which categorizes highways as follows.

Rural principal arterial system. A connected rural network of continuous routes having the following characteristics:

1. Serves corridor movements having trip length and travel density characteristics indicative of substantial statewide or interstate travel
2. Serves all, or virtually all, urban areas of 50,000 and over population, and a large majority of those with population of 25,000 and over
3. Provides an integrated network without stub connections except where there are unusual geographic or traffic flow conditions

The principal arterial system is further divided into interstate and other principal arterials.

Rural minor arterial road system. Routes whose design should be expected to provide relatively high overall travel speeds, with minimum interference to through movement, and all remaining routes on the rural network not included as "rural principal" having the following characteristics:

1. Links cities and larger towns and other traffic generators, for example, major resort areas, that are capable of attracting travel over long distances and forms an integrated network providing interstate and intercounty service
2. Spaced at such intervals, consistent with population density, that all developed areas of the state are within reasonable distances of an arterial highway
3. Provides service to corridors with trip lengths and travel density greater than those predominantly served by rural collectors or local systems

Rural collector road system. Routes generally serving intracounty rather than statewide travel and on which average trip lengths and travel speeds are less than on arterial routes. The rural collector road system is subdivided into major collector roads and minor collector roads.

Rural local road system. Routes generally serving to provide access to adjacent land and provide service to travel over relatively short distances as compared to collectors or other higher systems and not classified in a higher system.

Urban principal arterial system. Streets and highways serving major metropolitan activity centers, the highest traffic volume corridors, the longest trip desires, and a high proportion of total urban area travel on a minimum of mileage. Service to abutting land should be subordinate to the provision of travel service to major traffic movements. This system carries the major portion of trips entering and leaving an urban area, as well as the majority of through movements desiring to bypass the central city, and normally will carry important intraurban as well as intercity bus routes.

---

[3] *National Highway Functional Classification Study Manual*, FEDERAL HIGHWAY ADMINISTRATION, U.S. DEPARTMENT OF TRANSPORTATION (Washington, D.C.: Government Printing Office, 1969), pp. II-9–II-15.

Urban minor arterial street system. Streets and highways interconnecting with and augmenting the urban principal arterial system and providing service to trips of moderate length at a somewhat lower level of travel mobility. The system places more emphasis on land access and distributes travel to geographic areas smaller than those identified with the higher system. It includes all arterials not classified as principal.

Urban collector street system. Streets penetrating neighborhoods, collecting traffic from local streets in the neighborhoods, and channeling it into the arterial systems. A minor amount of through traffic may be carried on collector streets, but the system primarily provides land access service and carries local traffic movements within residential neighborhoods, commercial, and industrial areas. It may also serve local bus routes.

Urban local street system. Streets not classified in a higher system, primarily providing direct access to abutting land and access to the higher systems. They offer the lowest level of mobility and usually carry no bus routes. Service to through traffic is deliberately discouraged.

DESIGN TYPES OF HIGHWAYS

The most widely accepted design criteria in the United States are those developed by AASHTO. Although every state, and many counties, cities, and other governmental bodies, has developed its own standards, they are based largely on AASHTO standards for two reasons: (1) AASHTO design policies, standards, and guides have been developed and approved by every state and (2) the Federal Highway Administration has made them the applicable standards for design and construction of federal-aid highways.

General highway design standards, policies, and guides have been developed separately for (1) freeways, sometimes divided into interstate freeways and other freeways, (2) arterial highways other than freeways, and (3) collector roads (secondary roads) and local roads and streets. These three general types of highways (and their rural-urban and sometimes suburban subdivisions) can be considered a form of system classification originating with federal-aid funding distinctions, but more properly they are groupings of streets and highways for which geometric design values can be clearly separated. A brief description of the highways included in each type and the principal documents that are available for design guidance follow:

Freeways. Includes interstate highways and all other fully controlled access facilities regardless of the system of which they are a part. Design standards and criteria are found in three publications: *A Policy on Geometric Design of Rural Highways*[4] furnishes basic design criteria and values for all types of rural highways including freeways, described for a range of conditions, sometimes with both minimum and desirable values for general alinement, profile, sight distance, and other design values. *A Policy on Design of Urban Highways and Arterial Streets*[5] supplements the

---

[4] *A Policy on Geometric Design of Rural Highways* (Washington, D.C.: American Association of State Highway Officials, 1965).

[5] *A Policy on Design of Urban Highways and Arterial Streets* (Washington, D.C.: American Association of State Highway Officials, 1973).

rural policy and provides separate urban values for freeways, expressways-at-grade, arterial highways, and major streets. *Design Standards for the National System of Interstate and Defense Highways*[6] was specifically developed for interstate system highways, but it may also be used for design of other freeways.

Arterial highways. Includes the more important highways other than freeways. Design standards and criteria are found in the two AASHTO design policies cited above and in *Geometric Design Standards for Highways Other Than Freeways*[7] which is a short form of the main design controls for highways without control of access, largely expressed as minimum values, and having separate rural and urban controls.

Local roads and streets. Includes all other highways. Design standards and criteria also are found in the *Policy on Geometric Design of Rural Highways* cited above and in *Geometric Design Guide for Local Roads and Streets*[8] which is applicable for both rural and urban design. *Recommended Practices for Subdivision Streets*[9] also contains standards and criteria for design of such streets.

DESIGN CONTROLS AND CRITERIA

In geometric design, various controls and criteria are employed to ensure that the facility will accommodate the expected traffic requirements and to encourage consistency and uniformity in operation. To some degree these controls and criteria are applicable to all streets and highways.

CONTROL OF ACCESS

Definition. Control of access is the condition in which the right of owners or occupants of abutting land or other persons to access, light, air, or view in connection with a highway is fully or partially controlled by public authority.[10] There are two degrees of access control:

1. *Full control of access.* Where the authority to control access is exercised to give preference to through traffic by providing access connections with selected public roads only by prohibiting crossings at grade or direct private driveway connections.
2. *Partial control of access.* Where the authority to control access is exercised to give preference to through traffic to a degree that, in addition to access connections with selected public roads, there may be some crossings at grade and some private driveway connections.

---

[6] *Design Standards for the National System of Interstate and Defense* (Washington, D.C.: American Association of State Highway Officials, 1967).

[7] *Geometric Design Standards for Highways Other Than Freeways* (Washington, D.C.: American Association of State Highway Officials, 1969).

[8] *Geometric Design Guide for Local Roads and Streets* (Washington, D.C.: American Association of State Highway Officials, 1970).

[9] *Recommended Practices for Subdivision Streets* (Washington, D.C.: Institute of Traffic Engineers, 1967).

[10] Unless otherwise noted, all definitions of highway terms in this chapter are taken from *AASHTO Highway Definitions* (Washington, D.C.: American Association of State Highway Officials, 1968).

Application. Access control is accomplished either by obtaining access rights (usually when right of way is purchased) or by using frontage roads. When a freeway is developed on new location, of course, adjoining property owners have no inherent right to direct access to a highway that does not exist.

Highway function should determine the degree of access control needed. Freeways whose main function is to provide mobility should have full control of access. Urban arterial streets and major rural highways should have as great a degree of access control as is feasible.

The principal operational difference between a highway with and without control of access is in the amount of interference with through traffic by other vehicles and pedestrians. With access control, entrances and exits are located at points best suited to enable vehicles to enter and leave safely without interfering with through traffic. Vehicles are prevented from entering or leaving elsewhere so that regardless of the type and intensity of development of the roadside areas, the efficiency and capacity of the highway are maintained at a high level and the accident hazard remains low.

When there is no control of access on some new or improved arterial streets, such as a bypass route, the concentrated flow of traffic can be expected to attract roadside businesses with resultant traffic interference, reduction in capacity of the highway, and increase in accident hazard. As traffic increases and roadside businesses multiply, the hazard increases and considerable congestion may develop. Any remedial treatment may be frustrated by the high value of the roadsides. Therefore, in designing arterial streets and highways without access control, initial allowance should be made for potential roadside commercial development and its effects on the operation of the facility.

In summary, points of access on all streets and highways should be carefully planned at the design stage. Access should not be allowed at locations where entering and leaving vehicles will create a hazard (such as at locations where sight distance is limited or at a point too close to another intersection). On other than freeways, location of access points should be controlled by curb-cut regulations, driveway permits, or frontage roads.

DESIGN SPEED

Definition. A speed determined for design and correlation of the physical features of a highway that influence vehicle operation: the maximum safe speed maintainable over a specified section of highway when conditions permit design features to govern.

Application. Some features, for example, curvature, superelevation, sight distance, and gradient, are directly related to and vary appreciably with design speed. Pavement and shoulder widths and clearances to walls and rails are less directly related to design speed, but because they can affect vehicle speeds, higher standards for them should be used on highways with higher design speeds. Thus, nearly all geometric design elements of the highway are affected by the selected design speed.

Basis for selection. Design speed selection is influenced by the character of terrain, the density and character of the land use, the classification and function of the highway, the traffic volumes expected to use the highway, and by economic and environmental considerations. Usually, a highway in level terrain warrants a higher design speed than one in mountainous terrain; one in a rural area, a higher design speed than one in an urban area; an arterial highway, a higher design speed than a local road; and a high-volume highway, a higher design speed than one carrying low traffic volumes. Except for local roads and streets in urban areas, the design speed should be as high as possible commensurate with economic and environmental considerations.

Table 14.1 shows typical minimum design speeds for various highway classifications, types of terrain, and various volumes of traffic. These are minimum design speeds for the various conditions of terrain and traffic volumes. Higher design speeds

TABLE 14.1
Typical Minimum Design Speeds for Various Types
of Highways in MPH and KPH

Freeways		
Terrain	Rural	Urban
Level	70 (113)	50 (80)
Rolling	60 (97)	50 (80)
Mountainous	50 (80)	50 (80)

Arterial Highways—Rural		
Terrain	Current ADT 50–750 DHV Less Than 200	DHV 200 and Over
Level	50 (80)	70 (113)
Rolling	40 (64)	60 (97)
Mountainous	30 (48)	40 (64)

Arterial Highways	
Urban	Suburban
30–40 mph (48–64 kph) for all types of terrain and for all traffic volumes	40–50 mph (64–80 kph) for all types of terrain and for all traffic volumes

Local Roads and Streets—Rural			
Terrain	Current ADT Less Than 250	Current ADT 250–400	Current ADT Greater Than 400 DHV Greater Than 100
Level	40 (64)	50 (80)	50 (80)
Rolling	30 (48)	40 (64)	40 (64)
Mountainous	20 (32)	20 (32)	30 (48)

Local Roads and Streets—Urban	
Collector Streets	Local Streets
30–40 mph (32–64 kph) for all types of terrain and for all traffic volumes	20–30 mph (32–48 kph) for all types of terrain and for all traffic volumes

are used, up to 90 mph (145 kph) in the United States and 140 kph (87 mph) in some European countries.

OTHER SPEEDS USED AS A BASIS FOR DESIGN

Average running speed. For all traffic or component thereof, the summation of distance divided by the summation of running times.

Because 50 percent of all vehicles travel at speeds very close to the average running speeds shown in Table 14.2, they are used as a basis for design of geometric features such as intersection radii, speed-change lanes, and superelevation of above-minimum curves.

TABLE 14.2
Relationship of Average Running Speed to Design Speed

Design Speed mph (kph)	Average Running Speed		
	Low Volume	Intermediate Volume	Approaching Capacity
30 (48)	28 (45)	26 (42)	—
40 (64)	36 (58)	34 (55)	25 (40)
50 (80)	44 (71)	41 (66)	31 (50)
60 (97)	52 (84)	47 (76)	35 (56)
65 (105)	55 (88)	50 (80)	37 (60)
70 (113)	58 (93)	54 (87)	—
75 (121)	61 (98)	56 (90)	—
80 (129)	64 (103)	59 (95)	—

Operating speed. The highest overall speed at which a driver can travel on a given highway under favorable weather conditions and under prevailing traffic conditions without at any time exceeding the safe speed as determined by the design speed on a section-by-section basis.

Operating speed is usually about 5 mph higher than the average running speed for low-volume conditions on free-flow facilities. It is used as a measure of level of service for those highways that provide uninterrupted flow conditions (usually rural only) for vehicular travel and is therefore useful in determining the level of service provided by a specific traveled way.[11]

Average overall travel speed. The summation of distances traveled by all vehicles over a given section of highway during a specified period of time, divided by the summation of overall travel times.

Average overall travel speed is used as a measure of level of service for interrupted flow conditions (urban arterial and downtown streets) and is therefore useful in determining the level of service of a specific traveled way in urban areas.

DESIGN VOLUME

Definition. A volume determined for design, representing traffic expected to use the highway. Current average daily traffic (ADT) may be used for designing local roads and streets. For more important two-lane highways, the design hourly volumes

---

[11] The traveled way (carriageway) is that portion of the roadway for movement of vehicles exclusive of shoulders and auxiliary lanes. It differs from the roadway in that the roadway includes shoulders.

(DHV) concept is used—usually the 30th highest hourly volume that is expected to occur in some future design year. For multilane highways, use is made of the directional design hourly volumes (DDHV) for some future design year.

Application. The design volume represents the "load" that the highway must accommodate and it determines to a large degree the type of facility, required pavement widths, as well as other geometric features.

Determination. Determination of the design volume begins with the current ADT. For all except local roads and streets, the ADT is then projected to some future year, usually from 5 to 20 years beyond the estimated construction year. The DHV is then established by multiplying this future ADT by a *K* factor.

The *K* factor is the ratio of the DHV to the ADT. On the average main rural highway the *K* factor is about 15 percent[12] and for urban arterial streets about 11 percent of the ADT. A fortunate characteristic of the *K* factor is that for a particular facility it decreases only slightly with variations in the ADT. Therefore, the *K* factor determined for current traffic volumes needs only be adjusted slightly, if at all, when used to determine future design hourly volumes. For two-lane highways, the DHV is used for design.

The directional design hourly volume (DDHV) is determined by multiplying the DHV by a directional factor (*D*). The *D* factor is the percentage of traffic in the dominant direction of flow on highways with more than two lanes, and on two-lane roads with important intersections directional traffic estimates are essential. Traffic distribution by direction during peak hours is usually considered to remain unchanged throughout the week and year. Traffic data for rural and outlying urban highways show about 60 to 80 percent of peak-hour traffic in one direction, with an average *D* value of about 67 percent. In and near central business districts, particularly in large cities, *D* values approach 50 percent. Representative *D* values in large cities are: in downtown areas, about 55 percent on radial routes and inner belt lines; about 60 percent in intermediate areas; and 65–75 percent on radial routes in outlying areas.

The percentage of truck and bus traffic during the design hour, the *T* factor, should also be estimated so that allowances in the geometric design criteria can be made.[13]

Vehicles of different sizes and weights have different operating characteristics, and the effect on traffic operation of one truck or bus is often equivalent to several passenger vehicles, depending on the gradient and the operating characteristics of the trucks. On rural highways *T* values range up to 7 to 9 percent.[14] On principal urban highways *T* values may reach 10 percent, particularly where local bus lines utilize the same route. On other urban routes which carry traffic primarily between the suburbs and the central business district (CBD), *T* values do not generally exceed 5 percent. A *T* value established on the basis of current traffic generally can be assumed to be applicable to future traffic volumes. The effects of trucks and buses in the traffic stream are allowed for by either decreasing the service volumes or changing the volumes to passenger car equivalents (as discussed in Chapter 8).

---

[12] *A Policy on Geometric Design of Rural Highways* (Washington, D.C.: American Association of State Highway Officials, 1965), p. 56.

[13] In Europe, use is made of the passenger car unit (pcu) which corresponds to a private car. For other vehicles, a pcu equivalent is applied.

[14] *A Policy on Geometric Design of Rural Highways* (Washington, D.C.: American Association of State Highway Officials, 1965), p. 70.

In summary, it is necessary to know the following traffic elements for design purposes for various type highways:

Type Highway		Traffic Elements Required for Design
Local roads and streets	ADT_c	Current average daily traffic
Two-lane arterial highways	ADT_c	
	ADT	ADT for some future design year
	DHV	Design hourly volume for some future year (DHV)
	T	Percentage of trucks during design hour (T)
Multilane arterial highways and freeways	ADT_c	
	ADT	
	DDHV	Directional design hourly volume (DDHV)
	T	

DESIGN VEHICLES

Definition. A selected motor vehicle, the weight, dimensions, and operating characteristics of which are used in highway design. For purposes of geometric design, the selected design vehicle should be one with dimensions and minimum turning radius as large as almost all vehicles in its class expected to use the highway.

Application. Design vehicle characteristics are used to develop sight distance, intersection design, cross section, and other geometric design criteria. Six design vehicles are used for purposes of geometric design, one of which is selected with dimensions and turning characteristics equal to or greater than those of the largest vehicles expected in appreciable numbers (see Table 14.3). Similar data for vehicles of local or regional significance may be used to establish other design vehicles.

TABLE 14.3
Design Vehicle Dimensions

Design Vehicle		Dimensions in Ft					
Type	Symbol	Wheelbase	Front Overhang	Rear Overhang	Overall Length	Overall Width	Height
Passenger car	P	11	3	5	19	7	—
Single-unit truck	SU	20	4	6	30	8.5	13.5
Single-unit bus	BUS	25	7	8	40	8.5	13.5
Semitrailer— combination intermediate	WB-40	$13 + 27 = 40$	4	6	50	8.5	13.5
Semitrailer— combination large	WB-50	$20 + 30 = 50$	3	2	55	8.5	13.5
Semitrailer— fulltrailer combination	WB-60	$9.7 + 20.0 +$ $9.4* + 20.9 = 60$	2	3	65	8.5	13.5

*Distance between rear wheels of front trailer and front wheels of rear trailer.

The minimum turning path for the specific design vehicle is particularly important. The governing paths are the outer front overhang and the inner rear wheel. The outer front wheel is assumed to follow a circular arc which is the minimum turning radius as determined by the vehicle steering mechanism. A typical turning path of a WB-50 design vehicle is shown in Figure 14.2. For others, see cited reference.[15] The minimum outside and inside wheel paths generally used are as follows:

Design vehicle type	P	SU	BUS	WB-40	WB-50	WB-60
Minimum turning radius (feet)	24	42	42	40	45	45
Minimum inside radius (feet)	15.3	28.4	20.3	19.9	19.8	22.5

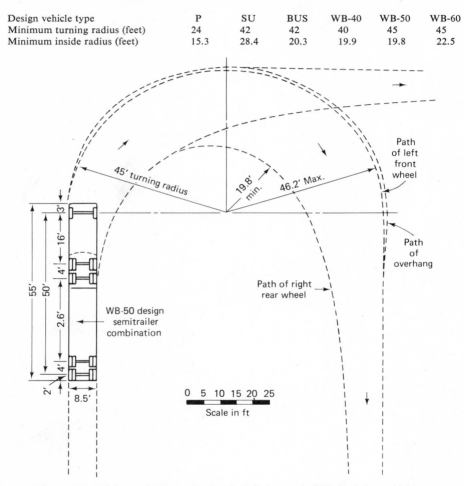

Figure 14.2. Minimum design path of typical design truck (WB-50 design vehicle). [Source: *A Policy on Geometric Design of Rural Highways* (Washington, D.C.: American Association of State Highway Officials, 1965), p. 84.]

DESIGN DESIGNATION

Design designation indicates the major controls for which a given highway is designed. It is largely independent of highway systems and of the highway type. The following tabulation is an example of a design designation.

---

[15] *A Policy on Design of Urban Highways and Arterial Streets* (AASHTO) 1973 pp. 268–276.

Design year = 1995
Average daily traffic (current year) = 20,100
Average daily traffic (design year) = 39,600
Design hourly volume = 4,400
Directional distribution of traffic = 67%
Trucks = 5%
Design speed = 60 mph
Control of access = full
Design level of service = C

Other necessary information for geometric design includes:

1. Pedestrian volumes and locations of crossings
2. Type, location, and nature of parking, if any required
3. Transit operation
4. Applicable design vehicle

DESIGN ELEMENTS

A number of design elements are common to all types of streets and highways. Reference to these elements will be made throughout the chapter.

SIGHT DISTANCE

Stopping sight distance. Sight distance is the length of highway visible to the driver. Sight distance everywhere along a highway should be adequate for all but a few drivers traveling at or near the probable top speed to come to a safe stop before reaching an object. Stopping sight distance used for design is the sum of two distances: (1) the distance a vehicle travels after the driver sights an object and before braking and (2) the distance it travels after braking.

The stopping sight distance (SSD) in feet is determined from the formula:

$$SSD = 1.47PV + \frac{V^2}{30(f \pm g)} \tag{14.1}$$

where V = speed from which stop is made in miles per hour,
 P = perception-reaction time in seconds,
 f = coefficient of friction (for wet pavement),
 g = percent of grade divided by 100 (added for upgrade and subtracted for downgrade).

If vehicle speed is in kilometers per hour, the stopping sight distance in meters is:

$$SSD = 0.278PV + \frac{V^2}{255(f \pm g)} \tag{14.2}$$

The minimum and desirable stopping sight distance for highways having various design speeds are shown in Table 14.4. Minimum distances assume that the vehicle is traveling at less than the design speed (the assumed speeds on which the minimum stopping distances are based are shown in Table 14.4). Desirable distances assume that the vehicle is traveling at the design speed.

Stopping sight distance is measured from a "seeing" height of 3.75 ft (1.1 m) to an object height of 0.5 ft (15 cm). Desirable stopping sight distance values should be used for design whenever possible. Stopping sight distance values less than the minimum should never be considered.

Passing sight distance. Passing sight distance is applicable only on two-lane, two-way highways. Passing sight distance is the length of highway ahead necessary for one vehicle to pass another before meeting an opposing vehicle which might appear after the pass began. Passing sight distances used for design, given in Table 14.5, are based on various traffic behavior assumptions found in the cited reference.[16]

Passing sight distances for purposes of pavement marking are also given in Table 14.5. No-passing zone markings, given in the *Manual on Uniform Traffic Control*

TABLE 14.4
Minimum and Desirable Stopping Sight Distances

Design Speed		Assumed Speed for Condition	Perception and Brake Reaction Distance*	Coefficient of Friction Wet Pavement	Braking Distance	Stopping Sight Distance (Rounded for Design)	
mph	kph	mph	ft	f	ft	ft	m
30	(48)	Minimum 28	103	.36	73	200	(61)
		Desirable 30	110	.35	86	200	(61)
40	(64)	Minimum 36	132	.33	131	275	(84)
		Desirable 40	147	.32	167	300	(92)
50	(80)	Minimum 44	161	.31	208	350	(107)
		Desirable 50	183	.30	278	450	(137)
60	(97)	Minimum 52	191	.30	300	475	(145)
		Desirable 60	220	.29	414	650	(198)
65	(105)	Minimum 55	202	.30	336	550	(168)
		Desirable 65	238	.29	485	750	(229)
70	(113)	Minimum 58	213	.29	387	600	(183)
		Desirable 70	257	.28	584	850	(259)
75	(121)	Minimum 61	224	.28	443	675	(206)
		Desirable 75	275	.28	670	950	(290)
80	(129)	Minimum 64	235	.27	506	750	(229)
		Desirable 80	293	.27	790	1,050	(320)

*Based on perception-reaction time of 2.5 seconds.

TABLE 14.5
Minimum Passing Sight Distances

Design Speed		Used for Design: Minimum Passing Sight Distance for Design		Used for Pavement Marking: 85th Percentile Speed		Minimum Passing Sight Distance for Pavement Marking	
mph	kph	ft	m	mph	kph	ft	m
30	(48)	1,100	(335)	30	(48)	500	(152)
40	(64)	1,500	(457)	40	(64)	600	(183)
50	(80)	1,800	(549)	50	(80)	800	(244)
60	(97)	2,100	(640)	60	(97)	1,000	(305)
65	(105)	2,300	(701)	—	—	—	—
70	(113)	2,500	(762)	70	(113)	1,200	(366)
75	(121)	2,600	(793)	—	—	—	—
80	(129)	2,700	(823)	—	—	—	—

[16] *Ibid.* pp. 140–145.

Devices,[17] are based on different assumptions which result in lower values. No-passing zones are based on the 85th percentile speed during low-volume conditions, which is slightly less than the design speed.

Sight distance adequate for passing should be provided frequently in design of two-lane highways, and each passing section should be as long as feasible. Although the frequency and lengths of such passing sections depend on physical and cost considerations and cannot be reduced to a standard, the importance of providing passing opportunities on as much of the length of a two-lane highway as possible cannot be overemphasized. The percentage of the highway where passing can take place affects not only capacity (see Chapter 8) but also the safety, comfort, and convenience of all highway users.

For purposes of design, passing sight distance for both horizontal and vertical restrictions is measured from a "seeing" height of 3.75 ft (1.1 m) to an object height of 4.5 ft (1.4 m). For purposes of marking pavement, it is measured from a "seeing" height of 3.75 ft (1.1 m) to an object height of 3.75 ft (1.1 m).

Intersection sight distance. Intersections should be planned and located to provide as much sight distance as possible. In achieving a safe highway design, as a minimum, there should be sufficient sight distance for the driver on the minor highway to cross the major highway without requiring approaching traffic to reduce speed. Minimums for different design speeds are shown in Table 14.6. Stop controls are assumed; other forms of traffic control have different intersection sight distance requirements.

TABLE 14.6
Suggested Corner Sight Distance at Intersections*

Design speed mph (kph)	20 (32)	30 (48)	40 (64)	50 (80)	60 (97)
Minimum corner intersection sight distance* ft (m)	200 (61)	300 (91)	400 (122)	500 (152)	600 (183)

*Corner sight distance measured from a point of the minor road at least 15 ft (4.6 m) from the edge of the major road pavement and measured from a height of eye of 3.75 ft (1.1 m) on the minor road to a height of object of 4.5 ft (1.4 m) on the major road.

Procedures for checking plans. It is often desirable during the preliminary design stage to determine graphically the sight distances and record them at frequent intervals. Methods for scaling sight distances and a typical sight distance record which should be shown on final plans are shown in Figure 14.3. For two-lane highways, passing sight distance, in addition to stopping sight distance, should be shown.

Horizontal sight distance on the inside of curves may be limited by obstructions such as buildings, plant growth, or cut slope. Horizontal sight distance is measured along a straight edge, as indicated in the upper left in Figure 14.3. The cut slope obstruction is shown on the work sheets by a line representing the proposed excavation slope at a point about 2.0 ft (0.6 m) (approximate average of 3.75 ft and 0.5 ft)

---

[17] *Manual on Uniform Traffic Control Devices for Streets and Highways,* FEDERAL HIGHWAY ADMINISTRATION, U.S. DEPARTMENT OF TRANSPORTATION (Washington, D.C.: Government Printing Office, 1971), p. 190.

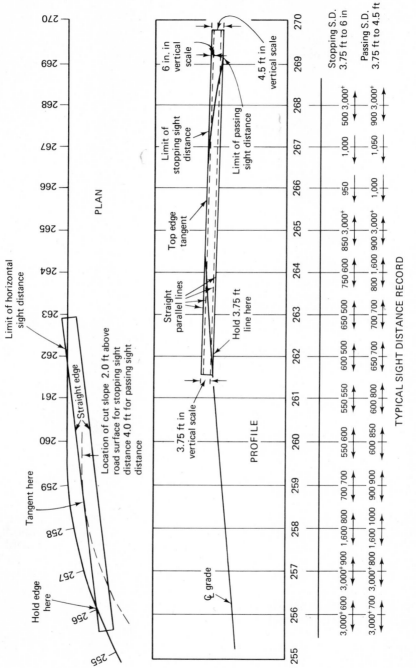

Figure 14.3. Scaling and recording sight distances on plans. (Source: *A Policy on Geometric Design of Rural Highways*, p. 150.)

614

above the road surface for stopping sight distance and about 4.0 ft (1.2 m) for passing sight distance. The stopping sight distance should be measured from center-to-center of the same traffic lane; the passing sight distance should be measured from the center of inside lane to center of outside lane.

Vertical sight distance may be scaled also as shown in Figure 14.3. A transparent straight edge is used with a scaled width of 4.5 ft (1.4 m). Lines are drawn 6 in. (15 cm) and 3.75 ft (1.1 m) from the upper edge, in accordance with the vertical scale. The 3.75-ft line is placed on the station from which the vertical sight distance is desired, and the strip is pivoted about this point until the upper edge is tangent to the profile. The distance between the initial station and the station on the profile intersected by the 6-in. line is the stopping sight distance. The distance between the initial station and the 4.5-in. line is the passing sight distance.

Sight distance design records are useful on two-lane highways for determining the percentage of highway length on which sight distance is restricted to less than the passing minimum—an important criterion in evaluating the overall design and the capacity.

VERTICAL ALINEMENT

Vertical alinement consists of tangent grades and vertical curves. Although there are no exact relationships, maximum grades and curvatures are generally related to design speed. Widely accepted criteria follow.

Vertical curves. In United States practice, vertical curves are parabolic. In Europe, circular curves are used. Parabolic curves are identified by their lengths and the algebraic difference of the grades they connect. The minimum length of vertical curve may be computed from the formula:

$$L = KA \tag{14.3}$$

where L = the length of vertical curve in feet,

$\quad\quad K$ = a constant for design,

$\quad\quad A$ = the algebraic difference in grades in percent,

or
$$L = KA/3.28 \text{ if } L \text{ is in meters.} \tag{14.4}$$

K is constant for each design speed, and its selection for crest vertical curves is based on sight distance requirements. For sag vertical curves, K is based on headlight sight distance, based on a light beam emanating from a source 2.0 ft (0.6 m) above the pavement on a 1° upward divergence from the longitudinal axis of the vehicle to where the beam intersects the roadway surface. The formula computes minimum lengths of vertical curve in each case, but K values vary for each design speed and condition. Table 14.7 gives K values to be used for the three controls.

Grades. Maximum grades recommended for road design are mostly governed by their influence on truck speeds. Table 14.8 summarizes typical maximum grade controls for various classes of highways.

General controls. In addition to grade and vertical curvature controls, additional considerations in designing vertical alinement include:

1. Smooth grade lines with gradual changes, consistent with the class of highway

TABLE 14.7
Design Controls for Vertical Curves: _K_ Values

Design Speed		Crest Curves			Sag Curves	
		Stopping Sight Distance		Passing Sight Distance	Stopping Sight Distance	
mph	kph	Minimum	Desirable		Minimum	Desirable
30	(48)	28	28	365	35	35
40	(64)	55	65	686	55	60
50	(80)	85	145	985	75	100
60	(97)	160	300	1,340	105	155
65	(105)	215	415	1,605	130	185
70	(113)	255	515	1,895	145	215
75	(121)	325	645	2,050	160	240
80	(129)	400	780	2,210	185	270

TABLE 14.8
Maximum Grades in Percent

Design Speed		Maximum Gradient	
		Freeways	
mph	kph	Other Than in Mountainous Terrain	Mountainous Terrain
50	(80)	5	7
60	(97)	4	6
70	(113)	3	5

		Arterial Highways*		
mph	kph	Flat	Rolling	Mountainous
30	(48)	6	7	9
40	(64)	5	6	8
50	(80)	4	5	7
60	(97)	3	4	6
65	(105)	3	4	6
70	(113)	3	4	5
75	(121)	3	4	—
80	(129)	3	4	—

		Local Roads and Streets†		
mph	kph	Flat	Rolling	Mountainous
20	(32)	7	10	12
30	(48)	7	9	10
40	(64)	7	8	10
50	(80)	6	7	9
60	(97)	5	6	—

*Short grades less than 500 ft in length and one-way downgrades may be 1 percent steeper. For extreme cases, in urban areas and at some underpasses and bridge approaches, steeper grades for relatively short lengths may be considered. For low-volume rural highways, grades may be 2 percent steeper.
†For highways with ADT's below 250, grades of relatively short lengths may be increased to 150 percent of value shown.

and character of terrain, should be provided. Figure. 14.4(a) shows an undesirable design in which short vertical curves are used in conjunction with relatively short tangent gradients. Figure 14.4(b) shows a much better alinement where long sweeping vertical curves are used in spite of the rugged terrain conditions.

2. "Roller-coaster" or "hidden-dip" profiles should be avoided. The profile in Figure 14.5(a) is aesthetically unpleasant and likely hazardous. Figure 14.5(b) shows the preferable use of a long, rolling grade line.

3. Broken-back grade lines (two vertical curves in the same direction separated by a short piece of tangent) should be avoided. Figure 14.6(a) shows an undesirable design of this type; Figure 14.6(b) illustrates a design that avoids a broken-back grade line.

4. Grades through at-grade intersections on highways with moderate or steep grades should be reduced whenever possible.

(a)

(b)

Figure 14.4. Examples of highways showing the visual effects of (a) using short vertical curves and (b) the more desirable practice of using a smooth grade line with gradual changes. [Sources: (a) Bob L. Smith, Professor of Civil Engineering, Kansas State University. (b) U.S. Department of Transportation, Federal Highway Administration.]

(a)

(b)

Figure 14.5. Examples of highways with (a) a "roller-coaster" grade line and (b) the much more desirable profile with a long vertical curve. [Sources: (a) Bob L. Smith, Professor of Civil Engineering, Kansas State University. (b) U.S. Department of Transportation, Federal Highway Administration.]

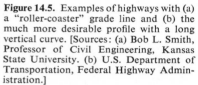

(a)

(b)

Figure 14.6. Example of (a) a vertical alinement with a "broken-back" grade line and (b) a much more desirable alinement in which a broken alinement has been avoided. [Sources: (a) Bob L. Smith, Professor of Civil Engineering, Kansas State University. (b) U.S. Department of Transportation, Federal Highway Administration.]

HORIZONTAL ALINEMENT

Horizontal alinement consists of tangents and horizontal curves. In United States practice, horizontal curves are circular curves with a constant radius, often connected to the tangents with transitional (spiral) curves. Criteria for determining maximum curvature are based on the laws of mechanics, design values depending on practical limits for superelevation and friction factors representative of pavement surfaces. The basic formula for determining horizonal alinement is:

$$e + f = \frac{V^2}{15R} \tag{14.5}$$

where e = rate of roadway superelevation, foot per foot,
 f = side friction factor,
 V = vehicle speed, mph,
 R = radius of curve in feet,
or if V is in kph and R is in meters, the formula is:

$$e + f = \frac{V^2}{127R} \tag{14.6}$$

Maximum superelevation. Maximum superelevation rates are controlled by several factors which may vary widely: (1) frequency and amount of snow and ice, (2) type of area, whether rural or urban, and (3) frequency of slow-moving vehicles. Maximum e values range upward to 0.12 for highways in rural areas where there is no snow and ice. Where snow and ice conditions prevail, e values normally range from 0.08 to 0.10. Although uncommon practice, much higher e values could probably be used at special locations (on a one-way, downgrade ramp, for example).

The most commonly used e value is 0.08 or 0.10 for all highway classifications in rural areas and for freeways in developed areas. For other highways in urban areas, superelevation is not generally used, although e values up to 0.06 are sometimes applicable. In other countries, superelevation is expressed as a percentage and the maximum used is most commonly 7 percent.

Maximum degree of curvature. The other variable that influences maximum curvature is the side friction factor (f) which is assumed to be 0.16 for 30 mph and 0.01 less than this value for each 10 mph increase in design speed. Based on these assumptions, the maximum radii and degree of curvature for the various design speeds for maximum e values of 0.06 to 0.12 are computed from the basic horizontal formula given above, and the values obtained are shown in Table 14.9.

In addition to the controls on maximum curvature to satisfy the above requirements, the need to provide minimum stopping sight distance around curves may control when sight distance cannot be otherwise provided by removing the sight obstruction. Figure 14.7 shows the required middle ordinates for clear sight areas for various degrees of curve. The chart applies only when the horizontal curve is longer than the sight distance required. For other cases (including a cut slope obstruction), either a graphic check as described previously or a more detailed analytical check should be made.

Superelevation rates. For curves of less than the maximum degree of curvature, the rate of superelevation is proportioned between the maximum curve for which superelevation is not required and the maximum curve for the governing maximum rate of superelevation. The superelevation rates (e) for various degrees of curvature

TABLE 14.9
Maximum Degree of Curve and Minimum Radius for Design

Maximum e	Design Speed, mph (kph)							
	30 (48)	40 (64)	50 (80)	50 (96)	65 (104)	70 (112)	75 (120)	80 (128)
	Minimum Radius ft (m)							
.04	300 (91)	561 (171)	926 (282)	1,412 (429)	1,657 (504)	2,042 (621)	2,500 (760)	2,844 (865)
.06	273 (83)	508 (155)	833 (254)	1,263 (384)	1,483 (451)	1,815 (551)	2,206 (670)	2,510 (763)
.08	250 (76)	464 (141)	758 (281)	1,143 (434)	1,341 (408)	1,633 (497)	1,974 (600)	2,246 (682)
.10	231 (70)	427 (130)	694 (211)	1,043 (317)	1,225 (372)	1,485 (451)	1,786 (543)	2,032 (615)
.12	214 (65)	395 (121)	641 (195)	960 (292)	1,127 (342)	1,361 (414)	1,630 (495)	1,855 (564)
	Maximum Degree of Curve, Rounded							
.04	19.0	10.0	6.0	4.0	3.5	3.0	2.5	2.0
.06	21.0	11.5	7.0	4.5	4.0	3.0	2.5	2.5
.08	23.0	12.5	7.5	5.0	4.5	3.5	3.0	2.5
.10	25.0	13.5	8.5	5.5	4.5	4.0	3.0	3.0
.12	26.5	14.5	9.0	6.0	5.0	4.0	3.5	3.0

for a maximum *e* value of 0.08 are presented in Table 14.10. Similar tables for other maximum *e* values are found in the cited reference.[18]

Superelevation runoff and transition curves. Transition curves used in the United States are either a series of compounded circular curves or a spiral curve. Spirals provide a natural, easy-to-follow driving path, can enhance highway appearance, and provide a desirable arrangement for superelevation runoff.

Spirals are identified by their length, which is usually governed by the distance required for superelevation runoff. Lengths required for superelevation runoff are from 100 to 250 ft, and consequently spirals are usually in this range—spiral lengths increasing with higher design speeds. Table 14.10 also presents recommended lengths of superelevation runoff or spiral curves for a maximum *e* value of 0.08.

Without spirals, a large portion of the superelevation runoff must be placed on the approach tangents. In general, from 50 to 100 percent of the length of superelevation runoff can be considered suitable. For more precise control, from 60 to 80 percent of the runoff should be located on the approach tangents.

General controls for horizontal alinement. In addition to the specific design elements for horizontal alinement, there are several other general controls:

1. Consistent with topography, alinements should be as direct as possible. Two-lane road alinements should provide as many safe passing sections as possible.

[18] *A Policy on Geometric Design of Rural Highways* (Washington, D.C.: American Association of State Highway Officials, 1965), pp. 168–71.

TABLE 14.10
Values for Design Elements Related to Design Speed and Horizontal Curvature

$e_{max} = 0.08$

D	R	V = 30 mph			V = 40 mph			V = 50 mph			V = 60 mph			V = 65 mph			V = 70 mph			V = 75 mph			V = 80 mph		
		e	L 2-lane	L 4-lane	e	L 2-lane	L 4-lane	e	L 2-lane	L 4-lane	e	L 2-lane	L 4-lane	e	L 2-lane	L 4-lane	e	L 2-lane	L 4-lane	e	L 2-lane	L 4-lane	e	L 2-lane	L 4-lane
0°15'	22918'	NC	0	0	NC	0	0	NC	0	0	NC	0	0	NC	0	0	NC	0	0	NC	0	0	RC	240	240
0°30'	11459'	NC	0	0	NC	0	0	NC	0	0	RC	175	175	RC	190	190	RC	200	200	RC	220	220	.024	240	240
0°45'	7639'	NC	0	0	NC	0	0	RC	150	150	.022	175	175	.025	190	190	.029	200	200	.022	220	220	.036	240	240
1°00'	5730'	NC	0	0	RC	125	125	.021	150	150	.029	175	175	.033	190	190	.038	200	200	.032	220	220	.047	240	240
1°30'	3820'	RC	100	100	.021	125	125	.030	150	150	.040	175	175	.046	190	200	.053	200	240	.043	220	290	.065	240	320
2°00'	2865'	RC	100	100	.027	125	125	.038	150	170	.051	175	210	.057	190	250	.065	200	290	.060	220	340	.076	240	380
2°30'	2292'	.024	100	100	.033	125	125	.046	150	190	.060	175	240	.066	190	290	.073	220	330	.072	230	370	.080	250	400
3°00'	1910'	.025	100	100	.038	125	125	.053	150	210	.067	180	270	.073	210	320	.078	230	350	.078	250	380			
3°30'	1637'	.028	100	100	.043	125	140	.058	150	230	.073	200	300	.077	220	330	.080	240	360						
4°00'	1432'	.032	100	100	.047	125	150	.063	150	260	.077	210	310	.079	230	340									
5°00'	1146'	.038	100	100	.055	125	170	.071	170	280	.080	220	320												
6°00'	955'	.043	100	120	.061	130	190	.077	180	280															
7°00'	819'	.048	100	130	.067	140	210	.079	190	290															
8°00'	716'	.052	100	140	.071	150	220																		
9°00'	637'	.056	100	150	.075	160	240																		
10°00'	573'	.059	110	160	.077	160	240																		
11°00'	521'	.063	110	170	.079	170	250																		
12°00'	477'	.066	120	180	.080	170	250																		
13°00'	441'	.068	120	180																					
14°00'	409'	.070	130	190																					
16°00'	358'	.074	130	200																					
18°00'	318'	.077	140	210																					
20°00'	286'	.079	140	210																					
22°00'	260'	.080	140	220																					
D max		.080	140	220	.080	170	250	.080	190	290	.080	220	320	.080	230	350	.080	240	360	.080	250	380	.080	260	400

D max: V = 30 → 23.0°; V = 40 → 12.5°; V = 50 → 7.5°; V = 60 → 5.0°; V = 65 → 4.5°; V = 70 → 3.5°; V = 75 → 3.0°; V = 80 → 2.5°

D—Degree of curve
R—Radius of curve
V—Assumed design speed
e—Rate of superelevation
L—Minimum length of runoff of spiral curve
NC—Normal crown section
RC—Remove adverse crown, superelevate at normal crown slope
Spirals desirable but not as essential above heavy line.
Lengths rounded in multiples of 25 or 50 feet permit simpler calculations.

Source: *A Policy on Geometric Design of Rural Highways*, AASHO, 1965, p. 169.

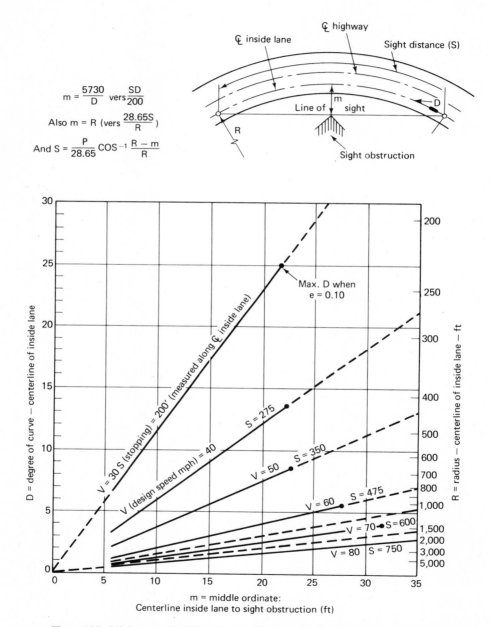

$$m = \frac{5730}{D} \text{ vers} \frac{SD}{200}$$

$$\text{Also } m = R \left(\text{vers} \frac{28.65S}{R} \right)$$

$$\text{And } S = \frac{P}{28.65} \text{ COS}^{-1} \frac{R - m}{R}$$

Figure 14.7. Minimum on 14.7(1) and on 14.7(2) Desirable Stopping sight distance on horizontal curves. 14.7(1) Minimum, 14.7(2) Desirable. (Source: *A Policy on Geometric Design of Rural Highways*, p. 188.)

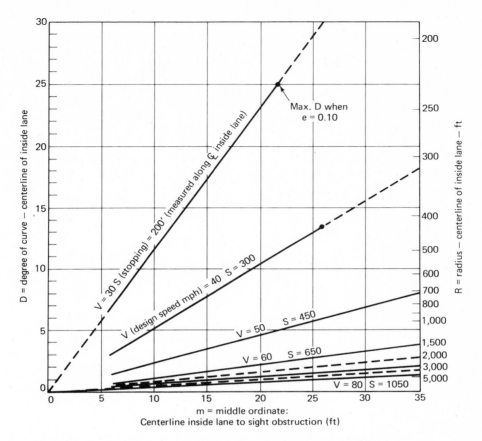

2. Maximum curvatures should be avoided whenever possible.
3. Consistent alinement should be sought. Sharp curves at the end of long tangents or at the end of long flat curves should be avoided.
4. Short lengths of curves should be avoided even for very small deflection angles.
5. Flat curvatures should be provided on long fills.
6. Compound circular curves with large differences in radii should be avoided.
7. Direct reverse curves should be avoided: A tangent length should be used between them.
8. "Broken-back curves" (two curves in the same direction on either side of a short tangent or large radius curve) should be avoided.

CROSS SECTION

The highway cross section is made up of traveled ways, auxiliary lanes, shoulders, medians, and roadsides. The dimensions of each vary with the type of highway. Each is discussed separately below.

Traveled way. The traveled way is that portion of the roadway available for movement of vehicles exclusive of shoulders and auxiliary lanes. It is normally comprised of two or more traffic lanes. On freeways where eight or more lanes are provided, two or more reversible lanes may be included in the cross section; sometimes express lanes are separated from lanes serving shorter distance travel resulting in a "dual-dual" facility.

On all freeways, expressways, and other arterial highways traffic lanes should be at least 12 ft wide. For local roads and streets, the desirable minimum width of a traffic lane is 11 ft, although 10-ft or even 9-ft lanes are adequate for low traffic volumes with few trucks. The full traveled way width for the approaches should be carried across all structures.

Undivided highways on tangents or flat curves have a crown or high point in the middle and slope downward toward both edges. The downward cross slope (crossfall) may be a plane or curved section or a combination of both. Divided highways may have each traveled way crowned or may have a straight slope across the entire width of each traveled way. Table 14.11 shows normal cross slopes for various type highway surfaces.

When two or more lanes are sloped in the same direction on multilane pavement, each successive lane outward from the crown line should have an increased slope. The lane adjacent to the crown line should be sloped at the normal minimum slope, and the slope on each successive lane should be increased by $\frac{1}{16}$ in. per ft. When curbs are used along high-type pavements, a cross slope of $\frac{3}{16}$ in. per ft on the outer lane is the practical minimum to reduce water sheeting on the traffic lane adjacent to the curb.

TABLE 14.11
Normal Cross Slopes

Pavement Surface Type	in./ft	ft/ft	cm/m
High	1/8–1/4	.01 –.02	1.0–2
Intermediate	3/16–3/8	.015–.03	1.5–3
Low	1/4–1/2	.02 –.04	2.0–4

Auxiliary lanes. An auxiliary lane is that portion of the roadway adjoining the traveled way for parking, speed change, turning, storage for turning, weaving, truck climbing, or for other purposes supplementary to through traffic movement.

When parking is permitted adjacent to a traffic lane, an additional pavement width of 10 or 12 ft (3 to 4 m) should be provided (see Chapter 15). The parking spaces should then generally be marked 8 ft from the curb regardless of the actual pavement width available, as shown in the *Manual on Uniform Traffic Control Devices*[19] to ensure proper use of the parking lane. The additional pavement width provided is necessary for proper operation of the adjacent traffic lane. Also, a parking lane of 10 ft or wider can be used as a through traffic lane during peak hours or converted into a storage lane, a turning lane, or a permanent through traffic lane if necessary.

A speed change lane is an auxiliary lane, including tapered area, primarily for the acceleration or deceleration of vehicles entering or leaving through traffic lanes (the design of speed change and turning lanes is discussed later).

---

[19] *Manual on Uniform Traffic Control Devices*, FEDERAL HIGHWAY ADMINISTRATION, U.S. DEPARTMENT OF TRANSPORTATION (Washington, D.C.: Government Printing Office, 1971), p. 203.

A weaving section is a length of one-way road where parallel traffic streams cross by merging and then diverging. The length and width of a weaving section is determined by capacity demand analysis (discussed later and in Chapter 8). Any required additional width is usually provided in the form of additional lanes through the weaving section. These additional lanes for weaving are a form of auxiliary lane.

Appreciable volumes of trucks operating on steep, sustained grades may significantly reduce the safety and capacity of a highway unless auxiliary lanes, termed climbing lanes, are provided. These lanes should be considered when an appreciable number of trucks travel at speeds significantly below that of passenger cars. Auxiliary downhill lanes for trucks are not generally required. When for safety reasons trucks *must* descend at speeds significantly below that of cars, an added downhill truck lane should be considered also.

Climbing lanes should be 12 ft wide. They should have delineation contrast for immediate recognition as extra lanes. The cross slope design should be the same as for the through traffic lanes. In general, full shoulders are not needed on the outer edge of climbing lanes since the climbing lane itself may be used for infrequent emergency stopping clear of the through traffic lanes. A paved shoulder from 4 to 6 ft wide is adequate in most cases.

The start of the climbing lane is determined by the speed reduction characteristics of a typical truck, usually taken as one with a 400 lb per hp weight-power ratio. The climbing lane should start when the speed of such a truck would be reduced to 30 mph. This can be determined from the speed of the truck as it enters the grade and the gradient in percent.[20] The full lane width should be preceded by a taper 150 ft long. The climbing lane should end at a point beyond the crest where a typical truck would attain a speed of 30 mph. An extension taper of 200 ft or more should be provided.

Shoulders. The term *shoulder* has several meanings depending on its modifying adjective: (1) The *graded shoulder* is the width from the edge of the traffic lane to the intersection of the shoulder slope and side slope planes. (2) The *surfaced shoulder* is the width outside the through traffic lane. It has an all-weather surface, including gravel, shell, or crushed rock surfaces, stabilization with mineral or chemical additives, bituminous treatments and various forms of asphaltic or concrete pavement. (3) The *usable shoulder* is the actual width usable for an emergency stop, including some of the rounding at the top of the flat earth slope. In Europe, the term *verge* or *outer verge* is used. The verge is comparable to the graded shoulder. Sometimes special paths reserved for pedestrians, cyclists, or similar traffic are provided on these verges.

Well-designed and properly maintained shoulders are necessary on rural highways that have any appreciable volume of traffic. They are desirable in urban areas also but cannot always be provided because of space limitations. Their functions are multifold:

1. Space is provided for emergency stops off the traveled way. Vehicles stopping on the traveled way introduce high accident potential.
2. Space is provided for drivers to recover safely should they lose vehicle control.
3. A sense of openness contributes to driving ease.
4. Sight distance is improved.

---

[20] *A Policy on Geometric Design of Rural Highways* (Washington, D.C.: American Association of State Highway Officials, 1965), p. 269.

5. Highway capacity is improved.
6. Space is provided for maintenance operations.
7. Structural support of the through traffic pavement is enhanced.

TABLE 14.12
Minimum Usable Shoulder Widths

	Minimum Usable Shoulder Width in Ft	
	Right Shoulder (ft)	Left Shoulder (ft)
Freeways		
Four-lane other than in mountainous terrain and where there are fewer than 250 trucks per hour in one direction in design year	10	6
Six or more lanes with fewer than 250 trucks per hour in one direction in the design year	10	10
Freeways in mountainous terrain	6	6
Four lanes with more than 250 trucks per hour in one direction in the design year	12	6
Six or more lanes with more than 250 trucks per hour in one direction in the design year	12	10

Arterial Highways Other Than Freeways		
Current ADT	DHV	
50–400	—	4
400–750	100–200	6
	Over 200	8

Local Roads		
Current ADT	DHV	
Less than 50	—	2
50–400	—	4
	Over 200	8

Shoulder widths vary depending on the type of highway. Table 14.12 shows minimum shoulder widths normally provided on various type highways. The full shoulder width should be carried across all structures except for major long-span structures, which should be analyzed individually. On highways other than freeways with current ADT less than 750 and design speed less than 50 mph, shoulder widths may be reduced somewhat. Minimum clearance from edge of traveled way to parapet or rail should be not less than 4 ft for arterial highways other than freeways and 2 ft for local roads.

Normally, shoulders are sloped to drain away from the traveled way. On super-elevated sections, however, it is sometimes necessary to slope the shoulder toward the through lane pavement in order to avoid too sharp a difference between through pavement cross slope and shoulder cross slope. The maximum algebraic difference in the pavement and shoulder grades should usually be about 0.07 ft per ft. Table 14.13 shows shoulder cross slopes used under normal conditions.

TABLE 14.13
Shoulder Cross Slopes

	Shoulder Cross Slope	
Type of Surface	in./ft	ft/ft
Bituminous	3/8–5/8	0.03–0.05
Gravel crushed stone	1/2–3/4	0.04–0.06
Turf	1	0.08

Medians (*central reserves*). Medians are desirable on highways that have four or more lanes. Median widths vary from 2 ft to over 100 ft. Rural freeway medians are usually at least 36 ft wide, but more often they are wider. When independent roadway design is used, medians can be of continually varying widths. Urban freeway medians are not less than 16 ft wide, and medians of this width should be provided with some form of barrier.

For highways other than freeways, medians can be categorized as follows:

1. Paint-striped separations, from 2 to 4 ft wide
2. Narrow, raised, or curbed sections, from 4 to 6 ft wide
3. Painted-striped or curbed sections providing space for separate left-turn lanes, from 14 to 18 ft wide
4. Traversible or curbed sections providing space for shielding protection of a vehicle crossing at an intersection and/or space for parkway landscape treatment, from 20 to 40 ft wide

Border areas. The border is the area between the roadway edge (or frontage road, if one is provided) and the right-of-way line. It does not include shoulders. It should be wide enough to accommodate any necessary cut or fill slopes, ditches, walls, bikeways, or sidewalks, and a landscaped buffer area between the highway and the surrounding land uses. Along city streets it might include an area for placement of utilities.

Borders should be at least from 30 to 60 ft wide for rural freeways and be flat or rounded to permit vehicle traverse without turnover. For rural highways other than freeways, desirably, a border should be wide enough for the basic ditch section and slopes, usually from 15 to 25 ft plus outer slope width. Desirably, the inner border area should be free of obstacles for at least from 10 to 20 ft and be flat or rounded. For urban roads and streets, a border width of from 4 to 8 ft plus sidewalk width should be provided.

Other cross section elements. The cross section may include a frontage road and outer separation between the frontage road and through roadway. Outer separations should be at least 30 ft wide in rural areas but may be as narrow as 4 ft with a suitable barrier in urban areas. Frontage roads should be at least 20 ft wide plus shoulder width or parking lane width as required; they are local roads and should be designed accordingly.

Right-of-way width. The total width of right of way required is the sum of the various elements described above. Right-of-way widths vary from a minimum 50 ft in urban areas to over 350 ft for freeways in open rural areas.

DESIGN PROCEDURES

The design of major highway improvements must be based on extensive investigations. A summary of these investigations is needed for joint consideration by the concerned organizations, as well as by the official responsible for deciding how and when the project should be undertaken.

PRELIMINARY ENGINEERING REPORT

A preliminary engineering report is needed for major highway projects in order to document the various factors considered in the project evaluation. The more complex and expensive the proposed highway, the more elaborate the report must be. Freeway development in urban areas requires the most detail. The following engineering report section headings highlight the major factors to be considered:

1. General description of the proposed improvement, need for improvement, and relationship of proposed highway improvement to transportation systems
2. Description and discussion of alternative locations and designs
3. Projected traffic volumes
4. Estimated total costs
5. Economic evaluation including user costs
6. Evaluation of consequences to natural environment, including effect on land values, employment, community values and services—this evaluation may be partly qualitative
7. Evaluation of the sociological effect on the community, including relocation of residences and businesses, and accessibility to community services—this evaluation may be mainly qualitative
8. A recommended scheme for the proposed highway improvement

PUBLIC HEARINGS

Public hearings should be held to ensure, to the maximum extent practicable, that highway locations and designs reflect and are consistent with local goals. Public hearings provide a medium for free and open discussion and are intended to encourage early and amiable resolution of controversial issues. They are also a way of keeping the public fully informed about specific details of highway location and design.

In the United States, major highway projects, particularly those involving federal funds, require at least two public hearings: a *corridor public hearing* and a *design public hearing*. A corridor public hearing is held before a route location is approved and before the agency responsible for determining the final corridor location is committed to a specific route. This hearing provides an opportunity for interested persons to participate in determining the need for, and the location of, a highway and for presenting views on the proposed alternatives as well as on the social, economic, and environmental effects of those alternatives. A design public hearing is held after the corridor public hearing and after the route location is approved but before the responsible agency is committed to a specific design. This hearing provides an opportunity for participation by interested persons in determining the specific location and major design features of a highway—an opportunity for presenting views on major design features including the social, economic, and environmental effects of alternate designs. Additional hearings or informal meetings may be desirable, most particularly at the systems planning stage prior to plan adoption.

The information sought from all public hearings includes the probable benefits or losses to the community as the result of constructing the highway, at least including the effects of the following:

1. Fast, safe, and efficient transportation
2. Economic activity
3. Employment
4. Recreation and parks
5. Fire protection
6. Aesthetics
7. Public utilities
8. Public health and safety
9. Residential and neighborhood character and location
10. Religious institutions
11. Local tax base and social service costs
12. Conservation including erosion, sedimentation, and wildlife
13. Natural and historical landmarks
14. Noise, air, and water pollution
15. Property values
16. Multiple use of space
17. Replacement housing
18. Disruption of schools
19. Displacement of families and businesses
20. Operation and use of existing highway facilities and other transportation facilities during construction and after completion

This information should be used to arrive at a design and location minimizing any adverse effects on the surrounding community and maximizing utility for the highway user.

ENVIRONMENTAL STATEMENT

In accord with the National Environmental Policy Act of 1969, in the U.S.A. it is a national policy that all federal agencies promote efforts for improving the relationship between man and his environment and to make special effort for preserving the natural beauty of the countryside and public park and recreational lands, wildlife and waterfowl refuges, and historic sites. It is also national policy that federal agencies consult with other appropriate federal, state, and local agencies; assess in detail the potential environmental impact in order that adverse effects are avoided and environmental quality is restored or enhanced to the fullest extent practicable; and utilize a systematic, interdisciplinary approach which will insure the integrated use of the natural and social sciences and the environmental design arts in planning and decision making which may have an impact on man's environment. The environmental assessments include the broad range of both beneficial and detrimental effects.

For these purposes an environmental statement is prepared and incorporated as a part of the design hearing and approval process. An environmental statement is a written statement containing an assessment of the anticipated significant beneficial and detrimental effects which the agency decision may have upon the quality of the human environment for the purposes of (1) assuring that careful attention is given to environmental matters, (2) providing a vehicle for implementing all applicable environmental requirements, and (3) insuring that the environmental impact is taken into account in the agency decision.

DESIGN CONCEPT TEAMS

Design concept teams have been formed primarily to solve the difficult problems inherent in locating new highways in heavily built-up urban areas. Interdisciplinary design teams normally include urban planners, highway and traffic engineers, architects, and sociologists, but they could also include real estate specialists, lawyers, economists, educators, and health planners. Such teams may have responsibility for (1) developing greater community participation in the design process, (2) evaluating the total environment of the highway corridor, (3) focusing greater attention on social problems, (4) suggesting changes in surrounding land uses when they conflict with the highway, and (5) planning possible community improvements in conjunction with the highway development (see Chapter 12).

MULTIPLE USE AND JOINT DEVELOPMENT

Multiple use of highway rights of way is the use of the airspace or other portion of the right of way for a nonhighway purpose such as buildings, recreational facilities, or parking areas. Joint development is the broader concept of satisfying not only a highway transportation need but also optimizing the use of land in the highway corridor; joint development usually takes place on land outside the normal highway right of way as well as inside it (Figure 14.8 shows shops built *under* an elevated freeway in Tokyo). Both multiple use and joint development projects must be considered at the *earliest* possible planning and design stages (see Chapter 12).

SAFETY CONSIDERATIONS IN DESIGN

Designing safety into highways is one of the main objectives of geometric design. This goal requires something more than the standards and principles thus far enumerated. Designs should anticipate and allow for driver error, should eliminate inconsistencies that cause driver confusion, and should promote the highway use intended.

Specific things that can be done to achieve safe designs are listed throughout the text. Some additional details that should be considered are related to structural design as well as geometric design. The extent to which they are applicable to the design of a particular facility depends on the highway type, the numbers and speeds of vehicles, and economic considerations. Such additional details include:

1. Roadside slopes should be made as flat as feasible, desirably 4:1 or flatter. The roadside area should be well rounded where slope planes intersect.
2. Sign and lighting supports either should be located far enough from the roadway to make them unlikely to be struck by an out-of-control vehicle or they should have breakaway capability.
3. All drainage structures should be designed so that out-of-control vehicles either can pass over them or can be safely deflected.

Figure 14.8. Example of multiple use of highway right of way—stores under elevated freeway. (Source: U.S. Department of Transportation, Federal Highway Administration.)

4. Guardrails should be considered when fill slopes 4:1 or flatter are unfeasible. The guardrail should be capable of safely stopping an out-of-control vehicle impacting it. For details on guardrail design, see cited reference.[21]

5. A suitably designed guardrail should be provided at approaches to bridges and at piers at undercrossing structures where there is likelihood they may be struck by an out-of-control vehicle.

6. On high-speed or high-volume divided highways with narrow medians, some form of median barrier should be considered to prevent out-of-control vehicles from crossing into opposing lanes.

7. Pavement cross slopes should be designed so that ponding of water will not occur.

8. When obstructions cannot be moved or protected by guardrails, such as at ramp gores on elevated structures, some form of impact attenuating device should be considered. The design of the gore area preceding the obstruction should allow for installing such devices.

9. Lanes should be dropped only at locations with excellent sight distances, and a recovery area should be provided to allow trapped vehicles to move safely to the adjacent lane.

10. Highway signing and marking is an integral part of design and should be developed progressively with geometric design. The design should allow for the latest traffic control devices as specified in the *Manual on Uniform Traffic Control Devices.*

---

[21] J. D. MICHIE and M. F. BRONSTAD, *Location, Selection and Maintenance of Highway Traffic Barriers,* National Cooperative Highway Research Program Report 118 (Washington, D.C.: Highway Research Board, 1971).

11. Design should not overlook problems caused by darkness, rain, and other nonoptimum operating conditions.

ARTERIAL HIGHWAYS OTHER THAN FREEWAYS

The design of the elements pertaining to arterial highways other than freeways were discussed previously, with the exception of those related to intersections, which will be discussed later in this section.

ROADWAY CROSS SECTION DESIGN

Rural roads other than freeways are most frequently two lanes wide with shoulders on both sides. Some degree of access control is desirable to limit points of access to locations where egress and ingress can be accomplished safety. Figure 14.9 shows typical cross sections of two-lane rural roads; Figure 14.10 shows them for rural divided highways. Typical cross sections used in Europe are shown in Figure 14.11. The bottom cross section is for a four-lane divided motorway having a design speed of from 80 to 100 kph. The top two are for a design speed of from 100 to 140 kph for four-and-six lane motorways. Particular care must be exercised in the location of all intersections to ensure sufficient sight distance, but above minimum sight distances should be provided at intersections-at-grade where located along divided highways. Unless provided with above minimum geometric features, these highways will produce higher than average accident rates.

Urban. Urban arterials most typically have two- or four-lane roadways, with a parking lane on either side and curb and gutter on the outside. Lane widths have already been discussed. Sometimes on urban arterials other than freeways it is desirable to convert the median to a continuous lane from which turns can be made from either direction. Special signing and pavement marking are required to ensure proper use of the lane. Minimum widths of this type lane should be comparable to a normal through traffic lane.

Curbs are used extensively on urban streets to control drainage, to discourage vehicles from leaving the pavement, to protect pedestrians, and to promote orderly roadside development. The two general classes of curbs are barrier and mountable. Either may be designed with a gutter to form a curb and gutter section. When the gutter contrasts in color with the through pavement, the width of gutter should not be considered part of the traffic lane width. When the gutter is the same color and texture as the through pavement, it may be considered as part of the through lane width.

Barrier curbs are from 6 to 10 in. (15 to 25 cm) or higher and have a steep face. Figures 14.12(a) and 14.12(b) show examples of typical barrier curbs. They have application only when vehicle speeds are relatively slow, and vehicle encroachment should be strongly discouraged. They are commonly used along streets and for refuge islands. Barrier curbs should be offset from 2 to 3 ft (0.6 to 0.9 m) from the through traffic lanes.

Mountable curbs are designed so that vehicles can cross them. They are 6 in. or lower and have a relatively flat sloping face. Figure 14.12(c) through 14.12(h) show typical examples of mountable curbs. Mountable curbs can be of bituminous mixtures or concrete.

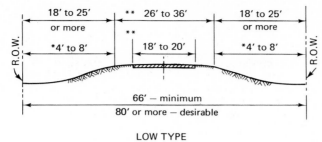

LOW TYPE

(a)

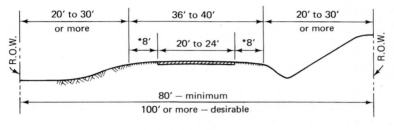

INTERMEDIATE TYPE

(b)

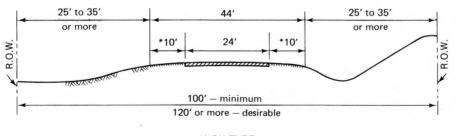

HIGH TYPE

(c)

* Usable shoulder width
** For low-volume roads with few trucks

Figure 14.9. Cross sections and right-of-way widths for two-lane rural highways. (Source: *A Policy on Geometric Design of Rural Highways*, p. 263.)

Although both barrier and mountable curbs can be used along medians, or at the edge of through traffic lanes, barrier curbs should be offset from through traffic lanes. Curbs of various types are also used at the edge of shoulders as a part of the drainage system.

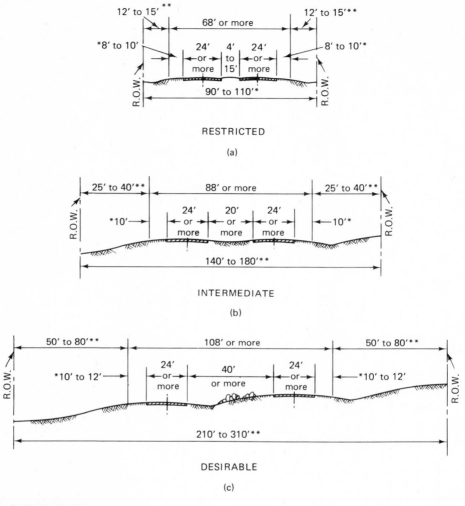

Figure 14.10. Cross sections and right-of-way widths for multilane divided highways. (Source: *A Policy on Geometric Design of Rural Highways*, p. 293.)

BORDER AREAS

Border areas along rural arterials should allow a driver who inadvertently leaves the through traffic lanes to recover without serious injury. This normally requires that border areas 10 to 20 ft beyond the shoulder edge be free of unyielding obstacles such as trees, large rocks, utility poles, and rigid sign posts. Ditches should be well-rounded in this area and side slopes should be 4:1 or flatter. When steeper slopes are required, as at approaches to overcrossing structures or at other locations involving

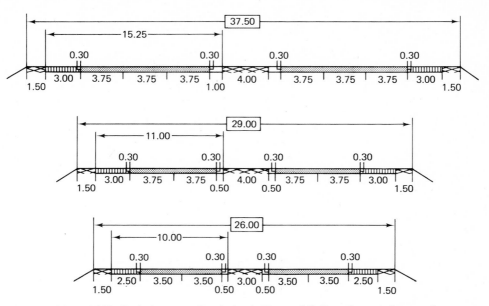

Figure 14.11. Typical cross-section design in Europe (all dimensions are in meters). [Source: U.S. Department of Transportation, Federal Highway Administration, *1972 World Survey of Current Research and Development on Roads and Road Transport* (Washington, D.C.: International Road Federation 1972, p. 321.)]

unmovable obstacles, guardrails should be provided; these are commonly located at the edge or 2 ft outside the edge of shoulder.

Border areas along urban arterial streets should be as wide as possible, but as a minimum, they should be sufficient to accommodate a sidewalk and a grass or utility strip. Sidewalks in built-up areas should be at least 8 ft wide, but in suburban residential areas 4 ft is sufficient. A grass strip is desirable between the sidewalks and curb, but in heavily built-up areas this is usually omitted. The grass strip, when used, can accommodate underground utilities.

INTERSECTIONS AT GRADE

Design principles. Individual maneuver areas are the smallest unit of intersection design. They may be combined variously to produce alternate geometric designs for any intersection. To a considerable extent their arrangement is governed by traffic demands, topography, land use, and economic and environmental considerations; the proper compromise is a decision to be made by the individual designer. Intersection design should consider the ten following fundamental principles:

Principle 1. Reduce number of conflict points. The number of conflict points among vehicular movements increases significantly as the number of intersection legs increases. For example, an intersection with four two-way legs has 32 total conflict points, but an intersection with six two-way legs has 172 conflict points.[22] Intersections with more than four two-way legs should be avoided wherever possible.

---

[22] T. M. MATSON, W. S. SMITH, and F. W. HURD, *Traffic Engineering* (New York: McGraw Hill Book Company, 1955), p. 505.

Barrier curns

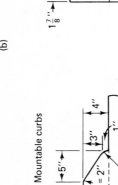
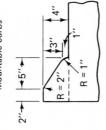

(a)

(b)

Mountable curbs

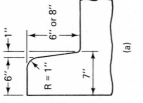
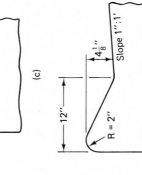

(c)

(d)

(e)

(f)

(g)

(h)

Figure 14.12. Typical highway curbs. (Source: *A Policy on Geometric Design of Rural Highways*, p. 228.)

Principle 2. Control relative speed. Relative speed is the rate of convergence or divergence of vehicles in intersection flow. A small difference (from 0 to 15 mph) in the speeds of intersecting vehicles and a small angle (less than 30°) between converging paths allow intersecting vehicular flows to operate continuously (uninterrupted flow). High relative speeds occur when there is either a large difference in vehicular speeds or a large angle of convergence. Since interrupted flow usually occurs under these conditions, traffic should be controlled by traffic control devices. Any intersection can be designed for either condition. Low relative speeds require elimination of both speed differences and large angles between intersection flows through design. Intersection of flows at high relative speeds should be as close as possible to 90°.

Principle 3. Coordinate design and traffic control. Maneuvers at intersections accomplished at low relative speeds require a minimum of traffic control devices. Maneuvers accomplished at high relative speeds are unsafe unless traffic controls such as stop signs and traffic signals are provided. Designs should physically divert or block the path of vehicles making dangerous movements. Intersection design should be accomplished simultaneously with the development of traffic control plans.

Principle 4. Use highest feasible crossing method. Vehicle crossing maneuvers can be accomplished in four ways: (1) uncontrolled crossing at-grade, (2) traffic sign or signal-controlled crossing at-grade, (3) weaving, and (4) grade separation. In general, both operational efficiency and construction cost increase in this order. The highest type should be used consistent with the numbers and types of vehicles using the intersection.

Principle 5. Substitute turning path. The method of providing turns can be changed. Separate roadways can be provided both for right- and left-turning vehicles, thereby reducing conflicts in the intersection area. For example, a direct connection can be provided to accommodate right turns at an intersection.

Principle 6. Avoid multiple and compound merging and diverging maneuvers. Multiple merging or diverging requires complex driver decisions and creates additional conflicts.

Principle 7. Separate conflict points. Intersection hazards and delays are increased when intersection maneuver areas are too close together or when they overlap. These conflicts may be separated to provide drivers with sufficient times (and distance) between successive maneuvers for them to cope with the traffic situation.

Principle 8. Favor the heaviest and fastest flows. The heaviest and fastest flows should be given preference in intersection design to minimize hazard and delay.

Principle 9. Reduce area of conflict. Excessive intersection area causes driver confusion and inefficient operations. Large areas are inherent in skewed and multiple-approach intersections. When intersections have excessive areas of conflict, channelization should be employed.

Principle 10. Segregate nonhomogeneous flows. Separate lanes should be provided at intersections when there are appreciable volumes of traffic traveling at different speeds. For example, separate turning lanes should be provided for high volumes of turning vehicles. When there are large numbers of pedestrians crossing wide streets, refuge islands should be provided so that more than three lanes do not have to be crossed at a time.

Provision for turning movements. Except for provision of turning movements, and sometimes widening through the intersection to accommodate through traffic, the horizontal design of a roadway in the area of an intersection-at-grade is the same as for the approach roadway. The layout to accommodate turning movements and the necessary through lanes determines the design of an intersection.

The design of the curb or radius return depends on the turning paths of the vehicles using the intersection. The intersection should be able to accommodate the selected design vehicles without requiring backing-up to complete the maneuver. Table 14.14 gives minimum edge of pavement designs for various design vehicles and angles of turn. Figure 14.13 shows a typical design of a radius return with a three-centered compound curve to handle semitrailer vehicles.

Separate parallel turning lanes. Both the safety and capacity of an intersection-at-grade can be improved significantly in many cases by providing added parallel turning or speed-change lanes on either the right or left or both sides. They are particularly advantageous in both rural and urban areas where through speeds are relatively high and there are appreciable volumes of either right- or left-turning traffic. The primary purpose of these turning lanes is to provide storage for vehicles. A secondary purpose is to provide space for turning vehicles to decelerate in advance of the intersection or to accelerate beyond it.

These turning lanes should be at least 10 ft and preferably 12 ft wide. The length of exit turning lanes consists of three components: (1) deceleration length, (2) storage length, and (3) entering taper.

Provision for deceleration clear of through traffic lanes is an important element on high-speed arterial streets and should be incorporated into their design whenever feasible. Lengths needed for deceleration (based on average running speeds) are as follows:

Average Running Speed		Deceleration Length	
mph	kph	feet	meters
20	35	160	49
30	50	250	76
40	65	370	113
50	80	500	155

On many highways the full length for deceleration plus storage and taper cannot feasibly be provided. Deceleration must then be partially accomplished before entering the turning lane. The length of exit turning lane necessary for storage should be sufficient to accommodate twice the average number of vehicles stored per traffic signal cycle. At intersections without traffic signals, the length may be based on the number of vehicles wishing to turn right or left in 2 min; vehicle length may be assumed to be 25 ft. The entering taper of a parallel turning lane should be at least 40 ft and preferably 100 ft.

Separate turning roadways. At intersections where larger semitrailer combinations must be accommodated and where passenger cars are to be allowed to turn at speeds of 15 mph or more, separate turning roadways should be provided. They should have the maximum radius possible. At urban intersections minimum type design is usually all that can be provided. The principal controls for minimum designs

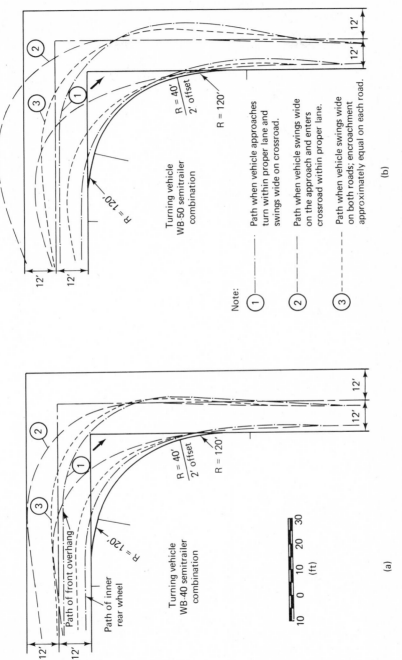

Figure 14.13. Minimum edge of pavement design for single-unit trucks and buses and necessary paths of larger vehicles. (Source: *A Policy on Geometric Design of Rural Highways*, p. 316.)

Turning vehicle WB-50 semitrailer combination

Note:

(1) Path when vehicle approaches turn within proper lane and swings wide on crossroad.

(2) Path when vehicle swings wide on the approach and enters crossroad within proper lane.

(3) Path when vehicle swings wide on both roads; encroachment approximately equal on each road.

(b)

Turning vehicle WB-40 semitrailer combination

Path of front overhang

Path of inner rear wheel

(a)

TABLE 14.14
Minimum Edge of Pavement Designs for Turns at Intersection
With No Encroachment on Adjacent Lanes

Design Vehicle	Angle of Turn	Simple Curve Radius	3-Centered Compound* Curve, Symmetric		3-Centered Compound Curve, Asymmetric	
			Radii	Offset	Radii	Offset
	degrees	feet	feet	feet	feet	feet
P	30	60	—	—	—	—
SU		100	—	—	—	—
BUS		140	—	—	—	—
WB-40		150	—	—	—	—
WB-50		200	—	—	—	—
P	45	50	—	—	—	—
SU		75	—	—	—	—
BUS		80	—	—	—	—
WB-40		170	—	—	—	—
WB-50		170	200-100-200	3.0	—	—
P	60	40	—	—	—	—
SU		60	—	—	—	—
BUS		70	—	—	—	—
WB-40		90	—	—	—	—
WB-50		—	200-75-200	5.5	200-75-275	2.0-6.0
P	75	35	100-25-100	2.0	—	—
SU		55	120-45-120	2.0	—	—
BUS		80	120-45-120	3.0		
WB-40		85	120-45-120	5.0	120-45-200	2.0-6.5
WB-50		—	150-50-150	6.0	150-50-225	2.0-10.0
P	90	30	100-20-100	2.5	—	—
SU		50	120-40-120	2.0	—	—
BUS		85	120-40-120	4.0		
WB-40		—	120-40-120	5.0	120-40-200	2.0-6.0
WB-50		—	180-60-180	6.0	120-40-200	2.0-10.0
P	105	—	100-20-100	2.5	—	—
SU		—	100-35-100	3.0	—	—
BUS		—	100-35-100	4.0		
WB-40		—	100-35-100	5.0	100-35-200	2.0-3.0
WB-50		—	180-45-180	8.0	150-40-210	2.0-10.0
P	120	—	100-20-100	2.0	—	—
SU		—	100-30-100	3.0	—	—
BUS		—	100-30-100	4.0		
WB-40		—	120-30-120	6.0	100-30-180	2.0-9.0
WB-50		—	180-40-180	8.5	150-35-220	2.0-12.0
P	135	—	100-20-100	1.5	—	—
SU		—	100-30-100	4.0	—	—
BUS		—	100-30-100	5.0		
WB-40		—	120-30-120	6.5	100-25-180	3.0-13.0
WB-50		—	160-35-160	9.0	130-30-185	3.0-14.0
P	150	—	75-18-75	2.0	—	—
SU		—	100-30-100	4.0	—	—
BUS		—	100-30-100	5.0		
WB-40		—	100-30-100	5.0	90-25-160	3.0-11.0
WB-50		—	160-35-160	7.0	120-30-180	3.0-14.0
P	180	—	50-15-50	0.5	—	—
SU	U-Turn	—	100-30-100	1.5	—	—
BUS		—	130-25-130	8.0		
WB-40		—	100-20-100	9.5	85-20-150	6.0-13.0
WB-50		—	130-25-130	9.5	100-25-180	6.0-13.0

*A simple curve with the smaller radius offset as shown above and connected to the normal edge of pavement with approximately 15:1 tapers on either end may be substituted for the three-centered compound curve.

Source: *A Policy on Design of Urban Highways and Arterial Streets*, p. 682.

of the turning roadways are the alinement of the inner edge of pavement and the width of roadway channel to accommodate the design vehicle.

Turning roadways are usually separated from through traffic lanes by islands in order to properly direct turns and reduce pavement areas (islands smaller than 50 sq ft should be avoided). Table 14.15 shows typical minimum design dimensions for three design classifications and shows approximate island sizes for various angles of turn. Space permitting, separate turning roadways with much larger radii are desirable and should be provided. Turning roadway layouts can sometimes approach those of freeway ramps.

Median opening. Design of median openings and median ends should be based on traffic volumes and types of turning vehicles. An important factor is the path of each design vehicle making a minimum left turn at low speed. Figure 14.14 shows a minimum median opening design to accommodate a WB-40 semitrailer design vehicle. However, as shown by the dashed line, a WB-50 design vehicle can also be accommodated (with slight encroachment on the adjacent lane). The intersection shown is at right angles; for the same control at skew intersections, the length of median opening will increase with an increase in the skew angle.

Channelization. Channelization involves the use of islands at intersections to guide and protect traffic. Many intersection design principles depend on it. Channelization also provides reference points within the intersection which enable drivers to better predict the path and speed of other drivers. This increases their ability to avoid accidents and congestion. Several rules govern the application of channelization:

1. Islands should be arranged so that the driving path seems natural and convenient.
2. There should be only one path for the same intersectional movement.
3. A few well-placed, large islands are better than a confusion of small islands.
4. Islands should be offset 2 ft or more from the edge of normal traveled way.
5. Adequate approach-end treatment is required to warn drivers and to permit gradual changes in speed and path.
6. Curving roadways should have radii and width adequate for the governing design vehicle.
7. Adequate visibility should be provided drivers approaching the intersection. There should be no hidden obstructions, and islands should be well-defined. In many cases some form of illumination will be necessary, depending on the type curb used to outline the island.

Channelizing islands may be traversable, deterring, or barrier types. The choice of type of border curb is often a compromise between the hazard caused by the barrier and the importance of positive control of vehicle speeds and paths. Islands themselves can be either paved or have a low-growing plant cover. Those used by pedestrians and small or narrow islands should be paved. Inasmuch as many factors affect the design of channelization, temporary installations are sometimes recommended for a trial period before permanent islands are installed.

Sight distance and vertical alinement. Minimum sight distance criteria at intersections were discussed previously. Much longer distances are desirable. The vertical alinement of the highway in the intersection area should be as flat as possible; the maximum grade on any approach leg should be about 5 percent.

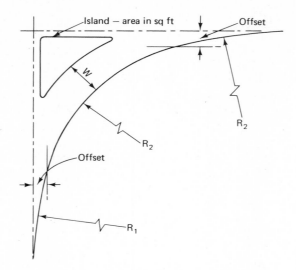

TABLE 14.15
Minimum Designs for Turning Roadways

Angle of Turn	Design Classification*	3-Centered Compound Curve		Width of Lane	Approxi- mate Island Size
		Radii	Offset		
Degrees		ft	ft	ft	sq ft
	A	150-75-150	3.5	14	60
75	B	150-75-150	5.0	18	50
	C	180-90-190	3.5	20	50
	A	150-50-150	3.0	14	50
90	B	150-50-150	5.0	18	80
	C	180-65-180	6.0	20	125
	A	120-40-120	2.0	15	70
105	B	100-35-100	5.0	22	50
	C	180-45-180	8.0	30	60
	A	100-30-100	2.5	16	120
120	B	100-30-100	3.0	24	90
	C	180-40-180	8.5	34	220
	A	100-30-100	2.5	16	460
135	B	100-30-100	4.0	26	370
	C	160-35-160	7.0	35	640
	A	100-30-100	2.5	16	1400
150	B	100-30-100	4.0	30	1170
	C	160-35-160	7.0	38	1720

*A—Primarily passenger vehicles; permits occasional design single-unit truck to turn with restricted clearances.
  B—Provides adequately for SU; permits occasional bus and WB-50 to turn with slight encroachment on adjacent traffic lanes.
  C—Provides fully for all design vehicles.
Note: Asymmetric 3-centered compound curves and straight tapers with a simple curve can also be used without significantly altering the width of pavement or corner island size.

Source: *A Policy on Geometric Design of Urban Highways and Arterial Streets*, p. 687.

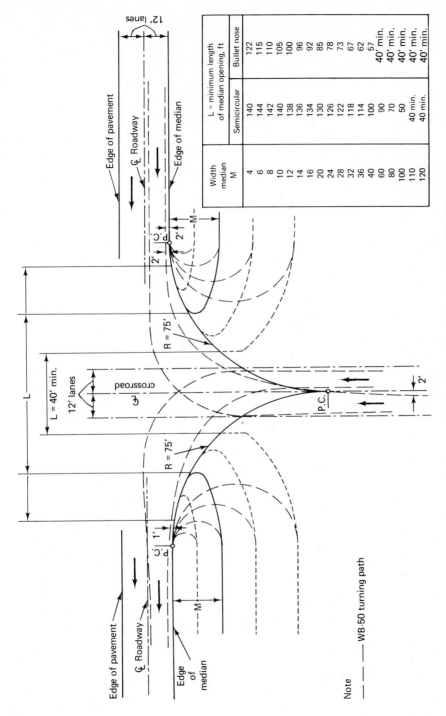

Figure 14.14. Minimum design of median openings to accommodate WB-40 design vehicles (control radius of 75 ft). (Source: *A Policy on Geometric Design of Rural Highways*, p. 415.)

Width median M	L = minimum length of median opening, ft	
	Semicircular	Bullet nose
4	140	122
6	144	115
8	142	110
10	140	105
12	138	100
14	136	96
16	134	92
20	130	85
24	126	78
28	122	73
32	118	67
36	114	62
40	100	57
60	90	40' min.
80	70	40' min.
100	50	40' min.
110	40 min.	40' min.
120	40 min.	40' min.

Note

——— WB-50 turning path

BICYCLE CROSSINGS[23]

Bicycle/motor vehicle crossing conflicts can best be eliminated by grade separation. A bikeway underpass is more desirable than an overpass because vertical clearance requirements are lower and the bicyclist can often gain sufficient speed on the downgrade approach to facilitate the subsequent upgrade pedaling. In most cases, the cost of an underpass and possible security and drainage problems rule out its use. Bicycle overpasses are also a possibility, but right-of-way restrictions frequently cause the approach ramps to be too steep.

Bicycle grade separations should be considered whenever heavily traveled bikeways must cross arterial routes with large traffic volumes. Grade separations are also desirable whenever high-volume bikeways must cross heavily traveled intersections and whenever significant bicycle traffic might disrupt motor vehicle and/or pedestrian movement.

At unchannelized at-grade intersections a bicyclist can follow a wide variety of paths to cross or turn at the intersection, thus increasing the potential for bicycle/motor vehicle/pedestrian conflicts. Therefore, whenever bikeways approach at-grade intersections, intersection channelization for bicycles should be seriously considered unless low motor vehicle and low bicycle volumes exist, motor vehicle speeds are low, and only a small percentage of the vehicles make a nearside turn across the projected bikeway alignment.

In Germany and the Netherlands the conflict between nearside turning motor vehicles and through bicycle movements is reduced by offsetting the bikeway crossing 16-33 ft (5-10 m) from the intersection. This type of crossing offset facilitates the sight distance of both motorists and cyclists and it allows a queue area for motor vehicles on the cross street so that through vehicles are not blocked on the main street. The Dutch also specify that a traffic island crossed by a bikeway should be at least 10 ft (3 m) wide and of a different surface color, thereby creating a safe queue area for the cyclists. Additional design alternatives are discussed in the reference document.

PROVISION FOR MASS TRANSIT

Urban streets that carry buses should incorporate various features in order to reduce conflicts between buses and passenger cars. Mass transit requirements should be considered early in the development of an urban highway: (1) bus stop spacing and location (near side, far side, or mid-block) must be selected; (2) bus stop design must be determined; (3) reservation of lanes, if advantageous, must be decided upon; and (4) any special traffic control measures required must be defined (see Chapter 18 for discussion of exclusive bus lanes).

The general location of bus stops is largely determined by patronage and intersecting bus routes. Highway and transit engineers should cooperate in these determinations. The specific location is influenced by convenience to patrons and operational considerations. No rigid criteria can be established. Everything being equal,

---

[23] This section summarized from the following publication which discusses numerous alternative at-grade street bikeway intersection designs: *Bikeway Planning Criteria and Guidelines* (Los Angeles: University of California, 1972), pp. 86–108.

however, far-side stops usually create less interference between buses and other vehicles because the buses can more readily reenter the traffic stream.[24]

The interference of buses with other traffic can be reduced considerably by providing added lane stops clear of through traffic lanes. To be fully effective, bus turnouts should incorporate: (1) a taper to permit easy entrance to the loading areas, (2) a standing space long enough to accommodate the most buses expected at the stop at any one time, and (3) a merging taper.

The turnout approach should be tapered at least 5:1 to encourage proper use of the turnout. The loading area should provide about 40 ft per bus and should be at least 10 and preferably 12 ft wide. The merging taper should be at least 3:1. The total length of bus turnout should then be about 180 ft for a mid-block location, 150 ft for a nearside location where a merging taper is not required, and 130 ft for a far-side location where a diverging taper is not required.

ROTARY INTERSECTIONS

A rotary intersection is one through which traffic passes by entering and leaving a one-way roadway connecting all intersection approach legs and running continuously around a central island. This specialized form of at-grade intersection is not well-suited for high-speed traffic or for accommodating high volumes on the approach legs. Rotary intersections are commonly called *traffic circles*, but proper design can result in central islands of various rounded shapes. Wherever British terminology prevails, they are quite properly known as *roundabouts*. Traffic flow around the central island is counterclockwise in those countries in which vehicles operate to the right of centerline.

Rotary intersections are best utilized at locations without grades and with unrestricted rights of way, where traffic volumes on all intersection legs are approximately equal but total less than 3,000 vph, and where the turning volumes approach or exceed the through traffic volumes.

When traffic volumes exceed the capacity of the weaving sections in the rotary, traffic signals must be installed. At this point, a normal at-grade intersection usually provides much better traffic service and control, and many rotaries have, in fact, been replaced by normal at-grade intersections. The choice between a rotary or normal at-grade intersection should depend on expected future traffic volumes. Figure 14.15 depicts the terms and elements commonly used in the design of rotary intersections. For additional details of rotary design, see cited reference.[25]

FREEWAYS

The highest type highway is the fully access-controlled freeway (motorway). Essential freeway elements include medians, grade separations, and ramp connections.

TYPES

Urban freeways may be constructed at ground level, be depressed or elevated, or be a combination of types. Rural freeways are generally constructed at ground level.

---

[24] *A Recommended Practice for the Proper Location of Bus Stops* (Washington, D.C.: Institute of Traffic Engineers, 1967).

[25] *A Policy on Geometric Design of Rural Highways* (Washington, D.C.: American Association of State Highway Officials, 1965), pp. 478–91.

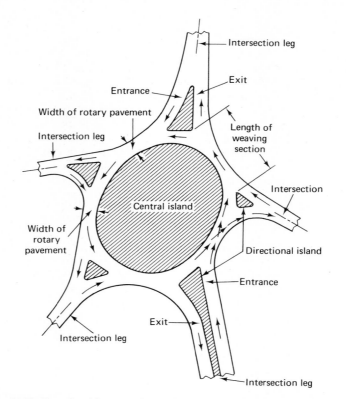

Figure 14.15. Terms used in rotary design. (Source: *A Policy on Geometric Design of Rural Highways*, p. 479.)

ALINEMENT

Horizontal alinement for freeways should safely accommodate high volumes at high speeds. Also, an alinement should be selected that will have the least effect on the landscape. In rolling terrain in rural areas an alinement with these attributes is best achieved by using a spline to develop alinement fitted to basic controls. The natural bending of the spline produces a smooth, flowing alinement without marked distortion. Because a freeway consists of two separated roadways, use can be made of independent roadway design and a variable width median to provide a superior facility at less cost. Smooth, flowing alinements for each one-way roadway should be the goal in designing a divided highway. Figure 14.16 shows a highway with an alinement as described above.

CROSS SECTION

Figure 14.17 shows a typical cross section of a freeway in rural areas. Figures 14.18, 14.19, and 14.20 show typical cross sections of urban at-grade, elevated, and depressed freeways, respectively. Figure 14.21 shows a freeway cross section in a tunnel.

Figure 14.16. Freeway with curvilinear alinement and variable width median. (Source: U.S. Department of Transportation, Federal Highway Administration.)

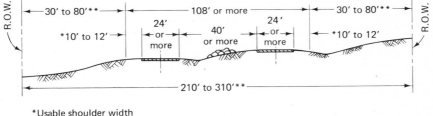

*Usable shoulder width
**Preferably wider

Figure 14.17. Cross sections and right-of-way widths for freeways in rural areas.

It is desirable that rural freeways and, where feasible, urban freeways have clear recovery area provided outside the edge of the shoulder. This area should be at least 20 ft and preferably greater, and it should contain no unyielding obstacles that could seriously damage an out-of-control vehicle. This includes large trees (over 4 in. in diameter), rocks, and unyielding sign and lampposts. Ditches should be well-rounded in this area, and fill slopes should be 4:1 or flatter. When it is unfeasible to remove obstructions or provide such flat fill slopes, suitable guardrails should be provided. Guardrails or other suitable barriers should also be provided in medians less than 30 ft wide and along piers and columns in both the median and on the outside where they are within about 20 ft of the edge of the through traffic lane.

GRADE SEPARATIONS

A grade separation is a crossing of two highways, a highway and a railroad, or a pedestrian walkway and a highway at different levels. An overpass is a highway passing over an intersecting street, railroad, or pedestrian facility. An underpass is a highway passing under an intersecting street, railroad, bicycle, or pedestrian facility. The type structure for a freeway underpass should be determined for the load,

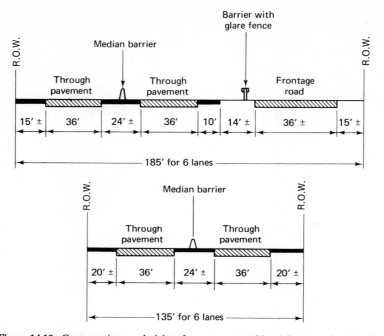

Figure 14.18. Cross sections and right of way on ground-level freeways (restricted). Note: Dimensions show for six-lane freeway; for other than six-lane freeways, similar dimensions are used except for through pavement width.

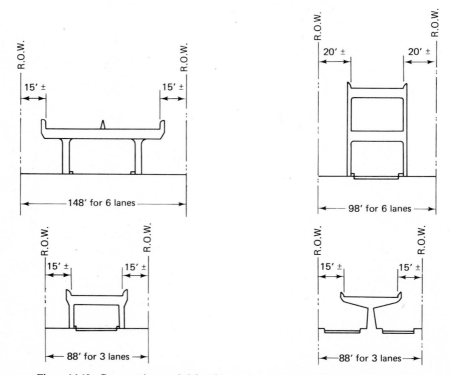

Figure 14.19. Cross sections and right of way; elevated freeways on structures without ramps. Note: Dimensions shown for six-lane freeways; for other than six lanes, similar dimensions are used except for through pavement widths.

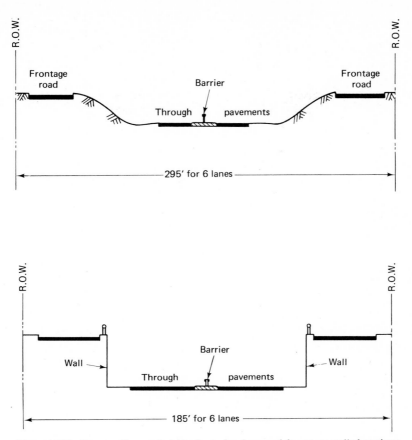

Figure 14.20. Cross sections and right of way for depressed freeways; walled sections without ramps. Note: Dimensions shown for six-lane freeways; for other than six-lane freeways, similar dimensions are used except for through pavement width.

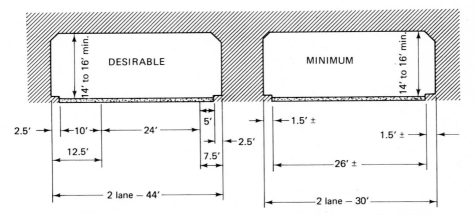

Figure 14.21. Typical section for tunnel. Note: Dimensions shown for four-lane freeway; for other than four lanes, dimensions are similar except for through lane pavement width.

foundation, and general site requirements for each case. The widths of traveled way and shoulders for bridges other than grade separations are essentially the same as for overpasses except for long-span structures.

Minimum clearances at underpasses from the edge of traffic lanes to piers, abutments, or columns are desirably the same as the clearances to obstructions on the remainder of the highway, generally 30 ft from the edge of the traffic lane. When this is unfeasible, the minimum clearances should be the normal shoulder width plus sufficient room to allow for protective devices between the edge of shoulder and the obstruction. Vertical clearance at underpasses in urban areas may be as low as 14.5 ft, but a 16-ft clearance is desired. In rural areas the minimum clearance is 16 ft.

On overpasses, the entire approach roadway width (including shoulders) should be carried across the structure. When a median is no wider than 30 ft, it is preferable, for safety, to deck over the entire median rather than to have separate structures for each one-way roadway.

Pedestrian and bicycle separations should be at least 8 ft wide. When they are located over freeways, there should be a vertical clearance of 16 ft; when they are located under freeways, they should have a clear height of 10 ft.

INTERCHANGES

Types. Numerous interchange configurations are now used, and many other logical arrangements could be constructed. Where there are four approach legs, the most common types are:

Cloverleaf. A four-leg interchange with loop ramps for some or all of the left turns. A full cloverleaf has ramps for two turning movements in each quadrant. Typical cloverleaf patterns are illustrated in Figure 14.22.

Diamond. A four-leg interchange with a single one-way ramp in each quadrant. All left turns are made directly on the minor highway. The regular diamond, diamond with "slip" ramps to frontage road, and "split diamond" are shown in Figure 14.23.

Directional. An interchange generally having more than one highway grade separation with direct or semidirect connections for the major left-turning movements. Figure 14.24 shows four of the most common types; Figures 14.24(a), (b), (d) show complete directional patterns; Figure 14.24(c) is a partial directional pattern with three loop ramps.

When there are three approach legs, T, Y, or trumpet interchanges are most commonly used. Figure 14.25(a) and 14.25(b) are examples of trumpet interchanges; Figure 14.25(c) and 14.25(d) illustrate Y interchanges, and Figure 14.25(e) and 14.25(f) are examples of T interchanges.

Interchange designs may vary greatly with physical controls (topography and culture), traffic patterns, and types of intersecting highways. A single route (or even an entire area) should, however, employ as few types as possible. Regardless of type, the exit terminal location, layout, and general appearance should be consistent. Except for special cases, complete interchanges usually should be constructed; partial ramp complements tend to confuse drivers and are frequently deficient in service.

Spacing. Interchanges should be located as needed to discharge and receive local traffic effectively. The spacing of major arterial crossroads usually governs and may range from less than one mile to many miles. Proper circulation of traffic between the

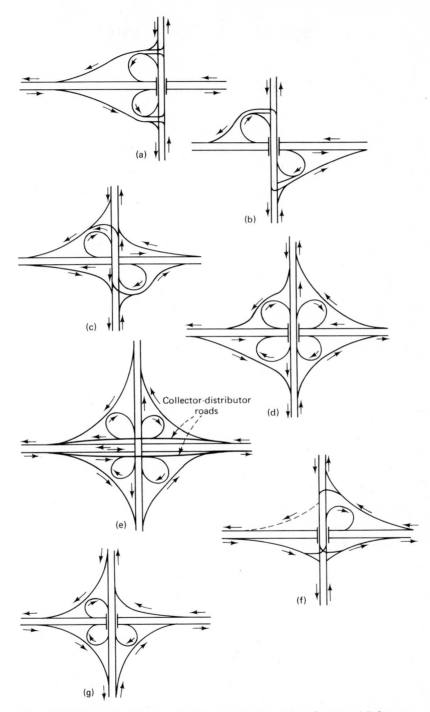

Figure 14.22. Sketches depicting cloverleaf interchange designs. (Source: *A Policy on Design of Urban Highways and Arterial Streets*, p. 594.)

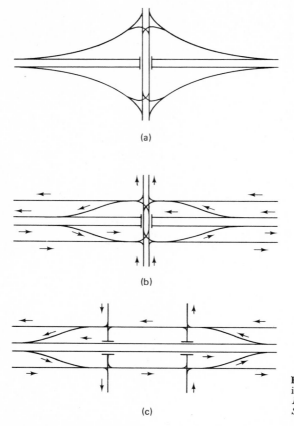

(a)

(b)

(c)

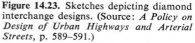

Figure 14.23. Sketches depicting diamond interchange designs. (Source: *A Policy on Design of Urban Highways and Arterial Streets*, p. 589–591.)

interchange crossroads, the local streets, and freeway systems must be assured. Interchange traffic may otherwise be concentrated at one or more crossroads in such volumes that serious disruption of traffic on both the local roads and the freeway result.

To provide optimum freeway operation with adequate weaving distance and sign placement, the average spacing of urban interchanges should be not less than 2 mi, in suburban areas not less than 4 mi, and in rural sections not less than 8 mi. However, individual spacings of adjacent interchanges may vary considerably. In urban and suburban areas the minimum distance between adjacent interchanges should desirably not be less than 1 mi, and never less than $\frac{1}{2}$ mi. In rural areas interchanges should be spaced not less than 3 mi.

Ramps. The term *ramp* includes all types, arrangements, and sizes of turning roadways that connect two or more legs at an interchange. The components of a ramp include a terminal at each leg and a central connecting roadway.

Ramps can be classed broadly as one of the five types shown in Figure 14.26. Each ramp generally is a one-way roadway. Although shown as a continuous curve in Figure 14.26(a), a diagonal ramp may be largely tangent or wishbone shaped, depending on the angle of intercepting roadways. A diamond type interchange gen-

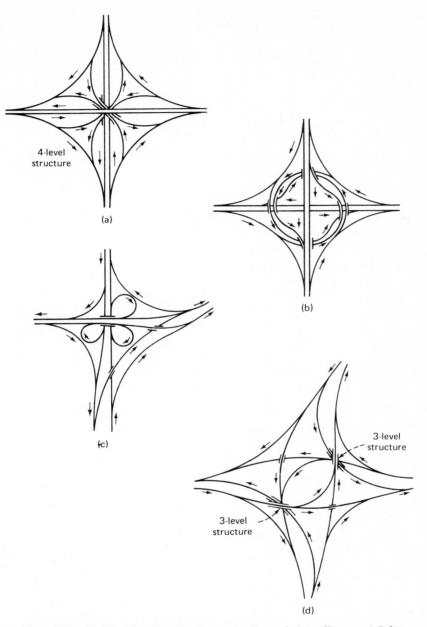

Figure 14.24. Sketches depicting directional interchange designs. (Source: *A Policy on Design of Urban Highways and Arterial Streets*, p. 596–600.)

erally has four diagonal ramps. With the loop pattern shown in Figure 14.26(b), the left turning movement is made without an at-grade crossing of the opposing through traffic. Instead, drivers making a left turn travel beyond the highway separation,

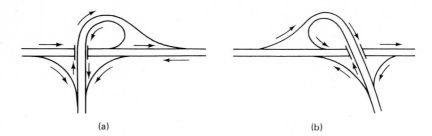

(a)   (b)

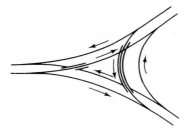 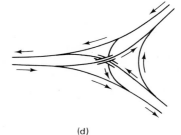

(c)   (d)

(e)   (f)

Figure 14.25. Sketches depicting three-leg interchange designs. (Source: *A Policy on Design of Urban Highways and Arterial Streets*, p. 587.)

swing right, and turn through approximately 270° to enter the other highway. With a semidirect connection [Figure 14.26(d)] drivers making a left turn first swing away from the intended direction, gradually reverse, and then follow directly around and enter the other road on the right (see solid line). In another semidirect connection, [Figure 14.26(d)] drivers exit to the left, loop around to the left, then reverse to enter on the right (see broken line). This is less desirable because of the left-hand off ramp. In the direct-connection left-turn movement, Figure 14.26(e), drivers leave on the left, turn directly toward the left, and enter on the left. Diagonal and outer connections without reverse alinement are direct connections for right turn movements.

Each ramp type can have a different shape according to traffic pattern, traffic volume, design speed, topography, culture, intersection angle, and type of ramp terminal. Several forms may be used for the loop and outer connection of a cloverleaf, as shown in Figures 14.27(a) and 14.27(b). The loop, except for its terminals, may

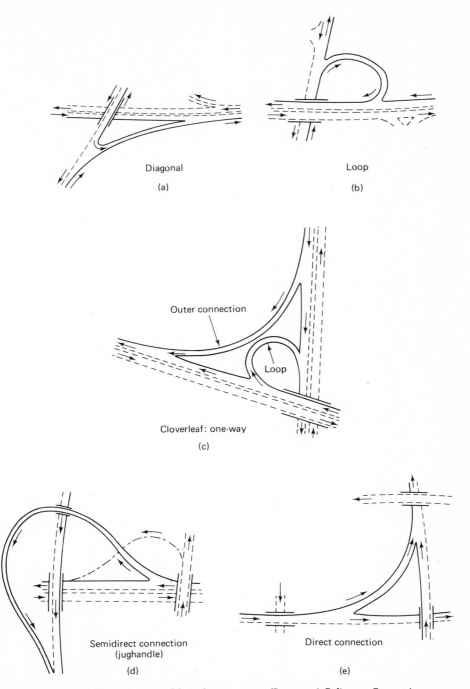

Diagonal
(a)

Loop
(b)

Outer connection

Loop

Cloverleaf: one-way
(c)

Semidirect connection
(jughandle)
(d)

Direct connection
(e)

Figure 14.26. General types of interchange ramps. (Source: *A Policy on Geometric Rural Highways*, p. 527.)

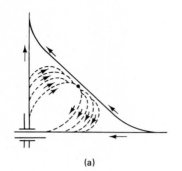

(a)

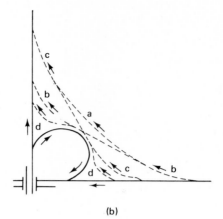

(b)

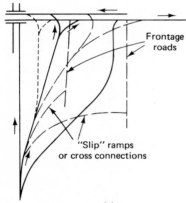

Frontage roads

"Slip" ramps or cross connections

(c)

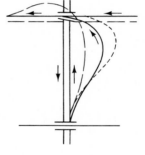

(d)

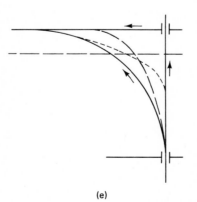

(e)

Figure 14.27. Ramp shapes. (Source: *A Policy on Geometric Design on Rural Highways*, p. 533.)

be a circular arc or some other symmetrical shape or nearly symmetrical form, as indicated by the group of dashed lines in Figure 14.27(a); or its shape may be asymmetric, a combination of any two dashed lines on opposite sides of the central common point in the diagram.

A desirable alinement for an outer connection is a continuous curve, as shown at *a* in Figure 14.27(b). This arrangement, however, may involve questionably extensive right of way. Another desirable arrangement has a central tangent and terminal curves, lines *b* and *c*. There are numerous choices for the angle and alinement of the outer connection. When the loop is more important than the outer connection, reverse alinement on the outer connection may be used in order to reduce the area of right of way as shown by line *d*. Any combination of *b*, *c*, and *d* may also be used.

Ramps forming a diamond interchange may assume a variety of shapes, depending on the pattern of turning traffic and right-of-way limitations. As shown by solid lines in Figure 14.27(c), the ramp may be a diagonal tangent with connecting curves. To favor a right turning movement, the ramp may be a continuous curve to the right with a spur to the left for left turns. On restricted right of way along the major highway, reverse alinements (with a portion of the ramp parallel to the through roadway) may be required, as shown by the short dashed line.

Diamond ramps of a type sometimes called *slip ramps* or *cross connections*, shown dashed in Figure 14.27(c), may also connect with parallel frontage roads. When either is employed, safety dictates one-way frontage roads. Cross connections to two-way frontage roads introduce the possibility of wrong-way entry onto the freeway and require extensive channelization to prevent it as well as to provide for proper operation for traffic leaving the freeway.

The shape of a semidirect connection, Figure 14.27(d), may be influenced by the extent of lateral separation of the one-way through pavements, the location of the terminals with respect to the structures, and the extent to which the structure pavements are widened or flared. The curve radii necessary to maintain a desired turning speed for an important left-turning movement sometimes determine the required lateral separation of the through roadway. When the semidirect connection leaves on the right and enters on the right, as dashed line in Figure 14.27(d), the through lanes can be closely spaced. The same applies to direct connections, Figure. 14.27(e).

It is rarely feasible to provide ramps on which turning traffic can travel in the same range of speeds as on the through roads. Nevertheless, there should not be a great difference between the design speed of the highway and the design speed of the connecting ramp. Table 14.16 shows minimum and desirable guide values for ramp design speeds as related to highway design speed.

TABLE 14.16
Guide Values for Ramp Design Speed as Related to Highway Design Speed

Highway Design Speed, mph (kph)	30(48)	40(64)	50(80)	60(97)	65(105)	70(113)	75(121)	80(129)
Damp design speed, mph (kph)								
Desirable	25(40)	35(56)	45(72)	50(80)	55(89)	60(97)	60(97)	65(105)
Minimum	15(24)	20(32)	25(40)	30(48)	30(48)	30(48)	35(56)	40(64)
Corresponding minimum radius, ft(m)								
Desirable		150(46)	300(91)	550(168)	690(210)	840(256)	1040(316)	1040(316)	1260(383)
Minimum		50(15)	90(27)	150(46)	230(70)	230(70)	230(70)	300(91)	430(131)

Direct connections should be designed for the desirable design speed as a minimum. The design speed of loops, on the other hand, usually must be near the minimum.

The minimum stopping sight distance values summarized in Table 14.4 apply directly to ramps. Longer sight distances should be considered whenever feasible. Table 14.7, showing K values for determining lengths of vertical curves in relation to the algebraic differences in grade, and Figure 14.4, showing required lateral clearances to objects in relation to radii of horizontal curves, are based on minimum and desirable sight distances. They apply to ramp design; usually, it is desirable to employ scaled drawings in lieu of these figures to check sight distances at vertical and horizontal curves.

Ramp grades should be as flat as possible in order to minimize driving effort in maneuvering from one road to another. The maximum grades shown in Table 14.8 for arterial highways other than freeways for various design speeds are generally applicable to ramps, but a precise relation has not been established.

Table 14.17 shows design widths of pavements for ramps. Desirably, on high-type ramps an 8 to 10 ft shoulder is provided on the right and 6 ft on the left along the length of the ramps. As a minimum, 6 ft should be provided on the right and 4 ft on the left. Curbs can be provided along the edge of ramp proper or on the outside of the shoulder. When they are provided on the inside of the shoulder, they should be the mountable type.

Cross slope on ramps should be the same as for through roadways, as given in Table 14.11. Superelevation on ramps should be the same as for through roadways; see Table 14.10 values for a maximum e value of 0.08. Higher maximum rates should be considered. A common direction of cross slope on a ramp should be provided both on tangent and superelevated sections.

The terminal of a ramp is that portion adjacent to the through traveled way, including speed-change lane, taper, approach nose, merging end, and island. Ramp terminals may be at-grade intersections, as at diamond or partial cloverleaf interchanges, or directional, where ramp traffic merges with or diverges from through traffic at flat angles.

Deceleration and acceleration lanes can be either the straight line taper or the parallel type. The straight line taper is most widely used. Figure 14.28(a) illustrates a straight line taper or direct type of deceleration lane. In the design of deceleration lanes of this type, the terminal edge departs from the through lane edge abruptly at an angle of approximately 3° or 5°. The taper for acceleration lanes is much flatter, as shown in Figure 14.28(b). A 50:1 to 70:1 rate of convergence between the outer edge of the acceleration lane and the freeway lane provides reasonable acceleration length and prescribes a proper path for an entering vehicle. As shown in Figure 14.28(a), the deceleration approach nose should be offset from the through lanes at least 12 ft, and the gore area between the through lane and the ramp should be surfaced. The surfaced area should be tapered from the nose to the edge of the traffic lane downstream from the nose. The nose should also be offset from the traveled way of the ramp a distance equal to the width of shoulder of the ramp.

Figure 14.28(b) illustrates an entrance terminal. Near the merging point the roadway entrance pavement should be aimed nearly parallel to the through roadway. The physical nose should be offset the full shoulder width from the edge of through lanes and the left shoulder width from the edge of the ramp.

TABLE 14.17
Design Widths of Pavements for Turning Roadways

R	Case I			Case II			Case III		
	1-Lane, One-way Operation—No Provision for Passing			1-Lane, One-way Operation—With Provision for Passing a Stalled Vehicle			2-Lane Operation —Either One-way or Two-way		
Radius on Inner Edge of Pavement, ft	Design Traffic Condition								
	A	B	C	A	B	C	A	B	C
50	18	18	23	23	25	29	31	25	42
75	16	17	19	21	23	27	29	33	37
100	15	16	18	20	22	25	28	31	35
150	14	16	17	19	21	24	27	30	33
200	13	16	16	19	21	23	27	29	31
300	13	15	16	18	20	22	26	28	30
400	13	15	16	18	20	22	26	28	29
500	12	15	15	18	20	22	26	28	29
Tangent	12	15	15	17	19	21	25	27	27

Pavement width in feet for:

Width Modification Regarding Edge of Pavement Treatment:

No stabilized shoulder	None	None	None
Mountable curb	None	None	None
Barrier curb: one side two sides	add 1' add 2'	None add 1'	add 1' add 2'
Stabilized shoulder, one or both sides	None	Deduct shoulder width; minimum pavement width as under Case I	Deduct 2' where shoulder is 4' or wider

Traffic Condition A—Predominantly P vehicles, but some consideration for SU trucks.
Traffic Condition B—Sufficient SU vehicles to govern design, but some consideration for semitrailer vehicles.
Traffic Condition C—Sufficient semitrailer, WB-40 or WB-50 vehicles to govern design.

Source: *A Policy on Geometric Design of Rural Highways*, p. 338.

Design of two-lane exit and entrance ramps is generally similar to single lane exit and entrance ramps, except for their additional width and length of speed-change lanes. These usually are special designs that entail a length of full added lane. For more complete information, see cited reference.[26]

Distance between successive ramp terminals. Urban freeways to serve the numerous traffic generators along a route may result in frequent ramp terminals in close suc-

---

[26] *Two-Lane Entrance Ramps,* ITE Informational Report (Washington, D.C.: Institute of Traffic Engineers, 1968).

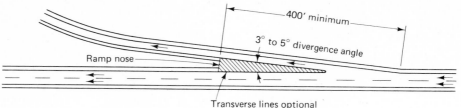

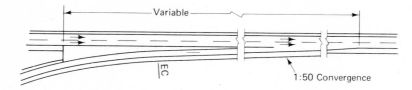

Figure 14.28. Designs for taper-type, single-lane exits and entrances.

cession. In order to provide sufficient maneuvering length and adequate space for signing, a reasonable distance is required between successive terminals. The minimum distances required for satisfactory freeway operation are 1,000 ft between successive exits on a freeway and 800 ft between a freeway exit and an exit on a collector-distributor road or a split in the ramp.

When the distance between an entrance ramp followed by an exit ramp is less than 2,000 ft, the speed-change lanes should be connected. When terminals are farther than 2,000 ft apart, a continuous operational lane should be provided wherever the weaving computations indicate the need (see Chapter 8).

When an exit ramp is followed by an entrance ramp, there should be 500 ft between the terminals so that drivers are not confronted with merging vehicles too soon after passing an exit area.

Control of access. Access should be controlled throughout all portions of interchanges, just as for the highway between the interchanges. When freeways interchange with noncontrolled access highways, it is desirable to extend the control of access at least 100 ft along the crossroad beyond the ramp terminal in urban areas and at least 300 ft in rural areas.

PROVISION FOR MASS TRANSIT

Freeways are sometimes combined effectively with mass transit facilities in large cities. Selected ramps and lanes can be reserved for buses; a rail transit line can be incorporated into a wide median.

Many metropolitan areas have freeway express buses. Almost all operate nonstop from suburban pickup points near the freeway into the CBD. Others utilize special bus stops at intersecting streets along the freeway where passengers transfer to or from other lines or automobiles. Freeway bus stops should be located only where site conditions are favorable and, if possible, where acceleration lanes are flat or down-grade. Bus stops may be provided at the freeway level for which stairs, ramps, or escalators are necessary, or they may be located at the street level that buses reach by way of interchange ramps.

Bus turnouts should be designed so that deceleration, standing, and acceleration is effected clear of and separated from the through traffic lanes. Speed-change lanes should be long enough to enable buses to leave and enter through traffic lanes at approximately the average running speed of the highway. The length of acceleration lanes from bus turnouts should be well above the normal minimum values because buses start from a standing position. Normal length deceleration lanes are suitable. Speed-change lanes, including the shoulder, should be 20 ft wide to permit passing a stalled bus. The dividing area between the outer edge of the freeway shoulder and the edge of the bus lane should be as wide as possible, preferably 20 ft or more, with 4 ft as an absolute minimum. Pedestrian loading platforms should not be less than 5 ft wide and should preferably be from 6 to 10 ft wide. Figures 14.29 and 14.30 show various arrangements for bus stop locations at cloverleaf and diamond-type inter-changes. Additional arrangements are shown in the cited reference.[27]

Rail transit can be located in the median, on either side, on top, or below the freeway. The most common arrangement is to place the transit line within the median of a depressed or ground-level freeway. When it is so located, a minimum median width of 64 ft is required between the two traveled ways, and when stations are also located in the median, a minimum width of 80 ft is usually required.

LOCAL ROADS AND STREETS

The design standards for local roads and streets vary considerably depending on the type of area served (subdivision, rural, commercial, industrial, etc.), traffic vol-umes, terrain, and the governmental body responsible for their design.

CROSS SECTION DESIGN

Typical cross sections for rural and urban local roads and streets are shown in Figure 14.31. In rural areas a shoulder is normally included in the cross section. A border area for errant vehicle recovery is desirable. It should have slopes desirably 4:1 or flatter and be clear of obstructions. In urban areas there may be sections with shoulders, but more commonly a plain curb (or curb and gutter) is included in the cross section at the edge of the outer parking lane. The curb should have a vertical or roll-type face and be not more than 6 in. high. The gutter usually is from 1 to 2 ft wide and can be separate or integral with the curb.

Sidewalks should be provided along streets used for pedestrian access to schools, parks, shopping areas, and transit stops. Minimum sidewalk widths should be 4 ft; widths of 8 ft or greater are required in commercial areas. Sidewalks may be located next to the curb, but, desirably, they should be at least 5 ft and preferably from 12 to 15 ft from the edge of curb. Borders should be made as wide as conditions permit.

---

[27] *Bus Stops for Freeway Operations* (Washington, D.C.: Institute of Traffic Engineers, 1971).

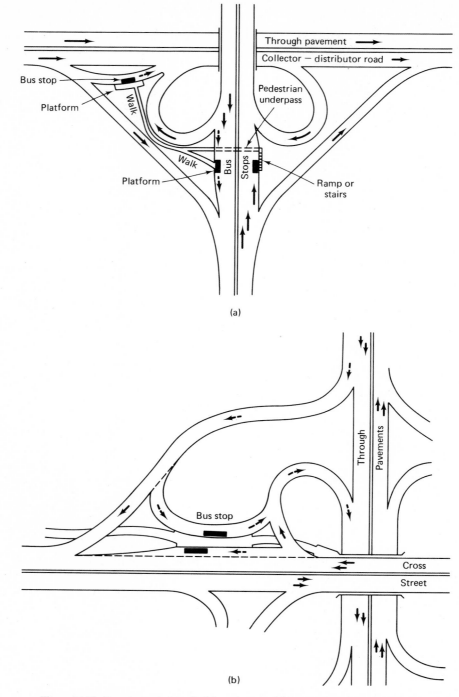

Figure 14.29. Bus stops at cloverleaf interchanges. (a) Bus stop at freeway level. (b) Bus stop at street level. (Source: *Bus Stops for Freeway Operations* (Washington D.C.: Institute of Traffic Engineers, 1971) p. 6.]

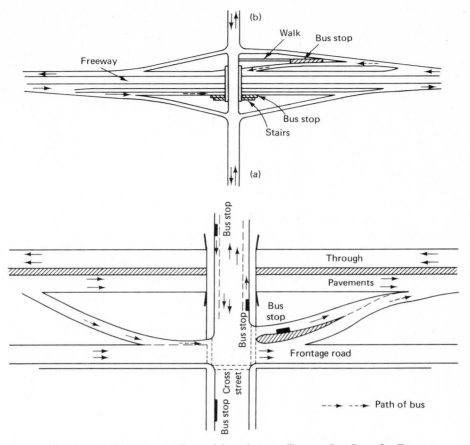

Figure 14.30. Bus stops at diamond interchanges. (Source: *Bus Stops for Freeway Operations* (Washington, D.C.: Institute of Traffic Engineers, 1971) p. 7.]

CRITERIA APPLICABLE TO SUBDIVISION STREETS

The primary objective of subdivision design is to provide maximum liveability. Driving convenience is secondary. Street alinement should fit topography closely enough to minimize the need for cuts or fills and at the same time to discourage high-speed through traffic. See cited reference.[28]

DRIVEWAYS

Typical residential driveways should generally be designed according to the guidelines shown in Table 14.18. Figure 14.32 shows a minimum design for a driveway along a residential street. Industrial driveways should generally be designed according

---

[28] *Recommended Practices for Subdivision Streets* (Washington, D.C.: Institute of Traffic Engineers, 1967).

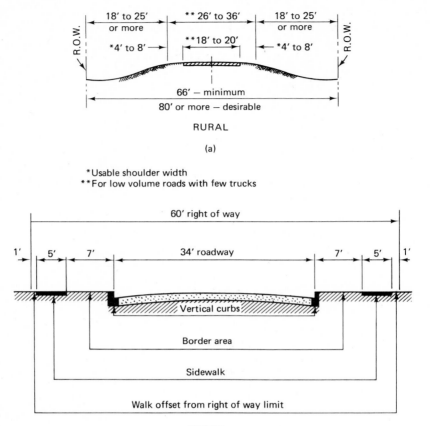

*Usable shoulder width
**For low volume roads with few trucks

Figure 14.31. Typical cross section of local roads and streets. [Sources: (a) *A Policy on Geometric Design for Rural Highways*, p. 263. (b) *Recommended Practices for Subdivision Streets* (Washington, D.C.: Institute of Traffic Engineers, 1967), p. 2.]

to the guidelines shown in Table 14.19. For additional details of driveway design, such as maximum grades, maximum change in grades, sight distance at driveways, and design of curbs in conjunction with spacing, see cited references.[29, 30]

BIKEWAYS[31]

The growing use of bicycles has led to the development of preliminary design standards for the three main classes of bikeway:

[29] *An Informational Guide for Preparing Private Driveway Regulations for Major Highways* (Washington, D.C.: American Association of State Highway Officials, 1960).

[30] *Guidelines for Driveway Design and Location: Tentative Recommended Practice* (Washington, D.C.: Institute of Traffic Engineers, 1973).

[31] The material in this section is summarized from *Bikeway Planning Criteria and Guidelines* (Los Angeles: Institute of Transportation and Traffic Engineering, University of California, 1972) pp. 19–140.

Class I: A completely separated right of way designated for the exclusive use of bicycles

Class II: A restricted right of way designated for the exclusive or semiexclusive use of bicycles

Class III: A shared right of way designated as such by signs placed on vertical posts or stenciled on the pavement

TABLE 14.18
Recommended Basic Guidelines for Residential Driveways

| Type of Development Served | Urban | | | | Rural | |
| | High Pedestrian Activity* | | All Other† | | | |
	Major	Secondary	Major	Secondary	Major	Secondary
Width‡						
Minimum ft (m)	10(3)	10(3)	10(3)	10(3)	10(3)	10(3)
Maximum ft (m)	20(6)	20(6)	30(9)	30(9)	30(9)	30(9)
Right turn radius of flare¶						
Minimum ft (m)	5(1.5)	5(1.5)	5(1.5)	5(1.5)	10(3)	10(3)
Maximum ft (m)	10(3)	10(3)	15(5)	15(5)	25(8)	25(8)
Angle‖	75°	60°	45°	45°	45°	45°

*As in central business areas, or in same block with auditoriums, schools, libraries.
†The remaining city streets including neighborhood business, residential, industrial.
‡Measured along right-of-way line, at inner limit of curbed radius sweep, or between radius and rear edge of curbed island at least 50 sq ft in area.
¶On side of driveway exposed to entry or exit by right-turning vehicles.
‖Minimum acute angle measured from edge of pavement.

Source: *Guidelines for Driveway Design and Location: Tentative Recommended Practice* (Washington, D.C.: Institute of Traffic Engineers, 1973).

TABLE 14.19
Recommended Basic Guidelines for Industrial Driveways

| Type of Development Served | Urban | | | | Rural | |
| | High Pedestrian Activity* | | All Other† | | | |
	Major	Secondary	Major	Secondary	Major	Secondary
Width‡						
Minimum ft (m)	20(6)	20(6)	25(8)	25(8)	35(11)	35(11)
Maximum ft (m)	35(11)	35(11)	40(12)	40(12)	40(12)	40(12)
Right turn radius of flare¶						
Minimum ft (m)	10(3)	15(5)	15(5)	15(5)	25(8)	25(8)
Maximum ft (m)	15(5)	20(6)	25(8)	25(8)	50(15)	50(15)
Angle‖	75°	60°	45°	45°	45°	45°

*As in central business areas, or in same block with auditoriums, schools, libraries.
†The remaining city streets including neighborhood business, residential, industrial.
‡Measured along right-of-way line, at inner limit of curbed radius sweep or between radius and near edge of curbed island at least 50 sq ft in area.
¶On side of driveway exposed to entry or exit by right-turning vehicles.
‖Minimum acute angle measured from edge of pavement.

Source: *Guidelines for Driveway Design and Location: Tentative Recommended Practice.*

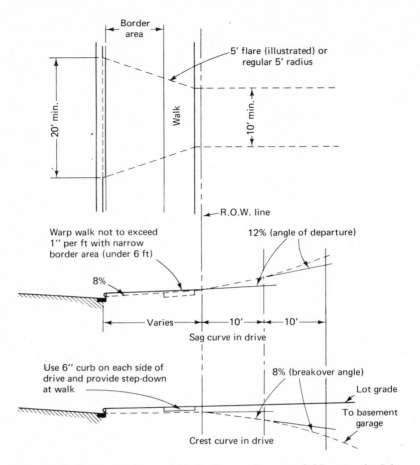

Figure 14.32. Residential driveway details. [Source: *Recommended Practices for Subdivision Streets* (1967), p. 8.]

Bikeway design controls and criteria include bicycle and cyclist dimensions, design speed, bikeway width and clearances, grade, radius of curvature, bikeway surface and drainage, and capacity.

BICYCLE AND CYCLIST DIMENSIONS

The mean bicycle and cyclist dimensions employed in European design are:

Handle bar width	1.96 ft (0.6 m)
Bicycle length	5.75 ft (1.75 m)
Pedal clearance	0.5 ft (0.15 m)
Vertical space occupied by cycle/cyclist	7.4 ft (2.5 m)

BIKEWAY DESIGN SPEED

Many factors, including the type of bicycle and gearing, pavement grade and surface, direction and magnitude of wind and air resistance, and physical condition of the cyclist, will influence bicycle operating speeds. Although bike riders have recorded speeds in excess of 30 mph (48 kph) most persons travel at about one-third of this rate.

For bikeway design purposes, a speed of 10 mph (16 kph) is a conservative value to use in setting criteria for minimum widths and radii of curvature on level bikeways. A more liberal standard of 15 mph (24 kph) has been specified in the USSR.[32]

Design speeds on upgrades will be influenced by the length and amount of grade as it influences energy expenditure as a function of desired speed and gear ratio. The increase of downgrade speeds will also be influenced by the length and amount of grade.

BIKEWAY WIDTHS, LOCATION, AND CLEARANCES

Bikeway width is one of the primary considerations in bikeway design. The minimum effective widths for Class I and Class II bikeways are shown in Table 14.20. The actual width selected will also be influenced by such factors as desired level of service, available space, and width of construction and maintenance equipment (paving material spreaders, sweeper brushes, snow plow blades, etc.).

TABLE 14.20
Minimum Effective Width for Class I and Class II
Bikeways as a Function of Number of Bikeway Lanes

Number of Lanes (One way)	Minimum Effective Width (ft)	
	German Specifications	Modified German Specifications Based Upon a Comfortable Maneuvering Allowance at a 10 mph Design Speed
1	3.3 (1 m)	3.3 (1 m)
2	5.3 (1.6 m)	6.4 (2 m)
3	8.5 (2.6 m)	10.9 (3.3 m)
4	11.8 (3.6 m)	15.3 (4.7 m)

Source: Adapted from *Bikeway Planning Criteria and Guidelines* (Los Angeles: University of California, 1972), p. 27.

Because there are serious problems in interfacing a two-way Class II bikeway with motorized and pedestrian traffic at urban intersections, two-way designs, in general, should be limited to Class I bikeways.

The minimum number of lanes provided on Class II bikeways depends on where the bikeway is located in relation to other street cross section elements. Various alternative schemes can be developed for incorporating a Class II or Class III bikeway on an existing street.

Among these alternate possibilities are combining a Class II bikeway with a sidewalk (if the bikeway is delineated by means of pavement markings and signing)

---

[32] V. F. BABKOV, ET AL., "The Road in Relation to Traffic Requirements, Question IV," *14th World Congress—Prague, 1971* (Paris: Permanent International Association of Road Congresses (IV-22)), p. 22.

or requiring pedestrians to share an unmarked sidewalk with a Class III bikeway. The feasibility of sharing a paved sidewalk (with or without markings and signing) varies directly with the width of the sidewalk and inversely with the volume of pedestrians and bicycles. In general, sidewalk-bikeway combinations should be avoided except in limited locations where sufficiently wide sidewalks exist in residential areas consisting of single-family dwellings. Almost all local ordinances prohibit bicycle traffic on sidewalks and, therefore, would have to be modified if sidewalk sharing is to take place.

Incorporating Class II or Class III bikeways with the roadway provides more alternatives than incorporating them with the sidewalk, but it also requires more careful planning and design because the conflicts with parked and moving vehicles pose more serious consequences to the cyclist than with standing or moving pedestrians. Parking density and turnover rate, motor vehicle volume in the outer lane, traffic composition and operating speed, and anticipated bicycle volume are major factors in determining the feasibility of separating the bicycles by varying degrees from the motor vehicles (as in a Class II bikeway) or sharing the roadway (as in a Class III bikeway).

Four alternatives for placing one-way Class II bikeways (which by definition have an exclusive or semiexclusive right of way) are shown in Figure 14.33. It should be noted that a symbolic barrier, e.g., striping, may be easily encroached either voluntarily or involuntarily by both bicycles and motor vehicles. Physical barriers, e.g., berms, median barriers, islands, fences, serve to delineate the edge of the bikeway right of way and minimize encroachments. Physical barriers are recommended for all bikeways that are immediately adjacent to traffic lanes.

Figure 14.34 shows alternative one-way Class III bikeway/roadway locations which have limited applications in urban areas.

Two lanes should be provided as a recommended minimum on a Class I bikeway in order to provide passing opportunities without leaving the bikeway. Whenever possible, one-way Class II bikeways should also be wide enough to provide passing maneuvers.

The minimum recommended horizontal clearances to vertical obstructions and other hazards are shown in Table 14.21. These values vary among countries, bikeway class, and situation encountered.

TABLE 14.21
Minimum Recommended Lateral Clearance From Edge of Bikeway
to Obstructions and Other Hazards

Description	Minimum Clearance
Horizontal clearance to obstructions	0.8–1.6 ft (0.2–0.5 m)
Class II bikeway at grade of sidewalk: clearance to curb drop-off	1.5–2.3 (0.5–0.7)
Class II bikeway at grade of roadway: clearance to raised curb	1.6 (0.5)
Shoulder clearance to edge of sloped drop-off (i.e., bikeway on an embankment, with less than 2:1 slope)	1.0 (0.3)
Soft shoulder: increase minimum width of bikeway by	1.6 (0.5)

Source: Derived from *Bikeway Planning Criteria and Guidelines*, p. 28.

Figure 14.33. Alternate one-way Class II bikeway location on roadway.

Alternative	Description	Pre-bikeway arrangement	Bikeway arrangement	Comments
IIA	Bicyclists have semi-exclusive right of way (no parking allowed or parking removed)	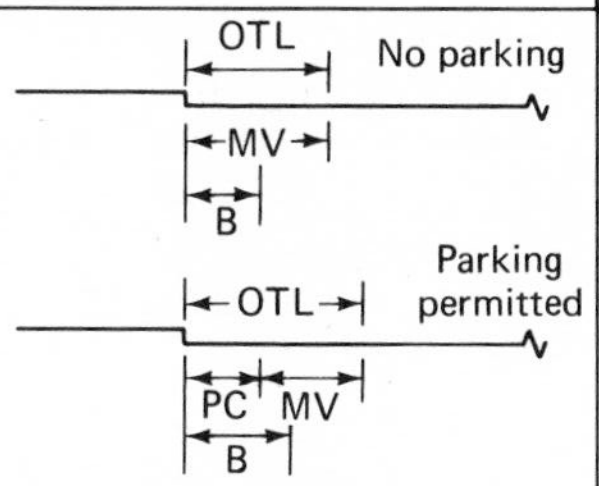 *(diagram: No parking / Parking permitted; OTL, MV, B, PC)*	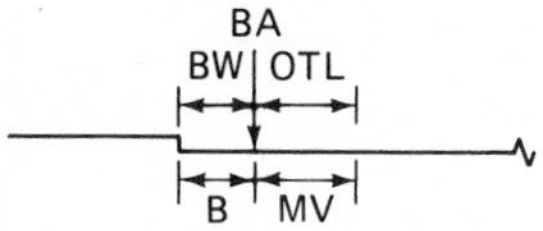 *(diagram: BA, BW, OTL, B, MV)*	Striping barrier not a positive means of separating bicycles and motor vehicles and local ordinances must be passed to prohibit use of the bikeway as a parking lane. A physical barrier (island, curb, berm, etc.) will minimize cross conflicts and is recommended. Elimination of curb parking may cause local problems and opposition.
IIB	Bikeway on roadway between curb and parked cars	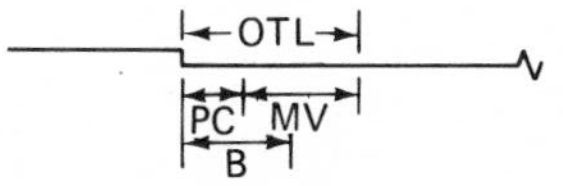 *(diagram: OTL, PC, MV, B)*	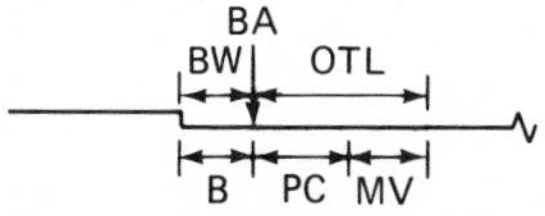 *(diagram: BA, BW, OTL, B, PC, MV)*	Parked vehicles eliminate moving vehicle bicycle conflicts. A curb barrier is recommended to keep parked cars from encroaching on bikeway. This alternative becomes ineffective with a high parking turnover rate which generates heavy pedestrian cross flow.
IIC	Parking permitted between bikeway and curb.	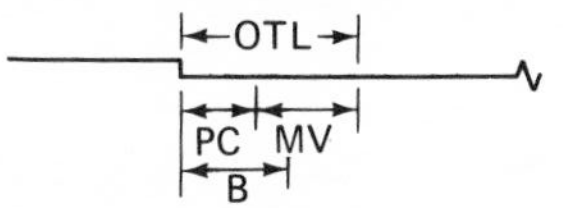 *(diagram: OTL, PC, MV, B)*	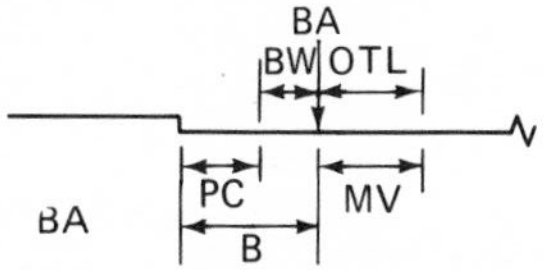 *(diagram: BA, BW, OTL, PC, B, MV)*	The bikeway cannot be protected by physical barriers because motor vehicles must have access and egress from the parking lane. Potential motor vehicle door opening conflicts with moving bicycles.
IID	Parking may be restricted during periods of peak bicycle flow	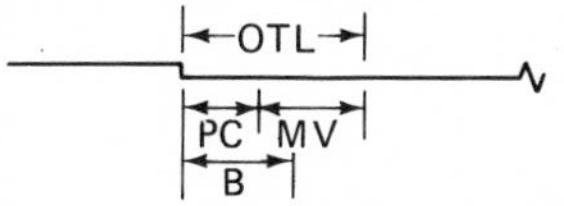 *(diagram: OTL, PC, MV, B)*	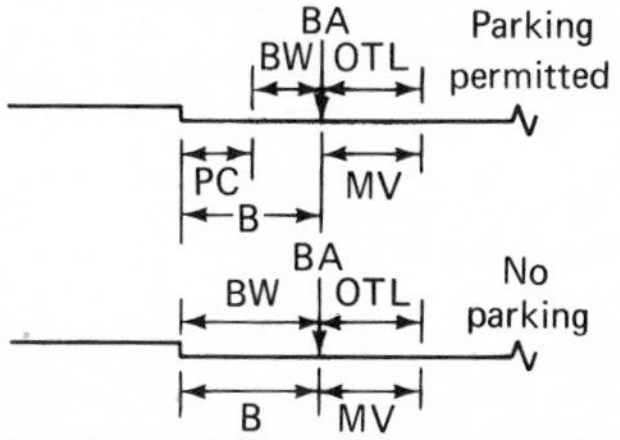 *(diagram: Parking permitted — BA, BW, OTL, PC, B, MV; No parking — BA, BW, OTL, B, MV)*	This alternative is most effective with parking restrictions in force during periods of peak bicycle flow thereby increasing the effective clearance to the outside traffic lane. Physical barriers cannot be used. High off-peak motor vehicle volumes coupled with heavy parking density and turnover can result in hazardous biking conditions.

Description of symbols

BW = Class II bikeway
B = Bicycle
PC = Parked car (static)
OTL = Outside traffic lane (parking may or may not be allowed next to curb)
MV = Motor vehicle (moving)
BA = Barrier

Figure 14.33. Alternate one-way Class II bikeway location on roadway. [Source: Adapted from *Bikeway Planning Criteria and Guidelines* (Los Angeles: University of California, 1972), pp. 68–74.]

Alternative	Description	Pre-bikeway arrangement	Bikeway arrangement	Comments
III A	Bicyclists share right of way with parked and moving vehicles	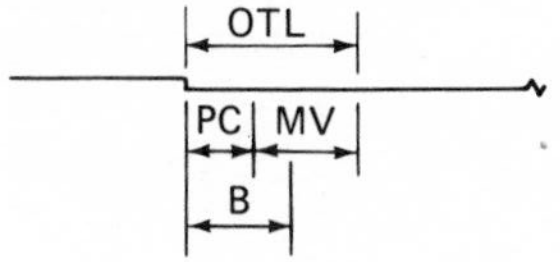 OTL, PC|MV, B	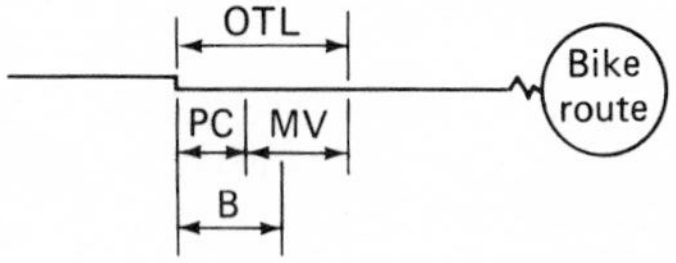 OTL, PC|MV, B — Bike route	Provision of signs and stencilled pavement messages provide only marginal protection to cyclists. Potential conflicts from doors of parked vehicles on one side and moving vehicles on the other side make this a viable alternative only with low-bicycle and motor vehicle volumes, low parking turnover, and low vehicle speeds in the outer lane.
III B	Bicyclists share right of way with moving vehicles (no parking allowed or parking removed) MV	No parking: OTL, MV, B / Parking permitted: OTL, PC|MV, B	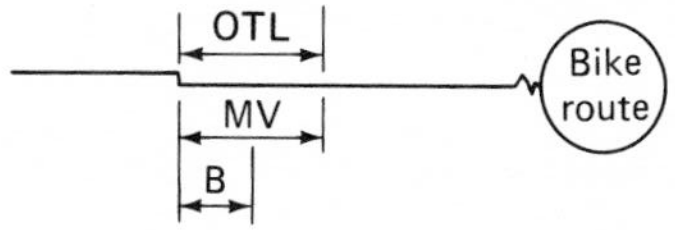 OTL, MV, B — Bike route	This alternative is essentially similar to the above alternative with the exception that parking is eliminated to afford space for bicycle traffic–without specifically designating a bicycle lane. Elimination of curb parking may cause local opposition and problems.

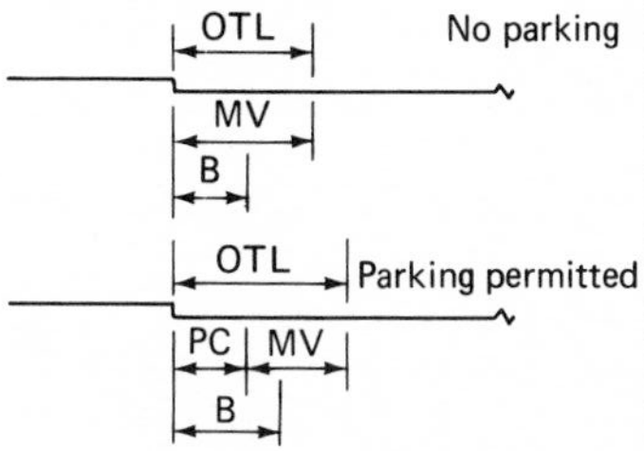

Description of symbols
B   = Bicycle
PC = Parked car (static)
OTL = Outside traffic lane (parking may or may not be allowed next to curb)
MV = Motor vehicle (moving)
BA = Barrier

Figure 14.34. Alternate one-way Class III bikeway location on roadway. [Source: Adapted from *Bikeway Planning Criteria and Guidelines*, pp. 62–67.)

TABLE 14.22
Minimum Space for Class II and Class III Bikeway Alternatives
(Given for One Side of Street Only)

Note: Dimensions in Parentheses are Based on Liberal Maneuvering Allowance

Bikeway Alternative (See Figures 14-33 and 14-34)	Minimum Bikeway Width in ft		Recommended Minimum in ft Space Required	Remarks
	Effective	Actual		
II A	3.3′	4.1′	4.1′ bikeway, outside traffic lane should meet suggested widths for functional classification	Painted strips with/without pavement markers
	5.3′ (6.4′)	6.8′ (7.9′)	7.5′ (8.6′) including curb barrier, outside traffic lane should provide 1.0′ minimum clearance	Curb barrier, 2-lane minimum to allow for passing cyclists
II B	5.3 (6.4′)	6.8′ (7.9′)	7.51′ (8.6′)	Curb barrier, no door-opening allowance given
II C	3.3′	3.3′ to 5.3′	13.3′ from curb to outer edge of bikeway 11.3′ from curb to outside stripe of bikeway	Medium to high parking density Low parking density and turnover
II D	3.3′	5.3′ to 13.3′	13.3′ from curb to outer edge of bikeway 11.3′ from curb to outer edge of bikeway	Medium to high parking density off-peak, low turnover Low parking density off-peak
III A			14.1′ for outside traffic lane	Low parking density, through motor vehicle traffic restricted
			22.1′ for outside traffic lane	Medium to high density parking
III B			14.1′ for outside traffic lane	Low motor vehicle volume and speed

Source: Derived from *Bikeway Planning Criteria and Guidelines*, p. 79.

Minimum space requirements for Class II and Class III bikeways are summarized in Table 14.22.

GRADES

The physical characteristics of the cylist and the bicycle, wind velocity, air resistance, and road surface are the major factors that determine maximum acceptable bikeway grades and the optimum lengths of these grades. Although additional research is still necessary, the current standards from three countries are shown in Table 14.23. In the design of Class I bikeways, the introduction of horizontal sections, rest stops, or low-grade switch backs may be desired when maximum lengths are inconsistent with maximum grades. In considering grade design, attention must be directed toward downgrade as well as upgrade cyclists.

TABLE 14.23
Grade and Grade Length Criteria

| Bikeway Gradient Percent | Desirable Length* ft (m) | Netherlands | | Denmark | India |
		Normal Length† ft (m)	Maximum Length‡ ft (m)	Maximum Length‡ ft (m)	Maximum Length‡ ft (m)
10	Not recommended	33 (10)	66 (20)	—	—
5	Not recommended	131 (40)	262 (80)	164 (50)	66 (20)
4.5	82 (25)	167 (51)	334 (102)	328 (100)	—
4	102 (31)	203 (62)	410 (125)	656 (200)	164 (50)
3.5	148 (45)	295 (90)	590 (180)	984 (300)	—
3.3	148 (45)	295 (90)	590 (180)	—	Unrestricted
2.9	200 (61)	400 (122)	800 (244)	1,640 (500)	—
2.5	262 (80)	525 (160)	1,050 (320)	—	—
2	410 (125)	820 (250)	1,640 (500)	—	—
1.7	590 (180)	1,180 (360)	—	—	—
1.4	—	1,610 (490)	—	—	—
1.3	—	2,100 (640)	—	—	—

*"Desirable" lengths include consideration of possible high wind conditions.
†"Normal" lengths represent judged acceptable gradient lengths.
‡"Maximum" recommended lengths of grade should not be exceeded.

Source: Koninklijke Nederlandsche Toeristenbond, *Fietspaden en oversteekplaatsen* (*Bicycle Paths and Cycle-Crossings*), Verkeersmemorandum No. 4 (*Trafic Memorandum No. 4*), Van de Verkeersafdeling van de ANWB (Traffic Division of the Algemene Nederlandse Wielrijders Bond), 2nd rev. ed., Amsterdam, Holland, December, 1967, p. 44. Ernst Renstrup, Assistant Chief Engineer, Vejdirektoratet, Copenhagen, Denmark. B. H. Subbarju *et al.*, "Urban Road Network, Question V," *14th World Congress—Prague, 1971*, Permanent International Association of Road Congresses (Paris, France), (V-11), pp. 1–9. As quoted in *Bikeway Planning Criteria and Guidelines*, p. 30.

RADIUS OF CURVATURE

Although a wide range of horizontal curve radii may be found on existing bikeways, the following empirical relationship between radius of curvature and bicycle velocity has been developed:

$$R = 1.25 \, V + 1.4 \tag{14.7}$$

where R = the unbraked radius of curvature in feet,
  V = bicycle design speed in mph.
If bicycle speed is in kph, the unbraked radius in meters is

$$R = 0.238 \, V + 0.41 \tag{14.8}$$

From the equation, for a Class I bikeway design speed of 10 mph (16.1 kph), the comfortable unbraked radius of curvature is 13.9 ft (4.2 m).

Since Class II bikeways in urban areas generally follow the alinement of existing streets, if the horizontal curvature can accommodate motor vehicles, it should be more than adequate for bicycles. Horizontal curves on downgrades should be checked because bike speeds easily and frequently exceed 10 mph (16.1 kph). If the expected or actual bicycle speeds exceed the available radius of curvature, traffic control devices should be employed to warn cyclists to regulate their speed.

BIKEWAY CAPACITY

A review of the available international literature has disclosed conflicting estimates for the capacities of single- and multiple-lane bikeways. Reported ranges in hourly bikeway capacity for one- and two-way movements are indicated in Table 14.24. Because the capacity of a bikeway is highly sensitive to climatic conditions, grades, proximity of barriers, and downstream bottlenecks (such as intersections), the capacity estimates shown in Table 14.24 should be considered as approximate ideal upper limits and should not be used for design purposes. Level of service considerations rather than capacity should be primary in deciding whether or not a multilane bikeway is required because the ideal capacity of even a one-lane bikeway is sufficient to accommodate all but the most extreme future demands.

TABLE 14.24
Range of Reported One- and Two-way Capacity of Bikeways
as a Function of Number of Lanes

Traffic Direction	Number of Lanes	Range of Estimated Capacity (Bicycles/Hour)
One-way	1	1,700– 2,530
One-way	2	2,000– 5,000
One-way	3	3,500– 5,000+
Two-way	1	850– 1,000
Two-way	2	500– 2,000
Two-way	3	1,700– 5,000
Two-way	4	4,000–10,000

Source: Adapted from *Bikeway Planning Criteria and Guidelines*, p. 37.

REFERENCES FOR FURTHER READING

A Design Guide for Local Roads and Streets (1970)
A Policy on Arterial Highways in Urban Areas (1972)
A Policy on Geometric Design of Rural Highways (1965)
Policy on Design of Urban Highways and Arterial Streets, 1973
Geomeric Design Standards for Highways Other Than Freeways (1969)
Guide for Bicycle Routes
Highway Design and Operational Practices Related to Highway Safety (1967)
American Association of State Highway Officials
341 National Press Building
Washington, D.C.

Traffic Control and Roadway Elements—Their Relationship to Highway Safety (1971)
Highway Users Federation for Safety and Mobility
Washington, D.C.

Recommended Practices for Subdivision Streets (1965)
Tentative Recommended Practice: *Guidelines for Driveway Design and Location* (1973)
Institute of Traffic Engineers

The Freeway in the City (1968)
U.S. Government Printing Office
Washington, D.C.

A Survey of Urban Arterial Design Standards (1969)
American Public Works Association
1313 East 60th Street
Chicago, Illinois

Designing Operational Flexibility into Urban Freeways (1963)
by J. R. Leisch
Proceedings
Institute of Traffic Engineers

"New Design Concepts for Urban Freeways and Rural Expressways" (1969)
by D. W. Loutzenheiser
International Road Federation Road Seminar, pp. 4–19
International Road Federation
Washington, D.C.

1972 World Survey of Current Research and Development on Roads and Road Transport, U.S. Department of Transportation, Federal Highway Administration (Washington, D.C., 1972), p. 321.

Chapter 15

PARKING, LOADING, AND TERMINAL FACILITIES

JAMES M. HUNNICUTT. Partner, Hunnicutt and Neale, Washington, D.C.

The provision of adequate parking is an essential element in the transportation system in the cities of the world today. This chapter covers the provision and operation of parking facilities, both on-street and off-street, and either in surface parking lots or in parking structures. The location, design, and financing of parking facilities is covered as well as parking related to zoning and to special purpose land uses. Off-street truck loading facilities are discussed as are transit terminal facilities.

DEFINITIONS

The following terms often appear in a discussion of parking.

Parking Supply. The number of legal parking spaces available in a given area.

Parking Inventory. The number of parking spaces available in a given area categorized by curb or off-street spaces, public or private use, or by other classifications.

Private Parking Supply. Parking spaces provided for employees or customers of a business and not available to the general public.

Public Parking Supply. Parking spaces available to the general public either free of charge or for a fee.

Parking Demand. The number of drivers desiring to park in a given area during a specified time period—often expressed as the number during the peak hour of the day.

Short-term Demand. Parking demand with a duration of less than three to four hours.

Long-term Demand. Parking demand with a duration exceeding three to four hours.

Parking Surplus. The extent to which the parking supply exceeds the demand of spaces.

Parking Deficiency. The extent to which the parking demand exceeds the supply of spaces.

Parking Accumulation. The total number of cars parked in a given area at a given time.

Space-hour. A single parking space occupied for one hour by a vehicle.

Parking Duration. The length of time a given vehicle remains in a specific space.

Turnover. The number of different vehicles that park in a given space during an average day.

Occupancy. The portion of time a vehicle is parked in a given space during the day.

Walking Distance. Distance on a normal walking path with crossings at intersections from the driver's parking space to the nearest door of his destination.

Parking Fee. An amount paid for parking for a specific period of time.

Blue Zone. An area in which parking is permitted. Use of these zones is common in Europe and they are designated by a circular blue sign with a red border.

Trip Purpose. The reason a person goes somewhere. Categories of trip purpose include work, shop, business, social-recreational, sales-service, and miscellaneous.

PARKING CHARACTERISTICS

Specific parking characteristics refer to either the amount of parking provided or to the manner in which that parking is used. Parking characteristics normally are influenced by the size of the city and by other factors pertinent to the city such as relative use of other available transportation modes and the size and relative importance of the CBD (central business district).

As discussed in detail in Chapter 4, General Traffic Characteristics, typical parking characteristics include information on the supply and utilization of parking facilities, the distance walked by parkers, and the accumulation of parked vehicles in an area by time of day.

PARKING METERS

A parking meter is a mechanical time-measuring device that continually indicates the available time remaining for a parked vehicle.[1] The parking meter was developed in 1935 and, in proper application, can greatly simplify the problem of enforcing parking regulations. In addition, parking turnover is increased.

The two general kinds of meters are the manual and the automatic meter. The manual parking meter is one in which the parker inserts a coin and then turns a handle that winds the clock and actuates the meter for a time period determined by the coin inserted and the duration that the meter allows. In the automatic parking meter, a coin is inserted and the time automatically registers for that coin. However, the clock mechanism of the automatic meter must be wound periodically by maintenance personnel. In practical use, the two meters are interchangeable, with the same time limits, choice of coins, etc. available for the two kinds.

In addition to the two basic meters, refinements are available for special applications. Within the meter, there are two coin boxes available: the open coin box and the sealed box. The open coin box can be emptied directly into the collection unit and replaced in the meter. The sealed box is removed from the meter, placed in a tray, and a new box placed in the meter. The sealed box is then opened only at some central

---

[1] J. S. BAKER and W. R. STEBBINS, JR., *Dictionary of Highway Traffic* (Evanston, Ill.: Traffic Institute, Northwestern University, 1960), p. 161.

location. The sealed box may also be placed on a portable collecting unit that is so keyed that when the sealed unit is unlocked, it cannot be taken from the collecting unit. Vandal-resistant coin receptacles are available for use with meters in high crime areas. There is also available a validating parking meter that dispenses a disc or token on insertion of a coin which can then be redeemed under a parker validation program by merchants.

Parking meters may be installed at either curb or off-street locations. For curb locations, the meters are mounted on a pipe generally placed about 18 in. (0.46 m) from the curb and about 2 ft (0.61 m) from the front edge of the parking stall. In some instances, two meter heads are mounted atop a single post. This can be done effectively in curb locations with "paired" parking in which one post (with two meter heads) serves the parking stalls immediately ahead and behind the meters or in off-street facilities where two parking spaces face each other across an island.

TABLE 15.1
Use of Color Coding of Meters by U.S. Cities

City Size	Color Coding Used		
	Yes	No	% Yes
Less than 2,500	56	11	84
2,500–5,000	119	47	72
5,000–10,000	168	71	70
10,000–25,000	184	67	73
25,000–50,000	105	33	76
50,000–100,000	76	23	77
100,000–250,000	51	21	71
250,000–500,000	14	9	61
500,000–1,000,000	10	10	50
Over 1,000,000	4	2	67
Total	787	294	73

Source: W. D. Heath, J. M. Hunnicutt, M. A. Neale, and L. A. Williams, *Parking in the United States—A Survey of Local Government Action* (Washington, D.C.: National League of Cities, Department of Urban Studies, 1967), pp. 66–87.

TABLE 15.2
Average Number of Meters in U.S. Cities

City Population	Number Cities Reporting	On-street Meters		Off-street Meters	Total Number of Meters
		In CBD	Outside CBD		
Less than 2,500	68	105	12	3	120
2,500–5,000	171	166	13	14	193
5,000–10,000	242	252	24	43	319
10,000–25,000	253	382	35	128	545
25,000–50,000	140	537	108	300	945
50,000–100,000	100	680	150	433	1,263
100,000–250,000	74	1,083	313	300	1,696
250,000–500,000	23	1,995	1,200	643	3,838
500,000–1,000,000	20	2,301	2,893	234	5,428
Over 1,000,000	6	2,204	19,153	1,957	23,314

Source: Heath, Hunnicutt, Neale, and Williams, pp. 66–87.

Many cities in the U.S. have adopted the use of color coding meters as an additional means of advising the parker of the time limit at a particular location. Table 15.1 summarizes information on the use or nonuse of color coding by more than 1,000 U.S. cities in all population ranges. Nearly three-quarters of all cities reporting indicated that they were using this means of identifying time limits.

The number of parking meters found in a city is directly related to city size. Table 15.2 presents the average number of on-street and off-street meters found in cities of all sizes.

The amount of revenue that can be expected from the average parking meter, along with average maintenance and collection costs, is shown in Table 15.3. Revenues and maintenance costs are directly related to population, but collection costs, although varied, are not strongly related to population.

TABLE 15.3
Average Meter Revenue, Maintenance Cost and Collection Cost in U.S. Cities in 1966

City Size	Average Revenue	Average Maintenance Cost	Average Collection Cost
Under 2,500	$44.69	$3.23	$3.54
2,500–5,000	43.72	4.09	5.79
5,000–10,000	51.52	4.78	7.34
10,000–25,000	59.92	7.02	7.33
25,000–50,000	59.95	8.66	5.98
50,000–100,000	75.92	8.16	5.57
100,000–250,000	82.13	12.04	5.92
250,000–500,000	105.13	11.63	5.78
500,000–1,000,000	104.36	14.73	6.01
Over 1,000,000	111.75	17.81	7.59

Source: Heath, Hunnicutt, Neale, and Williams, p. 64.

LOCATION OF PARKING FACILITIES

In order to establish the location of parking facilities, appropriate field studies are first conducted (see Chapter 10) to determine the areas of concentration of parking demand and the needs left unsatisfied in terms of existing supply. Only with data from these studies is it possible to properly locate a parking facility.

Many factors influence the location of parking facilities. These include parking shortages in each area, origin and destination of parkers, pedestrian walking distance and ease of movement, vehicular access, street capacities, kind of generators in the area, future development, economic factors, and the relationship to the overall CBD plan.

The kind of parking generators to be served influences the facility location. Short-term parking generators (department stores, other retail stores, banks, etc.) require greater proximity to parking spaces than do long-term generators (offices) because long-term parkers will accept a longer walking distance than will short-term parkers. In considering pedestrian access, ease of movement should be considered as well as actual walking distance. The longer of two routes may psychologically appear shorter because of pedestrian amenities, less conflict with vehicles, and other factors that tend to make one walking trip more acceptable than another.

Vehicular access is also an important factor in locating parking facilities. As much as practical, driving time to the facility by patrons should be minimized. If parker origins are concentrated in one direction because of physical or other barriers to travel, it is desirable to locate the parking facility between the driver's origin and his final destination. To locate the parking facility on the far side of the generators from the origins would require drivers to pass the generator on their way to the parking facility, thereby increasing travel distance and volume on CBD streets. The CBD facility should be located so that there is a minimum adverse effect on the CBD street system. Whenever possible, it is desirable to locate a parking facility near a major arterial or expressway, thereby minimizing travel on local streets and also providing quick and easy access to the major street and freeway systems.

Capacity on streets surrounding the proposed site should be adequate to handle the additional traffic attracted by the parking facility. In addition, entrances and exits to the facility should be placed as far as possible from intersections in order to provide maximum storage and maneuver space.

The location of a proposed parking facility should be consistent with the overall plan for the CBD. The location of future generators should be considered as well as the location of any new streets, freeway access points, or changes in the existing street system.

Economic considerations enter into the location of parking facilities. Land costs vary from one site to another. Land costs lead to a consideration of whether a garage or a surface lot should be constructed. When garages are being considered, underground facilities should be suggested if the economic costs can be borne. Although underground garages cost from 1.5 to 2 or more times per car space than aboveground facilities, they offer the advantage of preserving open space in the CBD. If an underground facility is built under an existing public park, the land costs may be eliminated. Even then, the additional cost of underground construction may negate the advantage of land cost.

The use of air rights over and under freeways, other streets, railroads, rivers, etc. for the location of parking facilities is a means of reducing or eliminating land cost related to parking facilities. Multipurpose buildings are also being constructed with increasing frequency because parking can be combined in the same structure with other land uses such as office, apartment, or institutional use. This greatly reduces the land costs chargeable to the parking portion of the development.

DESIGN OF OFF-STREET FACILITIES

ELEMENTS OF GOOD DESIGN

In designing any off-street parking facility, the elements of customer service and convenience and minimum interference with street traffic flow must receive first priority. Basically, a driver would like to park his vehicle immediately adjacent to his destination. The accessibility of the facility, the ease of entering, circulating, parking, unparking, and exiting are important factors in both the location and design of off-street parking facilities.

SURFACE LOT DESIGN

Site characteristics. The characteristics of the proposed site are an important design consideration. Factors such as site dimensions, topography, and profiles affect the design of off-street parking. Site dimensions may be such that angle parking may provide the only feasible alternative for effective utilization of the site. The relationship of the site characteristics to the surrounding street system will affect the location of entry and exit points as well as the internal circulation pattern.

Access points. External factors such as pedestrians, traffic controls, turning restrictions, and volumes of traffic on adjacent streets will affect the design of a parking facility, particularly in the location of entry and exit points. It is desirable to avoid locating entry-exit points where vehicles entering or leaving the site would conflict with large numbers of pedestrians. Likewise, street traffic volumes, turning restrictions, and other traffic controls may limit points at which entrances and exits can logically be placed. It is important to investigate these factors before making final decisions on entrances and exits.

Entry and exit points should be located in order to provide maximum storage space and maximum distance from intersections. Combined entry-exit points should preferably be located at mid-block. When entrances and exits are separated, the exit should preferably be placed in the downstream portion of the block, and the entrance should be placed as far upstream in the block face as possible.

Traffic circulation. The ideal movement into a parking facility is a left-hand turn from a one-way street. This places the driver in a left-hand turn pattern which is desirable on the parking site. The driver position is on the inside of the turn and it allows better visibility and more accurate judgment of the placement of the vehicle. A driver has more difficulty in judging his vehicle placement in a right-hand turn.

Vehicle circulation on the site may be either two-way or one-way, depending on site dimensions and the angle of the parking stall. Two-way circulation is generally allowed with 90° stalls, and one-way circulation is generally used with stall angles less than 90°. In any event, it is desirable to attempt to minimize traffic conflict points on site so that accident potential and congestion are minimized during peak hours.

Major use of facility. A factor for consideration in designing a parking facility is the use of the facility either in terms of kind of parker or kind of generator being served. Kind of parker can be identified by average duration with either short-term or long-term parkers, or a combination of the two. Design dimensions are often liberalized in facilities used by short-term parkers because of the high turnover rate and the desire to provide easy access, circulation, and parking-unparking.

A parking facility may serve any and all parking generators, particularly in a downtown area. Parking facilities that serve special events such as sports stadia, auditoriums, or other similar uses, require special design considerations because of the parker characteristics. People attending special events generally arrive over a short-time period and may all wish to leave at once upon conclusion of the event. This can place a severe strain on entrance and exit facilities and the internal circulation system. Special consideration must be given to providing adequate entrance and exit capacity at such facilities.

Landscaping. Landscaping parking facilities is desirable but should be limited to kinds that will not interfere with the parking function. Care should be taken to use shrubs, plants, and trees that can withstand auto fumes and the concentrated heat arising from a large, paved surface. Landscaping can be an effective means in controlling pedestrian paths. Hedges can serve to funnel pedestrians into desired walk patterns within the site.

Sufficient setback must be provided for all plants so that the front or rear overhang of cars does not destroy them. Extreme care should be exercised in placing shrubbery or other plants near entrances and exits so that sight distances are not restricted. This requires that the growth pattern of the plant be considered so that the small plant of today will not develop into a major sight restriction in future years.

Lighting. Adequate lighting of the parking site is very important. Mounting height and spacing of luminaires should be sufficient to distribute the desired lighting intensity to the entire facility. A normal lighting level is from 1.0 to 2.0 f.c. with a uniformity ratio (average illumination divided by the lowest level) not more than six to one.[2] The luminaire units should be placed so that vehicle movement and parking are not obstructed. If raised islands are used to separate adjacent parking stalls, the poles can logically be placed on the island. In any event, they should be placed between adjacent stalls and at the ends of the parking rows. Care should be taken to prevent excessive light spillover into adjacent residential areas.

Parking dimensions and layout. The long-term trend in American automobile design has been toward longer and wider vehicles. In the 1973 model year, American cars range up to 19.1 ft (5.3 m) in length and 6.66 ft (2.03 m) in width. The dimensions of parking aisle widths and stall sizes have necessarily increased to keep pace with increases in auto size. Typical parking dimensions vary with the angle at which the stall is arranged in relation to the aisle (see Table 15.4). Stall widths (measured perpendicular to the vehicle when parked) range from 8.5 to 9.5 ft (2.59 to 2.90 m). In attendant parking facilities, attendants can park standard-sized cars in spaces as narrow as 8.0 ft (2.44 m). However, the minimum stall size is recommended at 8.5 ft (2.59 m) for self-parking of long-term duration. For higher turnover self-parking, a stall width of 9.0 ft (2.74 m) is recommended. Stall widths at supermarket and other similar parking facilities, where shoppers generally have large packages, should desirably be 9.5 or even 10.0 ft (2.90 to 3.05 m) in width.

Substandard stall and aisle widths are a false economy. Although they permit the marking of more stalls per given length, vehicles tend to encroach upon adjacent stalls so that one or more spaces are unavailable for use. The end result is no gain in actual space usage but is a parking condition surrounded by confusion.

Table 15.4 is based on a stall length of 18.5 ft (5.64 m). The stall length should be sufficient to accommodate the length of almost all cars expecting to use the space. Many of the luxury American cars, however, exceed 18.5 ft (5.64 m) in 1973.

Aisle width is a function of the parking angle and stall width. One-way aisles are generally used with angle parking, and two-way circulation is generally used with 90° parking.

In designing parking facilities, a common unit of measure is the parking module. A module consists of the width of the aisle, plus the depth of the parking stalls (measured perpendicular to the aisle) on each side of the aisle. In many instances, parking modules are completely separated from each other. Such modules are represented by the wall-to-wall dimensions shown in Table 15.4. Another available module for angle parking is the interlocking module. The most common, and preferred, interlocking

---

[2] *Parking Principles*, Highway Research Board Special Report No. 125 (Washington, D.C.: Highway Research Board, 1971), p. 107.

TABLE 15.4
Typical Parking Dimensions in Ft

Parking Angle	Stall Width Parallel to Aisle	Stall Depth to Wall	Stall Depth to Interlock	Aisle* Width	Modules†	
					Wall to Wall	Interlock to Interlock
45°						
8.5-ft stall	12.0	17.5	15.3	13	48	44
9.0-ft stall	12.7	17.5	15.3	12	47	43
9.5-ft stall	13.4	17.5	15.3	11	46	42
60°						
8.5-ft stall	9.8	19.0	17.5	18	56	53
9.0-ft stall	10.4	19.0	17.5	16	54	51
9.5-ft stall	11.0	19.0	17.5	15	43	50
75°						
8.5-ft stall	8.3	19.5	18.8	25	64	63
9.0-ft stall	9.3	19.5	18.8	23	62	61
9.5-ft stall	9.8	19.5	18.8	22	61	60
90°‡						
8.5-ft stall	8.5	18.5	18.5	28	65	65
9.0-ft stall	9.0	18.5	18.5	26	63	63
9.5-ft stall	9.5	18.5	18.5	25	62	62

*Measured between ends of stall lines.
†Rounded to nearest ft.
‡For back-in parking, aisle width may be reduced 4.0 ft.
Note: These dimensions are for 18.5-ft length stalls, measured parallel to the vehicle and are based on results of a special study to evaluate the effects of varied aisle and stall width for the different parking angles shown. The study was conducted in December 1970 by the Federal Highway Administration and Paul C. Box and Associates.

Source: *Parking Principles*, Highway Research Board, Special Report No. 125, 1971, p. 101.

module is the one that places the bumpers of vehicles in adjacent stalls next to one another. This layout is illustrated in Figure 15.1, along with parking dimensions for various angles of parking. At 45°, a nested interlock is possible when adjacent aisles have one-way movement in the same direction. This places the bumper of one car adjacent to the front fender of another car and is not recommended, for the likelihood of damaged fenders is much greater than with other parking layouts.

TABLE 15.5
Parking Dimensions in Ft for Import-size Vehicles
(15 Ft Length)

Parking Angle	Stall Width	Aisle Length per Stall	Depth of Stalls at Right Angle to Aisle	Aisle Width	Wall-to-wall Module
45°	7.5	10.5	16.0	11.0	43.0
60°	7.5	8.7	16.7	14.0	47.4
75°	7.5	7.8	16.3	17.4	50.0
90°	7.5	7.5	15.0	20.0	50.0

Note: These measurements are *inadequate* for average American compacts. Each stall depth should be increased about 1 ft (2 ft total for the module) to accommodate the usual range of compact sizes.

Source: *Parking Principles*, p. 102.

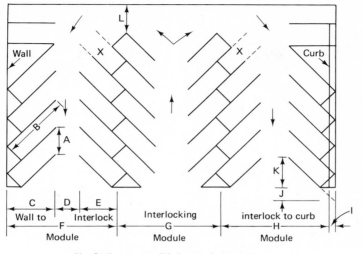

X = Stall not accessible in certain layouts

Parking layout dimensions (in ft) for 9-ft stalls
at various angles

Dimension	On diagram	Angle			
		45°	60°	75°	90°
Stall width, parallel to aisle	A	12.7	10.4	9.3	9.0
Stall length of line	B	25.0	22.0	20.0	18.5
Stall depth to wall	C	17.5	19.0	19.5	18.5
Aisle width between stall lines	D	12.0	16.0	23.0	26.0
Stall depth, interlock	E	15.3	17.5	18.8	18.5
Module, wall to interlock	F	44.8	52.5	61.3	63.0
Module, interlocking	G	42.6	51.0	61.0	63.0
Module, interlock to curb face	H	42.8	50.2	58.8	60.5
Bumper overhang (typical)	I	2.0	2.3	2.5	2.5
Offset	J	6.3	2.7	0.5	0.0
Setback	K	11.0	8.3	5.0	0.0
Cross aisle, one-way	L	14.0	14.0	14.0	14.0
Cross aisle, two-way	—	24.0	24.0	24.0	24.0

Figure 15.1. Stall layout elements. (Source: *Parking Principles*, Highway Research Board, Special Report No. 125, 1971, p. 99.)

Recommended parking dimensions for imported cars—15 ft (4.57 m) in length—differ from recommendations for standard U.S. cars (see Table 15.5). Stall lengths and widths are recommended at 15 ft (4.57 m) and 7.5 ft (2.29 m), respectively. If a number of these smaller-sized spaces are to be included in a facility, they should be placed together in a prime location to encourage their use. If these spaces are not convenient, small car drivers will park in the standard-sized spaces. Because of difficulties in predicting the amount of usage and in controlling the spaces, most American parking facilities are being designed with all spaces of sufficient size for standard American cars.

In the actual layout of parking stalls and circulation aisles, it is always desirable to have a row of parking on each side of the aisle. This gives the most efficient design. In addition, the greatest efficiency can generally be obtained by placing aisles and rows of parking parallel to the long dimension of the site. Greatest parking efficiency can usually be aided by placing a row of parking completely around the perimeter of the site. With adequate site dimensions, this places parking stalls on both sides of the aisle, including end aisles.

When pedestrian walks are used in parking facilities, they should direct pedestrians toward the major parking generators. Raised sidewalks can be used in larger facilities between rows of cars in order to aid pedestrian flow. Many pedestrians, however, will still use the aisles and the need for raised pedestrian walks is debatable.

Relative efficiency factors. Relative efficiency factors can be calculated for various parking angles and stall widths (see Table 15.6). The figures represent the number of square feet per stall plus one-half of the aisle width for a distance equal to the stall width measured parallel to the aisle. These dimensions were obtained from Table 15.4. Stalls arranged at 90° to the aisles provide the most efficient design and the efficiency decreases as the parking angle decreases.

TABLE 15.6
Relative Efficiency Factors (sq ft per stall area plus one-half aisle area)

| Parking Angle | Width of Stall | | |
	8.5 ft	9.0 ft	9.5 ft
45°	308	324	340
60°	288	296	310
75°	286	295	308
90°	276	284	295

Notes: (1) Does not include end aisle circulation areas or unusable area at ends of parallel parking rows. (2) Based on data included in Table 15.4.

Drainage. Adequate slope should be provided to surface lots in order to minimize the possibility of low or flat spots. Ponding of water in a lot is undesirable both for vehicle and pedestrian movement. This is particularly true in cold climates where freezing may create icy spots. Recommended minimum grades are one percent for asphalt surfaces and one-half percent for portland cement concrete surfaces.[3]

SURFACE VS. STRUCTURE PARKING

Development costs for surface parking lots normally are from $1.00 to $2.00 per sq ft exclusive of land, with an average figure approximating $1.50 per sq ft. This includes all improvement costs such as grading, paving, lighting, drainage, signing, and marking, etc. Land cost is often the factor that determines economic viability of surface parking in the downtown area. As land costs increase, it often becomes economically justifiable to expand parking vertically in a parking structure instead of expanding horizontally. Figure 15.2 illustrates comparative total cost per car space

---

[3] *Ibid.*

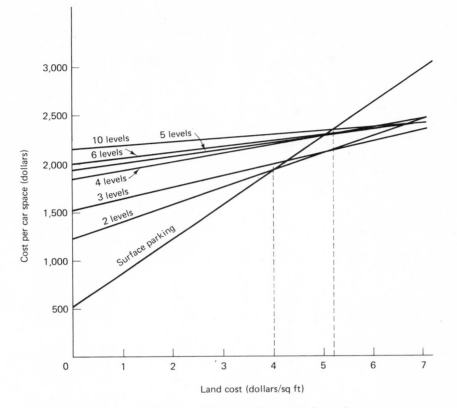

Figure 15.2. Surface vs. structure parking costs related to land cost. Area per car space equals 350 sq ft. [Source: Square foot construction costs obtained from R. F. Roti, *Square Foot Cost Averaging Principle for Parking Structures*, (Washington D.C.: 1971), National Parking Association, pp. 1–4.]

for surface and structure parking as land cost is varied. Since construction costs of parking structures have been shown to vary with the number of parking levels,[4] individual curves are shown for various levels of parking. The total cost per car space for a four- to six-level structure (commonly used number of levels) is approximately the same as surface parking costs per car space, when land costs reach $5.00 per sq ft (see Figure 15.2). Thus, for land costs above approximately $5.00 per sq ft, structured parking is often more economical than the same number of spaces in a surface lot. Decking of a surface lot to give two parking levels can be more economical than expanding surface parking when land costs exceed $4.00 per sq ft. These costs are for comparative purposes only. There are many factors affecting garage costs that can cause per space costs to vary widely. Further discussions of garage cost will be presented in a later section.

---

[4] R. F. ROTI, *Square Foot Cost Averaging Principle for Parking Structures* (Washington, D.C.: National Parking Association, 1971), pp. 1–3.

PARKING GARAGE DESIGN

Site characteristics. Many of the factors that affect the location and design of surface lots also affect the design of parking garages. Site characteristics such as size, shape, and topography are important factors in garage design. The topography of a

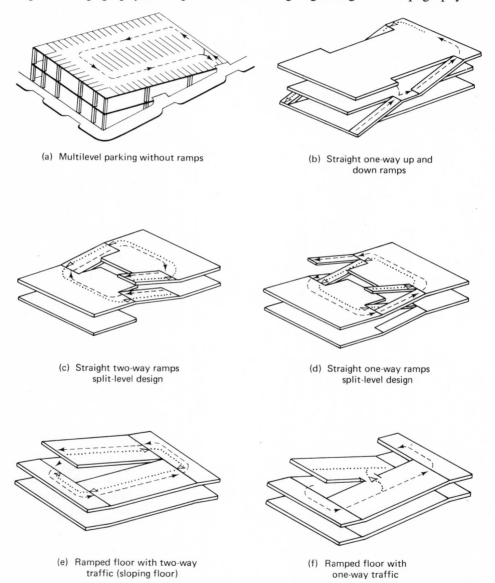

(a) Multilevel parking without ramps

(b) Straight one-way up and
down ramps

(c) Straight two-way ramps
split-level design

(d) Straight one-way ramps
split-level design

(e) Ramped floor with two-way
traffic (sloping floor)

(f) Ramped floor with
one-way traffic

Figure 15.3. Ramp systems.[Sources: Roti, R. F., (a, c, f, g, h) *Square Foot Cost Averaging Principle for Parking Structures*, National Parking Association, pp. 7–9, and Klose, D., (b, d, c, i, j) *Metropolitan Parking Structures*, (New York, N.Y.: Frederick A. Praeger, 1965), pp. 30–31.]

site may allow direct entry to more than one level of the garage. This will affect entry and exit locations as well as the interfloor travel system within the structure.

Access points. The location of entry and exit points is even more critical in garage design than in surface lot design because of the increased number of spaces available in structured parking. Street capacities, location of traffic controls, and other external factors must be carefully analyzed to assure a design that is compatible with the surrounding street system.

Major use of facility. As in surface lots, the dominant use, whether by short-term or long-term parkers or whether or not the facility will serve special events, will influence the design. When vehicles enter and exit in short spans of time, consideration should be given to express ramps, particularly for exiting. The "dump time" of the facility (time required to empty the facility when filled to capacity) is an important factor. This time should be kept to a minimum, and 30–45 minutes is generally considered acceptable for almost all facilities. A lesser time for emptying the facility is desirable for special-event use such as sporting events or concerts.

Interfloor travel systems. Interfloor travel systems may consist of either ramps or sloping floors, or various combinations of the two. Only on a sloping site that permits direct access to each level are ramps unnecessary, but they may be desirable for internal circulation. In ramped or sloping floor designs, the floors serve both as aisles and parking bays (see Figure 15.3). The ramped section is generally one or more

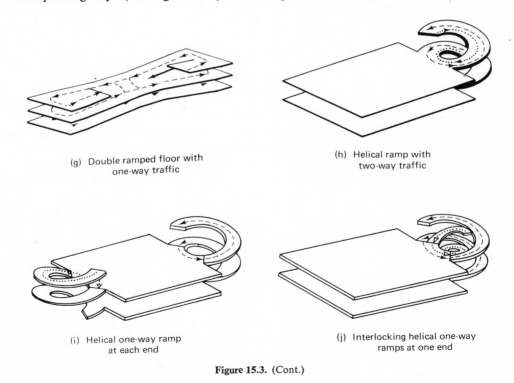

(g) Double ramped floor with
one-way traffic

(h) Helical ramp with
two-way traffic

(i) Helical one-way ramp
at each end

(j) Interlocking helical one-way
ramps at one end

Figure 15.3. (Cont.)

parking modules in width. In other designs, the ramps are used exclusively for travel between floors. Combinations of the various ramps are possible. An express exit ramp may be incorporated into a sloping floor design. This ramp may be a straight ramp along the side of the structure, or it may be a helical ramp.

Other kinds of ramps are used in addition to those shown in Figure 15.3. An additional helical ramp variation is the double helix. This design consists of two independent interwoven ramps that drop two levels in one complete 360° turn. One ramp then serves the odd-numbered floors and the other ramp serves the even-numbered floors. This design is useful for tall structures because it reduces the number of turns required in exiting or entering from the upper floors. It is desirable to keep the number of full turns to a maximum of five for self-parking facilities. With this criterion as a guide, from 10 to 12 levels are possible with a double helical ramp system. A single helix ramp, however, should be limited to about six levels.

For ramped or sloping floor designs, the number of levels should be limited to a maximum of six because of the amount of turning required and the number of spaces a driver must pass.

Another factor for consideration in determining the number of floors is the relative height of adjacent buildings. Many drivers develop acrophobia in taller garages, particularly when they are driving at a level about the rooftops of adjacent buildings. Proper design of the parapet walls will reduce this effect by limiting the driver's view of his surroundings while he is seated in his automobile.

The layout of parking aisles and stalls for garages is similar to that used for surface lots. Stall and aisle dimensions remain the same.

Ramp grades and dimensions affect the ease of circulation within the structure. Sloping or ramped floor grades should not exceed 3 to 4 percent. The parking angle on these sloping floors should be at least 60° in order to minimize the possibility of vehicles rolling out of the parking space and down the ramp. For 90° parking, sloping floor grades should not exceed 5 percent. Ramps without parking should be limited to about 10 percent, with grades up to 15 or 20 percent allowable in attendant parking structures.

Driving ramps should be from 14 to 18 ft (4.3 to 5.5 m) wide with 12 ft (3.7 m) sufficient for longer straight runs. Helical ramps should have a minimum outside radius of 32 ft (9.8 m) with a desirable radius of 35 to 37 ft (10.7 to 11.3 m).

Self-parking ramp capacities normally range between 500 and 600 cars per hour per lane. Typical capacity used for design purposes is often 400 cars per hour per lane. This capacity is reduced greatly to 150 to 200 cars per hour per lane when vehicles must stop at a cashier's booth on exiting.

Vertical clearance should approximate 7 ft (2.13 m) which normally results in floor-to-floor heights of about 10 ft (3.05 m).

Structural systems. Normally, there are four kinds of structural systems that should be investigated for any garage: structural steel, poured-in-place concrete, precast concrete, and post-tensioned concrete. Many factors should be considered in analyzing the various structural systems because they affect the relative economy and adaptibility of the systems. A partial listing of factors includes:

1. Building code requirements
2. Maintenance
3. Availability of materials and precast concrete fabricators

4. Shipping distance and costs
5. Availability of contractors experienced in each structural system
6. Environmental and atmospheric conditions

Building codes may affect the structural system because of code restrictions on kind of structural system allowed. Other building codes do not allow exposed steel design or may require fireproofing of the external columns only. Fireproofing of steel columns, beams, and girders can add greatly to the cost of this kind of construction.

Certain materials may not be readily available in particular areas and freight costs for them can be prohibitive. Similarly, the availability of quality precast concrete fabricators in some areas precludes this structural system. The post-tensioning of floor slabs of precast concrete structures requires contractors experienced in this work.

The relative maintenance of the various structural systems must be considered. Exposed steel requires periodic painting and care must be taken to specify a sufficient number of shop and field coats of quality paint to retard deterioration of the steel. Atmospheric and environmental conditions can greatly influence the amount of maintenance required, particularly in heavy industrial areas and areas near bodies of salt water. Weathering steels that develop a hard coat of rust do not require painting but they do cause difficult maintenance problems for approximately three years until they develop their hard coating.

Precast structural systems can be advantageous when many structural members are duplicated throughout the structure. Once the forms are available, a great number of members can be cast quickly and economically, but this advantage is lost when many different sizes of columns, beams, and girders are required such as in a garage on an odd-shaped site.

Erection time at the site is also favorable for precast or steel structural systems. The construction of forms for poured-in-place concrete requires time not needed for the other systems. Experience has shown, however, that the total construction time from awarding of contract to occupancy of the building is practically the same for all structural systems when time required to fabricate and ship the precast or steel members is considered.

It is not possible to say that one structural system is better or more economical in all locations and under all conditions. A comparative analysis of the systems should be made for each facility and all the influencing factors should be considered. Only then should a decision be made on the structural system.

Short- vs. long-span construction. Functional and operational considerations generally dictate that long- or clear-span construction be provided in most garages. There are certainly many advantages to this kind of construction, particularly in a free-standing garage. When an additional structure (office, apartments, or other use) is planned above the parking garage, short-span construction may be advantageous because of the additional loads imposed. Although functional considerations in the garage area favor long-span construction, consideration of the entire structure may require that short-span construction be used. Long-span construction usually costs in the vicinity of from 5 to 10 percent more than short-span, but this increase is offset by more parking spaces and better circulation. A comparison of short- and long-span construction is shown in Figure 15.4.

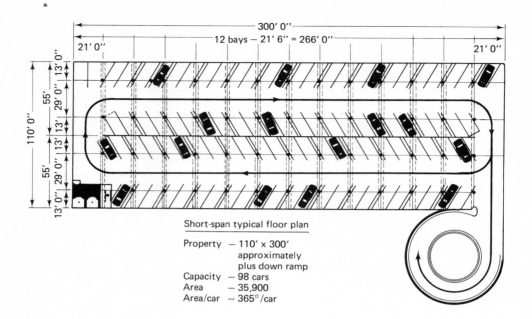

Short-span typical floor plan

Property   — 110′ x 300′
             approximately
             plus down ramp
Capacity — 98 cars
Area   — 35,900
Area/car — 365$^\square$/car

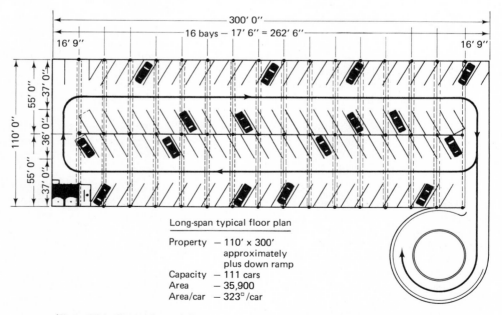

Long-span typical floor plan

Property   — 110′ x 300′
             approximately
             plus down ramp
Capacity — 111 cars
Area   — 35,900
Area/car — 323$^\square$/car

Figure 15.4. Comparison of short-span and long-span construction. (Source: Richard C. Rich, *Methods of Construction and Construction Costs*, National Parking Association, 1966, pp. 2, 4.)

A list of advantages of long- or clear-span construction in free-standing garages would include the following:

1. Column-free floors
2. Fewer columns and foundations
3. More parking space
4. Maximum operational efficiency and flexibility
5. Faster and easier parking and unparking manuevers
6. Maximum flexibility to change stall sizes or parking angle
7. Easy floor maintenance
8. Unrestricted sight distance
9. Fewer damaged cars
10. Greater driver acceptance
11. Minimum problem in locating ventilation and lighting equipment

Disadvantages of long spans include:

1. Deeper floor construction
2. Greater floor-to-floor heights
3. Higher construction costs

Lighting and electrical systems. Lighting within the garage is necessary to aid safety of movement and to discourage vandalism or acts of violence. Table 15.7 presents minimum and desirable lighting for garages. A high light level is needed at the entrance where a driver proceeds from bright sunshine into the garage.

TABLE 15.7
Garage Lighting Standards

Area	Light Intensity (footcandles)	
	Minimum	Desirable
Entrance	50–80	80–100
Driving aisles	8–10	10–15
Over-parked vehicles	3–5	5–10

It is desirable to have a central control panel from which all electrical and mechanical fixtures can be actuated. This control panel should be located in the manager's office. This allows a person to turn on the lights on the roof level without physically going up to that level and turning a switch. In addition, to facilitate operating economies, light circuits should be designed to allow a portion of the lights in a given area to be turned on and off independently of the remaining lights.

Interior garage signing. In signing a parking garage, it should be remembered that a parking facility is merely an extension of the street system. Directional and informational signs are generally needed and should conform as nearly as possible to standard street signs. Many signs in garages can be painted on walls or other areas that have good target value. The use of illuminated signs may be justified in some instances, but these are expensive. Well-placed, nonilluminated signs are often sufficient. Signs in a garage should direct motorists to parking spaces and exits and should inform the driver of such traffic conditions as one-way aisles and ramps. In addition, signs should direct the pedestrian to exits, stairs, and elevators.

It is difficult to determine a signing plan for a garage from a set of construction plans. It is recommended that the signing be postponed until the garage is nearly completed. The recommended approach, when possible, is to drive into the nearly completed garage in an auto and determine sign locations from the vehicle. This gives proper perspective to sign location in terms of sight lines, locations of parked cars, and sight obstructions such as beams and columns.

Drainage and waterproofing. Many people knowledgeable in garage design and operation consider water leakage through the floor slab to be the main unsolved long-term maintenance problem in parking structures. This leakage does not appear to be a problem during the early life of the garage. The water, mixed with salts in colder climates, seeps down through hairline cracks and attacks the reinforcing bars. These bars rust and expand resulting in scaling of the concrete. Over a period of years this action can cause structural damage. Without proper maintenance, a garage can be at the point of structural failure within 10 to 15 years because of water leakage.

The best waterproofing system is one that gets the water off the floor quickly and into the drains. This is particularly necessary for the roof level. The riser system should be designed to adequately accept a 10-year design rainfall. The floor of the garage should be sloped toward the drains with a minimum slope of $\frac{1}{8}$ in. per ft (10.4 mm/m). The minimum spacing of floor drains should be 60 ft (18.3 m) from center to center.

There are numerous waterproofing systems. One common system consists of a mastic asphaltic material placed over the floor slab to a thickness of about $\frac{1}{16}$ in. (1.6 mm). This material seeps and permeates down into the floor slab and forms an elastic, rubbery surface that seals small cracks. A wearing surface of from $\frac{1}{2}$ to 1 in. (from 1.3 to 2.5 cm) of asphaltic concrete is then placed.

A second waterproofing system consists of a plastic material that is usually sprayed onto the floor slab and seeps into the slab. For a wearing surface, a strong, tough, rubberized plastic is then sprayed on the deck in several layers and built up to a thickness of from .015 to .020 in. (from .38 to .51 mm). Flint chips or small gravel are often imbedded in the wearing surface to aid traction.

A third waterproofing system is a membrane surface of sheet rubber, placed over a concrete slab. A second slab is then poured over the membrane. Unfortunately, some of these membranes begin to distintegrate after about 10 years. Also, it is nearly impossible to locate or repair a rupture in the membrane because the membrane is embedded between two slabs of concrete.

There are many other waterproofing compounds. These are generally spray-on plastic types and have some waterproofing qualities, but they are of no value in preventing leakage when a hairline crack opens.

Traffic and revenue control systems. In a large garage where large amounts of money are being handled, there must be a revenue control system. This is particularly true for parking facilities at airports, convention centers, and at major CBD locations. A revenue control system is a system designed to record all entering and exiting vehicles and also includes a record of tickets, cashiers on duty, and the amount of money handled. The system provides the information needed to check for revenue losses and thefts.

The six basic elements of a revenue control system needed to keep an accurate record of parking activity and revenues are the following:[5]

1. An extremely accurate count of all entering and exiting vehicles.
2. No one must be allowed to enter the parking area without taking a ticket or exit without leaving a ticket.
3. All clocks within the system must be operated accurately and in conjunction with each other and must be designed so that cashiers cannot reset them.
4. There must be some type of validating device or cash register—not a time stamp—to record the outbound transactions. The tickets must be stamped with the appropriate information so individual cashiers may be held responsible for their actions.
5. The ticket-issuing machines and entry lanes must be operated in such a way that it is not possible for anyone to get more than one ticket from the machine. Future supplies of tickets also must be kept in a safe place.
6. Parking gates or other devices must be used to prevent cars entering through an exit or leaving through an entrance lane.

Safety and surveillance equipment. Any large garage is a potential source of problems of loitering, vandalism, thefts, and crimes against persons. Garages often become havens for drunks and derelicts. Within the garage, almost all crimes occur in the elevators and stairwells. For this reason, these areas should be well-lighted and should be closable at night. In addition, glass-enclosed stairwells should be used instead of masonry whenever possible. Glass-enclosed stairwells may conflict with fire codes but many cities have granted variances when the unique problems involved are explained to building officials and fire marshals.

In garages that do not remain open 24 hours a day, there should be positive ways of closing off the garage, such as roll-down doors at entry-exit points. In addition, the garage architecture should be such that large openings around the building do not exist to give access to the structure at other than designated points.

Sound and television monitoring systems are the two kinds of internal surveillance systems found in garages. The sound system monitors all sounds in the garage and in all elevators and stairwells. A speaker in the manager's office allows the monitoring of all sounds and the detection of suspected trouble areas.

In television monitoring, television cameras are placed throughout the garage. This system is not widely used because of the necessity of having someone continually looking at monitors and because the light intensity in the garage is often insufficient to provide a good picture contrast on the monitors. In addition, it is difficult to cover a large area with cameras and the cameras can be made inoperative by vandals.

In addition to the above equipment, regular security patrols are often used within garages to discourage vandalism and acts of violence.

Self vs. attendant parking. Because of the labor cost and the difficulty of retaining lower salaried parking attendants, almost all parking is now operated as self-parking.

---

[5] J. M. HUNNICUTT, "Safeguarding Your Parking Revenues," 1970 Annual Conference (Las Vegas: American Association of Airport Executives, 1971), p. 52.

When given the choice, the parker almost invariably prefers to park his own car. Reasons for this include:

1. The parker takes better care of his car than does the average attendant.
2. The self-parker can lock his car and take his keys with him.
3. The self-parker avoids the long delay often associated with attendant retrieval of cars.

Under normal conditions, however, self-parking results in from 10 to 15 percent fewer parking spaces for the same area as compared to attendant parking.

Attendant parking may be justified as a convenience service, and it is popular at restaurants and hotels. It may also be justified in the downtown areas of major cities where rates and parking demand are very high. The additional capacity available by attendant parking, and by double and triple stacking of cars, may produce sufficient additional revenue to offset the higher cost of attendant parking. Labor costs for attendants may add as much as 40 percent to the parking cost.

Central vs. exit cashiering. In central cashiering, the parker returns to a central location in the garage, pays his parking fee and receives a receipt, proceeds to his car, and on exiting the facility, hands the receipt to an attendant. In exit cashiering, the parker returns directly to his car and drives to the exit at which point he stops and pays his parking fee. In addition, under special conditions such as at sporting events, precashiering may be used. Under this system, the parker generally pays a flat fee on entering the facility and therefore can exit without stopping.

All alternatives should be considered before selecting a cashiering method. The design and use of the facility can influence the choice. For example, in a garage serving a department store or similar use where parkers use a common path in returning to the garage, central cashiering may be a good choice. Exit time is also reduced when central cashiering is employed. With central cashiering, when a receipt only is picked up on exiting, vehicles can exit at a rate per exit lane of one vehicle every 6 to 10 sec. With exit cashiering, the rate per lane would be approximately one vehicle every 20 sec. Central cashiering then increases exit lane capacity by a factor of two or three.

Aesthetics. A parking garage need not be an ugly structure, although some simply give the appearance of a concrete slab sitting up in the air. Thought should be given to making the garage an attractive building. This function normally falls to the architect, and the building is only as aesthetically pleasing as the architect's ingenuity makes it.

The exterior of the building can be improved by the use of such exterior sidings as aluminum, anodized aluminum in colors, vertical fins, and precast concrete. Judicious use of lighting and landscaping are also helpful.

Within the garage, columns, beams, and girders can be painted as well as concrete masonry units or cinder block walls. A form of color coding is possible by painting each floor a different color. The parker can then remember the floor on which he parked by its color.

By their nature, garages lend themselves to bright colors. This brightens the interior of the garage. Since the parker does not remain in the garage for any length of time, bland colors normally found in living and working areas are not necessary.

Fire protection. Many building codes require an inordinate amount of fire protection in garages. These requirements are based on old approaches to fire fighting which were generally incorporated into building codes many years ago. Tests of auto fires have proven the futility of trying to ignite an adjacent auto when parked beside a burning car.

The combustible materials in a car consist of the gasoline, upholstery, and paint. The fire loading in a garage, the number of BTU's of combustible material per square foot of floor area, is extremely low compared to that of office buildings, apartments, and private houses.

In underground garages, almost all building codes require sprinkler systems. Auto fires generally consist of smouldering upholstery, electrical fires under the hood, or the remote chance of a gasoline fire. Because the sprinkling of water or foam on the car is of little or no help, many cities now give variances to eliminate sprinklers.

Other fire detection equipment includes heat-rise indicators mounted in the ceiling of the garage. These indicators monitor temperatures in an area and are usually designed to sound an alarm when the temperature in an area rises 15° in one minute.

Placing large numbers of fire extinguishers in a garage has proven to be useless because of their thefts. Instead of small sprinklers, large extinguisher systems can be placed on dollies and situated at central locations in the garage. Although they are too large to be stolen easily, they can easily be rolled to the source of any fire.

Single-purpose vs. multi use garages. A single-purpose garage is a free-standing structure for parking vehicles with little or no area devoted to other uses. A multiuse garage is one in which the garage is a part of an overall complex consisting of more than one land use. In urban areas, there is a trend toward multipurpose garages because of the need to utilize land more effectively. Scarcity of land and its high cost often make a single-purpose garage impractical and financially unfeasible.

Parking garages may be included with many other land uses. They are often found in conjunction with office buildings, apartment complexes, and retail developments. The parking may be provided either above or below the other land uses.

The structural design of a multiuse facility is much more difficult than for a single-purpose garage. The major difficulty arises in the column spacing. The best and most efficient column spacing for a parking garage is not best for an office building or other land use above. Similarly, an ideal column spacing for an office or other building may render the parking garage inoperable. A compromise is usually required with the result that the garage will often operate less efficiently because of the compromise in column location.

Prefabricated garage systems. In the past few years a number of systems of patented construction have been introduced. Since these systems of construction are relatively new, long-range experience is unavailable on their success or failure. The kinds of prefabricated systems include:

1. Steel-frame structure with precast panels
2. Steel-frame structure with poured-in-place composite type design
3. Freestanding precast sections with concrete slab poured between the sections

Prefabricated garages are another tool available to solve the parking problem. Nevertheless, their apparent advantages and disadvantages should be studied closely before choosing the prefabricated design. The prefabricated design may have a cost advantage over conventional construction, and it may have utility when land will be available only for a limited time period before being converted to another use (such as in an urban renewal area). If, however, structural relocation is a consideration, the capability of being disassembled, moved, and reassembled in another location should be determined. This feature may be of doubtful value because of the difficulty of matching an existing garage to a new site. It is also doubtful if mechanical, electrical, and other systems could be salvaged.

An inherent feature of prefabricated garages is the large number of floor slab joints in the structure. There has been some difficulty in adequately sealing some of these garages and making the joints watertight. This is a problem that can be especially troublesome in areas of from moderate to heavy rainfall and snow, particularly as the structure ages.

Mechanical or elevator garages. In mechanical or elevator garages, the vehicle is not moved to the parking space under its own power but is moved there mechanically. There are many kinds of mechanical garages in existence, but their use worldwide has been declining in recent years, mainly because of maintenance problems associated with the mechanical equipment. In the average mechanical garage, the equipment has tended to wear out in about seven years—long before the garage has paid for itself. In addition, equipment failure can shut down the entire garage or at least a portion of it, creating poor customer relations when cars are stranded.

An additional problem with mechanical garages is their inability to accept surges of inbound or outbound traffic. Limited inbound capacity requires a large reservoir area to keep the cars off the streets. Low discharge capacity often creates customer irritation when the customers have to wait long periods of time for the return of their cars.

Mechanical garages generally provide satisfactory service only when the parking demand is relatively uniform throughout the day. In CBD's, they have proven a useful tool for serving a motel or hotel.

Underground garages. The major difference between above-ground and underground garages is the construction cost. An underground garage will usually cost from 1.5 to 2 or more times what an equivalent above-ground garage will cost per car space. This cost differential is caused by the more substantial lighting system required, as well as the additional costs of mechanical systems and air circulation and exhaust systems.

A high-water table may require increased size and depth of foundation in order to offset the effect of hydrostatic pressure. These factors generally tend to restrict the number of levels to three or four in an underground garage. Increasing construction costs and potential buildup of fumes and the problem of exhausting them generally weigh against providing more levels.

A major problem with underground garages is water seepage through the walls or through the floor joints because of hydrostatic pressure. Leakage through the roof may also become a problem if the garage is located under a park.

Design and operating characteristics of underground garages do not differ from above-ground garages, but additional emphasis should be placed on directional signing and markings because of the lack of orientation underground. It is important to keep the driver and pedestrian oriented as to streets and major land uses.

Garage construction costs. Many factors affect the total cost of a parking garage. A conventional ramp above-ground garage can usually be built for from $6.00 to $8.00 per sq ft or from $2,000 to $3,000 per car space. (Surface lots normally cost from $1.00 to $2.00 per sq ft for all improvements.)

Such factors as kind of elevator used (hydraulic vs. electric), extent of building facade, complexity of revenue control equipment, and requirement for fireproofing and water sprinklers can greatly alter the cost of a parking garage.

Underground garage construction will cost from $4,000 to $7,000 or more per car space. Mechanical garages can be expected to cost from $4,000 to $5,000 per space.

SPECIAL PURPOSE AND SPECIAL EVENT PARKING

SPORTS STADIA, CONVENTION CENTERS, AUDITORIUMS, AND EXHIBIT HALLS

Parking needs at sports stadia, convention centers, and other similar activities depend on many factors. Parking needs in a CBD will be less than for a suburban location for the same facility because of better public transportation and the ability to use existing parking facilities to meet a portion of the demand. Early analyses should be made of the stadium (or other facility) in order to determine the number and kind of events expected in the facility. Peak and average crowds should be estimated from which a design crowd can be determined.

Crowds for different events exhibit different characteristics. For example, car occupancy for baseball games may average 2.5 persons per car while car occupancy for football games may average 3.5 persons per car. Studies have also shown that the football fan will walk farther than will the baseball fan. Therefore, it is necessary to analyze the parking characteristics of each kind of event in designing the parking for the facility.

The *dump time* of the parking facilities is especially critical for special event or sports parking. This is the length of time from the end of the event until all of the parking facilities have been emptied and the last car has left the area. This value is usually between 30 and 60 min and efforts should be taken in design to minimize this time. One aid to reducing dump time is to use precashiering or collection of the parking fee on entry. The dump time then depends on the internal design of the parking facility, its ramps and exits, and the surrounding street system. Separation of pedestrian and autos should be achieved whenever possible because of the large volume of pedestrians. This is especially true at vehicle exit driveways. Pedestrian ways including overpasses or tunnels should be considered to aid in this separation.

AIRPORTS

The volume of enplaned and deplaned passengers is the most important factor affecting parking at an airport. Parking for employees, rental cars, casual visitors, and others is generally proportional to the amount of passenger traffic. For this reason, a good estimate of passenger travel is a necessity in planning parking facilities for airports.

The kind of parking used by employees and all other persons at selected U.S. airports is shown in Table 15.8. As might be expected, employees of the airport are more likely to park free than are other parkers.

TABLE 15.8
Parking Characteristics at Selected Airports*

Airport	Work Trips (%) Parking Lot				All Other Trips (%) Parking Lot			
	Paid	Free	On-Street	Not Parked†	Paid	Free	On-Street	Not Parked†
Atlanta	4	94	1	2	30	51	6	13
Buffalo	17	66	14	3	39	27	15	20
Chicago (Midway)	6	77	17	1	31	43	11	15
Minneapolis-St. Paul	12	81	3	3	16	45	21	18
Philadelphia	9	80	11	—	35	38	19	9
Providence	7	84	7	3	9	81	—	10
San Diego	20	80	—	—	7	29	43	21
Seattle-Tacoma	13	86	1	—	23	35	32	9
Washington (National)	4	91	4	1	23	46	15	15

*From transportation study data for various cities.
†Left for service or repairs, cruised, or otherwise not parked.

Source: L. E. Keefer, "Urban Travel Patterns for Airports, Shopping Centers, and Industrial Plants," NCHRP Report 24, Highway Research Board, 1966, p. 19.

Parking durations at airports range from a few minutes to more than one week. In a public parking facility at the Seattle-Tacoma International Airport, a study of parking durations showed nearly 68 percent of the total parkers stayed less than 4 hr while 19 percent stayed more than 1 day (see Table 15.9). Detailed analysis of those parked less than 4 hr showed an average duration of 57 min.

Airports are major generators of traffic and parking. In a metropolitan area, the airport is generally the largest single traffic and parking generator except for the CBD. For this reason, a rule of thumb for design of airport facilities is inappropriate. Each airport should be analyzed individually. The difference in mode of travel of passengers

TABLE 15.9
Parking Duration—Public Parking Lot Seattle-Tacoma International Airport

	Number	%	Cumulative %
Under 4 hr	514	67.8	67.8
4–8 hr	31	4.1	71.9
8–12 hr	36	4.7	76.6
12–16 hr	13	1.7	78.3
16–20 hr	8	1.1	79.4
20–24 hr	11	1.4	80.8
1 day	59	7.8	88.6
2 days	33	4.3	92.9
3 days	18	2.4	95.3
4 days	13	1.7	97.0
5 days	8	1.1	98.1
6 days	6	0.8	98.9
7 days and over	8	1.1	100.0
Total	758	100.0	

Source: James Madison Hunnicutt and Associates, *Future Traffic and Parking Requirements and Parking Financial Analysis—Seattle-Tacoma International Airport*, 1968, p. 12.

at the three New York airports is shown in Table 15.10. Within the metropolitan area, the auto usage by air passengers to the different airports ranges from 37 to 60 percent.

The number of cars generated at several airports per 1,000 departing air passengers varies widely (see Table 15.11). Los Angeles International Airport has a rate more than three times greater than O'Hare Airport in Chicago because of the different passenger characteristics of the two airports. Detailed study is then required at each airport to determine its individual characteristics before attempting to plan or design facilities.

TABLE 15.10
Mode of Travel to New York Airports

Mode	La Guardia Airport	Newark Airport	J. F. Kennedy Inter'n'l	Total
Auto	37%	60%	44%	45%
Taxi	46	12	31	32
Airport bus	9	13	12	11
Bus	2	10	3	4
Suburban limo	5	4	8	6
Other	1	1	2	2
	100%	100%	100%	100%

Source: L. E. Bender, "Airports and Their Surface Traffic Demands and Designs for 1970–1980," presented at the 1969 Annual Meeting of the Institute of Traffic Engineers, August 1969, p. 3.

TABLE 15.11
Motor Vehicle Generation Related to Passenger Departures

Airport	% Transferring from Other Planes	% Using Car or Taxi	Motor Vehicles per 1,000 Departing Air Passengers
La Guardia (New York)	11	83	530
O'Hare (Chicago)	70	81	180
Atlanta	70	93	200
Los Angeles	15	93	630

Source: Bender, p. 6.

CHANGE OF MODE FACILITIES

A change of mode facility is one in which people change from one form of transportation to a second form. In a more limited sense, change of mode terminals are often considered as the point of change from auto to mass transit. Because of the congestion in CBD areas, fringe parking areas are being used in many instances as parking facilities for commuters who then transfer to bus or rail transit.

Table 15.12 illustrates the travel mode to two change of mode stations in Milwaukee, Wisconsin. Remote sections of two shopping center parking lots were made available for commuter parking and these areas were served by express bus service to the CBD via freeways. In each instance, less than 40 percent of the commuters parked a vehicle at the change of mode station.

Table 15.13 shows the mode of travel of the same commuters prior to initiation of the shopping center-express bus service. More than 40 percent of those using the

TABLE 15.12
Mode of Travel to Shopping Center
Change of Mode Stations in Milwaukee, Wisconsin

Mode	Mayfair		Bayshore	
	No.	%	No.	%
Auto driver	157	38.9	119	38.4
Auto passenger	139	34.4	88	28.4
Drop-off	22	5.4	30	9.7
Walk	66	16.3	51	16.4
Another bus	20	5.0	22	7.1
Total	404	100.0	310	100.0

Source: H.M. Mayer, "Change of Mode Commuter Transportation in Metropolitan Milwaukee," Highway Research Circular No. 83, Highway Research Board, 1968, p. 9.

TABLE 15.13
Mode of Travel of Commuters Prior to Express Service
from Shopping Centers—Milwaukee, Wisconsin

Mode	Mayfair		Bayshore	
	No.	%	No.	%
Auto driver	182	41.2	141	44.4
Auto passenger	37	8.4	43	13.5
Bus	215	48.6	132	41.5
Train or taxi	8	1.8	2	0.6
Total	442*	100.0	318†	100.0

*Does not include 109 respondents who did not make the trip before the service began and 6 who failed to answer the question.
†Does not include 57 respondents who did not make the trip before the service began and 10 who failed to answer the question.

Source: Mayer, p. 11.

change of mode facilities had previously driven a car to work while most of the other commuters formerly had been bus passengers.

To be successful, a change of mode operation must present an attractive alternate to the auto. Frequent service from convenient locations is required. The combined cost of parking and transit fare should be less than CBD parking for a similar period and transit travel time should be competitive with auto time.

In addition to parking spaces, change of mode facilities should also provide adequate and easy access and egress location for those vehicles which deliver and pickup passengers.

SHOPPING CENTERS

Shopping centers are major generators of traffic which reach their peak in the month before Christmas. The peak day at a shopping center is generally the Saturday before Christmas. The parking accumulation on this peak day will generally occur in mid-afternoon. Typically, the peak parking accumulation on other days occurs in early evening, normally between 7:00 and 8:00 p.m. Figure 15.5 shows typical accumulation curves for the peak day and for a typical day during the pre-Christmas shopping period.

Figure 15.6 presents daily parking requirements for shopping centers presented in the form of number of days (or hours) in the year that parking demand exceeds a given figure. Studies concluded that the 30th highest hour is equivalent to the 6th highest day and that the 10th highest hour is equivalent to the 3rd highest day. Design at the 10th highest hour (3rd highest day) is recommended with a design standard of 5.5 parking spaces per 1,000 sq ft of gross leasable area. This is a reasonable figure for the average shopping center. Individual centers, however, often exhibit characteristics that result in a parking demand much more, or less, than the standard.

Actual practice in shopping center planning has been to provide spaces generally in excess of the above standard. In fact, in some instances parking spaces provided have exceeded the above standard by 100 percent or more.

The practice in some shopping centers today is to include some uses other than retail, for example, theaters and office buildings. The peaks for those other uses do not occur at the same time as the peak for retail shopping. Therefore, it has been found that up to 20 percent of the gross leasable area can be added to a shopping center as office space without affecting the peak parking demand.

INDUSTRIAL PLANTS

The parking characteristics of industrial plants are affected by a wide variety of factors. Plant location is a prime factor, particularly since it relates to suburban availability of mass transit. The amount of shift work, kind of industry, seasonal

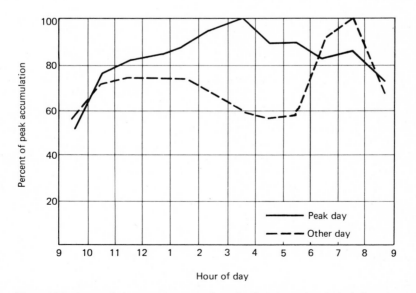

Figure 15.5. Accumulation pattern on peak and other days at a typical shopping center. (Source: A. M. Voorhees and C. I. Crow, "Shopping Center Parking Requirements," *Shopping Centers and Parking*, Highway Research Board, Record No. 130, 1966, p. 22.)

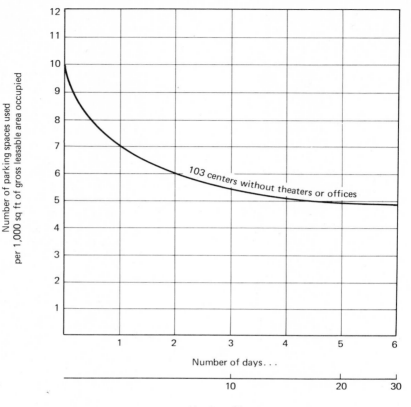

Figure 15.6. Shopping center daily parking requirements. (Source: Voorhees and Crow, p. 24.)

variations, sex of workers, and income level are additional factors that affect parking demand.

Figure 15.7 illustrates a typical vehicle accumulation curve for an auto assembly plant. The shift overlap peaks are quite distinct and, in this instance, nearly equal in magnitude.

The kind of parking facility used by employees of industrial plants varies with plant employment (see Table 15.14). The percentage of the employees using free off-street lots decreases as plant employment increases. Conversely, the number of employees parking on-street or in other facilities (generally off-street pay parking) increases with plant employment. It can be concluded that demand is more likely to exceed parking supply as the size of the plant increases.

Table 15.15 summarizes parking supply and demand per employee for various sized industrial plants. For multishift plants, the maximum shift employment is used.

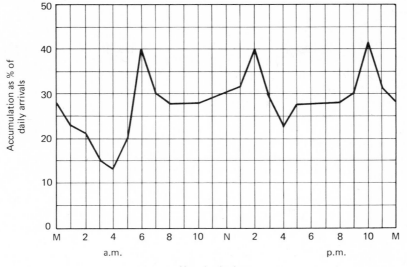

Figure 15.7. Vehicle accumulation vs. time of day at an industrial plant. [Source: *Parking Facilities for Industrial Plants* (Washington, D.C.: Institute of Traffic Engineers, 1969), p. 8.]

TABLE 15.14
Kind of Parking Used by Employees, by Plant Size

Plant Employment	Free Off-Street Lots	Percent Using On-Street Parking	Other	Total
0–2,500	95	3	2	100
2,500–5,000	94	4	2	100
5,000–10,000	77	19	4	100
10,000 +	67	14	19	100
Overall	89	6	5	100

Source: *Parking Facilities for Industrial Plants*, Institute of Traffic Engineers, Washington, D. C., 1969, p. 12.

TABLE 15.15
Parking Demand and Supply Related to Number of Employees, by Plant Size

Total Plant Employment	Parking Demand		Parking Supply	
	Number of Samples	Number of Occupied Spaces/Employee	Number of Samples	Number of Spaces Provided/Employee
0–500	24	.63	34	.81
500–1,000	11	.76	7	.70
1,000–5,000	18	.60	21	.72
5,000 +	8	.62	12	.80

Source: *Parking Facilities for Industrial Plants*, p. 13.

The parking supply per employee exceeds the corresponding demand because of the following factors:

1. Supply must exceed demand by from 5 to 10 percent in order to reduce searching for a space and possible illegal parking.
2. Assigned and reserved spaces are not always fully utilized.
3. Distant lots are often under-utilized while closer lots may be over capacity.
4. Problems of shift overlaps may require more spaces than a one-time demand count would indicate.

COLLEGES AND UNIVERSITIES

Campus parking often proves to be a major problem for college and university administrations. A survey of U.S. colleges and universities by the International Municipal Parking Congress (IMPC) showed that slightly more than 20 percent of the responding institutions had more than 10,000 student vehicle registrations while 23 percent of the respondents had faculty-staff registrations exceeding 3,000 (see Table 15.16).

The majority of campus parking is in off-street lots. Of the institutions responding to the IMPC survey, 3 percent indicated that they provide more than 10,000 off-street spaces in surface lots. This represents a land area of approximately 80 acres (32.4 hectares) devoted to surface parking. Only 8.7 percent of the responding institutions provide more than 1,000 parking spaces in garages (see Table 15.17).

The kind of parking facility used by parkers varies with campus population as determined from a study of 38 U.S. colleges and universities (see Table 15.18). As the campus population increases, the availability of free parking decreases. Less than 3 percent of the work trip parkers paid to park on campuses having less than 10,000 population. However, this increases to over 50 percent paying for parking when campus populations exceed 20,000.

TABLE 15.16
Number of University and College Vehicle Registrations

No. of Vehicle Registrations	Percentage of Responses	
	Student	Faculty-Staff
Less than 1,000	8.7	36.0
1,000–2,000		27.0
2,000–3,000		14.0
1,000–3,000	19.4	
3,000–5,000	18.5	12.0
Over 5,000		11.0
5,000–7,000	12.6	
7,000–10,000	20.4	
10,000–15,000	14.6	
Over 15,000	5.8	
Total	100.0	100.0

Source: *College and University Parking Survey*, International Municipal Parking Congress, Arlington, Va., 1971, p. 6.

TABLE 15.17
Kind of Parking Facilities Provided by Colleges and Universities

	Percentage of Responses			
No. of Spaces	Garage	Lot	On-Campus Street	Leased
Fewer than 100			49.5	95.5
100–500			25.3	3.6
Fewer than 500	83.5	11.8		
500–1,000	7.8	12.6	11.2	0.9
Over 1,000			14.0	0.0
1,000–2,000	6.1	16.5		
2,000–3,000	0.9	15.8		
Over 3,000	1.7			
3,000–5,000		22.8		
5,000–7,000		8.7		
7,000–10,000		8.7		
Over 10,000		3.1		
Total	100.0	100.0	100.0	100.0

Source: *College and University Parking Survey*, p. 6.

TABLE 15.18
Type of Parking Facility Used at Universities

	Work Trips				All Other Trips			
	Off-Street		On Street†	Not Parked‡	Off-Street		On Street†	Not Parked‡
Campus Population*	Free	Fee Paid			Free	Fee Paid		
Under 1,000	92%	—	8%	—	64%	5%	6%	25%
1,000–5,000	72	3%	25	—	45	12	28	15
5,000–10,000	91	2	6	—	60	8	24	8
10,000–20,000	63	21	15	1%	50	10	28	12
Over 20,000	38	52	9	1	22	32	27	19
Average	66	20	13	1	46	14	27	13

*Includes faculty, staff, and total student enrollment.
†Includes parking on residential property.
‡Passenger drop off, auto left for service or repairs, cruised, or otherwise not parked.

Source: L. E. Keefer and D. K. Witheford, "Urban Travel Patterns for Hospitals, Universities, Office Buildings, and Capitals," NCHRP Report No. 62, Highway Research Board, 1969, p. 77.

Average car occupancy differs more with trip purpose than with university size.[6] Car occupancy for work trips averages 1.17 persons per car, school trips average 1.22 persons per car, and all other trip purposes average 1.69 persons per car. The overall average for all trip purposes was 1.39 persons per car.

---

[6] L. E. Keefer and D. K. Witheford, "Urban Travel Patterns for Hospitals, Universities, Office Buildings, and Capitals," National Cooperative Highway Research Program Report No. 62 (Washington, D.C.: Highway Research Board, 1969), pp. 77–78.

URBAN RESIDENTIAL DISTRICTS

Parking in urban residential areas exhibits characteristics unlike that in any other area. Residential parking peaks at night and is a function of the car ownership characteristics of the residents of the area. Auto ownership is influenced by such factors as family income, age, and size, transit availability, kind of housing, and location. Table 15.19 relates auto ownership to age and income of the family and to kind of housing. As density of housing increases (low-rise vs. high-rise), auto ownership decreases. Auto ownership increases with income but decreases greatly for the elderly as compared to other age groups.

Parking for single-family residential areas is generally adequate because many single-family residences have driveways and generally some amount of on-street parking. In multifamily areas, however, particularly in older sections of the city, off-street parking space may be practically nonexistent and on-street space may be inadequate to meet the demand. A course of action adopted years ago by Milwaukee, Wisconsin, has helped solve the problem in that city. Milwaukee charges a fee for on-street parking in high-parking demand residential districts. This money then goes into a fund to build off-street parking facilities in these residential areas.

TABLE 15.19
Auto Ownership Related to Resident Characteristics, Washington, D. C.

Kind of Housing	Characteristics of Residents	Cars per Family
High-rise apartments	High income — elderly	0.33:1
High-rise ″	High income — other	1.30:1
High-rise ″	Middle income — elderly	0.20:1
High-rise ″	Middle income — other	1.10:1
High-rise ″	Low income — elderly	0.10:1
High-rise ″	Low income — other	0.20:1
Low-rise ″	High income	1.50:1
Low-rise ″	Middle income	1.30:1
Low-rise ″	Low income	0.40:1

Source: Alan M. Voorhees and Associates, Inc; *Transportation Planning for New Towns*, McLean, Virginia, p. 18.

HOSPITALS

Parking requirements for hospitals are quite varied depending on such factors as location (urban vs. suburban), number of doctors and employees, number of beds, kind of hospital, and outpatient load. Nevertheless, the parking accumulation characteristics of almost all hospitals are similar because they all require three shifts of employees. An accumulation curve for Detroit General Hospital (600 beds and 1,760 employees, including 200 part-time) shows that the peak accumulation occurs at the shift change in mid-afternoon (see Figure 15.8).

Table 15.20 shows the kind of parking facility used by parkers at hospitals. Almost all off-street facilities at hospitals are free to the parker and only a small percent pay to park. A higher percent of employees park off-street than do other persons coming to the hospital.

The basis for providing hospital parking varies considerably, as shown in parking ratios for hospitals (see Table 15.21). When employee lots are segregated from visitor

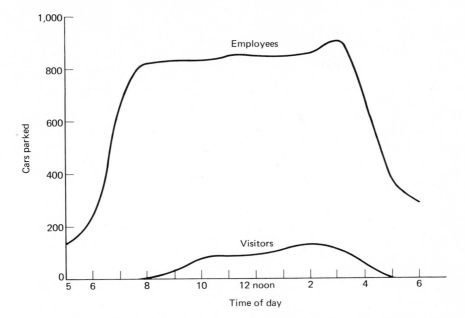

Figure 15.8. Accumulation curve, Detroit General Hospital. [Source: James Madison Hunnicutt and Associates, *Parking Study and Feasibility Report on Proposed Parking Garage*, (Detroit, Mich., 1967), p. 16.]

TABLE 15.20
Kind of Parking Facilities Used at Hospitals (%)

Kind of Hospital	Work Trips				All Other Trips			
	Off-Street		On-Street	Not Parked*	Off-Street		On-Street	Not Parked*
	Paid	Free			Paid	Free		
General:								
Under 300 beds	1	81	18	—	3	49	30	18
300–399 beds	1	91	8	—	2	54	25	19
400–499 beds	6	57	37	—	8	29	46	17
500–599 beds	—	83	17	—	—	44	31	25
600–999 beds	3	84	12	1	8	43	27	23
1,000 and over	4	84	11	1	8	47	23	22
Average	3	80	16	1	5	45	30	20
Veterans	6	84	10	—	6	56	25	13
Mental	—	97	3	—	—	62	14	24
University	25	67	7	1	7	39	12	42

*Auto left for service or repairs, cruised, or otherwise not parked.

Source: Keefer and Witheford, p. 31.

lots, general hospitals provide only 20 spaces per 100 personnel plus 32 spaces per 100 beds. When all parkers use the same lots, general hospitals provide 32 spaces per 100 beds and personnel combined. The ratio is smaller at larger hospitals than at smaller ones, indicating the probability of greater parking difficulty at larger hospitals.

TABLE 15.21
Reported Parking Ratios at Selected Hospitals Having Segregated Employee and
Visitor Parking Lots vs. Selected Hospitals Having Unsegregated Parking Lots

Kind of Hospital	Segregated Lots				Unsegregated Lots		
	No. of Hospitals	Employee Spaces per 100 Personnel	Visitor Spaces per 100 Beds	Total Spaces per 100 Beds + personnel	No. of Hospitals	Total Spaces per 100 Beds + Personnel	Grand Total Spaces per 100 Beds + Personnel
General:							
Up to 500 beds	22	21	52	31	9	41	32
Over 500 beds	6	18	8	14	11	29	20
All	28	20	32	24	20	32	26
Veterans	5	45	29	39	5	38	39
Mental	2	32	7	14	—	—	14

Source: Keefer and Witheford, p. 31.

Generally, the kind of hospital has a decided effect on the parking. Charity hospitals, those with heavy outpatient loads, specialized hospitals such as psychiatric or veterans hospitals, and those with medical schools all exhibit widely varying parking characteristics.

Car occupancy factors vary primarily with trip purpose. Work trips have an average car occupancy of about 1.1 and medical and visitor trips average 1.6 persons per car. Other trip purposes have an average car occupancy of 1.8 or 1.9 giving an overall average of 1.4 persons per car (see Table 15.22).

TABLE 15.22
Car Occupancy Factors for Hospitals

Kind of Hospital	Trip Purpose				
	Work	Medical	Visitor	Other	All
General:					
Under 300 beds	1.13	1.48	1.49	1.98	1.44
300–399 beds	1.11	1.61	1.56	1.90	1.43
400–499 beds	1.09	1.60	1.61	1.82	1.41
500–599 beds	1.08	1.77	1.44	1.78	1.44
600–999 beds	1.10	1.70	1.57	2.04	1.50
1,000 and over	1.16	1.48	1.58	1.62	1.38
Average	1.11	1.59	1.56	1.85	1.43
Veterans	1.10	1.32	1.66	1.70	1.29
Mental	1.15	1.31	1.60	1.71	1.32
University	1.17	1.46	1.63	1.84	1.44

Source: Keefer and Witheford, p. 32.

ZONING FOR PARKING

Zoning ordinance provisions are a means of ensuring an adequate supply of off-street parking in all new developments. The zoning requires that a specific number of parking spaces be provided with all new construction or major building modification.

TABLE 15.23
Zoning and Planning Standard Guidelines for Parking

Land Use	Unit	Parking Spaces/Indicated Unit			
		Zoning Requirements	Parking Space Needs	Planning Standard	Recommended Minimum Standard
Residential					
Single-family	Dwelling	1–2	0.5–2.0	1–2+	2
Multifamily	"	0.4–0.5 and up	0.3–2.0	0.7–2.0	
Efficiency					1.0
1 and 2 bedrooms					1.5
3 or more bedrooms					2.0
Hospital	Bed	0.25–1.40	0.60–1.40	1.0–1.4	1.2
Auditorium, theater, or stadium	Seat	0.08–0.25	0.08–0.50	0.25–0.33	0.3
Restaurant	Seat	Variable	N.A.*	0.33–0.50	0.3
Industrial	Employee	"	Variable	0.33–0.50	0.6
Church	Seat	0.10–0.33	N.A.*	0.20–0.33	0.3
College-university	Student	Variable	0.4–0.6	0.5–0.7	0.5†
Retail	1,000 sq ft (gross floor area)	1.5–3.0	1.5–8.0	2.0–8.0	4.0
Office	1,000 sq ft (gross floor area)	Variable	2.9–4.0	2.0–5.0	3.3
Shopping center	1,000 sq ft (gross leasable area)				5.5
Hotels, motels	Rooms and employees				1.0 per room and 0.5 per employee
Elementary–Junior High School	Classroom				1.0
Senior High School	Student and staff				0.2 per student and 1.0 per staff

*Not available.
†With auto access only (0.2 with good transit access).

Sources: Wilbur Smith and Associates, under commission from the Automobile Manufacturers Association, "*Parking in the City Center*, 1965, pp. 64–67. *Parking Principles*, Highway Research Board, Special Report No. 125, 1971, pp. 34–39.

It therefore represents a long-range approach to reducing parking problems in cities. An adequate supply of off-street spaces will allow on-street parking to be limited or prohibited. The streets can then better serve their primary function of moving traffic.

Parking demand not only varies according to land use, but it also varies within parts of the city for the same land use. Factors which affect parking demand include the availability of mass transit and the economic levels, local policies and customs of the city or portion of the city. Use of an average value as a zoning requirement will obviously result in inadequate parking facilities in some instances and in overdesign with attendant economic waste in others.

Local parking generation studies are useful in determining parking needs, and their results should be used to update zoning ordinances. Zoning ordinances can also specify parking requirements according to districts within the city. Zoning requirements for a specific land use could vary from district to district according to the character of the district. An appeals board could then handle any further request for variation from zoning requirements.

Planned Unit Development (PUD) is a relatively new tool in zoning. Under this zoning, several different land uses may be incorporated on a tract of land. The parking demand for the development is based on the characteristics of the land uses being planned. Almost all zoning ordinances require that the appropriate factor be applied to each land use in the development to determine the parking requirements for the individual land uses. These requirements are then added to determine the total parking requirements for the development. This total requirement, however, may be increased or reduced upon review of plans for the development, depending on the mix of land uses and the location of the project.

Table 15.23 summarizes zoning and planning standards for parking for various land uses. The data represented are from two different reference sources with the "minimum standard" figures taken from one source and the remainder from another source. In almost all instances, there is fairly good agreement between the two sources. Nevertheless, some discrepancies occur, for example, for industrial land uses; one source shows a planning standard of from 0.33 to 0.50 spaces per employee, and the other source recommends a minimum standard of 0.6 spaces per employee. The table clearly shows the variability of parking needs and requirements for the same land use.

MUNICIPAL PARKING PROGRAMS AND FINANCING

A variety of organizational structures and financing techniques has been used by cities in the effort to solve their parking problems. Each method has its advantages and disadvantages depending on city size, governmental organization, and other factors.

MUNICIPAL PARKING ADMINISTRATION

The parking function can be delegated to an existing department within the city government, for example, the Public Works Department or the Traffic Engineering Department. This is the simplest method of establishing a parking program, and it is often used in smaller cities. In this arrangement, however, parking needs often do not receive the necessary priority, and the parking program may be less active than it should be.

A separate parking department can give the parking program the required attention to solve parking problems. With a full-time staff devoted to parking, capable and experienced personnel are available, and the parking program is likely to be more effective than when parking is delegated to an existing department. There is, however, a need for close coordination between separate departments, such as between the parking department and the traffic engineering department.

A parking board, commission, or authority generally consists of a citizens board of advisors which makes recommendations to the city parking department. The members of this board of advisors or commission are appointed. This organization provides an opportunity for citizen participation in improving parking conditions.

The parking authority is a separate autonomous entity whose sole purpose is to carry out the parking program of the city. The parking authority is governed by an appointed parking board. It has its own staff and budget and can plan and carry out a parking program suitable to meet the needs of the city. Important features of parking authorities are that they can condemn property for parking development and they can issue nontaxable revenue bonds for financing parking developments.

FINANCING OF PARKING PROGRAMS

There are many kinds of arrangements available for financing municipal parking programs. One method primarily used in smaller cities is to finance parking improvements from the city's operating or capital improvements budget. This often results from an adopted policy in some jurisdictions that parking is a government service similar to the provision of streets or sanitary systems.

Parking revenues are often the source of funds for providing off-street parking facilities. The revenues may come from curb meters and existing off-street facilities as well as from parking tickets. These revenues are placed in a reserve fund for development of off-street parking facilities. The disadvantage of this pay-as-you-go system is that funds often do not accumulate fast enough to meet the parking demand.

A user benefit assessment is a third form of financing available to municipalities. Under this arrangement, when a parking improvement is planned, all properties within an assessment district are assessed according to a formula which apportions the improvement cost according to the benefits received. The formula for apportionment may consider such factors as number of parkers, distance from the proposed facility, and proportionate relation of gross sales of the businesses within the district. The principal disadvantage of this assessment is that disagreements over assessment rates and objectors to the system may delay the financing of the facility.

A fourth method of financing parking programs is through a parking tax. Within a parking district, a business has the option of providing the parking required by zoning ordinance or, in lieu of providing the parking, it can pay a parking tax. The parking tax in each district, plus funds from parking meters and fines, is used to provide sufficient off-street parking to meet the parking needs in that district.

Parking revenue bonds are a popular financing tool in many cities. For revenue bonds, only parking system revenues are pledged to the retirement of the bonds and the borrowing power of the city is unaffected. Because of this limitation, interest rates are generally higher than for general obligation bonds, and a debt coverage factor of 1.5 or greater is often necessary.

General obligation bonds are backed by the full faith and credit of the city and, therefore, result in a lower interest rate than revenue bond issues. A debt coverage factor of 1.1 or 1.2 is generally adequate for general obligation bonds. The borrowing power of the city, however, is reduced by the amount of the general obligation bond issue for the parking facilities.

A combination of private and public participation is possible in CBD developments. A parking authority or city may use its financing capability to aid in the development. This procedure normally requires the establishment of a private nonprofit corporation in which the parking authority has an active interest. After the parking facilities are built, they are leased to the nonprofit corporation to operate, and the corporation guarantees to pay the bonded indebtedness.

CURB PARKING

The primary use of the streets in any city is for the movement of vehicles. Parking on these streets (curb parking) must be considered a secondary use of street space as should other such uses as truck-loading zones. When parking or other secondary uses of the streets conflict with the movement of traffic, those uses should be removed so that the streets can best perform the function of moving traffic.

TYPES OF CURB SPACE

The curb space along streets in any city will have several categories of use, including, curb parking, truck-loading zones, no-parking zones, bus and taxi zones, passenger-loading zones, and others.

The amount of parking allowed at the curb varies with city size. In larger cities, curb parking may be severely restricted to aid the movement of traffic. In the Nashville, Tennessee CBD, only about 40 percent of the total lineal feet was available for public parking (see Table 15.24). Slightly more than 20 percent was devoted to special curb uses, and the remainder was in areas in which parking was prohibited.

TABLE 15.24
Use of Curb Space, Nashville CBD

Use	Lineal Feet	Percent	
Public parking			
Metered	29,948	20.8	
Unmetered	28,862	20.0	
			40.8
Parking Prohibited	54,023		37.5
Special uses			
Truck-loading zones	6,570	4.6	
Passenger-loading zones	550	0.4	
Bus zones	2,159	1.5	
Taxi zones	832	0.6	
Driveways	14,122	9.7	
Fire hydrants	4,698	3.3	
Alleys	1,804	1.2	
Mail boxes	505	0.4	
			21.7
Total	144,073		100.0

Source: *Nashville Metropolitan Area Transportation Study—Downtown Parking*, Nashville Parking Board, 1960, p. 24.

PARKING PROHIBITIONS

Parking prohibitions can be warranted on the basis of statutes, traffic capacity, or accident hazard. The statutory warrants of the *Model Traffic Ordinance*[7] authorize full-time prohibitions on both sides of roadways not exceeding 20 ft (6.1 m) in width and on one side for those not over 30 ft (9.1 m) in width. Studies of street capacity have found that parking reduces capacity by one-quarter to one-third or more on typical streets. When capacity problems exist, street parking should be removed if possible. Table 15.25 illustrates a recommended warrant for prohibiting parking because of traffic conditions.

A location where curb parking has proven to be an accident hazard could warrant a parking prohibition. Analysis of accident records would indicate locations where accidents are involving curb parkers.

TABLE 15.25
Parking Prohibition Criteria

Kind of Prohibition	Maximum Vehicles per hr per Lane When Parking Allowed (One Direction of Flow)	
	1 lane	2 or more lanes
Mid-block prohibition for entire street	400	600
Intersect prohibition up to 150 ft on approach and departure	300	500

Source: *Parking Principles*, p. 176.

REGULATIONS OF CURB PARKING

Time restrictions are often posted on curb parking spaces to increase the turnover of parkers. Time restrictions may be posted on signs only or parking meters may be used in conjunction with signs.

In addition to the parking meter, other methods are being used in Europe to regulate curb parking. Principal among these is the disc system (see Figure 15.9). Upon parking, the motorist sets the initial pointer to his arrival time (5:15 as illustrated in Figure 15.9). As the space between the pointers is fixed for each disc, the allowable duration is shown and the second pointer shows the time when the maximum allowable duration has been reached (6:45 in the illustration). The disc is displayed inside the vehicle and enforcement officers can readily check for violations.

The second method used in Europe is known as Park System. This system consists of color-coded parking tickets in a packet graduated by price and/or time limit which are purchased by parkers (see Figure 15.10). On parking, the motorist selects a ticket from the packet of the appropriate price or duration and punches the month, date, hour, and minute of arrival and displays the ticket in a window on his vehicle.

Very short limits (up to 15 min maximum) are used in areas where very high

---

[7] "Model Traffic Ordinance," Uniform Vehicle Code and Model Traffic Ordinance (Washington, D.C.: National Committee on Uniform Traffic Laws and Ordinances, revised 1968), p. 27.

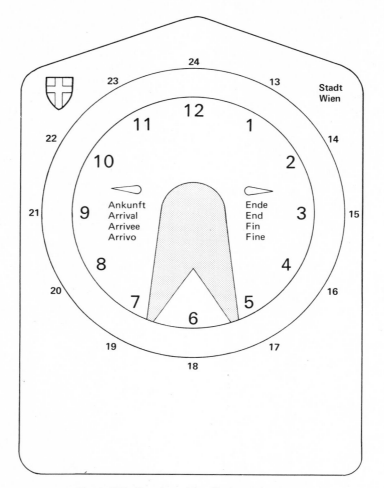

Figure 15.9. Sample parking disc in use in Europe.

turnover may be expected, such as near banks and post offices, and are often associated with "errand" parking. Average time limits are in the range of from $\frac{1}{2}$ hour to 2 hours and are typically found in retail areas, office areas, etc. Long-term limits are in the range of from 3 to 10 hr and are often used in fringe residential areas to discourage employee parking.

CURB FACILITY DIMENSIONS

Three kinds of stalls must be considered in dimensioning curb parking facilities: end stalls, interior stalls, and the "paired parking stall." Since a vehicle can either be driven directly into or out of it, the end stall need only be long enough to accom-

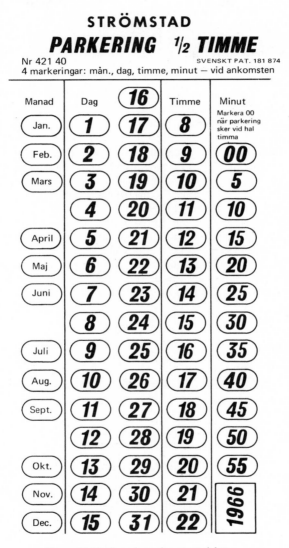

Figure 15.10. Sample park system ticket.

modate a parked vehicle. A length of 20 ft (6.1 m) is sufficient and often used today. Interior stalls must allow room for maneuvering, and a 23 to 26 ft (7.0 to 7.9 m) stall that allows for a 19-ft (5.8-m) vehicle is recommended.

"Paired" parking has stall layouts so that two vehicles are parked bumper to bumper, with the pairs of stalls separated by maneuver areas. Stall lengths of from 18 to 20 ft (5.5 to 6.1 m) are recommended with a well-defined maneuver area of from 8 to 10 ft (2.4 to 3.0 m) (see Figure 15.11).

Curb spaces marked for import size cars can be smaller than the standard sizes shown above. A suitable design vehicle would be 15 ft long (4.6 m). Interior parking stalls at the curb are recommended to be 19 ft long (5.8 m) as opposed to the 23 ft (7.0 m) required for standard American cars.

The parking stalls should be defined by white lines extending perpendicular from the curb for a distance generally recommended at 7 ft (2.1 m). The end stall line is generally marked with an *L* and interior lines have a *T* shape.

Truck-loading zones are generally from 30 to 60 ft (9.1 to 18.3 m) in length. Placing truck zones adjacent to no-parking areas allows additional maneuvering space for trucks.

Taxi zones require about 20 ft (6.1 m) for each stall included, plus an additional 5 ft (1.5 m) on each end for maneuver area.

Bus-loading zones range in length from 50 to 145 ft (15.2 to 44.2 m), depending on size of bus, number of buses loading at one time, location of stop (nearside, far side, or mid-block), and ease of access to the zone.

Passenger-loading zones should allow for the pickup of passengers without any backing maneuver required by the driver. This requires about 50 ft (15.2 m) for the initial space; an additional 25 ft (7.6 m) should be allowed for each additional space.

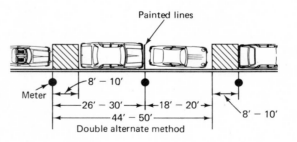

Figure 15.11. Paired parking layout.

TRUCK PARKING CHARACTERISTICS

Trucks are an integral part of the transportation system and a knowledge of truck parking characteristics is essential for a complete understanding of the transportation system in any city.

The distribution of truck trips by land use in several metropolitan areas is shown in Table 15.26 along with population data for each city at the time the study was undertaken. Residential and commercial land uses generate the bulk of the truck trips in each city. When the data are presented on the basis of truck trips generated per sq mi of land use, industrial land generates the largest number of truck trips and commercial land uses the second largest (see Table 15.27.)

Although trucks represent from 15 to 20 percent of the total CBD trips, they account for only approximately 7 to 8 percent of the total parkers in the downtown area. At the time of maximum parker accumulation, the truck influence is even less, with trucks accounting for only 2 to 3 percent of the maximum accumulation (see Table 15.28).

Truck parking in the CBD takes place predominantly at the curb, often in illegal

TABLE 15.26
Motor Truck Trip Generation in Relation to Various Land Uses

Item	Urban Area				
	Chicago	Philadelphia	Minneapolis	Nashville	Tucson
Population	5,169,663	4,342,897	1,376,875	357,585	244,495
Truck trips	828,000	1,089,786	390,559	91,605	57,000*
Land Use	Percent of Total Truck Trips				
Residential	49	35	37	39	25
Commercial	29	28	34	37	45
Retail	—	20	25	27	—
Service and wholesale	—	8	9	10	—
Public land and buildings	9	2	7	8	5
Industrial	9	9	17	7	20
Transportation	4	14	—†	6	—†
Other	—	12	5	3	5
Total	100	100	100	100	100

*Indicated distribution is based on 41,010 truck trips.
†Included with industrial land.
Data from Nashville, Penn-Jersey, Twin Cities, Tucson, Chicago area transportation studies.

Source: Wilbur Smith and Associates, under commission from the Automobile Manufacturers Association, *Transportation and Parking for Tomorrow's Cities*, 1966, p. 374.

TABLE 15.27
Truck Trip Generation Factors for Various Land Uses, Nashville, Tenn., 1959

Land Use	Truck Destinations per sq mi of Given Land Use		
	Light Trucks	Heavy Trucks	All Trucks
Residential	471	116	587
Industrial	808	801	1,609
Transportation, communication, utilities	310	308	618
Commercial	647	300	947
Public land and buildings	174	157	331
Other	10	11	21
All land uses	269	121	390

Source: Wilbur Smith and Associates, p. 251.

curb space. For trucks loading and unloading in the CBD, more than 95 percent use curb spaces (see Table 15.29).

Table 15.30 shows the trip purpose of truck parkers in the CBD. Loading and unloading of goods is the major purpose, although service and business also account for large numbers of truck parkers.

The average parking duration for commercial vehicles in Pittsburgh is related to trip purpose (Table 15.31) and to the land use at the destination (Table 15.32). Taxis

TABLE 15.28
Commercial Vehicle Parking in the CBD

	Chattanooga	Nashville	New Orleans	Cities Combined
Total parkers				
All vehicles	18,491	30,110	31,661	80,662
Trucks	1,767	1,942	2,182	5,891
Truck % of total	9.6	6.4	6.9	7.3
Maximum accumulation				
All vehicles	5,212	11,182	12,167	28,561
Trucks	167	221	324	712
Truck % of total	3.2	2.0	2.7	2.5
Maximum accumulation as % of total accumulation				
All vehicles	28.2	37.1	38.4	35.4
Trucks	9.4	11.4	14.8	12.1

Source: Wilbur Smith and Associates, p. 259.

TABLE 15.29
CBD Truck Parking by Kind of Facility

Kind of Facility	Chattanooga	Nashville	New Orleans
	All Trucks—Percentage Distribution		
Loading zone	50.0	35.7	16.7
Other legal curb	26.2	42.2	21.6
Illegal curb	13.9	8.8	47.3
Off-street	9.9	10.5	14.4
Off-street loading docks	N.A.*	2.8	N.A.*
Total	100.0	100.0	100.0
	Trucks Loading and Unloading Merchandise—Percentage Distribution		
Loading zone	73.5	55.9	17.6
Other legal curb	10.1	29.7	6.3
Illegal curb	15.7	10.7	75.0
Off-street	0.7	3.7	1.1
Total	100.0	100.0	100.0

*Not available.

Source: Wilbur Smith and Associates, p. 261.

were included as a part of the study. Heavy trucks exhibited much longer parking durations than did light or medium trucks. Handling of goods normally produces shorter parking durations than do other purposes, for example service, personal business, or parking at the base of operation. Parking durations at various land uses generally averaged less than one hour when all vehicles were considered together. Manufacturing land uses produced the longest average duration.

TABLE 15.30
Trip Purpose of CBD Truck Parkers

Trip Purpose	Chattanooga	Nashville	New Orleans
	All Trucks—Percentage Distribution		
Loading-unloading	30.2	36.0	43.7
Sales-service	24.7	13.7	8.0
Work	8.9	8.2	16.0
Business	21.6	30.4	16.0
Shopping	8.1	4.7	2.2
Other	6.5	7.0	14.1
Total	100.0	100.0	100.0
	Trucks in Loading Zones—Percentage Distribution		
Loading-unloading	44.3	56.3	45.2
Sales-service	27.4	8.2	17.0
Work	7.7	4.3	17.3
Business	12.0	20.5	15.3
Shopping	4.2	4.6	2.2
Other	4.4	6.1	3.0
Total	100.0	100.0	100.0

Source: Wilbur Smith and Associates, p. 376.

TABLE 15.31
Average Parking Duration of Commercial Vehicles
by Trip Purpose, Pittsburgh

Purpose	Average Duration (min)				
	Taxi	Light Truck	Medium Truck	Heavy Truck	Average
Pick up goods	—	29	51	110	36
Deliver goods	—	15	27	127	19
Pick up and deliver	—	14	25	50	17
Service	6	122	91	—	58
To base of operations	63	79	55	131	73
Personal business	—	90	47	—	82
Average	7	37	38	113	34

Source: L. E. Keefer, "Trucks at Rest," *Origin and Destination Techniques and Evaluations*, Highway Research Board, Record No. 41, 1963, p. 32.

TABLE 15.32
Average Parking Duration of Commercial
Vehicles by Land Use at Destination, Pittsburgh

Land Use	Average Duration (min)				
	Taxi	Light Truck	Medium Truck	Heavy Truck	Average
Residential	6	30	33	306	27
Retail	6	33	25	136	30
Services	6	46	35	33	37
Wholesale	5	59	48	69	53
Manufacturing	4	83	53	130	67
Transportation, utilities, communication	14	76	56	104	58
Public buildings	4	63	30	30	39
Public open space	2	68	50	—	51
Airports, streets, and railroad land	8	27	51	24	26
Average	7	37	38	113	34

Source: Keefer, p. 35.

OFF-STREET TRUCK FACILITIES

LOCATIONS AND DESIGN

The delivery and shipping of raw materials and finished goods are accomplished to a large extent by the use of trucks of all sizes. These trucks use the same streets for access to the industrial, retail, and other areas as do other vehicles but special provisions are required to meet the specialized needs of trucks on the site.

In general, trucks use the same entrances to a site as do employee vehicles and other traffic. The entrances and exits must be designed to accommodate the largest truck expected to visit the site. All of the United States now allow 55 ft (16.8 m) tractor-trailer lengths. Some states allow units of 60 and 65 ft (18.3 and 19.8 m). Figure 15.12 illustrates the entrance width and flare length required for the WB-50 design vehicle [overall length of 55 ft (16.8 m)] under the set of conditions shown [no parking, vehicle turning from curb lane, property line set back 15 ft (4.6 m) from curb]. If parking is allowed at the curb on the approach street, the vehicle path will be moved farther from the curb and result in a decreased entrance width and flare length. Adjustment of the property line location will likewise change the entrance dimensions. Ease of turning into the site may be accomplished by use of "Y" or angle approach. This may be particularly useful for access to and from a one-way street.

The width of roadway required at gates is generally recommended at 16 ft (4.9 m) for one-way operation, 28 ft (8.5 m) for two-way operation, and 34 ft (10.4 m) when pedestrian traffic is involved. If inbound trucks are stopped at the gate, it will be necessary to recess the gates so that sufficient storage space will be available for one, and preferably two trucks, without encroaching on the approach street.

Service roads within the property should be 24 ft (7.3 m) in width for two-way operation. Whenever possible, truck traffic should circulate counterclockwise because the left turn is easier with large commercial vehicles since the driver's position is on

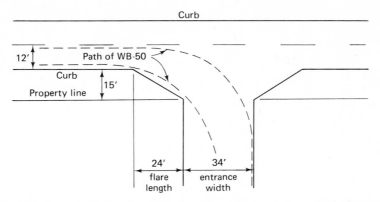

Figure 15.12. Driveway requirements to accommodate WB-50 design vehicle. [Source: WB-50 path from *A Policy on Geometric Design of Rural Highways* (Washington, D.C.: American Association of State Highway Officials, 1965), p. 84.]

the left side of the vehicle. Likewise, this places the truck in the most favorable position for backing into the dock. Care should be taken to prohibit parking where it may conflict with truck circulation or maneuvering into the truck dock areas.

A waiting or holding area for trucks is required next to the docks in order to accommodate trucks waiting for a dock space. The size of this area should be sufficient to provide space for the maximum number of trucks expected on the site less the number of dock spaces provided.

The maneuver area required in front of docks depends on the overall length of trucks using the facility, turning radii of the trucks, direction of traffic circulation, and the width of berths. A maneuver length of not less than twice the overall length of the longest vehicle using the facility has been recommended.[8]

Another recommendation is for a maneuver area of 105 ft (32.0 m) with counter-clockwise circulation and 165 ft (50.3 m) with clockwise circulation to avoid blind right-hand backing maneuvers.[9]

Individual berths should be from 12 to 14 ft (3.7 to 4.3 m) in width. Berths less than 12 ft (3.7 m) are generally not acceptable. Minimum overhead clearance generally should not be less than 15 ft (4.6 m). Height of dock for servicing all trucks should be from 48 to 50 in. (1.22 to 1.27 m). Almost all docks built today place the face of the dock flush with the outside wall of the building. The flush type dock can offer a covered, heated, closed dock operation without enclosing the trucks.

Enclosed docks completely enclose the truck when it is parked at the dock. They offer complete protection of goods, control over pilferage, ability to erect crane systems or other overhead systems for loading and unloading of unenclosed trucks, and an operating atmosphere not affected by outside weather conditions. The totally enclosed dock includes the maneuver area; trucks pull into the area, back into the dockspace, and generally leave by another door. A second kind of enclosed dock does not enclose the maneuver space; trucks back through a door and straight into the loading dock. Care must be taken to eliminate the accumulation of vehicle exhausts in the enclosed area.

---

[8] *Parking Facilities for Industrial Plants* (Washington, D.C.: Institute of Traffic Engineers, 1969), p. 26.

[9] *Modern Dock Design* (Milwaukee, Wisc.: Kelly Company, Inc., 1968), p. 2.

Modern industrial construction has often eliminated basements and dock-level buildings. Depressed approaches to docks can be used to create the elevation differential needed for the loading-unloading operation. Approach grades should not exceed 10 percent.

With a depressed approach, the top of the truck may contact the building wall before the truck bed contacts the bumpers if the dock is flush with the wall. To eliminate this condition, the approach grade can be reduced, the building wall recessed, or the dock face extended. Another solution is the use of a 15 to 25 ft (4.6 to 7.6 m) level section adjacent to the dock before beginning the approach grade.

When maneuver space is insufficient for trucks to use docks situated parallel to the building wall, a sawtooth arrangement may be used to reduce the maneuver area needed. The number of berths accommodated in a given dock length will be fewer with angular or sawtooth arrangement than with trucks parking at 90°. Berths at angles of 30° or less use twice as much dock space per berth as do berths at 90°. Truck circulation must be such that vehicles approach and depart with the angle of the dock so that maneuvering into the berth involves an angle less than 90°. Whenever possible, circulation should be counterclockwise in order to avoid blind backing maneuvers.

ZONING FOR LOADING FACILITIES

Truck activity varies greatly not only between land uses but also among similar land uses. Similarly, the number of off-street loading spaces varies widely. Table 15.33 summarizes the number of loading spaces provided for various industrial uses observed. The number of berths per 10,000 sq ft of heavy industrial use ranged from 0.09 to 0.98. Loading dock requirements are obviously dependent on the individual industrial operation.

Research by the ENO Foundation[10] found that 84 percent of the cities over 100,000, 66 percent of the cities between 50,000 and 100,000, and 64 percent of the cities under 50,000 have off-street loading controls. These are most often related to the gross floor area of the activity, but specifications varied greatly. The findings revealed that many cities do not require loading space for activities occupying fewer than 5,000 sq ft (464.5 sq m). Table 15.34 summarizes the range of requirements on the provision of the first and second loading space for various land uses. In contrast to these requirements, Table 15.35 presents recommended truck berth criteria for commercial and industrial land uses. One space per 100,000 sq ft (9,290 sq m) of floor area is reported as adequate for residential zones, hotels, and theaters.

TRUCK TERMINALS

The truck terminal, along with other freight terminals, forms an important link in the movement of goods in an urban area. There has been little change in the number of truck terminals in most cities, but there has been a trend toward increases in average terminal size. In the New York metropolitan area, there are 299 Class I and II intercity carriers operating out of 373 terminals[11] (Class I carriers have over $1

---

[10] G. E. KANAAN and D. K. WITHEFORD, "Zoning and Loading Controls," *Traffic Quarterly*, XXV, No. 3 (1971), 449.

[11] WILBUR SMITH and ASSOCIATES, under commission from the Automobile Manufacturers Association, *Transportation and Parking for Tomorrow's Cities* (New Haven, Conn.: 1966), p. 262.

TABLE 15.33
Relationship of Land Area and Building Size to Loading Facilities
Classified by Kind of Industrial Operation

Description	Land Area (sq ft)	Building Area (sq ft)	Number of Truck Docks	Number of Berths per 10,000 sq ft Land Area	Number of Berths per 10,000 sq ft Building Area
Bakery	600,240	259,890	24	.40	.92
Office and	62,530	30,220	3	.48	.99
research	74,330	35,160	4	.54	1.14
with	190,640*	60,000	1	.05†	.17†
warehouse				Average .51	1.07
Truck storage and terminal	47,720	7,730	12	2.52	15.52
Printing	78,190	43,250	2	.26	.46
	156,130	104,940	6	.38	.57
				Average .32	.52
Medium	45,610	18,540	2	.44	1.08
manufacturing	862,340	296,350	5	.06	.17
and				Average .25	.62
fabrication					
(electrical com-					
ponents, etc.)					
Heavy	20,440	8,425	2	.98	2.38
manufacturing	36,945	20,240	2	.54	.99
and	74,630	34,650	3	.40	.87
fabricating	89,900	50,230	3	.33	.60
(machine, tool	155,870	59,860	2	.13	.33
and die, metal	156,880	76,020	5	.32	.66
stamp, etc.)	296,760	187,260	6	.20	.32
	364,813	161,850	9	.25	.56
	690,990	321,590	7	.10	.22
	831,850	306,980	8	.09	.26
				Average .33	.62

*Office and research only.
†Not included in average.

Source: W. A. Alroth, "Parking and Traffic Characteristics of Suburban Industrial Developments," *Parking,* Highway Research Board, Record No. 237, 1968, p. 10.

million in gross revenue and Class II carriers $200,000 to $1 million). Although the average size is increasing, almost all terminals are still of moderate size with 75 percent having fewer than 20 spaces.

There has not been any large trend toward consolidation of terminals. Terminal consolidation can lower the vehicle miles of truck travel and time spent en route thereby reducing the cost of transport of merchandise. Consolidated terminals have been estimated to halve the amount of required truck travel.

The location of truck terminals depends on the road system, particularly the freeway system. Truck terminals seek to maximize access to major routes both to the CBD and to the region as a whole. A balance must also be sought between the advantages of a suburban location and those found in the high-density CBD. Figure 15.13 shows a generalized location in a metropolitan area which balances accessibility, density, and central location. Ideally, a location on the fringe of the CBD, with easy access to major freeways, is desired.

It is difficult to locate all truck terminals in this peripheral area because of differing

TABLE 15.34
Summary of Loading Space Requirements

Land Use	Total Floor Area (1,000 sq ft) requiring:	
	First Loading Space	Second Loading Space
Commercial, industrial	7–10	27–35
Institutional	12–15	75–80
Hotels, office, residential	20–30	80–113

Source: G. E. Kanaan and D. K. Witheford, "Zoning and Loading Controls," *Traffic Quarterly*, XXV, No. 3, ENO Foundation for Transportation, Saugatuck, Connecticut, 1971, p. 452.

TABLE 15.35
Suggested Truck Berth Criteria for Commercial and Industrial Land Uses

Floor Area (sq ft)	Truck Berths Number of
Under 8,000	1
8,000–25,000	2
25,000–50,000	3
50,000–100,000	4
100,000–250,000	5
Each additional 200,000	1

Source: Wilbur Smith and Associates, p. 265.

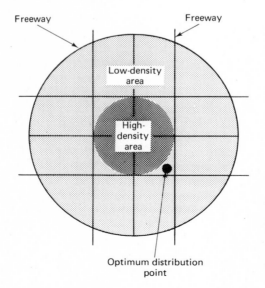

Figure 15.13. Generalized location of optimum distribution point for urban motor freight terminals (theoretical). (Source: Wilbur Smith and Associates, under commission from the Automobile Manufacturers Association, *Transportation and Parking for Tomorrow's Cities*, 1966, p. 264.)

functions among trucking companies. Intercity trucking companies normally do locate in the fringe areas near major highways or freeways, but local service trucking companies normally locate nearer the CBD area.

Terminal location and design are not static processes, but change with changes in technology in goods movement. Advances in vehicle design, expansion of freeway systems, improved loading-unloading facilities, and the concept of containerization of freight have all affected terminal location and design.

TRANSIT LOADING FACILITIES[12]

CURB LOADING ZONES—LOCATION

The location of bus-loading zones should receive priority over truck-loading zones or parking zones at the curb. In the location of bus-loading zones, factors that should be considered include:

1. Safety
2. Bus routing
3. Number of buses
4. Traffic and pedestrian volumes and direction of movement
5. Origin and destination of bus patrons
6. Transfer between lines
7. Location and operation of traffic control devices

Bus zones can be located on either the near or far side of the intersection or at a mid-block location as local factors dictate.[13] A particular location is often desirable as noted in the following listing:

1. When route direction changes require a left turn, a far side location is preferred.
2. When route direction changes require a right turn and the curb radius is short, a mid-block location is preferred.
3. If there is a high percentage of right turns at the intersection, bus stops should be located on the far side of the intersection.
4. If accumulation of buses occasionally exceeds the length of bus zones, far side stops should be avoided and the zone placed on the nearside.
5. At complicated intersections, a far side location is often advantageous.
6. At transfer points of two crossing lines, placing one stop on the nearside and the stop for the crossing line on the far side is an advantageous arrangement. This places both stops on the same corner and minimizes street crossings by transfer passengers.
7. When a large percentage of bus passengers using a stop are destined to a single large generator, the bus stop should be located so that pedestrian traffic is minimized in the intersection. The proper bus stop location could be either nearside or far side.

---

[12] Also see Chapter 6, Mass Transportation Characteristics.

[13] For further information, see *A Recommended Practice for the Proper Location of Bus Stops* (Washington, D.C.: Institute of Traffic Engineers, 1967).

CURB ZONES—DIMENSIONS

The recommended minimum lengths of curb loading zones are based on a 40 ft (12.2 m)-length bus and frequent service.[14] For longer or shorter buses, the zone length should be adjusted accordingly. A nearside bus stop for a single bus should be 105 ft (32 m) in length (measured from the front of the stopped bus to the front of the preceding parking stall). A farside bus stop should be 80 ft (24 m) in length for a single bus (measured from the rear of the stopped bus to the end of the first parking stall). A farside bus stop after a right turn should be 140 ft (43 m) in length for a single bus (measured from the edge of the lane from which the bus is turning to the end of the first parking stall. A mid-block bus stop for a single bus should be 140 ft (43 m) in length (measured from the front of the preceding parking stall to the rear of the next parking stall). All bus stop dimensions have been for a single bus. An additional 45 ft (13.7 m) should be added for each additional bus parked simultaneously. This allows for the 40 ft (12.2 m) length of bus and 5 ft (1.5 m) between buses.

The service volume of a bus route is dependent on the number of persons boarding or alighting at each bus stop. Each bus requires a certain length of stop that can be calculated if certain conditions about the fare system and the amount of baggage carried by patrons (see Table 6.7) are known. The average stop time per bus can then be determined.

Bus zones, when closely spaced, result in poor operating efficiency because of time lost in acceleration, deceleration, and in maneuvering into and out of bus zones. Spacing may range from one stop per block when city blocks are 500 ft (152 m) or more in length to stops in alternate blocks when city blocks are shorter. Recommended spacing for local transit service stops is between 850 and 950 ft. A closer spacing may be necessary in business areas.

CAPACITY CONSIDERATIONS FOR CURB LOADING

The effect of local buses on street capacity is a function of the area of the city, street width, parking conditions, number of buses, and bus stop locations.[15] Generally, when bus volumes are appreciable, nearside bus stops will reduce street capacity to a greater extent than will farside stops. Both right turn and through movements will be affected by nearside stops when parking is prohibited, Nearside stops, however, often allow the bus to combine loading-unloading delays with delays caused by the traffic signal. When parking is prohibited, the effect of farside stops depends partially on the percent of turning traffic. In some cases, the reduction in traffic caused by turns may nullify any adverse effect of the farside stop.

When parking is permitted, the effect of bus stop location depends on the limits of the parking zone that would be permitted if the bus zone did not exist. If parking would otherwise be permitted near the intersection, the nearside bus stop provides additional capacity for right turns when not in use. The farside stop will have a minor adverse effect primarily caused by vehicles pulling back into moving lanes.

---

[14] *Ibid.*
[15] *Highway Capacity Manual*, Highway Research Board Special Report No. 87 (Washington D.C.: Highway Research Board, 1965), pp. 142–45.

SIGNING AND MARKING FOR CURB LOADING

Curb bus zones should be adequately marked to prevent illegal parking within the limits of the zone. "No Parking—Bus Zone" signs of standard size and shape should be posted. In addition, the curb face is often painted solid yellow or some other specially designated color to discourage parkers. Painted pavement markings designating the limits of the zone are also frequently used.

OFF-STREET LOCAL SERVICE TRANSIT FACILITIES

Off-street transit facilities for local service are generally small in size and operation. They are often located at the end of a transit line or at a major transfer point providing off-street loading area and maneuvering space for the bus. However, the end-of-line facility is often being replaced, and the bus accomplishes its turnaround by circulating around a block.

The off-street facility often consists of a loop from 15 to 20 ft (4.6 to 6.1 m) in width. The bus pulls onto the loop from the street, stops for loading and unloading passengers and for any required layover if an end-of-line station, and then proceeds around the loop and back onto the street system. The length of the loop depends on the turning radius of the buses using the loop and the maximum number of buses using the loop at any one time.

Variations on a basic design are possible. Scheduling of buses may require that the loop be widened to allow one bus to pass another within the loop. A double loop is also possible when the loading platform is placed between the two loops and the outside loop operates clockwise while the inside loop operates counterclockwise (see Figure 15.14).

RAPID TRANSIT STATIONS

At rapid transit stations, particularly in suburban areas, access may be by a number of modes. The design of the station must then consider the access and circulation of each mode using the station, provision of easy circulation by pedestrians (separated from vehicular traffic whenever possible), and the geometrics of loading and unloading areas, access roads, and parking facilities.

Auto traffic at the transit station may be for varying purposes. Some drivers will park their cars and change their travel mode to transit ("park and ride") and others will be dropped off at the station and the vehicle will leave the station ("kiss and ride"). Separate facilities should be provided for these two kinds of auto arrivals. In addition, circulation and loading-unloading space must be provided for local or feeder bus service.

Parking areas for long-term parkers (park and ride) can be designed in much the same manner as other parking facilities. Standard dimensions of parking stalls are recommended. Short-term (kiss and ride) spaces will also be required. Turnover in these spaces will be very high, but they are particularly needed to park vehicles waiting to pick up passengers arriving at the station.

Loading-unloading areas for buses should be designed to provide continuous circulation without requiring backing maneuvers. A sawtooth pattern with bus spaces provided at about 65 ft (19.8 m) intervals will allow buses to pull into the space, load

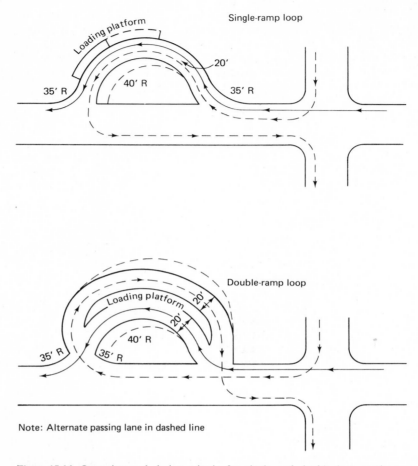

Figure 15.14. Operating and design criteria for single and double loop sections. [Source: *Traffic Engineering Handbook*, *Third Edition*, Institute of Traffic Engineers (Washington, D.C.: 1965), p. 516.]

or unload passengers, and pull out of the space without backing up. Since the transit station is often a terminal point for local buses, sufficient storage or layover area should be provided.

FREEWAY DESIGN AND MASS TRANSIT

Mass transit incorporated in controlled access highway rights of way may be either rail or bus facilities. Rail service should generally be provided in the median including loading and unloading facilities. Bus facilities should be located in order to provide quick and easy access to and from the through lanes of the freeway with a minimum of interference to through traffic. Acceleration and deceleration lanes are generally required as are loading-unloading areas separated from through traffic. For additional details, see Chapter 14, Geometric Design.

BICYCLE STORAGE FACILITIES

With the growing use of bicycles, it is important that consideration be given to adequate storage facilities at those strategic locations that generate large bicycle concentrations, for example, recreational centers, scenic sites, classroom buildings, dormitories, transit passenger terminals, fringe area parking lots, libraries, and other public buildings. In some central business districts bike racks have also been placed in public and private automobile parking facilities.

In designing bicycle storage facilities, consideration should be given to location and proximity to bikeways and user destination, storage lot identification and accessibility, orientation of storage spaces, kind of storage racks, lot boundary screening and protection, and provision of locking devices to eliminate casual and professional theft.

The previously described location and design principles for off-street automobile parking lots also apply to off-street bicycle storage lots.

The lot layout normally consists of stalls 2 ft $\times$ 6 ft (0.6 $\times$ 1.8 m) at 90° to aisles of a minimum width of 5 ft (1.5 m).

Various kinds of parking racks have been developed to support the bikes in a vertical position. The racks should be securely fastened to the lot surface, they should accommodate various wheel sizes, and they should be of sufficient structure strength to resist vandalism because they usually are the primary means of securing the bicycle from theft. Some lots have single- or double-wheel wells instead of vertical racks in the lot surface. These are more aesthetically pleasing, but cost, immobility, water, trash, and leaf removal constitute serious disadvantages.

Generally, the most successful method of theft prevention consists of weaving a case-hardened steel link chain [($\frac{5}{16}$ to $\frac{3}{8}$ in. (7.9 to 9.5 mm) minimum diameter] through the frame and both wheels and then attaching it to the loop, eyebolt, or structure of the rack by means of a durable lock. Because a chain of this size and adequate length may weigh as much as 12 lb (5.4 kg), some authorities provide the anchored chains as part of the storage facility and the cyclist only has to provide his own lock.

Smaller, compact bike storage lots are preferred over large spacious lots because cyclists tend to cluster around the most desirable locations and tend to use aisles and fencing instead of moving past loaded racks to more distant racks.

Three foot (0.9 m)-high dense hedging or fencing has been found effective in deterring improper movements and provides some screening while still providing lot visibility.

REFERENCES FOR FURTHER READING

Schemes for the Provision of Parking Space in Town Centres, Theme 6, Eighth International Study Week in Traffic Engineering, OTA. London, England, 1966.

HEATH, W. D., J. M. HUNNICUTT, M. A. NEALE, and L. A. WILLIAMS, *Parking in the United States—A Survey of Local Government Action.* Washington, D.C.: National League of Cities, Department of Urban Studies, 1967.

Parking Principles, Highway Research Board, Special Report No. 125. Washington, D.C.: 1971.

KLOSE, D., *Metropolitan Parking Structures.* New York: Frederick A. Praeger, 1965.

WILBUR SMITH and ASSOCIATES, under commission from the Automobile Manufacturers Association, *Parking in the City Center*. New Haven, Conn., 1965.

STOUT, R. W., *A Report on CBD Parking*, Highway Planning Technical Report No. 23. Washington, D.C.: U.S. Department of Transportation, Federal Highway Administration, 1971.

Parking Facilities for Industrial Plants. Washington, D.C.: Institute of Traffic Engineers, 1969.

A Recommended Practice for the Proper Location of Bus Stops. Washington, D.C.: Institute of Traffic Engineers, 1967.

Bus Stops for Freeway Operations, ITE Recommended Practice. Washington, D.C.: Institute of Traffic Engineers, 1971.

SEYMER, N., "Design of Parking Garages for European Needs," *International Road Safety and Traffic Review*. World Touring and Automobile Organisation, London, England, 1966.

Change of Mode Parking, an informational report. Arlington, Va: Institute of Traffic Engineers, 1973.

Chapter 16

TRAFFIC SIGNS AND MARKINGS

J. Robert Doughty, Director, Bureau of Traffic Engineering, Pennsylvania Department of Transportation, Harrisburg, Pennsylvania.

This chapter includes general information on the need for and use of traffic signs and markings. It does not serve as a standard or replace the *Manual on Uniform Traffic Control Devices for Streets and Highways* (hereafter, simply called the *Uniform Manual*),[1] but it does provide information and guidelines for: (1) kinds of design and elementary layouts; (2) warrants and requisite studies; (3) materials, maintenance, inventory, and schedules; and (4) equipment and shop requirements.

The standards recommended and frequently referenced herein are those prescribed in the *Uniform Manual*.[2] In the interest of uniformity throughout the United States, all signs should follow these accepted standards. A considerable number of states have issued their own sign manuals, patterned closely after the *Uniform Manual*, or have adopted the *Uniform Manual* by reference. Local authorities, in the design and application of signs, should follow the standards of their state. Practicing traffic engineers in other countries should follow their appropriate manual (Pan American, Canadian, Australian, European, etc.).

GENERAL PROVISIONS

Traffic control devices, such as signs and markings, are the primary means of regulating, warning, or guiding traffic on all streets and highways. The need for well-designed, adequately maintained devices grows in proportion to the density of traffic, speed of operation, and complexity of maneuvering areas on highways and at intersections.

Both signs and markings have the function of regulating, warning, guiding and/or channelizing traffic. To be effective, however, the installation of each device should: (1) fulfill a need; (2) command attention; (3) convey a clear, simple meaning; (4) command respect of road users; and (5) give adequate time for proper response.

The traffic engineer must employ five basic considerations to ensure that these requirements are met. Although detailed in the *Uniform Manual*,[3] the following help establish a relationship with the basic requirements above:

1. *Design:* the combination of physical features such as size, colors, and shape to command attention and convey a message.

---

[1] *Manual on Uniform Traffic Control Devices for Streets and Highways*, U.S. Department of Transportation, Federal Highway Administration (Washington, D.C.: U.S. Government Printing Office, 1971).

[2] *Ibid.*

[3] *Ibid.*

2. *Placement:* the installation of devices so that they are within the cone of vision of the user and thus command attention and give time for response.
3. *Operation:* the application of devices so that they meet the traffic requirements in a uniform and consistent manner, fulfill a need, command respect, and give time for response.
4. *Maintenance:* the upkeep of devices in order to retain legibility and visibility; the removal of devices if not needed in order to aid in commanding respect and attention while fulfilling the needs of the users.
5. *Uniformity:* the uniform application of similar devices for similar situations so that they fulfill the need of the user and command his respect.

The traffic engineer must also see that control devices supplement each other in terms of providing a meaningful message to motorists. Every effort should be made to remove obsolete or unnecessary signs or roadway markings and minimize any possible confusion.

STANDARDIZATION

The most outstanding characteristic of signs and markings in the United States, Canada, Europe, and other parts of the world is the high degree of standardization that has been achieved and that will be further implemented upon adoption of new manuals.

The adoption of the 1971 *Uniform Manual* provides for more extensive use of symbols on regulatory and warning signs. The recent changes in pavement markings, however, will still keep the United States apart from other countries because the use of yellow markings has not been uniformly adopted internationally for center lines or for no-passing zones.

LIMITATIONS AND EFFECTIVENESS

Traffic signs and roadway markings are inherently subject to becoming damaged or defaced and can become dirty and ineffective. Moreover, roadway markings have a relatively short durability and can also become obliterated by snow, weather conditions, and heavy concentrations of vehicles. A proper maintenance program, in addition to continued effort into research of materials and their application, will help minimize the many problems in these areas.

The effectiveness of signs and markings or the need for additional control can be measured by the amount of delay at a site or by the kind and number of accidents that develop over a period of time. When a problem develops in traffic control, an engineering study should be made in order to determine the primary causes. A driver or pedestrian observance study can provide an objective, quantitative evaluation of the existing traffic control devices and give guidance to corrective action.

Observance studies can indicate needed changes in regulations, traffic devices, education, or enforcement. If a study at one location indicates that motorist observance of a particular control device is lacking, this particular device should be carefully studied at other locations. If a pattern is evident, a good public relations program should be developed and disseminated to the motorists. If there is something inherently wrong with the design or application, corrective action must be taken.

DRIVER INFORMATION NEEDS

Research has identified the information needs of the driver and the interaction among these needs.[4] Information needs occur throughout the entire driving task and they fall into a hierarchy relative to satisfying those needs. The highest order of needs are those associated with the two main tasks of tracking and speed control, followed by the needs for obstacle avoidance and maintenance of the most efficient and safe course in the traffic stream. The lowest order of needs are those associated with trip preparation and direction finding.

The use of this hierarchy of information needs is of prime importance in developing and installing an information system for the drivers. That is, in areas where drivers will be busy with speed control or obstacle avoidance, they should not be overly burdened with directional signing. Directional information should be planned and installed in areas in which there are only "simple" steering and speed control maneuvers. Drivers should not be overwhelmed with complex or unexpected events on the roadway.

FEDERAL SAFETY STANDARDS

In the United States, a series of Highway Safety Program Standards has been issued by the U. S. Department of Transportation.[5] These program standards are designed to help identify areas in need of new or revised traffic control devices and to establish a means of scheduling adequate maintenance. Two of the programs that relate directly to signs and markings are:

1. *Number 10, Traffic Records:* the collection and maintenance of records, including data on drivers, vehicles, accidents, and highways. (Data pertaining to traffic control devices are listed under highways.)
2. *Number 13, Traffic Engineering Services:* the implementation of control device improvements that bear directly on reducing accidents.

Compliance with these acts and the adoption of similar programs in other countries will help reduce traffic accidents and deaths, injuries, and property damage on all streets and highways.

KINDS OF SIGNS

FUNCTIONAL USAGE CLASSIFICATIONS

In general, traffic signs fall into three broad functional classes:

1. *Regulatory signs:* used to impose legal restrictions applicable to particular locations and unenforceable in the absence of such signs.
2. *Warning signs:* used to call attention to hazardous conditions, actual or potential, otherwise not readily apparent.

---

[4] G. F. KING and H. LUNENFELD, "Development of Information Requirements and Transmission Techniques for Highway Users," *National Cooperative Highway Research Program Report 123* (Washington, D.C.: Highway Research Board, 1971).

[5] *Highway Safety Program Standards* (Washington, D.C.: U.S. Department of Transportation, National Highway Traffic Safety Administration, Federal Highway Administration, 1974).

3. *Guide or informational signs:* used to provide directions to motorists, including route designations, destinations, available services, points of interest and other geographic, recreational, or cultural sites.

TEMPORARY SIGNS

Almost all signing is permanent in nature and will be applicable to conditions that exist all day, every day. One exception is the use of temporary signs for construction or maintenance projects when signs are needed only during periods when there is activity on or near the roadway.

The main use of temporary signing is to warn motorists of the construction or maintenance operation and to guide them through or around the work area. These signs should be set up immediately before the start of work, moved and revised concurrently with progress of the work, and removed immediately when no longer needed.

Because of the special nature of the applications, the signs must always be placed where they will convey messages most effectively and will not unduly restrict lateral clearance or sight distance. Temporary construction signs generally should be mounted on the driver's side of the roadway, from 6 to 12 ft (1.8 to 3.6 m) from the edge of the traveled way or 2 ft (0.6 m) beyond a curb. Often they must be placed on barricades within the roadway: sometimes supplemental or duplicate signs are required on the opposite side of the roadway.

Construction warning signs must be placed in advance of the working area. Local conditions will dictate placement: a practical guide is 10 ft (3.0 m) of advance distance for each mile per hour of approach speed up to 35 mph (56 km/h) in urban areas; a considerably greater distance, up to as much as 1,500 ft (457 m) for rural highway conditions; and up to one-half mile (0.8 km) or more on high-speed freeway facilities.

Construction approach warning signs should be repeated wherever serious hazards exist. Standard practice is to place these signs 500, 1,000, and 1,500 ft (152, 305, and 457 m) in advance of heavy construction, restricted traffic movement, or detours in rural areas, and at correspondingly lesser distances in urban areas.

Regulatory signs should be placed only when officially authorized, and they should conform with standards of the authority having responsible jurisdiction.

Advisory speed signs generally are used in connection with construction zone warning signs. These should be mounted together. Warning, regulatory, and guide signs for construction and maintenance work generally should be mounted on two posts or on the larger barricades because of the large sign sizes involved. Detailed applications are shown in Part VI, *Uniform Manual.*

Because the problem of signing and marking these work areas may involve many governmental agencies, as well as large and small contractors and utility companies, a number of states and provinces have developed separate manuals or handbooks on the proper control and use of traffic control devices for construction and/or maintenance. Moreover, special preconstruction meetings are often required to review all phases of the construction project, including signing and traffic control. Individual questions on these matters should be referred to the proper state or provincial traffic engineers.

VARIABLE MESSAGE SIGNS

Variable message signs are used to inform drivers of regulations or instructions that are applicable only during certain periods of the day or under certain traffic conditions. The need for and use of variable message signs have increased considerably over the past several years. Moreover, the trend should quicken as more funds are made available for operational needs and as technological developments bring new equipment. These variable message signs, which can be changed manually, by remote control, or by automatic controls that can "sense" the conditions that require special sign messages, have applications in each of the functional usage classifications. Examples include:

1. *Regulatory:* NO LEFT TURN during peak hours, SPEED LIMIT _____ during school periods, and reversible ONE WAY operation.
2. *Warning:* ACCIDENT AHEAD and ICE AHEAD.
3. *Guide:* ALTERNATE ROUTE and STADIUM ACCESS.
4. *Information:* CONGESTION AHEAD and TRUCK SCALE OPEN.

Examples of these signs, which should conform to the same shapes and colors and be of the same dimensions as standard signs, are:

1. Blank-out signs, consisting of a fluorescent grid of illumination behind cut-out letters or symbols, which are effective only when the grid is illuminated.
2. Signs that change their message by means of a curtain drawn between a clear sign front and the electrical illumination provided behind the curtain or by means of a rotating drum.
3. Signs that change their message by illumination of incandescent lights in appropriate patterns, by selection of the light units that provide the desired letter, numeral, or symbol.
4. Signs that change their message by disks that are reflective white on one side and background color on the reverse side, and that rotate to form messages of alphabetical and numerical characters.

Each kind of unit will not necessarily serve all functional classifications of sign usage, but at least one of the designs can be adapted to a particular problem.

SIGNING ELEMENTS

GENERAL SPECIFICATIONS

The most important qualities of traffic signs are their attention value, legibility, and recognition:

1. *Attention value:* the characteristic that commands attention:
   a. *Target value:* the quality that makes a sign or group of signs stand out from the background.
   b. *Priority value:* the quality that makes it possible for a sign to be read first in preference to other signs.
2. *Legibility:* the characteristic of being read:
   a. *Pure legibility:* the distance at which a sign can be read in an unlimited time.
   b. *Glance legibility:* the distance at which a sign can be read at a glance, usually from 0.5 to 1.4 sec within a cone of approximately 5-ft (1.5 m) diameter at 100 ft (30.5 m) distance.

3. *Recognition:* the attribute of being recognized and understood by the utilization of standardized colors, shapes, and legends.

Other qualities are also important. For example, experiments indicate that signs located over the highway are more likely to be seen before those located on either side of the highway. The most important factors are brightness contrast of letters to sign and sign to background.[6]

SIGN DESIGN

The 1971 *Uniform Manual* retains the basic sign philosophy in shape, color, and message as the earlier editions, but it has adopted a larger number of "symbol" signs, consistent with those in Canada, Mexico, and European countries. Some of these changes and a comparison with Canadian signs and those reviewed at the 1968 United Nations Vienna Conference on Road Traffic are shown in Figure 16.1.

Sign shapes in the United States generally remain the same as approved in the 1961 *Uniform Manual* with the additions of a pennant-shaped warning sign and a pentagon-shaped school sign. The standard sign shapes are:

1. *Octagon:* reserved exclusively for the STOP sign.
2. *Equilateral triangle, with one point downward:* reserved exclusively for the YIELD sign.
3. *Round:* used for the advance warning of a railroad crossing.
4. *Pennant, an isosceles triangle with its longest axis horizontal:* used to warn of no passing.
5. *Diamond:* used only to warn of existing or possible hazards.
6. *Rectangle, ordinarily with longer dimension vertical:* used for regulatory, with the exception of STOP and YIELD signs.
7. *Rectangle, ordinarily with longer dimension horizontal:* used for guide and informational signs, with the exception of route markers and recreational area guide signs.
8. *Trapezoid:* may be used for recreational area signs.
9. *Pentagon, point up:* used for school advance and school crossing signs.
10. *Miscellaneous:* reserved for special purpose, such as the shield for route markers and the crossbuck for railroad crossings.

COLOR

The U.S. standard colors are also quite similar to earlier practices with the further identification of purple, light blue, coral, and strong yellow-green for future use and the colors orange and brown specified for immediate use. The standard colors[7] and their uses include:

1. *Red:* used as a background color for STOP, DO NOT ENTER, and WRONG WAY signs and supplemental plates on stop intersections; as a legend color

---

[6] T. W. FORBES, "Factors in Highway Sign Visibility," *Traffic Engineering*, XXXIX, No. 12 (1969), 20–27.

[7] Color cards showing the correct colors for highway signs are available on request from the Federal Highway Administration, Washington, D.C.

U.S.A.
1961

U.S.A.
1971

Canada
1966

Background — Yellow
Legend and Border — Black

Background — White
Legend and Border — Red

Background — Yellow
Legend and Border — Black

Background — White
Border — Red

No
Standard

Background — White
Legend and Border — Black

Background and Legend — White
Circle — Red

Background — Red
Bar — White

Background — White
Legend and Border — Black

Background — White
Legend
Border and Arrow — Black
Circle and Slash — Red

Background — White
Legend
Border and Arrow — Black
Circle — Green

Background — White
Legend
Border and Arrow — Black
Circle and Slash — Red

No
Standard

Background — White
Legend and Border — Black

Background — White
Legend and Border — Black

Background — White
Legend and Border — Black

Figure 16.1. Comparison of "old" and "new" U.S. signs with Canadian signs (1966)
and those detailed for the United Nations Vienna Conference on Road Traffic (1968).

for parking prohibition signs, and the circular outline and diagonal bar for prohibitory symbols; and used for the border and message for YIELD signs.

2. *Black:* used as a background for certain ONE WAY and WEIGH STATION signs as well as for specified night SPEED LIMIT signs. It is also used as a legend color for signs with white, yellow, and orange backgrounds.

3. *White:* used as a background for route markers, guide signs, fallout shelter directional signs, and regulatory signs except for the STOP sign; and used as the legend color on brown, green, blue, black, and red signs.

4. *Orange:* used as a background color for construction and maintenance signs.

5. *Yellow:* used as a background color for school signs, lane drop signs on expressways, and warning signs, except when orange is specified.

6. *Brown:* used as a background color for guide and information signs related to points of recreational, scenic, or cultural interest.

7. *Green:* used as a background color for milepost and guide signs, when the guide signs are not brown and white; also used as a legend color on white background for permissive parking regulations.

8. *Blue:* used as a background color for information signs related to motorist services, rest areas, and the evacuation route markers.

SIZE

Standard sizes have been established for regulatory and warning signs and for certain guide signs such as route markers. Larger sizes are used for high-speed highways, for locations involving impaired visibility, highways of four or more lanes, or other locations where added emphasis is desirable.

Standard sign sizes include:

1. STOP: 30 in. by 30 in. (76 cm by 76 cm), except that a sign 24 in. by 24 in. (61 cm by 61 cm) may be used on minor roads and secondary streets.

2. YIELD: side dimensions 36 in. (91 cm).

3. NO _____ TURN (symbol): 24 in. by 24 in. (61 cm by 61 cm) plus plaque 24 in. by 18 in. (61 cm by 46 cm).

4. DO NOT ENTER: 30 in. by 30 in. (76 cm by 76 cm).

5. Urban parking signs: 12 in. by 18 in. (31 cm by 46 cm), except when the message requires additional space.

6. Other regulatory signs: 24 in. by 30 in. (61 cm by 76 cm), except that some signs 18 in. by 24 in. (46 cm by 61 cm) may be used on minor roads and secondary streets.

7. Guide signs: sufficient to accommodate minimum lettering 6 in. (15 cm) high on major routes in rural districts, 4 in. (10 cm) high on less important rural roads and urban streets, and 8 in. (20 cm) or more on expressways.

The overall sizes of guide signs for expressways and the Interstate system are based on the kind of mounting, the kind of exit, the kind of information (supplemental, gore, advance guide), and the message itself. Minimum letter and numeral sizes have been established in the *Uniform Manual* and computer programs have been developed for custom designing these signs.[8]

---

[8] R. L. BLEYL and H. B. BOUTWELL, "A Computer Program for Guide Sign Designing and Drafting," *Traffic Engineering,* XXXVIII, No. 6 (1968), pp. 22–26.

LEGEND

In general, both sign and letter size have been established for all regulatory and warning signs. Moreover, the *Uniform Manual* sets forth criteria establishing the series of letters to be used and the spacing between letters for these signs. Series A and B alphabets are restricted to street names and parking signs because of the limited breadth and stroke width.

Only minimum sizes have been established for guide signs. The letter size needed to give motorists ample opportunity to read a sign easily at normal approach speed will, in general, determine the size of sign needed. In selecting proper letter size, the following factors must be considered: (1) speed of the approach vehicle; (2) location of the sign; (3) width and kind of letters; (4) illumination or reflectorization; (5) necessary warning time (perception-reaction time and stopping or deceleration time for the necessary maneuver); and (6) amount of sign copy.

The standard lettering for conventional highway signs is a rounded style of upper-case letters. When letter height exceeds 8 in. (20 cm), however, place names on guide signs should be composed of upper- and lower-case letters. The initial upper-case letters shall be $1\frac{1}{3}$ times the "loop" height of the lower-case letters.[9]

The style of lettering to be used on freeway guide signs shall be one of the following: (1) upper-case letters for all word legends; (2) lower-case letters with initial upper-case letters for all names of places, streets, and highways and upper-case letters for other word legends.

Word messages in the legend of freeway guide signs shall be in letters at least 8 in. (20 cm) high. Larger lettering is necessary for major guide signs at and in advance of interchanges and for all overhead signs. Recommended numeral and letter sizes according to interchange classification, kind of sign, and component of sign legend have been specified in the *Uniform Manual*.[10]

Appropriate descriptive symbols instead of words on highway signs provide a great advantage in conveying the message. Except for symbolism in sign shapes, the principal symbol used in the United States is the arrow, which shows route turns and directions, changes in alignment, and in some urban areas, turns permitted from a given lane of multilane roadways. Arrows are also used in conjunction with a brief message to denote regulations applying to a certain lane, for example, a reversible-lane section or a restriction on the kind of vehicles permitted in that lane. The arrow is used only on overhead guide signs that restrict the use of specific lanes to traffic bound for the destinations and routes indicated by these arrows.

Diagrammatic signing should use arrows that approximate the intersection geometrics, or the necessary part of it, in a clear, understandable manner that will impart a glance-legible message as shown in Figure 16.2(a) and Figure 16.2(b).

Symbols showing pictures of various kinds of hazards and traffic interference are used extensively in Europe because of the language differences on the European continent.[11] These symbols are used to some extent in Canada and have now been added to the 1971 *Uniform Manual*.

An example of computer programming to determine legend positioning on highway signs may be found in Chapter 11, Computer Applications.

---

[9] Recommended designs for the upper-and lower-case letters, together with tables of spacing are available from Federal Highway Administration, Washington, D.C.

[10] *Manual on Uniform Traffic Control Devices for Streets and Highways*, pp. 142–44.

[11] J. M. ZUNIGA, "International Effort Toward Uniformity on Signs, Signals, and Markings," *Traffic Engineering*, XXXIX, No. 8 (1969), pp. 32–39.

Figure 16.2. (a) Sample diagrammatic sign for cloverleaf interchange. (b) Sample diagrammatic sign for directional interchange. [Source: *Manual on Uniform Traffic Control* (Washington, D.C.: U.S. Department of Transportation, Federal Highway Administration, 1971), p. 139.]

REFLECTORIZATION AND ILLUMINATION

Signs that carry regulatory messages, warning, or essential directional information must be as legible at night as in the day. They must be reflectorized or illuminated unless they are not applicable at night or are the urban parking series and are in an area of sufficient level of illumination to provide nighttime visibility.

The combination of improved automobile headlighting and more effective reflecting materials during the past decade has made reflectorized signs as effective as certain kinds of illuminated signs. Reflectorized signs, therefore, provide satisfactory visibility under almost all driving conditions. Practical financial considerations demand the use of reflectorized signs instead of illuminated signs for almost all applications on both rural and urban roads, particularly those mounted at the side of the roadway.

Sign reflectorization is accomplished by (1) reflectorizing the message, lettering, symbols, and the border, (2) reflectorizing the background, or (3) reflectorizing both background and message.

When only the message is reflectorized, the lettering or symbol stroke must be relatively wide and the reflectorization must have a high intensity so that adequate target value is achieved. When both background and letters or symbols are reflectorized, the message should be provided with a higher brightness value than the background material.

The desirable physical characteristics of a reflectorized traffic sign should assure that the sign provides the following:

1. Sufficient reflectivity over a normal approach distance so that good target value and legibility are provided from the maximum to the minimum approach distance.
2. Lack of glare from the reflectorized portion of the sign that would tend to blot out the legibility of the message.
3. Seventy-five percent of its dry reflectivity when subjected to adverse conditions of rainfall and fog.
4. The same general night and day appearance and message unless the sign is designed to change its message after nightfall.
5. A self-cleaning surface so that there will not be a heavy accumulation of dirt on the reflective surface.
6. A durable surface capable of replacement by normal maintenance procedures.
7. A durable surface which is resistant to vandalism and capable of replacement by normal maintenance procedures.

Reflective elements are of two principal kinds: reflector buttons and reflecting coatings. They both possess the quality of reflex or retrodirective reflections, that is, the ability to reflect the incident light directly back toward its source.

Reflector buttons are small reflecting units set on or in the face of the sign in a pattern to form a symbol or a series of letters. They generally consist of glass or transparent buttons having rear inner faces made up of prismatic reflectors.

Reflective coatings are materials intended to cover areas of any shape such as the entire background of a sign or to be used for complete letters or symbols. The type in common use consists of flexible adhesive coated sheeting with very small glass spheres embedded in a plastic surface. Each sphere acts as a miniature reflector.

The qualities of reflective materials that affect their uses include:

1. The amount of incident light from the car headlamps back to the driver's eyes under normal viewing distances
2. The maximum entrance angle through which effective reflection is possible
3. Resistance to weather deterioration and vandalism
4. Ease of application and replacement
5. The sign cost per year of useful life

The divergence angle, which is the angle between the incident beam and the light reflected to the observer, must be wide enough to include the eyes of the motor vehicle driver. (This varies from less than 0.2° at great distances to 1.5° at 150 ft (46 m) for drivers of some trucks.) Therefore, the maximum divergence angle is required when a sign is to be read from a vehicle about to pass the side of the sign and for truck drivers whose eyes are a substantial distance above his headlights.

The entrance angle is the angle between the incident beam and a perpendicular to the face of the sign. Reflective materials used must have a wide-angle response to reflect light even when the approaching vehicle headlights are well to one side of the sign. The wide-angle materials available to fulfill this requirement give good reflectivity up to 40°. Good reflectivity at wide-entrance angles is required to:

1. Secure good visibility over a long range of approach distances, including close viewing distance
2. Offset the effect of inaccuracies in sign installations
3. Compensate for any temporary damage to supporting structure
4. Provide legibility on multilane roads where the lateral distance between the vehicle path and sign placement may be considerable (This is particularly true today with safety programs to set back signs from edge of highway in order to minimize points of conflict.)

When reflectorized signs are not likely to give effective results, such as at overhead locations, sign illumination is mandatory in order to provide adequate contrast and visibility. Illumination is normally used on all overhead signs of the Interstate Highway System, on high-speed turnpikes, on urban expressway sections, and in areas in which rapid decisions and response are required of the motorist.

Illumination may be provided by:

1. Directing external light upon the surface of the sign itself so that it is illuminated from the direction of approach. Because of the possibility of lighting failure, the message must be reflectorized and the background may be either reflectorized or nonreflectorized, but the mixing of reflectorized and nonreflectorized signs in the same general area should be avoided. Depending on local policy, lighting may be placed either above or below the sign surface. Illumination from below, however, avoids undesirable daytime shadows on the face of the sign. Illumination may be provided by fluorescent tubes or by mercury vapor luminaires. The latter tend to be favored because of easier mounting, easier maintenance (longer life), and better light distribution.
2. Light produced by incandescent or fluorescent tubes shining through a translucent background upon which the sign message is lettered.

In a modified form of sign illumination, flashing lights are used in order to improve target value of some of the more critical signs. Red flashing lights can be used in connection with STOP signs and yellow flashing lights can be used in connection with various warning signs and school crossing signs. In one of the latter forms, the speed limit prevailing through a school zone or school crossing can be indicated in illuminated letters during the period when the speed regulation is in effect.

SIGN LOCATION, ERECTION, AND SUPPORTS

Standardization of position cannot always be attained. The general rule, however, is to locate signs on the driver's side of the roadway where he would be looking for them. Signs in other locations, except for those mounted overhead, should be considered as supplementary to signs in their normal locations.

In general, signs should be located to optimize nighttime visibility and minimize the effects of mud spatter. Signs should be located so that they do not obscure each other or are hidden from view by other roadside objects.

Post-mounted signs should be placed at an angle in order to minimize glare from their surfaces. Reflectorized signs may be aimed slightly toward the roadway facing traffic because this position gives maximum brilliance over the greatest possible approach distance. But a slight rotation away from the roadway is preferred for it reduces mirror reflection and does not unduly reduce readability. Signs should be placed with regard to the alignment of approaching traffic, not necessarily at a predetermined angle with the roadway edge. It is sometimes desirable, in the case of reflectorized signs on grades, to tilt signs forward or backward in their mountings so that they face approaching traffic more squarely or to raise them above normal mounting heights so that they are within the range of visibility of motorists driving uphill.

Whenever possible, each traffic sign should be mounted on an individual post in order to provide a minimum of competition among groups of signs. Speed advisory signs are an exception because they should serve as a supplementary message to a warning sign. Advertising signs should not be permitted within a group of traffic signs or on the highway right of way.

The lateral clearance for regulatory and warning signs or the smaller directional signs should be from 6 to 12 ft (2 to 4 m) from the edge of the pavement or traveled way in rural areas, as shown in Figure 16.3. The larger directional signs should retain a 30 ft (9 m) clearance on high type highways.

In urban areas, signs generally are mounted alongside the roadway in the space between the curb and the sidewalk, or above the sidewalk itself, preferably in such a location that the signs will not have to compete with advertising signs on adjacent buildings. The recommended minimum urban mounting height of 7 ft (2 m) should always be followed in order to ensure uniformity in sign placement, to be sure that the sign is not obscured by parked or standing vehicles, and to prevent the sign from being a hazard to pedestrians. When practicable, a sign should not be less than 10 ft (3 m) from the edge of the nearest traffic lane. Large guide signs especially should be farther removed, preferably 30 ft (9 m) or more from the nearest traffic lane. Lesser clearance, but not generally less than 6 ft (92 m) may be used on connecting roadways or ramps at interchanges.

On wide expressways, or where some degree of lane-use control is desirable, or where space is not available at the roadside, overhead signs are often necessary. Other factors that may justify the erection of overhead signs include:

1. Traffic volume at or near capacity
2. Complex design
3. Three or more lanes in each direction
4. Restricted sight distance
5. Closely spaced interchanges or intersections
6. Multilane exits or turns
7. Large percentage of trucks

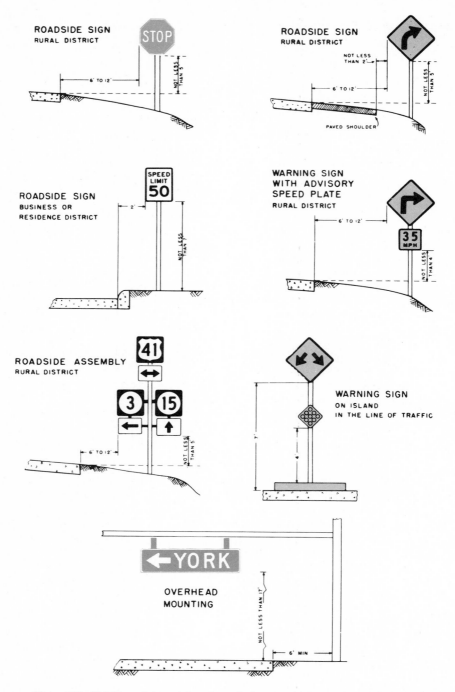

Figure 16.3. Height and lateral location of signs—typical installations. (Source: *Manual on Uniform Traffic Control Devices*, p. 18.)

8. Background of street lighting
9. High-speed traffic
10. Consistency of sign message locations
11. Insufficient space for ground signs
12. Located near signal heads so that a motorist can see all controls within a confined area

In general, the longitudinal placement of signs calls for regulatory signs to be installed where the mandate or prohibition applies or begins, warning signs to be placed in advance of the conditions to which they call attention, and guide signs to be placed as needed to keep drivers adequately informed. More specifically, the following guidelines should be followed on the conventional streets and highways:

1. STOP and YIELD signs should be placed at the point at which compliance is to be made. Even in open rural areas, STOP signs should not be placed more than 50 ft (15 m) from the intersecting roadway. A suitable stop line or other marking device should be placed in the roadway at the intended point of compliance to supplement the STOP sign if the sign is not at or near the point at which the vehicle should stop. Channelization may be necessary in order to provide a suitable location for the STOP sign at complicated intersections or where physical conditions make it difficult to provide proper location.

2. Warning signs are generally placed slightly in advance of the point of compliance. In urban areas where speeds are relatively low, warning signs should be posted about 250 ft (76 m) in advance of the hazardous condition to which they are directing attention. In rural areas and on higher-speed roadways, warning signs should be posted at a distance from 750 to 1,500 ft (229 to 457 m) in advance of the hazard. These distances are necessary in order to permit the driver to make the necessary response to comply with the regulation.

3. Guide signs are posted in advance of intersections and within the intersections themselves. A statewide or national uniform plan should be followed so that motorists, once accustomed to the plan, will be able to find signs in uniform locations. Junction signs and advance turn arrows in rural areas should be erected not less than 400 ft (122 m) in advance of the intersections, as shown in Figure 16.4; in built-up areas they should be located approximately mid-block preceding the intersection and generally not more than 300 ft (91 m) in advance of the intersection.

4. Destination signs in rural areas are to be located not less than 200 ft (61 m) or more than 300 ft (91 m) in advance of the intersection; in urban areas, shorter distances are permissible. Directional route markers should be at the intersection. Confirmatory route markers are placed from 25 to 200 ft (8 to 61 m) beyond the intersection in rural areas and from 10 to 80 ft (3 to 24 m) beyond it in urban areas.

Expressway interchange signs generally combine the functions of route markers and destination signs. They require greater advance warning distances than signs on conventional highways. For each interchange there should be a minimum of three guide signs and a maximum of five, not including any signs placed as supplements to the regular signs. In all cases, major guide signs placed in advance of deceleration lanes should be at least 800 ft (244 m) apart. Figure 16.5 shows the typical signing sequence for a cloverleaf interchange.

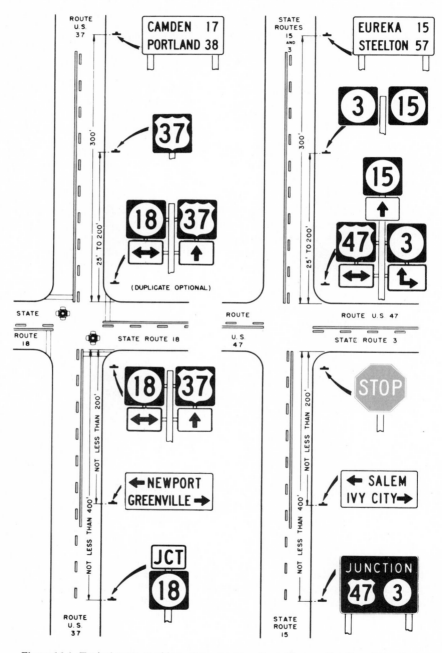

Figure 16.4. Typical route markings at rural intersections (for one direction of travel only). (Source: *Manual on Uniform Traffic Control Devices*, p. 26.)

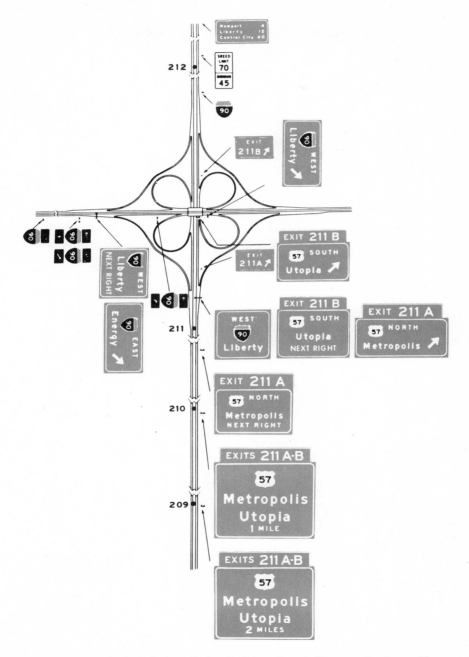

Figure 16.5. Typical signing of cloverleaf interchanges (for one direction only). (Source: *Manual on Uniform Traffic Control Devices*, p. 152.)

Signs erected at the side of the road in rural areas shall be mounted at a height of at least 5 ft (1.5 m), measured from the bottom of the sign to the near edge of the pavement. In business, commercial, and residential districts where parking and/or pedestrian movement is likely to occur or where there are other obstructions to view, the clearance to the bottom of the sign shall be at least 7 ft (2.1 m).

The height requirements for ground installations on expressways vary from those on conventional streets and highways. Directional signs on expressways shall be erected at a minimum height of 7 ft (2.1 m) above the near edge of the pavement to the bottom of the sign. If, however, a secondary sign is mounted below another sign, the major sign shall be at least 8 ft (2.4 m) and the secondary sign at least 5 ft (1.5 m) above the level of the pavement edge. All route markers and warning and regulatory signs on expressways shall be at least 6 ft (1.8 m) above the level of the pavement edge. If guide signs are placed 30 ft (9 m) or more from the edge of the nearest traffic lane for increased roadside safety, the height to the bottom of such signs may be 5 ft (1.5 m).

Overhead signs shall provide a vertical clearance of not less than 17 ft (5 m) over the entire width of the pavement and shoulders except when a lesser vertical clearance is used for the design of other structures. The vertical clearance to overhead sign structures or supports need not be greater than 1 ft (30 cm) in excess of the minimum design clearance of other structures.

POSTS AND MOUNTINGS

Signs can be correctly placed on existing supports used for other purposes, such as traffic signal, street light, and public utility poles when permitted, thereby saving expense and minimizing sidewalk obstruction. These signs should not be installed on wooden utility poles where utility crews could be hurt while working on the poles.

Other sign posts and their foundations and sign mountings shall be so constructed that they hold signs in a proper and permanent position and resist swaying in the wind or displacement by vandalism. Yet, they should be so constructed and installed that they yield or break away upon contact. The only exceptions to these safety features should be (1) overhead signs or other signs that are protected by curbs in urban areas or guardrails in rural areas and (2) sign installations that are a sufficient distance from the pavement edge to allow an errant driver time to recover.

In noncurbed or other unprotected areas, regulatory and warning signs can normally be mounted on the driver's side of the road on square or U-shaped posts that break away at pivot points or yield to any vehicle that strikes them. In these cases, signs that do not exceed 36 in. (91 cm) in width or 10 sq ft (0.9 sq m) in area can be installed on one post. Slightly larger signs can be installed on wood posts 4 in. by 4 in. (10 cm by 10 cm).

Larger signs, including guide or destination signs, with a width of 48 in. (122 cm) should be mounted on two supports that yield or break away. Tests in Pennsylvania[12] have indicated that:

1. Posts of Douglas fir or structural pine [4 in. by 4 in. (10 cm by 10 cm) and 4 in. by 6 in. (10 cm by 15 cm)] will break upon contact.

---

[12] "Dynamic Field Test of Wood Post Sign Supports," *Research Project No. 68–36,* Bureau of Materials, Testing and Research, Pennsylvania Department of Highways, October 1968.

2. Sign posts 6 in. by 6 in. (15 cm by 15 cm) should have two holes $1\frac{1}{2}$ in. (3.8 cm) in diameter drilled in the center of the post, parallel to the face of the sign; one hole should be at a height of 6 in. (15 cm) and the other at 18 in. (46 cm) above the ground line.
3. Wood posts 6 in. by 8 in. (15 cm by 20 cm) should have two holes $2\frac{1}{2}$ in. (5.8 cm) in diameter drilled as above.

Large directional signs must have specially designed breakaway supports that can be either of wood or steel. Steel breakaway supports include:

1. The Texas design, with a slip joint feature at the base and a hinge joint below the sign to allow the post, after impact, to slip off the foundation and swing up and away from the vehicle.[13]
2. The New Jersey design, with a load concentrated breakaway coupling (LCBC) at the base and a separate breakaway feature between sign and post to allow the post to shear from the foundation upon impact and to swing away from the vehicle to avoid a secondary collision between post and vehicle.[14]

From a safety and an aesthetic standpoint overhead signs should be mounted on overhead bridge structures whenever possible. Since, however, such structures are not normally in the desired location, the engineer must design a sign support that will handle the dead load and wind load of the completed sign and lighting system. The design features must further provide for the proper vertical and horizontal clearances in order to meet the minimum standards for that highway.

Experimentation is taking place on overhead breakaway structures that provide for a dual support system so that one support can be knocked away without allowing the span and sign to fall onto the highway.[15] Future technology may allow a wide-spread use of these new structures, but the engineers of today will generally have to use current designs and see that they are properly protected.

SIGN WARRANTS AND REQUISITE STUDIES

Regulatory and warning signs should be used only when needed and warranted so that their effectiveness will not be destroyed by excessive frequency. Guide signs, however, should be used whenever they can contribute to the convenience and facilitation of traffic movement.

UNIFORM MANUAL SIGN WARRANTS[16]

Specific warrants for the use of some regulatory signs are provided as general policy statements rather than as absolute warrants. Warrants provide a guide to sound sign application and serve as an aid in preventing the overuse of regulatory signs.

---

[13] "Impact Behavior of Sign Support—11," *Staff Report 68–2* (College Station, Texas: Texas Transportation Institute, 1965).

[14] "Full-Scale Dynamic Testing of the New Jersey Breakaway Sign Structure," *Cal. Report VJ-2955-V-1* (Buffalo, N.Y.: Cornell Aeronautical Laboratory, Inc., 1970).

[15] D. L. IVEY, R. M. OLSON, C. E. BUTH, T. J. HIRSCH, and D. L. HAWKINS, "Breaking Overhead Sign Bridges, Crash Testing," *Highway Research Record No. 346* (Washington, D.C.: Highway Research Board 1971).

[16] *Manual on Uniform Traffic Control Devices for Streets and Highways.*

As an example, a STOP sign may be warranted at an intersection when one or more of the following conditions exist:

1. Intersection of a less important road with a main road where application of the normal right-of-way rule is unduly hazardous.
2. Intersection of a county road, city street, or township road with a state highway.
3. Street entering a through highway or street.
4. Unsignalized intersection where a combination of high speed, restricted view, and serious accident records indicate a need for control by the STOP sign.

STOP signs cannot be erected at intersections where traffic control signals are present because the signals should be operated continuously, either in normal or flashing operation.

The "multiway" (4-WAY or ALL WAY) STOP installations can be used as a safety measure at some locations where the volume on the intersecting roads is approximately equal and the following conditions have been established.[17]

1. Where traffic signals are warranted and urgently needed, the multiway is an interim measure that can be installed quickly in order to control traffic while arrangements are being made for the signal installation.
2. An accident problem as indicated by five or more reported accidents in a 12-month period of a kind susceptible to correction by a multiway STOP installation.
3. Minimum traffic volume:
   a. The total vehicular volume entering the intersection from all approaches must average at least 500 vph for any 8 hr of an average day, and
   b. the combined vehicular and pedestrian volume from the minor street or highway must average at least 200 units per hr for the same 8 hr, with an average delay to minor street vehicular traffic of at least 30 sec per vehicle during the maximum hour, but
   c. when the 85th percentile approach speed of the major street traffic exceeds 40 mph (64 km/h), the minimum vehicular volume warrant is 70 percent of the above requirements.

YIELD sign warrants are established as follows:[18]

1. On a minor road at the entrance to an intersection where it is necessary to assign right of way to the major road, but where a stop is not necessary at all times, and where the safe approach speed on the minor road exceeds 10 mph (16 km/h).
2. On the entrance ramp to an expressway where an adequate acceleration lane is not provided.
3. Within an intersection with a divided highway, where a STOP sign is present at the entrance to the first roadway and further control is necessary at the entrance to the second roadway and also where the median width between the two roadways exceeds 30 ft (9 m).

---

[17] *Ibid.*
[18] *Ibid.*

4. Where there is a separate or channelized right-turn lane, without an adequate acceleration lane.
5. At any intersection where a special problem exists and where an engineering study indicates that the problem is susceptible to correction by the use of the YIELD sign.

YIELD signs should not be used to control the major flow of traffic, approaches of more than one of the intersecting streets, or at intersections where there are STOP signs on one or more approaches.

Criteria for installation of lane-use control signs at intersections have been established and are as follows:

1. Lane-use control signs at intersections shall be used whenever it is desired to require vehicles in certain lanes to turn or to permit turns from an adjacent lane.
2. Lane-use controls permitting left (or right) turns from two (or more) lanes normally are warranted whenever the turning volume exceeds the capacity of one turning lane and when all movements can be accommodated in the lanes available to them.

When an engineering and traffic investigation indicates that attention must be called to an actual or potential hazardous condition, warning signs should be installed. Typical locations that may warrant such signs include:

1. Turns, curves, or intersections
2. Advance warning for STOP signs, signals, or railroads
3. Grades, drops, or bumps
4. Narrow roadways, bridges, or other points of limited clearance
5. Advisory speed signs on curves or through work areas

Adequate signing at intersections to show route directions and destinations includes advance notice of route junctions and turns, directional route markings at the intersection, and destination signs showing the names of important cities and towns with their directions and distances. Confirmatory route markers should be placed on numbered routes just beyond every marked intersection and occasional reassurance markers should be placed between intersections.

Although the *Uniform Manual* gives detailed specifications for the application and placement of route markers at intersections, many intersections require individual engineering treatment because of their particular design and other unusual physical characteristics.

SPECIAL SIGNS

Signing needed to direct traffic to reach a freeway may be required in a relatively narrow corridor along the facility. When used, the corridor should extend to the nearest important intersection of a major street leading to the freeway and, in some cases, to the nearest major street paralleling the facility. In other cases, it should extend to a highway route replaced by the freeway. Since the width of the band should be determined by the street network in the freeway vicinity, it therefore cannot be preestablished.

Within this corridor the signing must be custom-designed to the conditions. The sign locations and messages depend on both the movements required and on the street and freeway layout. The following rules, which elaborate on but do not supersede the basic principles, have been developed to govern location and message content of this kind of signing:

1. Access to the freeway can be provided only at widely spaced locations in terms of city blocks. Many motorists approach the general vicinity of the freeway with only a vague knowledge of its specific location, and having arrived in the narrow band described above, they start groping for the nearest entrance in the proper direction. Therefore, the signing to the entrances must be continuous along this band, especially wherever the freeway can actually be seen from the intersecting surface streets.
2. The proper lane for each movement should be indicated in advance of the point where the turn must be made.
3. Advance notice signs should clearly state what motorists must do to reach the entrance.
4. The signs at the entrance should be positioned uniformly with respect to the point where the turn from the street to the entrance must be made.

At locations where motorists must use a street other than the one they are on in order to reach the freeway entrance, special emphasis is necessary to impart the information that the freeway can be reached only by turning onto the other street. At locations where motorists must make a movement that seems illogical, the signing should be particularly clear and well-positioned, both in advance of the actual turning point and at the point where the turn is to be made.

Other special informational and directional signs are warranted to direct motorists to destinations that generate a large number of personal trips. Local pressure groups can make requests and/or demands for all kinds of directional signs at every conceivable location, but extreme care must be exercised to properly identify and sign only such destinations that are most needed by the motoring public.

LIABILITY

Missing, damaged, or deteriorated traffic control devices or those that do not conform with national or state standards can cause accidents. An improper installation, the failure to replace or repair a damaged or missing control device, or failure to conform to standards outlined in the *Uniform Manual* may lead to liability on the part of a governmental agency or the responsible governmental officials themselves. When and to what extent the governmental agency and its employees can be held legally liable for their acts or failure to act vary from state to state.

In a number of states the liability of governments, governmental agencies, and employees is limited by the doctrine of sovereign immunity, that is, the inherent immunity of the state against legal actions brought by its citizens. When sovereign immunity exists, it generally extends to employees to the extent that they are not liable for damages or injuries resulting from the exercise of discretion while performing governmental functions.

In many states sovereign immunity was adopted in the original constitutions and thus cannot be changed without a constitutional amendment. In some states sovereign immunity has been waived to various degrees and courts of claims have been established by the legislatures to hear and decide claims against the states or governmental agencies. In these cases the governmental agency permits itself to be subject to claims for damages or injuries caused by its agents or employees.

In many states, too, the doctrine has been restricted to the state government, thus leaving local governments open to law suits. In many situations, application of the doctrine is subject to court interpretation on a case-by-case basis. Because governmental employees are becoming more and more vulnerable to lawsuits relating to their official duties, it is highly desirable that the state legislature adopt legislation providing that there will be no liability for discretionary acts, and, more specifically, that failure to initially install traffic control devices shall not be the basis of liability. A further means of protection for governmental employees is for the public bodies to carry liability insurance to cover any possible negligence. (When sovereign immunity does not exist, such insurance is almost mandatory, especially for smaller municipalities.)

When states and local governmental units are subject to liability for failure to maintain control devices or failure to repair or warn of hazardous conditions, some notice or a reasonable opportunity to learn of the problem and make corrections is required.

In all cases, regardless of the extent of sovereign immunity or liability, each governmental agency should see that adequate programs are developed for the proper installation, observation, and maintenance of all control devices and repair or warning of hazardous conditions.

SIGNING MATERIALS, MAINTENANCE AND INVENTORY PRACTICES

SIGN BACKING MATERIALS

Materials in general use for traffic signs include outdoor plywood, aluminum and steel sheeting, fiberglass, and plastic. Regular plastic and fiberglass signs are often used for the construction and maintenance series because they are relatively light and do not develop metal slivers which can cut the workmen if the signs are roughly handled. Translucent plastic sign faces are generally used for internally illuminated signs and provide excellent daytime visibility and legibility.

Exterior high-density overlay plywood is preferable to the older wooden signs with batten reinforcements. High-density overlay plywood is available in thicknesses from $\frac{1}{4}$ in. to 1 in. (0.6 cm to 2.5 cm), although thicknesses in the range of $\frac{1}{2}$ in. to $\frac{3}{4}$ in. (1.3 cm to 1.8 cm) are customarily used for larger sized signs so that no additional support is needed.

Commercial sheet steel, coated or rust proofed by one of the various chemical treatments recommended by manufacturers, is suitable for traffic signs. Eighteen-gauge steel is adequate for smaller signs up to 24 in. (61 cm) in their largest dimensions, but 16-gauge steel is recommended for larger signs.

Aluminum sign blanks should conform to the American Society for Testing and Materials (ASTM) Specification B-209 Alloy 6061-T6 or 5154-H38. They are easily fabricated into any of the standard shapes and sizes. A chemical conversion treatment coating conforming to ASTM Specification B449-67, class 2 (10-25 mg/ft^2), which produces a light, tight, iridescent coating, provides a satisfactory basis for reflective or nonreflective sheeting and paint, and also imparts increased corrosion resistance.

For larger signs mounted on structures over the roadway, extruded or laminated aluminum panels are required. These are bolted on the support structures with special clips manufactured for that purpose. The extruded or laminated aluminum panels are better for the larger signs in that extra support is minimal. Experimentation in Illinois has shown that a louvered sign can be effectively installed that meets the visibility and legibility needs of the motorist, but it reduces the wind load considerably so that lighter supports can be used or larger signs can be mounted on existing supports.[19]

SIGN SURFACE MATERIALS

The material used for sign faces can include paint, adhesive-coated plastic film, porcelain, reflective sheeting, and reflective coatings of beads and a binder. Cut-out letters, numerals, and symbols can be made of plastic reflective buttons in a porcelain enameled frame or reflective sheeting.

Paint, nonreflective plastic film, and porcelain materials should only be used on signs that are used exclusively for daylight operations or are adequately illuminated for nighttime visibility.

In developing specifications and purchasing legend materials, the engineer must make certain that:

1. The colors are within the allowable tolerance under both daylight and night lighting conditions
2. The material is weather resistant
3. Minimum brightness levels are met for reflective signs
4. There is adequate durability
5. Materials are packaged and shipped so that there is minimal damage

When local governmental agencies do not have the necessary expertise to finalize these specifications, they should check with their state or provincial counterparts.

SIGN MAINTENANCE

Adequate maintenance of the traffic sign system is of equal importance with adherence to proper warrants and good installation practices in the original installation. Inspection of the system should be made at least twice during daylight hours and once during darkness every year to ensure proper visibility and legibility.

To ensure adequate maintenance, a suitable schedule for inspection, cleaning, and replacement of signs should be established. Employees of the highway department, police, and other governmental employees whose duties require that they travel on the highways should be encouraged to report any damaged or obscured signs immediately.

Maintenance should include washing signs with a good soap or detergent as often as local conditions demand or cleaning by means of a steam generator unit (see Figure 16.6). The cleaning cycle may range from once every two years for self-cleaning surfaces in clean residential areas to once every few months in industrial areas. Special attention and necessary action should be taken to see that weeds, trees, shrubbery, and construction materials do not obscure the face of any sign.

---

[19] "Evaluation of a Louvered Panel for Use as a Freeway Sign Background" (Springfield, Ill.: Illinois Department of Transportation, Bureau of Traffic, 1971).

Figure 16.6. Typical sign washing equipment and operation. (Source: Pennsylvania Department of Transportation, Bureau of Traffic Engineering, Harrisburg, Pennsylvania.)

Illuminated signs demand greater attention. Because they are generally of larger size, an occasional check for weather damage to sign and mountings is essential.

A regular cycle for lamp inspection and replacement should be established and lamps should be replaced when the level of illumination goes below predetermined standards. This is very important in cold weather because old lamps tend to become inefficient at low temperatures.

Large freeway signs, both overhead and side mounted, provide special maintenance problems because of their size and locations. They should be washed once or twice a year and have the background and legend refurbished as needed. The refurbishing, when routine maintenance is inadequate, often calls for the removing of the legend and the application of an overlay panel with a new background upon which a new legend of reflective buttons or reflective sheeting is applied. Steel sign supports should be repainted on about four or five year cycles.

INVENTORY

Traffic signs constantly require maintenance and modernization. Before establishing a schedule for such a program a sign inventory should be undertaken to establish the requirements and provide the basic planning information. A systematic sign survey, which can pinpoint deficiencies or damaged signs, should be planned in detail so that the necessary forms are available for the initial inventory and future updates. These forms must be convenient for field use and easily compiled in the office so that the information is in a usable form.

Objectives of the inventory are:

1. Classification of all traffic signs by size, location, condition of sign and post by day, and reflectivity at night.
2. Discovery of conditions requiring change in design or size of signs, refurbishing, or reflectorization.
3. Establishment of existing and future sign needs and a plan to upgrade signs systematically to current standards.

The field inventory procedure will depend on the governmental agency responsible for the control devices and the availability of computer processing equipment (see discussion in Chapter 11, Computer Applications). A municipal agency would be interested in obtaining and maintaining data for each block face;[20] a larger agency would be more interested in longer segments, for example, a traffic route.[21] In any case, the minimum data collected should include:

1. Block number or route number
2. Side of roadway (North, South, etc.)
3. Direction sign is facing
4. Sign code
5. Kind of post
6. Distance from intersection or other control points
7. Offset distance
8. Installation date
9. Condition of sign
10. Other signs on same post
11. Condition of other signs
12. Visibility factor

A second inventory that should be maintained by any agency responsible for traffic control devices is that of all completed signs in the sign shop and storage buildings plus an inventory of sign supports and raw material for the manufacture of signs. Such a program, if kept current as signs are removed for installation, will aid in the schedule of production for signs, aid in monitoring the work of the field crews, and be used as a basis for budget forecasting.

SIGN SHOP REQUIREMENTS AND PRACTICES

The size and scope of sign shop operations in a given jurisdiction is dependent primarily on decisions of a local nature, such as local sign fabrication vs. purchasing of manufactured signs. The decisions involve these factors:

---

[20] H. A. SWANSON, "An Urban Sign Inventory Procedure," *Traffic Engineering*, XL, No. 8 (1970), pp. 44–46.

[21] J. A. VASCONCELLES, "A Sign Inventory Procedure," *Traffic Engineering*, XL, No. 1 (1969), pp. 36–41.

1. Number of signs to be used
2. Cost of labor (can exterior workers be used in offseason as shop men?)
3. Time factor (can emergency needs be filled and are warehousing facilities available to stock needs in advance?)
4. Local policies on purchasing

EQUIPMENT NEEDS

Sign production is greatly simplified if sign backings are purchased in proper blank sizes and have been processed so that nothing further is required except for painting or the application of sheeting. When further processing is needed, a separate well-ventilated area should be provided and there should also be a spray booth for high production. (Often the maintenance section will have painting facilities that can be used jointly.)

Although some or almost all blanks are purchased to size, facilities are generally needed for fabricating sign blanks. Large shears can be used to cut sheet steel or aluminum for original sign blanks and also to salvage damaged signs by making smaller blanks. Power saws are a necessity in fabricating plywood signs.

Additional equipment, depending on the size and operation of the sign shop, will include:

1. Equipment to paint and/or coat the blanks and for the application of reflective sheeting, plastic film, or beads
2. Equipment to silk-screen and stencil messages
3. Ovens or infrared drying units with drying racks
4. Storage racks, bins, or shelves

In addition, some sign shops have facilities for stripping old finishes from reusable blanks. In recent years, however, this specialized service has been furnished by commercial companies that supply the aluminum blanks. Pennsylvania and West Virginia have been using this service and have found that the cost is approximately one-half of the original cost of the blank.

Generally, facilities for sign layout, letter and symbol design, and letter spacing will be needed as a close adjunct to the sign shop. Means for servicing illumined signs must also be available.

PRACTICES

Sign shop practices vary from state to state and from municipality to municipality as dictated by policy decisions. In some states, prison labor may cut and furnish sign blanks from sheet metal and in other states they may furnish the completed sign to the highway department. When highway or transportation departments completely manufacture their own signs, there are two common practices: (1) all signs are manufactured and supplied from one central sign shop or (2) signs are manufactured at the district level.

Large signs for freeways are generally bought as completed signs, but some manufacturing is necessary at sign shops for temporary signs, replacements, or for overlay panels.

Many municipalities having over 50,000 population have sign shops that manufacture all or a part of their signs. The amount of work varies depending on the size of the municipality and its policies. Some buy completed faces and just apply them to the sign blanks; others apply a sheeting to the sign blank and screen a message or else do all signs that do not need a reflective background.

A governmental agency should continue to review its policies of materials being used and its purchasing procedures. It can obtain guidance for its local requirements from manufacturers of shop equipment, manufacturers of sign materials, and by visiting sign shops already in operation by similar state highway departments, municipalities, or counties.

KINDS OF MARKINGS

Traffic markings include all traffic lines (both longitudinal and transverse), symbols, words, object markers, delineators, cones, or other devices, except signs, that are applied or attached to the pavement or mounted at the side of the roadway to guide traffic or warn of an obstruction.

FUNCTIONS OF MARKERS

Pavement markings have definite functions to perform in the proper control of vehicular and pedestrian traffic. In some instances they serve to regulate, to guide, to channelize traffic into the proper position on the street or highway, and to supplement the regulations or warnings of other traffic control devices. In other instances they serve as a psychological barrier for opposing streams of traffic and as a warning device for restricted sight and passing distances; they also give information for turning movements, special zones, etc. As an aid to pedestrians, they channelize movement into locations of safest crossing and, in effect, provide for an extension of the sidewalk across the roadway. Under favorable conditions, pavement markings aid the vehicle driver in many ways without diverting his attention from the roadway.

More permanent marking materials having increasingly improved visibility and durability under all driving conditions have been and are continually being developed.

TEMPORARY MARKINGS

Almost all applications of markings are in areas in which it would be desirable to have a permanent or long-life installation as opposed to a temporary or short-life installation. Permanent traffic markings generally do not have the same life as normal traffic signs, and they must be replaced more often.

There is definite need and use for temporary markings in construction areas and at locations where a temporary hazard must be properly marked until the necessary repairs or improvements can be made. Provisions must be made so that the temporary installations are removed when no longer needed and that motorists will not be confused by an application of a permanent device in the vicinity of temporary markings.

MARKING ELEMENTS

LONGITUDINAL PAVEMENT MARKING DESIGN

Longitudinal lines are used to organize traffic into proper channels, advise motorists where passing is prohibited, and to supplement other warning devices when there are hazards within the roadway.

In the United States, the color of longitudinal markings can be white, yellow, or red. Black can also be used in combination with the three primary colors when the pavement does not provide sufficient contrast. Application and basic concepts of the standard colors have been established in the *Uniform Manual*[22] and include:

1. Yellow lines delineate the separation of traffic flows in opposing directions.
2. White lines delineate the separation of traffic flows in the same direction.
3. Red markings delineate a roadway that shall not be entered or used by the viewer of that line.
4. Broken yellow and white lines are permissive in character. These are normally formed by segments and gaps in a ratio of from 3 to 5, such as from 15 ft to 25 ft (4.6 to 7.6 m) or from 9 ft to 15 ft (2.7 to 4.6 m).
5. Solid lines are restrictive in character.
6. Width of line indicates the degree of emphasis. A line from 4 in. to 6 in. (10 to 15 cm) wide is a normal width line.
7. Double lines indicate maximum restrictions.

Some of the applications of these principles are shown in Figure 16.7, but a brief summary includes:

1. Centerlines vary depending on the width of the traveled roadway and the area. A single, broken yellow line, which is used on two-lane, two-way streets, indicates that passing is permitted from either direction. (A solid yellow centerline on either side of the broken line indicates that passing is prohibited from the direction that is immediately adjacent to the solid line.) Double, solid yellow center lines should be used on two-way streets of four or more lanes, in areas where the centerline is offset, and in areas where passing is prohibited in both directions.
2. Lane lines are broken white lines with a standard spacing of 15 ft (4.6 m) painted and a gap of 25 ft (7.6 m). Some urban areas use a pattern of 9 ft (2.7 m) painted and a gap of 15 ft (4.6 m). Solid lane lines can be used on approaches to intersections and in other areas where lane changes should be restricted. A dotted white line, normally 2 ft (0.6 m) in length with 4 ft (1.2 m) gaps can be provided to delineate the extension of a line through intersections or across deceleration lane openings where special problems indicate that motorists can be confused.
3. Reversible lane markings are broken, double yellow lines with a standard spacing of 15 ft (4.6 m) painted and a gap of 25 ft (7.6 m). Reverse lane signs or a signal system is required to ensure that the lane is operating in one direction only for each time period.
4. Pavement edge lines, which should be neither less than 2 in. (5.1 cm) nor more than 4 in. (10.2 cm) wide, include white lines on each side of the traveled roadway, except on divided highways where there is an inadequate clear space to the left of the line for emergency refuge; the markings adjacent to these medians should be yellow.
5. Channelizing lines and median islands should be wide or double lines and can be either white or yellow depending on the traffic patterns. If the lines are to form traffic islands where travel in the same direction is permitted on both

---

[22] *Manual on Uniform Traffic Control Devices for Streets and Highways.*

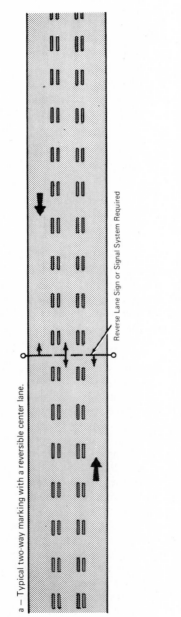

a – Typical two-way marking with a reversible center lane.

Reverse Lane Sign or Signal System Required

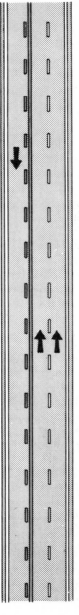

b – Typical two-way marking where motorists in a single lane are permitted to pass.

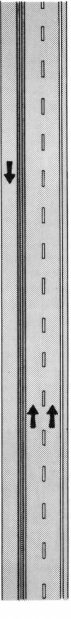

c – Typical two-way marking where motorists in a single lane are not permitted to pass.

Figure 16.7. Typical two-way marking applications. (Source: *Manual on Uniform Traffic Control Devices*, p. 182.)

sides, the markings are to be white. If islands are formed to separate travel in opposite directions, the markings are to be yellow. In either case, any cross-hatching should be of the same color as the longitudinal lines.

TRANSVERSE MARKING DESIGN

Transverse markings, either by themselves or in combination with traffic signs, are used to convey a special message to the motorist. They include shoulder markings, word and symbol markings, stop lines, crosswalk lines, and parking space markings. All of these markings are to be white except that they are to be yellow when they are a part of a median marking and they are to be red when they are visible only to traffic proceeding in the wrong direction on a one-way roadway.

Because of the low-approach angle at which pavement markings are viewed, all transverse lines must be proportioned in order to give visibility equal to that of longitudinal lines and to avoid apparent distortion where longitudinal and transverse lines combine in symbols or lettering.

Detailed applications for these markings are provided in the *Uniform Manual.*[23] A brief summary includes:

1. Crosswalk markings, which are a minimum of 6 in. (15 cm) in width and at least 6 ft (1.8 m) apart, are used to guide pedestrians in the proper paths and to serve to warn motorists of a pedestrian crossing point. The solid, white boundary lines can be up to 24 in. (61 cm) in width when vehicle speeds are over 35 mph (56 km/h) or when crosswalks are unexpected. Moreover, the area of the crosswalk may be marked with white diagonal lines at a 45° angle or longitudinal lines at a 90° angle (see Figure 16.8).

2. Stop lines, which are normally from 12 to 24 in. (31 to 61 cm) wide, should extend across all approach lanes. They should be placed at the desired stopping point, in no case more than 30 ft (9 m) or less than 4 ft (1.2 m) from the nearest edge of the intersecting roadway. If crosswalks have been installed, the stop line should ordinarily be placed at 4 ft (1.2 m) in advance of the nearest crosswalk line.

3. Railroad crossing markings are placed in advance of a railroad crossing and are to consist of an *X*, the letters *RR*, a no-passing marking, and certain transverse lines. They should be placed on all paved approaches to railroad crossings and elongated to allow for the low angle at which they are viewed (see Figure 16.9).

4. Parking space markings encourage more orderly and efficient use of parking space when parking turnover is substantial and they tend to prevent encroachment on fire-hydrant zones, bus stops, loading zones, approaches to corners, clearance spaces for islands, and other zones where parking is prohibited (see Figure 16.10).

5. Word and symbol markings, which are white, may be used for guiding, warning, or regulating traffic. They should be limited to not more than a total of three lines of words and/or symbols. They can serve as regulatory (STOP, RIGHT TURN ONLY, etc.), warning (STOP AHEAD, SCHOOL, etc.), or guide (US 30, ROUTE 123, etc.), but when used as a regulatory device, they must

---

[23] *Ibid.*

a — Standard crosswalk marking.

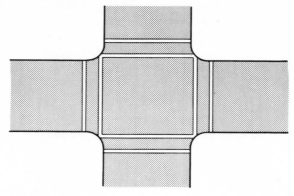

b — Crosswalk marking with diagonal lines for added visibility.

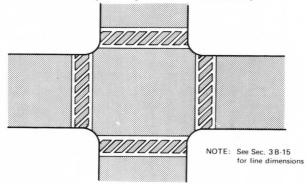

NOTE: See Sec. 3B-15 for line dimensions

c — Crosswalk marking with longitudinal lines for added visibility.

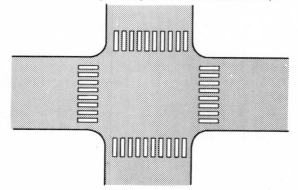

Figure 16.8. Typical crosswalk markings. (Source: *Manual on Uniform Traffic Control Devices*, p. 200.)

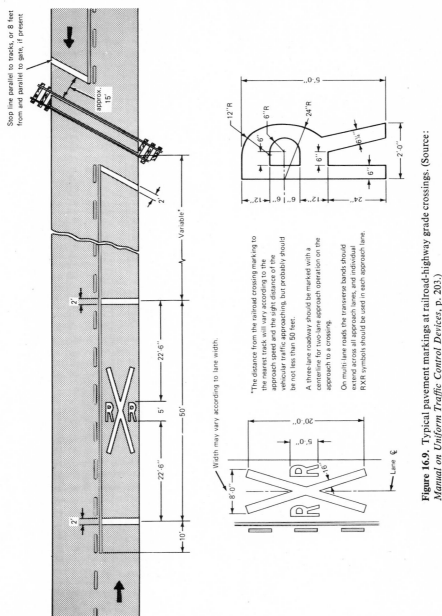

Figure 16.9. Typical pavement markings at railroad-highway grade crossings. (Source: *Manual on Uniform Traffic Control Devices*, p. 203.)

763

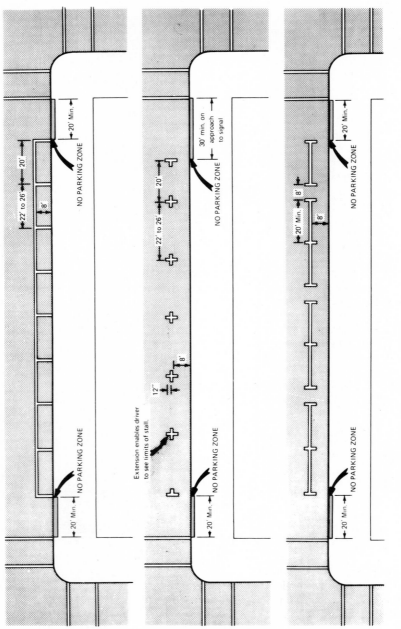

Figure 16.10. Typical parking space-limit markings. (Source: *Manual on Uniform Traffic Control Devices*, p. 204.)

supplement necessary signs. Symbols are preferable to words, and in all cases both must be elongated in the direction of traffic movement. Large letters, symbols, and numerals [8 ft (2.4 m) or more in height] should be used. If the message consists of more than one word, it should read UP, i.e., the first word should be nearest to the driver.

6. Curb markings can be either roadway delineation or parking regulations. When used as roadway delineations, the color should follow the guidelines for edge lines or channelizations. When used as parking regulations, they should supplement standard signs and can be a special color as prescribed by local authorities.

OBJECT MARKERS AND DELINEATORS

Physical obstruction in or near a roadway, including installations designed for the control of traffic, constitute serious hazards to safe traffic movement and should be adequately marked. Every effort, however, should be made to remove the obstruction or to minimize the potential danger.

In United States practice, when an obstruction within or adjacent to the roadway requires marking, the marker shall consist of an arrangement of one or more of the following designs:

1. A Type 1 marker is a nine-unit marker consisting of nine reflectors having a minimum dimension of approximately 3 in. (7.6 cm), yellow in color, mounted symmetrically on an 18 in.- (46 cm) diamond background, either black or yellow in color; or an 18 in.- (46 cm) diamond, all yellow reflector.
2. A Type 2 marker is (1) a three-unit marker consisting of three yellow reflectors having a minimum dimension of approximately 3 in. (7.6 cm) arranged either horizontally or vertically or (2) a rectangular yellow reflector 6 in. by 12 in. (15.2 cm by 30.5 cm).
3. A Type 3 marker is a striped marker consisting of a vertical rectangle approximately 1 ft by 3 ft (30.5 cm by 91 cm) in size with alternating black and reflectorized yellow and white stripes sloping downward at an angle of 45° toward the side of the obstruction on which traffic is to pass. The minimum width of the yellow or white stripe shall be 3 in. (7.6 cm). A better appearance can be achieved if the black stripes are wider than the yellow or white stripes. The black-on-yellow marker shall be used to the left of the direction of travel.

The application of these object markers includes:

1. Objects in the roadway are to be marked with a Type 1 or 3 marker. In addition, large surfaces can be painted with diagonal stripes, 12 in. (30.5 cm) or greater in width, similar in design to the Type 3 object marker. The alternating black and reflectorized yellow stripes shall be sloping down at an angle of 45° toward the side of the obstruction which traffic is to pass.
2. Objects adjacent to the roadway, which are so close to the road that they need a marker, are to be marked with Type 2 or 3 object markers. The inside edge of the marker is to be in line with the inner edge of the obstruction.
3. End of roadway is to be marked with (1) a nine-unit marker consisting of nine reflex reflectors having a minimum dimension of approximately 3 in. (7.6 cm),

red in color, mounted symmetrically on an 18-in. (46 cm) square background, either red or black in color or (2) an 18 in.- (46 cm) square all red reflector.

Road delineation markers, with a minimum dimension of about 3 in (7.6 cm), are to be considered as guide markings instead of warning devices. Delineators may be used on long, continuous sections of highway or through short stretches where there are pavement width transitions or other changes in horizontal alignment, particularly where the alignment might be confusing.

When curbs of islands are located in or near the line of traffic flow, they should be marked with white delineators if on the right or if traffic in the same direction passes on both sides. If the curbed island separates traffic in opposing directions, it should have yellow delineators.

Other applications of delineators include:

1. Delineators used on through two-lane, two-way roadways shall be single white reflector units on the right side. Single white reflector units may be placed on the left side of two-way roadways, particularly at sharp right-hand curves.
2. On through roadways of expressway type facilities, single white delineators are to be placed continuously along the right side. If used on the left side for additional guidance, they are to be white also.
3. Double or vertically elongated yellow delineators are to be spaced at 100 ft (30.5 m) intervals along acceleration and deceleration lanes along the outside or both sides of tangent portions of interchange ramps. When delineators are used only on one side of an interchange ramp and the ramp is curved, the delineators shall be placed on the outside of the curve.
4. Red delineators may be used on the reverse side of any delineator whenever it would be viewed by a motorist traveling in the wrong direction.

COLORED PAVEMENT

Colored pavement, which significantly contrasts with adjoining paved areas, can be used to help guide or regulate traffic. The applications to date have shown better contrasts in daylight operations because the colors have not been satisfactorily reflectorized and thus blend in with the abutting pavement at night.

When used, the colors are to be limited to the following:

1. Red is to be used only on the approaches to a stop sign that is in use 24 hours a day.
2. Yellow is to be used only for medians separating traffic flows in opposite directions.
3. White is to be used only for delineation in order to provide contrast with other colors such as for right-hand shoulders or for channelizing islands for traffic flows passing on both sides, etc.

MARKING WARRANTS AND REQUISITE STUDIES

Although traffic markings have definite limitations, their advantages of aiding the motorists without diverting his attention from the roadway and their acceptance or demand by the motoring public make it necessary to establish criteria for proper installations.

UNIFORM MANUAL PROVISIONS

Specific guidelines for the use of markings and the requisite studies in the United States are provided in the *Uniform Manual*.[24] Some of these guidelines and studies include:

1. Center lines, which separate traffic traveling in opposite directions, are desirable on paved highways when:
   a. The two-way roadway in rural districts is 16 ft (5 m) or more in width and speeds are in excess of 35 mph (56 km/h).
   b. The highways are in residential or business districts and where there are significant volumes of traffic.
   c. The undivided highway has four or more lanes.
   d. The highway is wide enough for three lanes of traffic. (Here, the highway should be marked for two lanes in one direction and one lane in the opposing direction.)
2. Lane lines should be installed on multilane highways when an engineering study indicates that the roadway will accommodate more lanes than when unpainted. In addition, lane lines should be installed:
   a. On one-way streets where maximum efficiency in use of the street width is desired.
   b. At the approaches to important intersections and in dangerous locations where the lines would organize the traffic for better roadway use.
   c. On rural highways that have an odd number of traffic lanes.
3. A no-passing zone at a horizontal or vertical curve is warranted when the sight distance is less than the minimum necessary for safe passing at the prevailing speed of traffic. The conditions that justify such zones are shown in Table 16.1.

TABLE 16.1
Minimum Passing Sight Distance vs. Speed of Traffic

85 Percentile	Speed	Minimum Passing Sight Distance*	
(mph)	(km/h)	(ft)	(m)
30	48	500	152
40	64	600	183
50	81	800	244
60	97	1000	305
70	113	1200	366

*Passing sight distance on a vertical curve is the distance at which an object 3.75 ft (1.1 m) above the pavement can be seen from a point 3.75 ft (1.1 m) above the pavement (see Figure 16.11). Similarly, passing sight distance on a horizontal curve is the distance measured along the center line (or the lane line of a three-lane highway) between two points 3.75 ft (1.1 m) above the pavement on a line tangent to an obstruction that cuts off the view on the inside of the curve.

Source: *Manual on Uniform Traffic Control Devices* (Washington, D.C.: U.S. Department of Transportation, Federal Highway Administration, 1971), p. 190.

---

[24] *Ibid.*

VERTICAL CURVE

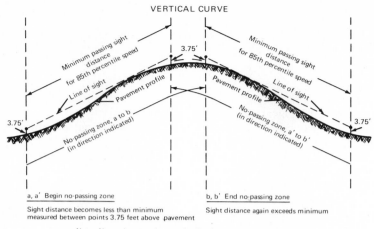

a, a' Begin no-passing zone

Sight distance becomes less than minimum
measured between points 3.75 feet above pavement

b, b' End no-passing zone

Sight distance again exceeds minimum

Note: No-passing zones in opposite directions may or may not overlap,
depending on alignment.

HORIZONTAL CURVE

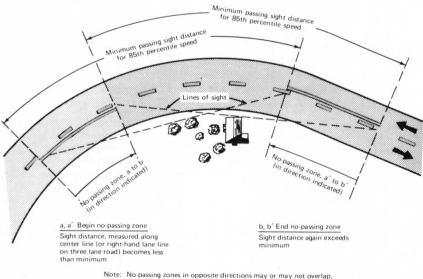

a, a' Begin no-passing zone

Sight distance, measured along
center line (or right-hand lane line
on three lane road) becomes less
than minimum

b, b' End no-passing zone

Sight distance again exceeds
minimum

Note: No-passing zones in opposite directions may or may not overlap,
depending on alignment.

Figure 16.11. Method of locating and determining the limits of no-passing zones at vertical and horizontal curves. (Source: *Manual on Uniform Traffic Control Devices*, p. 189.)

4. Markings for railroad crossings are to be placed where railroad highway grade crossing signals or automatic gates are operating and at all other crossings where the prevailing speed of highway traffic is 40 mph (64 km/h) or more.
5. Pavement edge markings are to be placed on all interstate highways and should

be placed where edge delineations are desirable in order to reduce driving on paved shoulders or refuge areas.

6. Crosswalks should be marked wherever studies show that there is a substantial conflict between vehicles and pedestrian movements. They should also be installed at points of pedestrian concentration and where pedestrians cannot recognize the best place to cross.

SPECIAL MARKINGS

The safe operation and needed capacity of many streets and highways quite often depends on special applications of traffic markings. These special markings which utilize pavement markers as well as signs, object markers, and delineators can be identified as:

1. *Lane reduction transitions.* Pavement line markings can be effectively used to supplement the standard signs that guide traffic where the pavement width reduces to a lesser number of lanes (see Figure 16.12). Many variations are possible, depending on which lanes must be offset or eliminated and on the amount of the offset. One or more lane lines must be discontinued and the remaining center and lane lines must be connected in such a way that traffic safely merges into the reduced number of lanes.

    Lines marking pavement width transitions should be the standard design for center, lane, or barrier lines. Converging lines should have a length of not less than that determined by the formula $L = S \times W$, where L equals the length in feet, S the off-peak 85-percentile speed in miles per hour, and W the offset distance in feet.

2. *Obstruction approach markings.* Pavement markings are frequently used to supplement standard signs in order to guide traffic approaching a fixed obstruction within a paved roadway (see Figure 16.13). If the obstruction is in the center of the roadway, all traffic is usually directed to drive to the right of it. Sometimes the obstruction may be between two lanes of traffic moving in the same direction. The use of obstruction approach markings and signs does not eliminate the need for adequate object markings on the obstruction itself. These markings normally consist of a diagonal line (or lines) extending from the center or lane line to a point from 12 to 24 in. (31 to 61 cm) to the right side (or to both sides) of the approach end of the obstruction. The length of the diagonal markings can be determined by the formula $L = S \times W$, where L equals the length in feet, S the off-peak 85-percentile speed in miles per hour, and W the width of the obstruction in feet. The diagonal line should never be less than 200 ft (61 m) in length in rural areas or 100 ft (30 m) in urban areas.

3. *Reversible lane markings.* The capacity of many urban and suburban arterial streets and highways has been effectively increased by the use of reversible lanes, i.e., lanes that are assigned to opposite directions of traffic movement at different times of the day. The proper use of the assigned lanes can be achieved by lane-direction traffic signals. In addition, double, broken yellow center and lane lines are to be used on each side of the dual-usage reversible lanes.

4. *Two-way left turn lanes.* Many urban streets and highways that are wide enough to create an odd number of lanes can have their capacity increased by making the center lane a two-way left-turn lane. (Vehicles from both directions turn

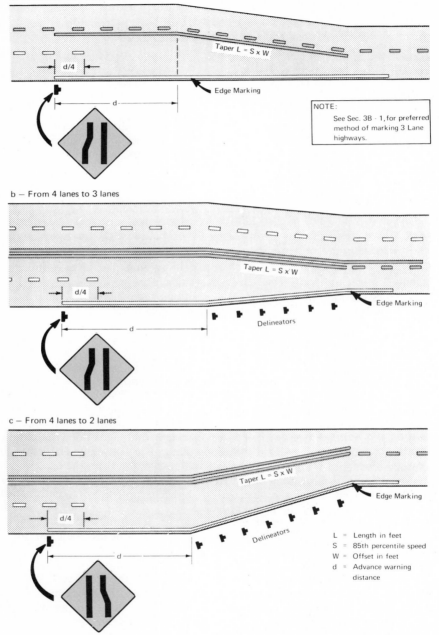

a — From 3 lanes to 2 lanes

Taper L = S x W

Edge Marking

d/4

d

NOTE:

See Sec. 3B - 1, for preferred method of marking 3 Lane highways.

b — From 4 lanes to 3 lanes

Taper L = S x W

d/4

Edge Marking

d

Delineators

c — From 4 lanes to 2 lanes

Taper L = S x W

Edge Marking

d/4

d

Delineators

L = Length in feet
S = 85th percentile speed
W = Offset in feet
d = Advance warning
        distance

Figure 16.12. Typical pavement-width transition markings and signs. (Source: *Manual on Uniform Traffic Control Devices*, p. 193.)

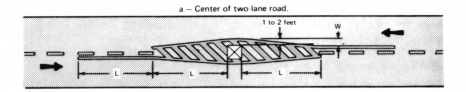

a – Center of two-lane road.

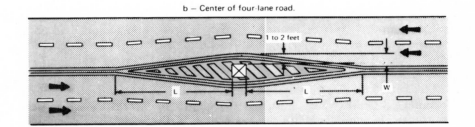

b – Center of four-lane road.

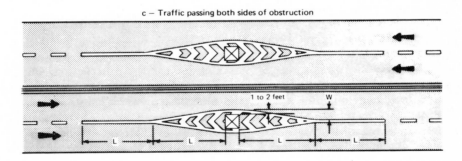

c – Traffic passing both sides of obstruction

L = S x W

S = 85th percentile speed in miles per hour
W = Offset distance in feet

Minimum length of: L = 100 feet in urban areas
                    L = 200 feet in rural areas

Length "L" should be extended as required by
sight distance conditions.

Figure 16.13. Typical approach markings for obstructions in the roadway. (Source: *Manual on Uniform Traffic Control Devices*, p. 197.)

left into driveways and side streets from the same lane.) The two-way left-turn lane must be well-defined and marked with a yellow, no-passing solid line on each edge of the lane.

5. *Channelization.* Painted channelization can be used to increase efficiency and safety and has the advantage of easy modification when warranted by driver behavior. If a more positive barrier is required, curbs and islands may be constructed, but the paint channelization may well serve initially to establish the best layout arrangement before permanent construction is established.

MARKING MATERIALS, MAINTENANCE PROCEDURES, AND SCHEDULES

Although traditionally the most common materials used for pavement and curb markings have been paint and glass beads, newer materials that are more durable and sometimes more effective during inclement weather are now available. Although they may have a much higher initial cost, lower maintenance costs, less interruption of traffic, better visibility and legibility through winter months, etc. can justify their use.

PAINTED TRAFFIC LINES

Technological improvements also continue to be made in traffic paints materials, beads and their gradations, and methods of applications. Research continues to improve the final product—a reflectorized line that has been installed at a minimum cost and will serve for six months or, more desirably, one year on the heavily traveled streets and highways. In 1965 in almost all state highway departments in the United States, the predominant wet film thickness of traffic paint was 15 mils, and the beads, which were applied at a rate of about 6 lb of drop-on beads per gallon of paint, had a standard refractive index of 1.50+ as opposed to 1.65 or more.[25]

One of the biggest improvements in paint has been the development of "rapid-dry" binders that can be applied with slightly modified, existing equipment at "low" heat or with specially constructed "high"-heat equipment. Their main advantage is less disruption to traffic flow. As the material costs are further reduced, there will be very little difference in applied costs because the amount of equipment can be reduced and the need for cones or other devices during the drying period will be eliminated.

There are several methods used for selection of a traffic paint or binder. The final selection and purchase, however, is generally based on price. A brief summary of established specifications include:

1. A performance specification with a laboratory and service test procedure to be used to rate the submitted samples. Usually, a committee of representatives of traffic engineering, materials testing, and purchasing departments evaluate paint performance. Qualities evaluated may include general daylight appearance, color, film condition, bead retention, and reflectance. Various rating methods have been used to evaluate these qualities.[26]

---

[25] "1965 Usage of Pavement Marking Materials by Government Agencies in the United States," *Highway Research Circular No. 79* (Washington, D.C.: Highway Research Board, 1968).

[26] "A Model Performance Specification for the Purchase of Pavement Marking Paints," a Tentative Revised ITE Standard, *Traffic Engineering*, XLII, No. 6 (1972), pp. 18–24.

2. A formulation specification based on rigid analysis of the ingredients and generally subject to definite laboratory tests and, occasionally, service tests. Since it is impossible to determine all characteristics of a paint on the basis of chemical analysis, the specifications should include both laboratory and service tests for best results. Usually, the buyer reserves the right to inspect the manufacturing facility. This specification should guarantee the buyer a paint well-suited to his requirements, provided the buyer has the means of determining the paint formulae best suited for the climate and road surface conditions of his particular area. Nevertheless, there are several disadvantages in this method:

   a. The buyer will probably not be taking advantage of the new developments in paint formulation unless he has available a paint chemist whose own product development keeps pace with those of the paint manufacturers.

   b. Some paint manufacturers claim this specification tends to stifle their research and development.

   c. It may be impossible to analyze quantitatively all ingredients in a paint because of chemical changes during processing.

3. A specification which is by brand product or equal. This method is not generally advisable because it subjects the buyer to complaints of favoritism or discrimination. Also, this method is complex because of the many factors that must be considered in order to determine the equality of two products.

The maintenance procedures and schedules are generally dependent on the kind of road, traffic volumes, and lane widths. Schedules should be established that will provide for a usable line at all times. This scheduling can include many miles of streets and highways that only require an application of paint once a year. Major arterials, however, may require paint applications in the spring and fall, or more often if weather permits. The maintenance procedure calls for the repainting of a new line directly over an earlier application. Wherever needed, the old line should be cleaned if covered with dirt or spotted by applying temporary marks on the highway where the previous lines have been worn away or where there has been resurfacing. (All resurfaced roads or new construction should be painted before opening to traffic.)

THERMOPLASTIC TRAFFIC LINES

Sprayed, extruded, and cold-rolled thermoplastic markings have been successfully applied in numerous countries throughout the world. Since technological advances continue, all users or potential users must keep in contact with the numerous suppliers and contractors.

The use of cold-rolled or glue-down plastic stripes, which have an adhesive backing, has primarily been for crosswalks and stop lines on bituminous pavements in high-density urban areas. Other applications have included lane and center lines on new bituminous pavement where the plastic stripes can be rolled into the surface with the last rolls of the roller. There applications, however, should be in areas where there are good street lights because the reflectivity of the beaded line will not be maintained at a desirable level.

A recent survey and report has made a comparison between thermoplastic striping materials and conventional paint.[27] The report indicated that:

---

[27] BERNARD CHAIKEN, "Comparison of the Performance and Economy of Hot-Extruded Thermoplastic Highway Striping Materials and Conventional Paint Striping," *Public Roads*, XXXV, No. 6 (1969), pp. 135–56.

1. Thermoplastics are much more durable on bituminous pavements than on portland cement concrete pavements.
2. The newer the concrete surface, the poorer the adhesion.
3. Thermoplastics are subject to snowplow damage.
4. Thermoplastics may be more economical than conventional paint striping only when high-traffic volumes are prevalent.

Economics aside, Pennsylvania, like other northern states, has several freeways on which it has been impossible to maintain a painted line throughout the winter months because of high-traffic density, studded tires, snowplows, and the use of large quantities of abrasive materials on ice and snow. Here thermoplastic lines appear to be the only answer.

The application to sprayed or extruded thermoplastic materials with glass beads requires special equipment, and it also generally requires a service-purchase contract. Such contracts should include specifications for all materials, construction requirements, and warranty provisions.[28]

Maintenance procedures and schedules call for reapplication of thermoplastic markings on compatible material when the earlier line has worn out or lost reflectorization. Schedules should be established so that all needed markings are applied before freezing weather.

PREFABRICATED TAPE MARKINGS

Prefabricated tape markings with an adhesive backing that is designed to conform closely to the texture of the roadway surface have found wide use for temporary markings for routing traffic during construction and for semipermanent markings such as pavement symbols, parking stalls, and parking lot markings. More recent materials give a measure of wet night reflection as well. When longer service is needed for this kind of marking, a primer should be applied to the road surface prior to the placing of the tape.

RAISED MARKERS

Raised markers, which form a semipermanent marking in areas where there is little or no snow plowing or studded tires, improve visibility under nighttime, wet weather conditions. These markers, which can be from $\frac{1}{4}$ in. (0.62 cm) to less than 1 in. (2.5 cm) high, can be used singularly as a reflective unit to supplement painted or thermoplastic lines or they can be used instead of other lines.

In snow-free areas, raised markers have been reported to be superior to standard traffic paint in terms of durability, driver preference, and night-wet visibility, and their cost for a 10-year period has been claimed to be compatible with the cost of standard paint striping.[29]

Because there is still a lot of development in this area, it is best to keep abreast of latest developments through literature, suppliers, or other governmental agencies prior to making an installation.

---

[28] "A Model Performance Specification for the Purchase of Thermoplastic Pavement Marking Materials," a Tentative Revised ITE Standard, *Traffic Engineering*, XLII, No. 6 (1972), pp. 26–36.

[29] BERNARD CHAIKEN, "Traffic Marking Materials—Summary of Research and Development," *Public Roads*, XXXV, No. 11 (1969), pp. 251–64.

Any schedule for maintenance should include necessary replacements prior to periods of prolonged inclement weather.

REFLECTORIZED STRIPING POWDER

Recent developments have provided a reflectorized striping powder that can provide "instant" track-free marking of crosswalks, school zones, and other legends. The material has glass beads distributed throughout and is applied to the pavement by a special striper with a propane flame that melts the powder in air just above the surface and then bonds it to the pavement. To date, the material does not have a life expectancy any longer than paint, but its greatest advantage is its minimum disruption of traffic.

GLASS BEADS

Glass spheres have been used with numerous binders to provide the night visibility where continuous roadway lighting is not provided. These beads can be premixed with the binder (paint, thermoplastic, striping powder, etc.) prior to application, dropped on by the gravity method, or applied by a combination of premix and drop on. The advantages of premix or beads-in-paint are their convenience in use and better distribution of the beads in the paint film. However, a further application of 2 lb of coarser beads per gal of paint dropped on the wet line provides an immediate reflective line and eliminates the need for traffic to wear off the thin paint film on the top beads.

Specifications for glass beads are written primarily for the bulk purchase of beads for the "drop-on" application method. Fewer specifications have been written to cover the premixed paint application method.

Essential requirements include the following major specifications: (1) gradation, (2) color and shape, (3) imperfections, (4) crushing strength, (5) index of refraction, (6) silica content, (7) chemical stability, (8) reflectivity, and (9) packaging.

OBJECT MARKERS AND DELINEATORS

Object markers, as discussed under "Marking Elements," can consist of yellow 3-in. (7.5 cm) reflectors mounted on yellow or black panels or can be yellow or white reflectorized stripes on a black background. These yellow reflectors are to be similar to the yellow and white delineators which are capable of clearly reflecting light under normal atmospheric conditions from a distance of 1,000 ft (305 m) when illuminated by the upper beam of standard automobile headlights.

Maintenance procedures and schedules for these devices are quite similar to those for traffic signs. Any installation in splash areas, such as median or edge of the roadway, should be scheduled for washing at least twice a year. Because of their location, they should be kept under constant observation for damage caused by vehicular hits and should be scheduled for replacement as soon as damage is detected.

EQUIPMENT AND OPERATION

PAINT MARKINGS

Traffic engineering departments use a large variety of paint application equipment varying from small gravity-fed, manually operated devices to large, elaborate machines designed for precision, high-output striping work. Two kinds are most common: manually propelled machines and self-propelled machines. Basic application problems make the smaller, manually propelled piece of equipment more efficient for painting word messages, stop lines and other transverse markings, and object markings. Supplemental equipment in use on the larger self-propelled machines includes bead bins and dispensers to apply reflecting glass spheres, skipping devices to automatically produce broken lines, and paint pumps to load paint tanks. Some machines have all this equipment, in addition to both inboard and outboard riggers for maximum longitudinal marking flexibility (see Figure 16.14). These machines can paint several lines at the same time in various changing patterns at operating speeds from 10 to 15 mph (16 to 24 km/h) on anything from a multilane Interstate highway to a 16-ft (5 m) rural secondary road. Special features can include two 250-gal paint tanks, five paint guns and bead dispensers, paint heaters, paint pumps, dual steering, two individually steerable paint carriages (inboard and outboard), storage space for the wet line guards, and a complete intercommunication system.

The speed of operation varies considerably with the kind of machine used and the kind of work done. Under ideal conditions, between 15 and 30 machine miles (32 to 48 km) of striping can be applied per 8-hr day.

On sections where there is no previous stripe available for guidance, a construction joint can be followed or a manual layout provided by placing spots of paint along the roadway or by stretching a rope between control points and painting along the rope either in spots or in a solid line. [Six hundred-ft (183 m) lengths of $\frac{3}{8}$ in.- (1 cm) manila rope or $\frac{3}{16}$ in.- (0.5 cm) diamond-braided nylon rope are most effective for this purpose.]

On tangents, control points should not be in excess of 600 ft (183 m). On curves, control points should be placed at such intervals as will ensure the accurate location of the line. On multilane highways, where many lines are parallel, offset attachments on many striping machines allow the painting of all parallel lines from only one layout line.

Figure 16.14. Modern high-capacity, high-speed pavement marking machine. (Source: Pennsylvania Department of Transportation, Bureau of Traffic Engineering, Harrisburg, Pennsylvania.)

Continuing experimentation by both operating agencies and manufacturers of paint and painting equipment is under way in order to improve application characteristics and durability of marking materials. Common causes of premature paint failures include:

1. Insufficient cleaning of pavement
2. Over-thinning of paint
3. Damp or wet pavement
4. Applying on windy days or when the temperature is below 40°F
5. Presence of limestone or other alkaline materials that tend to break down the paint and cause it to be washed away
6. Insufficient paint film applied

Glass beads can be introduced separately into the paint with the aid of a special dispenser attached to a standard striping machine. When the beads are premixed, however, the paint may be sprayed directly with conventional equipment.

THERMOPLASTIC LINES

Thermoplastic traffic lines can be installed by manually propelled machines for small applications such as crosswalks and stop lines or they can include large self-propelled machines with large tanks for heating the reflectorized thermoplastic material. In either case, the machine, which must heat the material to approximately 400°F, needs supplemental equipment such as bead bins and dispensers for spraying reflecting glass spheres onto the hot line. Skipping devices are a necessity for lane line and center line applications if any volume of material is to be applied.

Because the equipment is specialized in nature and the amount of material applied in a year by any one governmental agency is relatively small, these machines are generally owned by a vendor or contractor. Contracts for their use should specify:

1. The material to be used
2. The amount of beads in the compound
3. The amount of beads to be applied as top dressing
4. The width of line and the total length of the project
5. A minimum application rate
6. Warranty, if desired

RAISED MARKERS

Almost all raised marker applications in the United States use unsophisticated equipment. The individual units have merely been epoxied to the pavement following the layout by tape and marker. Units with a steel casing and two keels have been experimentally installed in snowplow areas. The keels have been cemented with epoxy resin into grooves cut into the pavement.

A prototype machine was developed, as a part of an NCHRP project, to clean the pavement, drop an epoxy resin onto the pavement, and drop 0.25 in.- (0.62 cm) glass beads into the epoxy.[30] This machine installed one formed-in-place marker and then moved down the roadway to the location of the next unit.

---

[30] JOHN DALE, "Development of Formed-in-Place Wet Reflective Markers," *National Cooperative Highway Research Program Report 85* (Washington, D.C.: Highway Research Board, 1970).

OTHER MARKERS

Objects within the roadway and hazards adjacent to the edgeline can be painted with diagonal stripes if additional emphasis is needed. When signs or markers are used, they can be mounted on the object and, if needed, can be post mounted on the approaches. The best equipment for driving posts, prior to mounting delineators or object markers, is an air-operated post driver. Various models are available from light-duty post drivers for ground rods, stakes, and small posts to large capacity, heavy-duty post drivers for larger posts. A truck mounted air compressor is needed for the post driver and a post extractor is quite useful for pulling out damaged posts.

MISCELLANEOUS TRAFFIC CONTROL DEVICES

Standards and/or specifications have been developed in the United States for additional traffic control devices that are not included in previous sections of this chapter. These additional devices are used to guide traffic in and around work areas, alert traffic to hazards ahead, and provide a means of identifying specific locations on the street or highway.

BARRICADES

Temporary devices are used to warn and alert drivers of hazards created by construction or maintenance activites in or near the traveled way and to guide and direct drivers safely past the hazards. Barricades shall be one of three types: Type I, II, or III. The characteristics of these types are shown in Figure 16.15.

Markings for barricade rails shall be alternate orange and white (sloping downward at an angle of 45° in the direction that traffic is to pass). For nighttime use, they shall be reflectorized and/or equipped with lighting devices for maximum visibility.

TRAFFIC CONES

Traffic cones are portable, temporary devices used to guide drivers through or past an obstacle. Traffic cones and tubular markers of various configurations are available. They should be a minimum of 18 in. (46 cm) in height with a broadened base and may be made of various materials to withstand impact without damage to themselves or to vehicles. Larger sized cones should be used whenever speeds are relatively high or whenever more conspicuous guidance is needed. The predominant color of cones is orange. They should be kept clean and bright for maximum target value. For nighttime use, they shall be reflectorized or equipped with lighting devices for maximum visibility.

BARRICADE WARNING LIGHTS

Portable, lens-directed, enclosed lights are needed for mounting on barricades to further warn traffic of hazards ahead. The color of the light emitted shall be yellow. They may be used in either a steady burn or flashing mode. Barricade warning lights should be in accordance with the requirements of the "ITE Standard for Flashing and Steady-burn Barricade Warning Lights," as summarized in Table 16.2.

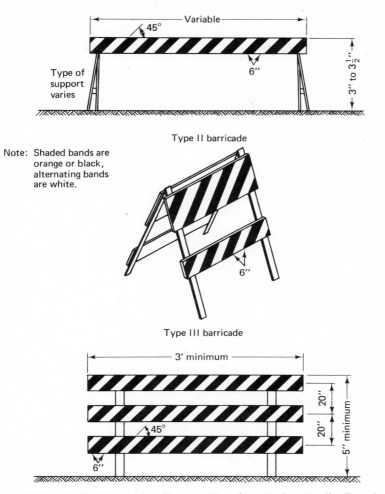

Figure 16.15. Standard barricades. (Source: *Manual on Uniform Traffic Control Devices,* p. 299.)

TABLE 16.2
Barricade Warning Lights

	Type A Low-Intensity	Type B High-Intensity	Type C Steady-Burn
Lens-directional faces	1 or 2	1	1 or 2
Flash rate per minute	55 to 75	55 to 75	Constant
On-time	10%	8%	Constant
Minimum effective intensity	4.0 Candelas	35 Candelas	
Minimum beam candle power			2 Candles
Hours of operation	Dusk to dawn	24 hr/day	Dusk to dawn

Source: "ITE Standard for Flashing and Steady-burn Barricade Warning Lights," *Traffic Engineering,* XLI, No. 11 (1971), pp. 42–43.

Type A low-intensity flashing warning lights are most commonly mounted on separate portable supports, on Type I or Type II barricades, or on vertical channelizing devices and are intended to continually warn the driver that he is proceeding in a hazardous area.

Type B high-intensity flashing warning lights are normally mounted on the advance warning signs or on independent supports. Extremely hazardous site conditions within the construction area may require that the lights be mounted on Type I barricades, signs, or other supports. Since these lights are effective in daylight as well as dark, they are designed to operate 24 hours per day.

Type C steady-burn lights are to be used to delineate the edge of the traveled way, on lane changes, on lane closures, and on other similar conditions.

RUMBLE STRIPS

A rumble strip is a device used to alert the driver that there is a change of conditions ahead. It can consist of sawed grooves in the pavement, a series of transverse sprayed thermoplastic strips, or some other means of creating the "rumble" effect. Successful applications have been made on sharp curves where an accident problem exists, on approaches to STOP signs, and as an outline in taper areas where there is a reduction in pavement width. Some work has been completed on the development of effective patterns for these rumble strips.[31]

MILEPOSTS

Properly installed milepost markers can assist the driver in estimating his progress, provide a means for identifying the location of emergency incidents, and aid in highway maintenance and servicing. Mileposts may be erected along any section of a numbered highway route, but zero mileage in the United States should begin at the south and west state lines or junctions where routes begin. Milepost signs shall be vertical panels with 6 in.- (15.2 cm) white numerals, a border, and the word MILE in 4 in.- (10.2 cm) letters on a green background and shall be reflectorized. Milepost signs shall be mounted at a minimum height and lateral placement equal to that used for delineators. To further enhance the usefulness of mileposting, delineators may be spaced at a distance of $\frac{1}{10}$ or $\frac{1}{20}$ of a mile or kilometer and can be marked in fractions of a mile or kilometer by a stencil on the back of the delineator or post or by a small plate on the delineator post.

REFERENCES FOR FURTHER READING

Manual on Uniform Traffic Control Devices for Streets and Highways, U.S. Department of Transportation, Federal Highway Administration. Washington, D.C.: U.S. Government Printing Office, 1971.

Manual of Uniform Traffic Control Devices for Canada, Road and Transportation Association of Canada. Ottawa, Ontario, 1966.

Australian Standard Rules for the Design, Location, Erection and Use of Road Traffic Signs and Signals, The Standards Association of Australia. Sydney, Australia, 1960.

[31] W. R. BELLIS, "Development of an Effective Rumble Strip Pattern," *Traffic Engineering*, XXXIX, No. 7 (1969), 22–25.

KIRCHNER, SIEGFRIED, "Traffic Signs and Markings in the German Democratic Republic," *Traffic Engineering and Control*, XII, No. 6 (1970), 316–18.

CAL Y MAYOR, RAFAEL, "Mexico's New Traffic Signs," *Traffic Engineering*, XLI, No. 2 (1970), 60–67.

KING, G. F. and H. LUNENFELD, "Development of Information Requirements and Transmission Techniques for Highway Users," *National Cooperative Highway Research Program Report 123*. Washington, D.C.: Highway Research Board, 1971.

ALLINGTON, R. W., "Criteria for Longitudinal Placement of Warning Signs," *Traffic Engineering*, XL, No. 11 (1970), 54–56.

ROWAN, N. J. and R. M. OLSON, "The Development of Safer Highway Sign Supports," *Traffic Engineering*, XXXVIII, No. 2 (1967), 46–50.

ROBINSON, C. C., "Color in Traffic Control," *Traffic Engineering*, XXXVII, No. 6 (1967), 25–29.

PARISI, R. P., "Cheap Paint Is Costly," *Traffic Engineering*, XXXVIII, No. 3 (1967), 10–13.

Chapter 17

TRAFFIC SIGNALS

SAMUEL CASS, Commissioner of Roads and Traffic, The Municipality of Metropolitan Toronto, Toronto, Canada.

The technical literature on traffic engineering devices provides many definitions of a traffic signal. All sources agree that traffic control signals are a "power-operated traffic control device which alternately directs traffic to stop and to proceed." Many sources also refer to traffic signals as "any power-operated traffic control device by which traffic is warned or directed to take some specific action." This would include flashing beacons, speed control signals, lane use control signals, railroad crossing signals, and several other similar devices.

This chapter will emphasize traffic control signals or those devices which alternately direct traffic to stop and to proceed. Other signals, not specifically used to control the intersection of two flows of traffic, will also be discussed.

AUTHORITY

Traffic signals must always be easily and readily recognized by the road user and their meaning must be clearly understood. To achieve these objectives, traffic signals should be uniform; the authority for their installation should be unimpeachable; and compliance with the requirement for obedience to the signal indications should be legally enforceable.

The installation should be as prescribed in the *Manual on Uniform Traffic Control Devices for Streets and Highways*,[1] or similar national standards, and the actions required of motorists and pedestrians should be specified by statute or by local ordinance or resolution, consistent with national standards. Suitable legislation establishing the authority for the installation, the meaning of the signal indications, and the required obedience to these indications by the road user are outlined in the Uniform Vehicle Code[2] and in the Model Traffic Ordinance.[3]

GROWTH OF TRAFFIC SIGNALS

Traffic control signals for street traffic were first used a little more than a century ago. An account of the history of traffic signals has been prepared by Edward A. Mueller.[4]

---

[1] *Manual on Uniform Traffic Control Devices for Streets and Highways*, U.S. Department of Transportation, Federal Highway Administration (Washington, D.C.: U.S. Government Printing Office, 1971).

[2] *Uniform Vehicle Code* (Washington, D.C.: National Committee on Uniform Traffic Laws and Ordinances, 955 North L'Enfant Plaza, S.W., Revised 1968 and Supplement I, 1972), Sections 11–201 through 11–206, 15–102, and 15–106.

[3] *Model Traffic Ordinance* (Washington, D.C.: National Committee on Uniform Traffic Laws and Ordinances, 955 North L'Enfant Plaza, S.W., Revised 1968 and Supplement I, 1972), Section 4–1 through 4–9.

[4] EDWARD A. MUELLER, "Aspects of the History of Traffic Signals," *IEEE, Transactions on Vehicular Technology*, VT–19, No. 1 (1970).

Almost all traffic signals are installed under the authority of government agencies. Some are under the jurisdiction of nongovernmental agencies for such locations as university campuses, medical centers, and industrial or commercial complexes.

TRAFFIC SIGNALS IN USE

In the United States in 1965, there were about 159,000 signalized intersections in rural areas: about one for every 900 persons in urban areas and one for every 5,200 persons in rural areas, or overall about one for every 1,160 persons.[5] The relationship between urban population and signalized intersections in the United States and Canada is presented in Figure 17.1. In Canada, there is approximately one signal for every 2,300 persons. By comparison, Hamburg, Germany, was reported to have 900 traffic signal installations in 1970 when the population was 1.8 million. This is approximately one signal installation for every 2,000 persons.

PROJECTION OF FUTURE USE

Annual installation of new signals based on a sampling of some cities and state highway departments in the United States totals about 5,000 per year, and the number of signalized intersections is increasing at a rate of about 3 percent annually. Compared to this, in the Canadian sample and in the Hamburg example, the rate of increase seems to be about 6 percent annually. The steps leading to the installation of a traffic signal, and its subsequent operation and maintenance, are illustrated in Figure 17.2. These steps provide a convenient outline for the remainder of this chapter.

WARRANTS: CRITERIA FOR SIGNAL CONTROL

Because the principal function of traffic control signals is to permit crossing streams of traffic to share the same intersection by means of time separation, the major criterion for signal control is the volume of traffic entering the intersection. Generally, traffic control signals are not needed if traffic is sufficiently light that adequate gaps appear at frequent enough intervals to permit all intersecting traffic to enter and cross the intersection with little delay. The need for a traffic control signal at any particular location must, however, be carefully evaluated in relation to several warrants (discussed in detail in the *Manual on Uniform Traffic Control Devices*).

Warrant 1   Minimum vehicular volume
Warrant 2   Interruption of continuous traffic
Warrant 3   Minimum pedestrian volume
Warrant 4   School crossings
Warrant 5   Progressive movement
Warrant 6   Accident experience
Warrant 7   Systems
Warrant 8   Combination of warrants

Warrants should be considered as a guide in the determination of the need for traffic control signals instead of absolute criteria. For example, such factors as physical

---

[5] WILBUR S. SMITH, "Marketing Study, Traffic Signal Equipment," prepared for the 3M Company, 1966.

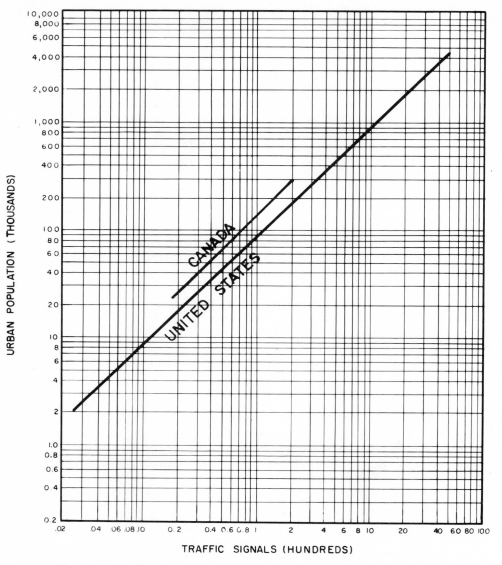

Figure 17.1. Relationship of urban area population and signalized intersections in the United States and Canada. [Source: From data provided by the National Safety Council (U.S.) and from data collected by the Department of Roads and Traffic, Municipality of Metropolitan Toronto (Canada).]

roadway features, age of pedestrians, effect of adjacent signalized intersections, etc., may modify a decision based on the warrants. Warrants must be used in conjunction with professional judgment based on experience and consideration of all related factors.

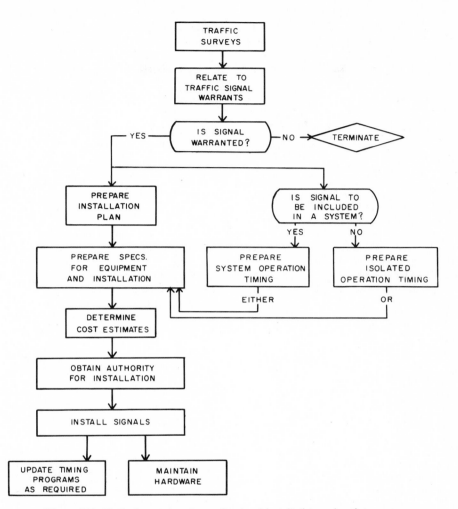

Figure 17.2. Typical procedure for traffic signal installation and maintenance.

WARRANT 1. MINIMUM VEHICULAR VOLUME

At an intersection at grade the intersecting flows must clear a common roadway area within the intersection. Time is the only separation. Prior to signalization the traffic entering the intersection from the major street can be assumed to enter at normal speed while traffic entering from the minor street must be assumed to enter after a full stop. The average delay per vehicle to the minor street traffic is, therefore, greater than that of the major street traffic. In order to effect an improvement following signalization, the delay per vehicle averaged for all approaches should be reduced.

The warrant is satisfied when, for each of any eight hours of an average day, the traffic volumes given in Table 17.1 exist on the major street and on the higher-volume minor street approach to the intersection.

TABLE 17.1
Minimum Vehicular Volume Warrant

Number of Lanes for Moving Traffic on Each Approach		Vehicles per Hour on Major Street (Total of Both Approaches)	Vehicles per Hour on Higher-volume Minor Street Approach (One Direction Only)
Major Street	Minor Street		
1	1	500	150
2 or more	1	600	150
2 or more	2 or more	600	200
1	2 or more	500	200

Source: *Manual on Uniform Traffic Control Devices*, U.S. Department of Transportation, Federal Highway Administration, U.S. Government Printing Office, Washington, D.C., 1971, p. 236.

These major street and minor street volumes are for the same eight hours. During each of those eight hours, the higher approach volume on the minor street is considered, regardless of its direction.

When the 85 percentile speed of major street traffic exceeds 40 mph (64 kph) or when the intersection lies within the built-up area of an isolated community having a population less than 10,000, the minimum vehicular volume warrant is 70 percent of the requirements above, in recognition of differences in the nature and operational characteristics of traffic in urban and rural environments and smaller municipalities.

WARRANT 2. INTERRUPTION OF CONTINUOUS TRAFFIC

Warrant 2 applies whenever unusually long delays are experienced by a relatively few motorists because of the absence of acceptable gaps at reasonable frequency. Although a signal installed under these conditions will usually result in greater delay per vehicle averaged for all approaches to the intersection, the installation can be justified by the reduction in potential hazard which existed prior to signalization.

The warrant is satisfied when, for each of any eight hours of an average day, the traffic volumes given in Table 17.2 exist on the major street and on the higher-volume minor street approach to the intersection and where the signal installation will not seriously disrupt progressive traffic flow. These major street and minor street volumes are for

TABLE 17.2
Minimum Vehicular Volumes for Warrant 2

Number of Lanes for Moving Traffic on Each Approach		Vehicles per Hour on Major Street (Total of Both Approaches)	Vehicles per Hour on Higher-volume Minor Street Approach (One Direction Only)
Major Street	Minor Street		
1	1	750	75
2 or more	1	900	75
2 or more	2 or more	900	100
1	2 or more	750	100

Source: *Manual on Uniform Traffic Control Devices*, p. 237.

the same eight hours. During each of those eight hours, the higher volume on the minor street is considered, regardless of its direction.

A reduced volume, similar to that described under Warrant 1, can be used in place of those shown in Table 17.2 on higher-speed roads or in smaller communities.

WARRANT 3. MINIMUM PEDESTRIAN VOLUME

The minimum pedestrian volume warrant reflects the adequacy of crossing gaps and the frequency of such crossing gaps in relation to the number of pedestrians who may desire to cross. In conjunction with the volume warrants, careful consideration should be given to modifying influences such as speed of approaching traffic, the age group of the pedestrians, the physical characteristics of the intersection, and the distribution of the times of the day during which pedestrians wish to cross at the location in question.

The minimum pedestrian volume warrant is satisfied when, for each of any eight hours of an average day, the following traffic volumes exist:

1. On the major street, 600 or more vph enter the intersection (total of both approaches) or 1,000 or more vph enter the intersection (total of both approaches) on the major street where there is a raised median island 4 ft or more in width, and
2. During the same eight hours as in paragraph (1) there are 150 or more pedestrians per hour on the highest-volume crosswalk crossing the major street (see Table 17.3).

TABLE 17.3

Vehicles per Hour on Major Street (Total of Both Approaches)		Number of Pedestrians on Highest-Volume Crosswalk
With Median	Without Median	
1,000	600	150
		150

These warrants may be reduced, as described under Warrant 1, when applied to higher speed roads or smaller communities.

WARRANT 4. SCHOOL CROSSINGS

The fourth warrant recognizes the unique problems related to children crossing a major artery on the way to and from school, particularly near the school, and may be considered as a special case of the pedestrian warrant.

A traffic control signal may be warranted at an established school crossing when a traffic engineering study of the frequency and adequacy of gaps in the vehicular traffic stream (see Chapter 3), as related to the number and size of groups of school children at the school crossing, shows that the number of adequate gaps in the traffic stream during the period when the children are using the crossing is less than the number of minutes in the same period.[6]

---

[6] *Manual on Uniform Traffic Control Devices for Streets and Highways,* Section 7A–3.

WARRANT 5. PROGRESSIVE MOVEMENT

The progressive movement warrant relates to the desirability of holding traffic in compact platoons that will arrive at each successive signalized intersection at the beginning of the green interval. This reduces the number of stops and delays and also provides longer gaps along the route so that crossings may be made at intermediate points without the assistance of traffic control devices. Traffic signals are installed at carefully designed spacings to permit the relationship of normal vehicle speed and signal cycle time to fit into a planned progressive movement of traffic along the route.

The progressive movement warrant is satisfied when:

1. On a one-way street or a street that has predominantly unidirectional traffic, the adjacent signals are so far apart that they do not provide the necessary degree of vehicle platooning and speed control.
2. On a two-way street, adjacent signals do not provide the necessary degree of platooning and speed control and the proposed and adjacent signals could constitute a progressive signal system.

The installation of a signal according to this warrant should be based on the 85 percentile speed unless an engineering study indicates that another speed is more desirable. According to this warrant, the installation of a signal should not be considered whenever the resultant signal spacing would be less than 1,000 ft (300 m).

WARRANT 6. ACCIDENT EXPERIENCE

This warrant must be used with caution, not because of lack of concern for traffic accidents, but because experience has indicated that the traffic signal does not always succeed as a safety device. Under certain conditions, a carefully designed traffic signal will materially reduce a right-angle accident collision pattern. This does not always happen, however, and often a rear-end collision accident pattern develops which far exceeds the original accident frequency.

The accident experience warrant is satisfied when:

1. An adequate trial of less restrictive remedies with satisfactory observance and enforcement has failed to reduce the accident frequency and
2. Five or more reported accidents of types susceptible to correction by traffic signal control have occurred within a twelve-month period, each accident involving personal injury or property damage to an apparent extent of $100.00 or more.
3. There exists a volume of vehicular traffic not less than 80 percent of the requirements specified in the minimum vehicular volume warrant; the interruption of continuous traffic warrant; or the minimum pedestrian volume warrant and
4. The signal installation will not seriously disrupt progressive traffic flow.

WARRANT 7. SYSTEMS

A traffic signal installation at some intersections may be warranted to encourage the concentration and organization of traffic. The systems warrant is applicable when the common intersection of two or more major routes has a total existing or immediately projected entering volume of at least 800 vehicles during the peak hour of a typical weekday or each of any five hours of a Saturday and/or Sunday.

A major route as used in the above warrant has one or more of the following characteristics:

1. It is part of the street or highway system that serves as the principal network for through traffic flow.
2. It connects areas of principal traffic generation.
3. It includes rural or suburban highways outside of, entering, or traversing a city.
4. It has surface street freeway or expressway ramp terminals.
5. It appears as a major route on an official plan such as a major street plan in an urban area traffic transportation study.

WARRANT 8. COMBINATION OF WARRANTS

In exceptional cases, signals may occasionally be justified when no single warrant is satisfied but when any combination of warrants 1, 2, and 3 is satisfied to the extent of 80 percent or more of the stated values. Adequate trial of other remedial measures which cause less delay and inconvenience to traffic should precede installation of signals under this warrant.

WARRANTS IN OTHER JURISDICTIONS

The warrants as described in the *Manual on Uniform Traffic Control Devices* for Canada are identical to those described in the United States. Traffic authorities in Australia also rely on the warrants in the United States Manual for guidance.

In Great Britain,[7] the Ministry of Transport does not lay down set conditions which must be attained before signals can be considered. Each case is judged on its merits. Broadly speaking, the three primary aims of signal control are:

1. To reduce traffic conflicts and delay. A total of approximately 300 vph, with at least 100 vph on the side road, averaged over 16 hours of the day would be required as the minimum justification for signal control. This would be equivalent to a peak-hour value (taken as 10 percent of the 16-hour total) of 480 vph. Many intersections with flows substantially in excess of this volume work quite satisfactorily with no control at all, and some examples of hypothetical intersections showed that about twice this volume was required to make the average delay the same with and without traffic signals.
2. To reduce accidents.
3. To minimize use of police time.

SIGNAL EQUIPMENT

A traffic signal installation is the result of grouping various pieces of traffic control equipment into a working system. In any installation, certain fundamental items such as controllers, signal heads, detectors, interconnecting cables, and associated hardware are required. The equipment required in any installation will be determined by the

---

[7] *Research on Road Traffic*, Department of Scientific and Industrial Research, Road Research Laboratory (London, England: Her Majesty's Stationery Office, 1965), p. 292.

degree of sophistication desired and the particular application of the control device. Many variations are found in the installation of vehicle signals, drawbridge signals, railway signals, pedestrian signals, beacons, and lane control signals. Specifications for the basic items have been written and approved by the Institute of Traffic Engineers and other agencies to ensure uniformity and standardization. Basic references include:

1. *Revised Standard for Adjustable Face Vehicle Traffic Control Signal Heads*, 1970.
2. *Adjustable Face Pedestrian Signal Head Standard*, 1963.
3. *A Standard for Traffic Signal Lamps*, Revised, 1967.
4. *Pre-Timed, Fixed Cycle Traffic Signal Controllers*, Revised, 1958.
5. *Traffic Actuated Traffic Controllers and Detectors*, Revised, 1958.
6. Colors of Signal Lights, U.S. Department of Commerce, National Bureau of Standards (NBS Monograph 75).

TRAFFIC SIGNAL CONTROLLERS

The function of a controller, whether it be electromechanical or electronic, is to switch the signal indications on and off, according to a fixed or alterable plan to correctly and safely assign the right of way at a given location. Two basic kinds of controllers are used: pretimed and traffic actuated. Both come in a variety of models to effect control of practically any given traffic situation. The kind to use will be determined by an engineerng study of local conditions. Figure 17.3 shows the schematic design of a basic pretimed controller.

Pretimed controllers. A pretimed controller operates on the basis of predetermined cycle lengths and phase intervals. It is frequently used when there are predictable and stable traffic volumes. It provides a simple, economical means of traffic control, and because of its simplicity, it is very reliable and relatively easy to maintain. The expansion of this kind of controller to provide additional timing programs is often limited because of the timing and logic circuitry required.

Traffic actuated controllers. A traffic-actuated controller operates on the basis of traffic demands as registered by the actuation of vehicle and/or pedestrian detectors. There are several kinds of traffic-actuated controllers, but their basic advantage is their ability to continuously adjust cycle lengths and phase intervals. Successful operation of these controllers is dependent primarily on the reliability of the detectors employed. An improperly adjusted or inoperative detector will negate almost all the advantages of using a traffic-actuated controller.

Semiactuated controllers. Semiactuated controllers provide a continuous green indication on the major street except when crossing time is required for the minor street. Vehicle detectors are placed on the minor street at the intersection approach. This kind of controller is most useful when minor street traffic is light or intermittent because it minimizes delays on the major street. Although many semiactuated controllers respond to both vehicle and pedestrian actuations, they may also be used advantageously to permit pedestrian crossings at mid-block locations. The use of controllers solely for the purpose of providing pedestrian crossing generally requires that the user know where the push buttons are and how to operate them. This can be accomplished by the use of signs. The user should also receive assurance that his call

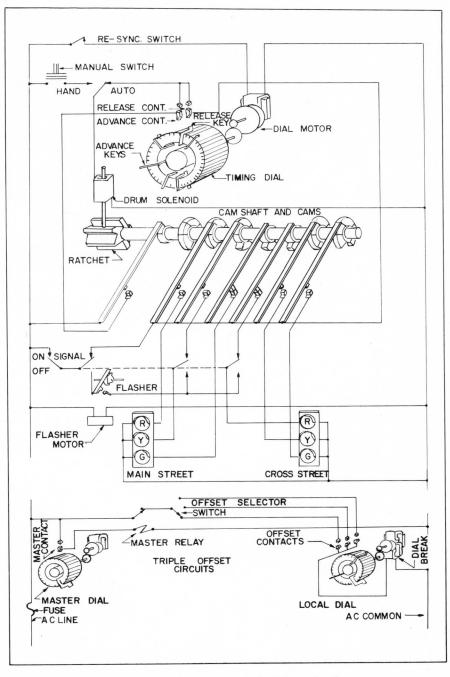

Figure 17.3. Traffic signal basic fixed time controller.

has been registered. This can be accomplished by an illuminated push button indicator.

Although many electronic or solid-state controllers are now used, the proper equipment for a particular application should be carefully assessed. Some controllers are completely solid-state devices and others use solid-state components in the timing circuitry only. The greatest advantage of the solid-state unit lies in its modular construction which permits selection of a basic controller and the expansion of its facilities merely by adding additional modules. Although the initial cost of the modular solid-state controller is more than the electromechanical kinds, the expansible feature for future sophistication of a system may well justify the added cost. The integration of electronic and electromechanical controllers in a coordinated system may present a problem because their circuitry is unlikely to be compatible. A modern solid-state electronic controller is shown in Figure 17.4.

Fully actuated controllers. Fully actuated controllers provide variable cycle lengths and phase intervals based on the relationship of the approach volumes on intersecting streets. This controller is most useful when approach volumes vary and when frequent timing changes are necessary. Vehicle detectors are placed at all approaches to the intersection. The completely solid-state electronic controllers developed recently for fully actuated operation provide the greatest flexibility and variation possible. The timing circuits permit adjusting each interval for maximum efficiency. Facilities to call up or omit extra phases, introduce protected movements, vary clearance intervals, and set maximum and minimum limits for all phases are some of the many features that the fully actuated controller can provide.

A specialized fully actuated controller is available in the form of a volume-density controller which permits adjustment of green time in accordance with traffic density as well as volume.

Master controller. A master or supervisory controller is often used to control the operation of one or more "local" controllers, which control individual intersections. Standard pretimed controllers may be used for simple coordination when traffic patterns are basically predictable and repetitive, day by day, and week by week. When traffic patterns vary widely and when cycle lengths, splits, and offsets must be changed

Figure 17.4. Solid-state electronic controller. (Source: Canadian General Electric Company.)

frequently, some form of sampling and computing equipment may be necessary. Although the cost of this equipment may seem high, it may well be justified because of increased efficiency in traffic operations.

Computers that function as a master or supervisory controller are available in many types. In its simplest form, a computer can select an almost infinite number of cycle lengths, phase intervals, and offsets. In its most complex form, it can digest and compare traffic data input, make an instantaneous decision, and directly control signals based on the decisions reached. For maximum efficiency, a computer must be supplied with continuously updated data from the street or streets.

Ancillary equipment. Controllers are normally designed and built for operation in a specific number of phases, each having standard indications. When unusual conditions make the use of additional phases or special operations necessary, ancillary equipment must be used. The following components are commonly used for the applications noted:

1. *Flasher:* to provide flashing operation.
2. *Relays:* to handle additional current loads when the normal controller load switches are inadequate.
3. *Pre-emptor:* to provide a special indication on actuation from a fire hall, train approach signal, etc.
4. *Time switch:* to select locally a particular operation. Time clocks should have a mechanical carry-over feature in case of power failure.
5. *Manual control panel:* to permit manual operation; a separate panel accessible through a special door should permit selection of OFF, AUTO, FLASH, or MANUAL operation in emergencies.

Although almost all electromechanical controllers may be modified by the addition of the accessories indicated above, these features should be incorporated in electronic controllers at the time of manufacture.

Controller location.[8] The signal controller may be attached to any convenient pole or, if a console cabinet is used, it may be placed wherever desired, provided that in either case the location chosen satisfies that:

1. A power supply can be conveniently obtained.
2. There will be an unobstructed view of all approaches to the intersection in the event of manual operation. When this condition cannot be satisfied and manual operation is frequently required, it may be desirable to install a special remote unit at a more favorable position with its switches in parallel with those of the controller proper.
3. The cabinet does not unduly obstruct the pedestrian right of way.
4. The cabinet will not be unduly exposed to accidental damage caused by passing traffic.

---

[8] J. T. HEWTON, "Traffic Signals Manual," *Ontario Traffic Conference*, 1965.

SIGNAL HEADS AND OPTICAL UNITS

Definitions[9]

1. *Signal head:* an assembly containing one or more signal faces that may be designated accordingly as one-way, two-way, etc.
2. *Signal indication:* the illumination of a traffic signal lens or equivalent device or a combination of several lenses or equivalent devices at the same time.
3. *Signal face:* that part of a signal head provided for controlling traffic in a single direction. Turning indications may be included in a signal face.

Number of signal faces and location.

To be effective, a traffic signal must be easily seen and identified. Uniformity in the location of signal faces is desirable for safety and efficiency. It has been demonstrated that almost all observers are able to detect a target in an expected location much better than in an unexpected location.

Critical elements in the location of signal faces are lateral and vertical angles of sight toward a signal face as determined by typical driver eye position. The distance of unobstructed view should vary with the 85 percentile approach speed as shown in Table 17.4.

The drawings shown in Figure 17.5 indicate the cone of perception of the average motorist without turning his head. Good design would ensure that at least one, and preferably both, signal faces would be located within this perception cone. The *Manual of Uniform Traffic Control Devices* shows the desirable location of signal faces within an angle of approximately 20° either side of the driver's straight-ahead line of sight when the vehicle is at the stop line.

A minimum of two signal faces is required on the far side of the intersection for each approach. If not mounted over the traveled portion of the road, the signal face should be neither less than 8 ft (2.5 m) nor more than 15 ft (4.5 m) above the sidewalk or the grade of the road. If mounted over the traveled portion of the road, the signal face should be neither less than 15 ft (4.5 m) nor more than 19 ft (6 m) above the grade of the road.

In many instances it is desirable to mount the far right-side signal face on a mast arm so that its visibility is unobstructed. Supplemental signal faces should be employed wherever necessary in order to obtain complete unobstructed intersection visibility. If nearside signals are used, they should be located as near as practicable to the stop line. A single signal face is permissible for the control of an exclusive turn lane. Insofar as possible, such a signal face should be located directly in front of the controlled driver. If used, the single signal face shall be in addition to the minimum of two signal

TABLE 17.4
Minimum Signal Visibility Distance for Varying Approach Speed

85 Percentile Speed	mph (kph)	20 (37)	25 (40)	30 (48)	35 (56)	40 (64)	45 (72)	50 (81)	55 (89)	60 (97)
Minimum Visibility Distance	Feet (Meters)	100 (30)	175 (53)	250 (76)	325 (99)	400 (122)	475 (145)	550 (168)	625 (191)	700 (213)

Source: *Manual on Uniform Traffic Control Devices*, p. 225.

[9] "Adjustable Face Vehicle Traffic Control Signal Heads," *ITE Technical Report No. 1*, 1970.

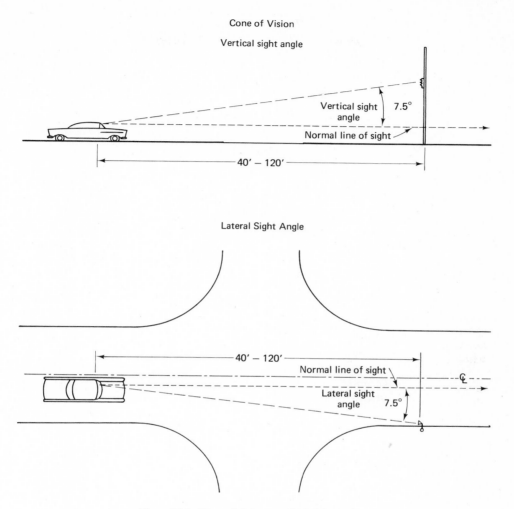

Figure 17.5. Cone of vision—vertical sight angle.

faces for through traffic. Unless physically impractical because of the size of an intersection, signal faces should be located neither less than 40 ft (12 m) nor more than 120 ft (36.5 m) from the stop line.

All signal faces should be equipped with visors in order to reduce reflected glare from the sun and other illumination and to direct the signal indication to approaching traffic.

A typical signal layout is shown in Figure 17.6. More detailed information on the location of signal faces is contained in the *Manual on Uniform Traffic Control Devices* (4B-12).

Number of lenses per signal face. The number of lenses per signal face will vary with the number of movements controlled at a given intersection, but except in the

Typical Signal Layout

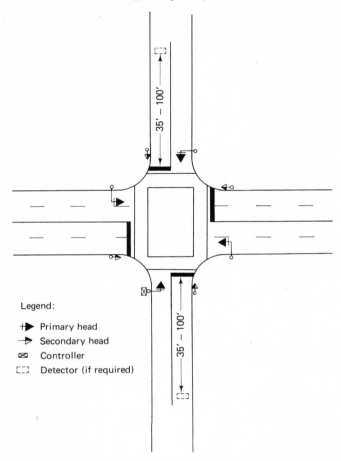

Figure 17.6. Typical sign layout.

case of pedestrian signal faces, they shall consist of at least three lenses. The lenses shall be red, yellow, or green.

When separation of vehicular movements is required at more complex intersections, arrow lenses may be added. Physical arrangement of signal lenses is shown in Figure 17.7, and it should be noted that both vertical and horizontal mountings are acceptable.

The color of signal lenses is closely controlled both in glass and plastic, and specifications have been prepared[10] for ensuring that the basic colors of red, yellow, and green are true and provide the correct uniform indications. Plastic lenses are preferred to glass because of their shatterproof characteristics and polycarbonate plastics are preferred to acrylic plastics because of their higher temperature characteristics.

---

[10] *Revised Standard for Adjustable Face Vehicle Traffic Signal Heads* (Washington, D.C.: Institute of Traffic Engineers, 1970).

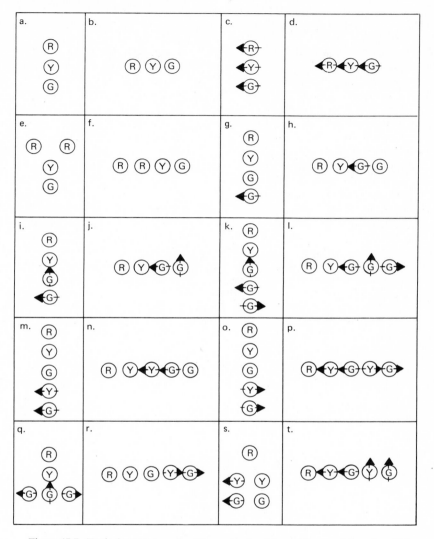

Figure 17.7. Typical arrangements of lenses in signal faces. [Source: *Manual on Uniform Traffic Control Devices for Streets and Highways*, U.S. Department of Transportation, Federal Highway Administration (Washington, D.C.: U.S. Government Printing Office, 1971), p. 222.]

Arrow lenses. Arrow lenses are normally used instead of circular lenses where exclusive turning movements are provided, where certain movements are prohibited or physically impossible, or at locations where all movements from one approach do not begin or end at the same time. Red, yellow, and green arrow lenses may be used in the latter application. When left turns are restricted to a specific interval within the signal phase, a green arrow can be used to indicate when left turns are permitted.

When the left-turn interval terminates before the end of the green phase, there is no standard provision to advise the motorists of its termination. To overcome this difficulty, some use has been made of a signal face containing green, yellow, and red left-turn arrows. By using a special signal face for the left-turn movement only, the green arrow can be followed by the yellow warning arrow, which is then followed by the red arrow signifying the termination or the prohibition of any further left-turn movement during this phase (see Figure 17.8.) It is important to note, however, that the *Manual on Uniform Traffic Control Devices* requires that a red arrow indication shall be used only in a separate signal face which also contains yellow arrow and green arrow indications. This system of three color arrows can be used for a right-turn interval as well as for a left-turn interval. On a one-way street it may be used instead of the corresponding circular indications.

In some countries, a standard traffic signal head using the circular red, yellow, and green lenses is used to control exclusive unidirectional movements by signing the signal face for that purpose. The sign can advise that the signal either applies to left turns only or to right turns only (see Figure 17.9). Arrow lenses do not have the same visibility as the corresponding circular lenses. To compensate, consideration should be given to the use of 12 in. (30 cm) arrow lenses for improved visibility.

Optically directed lenses. The problem of directing the visibility of signal indications toward certain movements only may be accomplished by using various visors or optically directed lenses. Signals using these features provide a sharp optical cut-off of the indication, both vertically and horizontally, which cannot be duplicated by the use of visors. Because the need to restrict the view of a signal indication may not apply to all the indications in a signal face, careful consideration should be given to the number of optically directed lenses actually required. Because the satisfactory operation of optically directed signals depends on their correct alignment, rigid mounting of the signal face is required.

Pedestrian signals. Pedestrian signal indications are a special kind of traffic signal indication intended for the exclusive purpose of controlling pedestrian traffic. They

Figure 17.8. Application of red, yellow, and green arrows.

Figure 17.9. Signal heads for designated movements.

consist of two lenses with the legends WALK and DON'T WALK. Portland orange DON'T WALK and lunar white WALK legends are specified in the *Manual of Uniform Traffic Control Devices*. Symbol indications are used in many other countries. One such symbol for the DON'T WALK indication is a portland orange silhouette hand. Another symbol that is often used is the silhouette of a standing man. A silhouette of a walking man is used to represent the WALK message.

The legends may be displayed in either the steady or flashing state. The DON'T WALK flashing state, in most cases, corresponds to the yellow clearance interval but may not necessarily be of the same time duration.

A new concept employing the dynamic display of pedestrian indications with a single optically directed lens has been recently developed. This method makes it possible to display during the clearance interval simultaneous WALK and DONT WALK indications to separate portions of a pedestrian crosswalk. The indications are dynamically moved across the crosswalk at a predetermined rate of speed ensuring adequate pedestrian clearance time.

Pedestrian signal faces should be located at each end of a controlled crosswalk. Pedestrian signal faces should be physically separated from vehicular signal faces, and they should be mounted so that they are clearly visible to the pedestrian throughout his entire passage in the crosswalk. Detailed information on the meaning, application, design, and warrants for pedestrian signals is presented in the *Manual of Uniform Traffic Control Devices*.

Lamps. Traffic signal lamps[11] require that the filament be supported to withstand excessive vibration. These lamps form a part of a complete optical system, consisting of light source, reflector, and lens. The critical factors are light center length, wattage, and rated lumen output. Lamps should be selected for the size of indication required, but when 12-in. (30 cm) signal lenses having 150 watt lamps are operated as a flashing yellow indication, especially at night, an automatic dimming device should be used to reduce its brilliance. The life of signal lamps is greatly reduced by abnormal high-voltage conditions and excessive vibration.

Hardware and mounting. Signal faces may be mounted on posts or poles at the sides of the roadway or on islands in the roadway, or they may be suspended from span wires, mast arms, or signal bridges over the roadway. The method of mounting will vary according to local conditions, street widths, intersection layouts, approaching traffic volumes, speeds, and visibility considerations. Signal faces should be mounted so that they provide a clear and unmistakable indication of right-of-way assignment to approaching motorists and to motorists and pedestrians as they enter the intersection. Over-road clearance and adequate physical separation of both vehicle and pedestrian signal faces are essential.

Because signal equipment is exposed to the elements, special attention must be given to the selection of materials to ensure that the effects of weather and corrosion are kept to a minimum. Aluminum, stainless steel, and plated hardware are widely used, but combinations of these materials may result in electrolytic action in damp environments. The effects of wind loading on equipment vary with location, and mounting hardware should be capable of withstanding the prevailing local weather conditions.

––––––––––

[11] *A Standard for Traffic Signal Lamps* (Washington, D.C.: Institute of Traffic Engineers, 1967).

Push buttons used by pedestrians should be placed at a convenient height, and care should be taken to ensure that pedestrians will not injure themselves on traffic signals mounted too low or on poles and controllers located in crosswalks. Typical mounting arrangements are shown in Figure 17.10.

DETECTORS

Whenever a traffic-actuated controller is used or whenever it is desired to gather volume data, some form of detector is required to actuate one or more of the signal phases or a counter. The simplest form of detector is the push button provided for pedestrians; a pneumatic tube stretched across a roadway is the simplest vehicle detector. Detectors can be grouped into four main categories:

1. *Over-pavement detectors* are either mounted on poles at the side of the roadway or are suspended over the roadway on span wires. They are nondirectional and may be either presence-sensitive or movement-sensitive as desired. Some are capable of measuring speed by application of the Doppler effect (by means of appropriate radar equipment).
2. *On-pavement detectors* are the simplest, most inexpensive, and easiest to install, but their use should be confined to locations where ideal weather conditions exist because they are also the most easily damaged.
3. *In-pavement detectors* are embedded in the roadway and consist of a pressure-sensitive pad of a variety of forms. Apart from the damage caused by normal wear, they are rendered completely inoperative by repaving or other roadwork and occasionally by snow or ice conditions.
4. *Under-pavement detectors* react to the disturbance caused to an electrical field by the presence of a mass of ferrous metal. They are highly accurate, reliable, versatile, and the least susceptible to damage (other than by a complete reconstruction of a roadway).

Methods of operation. The method of operation of various detectors is described below:

1. *Passage detector:* Responds only when a vehicle passes a point on the roadway (also: dynamic, movement, or motion detector).
2. *Limited-presence detector:* Responds to a stationary vehicle within a selected area for only a limited period, and then it resets itself so that the entry of a further vehicle may be detected.
3. *Continuous-presence detector:* Gives a continuous response when a stationary or moving vehicle, or vehicles, is within a selected area of roadway. It also maintains that response indefinitely for as long as the selected area is so occupied. The detector may or may not detect the number of vehicles within the area of detection. This detector is also called an *occupancy detector*.
4. *Dynamic-presence detector:* A presence detector gives an output signal only when a vehicle is moving above a critical minimum speed within a selected area of roadway. No output signal is provided by a stationary vehicle.
5. *Speed detector:* This may consist of two passage detectors in close sequence with appropriate timing circuitry or Doppler speed detectors.

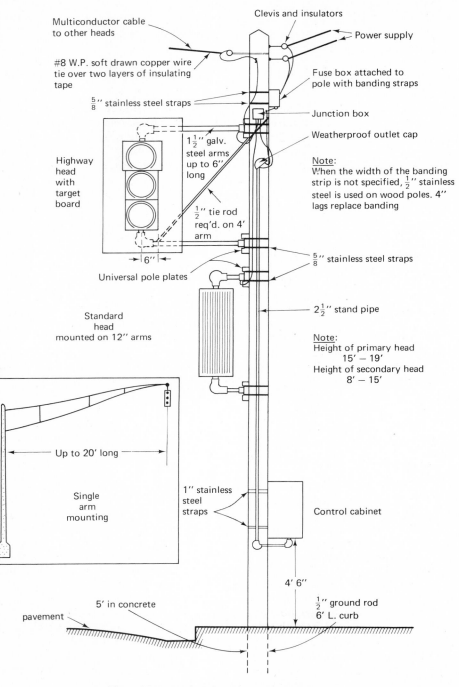

Multiconductor cable
to other heads

#8 W.P. soft drawn copper wire
tie over two layers of insulating
tape

$\frac{5}{8}''$ stainless steel straps

Highway
head
with
target
board

$1\frac{1}{2}''$ galv.
steel arms
up to 6''
long

$\frac{1}{2}''$ tie rod
req'd. on 4'
arm

|← 6'' →|

Standard
head
mounted on 12'' arms

Universal pole plates

Clevis and insulators

Power supply

Fuse box attached to
pole with banding straps

Junction box

Weatherproof outlet cap

Note:
When the width of the banding
strip is not specified, $\frac{1}{2}''$ stainless
steel is used on wood poles. 4''
lags replace banding

$\frac{5}{8}''$ stainless steel straps

$2\frac{1}{2}''$ stand pipe

Note:
Height of primary head
15' − 19'
Height of secondary head
8' − 15'

— Up to 20' long —

Single
arm
mounting

1'' stainless
steel
straps

Control cabinet

4' 6''

5' in concrete

pavement

$\frac{1}{2}''$ ground rod
6' L. curb

Figure 17.10. Typical signal mounting methods.

Principles of detection. The fundamental methods of vehicle detection are described below and nearly all have been employed, at least experimentally. Several methods depend on pressure exerted by the tires of passing vehicles.

1. *Contact:* Pressure closes an electrical contact.
2. *Pneumatic:* Pressure compresses an air-filled flexible tube, and the air-pressure wave operates a diaphragm to make or break an electrical switch.
3. *Hydraulic:* Pressure compresses a liquid-filled flexible tube and the movement of the fluid operates a pressure-sensitive switch.
4. *Capacitive (mechanical):* Pressure alters the separation of two flexibly mounted metallic surfaces, thus changing the electrostatic coupling between them. This change of capacitance is then detected by a suitable electrical circuit.
5. *Capacitive (nonmechanical):* The change of capacitance between two electrodes by the intrusion of the metallic mass of the vehicle (or between one electrode and the vehicle itself) is detected in a similar manner as above.
6. *Piezoelectric:* Pressure is used, by suitable mechanical linkage, to stress a piezoelectric element to give a voltage output.
7. *Triboelectric:* Pressure rubs two insulators together to create a static charge which is converted to a voltage and is measured.
8. *Seismic:* Vibrations set up in a metal bar by the passage of a vehicle wheel are detected by an accelerometer.
9. *Magnetic:* The distortions of a magnetic field by the ferrous metallic mass of a vehicle brings about a change of flux in a sensing element which may be readily detected. The magnetic field may be artificially produced or use may be made of the natural magnetism of the earth.
10. *Photoelectric (interrupted beam):* A beam of light is projected across the roadway onto a light-sensitive cell. The interruption of the light beam by a vehicle is detected by the cell.
11. *Photoelectric (reflected beam):* A beam of light is reflected from the road surface onto a light-sensitive cell; the sudden change in the reflectance produced by the intrusion of a vehicle is then detected by the cell.
12. *Infrared (active):* This detector operates the same way the photoelectric kind does, but it uses a semiconductor infrared generator (a gallium arsenide diode) as the transmitter.
13. *Infrared (passive):* The change from the level of the emission of infrared energy from the surface of the road to that which is radiated by a vehicle is detected by suitable infrared receiver.
14. *Acoustic:* The sound produced by a vehicle is detected; usually, it is the sound produced by a vehicle wheel passing over a resonant cavity.
15. *Ultrasonic (pulsed):* Transmitted pulses of ultrasonic energy are reflected from the road surface to a suitable receiver. The reduction in the transit time of the reflected pulses created by the presence of a vehicle is then detected.
16. *Ultrasonic (continuous wave):* A beam of continuous ultrasonic energy is directed toward possible approaching vehicles. The frequency shift of the energy reflected from the approaching vehicle caused by the Doppler effect is then detected.
17. *Radar (pulsed):* Operates the same way the pulsed ultrasonic detector does, but it makes use of centimetric X-band radio-frequency energy instead of ultrasonic energy.

18. *Radar (continuous wave):* Operates the same way the continuous ultrasonic detector does, but it makes use of centimetric X-band radio-frequency energy.
19. *Radar (guided):* Radio frequency is fed into a buried transmission line which is terminated by its correct characteristic impedance. The electric mismatch caused by a vehicle crossing the transmission line is then detected.
20. *Inductive loop:* The metallic mass of a vehicle alters the mutual or self-inductance of a conducting loop or loops normally placed under or in the road surface. Two common loop detectors are:
    a. *Inductive loop (resonant circuit):* Almost all loop detectors are inductive loops which consist of a loop and a variable condensor in an oscillating circuit. The frequency of the circuit is adjusted so that the circuit is in resonance with an oscillator of constant frequency. A vehicle on the detector alters the inductance of the loop, causing an alteration in frequency of the oscillating circuit. The difference in frequency between the constant oscillator and variable frequency oscillation circuit actuates another circuit which closes a relay. It is sometimes difficult to balance the oscillating circuit and the constant frequency oscillator so that they are in resonance.
    b. *Inductive loop (phase displacement):* Alteration of inductance of the loop caused by a vehicle is determined by measuring the phase shift between voltage and current at a complex resistor of which the loop is a component. An oscillator feeds the complex resistor and thereby the loop. Displacement of the high-frequency loop current against the applied voltage of the oscillator is determined by a phase measuring rectifier. A measure of the loop detuning is the voltage occurring at the exit of the rectifier. A differentiating circuit is used to avoid slow detuning of the loop, which may be caused by temperature, moisture, and aging.

An important advantage of detectors acting on this principle is that the loop is not a component of an oscillating circuit, and the system does not act upon the difference of frequencies. Therefore, difficult adjustment of detectors is not normally necessary.

Detector installation. The effectiveness and efficiency of signal equipment are greatly affected by proper detector installation. The frequency of repair, either mechanical or electrical, depends on the initial installation. Detectors installed in roadways are more susceptible to damage than those mounted overhead.

Although the physical mounting or location of a detector may be dictated by roadway characteristics and technical limitations, its placement in relation to a controlled intersection determines the effectiveness of signal operation. Distance of the detector from the stop line for simple traffic-actuated signal operation may be as little as one car length. For more sophisticated traffic signal control operation, the distance may vary from one car length up to 500 ft (150 m).

Such detectors as radar, sonic, infrared, and photoelectric are usually mounted on mast arms or span wires over the lane or lanes of the roadway on which detection is desired. The necessity for providing adequate height for clearance of high loads may result in loss of sensitivity or "over-shoot" to other lanes. Because many kinds of overhead detectors utilize the Doppler principle, directional characteristics are difficult to achieve unless narrow beam widths are used and this may result in a loss of sensitivity.

On-pavement detectors are the most simple and inexpensive to install because they have only to be cemented, nailed, or taped to the roadway pavement. Examples are the pneumatic tube and portable loop.

The most efficient and reliable are the under-pavement magnetic and loop detectors. Installation of a magnetic detector requires that a hole, slightly larger than the circular detector, be drilled in the pavement for each detector and the necessary interconnecting sawcuts made for the cables. With epoxy and/or bituminous sealants, this installation is likely to remain operational provided that no physical damage occurs.

To install loop detectors, a rectangular sawcut is made in the pavement from $1\frac{1}{2}$ in. to 2 in. (3.8 cm to 5 cm) deep. The size of this rectangular sawcut varies in length in accordance with the application of the detector but its width is determined by the lane width. Three to four turns of wire are then laid into this sawcut and the two ends of the loop are joined to a coaxial cable. The coaxial cable has a constant impedance and its use thereby eliminates one of the variables in tuning to frequency as well as reducing the possibility of interference pickup of stray electrical fields by the feeder. The sawcuts are then sealed with a bituminous sealant or soft epoxy because the detector must be free to move in the pavement without breaking. Hard epoxies will not permit this movement.

Once the loop and feeder have been installed, the detector must be tuned unless it is self-tuning. Loop detectors are quite stable and little test equipment is necessary. Some manufacturers provide a special test set with their detectors to assist in determining the values of components which must be added or changed.

CONTROLLER INTERCONNECTION

Interconnection may involve two adjacent controllers, a master controller and several slave controllers on a single street, or a master controller and all controllers in an area system. It provides positive time relationships between signals and thereby ensures an orderly progression of traffic. Interconnection may also be necessary for the purpose of preemption of traffic signals at fire halls, railways, and drawbridges.

The simplest form of interconnection is that provided by the power lines which feed the individual controllers. If simple controllers use a common cycle length, and if the supply frequency is stable, visual progressions may be set up that will retain the time relationships between intersections unless a power interruption occurs.

Relying on this method of interconnection for more than two or three adjacent intersections is not recommended because the fine tuning is difficult, time consuming, and does not lend itself to complex equipment and modes.

Kinds of interconnection. Interconnection may be satisfactorily achieved by any of the three following methods:

1. *Cable:* A multiconductor cable, usually having from Number 14 to Number 10 AWG wire size may be run between a master and a number of slave controllers. The distance between the power source and the controlled relays will determine the size of conductors because ac and dc voltage drops must be minimized in order to effect reliable operation. The cables may be installed either overhead or underground, but an underground installation provides the most reliable interconnection.
2. *Radio:* Radio control, in connection with a reliable telemetry system, may be

used when a central transmitter can be combined with a high antenna in order to provide adequate signals to the controller receivers. This system requires a separate radio receiver and suitable decoding and actuating equipment at each controlled intersection. Interference from random electrical disturbances and from other radio channels may disable the system, and maintenance is a high-cost factor. Unless special fail-safe circuits are built into each controller, it is possible that interference could result in abnormal signal indications being displayed. Nevertheless, the overall reliability of a properly designed, installed, and maintained radio system is comparable with other methods of interconnection. Although not specifically in the radio field, experiments using laser beams have been conducted to determine the feasibility of their use for traffic controller interconnection.

3. *Leased telephone lines:* Leased telephone lines and low-level telemetry equipment are used extensively for interconnection of large systems. They offer the advantage of reliability, flexibility, economy, and predictable cost. A single pair of voice level service telephone lines used in conjunction with multiplex equipment can provide sufficient channels for all required functions as dictated by a master controller as well as provide such monitor data as detector actuations for vehicle counting and the type and duration of signal indications at an intersection. This interconnection requires a minimum of maintenance and lends itself to computer controlled systems.

SIGNAL TIMING

CYCLE LENGTHS—PHASING, GENERAL

A number of terms in signal timing require definition:

1. *Cycle length* (*cycle*): The total time required for the complete sequence of phases.
2. *Signal phase* (*phase*): The sequence of conditions applied to one or more streams of traffic which, during the cycle, receive simultaneous right of way during one or more intervals.
3. *Interval:* An exclusive division within a cycle in which the traffic signal indication does not change.
4. *Offset:* The number of seconds or percent of the cycle length that the green indication appears at a given traffic control signal after a certain instant used as a time reference base.

The objective of signal timing is to alternate the right of way between traffic streams so that the average delay to all vehicles and pedestrians, the total delay to any single group of vehicles or pedestrians, and the possibility of accident-producing conflicts are minimized. These criteria frequently conflict and require compromise based on engineering judgment. For example, to minimize total delay, the number of distinct phases should be kept to a minimum consistent with safety and traffic demands. The usual procedure is to implement two-phase operation whenever possible, such as at normal right angled intersections, at T intersections, or at mid-block locations. However, at irregular multileg intersections or where interphase conflicts are present, multiphase (i.e., three-phase, four-phase, etc.) operation may become unavoidable in order to resolve vehicular and/or pedestrian conflicts. In the same way, when the need to accommodate exclusive or irregular movements is indicated, the green phase is sometimes split (this application of phasing is termed *split phase*). Illustrative examples are shown in Figure 17.11.

Studies in the United States[12] and elsewhere[13] have demonstrated that longer cycles will accommodate more vehicles per hour, and cycles longer than necessary to accommodate the traffic volume present will produce higher average delays. The best rule is to use the shortest practical cycle length that will serve the traffic demand.

Vehicles stopped at a traffic signal do not instantaneously enter the intersection at minimum headways following display of a green indication. This starting delay has been measured in a number of studies. Greenshields[14] concluded that the first five vehicles in a lane require 14.2 sec to enter, whereas succeeding vehicles require 2.1 sec each. Thus, a starting delay of 3.7 sec is inferred. Capelle,[15] measuring somewhat differently, arrived at a value of 5.8 sec for the first two vehicles, followed by a mean time headway of 2.1 sec; this yields a starting delay of 1.6 sec. A Traffic Research Corporation study[16] indicated that a motion wave in a queue moves back from the stop line at the rate of 100 ft (30.5 m) in 6 sec. Once the motion wave reaches the detector, located approximately 300 ft (91.5 m) back from the stop line, detections occur at the rate of one every 2 to $2\frac{1}{2}$ sec (in each lane) until either the end of the queue clears the detector or the movement is stopped again by a red light and the queue has closed up.

With the known influence of geometrics on intersection capacity, the variation in the above figures is not surprising. For purposes of general calculation, a starting delay of 2.5 sec per phase is a reasonable value, but for more accurate calculation, measurement at the specific intersection is required.

In addition, there is in each phase a yellow clearance interval (and possibly an all-red interval) that is not efficiently used for traffic movement. Thus, over any time period, a short cycle with a larger number of starting delays and yellow intervals will produce more lost time and accommodate fewer total vehicles.

For a long cycle to be efficient, there must be a constant demand during the entire green period on each approach. As the cycle and green periods lengthen, there is less probability that demand will be present during the latter portions of the green phase, while a higher average delay must occur for all vehicles waiting on the corresponding red phase. This produces a higher total delay. The selection of an appropriate cycle length is, then, an effort to select the shortest cycle that will accommodate the demand present and at the same time minimize the average delay.

ISOLATED SIGNALS

An isolated signal is one at which the timing is independent of any other signal control in the vicinity. It is generally at an intersection too remote from other signal locations to require coordination with them.

---

[12] J. H. KELL, "Results of Computer Simulation Studies as Related to Traffic Signal Operation," *Proceedings, Institute of Traffic Engineers*, 1963, pp. 70–107.

[13] F. V. WEBSTER, "Traffic Signal Settings," *Road Research Technical Paper 39* (London, England: Her Majesty's Stationery Office, 1958).

[14] B. C. GREENSHIELDS, D. SCHAPIRO, and E. L. ERICKSON, "Traffic Performance at Urban Street Intersections," (New Haven, Conn.: Yale University Bureau of Highway Traffic, 1947.)

[15] D. G. CAPELLE and C. PENNEL, "Capacity Studies of Signalized Diamond Interchanges," *Freeway Design and Operations, Bulletin 291* (Washington, D.C.: Highway Research Board, 1961), pp. 1–25.

[16] L. CASCIATO and S. CASS, "Pilot Study of the Automatic Control of Traffic Signals by a General Purpose Electronic Computer," *Electronics in Traffic Operations, Bulletin 338* (Washington, D.C.: Highway Research Board, 1962).

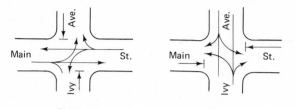

Phase A　　　　　　　　Phase B

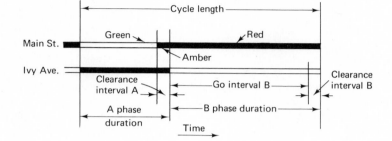

Usual phasing of traffic flow at ordinary
right-angle intersection. Two, two-way traffic streams

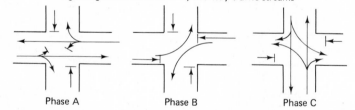

Phase A　　　　Phase B　　　　Phase C

Use of three phases at ordinary, "right-angle"
intersection. Two, two-way traffic streams with
separate phase provided for left turns

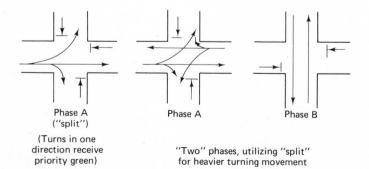

Phase A　　　　　Phase A　　　　　Phase B
("split")

(Turns in one　　　"Two" phases, utilizing "split"
direction receive　　for heavier turning movement
priority green)

Figure 17.11. Phasing of traffic flows. [Source: Theodore Matson, Wilbur S. Smith and Frederick W. Hurd, *Traffic Engineering* (New York: McGraw-Hill Book Company, 1955), p. 328.]

Studies required. Before attempting to set the timing of any traffic signal, there must be available adequate information on the pedestrian and vehicular traffic using the intersection. These studies include:

1. *Location plan:* A scale plan of the location in question, which includes all relevant physical details within 400 ft (122 m) of the intersection, should be prepared. This includes such features as intersectional geometrics, channelization, grades, sight distance restrictions, bus stops and routings, parking conditions, pavement markings, street lighting, driveways, location of nearby railway crossings, distance to the nearest signals, utility poles and fixtures, and adjacent buildings.
2. *Accident record:* The number, kind, time of occurrence, weather condition, road condition, and reason, if known, for all accidents that have occurred at the location in question for at least three years should be obtained and shown on a schematic diagram.
3. *Speed study:* The 85 percentile speed of vehicles on each approach to the intersection should be determined at points at least 150 ft (46 m) from the intersection. On roadways where normal vehicle speeds are in the 15 to 25 mph (24 to 40 kph) range, the speed check would be made from 150 to 200 ft (46 to 61 m) before the intersection; on high-speed highways (50 to 70 mph or 80 to 113 kph), the checkpoint should be between 800 and 1,200 ft (244 and 366 m) ahead of the signal.
4. *Turning movement study:* A manual count of the number of vehicles and pedestrians during 15-minute periods for each traffic movement from each approach should be undertaken. The study should be carried out for eight, not necessarily consecutive hours on a normal day, with the counts covering the following traffic conditions: (1) the morning rush hour, (2) mid-morning period, (3) mid-afternoon period, and (4) the evening rush hour.

To allow for ready comparison between locations and to provide standardized design criteria, the results of the movement study are generally expressed in the form of *equivalent volumes* in order to compensate for the number of traffic lanes actually in use and for the effect of turning movements and heavy vehicles. The equivalent volume per lane for any approach will be:

$$V_e = \frac{V + 0.5\,H + 0.6\,L + 0.4\,R}{n}$$

where V_e = equivalent volume per lane,
V = actual total volume for the approach,
H = number of heavy vehicles for the approach,
L = number of left turns for the approach,
R = number of right turns for the approach,
n = number of usable traffic lanes for the approach.

Signal timing for pretimed signals. Numerous methods are used for estimating the appropriate green time for each phase. Although the methods described in this section may seem to apply to two-phase operation, these same techniques may be used to calculate timing for any desired number of phases. For example, separate turn phases should be treated as an additional phase in the determination of a leading or lagging green requirement.

Whenever pedestrians are present, the minimum green for each phase is normally set by pedestrian requirements. In other words, the phase time should provide sufficient walking time for a pedestrian to clear the conflict zone prior to the release of opposing vehicles. A walking speed of 4 ft per sec (1.2 m per sec) is a value frequently used to calculate the minimum green requirement; in general terms, however, the relationship between minimum green or walk time and pedestrian walking speed is:

$$G = \frac{D}{v_p}$$

where G = minimum green time in sec,

D = length in ft (or m) of the longest crosswalk in use during the phase,

v_p = pedestrian walking speed in ft per sec (or m per sec).

On wide, divided roadways with median widths of at least 6 ft (1.8 m), it is sometimes advantageous to provide sufficient crossing time to reach only the median island. Here pedestrians would cross in two cycles.

When pedestrians are not present, the minimum green time may be set by a practical vehicle operational minimum, usually taken to be 12 sec. In some places, however, vehicle minimums as low as 7 or 8 sec have been used successfully.

The determination of green time above these minimum values is based on (1) proportioning the green time among the approaches so that the ratio of capacity to demand on each of the critical approaches is approximately equal and (2) having each green interval of sufficient length so that almost all phases during the peak demand periods are long enough to accommodate all traffic that has accumulated prior to the beginning of the green phase and the traffic arriving during the green phase.

Examples of some of the approaches to optimizing traffic signal timing are outlined below.

1. *Webster method:* Webster developed a model for computing the approximate cycle length that will minimize the total intersection delay as well as the effective green time for each approach.[17] Webster's formula is represented by:

$$C_o = \frac{1.5L + 5}{1 - y_1 - y_2 \dots y_n}$$

where C_o = optimum cycle length in sec,

L = total lost time per cycle, generally taken as the sum of the total yellow and all red clearance per cycle in sec,

y = volume/saturation flow[18] for the critical approach in each phase,

and for a two-phase signal the effective green time on each approach is represented by:

$$\frac{g_1}{g_2} = \frac{y_1}{y_2}$$

where g_1 and g_2 are the effective green times of phase 1 and 2, respectively.

---

[17] F. V. WEBSTER and B. M. COBBE, "Traffic Signals," *Road Research Technical Paper No. 56* (London, England: Her Majesty's Stationery Office, 1966) pp. 57–60.

[18] Saturation flow is defined as the flow that would be obtained if there were a continuous queue of vehicles and they were given a 100 percent green time. It is generally expressed in vehicles per hour of green time.

Examples of the variations of delay with cycle length are shown in Figure 17.12. It has been found that for cycle lengths within the range from $\frac{3}{4}$ to $1\frac{1}{2}$ times the optimum value, the delay is never more than from 10 to 20 percent above that given by the optimum cycle.

2. *Davidson method:* The technique of signal phasing as developed by Davidson can be accomplished as follows:[19]

   a. Select a tentative phasing for the intersection.

   b. Compute necessary clearance periods for each approach and select the longest required for each phase.

   c. Compute the minimum green required for each phase accommodating pedestrians.

   d. Compute the capacity of each approach on the basis of a full hour of green time.

   e. From traffic counts, determine the average hourly volume on each approach for the period under consideration (usually the peak hour of a typical weekday).

   f. Compute the ratio of volume to capacity for each approach and determine the critical (highest) ratio for each phase.

   g. Assign to the phase having the lowest critical ratio the green time determined in Step c (but not less than 15 sec).

   h. Assign to each additional phase a green time in the same proportion as its

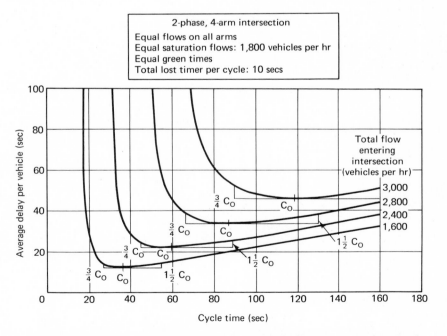

Figure 17.12. Effect on delay of variation of cycle length. [Source: *Research on Road Traffic* (London, England: Her Majesty's Stationery Office, 1965) p. 306.]

---

[19] B. M. DAVIDSON, "Traffic Signal Timing Utilizing Probability Curves," *Traffic Engineering*, XXXII, No. 2 (1961), pp. 48–49.

critical ratio bears to the critical ratio of the first phase (but not less than its minimum green determined in Step c).

i. Sum the clearance intervals determined in Step a, and sum the green intervals determined in Steps g and h. The values of each should then be adjusted so that the sum equals a multiple of five, and each interval is an integer percentage of the sum. (This step is required to accommodate the mechanical limitation of existing control equipment.) The result of these nine steps is the minimum cycle length that meets the first objective for a period under study. It is necessary, however, to determine whether or not a number of phases during the period will fail to accommodate traffic arriving during that phase. The number of vehicles arriving during any particular phase will usually vary in a random manner. It is possible to determine the probability of arrival of any particular number of vehicles from Poisson's distribution formula. Figure 17.13 is useful here.

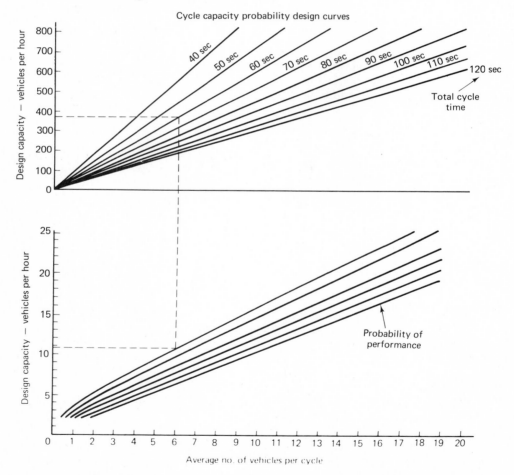

Figure 17.13. Cycle capacity probability design curves for use in traffic signal timing. [Source: Bruce M. Davidson, "Traffic Signal Timing Utilizing Probability Curves," *Traffic Engineering*, XXXII, No. 2 (1961), p. 49.

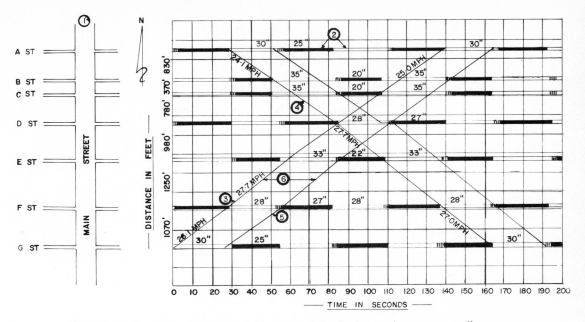

Figure 17.14. Typical time-space diagram. Note: The following items refer to corresponding numbers in Figure 17.14. (1) Plan of street drawn to scale. (2) "Go" intervals for Main Street designated by open space; "stop" intervals for Main Street designated by solid line; "yellow" intervals for Main Street designated by crosshatching. (3) Slope of this line indicates the speed and represents the first vehicle of a group or platoon moving progressively through the system from G Street to A Street. (4) As in 3 above, this line represents the first vehicle moving from A Street to G Street in the southbound direction. (5) This line is roughly parallel to line 3 and represents the last northbound vehicle in a group to go through all intersections. (6) The space between lines 3 and 5 on the time scale is the width of the "through band" in seconds; this width may vary over different sections of street. [Source: *Traffic Engineering Handbook*, 3rd ed. (Washington, D.C.: Institute of Traffic Engineers, 1965) p. 413.]

j. Determine the level of performance it is desired to achieve, i.e., 95 percent of cycles serving the demand present. Enter the upper portion of Figure 17.14 with the critical hourly volume for each phase (example: 360). From the intersection of this horizontal line with the trial cycle length (example: 60 sec), trace vertically to the desired probability level (example: 95 percent) on the lower portion and then horizontally to the left scale for the solution (example: 11 vehicles). To adequately serve 95 percent of the cycles with this example volume and cycle, the green phase must be sufficient for 11 vehicles.

k. Determining the minimum green phases required to accommodate the value from Step j can proceed two ways: (1) average time values developed earlier in this chapter for each vehicle may be used or (2) the computed capacity per hour of green for the approach may be assumed to be proportionally available in each cycle (hourly capacity of 1,200 = $\frac{1}{3}$ vehicle per sec or 3 sec per vehicle). Both methods are approximations. By the method selected, determine whether or not the selected level of performance is achieved for each phase.

If this test is positive, the signal timing is completed for the selected time period. It is desirable to repeat the process for other time periods in which the signal will be in operation in order to determine if better performance could be obtained by altering the timing (using multidial equipment) or by a compromise in the timing schedule. Should the test in Step k give negative results, one of several courses is available: (1) Select a longer cycle length and recalculate Steps h through k with adjusted green periods, or (2) select a different phasing pattern (i.e., provide a left-turn phase, reduce from three to two phases, etc.) and recompute, or (3) make physical or operational changes in the intersection or approach roads to increase capacity (or in some cases decrease demand), or (4) select a lower level of performance in Step j and recompute.

3. *Canadian Manual on Traffic Control Devices Method:*[20] By way of contrast, the *Manual on Uniform Traffic Control Devices for Canada* presents a table of phase time which is the sum of the green and yellow intervals required to provide a 95 percent probability that all vehicles arriving at that approach in question during a complete signal cycle will be able to clear during the next green interval. If the equivalent hourly volume in the heavy direction for each signal phase is known, the required time interval for any phase can be found in Table 17.5 by equating the volume for that phase against the sum of the volumes for all other phases.

TABLE 17.5
Phase Time Requirements

Sum of the Equivalent Hourly Volume in the Heavy Direction on All Other Phases	Phase Time in Sec Required When the Equivalent Hourly Approach Volume Is											
	50	100	150	200	250	300	350	400	450	500	550	600
200	11	12	13	14	15	16	17	19	21	23	26	29
250	11	12	13	14	16	17	18	20	22	24	27	30
300	11	12	13	14	16	17	19	20	22	24	29	32
350	11	12	13	14	16	17	19	21	23	26	30	34
400	11	12	13	15	16	18	20	22	25	28	32	37
450	11	12	14	15	17	19	21	23	26	30	35	41
500	11	12	14	15	17	19	22	25	28	33	38	45
550	11	12	14	15	17	20	23	26	31	36	42	51
600	11	13	14	16	18	21	24	28	33	39	48	60
650	11	13	14	16	19	23	26	31	37	44	56	73
700	11	13	15	17	20	24	28	34	41	52	68	97
750	11	14	15	17	21	26	31	38	48	63	88	143
800	12	14	16	19	23	28	34	43	57	81	132	—
850	12	14	17	20	25	31	40	52	74	121	—	—
900	12	14	18	22	27	35	47	67	110	—	—	—
950	12	15	19	24	31	42	60	99	—	—	—	—
1000	13	16	20	27	36	53	88	210	—	—	—	—

Source: *Manual on Uniform Traffic Control Devices for Canada*, 2nd ed. rev. 1971, Roads and Transportation Association of Canada, Table 4B. 4.09.

Yellow clearance interval. The purpose of a yellow clearance interval is twofold: (1) to advise the motorists that the red interval is about to commence[21] and to permit the motorists to come to a safe stop and (2) to allow vehicles that have entered the intersection legally sufficient time to clear the point of conflict prior to the release of opposing pedestrians or vehicles. Thus, the duration of the clearance period is a func-

---

[20] "Traffic Control Signal Timing," *Manual on Uniform Traffic Control Devices for Canada*, 2nd ed. (Reinge Press Limited, Ottawa, Canada, 1966).
[21] In Great Britain it is the practice to use the yellow clearance interval before the beginning of green as well as before the beginning of red. This is not permitted in the *Uniform Manual on Traffic Control Devices*.

tion of approaching speed. To satisfy the first purpose, the minimum duration of the yellow period is:

$$y_1 = t + \frac{1}{2} \frac{v}{a}$$

where y_1 = yellow interval in sec,
t = perception-reaction time of driver in sec (the value for t can be taken as 1 sec),
v = approach speed in ft per sec (or m per sec)
a = deceleration rate in ft per sec per sec (or m per sec per sec).

An incorrect choice for the length of yellow period, however, can lead to the creation of a *dilemma zone*. This is an area close to an intersection in which a vehicle can neither stop safely nor clear the intersection before the beginning of the red interval without speeding. Olson and Rothery[22] have suggested that the yellow period should be such that a driver could just stop his vehicle on seeing the yellow light before entering the intersection or he could continue at uniform speed and cross the intersection before the beginning of the red interval. The nondilemma yellow period to satisfy both conditions is:

$$y_2 = t + \frac{1}{2} \frac{v}{a} + \frac{w + l}{v}$$

where y_2 = nondilemma yellow interval in sec,
w = width of intersection,
l = length of vehicle.

When reasonable limiting values of $t = 1$ sec, $a = 15$ ft per sec per sec (4.6 m per sec per sec), and $l = 20$ ft (6.1 m) are used, the values of y_1 and y_2 for various approach speeds may be calculated as shown in Table 17.6.

The yellow clearance interval should equal or exceed the values of y_1 for the approach speed selected. On the general assumption that excessively short or long yellow intervals encourage driver disrespect, common practice sets yellow periods between 3

TABLE 17.6
Theoretical Minimum Clearance Intervals* for Different Approach
Speeds and Crossing Street Widths

Approach Speed mph	Minimum Time to Stop (y_1) (sec)	Street Width†				
		$w = 30$	$w = 50$	$w = 70$	$w = 90$	$w = 110$
20	2.0	3.8	4.4	5.6	5.7	6.4
30	2.5	3.6	4.1	4.5	5.0	5.5
40	3.0	3.9	4.2	4.5	4.9	5.2
50	3.4	4.1	4.4	4.7	5.0	5.2
60	3.9	4.5	4.7	4.9	5.1	5.4

*Obtained from the formulas:

$y_1 = t + \frac{v}{2a}$ and $y_2 = t + \frac{1}{2} \frac{v}{a} + \frac{(w + l)}{v}$, when $t = 1$ sec, $a = 15$ ft per sec, and $l = 20$ ft.

†Crossing street width in ft.

22 P. L. OLSON and R. W. ROTHERY, "Driver Response to the Amber Phase of Traffic Signals," *Traffic Engineering*, XXXII, No. 5 (1962) pp. 17–20, 29.

and 5 sec. When y_2 exceeds the value selected for the yellow interval and when hazardous conflict is likely, an all-red clearance interval could be used for 2 to 3 sec between the yellow interval and the start of green for opposing traffic. The total time of the yellow and all-red intervals should be held to the minimum necessary to clear the intersection.

SYSTEM TIMING FOR ARTERIAL ROUTES

A signal system consists of two or more individual signal installations operated in coordination, i.e., having a fixed time relationship to each other. All signals in the system must operate with the same (common) cycle length. In rare instances some intersections within the system may operate at double or half the cycle length of the system. In the case of an actuated signal within a system, the usual practice is to provide a common cycle length as a background cycle with the appropriate main street offset. At individual intersections, the intervals (red, green, and yellow) may vary depending on traffic conditions; however, it is desirable that the coordinated street have a green plus yellow period equivalent to at least 50 percent of the cycle length.

The methods of linking signals into a system network vary widely. An extensively used method involves an interconnecting cable joining the traffic signal control units that are to be linked. One of the units serves as a master, and the balance are local or slaves. The master merely maintains the established synchronization among each of the units.

Many modern systems involve centralized control by means of an analog or digital computer. Central control generally provides for greater flexibility by permitting more timing cycles than can normally be accommodated by the intersection control units. Detectors that measure traffic flow at selected locations also permit the selection of various signal timing arrangements in real time through central control.

Advantages of system synchronization. Some of the advantages of providing coordination among signals are:

1. A higher level of traffic service is provided in terms of higher overall speed and reduced number of stops.
2. Traffic should flow more smoothly, often with an improvement in capacity.
3. Vehicle speeds should be more uniform because there will be no incentive to travel at excessively high speed to reach a signalized intersection within a green interval that is not in step. Also, the slow driver is encouraged to speed up in order to avoid having to stop for a red light.
4. There should be fewer accidents because the groups of platoons of vehicles will arrive at each signal when it is green, thereby reducing the possibility of red signal violations or rear-end collisions. Naturally, if there are fewer red intervals displayed to the majority of motorists, there is less likely to be trouble because of driver inattention, brake failure, slippery pavement, etc.
5. Greater obedience to the signal commands should be obtained from both motorists and pedestrians because the motorist will try to keep within the green interval and the pedestrian will stay at the curb because the vehicles will be closer spaced.
6. Through traffic will tend to stay on the arterial street instead of on parallel minor streets.

Applications of system timing. In a discussion of the two-way and one-way street applications of system timing, the following terms are frequently used:

1. *Through band:* the time in seconds elapsed between the passing of the first and the last possible vehicle in a group or platoon of vehicles moving in accordance with the designed speed of a progressive signal system.
2. *Progression:* a time relationship between adjacent signals permitting continuous flow of groups of vehicles at a planned rate of speed.
3. *Offset:* the number of seconds or percent of the cycle that the start of the green interval appears at a given traffic control signal after a certain instant used as a time reference base.

One-Way Street. The simplest form of coordinating signals is along a one-way street or to favor one direction of traffic on a two-way street that contains highly directional traffic flows. Essentially, the mathematical relationship between the progression speed S and the offset L can be described as:

$$S \text{ mph} = \frac{D(\text{ft})}{1.47L} \quad \text{or} \quad S \text{ kph} = \frac{D(\text{m})}{0.278L}$$

where S = speed of progression,
$\quad D$ = spacing of signals,
$\quad L$ = offset, sec.

Two-Way Street. For a two-way movement, four general progressive signal systems are possible: (1) simultaneous, (2) alternate, (3) limited (simple) progressive, and (4) flexible progressive. The relative efficiency of any of these systems is dependent on the distances between signalized intersections, the speed of traffic, the cycle length, the roadway capacity, and the amount of friction caused by turning vehicles, parking and unparking maneuvers, improper or illegal parking or loading, and pedestrians. In general, a two-way progression with maximum band widths can only be achieved if the signal spacings are such that vehicular travel times between signals are a multiple of one-half the common cycle length; otherwise, inevitable compromises have to be made in the progression design. A discussion of the four general progressive signal systems follows:

1. *Simultaneous system:* All signals along a given street operate with the same cycle length and display the green indication at the same time. Under this system, all traffic moves at one time, and a short time later all traffic stops at the nearest signalized intersection to allow cross street traffic to move.

   The mathematical relationship between the progression speed (in both directions) and signal spacing in a simultaneous system can be described as follows:

$$S \text{ mph} = \frac{D(\text{ft})}{1.47C} \quad \text{or} \quad S \text{ kph} = \frac{D(\text{m})}{0.278C}$$

   where S = speed of progression,
   $\quad D$ = spacing of signals,
   $\quad C$ = cycle length, sec.

   For example, a system of signalized intersections at one-half mile or one-half kilometer spacing could have a progressive speed in a simultaneous system of

30 mph or 30 kph, respectively, with a 60 sec cycle. With closely spaced intersections, however, a simultaneous system may encourage excessive speeds as drivers attempt to travel through a maximum number of intersections during the green interval.

2. *Alternate system:* Each successive signal or group of signals shows opposite indications to that of the next signal or group. If each signal alternates with those immediately adjacent, the system is called *single alternate*. If pairs of signals alternate with adjacent pairs, the system is termed *double alternate*, etc. The progressive speed in a single alternate system is:

$$S \, \text{mph} = \frac{D(\text{ft})}{0.735C} \quad \text{or} \quad S \, \text{kph} = \frac{D(\text{m})}{0.139C}$$

where S = speed of progression,
   D = spacing of signals,
   C = cycle length, sec.

In a double alternate system, the progression speed is determined by the same formula with D being the distance between the mid-points of adjacent pairs. Generally speaking, the alternate system may provide excellent traffic service, depending on the distances between signals and the cycle length. Equal distances provide the best results.

3. *Limited progressive system:* Also known as a *simple progressive system*, this system uses a common cycle length, and the various signal faces controlling a given street provide green indications in accordance with a time schedule to permit continuous operation of platoons along the street at a designed rate of speed. The speed may vary within different parts of the system.

4. *Flexible progressive system:* This system is a refinement of the limited progressive system in that the signal offsets, splits, and/or cycle length of the common cycle are changed to suit the needs of traffic throughout the day. For example, an inbound progression toward the CBD during the morning peak, and an outbound progression during the remainder of the day, can be provided by merely changing the signal offsets. Or else a longer cycle length can be provided during the morning and evening peak hours in order to provide greater capacity than would normally be provided during the off-peak period.

Progressive signal system design[23]

Selection of a Cycle Length. Ideally, a coordinated signal system will provide progression speeds at or near the mean operating speed of vehicles on the street. This criterion frequently exists in the selection of a trial cycle length. If the spacing of signals in the system is fairly regular, the equations presented in the preceding section may be solved for C (cycle length) by using the measured operating speed for S and the typical distance between proposed signals for D. Either of the resultant cycle lengths falling in a usable range should be compared with the cycle length computed for each individual intersection. If it approximates or slightly exceeds the cycle length computed for a majority of individual intersections, it should be selected on a trial basis. First, however, each individual intersection must be reexamined to assure that it can operate effectively with the selected cycle. Sometimes rephasing or geometric and/

---

[23] JAMES H. KELL, "Course Notes," *University of California, Division of Transportation Engineering.*

or operational improvements at an intersection will be required. If such changes are not feasible, and the operation with this cycle at one or more intersections would be seriously impaired, a new trial cycle length should be selected. In practice, the cycle length already established for signal systems intersecting or closely adjacent to the system under study will frequently dictate the cycle length to be used.

Design of a Coordinated System for Two-directional Flow with Uniform Block Spacing. This method can generally be used only when signals can be spaced approximately uniformly, even though, in fact, some of these signals may not be installed. Signals installations at every intersection must be assumed.

CASE 1. CYCLE LENGTH NOT PREDETERMINED

This method applies to a street that is not part of any other system and when the cycle length is only restricted by the traffic requirements at individual intersections along the route.

1. Select a desired speed of progression for the system.
2. Compute the time required to travel one block at the desired speed.
3. Select a single, double, or triple alternate system on the basis of the time required for a round trip from the first intersection to the second, third, or fourth intersection. If a round trip to the second intersection results in an acceptable cycle length that satisfies the traffic requirements at all intersections, use the single alternate system; if the trip to the third intersection and back gives a good cycle length, use the double alternate system; if the round trip to the fourth intersection gives a better cycle length, use the triple alternate system.

   Example

   Uniform block spacing of 400 ft (122 m).
   Desired speed of 25 mph (40 kph).
   25 mph $=$ 36.7 ft per sec (11.2 m per sec).

   $$\text{Travel time per block} = \frac{400 \text{ ft}(122 \text{ m})}{36.7 \text{ ft per sec}(11.2 \text{ m per sec})} = 10.9 \text{ sec.}$$

   Round trip to second intersection $=$ 21.8 sec.
   Round trip to third intersection $=$ 43.6 sec.
   Round trip to fourth intersection $=$ 65.4 sec.
   In this example, a double alternate system with a 45 sec cycle length would be used if the 45 sec cycle satisfies the traffic conditions at the individual intersections.
4. The offsets for all signals would be either zero or one-half the cycle length (for example, in a double alternate system with a 45 sec cycle, the first two intersections would have zero offset, the next pair 22.5 sec offsets, the next pair zero, etc.). Unsignalized intersections are included when determining offsets.
5. The division of the cycle length, i.e., green, yellow, and red intervals, for individual intersections is obtained by analyzing each case. Thus, although the beginning of the green interval is synchronized to provide coordinated flow, the end of the green interval may present a slightly irregular pattern.
6. The through band width (the time interval that determines the size of each platoon progressing through the system) depends on the system that has been selected. For a single alternate system, the width of the through band is equal to the shortest green plus yellow period; for a double alternate, the width is one-

half the green plus yellow; and for the triple alternate, the width is one-third the green plus yellow. The triple alternate should be sparingly used because of the reduction in the efficiency of the system.

CASE 2. CYCLE LENGTH PREDETERMINED

This method applies to a street containing equal block spacing and when the cycle length is predetermined, e.g., one intersection may be part of an intersecting coordinated system.

1. Obtain block spacing and cycle length.
2. Determine speed of progression by dividing the block spacing by one-half, one-fourth, and one-sixth of the cycle length, respectively, for single, double, or triple alternate systems.

Example

Uniform block spacing of 400 ft (122 m).
Cycle length of 50 sec.

Single alternate $\dfrac{400 \text{ ft}}{\frac{1}{2}(50 \text{ sec})} = 16$ ft per sec or 10.9 mph

or $\dfrac{122 \text{ m}}{\frac{1}{2}(50 \text{ sec})} = 4.9$ m/sec or 17.5 kph

Double alternate $\dfrac{400 \text{ ft}}{\frac{1}{4}(50 \text{ sec})} = 32$ ft per sec or 28.0 mph

or $\dfrac{122 \text{ m}}{\frac{1}{4}(50 \text{ sec})} = 98$m/sec or 35 kph

Triple alternate $\dfrac{400 \text{ ft}}{\frac{1}{6}(50 \text{ sec})} = 48$ ft per sec or 32.7 mph

or $\dfrac{122 \text{ m}}{\frac{1}{6}(50 \text{ sec})} = 147$ m/sec or 52.5 kph

In this example, a double or possibly a triple alternate system would be used, depending on the desired speed.

Design of a Coordinated System for Two-directional Flow with Nonuniform Block Spacing. The graphical construction of a time-space diagram is a method that can be used to coordinate any series of signalized intersections regardless of the spacing and, if properly applied, should result in the most efficient timing plan possible. The principle involved in the construction of this diagram is based on the fact that, in general, ideal progression with maximum band widths can be achieved in both directions if the spacings are such that travel times between signals are multiples of one-half the cycle length. Figure 17.14 shows a typical time-space diagram.

To illustrate the procedures involved, an example system is constructed as follows:

1. A time-space diagram is constructed in the following manner:
   a. A two-dimensional graph is established, generally with distance on the horizontal scale and time on the vertical scale. The intersections are then located on the distance scale and vertical lines are constructed at the centerline of all signalized intersections (see Figure 17.15).
   b. A horizontal working line is drawn across the diagram. The use of this line is discussed below.
   c. A cycle length scale is then established. This is usually an arbitrary scale having no numerical value. The actual cycle length is determined after completing the time-space diagram. (In systems in which the cycle length is predetermined, that cycle length must be used, possibly resulting in a less efficient coordination.) The scale of one cycle length should be approximately one-fourth of the vertical dimension of the diagram.
   d. The division of the cycle into red, green, and yellow intervals is determined for each intersection. (Green only should be used in jurisdictions in which it is illegal to utilize the yellow portion of the cycle.) It is usually convenient to use an average cycle length in the construction of the time-space diagram and then adjust the individual phases. For our example, 50 percent green plus yellow (45 percent green and 5 percent yellow) cycle split used.
   e. The construction of the time-space diagram is continued in the following manner. The signal phases are laid out on the vertical lines representing the intersections according to the procedure explained below. These signal phases *must* be drawn so that the mid-point of either a green plus yellow or a red portion of the cycle is centered on the horizontal working line. Placing the mid-point of a phase on the working line must be followed rigidly at each intersection to ensure equal treatment of each direction of traffic.
2. The construction of the signal phases is started at the first signalized intersection at the left edge of the diagram. Either a green plus yellow or a red phase is centered on the working line. The entire vertical line representing the intersection is then divided into red, green, and yellow intervals. In the example system, a green plus yellow phase is centered on the working line (see Figure 17.16).
3. A temporary construction line is then drawn through the beginning of the first green interval with a slope equal to one-half the cycle length per 1,000 ft (304.8 m). This line is shown as *A* in Figure 17.17.
4. The signal phases are then constructed at the second and third signalized intersections, centering either a green plus yellow or a red phase on the working line so that the beginning of a green interval is placed as near line *A* as possible. For the example system, this is shown in Figure 17.17. A green plus yellow phase is centered on the working line for the intersection of *B* Street and a red phase is centered at *D* Street. This places the beginning of green phases at *B* Street and *D* Street as near line *A* as possible.
   a. If the first and second intersections have the same phase centered on the working line, and the third intersection has the opposite phase centered (as is the case in the example system), a new construction line is drawn connecting the beginning of green phases at the first and third intersections and intersecting the fourth vertical line. This new line should have approximately the same slope as line *A* (see line *B* in Figure 17.17).

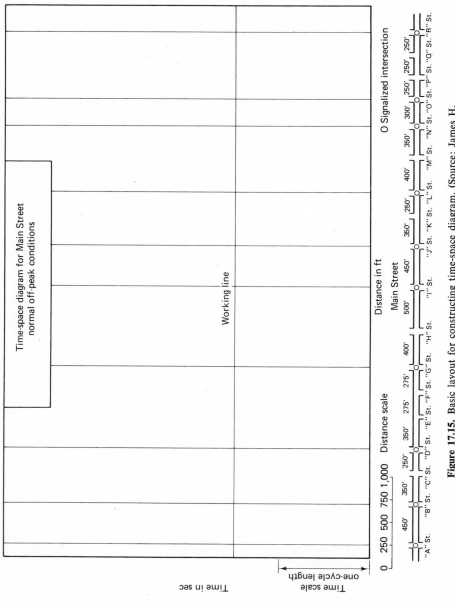

Figure 17.15. Basic layout for constructing time-space diagram. (Source: James H. Kell, *Course Notes*, University of California, Division of Transportation Engineering.)

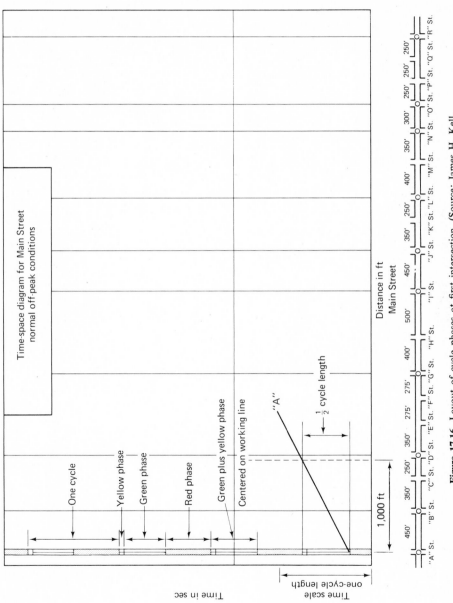

Time-space diagram for Main Street
normal off-peak conditions

One cycle

Yellow phase

Green phase

Red phase

Green plus yellow phase

Centered on working line

"A"

$\frac{1}{2}$ cycle length

Time in sec

Time scale
one-cycle length

1,000 ft

Distance in ft
Main Street

"A" St. | "B" St. | "C" St. | "D" St. | "E" St. | "F" St. | "G" St. | "H" St. | "I" St. | "J" St. | "K" St. | "L" St. | "M" St. | "N" St. | "O" St. | "P" St. | "Q" St. | "R" St.

450' | 350' | 250' | 350' | 275' | 275' | 400' | 500' | 450' | 350' | 250' | 450' | 400' | 350' | 300' | 250' | 250' | 250'

Figure 17.16. Layout of cycle phases at first intersection. (Source: James H. Kell, *Course Notes*.)

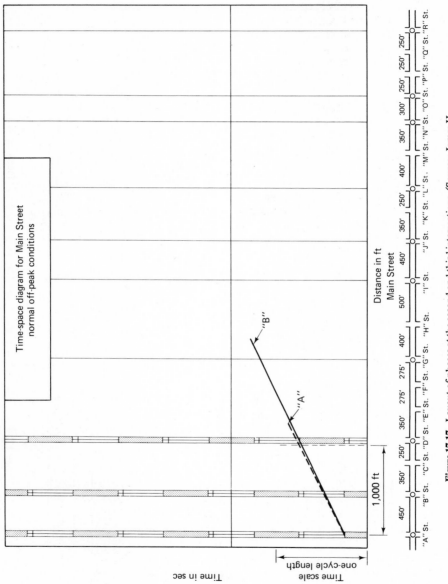

Time-space diagram for Main Street
normal off-peak conditions

"A"

"B"

Distance in ft
Main Street

1,000 ft

Time in sec

Time scale
one-cycle length

"A" St. "B" St. "C" St. "D" St. "E" St. "F" St. "G" St. "H" St. "I" St. "J" St. "K" St. "L" St. "M" St. "N" St. "O" St. "P" St. "Q" St. "R" St.

450' 350' 250' 350' 275' 275' 400' 500' 450' 350' 250' 400' 350' 300' 350' 250' 250'

Figure 17.17. Layout of phases at the second and third intersections. (Source: James H. Kell, *Course Notes.*)

b. Under any other combination than above, line *A* should be extended to the fourth signal and the proper phase at this intersection centered on the working line so that the beginning of a green phase is as near line *A* as possible. A new construction line should then be drawn through the beginnings of the green phases of two intersections so that the line intersects the green phase at the other signals.

5. The fourth signal is then examined. The proper phase is centered on the working line so that the beginning of the green is as near the construction line as possible. The construction line is adjusted (see Figure 17.18) and the process is repeated at each succeeding signal through to *R* Street. Throughout, the construction line indicates the speed of progression through the system. The slope of this line should remain approximately equal to the original line *A*.

6. When the layout of the signal phases is completed, the final adjusted construction line becomes the lower line of the through band. It is now desirable to establish the cycle length (if not predetermined) to provide a reasonable speed of progression. As mentioned previously, the slope of the construction line indicates the speed of progression. In the example system, the distance between the first signal (*A* Street) and the last signal (*R* Street) is 5700 ft (1,727 m). It takes 3.06 cycles to traverse this distance. The speed of progression for various cycle lengths is shown in Table 17.7. If the speed limit in this area is 25 mph (40.2 kph), and the individual intersection requirements are such that a 50 sec cycle is acceptable, this cycle length should be selected.

TABLE 17.7

Cycle Lengths (sec)	Speed of Progression	
40	31.8 mph	51.2 kph
45	28.3 mph	45.5 kph
50	25.4 mph	40.9 kph
55	23.0 mph	37.0 kph
60	21.2 mph	34.1 kph

7. The top line of the through band is constructed parallel to the final construction line (line *E* in Figure 17.19) so that this line intersects all signals in the green or yellow interval of the cycle. (*Note:* In some areas where it is illegal to enter the intersection on the yellow, this line can only intersect the green interval of the cycle, thereby reducing the width of the through band.) In the example system (see Figure 17.19), the slope of the lower line of the through band is controlled by the first signal (*A* Street) and the eighth signal (*N* Street). The top line and the band width is controlled by the sixth signal (*J* Street). With a 50-sec cycle this results in a through band that is 11.2 sec wide at a speed of 25.4 mph (40.9 kph).

8. The through band for the opposing direction is then constructed. This through band is approximately equal in width and slope to the first one. In the example system (see Figure 17.19), the lower line of the band is controlled by the second signal (*B* Street) and the sixth signal (*J* Street). The top line is controlled by the eighth signal (*N* Street). This through band is 10.5 sec wide at a speed of 25.9 mph (41.7 kph).

9. The completed time-space diagram showing all bands for the example system is illustrated in Figure 17.20. The offsets of the individual signals are either zero

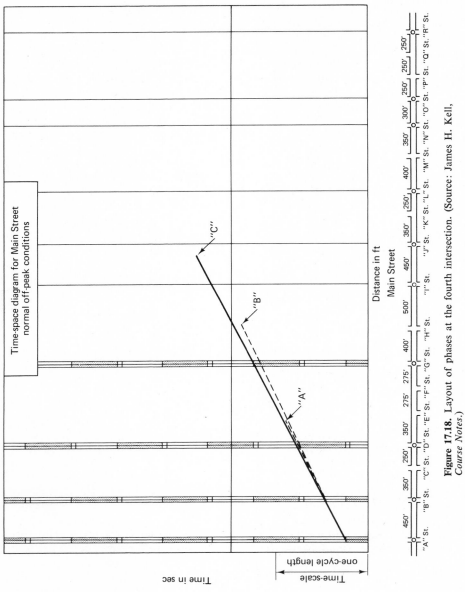

Figure 17.18. Layout of phases at the fourth intersection. (Source: James H. Kell, *Course Notes*.)

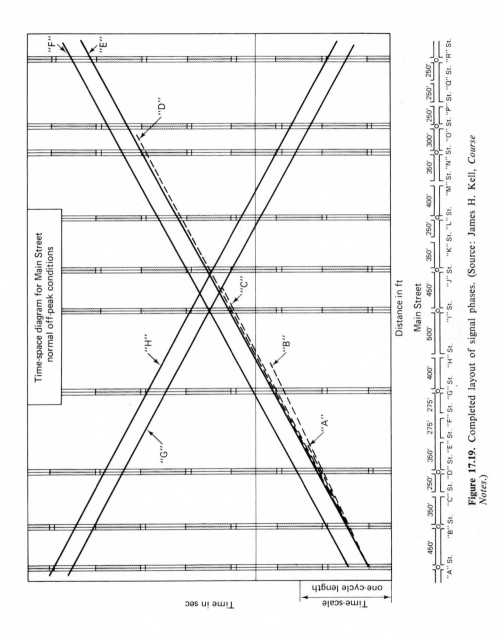

Figure 17.19. Completed layout of signal phases. (Source: James H. Kell, *Course Notes.*)

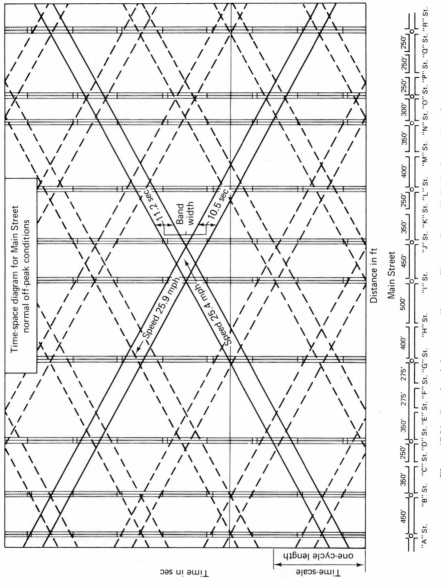

Figure 17.20. Completed time-space diagram. (Source: James H. Kell, *Course Notes*.)

or one-half cycle. In the example system, assuming *A* Street to be the master controller with a zero offset, *A, B, G, L,* and *R* Streets would have zero offsets and *D, I, L, N,* and *O* Streets would have 25 sec (one-half cycle) offsets.

Design of a Coordinated System for One-directional Flow with Nonuniform Block Spacing. The method for developing one-directional flow progression plans is described below:

1. The cycle length for the system must be determined. For peak-hour flows this may be longer than off-peak cycles to increase the band width for the peak flow.
2. The desired speed of progression is selected.
3. A basic layout for a time-space diagram is prepared with all signalized intersections located along the horizontal scale.
4. A construction line is drawn across the diagram with a slope equal to the desired speed of progression. This line is the bottom line of the through band.
5. The phases are then constructed at each intersection so that the beginning of a green phase is placed on the construction line at each intersection.
6. The top line of the through band is placed parallel to the bottom line. If all signals have the same phase length, then the through band width is equal to the green plus yellow portion of the cycle. If the phase is not the same at all signals, the through band width is equal to the shortest green plus yellow period in the system.
7. Offsets are determined by measuring the displacement of the beginning of the green interval at individual intersections from the beginning of the green interval at the master station.
8. For the example system, assume a cycle length of 60 sec, a speed of progression of 25 mph (40.2 kph), and direction of progression from *A* Street to *R* Street.
   a. Line *A* (see Figure 17.21) is first constructed with a slope equal to 25 mph (40.2 kph) with a 60 sec cycle.
   b. Signal phases are laid out at each intersection with the beginning of green placed on line *A*.
   c. The top line of the through band is then drawn. Since in the example system there is a uniform split of 50 percent, the through band is equal to the green plus yellow period of 30 sec.
   d. Assuming *A* Street to be the master intersection with zero offset, the individual intersection offsets are shown in Figure 17.21.
   e. It should be noted that although no recognizable through band exists in the opposite direction, opposing traffic may still travel through the system, but it will be stopped at one or more signals.

Computer timing models for arterial route systems. In recent years, the above procedures have been reduced to mathematical models suitable for computer solution as discussed in Chapter 11, Computer Applications. (Obviously, the manual or graphical solutions are not only subject to human error, but also require extensive man-hours of scarce traffic engineering staff time.) As an example, the mathematical model SIGART[24] developed for use in the Toronto Traffic Control System is described below.

---

[24] *Program Documentation Manual,* Metropolitan Toronto Traffic Control System, Metropolitan Toronto Department of Roads and Traffic, City Hall, Toronto, Canada, April 1965.

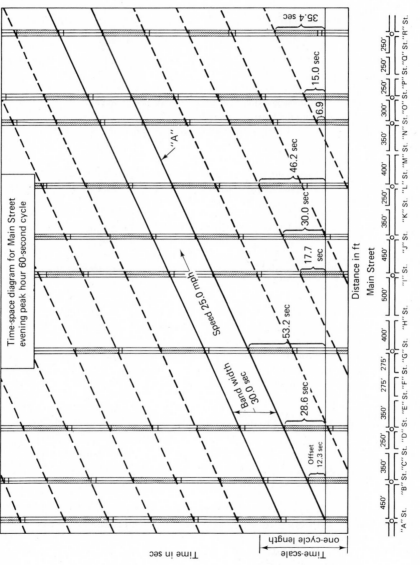

Figure 17.21. Completed time-space diagram favoring one direction of flow. (Source: James H. Kell, *Course Notes.*)

The *SIGART program* produces all possible through bands (wider than some desired minimum) on an arterial route within a specified progression speed range for all feasible cycle lengths. The traffic engineer can then evaluate these through bands in relation to the specific arterial and select the appropriate band that is most desirable for the existing conditions. Input to the program consists generally of: (1) the number of signalized intersections along the section under consideration; (2) the number of cycles to be tested; (3) three speed values (in mph or fps)—the desired speed, the maximum speed to be considered, and the minimum speed that would allow a minimum band width considered acceptable; (4) the cycle lengths in seconds, (5) the percent of cycle that is red at each individual intersection, (6) and the centerline to centerline distances between signalized intersections.

The program computes every possible through band within the specified speed range for each requested cycle length. As each feasible band is found, the program outputs the appropriate data (see Figure 17.22), including the progression speed, the through band width, the various offsets, and the critical intersections that delimit the band. The program's terminal phase is to print a summary of the feasible bands found for each of the requested cycle lengths. For each band, the summary contains the speed of progression, the through band width, and the relative band efficiency, i.e., the band width expressed in terms of percent of cycle.

SIGNAL NETWORK

The time-space relationships in signal coordination thus far developed have been primarily concerned with a single route. When two or more routes cross at a common intersection, the result becomes a signal network that may be classed as either open or closed.

Open network. An open network contains only one common intersection throughout its entire length. Examples of open networks are illustrated in Figure 17.23. In general, the cycle length for the network is fixed by the requirements at the common intersection, and the phase lengths at the common intersection are fixed as well. Subject to the limitations imposed at the common intersection, the design of the signal timing for each route may proceed independently.

Closed network. A closed network contains two or more common intersections as illustrated in Figure 17.24. All signals forming the closed network must have the same cycle length, which is determined by the requirements at the critical intersection, i.e., the intersection with the longest cycle length requirement. Once the cycle length is determined, the timing of each route should be considered separately and adjustments made to the offsets and/or green and yellow times of those signals within the closed network in order to achieve a balance. In other words, the sum of the offsets plus the green and yellow times taken in sequence around a closed network must be equal to the network cycle length or a multiple thereof.

Given an example of a simple closed loop with traffic signals located in a clockwise direction at points A, B, C, and D, the coordination of these four signals to provide a clockwise progression is given by the following equation:

$$O_{B(AB)} + G_{B(AB)} + O_{C(BC)} + G_{C(BC)} + O_{D(CD)} + G_{D(CD)} + O_{A(DA)} + G_{A(DA)} = NC$$

where $O_{B(AB)}$ = offset in sec of B from A to B,

$\quad\quad G_{B(AB)}$ = green plus yellow time in sec at B in the direction of A to B,

$\quad\quad\quad\quad C$ = cycle length in sec for the closed network,

$\quad\quad\quad\quad N$ = an integer (i.e. 1, 2, 3, . . .).

```
DUFFERIN STREET -- PRELIMINARY STUDY TO SELECT CYCLE LENGTH     22/4/68

ALL POSSIBLE BANDS GREATER THAN 10.00 SECONDS  FOR  A  70  SECOND  CYCLE

                                                              BAND NUMBER  70-1
SPEED IS  32.72 MPH (48.00 FPS)

BAND WIDTH IS  12.36 SECONDS ( .177 CYCLE)

OFFSETS ARE  .064 .494 .994 .539 .964 .474 .954 .459 .964 .989

CRITICAL SIGNALS ARE  2,  9, AND 10
                                                              BAND NUMBER  70-2
SPEED IS  31.28 MPH (45.88 FPS)

BAND WIDTH IS  11.70  SECONDS ( .167 CYCLE)

OFFSETS ARE  .064 .494 .994 .539 .964 .474 .954 .459 .964 .489

CRITICAL SIGNALS ARE  2,  10, AND 4


        DUFFERIN STREET -- PRELIMINARY STUDY TO SELECT CYCLE LENGTH     22/4/68

 •••••• SUMMARY OF FEASIBLE BANDS ••••••     (CONTINUED)

BANDS FOR A 70 SECOND CYCLE

SPEED (MPH) BAND (SECS) EFFICIENCY BAND NUMBER     DESIRED SPEED BAND

   30.00      7.88        11.3        70-0

   32.73     12.36        17.7        70-1
   31.28     11.70        16.7        70-2
```

Figure 17.22. Sample output of SIGART Program.

832

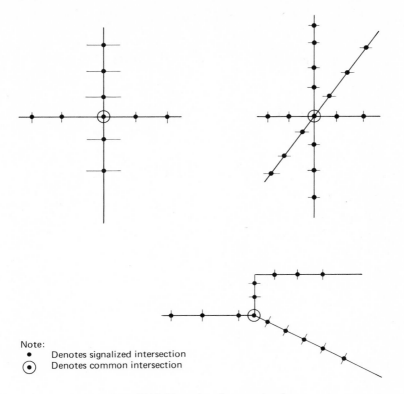

Note:
● Denotes signalized intersection
⊙ Denotes common intersection

Figure 17.23. Examples of open networks.

Computer timing models for closed systems. To overcome the often difficult manual task of optimizing traffic signal patterns within a network, mathematical models have been developed, such as SIGRID[25] and SIGOP,[26] which are suitable for computer solution. The SIGRID program developed for use in the Toronto Traffic Control System is one example:

The *SIGRID program* determines the optimum offsets for pretimed and vehicle-acuated interconnected traffic signal systems. In addition, a rating system is presented as program output for both the existing offsets (if any) and the optimized system for each cycle length studied. This permits the evaluation and comparison of the effects of altering any of the variables describing the system, e.g., cycle length, location and number of signals, and coordination priorities. In its present form, the program can handle up to 300 signals and up to 1,000 directional street sections (links) for a grid network of any configuration. Figure 17.25 illustrates a sample output from the SIGRID program. In addition to a description of the street network, the input data required for this program includes the following:

1. Locations of all signalized intersections.
2. Representative volume counts for the time period of traffic flow requiring separate signal settings.

[25] *Program Documentation Manual,* Metropolitan Toronto Traffic Control System, April, 1965.

[26] *SIGOP:* "Traffic Signal Optimization Program," prepared for the Bureau of Public Roads, U.S. Department of Commerce by Traffic Research Corporation, September, 1966.

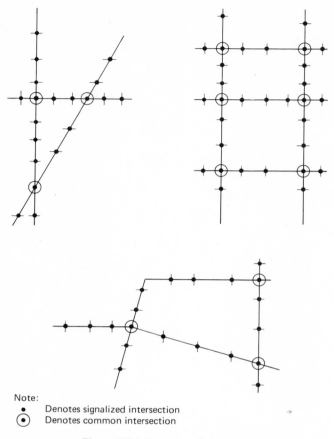

Note:
- Denotes signalized intersection
⊙ Denotes common intersection

Figure 17.24. Examples of closed networks.

3. A relative importance factor for each street section based on the usefulness and importance of signal coordination during the time period under study. For example, streets with excessive marginal frictions should have smaller importance factors than streets without marginal frictions even though the volumes carried by both streets were the same.
4. Desired differences in offsets between adjacent intersections. The traffic engineer may specify these directly in seconds or they may be calculated by the program. If they are to be calculated, the following are necessary:
   a. Distances between intersections
   b. Desirable travel speed for each traffic flow study, taking into account volumes and marginal frictions
   c. Special adjustments for the release of queues caused by turning movements
5. Suitable phase lengths for each signal. The traffic engineer may specify these as a percentage of cycle length or the phases may be specified as fixed durations in sec for all cycle lengths.
6. Current traffic signal settings to be used for comparison with the new settings (this is, of course, optional and not required for the optimization).

SIGRID SAMPLE RUN

P = .90

CYCLE LENGTH 70 SECONDS

SIGNAL NO.	SPLIT (N-S)	ORIGINAL OFFSET	CHANGE IN OFFSET	NEW OFFSET
1	.40	-.00	.47	.47
2	.30	-.00	.02	.02
3	.40	-.00	.43	.43
4	.40	-.00	.08	.08
5	.30	-.00	.68	.68
6	.40	-.00	.00	.00

FROM SIGNAL	TO SIGNAL	DIREC-TION	DIFFERENCE IN OFFSET			DELAY PROPENSITY FACTOR		VOLUME FACTOR	IMPORTANCE FACTOR
			DESIRED	ORIGINAL	NEW	ORIGINAL	NEW		
1	2	EBND	27	-7	32	14	5	600	20.0
2	3	EBND	34	7	36	17	2	600	20.0
3	4	NBND	20	0	46	29	26	240	10.0
4	5	WBND	34	63	35	29	1	1200	20.0
5	6	WBND	27	7	29	12	2	1200	20.0
6	1	SBND	20	0	33	29	13	600	15.0
1	6	NBND	20	0	37	29	17	350	10.0
5	2	SBND	20	0	24	46	4	300	15.0
2	6	NBND	20	0	46	46	26	200	10.0
4	3	SBND	20	0	24	29	4	750	15.0

WEIGHTED AVERAGE DELAY PROPENSITY FACTOR

ORIGINAL SYSTEM	NEW SYSTEM
24.4	6.2

Figure 17.25. Sample output of SIGRID Program.

TRAFFIC-ACTUATED SIGNALS

Actuated signals may be pedestrian-actuated, semiactuated, fully actuated, or volume-density. The procedures involved in timing an actuated signal are basically the same as for pretimed signals. The major difference lies in the refinements which the flexibility of actuated equipment permits. With the aid of vehicle detectors and/or pedestrian push buttons, the actuated controller is capable of varying the cycle length and/or intervals within the cycle in accordance with traffic demand. Its greatest functional advantage is its automatic response and adjustment not only to the daily, weekly, and seasonal variation in traffic flow, but also to the momentary surges that are inherent in traffic demand. In this way, delay can be held to a minimum, and maximum intersection capacity may be achieved. When network coordination is required, the normal practice is to equip the controller with a constant cycle length (known as a background cycle). Each kind and make of controller varies as to the exact adjustments provided and the setting-up procedure to be used, but the following may be considered as a general indication of the basic intervals

1. *Minimum green:* The minimum green that can be given to any phase must be the pedestrian requirement to cross the roadway at a reasonable speed, usually 4 ft per sec (1.2 m per sec). Under conditions of light traffic, this time may be excessively long, but it can only be reduced in locations where no pedestrians are present or if a separate pedestrian-actuated interval timer is installed together with pedestrian signal heads.
2. *Vehicle interval:* Each time an actuation is received, it causes the green time to be extended beyond the minimum green or initial interval by a time equal to a prespecified amount. A vehicle interval of from 3 to 5 sec is normally used.
3. *Maximum interval:* The maximum interval terminates the green period for that phase. The value of the maximum interval may be determined by calculating the phase time required for operation as a pretimed controller.

COMPUTER CONTROLLED SIGNAL TIMING

Digital computer traffic control systems have been installed in many cities; many benefits have accrued and advantages are cited. One striking potential advantage of digital computer control is flexibility. Although the computer does not have built-in traffic control functions, *per se*, it can be programmed for various modes of network traffic control. Thus, providing that adequate and reliable sensor information is available, a number of control methods can be implemented by a computer.

The potential advantages of this flexibility are still largely unexplored. The traffic control logic used in almost all operational computer controlled traffic systems provides limited adjustment of traffic signal timings to traffic conditions. Often, a library of preengineered, pretimed signal plans is stored in the computer memory, and control is implemented by table look-up procedures operating on the stored signal plans. The appropriate signal plan is selected either by time of day or by special logic that responds to fluctuations of traffic parameters at selected detector locations. In some cities, such pretimed plan selection techniques are augmented with logic for individual intersection control, similar to conventional (on-the-street) vehicle-actuated controllers. Table 17.8 presents a summary of digital computer control experience in some 13 cities in 1970.

Considerable research effort is presently directed toward developing new control

TABLE 17.8
Area Traffic Control Installations

City	Status	Control Method	No. of Inter-sections	Area (sq mi)	Length of Streets (mi)	Kind of Area	Computer	Data Transmission	Number of Detectors	Kind of Detectors	Display	Control Modes Run	Control Modes Proposed	Present Conclusions	Remarks
Canada Toronto	Operational	Digital computer (2) (5) (6)	910	240			Univac 1107 and 418	Telephone cable	500	Loop detectors	Aerial photograph plus 2 lights for intersection, showing whether under computer control	Fixed-time preprogrammed splits vary with intersection demand. Volume-density mode during peaks, 3 other kinds of actuated control, up to 10 plans for each control area.	More self-determination on part of computer with computer control morning peak; avg. speeds increased about 15 to 20%; stops decreased 40–50% with flow up 10%.		Failure rate on electronic equipment telemetering, detectors, logic cards, relays 3% per year.
Germany Aachen (9) (10)	Operational	Digital computer	10		1	Adjacent to centers (two intersecting streets)	Siemens VSR–16000	Cables	26	Inductive detectors (including 12 with "analog digital-converters" for speed) remainder measure presence. Inductive detectors (including several "analog digital converters" for speed)	Green indications of three main intersections.	Fixed time changed by (a) time of day and (b) traffic demand. Also prediction of and minimization of delay at two critical intersections.			Computer may control up to 160 junctions but several computers may be coordinated.
Hamburg	Operational	2 Digital computer	54+ 50–60		4		Siemens VSR–16000	Cables + voice frequency multiplexing	60			8 programs changed according to traffic demand.			

TABLE 17.8 (*Continued*)
Area Traffic Control Installations

City	Status	Control Method	No. of Inter-Sections	Area (sq mi)	Length of Streets (mi)	Kind of Area	Com-puter	Data Trans-mission	Number of Detec-tors	Kind of Detec-tors	Display	Control Modes Run	Control Modes Proposed	Present Con-clusions	Remarks
Spain Madrid	Opera-ational	Digital com-puter plus 8 sub-masters	90	1.5	55	City center	Elliot 903	Cables	76	Loop detectors	Console Offset, split, cycle, count-ing, head-way and queue dura-tion can be seen (one per time and in one location or intersection per time.)	Dynamic signal plan genera-tive system. Takes into consideration volume, satura-tion degree, platoon speed, and queues. Maximum and minimum cycle may be fixed but otherwise is completely flexible. Fixed time programs are available by time of day or manual choice. Fixed programs are available in emergency without com-puter control.			Computer con-trols cycle length; sub-masters con-trol splits and offsets for a group of 10 or 12 intersections. Similar systems in Lisbon and Barcelona.
United Kingdom Glasgow	Opera-tional	Digital com-puter (3)	80	1	10	City centre	Marconi Myriad; Failure rate once in 3000 hours 16 k core 80 k drum	15 pair cable to 70 inter-sec-tions ditto over 3 conduc-tors GPO multi-plex to 10 inter-	250	Pneumatic 8% work-ing at any one time.	Network display of green indi-cations.	Existing (vehi-cle actuated) with some linking; three types of fixed time; combi-nation method, TRANSYT and SIGOP, two types of fixed time with varying splits (FLEXI		(1) Opti-mized fixed time linking on area basis provided substantial improve-ments (2) No scheme has yet been shown to	Simulation be-ing pursued in addition to simplify field testing and to generalise results.

Location	Status	Computer	Intersections			Computer location	Computer	Communication	Detectors		Display	Control method	Results	Remarks
(continued)								sections. Voice frequency multiplexing.				PROG and EQUISAT Platoon identification (PLIDENT).	be superior from optimized fixed time.	
West London	Operational	2 Digital computer (3)	70 + 30 pedestrian crossings	7	30	Adjacent to center	Plessey XL9 16k core; 83k drum	2 pairs + time division multiplex		Pneumatic double-loop queue detectors; double-loop speed detectors multiloop counting detectors.	Network display of main road greens and queue detectors; detailed display of any 3 intersections at a time.	FLEXI RPOG and combination method and TRANSYT.	Improvement from combination method consistent with Glasgow.	TV surveillance of major intersections. Control area being extended to include further 300 intersections.
Liverpool	Operational	Digital computer (4)	47		30 10	Part of central area adjacent to tunnel entrances	Plessey XL9	Cable		Pneumatic double-loop queue detectors.	Similar to West London	Basic purpose to prevent long approach queues to tunnel backing through intersections.	Subject assessment; system satisfactory.	
United States San Jose, California	Operational	Digital computer (1) (2)	60	1	12		IBM 1800	4 wires plus 1 per detector	400	320-loop 80 pressure flow speed, lane-occupancy.	Network display green for major street.	One and three dial fixed adjusting systems tested.	3 dial better than 1 dial. Green adjusting at critical intersections and preplanned program based on simulation better still. Evaluation based on data from flow, speed, and occupancy detectors.	Phase skipping not permitted.

TABLE 17.8 (*Continued*)
Area Traffic Control Installations

City	Status	Control Method	No. of Inter-Sections	Area (sq mi)	Length of Streets (mi)	Kind of Area	Computer	Data Transmission	Number of Detectors	Kind of Detectors	Display	Control Modes Run	Control Modes Proposed	Present Conclusions	Remarks
Wichita Falls, Texas	Operational	Digital computer (1) (2)	77	1	14	Downtown and fringes	IBM 1800 16k core card input output and typewriter 512k disc	4 wires plus 1 per detector	51	32 loop, 19 pressure. Measure flow, speed, lane occupancy. 26 different locations.	Network display, red, yellow, green signals for each street, plus blue to show computer is controlling signals.			Reduced delays 31% stops 16%.	Similar systems are proposed in Portland, Ore; Austin, Texas; Baltimore County (not city) Md.; Queens, New York. This type of computer may control up to 500 inter-sections.
Washington, D.C.	Planned	Digital computer	112	1.5	23	Downtown	Dual Sigma 5 65k core	Leased telephone lines FSK	517 Vehicle 144 Bus	Vehicle induction loop. Special bus transmitters and antennas.	Wall map display for signals and traffic parameters. CRT operational status display.		First-generation precomputed patterns. Second generation online optimization. Future generations.		Operational in summer 1972 as test facility for control strategies.
Fort Lauderdale Florida	Operational	Digital computer operating on analog principles (2)	83						83	Primarily lane occupancy.					

Baltimore, Maryland	Operational	Analog master controllers. some vehicle actuated.	1,050 (8 systems)	50	100	Entire city	Automatic Signal PR Controllers.	100	Volume and radar speed (density calculated).	Negligible		Philadelphia, Los Angeles, Fort Worth and, 150 other cities have similar although smaller installations. Some evaluations have been made.
Washington, D.C.	Operational	Three-dial fixed time system.	1,000	80	100	Entire city	Radio and cable	Very few		None		Many other cities including New York have fixed-time, interconnected systems of various kinds.
Charleston, N.C.	Under construction	Hybrid	85	1.5	11	Entire city	PDP-81 plus automatic signal modular solid-state master.	25 sampling 60 counting 20 local	Induction loop	Real time R and G for each intersection on map. Indication of unduly noncycling controllers. 5 c.c. TV cameras. Printout of discrepancies between time-space diagram speeds and actual ave. speed.	Common cycle (variable) choice of 2 independent offsets for the 2 subsystems in which area is divided (each of the 2 offsets independently variable)	

Source: *Area Traffic Control Systems*, Road Research Group, Organization for Economic Cooperation and Development (Paris, France, 1972), pp. 20–22.

logic that begins to utilize the capabilities of the digital computer for data processing, decision making, and control command functions. These projects will provide valuable knowledge in both the hardware and the software areas of operating computer traffic control systems. Some of the areas of research being pursued at this time are:

1. Exploration of the feasibility for mathematical formulation and analytical description of urban traffic flow.
2. Development of the criteria and configuration for an instrumented surveillance system, including the definition and evaluation of traffic parameters, data measurement, and processing techniques.
3. Development of system plans and specifications for a digital computer controlled traffic signal system to serve as a laboratory or a test site for the overall effort.
4. Development, test, and validation of traffic simulation models, toward developing a workable model on which new control strategies can be tested prior to actual field application.
5. Development of advanced control software.

SPECIAL SIGNALS

LANE-USE CONTROL SIGNALS

Lane-use control signals are special overhead signals having indications to permit or prohibit the use of specific lanes of a street or highway (see, also, Chapter 18). Such signals may consist of either a red X on an opaque background or a green arrow on an opaque background, with the arrowhead pointing downward. In addition to these two symbols, a yellow X may be used. It will be described later.

Lane-use control signals are most commonly used for reversible lane control; but they may also be used to (1) clear a freeway lane(s) when this is deemed necessary at any time, (2) indicate the termination of a freeway lane, (3) indicate the blockage (caused by an accident or a hazard) of a lane ahead, or (4) permanently operate a two-way street with an unequal lane distribution.

On a two-way street, road users are accustomed to the traffic in each direction being restricted to one-half the roadway width as indicated by an appropriate center lane line. In special cases, however, it may be desirable to operate the two-way street with more lanes in the one direction than in the opposing direction (see Figure 17.26). Such treatment is normally, although not always, restricted to a roadway with an odd number of marked lanes.

When the distribution of the lanes is permanent, there should be lane-use control signals mounted from 15 to 19 ft (4.5 to 5.8 m) over the center of each lane across the width of the roadway. When lane direction control signals are used to reverse the direction of flow on one or two lanes, a *red X, yellow X* (optional), and *green arrow* signal (in that order, from left to right) must be suspended over the center of the lane(s) to be reversed, and a *red X* signal or *green arrow* signal must be suspended over the adjacent lanes. Overhead signals should be placed at sufficiently close intervals so that at least one, and preferably two, lane-use control signals can always be seen.

Care must be taken to ensure that when the signal indications are reversed, there is no traffic proceeding in the wrong direction. This can be accomplished by providing

Figure 17.26. Application of lane use control signals. (Source: Bureau of Traffic Engineering and Electrical Services, City of Milwaukee.)

an additional signal head having a *yellow X* which could be continuously illuminated during the period immediately prior to the introduction of the *red X*, thereby giving motorists using the lane a warning indication to get off the lane before the change occurs. Or, alternatively, it can be achieved by having the *red X* appear in both directions on the lane for a sufficient clearance period to permit all traffic to vacate the center lane.

Where a *yellow X* is used, such a lane may be used by motorists for a left turn or for passing in both directions by flashing the *yellow X* for both approaches, which would signify that the lane may be used only with caution.

To prevent any confusion between lane-use control signals and the signals at an intersection on the same street, it is advisable to locate the lane-use control signals at least 200 ft (60 m) from any signalized intersection, and care should be taken to ensure that the placement of lane-use control signals does not interfere with the motorists' view of intersection traffic control signals.

When the division of the lanes is maintained permanently, the lane-use control signals can be illuminated directly from an adjacent power supply. When a reversible lane system is being applied, the signals must be controlled through a control unit that would ensure the *green arrows* are not shown on the same lane in both directions and that would provide a clearance interval when a lane change is imminent. The lane-use control signals may be switched through a time clock or some other centrally controlled device. Even so, it is desirable to have a manual override switch available.

RAILROAD-HIGHWAY GRADE CROSSING SIGNALS

There are approximately 220,000 highway-railroad grade crossings in the United States and approximately 44,000 have been provided with special protection of some sort—gates, flashing lights, flagmen, traffic signals, or bells. The remaining 176,000 have no special protection except signs. Approximately 3,200 accidents per year involve trains. Accidents involving trains have much greater severity than other traffic accidents; they result in approximately 1,200 deaths and 3,400 persons injured each year.[27]

Train approach signals are used at highway-railroad grade crossings to warn vehicular traffic that a train is about to pass. The signal may be fitted with a bell and may be used in conjunction with gates, if desired, all of which may be operated either automatically or manually. The latter is preferable at locations where the volume of rail traffic is large because less time will be lost to road users.

Design and specification for the installation of signals at highway-railroad grade crossings are described in Bulletin No. 6 of the Association of American Railroads, called "Railroad-Highway Grade Crossing Protection—Recommended Practices." In other countries, the appropriate authorities have established the required specifications.

When the grade crossing signal is operated automatically, track switches must be arranged so that the signal will operate for at least 20 sec before the fastest train reaches the crossing and will continue to operate until the last coach or truck of the train has passed. On single tracks, where trains move in both directions, suitable switching circuits must be provided so that the signals will operate on the approach of a train from either direction.

In order to reduce vehicular delay at locations where train speeds vary considerably and where vehicular volumes are large, a special form of control may be necessary to ensure a uniform minimum advance warning time irrespective of the speed of the train. At locations where shunting or switching operations adjacent to but not on the crossing could cause prolonged and unnecessary operation of the signals, a manual or automatic cut-out should be provided; however, arrangements must be made so that actuation of the signals by through trains (other than the one that is stopped or shunting) will still be possible.

One kind of flashing light signal now considered standard for all highway-railroad grade crossings is shown in Figure 17.27. The lenses are illuminated alternately for equal time intervals with each flashing on and off not less than 30 or more than 40 times per min. In almost all cases, a bell sounds continuously when the signal is operating, and a gate is often used in conjunction with the signals.

When a signalized intersection is located within approximately 200 ft (60 m) of a highway-railroad grade crossing that is controlled by train approach signals, the traffic control signals should be interconnected with the train approach signals so that conflicting aspects of the traffic signal and the train approach signal are avoided.

FLASHING BEACONS

A flashing beacon commonly consists of one or more sections of a standard traffic signal, displaying yellow or red lenses. They are mounted either as single or twin

---

[27] *Factors Influencing Safety at Highway-Rail Grade Crossings*, National Cooperative Highway Research Program, Report 50 (Washington, D.C.: Highway Research Board 1968), p. 5.

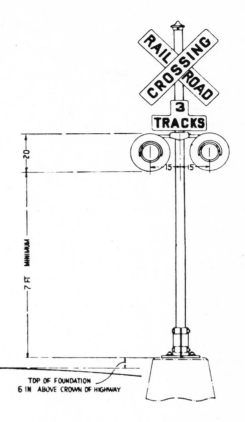

Figure 17.27. Highway grade-crossing signal of the flashing light type. [Source: *Railroad-Highway Grade Crossing Protection—Recommended Practices*, Bulletin No. 6 (Washington, D.C.: Association of American Railroads, 1966), p. 9.]

units, with the indication illuminated alternatively on and off at a rate of 50 or 60 flashes per min. The illuminated period is usually the same or slightly longer than the off period.

Intersection control beacons. Flashing beacons are used at locations where full traffic control signals are not warranted but where, because of lack of visibility or other hazard, signs alone are not sufficient.

Intersection control beacons are suspended over the roadway to draw attention to a hazardous location, and they must show a separate indication for each direction of movement or each approach. Typically, such beacons will have flashing yellow indications facing the major street approaches and flashing red indications facing the minor street approaches. In special circumstances, flashing red beacons may be used in all four directions to augment a four-way stop regulation. Stop signs should be in place at intersections where flashing red beacons are used.

Hazard identification beacon. Flashing yellow beacons may be located on the approaches to narrow bridges, underpasses, or other obstructions or hazards that require special attention by the motorist. Or, they may be used to supplement an appropriate warning or regulatory sign or marker. In this application the beacon shall not be included within the border of the sign.

As a general rule, beacons or any other device having flashing indications should not be installed too close to the approach to a signalized intersection because they may very easily distract the attention of the driver from the signal indications and cause accidents.

Speed limit sign beacon. A single beacon or a twin beacon showing two alternately flashing yellow indications, vertically aligned, may be used in conjunction with a fixed or variable speed limit sign. This application is particularly appropriate for school zones where lower speed limits apply during specified times and on specified days. When the beacons are flashing, they indicate that the school zone speed limit is in effect.

Stop sign beacon. A single or twin beacon with red lenses can be used to augment a stop sign. The beacon should be mounted from 12 to 24 in (0.3 to 0.6 m) above the stop sign. If a twin beacon is used in this application, it should be aligned horizontally and flashed simultaneously.

TRAFFIC SIGNALS AT DRAWBRIDGES

When a roadway crosses a drawbridge, traffic control signals are used to stop vehicular traffic from approaching the bridge when it is open or when it is about to be opened. Signal heads are installed at both approaches to the bridge, often in conjunction with gates or other forms of protection. The signal control mechanism should be interconnected with the bridge control and arranged so that an adequate warning indication will be shown sufficiently in advance of the bridge opening to ensure that the bridge will be clear of all traffic.

When standard intersection traffic control signals are located within approximately 500 ft (150 m) of the bridge ends, they should be fitted with preemptors interconnected with the bridge signals so that the intersection itself can be kept unobstructed by the traffic stopped at the drawbridge signal. Similarly, when the bridge is close to a railway grade crossing, special advance signals may be necessary to ensure that vehicles will not be stopped on the tracks.

EMERGENCY VEHICLE ACCESS

Traffic signals are sometimes installed to ensure safe and quick access for emergency vehicles from fire stations or ambulance stations to a street where such access might otherwise be difficult. In these instances, the warrants prescribed for traffic signal installation will not normally be satisfied. The operation of such signals provides continual right of way to the through movement with a preemptive control in the station to permit the emergency vehicle to receive the right of way and enter at such time as it is required.

SPEED CONTROL

Traffic control signals are sometimes used simply for speed control. When so used, the signal is not placed at a crossing or intersection, and is almost exclusively used on one-way roadways. Its steady state is normally red. A vehicle detector is installed some distance in advance of the signal, and when a motor vehicle is driven across the vehicle detector, the timing mechanism for the traffic signal will turn the indication to green in time to allow the motorist to proceed uninterrupted if he is traveling at or below the posted speed.

Each actuation received while the green is showing resets the timing circuit causing the green to remain for a further minimum period and thus the red indication will only reappear when there is a gap greater than the minimum green interval in the traffic flow crossing the detector. Thus, the resulting speed control will apply only to the first vehicle in a group and if the traffic flow is continuous, the speed control becomes ineffective.

SIGNAL CONTROL BY EMERGENCY VEHICLES

Preemption of the right of way through signalized intersections by emergency vehicles (usually fire-fighting vehicles) has been achieved by means of directed transmissions by radio or light beam from a transmitter in the emergency vehicle to a receiver at the signalized intersection.

When the receiver detects the transmission from the oncoming emergency vehicle, the phase selector determines whether or not the signal is green in the direction of the approaching emergency vehicle. If the signal shows green, it is maintained in that state until the emergency vehicle has passed and/or the transmissions have ceased to be received by the intersection receiver. If, on the other hand, the traffic signal is not green in the direction of the oncoming emergency vehicle when the transmissions are first received, the controller is advanced in accordance with timing adjustments which are set into the phase selector so that the green will show in the direction of the oncoming emergency vehicle. As soon as the emergency vehicle has cleared the intersection, the phase selector returns the traffic signal to its standard operation.

MAINTENANCE AND MODIFICATION OF EQUIPMENT

A poorly maintained signal system results in frequent breakdowns and brings discredit to the responsible authority. Because traffic control devices are designed to provide for the safe and efficient movement of vehicles and pedestrians, it is essential that the devices operate reliably and continuously. Emergency maintenance or repair can be very costly, especially if it must be done outside normal working hours. It is much better to *choose* the time when maintenance is done.

ROUTINE MAINTENANCE

Routine maintenance, according to a planned program, can be made to effect definite economies in the operation of any signal system. A simple preventive maintenance program requires: (1) regular replacement of lamps before the end of their rated life, (2) overhaul of the moving parts in controllers, (3) repainting equipment and hardware and routine checks of voltage, and (4) inspection of the condition of equipment.

MODIFICATIONS

When new equipment is being acquired, purchasing facilities for future expansion will reduce the cost of modifications. It makes little sense, for example, to purchase a pretimed single dial controller when future requirements may dictate using an expansible controller. The use of plug-in components, whether electrical or electronic, makes for ease of maintenance when personnel is semiskilled, because parts may be substituted directly in a minimum of time.

If traffic and accident studies indicate shortcomings in a traffic control installation, it is essential that the necessary modifications be carried out with a minimum of delay. Frequently, an overall upgrading of an installation may be done during modification. Here the value of having anticipated future needs becomes apparent. Underground conduits may be too small to accommodate extra cables; poles may not have sufficient strength to support longer mast arms or signal heads; and cables may have insufficient conductors of the correct gauge to permit modifications without extensive updating of existing equipment.

ACCIDENTAL DAMAGE

A signal installation is exposed to the elements and in some cases to the risk of damage by vehicles. Considerable accidental damage can occur. Poles and controllers are most susceptible to being hit by vehicles, and high wind loads may twist and damage signal heads. Overhead cables are particularly vulnerable to damage and their failure results in loss of some, if not all, signal indications. The importance of locating equipment so that accidental damage is minimized cannot be too strongly stressed.

When damage to a signal installation occurs, it should be repaired immediately so that orderly control of traffic is restored. Traffic control devices are legally installed to assign right of way and failure to keep them operative in a proper manner may be construed as negligence on the part of the traffic authority. In order to minimize downtime caused by accidents, spare controllers, temporary stands for signal heads, emergency flashers, and a fully equipped maintenance vehicle should always be ready for emergencies. If an installation is going to be out of service for some time, police assistance should be sought and consideration given to the erection of stop signs to provide control.

RECORDS

Detailed maintenance records on individual installations should be kept to determine the time interval for routine maintenance and cost analysis. These records will provide information on equipment, reliability, manufacturers' service, susceptible component failure and will also provide a documented history in the event of litigation involving the responsible authority.

EFFECTIVENESS OF SIGNALS

DRIVER OBSERVANCE

If traffic control and other regulatory signals are to be effective, their indications must be strictly observed and legislation must be enacted specifying driver and pedestrian requirements and penalties for failure to comply. The enforcement of this legislation must be a police function, and, therefore, can only be successfully undertaken when the signals are installed in conformity with certain minimum standards and by a duly authorized public body.

This requires that signals be installed only when they have been authorized by the appropriate legislative authority and in accordance with legislation established to regulate their design, placement, and maintenance. Greatest observance is achieved by uniformity in the application and appearance of traffic signals throughout the whole country; thus, best results will be achieved when the requirements are in conformity with the recommendations contained in the *Manual on Uniform Traffic Control Devices for Streets and Highways.*

ENFORCEMENT

The primary function of a traffic signal is to promote the safe, orderly, and efficient movement of traffic. This is also a police responsibility, and, therefore, a most important reason for installing the signal is to assist police in carrying out their duties and to relieve them of the onerous and often hazardous task of manually directing traffic. To be of the utmost value, the signal must be operating in the most efficient manner possible and it is quite definitely in the interest of the police to ensure that it is. Therefore, they should be prepared to devote time and thought to the subject and to cooperate with the traffic engineering department or other agency responsible for installing and maintaining the signals by providing them with information on both physical and operational defects.

Police enforcement can materially determine the effectiveness of the signals. Signals installed in areas where police enforcement is known to be persistent will usually result in greater safety and efficiency of traffic operations than in areas where enforcement is lax.

STANDARDIZATION AND INNOVATION

On the subject of uniformity and standardization, the *Manual on Uniform Traffic Control Devices* says:

> Uniformity of traffic control devices simplifies the task of a road user because it aids in recognition and understanding. It aids road users, police officers and traffic courts by giving everyone the same interpretation. It aids public highway and traffic officials through economy in manufacture, installation, maintenance and administration.
>
> Simply stated, uniformity means treating similar situations in the same way. The use of uniform traffic control devices does not, in itself, constitute uniformity. A standard device used where it is not appropriate is as objectionable as a nonstandard device; in fact, this may be worse, in that such misuses may result in disrespect of those locations where the device is needed.[28]

The need for uniform standards was recognized long ago and has resulted in the creation of national and international guides on uniform standards such as those contained in the *Manual on Uniform Traffic Control Devices*, the *Uniform Vehicle Code*, and the *Model Traffic Ordinance* (see also Chapter 22). It is important, however, to differentiate carefully between standardization and stagnation. There must be room for innovation, experimentation, and possible improvements that will lead to the revision of standards from time to time.

Experimentation should be carefully controlled to ensure that it does not constitute an attempt to introduce change merely for the sake of change. Experimentation should

---

[28] *Manual on Uniform Traffic Control Devices for Streets and Highways*, p. 4.

normally arise from a conclusion that the existing standard or application fails to meet all necessary requirements. The following procedures are outlined in the *Manual on Uniform Traffic Control Devices* for handling of interpretations, experimentation, and changes to the Manual:

1. The request for any clarification, permission to experiment or change in the Manual should be addressed to one of the appropriate organizations of the National Joint Committee who will review the request and advise the Federal Highway Administration. These organizations include the American Association of State Highway Officials; the Institute of Traffic Engineers; the National Association of Counties; the National Committee on Uniform Traffic Laws and Ordinances, and the National League of Cities. The action on the request will be by the Federal Highway Administration.
2. The Manual outlines further the kind of information which should be contained in the request, and the form in which rulings may be given.

The following few examples are submitted to show how experimentation and innovation have been tried.

Split-phase signalling (advanced or extended green interval). In an attempt to indicate more clearly to approaching motorists the portion of a green phase that is advanced or extended beyond that shown to the opposing traffic, the green interval during that portion of the phase which is exclusive has been flashed. By doing so, it clearly indicates that the green signal and its meaning still apply; however, because it is being flashed it is differentiated from the steady green signal. In addition, the flashing interval indicates clearly when the extended or advanced portion of the green phase commences and terminates. This system is now commonly used in the Province of Ontario, Canada, but it is not authorized under the provision of the *Manual on Uniform Traffic Control Devices.*

Time-elapsed signals. There have been persistent attempts to indicate in some manner the amount of green that still remains. Some experiments have used a clock method showing a dial with a moving arm that rotates completing a 360° sweep in the interval allotted to the green phase. Another method that has been tried shows a dark shadow crossing the green lens so that it would complete the sweep from one side to the other at the conclusion of the green phase. None of these methods has met with sufficient success for introduction in the standards.

Green wave (pacer) system using speed indications. So many of the signals today on streets and highways are interconnected in order to provide for progressive movement of traffic that there has evolved (principally in Europe) an innovation that uses a standard signal head placed in advance of the intersection signals with the standard red, yellow, and green lenses replaced by three lenses indicating appropriate speeds. These are illuminated at the appropriate moment in time so that the oncoming motorist is advised of the speed required to permit him to proceed through the intersection without interruption on a green signal.

The Green Wave System was first introduced by Dr. Von Stein in Dusseldorf, Germany in 1954 under the name of *Dusseldorf Traffic Funnel* (see Figure 17.28). Stemming from this development and following a presentation made by Dr. Von Stein

Figure 17.28. Speed indications, Dusseldorf Traffic Funnel. (Source: Herr Von Stein, Landeshaupt Stadt Dusseldorf.)

in Detroit in 1959, an experiment of this system (*Traffic Pacer System*) was introduced on Mound Road in Macomb County, Michigan in July 1961. The results of this experiment did not indicate any substantial improvement in traffic flow. This operation is not authorized under the provision of the *Manual on Uniform Traffic Control Devices.*

ACCIDENT REDUCTION

The common acceptance of traffic signals as a solution to all right-of-way problems has perhaps naively led many people to the conclusion that by simply introducing a traffic signal one can prevent or eliminate all intersectional accidents. Traffic studies in urban areas indicated that in a majority of locations studied accidents increased subsequent to the installation of traffic control signals. It is not the purpose at this time to examine this statement rigorously. In Special Report 93 of the Highway Research Board,[29] however, there is reported a study undertaken in 1959 in Cincinnati involving all signals installed between the years 1950 and 1958. This study indicated that accidents increased at 102 intersections, decreased at 23, and that no significant change was noted at 27 intersections. Another study in Cinicinnati in 1967 involving all signals installed between the years 1959 and 1964 showed that accidents increased at 10 intersections and decreased at 22. The exact reason for the change is not clearly understood, but it can be hypothesized that more careful studies, better adherence to the warrants, and more careful design of the physical aspects of the installation are tending to improved accident results following traffic signalization.

Despite all this, however, it is still clear that in a substantial number of cases there is an increase in accidents following the installation of traffic signals; therefore, it is very important that in answer to a request to install traffic signals for the primary purpose of reducing accidents, particular care must be taken with the traffic surveys

---

[29] *Improved Street Utilization Through Traffic Engineering,* Highway Research Board Special Report No. 93 (Washington, D.C.: Highway Research Board, 1967), pp. 84–87.

and examination to determine the probability that accidents will be reduced. If a decision is then made to install signals, extra care must be taken in the physical design and installation of signals.

REFERENCES FOR FURTHER READING

Manual on Uniform Traffic Control Devices for Streets and Highways, U.S. Department of Transportation. Washington, D.C., 1971, Part IV Signals.

Manual on Uniform Traffic Control Devices for Canada (2nd ed. rev.) Ottawa, Canada, Runge Press, Ltd. Roads and Transportation Association of Canada, 1971, Part B, Traffic Signals.

"Colors of Signal Lights," *NBS Monograph* **75**, U.S. Department of Commerce, 1967.

"Intersections," *Traffic Control and Roadway Elements*. Washington, D.C.: Highway Users' Federation for Safety and Mobility, 1970, Chapter 4.

HUPPERT, WILLIAM W., "Familiarity with Loop Detectors," *Traffic Engineering*, XXXV, No. 11 (1965), p. 21.

"Traffic Signal Visibility," New York Metropolitan Section, Institute of Traffic Engineers, Section Technical Report, July, 1970.

MATSON, T. M., W. S. SMITH, and F. W. HURD, *Traffic Engineering*, Civil Engineering Series. New York: McGraw-Hill Book Company, 1955, pp. 324–54.

WEBSTER, F. V., and B. M. COBBE, "Traffic Signals," Road Research Technical Paper No. 56. London, England: Her Majesty's Stationery Office, 1966.

BARTLE, R. M., VAL SKORO, and D. C. GERLOUGH, "Starting Delay and Time Spacing of Vehicles Entering Signalized Intersections," *Effects of Traffic Control on Street Capacity*, Bulletin 112. Washington, D.C.: Highway Research Board, 1956, pp. 33–41.

HEWTON, J. T., *Traffic Signals*, Traffic Training Course. Ontario Traffic Conference; King Edward-Sheraton Hotel, 37 King St. E., Toronto, Ontario, Canada M5CIE9.

WOHL, M., and B. V. MARTIN, *Traffic System Analysis for Engineers and Planners*. New York: McGraw-Hill Book Company, 1967, pp. 426–95.

BONE, A. J., B. V. MARTIN, and T. N. HARVEY, "The Selection of a Cycle Length for Fixed Time Signals," Department of Civil Engineering Research Report R 62–37. Cambridge, Mass.: Massachusetts Institute of Technology 1962.

WEBSTER, F. V., "Traffic Signal Settings," Road Research Technical Paper 39. London, England: Road Research Laboratory, 1958.

MORGAN, J. T., and J.D.C. LITTLE, "Synchronizing Traffic Signals for Maximal Bandwidth," Operations Research Volume 12, No. 6, 1964.

Improved Street Utilization Through Traffic Engineering, Special Report No. 93. Washington, D.C.: Highway Research Board, 1967, pp. 84–143.

An Introduction to Highway Transportation Engineering. Washington, D.C.: Institute of Traffic Engineers, 1968, pp. 121–33.

Chapter 18

SPEED REGULATIONS AND
OTHER OPERATIONAL CONTROLS

ROBERT E. TITUS, Head of Environmental Review Unit, West Virginia Department of Highways, Charleston, West Virginia

This chapter discusses speed controls, both regulatory and advisory, and miscellaneous operational controls such as one-way streets, reversible lanes and roadways, exclusive bus lanes, turn regulations, barriers and malls, and similar controls that the traffic engineer uses to improve traffic flow.

The sections on speed control present factors that should be considered in the development and application of regulatory speed limits and of advisory speed indications. The public attitude toward speed controls is discussed; relationships between speed and accident frequency and severity are presented, as is the matter of safe speeds in relation to kinds of roadway, location, and condition. Almost all aspects of establishing speed limits are covered, including legal requirements and authority, the different kinds of speed limits, studies prerequisite to the establishment of limits, criteria for setting speed limits, and the factors involved in the actual determination and posting of the numerical limits.

The use of advisory speed indications is explored. In addition, special problems involving regulatory speed limits, day vs. night, weather conditions, and others are considered. Finally, the effectiveness of the speed limits and several aspects of enforcement are mentioned.

The sections on miscellaneous operational controls cover a wide range of special operational controls that are used primarily in urban areas to increase street capacity, improve vehicular and pedestrian safety, and generally to increase efficiency of street and highway usage.

SPEED REGULATION, ZONING, AND ENFORCEMENT

Speed regulations and speed limits restrain a driver's freedom to drive at any speed he wishes. Speed limits should be imposed, then, only when they will promote better traffic flow or increase safety. If drivers do not recognize particular speed limits as being reasonable, the limits will be disrespected and ineffective. Speed regulations imposed only where and when necessary will be readily accepted.

In North American countries the practice of posting different speed limits for varying roadway and environmental conditions has been traditional. This general method requires careful analysis of each section of highway to determine the appropriate speed limit to be posted. A highway of considerable length may require a wide range of speed limits.

Many European and other countries have tended historically either to enforce speed limits only in hazardous rural areas or in urban areas or to enact single speed limits by statute to cover all highways of a certain kind in a given area. The practice of setting special speed limits for specific roadway conditions is relatively new. In some cases (Route M-1 in England, for example) highways were or are operated with no speed limit whatever. This total lack of speed control has generally proven unsafe.

SPEED REGULATIONS

FACTORS AFFECTING SPEED REGULATIONS

Public attitude. The traffic engineer will receive many requests for establishing new speed limits or for altering existing limits upward or downward. Such requests often reflect citizen opinion that something is wrong with a particular section of highway or with the operation of traffic thereon. A request for a revised speed limit, usually lower than the limit posted, is sometimes the only immediate solution that the public can offer. Such requests often are based on the misconception that almost all motorists will automatically exceed the posted limit by 5 or 10 mph and that the only way to reduce speeds is to reduce the speed limit. Citizens, acting as individuals or groups, will frequently request lower speed limits for their own neighborhood streets than they, as drivers, would consider reasonable in similar neighborhoods elsewhere.

The consensus of traffic engineers in the United States is that motorists usually adjust their speeds according to conditions on the road and not necessarily to posted speed limits. Hence, if unreasonably low limits are posted, the limit will be violated by large numbers of drivers. This leads to disrespect of other posted limits as well.

Studies of speed in Europe have shown, almost without exception, that the speed of vehicles can be considerably reduced by installing a speed limit. Experience in the United States indicates that drivers do not drastically alter speed patterns with changes in speed limits. One possible explanation is that European experience generally deals with the application of speed limits for the first time, but the United States experience usually deals with revision of existing speed limits.[1]

Public reaction to the imposition of speed limits varies. In 1971 West Germany proposed the imposition of a 100 kph (62 mph) speed limit on two-lane rural roads where previously no speed limit had been posted. The purpose was to reduce West Germany's high accident rate. The general public reaction was one of anger.[2] In other instances, speed limits have been welcomed.

Accident frequency and severity vs. speed. Various safety campaigns aimed at drivers have attempted to persuade them that speed is the cause of almost all accidents, and that if speed can be controlled, accidents will be prevented or reduced. Although excessive speed has often been listed in police reports as the cause or major contributing factor in accidents, the real problem is driving too fast for prevailing conditions.

Statistics have generally shown that the imposition of a speed limit in an urban

---

[1] DONALD C. CLEVELAND, "Speed and Speed Control," *Traffic Control and Roadway Elements—Their Relationship to Highway Safety/Revised*, Chapter 6 (Washington, D.C.: Highway Users Federation for Safety and Mobility, 1970), p. 6.

[2] ALICE SIEGERT, "Speed Limits Irk Germans," *Chicago Tribune*, October 11, 1971, Sec. 1-A, p. 3.

area leads to a reduction in serious injury rate and in the overall accident rate on a specific highway section. The most marked general effect of the imposition of speed limits in urban areas in several European countries has been a reduction in fatal accidents. The effect on slight-injury or property-damage-only accidents is much smaller.[3] In Kent County, England, in a new study of 40 sites where the speed limit was raised from 30 mph to 40 mph, accidents were reduced by about 20 percent. The 85-percentile speed decreased at 20 of the 40 sites and increased at eight.

Figure 18.1, taken from a study made by the Bureau of Public Roads (now Federal Highway Administration), reveals some interesting findings regarding accident involvement vs. speed on main rural highways, not including freeways. Accident-involvement rates are the highest at very low speeds, are lowest at about the average speeds, and increase again at very high speeds. A principal conclusion is that the more a driver deviates from the average speed of traffic, the greater his chance of being involved in an accident.[4]

In England, where in urban areas when the 30 mph limit was poorly observed, there was no significant change in accident experience when a 40 mph limit was installed.[5]

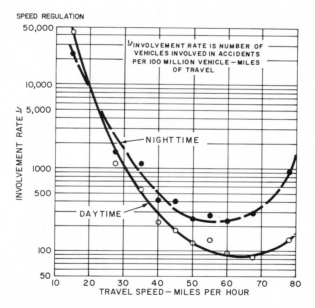

Figure 18.1. Involvement rate by travel speed, day and night. [Source: David Solomon, *Accidents on Main Rural Highways Related to Speed, Driver, and Vehicle*, U.S. Department of Commerce, Bureau of Public Roads (Washington, D.C.: U.S. Government Printing Office, 1964), p. 1.]

---

[3] *The State of the Art of Traffic Safety* (Cambridge, Mass.: Arthur D. Little, Inc., 1966), p. 171.

[4] DAVID SOLOMON, *Accidents on Main Rural Highways Related to Speed, Driver, and Vehicle*, U.S. Department of Commerce, Bureau of Public Roads (Washington, D.C.: U.S. Government Printing Office, July 1964), p. 1.

[5] *How Fast? A Paper for Discussion*, Ministry of Transport (London: Her Majesty's Stationery Office, 1968), p. 10.

"SAFE" SPEEDS FOR VARIOUS CONDITIONS

Although much information directed at the driver involves use of the term "safe speed," the term is purely relative and depends on many conditions and the situation involved. A "safe speed" in one location may not be safe in another, and a "safe speed" at a specific time at one location may not be safe under other conditions at the same location.

Safe speeds may be recommended at spot locations through the use of advisory-speed plates. Such plates are often used in conjunction with curve warning signs (described later). Advisory plates are also used to indicate "safe," or reasonable, speeds through intersections where sight distance is restricted, through rough road areas, and for other similar conditions where hazardous conditions exist.

Roadway type and condition. Higher speeds are relatively safer on roadways with high design standards—wide lanes, absence of sharp curves, adequate sight distance, absence of driveways and intersecting crossroads, and clear roadsides—conditions such as exist on freeways. Average speeds for various kinds of vehicles by highway type in the United States are shown in Table 18.1. Roadway surface conditions are also a significant factor affecting "safe" speed, especially surface characteristics which makes one surface more slippery than another when wet.

TABLE 18.1
Average Speed of Vehicles by Kind of Highway, 1971 (U.S.)

Kind of Highway	All Vehicles	Average Speed Passenger Car	Truck	Bus
Main rural	60.6	62.0	56.1	60.2
Rural insterstate	64.7	66.3	59.4	64.5
Rural primary	57.7	58.9	54.0	57.5
Urban interstate	56.0	57.2	52.6	55.7

Source: "Traffic Speed Trends" (Washington, D.C.: U.S. Department of Transportation, Federal Highway Administration, 1972).

Urban and rural locations. "Safe" driving speeds are also affected significantly by urban vs. rural conditions. "Safe" speeds on urban streets tend to be much lower than on rural highways because of the houses, businesses, and other kinds of development and increased marginal friction that slows traffic.

Weather conditions. Weather is also an important factor affecting "safe" speed. The most significant condition is the presence of snow and ice on the pavement. Rain and fog appear to have less influence. Data obtained at selected sites on California freeways and expressways in both day and night conditions indicate that the effects of fog on traffic flow are not large, mean speeds being reduced only from 5 to 8 mph.[6] In extremely dense fogs, of course, traffic may be slowed to crawl speeds. Even heavy rain does not appear to have the same influence because sight distance is not reduced too much. Heavy rains, however, have caused drivers to pull off the roadway and stop because of the hazardous driving conditions.

---

[6] *Highway Fog*, National Cooperative Highway Research Project Report 95 (Washington, D.C.: Highway Research Board, 1970), p. 3.

SPEED ZONING

DEFINITION

The *Uniform Vehicle Code* contains the following provision:

> Whenever the (State highway commission) shall determine upon the basis of an engineering and traffic investigation that any maximum speed herein before set forth is greater or less than is reasonable or safe under the conditions found to exist at any intersection or other place or upon any part of the State highway system, said (commission) may determine and declare a reasonable and safe maximum limit thereat, which shall be effective when appropriate signs giving notice thereof are erected.[7]

Almost all states in the United States under this or similar authority permit state or local officials to establish speed regulations at specific locations. This is usually done on the basis of a traffic engineering investigation, and the revision usually takes the form of modifying the basic state speed limits. The establishment of the regulations at specific locations is termed *speed zoning*.

KINDS OF SPEED CONTROL

There are two basic speed controls: (1) regulatory limits that have the effect of law and are enforceable and (2) advisory limits that are not enforceable but warn motorists of suggested safe speeds for specific conditions at a specific location.

Regulatory controls. Speed regulation may be classified as (1) statewide regulations established by legislative authority and generally applicable throughout the state and all political subdivisions and (2) zoned speed regulations for specific locations established by administrative action.

Each state has some form of the basic rule which, in principle, defines the maximum safe speed. Fundamentally, the basic rule states that:

> No person shall drive a vehicle at a speed greater than is reasonable and prudent under the conditions and having regard to the actual and potential hazards then existing. Consistent with the foregoing, every person shall drive at a safe and appropriate speed when approaching and crossing an intersection or railroad grade crossing, when approaching and going around a curve, when approaching a hill crest, when traveling upon any narrow or winding roadway, and when special hazards exist with respect to pedestrians or other traffic or by reason of weather or highway conditions.[8]

A variation of this definition is that "no person shall drive a vehicle in or upon any highway at a greater speed than will permit him to bring it to a safe stop within the assured clear distance ahead."

There are two basically different numerical maximum speed limits: (1) maximum lawful or *absolute* limit and (2) prima facie limit. An *absolute* speed limit is a limit above which it is unlawful to drive regardless of roadway conditions, the amount of traffic, or other influencing factors. A prima facie speed limit is a limit above which the driver is presumed to be driving unlawfully but where, if charged with a violation, he may contend that his speed was safe for conditions existing on the roadway at that

---

[7] *Uniform Vehicle Code, Revised—1968*, National Committee on Uniform Traffic Laws and Ordinance (Charlottesville, Va.: The Michie Company, 1968), p. 156.
[8] *Uniform Vehicle Code*, p. 155.

time and, therefore, was not guilty of a speed violation. Enforcement officials prefer the absolute limit because it is much easier to enforce.

The *Uniform Vehicle Code* presently provides the following maximum speed limits:

1. Thirty miles per hour in any urban district.
2. Sixty miles per hour in other locations during the daytime.
3. Fifty-five miles per hour in such other locations during the nighttime.[9]

In many locations minimum speed limits are also posted. A minimum speed limit prohibits driving "at such a slow speed as to impede the normal and reasonable movement of traffic except when reduced speed is necessary for safe operation or in compliance with law."[10] On freeways, the minimum speed limit is often 40 mph.

Advisory controls. Advisory speed limits are not enforceable but merely warn motorists of suggested safe speeds for specific conditions on a highway. They may be posted in the form of advisory speed plates (see Figure 18.2) generally used as a supplementary panel with a warning sign. Advisory speed limits are posted in multiples of 5 mph. In some court jurisdictions, violation of advisory speeds may be admitted as evidence that the driver was operating in a reckless manner.

Figure 18.2. A typical application of the advisory speed plate. (Source: West Virginia Department of Highways.)

Should drivers find that some curves may be negotiated safely, particularly in dry weather conditions, at speeds somewhat above the posted advisory speed, they may tend to exceed the posted advisory limit on all curves. It is important, therefore, that all curves be posted consistently so that those drivers do not lose control on any particular curve.

REGULATORY SPEED LIMITS

Initiating speed regulations. Speed limits must be based on sufficient engineering and traffic data to justify their reasonableness. Traffic officials are often called upon to testify in court cases regarding speed limits which have been enacted and they must support their testimonies with data accumulated prior to the establishment of

---

[9] *Uniform Vehicle Code*, pp. 155–56.
[10] *Uniform Vehicle Code*, p. 157.

specific speed limits. This information should be gathered both in a qualitative and quantitative manner which will justify the actions taken.

Requisite studies. The *Uniform Vehicle Code* requires that an engineering and traffic investigation shall be the basis for altering any maximum speed limit set forth in the Code. Almost all state laws contain a similar provision. What constitutes "an engineering and traffic investigation" is not described in the statute or in the *Uniform Vehicle Code*. Judgment must be used to select the pertinent data. Because posted speed limits apply to normal roadway conditions (dry pavement, good visibility, roadway uninhibited by traffic congestion or accidents) those data should be collected so that they truly indicate what would be considered normal maximum speed under such conditions.

The following factors should be considered, and appropriate data gathered, in establishing speed limitations:

1. Prevailing vehicle speeds
   a. 85-percentile speed
   b. Pace
   c. Average test run speeds
   d. Speed distribution data
2. Physical features
   a. Design speed
   b. Measurable physical features
      1. Maximum comfortable speed on curves
      2. Spacing of intersections
      3. Number of roadside businesses per mile
   c. Roadway surface characteristics and conditions
      1. Slipperiness of pavement
      2. Roughness of pavement
      3. Presence of transverse dips and bumps
      4. Presence and condition of shoulders
      5. Presence and width of median .
3. Accident experience
4. Traffic characteristics and control
   a. Traffic volumes
   b. Parking and loading vehicles
   c. Commercial vehicles
   d. Turn movements and control
   e. Traffic signals and other traffic control devices that affect or are affected by vehicle speeds
   f. Vehicle-pedestrian conflicts[11]

The spot speed check whether only free-moving vehicles or whether all vehicles were recorded. A free-moving vehicle is one in which the driver is not restricted by other vehicles in selecting his speed. Observations should be restricted to those vehicles having at least from 6- to 9-sec headways from those ahead and making no apparent effort to overtake and pass them.

---

[11] For additional information on these factors and their application to speed zoning, see "An Information Report on Speed Zoning," *Traffic Engineering*, XXXI, No. 10 (1961), pp. 39–44.

The 85-percentile speed as determined by spot speed studies is the principal factor generally used by traffic engineers to determine speed limits. Although this method is highly satisfactory on streets and highways carrying moderate to heavy volumes of traffic, it is difficult to apply on low-volume roads because of the time consumed in gathering the necessary number of observations. In such cases, trial runs can serve as a satisfactory substitute.

Criteria for establishing speed limits. The Traffic Committee for the American Association of State Highway Officials adopted in 1970 the following policy statement for the establishment of speed zones:

> The 85th percentile speed is to be given primary consideration in speed zones below 50 miles per hour, and the 90th percentile speed is to be given primary consideration in establishing speed zones of 50 miles per hour or above. To achieve the optimum in safety, it is desirable to secure a speed distribution with a skewness index approaching unity.

Signing for speed limits. Signing for speed limits should be consistent with the appropriate sections of the latest edition of a manual on uniform traffic control devices, or an equivalent, used in each country (see Chapter 16, Traffic Signs and Markings).

Signs for speed limits are erected at varying intervals, depending on highway type and general location. In urban areas, speed limit signs are usually erected at intervals not exceeding one-half mile if the speed limit is 40 mph or less. On freeways and in rural areas, frequency of signing varies considerably, with intervals between signs usually ranging from one to five miles.

DETERMINATION OF ADVISORY SPEED INDICATIONS

Two basically different methods are available for determining advisory speed limits on horizontal curves: (1) by trial speed runs with a test vehicle or (2) by office calculation. Either method is satisfactory, but field runs to check the office calculations are desirable in any event.

The trial speed runs method involves using a vehicle equipped with a ball-bank indicator to show the combined effect of the body roll angle, the centrifugal force angle, and the superelevation angle. Safe speeds on curves are indicated by ball-bank readings of 14° for speeds below 20 mph, of 12° for speeds between 20 and 35 mph, and of 10° for speeds of 35 mph and higher. Also, 10° is safe for 50 mph and even 60 mph, but for higher speeds a smaller reading should be used.[12]

In using the office method for determination of advisory speed, the advisory speed indication for a curve may be calculated by the following formula:

$$V = \sqrt{\frac{(e+f)R}{0.067}} = \sqrt{15(e+f)R}$$

where V = advisory speed of vehicle in mph,
$\quad e$ = superelevation in ft per ft of horizontal width,
$\quad f$ = transverse coefficient of friction,
$\quad R$ = radius of curvature in ft.

---

[12] *A Policy on Geometric Design of Rural Highways* (Washington, D.C.: American Association of State Highway Officials, 1965), p. 154.

The formula is solved for the advisory speed on the curve. The resulting speed should be rounded to the nearest 5 mph for signing in the field.

If D is used as the degree of circular curve, arc definition, $D = \dfrac{5730}{R}$ and the standard formula becomes:

$$V = \sqrt{\frac{85{,}900(e + f)}{D}}$$

Using this formula, and with given values for e and f, the minimum radius and the maximum degrees of curvature for the designated design speeds have been computed as shown in Table 18.2.

TABLE 18.2
Maximum Degree of Curve and Minimum Radius Determined
for Limiting Values of e and f

Design Speed	Maximum e	Maximum f	Total $(e + f)$	Minimum Radius	Max. Degree of Curve	Max. Degree of Curve, Rounded
30	.06	.16	.22	273	21.0	21.0
40	.06	.15	.21	508	11.3	11.5
50	.06	.14	.20	833	6.9	7.0
60	.06	.13	.19	1263	4.5	4.5
65	.06	.13	.19	1483	3.9	4.0
70	.06	.12	.18	1815	3.2	3.0
75	.06	.11	.17	2206	2.6	2.5
80	.06	.11	.17	2510	2.3	2.5
30	.08	.16	.24	250	22.9	23.0
40	.08	.15	.23	464	12.4	12.5
50	.08	.14	.22	758	7.6	7.5
60	.08	.13	.21	1143	5.0	5.0
65	.08	.13	.21	1341	4.3	4.5
70	.08	.12	.20	1633	3.5	3.5
75	.08	.11	.19	1974	2.9	3.0
80	.08	.11	.19	2246	2.5	2.5
30	.10	.16	.26	231	24.8	25.0
40	.10	.15	.25	427	13.4	13.5
50	.10	.14	.24	694	8.3	8.5
60	.10	.13	.23	1043	5.5	5.5
65	.10	.13	.23	1225	4.7	4.5
70	.10	.12	.22	1485	3.9	4.0
75	.10	.11	.21	1786	3.2	3.0
80	.10	.11	.21	2032	2.8	3.0
30	.12	.16	.28	214	26.7	26.5
40	.12	.15	.27	395	14.5	14.5
50	.12	.14	.26	641	8.9	9.0
60	.12	.13	.25	960	6.0	6.0
65	.12	.13	.25	1127	5.1	5.0
70	.12	.12	.24	1361	4.2	4.0
75	.12	.11	.23	1630	3.5	3.5
80	.12	.11	.23	1855	3.1	3.0

Source: *A Policy on Geometric Design of Rural Highways* (Washington, D.C.: American Association of State Highway Officials, 1965), p. 158.

Safe speeds determined by these methods may need to be modified by other factors. For example, the safe-stopping sight distance around the curve may require a more restrictive speed than the curvature itself. In this case, it would be advisable to post the advisory speed at the lesser speed (see Chapter 14, Geometric Design).

SPECIAL PROBLEMS

Day vs. night speed limits. At least 17 states in the United States use reduced maximum speed limits at night.[13] For example, a 60-mph limit during daylight might be 50 mph at night. Reduced sight distance after sundown is apparently the basis for this speed limit reduction.

No specific warrants exist for use by the traffic engineer in determining whether or not a nighttime speed reduction is desirable. In the United States there is no uniformity in application of such limits. It appears generally to be a matter of practice, or law, in each state.

Differential limits by kind of vehicle. Some jurisdictions have laws or follow the practice of posting different speed limits for different kinds of vehicles. Differential limits are most common for (1) passenger cars, (2) trucks, and (3) buses. Some jurisdictions also post a reduced limit for towed vehicles, such as trailers, wrecked vehicles, or race cars.

Witheford's study reports that 76 percent of the states responding to a questionnaire post different limits for cars and trucks. Differentials are more likely to occur on at-grade rural highways than on freeways and urban streets. The merits of differential speed limits are still debated. Proponents contend that reduced speed is desirable for larger vehicles because their operating characteristics, e.g., stopping distance, are not as good as for passenger cars. Opponents, on the other hand, argue that a differential limit creates a built-in hazardous condition. Such variance in speed is apparently undesirable as is evidenced by the results of the study by the Federal Highway Administration (see Figure 18.1).

Speed limits for adverse weather conditions. The basic speed law requires the driver to adjust his speed to existing road conditions. The primary responsibility for accommodating to adverse weather conditions thus rests with the driver. Nevertheless, some jurisdictions, e.g., New Jersey, have found it desirable primarily for safety reasons to reduce speed limits at specific locations during adverse weather conditions by means of internally illuminated signs. Such practice is generally limited to freeways or expressways.

Variable speed limits by lanes on freeways. In order to improve the quality and safety of traffic flow, the use of different speed limits for various lanes of a highway has been tried, principally on freeways or expressways. Generally, the practice is to post the higher limits on lanes closer to the median. Figure 18.3 illustrates a minimum speed limit in the left lane only on a freeway.

One study reports that using changeable speed limit signs during the off-peak

---

[13] DAVID K. WITHEFORD, *Speed Enforcement Policies and Practice* (Saugatuck, Conn.: ENO Foundation for Transportation, 1970), p. 18.

Figure 18.3. Illustration of the posting of a minimum speed limit in the left lane only on a freeway. (Source: Tennessee Department of Transportation.)

period produced negligible benefits and that the use of these devices during the peak period was judged to have produced essentially no effect.[14]

An underpass in Aalborg, Denmark has been equipped with illuminated speed indicators with changeable figures. Normally, there are no speed limits in the underpass itself or on its approaches, but they can be applied by using the illuminated signs in case of emergencies, for example, when a lane is blocked by a stalled car.[15]

Schools. *The Uniform Vehicle Code* and the *Model Traffic Ordinance* have not contained provisions for regulating vehicle speeds in school zones since 1934. However, numerical speed limits are mentioned in the state codes of at least eleven states, and judging from Table 18.3, may appear in many others.

A West Virginia study indicated that the most significant factors influencing speeds in school zones were the approach speed limit, the distance of school buildings from the roadway edge, traffic volumes, and the length of the school zone.[16]

TABLE 18.3
Urban School Zone Speed Limits

Group	Percent Reporting Indicated Limit (mph)					
	15	20	25	30	Other*	Total
States	42	15	12	0	31	100
Cities over 100M	30	31	27	2	10	100
Cities 50–100M	26	27	36	1	10	100
Cities 25–50M	33	33	24	3	7	100

*Other: School zone speed limits not applicable or not specified.

Source: David K. Witheford, *Speed Enforcement Policies and Practice* (Saugatuck, Conn.: ENO Foundation for Transportation, 1970), p. 26.

[14] GERALD C. HOFF, "A Comparison Between Selected Traffic Information Devices," Chicago Area Expressway Surveillance Project, Report No. 22 (Illinois Division of Highways: Chicago, 1969), p. 7.
[15] NIELSEN, B. P., "Speed and Lane Signals on Danish Motorway," *Traffic Engineering and Control*, XII, No. 1 (1970), p. 34.
[16] *Establishing Criteria for Speed Limits in School Zones* (Morgantown, Engineering Experiment Station, West Virginia University, 1967), p. 61.

SPEED ENFORCEMENT

Almost all enforcement agencies regard their speed enforcement role as contributing primarily to improved highway safety, even if the immediate objective of enforcement is regulatory or punitive.

Various methods are used in speed enforcement. The predominant technique involves some form of electronic device, radar, for example. The oldest method involves trailing offenders by car. The newest method involves surveillance by aircraft. The latter requires timing a vehicle through a marked section of highway of known length and calculating the speed of the vehicle from the passage time.

Driver reaction to posted speed limits is influenced by many factors. Among these are the reasonableness of a speed limit, the degree of enforcement in a particular area, and the tolerance permitted by enforcement agencies before making arrests. Degree of enforcement varies widely from place to place. Some localities are known for strict enforcement of speed limits, and drivers in those localities tend to observe the speed limit better than in places where enforcement is lax. Such variations cause some locations to be called "speed traps." The term implies that speed enforcement is overly strict. Because strict or good enforcement of speed limits is usually desirable, the term speed trap properly should be applied only when strict enforcement is coupled with a section of highway having an unreasonably low speed limit. Variable tolerances above posted speed limits allowed by police officials is a related problem. Some enforcement agencies allow tolerances of up to 15 mph before making arrests. The unfamiliar driver does not know the amount of tolerance allowed.

A carefully controlled study on the effect of different degrees of enforcement on accidents, traffic diversions, and speed indicated that although there was no significant change between test and control routes in average speed or percentage of drivers exceeding the speed limit, there was a significant reduction in the variance of speeds, directly related to the enforcement level. However, no effects were found on traffic diversion or accidents.[17]

Extensive information on the subject of speed enforcement is contained in the publication, *Speed Enforcement Policies and Practice*, published by the ENO Foundation for Transportation, Saugatuck, Connecticut, 1970.

ONE-WAY STREETS

Nearly all streets and highways were originally designed for use as two-way streets and were used as such. With increasing traffic volumes, particularly in urban areas, various means were sought to improve existing street usage without having to construct entirely new streets or street systems. One of those means was the conversion of two-way streets to one-way traffic operations.

Because one-way streets can handle traffic in only one direction, they are commonly operated one in each direction. The two streets are generally located parallel to and as near to each other as possible in order to reduce confusion to motorists and to facilitate traffic flow. A pair of one-way streets resembles a divided highway with a wide median strip, the wide median being the block or series of blocks separating the two parallel one-way streets.

---

[17] Richard M. Michaels, "The Effect of Enforcement on Traffic Behavior," *Public Roads*, XXXI, No. 5 (1960), pp. 109–13, 124.

ONE-WAY STREET OPERATIONS

One-way streets are generally operated in one of three ways:

1. A street on which traffic moves in one direction at all times.
2. A street that is normally one-way but at certain times may be operated in the reverse direction in order to provide additional capacity in the predominant direction of flow.
3. A street that normally carries two-way traffic but which during peak traffic hours may be operated as a one-way street, usually in the heavier direction of flow. Such a street may be operated in one direction during the morning peak hour and in the opposite direction during the evening peak hour, with two-way traffic during all other hours.

ADVANTAGES AND DISADVANTAGES OF ONE-WAY STREETS

One-way streets are usually installed to reduce congestion and to increase the capacity of an existing two-way street. One-way streets also have important effects on highway safety, traffic operation, and economic conditions.

Effect on capacity. Intersection conflicts and delays are a principal cause of congestion and reduced travel time on two-way urban streets. On one-way streets, left-turning movements are not delayed by the oncoming traffic present on two-way streets, and thus, intersection delay is greatly reduced. In addition, full use may be made of streets that have an odd number of lanes. The capacity of a one-way street may increase from 20 to 50 percent over its capacity as a two-way street, with the greater advantage occurring on the more narrow streets (see Chapter 8, Highway Capacity and Levels of Service).

With the increased highway capacity afforded by one-way streets, it may be feasible to permit parking, either part or full time, on streets that would require removal of parking if operated as two-way streets.

Effect on highway safety. Traffic safety is generally increased by one-way streets because they provide a divided highway effect. Vehicles also platoon better with signals progressed for one-way operation. Such grouping provides gaps in traffic for safer movement of vehicles and pedestrians from cross streets across the main one-way street. In addition, drivers and pedestrians crossing the street need look in only one direction to observe traffic.

One common problem with one-way street design and signal control is that traffic signals are erected for movement of traffic on the one-way and the cross street with little attention to the needs of pedestrians crossing the one-way street. Traffic and pedestrian signals must be located so that all pedestrians have a clear indication of when they can safely cross both the one-way street and the cross street.

Numerous studies have shown, however, that the conversion of two-way streets to one-way operation reduces total accidents on the order of from 10 to 50 percent. In some cases, specific kinds of accidents are reduced even more.[18] Minor mid-block

---

[18] Much detailed information on before and after studies is contained in the report entitled, "One-Way Streets and Parking," *Traffic Control and Roadway Elements—Their Relationship to Highway Safety/Revised,* Chapter 10.

traffic collisions may increase as a result of improper weaving by drivers to position themselves for an available parking space or to get in the proper lane for a turn. Transition areas between one-way and two-way operations are frequently hazardous and require special traffic control treatment.

Effect on traffic operation. The prime purpose of one-way streets is to improve traffic operations and to reduce congestion. The degree of improvement in operating conditions, travel speed, and safety depends, of course, on the previous situation. Results on several streets in London, as shown in Table 18.4, are fairly typical of those reported elsewhere. Generally, travel times were reduced from 10 to 50 percent and accidents from 10 to 40 percent, despite a slight increase in overall traffic volumes.

Such general improvement in traffic operations must be balanced against the following negative aspects:

1. Some vehicles must travel extra distances to reach their destination.
2. Strangers may become confused with the one-way street pattern, particularly if the two streets in the one-way system are not close together and markings and signal indications are not clearly provided.
3. Transit operations may be adversely affected (this factor must be taken into consideration before the one-way street system is implemented).
4. Emergency vehicles, such as fire trucks, may need to take a more circuitous route to reach their destinations. This may be alleviated somewhat, however, by providing signal controls that will allow emergency vehicles to move against the flow of traffic on one-way streets by holding vehicles at the next intersection before the emergency vehicle will enter the one-way system.

Effect on economic conditions. Improved traffic movement and increased safety generally produce broad economic benefits both to adjacent land users and to the general public. Nevertheless, when planning the one-way system, especially one involv-

TABLE 18.4
Accident Changes and Traffic Characteristics on One-way Streets
(London, England)

Street	Percentage Change in Traffic (Avg. Weekday)			Travel Time (% Change)				Accidents (% Change)	
	Mileage	Volume	Vehicle Miles	Off Peak Each Direction		PM Peak Each Direction		Injury	Pedestrian
1. Tottenham Ct. Rd.*	5.1	+4	+8	−49	−34	−43	−14	−21	−33
2. Baker St.*	2.1	+2	+3	−48	−35	−65	−55	+4	−8
3. Earls Ct. Rd.†	6.3	+10	+12	−33	−15	−27	−16	−27	−18
4. Kings X*	2.5	−2	+18	−28	0	−27	+40	−33	−40
5. Bond St.†	1.3	+9	+14	−26	−38	−15	−38	0	0
6. Piccadilly*	1.3	−4	0	−19	−12	−5	−12	−14	−38

*6-months before and after.
†3-months before and after.

Source: J. T. Duff, "Traffic Management," Conference on Engineering for Traffic, p. 49, 1963.

ing commercial streets, traffic engineers should expect objections by affected business owners who contend that one-way streets will adversely affect their trade.

Yet studies made in various parts of the United States have generally tended to disprove such claims. Moreover, where one-way systems have once been implemented, many businessmen formerly opposed to the one-way street plan have become their staunchest supporters. Although statistics are not readily available, one-way streets are rarely reverted to two-way operation unless the construction of major new highway facilities makes the continued operation of the one-way street system unnecessary.

Although the economic impact of converting to a one-way street system will undoubtedly vary from one place to another, the results of an in-depth study by the Michigan Department of State Highways reveals some interesting results:[19]

1. The greatest degree of environmental dissatisfaction existed with the residents who resided adjacent to the one-way pairs where a majority expressed dissatisfaction.
2. This attitude or feeling diminished in the study area which was at least one residential lot removed from the one-way pair.
3. Residents felt that the one-way conversions would cause a loss in property value and make the area less desirable from an environmental viewpoint. However, results of the market analysis showed that the greatest residential property value increase occurred on the low traffic volume converted residential streets where the greatest degree of environmental dissatisfaction occurred.
4. There was no indication of adverse economic influence on business activity within the one-way corridor. The number of business failures was reduced substantially after one-way conversion.

TRENDS IN ONE-WAY STREET USAGE

The number and total length of one-way streets have increased significantly in recent years. In 32 European towns, the total length of one-way streets has increased from 225 to 575 km in 10 years.[20] Figures are not readily available for the United States, but general observation suggests a similar trend.

CRITERIA FOR APPLICATION

General requirements. With exceptions, two-way streets should be made one-way only when:

1. It can be shown that a specific traffic problem will be relieved.
2. A one-way operation is more desirable than alternate solutions.
3. Parallel streets of suitable capacity, preferably not more than a block apart, are available.
4. Such streets provide adequate traffic service to the area traversed and carry traffic through and beyond the congested area.
5. Safe transition to two-way operation can be provided at the end points of the one-way sections.

---

[19] *The Economic and Environmental Effects of One-Way Streets in Residential Areas* (Lansing: Michigan Department of State Highways, 1969).

[20] E. NIELSEN, "Experience from 10 Years Fight Against Traffic Congestion," *XXXVIth International Congress* (Brussels, Belgium: International Union of Public Transport, 1965), p. 15.

6. Adequate transit service can be maintained.
7. Such streets are consistent with the master street plan.
8. Thorough study shows that the total advantages significantly outweigh the total disadvantages.

Increased capacity. One-way streets may be warranted if they will:

1. Reduce intersection delays caused by vehicle turning-movement conflicts and pedestrian-vehicle conflicts.
2. Allow lane-width adjustments which increase the capacity of existing lanes or actually provide an additional lane.

Increased safety. One-way streets may be warranted if pedestrian and traffic safety will be increased substantially by:

1. Reducing vehicle-pedestrian conflicts at intersections.
2. Preventing pedestrian entrapment between opposing traffic streams.
3. Permitting more effective spacing of traffic signals and thus better traffic operation.
4. Improving drivers' fields of vision at intersection approaches.

Improved operation. One-way streets may be warranted if traffic operations will be improved substantially by:

1. Reducing travel time.
2. Permitting improvements in public transit operations, such as routings without turnback loops (out on one street and return on the parallel streets).
3. Permitting turns from more than one lane and doing so at more intersections than would be possible with two-way operation.
4. Redistributing traffic to relieve congestion on adjacent streets.
5. Simplifying traffic signal timing by:
   a. Permitting wider range of offsets for progressive movement of traffic.
   b. Permitting offsets to achieve wider through bands.
   c. Reducing multiphase requirements by making minor streets one-way away from complex areas or intersections.

Improved economy. One-way streets may be warranted if definite economic gains will be realized by:

1. Providing additional capacity to satisfy traffic requirements for a substantial period of time without large capital expenditures.
2. Permitting stage development of a master plan.
3. Meeting changing traffic patterns almost immediately and at negligible cost.
4. Saving sidewalks, trees, and other valuable frontage assets that would otherwise be lost because of the widening of existing two-way streets.

PLANNING FOR ONE-WAY STREETS

Planning for a one-way street system should begin before the existing street system has become seriously overloaded. The amount of planning for the new system will depend largely on the size and complexity of the one-way system under consideration. The following questions should be considered:

1. Is the present street system laid out so that one or more pairs of one-way streets can be implemented on a practical basis?
2. What effect would the proposed one-way street(s) have on existing transit operations?
3. Are the streets of adequate width and must parking be restricted in certain areas to provide adequate street width?
4. What changes need to be made in signs, markings, traffic signals, and other traffic control devices?
5. Are there marked federal and/or state routes to be considered?
6. Are there major traffic generators on the streets to be considered for one-way operation, and what, if any, effect would there be on such generators?
7. Do the street sections proposed for one-way traffic operation terminate so that transition to two-way traffic or the termination at a street intersection would not cause traffic problems?

After the system has been thoroughly planned and designed, the start date for one-way operation should be established. A Sunday or other low-traffic period is usually desirable.

Other details concerning the change must also be considered:

1. Preparation of the proper ordinances to validate the one-way system, to revise parking restrictions, and, if necessary, to alter speed limits.
2. Changes in all traffic control devices, including route markings, changes in traffic-signal timing, and bus stops.
3. Relocation of parking meters where necessary.
4. Arrangements with local police for adequate field supervision, particularly during the first days of operation.
5. Arrangements using all news media to provide adequate publication of the impending change.

DESIGN OF SYSTEM

Roadway requirements. Although all one-way systems will probably differ from one another in details, there are certain basic conditions to consider in designing a specific system:

1. The capacity of the street(s) in one direction should approximately balance the capacity of the street(s) in the opposite direction. If capacities cannot be balanced, the street having the lower capacity must at least have adequate capacity for current traffic and, preferably, for at least a few future years.
2. The one-way pair should preferably be adjacent streets (although systems are operating satisfactorily where there are intervening parallel streets).

Problems at termini. Some street patterns readily lend themselves to good traffic operations at system termini: for example, when two streets join in a Y-pattern to become one. In a gridiron pattern, however, the one-way system must end at a typical four-way intersection. When the one-way system would normally terminate at a major cross arterial, it is usually desirable to extend the system one block beyond that point. This is particularly true of the one-way street carrying traffic toward the crossing arterial.

REVERSIBLE LANES AND ROADWAYS

Arterial routes that are normally operated as two-way streets, particularly those in urban areas, usually experience much heavier peak-hour traffic volumes in one direction than in the other. This condition typically wastes street capacity. Reversible lanes can provide for better utilization of such streets.

Under the reversible-lane system, one or more lanes are designated for movement one way during part of the day and in the opposite way during another part of the day. On a three-lane road, for example, the center lane might normally operate as a two-way left-turn lane, but during the peak hour it operates as a one-way lane in the direction of heaviest flow. The purpose of the reversible-lane system, obviously, is to provide an extra lane or lanes for use by the dominant direction of flow (see Figure 18.4). An increasingly used method is to reverse the flow of an entire street during peak-hour periods or to make the two-way street operate one-way during that period.

A new, seven-lane, single-carriageway motorway was opened to traffic in Birmingham, England in late 1971 with signal control to produce directional splits of up to five lanes to two. In the United States one of the outstanding examples of multiple reversible lanes is the eight-lane Outer Drive in Chicago which operates on a split of six lanes to two when needed.

Figure 18.4. Reversible lanes on a three-lane roadway with overhead lane control signals. (Source: City of Atlanta, Office of City Traffic Engineer.)

WARRANTS AND APPLICATION CRITERIA

Although no definite warrants have been established for reversible lanes, the concept, whether involving one or more lanes or entire streets, has as its objective the provision of adequate capacity for traffic in the direction of heavier flow. Care must be taken to ensure that adequate capacity is also available in the direction of lighter flow.

ADVANTAGES AND DISADVANTAGES

A reversible lane system is clearly one of the most efficient methods of increasing rush-period capacity of existing streets. With minimal capital costs, it takes advantage of unused capacity in the direction of lighter traffic flow by making one or more of those lanes available to traffic in the opposite direction—the heavier flow. The result is that all lanes are better used. The system is particularly effective on bridges such as the Liberty Bridge in Pittsburgh, Pennsylvania and in tunnels where the cost to provide additional capacity would be high and, perhaps, impossible. Some disadvantages are:

1. Capacity may be reduced for minor flows during peak periods.
2. Reversible lanes frequently result in operational problems at their terminals.
3. Driver behavior can require concentrated control efforts.

SYSTEM CONSIDERATIONS

There are several factors to be considered in determining whether or not reversible lanes are justified:[21]

Evidence of congestion. A street or highway is congested if the average speed of vehicles during certain periods decreases by at least 25 percent over normal period experience, or if there is a noticeable back-up at signalized intersections so that most vehicles miss one or more green signal intervals. Under these conditions, there is evidently a traffic demand in excess of actual capacity.

Periodicity of congestion. The periods during which congestion occurs should be periodic and predictable because traffic lanes are usually reversed at a fixed time each day.

Ratio of directional traffic volumes. Lane reversal requires that the additional capacity for the heavier direction be taken from the traffic moving in the opposite direction. Traffic counts by lane will determine whether or not the number of lanes in the counter direction can be reduced, how many lanes should be allocated to each direction, and when the reversal should begin and end. The ratio of major to minor movements should be at least 2 : 1 and preferably 3 : 1.

Capacity at access points. There must be adequate capacity at the end points of the reversible-lane system, with an easy transition of vehicles between the normal and

---

[21] *Manual on Uniform Traffic Control Devices for Streets and Highways*, U.S. Department of Commerce, Bureau of Public Roads (Washington, D.C.: U.S. Government Printing Office, 1961), pp. 226–27.

reversed-lane conditions. Installation of a reversible-lane system with insufficient end-point capacity may simply aggravate the traffic problem.

Lack of alternate improvements. Design controls and right-of-way limitations preclude widening the existing roadway or providing a divided highway or two or more parallel facilities on separate rights of way.

Once a reversible system is shown necessary and feasible, the method of designing lanes to be reversed and the direction of flow must be selected. Three general methods are used: (1) special traffic signals suspended over each lane, (2) permanent signs advising motorists of the changes in traffic regulations and the hours they are in effect, and (3) various physical barriers, such as traffic cones, signs on portable pedestals, and movable divisional medians.

Although reversible-lane operation is principally used on existing streets and roa ways, it can also be designed into new freeways and expressways, as for example, on the Northwest Freeway in Chicago. Applications to older limited access facilities is difficult because at most all such roadways have fixed medians separating the two directions of traffic. By constructing special median crossing locations and by properly using traffic control devices, however, even these facilities can be used in a reversible manner. Obviously, extreme care must be exercised to maintain a safe operation.

OPERATIONAL METHODS

Reversible lanes may be controlled in several different ways, all of which involve the use of traffic control devices. The reversible-lane situation must be emphatically conveyed to the driver because of the possible disastrous consequences of drivers using these lanes improperly. Such methods include:

1. Signs and markings
2. Lane control traffic signals
3. Traffic cones, barricades, and other similar devices
4. Physical barriers

Details on the use of traffic control devices are found in the *Manual on Uniform Traffic Control Devices*, 1971 edition[22] (see also Chapters 16 and 17).

SPECIAL ROUTINGS

One means for improving traffic movement has been to segregate the various kinds of vehicles comprising the traffic stream. Several methods are used:

1. Provision of special routings for trucks, particularly in urban areas.
2. Devotion of special lanes on streets for use by transit vehicles only.
3. Provision for handling traffic in general during special events, such as parades, athletic events, and other events attracting large volumes of traffic.

---

[22] *Manual on Uniform Traffic Control Devices*, U.S. Department of Transportation, Federal Highway Administration (Washington, D.C.: U.S. Government Printing Office, 1971), pp. 182, 194, 197, 249–52.

TRUCK ROUTES

Trucks in congested areas may make an already difficult or serious traffic problem worse. It is often helpful to divert through truck trips around the business or congested areas of a city. Truckers usually wish to avoid these areas anyway because of the delays. Consequently, it may be desirable to establish signed truck routes through less congested areas of a city or, perhaps, entirely around the city.

In locating truck routes, several factors should be considered:

1. The effect of trucks on the area through which the truck route is located. It would, for example, be undesirable to sign a truck route through a residential area if another area such as a warehouse or semiindustrial area may be found.
2. The particular needs of local trucking companies or truck-dependent industries.
3. Street widths, parking, number of turns, turn radii at intersections, and other factors affecting truck operation along the route selected and the problems that might be caused by trucks on narrow streets.

All established truck routes should be clearly marked to distinguish them from normally signed routes for other vehicles.

EXCLUSIVE BUS LANES AND BUS ROADWAYS

One means of improving bus speeds and thereby inducing drivers to divert to buses, thus reducing traffic congestion, has been to reduce the travel time of buses compared with cars over the same route by providing a lane exclusively for buses. Examples of exclusive bus lanes are shown in Figures 18.5 through 18.7. Figure 18.5 illustrates an

Figure 18.5. A combination of plastic traffic posts, overhead lane directional signals, and changeable signs are used to separate eastbound bus lane from westbound traffic flow. (Source: GMC Truck and Coach Division, General Motors Corporation.)

Figure 18.6. Reserved lane for buses in Brixton Road in London. (Source: International Union of Public Transport, Brussels, Belgium.)

exclusive bus lane operated on a freeway. Note that here the buses travel in an exclusive lane opposing the main flow of traffic. The lane is marked by plastic safety posts and is on the opposite side of the median from which the buses would usually travel. Figure 18.6 is a reserved bus lane for buses in Brixton Road in London, and Figure 18.7 shows a reserved lane for buses in the Prinzenstrasse in Hanover, Germany.

Exclusive bus lanes are normally justified only in and around large urban areas

Figure 18.7. Reserved lane for buses in Prinzenstrasse in Hanover. (Source: International Union of Public Transport.)

where there is heavy traffic in a particular corridor and the potential exists for diverting many people from cars to buses.

Consideration for exclusive bus lane. The general warrant for an exclusive bus lane, whether on a freeway or surface arterial, is that more people will be accommodated than by the use of the lane by general traffic. Bus loading problems and their solutions are discussed in Chapter 15, Parking, Loading, and Terminal Facilities.

An engineering analysis of the actual vehicular capacity at a desirable level of service should be made and the existing vehicle occupancy applied to determine the person capacity of the lane under consideration for the peak period. Care should be exercised in the selection of the automobile occupancy figures for they vary considerably by time of day. Rarely do peak-period average automobile occupancies exceed 1.3 persons; some are as low as 1.1. Off-peak occupancy values are generally higher. From a person-moving standpoint, the difference between using a value of 1.1 as against 1.3 can produce a 25 percent variation in the number of persons accommodated.

Considering this, we can apply the general warrant to determine whether or not more people would be accommodated with an exclusive bus lane. Other factors that should be considered are: the effect on the operation of the remaining lanes if one is taken away for bus use, the benefits to bus users, the disbenefits to automobile users, the expected shift of automobile users to buses as a result of the lane reservation, the amount of reserve capacity that the lane might provide, the effect on abutting land use, and the benefit to the community of encouraging a reduction in the number of automobiles.

Consideration for exclusive bus roadways. Special roadways to serve buses exclusively have been considered in St. Louis, Milwaukee, Pittsburgh, and elsewhere. Pittsburgh's "East PATway" illustrates the concept of a two-lane roadway to be built along an available railroad right of way, not part of a joint transit-highway corridor, serving conventional rubber-tired buses, and connecting downtown Pittsburgh with an outlying, secondary business district and surrounding residential areas (a "South PATway" is already under construction). Such exclusive bus roadways have several potential advantages:

1. At a reasonable cost they can provide express bus service between areas that cannot otherwise be served by exclusive bus lanes on new or existing expressways.
2. Taking advantage of abandoned or underutilized railroad or other narrow rights of way, they can be fitted into the urban landscape with a minimum of land taking, often into corridors too confined for a full expressway.
3. They can be converted to rail rapid transit should ridership demand prove subsequently warranted (which is the expectation and plan in Pittsburgh).

Specific warrants for exclusive bus roadways do not exist, and such proposals must be fully evaluated through the methodology of the long-range urban transportation planning process (see Chapter 12, Urban Transportation Planning).

Consideration for preferential bus treatment. Some of the many ways of providing preferential treatment to buses include: metering vehicles entering freeways with provision for buses to bypass the metering, closing certain ramps to all vehicles except buses and emergency vehicles, reserving portions of a facility for buses and selected other vehicles, such as car pools, designating curbside lanes for buses and right-turning vehicles, and bus-actuated traffic signals.[23]

Operational results. Exclusive bus lanes are being used in many cities throughout the world. Results in several United States cities have shown reductions in travel times for buses of from 10 to 30 percent, often with an accompanying decrease in travel times for cars.

In Reading, England a special bus lane runs the buses on a one-way street opposite the normal flow of traffic on the street. Peak-hour travel times were reduced nearly 50 percent.[24] Exclusive lanes have also been established in many other European cities and have had similar results.

Provision of exclusive bus lanes on freeways has been extremely beneficial. On the Shirley Highway (I-95) in Virginia, near Washington, D.C., bus patronage increased 70 percent and travel time savings ranged from 12 to 18 min. It is estimated that buses will ultimately save up to 30 min upon completion of the entire bus lane. On I-495

---

[23] "Warrants for Exclusive Bus Facilities," *Instructional Memorandum 21–13–67 (1)* (Washington, D.C.: U.S. Department of Transportation, Federal Highway Administration, 1970), pp. 2–3.

[24] TERRENCE BENDIXSON, "Results of Reading Bus Lane Experiment," *Traffic Engineering and Control*, XII, No. 4 (1970), p. 188.

in New Jersey, near New York City, a similar lane has reduced travel time from 25 min to 10 min between trip origin and destination.[25]

Methods of implementing marking and signing for exclusive bus lanes are shown in Figs. 18.5 through 18.7.

PARADES, ATHLETIC, AND OTHER SPECIAL EVENTS

There are many events (parades, baseball, football, or soccer games, auto races, inaugurations, etc.) that attract so much pedestrian and vehicular traffic that special provisions must be made for handling traffic. Probably the most important aspects of handling such traffic successfully are (1) detailed preevent planning and (2) coordination of all agencies that would in any way be involved in handling traffic, caring for emergencies, and other related needs.

The following factors should be considered:

1. Determine the area and streets affected and how to handle traffic in or around the area.
2. When streets must be blocked off, determine barricade and detour sign and warning device needs.
3. Assure that no roadway, utility, or other work in the roadway will occur during the affected time period.
4. Effect necessary parking restrictions and other vehicular controls.
5. Provide for handling emergencies such as accidents.
6. Obtain necessary temporary manpower assistance for traffic routing and control, handling of parking, medical needs, and other related items.

PLANNING AND ROUTING FOR EMERGENCY SITUATIONS

Events frequently occur on a particular highway (an accident) or in an entire city or area (snow) which create emergency traffic situations that must be handled immediately. If there is snow or a similar weather emergency, advance preparations can often be based on weather forecasts. Nevertheless, accidents happen spontaneously and advance knowledge is impossible. To ensure that traffic is well-handled during these periods, considerable preplanning is desirable and should include:

1. Arranging with police for ambulance and/or wrecker service.
2. Determining alternate streets to be used in case a street is closed, making police assignments for directing traffic, and having available temporary signs and flashers.
3. Designing signals and signal systems to let emergency vehicles proceed with minimal delay.
4. Installing variable message highway signs (see Chapter 16, Traffic Signs and Markings) to warn motorists of roadway conditions ahead, of alternate routes to use, and similar information. Use of such signs is particularly important on freeways and other high-volume, high-speed highways. Figure 18.8 illustrates a sign currently in use.

---

[25] "Preferential Treatment of Buses," FHWA Notice HP-24 (Washington, D.C.: U.S. Department of Transportation, Federal Highway Administration, 1971).

Figure 18.8. Sign currently in use to warn motorists of roadway conditions ahead. (Source: Winko-Matic Signal Company, Avon Lake, Ohio.)

To meet weather emergencies, additional preplanning should include:

1. Enacting appropriate legislation or directives to cover action to be taken during the emergency. Cleveland's "snow" ordinance is as follows:

CITY OF CLEVELAND'S "SNOW" ORDINANCE

Section 9.090301. Emergency During Heavy Snow Storms.

Whenever, during any period of twenty-four hours or less, snow falls in the City of Cleveland or in a section or sections thereof to a depth of two inches or more, an emergency is declared to exist in that such a heavy snow storm constitutes a serious public hazard impairing transportation, the movement of food and fuel supplies, medical care, fire, health and police protection, and other vital facilities of the city. Said emergency shall continue until an announcement by the Director of Public Safety that snow plowing operations have been completed, which announcement shall be made in the same manner as outlined in Section 9.090302.

Section 9.090302. Parking Regulations During Emergency.

Whenever such an emergency exists the Director of Public Safety shall request the cooperation of the local press and radio and television stations to announce the emergency and the time that emergency parking regulations will become effective, which time shall be no sooner than one hour after the first announcement, and such announcement by two local radio stations, or two local television stations, or in a daily newspaper of general circulation published in the City of Cleveland shall constitute notice to the general public of the existence of the emergency. However, the owners and operators of motor vehicles shall have full responsibility to determine existing weather conditions and to comply with the emergency parking regulations.

During the period of the emergency, the Director of Public Safety may prohibit the parking of any vehicles upon any or all of the city streets designated as a through highway by Section 9.1102 of the Codified Ordinances of the City of Cleveland. During the emergency, it shall be unlawful for any person to park, or cause to be parked or permit to be parked or permit to remain parked, or to abandon or to leave unattended, any vehicle of any kind or description upon such specified streets; provided, however, that vehicles may be parked for a period of not longer than three minutes for actual loading or unloading of passengers or thirty minutes for actual loading or unloading of property; provided, further, that no other ordinance restricting parking as to place or time is violated thereby.

Penalties: The penalties for violation of the above sections are the same as the "rush hour" penalties. Violation may be handled on a waver basis on the payment of $5 within 48 hours after the citation or $10 within 72 hours after citation. If not wavered, the violator must appear in court and could be fined not less than $15 nor more than $25 for first offense. In addition, charges for towing and impounding illegally parked vehicles must be borne by the owner.

2. Erecting signs designating an important street or highway as a SNOW ROUTE and other appropriate signs such as NO PARKING WHEN ROAD IS COVERED WITH SNOW.

TURN REGULATIONS

AVOIDING VEHICLE-TO-VEHICLE CONFLICTS

A leading cause of congestion and traffic problems in urban areas is the turning movement made by a vehicle at an intersection. These turning movements not only delay vehicles in the traffic stream, but they also create conflicts with pedestrians in crosswalks. Turn restrictions at intersections, or in mid-block areas, are a means of reducing such delays and conflicts. Because of the many differences from city to city, there is little uniformity and no specific warrant for the application of turn restrictions. Each situation must be studied independently.

Left-turn movements in the face of opposing traffic create a major problem. When opposing traffic is heavy, left turns may be very difficult and, perhaps, impossible to make. This may impede through traffic trapped behind the left-turning vehicle. Elimination of such left turns may add considerably to the intersection capacity and result in reduced congestion and delay. Factors that should be considered before initiating a left-turn prohibition include:

1. The possibility of alternate solutions, such as channelization with the provision of a separate left-turn lane.
2. A provision in the timing of the traffic signal of a "leading" or "lagging" green signal phase providing for left-turn movements. In some situations, if separate turning lanes are used, the use of special vehicle-actuated detectors can be provided to permit separate turn phases only when vehicles are waiting in the turn lane.
3. The seriousness of the problem indicated by the amount of traffic congestion as indicated by queuing, traffic counts, and other indicators such as time-delay runs.
4. The availability of alternate turn points for vehicles that are prohibited from making a left-turn movement at a particular location.

Full-time prohibitions are not always necessary; part-time prohibitions can be used effectively during problem hours only. When part-time prohibitions are used, internally illuminated signs rather than ordinary signs should be used so that the regulation in effect at any particular time will be noticeable. Such signs are usually suspended over the roadway in a readily visible location and are illuminated only when the special regulation is in effect.

Before turn regulations are imposed, consideration should be given to the possible adverse effects resulting therefrom, such as (1) creating additional travel and increased turning movements at other locations, particularly when left turns are prohibited, (2) routing of traffic following national or state signed routes if junctions of such routes are involved, (3) causing severe U-turn problems along a street or highway, particularly where around-the-block travel is lengthy or perhaps impossible.

AVOIDING VEHICLE-TO-PEDESTRIAN CONFLICTS

Prohibition of turning movements is sometimes used to avoid conflict between vehicles and pedestrians, particularly when a crosswalk carries heavy pedestrian movement throughout the green period. This may not only eliminate the pedestrian-vehicle conflict, but it may also add capacity to the intersection by precluding vehicles queuing and thus blocking it while waiting for a gap in the pedestrian flow (also see section on "Pedestrian Controls and Facilities" for more information on this subject).

Another approach is to delay vehicles in the right lane from moving when the green light appears until pedestrians have had a chance to move or to clear the crosswalk. The right-turn movement is then permitted for the duration of the green period. The same techniques can be applied to left-turns on one-way streets.

SPECIAL TURNING PROVISIONS

Multiphase signal timing. Added capacity and safety can be gained on certain intersection approaches by signalized control of turning movements. Actuated traffic signals are particularly conducive to this method. One situation where this is possible is on a multilane highway with left-turn lanes existing in both directions on one street at an intersection. If no vehicle is present in one of the turn lanes, the traffic signals can be operated to permit the opposing through movements to proceed while the left-turning vehicles from this direction are also moving (see Chapter 17, Traffic Signals).

Continuous left-turn lanes. Some sections of road have roadside development that induces a significant demand for virtually continuous left turns and not merely at intersections. This is particularly true on "string development roads." If the roadway has adequate width, usually five lanes or more and no median, a continuous left-turn lane can be created in the center lane. All vehicles turning left off the main roadway make their turns from this lane. No straight-through traffic may use the lane. Devices for control of traffic in such lanes are described in Chapter 16, Traffic Signs and Markings.

BARRIERS, MALLS, AND STREET CLOSURES

The erection of physical barriers or grade separation structures is often desirable to increase the safety of vehicles or pedestrians or to prevent vehicles from using certain areas.

VEHICLE CONTROLS

Median barriers and controls. Various permanent and temporary barriers are used to separate vehicles moving in opposite directions on a highway. Permanent barriers used in median of highways include guardrails, concrete median barriers and curbs, various kinds of fence, thick plantings of shrubbery, cones of metal, plastic, rubber, or other material, or other temporary devices placed between opposing lanes.

One of the latest median barriers is illustrated by the cross section shown in Figure 18.9. The design tends to deflect a vehicle striking or brushing it back into its own lane of travel and overcomes the vaulting effect found in some older designs with a 6- or 8-in. curb at the outer edges of the barrier. Double-beam, W-section, or box-section guardrail is also commonly used as a median barrier.

Erection of a barrier median is not feasible when it is necessary or desirable to permit vehicles to cross the median more or less continuously. In such cases, mountable medians may consist of a raised 3- to 6-in. curb with sloping sides which may be easily and safely mounted by vehicles; or they may be designed flush with the pavement but grooved or roughened to make them obvious to an encroaching driver.

Glare fence is often used on narrow medians to restrict glare from opposing headlights at night. Such *glare screens* are shown in Figures 18.10 through 18.12.

Shoulder edge barriers and controls. Although median barriers may also be used on the shoulder edge of the highway when conditions warrant, regular steel guardrail is the most commonly used shoulder edge barrier. Latest AASHO designs or standard guardrail used on high-speed highways require a post spacing of 6.25 ft instead of the previously used 12.5 ft, and the steel rail is usually blocked out from the posts to prevent vehicles' wheels from being entrapped by the posts.

Impact attenuating devices. The designs of many freeways and expressways have created raised concrete obstructions that may be extremely hazardous when struck. The gore of an exit ramp on a bridge where the parapet walls and curbs of the bridges on the ramp and the main roadway join is an example. Numerous fatalities have occurred from vehicles striking such obstacles. Impact attenuating devices are often installed to lessen the consequences of collision. Such devices include barrels, sand-or water-filled containers, and mechanical devices that absorb the shock when struck by a vehicle. A device using barrels is illustrated in Figure 18.13.

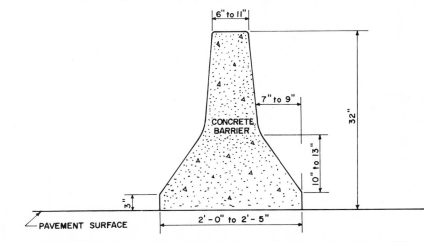

Figure 18.9. Cross section of one kind of median barrier in current use. (Source: West Virginia Department of Highways.)

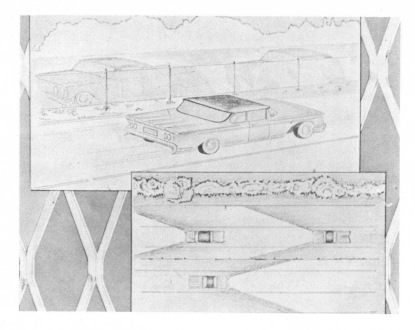

Figure 18.10. A typical use of glare screen. (Source: Wheeling Corrugating Company, A Division of Wheeling-Pittsburgh Corporation.)

Figure 18.11. A view of the installation of glare screen on median guardrail. (Source: Wheeling Corrugating Company.)

MALLS AND STREET CLOSURES

Pedestrian malls are considered a means of revitalizing central business districts. Malls represent an attempt by downtown interests to incorporate one of the basic features of successful suburban shopping centers, namely, the separation of pedestrians from the more serious traffic conflicts within the retail environment, thereby improving the shopping atmosphere.

Although malls may be of many types, the one with which the traffic engineer is most concerned involves the closing of one or more blocks of street to vehicular traffic.

Figure 18.12. Glare fence in median along highway paralleling the Rhine River in West Germany. (Source: John E. Baerwald.)

Figure 18.13. A cushion type barrel barrier installation at elevated gore. (Source: U.S. Department of Transportation, Washington, D.C.)

Usually, the entire sidewalk and former street area is developed with benches, plants, trees, and decorations (see Figure 18.14 and 18.15). In some cases, roadway space is left through the mall area for use by emergency vehicles. In other cases, special bus lanes are provided through the area.

Extensive information on malls may be found in a special publication by the Institute of Traffic Engineers.[26]

PEDESTRIAN CONTROLS AND FACILITIES

Pedestrians are involved in a significant percentage of urban traffic accidents. Allowing pedestrians to cross streets uncontrolled poses a serious traffic hazard. Even at intersections with crosswalks and traffic signals, there is significant hazard for the pedestrian because of the conflict with motor vehicles. To alleviate such hazards it is often desirable to erect physical barriers to control the movement of pedestrians

---

[26] *Traffic Planning and Other Considerations for Pedestrian Malls* (Washington, D.C.: Institute of Traffic Engineers, 1966).

Figure 18.14. A view of the Burdick Street Mall in Kalamazoo, Michigan. (Source: Courtesy of *Kalamazoo Gazette.*)

Figure 18.15. Another view of the Burdick Street Mall in Kalamazoo. (Source: Courtesy of *Kalamazoo Gazette.*)

across the street and sometimes to erect grade separations to separate physically the movement of vehicles and pedestrians.

Two distinct kinds of barriers or controls may be used: (1) grade separation structures, either overpasses or underpasses that physically separate vehicles and pedestrians (see Figures 18.16 through 18.19) and (2) fences, chains, or other similar devices

Figure 18.16. Pedestrian overpass in Iowa City, Iowa. (Source: John E. Baerwald.)

Figure 18.17. Formed concrete pedestrian underpass in Montreal, Canada. (Source: John E. Baerwald.)

Figure 18.18. Pedestrian bridge in Freiburg, West Germany. (Source: John E. Baerwald.)

Figure 18.19. Entrance to underground pedestrian crossing and subway station in central business district of Hamburg, West Germany. (Source: John E. Baerwald.)

which, while allowing pedestrians to cross streets at grade, do control the locations at which such crossings may be made (see Figures 18.20 through 18.23).

When pedestrian grade separations are used, it is usually necessary to lead or herd people to the facility so that they may not cross the street or highway at grade. When grade separations are used on freeways, this is usually not a problem.

Whether to provide a pedestrian overpass or underpass must be determined for each location. There are advantages and disadvantages for each. The following factors

Figure 18.20. Use of posts and chain to separate pedestrian-vehicle traffic. (Source: Department of Streets and Sanitation, Chicago.)

Figure 18.21. Use of pedestrian control to regulate the location where pedestrians may cross the street. (Source: Department of Streets and Sanitation, Chicago.)

Figure 18.22. Pedestrian barriers in the streets surrounding the main railroad station in Hamburg, West Germany. (Source: John E. Baerwald.)

Figure 18.23. Use of post-bar type pedestrian barrier in Washington, D.C. (Source: Department of Highways and Traffic, District of Columbia.)

should be considered: (1) cost, (2) security, (3) vandalism, (4) aesthetics, (5) possible conflict with utilities, and (6) terrain condition.

Geographic conditions and presence of utilities, both above and below ground, are major factors in determining if an overpass or underpass is more suitable. A particular problem with underpasses, especially at night, is personal safety. Unless they can be made secure from crime, underpasses simply will not be used.

OTHER OPERATIONAL CONTROLS

RUMBLE STRIPS

When unusual alertness is required of drivers, and standard traffic control devices such as signs and/or flashers have apparently not proven to give adequate warning, *rumble strips* have sometimes been used. These are slightly raised strips of asphalt, plastic, or other suitable material across the lane(s) of pavement approaching the hazard ahead. Each strip is usually approximately from 3 to 6 in. wide and $\frac{1}{2}$ to 1 in. high. Several strips are placed laterally across the pavement to cause a rumble or bumpy motion that when traversed by a vehicle will alert the driver (see Figure 18.24).

Typical locations for use of rumble strips are approaches to toll gates and to stop signs hidden by horizontal or vertical curves. Proximity of the rumble strip to the hazard is important. If located too close, sufficient reaction time is not given the driver, and if they are located too far away, the driver may not relate the rumble strip to the hazard.

Use of rumble strips, bumps, or similar devices to control speeding is definitely not recommended.

An extensive study on rumble strips was conducted by the Illinois Department of Public Works. One significant conclusion was " . . . rumble strips, like many other

Figure 18.24. An asphalt rumble strip. (Source: Ohio Department of Transportation, Columbus, Ohio.)

nonstandard traffic control devices, are effective only as long as they are startlingly different . . ."[27]

FOGGY WEATHER CONTROLS

Fog, particularly on high-speed highways, has resulted in many spectacular accidents, often involving tens of vehicles and causing many deaths and injuries. Various methods and devices that have been used to reduce the incidence of accidents in fog include reduced speed limits and illuminated warning signs. To date no really satisfactory solution to the fog problem has been found.

STAGGERED WORKING HOURS

Use of traditional traffic engineering methods often proves inadequate to solve peak-hour traffic problems. When highways operate at or above capacity levels and no other alternative solution exists, only a reduction in traffic demand will help. One method of reducing demand is to encourage major employers to stagger their working hours in order to spread the peak traffic load. This can be particularly effective at large industrial plants and government office complexes. Although many successful examples can be cited, the practice has yet to be completely accepted. Some of the drawbacks include the interruption of car-pooling arrangements, the interruption of office or plant routines because of the difference in working hours, and the possible difficulty in controlling employee working times when employees work on different schedules.

---

[27] *Accident Study Report No. 102, Rumble Strips Used as a Traffic Control Device, an Engineering Analysis* (Springfield, Ill.: Illinois Department of Public Works and Buildings, Division of Highways, Bureau of Traffic, 1970), p. 2.

REFERENCES FOR FURTHER READING

WITHEFORD, DAVID K., *Speed Enforcement Policies and Practice.* Saugatuck, Conn.: ENO Foundation for Transportation, 1970.

Uniform Vehicle Code and Model Traffic Ordinance. Washington, D.C.: National Committee on Uniform Traffic Laws and Ordinances, 1968 and Supplement I, 1972.

A Policy on Geometric Design of Rural Highways, Washington, D.C.: American Association of State Highway Officials, 1965.

Manual on Uniform Traffic Control Devices for Streets and Highways, U.S. Department of Transportation, Federal Highway Administration. Washington, D.C.: U.S. Government Printing Office, 1971.

The Federal Role in Highway Safety, House Document No. 93, 86th Cong., 1st sess., 1959. Washington, D.C.: U.S. Government Printing Office, 1959.

Improved Street Utilization Through Traffic Engineering, Special Report 93. Washington, D.C.: Highway Research Board, 1967.

"Speed and Speed Control," and "One-Way Streets and Parking," *Traffic Control and Roadway Elements— Their Relationship to Highway Safety/Revised.* Chapters 6 and 10 (1970 and 1971). Washington, D.C.: Highway Users Federation for Safety and Mobility.

Chapter 19

TRAFFIC SURVEILLANCE AND CONTROL

Patrick J. Athol, Adjunct Associate, University of Pittsburgh, Pittsburgh, Pennsylvania

A traffic surveillance and control system reduces freeway congestion by controlling the rate at which vehicles enter the freeway, thereby tending to balance demand to capacity. The entry rate is based on the ability of the freeway to carry the additional ramp volume without breaking down in congestion. The controlled freeway can carry almost the same volumes as the uncontrolled freeway, but without congestion. Ramp volumes in excess of the allowed entry rate are distributed proportionately to arterial alternates to make use of available capacity on those roadways.

A traffic surveillance and control system accomplishes its management function in several steps. Electronic surveillance provides data on prevailing traffic conditions. These data are analyzed and used to make rational decisions as to any actions required to prevent congestion. The resulting impact of these actions on traffic conditions is then monitored by the surveillance system. In this way, the loop of information, decisions, control, and impact is said to be closed. The result of previous surveillance and control projects has been not only a smooth-running freeway, but a marked reduction in overall total travel time for corridor drivers, and a significant decrease in peak period freeway accidents.

Freeways have introduced the motoring public to safer and faster travel standards. High levels of service result in the heavy loading of freeways, not only with the longer trips for which they are best suited, but also with numerous short trips. A new freeway will absorb, like a blotter, traffic from the arterial street system. The corridor communities are thereby relieved of traffic on their arterials and the community residents enjoy the benefits of access to the freeway. The result is likely to be congestion on the freeway.

Surveillance and control are used as remedial devices to cut down congestion by controlling an excessive traffic shift from arterials to freeways. Early control prevents the need to reallocate excess freeway traffic back to the arterials.

In the transportation planning process, the term *surveillance*, or monitoring traffic growth and urban developments, has now generally been accepted. In most urban areas, however, there has not been a corresponding awareness of the need to know and record the operating performance of a highway corridor.

An important aspect of surveillance is measuring the effect of various controls or information techniques on the system. Most urban travel corridors are so complex that traditional traffic engineering techniques are inadequate to handle the necessary data collection. It is important to have measurements on a systems basis so that hypotheses can be rationally developed on the interaction among all affected highways in the network. When one link is influenced by control, or other techniques, the performance of the remaining links, at the same instant in time, can be recorded.

HISTORICAL BACKGROUND

Traffic surveillance and control first gained prominence in the U.S. in two major research projects sponsored by state, county, and city governments, in conjunction with the Federal government. These two early projects, in Detroit and Chicago, were designed to provide different end results.

The Detroit Surveillance System[1] pioneered the use of television monitoring on the John C. Lodge Freeway. Manually set lane control signals and advisory speed messages were the major forms of control and driver information. In 1966, ramp metering was initiated and a digital computer added for data logging.[2]

In contrast, the Chicago Expressway Surveillance Project[3] pioneered, in 1961, the instrumentation of automatic detection systems monitoring the Eisenhower Freeway through the use of sensors along the roadway. This system exercised control, with a central digital computer, through ramp metering. It has since evolved into a fully operational system of sophisticated technology, monitoring, and control involving many of the freeways in metropolitan Chicago.

Extensions of this work were then used in other cities. Notable in this area was the work done in Houston, Texas, on the Gulf Freeway, where television and ramp control, emphasizing gap acceptance, were combined.[4] In the Los Angeles area the automatic system was extended, utilizing ramp control, to coordinate with freeway improvements and trouble incident detection.[5] The Dallas project is noteworthy because it coordinates the arterial signal control with the freeway ramp control and selective television monitoring.

Other systems, developing in most parts of the U.S., each have some different features. Some, such as the Boston system, attempt to incorporate the surveillance and control plan in their highway construction program. Other areas, such as Seattle, Minneapolis, Atlanta, and Phoenix, are attempting to utilize traffic surveillance and control techniques at various levels to relieve existing congestion patterns.

There has been a growing interest by other nations in studying the results of work done in the U.S. and applying, where practical, comparable techniques to comparable problems.[6] In Japan, traffic officials have implemented a traffic surveillance and control system using the automatic monitoring approach, with ramp access control techniques. They have also implemented a system on some facilities whereby some access control is provided by toll booth collectors who modify the rate of toll collection to influence access to the expressway. The United Kingdom has undertaken an extensive information and lane control system on a major portion of its motorways. This is somewhat analogous to the initial approach taken by the Detroit study. These systems, both in the United Kingdom and in Japan, are operational and were built as such.

[1] F. DeRose, "Lodge Freeway Traffic Surveillance and Control Project," *Freeway Operations—7 Reports*, Highway Research Record No. 21 (Washington, D.C.: Highway Research Board, 1963), pp. 69–89.

[2] E. F. Gervais, "Optimization of Freeway Traffic by Ramp Control," *Freeway Operations—7 Reports*, Highway Research Record No. 59 (Washington, D.C.: Highway Research Board, 1964).

[3] J. M. McDermott, "Automatic Evaluation of Urban Freeway Operations," *Traffic Engineering*, XXXVIII, No. 4 (January, 1968).

[4] C. Pinnel et al., "Evaluation of Ramp Control on a Six-Mile Freeway Section (Houston)," *Freeway Traffic Characteristics and Control—5 Reports*, Highway Research Record No. 157 (Washington, D.C.: Highway Research Board, 1967), pp. 22–76.

[5] J. T. West, "California Makes Its Move," *Traffic Engineering*, XLI, No. 4 (January, 1971).

[6] J. Duff, "Accomplishments in Freeway Operations Outside the U.S.," *Accomplishments in Freeway Operations—5 Reports*, Highway Research Record No. 368 (Washington, D.C.: Highway Research Board, 1971).

TRAFFIC SURVEILLANCE

Surveillance is commonly characterized by the use of traffic sensors, a transmission network, and a central digital computer with peripheral devices and consoles. Their functions are to collect traffic data, to indicate prevailing traffic conditions, and to anticipate developing congestion problems.

VISUAL DETECTION

Visual detection techniques include photographic, human observers, and closed circuit television (CCTV). All these techniques require "observers" at the scene of the traffic problem, incident, or area of data collection.

Photographic techniques can be used in the evaluation phases of a system by recording data at areas where only occasional information is required and at locations where it is inadvisable to install permanent record-keeping equipment. Aerial photography has gained wide usage when a great deal of information is required at a location over a short period of time and when there is a small amount of traffic. It is also used in flying over a wide area for which an occasional qualitative review is needed. For an entire traffic surveillance and control system, however, there is no foreseeable use of photographic techniques.

The human observer may have a role to play in traffic surveillance and control by being able to report the nature and character of certain incidents occurring along the freeway. It would be expensive, of course, systematically to station people along the freeway to gather this information. Human observers do exist, however, in the form of police forces, highway department personnel, and private traffic reporting agencies who travel the freeway and assist by calling in incidents or problems to the control center.

CCTV should not be used as a data collection tool but should be considered supplementary to automatic sensors. The primary advantage of CCTV is providing "on-the-spot" visual surveillance.

NONVISUAL TECHNIQUES

The freeway introduced the special problem of counting adjacent lanes of traffic. Accurate counts are required for each lane. Induction loop detectors are most often used for this purpose, supplemented as required, by overhead ultrasonic detectors and magnetometers in the pavement. Magnetometers and ultrasonic and radar detectors can also meet the freeway counting requirements, but they are less suited to lane occupancy measurement.

Detector accuracy. There is a lack of widely accepted levels of accuracy to be expected from various detectors. The following comments may be helpful.

Volume accuracy should be within 1 percent of the actual volume in a multilane freeway environment. Detector performance to within 1 percent of the actual volume for a multilane freeway station should be an accuracy that can be achieved and sustained year round without adjustment. Accurate counting on a single test lane installation is not adequate in detector elvaluation.

Errors in direct measurements of speed should not exceed a 1 mph variance on individual vehicles in a multilane environment. Such accuracy is very good for most surveillance applications.

Lane occupancy is the parameter most neglected in determining the accuracy of detectors. It is often assumed that total volume accuracy yields corresponding accuracy in lane occupancy measurements, but this has failed to be the case. Occupancy accuracy depends on the detector's area of coverage and the effective length of the vehicle (the effective length of a vehicle depends on the pickup and drop-out characteristics of the vehicle entering the detection zone). For an individual vehicle, occupancy should be accurate to 5 milliseconds in timing the individual detector output pulse. Sample time, percentage occupancy, tends to be more accurate at the lower speed levels than at the very high speed levels.

INCIDENT DETECTION

Perhaps the most important aspect of surveillance is its ability to detect incidents within the highway system. Stalled vehicles, disabled vehicles, accidents, and other events may diminish highway capacity by half or greater for a relatively large number of days in the year. Major incidents may require rescheduling and reallocating travel demand within the corridor.

The extent of incidents occurring is very difficult to assess without a surveillance system. Incident detection allows a measure of these problems within the system and later on it may become very important in providing additional capacity through the quick correction of such problems.

MEASURING PERFORMANCE

Establishing the effectiveness of various control techniques is also an important function of surveillance. It relates the control techniques to the amount of traffic service provided and the level at which that service is provided. Measures of performance also allow comparisons between geometric design features. The measure of performance indicates the potential of control in those cases in which remedial action is needed. Measures of performance also are key indicators of developing problems within the system, and these indicators can provide lead time to prevent congestion.

DATA TRANSMISSION

Data transmission requirements of a surveillance and control system consist of (1) a widely spread out system of detection and control which has relatively slow speed requirements and relatively few gathering points at which to concentrate data and take advantage of high speed data links and (2) a video or closed circuit television network with a large band-width requirement laid out along a line with hook-ups at various points along the line.

Between each of the points in the traffic surveillance and control system, the various lengths in the communication system can be considered as channels. Each channel will have a frequency band width to carry information. For example, a voice grade telephone line would leave 2,400 hertz bandwidth and an FM radio station a 15,000 hertz bandwidth, but television uses 5,000,000–10,000,000 hertz per channel.

The channel capacity of a communication system may not be fully utilized if channels are provided for individual functions. Multiplexing is a technique in which various signals are combined on the same communication channel. In this manner, the number of communication channels may be reduced and the utilization of those channels increased. To make efficient use of individual channels, two multiplexing techniques are used.

TIME- OR FREQUENCY-DIVISION MULTIPLEXING

Time division is the technique in which the full frequency band width of the channel is utilized. This technique transmits information from the various functions in serial form. For example, in sending the status of twenty detectors along a line in a time-division multiplex system, the system will send a "0" or "1" for each detector, number 1 being first, followed by 2, and so on until 20 detector outputs are transmitted. The status of the first detector can be repeated only after the completion of the twentieth transmission.

In the *frequency-division* multiplex system, the band width of the channel is subdivided into subchannels of smaller band widths. The information is sent with one function on each subchannel. In such a situation, the condition of twenty detectors can be sent simultaneously along each subchannel having about 1/20th of the original band width.

A time-division multiplex system is more cost effective when larger quantities of data can be gathered onto a single channel. A frequency-division multiplex system is more cost effective for lower data rates. Local pricing determines the break point between time-division and frequency-division multiplexing. However, by the time the number of signals or functions reaches ten, consideration should be given to time-division multiplexing.

DATA COMPILATION

The detector signals coming from the field, transmitted over the data transmission network to the control center, must be reduced in form to produce traffic measures. These are, namely, volume and occupancy supported by speed, density, or other measures as needed. The cost of providing special equipment for each detector to produce these measures of volume and occupancy generally exceeds the cost of a central computer system that will perform all the data functions, data logging, and other system requirements.

Some of the features of the digital computer that are particularly suited to surveillance and control are reflected in its ability to:

1. Perform various functions according to a pre-established priority order.
2. Gather information, calculate parameters, make decisions, and output control commands in the same system at high speeds.
3. Operate in real-time so that events, as they occur, initiate pre-established priority level programs.
4. Operate automatically and without human intervention on a seven day, twenty-four-hour-day basis.
5. Record system performance and output the performance in printed form for permanent records as well as in display form to allow operator interaction with the system.
6. Control various kinds of systems, including signing and television subsystems,

by turning on, turning off, or selecting equipment, based on control and decision logic programmed into the computer.

7. Perform technical analysis in the free time of the control system.

AUDIO COMMUNICATIONS

A Motorist Aid System (MAS) helps to meet the needs of the traveling public by assisting authorities in making them aware of a disabled motorist, offering a means by which his needs are communicated and providing the appropriate aid response.

EXISTING AND PROPOSED SYSTEMS

A survey reported by the National Cooperative Highway Research Program indicates that there are at least 19 operating MAS (in 12 states), each with more than 25 units, that provide direct communication with a control center.[7] There are at least a dozen additional small-scale systems (25 or less roadside units each) currently in operation. These systems are summarized in Table 19.1.

The most extensive operation to date is the Southern California System which consists of 2,103 telephone installations on 12 major urban freeways in the greater Los Angeles area. All phones are spaced at one-half mile intervals. The 275 miles of freeway handle more than a million tourist and commuter traffic vehicles each day. The total 2,103 unit system cost was approximately \$965,000.00 to install, plus \$225,000.00 per year rental and maintenance charges. All calls are received by the Highway Patrol and required emergency services are dispatched by the Patrol. The system averages 25,000 calls per month or about 1.2 calls per phone per month. The conclusion of studies on the operation of this system indicated that although the total system is relatively expensive, its cost is apparently justified based on the number of persons served.

COMMUNICATING WITH THE DRIVER

The most critical need of a MAS is to provide help to the stranded motorist. The two main methods of achieving this are by the two-way, telephone type voice installation and by the nonverbal push-button installation discussed in the following paragraphs.

Handset telephones are the most common installation in Motorist Aid Systems. Figure 19.1 shows one of these telephones in use. The telephones are located on the shoulder of the roadway where shoulder clearance is available. They can be either drive-up or walk-up installations and are usually activated merely by removing the handset from its cradle. Some of the advantages of this method are:

1. The agency receiving the call can determine reasonably well that the aid call is authentic.

2. Radio handsets could be installed at isolated problem locations without having to install cable and conduit or to lease transmission facilities.

---

[7] *Motorist Aid Systems*, Synthesis of Highway Practice No. 7, National Cooperative Highway Research Program Report (Washington, D.C.: Highway Research Board, 1971).

TABLE 19.1
Existing Motorist Aid Systems of 25 or More Units

Location	Land Use	Number Locations	Spacing Miles	Kind
California				
Southern California (L.A.)	Urban	2,103	$\frac{1}{4}$	Telephone
San Francisco—Oakland Bay Bridge	Urban	250	$\frac{1}{10}$	Coded radio
Richmond—San Rafael Bridge	Urban	100	$\frac{1}{5}$	Coded radio
San Mateo Bridge	Urban	134	$\frac{1}{10}$	Coded radio
Connecticut				
Waterbury Viaduct		36	$\frac{1}{10}$	Telephone
Illinois				
I–80	Rural	302	1	Telephone
Kentucky				
Louisville freeways	Urban	110	$\frac{1}{2}$	Telephone
Maryland				
Harbor Tunnel Thruway		44	$\frac{1}{2}$	Telephone
I–495 (Capital Beltway)		324	$\frac{1}{8}-\frac{1}{2}$	Coded radio
Michigan				
I–94 (Jackson–Battle Creek)	Rural	62	$\frac{3}{5}-1$	Telephone
New Jersey				
Atlantic City Expressway	Urban	100	1	Coded radio
I–287	Urban	425	$\frac{1}{25}$	Push-button wire
New York				
I–87 Northway	Rural	712	$\frac{1}{2}$	Telephone
Pennsylvania				
I–80		370	$\frac{1}{2}$	Telephone
I–95		56	Varies	Telephone
Texas				
I–45 (Houston)	Urban	145	$\frac{1}{4}$	Coded radio
Virginia				
Chesapeake Bay Bridge–Tunnel		118	$\frac{1}{2}$	Telephone
Washington				
Alaskan Way Viaduct	Urban	35	$\frac{1}{18}-\frac{1}{4}$	Telephone
Evergreen Point Bridge	Urban	36	$\frac{1}{5}$	Telephone

3. Valuable time is saved by being able to dispatch the correct emergency equipment immediately.
4. The motorist has more confidence in the system because he receives instant acknowledgment and reassurance that help is on the way.

Figure 19.1. Roadside telephone.

Disadvantages are:

1. Limited shoulder space prohibits installation of telephones at some locations.
2. Stopped motorists can create a tie-up because of curious drivers slowing down to see what has happened. Once the habitual (commuter) driver becomes accustomed to the MAS, this slow-down tends to moderate, if not disappear completely.

The reliability, information capacity, and user acceptance of the telephone method has made this kind of installation the basic component of most existing motorist aid systems.

Push-button speakers have not been fully developed, but they are similar in character to the handset installation. These units are likewise located on the shoulder of the roadway and can be either of the drive-up or walk-up kind. The installation is activated by the user pushing a button on the face of the panel, thereby initiating two-way conversation with the monitoring authority. The main advantage of this system is that it is less susceptible to vandalism. The disadvantages are the same as those for the handset installation, plus the fact that the speakers must be far enough removed from the roadway so that they are unaffected by road noise. This is often impossible on a busy highway.

Push-button voiceless communication call boxes can be installed to transmit either by wire or by a radio wave transmitter at each installation. Most boxes have one or more buttons to push for police, fire, service, or ambulance, depending on the needs of the stranded motorist. Most boxes are not equipped with some means to let the motorist know his call has been received. The chief advantage of the system is that the radio installations are a relatively inexpensive means of communication and can be installed at isolated locations. The chief disadvantage is that the one-way communication capability results in less information being transmitted. Past experience has shown that specialized services, such as fire and ambulance, may not respond to the call until a police car has reached the scene and confirmed the need. This means that most systems operate as a one-button system regardless of how many alternatives are provided.

AM-FM radio broadcasting is the most common means of communicating information to the traveling motorist. Many radio stations in larger cities have peak-hour traffic reports, but unfortunately, they are not always as effective as they could be. The principal problem is the time lapse between the occurrence of an incident on the freeway and the broadcast report of the incident. Nevertheless, there is great potential for incorporating commercial radio into the overall motorist information system.

ROADSIDE RADIO TRANSMITTERS

There is continued interest in a driver information system that uses standard car radios and small roadside transmitters. If the driver is tuned to a certain frequency, he will hear recorded messages at certain intervals along the freeway when he is near one of the transmitters. The advantages of this method are: (1) the messages are constantly available to the motorist without his having to wait for new programming or commercials to end before broadcasts are made and (2) the message can be updated continuously to get the information to the motorist quickly. That major disadvantage is that, although it shows promise, it is subject to Federal Communications Commission approval and radio station acceptance.

COMMUNICATIONS CENTER

In the past the most frequent recipient of motorist aid calls has been the responsible police agency. It in turn dispatches the proper emergency equipment. Development of traffic surveillance and control centers and increasing nonemergency use of the MAS suggest closer coordination of the two efforts. This would allow a more rapid exchange of information and expedite necessary services required by the motorist. It would also aid in determining the necessary control requirements resulting from an incident.

VISUAL COMMUNICATION

For the purpose of this discussion, visual communication (a driver information system) shall be considered distinct and separate from a motorist aid system, even though in practice they usually are not separated.

Driver information can be considered in two ways: (1) information that would allow a driver to alter his route through a network so that he could avoid problems or other undesirable situations or (2) information that would inform the driver, while traveling a particular route, of specific problems within the network in order to allay some of his concern, irritation, and discomfort without requiring or expecting him to change his routing.

Driver information advising the motorist of incidents or problems within the network can be provided by the use of visual signing such as changeable message signs. Changeable message signs, shown in Figure 19.2, can be considered for route guidance, incident and congestion warnings, lane use designation, weather conditions, and general traffic information (see Chapter 16, Traffic Signs and Markings).

Figure 19.2. Changeable message sign.

GUIDELINES FOR USE

An effective driver information system can produce savings to motorists by reducing delay, aggravation, and frustration. Delay can be reduced through voluntary diversion (if good alternate routes are available and known) and through more efficient traffic operations (if the information presented is able to prevent flows from "breaking down").

To be effective, visual communication must provide information that is timely, accurate, understandable, and specific enough to be meaningful. These requirements in themselves pose an operational problem for the manager. The more specific the message, and the faster it is presented, the greater are the chances that it will be incorrect. The effectiveness of the system will decrease in proportion to the number of times it is wrong, or obviously redundant.

Informational signs may be used for three major functions: to identify the location of an incident, to indicate alternate routing, and to indicate the extent of disruption to traffic. Messages are generally displayed on changeable message signing. The changeable message signs are of different kinds, including lamp matrix, dot matrix, roller blind, and rotating drum.

The changeable message matrix equipment has the advantage of wide flexibility in message selection. The rotating drum displays various messages mounted on the facets of a rotating drum, but it is limited in the number of messages that can be displayed. However, it does have the advantage of legibility and the use of colors in the message. The roller blind sign can also be used for this application. It has more messages than the drum, and it also has the flexibility of showing color.

Regardless of the technique used, changeable message signs essentially provide information on the location of an incident in the system. Surveillance is used to initiate the command for these signs when it detects an incident. Manual controls may be used when these signs are operated by local officials, police, or highway personnel.

Alternate routing plans can be instrumented using changeable message signs, or changing symbol signs, for choices between freeway and arterial street operations. The characteristics of most trips are such that drivers do not like to radically change their routes each and every day. There is a preference for a steady and regularly used travel pattern. It is, therefore, desirable to provide routing information that is compatible with this preference. The routing system should be stable to the extent that the driver is directed to one alternate route or another, instead of being continually rerouted at each and every step of his journey.

Fixed legend signs and symbols may be used on parallel arterial roads to indicate a local alternate route at those locations where ramps are controlled. These signs essentially direct the driver, particularly the one unfamiliar with the local area, to an alternate access point along the freeway. They do not attempt to divert traffic in large numbers through the adjacent arterial street system.

Alternating routing between the freeway and arterial streets should be treated with great caution. There are very few situations in which it is beneficial to divert traffic from the freeway onto the arterial street system. More information about incidents along the freeway will generally, in and of itself, cause some local diversion to the arterial street system. However, to deliberately encourage traffic to leave the freeway for the arterial street system should be strongly discouraged.

The occurrence of a major incident on the freeway, in which there is a radical loss in freeway capacity and for which no remedy of removal can be achieved in a reasonable time, warrants the consideration of local detours. Such detours need close coordination between the freeway and arterial street operating agencies. Pre-planning for such emergencies, and pre-arrangements for implementing these detours, will greatly assist in the execution of an efficient detour.

RAMP CONTROL

The primary role of ramp control is to allocate the demand for travel in the corridor to the various highways within the corridor. This allocation attempts to provide the best level of service on all of the various highway links in the system. In the planning of the control scheme, the projected levels of service are derived from the estimates of demand and capacity for the individual links of the network. The capacity of both the freeway sections and the intersection capacities along the arterial street system are calculated, and traffic demand is then allocated by link and a level of service derived for each link. It is impractical to operate a ramp control scheme at peak efficiency without information about current conditions on the expressway. This information is best obtained directly from the surveillance information.

METHODS OF RAMP CONTROL

The three major kinds of ramp control that can be used in an operational system are ramp closure, ramp metering, and merge control.

Ramp closure consists of physically closing the ramp to traffic. This can be done either on a short-term basis, such as during peak period critical times, or it may be a permanent closure. Ramp closure during a peak period condition may be justified when the ramp demands are such that the ramp can be closed and traffic still has alternate routes for access to the freeway. At times, total closure is the only solution. If a ramp volume is sufficiently large, any partial control, such as ramp metering, may simply create backups and intolerable congestion in nearby parts of the street system.

Ramp closure may also be introduced when other control techniques cannot be used. Closure should be considered at locations where geometrics are inadequate and cannot be improved, at short ramps with inadequate storage or acceleration lanes, and at those locations where the ramp, even when controlled, does not allow traffic to merge into the traffic stream without considerable hazard or disruption.

Ramp metering should be used when the control of entrance ramps allows the freeway to operate at a higher level of service. Ramp metering may also be utilized to control ramp volumes so that the demand on a critical geometric section is reduced sufficiently to allow a more efficient level of service. In cases of geometric improvements, it allows additional traffic to enter the freeway in the area of geometric improvement, but only to the extent that the additional flow is commensurate with maintaining the level of service. Ramp metering in a system should only be used when an overall system benefit is achieved. Figures 19.3 and 19.4 show a typical mainline ramp metering layout and typical ramp metering signal and signing.

The development of ramp metering controls has followed a general pattern of simple controls gradually increasing in refinement to include not only the local condition, but critical bottlenecks in the network, by controlling the inputs to the system at points upstream from the physical bottleneck. Ramp metering started with a traffic responsive system in which the allowable ramp volumes were predicated on the local traffic flow conditions adjacent to the ramp. In the first Chicago ramp metering con-

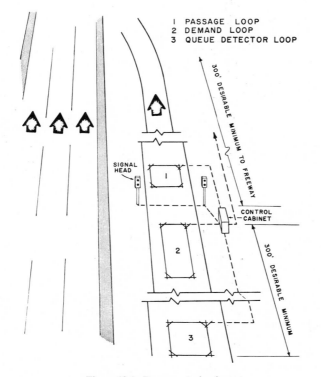

Figure 19.3. Ramp metering layout.

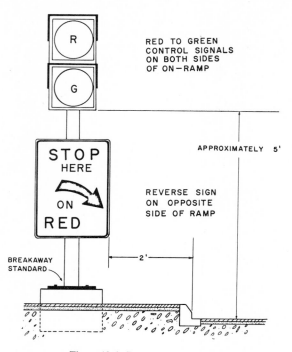

Figure 19.4. Ramp metering signal.

trol experiment, the allowable ramp volumes were derived from the lane occupancy measurements in the area of the controlled ramps.[8]

The experiments in Houston[9] and Los Angeles[10] used less equipment and started by pre-timing and pre-establishing allowable ramp volumes. The pre-timed system is based on historical data for traffic operations in that area. These systems, the pre-timed and the traffic responsive systems, are mutually exclusive and are not used together. Because the pre-timed system is based on historical data, it is allowed to continue in the same mode of operation until a new traffic survey is conducted on the system and until a reallocation of demand and capacity is made.

The traffic responsive system has generally undergone additional control refinement as the amount of instrumentation in the surveillance network has increased. The traffic responsive system also lends itself to the idea of multilevel control in which successive refinements in control parameters do not override the lower-level controls. For instance, the local occupancy control system should not be violated by any other control requirement. This stems from the fact that the allowable ramp volumes are based on avoiding turbulence at that location. It may be that the larger control system tying together a number of ramps upstream from a bottleneck would call for fewer ramp vehicles to enter than the local control logic would call for. Under these conditions the number of vehicles entering would be reduced. Should a network control system call for ramp volumes in excess of those volumes that can be handled at the local level, then the local control strategy would govern.

---

[8] McDermott, "Automatic Evaluation of Urban Freeway Operations."

[9] Pinnel et al., "Evaluation of Ramp Control."

[10] West, "California Makes Its Move."

The first major refinement in operating technique, over local occupancy control, was the coordination of ramp controls upstream of a bottleneck. In this control strategy, the bottleneck is identified, its operating volume established, and the flows upstream chosen so that traffic arriving at that bottleneck does not overload it. This control tends to be more restrictive on entering ramp volumes than the local control situation. It also is confusing in some degree to the motorist entering the freeway, because at the points at which he may be controlled, there is less congestion than at the critical bottleneck section. He may not relate the severity of the entrance ramp control to the unseen problem of congestion downstream at the bottleneck.

The control technique in this situation is the establishment of capacity, determined at the bottleneck from dynamic measurements, converted into ramp volumes entering upstream. The largest variable is the uncontrolled exit ramp traffic between the upstream ramp and the bottleneck. The amount of traffic reaching the bottleneck is the sum of the entering traffic less the exit ramp traffic. Through the exercise of control, the fluctuations in entrance ramp traffic are evened out and a steady flow pattern is achieved. By contrast, the exit ramp volumes do not smooth out and are subject to fluctuation. This fluctuation in exit ramp traffic feeds back and causes a readjustment to entrance ramp flow. Because the upstream ramps are removed in time from the operation of the bottleneck downstream, the control tends to be conservative in the maximum flow allowed to the bottleneck. If the occupancy is seen to increase at the bottleneck, the upstream controls are tightened. If storage is detected within the control section between the input to the section and the bottleneck, the controls are again increased to reduce the storage in the section and improve the level of service.

The control is primarily based on the level of service desired in the system. By controlling the operating level of service in the system, overall higher volumes for longer periods of time may be achieved throughout the length of the facility. However, at the output of the bottleneck, there is often a trade off between flow and level of service. For short bottlenecks, there may be a sizable loss in the output of the controlled section because short bottlenecks, with entrance ramps preceding them, have the ability to pack more traffic in congested conditions than do longer bottlenecks. Short bottlenecks are limited to a few hundred feet and in some cases they may be as short as from 20 ft to 30 ft. The control system here monitors level of service performance at all access points in the system and, from this measurement, establishes an allowable ramp flow. This is then compared to an allowable ramp flow developed from the downstream bottleneck location, and the lesser allowable flow is the one chosen in the control scheme. This control tends to be more conservative than the local control situation.

The system is then quite adaptable to incidents occurring within the control section. Should an incident occur away from a physical bottleneck, the surveillance subsystem detects this problem, develops a new control scheme for upstream, and returns downstream to a local control situation, which allows the ramp to replace the flow inhibited by the incident upstream.

Another form or method of entrance ramp control is called *merge control* or *gap acceptance*. For this technique, more instrumentation (and therefore greater expense) is required for successful operations than for ramp metering. A series of roadway detectors, installed in the right-hand lane of a roadway, are connected to a control center where a computer monitors the flow in that lane. The computer scans the sensors and locates a space (a gap) between any two successive vehicles. If this gap is of a pre-determined length or greater, a vehicle is released from an up-ramp holding point and allowed to enter the flow of the right-hand lane. Gap acceptance or merge control is, therefore, a method of merging on-ramp traffic into available spaces on the expressway. Merge control is most effective on ramps at which there is a turbulence generated by vehicles entering the main traffic stream for long periods of time. The turbulence should be persistent and not isolated to an individual minute or so. It should probably exist for at least half an hour in each peak period.

Merge control may also be considered when there are inadequate acceleration lanes. This control should not, however, be used as a substitute for providing adequate acceleration lanes. In a freeway situation, direct taper acceleration lanes of 300 ft and more can operate adequately as a ramp metering setup. Thus, merge control is to be favored in acceleration lanes of about 300 ft or less. This control operates effectively up to ramp volumes of about 500 vehicles per hour. Above this, ramp volumes and platooning on the ramps become excessive to the point that some form of ramp metering control may be needed. Merge control is more effective at the lower volumes. Another factor in merge control is the main line volume on the approach area to the merge zone. The volumes on the lane adjacent to the ramp should not exceed a value such that the right-hand lane, plus the ramp lane, exceeds 1,800 vehicles per hour.

Merge control schemes are very similar to ramp metering control schemes that are based on local conditions at each installation along the freeway. In other words, merge control is principally a local control technique. The allowable flow under the merge mode may be more, and is generally more, than the flow allowed under the bottleneck consideration. In such a situation, the merge control is modified so that the acceptable gap is lengthened, thereby rejecting the ramp flow. As the flow reaches the higher levels, the merge control becomes more subservient to the bottleneck considerations.

On urban freeways, during the peak period, the allowable flows (as calculated from a freeway capacity standpoint) on entrance ramps are generally sufficiently small such that with contemporary design standards, most entering vehicles have little difficulty in merging. For these low flows, one-at-a-time release methods, such as the traffic responsive system, do not experience any merge problems. During light freeway flow conditions, ramp flows are usually light. If a ramp has a high demand during this off-peak period, merge control may be effective in limiting the release of vehicles into platoons of vehicles on the freeway. Under such conditions, the entering traffic, even released one at a time, can adjust to the platooning on the freeway if the ramp controlled vehicle can attain freeway speeds within the acceleration lanes.

Most drivers have little trouble in merging under such a condition. Refinement in control can extend to the concept of treating the origins and destinations or traffic entering the freeway by exercising control in order to optimize flow with linear programming or other programming techniques. Some of these techniques become demanding on the computational system, and returns may be small once the bottleneck control system is implemented. Performance does not appear to be superior to less elaborate methods because of the necessary vehicle acceleration assumptions, overrides, and gap stability.

Another control refinement is the attempt to redistribute the queues at the individual ramps in some optimal manner. This may be used in a control scheme. Typically, when the ramps are queued, most of the ramps tend to be lined up, and there are no alternate locations to which one can shift control. The queue effect is rationalized on the basis of allowing longer queues at ramps where there is less interference with the cross streets or other traffic, thus keeping queues short at those locations where modest queues will interfere with cross-street traffic. This refinement should always be considered, and there may be times when it is effective. However, there is often a demand and there are queues at all ramps. The other measure to be considered is that when queues do interfere, thought should be given to actually closing the ramp for operational reasons.

In addition to the central computational system, there are additional controllers for ramp controls, signs, and traffic signals. These systems may be either special purpose equipment or incorporated within the central computation equipment. The ramp metering controller, in a traffic surveillance and control system, may be either located in the field at the ramp location or at a central location. When the device is located in the field, it gathers information locally, makes a control decision, and changes the signal on the spot. It may be overridden or modified in timing from a central system. When there is a communication link with the field, the ramp controllers themselves may be either programmed into the computer or be on control circuit cards located at the central location. The use of circuit cards allows the manual system to continue the operation even if the computer has failed. Generally speaking, the demands for ramp control are such that computer software controllers are not used because a system failure would cause loss of control for the ramps.

If the cost of a central control location exceeds the cost of a special-purpose computer in the field, consideration should be given to using special-purpose computers. If telemetry communication costs are not large, the information may be brought to a central control and command signals returned to the field. Locating equipment in the field means that maintenance would require some form of control or inspection process. Equipment in a central office allows maintenance personnel to replace equipment on the spot and restore the system. In either case, the computer itself is used to monitor the total traffic surveillance and control system for malfunctions or other equipment failures.

CLOSED-CIRCUIT TELEVISION

The most significant role of television in the system[11] is to aid and reassure the human operator of the need for special service at the scene of an incident. The automatic detection system, with continued use, builds up the operator's confidence in the ability of the system to detect incidents or other problems. When such detection initiates a standard service procedure, the automatic detection system is adequate. When the human operator wishes to instigate special service or take action on an incident, he needs some visual display of conditions in the system on which to justify this action. Another extension of this concept is the desire of the operating agency to manage incident removal procedures from a central location.

Closed-circuit television (CCTV) can perform three major functions:

1. As an adjunct to certain control actions
2. As an identification device for traffic incidents
3. As a supplement to a motorist aid telephone system in identifying the need for assistance in areas where it is not possible or practical to install telephones

Ramp closures or lane use regulations may require supplemental visual information in order to initiate proper control. To this end, television can be considered a sur-

[11] Freeway Operations Group, "Closed-Circuit Television Surveillance and Reversible Roadway Control," Washington Department of Highways, 1969.

Figure 19.5. Television camera.

veillance tool at persistent congestion or incident locations. Figure 19.5 shows a television camera installation.

Closed-circuit television might also be coordinated with detector surveillance so that incident alarms would automatically indicate the appropriate camera for viewing on the television surveillance monitor.

CCTV should not be considered as a primary surveillance tool outside of the limitations mentioned above. As a supplement it is excellent, but as a data-collection device it is inefficient.

The use of television can greatly assist the operating agency in the coordination of police efforts, maintenance crews, and service operators, all of whom may be at any one time active in a certain area of the freeway. The television system allows for coordinated management of operations at the scene of the incident, and it allows decisions to be made based on information from the scene, as viewed through the coordinator's eyes, and not just as reported by individual groups. The need for this action may be most prevalent on those sections of freeway where there is no shoulder or refuge for disabled and stalled vehicles. The detectors can locate an incident that causes a traffic reaction and television can assist in clearing the roadway of the incident. At critical locations it is important not simply to remove the incident itself, but also to rapidly remove all police, maintenance, or service activities from the critical section and again restore the much needed capacity to the freeway.

CONTROL CENTER

The control center should house the digital computer and related peripheral equipment, communication consoles, and network control stations, and it should have the equipment necessary to dispatch emergency and maintenance vehicles to problem locations. It receives information concerning traffic, analyzes and evaluates it, and makes control decisions. It should be capable of informational services not only to news media, but to the general public as well.

The control center is not only a focal point for communication and control equipment, but control center staff must also maintain good working relationships with other highway service agencies such as police, aid services, and maintenance crews. In addition, a high degree of cooperation must be exercised with traffic engineering offices of other governments through which the freeway system passes because the operation of freeways and arterial streets cannot be viewed as independent from one another.

SERVICE TO MOTORISTS

The last step, and a critically important one, in the chain of events following the occurrence of an incident is the rapid removal of the incident and conscientious assistance to anyone directly involved in that incident. The severity of delay resulting from a major incident is much higher than the delay expected from congestion or other overloading. In some cases, there is considerable difference between the delay caused, say by an accident, and the delay saved through the exercise of ramp control on a congested section of freeway. The return in removing an incident from the freeway is tenfold compared to the introduction of control on that facility. This realization of the impact of incidents within the freeway network has extended the need for remedial action beyond the exercise of ramp controls.

Surveillance is the key to detection of the incident. This is supplemented by the presence of police patrols in the area. However, the benefits of surveillance to locate incidents within the system have not been fully realized because of the lack of ability to respond quickly to detected incidents. The potential benefit in the rapid removal of incidents within the freeway system, and in the corridor, depends on an action program for patrolling and servicing incidents. This program, for a freeway environment, should ideally include legislation and service facilities which, upon location of an incident, allows the rapid removal of that incident completely off the freeway and out of sight of traffic in both directions. Incident removal also relies heavily on operational procedures by operating and maintenance personnel to remove themselves from a conspicuous situation and to settle paper work and other administrative requirements in a location off the freeway, thereby providing both safety to those involved and minimizing hazard to others.

Freeway operations can be sustained at lane volumes of approximately 1,800 vehicles per hour per lane averaged on all lanes. Design values are set at approximately 1,500 vehicles per hour per lane. For an accident, major congestion, or incident in the system, however, the lane volumes drop to approximately 1,200 to 1,300 vehicles per lane. Thus, for a four-lane freeway section in one direction having an incident in one lane attracting the interest of other motorists, the operating volumes drop from 7,200 vehicles per hour to 3,600 vehicles per hour, or a 50 percent loss in performance.

When traffic lane volumes exceed the 1,200 to 1,300 vehicles per hour value, major shoulder activities such as stalled cars, abandoned vehicles, or necessary police apprehensions can cause flow reductions and development of congestion queues or shock waves. One of the major contributions to secondary delays is the attention paid by motorists to incidents that have occurred in the opposite direction of travel.

GEOMETRIC CAPACITY IMPROVEMENTS

In order to sustain a balance between demand and capacity on the freeway, two approaches can be considered. One is to control demand through the use of ramp control techniques and the other is to improve freeway capacity to meet critical volume demands by making changes in geometrics.

In urban areas many original designs were compromised because of the lack of right of way or other facilities. Thus, geometric redesign is often impractical. There is considerable merit, however, in complete redevelopment of the freeway within the existing right of way. As an example, the combined cross section of pavements, shoulders, and median may be redeveloped to provide additional capacity in a particular section. Because traffic demands are continually changing and because geometric redesigns will contribute to changing demand patterns in the area, it is desirable to take a step-by-step approach to geometric redesign. On a freeway with shoulders and median it may be helpful to restripe the pavement and shoulders in order to gain extra traffic lanes. After this, medians can often be modified by introducing barrier type separators and, finally, the whole section can be rebuilt.

It is extremely important to be sure that additional capacity along the roadway will in fact improve the operations of the system. For example, widening one small section of the freeway may accomplish nothing more than providing a small increase in speed through the widened section and moving the traffic to another narrow section downstream. To avoid wasteful redevelopment, capacity improvements should be developed to work in directions counter to the flow of traffic. In attempting to improve freeway operations through the use of capacity improvements, they should be considered at downstream locations, or at the bottleneck first, and then progressively work back upstream against traffic flow, thereby assuring that the improvement will allow traffic to leave the problem area.

The obvious importance of the demand-capacity relationship is that geometric design depends very heavily on the ability to control demand in some manner. In some cases, when capacity improvements are made, it will be necessary to restrict additional demand for the freeway so that a high level of service can be maintained. By controlling this demand, the monies expended on the geometric redesign will bring a return in operational improvements.

EFFECTIVENESS OF SURVEILLANCE AND CONTROL

The first major experiments in surveillance and control,[12,13,14] conducted at very carefully selected sites, showed that predictable benefits could be achieved. These sites are categorized as corridors in which freeway congestion is caused by poor operational design and those corridors in which the travel demand caused congestion on the freeway.

The characteristic performance of these two categories of systems is substantially different. When geometric design is the major problem, significant reduction in accidents is achieved by surveillance and control. When traffic demand is the major problem, there is less likelihood of accident reduction. At geometric constraints, it is more difficult to prevent breakdown at the bottlenecks. In other situations, bottlenecks can operate at a higher level of service for sustained periods. The benefits in both categories of systems can be summarized as follows:

1. Total travel served: In the total corridor, the amount of travel, measured by vehicle miles, remains essentially unchanged within the accuracy of the measur-

[12] McDermott, "Automatic Evaluation of Urban Freeway Operations."
[13] Pinnel et al., "Evaluation of Ramp Control."
[14] Gervais, "Optimization of Freeway Traffic by Ramp Control."

ing equipment. Thus, the surveillance and control system serves the same travel demand, but at a different level of service.

2. Total travel time expended: Most experiments show a significant reduction in travel time in the corridor. Travel time for the corridor is measured accurately by recording travel times on the freeway, queue delays at entrance ramps, and moving vehicle estimates of travel time on the arterial streets. Changes in travel time on the arterial street are in most cases too small to be significantly differentiated from the previous uncontrolled travel times. The cumulative travel times on the arterials remain the same, but in contrast, the travel time on the freeway shows a significant reduction. In cases where the control is more stringent, and there is an absence of geometric bottlenecks, the reduction in freeway travel time is large. The freeway changes from a level of service "F" back up to a level of service about "D." Against this benefit there is an induced delay to traffic entering the freeway at metered entrance ramps and in some instances to traffic at the intersection adjacent to the controlled ramps. Counting the ramp queue lengths at fixed time intervals gives the ramp delay. Individual travel times for specific movements through the adjacent intersections yield measures of delay. The volume and delay for each movement give the total intersection delay caused by the change in traffic. In some situations, input-output counts estimate intersection delay. The time record of vehicles within the intersection area gives total travel time.

   The net reduction in travel time on the freeway must always exceed the delay to vehicles entering the freeway on the entrance ramps. On critical occasions the delay to intersection traffic or cross-street traffic may approach the freeway savings. A successful system should show a significant reduction in travel time for all travel in the corridor.

3. Accident and hazard rates: On freeways where geometric design elements cause congestion, surveillance and control significantly reduce the number of accidents on both the freeway and on the entrance ramps. The introduction of one-at-a-time metering smooth the merge operation and reduce the number of hazardous maneuvers in the merging areas. On highly substandard acceleration lanes the use of merge control reduces the incidence of hazardous maneuvers in the merge area.

   On freeways not deficient in geometric design, there tends not to be a reduction in accidents. In fact, increases in accidents can occur in the initial control period.

4. Incident delays: The surveillance system provides a record of the incidents and the delays occuring on the freeway. Early attempts to isolate the impact of different control strategies met with limited success because of the persistence of incidents within the control area. Although these incidents occur at frequent intervals, the time, location, and severity cannot be usefully predicted. The delay from a major incident is much higher than the delay expected from congestion caused by traffic overloading; in some cases this is a tenfold difference. Fast removal of an incident from the freeway can save much more delay than ramp control on the freeway. The potential benefit in the rapid removal of incidents within the corridor depends on an action program for service patrols. One of the major contributions to the excessive delay caused by incidents is the attention paid to the incident by motorists traveling in the opposite direction.

5. Developing congestion: The surveillance system allows the operating agency to

monitor the growth of traffic. As discussed previously, freeways can sustain lane volumes of 1,800 vehicles per hour (averaged on all lanes), even though the lane volumes are 1,500 vehicles per hour. An accident or major incident can cause a 50 percent decrease in lane volumes. When traffic lane volumes exceed the 1,200 vehicle per hour rate, major activities on the shoulder can cause congestion and flow reductions. Such volume levels warrant a surveillance system. When entering ramp volumes show fast growth, congestion can develop on the freeway. If ramp control is implemented early enough, and the controlled volumes equal the demand, there will be little or no adverse delay to traffic. Adequate surveillance provides the operating engineer with timely information so that he can prevent the growing congestion so often witnessed on uncontrolled freeways.

SYSTEM COSTS AND DESIGN

COSTS

Sensors. For large-scale systems the cost of freeway installations can vary from under $500 per detector up to $1,000 depending on the local conditions. In the arterial street system the cost of detector installation is much less. The higher cost of freeway installation includes the required traffic control provisions associated with freeway work.

Communications costs. It is difficult to balance the competing interests of leased and privately owned facilities. This is particularly acute in surveillance and control. As a general guide for estimating costs, communications within the highway right of way should be privately owned and communications away from the highway should be leased. In surveillance and control, frequency-division multiplexing has found wide application because of its cost advantage for low information rates. The time-division multiplexing is most useful and economical when a large number of signals can be gathered together at a point and then transmitted from that point to another such as a central location. For each detector using frequency-division multiplexing, the cost of a transmitter and receiver installed and wired in place would vary from $300 to $500. The corresponding cost of time-division multiplexing would show a price advantage for about 10 or more detectors on a single channel. Capital costs alone should not determine the kind of communications system used. Maintenance costs and the ability to recruit trained personnel are key factors in selecting the communications system.

Computational equipment. The role of surveillance and control should not depend on the development of special equipment. There is adequate hardware, operational and fully developed, to satisfy the needs of a surveillance and control system. Computational equipment chosen from a digital computer supplier should be made up entirely of standard products in the computer line. It should be unnecessary to call for special equipment. Software packages from the manufacturer should be utilized and a minimal amount of programming should be undertaken by the operating agency. Since most operating agencies have neither the time nor the expertise to develop software packages, the success of the surveillance and control system can be seriously jeopardized by inadequate programming. Development of new software should only be undertaken with the full understanding of the need and requirements of such a choice. The computer system costs vary from $200,000 to $400,000.

Facilities. A central facility requires space for the computational equipment, information display boards, control monitoring, communications equipment, and personnel. Space requirements can rapidly reach 5,000 sq ft for a major system and exceed 10,000 sq ft for an urban area. The cost of these facilities would follow prevailing building costs. The special requirements of the building can bring the cost of the building above common pricing. In an effort to conserve land, strenuous efforts are usually made to house the surveillance and control systems within available highway right of way. Installation of a surveillance and control center within leased property is very constraining unless adequate long-term commitments can be made.

Operation costs. Maintenance cost will vary, depending on the standard set by the operating agency, but for a large system it can be estimated at approximately 10 percent of the capital cost per year.

PLANNING AND DESIGN

In order to achieve flexibility certain design features should be included in the initial highway construction. Some of the major design factors are as follows:

1. Provide continuous routes in the travel corridor. Frontage roads, when used, should be continuous throughout the corridor. It is not sensible for traffic operations to invest heavily in frontage roads along portions of the highway and then abandon them at critical sections.
2. Locate entrance ramps as far away as possible from cross-street intersections. The farther the ramp is away from the cross-street intersection, the better the operation of ramp control.
3. Controlled ramps, operating as single lanes, should be only single-lane width (12 ft basic). Such ramps should have single-lane pavements for traffic and full width shoulders for emergency passing.
4. Provision to integrate bus or rapid transit with the controlled freeway traffic. Terminals with special transit entrance ramps to the freeway allow for preferential transit operations.
5. Shoulders at critical bottlenecks should have adequate structural strength to allow for restriping or widening at minimum cost.
6. Flexibility should be built into the system. The design must accommodate the traffic demand that eventually develops.
   Success depends on meeting actual traffic demand, not in meeting projections.

REFERENCES FOR FURTHER READING

EVERALL, PAUL F., *Urban Freeway Surveillance and Control.* Washington, D.C.: Federal Highway Administration, 1972.

MOSKOWITZ, K., *Analysis and Projections of Research on Traffic Surveillance, Communications, and Control,* National Cooperative Highway Research Program Report 84. Washington, D.C., 1970.

Chapter 20

LIGHTING OF TRAFFIC FACILITIES

NEILON J. ROWAN, P.E., Ph.D., Research Engineer and Professor, Texas A&M University, and NED E. WALTON, P.E., Ph.D., Assistant Research Engineer and Assistant Professor, Texas A&M University, College Station, Texas.

To the urban community, street lighting is a means of improving the urban environment through increased comfort, convenience, and safety of night-traffic operation and reduced crime and accidents. To the traffic engineer, street lighting is a tool that can be used to increase his ability to communicate with the driver. Lighting, in itself, cannot produce traffic safety, but it can significantly improve the efficiency of traffic operation and provide increased safety as a byproduct. The traffic engineer should thus approach the application of street lighting principally as a means of increasing his communication with the nighttime driver.

Public lighting should permit users of traffic facilities to move about at night with the greatest possible safety, comfort, and convenience. The driver must be able to see distinctly and locate with certainty and in time all significant details of the driving environment. The pedestrian must also be able to see distinctly his path and his relationship to vehicles and possible obstacles. Although public lighting must satisfy the informational needs of both drivers and pedestrians, in practice, the driver's requirements are the more stringent.

The importance of providing the driver with information needed for satisfactory performance has been pointed out by Cumming,[1] who states, "The road complex must provide for the operator a comprehensive display of information both in the formal sense of signs, signals, guidelines and edgeposts, and in the informal sense of clear visibility in all relevant directions." The purpose of information, whether through visual or other senses, is to reduce uncertainty. As long as uncertainty exists, possible alternate decisions cannot be fully evaluated.

Visual information needs of the driver are a direct function of what the driver does in performance of the driving task. The requirements of fixed lighting are based upon how the driver sees and what he needs in terms of information for satisfactory driving.

LIGHT AND SEEING

LIGHT

Light is radiant energy defined in terms of its capacity to produce visual sensation.[2] Radiant energy of the proper wavelength from a source of sufficient visual angle

---

[1] R. W. CUMMING, "Progress Report of Human Factors Committee," *Australian Road Research*, XX, No. 2. (1962).

[2] *IES Lighting Handbook*, 4th ed. (New York: Illuminating Engineering Society, 1966), p. 1–1.

becomes visible when the energy is emitted or reflected in sufficient quantities to activate the receptors in the eye.

CONTRAST

Contrast is one of the most important contributors to nighttime visual performance. To a great extent, the recognition of objects is based upon a discrimination of brightness (luminance) differences between an object and its background.[3] For night conditions, an obstacle may appear as a dark area against a bright background (discernment by silhouette) or it may appear as a bright area against a dark background (discernment by reverse silhouette). To enhance discernment by silhouette, brightness of the pavement and uniformity of brightness along and transverse to the roadway are essential. Discernment by reverse silhouette usually applies to the visibility of objects on areas adjacent to the roadway, projections above the pavement surface such as channelizing islands and abutments, and the upper portions of pedestrians or vehicles.

DISCERNMENT BY SURFACE DETAIL

Discernment by surface detail depends on a high order of direct illumination on the face of the object toward the driver.[4] The object is seen by variations in brightness or color over its own surface without general contrast with a background. Discernment by surface detail may be a principal method of seeing in heavy traffic when the complexity of the situation requires considerable visual detail.

VISUAL ACUITY

Contrast sensitivity, or the ability to distinguish luminance (brightness) differences, provides for the detection of objects, but the identification of most objects is accomplished by visual acuity. By definition, visual acuity is the ability of the eye to resolve small detail.[5] In driving, two kinds of visual acuity are of concern—static and dynamic visual acuity.

Static visual acuity occurs when both the driver and the object are stationary and is a function of background brightness, contrast, and time. With increasing illumination, visual acuity increases up to a background brightness of about 10 millilamberts, and then it remains constant despite further increases in illumination. Static visual acuity also increases with increasing contrast of the object. Optimal exposure time for a static visual acuity task is from 0.5 to 1.0 sec when other visual factors are held constant at some acceptable level.

When there is relative motion between the driver and an object, such as occurs in driving, the resolving ability of the eye is termed *dynamic visual acuity*. Dynamic visual acuity is more difficult than static visual acuity because eye movements are not generally capable of holding a steady image of the target on the retina. The image is blurred and, therefore, its contrast decreases. The conditions favorable for dynamic visual acuity are slow movement, long tracking time, and good illumination. These

---

[3] INGEBORG SCHMIDT, "Visual Considerations of Man, the Vehicle, and the Highway, Part I," Publication SP-279, Soc. of Automotive Engrs., Inc. (March, 1966), p. 7.

[4] *American Standard Practice for Roadway Lighting* (New York: Illuminating Engineering Society, 1963), p. 29.

[5] SCHMIDT, "Visual Considerations of Man, the Vehicle, and the Highway, Part I," p. 10.

are rarely found in the nighttime driving environment except in sign reading, an important dynamic visual acuity task.

GLARE

Two kinds of glare have a critical influence on driver visual performance:

1. Discomfort glare (psychological glare): ocular discomfort from a bright light source
2. Disability glare (physiological glare): stray light that reduces contrast sensitivity and, thus, produces a loss of visual efficiency

Glare is an especially disturbing influence when one is viewing a difficult visual task under low brightness conditions.

Discomfort glare effects can be diminished by reducing luminaire brightness, by increasing mounting height, and by increasing the background brightness in the observer's field of view. Luminaire brightness may be reduced by increasing the effective luminaire area and by decreasing the intensity of light at angles higher than that required for optimum pavement brightness (luminance). The background brightness is increased by raising the general level of illumination. Disability glare can be reduced by increasing mounting height, moving the luminaire from the line of sight, and increasing background brightness.

Current practices in light design, such as higher mounting heights, increased lateral set-back of supports, and restriction of light from the luminaire at vertical and horizontal angles where interference with driver visibility is most significant are effective in controlling discomfort and disability glare.

TIME AS RELATED TO SEEING

The time available to the driver to perform the visual task, whether recognition of an object or perception of a traffic situation, is very important. This time decreases as vehicular speed and situation complexity increase. The time factor is extremely critical in low-illumination areas or in high-brightness variation areas where the eye is in a continual state of adaptation.

DRIVER VISUAL INFORMATION NEEDS

Visual information needs associated with the driving task can be organized in accordance with three driver performance levels. These levels and the general information needs are as follows:

1. Positional performance: needs associated with routine steering and speed control. These needs are satisfied primarily through pavement markings, curb delineation and delineation of road edges, lane divisions, and roadside features.
2. Situational performance: needs associated with required changes in speed, direction of travel, or position on the roadway as a result of a change in the geometric, traffic, and/or environmental situation. These needs are as varied as the number and kinds of road and traffic situations encountered in driving.
3. Navigational performance: needs associated with selecting and following a route from an origin to a destination. These needs are satisfied primarily by

formal information sources (signs, etc.) and informal information sources (landmarks, etc.).

From the fixed lighting standpoint, situational information needs are most critical. Positional and navigational needs can usually be accommodated by the vehicle headlights because important signing, marking, and delineation are usually either retroreflective or externally illuminated.

Table 20.1 lists the more important elements of the nighttime driving visual environment and should provide insight into the selection and area of coverage for a lighting system. From the standpoint of fixed lighting, these elements become more complex and important as geometric, operational, and environmental complexity increases. The complexity of such conditions can be considered as warrants or minimum justification for the installation of a lighting system since they are the conditions

TABLE 20.1
Important Elements of the Night-Driving Visual Environment

Element	Kind of Information	Description
Roadway geometry	Positional Situational	Perception of the roadway alignment, topography, and cross section at a distance commensurate with travel speed
Intersection	Situational Navigational	Perception of intersecting roadway ahead commensurate with travel speed
Channelization	Positional Situational	Perception of markings, curbs, medians, etc., that indicate an assigned path
Lane markings	Positional Situational	Perception of lane lines, edge lines, center lines
Roadside and roadside objects	Positional Situational Navigational	Perception of the environment for dynamic appreciation and recognition of possible hazards
Curbs	Positional Situational	Perception of curb as an object and guide
Access drives	Situational	Recognition of curb break, pavement contrast, or other features indicating an access opening
Pedestrians	Situational	Perception of pedestrian on or adjacent to the roadway and recognition as a possible conflict.
Vehicles	Positional Situational	Perception of other vehicles on the facility, their location and intended directions in relation to own location and movement
Signs	Situational Navigational	Perception and recognition of signs and contents
Signals	Situational	Perception of color and/or orientation of signal heads indicating assignment of right of way.
Pavement edge	Positional	Perception of pavement boundaries, contrast between pavement and shoulder or roadside and edge lines
Delineation	Positional Situational	Perception of roadway delineation as indicative of roadway features
Special geometric features	Positional Situational	Perception of conflict points, ramp exits and entrances, merges, ramp configuration and direction
Roadway objects	Situational	Perception of hazardous objects on the roadway at a distance commensurate with the travel speed
Road condition	Situational	Perception of road surface indicating structural and climatic conditions
Special roadside features	Situational Navigational	Perception of signs, land marks, etc., indicating an intermediate or final destination

usually producing various informational needs. From the geometric aspect, the need to see roadway features, access drives, channelization, and other cross-section elements depends on the complexity of the geometrics. From an operational viewpoint, high speeds, high volumes, and frequent interactions and conflicts produce high visual information needs. Environmental conditions, such as extraneous lighting, roadside development, and service drives also create information needs and compete for the driver's attention.

PAVEMENTS

Under fixed lighting, drivers usually see objects within the roadway as dark silhouettes against a bright background formed by the roadway and its surroundings. Therefore, one of the important uses of fixed lighting is to brighten the roadway surface. In producing this brightness, the reflection of light from the surface and the surface illumination are equally important.

Efficient fixed-lighting systems should provide suitable light distribution in which the bright patches of light cover the roadway from any direction of viewing, and the surface should be as brightly and uniformly illuminated as possible without excessive glare. Luminaires do not usually produce a uniform intensity distribution, however, because more light is directed up and down the roadway in order to lengthen the bright patch and thereby reduce the number of luminaires required.

The way a roadway is brightened by street lighting depends on the exposed face or surface of the roadway. The reflection characteristics of a surface depend on a number of properties: (1) the surface texture (also influences skid resistance), (2) the material used, (3) the color and lightness, (4) the extent to which the surface has been polished by traffic, and (5) the degree of wetness or dryness. Lighter colored pavements of Portland cement concrete, synthetic aggregates, and certain surface treatments can be used economically where it is apparent that roadway lighting is justified.

Regardless of the paving material used, there are aspects of pavement texture and color that should be considered in the design of a lighting system. Reflectance factors developed by deBoer[6] can be used with adequate accuracy to predict the resulting luminance of pavements when illuminated. These factors can be applied to the design level of illumination to compensate for pavement surface properties. A simple pavement classification scheme can be used as follows:

Pavement Condition	Extremely Light Pavement	Above Average Pavement	Average Pavement	Below Average Pavement	Extremely Dark Pavement
Reflectance factors	.34	.31	.27	.23	.19

By assigning a unit value to the average pavement condition, an approximate weight, or multiplier for the pavement conditions, can be established. These multipliers are shown below:

Pavement Condition	Extremely Light Pavement	Above Average Pavement	Average Pavement	Below Average Pavement	Extremely Dark Pavement
Multiplier	.80	.90	1.0	1.2	1.4

[6] J. B. DEBOER, *Public Lighting* (Eindhoven, The Netherlands: Philips Technical Library, 1967), p. 197.

To use the multiplier, the design level of illumination (in foot-candles) is simply corrected by the multiplier. Although the technique is inexact, it responds reasonably to pavement factors that relate significantly to information needs and seeing.

TRAFFIC CRITERIA AND WARRANTING CONDITIONS

ROADWAY AND WALKWAY CLASSIFICATIONS

Most highway and street systems encompass several classes or types of roadways and walkways. At one extreme are high-speed, high-volume facilities carrying through traffic, with no attempt made to serve abutting property, pedestrians, or local trips. At the other extreme are local highways, streets, or roads that carry low volumes, at low speeds, with a primary function of land access instead of vehicular movement. A comprehensive lighting program requires that the road and street be classified on the basis of intended function.

The following classifications are those recommended by the Illuminating Engineering Society:[7]

1. *Major.* The part of the roadway system which serves as the principal network for through traffic flow. The routes connect areas of principal traffic generation and important rural highways entering the city.
2. *Collector.* The distributor and collector roadways serving traffic between major and local roadways. These are roadways used mainly for traffic movements within residential, commercial, and industrial areas.
3. *Local.* Roadway used primarily for direct access to residential, commercial, industrial, or other abutting property. They do not include roadways carrying through traffic. Long local roadways will generally be divided into short sections by collector roadway systems.
4. *Expressway.* A divided major arterial highway for through traffic with full or partial control of access and generally with interchanges at major crossroads. Expressways for non-commercial traffic within parks and park-like areas are generally known as parkways.
5. *Freeway.* A divided major highway with full control of access and with no crossings at grade.
6. *Alleys.* A narrow public way within a block, generally used for vehicular access to the rear of abutting property.
7. *Sidewalks.* Paved or otherwise improved areas for pedestrian use, located within public street rights-of-way, which also contain roadways for vehicular traffic.

AREA CLASSIFICATION

Although the above classifications normally reflect the geometric and operational conditions to be expected on a traffic facility, sometimes additional subclassifications

---

[7] *American National Standard Practice for Roadway Lighting* (New York: Illuminating Engineering Society, American National Standards Association, 1972), p. 9.

by area type and environmental conditions are desirable. The Illuminating Engineering Society[8] recommends the following area classifications:

1. *Commercial.* That portion of a municipality in a business development where ordinarily there are large numbers of pedestrians and a heavy demand for parking space during periods of peak traffic or a sustained high pedestrian volume and a continuously heavy demand for off-street parking space during business hours. This definition applies to densely developed business areas outside of, as well as those that are within, the central part of a municipality.
2. *Intermediate.* That portion of a municipality which is outside a downtown area but generally within the zone of a business or industrial development, often characterized by moderately heavy nighttime pedestrian traffic and a somewhat lower parking turnover than is found in a commercial area. This definition encompasses densely developed apartment areas, hospitals, public libraries, and neighborhood recreational centers.
3. *Residential.* A residential development or a mixture of residential and commercial establishments characterized by few pedestrians and a low parking demand or turnover at night. This definition includes areas with single family homes, townhouses, and/or small apartments. Regional parks, cemeteries, and vacant land are also included.

NIGHT TRAFFIC VOLUMES

Vehicular traffic volume must be considered as one warranting condition for the installation of a fixed lighting system. This is especially true in a comprehensive lighting program where priorities must be assigned for efficient use of public funds. Accidents increase demonstrably as the average daily traffic (ADT) increases. Traffic conflicts and interactions also increase proportionately with volume increases.

Volume counts can be made to determine average night volumes on a given traffic facility. The average night volume for higher type facilities, however, can be adequately estimated as 25 percent of the average daily traffic for most facilities.[9]

SPEEDS

Traffic speeds must also be considered a warranting condition for roadway lighting. As traffic speeds increase, less time is available for seeing. Every possible advantage, including roadway lighting, should be afforded the driver. Typical vehicle lighting provides for an operating speed of approximately 45 mph. When competition for driver attention and other complex situations exist, lower speeds often justify fixed lighting.

TRAFFIC MANEUVERS AND ACCESS CHARACTERISTICS

Traffic maneuvers and access characteristics must also be considered when determining warranting conditions for fixed lighting. In areas in which there are frequent turns, entrances and exits from abutting properties, and interchanging vehicles, road-

---

[8] IBID., p. 9.

[9] PAUL C. BOX, "Relationship Between Illumination and Freeway Accidents," Project 85–67 (New York: Illuminating Engineering Research Institute, April, 1970), p. 23.

way lighting can be critical. The need for lighting increases proportionately with the frequency and complexity of these maneuvers.

PEDESTRIANS

Pedestrian death rates are extremely high in unlighted urban areas. There is a definite need to light pedestrian walkways, sidewalks, and other locations (especially mid-block) where pedestrian activity is high.

VEHICLE ACCIDENTS

Vehicle-vehicle, vehicle-object, and vehicle-pedestrian accidents at night must continue to be a warranting condition for fixed street lighting. In areas where high night-to-day accident rates exist, roadway lighting must be given priority consideration. In addition, locations where accident potential is high (usually reflected in operational, geometric, and environmental conditions) should be considered for early priority lighting.

LIGHT SOURCES

From the torches and oil lamps of our ancestors, through the gas lamps, carbon arc lamps, Thomas Edison's filament lamps, to the highly efficient discharge lamps of today, lighting has come a long way. Like technology in other fields, the most phenomenal growth has occurred since World War II. The gaseous discharge lamp, the mercury vapor, was invented in the 1930's and has been developed over the years to be the "work horse" of street lighting.

Light sources are normally compared on the basis of their efficacy, the number of lumens[10] produced per watt of energy expended. The principle of operation of the mercury lamp is representative of all gaseous discharge lamps: an electric current generates light by its passage through a gaseous medium. The characteristics of the gaseous medium determine the color of the light and the efficacy of light output. For example, mercury vapor produces a blue-white light at approximately 55 lumens per watt; sodium vapor produces yellow monochromatic light at an efficacy of approximately 175 lumens per watt.

Mercury and sodium sources were the principle gaseous discharge sources until recent years, and each has its advantages and disadvantages. Mercury perhaps has the more acceptable color characteristics, particularly when the lamp is coated with a color-correcting phosphor material. Also, the blue-white appears to be more appealing in warm climates. Sodium vapor has a great advantage in operating efficacy, producing three times as much light for a given amount of expended energy. This is particularly appealing wherever energy costs are high. Also, the warm yellow color seems to appeal to the people of the cooler and wet climates of Europe. The disadvantages are characterized by the monochromacity, the lack of color rendition, and the physically larger lamp required to produce equal amounts of light.

Metal halide lamps, introduced in the 1960's, provide better color with greater efficacy than mercury through a combination of metallic vapors in the arc tube. Metal

---

[10] Lumen is defined on p. 924.

TABLE 20.2
Kinds of Light Sources

Kind	Wattage	Efficacy (Approx.) (Lumens/Watt)	Lamp Life (Approx.) (Hours)
Mercury vapor	175–1,000	55	24,000
Metal halide	175–1,000	90	12,000
High-pressure	400	110	16,000
sodium	1,000	130	16,000
Low-pressure	60–180	180	11,000
sodium			
Fluorescent	40–120	70	6,000

halide lamps provide 85–90 lumens per watt of energy expended. They are available in a number of sizes, as shown in Table 20.2 and are the lamps most generally used for high-mast lighting.

The most recent addition to the discharge lamp family is the high-pressure sodium lamp, with 110 lumens per watt, which is second only to the sodium vapor lamp. The 400-watt source was the largest high-pressure sodium lamp available in this country until a recent announcement of a 1,000-watt lamp which manufacturers claim produces 130 lumens per watt. High-pressure sodium, with a combination of gases in the arc tube, provides a soft, pinkish yellow light that is accepted readily by the driving public. Drivers say that high-pressure sodium light provides better visibility with less glare.

Fluorescent light sources have been used to some extent for sign lighting, particularly in some European countries. Mercury and metal halide lamps, however, have begun to replace these for the lighting of traffic signs. Fluorescent lamps have a fairly high efficacy (70 lumens per watt) and provide excellent color rendition, but the lamps are very large (up to 8 ft long), have a relatively short life, and are difficult to start in low temperatures.

There are many other kinds of light sources available today, including many "exotics," which are impractical for roadway lighting. For example, xenon sources have been manufactured in single-unit sources up to 20,000 and possibly even 50,000 watts. Metal halide lamps up to 10,000 watts have been used to light soccer stadiums in Europe. These larger sources may not be practical for roadway lighting, but they support the technological development necessary to improve roadway lighting sources. The staggering future demand for electrical energy will doubtless bring longer lamp lives and the continued improvement of light source efficacy.

LUMINAIRE DESIGN AND PLACEMENT

LUMINAIRE DESIGN AND TYPES

Webster defines "luminaire" as "any body that gives light," but in street lighting the term "luminaire" describes the complete lighting assembly, less the support assembly. The modern-day luminaire, as shown in Figure 20.1, consists of a weatherproof housing enclosing the light source, a reflector, and in many cases the electrical ballast for discharge type lamps. A refractor serves as the lower part of the enclosure and serves, with a reflector, to control the distribution of light on the roadway. The

Figure 20.1. Examples of typical roadway luminaire providing a Type III medium light distribution. (Source: Texas Transportation Institute.)

refractor is generally a molded glass element that provides prismatic control of light (Figure 20.1).

Luminaires are designed and identified primarily on the basis of the area of coverage, i.e., the width and length of the area to be lighted and the "allowable beam angle." The higher beam angles permit greater spacing of luminaires for uniform coverage, but higher beam angles mean more glare and reduced effectiveness of the lighting system.

To standardize luminaires for manufacture and design purposes, the Illuminating Engineering Society has assigned type numbers, Types I through V, to luminaires that produce different light distribution patterns used for various purposes. These pertain mainly to the street width and the location of the luminaire in relation to the roadway. A brief description with illustrative sketches of each luminaire type is given in Figure 20.2. A more detailed and technical description of luminaire types can be found in the proposed *American National Standard Practice for Roadway Lighting.*[11]

Luminaires are also classified on the basis of vertical light distribution, the ability to spread light along the length of the roadway. Short, medium, and long distributions are established on the basis of the distance from the luminaire where the light beam of maximum candlepower strikes the roadway surface, defined as follows:

1. *Short distribution.* The maximum candlepower beam strikes the roadway between 1.0 and 2.25 mounting heights from the luminaire.
2. *Medium distribution.* The maximum candlepower beam strikes the roadway at some point between 2.25 and 3.75 mounting heights from the luminaire.
3. *Long distribution.* The maximum candlepower beam strikes the roadway at a point between 3.75 and 6.0 mounting heights from the luminaire.

On the basis of the vertical light distributions, theoretical maximum spacings of luminaires are such that the maximum candlepower beams from adjacent luminaires are joined on the roadway surface. With this assumption, the maximum luminaire spacings would be 4.5 mounting heights for a short distribution, 7.5 mounting heights for a medium distribution, and 12.0 mounting heights for a long distribution. These spacings will not, however, satisfy the design criteria outlined later.

In practice, the medium distribution is most widely used, and the luminaire spacing

---

[11] *American National Standard Practice for Roadway Lighting*, p. 10.

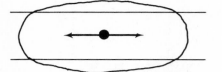

Type I - A luminaire designed
for center mounting over
streets up to 2.0 mounting
heights in width.

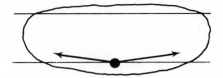

Type II - A luminaire designed
for mounting over the curb
line of street widths less
than 1.5 mounting heights.

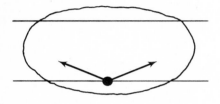

Type III - A luminaire
designed for mounting
over the curb line of
street widths up to 2.0
mounting heights.

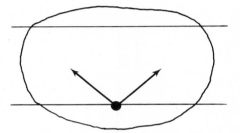

Type IV - A luminaire designed
for mounting over the curb line
of street widths greater than
2.0 mounting heights.

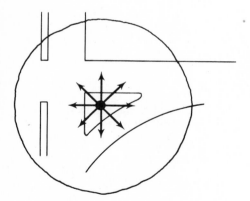

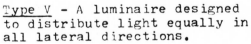

Type V - A luminaire designed
to distribute light equally in
all lateral directions.

Figure 20.2. Illustrations and descriptions of IES luminaire type nomenclature. [Source: *American National Standard Practice for Roadway Lighting* (New York: Illuminating Engineering Society, American National Standards Association, 1972), p. 19.]

normally does not exceed 5 mounting heights. Short distributions are seldom used for reasons of economy, and long distributions are not used to any great extent because the high beam angle of maximum candlepower produces excessive glare.

LUMINAIRE PLACEMENT

Luminaire placement is an integral part of the design of an effective lighting system. Luminaires are mounted at a given height above the roadway, depending on the lamp output, and at a given point in relation to the roadway, depending the character of the roadway to be lighted. For roadways not having medians, the luminaire is normally installed in a "house-side" location, which may be further described as a one-side system for narrow streets, a "staggered" system for medium width streets, and an "opposite" system for wide streets. For streets having wide medians, especially freeways, a "median lighting system," provides a very effective lighting system at less cost because of the saving in luminaire supports and electrical conductors. See Figure 20.3.

Mounting height is generally determined by the lamp output and the desired average illumination on the roadway and by the required uniformity of the distribution of light (factors to be covered more specifically later). By rule of thumb, however, light sources of 20,000 lumens or less should be mounted at heights of approximately 30 ft, light sources of from 20,000 to 45,000 lumens should be mounted at heights from 30 to 45 ft, and light sources of from 45,000 to 90,000 lumens should be mounted at heights from 45 to 60 ft.

Mounting heights from 100 to 175 ft have been utilized in a special roadway lighting technique called high-mast lighting. In high-mast lighting, which is used mainly to light large areas, for example, interchanges, intersections, toll plazas, and parks, floodlight units are arranged in combinations in order to provide a total system output of from 500,000 to 1,000,000 lumens distributed over a large area. The justifications for high-mast lighting pertain mainly to the removal of luminaire supports near the traffic area for safety reasons and to providing a complete view of an area that would be similar to daylight conditions.

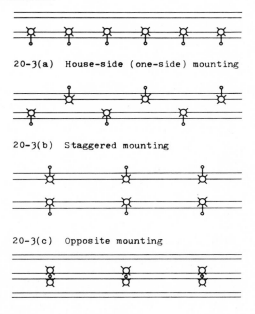

20-3(a) House-side (one-side) mounting

20-3(b) Staggered mounting

20-3(c) Opposite mounting

Figure 20.3. Examples of typical luminaire mounting arrangements. [Source: Antanas Ketvirtis, *Highway Lighting Engineering* (Toronto: Foundation of Canada Engineering Corporation Limited, 1967), p. 36.]

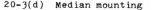

20-3(d) Median mounting

Figure 20.4. High-mast lighting installation in Dallas. (Source: Texas Highway Department.)

Luminaires for high-mast lighting include (1) a combination of floodlights that can be aimed to provide a rectangular or any other particular distribution to fit the area to be lighted, (2) the IES Type V luminaire that provides a circular distribution, and (3) the IES Type III luminaire that provides an asymmetric distribution. The diameter of the Type V distribution, and thus the illumination level, is varied by changing the vertical location of the lamp in relation to the reflector and the refractor of the luminaire. Examples of high-mast lighting systems are illustrated in Figure 20.4.

DESIGN OF LIGHTING SYSTEMS

DEFINITION OF TERMS

Lumen. The unit quantity of light output, analogous to cu ft per sec for water, or vehicles per hr for traffic volume. In specific terminology, it is defined as the amount of light that falls on an area of 1 sq ft, every point of which is 1 ft distant from a source of one candela.

Footcandle. The unit of illumination when the unit length is 1 ft; 1 lumen distributed uniformly over an area of 1 sq ft.

Lux. The unit of illumination when the meter is the unit length; 1 lumen distributed uniformly over an area of 1 sq m.

Horizontal footcandle. One lumen distributed uniformly over a *horizontal* surface 1 sq ft in area. Thus, horizontal footcandle is a measure of the light that strikes the pavement surface.

Vertical footcandle. One lumen distributed uniformly over a *vertical* surface 1 sq ft in area. Thus, vertical footcandle is a measure of the light that strikes vertical surfaces such as curbs, piers, or retaining walls.

Luminance. The luminous intensity of a surface in a given direction per unit of projected area of the surface as viewed from that direction (measured in footlamberts).

Footlambert. The unit of photometric brightness (luminance); one footlambert is equal to $1/\pi$ candela per sq ft.

Candela. The unit of luminous intensity; 1 lumen per unit solid angle (steradian).

GENERAL DESIGN CRITERIA

In the United States the principal criteria in the design of lighting systems are average intensity and uniformity of illumination. Average intensity of illumination, expressed in horizontal footcandles or lumens per sq ft on a horizontal surface, is a measure of the total illumination on the roadway surface. Uniformity, expressed in terms of average-to-minimum or maximum-to-minimum intensity ratios, describes how the total illumination is distributed on the roadway surface.

Average intensity of illumination is not necessarily directly related to the ability to see. The ability to see is primarily a function of the amount of light striking the pavement surface which is diffused and reflected toward the driver's eye. Seeing, therefore, is a function of pavement brightness or pavement luminance characteristics. The determination of pavement brightness is a very difficult task, both from a technical and a realistic standpoint. The measuring equipment is complex, and the procedure is only applicable to pavements already in service. Also, pavement reflectance characteristics change with pavement age, wear, and maintenance.

Realistic consideration of pavement brightness, as discussed previously, can be made in general terms as light, medium, and dark surfaces. Some authorities say that pavement brightness sometimes varies up to 100 percent, indicating that almost twice as much light is needed on dark pavements to have equal effectiveness in pavement brightness. Since current recommended values of average intensity of illumination are representative of average conditions, it is recommended that these values be increased up to 40 percent for extremely dark pavements.

The current values for average intensity of illumination as recommended by IES[12] are given in Table 20.3. These are recommended minimum values for average maintained horizontal illumination. At least one authority[13] feels that the values for freeways are too low and recommends 1.0, 0.8, and 0.8 footcandles (Lux) for commercial, intermediate, and residential areas, respectively.

TABLE 20.3
Recommendation for Average Maintained Horizontal Illumination

Roadway and Walkway Classification	Footcandles (Lux) Area Classification		
	Commercial	Intermediate	Residential
Vehicular roadways			
Freeway*	0.6 (6.)	0.6 (6.)	0.6 (6.)
Major and expressway*	2.0 (22.)	1.4 (15.)	1.0 (11.)
Collector	1.2 (13.)	0.9 (10.)	0.6 (6.)
Local	0.9 (10.)	0.6 (6.)	0.2 (4.)
Alleys	0.6 (6.)	0.4 (4.)	0.2 (2.)
Pedestrian walkways			
Sidewalks	0.9 (10.)	0.6 (6.)	0.2 (2.)
Pedestrian ways	2.0 (22.)	1.0 (11.)	0.5 (5.)

*Both main lanes and ramps.

Source: *American National Standard Practice for Roadway Lighting*, Illuminating Engineering Society, American National Standards Association, 1972, p. 16.

---

[12] IBID., p. 16.
[13] ANTANAS KETVIRTIS, *Highway Lighting Engineering* (Toronto: Foundation of Canada Engineering Corporation Limited, 1967), p. 25.

The IES recommends average-to-minimum uniformity ratios of 3 : 1 for all road-ways except local residential streets, which should have a ratio not exceeding 6 : 1.[14] The IES has not suggested specific maximum-to-minimum ratios, but Ketvirtis[15] has recommended ratios as given in Table 20.4. Many feel that the maximum-to-minimum ratio is the most realistic measure of uniformity because it represents the full range of illumination rather than generalized considerations.

TABLE 20.4
Recommended Maximum to Minimum Ratios on Road Pavement

Road Classification	Urban			Rural
	Downtown	Inter-mediate	Outlying	
Freeway	6 : 1	6 : 1	6 : 1	6 : 1
Expressway	6 : 1	6 : 1	6 : 1	6 : 1
Arterial	6 : 1	6 : 1	8 : 1	8 : 1
Collector	6 : 1	8 : 1	8 : 1	8 : 1
Local	8 : 1	8 : 1	10 : 1	10 : 1

Source: A. Ketvirtis, *Highway Lighting Engineering* (Toronto: Foundation of Canada Engineering Corporation Limited, 1967), p. 27.

DESIGN OF CONTINUOUS LIGHTING SYSTEMS

The illumination design process is one of evaluating the lighting needs, selecting appropriate illumination criteria for design, selecting proper equipment for the job, and establishing the geometry of the system to provide the most effective lighting system to satisfy the needs. The major steps of the design process are outlined below:

1. *Evaluation of Existing Conditions*
   Much emphasis should be placed on the evaluation of existing conditions which determine the need for lighting, and on the roadway, traffic, and environmental conditions which influence the selection of illumination levels. It is recom-mended that the designer develop a checklist and possibly a numerical rating scheme for assigning priorities to lighting jobs within his city or area of juris-diction. At least the following should be included in an evaluation checklist:
   A. Description of Roadway or Street
      a. Kind of facility
      b. Number of lanes and lane width
      c. Median type and width
      d. Curb type and degree of access
      e. Pavement surface type
      f. Geometric features (curves, grades, etc.)
   B. Traffic Conditions
      a. Night traffic volumes
      b. Speeds (85 percentile, 15 percentile)
      c. Percent of entering and exiting traffic

---

[14] *American National Standard Practice for Roadway Lighting*, p. 19.
[15] KETVIRTIS, *Highway Lighting Engineering*, p. 27.

    d. Accident rates (night vs. day)

    e. Accidents by type

    f. Actual or potential pedestrian activity

C. Environmental Conditions

    a. Type of land use (commercial, residential, and subcategories of each that relate to street use)

    b. Extent of commercial lighting, including signs, area lighting, novelty lighting, and degree of animation

    c. Pedestrian generation potential

    d. Traffic generation potential

    e. Possible influences of driver condition, principally in regard to alcoholic consumption

A rational method of rating geometrics, operational, and environmental conditions to determine warrants and priorities was recently published in a paper entitled, "A Total Design Process for Roadway Lighting," by Walton and Rowan, Highway Research Record No. 440 (pp. 1-19), Transportation Research Board, National Academy of Sciences, Washington, D. C.

2. *Selection of Illumination Level*

Using the data collected in the evaluation in Item 1 above, the designer should first classify the highway facility and then the area type according to the classification schemes presented previously. At this point, the *minimum* average intensity of illumination may be determined from Table 20.3. This should be recognized as a minimum value for average conditions as outlined in Item 1, and a higher average intensity should be used for design if the evaluation in Item 1 reveals several critical factors. Selection of higher than minimum average intensity values is a matter of judgment. Here a numerical evaluation scheme could be used effectively.

3. *System Characterization*

The designer should determine the location and mounting height of luminaires and the type and wattage of the light source to be used. If the system is to be placed on existing supports (not necessarily recommended), the spacing and mounting heights are fixed and will determine the other characteristics. For a new installation, the guidelines for an optimum system should be followed. The highest mounting heights and largest light sources practicable should be used in the general interest of safety, economy, and overall system efficiency. Tables 20.5 and 20.6 provide a guide to the selection of mounting heights based on the size of the light source and on the type and arrangement of luminaires related to street width and location of luminaires.

    Maximum spacing, consistent with good illumination design, should be emphasized. Luminaire supports are hazardous roadway objects and, for safety, the number should be minimized and they should be strategically located. Supports should be set back as far as practicable, with the luminaire mounted over or near the curb instead of extended over the roadway. Types II, III, and IV luminaires are designed to be mounted over the curb, and extension over the roadway normally contributes nothing more than glare and additional cost. Median mountings (twin mast arms on one support located in the median) should be used wherever practicable because they reduce the number of supports and provide a high-quality, more economical system.

TABLE 20.5
Guide for Luminaire Lateral Light Distribution Type Selection and Placement

Rectangular Roadway Area					
Side of the Roadway Mounting			Center of Roadway Mounting		
One side or staggered	Staggered or opposite	Grade intersections	Single roadway	Twin roadways (median mtg.)	Grade intersections
Width up to 1.5 M.H.	Width beyond 1.5 M.H.	Width up to 1.5 M.H.	Width up to 2.0 M.H.	Width up to 1.5 M.H. (each pavement)	Width up to 2.0 M.H.
Types II–III–IV	Types III–IV	Type II 4-way	Type I	Types II–III	Types I 4-way, V

Note: In all cases suggested *maximum* longitudinal spacings and associated vertical distribution classifications are:

Short distribution = 4.5 M.H.
Medium distribution = 7.5 M.H.
Long distribution = 12.0 M.H.

Source: *American National Standard Practice for Roadway Lighting*, 1972.

TABLE 20.6
General Guide to the Selection of Luminaire Mounting Heights

Lamp Lumens	Mounting Height (Ft)
≤ 20,000	≤ 35
20,000 to 45,000	35 to 45
45,000 to 90,000	45 to 60

4. *Acquisition of Luminaire Data*

Once the type of luminaire and size of light source have been selected, *the designer should obtain photometric data from prospective manufacturers for the specific luminaire and light source to be used.* These data should include:

A. Isofootcandle Diagram. An isofootcandle diagram, as illustrated in Figure 20.5, should be obtained from the manufacturer of the equipment to be used. The isofootcandle diagram should desirably be prepared from measurements in a full-scale field laboratory, but most manufacturers use photometric laboratory sources where precise control is achieved. Too frequently, field conditions cannot be controlled as well as photometric laboratory conditions, and field installation does not provide the luminaire efficiency indicated by the isofootcandle diagram.

B. Coefficient of Utilization Curve. The coefficient of utilization curve plots the percentage rated lamp lumens related to the width of roadway. An *example curve* is shown in Figure 20.6. The plot shows that 38.5 percent of the rated lamp lumens fall on a street 1.5 mounting heights wide when the luminaire is mounted over one curb line. Most coefficient of utilization plots provide one curve for the "street side" or front side of the luminaire and another curve for the "house side" or the back side of the luminaire. *The*

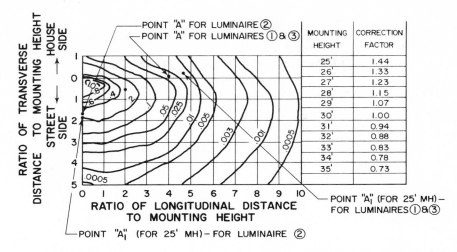

Figure 20.5. Example of an isofootcandle diagram of horizontal footcandles on pavement surface for a luminaire providing a Type III medium light distribution. [Source: *American National Standard Practice for Roadway Lighting* (New York: Illuminating Engineering Society, American National Standards Association, 1972), p. 26.]

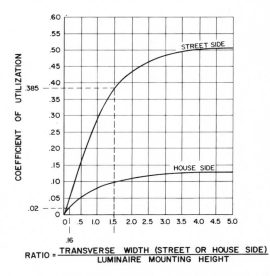

Figure 20.6. Example of coefficient of utilization curves for luminaire providing Type III medium light distribution. These curves are often plotted directly on the isofootcandle diagram. (Source: Proposed *American National Standard Practice for Roadway Lighting*, 1972), p. 25.

coefficient of utilization curve should be a curve for the specific luminaire and light source used instead of a typical curve for a given type of luminaire.

C. Lamp Lumen Depreciation Curve. A depreciation curve for the lamp to be used should be obtained from the manufacturer. This curve provides information about the output of the lamp related to length of service. This information is useful in design and maintenance.

D. Luminaire Maintenance Factor. Figure 20.7 provides a guide to the selection of a maintenance factor for design and scheduling of maintenance, depending on the type of area in which lighting is to be provided.

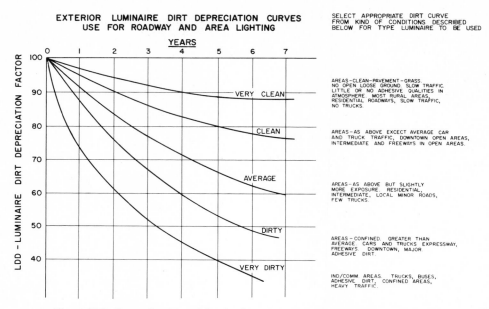

Figure 20.7. Curves for determining luminaire dirt depreciation. (Source: Proposed *American National Standard Practice for Roadway Lighting*), p. 24.

5. *Computation of Luminaire Spacing*

The computational procedure for the spacing of luminaires is based on the average intensity of illumination required, the light that can be utilized from the luminaire, and the area to be covered by the luminaire. The following formula includes all of these factors and provides a solution to luminaire spacing when all other factors are known:

$$\text{Spacing between luminaires (ft)} = \frac{\text{lamp lumens} \times \text{coefficient of utilization}}{\text{average intensity} \times \text{roadway width (ft)}}$$
$$\text{of illumination (horizontal ft-cd)}$$

In this form, the formula is only valid for initial conditions, with no provision for lamp depreciation and dirt accumulation on the luminaire. When these factors are introduced as a luminaire maintenance factor, the formula becomes:

$$\text{Spacing between luminaires (ft)} = \frac{\text{lamp lumens at} \times \text{coefficient of} \times \text{luminaire maintenance}}{\text{replacement time} \times \text{utilization} \times \text{factor}}$$
$$\text{average intensity} \times \text{width of roadway (ft)}$$
$$\text{of illumination (horizontal ft-cd)}$$

6. *Computation of Uniformity Ratio*

With luminaire spacing determined, the designer should then check the uniformity of illumination by computing the average-to-minimum ratio and comparing it with selected criteria. Minimum illumination is determined from the proper isofootcandle diagram (a typical diagram is illustrated in Figure 20.5). This is accomplished by selecting points on the layout (Figure 20.8) where it is anticipated that minimum illumination will occur and checking the illumination on the isofootcandle curve. The designer should include the contributions from all luminaires in determining the minimum illumination because they are additive. The uniformity ratio is as follows:

$$\text{Uniformity ratio} = \frac{\text{average intensity (maintained)}}{\text{minimum intensity (maintained)}}$$

The uniformity ratio should not exceed 3 to 1 except on residential streets where 6 to 1 is generally acceptable.

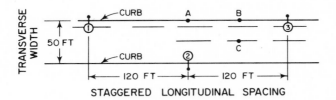

Figure 20.8. Layout of luminaire and roadway assumed for typical computation. (Source: Proposed *American National Standard Practice for Roadway Lighting*,) p. 26.

With a satisfactory check of uniformity, the computational procedure is complete. It may be necessary to return to Step 3 in order to alter the system characteristics if luminaire spacing is too short or the uniformity ratio is too large.

7. *Consideration of Transitional Lighting*
The need for transition to continuous lighting is more psychological than physiological. The eye normally adapts to light changes rapidly, achieving approximately 90 percent efficiency within seconds. When the lighting system terminates in an extremely dark area, the driver may want a transitional effect because of the rapid change in the flow of information. Accepted practice in the United States is to accomplish transition within 15 sec using lower wattage sources at the same spacing and mounting height used in the continuous lighting system.

The designer should pay particular attention to the terminal points of a lighting system to see that the terminal points do not correspond with significant changes in roadway geometry, traffic control, or environmental characteristics. An example is the upgraded section of highway that has been improved with additional lanes, a median, and high-quality fixed lighting. Too frequently, this new section connects directly to the old unlighted section. The desired treatment continues the lighting system through the transition to the old facility, and then it gradually reduces the lighting to the darkened condition.

8. *Selection of Safety Features*
Breakaway bases for luminaire supports are now commonly used on freeways and other major highways and have saved many lives. Collisions with fixed-base luminaire supports at speeds over 30 mph are likely to produce serious injury or death to the vehicle occupants. Therefore, breakaway bases should be used

any time the 85 percentile speed on the traffic facility exceeds 30 mph and when the luminaire support is not protected by a median barrier or guardrail. It is *not* feasible to install guardrail for the sole purpose of protecting luminaire supports. Guardrail used in this manner is more of a hazard than luminaire supports and should be used only when the hazard is otherwise greater than that produced by the installation of the guardrail.[16]

Questions have arisen concerning the safety of pedestrians and private property where breakaway luminaire supports are used. Breakaway devices are advisable where pedestrian activity is high or where property is developed right to the property line. Speeds of 30 mph or more are unlikely in these areas.

There are several types of breakaway luminaire supports available:

A. Frangible bases, characterized by the cast aluminum transformer base
B. Progressive shear bases
C. Slip bases

Additional information on the types of bases can be obtained from the Highway Research Board, the Federal Highway Administration, and various luminaire support manufacturers.

PARTIAL LIGHTING

Where roadways are not continuously lighted, it is often desirable to provide lighting at intersections, interchanges, and elsewhere as justified. Lighting installed only at key points is normally referred to as "partial" or "safety" lighting. The objective is to draw the driver's attention to unusual conditions and to convey necessary information. Sometimes it is sufficient to light only the major points of conflict with other traffic or the roadway features themselves, but generally it is desirable to illuminate the entire area, including the geometric and environmental features, as well as the major conflict points.

Because conditions related to partial or safety lighting are so variable, it is difficult to outline a specific design procedure. The following general points may, however, be helpful:

1. A single luminaire calls attention to a point on the roadway, but it has very little revealing power. Two or more luminaires, close together, are necessary to improve visibility significantly. For example, a single luminaire at the nose of an entrance ramp does little more than define the nose of the ramp.
2. Merge areas should be lighted so that mainstream traffic can judge the location and speed of entering traffic.
3. The main lanes before a merge point should be lighted so that entering traffic can judge the speed and distance of oncoming traffic.
4. Intersections should be lighted so that vehicles, curb faces, obstructions, signs, and other vertical surfaces are illuminated by direct light. This generally means placing the luminaire on the nearside corner of the intersection.
5. Intersections, pedestrian crosswalks, auxiliary lanes, and other special locations should be illuminated effectively.
6. Intersection approaches should be lighted whenever the driver on the inter-

---

[16] W. F. McFARLAND and N. E. WALTON, "Cost Effectiveness Analysis of Roadway Lighting Systems," Research Report 137–1 (College Station, Texas: Texas Transportation Institute, 1969).

secting roadway will need to identify the roadway for informational purposes and to judge the speed and distance of intersecting traffic. This is particularly true for complex intersections with turning roadways.

7. Considerable informational value is gained when the driver can see the entire intersection or interchange rather than only the specific part he uses. Thus, the designer should consider lighting the entire area when the information value justifies it. This particular point has contributed substantially to the extensive use of high-mast lighting of intersections and interchanges.

8. Specific consideration should be given to the hazards introduced by the installation of luminaire supports at intersections and interchanges. Because these are areas of conflict and decision, the probability of a collision with a luminaire support is much higher than for continuous lighting systems. Supports should be strategically located to lessen the probability of collision, and breakaway bases should be used except at the specific points where a fallen luminaire support and an errant vehicle would obviously result in more serious consequences. For example, it would be impractical to place a breakaway support next to a busy pedestrian walkway, but breakaway supports would be entirely appropriate for merge points, turning roadways, etc.

There are no specific criteria for illumination design of partial lighting, but the *American National Standard Practice for Roadway Lighting* [17] says:

> Intersecting, converging or diverging roadway areas require higher illumination. The illumination within these areas should at least be equal to the *sum* of the values recommended for each roadway which forms the intersection. Such areas include ramp divergences or connections with streets or freeway mainlines. They also include very high volume driveway connections to public streets and midblock pedestrian cross-walks.

This statement apparently pertains only to the conflict points and not to the entire area of large intersections and interchanges. In such instances, it is felt that the criteria for continuous lighting (Table 20.3) apply.

Design procedures presented earlier for continuous lighting generally apply for partial lighting also. Luminaire location for partial lighting is aided greatly, however, by making scale drawings of isofootcandle curves on transparent materials such as sheet acetate. The transparency permits the designer to strategically locate the luminaires so that critical points receive the greatest amount of illumination. By using two transparencies, the designer can establish realistic spacings of luminaires without immediately going through the computational procedure.

HIGH-MAST LIGHTING

High-mast Lighting [18] describes the application of the area lighting concept to highway interchanges, complex intersections, and highways having wide cross sections and many lanes. Although high-mast lighting is recognized as a "popular new concept," it is the application rather than the concept that is new. The concept dates back to the 1800's when tall masts were installed in several cities, including Philadelphia and Vancouver, to illuminate large areas and thus provide a pleasant nighttime envi-

---

[17] *American National Standard Practice for Roadway Lighting*, p. 21.

[18] N. E. WALTON and N. J. ROWAN, "High-mast Lighting," Research Report 75-12 (College Station, Texas: Texas Transportation Institute, 1969), p. 1.

ronment. The operation and maintenance of these installations proved very costly, and most were abandoned. But at least one is still used. About the turn of the century, the city of Austin, Texas, traded a narrowgauge railroad for a number of 150-ft towers and installed them at several points throughout the city where artificial "moonlight" was desired. Although the light sources have been changed with advancements in technology, most of the towers are still used effectively.

The first known application of high-mast lighting for modern highways was the Heerdter Triangle installation in Dusseldorf, Germany, in the late 1950's. It was followed by installations in other European countries, including Holland, France, Italy, and Great Britain. Interest in high-mast lighting in this country was stimulated by the successful applications in Europe and the increasing difficulty of lighting some highway interchanges by conventional methods.

The principal objective in the application of high-mast lighting to highway interchanges is to synthesize the visual advantage provided the driver by daylight. Thus, the driver can see all things pertinent to the decision-making process in time to assimilate the information and then plan and execute his maneuvers effectively. He can distinguish roadway geometry, obstructions, terrain, and other roadways, each in proper perspective.

Additional advantages of high-mast lighting are related to safety and aesthetics. When there are fewer poles, there are fewer opportunities for collision. The masts can be located farther from the roadway so that the possibility of a collision with the luminaire support is virtually eliminated. Daytime aesthetics are greatly improved by removing the "forest" of poles generally necessary to light complex interchanges and intersections with continuous lighting.

There are no established illumination criteria unique to the high-mast lighting concept. The criteria for continuous lighting (Table 20.3) are generally applicable. An evaluation procedure similar to that recommended for continuous lighting can be used, and increases in average intensity can be made whenever justified by roadway, traffic, or environmental conditions. The average intensity and uniformity values should be applicable only to traffic areas. In nontraffic areas, a minimum of 20 percent of the average intensity for traffic areas is suggested.

The design process for high-mast lighting differs from that for continuous lighting mainly in the system characterization arising from the great differences in illumination equipment and in the computational procedure. The following section is devoted primarily to these elements of the design process.

System characterization. System characterization consists of establishing a design configuration or alternative configurations to be analyzed comparatively in the design process. First, the type of luminaire and the mounting height must be selected. The designer may choose the Type V circular distribution, Type V asymmetric distribution (sometimes listed as IES Type III), or a system of floodlight-type units that can be aimed to achieve these patterns.

The mounting height of high-mast systems is variable, depending on the elevation of the light source in relation to the roadway to be lighted. At least a 100-ft difference in elevation of the light source and the roadway is desirable. Mounting heights of 150 ft are generally applicable.

Achieving a specified illumination level is a function of mounting height, type of light source, type of luminaire or floodlight, and number of luminaires or floodlights in a system (system refers to the assembly on a single mast). A trial number of system units can be estimated by a computational process involving total lamp lumens, a coefficient of luminaire efficiency, and the area to be covered. Experience helps, but the designer will more than likely determine by trial and error the number of units required to provide a given illumination level.

Preliminary spacing of masts in the interchange area should be made on a maximum spacing-to-mounting height ratio of 5:1 if the same cut-off characteristics that result from medium distribution luminaires in continuous lighting are desired. Spacing adjustments should be made on the basis of location criteria and computational procedures of illumination by the point-by-point method discussed later.

Certain considerations regarding mast locations are required in order to achieve the greatest effectiveness from the lighting system:[19]

1. Masts should be located so that the driver's line of sight is not directly toward the light source when he is within 1,500 ft of the mast. More specifically, the driver's line of sight should not be above the lower third point of the mast when he is within 1,500 ft of the mast. This rule is applicable when the driver's line of sight is within 10° on either side of the mast.
2. Masts should be located so that the light source will not at any time be in the direct line of sight with signs (especially overhead signs) and other visual communication media.
3. Masts should not be placed at the end of long tangents or other vulnerable locations where there is an appreciable probability of collision. If such a location is necessary, adequate impact attenuation protection should be provided.
4. Masts should be located so that the highest localized levels of illumination fall in the traffic conflict areas; for example, ramp terminals. Otherwise, masts should be located a sufficient distance from the roadway in order to position the greatest uniformity of illumination on the pavement surface. This is done by using plastic overlays of isofootcandle curves of the light units to be used. This will normally result in the masts being placed a sufficient distance from the roadway so that the probability of collision is virtually eliminated.

Illumination computational procedures. Following the preliminary location and spacing of masts and the initial selection of the number of units on each mast, it is necessary to check the distribution of illumination against the established criteria. The most rational approach to this check process is by computation of illumination using a point-by-point procedure. To facilitate computations, the entire interchange area is superimposed with grid lines at 25-ft to 50-ft intervals. Intervals of 25 ft are desirable if a computer is used, but 50-ft intervals are more appropriate for hand calculations.

The point-by-point computational process utilizes a candlepower distribution curve as illustrated in Figure 20.9. Candlepower distribution curves are normally developed for single-unit Type V luminaires, in which case they must be multiplied by the number of luminaires in the system. This curve can be developed for all symmetrical high-mast systems, whether they consist of Type V units or individual floodlights arranged in a symmetrical pattern.

---

[19] *Ibid.*, p. 57.

The illumination in horizontal footcandles at a grid point resulting from one high-mast assembly can be computed by using the formula:

$$E_H = \frac{CP \cos \theta}{d^2}$$

where $E_H =$ illumination at the point in horizontal footcandles,
$CP =$ candlepower at angle θ, lumens,
$\theta =$ the angle from the vertical axis through the system to the point in question (Figure 20.9), and
$d =$ the distance from the light source to the point in question (Figure 20.10).

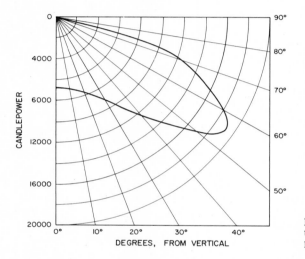

Figure 20.9. Example of candlepower distribution curve used for making point-by-point illumination calculations.

Then, the total illumination at each of the grid points is the sum of the contributions of illumination from the high-mast assemblies within an effective range of the point in question. This process is illustrated in Figure 20.10.

Once the amount of illumination is computed for all of the grid points, an iso-footcandle diagram may be drawn for the entire interchange area. This will facilitate an overall appraisal of the illumination design.

For a more specific appraisal, the designer should plot an illumination profile for each section of roadway in the interchange. For wider roadways, it may be necessary to plot two or more profiles in order to fully represent the traveled way. These profiles are plotted either by using the contour values and contour spacings along the roadway or by using interpolation of the grid matrix of illumination values at the grid points. If computer techniques are used, the latter is the more adaptable process. The average illumination values and uniformity ratios are computed from the illumination profile. By comparing the average illumination and uniformity ratios with the previously established criteria, the spacing of masts and/or number of units on each mast may be adjusted accordingly, and the computational process repeated until the desired criteria are achieved.

The major weakness in the point-by-point design process is the lack of reliability of the input photometric data, as in the case of computational procedures for continuous lighting. The candlepower distribution curve is developed under controlled

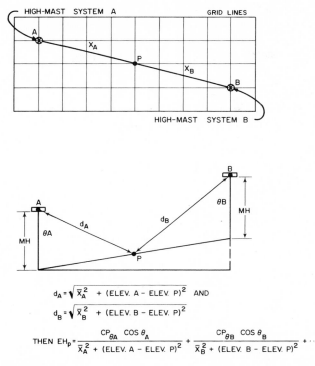

Figure 20.10. Illustration of the point-by-point process of illumination computations.

laboratory conditions. There is some preliminary indication from unpublished research that the loss from laboratory to field installation is approximately 25 percent. Therefore, it seems justified that the designer should consider reducing the candlepower values by 25 percent unless there is evidence that he can be assured of achieving the light output indicated by the candlepower distribution curve.

High-mast lighting hardware. Both poles and towers have been used on high-mast lighting installations with apparently equal functional capability. Manufacturers now offer either type in a "package deal" that includes lowering assemblies and other mounting hardware. The best choice depends on local factors instead of on any set guidelines.

There are three basic methods available for mounting high-mast light sources:[20]

1. Light sources fixed at top of mast, with provisions for climbing the mast for service and maintenance.
2. Light sources fixed at top of mast, with provisions for a motorized personnel carrier to transport service man to light sources for service and maintenance.
3. Light sources mounted on a lowering assembly, with a motorized winch to facilitate lowering and raising of lighting assembly for service and maintenance. Several versions of this method are available. Some have locking mechanisms

---

[20] *Ibid.*, p. 33.

and power connections at the top; others provide continuous cable suspension for the lighting assembly and a continuous power supply line connected at the base. Once again, the selection of the mounting method is largely dependent on local factors.

TUNNEL AND UNDERPASS LIGHTING

Since tunnels and underpasses involve different physiological characteristics of the driver, they deserve special consideration. Tunnels and underpasses of any substantial length involve eye adaptation to ambient light conditions, and transient adaptation becomes the critical factor in design. Lighting of tunnels in the United States generally provides high illumination intensities (30–60 footcandles) at the entrance and exit in order to accommodate the adaptation process. Modern European designs, however, have involved screening of the roadway area in advance of the portal, as well as transitional lighting to better provide for eye adaptation.

Lighting within the tunnel is required for any tunnel over 100 ft long. For short tunnels (from 100 to 500 ft), entrance illumination levels (30–60 footcandles) should be maintained throughout the length of the tunnel during the daytime; at night, illumination levels can be reduced to from 0.7 to 2.0 footcandles. For long tunnels (longer than 500 ft), illumination within the tunnel may be reduced to 5 to 10 footcandles in the daytime; night illumination requirements are the same as for short tunnels.

Although the illumination levels presented above are in conformance with *An Informational Guide for Roadway Lighting*,[21] there is a very strong feeling that the levels are too low. Every consideration should be given to exceeding these levels whenever practicable. For more realistic illumination levels for tunnels, the designer should look closely at current practices.

No additional illumination is required for an underpass less than 75 ft in length, provided roadway luminaires are positioned at each end of the underpass to light the area within. When the underpass exceeds 75 ft in length, supplementary lighting is generally used.

Tunnel and underpass lighting involves many special design features not common to the lighting of streets and highways. It is advisable to consult more specific references for the many considerations in design.[22,23] A good reference is the *IES Guide for the Lighting of Tunnels and Underpasses* published by the Illuminating Engineering Society, New York.

RURAL AT-GRADE INTERSECTION ILLUMINATION

For the past several years, researchers at the University of Illinois at Urbana-Champaign have undertaken an extensive investigation of rural at-grade intersection illumination. The following is recommended as the basis for illumination warrants:

> Rural intersections should be considered for lighting if the average number of nighttime accidents (N) per year exceeds the average number of day accidents (D) per year divided by three. All the accident

---

[21] *An Informational Guide for Roadway Lighting* (Washington, D.C.: American Association of State Highway Officials, 1969).

[22] "Lighting Traffic Tunnels and Underpasses," *Illuminating Engineering*, LII, No. 6 (1957), 325.

[23] J. B. DeBoer, *Public Lighting* (Eindhoven, The Netherlands: Philips Technical Library, 1967).

data available since the date of the last modification to the intersection should be used when calculating these averages. If N is greater than D/3 the likely average benefit should be taken as N — D/3 accidents/year.

The likely benefits of lighting new or modified intersections should be estimated from previous experience. It is recommended that illumination should be provided whenever an intersection is channelized.

The estimated cost of lighting the intersections, which show a benefit using the above criteria, should be computed. The lighting program should then be based on the resulting list of intersections ranked in priority order by means of the benefit/cost ratio (expressed as annual reduction in accidents/annual cost).[24]

The recommended warrant is designed to give the decision makers the most information possible based on current knowledge. It is implicitly assumed that the highway improvement budget is limited, and thus interest is focused on maximizing the benefits of a limited budget. For this reason, reductions in number of accidents rather than accident rates are used. One important implication of this approach is that the distribution of funds for lighting improvement tends to be directed into the areas of high traffic volumes. Thus, if intersections are ranked on a statewide listing, the distribution of the budget would not be the same as one distributed by listing intersections on a district basis. The latter would spread improvements more uniformly throughout a state, but at a lower overall benefit/cost ratio.

OPERATION AND MAINTENANCE

ELECTRIC POWER CONSIDERATIONS

Many lighting installations require enough electrical power to justify special distribution equipment and coordination with an electric power company. Usually, the operating line voltage is available in multiples of 120 volts, the upper limit determined by the breakdown voltage of the insulation used in each component of the system.

Lamps used for roadway lighting are of the gas-discharge type and require current limiting while burning. This is accomplished by placing a coil of inductance, called a ballast, in series with the lamp. Common practice incorporates an autotransformer with the ballast so that voltage changes (step up or step down) are easily facilitated. To minimize line losses, higher operating voltages (480 volts) are usually used.

BALLASTS

Three basic multiple mercury lamp ballasts are used in mercury lamp circuits: the reactor, the autotransformer (high reactance), and the regulated output, also referred to as constant wattage or stabilizing type.[25]

The reactor ballast in series with the lamp is recommended for use with circuits

---

[24] ROBERT H. WORTMAN, M. E. LIPINSKI, et al., "Interim Report—Development of Warrants for Rural At-Grade Intersection Illumination," Illinois Cooperative Highway Research Program Series No. 135, University of Illinois at Urbana-Champaign, 1972, p. 37.

[25] *IES Lighting Handbook*, pp. 8–32.

that provide good regulation and sufficient line voltage to reliably start the lamp. Voltage taps may be provided to match the ballast to the primary voltage. The ballast may also include a capacitor for power factor correction.

The high reactance ballast adds an autotransformer to the reactor in order to increase the line voltage to a value that will provide reliable lamp starting whenever the line voltage is low. This ballast may also be provided with line voltage taps to match the ballast to the line voltage. Power factor correction can be provided with a capacitor.

Regulated output (constant wattage or stabilizing) ballasts are used whenever relatively poor voltage regulation is experienced or whenever gradual voltage change may occur. It eliminates the necessity for matching a ballast voltage tap with the line voltage. This ballast allows the lamp to draw its rated current with an input voltage variation of plus or minus 10 to 13 percent.

When ballasts are being selected, careful consideration should be given to the various characteristics, including use with various lamp types and low-temperature starting characteristics.

DESIGN OF THE ELECTRICAL SYSTEM

The electrical design of a lighting system usually falls outside the traffic engineer's responsibility. An excellent text on lighting electrical design, *Highway Lighting Engineering*, by Antanas Ketvirtis,[26] provides very detailed descriptions, plans, and specifications for all electrical design associated with highway lighting.

MAINTENANCE

Because light reduction caused by dirt and lumen depreciation of the lamp cannot readily be observed, periodic cleaning and relamping are important maintenance requirements. Although, in practice, maintenance intervals range from six months to four years (and often lamp replacement and cleaning are performed only at lamp burn-out), it must be stressed that it is economically advantageous to follow a set maintenance schedule based on expected lamp life and dirt accumulation characteristics of the area. Generally, the benefits of maintenance parallel the initial benefits of illumination, and the most effective cleaning schedule is one that allows no more than a 20 percent light reduction caused by dirt. Lamp replacement should follow the same schedule. This usually amounts to from approximately 16,000 to 24,000 hours of service at rated conditions for most mercury, metal halide, and high-pressure sodium lamps, making the four-year interval tolerable.

One question that often arises in maintenance considerations is the economic advantages of higher mounting heights (from 40 to 50 ft). It has been demonstrated that even a small lighting program can achieve sufficient savings from higher mounting heights to purchase the equipment necessary to service the higher heights.[27] In many cases, effectual communication between municipalities and the electric companies is necessary to demonstrate the benefits (savings) of the higher heights.

Another problem is the cleaning of polished aluminum reflectors that are easily tarnished or damaged. Many trade journals advertise cleaning solutions made especially for polished aluminum. This not only makes the use of these reflectors more feasible, but it also provides for adequate maintenance.

---

[26] KETVIRTIS, *Highway Lighting Engineering*.
[27] McFARLAND and WALTON, "Cost-Effectiveness Analysis of Roadway Lighting Systems."

BENEFITS OF LIGHTING

The real value of lighting is finally and directly related to driver safety and comfort, and, consequently, to a reduction in the nighttime accident rate. Many factors have been cited as the cause of the disproportionate number of nighttime accidents. Among them are fatigue, alcohol, and reduced visibility. The best evidence that reduced visibility contributes to increased accidents are the many examples of accident reduction resulting from improved roadway illumination.[28]

The beneficial effects of fixed roadway lighting are not limited solely to high-speed freeways and interchanges. Many studies have shown a reduction in the number of accidents in urban, suburban, and rural areas, regardless of the class of roadway.

The effect of lighting on the accident rate is significant not only for continuous lengths of roadways, but for specific locations as well. Studies showing marked reduction in nighttime accident rates following illumination have been conducted in California, Connecticut, New York, and many other states and cities. Undoubtedly, modern lighting techniques reduce the number of fatal and nonfatal accidents, as well as the severity of those accidents that do occur despite improved lighting techniques.

The tangible benefits of roadway lighting include many social and economic gains, especially in downtown urban areas. Serious crime can be reduced and business can be improved. Many downtown areas, almost deserted after dark, have increased their aesthetic appeal and business activity after the addition of improved lighting. Architects and city planners consider better lighting a major source of economic stimulation and beautification for downtown areas. In fact, many cities across the country are reopening parks and recreation areas after dark because of the effects lighting has on the crime rate, business, and personal security. Police forces across the country have praised and actively pushed for improved lighting in order to make law enforcement more effective.

COST CONSIDERATIONS

The design and specification of a lighting system should be such that maximum returns are achieved at the lowest possible costs. Several designs providing a predetermined level of effectiveness should be developed and their cost compared. Design variables that usually are most influential to both effectiveness and costs are as follows:

1. Arrangement or placement of lighting units with respect to the roadway (median, one-side, staggered on opposite sides, opposite on two sides, and median plus on the side)
2. Mounting heights
3. Pole and base type
4. Lamp characteristics (type, ASA designation, initial lumens, life and lumen history, wattage)
5. Luminaire type
6. Light distribution type
7. Type of distribution system

[28] D. E. CLEVELAND, "Illumination," in *Traffic Control and Roadway Elements—Their Relationship to Highway Safety/Revised*, Chap. 3 (Washington, D.C.: Automotive Safety Foundation, 1969).

8. Burning hours per year
9. Type of ownership and maintenance

In addition to these design variables, there are roadway, traffic, and environmental variables influencing effectiveness and costs that should be considered in cost analysis. The following are included among the influential roadway variables:

1. Median width
2. Presence or absence (or plans for such) of a median barrier and its type
3. Number of traffic lanes
4. Width of roadway shoulders
5. Overall roadway cross section
6. Number and type of exit and entrance ramps
7. Number and type of intersections and intersecting roadways and streets

Among other important variables are the amount and type of traffic, the type of area through which the roadway passes, and weather conditions in the area.

From a general standpoint, Thompson and Fansler[29] have shown that higher mounting heights are least costly for continuous and intersection lighting. Cassel and Medville[30] state that the use of 700-watt and 1,000-watt units at mounting heights of 35 ft and higher greatly reduces costs because fewer lighting poles are required per mile. McFarland and Walton[31] recommend higher mounting heights, higher luminaire output, careful lateral placement of units (for safety reasons), and breakaway devices for reducing overall lighting system costs (initial, operation, maintenance, and accident).

CURRENT RESEARCH

Roadway lighting research is currently being conducted in the areas of lighting requirements, design guidelines, warrants, benefits, and cost-effectiveness under the National Cooperative Highway Research Program.[32] Application of the results of this research should substantially contribute to the state-of-the-art in roadway lighting technology and practice.

REFERENCES FOR FURTHER READING

DeBoer, J. B., *Public Lighting*. Eindhoven, The Netherlands: Philips Technical Library, 1967.

IES Lighting Handbook, 4th ed. (1966). Proposed *American National Standard Practice for Roadway Lighting* (1972). Illuminating Engineering Society, New York.

An Information Guide for Roadway Lighting. American Association of State Highway Officials, Washington, D.C., 1969.

---

[29] J. A. Thompson and B. I. Fansler, "Economic Study of Various Mounting Heights for Highway Lighting," HRR No. 179 (1967), pp. 1–15.

[30] A. Cassel and P. Medville, "Economic Study of Roadway Lighting," NCHRP Report 20 (1966).

[31] McFarland and Walton, "Cost-Effectiveness Analysis of Roadway Lighting Systems."

[32] *Warrants for Highway Lighting*, National Cooperative Highway Research Program, Report No 152, (Washington, D.C.): Transportation Research Board, 1974.

Ketvirtis, Antanas, *Highway Lighting Engineering*. Toronto: Foundation of Canada Engineering Corporation Limited, 1967.

Lipinski, M. E., G. C. Meador, A. L. Gilbronson, M. L. Traylor, W. D. Berg, C. L. Anderson, and R. H. Wortman, "Summary of Current Status of Knowledge on Rural Intersection Illumination," *Night Visibility and Driver Behavior*, Highway Research Record No. 336. Washington, D.C.: Highway Research Board, 1970.

"Tentative Recommended Practice for Selection and Placement of Lighting Poles," *Traffic Engineering*, (XL, No. 9, 1973), Institute of Traffic Engineers, Washington, D.C., pp. 24–31, 33.

Rex, Charles H., "Effectiveness Ratings for Roadway Lighting," *Illuminating Engineering*, LVIII, No. 7 (1963), 501–20.

Street Lighting Manual, Publication No. 68–72. Edison Electrical Institute, New York, 1969.

Chapter 21

ENVIRONMENTAL CONSIDERATIONS
IN TRAFFIC ENGINEERING

JOHN D. EDWARDS, JR., President, Traffic Planning Associates, Inc., Atlanta, Georgia.

The transportation system and particularly the streets and highways in the United States and other countries have been one of the most important instruments for achieving the high standard of living that the great majority of their citizens enjoy. Highways and streets coupled with the personal motor vehicle have made possible unprecedented mobility for many people, permitting them to enjoy the varied wonders of the world, contributing to social and commercial interaction, and binding the states and regions together.

In the early years of transportation development in the United States, the primary concern was to enable people to move themselves and their goods from one place to another. Little regard was paid to the side effects of transportation development on environment and natural resources. The recent development of street systems, freeways, and highways has not been without some compromises on the environment, especially in urban areas. The stage has now been reached where the complexity of the urban areas demands a broader consideration of highway impact. Other criteria— in addition to efficiency—have risen in importance in judging the value and the impact of a traffic improvement: safety, aesthetics, and effect on the social and physical environment. Realizing the tremendous impact that streets and highways have on each individual and on the communities within which they live, traffic engineers must be more concerned with the compatibility of streets and highways and the environment.

No traffic engineering improvement or a highway project is without both positive and negative effects on the environment. Only certain impacts, such as changes in air quality, visual appearance, noise levels, and overall effect on the urban environment, can be effectively quantified and evaluated. Others, such as aesthetics, effects on established neighborhoods, and social impacts, are more difficult to measure. Environmental considerations discussed in this chapter are limited to those elements and factors that have readily definable qualitative or quantitative limits. Clear and concise definitions are a prerequisite to understanding environmental considerations. Pollution and adverse impact are separated in the discussion because they represent different characteristics of the problem.

Pollution results from the addition of some man-made impurity to air or water or to some other element of nature. Pollution, as a chemical agent, can be quantified and factual criteria can be established to measure its level and effect. Air pollution is the obvious kind of pollution and readily available criteria can be used to measure the level of air pollution. For instance, pollution by automobile transportation can be seen and measured readily by accepted standards. Noise pollution is also readily measurable, although standards for measuring the nonacceptable levels are more subjective. Figure 21.1 illustrates a typical air pollution situation.

Figure 21.1. Air pollution in the urban area. (Source: Courtesy U.S. Environmental Protection Agency. EPA-DOCU-MERICA-*Photograph by Gene Daniels.*)

Adverse impact is much more subjective. Quantitative criteria are not generally applicable. Adverse impact includes such disparate elements as "visual pollution," depletion of natural resources, undesirable land uses, traffic congestion and accidents, and inefficient space utilization, which result in waste or spoilage of the natural endowment. Although they do not permanently damage the balance of nature, they do cause problems, displeasure, anxiety, and grief to society.

Adverse impact is measured on a scale that varies from individual to individual. What may be pleasure or profit to one is misery and loss to another. The broad area of concern and the lack of unanimity on degree of the adversity make even general subjective measures of the problem difficult. Figure 21.2 illustrates a typical example of visual pollution.

TRAFFIC AS A POLLUTER

Transportation pollution is primarily concentrated around the major metropolitan areas and is principally caused by urban traffic. Transportation-related pollution has accounted for an average of 60 percent of the total pollutants in the atmosphere for

Figure 21.2. Visual pollution. (Visual Survey & Design Plan, AIA, Georgia Chapter, p. 14. *Photograph by Dave Miller.*)

the major metropolitan areas in the United States. Of this 60 percent, the private automobile contributed from 90 to 95 percent of the air pollution. The private vehicle also accounted for roughly one-half of the air pollution in the major metropolitan areas within the United States.

Pollution levels and contributing factors are extremely variable. The above statistics would not, therefore, represent world-wide conditions. Because of low private vehicular usage the auto is not a major polluting influence within many still-developing countries.

AIR POLLUTION DEFINED

Air pollution, like other forms of pollution in the United States, is caused by the multiplicative effect of population growth and increasing per capita demands for goods and services. In the process of meeting these demands, millions of tons of gaseous and particulate pollutants are dumped into the atmosphere. These pollutants soil, produce corrosion, and may be harmful to both plants and animals. The potential damage of any pollutant may be inferred from the report of the Committee on Pollution of the National Academy of Science–National Research Council, which states that pollution is "an undesirable change in the physical, chemical, or biological characteristics of our air, land, and water that may or will harmfully affect human life or that of any other desirable species, or industrial process, living conditions, or cultural assets; or that may or will waste or deteriorate our raw material resources."[1] Air pollution includes those detrimental effects related to the atmosphere.

---

[1] *Report of the Committee on Pollution* (Washington, D.C.: National Academy of Science, National Research Council, 1970.

KINDS OF POLLUTANTS

There are several elements of air pollution, which include the following:

1. *Carbon monoxide.* This poisonous gas from the internal combustion engine drives out the oxygen in the bloodstream. A high concentration over a short exposure time can kill; a small amount can cause dizziness, headaches, fatigue, and slow driving reactions. Moderate concentrations often exist in tunnels, garages, and heavy traffic. It is especially dangerous for people with heart disease, asthma, anemia, etc.

2. *Sulphur oxides.* The primary source of these pollutants are factories and power plants burning coal or oil containing sulphur, which forms sulphur dioxide. This pollutant by itself is usually not harmful, but when it is mixed with other pollutants and with moisture, it irritates the eyes, nose, and throat, damages the lungs, kills plants, rusts metals, and reduces visibility.

3. *Oxides of nitrogen.* A result of burning fuels that convert nitrogen and oxygen to nitric oxide (NO) the greatest toxic potential is the tendency of nitric oxide (which is not an irritant) to oxidize to nitrogen dioxide (NO_2), which at high concentrations acts as a pulmonary irritant and is a major component of smog.

4. *Hydrocarbons.* These are unburned chemicals in combustion, such as car exhaust, which react in air to produce smog. Hydrocarbons have produced cancer in animals and may be a cancer-producing element in cigarette smoke. They are primarily of concern because of their role in the formation of photochemical oxidants.

5. *Particles.* Smoke, fly ash, dust, fumes, etc., are solid and liquid matter in air. They may settle to the ground or may stay suspended. They soil clothes, dirty window sills, scatter light, and carry poisonous gases to the lungs. They come from autos, fuels, smelters, building materials, fertilizers, etc.

6. *Photochemical smog.* Photochemical smog is a mixture of gases and particles oxidized by the sun from products of gasoline and other burning fuels. It irritates the eyes, nose, and throat, makes breathing difficult, and damages crops and materials.

Figure 21.3 illustrates vehicular contributions to air pollution.

EFFECTS OF AIR POLLUTION

The effects of air pollution cannot be quantified precisely even for a particular area. Recent estimates have placed the loss of crops and flowers generally in the range of $500 million per year in the United States alone. Recent studies by Dr. L. B. Lave and E. P. Seskin of the Carnegie-Mellon School of Industrial Management indicate that if air pollution were cut by 50 percent in the major cities of the U.S., (1) a newborn baby would have an additional three to five years' life expectancy, (2) deaths from lung diseases would be cut by 25 percent, (3) death and disease from heart and blood vessel disorders might be cut by 10–15 percent, and (4) all disease and death would be reduced by 4.5 percent yearly, and the annual saving to the nation would be at least $2 billion. It has been estimated that a 50 percent reduction in air pollution would save nearly as much money and life as a complete cure for cancer.[2] These are some of the benefits of reducing air pollution.

---

[2] L. B. LAVE and E. P. SESKIN, *The Health Costs of Air Pollution* (Pittsburgh: Carnegie-Mellon School of Industrial Management, 1970).

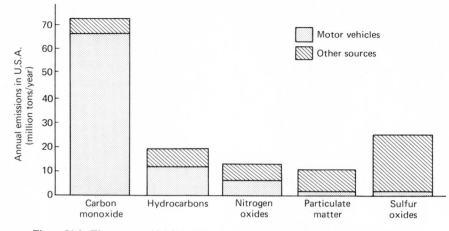

Figure 21.3. The motor vehicle's contribution to five major pollutants. [Source: M. D. Harmelink and W. J. Peck, *Transportation Air Pollution*, DHO Report #RR169 (Ontario: Department of Highways, 1971), p. 31.]

The major sources of pollution at any particular place depend on the relative kinds of activity. Examples of activities that influence pollution levels include agricultural production (crop dusting), manufacturing (smelting), mining (rock crushing), and transportation (vehicle emissions). Emissions from motor vehicles accounted for approximately 80 percent of Atlanta's air pollution in 1970 but accounted for less than 10 percent of the problem in Louisville, Kentucky.[3] This does not mean that automotive pollution in Louisville is not important, but that uncontrolled pollution from other sources is much greater.

Table 21.1 provides data on nationwide major sources of air pollution by type. This table indicates that transportation sources contribute 42 percent of the pollution.

TABLE 21.1
Estimated Nationwide Emissions from Major Sources, 1968

| Source | Emissions in Millions of Tons Annually | | | | | Percent by Source |
	CO	Particulates	SO$_2$	HC	NO$_2$	
Transportation	63.8	1.2	0.8	16.6	8.1	42.3
Fuel combustion in stationary sources	1.9	8.9	24.4	0.7	10.0	21.4
Industrial processes	9.7	7.5	7.3	4.6	0.2	13.7
Solid waste disposal	7.8	1.1	0.1	1.6	0.6	5.2
Miscellaneous	16.9	9.6	0.6	8.5	1.7	17.4
Total	100.1	28.3	33.2	32.0	20.6	100.0

Source: National Inventory of Air Pollutant Emissions, 1968, U.S. Department of Health, Education and Welfare, Public Health Service, National Air Pollution Control Administration, Raleigh, North Carolina, August, 1970.

_____

[3] *Maintaining a Quality Environment in Georgia*, Georgia Office of Comprehensive Health Planning, Georgia Department of Public Health (January, 1971), p. 49.

This compares to a 60 percent contribution in the major metropolitan areas. Transportation sources contribute the highest amounts of pollution in the form of carbon monoxide and hydrocarbons, with from 50 to 60 percent of the total amount coming from this source.

Two factors have been responsible for the national prominence of the air pollution problem: the increase in population in the urban areas and the increased per capita consumption of energy. Two of the major sources of pollution—motor vehicles and power production—are increasing much faster than the population because of the push–pull factors of increased population concentration and increased per capita consumption. Two-thirds of the increase in automobiles and 90 percent of the increase in power production are caused by increased per capita demands for the United States. The most important step in the reduction of pollution is the control of the combustion operations, which contribute the major portions of the carbon monoxide, unburned hydrocarbons, and oxides of nitrogen and sulphur.

The major thrust of this effort at air pollution reduction must be the control of vehicular emissions. Steps in this direction have and are being taken through state and national emission standards.

MEASURES OF AIR QUALITY

Several measures of air quality have been developed for use by the various governmental agencies in quantifying the characteristics of air pollution. Air quality as surveyed by the U.S. Public Health Service for metropolitan areas is stated in terms of parts per million of pollutants. Surveys have been conducted within urban communities at selected sites termed "Continuous Air Monitoring Program Site" (CAMP Site), and these surveys provide a basis for determining the overall quality of the atmosphere within a region using these measures.

The surveys at CAMP Sites provide background information on the quality of atmosphere of major urban areas in terms of carbon monoxide levels, nitrogen oxide levels, particulates, and sulphur oxide levels. Generally, the highest concentrations of all pollutants are found during the winter months. For example, in Washington, D.C., carbon monoxide levels of from 8.0 ppm to 12.0 ppm in winter compare to levels of from 5.0 ppm to 8.0 ppm in summer. Hydrocarbon levels range from 3.0 ppm in winter to 2.0 ppm in summer. These general trends allow the comparison of air pollution and health and mortality statistics to determine the general impact of air pollution.

Corridor pollution levels. There have been numerous surveys of carbon monoxide levels on expressways and major arterials in several major cities. As a general standard 80 ppm of carbon monoxide breathed over 8 hours for a person at rest is not considered serious. The carbon monoxide levels for traffic corridors as shown in Table 21.2 are generally within the above standard, even though they are considerably higher than for ambient levels within the community as a whole.

Carbon monoxide concentrations on expressways range from 12 ppm to 28 ppm, while those on arterial streets range from 20 ppm to 33 ppm. Downtown streets and arterial streets are essentially the same. Operating conditions on the major arterial facilities and the volume of traffic that was handled on the survey sections account for the range in peak values shown in the table.

TABLE 21.2
Carbon Monoxide in Traffic in Several American Cities
(concentration in ppm by volume)

City	Means of Averages from All Runs			Range of Peak Values
	Expressway	Downtown	Major Artery	
Atlanta, Georgia	28	25	33	37–96
Baltimore, Maryland	13	28	20	10–95
Chicago, Illinois	26	25	25	20–100+
Cincinnati, Ohio	12	28	22	10–85
Detroit, Michigan	29	28	30	13–120
Louisville, Kentucky	12	27	20	10–66
New York, New York	22	40	32	19–95

Source: U.S. Public Health Service, "Survey of Lead in the Atmosphere of Three Urban Communities," U.S. Department of Health, Education and Welfare, PHS Publication No. 999-AP-12, January, 1965.

Vehicular pollution levels as affected by time of day. Pollution levels vary considerably depending on the type of vehicle operation, the time of day, and the atmospheric conditions within the urban area. In some cities the peak condition for carbon monoxide concentration follows very closely the chronological sequence of peak-hour vehicular operations, but in other cities this relationship does not occur.

Figure 21.4 illustrates the hourly variation in traffic volumes entering and leaving the central business district of Los Angeles and the concentration of nitrogen oxide and carbon monoxide in the atmosphere for this area during comparable hours. The peak concentration for both carbon monoxide and nitrogen oxide occurs between 7:30 and 8:00 a.m., with concentration levels of from 15 ppm and 18 ppm, respectively. The pollution level drops during the late morning hours for nitrogen oxide to only 1 to 2 ppm. During this period sunlight tends to dissipate the nitrogen oxide. Nitrogen oxide levels increase again toward late afternoon. Carbon monoxide concentrations decrease to 10 ppm at approximately 2:00 p.m. and gradually rise to about 12 ppm where they level out throughout the evening with very little variation until the following morning peak hour.

Speed as a factor in emissions. There is a considerable difference in arterial and expressway carbon monoxide concentrations. A study[4] completed by the California Vehicle Pollution Laboratory indicates that the hydrocarbon level varies depending on the average speed of operation of the vehicle. Tests by the Vehicle Pollution Laboratory indicate that for average freeway operating conditions the hydrocarbon concentration was from 350 ppm to 400 ppm while for average operating conditions on the arterial streets the hydrocarbon concentration was from 580 ppm to 590 ppm.[5] The hydrocarbon emissions were 100 ppm for an engine idling, 540 ppm for acceleration, 485 ppm under cruising conditions, and 5,000 ppm during deceleration. Stopping and starting (for example, on a major signalized arterial street) have a very dramatic effect on hydrocarbon and carbon monoxide emissions from vehicles.

An increase in hydrocarbon emissions of from 34 to 40 percent is indicated on arterial facilities over freeway facilities.

---

[4] JOHN C. CHIPMAN, "Comparison of Auto Exhaust Emissions: Freeway vs. Average Type Driving" (Los Angeles: California Vehicle Pollution Laboratory, 1964), pp. 1–6.

[5] IBID., Tables III and IV, p. 6.

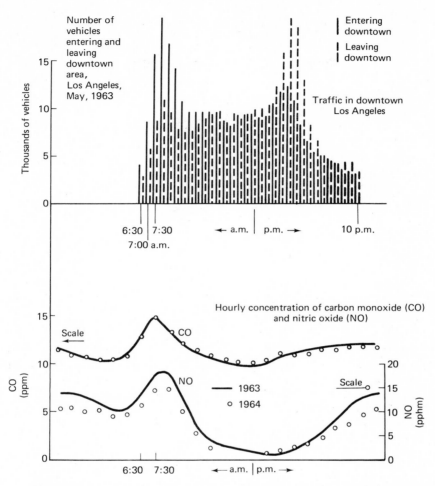

Figure 21.4. Concentrations of carbon monoxide and nitric oxide vs. traffic in down-town Los Angeles. Note that these data were collected before widespread usage of emission controls on motor vehicles and may not represent current numerical data. (Source: California Vehicle Pollution Laboratory, 1964.)

Figure 21.5 illustrates the emissions of hydrocarbons of an automobile for speeds of from 10 to 40 mph. At 40 mph the emission is approximately .009 pounds per mile, but at 10 mph it is approximately .025; that is, more than two and one-half times the emission at 40 mph.

Figure 21.6 illustrates the relationship between speed and carbon monoxide (CO) emissions. This curve is similar to that of hydrocarbon emissions with CO emissions at 10 mph, that is, approximately three and one-half times the emissions measured at 40 mph.

Emissions for nitrogen oxide increase with increasing speed, contrary to the declines in emissions for hydrocarbons and carbon monoxide. Nitrogen oxide emissions (see Figure 21.7) for cars equipped with no emission controls range from .0025 at 10

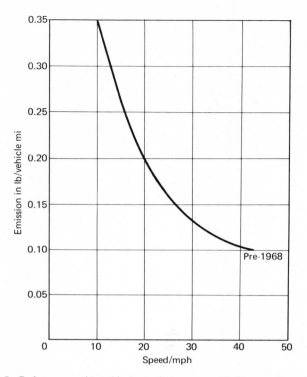

Figure 21.5. Carbon monoxide vehicular emission vs. speed. (Source: Second report, Secretary of HEW to U.S. Congress pursuant to P.L. 88–206 Clean Air Act, February 19, 1965, Table 1.)

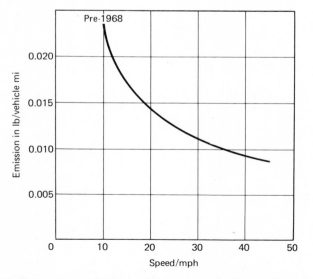

Figure 21.6. Hydrocarbon vehicular emission vs. speed. (Source: Second report, Secretary of HEW to U.S. Congress, Table 1.)

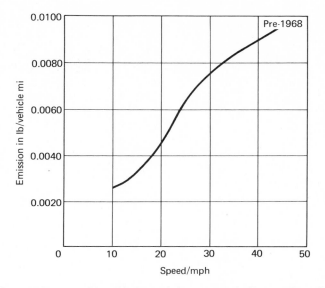

Figure 21.7. Nitrogen oxide vehicular emission vs. speed. (Source: Second report, Secretary of HEW to U.S. Congress, Table 1.)

mph up to .0090 at 40 mph. Higher operating speeds and improved operating conditions provided by freeways do not reduce emissions of nitrogen oxide. These emissions may be reduced by the recirculation of the exhaust gas through the carburetor. The results are reduced power and more expensive operating conditions.

Vehicle age and type as a factor in emissions. Vehicle age is a factor in the level of pollution primarily because of innovations in anti-pollution equipment rather than lack of maintenance. Hydrocarbon emissions come from several sources, including blowby, exhaust, and evaporative sources. Blowby emissions from improper carburetion, combustion, and crankcase fumes amount to approximately 7 percent of total hydrocarbon emissions. The installation of blowby equipment on all vehicles as required by California after 1961 has almost eliminated this element.

Surprisingly, studies and reports[6] by the Automobile Club of Southern California indicate that automobile maintenance has very little, if any, effect on hydrocarbon emissions. The studies indicate that although there may be some increase in emissions caused by improper functioning of spark plugs or carburetor, these malfunctions are generally promptly corrected by the owner because they adversely affect the operating characteristics of the vehicle.

Although trucks generally account for about 20 percent of the total vehicle miles of travel and for as much as 50 percent of the volume on truck routes, they play a relatively minor role in air pollution for several reasons. Most large trucks, which account for the largest vehicle miles of operation, are powered by diesel engines. The level of pollution produced by diesel engines is considerably less than that of gasoline. Figure 21.8 shows that a gasoline engine produces 13.7 times more emissions than a diesel engine of comparable size (piston displacement).

---

[6] Louis J. Bintz, *Report on the Status of Automotive Air Pollution in the Los Angeles Basin* (Los Angeles: Automobile Club of Southern California, January, 1967), p. 10.

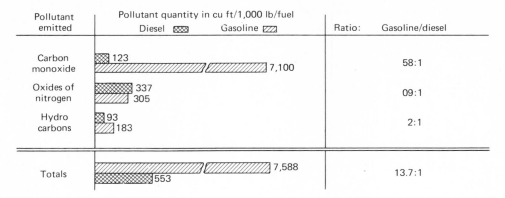

Figure 21.8. Comparison of diesel and gasoline engine emissions. (Source: Harmelink and Peck, *Transportation Air Pollution*, p. 31.)

Also, truck travel is more constant throughout the day, with less peaking occurring during normal morning and afternoon commuting hours. Since air pollution has time as well as emission dimensions, the time distribution of truck travel tends to lessen the impact of emissions by these vehicles.

Emissions as affected by anti-pollutant devices. As early as 1965, major efforts were underway to reduce the level of pollution created by the motor vehicle. These efforts were started in California in response to the smog conditions in the Los Angeles area in the early 1960's. Since then an extensive amount of work has been done nationally on air pollution devices that can reduce the emissions from motor vehicle exhaust, fuel evaporation, and blowby. (See References for Further Reading at the end of this chapter.)

In order to evaluate the impact that this technology will have on vehicle emissions, consider the vehicle emissions standards that have been established by the U.S. Department of Health, Education and Welfare and, more recently, the Environmental Protection Agency to obtain the air quality thought necessary for the protection of public health and welfare. The standards are discussed in detail on page 979, but the general effect of the standards can be seen in Figures 21.9, 21.10, and 21.11. If standards are met by 1975, carbon monoxide emissions will be approximately one-seventh of the emissions experienced in the period prior to 1968 for comparable speeds. The hydrocarbon emissions per vehicle mile after 1975 will be one-tenth of the emission of pre-1968 conditions for a comparable speed. Nitrogen oxide will be decreased from .0025 lb per vehicle mile to .00075 lb per vehicle mile, or about one-third of the pre-1968 level.

NOISE POLLUTION

Noise has often been described as unwanted sound, sound without value, or vibrational energy out of control. Noise has become a national concern and an increasing problem. Generally, noise has three sources: (1) transportation noise, (2) occupational or industrial noise, and (3) community background noise. Because any noise is capable of producing both physical and psychological damage, all sources must be controlled.

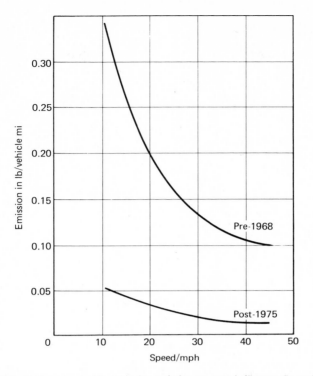

Figure 21.9. Carbon monoxide vehicular emission vs. speed. (Source: Second report, Secretary of HEW to U.S. Congress, Table 1.)

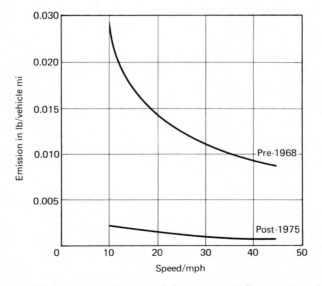

Figure 21.10. Hydrocarbon vehicular emission vs. speed. (Source: Second report, Secretary of HEW to U.S. Congress, Table 1.)

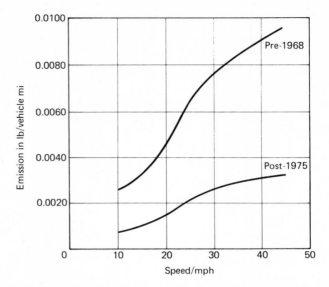

Figure 21.11. Nitrogen oxide vehicular emission vs. speed. (Source: Second report, Secretary of HEW to U.S. Congress, Table 1.)

Although aircraft noises may first come to mind when transportation noises are being discussed, noise from surface vehicles, such as automobiles, trains, trucks, and motorcycles, and industrial or occupational noise are more widespread and more significant. Most occupational or industrial noise results from metal-to-metal contact or from high-speed equipment (blowers, for example) and is a serious problem in steel, paper, textile, and petro-chemical industries. Occupations other than the heavy production industries are also affected.

Even in the confines of the home, one is bombarded with noise from the radio, record player, dishwasher, lawnmower, disposal unit, and ice crusher. These convenience tools are increasing to such an extent that the noise level in the home is approaching that of a mechanized industry. These home noises contribute to community background noise.

Typical sound sources. In this chapter noise pollution is primarily related to surface transportation noise and comparisons with community noise levels. Table 21.3 provides examples of typical sound sources. The sound levels illustrate the relative amounts of noise generated by selected sound sources as compared to human response and hearing effects.

Comparison of transportation noise sources. Transportation noise is one of the most prevailing sources of sound in the urban area. Much of the urban area background noise is composed of the aggregate sounds produced by the thousands of cars, trucks, airplanes, and other transportation modes in the urban area.

Measurements of various transportation sound sources provide the comparative data in Table 21.4. Many transportation noise sources are between 80 dBA and 100

TABLE 21.3
Sound Levels and Human Response

	Noise Level (dBA)	Response	Hearing Effects	Conversational Relationships
	150		CONTRIBUTION TO HEARING IMPAIRMENT BEGINS	
Carrier deck				
Jet operation	140			
		Painfully loud		
	130	Limit amplified Speech		
Jet takeoff (200 ft)	120			
Discotheque		Maximum vocal effort		
Auto horn (3 ft)				
Riveting machine	110			
Jet take-off (2,000 ft)				
Shout (0.5 ft)	100			Shouting in ear
N.Y. subway station		Very annoying		
Heavy truck (50 ft)	90	Hearing damage (8 hr)		Shouting at 2 ft
Pneumatic drill (50 ft)	80	Annoying		Very loud conversation, 2 ft
Freight train (50 ft)				
Freeway traffic (50 ft)	70	Telephone use Difficult Intrusive	↑	Loud conversation, 2 ft
Air conditioning unit (20 ft)	60			Loud conversation, 4 ft
Light auto traffic (100 ft)	50	Quiet		Normal conversation, 12 ft
Living room				
Bedroom	40			
Library				
Soft whisper (15 ft)	30	Very quiet		
Broadcasting studio	20			
	10	Just audible		
	0	Threshold of hearing		

Source: "Noise Pollution," U.S. Environmental Protection Agency (Washington, D.C.: Government Printing Office, August, 1972), p. 6.

dBA[7] as measured at the operator's ear and fall to significantly lower levels at the average observer's distance. Trucks and buses have noise levels varying from 81 dBA to 96 dBA. This compares to aircraft noises for the operator of from 73 dBA to 119 dBA and outside noise levels of in excess of 110 dBA measured from a 400 ft altitude. Sound levels in the immediate vicinity of the aircraft may easily reach 160 dBA.

Public transportation vehicles make noise, too. Noise levels for the public transportation facilities in Table 21.5 are based on current experience with equipment now in operation in Germany, England, Toronto, and Philadelphia. A range of noise levels from 78 dBA to 100 dBA, depending on the observer's location, the type of

[7] JAMES H. BOTSFORD, "Damage Risk," *Transportation Noises—A Symposium on Acceptability Criteria* (Seattle: University of Washington Press, 1970), pp. 110–13.

TABLE 21.4
Comparative Transportation Sound Levels

Miscellaneous Vehicles	Maximum Noise at Operator's Ear
Cranes	85–113 dBA*
Outboard motor	85 dBA
Street sweeper	96 dBA
Buses	82–96 dBA
Trucks (2000–2500 rpm)	81–92 dBA
Tractors	85–113 dBA
Road graders	97–100 dBA
Power boats, at seat nearest motor (cruising speed)	83–104 dBA
River barge two boat, 919 tons gross (engine room)	101–112 dBA
Jet airliner (take-off power)	140–160 dBA
Shovels, diesel	91–107 dBA
Shovels, electric	83–91 dBA
Bulldozers	102–106 dBA
Diesel locomotives	88–100 dBA

*A-Weighted Sound Level.

Source: James H. Botsford, "Damage Risk," *Transportation Noises—A Symposium on Acceptability Criteria* (Seattle: University of Washington Press, 1970), p. 111.

TABLE 21.5
Public Transportation Noise for Selected Rail Transportation Systems

Type and Location	Below Ground	Above Ground
ARWEG monorail		
Inside cars	—	81 dBA
Station platform (25 ft)	—	80 dBA
British Railroad—suburban train		
Inside cars	—	71 dBA
Station platform (25 ft)	—	88 dBA
London Transport—METRO		
Inside cars	82 dBA	78 dBA
Station platform (25 ft)	87 dBA	—
Toronto subway car		
Inside cars	78 dBA	—
Station platform (25 ft)	84 dBA	—
Philadelphia rail car		
Inside cars	82–95 dBA	78–90 dBA
Station platform	93–98 dBA	83–93 dBA
Philadelphia—trolley car		
Inside cars	74–87 dBA	65–75 dBA
Station platform	84–100 dBA	80–85 dBA

Sources: P. T. Lord, "Study of Rapid Transit Systems and Concepts," *Manchester Rapid Transit Study, Volume 2*, De Leuw, Cather and Partners, Hennessy Chadwick, O. Heocha and Partners, Ministry of Transport, City of Manchester and British Railways, August, 1967.

James H. Botsford, "Damage Risk," *Transportation Noises—A Symposium on Acceptability Criteria*, pp. 103–13.

equipment, and the location (above ground or below ground) have been recorded. These noise levels are representative of most of the noises associated with existing public transit systems. For comparison, noise levels on the Montreal Metro, a modern transit system utilizing rubber tires on concrete rail, are generally in the range of from 60 dBA to 80 dBA.[8]

Measurement of noise. The measurement of noise and the correlation of that noise to human annoyance is a very complex procedure. Extensive research has been completed to quantify the most significant measures of loudness. Two approaches have been utilized: (1) research on how an individual judges loudness and noise from a psychological viewpoint and (2) experiments on groups of people in order to determine how they judge the acceptability of various sounds on category judgment scales.

There are many scales for measuring sound (see References for Further Reading) that incorporate psychologically derived measures and physical measures. Among the former are *Loudness, Loudness Level, Perceived Noise Level (PNdB)*, and the *Speech Interference Level (SIL)*. Among the latter are *Overall Sound Pressure Level (OASPL)*, and *A-Weighted Sound Level (dBA)*.[9]

COMMUNITY RESPONSE TO NOISE

Although community response to noise, especially highway-related noise, must certainly be negative, a certain amount of transportation noise is unavoidable. Quantifying objectionable transportation noise levels is difficult because of the extreme variability of human response to noise and, even more importantly, human response to sound. Any evaluation of community response to noise must consider the following elements:

1. The attitudes of the sample upon which the response is based.
2. The socioeconomic background of the sample (collectively and individually).
3. The presence of other negative stimuli that could be associated with the sound source.
4. The characteristics of the individual with respect to his exposure to sound.

Reactions to freeway noise. One study of freeway noise conducted in 1962 found that persons judge freeways not so much by noise generation or other assumed adverse factors but by attributes such as accessibility, the advantages of a freeway location, and the appearance of the freeway.[10] Some factors that are generally assumed as being adverse, for example, the appearance of the freeway or the high lighting level associated with most freeways, were reacted to favorably on the part of many respondents—even by those who lived near the freeway. Significantly, the research study found that those who expressed annoyance at noise levels were doing so on the

[8] M. D. HARMELINK, *Noise and Vibration Control for Transportation Systems*, DHO Report No. RR168 (Ontario: Department of Highways, October, 1970), Table 191, p. 37.

[9] The term "level" is used to indicate a logarithmic instead of a linear scale. It so happens that the psychological response of the auditory system is essentially exponential; this is another reason for the decibel scale.

[10] W. J. GALLOWAY, W. E. CLARKE, and J. S. KERRICK, *Highway Noise-Measurement, Simulations and Mixed Reactions*, NCHRP Report No. 78 (Washington, D.C.: Highway Research Board, 1969).

basis of an average of only one dbA increase above the mean level of those who expressed no annoyance at freeway noise.

Community background noise. Sound sources, when viewed in proper perspective, must be related to the prevailing community background noise level. Extensive research indicates that for each community there is a noise signature that corresponds to the type and spatial arrangement of industrial activities, transportation corridors, and land-use patterns within the region as well as atmospheric conditions and the topography.

A study of Ottawa, Canada, attempted to develop a predictive equation to determine the background noise at any given point in the community.[11] Recordings were made for four days at a location within a residential area but not adjacent to a freeway or major street. Figure 21.12 shows the sound signature of the background noise levels for a Saturday, a Sunday, a Monday, and a Tuesday in 1969. The average noise level was found to be 48 dBA during the daylight hours and 38 dBA at night for the location shown. The signature shows extreme variations during the entire mid-day portion on Sunday—the highest noise level occurring as extreme peaks around 10:30 a.m. at 65 dBA. At approximately 9:00 p.m. and at 7:00 p.m. on Saturday the highest noise levels occurred at 50 dBA. Otherwise, very little variation in the noise level occurred.

For the weekday patterns, the 8:00 a.m. hour and the 5:00 p.m. peak hour are indicated with the a.m. peak showing more prominently. Instantaneous peaks of 65 dBA were common throughout the daylight hours during the week with the mean level between 50 dBA and 55 dBA. This is contrasted to a mean noise level of from 45 dBA to 50 dBA on Saturday and Sunday.

Predicting background noise levels. Further research suggests that the background noise level within a community is not greatly affected by an increase in city size.[12] Noise levels in a city 100 miles in diameter would only increase 4 dBA over a city of 10 miles in diameter.

Traffic movement in an urban area does not drastically change the background noise levels. Research studies indicate that doubling of the density of traffic would only increase the background noise by 3 dBA.

Auto noise levels as affected by traffic operations. Noise from motor vehicles in motion comes from two major sources: the engine-exhaust system and the tire-roadway system. Major engine-exhaust noise sources consist of intake noise at the carburetor, cooling fans, valve lifters, gear boxes, and exhaust noise. Exhaust noise, when properly muffled (as with factory-installed equipment), probably contributes no more than 10 or 15 percent to the overall noise signature of the vehicle.

Tire-roadway noise is present for all vehicles in motion and is the dominant source of noise under certain operating conditions. The tire-roadway interaction produces a sound signature primarily attributable to the pattern and depth of tire tread, roughness of the road surface, wetness, tire stiffeners, tire loading, and the suspension system of the vehicle.

---

[11] G. J. THIESSEN, "Community Noise Levels," *Transportation Noises—A Symposium on Acceptability Criteria* (Seattle: University of Washington Press, 1970), p. 24.
[12] IBID., p. 29.

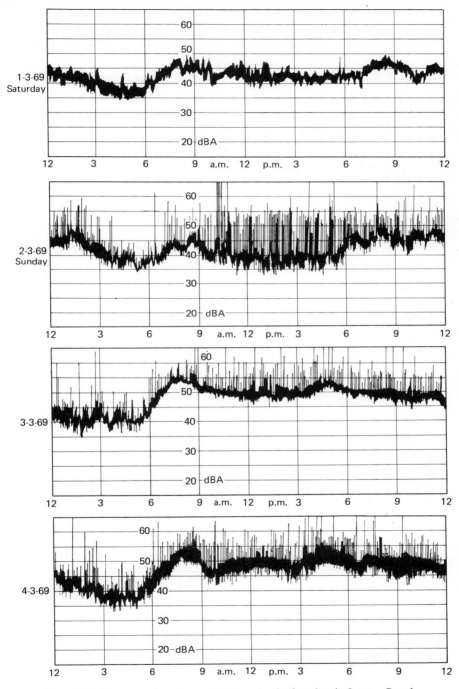

Figure 21.12. Sound signature for a selected location for four days in Ottawa, Canada. Continuous recording of the noise level with a writing speed of 1.0 db per sec. [Source: James D. Chalupnik, *Transportation Noises, A Symposium on Acceptability Criteria* (Seattle: University of Washington Press, 1970), p. 27.]

The importance that tire-roadway noise plays in the overall automobile noise signature is illustrated in Figure 21.13. The characteristic spectral shapes for passenger car noise were measured at a position 25 ft laterally from the vehicle. The narrow hatched band illustrates the difference in spectra measured for cars traveling at 65 mph and spectra for cars coasting with the engine disengaged for the upper and the lower edge of the band, respectively. The narrow band width indicates that the noise is primarily the tire-roadway interaction that is independent of the engine operation. Greater differences between the two curves in the left half of the spectrum show that engine-exhaust system noises are important in this low-frequency range at 65 mph during normal cruise conditions. The differences of from 3 to 5 decibels (db) suggest that one-half or more of the noise in the second, third, and fourth octave band comes from the engine-exhaust system.

The other banded area in Figure 21.13 shows the difference in the range of spectra from normal cruise conditions at 35 mph to maximum acceleration conditions (represented by the top edge). The great difference between these operating conditions suggests that engine-exhaust system noise predominates over the entire spectrum for conditions of maximum acceleration.

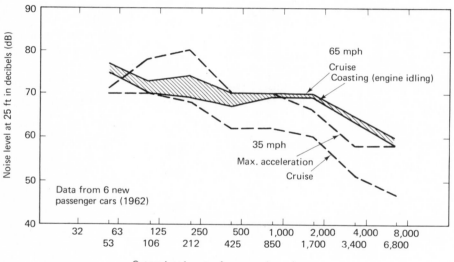

Figure 21.13. Elements of passenger car noise. [Source: Clarke, W.E. Galloway, W.J. and Kerruch, J.S., *Highway Noise, Measurement, Simulation, and Mixed Reactions,* National Cooperative Highway Research Program Report 78 (Washington, D.C.: Highway Research Board, 1969), p. 32.]

Speed and noise production. Speed produces variations in noise: the higher the speed, the louder the noise. Figure 21.14 illustrates the results of a controlled experiment.[13] When a vehicle is driven at speeds of 30, 40, 50, and 60 mph past the measurement point, the noise spectra illustrate a consistent increase in all octave bands of frequency for each increase in speed.

---

[13] GALLOWAY *et al.*, p. 35.

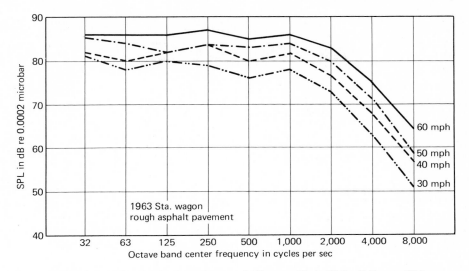

Figure 21.14. Noise spectra at varying speed. (Source: Clark, W.E., Galloway, W.J. and Kerrick, J.S., *Highway Noise, Measurement, Simulation, and Mixed Reactions,* p. 35.)

Vehicle age and noise production. Experiments comparing a new vehicle with a two-year-old vehicle of the same model indicates that differences in noise levels in relation to vehicle age are primarily caused by exhaust system wear and body rattles. The average difference in noise levels (over all ranges of speed) of from 3 to 5 db indicates a relatively small change in noise as related to age (see Figure 21.15). These data represent laboratory experiments of vehicles in good condition, i.e., no muffler defects and no exhaust system modifications.

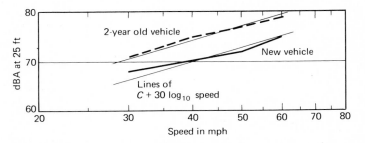

Figure 21.15. Noise production vs. vehicle age. (Source: Clark, W.E., Galloway, W.J. and Kerrick, J.S., *Highway Noise, Measurement, Simulation, and Mixed Reactions,* p. 36.)

Pavement surface and noise. Figure 21.16 illustrates data collected from random recordings of passenger vehicles over various road surfaces at a distance of 25 ft from the travel lane. Generally, the difference in sound level produced from the smooth to very rough pavement is approximately 10 dBA over the entire operating speed range. At 50 mph the sound levels from rough pavement were 10 dBA louder than from very smooth pavement.

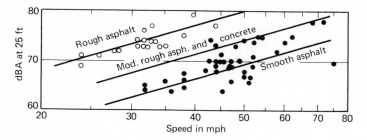

Figure 21.16. Noise vs. pavement surface. (Source: Clarke, W.E., Galloway, W. J. and Kerrick, J.S., *Highway Noise, Measurement, Simulation, and Mixed Reactions*, p. 36.)

Intersections and noise production. The loudest sounds are generated at intersection areas by stopping and starting vehicles. These noise sources consist of braking noises, including tire-roadway and brake system (primarily in deceleration and engine-exhaust noise) and tire-roadway interaction (primarily in acceleration). Generally, deceleration at intersections is controlled and without tire skid, therefore the acceleration phase of intersection operation produces the greater sound level.

Vehicle type and noise. There are significant variations in the noise signature of individual vehicle types. Trucks, especially large diesel units, although a relatively small proportion of the total traffic stream, contribute significantly to the noise produced by traffic. They are impressive sources of noise and are usually clearly audible in the mixture with automobiles in traffic. Diesel trucks are inherently much noisier than passenger cars, producing a noise level about 15 dBA higher in each type of vehicle operating condition. The noise source for large diesel trucks is a composite of contributions from the engine-exhaust system and the tire-roadway interaction, but the engine and exhaust noise tends to be dominant at low operating speeds and tire-roadway interaction dominates at high operating speeds.

The contribution of tire noise to the total noise signature from diesel trucks in motion can be observed in the lowest curve in Figure 21.17 where the truck is operating down-grade, engine at idle.

Engine and exhaust noises are illustrated by the high spectra associated with maximum acceleration. Comparison of passenger car noise with truck noise indicates that truck noise ranges from 10 to 15 dBA higher across the entire spectra at maximum acceleration (see Figure 21.18).

Motorcycles and sports cars are often cited as the prime violators of noise restrictions. The conclusions from the measurements are that no particular correlation exists between the amount of power developed (or the size of the vehicle) and the amount of noise output. Muffling the exhaust level is highly variable for each manufacturer's products and is much more variable in the hands of the users. In particular, the noise from acceleration of these vehicles is a problem because of the low standard of muffling of exhaust noise.

Highway noise abatement.[14] A number of measures may be used to reduce the undesirable noise from highway and street traffic. A rigid (fairly massive) barrier can

---

[14] "Transportation Noise and Its Control" (Washington, D.C.: U.S. Department of Transportation, June, 1972), pp. 14–16.

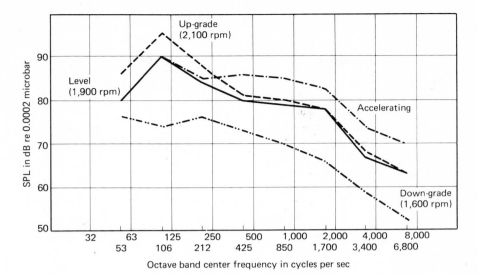

Figure 21.17. Effects of operating conditions. (Source: Clarke, W.E., Galloway, W.J. and Kerrick, J.S., *Highway Noise, Measurement, Simulation, and Mixed Reactions,* p. 39.)

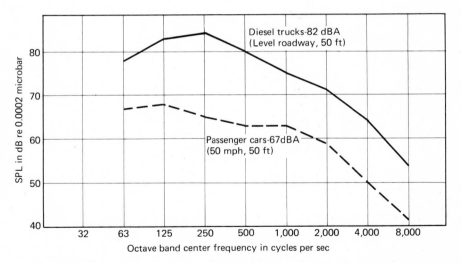

Figure 21.18. Comparison of truck and auto noise for acceleration. (Source: Clarke, W.E., Galloway, W.J., and Kerrick, J.S., *Highway Noise, Measurement, Simulation, and Mixed Reactions*, p. 42.)

be an effective means to reduce noise from highways depending on the relative heights of the barrier, the noise source, and the affected area, as well as the horizontal distance between the source and the barrier and between the barrier and the noise-affected area. In general, it has been found that the closer a barrier is constructed to either the noise source or to the noise-affected area, the greater the sound-level reduction. In addition, increasing the height of a barrier increases the amount of reduction in sound level.

Another technique is to depress or elevate the roadway. These differences in grade provide some shielding of traffic noise and can reduce the noise levels at adjacent properties (see Figure 21.19). Construction costs of major traffic ways are expensive, especially in urban areas, and designing to reduce detrimental noise levels may well increase the cost, as indicated in Table 21.6.

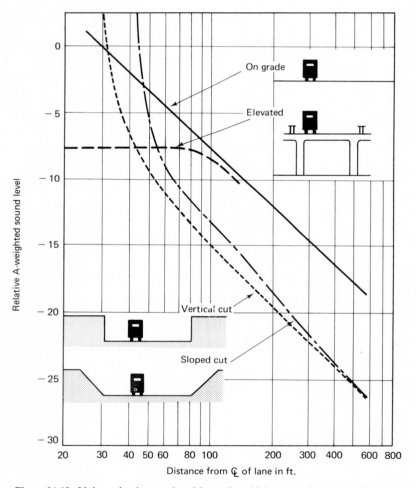

Figure 21.19. Noise reduction produced by various highway configurations. [Source: "Transportation Noise and Its Control," U.S. Department of Transportation (Washington, D.C.: Government Printing Office, 1972), p. 15.]

Landscaping (trees, bushes, shrubs, etc.) adjacent to a highway produces little physical reduction of noise level unless it is very dense and of significant depth. The results of recent investigations indicate that trees do reduce sound levels but not enough to justify a large expenditure for this purpose alone. Decorative plantings,

TABLE 21.6
Relative Construction Costs of Urban Motorways
(Excluding Land Costs, Etc.)

Form of Construction	Average Index Number	Range of Index Number
At ground level*	1	0.7–1.3
Depressed open cutting	1.5	—
Elevated on embankment	2	—
Elevated: retained	3	—
Depressed: retained	5.5	—
Elevated: viaduct	7.5	4–10
Tunnel: bored in good condition	13	10–16
Tunnel: immersed tube	25	—
Tunnel: cut and cover	14	5–30
Tunnel: bored under river	50	18–65

*Cost per lane kilometer at ground level taken as approximately £70,000 ($108,000) at 1969 prices.

Source: L. H. Watkins, *Urban Transport and Environmental Pollution*, TRRL Report LR 455 (Crowthorne, Berkshire, U.K.: Transport and Road Research Laboratory, 1972), p. 11.

although they do not significantly reduce the sound level, have reduced the number of complaints about noise.

OTHER TRANSPORTATION IMPACTS

Although the positive impacts of transportation facilities far outweigh the negative impacts, both must be considered and constantly evaluated.

There can be no question that highway transportation has played a major role in the economic development of the United States and other nations of the world. For the purposes of this discussion, however, greater attention is paid to the negative impacts as a reminder to the practicing traffic engineer that the public is often *most* sensitive to the negative impacts.

Several negative impacts may be described. Some are temporary, for example, the construction of a highway that may cause erosion and result in the silting of natural streams. Others are of a more permanent nature, for example, the day-to-day over-capacity use of existing streets and freeways that results in congestion, delay, or accident hazard. A secondary impact of such heavy traffic may be the stimulation of development patterns that create undesirable effects on adjacent land use as well as problems with traffic flow.

FREEWAYS AND MAJOR STREETS AS ACCIDENT ZONES

Accidents cannot be considered "pollution" in the normal sense, although automobile accidents are clearly destructive to society. It was estimated that in 1972 accidents on streets and highways in the United States cost $19.4 billion annually in property damage and loss of productive time. There were 56,600 persons killed in automobile accidents during that year as well. Much of this pain, suffering, and economic loss can be prevented through improved vehicular and highway design, safety programs, and other accident prevention techniques.

Accident frequency patterns are concentrated on freeways and major streets. In the United States it was estimated that in 1972, 56,600 out of a total of 117,000 accidents and deaths and 2.1 million out of a total of 11.5 million disability injuries were auto related.[15] This implies a tremendous challenge and opportunity to the traffic engineering profession. Improvements in the safety of these streets and highways can effectuate a major reduction in injury and loss of life. See also Chapter 9, Traffic Accident Analysis.

VISUAL IMPACT

The visual impact of the highway environment is the basis on which most of the driving public judges a "good" or "bad" road. There are many miles of beautiful scenic areas made accessible through the development of modern freeways, but the good areas tend to be forgotten when the motoring public is exposed to the bad areas along many arterial streets in most urban areas where inadequate land use and sign controls exist. For a comparison of good and bad visual impact in highway development, refer to Figure 21.20. Unfortunately, major highway improvements have not always been designed with their impact on the total urban environment in mind. None of the design professions have done a comprehensive job of considering the total improvement corridor when highway improvements have been made. Whatever the causes of adverse visual impact, most of the problems can be solved by good design, by the proper consideration of the total environment, and by the use of joint development proposals that complement the proposed facility.

(a)   (b)

Figure 21.20. Good and bad visual impact. [Source: W. R. Ewald and D. R. Mandelker, *Street Graphics* (Washington, D.C.: American Society of Landscape Architects Foundation, 1971), p. 7.]

Considerable research effort has been expended in evaluating visual impact and in developing techniques for improving the environment. One study by the North Georgia Chapter, American Institute of Architects, entitled "Improving the Mess We Live In," illustrates the problem in dramatic graphic style and suggests that sign control is a necessary part of any project improvement program for freeways and major arterial highways.

---

[15] *Accident Facts* (Chicago, Ill.: National Safety Council, 1973), pp. 3–5.

Quoting from another study, *Visual Survey and Design Plan*,[16]

> Our streets, especially commercial streets, are crowded corridors where automobiles
> and pedestrians are hopelessly intermingled and where ugly, unplanned, uncoordinated
> street furniture completes the visual confusion. The graphics of an area should be
> integrated with the architecture and landscape of the area so as to achieve visual unity
> and individual identity. Traffic signs, street signs, store signs, lettering outside and inside
> buildings as well as letters on traffic lights, sidewalks, and benches should all show a
> relationship of form and color. Planned graphics will not only be pleasing to the eye,
> but will guide the motorist and pedestrian through the area efficiently.

The efforts of design professionals to achieve some coordination in sign elements
includes the consideration of traffic signs as well. "Street Graphics,"[17] a report
published by the American Society of Landscape Architects, adds:

> While the street graphics controls proposed herein function within the performance
> standards of a legal framework to help produce a more pleasing esthetic result, it is
> hoped that this study will go beyond legalities to really interest sign manufacturers
> and their clients and to help strengthen decisions of our courts which are coming more
> and more to recognize amenity as a *legitimate basis* for public concern and regulation.

Recognition of the sign control problem has received attention from the Federal
Highway Administration in the allocation of a 5 percent additional funding for state
highway departments who exercise controls on billboards within 800 ft of the interstate system right-of-way.

Parking lots. The proper design of the large expanse of asphalt that provides the
parking so necessary to major traffic generators is a major area of concern. These
parking areas have traditionally been barren with hardly a sprig of green or any other
landscaping to make the parking area more attractive and more functional.

Landscaped islands and other channelization and orientation elements have real
functional uses in traffic operations. It is very difficult for a motorist driving through
a sea of parked cars to locate the access or egress point. The use of islands as a channelizing device helps to reduce high-speed diagonal maneuvers across parking areas,
provide direction for the motorist, and reduce parking lot accidents. Large parking
lots must be divided into series of smaller lots (generally not exceeding 400 or 500 cars
each) in order to obtain more efficient usage, safer operation, and better traffic circulation.

Traffic signs. In recent years the National Joint Committee on Uniform Traffic
Control Devices has been very active in establishing uniform signs and pavement
markings throughout the United States. These uniform standards were developed in
order to assure that every highway is signed and marked in accordance with an
overall plan.

Uniformity has received international support from a broad range of professional
groups. Some of the same groups have strongly objected to multiple signing or over-
signing. Harold Lewis Malt, a landscape architect, made the following comment:

> When three, four, five or ten traffic signals are required at one intersection to let the
> motorist know who has the right-of-way, then it is clear that something is wrong with
> the design of the signal, the design of the installation, or the design of the intersection.

---

[16] *Visual Survey and Design Plan*, Georgia Chapter, American Institute of Architects, p. 38.
[17] W. R. Ewald and D. R. Mandelker, *Street Graphics* (Washington, D.C.: American Society of
Landscape Architects Foundation, 1971), foreword.

Perhaps all three. This kind of design is deliberate. It results from the theory that communicating a message with more than one sign or signal helps minimize the background "noise" of all the competing stimuli on the urban scene. In other words, redundancy is considered desirable. Although usual engineering practice tolerates some over-design as a factor, redundancy may cause chaos and confusion in the hands of an unskillful engineer. Further, the acceptance of this theory tends to a hardened position that forecloses a number of other options and makes experimental progress much more difficult.[18]

In Buffalo, New York, there are estimated to be 45,000 traffic signs displayed on light standards and makeshift supports. In a single year, 1966, the city's Bureau of Signs and Meters rehabilitated 5,304 traffic signs, fabricated 6,600 new signs, installed 963 concrete foundations, and erected 853 street name signs. This pattern exists in other cities of the country. In a recent study in the District of Columbia alone, there was an inventory of 34,100 parking signs estimated to carry over 51,000 parking messages.[19]

The point the landscape architects and other design professionals are raising is that the sheer number of traffic signs present on many streets and highways constitutes visual pollution and contributes to confusion and, possibly, to accidents. At the same time, many of the signs are not necessarily contributing to the efficient flow of traffic. Additional research on better ways of communicating messages more effectively to the motorist is needed.

Freeways. The opportunities that freeways provide to enhance the environment are virtually limitless. There are problems with freeway visual appearance because of the overwhelming physical dimensions of freeways. The detrimental visual aspects of freeways can be divided into two basic areas:

1. The physical characteristics of the facility itself, such as ramps, structures, barriers, medians, retaining walls, and other elements which clash with the environment.
2. Land development patterns for which the freeway has served as a catalyst.

Examples of visually negative physical characteristics of freeways exist primarily along those segments that were constructed in the early days of freeway development. Some of these facilities were built without the benefit of extensive experience in freeway design. As a result, these facilities have design deficiencies that translate into negative visual forms. Such features as narrow, barrier type medians, high retaining walls, abutment type bridges, prolonged elevated structures, and heavily signed transition areas create a negative visual reaction on the part of the motorist. Conversely, many situations exist in which freeways have been built with wide medians, with long graceful bridges having guardrails that flow into the bridge structure, and with shrubbery and other landscaping on slopes in order to increase their attractiveness and create a positive visual impact. Generally, the visual impact created is the result of the designer's appreciation for visual amenities.

Land development patterns are dramatically influenced by the accessibility provided by a freeway. High-intensity development of many roadside businesses (see Figure 21.21) is frequently stimulated at interchanges with the freeway. This overuse

---

[18] Harold Lewis Malt, *Furnishing the City* (New York: McGraw-Hill Book Company, 1970), p. 178.
[19] Malt, *op. cit.*, p. 178.

Figure 21.21. High-intensity development. (Source: Ewald and Mandelker, *Street Graphics*, p. 2.)

of the land usually creates a negative visual impact. The ribbon development, coupled with the lack of sign control and the lack of access control, is characteristic of too many highway-oriented commercial districts. The lack of adequate sign and land use control along a proposed new freeway can result in adverse impact on the development of a quality environment.

Ribbon commercial development. The roadside-oriented business areas are generally called "ribbon" or "strip commercial development." The negative effect of ribbon development along a street or highway may be measured in visual, accident, and congestion impacts. An "affiliated commercial area" on either side of the highway that must be served by a network of streets usually develops. The traffic generated on the cross streets and at major access points enters the highways at a number of closely spaced intersections. The most significant detrimental impact is the reduction in capacity for through traffic caused by the number of turning movements generated by the highway commercial businesses. A study indicates that restaurants, snack bars, and commercial uses contribute the highest interference to through traffic flow.[20] The number of turns for the peak hour and the turns that involved interference with through traffic are shown by destination type in Table 21.7. These turning movements frequently result in a drastic reduction of capacity.

TABLE 21.7
Interference Study—Turning Movements into Roadside Establishments

Type	Number of Establishments	Number of Turns	Average Number of Turns per Establishment	Interference Turns	Percent Interference Turns
Restaurants	12	91	7.5	23	25.3
Snack bars	3	8	2.7	2	25.0
Residences	12	8	0.7	2	25.0
Commercial	27	110	4.1	28	25.4
Motels	12	49	4.1	11	22.4
Service stations	11	80	7.3	12	15.0
Total	77	346	4.5	78	22.5
Intersections*	3	89	29.7	6	6.7

*Intersections with county roads only (locations were chosen away from major intersections).

Source: R. I. Wolfe, "Effect of Ribbon Development on Traffic Flow," *Traffic Quarterly*, XVIII, No. 1 (1964), 116.

TRAFFIC AS A CATALYST FOR UNDESIRABLE DEVELOPMENT

There seems to be a close relationship between the improvement of a major route and subsequent roadside commercial development. This has been illustrated by a number of studies on traffic impact. The Federal Highway Administration conducted a study that involved interviews with leading officials of 52 business firms in Chicago, Hartford, and Pittsburgh. The study attempted to determine the important factors that motivated the selection of a particular site for new retail commercial locations.[21] Figure 21.22 shows the relative importance that each of the 52 firms placed on the dominant location factors. Retail firms generally place a greater emphasis on accessibility of vehicular traffic rather than on walking traffic. Good visibility from the highway is a form of advertising and is a major consideration in terms of prestige for many retail businesses. Highway exposure was found to be the dominant location criterion.

Retail establishments seek a location on a highway on which potential customers are most likely to travel. One retailer selling boats selected a location along a two-lane road because he felt prospective boat owners would travel that two-lane road.

[20] R. I. WOLFE, "Effect of Ribbon Development on Traffic Flow," *Traffic Quarterly*, XVIII, No. 1 (1964), 116.
[21] U.S. Department of Commerce, Bureau of Public Roads, *Highways and Economic and Social Changes* (Washington, D.C.: Government Printing Office, 1964), Figure 3-18, p. 84.

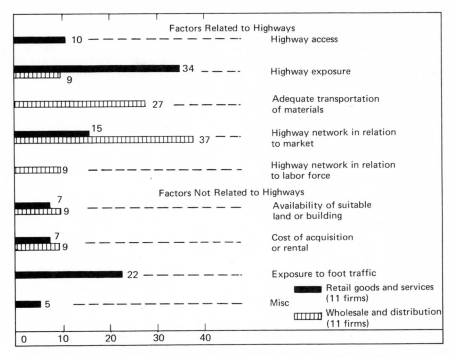

Figure 21.22. Dominant criteria in the selection of new locations. [Source: *Factors in the Selection of Commercial and Industrial Locations* (Chicago: Real Estate Research Corporation, 1958).]

Another firm, engaged in sales and service of business machines and data-processing equipment, endeavored to locate branch offices in prestige locations on highways traveled daily by business executives.

IMPACT OF TRAFFIC ON EXISTING DEVELOPMENT

Ribbon development, which may occur in response to accessibility or traffic volumes, adversely affects the existing development in adjacent areas. Ribbon development is highway oriented and requires visual exposure to the major arterial or the freeway corridor along which it locates. Sites that provide direct visual exposure are sought by most highway-oriented development establishments. Thus, the development corridor is relatively narrow, usually only one lot deep along the major arterial. There is little tendency for this development to spread laterally to adjacent areas.

Because of the linearity of ribbon development, these areas tend to extend for miles along the major arterial routes with their attendant negative aspects, including exposure of residential areas to commercial operations. This is especially harmful in the case of restaurants because of the long hours of operation and the odor of cooking food. This adverse influence has been used in many zoning cases as an excuse to extend the commercial zoning laterally into the residential areas, thereby posing a threat to the existing neighborhoods.

TRAFFIC AND URBAN SPRAWL

Many books and papers have been written about the adverse effects of urban sprawl. Many of these books indicate that urban sprawl is inefficient, is wasteful of our natural resources, and results in unattractive development. It is said that urban sprawl results in excessive expenditures for highway transportation facilities, requires a high investment in public services such as water and sewer facilities, and does not allow proper timing of land development, which, in turn, results in undesirable development patterns.

Some aspects of urban sprawl. Urban sprawl is a relatively recent phenomena, which, in general, has primarily occurred since World War II. This trend corresponds in time span with a rapid increase in the usage of the private automobile and a corresponding decline in mass transit patronage. Urban sprawl results in a lower density for the metropolitan area. Trends in density for 21 American cities are illustrated in Figure 21.23. In this study Smith found that density patterns reflect the city's age and the mode of intra-urban transport that dominated the transit picture during the city's formative years.[22]

The decrease in density of American cities reflects the results of sprawl. The changes in residential densities in the United States are related to a number of factors, which include:

1. The availability of mortgage money to substantially finance 90 percent of the entire cost of a single family house.
2. The income tax break by the federal government for those persons financing homes (through the write-off of interest cost).
3. The reduction of the local tax burden by living outside the incorporated limits of the larger cities.
4. Improved accessibility within the cities provided by freeways.
5. Increased vehicular ownership provided by higher wages.

There are other factors that play a role in this trend. As a result of the trend, however, the reliance on the private vehicle has become dominant.

In every land, pedestrian-oriented cities have always been tightly clustered, and cities built before mechanical transportation have been typified by high density. Thus, comparison of modern American cities with large cities throughout the world provides insight into the changes that new forms of transport have wrought on the patterns of urban living. Cities in Europe, Asia, and South America invariably exhibit much higher densities than North American cities of similar population.

Transportation costs. One major disadvantage of low-density development (urban sprawl) is the increased cost for transportation. An example of the differences in transportation costs between a high-density plan and an urban-sprawl plan is provided by the Southeastern Wisconsin Regional Transportation Study for the Milwaukee area. This high-density or "centralized" land-use plan provided an average population

---

[22] WILBUR SMITH and ASSOCIATES, *Transportation and Parking for Tomorrow's Cities* (Detroit: Automotive Manufacturers Association, 1966), pp. 15–18.

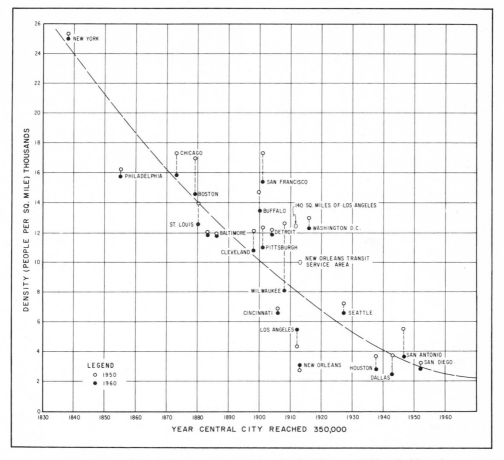

Figure 21.23. Effect of city age on population density. [Source: Wilbur Smith and Associates, *"Transportation and Parking for Tomorrow's Cities* (Detroit: Automotive Manufacturers Association, 1966), p. 15.]

density in the urbanized area of 4,400 persons per square mile as compared to the density in the unplanned land-use alternative of 2,700 persons per square mile. These data are shown in Table 21.8. Vehicle hours of operation and operating cost per day for the centralized or recommended land-use plan as compared to the unplanned land-use alternative were $20.3 billion and $22.31 billion, respectively. This represents a difference of approximately $2 billion per day, indicating that there are very real additional costs of transportation system operation with different densities and patterns of development. Interestingly, Figure 21.23 illustrates that densities in Milwaukee decreased from 12,500 persons per square mile in 1950 to 8,000 persons per square mile in 1960.

Land utilization. One of the interesting aspects of the two alternative land-use patterns for the Southeastern Wisconsin Region is the amount of land required for transportation. For the recommended land-use plan, which involves a mass-transpor-

TABLE 21.8
Comparison of the Transportation Operations of a Centralized Plan
and a Sprawl Land Use Pattern

	Recommended Land Use Plan	Unplanned Land Use Alternative
Vehicle hours of travel	926,000 hr/day	1,016,000 hr/day
Operating costs	$20.34 billion	$22.31 billion
Vehicle miles of travel	32.29 million	36.95 million
Land for transportation	66.5 sq mi	94.6 sq mi
Average population density	4400 persons/ sq mi	2700 persons/ sq mi
System costs	$2.10 billion	$2.23 billion

Source: SEWRPC Planning Report No. 7, *Land Use-Transportation Study*, Volume 3, Recommended Regional Land Use and Transportation Plans—1990, (Southeastern Wisconsin Regional Planning Commision, Waukesha, Wisconsin, 1966), pp. 69-71.

tation network, the total land for transportation was estimated at 66.5 square miles or approximately 30 percent of the total land area. For the unplanned land-use alternative, however, the land required for transportation was 94.6 square miles or approximately 40 percent of the total urbanized area.[23]

One basic argument contends that urban sprawl is inherently less efficient in terms of land use. Comparing the planned or unplanned land-use alternatives for Southeastern Wisconsin with land presently allocated to streets and transportation facilities (illustrated in Table 21.9), it is apparent that the percentage of land for transportation in the future with the planned alternative will not drastically increase but may well decrease. This suggests that arguments on the excessive use of land for transportation for the urban area as a whole are not valid.

Central business district area trends in parking. A characteristic of urban sprawl is the extreme reliance upon the private vehicle as the dominant mode of transportation. This dependence has implications for the amount of area devoted to transportation uses in the CBD as well as for the urban area as a whole.

Considerable discussion of the possible advantages and disadvantages of providing parking in the CBD has been presented in various professional publications for a number of years. There is yet no consensus on optimum amount of space to be allocated to parking and other transportation uses and there probably never will be because of the variability of land requirements for each individual urban area.

Table 21.10 illustrates the percent of the CBD for selected major urban areas. These percentages range from 35.0 percent for Chattanooga to 59.0 percent for Los Angeles. The vast majority of the cities fall within the 40 percent to 50 percent range, irrespective of size.

The proportion of ground area devoted to roadways in selected cities ranged from 21 percent to 44 percent—the latter figure applying to the L'Enfant plan for Washing-

---

[23] Source: SEWRPC Planning Report No. 7, *Land Use-Transportation Study*, Volume 3, *Recommended Regional Land Use and Transportation Plans—1990*, (Southeastern Wisconsin Regional Planning Commission, Waukesha, Wisconsin, 1966), pp. 69-71.

TABLE 21.9
Percent of Developed Land Devoted to Streets in Selected Urban Areas

Area	Year	Percent of Developed Land Devoted to Streets
New York City	1960	34.6
Detroit	1953	30.8
Minneapolis-St. Paul	1960	29.1
Tucson	1960	28.3
Washington, D.C.	1948	27.8
Chicago	1956	25.9
Pittsburgh	1958	25.0
Chattanooga	1960	23.9
53 central cities	1955	28.2*
33 satellite areas	1955	27.7*
11 urban areas	1955	27.6*

*H. Bartholomew, *Land Uses in American Cities* (Cambridge, Mass.: Harvard University Press, 1955), p. 121.

Source: Wilbur Smith and Associates, *Transportation and Parking for Tomorrow's Cities* (Detroit: Automotive Manufacturers Association, 1966), p. 29. Compiled from origin-destination and land-use studies in each urban area. Approximately one-third of the street reservation is used for parkway strips and sidewalks.

TABLE 21.10
Proportion of CBD Land Devoted to Streets and Parking

Central Business District	Year	Total Acres	Percent of CBD Land Devoted to		
			Streets	Parking	Streets and Parking
Los Angeles	1960	400.7	35.0	24.0	59.0
Chicago	1956	677.6*	31.0	9.7	40.7
Detroit	1953	690.0	38.5	11.0	49.5
Pittsburgh	1958	321.3*	38.2	†	†
Minneapolis	1958	580.2	34.6	13.7	48.3
St. Paul	1958	482.0	33.2	11.4	44.6
Cincinnati	1955	330.0	†	†	40.0
Dallas	1961				
Core area		344.3	34.5	18.1	52.6
Central district		1,362.0	28.5	12.9	41.4
Sacramento	1960	350.0	34.9	6.6	41.5
Columbus	1955	502.6	40.0	7.9	47.9
Nashville	1959	370.5	30.8	8.2	39.0
Tucson	1960	128.9	35.2	†	†
Charlotte	1958	473.0	28.7	9.7	38.4
Chattanooga	1960	246.0	21.8	13.2	35.0
Winston-Salem	1961	334.0	25.1	15.0	40.1

*Excludes undevelopable land.
†Not itemized.

Source: Wilbur Smith and Associates, *Transportation and Parking for Tomorrow's Cities* (Detroit: Automotive Manufacturers Association, 1966), p. 59.

ton, D.C. At that time there was negligible space provided for off-street parking areas. The selected cities in Table 21.10 reflect a time span of eight years. There is no discernible increase in percentage for those cities recently surveyed. Thus, no trend toward larger percentages of CBD land being devoted to parking is apparent.

IMPROVING THE ENVIRONMENT BY LEGAL AND ADMINISTRATIVE MEANS

Numerous efforts have been made to reduce pollution and thereby improve the environment by legal as well as administrative techniques. These attempts include legislation, establishment of standards, and attempts at the control and stimulation of desirable land development patterns, for example, the following:

1. Air pollution standards
2. Noise regulation and guidelines
3. Land development controls

Perhaps the most successful attempts to control vehicular pollution are those controls established in the area of air pollution.

AIR POLLUTION STANDARDS

Air pollution control standards have existed for a number of years, generally as part of the municipal code. One of the earlier efforts at control was an anti-pollution ordinance passed in 1958 by the City of Newark, New Jersey. The ordinance contained the following provisions:

1. No motor vehicle shall be operated which causes a nuisance by emitting obnoxious or excessive smoke, gases, vapors, or fumes while stationary or going for a distance of more than one hundred (100) feet anywhere within the municipal boundaries.
2. No gasoline- or diesel-fueled bus shall be permitted to operate while discharging polluting gases for more than three (3) minutes while stationary at a bus stop.
3. No gasoline-fueled bus picking up or discharging passengers at the curb shall be operated after one (1) year after the effective date of this ordinance, unless it is equipped with a device to minimize the deposit of pollutants on gasoline intake passages and the manifold of the engines while decelerating. Such deficiencies shall be approved by the Chief.
4. No motor vehicle, except as hereinafter provided, which uses gasoline or diesel fuel that discharges the exhaust caused by the combustion of such fuel into the air through a vertical exhaust pipe, shall be operated upon the streets and highways of the City after one (1) year following the effective date of this ordinance. Provided, however, that subject to approval by the Chief, exhaust pipes may be used on motor vehicles where the use of horizontal exhaust systems would endanger the safety of the operators or occupants of such motor vehicles.

Other attempts at municipal air pollution control were made by Toledo, Ohio, Chicago, Illinois, and elsewhere. Generally, the Toledo ordinance is similar to that for Newark, New Jersey, except that provision was made for a maximum fine of $100.

The problem with municipal ordinances is the difficulty of enforcement of the subjective standards stated in the ordinances.

Qualitative air pollution standards by legislation. California has generally led the way in air pollution control standards. Standards were adopted and modified over a period of several years in the early and middle 1960's in response to smog in the Los Angeles area. Criteria on smoke emissions were adopted as early as 1963. The California standards provided a progressively more restrictive standard for each year from 1966 to 1975. These standards are summarized in Table 21.11.

TABLE 21.11
Trends in Vehicle Emission Standards—California and U.S.

Pollutant	Amount of Pollutant (g/mi)						
	Prior to Control	California	U. S. or Federal				
		1966	1968	1970	1973	1975	1976
Hydrocarbons	11.0	3.4	—	4.1	0.41	0.41	0.41
Carbon monoxide	80.0	34.0	34.0	34.0	3.4	3.4	3.4
Nitrogen oxide	6.0	—	—	4.0*	3.0	3.0	0.4

*This standard was to be established in 1971 by the California Clear Air Act.

Source: *Maintaining a Quality Environment in Georgia*, Georgia Office of Comprehensive Health Planning, Georgia Department of Public Health, 1971.
Technical Advisors Committee, California Air Resources Board, *Environmental News*, Environmental Protection Agency, Washington, D.C., June 30, 1971.

Air pollution control standards were initiated in 1965 by the federal government under the authority of the Clear Air Act of 1965, as amended by the United States Congress. The motor vehicle exhaust standards were placed in effect in 1968 with the new model automobiles. Federal standards since that time have become progressively more restrictive. The goal is to reduce hydrocarbons and carbon monoxide emissions by 90 percent by 1975 over the standards established in 1970 and to reduce nitrogen oxides 90 percent in 1976 from the 1971 model cars which had no nitrogen oxide control system.

The California Clear Air Act (AB351) included a 1970 carbon monoxide standard for California of 23.0 grams per mile vs. 34.0 grams per mile for the United States. The 4.0 grams per mile standard established for nitrogen oxide in 1971 will not be applied until 1973.

Other legislative action on air pollution standards has been taken by the Province of Ontario, Canada, and by the Canadian government. Generally, the standards accepted by the Canadian government match the standards established for the United States.

Results of emission standards. Emission standards are expected to drastically reduce the amount of air pollution from all sources. The goal of emission standards is to reduce 97 percent of the hydrocarbons and 96 percent of all carbon monoxide from automobile exhausts by the 1975 model. By 1976, the standards requires a 92 percent reduction in nitrogen oxide.

As a result of emission controls on current models, 80 percent of the hydrocarbon emissions,[24] 70 percent of the carbon monoxide and, on cars sold in California, 25 percent of nitrogen oxide have been reduced in the United States. Because the motor vehicle creates a major portion of the pollution, the emission standards are expected to provide dramatic relief.

Figure 21.24 shows that as a result of legislation in the mid-1960's, the atmospheric pollution levels in the early 1970's will decline. The decline will not continue, however, unless the more stringent standards now proposed are implemented, because the number of motor vehicles will increase as will pollution from other sources.

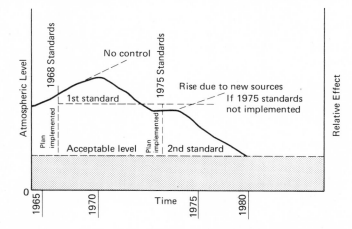

Figure 21.24. Relative effect of implementation of air quality standards in the United States. (Source: Harmelink and Peck, *Transportation Air Pollution*, p. 32.)

If the standards for vehicle emissions now proposed for 1975 are met, a further reduction in air pollution is possible. This is illustrated by the sharp drop in the curve beyond 1975. It is anticipated that by 1980 the rate of emissions will have declined to an acceptable level.

Noise regulation and legislation. The existing local legislation in the United States related to motor vehicle noise is subjective in character. A majority of states and cities utilize elements of the uniform vehicle code published by the National Committee on Uniform Traffic Laws as a basis for noise regulation, even though this code is relatively vague about vehicular noise. The code states:

> Every motor vehicle shall have at all times mufflers in good working order and in constant operation to prevent excessive or unusual noise and annoying smoke, and no person shall use a muffler cut-out bypass or similar device on a motor vehicle on the highway.[25]

---

[24] "Growing Battle Over Auto Pollution," *U.S. News and World Report*, LXXII, No. 24 (June 12, 1972), 10. Copyright 1972, U.S. News and World Report, Inc.

[25] W. J. GALLOWAY *et al.*, *op. cit.*, p. 11.

The passage implies that there is no objective measure of permissible noise. Enforcement depends to a great extent on the police officer's judgment of what is excessive or unusual noise. This leads to ambiguity in enforcement and application. As a result, in several localities the courts have declared that local noise regulations are unconstitutional.

The lack of an objective standard for excessive noise has been recognized by a number of enforcement agencies and has promoted some interest in revising legislation to specify an objective measurement of motor vehicle noise. Generally, these efforts center around the use of a sound level meter. Three cities in the United States have noise ordinances that utilize an objective sound limit for the measurement of noise. The city of Wilmington, Indiana, specifies a noise limit for all vehicles of 95 dBA. The measurement is to be made by a sound meter at a point 20 ft behind the vehicle at the right rear wheel. Similar standards have been established in other cities, for example, Cincinnati, which specifies a maximum sound level of 95 dBA at a distance of 20 ft, and Cleveland, which specifies 95 dBA measured not less than 5 ft from the source.[26]

Toledo and Milwaukee have had motor vehicle noise ordinances that included objective standards. In both cases the statutes have been repealed, primarily because the ordinances were either unenforceable or required excessive amounts of police personnel time. Trying to isolate the noise emitted from one particular vehicle out of many vehicles and applying such regulations on a city-wide basis present enforcement problems. Nonresidents driving through the area who may be unfamiliar with the regulations are particularly affected.

In 1965, the state of New York passed legislation that defines unacceptable noise as anything above 88 dBA. This state law included a 2 dBA tolerance with the test based on a measurement 50 ft from the center line of the lane of travel. This measurement was to be made with the vehicle operating at a speed of 35 mph. Because of the difficulty in enforcing these regulations, the New York legislation has never been utilized extensively. A member of the New York State Police indicated that 335 man hours of police time were required in making 16 arrests. This averages approximately 21 man hours of police time per arrest and is considered excessive for what is achieved.[27]

In 1970, Congress passed the Noise Pollution and Abatement Act (Pl 91–604) which provided the initial step toward establishing measures and standards for noise levels.

Because of the enforcement problems, it would appear that any efforts at vehicular noise control must be made on at least a state-wide basis. There is a need for promulgating objective standards that bear a direct relationship to goals and objectives on noise control.

Noise guidelines for housing. In another approach to improving the environment, guidelines for evaluating potential housing sites have been established by the U.S. Department of Housing and Urban Development. These guidelines cover air, highway, and rail transportation noise sources and include procedures for calculating noise levels. Presumably, no federal loan guarantees or other government assistance will

---

[26] *Ibid.*
[27] *Ibid.*

be available for housing sites that do not meet the standards. Figure 21.25 illustrates the application of site exposure to truck noise. Adjustments are made for road grade, sound barriers, and stop-and-go traffic.[28]

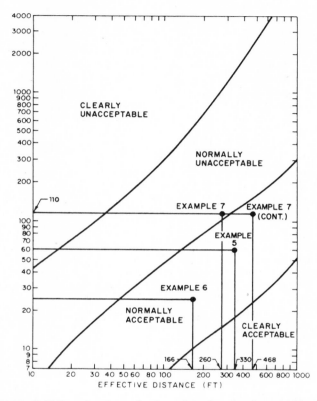

Figure 21.25. Guidelines for the evaluation of truck noise for a residential site. [Source: Bolt Berabek and Newman, Inc., *Noise Assessment Guidelines* U.S. Department of Housing and Urban Development (Washington, D.C.: U.S. Government Printing Office, 1971), p. 10.]

ADMINISTRATIVE REVIEWS OF HIGHWAY PROJECTS

Another means of improving the environment is provided by the establishment of an administrative procedure for the review of environmental impact for each highway project. The administrative mechanism was established nationally in the United States by the National Environmental Policy Act of 1969 (Pl 91–190). The purpose of the act is to provide a detailed statement on reports and proposals that affect the quality of the human environment. Such a statement shall include:

---

[28] Bolt Berabek and Newman Inc., *Noise Assessment Guidelines*, U.S. Department of Housing and Urban Development (Washington, D.C.: U.S. Government Printing Office, 1971), pp. 10–13.

(i) the environmental impact of the proposed action,[29]
(ii) any adverse environmental effects . . . ,
(iii) alternatives to the proposed action,
(iv) the relationship between local short-term uses of man's environment and the maintenance and enhancement of long-term productivity, and
(v) any irreversible irretrievable commitments of resources . . .

The processing of the environmental statement outlined above for a highway location is illustrated in Figure 21.26. The procedure provides for a location hearing and then a public hearing on design when the design stage of the project is reached.

Another administrative procedure relating to the United States Government Standards is the establishment of environmental review responsibilities at the state level. These responsibilities, commonly referred to as A-95 review, include consideration of each project by local, state, and federal agencies.

Responsibilities for public hearing and environmental review procedures have also been established at the state level by several reorganization acts establishing state departments of transportation. An example of this approach is illustrated by Act 120 (SB 408) of the General Assembly of the Commonwealth of Pennsylvania which established the Department of Transportation (Penn DOT) and specified that the Department:

Shall consider the following effects of the transportation route or program: (1) residential and neighborhood character and location; (2) conservation including air, erosion, sedimentation, wildlife and general ecology of the area; (3) noise, air and water pollution, (4) multiple use of space; (5) replacement housing; (6) displacement of families and businesses; (7) recreation and parks; (8) aesthetics; (9) public health and safety; (10) fast, safe and efficient transportation; (11) civil defense; (12) economic activity; (13) employment; (14) fire protection; (15) public utilities; (16) religious institutions; (17) conduct and financing of government including the effect on the local tax base and social service cost; (18) National and Historic landmarks; (19) property values; (20) education including the disruption of school district operations. . . .[30]

IMPROVING THE ENVIRONMENT THROUGH DESIGN AND PLANNING TECHNIQUES

Techniques other than legislation may be far more effective in reducing the adverse impacts of traffic. Transportation planning techniques that are applicable are: integration of land use and transportation planning to encourage transit usage, reduction in travel by the arrangement of traffic generators, and freeway design concepts that may help preserve the environment.

Some of the ideas discussed in this chapter have not been tested and are at the very forefront of an effort to better integrate the private vehicle with a desirable environment.

---

[29] "Guidelines for Implementing Section 102 (2) (C) of the National Environmental Policy Act of 1969, Section 1053 (f) of 49 U.S.C. . . . ," *Policy and Procedure Memorandum 90–1*, U.S. Department of Transportation, Transmittal 202, Federal Highway Administration (August 24, 1971), p. 1.

[30] Act No. 120 (SB 408), General Assembly of the Commonwealth of Pennsylvania (May 6, 1970), p. 13.

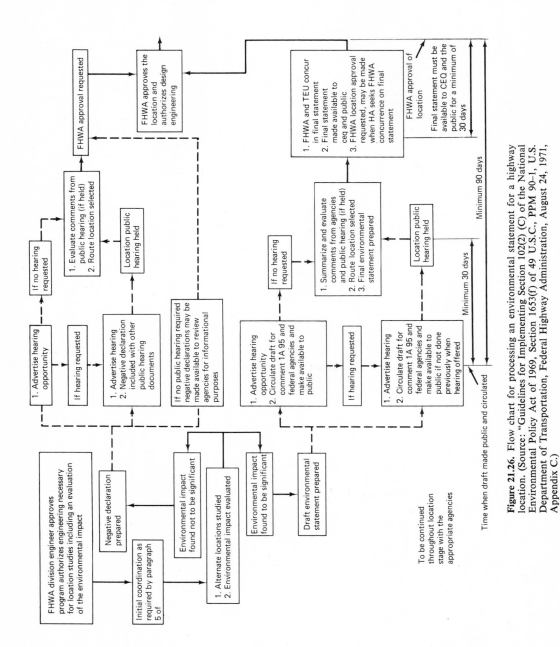

Figure 21.26. Flow chart for processing an environmental statement for a highway location. (Source: "Guidelines for Implementing Section 102(2) (C) of the National Environmental Policy Act of 1969, Section 1653(f) of 49 U.S.C., PPM 90–1, U.S. Department of Transportation, Federal Highway Administration, August 24, 1971, Appendix C.)

INTEGRATING LAND USE AND TRANSPORTATION PLANNING
AT THE LOCAL SCALE

Chapter 11, Urban Transportation Planning, and Chapter 12, Statewide and Regional Transportation Planning, contain discussions of land use considerations. Land use and transportation planning can be coordinated to reduce the adverse impact of traffic in several ways:

1. Land use patterns can be established to encourage transit usage and, thereby, reduce vehicular emissions.
2. The transportation network can be designed to minimize vehicular hours of travel, thereby reducing emissions and noise.
3. Land use plans can be coordinated with transportation plans to reduce the exposure of residential areas to heavy traffic movement.
4. Land use controls can be established to preserve capacity and minimize ribbon commercial development along major radial routes.

The 1962 Highway Act provided the basis for better coordination of land use and transportation planning by requiring that there be a "continuing, cooperative, comprehensive transportation planning process" in all urban areas of over 50,000 population.

Land use patterns to encourage transit. The development of land use plans to encourage transit usage generally involves higher densities along existing or proposed transit corridors. This increased density provides the market for transit ridership to the CBD. The result may be a higher proportion of transit ridership and a consequent reduction in the use of the private vehicle. This translates to lower air pollution and noise.

Transportation network to minimize travel. Systems planning usually seems to minimize vehicle miles and vehicle hours of travel. The proper balance of freeways, arterial streets, and public transit will result in a compromise situation in which vehicle miles and vehicle hours are minimized consistent with capital investment and operating cost for highways and transit. Reductions in vehicle hours of operation will reduce air pollution and noise.

Protection of residential areas. Transportation networks should be selected in such a way that existing residential areas are protected and that a framework for logical development of future residential areas is provided. The design of the network with adequate spacing allows the development of residential areas between the major segments. The network must have sufficient capacity, however, to eliminate through traffic in the residential area. This design minimizes transportation noise levels within neighborhoods and reduces the number of accidents.

Land use controls and the environment. Heavy traffic flow can serve as a catalyst for ribbon development and frequently reduces capacity on the major arterial facility. Land use controls, when properly applied, can restrict undesirable development and/or can provide design solutions such as control of access or driveway design that minimize the detrimental aspects of the development. The result is the protection of capacity on major arterial routes, the reduction of accident rates within these corridors, the improvement of visual appearance by sign controls, and the removal of "zoning pressure" on adjacent residential development.

REDUCING TRANSPORTATION POLLUTION AT
THE REGIONAL SCALE

Preparing a transportation–land use plan provides the opportunity to coordinate the development of road network or transit system with land-use policies. Hopefully, it results in maximizing the efficiency of the transportation system and, at the same time, minimizing the adverse influence on the environment, for example, air pollution emissions. One goal of the transportation planning process is to minimize the amount of travel required within the region and yet provide an acceptable level of service. The evaluation of alternative regional land use plans has frequently provided a basis for reducing the amount of pollution by simply reducing the amount of travel. One of the best examples of this approach to reducing adverse influence is exemplified by the transportation–land use plan prepared by the Southeastern Wisconsin Regional Planning Commission. This plan considered a number of alternative land use plans and recommended one in which vehicular travel and vehicular hours of operation were minimized. The plan also provided the lowest level of vehicular emissions.

Another approach used in evaluating transportation plans is illustrated by Table 21.12 which shows, for two alternative land use patterns and three highway networks, the reduction in various vehicular emissions over the base network. The table includes emissions based on pre-1968 emission standards and post-1975 emission standards. The best combination of plans includes the "A-Sprawl" land use coupled with four major radial freeways and inner loop. This plan results in a 35 percent decrease in carbon monoxide, a 14 percent decrease in hydrocarbons, and a 154 percent increase in nitrogen oxide. These are based upon 1968 emission rates for automobiles. If the standards adopted by the Environmental Protection Agency for 1975 are assumed, the emission rates are reduced considerably, as is shown. The table suggests that criteria for evaluating alternative regional transportation networks should include air pollution as a factor.

THE MULTIDISCIPLINARY DESIGN TEAM APPROACH
IN FACILITY PLANNING

The design of a transportation facility has great impacts on the immediate environment. A freeway brings definite transportation benefits and can serve as an opportunity to improve the environment in many ways. The result of new freeway construction has far-reaching geographic and temporal influences on individuals as well as on neighborhoods, cities, and entire regions.

In order to respond in a comprehensive manner to the impacts of freeways, use is often made of a multidisciplinary design team. The Federal Highway Administration states that the objective of the design team is "to make sure that adequate attention is given to the preservation and enhancement of the quality of the environment and related social and economic factors."

The design team approach provides the technical expertise for a broad consideration of impacts and opportunities associated with freeway design and construction.

TABLE 21.12

Changes in Air Pollution as Related to Hypothetical City Form and Highway Network Configuration

Urban Pattern	Highway System	Base Network	Reductions (−) for Different Highway Systems as Compared to a Base Highway Network*		
			Pre-1968 Emission Rates		
			CO	HC	NOx
A Sprawl	2 Freeways along major radials	1 Basic arterial grid	−13%	−4%	+82%
A Sprawl	4 Additional radial freeways and an inner beltway for maximum coverage	1 Basic arterial grid	−35%	−14%	+154%
B Moderate corridors	3 Major radial freeways with outer beltway and inner loop added	2 Freeways along major radials	−11%	+5%	+30%
			Post-1975 Emission Rates		
			CO	HC	NOx
A Sprawl	2 Freeways along major radials	1 Basic arterial grid	−9%	−29%	+52%
A Sprawl	4 Additional radial freeways and an inner beltway for maximum coverage	1 Basic arterial grid	−31%	−19%	+160%
B Moderate corridors	3 Major radial freeways with outer beltway and inner loop added	2 Freeways along major radials	+5%	+5%	+26%

Source: S. J. Bellomo, and Edward Edgerby, "Ways to Reduce Air Pollution Through Planning, Design and Operations," Highway Research Board (unpublished paper, 1971), Table 10, p. 24.

Some factors for consideration in design are:

1. Economic and fiscal structure, which can include either loss or gain in local tax revenue, employment gains and losses.
2. Housing displacement and replacement.
3. Community facilities, especially schools and school zones because these areas are directly affected by the barrier effect of freeways.
4. Circulation. Traditional traffic patterns may be altered drastically by the opening of the freeway (see Figure 21.27). Also to be considered are pedestrian circulation patterns.
5. Neighborhood environmental impacts, which may include increases in noise and air pollution (see Figure 21.28), noise alleviation by sound barriers, visual and aesthetic values.
6. Joint-use development. This concept of many design teams involves the use of air rights and areas under the freeway for development purposes (see Figure 21.29).

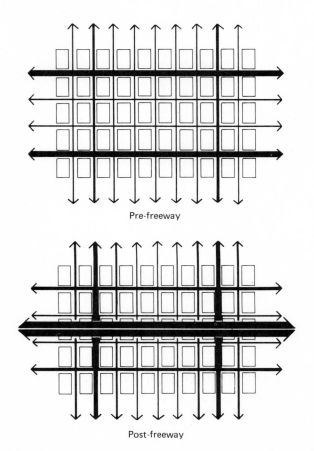

Figure 21.27. Changes in vehicle movement patterns with construction of a freeway. [Source: Gruen Associates, *Interstate 105 Freeway-Design Team Concepts* (Los Angeles: Prepared for the California Division of Highways, 1970), p. 28.]

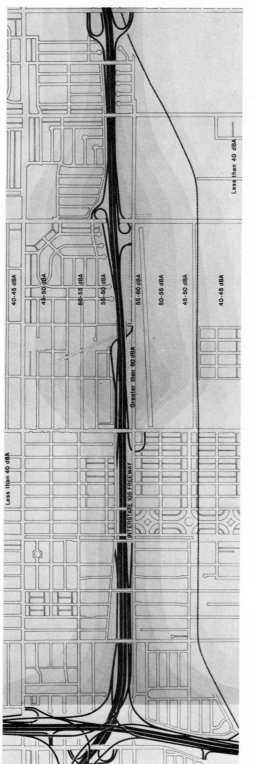

Figure 21.28. Freeway generated noise contours. [Source: Gruen Associates, *Interstate Freeway–Design Team Concept* (Los Angeles, Prepared for the California Division of Highways, 1970), p. 33.]

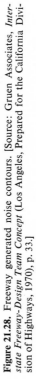

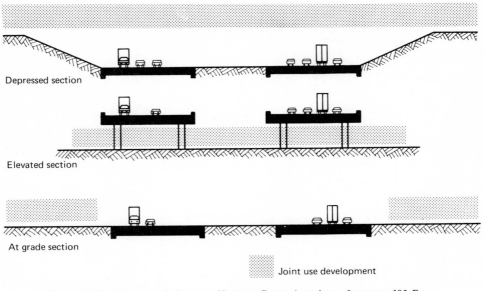

Depressed section

Elevated section

At grade section

Joint use development

Figure 21.29. Joint use of air space. [Source: Gruen Associates, *Interstate 105 Freeway-Design Team Concepts* (Los Angeles, Prepared for the California Division of Highways, 1970), p. 38.]

One of the most significant contributions of the design team is in pinpointing possibilities for the application of multiple federal programs such as fringe parking, rapid busway lanes, and joint-use development to maximize the return on highway construction programs. Also see Chapter 12, Urban Transportation Planning.

DESIGN APPROACHES TO ENVIRONMENTAL ENHANCEMENT

The section on visual pollution included a description of environmental problems relating primarily to signs and sign control. One major area of concern to the traffic engineer is traffic signs. Urban design is increasingly concerned with the coordination of information systems for the pedestrian and the vehicle resulting in a more systematic means of visual communication. This usually involves the design or selection of a graphics system, the combination of information systems (street name, directional signs, regulatory signs, parking control signs, etc.) within a standardization structure for simplicity and aesthetic reasons. Figure 21.30 illustrates one attempt at the design of an information system that incorporates a number of information sources. In this system, vehicular signals and pedestrian signals are combined within one element. This approach could reduce clutter, but it may present operational problems.

Another approach is illustrated in Figure 21.31. Here, lane signs and a standard traffic signal head are mounted within a space frame. The space frame is part of a combination lighting standard and sign system for parking regulation; it also has a trash receptacle. The objective is to combine many separate and uncoordinated street furniture elements into a unified whole.

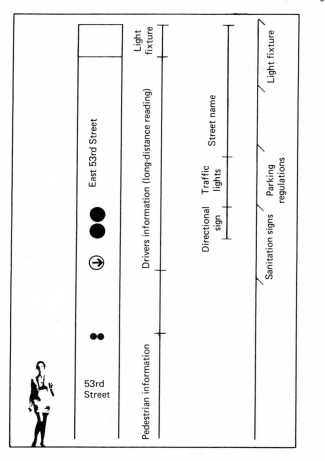

Figure 21.30. A combination of traffic control and information system. [Source: Western ITE, *Street Furniture*, XXC, No. 3 (1971), 3.]

IMPROVING THE ENVIRONMENT BY INNOVATIVE TRANSPORTATION MODES

Another possibility for improving the environment is through the application of new technical innovations in transportation. This primarily includes the use of new transportation modes and methods of propulsion. The validity of the new propulsion systems will depend on how attractive these systems are to the traveling public on a mass basis, since any propulsion system must be mass produced to be economically feasible. Table 21.13 contains a description of alternative propulsion systems in addition to the internal combustion gasoline engine.

All eleven alternative systems shown have been studied in some detail as a part of research undertaken by the various automobile manufacturers. The electric propulsion system is least polluting, but it is impractical for the private automobile at the present time because of the lack of efficient batteries. Perhaps the most practical

TABLE 21.13

Alternative Propulsion Systems and Their Applicability to Motor Vehicles

Alternative Propulsion System	Power Source	Principal Applications for System	Alternative Road Test Performance	Feasibility as Alternative to Gasoline Internal Combustion Engine	Air Pollution Problems Associated with System
Diesel	Gasoline, diesel fuel, kerosene	Trucks, busses, trains auxilary power for other systems (G.V.T.) boats	Engine has been fully tested by Mercedes Benz and Peugeot, poorer acceleration performance than gasoline engine	Economical in larger sizes, less so in auto, Mercedes Benz and Peugeot manufacturing auto diesel engines	Smoke and odor are principal concerns Low emissions of H_xC, CO NO_x Emission could be controlled by design
Modified diesel	Gasoline or diesel fuel	Alternative to gasoline I.C.E.	Ford Motor Co. and U.S. Army Tank Command testing model at present	Main reasons diesel and modified diesel not used for autos are because of noise, odor and costs. The modified diesel should overcome these problems	Same as above
Gas turbine	Gasoline and other petroleum fuels	Cars, trucks, busses, trains, aircrafts, TACV hovercraft, boats	Chrysler has field tested 50 turbine cars and found them reliable and operationally economical as new cars	Reasonable alternative in larger sizes, trucks, bus, trains but not yet economical for use in auto	Low emission of CO and H_xC NO_x could be controlled by design Future emission possibilities in gms/mile: H_xC 0.5–1.2, CO 3.0–7.0, NO_x 0.15–0.4
Natural gas propane engine (I.C.E. can be modified)	Natural gas fuel	Alternative to gasoline I.C.E.	Runs well with somewhat poorer acceleration performance than gasoline I.C.E. Refuelling less convenient than with gasoline engine Longer valve life	Initial cost probably higher, although could drop if production increased Cost of modifying gasoline I.C.E.: **$500** Propane supply likely insufficient for automobile population	Natural gas is one of the cleanest fuels CO can be considerably reduced No NO_x, virtually no H_xC or sulfur oxides Future emission possibilities in gms/mile H_xC 1.5 CO 3.0–5.0 NO_x 1.5

Battery electric	Electricity	Possible alternative to gasoline I.C.E.	From the experiments carried out to date it's estimated that it will be 20 years before these vehicles can compete economically on an equal performance basis with auto	Uneconomical at present because of short range, low speed, recharging constraints on present day lead batteries. Requires development of dependable batteries (high energy fuel cells may be the answer)	Virtually pollution free
Linear electric	Electricity	TACV, trains, hybrid vehicles	Efficiency approaches 85%	N/A for use in auto; main application is for new modes of mass transit	Virtually pollution free except for pollutants that may result from generating source
Rotary electric	Electricity	Cars, subways, trains, mini-cars, belted cable systems	Little available except on trains	Not feasible for private motor vehicles at present, and unlikely to be in future, because of ever-increasing demands for electric power generally	Same as above
Steam	Steam powered Rankine engine burns fuel	Cars, trains, boats	Lear of Lear Jet has devoted time and money into development of steam car, but has now stopped work on steam and is concentrating on gas turbine. Minto has developed a freon engine which operates like steam engine, has sold licensing rights in Japan	Research has proven that the steam car is an uneconomical alternative to auto at present time	Burns fuel more cleanly than I.C.E. thus emitting less pollutants. Future emission possibilities in gms/mile, H_xC 0.2–0.7, CO 1.0–4.0, NO_x 0.15–0.40
Hybrid	Combines attributes of internal and electrical combustion engine	Cars, busses, trains, TACV	Overall efficiency is estimated to equal present engines. Mercedes Benz has developed a workable system for buses	Initial costs of methods are high and could not hope to compete with auto unless technological breakthrough is made	Emits less pollutants than internal combustion engine

TABLE 21.13 (Contd.)

Alternative Propulsion System	Power Source	Principal Application for System	Alternative Road Test Performance	Feasibility as Alternative to Gasoline Internal Combustion Engine	Air Pollution Problems Associated with System
Wankel	Several fuel types	Cars, trucks	Powers several production cars in Europe and Japan Reliability not proven on mass production basis	Shape of engine parts, rubbing surface requiring special treatment, and many seals being required make this car inherently expensive to produce	Worst possible engine from the viewpoint of pollutant emission Future emission possibilities in gms/mile H_xC 1.8, CO 2.3, NO_x 2.2
Stirling	Hydrogen and helium	Cars, trucks, busses, boats	Philips Research Laboratory developer of modern Stirling engine Efficiency in neighbourhood of 60% A Stirling powered boat has been operating for the last 5 years. It is expected that a Stirling powered bus will be on the road in the near future	Higher material costs, more expensive manufacturing, low production and large size indicate selling price slightly higher than diesel Not economical at present for use in auto	Low emission of pollutants Future emission possibilities in gms/mile H_xC 0.006, CO 0.3, NO_x 2.2.

Source: M. D. Harmelink, and W. J. Peck, *Transportation Air Pollution*, DHO Report No. RR169 (Ontario: Department of Highways, 1971), p. 31.

Figure 21.31. Combination of traffic control and information system. (Source: Harold Lewis Malt, Harold Lewis Malt Associates, Inc., *Furnishing the City*, McGraw-Hill Book Co., New York, 1970, p. 178.)

low-emission propulsion is the modified diesel engine which uses gasoline or diesel fuel as a power source. The modified diesel emissions are relatively low in hydrocarbons, carbon monoxide, and nitrogen oxide. The major drawbacks to the modified diesel engine are odor, smoke, and noise.

REFERENCES FOR FURTHER READING

CURRY, DAVID A., and DUDLEY G. ANDERSON, *Procedures for Estimating Highway User Costs, Air Pollution, and Noise Effects*, National Cooperative Highway Research Program Report 133. Washington, D.C.: Highway Research Board, 1972.

Environmental Considerations in Planning, Design, and Construction, Special Report 138. Washington, D.C.: Highway Research Board, 1973.

"Transportation Noise and Its Control," U.S. Department of Transportation. Washington, D.C.: Government Printing Office, June, 1972.

WATKINS, L. H., *Urban Transport and Environmental Pollution*, TRRL Report LR 455. Crowthorne, Berkshire, U.K.: Transport and Road Research Laboratory, 1972.

Chapter 22

TRAFFIC ENGINEERING ADMINISTRATION

DANIEL J. HANSON, Sr., Executive Vice President, American Road Builders' Association.

This chapter covers the subjects of traffic engineering and transportation administration, functions, organization, personnel requirements, legislative authority, financing, decision making, public relations, and support activities. Considerable treatment is given to the aspects of traffic engineering dealing with the public, budgeting, and sound administrative procedures.

TRANSPORTATION FUNCTIONS

To properly evaluate all of the elements involved in traffic engineering and transportation administration, it is essential to consider all of the functions that are vital to a comprehensive street and highway program. A Yale University Bureau of Highway Traffic study, "Urban Transportation Administration,"[1] revealed that in the United States these functions are basically the same for cities, counties, and state governmental organizations. Because there may be less emphasis at the state level on aspects such as parking, pedestrian activities, transit operations, terminal facilities, and rush-hour traffic controls, most of the following discussion centers on urban transportation functions.

The functions and duties in the traffic engineering field were once centered around traffic control devices and fact finding surveys. Today's activities include many other responsibilities. They vary according to legislative authority, organizational structure, financial considerations, and numerous other resources of the individual traffic engineer. The following functions are typical:

1. Collection and analysis of factual data. Traffic volumes, speeds, origin and destination, accident analysis, and pedestrian studies are included in this group.
2. Traffic regulations such as parking and loading controls, one-way streets, and turn restrictions.
3. Traffic control devices, including traffic signs, signals, and pavement markings.
4. Traffic design such as alignment, grade problems, cross section, intersections, and subdivision regulations.
5. Traffic planning, including traffic projection and assignment, types of routes, location, and economic justification.
6. Street lighting, including luminaire selection and location.

---

[1] THOMAS J. SEBURN and BERNARD L. MARSH, *Urban Transportation Administration* (New Haven: Yale University, Bureau of Highway Traffic, 1959).

7. Transit operations.
8. Storage and retrieval of traffic engineering information.

A study of 44 cities representative of communities between 50,000 and 200,000 population in the United States and Canada concluded that activities relating to traffic operations, traffic controls, and parking had the highest degree of involvement by city traffic engineering units (see Table 22.1). Clearly, the major activities

TABLE 22.1
Functions Performed by Traffic Engineering Agencies
in Cities of 50,000 to 200,000 Population*

Function	Average Percentage of Cities Reporting Traffic Engineering Responsibility for the Function Listed
Surveys and studies related to traffic operations	79
Traffic control and driver aids	72
Parking and standing	72
Surveys and studies related to transportation planning	68
Transportation planning and programming	39
Design	30
Street use	23
Miscellaneous functions	29
Weighted average for the performances of all functions	52

*Based on a sample of 44 cities representative of communities in the United States and Canada.

Source: "Traffic Engineering Functions in Cities of 50,000 to 200,000 Population," *Traffic Engineering*, XXXIII, No. 10 (1963), 54–55, 57.

of traffic engineering units in medium-sized cities are delegated to functions in the area of traffic operations.[2]

ADMINISTRATION

The location of the traffic engineering function may vary within any governmental organization. All traffic control work was once considered strictly a police activity. The prime responsibility was to maintain order, investigate accidents, direct traffic, and provide for the public safety. As the traffic problem became more complex, many governmental agencies found that solutions were far more dependent on the application of sound traffic engineering techniques. The resulting trend has been to locate the function within an engineering department, with the specific location depending on:

1. Existing organizational structure
2. State constitution and/or other legal statutory provisions

---

[2] "Traffic Engineering Functions in Cities of 50,000 to 200,000 Population," *Traffic Engineering*, XXXIII, No. 10 (1963), 54–55, 57.

3. Elected public officials' relationship to traffic engineering function
4. Interest on the part of department heads

The traffic engineering function is perhaps best performed in a separate department. In smaller units of government, however, it may appropriately be placed within an engineering department whose director is familiar with the importance of sound traffic planning and operations. Regardless of organizational arrangement, of course, the traffic engineer should have authority fully commensurate with his responsibility.

STATE ORGANIZATIONS

With the advent of state departments of transportation (DOTS), the role of the state traffic engineer has been changed in some instances. Almost half of the states have DOTS responsible for all major transportation functions, including traffic engineering.

Whenever the traffic engineering activity has been maintained within the state highway department or department of public works, the role of the traffic engineer has remained unchanged. Within the state highway departments, all engineering functions are usually located in one central authority. Various methods of combining the functions from the standpoint of administration can be found. Some state highway departments employ a strong headquarters organization; others depend more on a strong district organization. In the former, practically all planning, traffic engineering, design, research, and similar functions are handled by the headquarters staff. The district personnel may include very few technical men. Most, or all, of the traffic engineering function may be carried out from the central office.

In the decentralized organization, district traffic engineers have considerable responsibility. Headquarters staff develops general policy, criteria, standards, and programs and also performs the overall staff function. Direct responsibility for application of basic policy is vested within the district engineer and his technical staff.

The size of the state and the distances between major population centers are important factors in determining which organizational pattern is more applicable. Within a large decentralized organization, such as the state traffic engineering unit in a populous state, most of the actual work must be performed at the local level. Therefore, the district traffic engineer and his staff become a first-line operating agency. The wide range in population density, area characteristics, and geography results in a wide variation in the size of the district organizations and the number of employees.

The work, of course, must be carried on in accordance with policies, standards, and methods set forth by the state traffic engineer and the headquarters staff. In the organization of the district traffic engineering staffs, there is wide latitude in meeting local conditions. The complexity of the area, the difference in workloads, and the variation in the traffic engineering services required all dictate a need for maximum flexibility. Typical organizational patterns, including the DOT concept, are shown in Figures 22.1 through 22.4.

The state traffic engineer is responsible for:

1. Activities to promote the safe, orderly, and expeditious movement of traffic on the state highway system.
2. Coordinating the work of and furnishing advice, guidance, and consultation

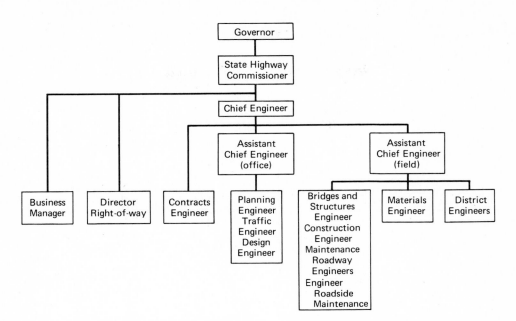

Figure 22.1. Typical organization of a state highway department with a strong headquarters organization.

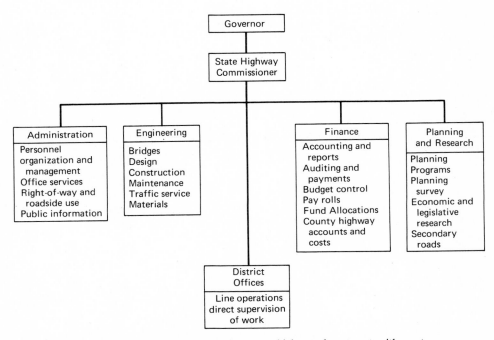

Figure 22.2. Typical organization of a state highway department with a strong district organization.

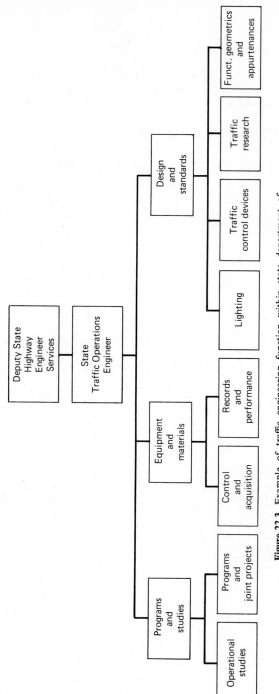

Figure 22.3. Example of traffic engineering function within state department of transportation. (Source: Edward A. Mueller, State of Florida Department of Transportation, 1971.)

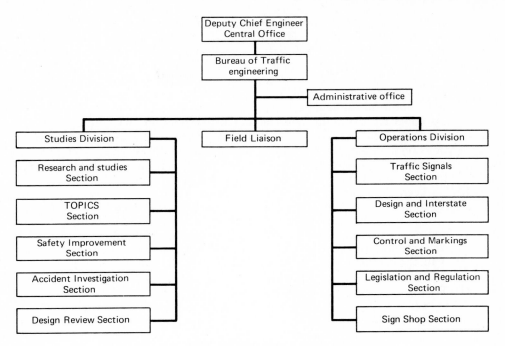

Figure 22.4. Pennsylvania Department of Transportation.

to the eleven district traffic departments and other district and headquarters departments concerned with traffic functions.

3. Active participation on state and national committees concerned with traffic engineering.

The major functions of the traffic department are:

1. To develop or assist in the development and administration of policies, standards, and practices which are primarily aimed at facilitating safe traffic movement. This includes the preparation of an up-to-date fully adequate manual of instructions for the guidance of the district traffic department personnel and all others concerned with traffic functions.
2. To carry on a continuous check of all state highways to locate points of accident concentration and areas of high accident frequency, make studies of such locations as required, and recommend or take immediate action to improve conditions.
3. From actual field observation, traffic volume, and delay studies, to make recommendations or take steps to relieve congested areas and facilitate the orderly flow of traffic.
4. To conduct an aggressive research and development program in order to constantly add to the existing body of knowledge on safe and efficient traffic flow.
5. To apply or cooperate in applying knowledge gained from operating experience and research to the planning and design of new facilities.
6. To make available basic traffic data for planning, budgeting, and design, in-

cluding both current and estimated future volumes, travel patterns, turning movements, accident data, and other traffic information as required.

7. To cooperate with local officials to establish a working relationship between them and the department as a basis for solution of common problems and provision of a high level of transportation service in local jurisdictions.

8. To maintain close working relationships and continuous cooperation with state and local traffic law enforcement agencies.

9. To cooperate with the construction department in the control of traffic through construction areas and to review, evaluate, and make recommendations on contract change orders which affect the operating characteristics of a facility. The traffic department should inspect new facilities before opening to check on safety and completeness.

10. To keep the public informed on activities. Although not a primary traffic function, it is of primary concern to the traffic engineer that the public be promptly and fully informed regarding all traffic regulations, control devices, and other measures which effect safe, expeditious traffic flow. It is a traffic responsibility to take the initiative in providing the public information unit with the factual data needed to keep the public fully informed on traffic matters.

11. To investigate requests and complaints. Very careful attention must be given to requests and complaints from organizations and individuals. They should be investigated and answered promptly with an understandable explanation. Frequently, worthwhile constructive suggestions are received. In many cases, a personal contact to talk over the problem or review it in the field is highly desirable. Public understanding of regulations and control devices and of what the division is doing and why will aid in securing compliance and support.[3]

In the National Safety Council's *Annual Traffic Inventory*, information has been obtained from state highway departments concerning their traffic engineering function.[4] Of 49 states reporting on this function in the last inventory available, 44 had traffic engineering divisions comparable in authority with the divisions of design, construction, and maintenance within the state highway departments. Separate traffic engineering divisions had been established in almost two-thirds of the states (see Table 22.2). The rest had traffic engineering functions combined with those of other divisions, most frequently with highway planning. Table 22.3 summarizes the diverse state traffic engineering activities undertaken.

COUNTY ORGANIZATIONS

Many counties in the United States, particularly those in which larger cities are located, have established separate traffic engineering units. Typical of these are Wayne County (Detroit), Michigan; Cook County (Chicago), Illinois; St. Louis County, Missouri; Montgomery County (adjacent to Washington, D.C.) in Maryland, and Los Angeles County (Los Angeles), California.

---

[3] GEORGE M. WEBB, "Organization and Administration of State Traffic Department," *Traffic Engineering*, Vol. 35, No. 5. (1965), pp. 13, 14, 37, 39, 44.

[4] *Annual Traffic Inventory* (Chicago: National Safety Council, 1964).

TABLE 22.2
State Agency Responsible for Traffic Engineering Function

Responsible Bureau or Division	Number of States Reporting
Traffic Engineering	15
Traffic	14
Traffic and Planning	7
Traffic Services	2
Highway Commission	2
Planning and Research	1
Planning	1
Traffic Safety Service	1
Traffic Safety Engineering	1
Traffic Control and Safety	1
Traffic Operations	1
Maintenance Operations	1
None designated	1
Not reported	1
Total	49

Source: *Annual Traffic Inventory*, National Safety Council, Chicago, 1964.

TABLE 22.3
Composite Summary of State Traffic Engineering
1963 and 1965 Traffic Inventory Activities

	1963		1965	
1. Administration and Personnel:				
A. Highway departments.	(49)*	43	(49)	44
B. Comparable with design, construction, and maintenance divisions.	(49)	44	(49)	44
C. Is the state divided into districts?	(49)	Yes 47	(49)	Yes 48
If yes, how many?	(47)	Total 392	(49)	Total 432
How many have a district traffic engineer?		Totel 212		Total 230
Percent with district traffic engineer.		54%		53%
D. Equivalent full-time persons assigned to traffic functions, traffic administration, surveys, and planning studies (reg. empl.).	(49)	4,677	(50)	5,259
Rate per billion miles of travel on main rural roads		15.5		16.6
Man-days on actual maintenance of signs, signals, and pavement markings.	(46)	1,988,192	(45)	2,179,635
E. Separate budget for traffic engineering.	(49)	Yes 44		
Expenditures for signal program.	(46)	$25,327,762	(48)	$30,577,087
Expenditures for sign program.	(48)	$72,158,524	(47)	$72,334,151
Expenditures for pavement marking program.	(48)	$39,279,358	(47)	$40,838,009
F. Was there a formal training program for new and/or in-service engineering employees conducted by the Highway Department?	(48)	Yes 47	(49)	Yes 45
Includes traffic engineering.	(47)	Yes 40	(47)	Yes 40
G. Does the Highway Department have authority to control parking on highways for which it is responsible for maintenance?				
In rural areas?	(49)	Yes 49	(49)	Yes 48
In incorporated areas?	(48)	Yes 42	(47)	Yes 41
To regulate angle parking?	(49)	Yes 44	(49)	Yes 43

TABLE 22.3 (*Cont.*)

	1963		1965	
H. Is the traffic engineering division consulted throughout planning, design, and construction of projects on highways for which the department is responsible for maintenance and operation by the:				
a. Design division on geometric design, safety, capacity, etc.?	(49)	Yes 47	(49)	Yes 46
b. Right-of-way division on crossovers and entrances, etc.?	(49)	Yes 46	(49)	Yes 47
c. Construction division on field changes in:				
Design?	(49)	Yes 45	(49)	Yes 47
Detours?	(48)	Yes 43	(48)	Yes 45
Construction signing?	(47)	Yes 44	(48)	Yes 47
I. Does the traffic engineering division have access to or itself maintain:				
Accident location file?	(49)	Yes 49	(48)	Yes 47
Spot map?	(48)	Yes 43	(49)	Yes 46
Strip map?	(49)	Yes 46	(48)	Yes 45
K.1. Are field investigations made by traffic engineering personnel of high accident locations along state highways?	(49)	Yes 49	(48)	Yes 48
a. Does an established number of accidents within a given period of time automatically require this investigation?	(48)	Yes 33	(48)	Yes 39
J. Does your state have a manual of signs, signals, and pavement markings?	(49)	Yes 48	(49)	Yes 48
J.1. Is it published and made available to all municipalities?	(48)	Yes 46	(49)	Yes 48
J.2. Does it conform to the Manual on Uniform Traffic Control Devices?	(48)	Yes 46	(47)	Yes 47
K. Does your state have standards for channelizing commercial driveway entrances?	(49)	Yes 47	(48)	Yes 47
a. Are permits for commercial driveway entrances reviewed by the traffic engineer?	(49)	Yes 43	(48)	Yes 46
L.1. Does the traffic engineer or traffic engineering department have the authority to initiate proposed legislation affecting traffic matters to be presented to the legislature through authorized departmental channels?	(49)	Yes 46	(48)	Yes 44
L.2. Does the traffic engineer review bills affecting traffic matters that have been presented to the state legislature and does he make recommendations for their support or rejection?	(49)	Yes 46	(48)	Yes 43
L.3. Does the traffic engineer appear before legislative committees and testify either in favor of or in opposition to traffic legislation bills?	(49)	Yes 37	(48)	Yes 39
L.4. Is the traffic engineering department called upon to submit supporting data on proposed traffic legislation?	(49)	Yes 43	(48)	Yes 45
L.5. Is the Uniform Vehicle Code considered a standard in recommending proposed legislation or when reviewing proposed bills?	(49)	Yes 48	(47)	Yes 47

*( ) indicates the number of states reporting.
†Rate travel: per billion vehicle-miles on main rural roads (302 billion in 1963 and 316 billion in 1965).

Source: *Annual Traffic Inventory*, 1966.

A typical county road department within a rather large organization is shown in Figure 22.5. Such departments can actually be larger than either a small state or city department. As suburban population grows, more and more counties may be expected to create specific traffic engineering departments. Figure 22.6 shows how this general organization actually operates in St. Louis County, Missouri with the Division of Traffic on the same level as other major divisions within the County Department of Highways and Traffic.

Figure 22.7 delineates a traffic engineering unit as it might be located within a county department of public works. In this case, the unit would represent a rather small division with its operations handled on a direct "in-line" basis. Figure 22.8 shows the actual operation of Maryland's Montgomery County Bureau of Traffic Engineering within the organizational structure of the Department of Public Works.

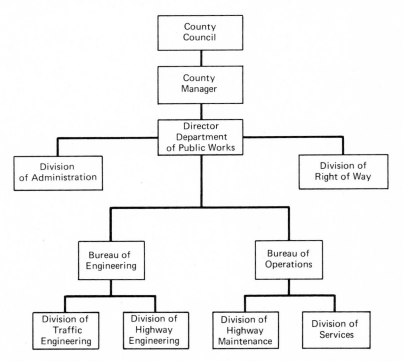

Figure 22.5. Typical organization of a county public works department showing location of traffic engineering funtion.

CITY ORGANIZATIONS

Because of the wide variation in population, geographical size, and area, numerous traffic engineering organizational patterns can be found within cities.

In the United States, there are basically three forms of city governments:

1. The "strong mayor" form in which the legislative activities are vested in the city council with the mayor serving as the chief administrative officer. Normally, the mayor has the power to veto matters of legislative importance.

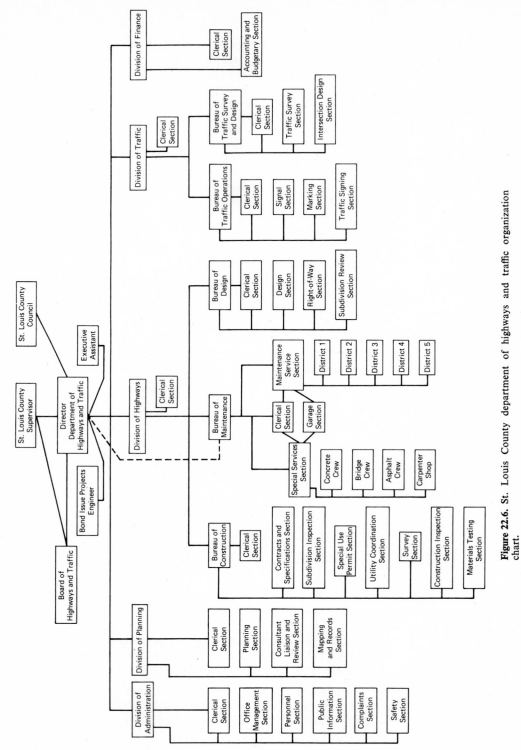

Figure 22.6. St. Louis County department of highways and traffic organization chart.

1006

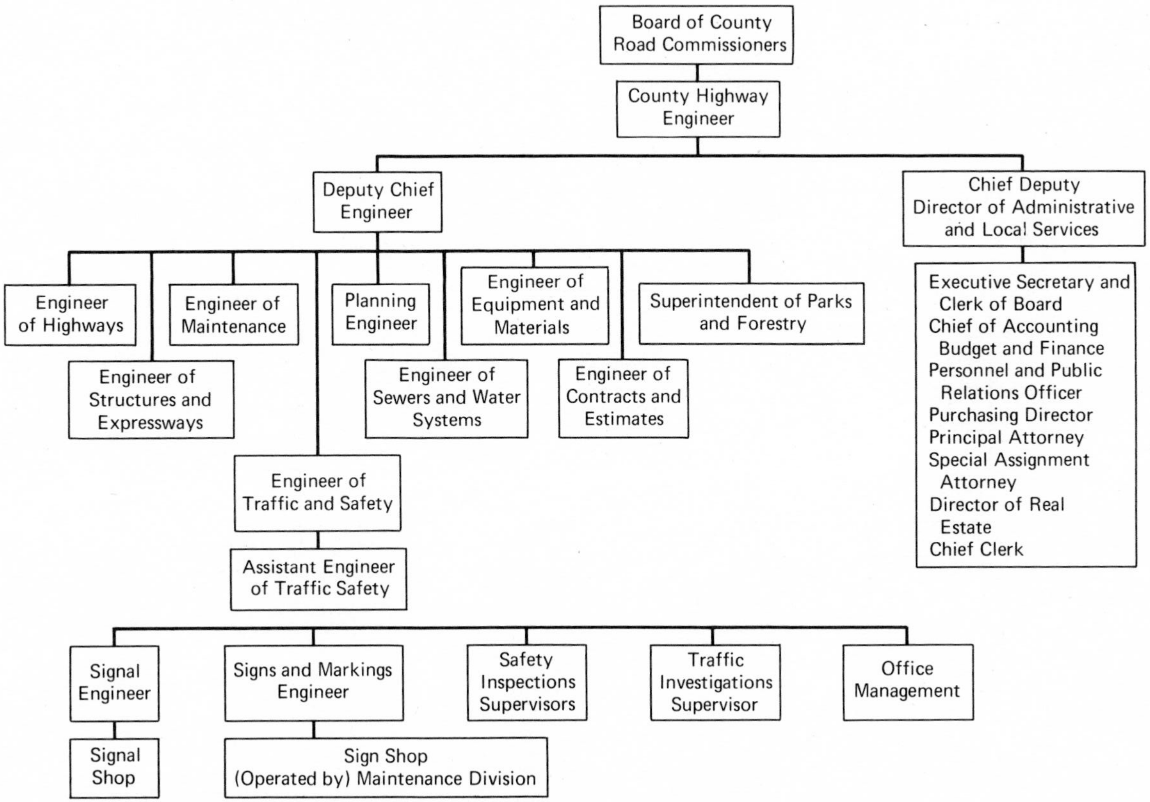

Figure 22.7. Typical organization of a county road department showing location and activities of traffic engineering function.

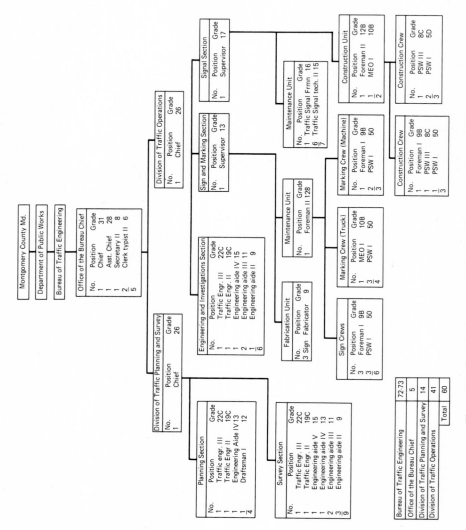

Figure 22.8. Organization chart, 1972–1973, Montgomery County, Md.

2. The council-manager form in which the mayor usually serves as a member of the city council and the city manager is chief administrator. The city manager is appointed by and serves at the pleasure of the council.
3. The commission form in which each elected commissioner generally serves as the head of an operating department. The commissioner, who may be elected at large or by districts, constitutes the legislative body and may choose one of its members to serve as chairman.

Figure 22.9 illustrates the strong mayor form with a separate traffic engineering department. The traffic engineering department has equal status with other important departments such as public works, fire, and police. It has complete authority and responsibility over all engineering aspects of traffic operations. Installation and maintenance of traffic signals and street lights are performed by the public works department on specifications and standards established by the traffic engineering department. Various combinations of this arrangement can be found in different cities. In some cases the traffic engineer performs the work with his own forces. Such a departmental organization is usually found in the larger cities.

Figure 22.10 shows the traffic engineering function as a division of the public works department. In this case, the division is usually responsible for traffic planning, geo-

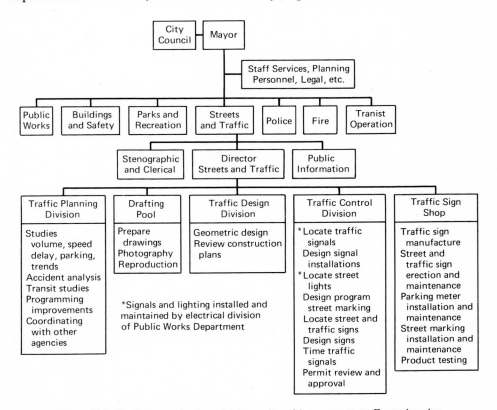

Figure 22.9. Typical organization of a large city with a separate traffic engineering department.

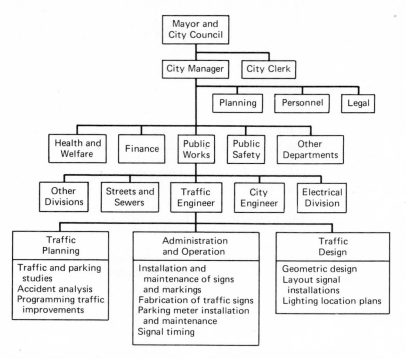

Figure 22.10. Typical organization of a smaller city with a traffic engineering division within the public works department.

metric design, and the installation and maintenance of traffic control devices. The traffic engineer also determines the need and prepares layouts for the location of signals. The timing of the traffic signal system is also performed by the traffic engineering division although the maintenance may be performed by another unit within the public works department.

Figure 22.11 shows the traffic engineering department as a unit of government under the commission form. Since the commission form of government is more prevalent in smaller cities, this organizational chart has been simplified. The traffic engineer and his staff perform all of the planning, design, operation, and maintenance operations within the department. The traffic engineering department is on the same organizational level as the police, fire, and public works departments and reports directly to one of the elected commissioners.

The determination of the proper organizational pattern depends on such factors as the size of the city, existing city charter, and other legislative provisions. The general character of the city can also provide for variations in traffic engineering administration because of the interrelationships with county, state, and metropolitan governments.

Several interesting statistics are available on traffic engineering staffing within various size cities. Table 22.4 indicates the number of cities reporting at least one traffic engineer on their staff, with 90 percent of the cities of over 100,000 population reporting at least one or more traffic engineers. Table 22.5 indicates which city official is responsible for the traffic engineering function in smaller cities. Only cities of 10,000

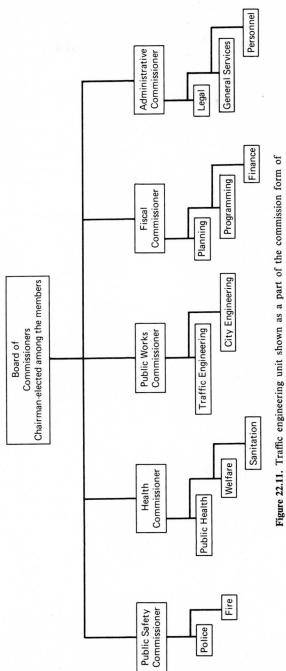

Figure 22.11. Traffic engineering unit shown as a part of the commission form of government.

TABLE 22.4
Traffic Engineering Staffing

Population Group	Number of Cities Eligible	Reporting	Number of Cities Having One or More Traffic Engineers
Over 500,000	25	24	23
350–500,000	11	11	11
200–350,000	28	26	24
100–200,000	79	60	51
50–100,000	205	121	60
25–50,000	409	189	23
10–25,000	1,124	351	12
Totals	1,881	782	204

Source: *Annual Traffic Inventory*, 1965.

TABLE 22.5
City Official Responsible for Traffic Engineering in Cities of 10,000 to 50,000 Population

City Population	10–25,000 population	25–50,000 population
Traffic Engineer	12	23
City Engineer or Director of Public Works	203	110
Mayor	2	0
City Manager	22	1
Police	54	35
Others	30	13
None	7	1
No report	21	6
Totals	351	189

Source: *Annual Traffic Inventory*, 1965.

to 50,000 population were included in the study. Only 7 percent of the cities reporting had a traffic engineer responsible for the traffic engineering function.

Table 22.6 shows the professional status of traffic engineers in various size cities. Well over 60 percent of all the traffic engineers responding indicated either professional

TABLE 22.6
Status of Professional Standing

Professional Standing	Registered Professional Engineer	Graduate Engineer	P.E. and G.E.	No Report
Over 500,000 population	8	2	10	4
350–500,000 ″	2	2	6	1
200–350,000 ″	9	2	7	8
100–200,000 ″	12	16	9	23
50–100,000 ″	30	23	20	48
25–50,000 ″	37	36	51	65
10–25,000 ″	112	37	58	144
Totals	210	118	161	293

Source: *Annual Traffic Inventory*, 1965.

registration, graduation as an engineer, or both. The actual percentage may be somewhat higher since a large number of cities did not respond to this specific inquiry.

METROPOLITAN GOVERNMENT

The rise of metropolitan governments in several parts of the United States has provided an interesting development in relation to traffic engineering organizations. Such governments currently exist in major metropolitan areas like Dade County–Miami, Florida; Davidson County–Nashville, Tennessee; Hennepin County–Minneapolis, St. Paul, Minnesota. Basically, the traffic engineering unit is found at the same organization level, and the organizational structure is similar to that of a large county or a major city traffic engineering department. Under this structure, the traffic engineering department is responsible for all planning, geometric design, operation, and maintenance functions for traffic control devices and programs.

LEGISLATIVE AUTHORITY

As a unit or subunit of government, any public traffic engineering agency must have a legal basis for its authority in order to perform its public functions, both administrative and regulatory.

UNIFORM VEHICLE CODE

The *Uniform Vehicle Code*[5] is a specimen set of motor vehicle laws that has been designed to be a comprehensive guide for state motor vehicle and traffic laws. It is based on actual experience of various state laws and reflects the highly desirable need for uniformity in traffic regulations throughout the United States. Chapter 15 deals specifically with the respective powers of state and local officials and indicates that traffic control provisions shall be applicable and uniform throughout a state and all its political subdivisions. No local authority can enact or enforce any traffic control ordinance unless expressly authorized. Some of the more significant powers, duties, and regulations found in Chapter 15 include:

§ 15-102—**Powers of local authorities**

(a) The provisions of this act shall not be deemed to prevent local authorities with respect to streets and highways under their jurisdiction and within the reasonable exercise of the police power from:

1. Regulating or prohibiting stopping, standing or parking;
2. Regulating traffic by means of police officers or official traffic-control devices;
3. Regulating or prohibiting processions or assemblages on the highways;
4. Designating particular highways or roadways for use by traffic moving in one direction as authorized by §11-308;
5. Establishing speed limits for vehicles in public parks notwithstanding the provisions of §11-803(a) 3;
6. Designating any highway as a through highway or designating any intersection or junction of roadways as a stop or yield intersection or junction; (Revised, 1971.)
7. Restricting the use of highways as authorized in §14-113;
8. Regulating the operation of bicycles and requiring the registration and inspection of same, including the requirement of a registration fee;

---

[5] *Uniform Vehicle Code*, National Committee on Uniform Traffic Laws and Ordinances (Charlottesville, Va.: The Michie Company, 1968; Supplement I, 1972).

9. Regulating or prohibiting the turning of vehicles or specified types of vehicles;

10. Altering or establishing speed limits as authorized in §11-803;

11. Requiring written accident reports as authorized in §10-115;

12. Designating no-passing zones as authorized in §11-307;

13. Prohibiting or regulating the use of controlled-access roadways by any class or kind of traffic as authorized in §11-313;

14. Prohibiting or regulating the use of heavily traveled streets by any class or kind of traffic found to be incompatible with the normal and safe movement of traffic;

15. Establishing minimum speed limits as authorized in §11-804(b);

16. Designating hazardous railroad grade crossings as authorized in §11-702;

17. Designating and regulating traffic on play streets;

18. Prohibiting pedestrians from crossing a roadway in a business district or any designated highway except in a crosswalk as authorized in §15-107;

19. Restricting pedestrian crossings at unmarked crosswalks as authorized in §15-108;

20. Regulating persons propelling push carts;

21. Regulating persons upon skates, coasters, sleds and other toy vehicles;

22. Adopting and enforcing such temporary or experimental regulations as may be necessary to cover emergencies or special conditions;

23. Adopting such other traffic regulations as are specifically authorized by this act.

(b) No local authority shall erect or maintain any official traffic-control device at any location so as to require the traffic on any State highway to stop before entering or crossing any intersecting highway unless approval in writing has first been obtained from the (State highway commission).

(c) No ordinance or regulation enacted under subdivisions (4), (5), (6), (7), (9), (10), (12), (13), (14), (16), (17) or (19) of paragraph (a) of this section shall be effective until official traffic-control devices giving notice of such local traffic regulations are erected upon or at the entrances to the highway or part thereof affected as may be most appropriate. (Section Revised, 1968.)

§ 15-103—Adoption by reference

Local authorities by ordinance may adopt by reference all or any part of the (name of State) Model Traffic Ordinance (include any further description of the ordinance that may be necessary) without publishing or posting in full the provisions thereof, provided that (the enacting ordinance is published and) not less than three copies are available for public use and examination in the office of the (clerk) (commencing at least _____ days prior to such adoption).[1] (New, 1968.)

§ 15-104—(State highway commission) to adopt sign manual

The (State highway commission) shall adopt a manual and specifications for a uniform system of traffic-control devices consistent with the provisions of this Act for use upon highways within this State. Such uniform system shall correlate with and so far as possible conform to the system set forth in the most recent edition of the *Manual on Uniform Traffic Control Devices for Streets and Highways* and other standards issued or endorsed by the Federal Highway Administrator.[2] (Revised, 1971.)

§ 15-105—(State highway commission) to sign all State (and county) highways

(a) The (State highway commission) shall place and maintain such traffic-control devices, conforming to its manual and specifications, upon all State (and county) high-

---

[1] This section should be considered together with existing constitutional and legal requirements concerning the adoption and publication of municipal ordinances. Also, many states already have laws relating to municipal adoption of codes by reference and they should also be consulted. Consideration should be given to whether subsequent changes in the model ordinance adopted by reference will be adopted automatically or separately. If a state does not have or contemplate having an official or unofficial model traffic ordinance for use by its municipalities, some consideration might be given to authorizing adoption by reference of a printed code of traffic ordinances compiled by a nationally recognized organization such as the *Model Traffic Ordinance* of the National Committee on Uniform Traffic Laws and Ordinances.

If the recommendation of the National Committee is followed and a model traffic ordinance is adopted by the state legislature, then this section should be included as a part of that enactment.

[2] As to the last paragraph in footnote 2 on page 233, the 1971 edition of the *Manual on Uniform Traffic Control Devices for Streets and Highways,* has recently been published. Copies may be obtained from the U.S. Government Printing Office, Washington, D.C. 20402 for $3.50 per copy.

ways as it shall deem necessary to indicate and to carry out the provisions of this act or to regulate, warn or guide traffic.

(b) No local authority shall place or maintain any traffic-control device upon any highway under the jurisdiction of the (State highway commission) except by the latter's permission.

§ 15-106—Local traffic-control devices

(a) Local authorities in their respective jurisdictions shall place and maintain such traffic-control devices upon highways under their jurisdiction as they may deem necessary to indicate and to carry out the provisions of this act or local traffic ordinances or to regulate, warn or guide traffic. All such traffic-control devices hereafter erected shall conform to the State manual and specifications.[3]

Optional (b) Local authorities in exercising those functions referred to in the preceding paragraph shall be subject to the direction and control of the (State highway commission).[4]

§ 15-107—Authority to restrict pedestrian crossings

Local authorities by ordinance, and the (State highway commission) by erecting appropriate official traffic-control devices, are hereby empowered within their respective jurisdictions to prohibit pedestrians from crossing any roadway in a business district or any designated highways except in a crosswalk. (Revised, 1968.)

§ 15-108—Authority to close unmarked crosswalks

The (State highway commission) and local authorities in their respective jurisdictions may after an engineering and traffic investigation designate unmarked crosswalk locations where pedestrian crossing is prohibited or where pedestrians must yield the right of way to vehicles. Such restrictions shall be effective only when official traffic-control devices indicating the restrictions are in place. (New, 1968.)

§ 15-109—Authority to designate through highways and stop and yield intersections

The (State highway commission) with reference to State (and county) highways and local authorities with reference to (other) highways under their jurisdiction may erect and maintain stop signs, yield signs, or other traffic control devices to designate through highways, or to designate intersections or other roadway junctions at which vehicular traffic on one or more of the roadways should yield or stop and yield before entering the intersection or junction. (Revised, 1971.)

---

[3] Section 15-106(a) leaves to local authorities complete jurisdiction to determine the number and location of all traffic-control devices upon highways under their jurisdiction, requiring only that all such devices shall conform to the State manual and specifications.

[4] Optional paragraph (b), if adopted, would vest in the (State highway commission) authority to direct and control where and what number of traffic-control devices might be erected by local authorities. This may be objectionable to some local authorities although it is recognized that in certain instances local authorities having a free hand in this matter have erected such numbers of regulatory signs and signals as to unduly delay traffic and invite disobedience by the motoring public.

MODEL TRAFFIC ORDINANCE

The *Model Traffic Ordinance*[6] is a specific set of motor vehicle ordinances for a municipality which is consistent with the provisions of state law as embodied in the *Uniform Vehicle Code*. Its provisions are designed as a guide for municipalities to follow. From the standpoint of the traffic administrator, the *Model Traffic Ordinance* can only be effective to the degree that essential functions are properly coordinated and authority is established by law.

---

[6] *Model Traffic Ordinance*, National Committee on Uniform Traffic Laws and Ordinances (Washington, D.C.: The Michie Company, Charlottesville, Va., 1968; Supplement I, 1972).

Non-uniform laws and ordinances are a source of inconvenience and hazard to both the pedestrian and motorist. In recognition of these problems, both the *Model Traffic Ordinance* for municipalities and the *Uniform Vehicle Code* for states have been developed for adoption by *all* governmental jurisdictions. Some of the more significant provisions of the *Model Traffic Ordinance* include:

§ 2-10—City traffic engineer

(a) The office of city traffic engineer is hereby established. The city traffic engineer shall be a qualified engineer and shall be appointed by _____ (under civil service) and he shall exercise the powers and duties as provided in this ordinance and in the traffic ordinances of this city.

Alternate (a) The office of city traffic engineer is hereby established. The (city engineer) shall serve as city traffic engineer in addition to his other functions, and shall exercise the powers and duties with respect to traffic as provided in this ordinance.

(b) It shall be the general duty of the city traffic engineer to determine the installation and proper timing and maintenance of traffic-control devices, to conduct engineering analyses of traffic accidents and to devise remedial measures, to conduct engineering investigations of traffic conditions, to plan the operation of traffic on the streets and highways of this city, and to cooperate with other city officials in the development of ways and means to improve traffic conditions, and to carry out the additional powers and duties imposed by ordinances of this city.

§ 2-11—Emergency and experimental regulations

(a) The chief of police by and with the approval of the city traffic engineer is hereby empowered to make regulations necessary to make effective the provisions of the traffic ordinances of this city and to make and enforce temporary or experimental regulations to cover emergencies or special conditions. No such temporary or experimental regulation shall remain in effect for more than 90 days.[4]

(b) The city traffic engineer may test traffic-control devices under actual conditions of traffic.

(c) The chief of police may authorize the temporary placing of official traffic-control devices when required by emergency. The chief of police shall notify the city traffic engineer of his action as soon thereafter as is practicable. (New, 1971.)

§ 4-1—Authority to install traffic-control devices

The (city traffic engineer) shall place and maintain official traffic-control devices when and as required under the traffic ordinances of this city to make effective the provisions of said ordinances, and may place and maintain such additional official traffic-control devices as he may deem necessary to regulate, warn or guide traffic under the traffic ordinances of this city or the State vehicle code. (Revised, 1968.)

§ 4-6—Authority to establish play streets

The city traffic engineer shall have authority to declare any street or part thereof a play street and to place appropriate signs or devices in the roadway indicating and helping to protect the same.

§ 4-8—City traffic engineer to designate crosswalks and establish safety zones

The city traffic engineer is hereby authorized:

1. To designate and maintain, by appropriate devices, marks, or lines upon the surface of the roadway, crosswalks at intersections where in his opinion there is particular danger to pedestrians crossing the roadway, and at such other places as he may deem necessary;

2. To establish safety zones of such kind and character and at such places as he may deem necessary for the protection of pedestrians.

----

[4] This subsection is authorized by UVC §15-102 (a) 22.

§ 4-9—Traffic lanes

The city traffic engineer is hereby authorized to mark traffic lanes upon the roadway of any street or highway where a regular alignment of traffic is necessary.

§ 15-1—City traffic engineer to designate loading zones

The city traffic engineer is hereby authorized to determine the location of loading zones and passenger loading zones and shall place and maintain appropriate signs indicating the same and stating the hours during which the provisions of this article are applicable. (Revised, 1971.)

§ 15-5—City traffic engineer to designate public carrier stops and stands

The city traffic engineer is hereby authorized and required to establish bus stops, bus stands, taxicab stands and stands for other passenger common-carrier motor vehicles on such public streets in such places and in such number as he shall determine to be of the greatest benefit and convenience to the public, and every such bus stop, bus stand, taxicab stand or other stand shall be designated by appropriate signs. (Amplified, 1952.)

LEGAL AUTHORITY IN STATES

State highway departments usually include all of the basic transportation functions of planning, design, construction, operations, and maintenance. Therefore, the chief administrative officer usually has considerable latitude in the distribution of traffic engineering activities if he has the essential broad powers. For example, traffic engineering operations are consolidated in some states with highway planning and the division head is the traffic and planning engineer.

The geometric design functions may be located within the design division with the work being performed by engineers who have had training in the basic principles of traffic operations. This can best be accomplished in those jurisdictions which enjoy broad statutory authority.

LEGAL AUTHORITY IN COUNTIES

Legal authority in counties may stem from several sources. One kind is represented by that of Wayne County (Detroit), Michigan. The Board of Supervisors, by constitutional provision, is the legislative body of the county and has such powers, both legislative and administrative, as may be conferred upon it by the legislature and the state constitution. Under authority of the *Public Acts* for the State of Michigan, as amended, the people of Wayne County adopted the County Road System and thereby established a Board of Wayne County Road Commissioners. This Board became a legal entity known as a "Public Corporate Body." The power to use road funds from state gasoline and vehicle weight taxes, as apportioned to the county, has been conferred upon the Board. The County Highway Department functions under this power.

Another kind of legal authority is represented in Montgomery County, Maryland (adjacent to Washington, D.C.) The Charter of Montgomery County, adopted in 1969, established the county executive form of government with the county council holding legislative powers and the executive as the chief administrative official. The charter specifically established several departments, bureaus, and divisions by local law. Under this authority, the Bureau of Traffic Engineering was established in the Department of Public Works.

LEGAL AUTHORITIES IN CITIES

The legal basis upon which the traffic engineering function is established has a large bearing on the effectiveness with which the activities are pursued. Three basic methods of establishing such "legality" in cities are by: (1) traffic ordinance, (2) charter amendment, and (3) administrative regulation or order.

Traffic ordinance. The most popular method in use in a majority of cities is the designation of traffic engineering responsibilities as a part of the traffic ordinance. This ordinance should comply as closely as possible with the provisions of the nationally recommended *Model Traffic Ordinance.*

Charter amendment. The most effective means of establishing the traffic engineering function in a city is by charter amendment. Often, existing charters make specific assignment of key activities to an existing agency and it is impossible to relocate the responsibility without repeal of such provisions. This results in divided authority over various aspects of the program. Although charter amendment is the more effective, more permanent method, it is sometimes the most difficult and time consuming to accomplish.

Administrative regulation or order. In smaller communities, a popular method of establishing the traffic engineering function is the executive or administrative order. The provisions recommended in the Model Traffic Ordinance may serve as a useful guide in prescribing the activities.

Delegation of Authority. In carrying out many of the functions of traffic engineering, it is necessary to adopt such regulations as parking controls, intersection controls, and speed controls. The nature of traffic operations is such that these should be dynamic rather than static regulations so that they can be revised as conditions change. Too often regulations are of a static, legislative nature and require the cumbersome procedure of acts of the legislative or city council to change.

Many governmental agencies are currently seeking methods of delegating regulatory powers to administrative officials. Of the many higher court decisions on this question, an excellent example is the opinion of the Court of Appeals of Maryland on a case in the city of Baltimore. In accordance with authority vested in him by ordinance, the Director of Traffic had adopted regulations establishing certain speed limits within the city. A taxpayer sought a declaratory judgment in which he attacked the provision on two grounds:

1. That the city ordinance unlawfully delegated functions to an administrative official; and
2. That it did not provide proper standards for the guidance of the Director in adopting his rules, regulations, orders, and directives.

The Circuit Court upheld the validity of the ordinance and an appeal was taken to the Court of Appeals of Maryland. With citations of numerous Supreme Court decisions in various states, the higher court affirmed the decree in this aspect with comments in part as follows:

> On account of the tremendous growth of traffic and the need for constant supervision
> of traffic control, it has become increasingly imperative for city councils in metro-

politan centers to delegate to traffic experts a reasonable amount of discretion in their administrative duties. New traffic problems are constantly arising, and therefore to require the enactment of an ordinance to cover each specific problem would be likely to result in widespread delays and even serious hazards. It is obvious that there is a practical necessity for expert and prompt judgment in the application of the concept of public safety to concrete situations, and that the standards for administrative officials in the domain of public safety should be at least as flexible as in the domain of public health.

In states in which court interpretations have not yet provided for such complete delegation of authority, counties and cities are seeking other methods of providing more direct service of the traffic engineering function.

FINANCING TRAFFIC ENGINEERING OPERATIONS

STATE TRAFFIC ENGINEERING BUDGETS

State traffic engineering budgets for traffic control devices have increased sharply in the last 25 years. A trend analysis by the National Safety Council reveals that in 1955 the states spent almost $70 million. In 1965, they spent almost $144 million.[7] By 1975, it is estimated they will spend close to $250 million.

These figures have been updated in connection with the more recent 1968 Federal Department of Transportation Needs Study. The figures included in the Department of Transportation survey relate to the amount of funds required to bring all traffic control devices up to the standards included in the *Manual on Uniform Traffic Control Devices for Streets and Highways*. The dollar values involved and a ten-year program to fulfill these requirements are shown in Figure 22.12.

CITY AND COUNTY BUDGETS

Any consideration of traffic engineering financing and budgets must take into account the wide variety of activities assigned to the traffic engineer. In some jurisdictions, for example, the traffic engineer may serve only in a staff capacity, with the actual work of installing and maintaining various traffic control devices performed by other departments having separate budgets; the traffic engineer's budget would not reflect the true costs of providing traffic engineering services. To get around this problem, Seburn and Marsh studied several cities in order to determine the actual expenditures for *all* traffic engineering activities carried out by *all* city departments. The results of this analysis for four cities with good performance budgeting practices are shown in Table 22.7. The data are not submitted as representing the optimum condition but they are accurately indicative of what some cities were actually spending at that time.

Figures 22.13 through 22.16 show a selection of traffic engineering budgets for several cities and counties of varying sizes. No attempt has been made to classify these budgets, and they serve only to illustrate the kinds of financial structures, cost items, and programming available in the cities and counties shown.

---

[7] "1965 Traffic Inventory—States" (Chicago: National Safety Council, 1966).

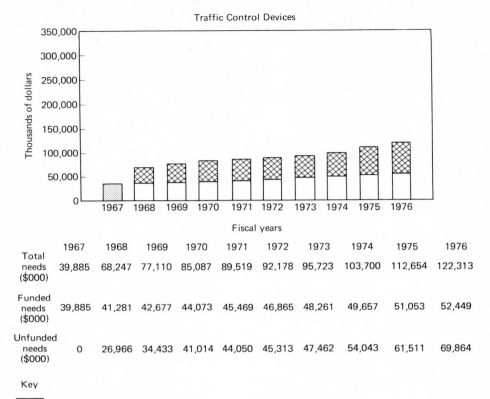

Traffic Control Devices

	1967	1968	1969	1970	1971	1972	1973	1974	1975	1976
Total needs ($000)	39,885	68,247	77,110	85,087	89,519	92,178	95,723	103,700	112,654	122,313
Funded needs ($000)	39,885	41,281	42,677	44,073	45,469	46,865	48,261	49,657	51,053	52,449
Unfunded needs ($000)	0	26,966	34,433	41,014	44,050	45,313	47,462	54,043	61,511	69,864

Key

 Actual expenditure in FY 1967 by state and local governments.

Funded needs: 1967 state and local expenditures projected to maintain 1967 program levels.

Unfunded needs are the computed additonal cost of expanding program levels and instituting new programs to the total level estimated by the states. "Unfunded needs" are thus the difference between "funded needs" and "total needs".

Figure 22.12. State and local highway safety programs needs estimate. (Source: U.S. Department of Transportation Needs Study, 1968.)

TABLE 22.7
Expenditures for Traffic Operations

City	Dollars Spent in Fiscal Year 1957–1958 per			
	Capita	Mi of Street	Sq Mi of City Area	Registered Vehicle
1	$3.92	$1,645	$18,188	$10.79
2	4.30	1,347	20,450	10.74
3	3.57	1,476	21,522	12.26
4	3.01	1,136	14,331	10.43
Average	3.70	1,401	18,623	11.05

Source: Thomas J. Seburn and Bernard L. Marsh, "Urban Transportation Administration" (New Haven: Yale University, Bureau of Highway Traffic, 1959), p. 40.

1610-020 EXPENDITURE SUMMARY				
Function	Department Transportation	Fund General	Account No. 1-37100-37500	
1. Character Classification	2. Actual Expenditures 1969-70	3. Current Budget 1970-71	4. Estimated Requirements 1971-72	5. Final Allowance
Personal Services	646,543	833,330	1,027,703	
Contractual Services	796,380	837,097	1,123,377	
Commodities	130,097	106,575	115,975	
Capital Outlay	12,325	13,375	6,400	
Total	1,585,345	1,790,377	2,273,455	
Summary By Activities				
Public Safety				
1-37100 Administration*	–	155,590	183,048	
1-37101 Transportation Planning †	397,441	195,598	336,114	
1-37102 Traffic Signals	603,297	716,798	757,584	
1-37500 Traffic Operations Programs (TOPICS)	–	–	138,900	
Streets and Highways				
1-37203 Street Signs and Markings	347,563	415,457	479,128	
1-37204 Tow Service	211,106	306,934	378,681	
1-37400 Special Transportation Facility	25,938	–	–	
Total	1,585,345	1,790,377	2,273,455	

*This activity transferred from Transportation Planning, Account No. 1-37101, effective May 1, 1970

† Certain activities in this account transferred to Administration, Account No. 1-37100, Traffic Signals, Account No. 1-37102, and Street Signs and Markings, Account No. 1-37203, effective May 1, 1970.

CITY OF KANSAS CITY, MISSOURI

Figure 22.13. Traffic engineering budget, Kansas City, Mo.

TRAFFIC ENGINEERING DEPARTMENT SUMMARY					
Department	Division				Account No.
Traffic Engineering	–				2800

Character Classification	1970-71		1971-72 Allowance		
	Adopted Budget	Estimated Expenditures	Current Operations	Service Improvements	Total
Personal Services	689,083	645,598	709,323	8,749	718,072
Contractual Services	237,031	281,601	266,064		266,064
Commodities	250,158	225,331	243,216		243,216
Capital Outlay	73,593	86,526	81,700	10,200	91,900
Debt Service					
Total	1,249,865	1,239,056	1,300,303	18,949	1,319,252
Less Work Order Credits	−18,000	−48,826	−20,900		−20,900
Total	1,231,865	1,190,230	1,279,403	18,949	1,298,352
Number of Positions	70.9	69	69	1	70

EXPENDITURES BY ACTIVITY					
2810 – Administration	94,698	97,624	97,929		97,929
2820 – Design and Planning	88,230	93,135	100,610	400	101,010
2830 – Traffic Safety	75,819	13,885	37,380	3,095	40,475
2840 – Operations	794,454	762,880	817,484	15,454	832,938
2850 – Traffic Signals	178,664	222,706	226,000		226,000
Total	1,231,865	1,190,230	1,279,403	18,949	1,298,352

PERSONNEL BY ACTIVITY					
2810 – Administration	7.9	7.2	7.2		7.2
2820 – Design and Planning	7	7.3	7.3		7.3
2830 – Traffic Safety	5	3	3		3
2840 – Operations	51	51.5	51.5	1	52.5
Total Positions	70.9	69	69	1	70

CITY OF PHOENIX, ARIZONA

Figure 22.14. Traffic engineering budget, Phoenix, Arizona.

FINAL 1971 BUDGET

(Including Wage Survey and Vehicle Depreciation in the Work Program Costs)
City and County of Denver, Colorado, Dept. of Public Works

054   Traffic Engineering Division

(No.)   (Appropriation Name)

Program Plan and Projection — Units Output Cost

Program	Program Measure Output (Units)	Actual 1969		Estimated 1970		Requested 1971		Projected 1972	
		Number of Units	Program Cost	Number of Units	Program Cost	Number of Units	Program Cost	Number of Units	Program Cost
(1)	(2)	(3)	(4)	(5)	(6)	(7)	(8)	(9)	(10)
00 Administration	-	-	7,128	-	7,744	-	9,366	-	8,855
10 Engineering	-	-	68,147	-	80,808	-	97,989	-	92,575
21 Signal Install, New	Locations	31	98,253	33	114,310	35	128,357	36	130,970
22 Signal Rebld to Std Design	Locations	42	183,410	29	98,371	30	122,265	31	112,725
23 Signal Rebld to Imprv Opr	Locations	66	37,167	25	14,335	26	16,731	27	16,388
24 Signal Relocate on Rebld	Locations	.18	27,610	240	84,750	250	90,531	260	97,133
25 Signal Repair, Accident	Locations	457	93,644	1,600	98,757	1,650	113,945	1,700	113,190
26 Signal Removal	Locations	8	4,093	6	4,437	5	4,936	5	5,042
27 Signal Maintenance Jobs	Locations	15,965	114,555	16,000	158,410	16,500	188,115	17,000	181,729
27B Signal Intercon Service	Locations	385	6,165	400	5,500	420	7,800	440	6,303
Total Signals, End Year	Locations	1,104	564,897	1,137	578,870	1,167	672,680	1,198	663,480
30 Signs Delivered to Others	-	-	-	-	40,000	-	16,049	-	45,827
31 Sign Install, New	Signs	5,142	63,253	5,400	69,779	5,600	78,517	5,800	79,939
32 Sign Rebld to Std Design	Signs	5,740	74,128	6,030	82,255	6,300	90,541	6,600	94,248
33 Sign Rebld, Chg Regulation	Signs	1,077	10,111	1,130	11,291	1,180	13,592	1,240	12,935
34 Sign Relocate for St Imprv	Signs	405	5,282	430	5,985	450	7,474	470	6,830
35 Sign Repair, Accident	Signs	9,122	93,750	9,580	102,633	10,000	116,257	10,500	117,790
36 Sign Removal	Signs	435	1,195	310	855	325	1,481	340	954
37 Sign Maintenance	Signs	1,861	18,079	1,950	20,141	2,050	22,331	2,150	23,047
Total Signs, End Year	Signs	126,473	265,798	131,563	332,939	136,838	346,242	142,298	381,570

Figure 22.15. Traffic engineering budget, Denver, Colorado.

1023

Program Plan and Projection – Units Output Cost

Traffic Engineering Division
(Appropriation Name)

054
(No.)

(1) Program	(2) Program Measure Output (Units)	Actual 19 69		Estimated 19 70		Requested 19 71		Projected 19 72	
		(3) Number of Units	(4) Program Cost	(5) Number of Units	(6) Program Cost	(7) Number of Units	(8) Program Cost	(9) Number of Units	(10) Program Cost
41 Meter Installation, New	Meters	149	13,488	155	15,921	160	17,030	165	18,230
42 Meter Rebld to Std Design	Meters	881	11,282	925	12,727	970	16,642	1,000	14,570
43 Meter Covering, Temp	Orders	250	1,195	265	1,406	275	1,787	285	1,609
44 Meter Relocate for St Imp	Meters	2	(112)	2	(106)	2	(42)	2	(106)
45 Meter Repair, Accident	Meters	554	29,671	580	35,322	600	37,656	620	40,454
46 Meter Removal	Meters	256	(14,900)	270	(18,174)	280	(19,209)	290	(20,807)
47 Meter Maintenance Jobs	Meters	11,306	30,814	11,870	33,941	12,400	43,125	13,000	41,845
47A Meter Replacement	Meters	1,648	5,618	1,241	2,604	595	3,259	–	–
Total Meters, End Year	Meters	3,719	77,056	3,604	83,641	3,484	100,248	3,359	95,795
52A Centerline, Paint	Miles	501	52,108	525	66,310	530	72,122	535	75,988
53A Lane Line, Paint	Miles	396	58,669	415	71,654	420	79,983	425	82,169
54A Xwalk & Stop Line, Paint	Locations	1,731	58,512	1,815	68,744	1,900	78,350	1,925	78,784
55A Pave Message, Paint	Locations	539	2,103	565	2,435	575	2,875	580	2,770
56 Removal, Cl or Lane Line	Miles	7	4,258	8	4,803	9	5,413	9	5,499
Total Markings, End Year	Miles	493	175,650	500	213,946	510	238,743	520	245,200
60 School Xings Supervised	Locations	23	23,609	50	90,287	50	62,385	50	103,362
70 Park Enforce, Walk Patrol	Citations	116,793	41,529	54,070	21,087	–	–	–	–
71 Park Enforce, Veh Patrol	Citations	71,593	26,685	216,680	86,843	–	–	–	–
Total Enforce, End Year	Citations	188,386	68,214	270,750	107,930	(Transferred to Police Department)			
89 Off-Street Parking, Supvsn							4,722		
99 Street Light Prgm, Supvsn							10,346		
100 Equip, Tools, Bldg.	–	–	12,945	–	22,635	–	32,979	–	19,160
Totals	–		1,263,444		1,518,800		1,575,700		1,610,000

Figure 22.15. contd.

1024

```
                MONTGOMERY COUNTY, MARYLAND
                 Bureau of Traffic Engineering
                      1972-1973 Budget

                Bureau's Total Budget Summary

Operating Budget

Salaries and Wages
   Regular Positions                $550,950
   Additional Positions               30,300
   Overtime                           13,950
      Total                                            $595,200

Operating Expense                                     1,397,720
Capital Outlay                                           17,150
   Total Operating Budget                            $2,010,070
   Total Capital Improvement Program Budget             572,000
   Grand Total                                       $2,582,070

Office of the Bureau Chief

Salaries and Wages                                      $69,770
   Communications                    $2,900
   Motor Pool Charges                 1,530
   Office Supplies                      400
   Misc. (Subscriptions, Membership Dues,
        Travel Conf.)                    560
      Total                          $5,390

   Total Operating Expense                              $5,390

Capital Outlay                                             180
   Total Operating Budget                              $75,340

Division of Traffic Planning and Survey

Salaries and Wages
   Regular Positions                $143,920
   Additional Positions               10,580
   Overtime                            7,200
      Total                                            $161,700

Operating Expense                                    $1,129,920
Capital Outlay                                            6,020
   Total Operating Budget                            $1,297,640

Administrative Section

Salaries and Wages
   Regular Positions                                    $20,970
Motor Pool Charges                    $770
Misc. (Subscriptions, Membership Dues,
     Travel Conf.)                     150
   Total                             $920
   Total Operating Expense                                 920

   Total Operating Budget                              $21,890
```

Figure 22.16. Traffic engineering budget, Montgomery County, Md.

Division of Traffic Planning and
Survey Planning Section

Salaries and Wages
  Regular Positions $30,940
  Additional Positions   10,580
    Total $41,520

Motor Pool Charges   900
Office Supplies   300
Professional and Technical Services   2,500
  Total   3,700
  Total Operating Expense $3,700

Permanent Continuous Traffic Count
Station, Office Furniture   6,020
  Total Capital Outlay $6,020
  Total Operating Budget $51,240

Survey Section

Salaries and Wages
  Regular Position $92,010
  Overtime   7,200
    Total $99,210

Uniforms   180
Motor Pool Charges   3,900
Office Supplies   400
Equipment, Supplies and Maintenance   1,600
Misc. (Subscriptions, Membership Dues)   50
Street Lighting   1,119,170
  Energy Cost O.H.:- 1,079,495
  Energy Cost U.G.:-   39,675

  Total $1,125,300
  Total Operating Expense $1,125,300
  Total Operating Budget $1,224,510

Division of Traffic Operations

Salaries and Wages
  Regular Positions $337,260
  Additional Positions   19,720
  Overtime   6,750
    Total $363,730
    Operating Expense   262,410
    Capital Outlay   10,950

    Total Operating Budget $637,090

Administrative Section

Salaries and Wages
  Regular Position $15,500

Motor Pool Charges $1,080
Office Supplies   20
Misc. (Subscription, Membership Dues)   50
  Total $1,150
  Total Operating Expense $1,150
  Total Operating Budget $16,650

Figure 22.16. contd.

```
Division of Traffic Operations. Con't.
Engineering and Investigations Section

Salaries and Wages
  Regular Positions                    $37,730
  Additional Positions                  16,180
  Overtime                               1,000
    Total                                            $54,910
Office Supplies                            100
    Total Operating Expense                              100
Office Furniture                           220
    Total Capital Outlay                                 220
    Total Operation Budget                           $55,230

Sign and Marking Section

Salaries and Wages
  Regular Positions                   $136,400
  Additional Positions                   3,540
  Overtime                                 750
    Total                                           $140,690

Uniforms                                 2,400
Janitorial Supplies                        100
Motor Pool Charges                      14,860
Office Supplies                            120
Equipment Supplies and Maintenance       2,250
Small Tools                                250
Traffic Paint:- 22,200
Bars and Special Marking:- 14,000
Pavement Arrows:- 2,400                 38,600
Street Name Sign (New and Replace)      23,400
Traffic Sign Blanks                     21,930
Prepared Sign Faces                      5,480
Rolled Material for Sign Faces          18,360
Traffic Sign Hardware and Posts         14,450
  Total                                $142,200
    Total Operating Expense                         $142,200

File Cabinet                                90
Shop Equipment (Bench Vice, Storage
  Cabinet, Shop Benches, 48 in. Squeeze
  Roller Applicator)                     7,310
    Total                                7,400
    Total Capital Outlay                                7,400
    Total Operating Budget                          $290,290

Signal Section

Salaries and Wages
  Regular Positions                   $147,630
  Overtime                               5,000
    Total                                           $152,630
Uniforms                                 2,100
Janitorial Supplies                        100
Motor Pool Charges                      18,750
Office Supplies                            110
Equipment Supplies and Maintenance       2,550
Small Tools                                250
Signal and Controller Parts             24,450
Traffic Signal Lamps                     1,180
Energy Charge for Traffic Signals       69,470
  Total                                $118,960
    Total Operating Expense                         $118,960
```

Figure 22.16. contd.

```
Division of Traffic Operations
Signal Section. (Con't.)

Desk, Chair and Adding Machine                          700
Shop Equipment (Bench Vice, Benches,
   Storage Cabinet)                                   2,630
   Total                                              3,330
   Total Capial Outlay                                              3,330
   Total Operating Budget                                        $274,920


                    Capital Improvement Program

Street Lighting
(For those areas served with underground utilities)
To install post top luminaires

1972-1973                                                         $7,000

Traffic Signals

1.  New Installations-20
2.  Modification of Existing-25
3.  Interconnection of existing signals on several
    arterial routes estimated total cost                        $175,000


Georgia Avenue

This project is the County share of a $1.1 million
M.S.H.A. project which provides that 22 intersections
along Georgia Avenue corridor between 16th Street and
Layhill Road will be interconnected and standardized
so as to provide a traffic responsive progressive
system.                                                          390,000


Total Capital Improvement Program                               $572,000
```

Figure 22.16. contd.

In the 1972 National Highway Needs Report it was estimated that the cost of traffic engineering improvements in all urban areas in the United States for the 1970–1990 period was $2.2 billion.[8]

FEDERAL FUNDING

The magnitude of the financial needs in the traffic control device field alone can be measured by the U.S. Department of Transportation 1968 Report to Congress entitled "Estimate of the Cost of Carrying Out the Provisions of the Highway Safety Act of 1966."[9]

---

[8] Secretary of Transportation, "1972 National Highway Needs Report," House Document No. 92–266, Part II, 92nd Cong., 2nd sess. (Washington, D.C.: Government Printing Office, 1972), p. IV-91.

[9] "Estimate of the Cost of Carrying Out the Provisions of the Highway Safety Act of 1966," A Report to the Congress from the Secretary of Transportation, U.S. Department of Transportation (Washington, D.C., October, 1968).

Traffic control devices. Federal funding has been made available for numerous studies in this area. For the most part, however, the cost of installing new or modernized traffic control devices has been a financial responsibility of the state or local governments. Some projects on federal-aid routes have been programmed for 50–50 matching funds, but in relation to the total estimated needs these projects represent only a small token of the total requirements during the next decade.

The following significant documents and texts have been extracted directly from the 1968 Department of Transportation report.

TRAFFIC CONTROL DEVICES

General Description:

This program area stipulates that traffic control devices (signs, signals, markings, etc.) on all streets and highways shall be uniform in design and application. Existing devices are to be upgraded, and where engineering studies determine that new devices are required, such devices are to be installed, all in conformance with standards for uniformity. The program area provides that there is to be a program of inspection, maintenance, and repair of traffic control devices. Furthermore, the program calls for a program of establishing speed zones based on engineering and traffic investigations.

Background:

"Traffic control devices, signs, and signals on all highways and streets should be uniform, and standards should be continually reviewed and upgraded."

> Report No. 1700, House of
> Representatives, 89th Congress,
> 2nd Session, July 15, 1966, p. 19.

"The value of uniformity is clear in such matters as uniform signs and signaling devices..."

> Report No. 1302, United States
> Senate, 89th Congress, 2nd Session,
> June 23, 1966, p. 5.

Purpose:

To assure the full and proper application of modern traffic engineering practice and uniform standards for traffic control devices to reduce the likelihood and severity of traffic accidents.

Program Activity:

Within this program area, States are working to implement such activities as:
- Establishing a system and procedures for:
  —Identifying needs and deficiencies of traffic control devices
  —Developing programs to maintain, upgrade, and install traffic control devices
- Formulating action programs to:
  —Upgrade existing traffic control devices on all streets and highways
  —Install new traffic control devices on all streets and highways, based on engineering studies
  —Maintain, repair, and inspect on a daytime and nighttime basis all traffic control devices
  —Establish fixed or variable speed zones, at least on expressways, main streets, and highways, and through streets and highways, based on engineering and traffic investigations

A typical project for this program would be directed toward conducting an onsite survey of existing traffic control devices. An inventory of signs, signals, and markings would be obtained specifying location, identification by type, and conformance to standards.

Title 23, U.S. Code, requires conformance with established traffic control standards on all Federally aided highways. This program area would extend that requirement to local roads and streets that constitute the approximately 75 percent of the nation's highways not on the Federal-Aid systems.

Expenditures in the States amounted to $39.9 million in 1967. The breakdown was $19.3 million at the State level and $20.6 million at the local level.

The TOPICS program. Recognizing the dilemma of the cities, in 1967 the Bureau of Public Roads (now the Federal Highway Administration) announced a program to focus attention on raising the efficiency of existing streets and highways in urban areas. The Traffic Operations Program to Increase Capacity and Safety, known as TOPICS, was established as a part of the 1968 Federal-aid Highway Act. This program provides federal funds on a (50–50) matching basis for improvements to relieve traffic bottlenecks and reduce hazards on roads in urban areas. Funds are allocated to the various states and administered to local areas by the state under the supervision of the Federal Highway administration.

The 1968 Federal-aid Highway Act[10] authorized the expenditure of $200 million specifically for TOPICS for each of the fiscal years 1970 and 1971, to be matched on a 50–50 basis by state and local funds. An amount of $100 million has been authorized for each of the fiscal years 1972 and 1973, in accordance with a specific provision of the 1970 Federal-aid Highway Act.[11]

The objective of the TOPICS program is basically the application of traffic engineering principles to improve traffic flow and increase safety. Major construction of new facilities is not within the scope of this program. The following kinds of improvements, generally within the existing right of way, are eligible for TOPICS funds:

1. Channelization of intersections
2. Providing additional traffic lanes on approaches to signalized intersections
3. Improvement of traffic signal systems
4. Establishment of reversible traffic lanes and their signal control system
5. Development of separate traffic lanes for loading and unloading transit passengers
6. Establishment of traffic surveillance systems
7. Additions and upgrading of highway lighting
8. Construction of pedestrian grade separations or highway-railroad grade crossings
9. Development of truck loading and unloading facilities

The following steps are generally necessary in order to obtain TOPICS funds:

1. TOPICS is a federal-aid program and not a direct grant to cities. Therefore, projects must be developed and schedules through the state highway department as in any other federal-aid project.
2. The execution of a TOPICS program begins with the development of an area-wide program for traffic operation improvements. This should include an inventory and analysis of all elements related to the plan.
3. In this plan, certain arterial highways and major streets not on the federal-aid systems are designated Type II, federal-aid primary routes.

---

[10] (Washington, D.C.: U.S. Government Printing Office, 1968).
[11] (Washington, D.C.: U.S. Government Printing Office, 1970).

4. Once the plan is completed and the Type II System designated, projects eligible for TOPICS may be programmed. These may be projects for not only the Type II System but also for the Type I Primary System and Secondary System.
5. Under TOPICS, projects in areas having over 5,000 population can become eligible for federal-aid. In areas having under 50,000 population, it will also be necessary for the state highway departments and the local communities to agree on procedures for carrying out the transportation planning process.

In order to implement a TOPICS program, additional qualified personnel may be required. A city may add to its existing staff or retain qualified traffic engineering consultants to perform this work. Consulting firms can provide an area transportation plan, an area-wide traffic operations plan, or a specified detailed project plan.

A great deal of administrative work is required to process each phase of any TOPICS improvement project, but the hours of work necessary to meet TOPICS regulations can result in a city or county's receiving up to several million dollars in TOPICS funds. This will allow the city and county to finance additional urgently needed highway and traffic services.

Other federal funds. There are many means of obtaining federal funds from other sources that should be mentioned. Within the U.S. Department of Transportation, the Federal Highway Administration, the National Highway Traffic Safety Administration, and the Urban Mass Transportation Administration are all potential providers of grants-in-aid and other federal matching funds.

Such monies can be used for purposes such as street improvements, safety projects, bus lanes and terminals, parking facilities, traffic control devices, railroad-highway grade crossings improvements, and other traffic engineering projects subject to federal requirements and regulations. Most of these federal funding programs are provided on a 50–50 percent matching basis. The local share can usually be made available for either state, county, or city funds.

In addition, federal funds are also available from the U.S. Department of Housing and Urban Development (HUD). In most cases, these monies must be applied to specific projects dealing with urban street improvements in connection with urban renewal projects and other community development programs. HUD funds may also be made available for parking facilities to relieve street congestion, signal systems to optimize intersection capacity, and mass-transportation projects to serve urban centers.

RELATING BUDGETS TO ACCIDENT TRENDS[12]

Urban traffic deaths account for over 30 percent of all traffic deaths in the United States. On a vehicle/mile basis, the urban rate of 3.6 deaths per 100 million vehicle/miles was far below the rural rate of 7.5. The significant fact, however, is the rising proportion of urban deaths over the past few years in comparison with rural deaths (see Table 22.8).

---

[12] DAVID M. BALDWIN, "Accident Trends in Cities and City Traffic Engineering Staff, Budgets and Responsibilities," *Improved Street Utilization Through Traffic Engineering*, Highway Research Board Special Report No. 93 (Washington, D.C.: Highway Research Board, 1967), pp. 178–81.

TABLE 22.8
Motor Vehicle Deaths and Mileage Death Rates (1955–1965)

Location of Accident	Deaths			Mileage Death Rate		
	1955	1965	Change (%)	1955	1965	Change (%)
Urban	9,390	15,000	+60	3.5	3.6	+3
Rural	29,030	34,000	+17	8.6	7.4	−14
Total	38,420	49,000	+28	6.4	5.6	−12

Source: National Safety Council. Rate is deaths per 100 million veh/mi., as reported by David M. Baldwin, "Accident Trends in Cities and City Traffic Engineering Staff, Budgets and Responsibilities," *Improved Street Utilization Through Traffic Engineering*, Highway Research Board Special Report No. 93 (Washington, D.C., 1967), p. 178.

Moreover, the kinds of accidents resulting in death changed markedly during the last decade (see Table 22.9). The significant points are the major differences in the changes, urban and rural, in two-vehicle and noncollision accidents. The increases in urban accidents of these two kinds far exceed the changes in rural areas and are responsible for the fact that the urban mileage death rate remained nearly constant during a period when the rural rate dropped more than 12 percent.

TABLE 22.9
Types of Fatal Motor Vehicle Accidents (1955–1965)

Type of Accidents	Urban Deaths			Rural Deaths		
	1955	1965	Change (%)	1955	1965	Change (%)
Pedestrian	5,200	5,700	+10	3,000	3,100	+3
Two-vehicle collision	1,900	4,700	+148	12,600	16,000	+27
Other collision	1,290	1,800	+40	2,330	2,800	+20
Noncollision	1,000	2,800	+180	11,100	12,100	+9
Total	9,390	15,000	+60	29,030	34,000	+17

Source: National Safety Council, as reported by David M. Baldwin, "Accident Trends in Cities and City Traffic Engineering Staff, Budgets and Responsibilities," p. 179.

Table 22.10, a comparison by size of city, indicates that the problem is not uniformly distributed. The comparison of death rates by size groups suggests that the largest cities of over 750,000 population have experienced the smallest increase in accident rates. The large city registration rate has actually decreased in the ten years from 1955 to 1965, although it still is higher than for smaller urban areas. The largest increases in rates, on the other hand, have taken place in the cities from 100,000 to 750,000 population. Cities below 100,000 population have experienced increases, but they are not as great.

Nonfatal injuries in urban accidents are greater in total number than those in rural accidents. Three kinds of accidents account for a large part of the total: pedestrian, two-vehicle, and noncollision. The differences between urban and rural experience are given in Table 22.11. The large number of injuries in urban two-vehicle accidents is immediately apparent. Similarly, although the totals are much smaller, injuries in

TABLE 22.10
Motor Vehicle Death Rates, Population and Registration (1955–1965)

Urban Population	Deaths per 100,000 Pop.			Deaths per 10,000 Veh. Reg.		
	1955	1965	Change (%)	1955	1965	Change (%)
Over 1,000,000	10.3	10.7	+4	3.9	3.4	−13
750,000–1,000,000	10.3	12.0	+16	3.3	3.0	−9
500,000–750,000	9.8	12.8	+30	2.6	3.0	+15
350,000–500,000	11.1	16.3	+47	2.7	3.1	+15
200,000–350,000	10.1	13.9	+38	2.2	2.5	+14
100,000–200,000	9.5	13.9	+46	2.1	2.5	+19
50,000–100,000	9.0	9.8	+9	2.0	1.9	−5
25,000–50,000	8.9	10.8	+22	1.9	2.0	+5
10,000–25,000	9.0	11.2	+24	1.6	1.9	+19
All cities	9.8	11.7	+19	2.4	2.5	+4

Note: Not all cities in the United States are included, but the samples are believed representative. Data for 1955 and 1965 are not necessarily based on identical cities, partly because some cities have moved from one population group to another during the 10 years.

Source: David M. Baldwin, "Accident Trends in Cities and City Traffic Engineering Staff, Budgets and Responsibilities," p. 179.

TABLE 22.11
Motor Vehicle Deaths and Non-Fatal Injuries by Type of Accident (1965)

Type of Accident	Deaths			Non-Fatal Injuries		
	Total	Urban	Rural	Total	Urban	Rural
Pedestrian	8,800	5,700	3,100	140,000	128,000	12,000
Two-vehicle	20,700	4,700	16,000	1,230,000	760,000	470,000
Non-collision	14,900	2,800	12,100	330,000	95,000	235,000
All others	4,600	1,800	2,800	100,000	67,000	33,000
Total	49,000	15,000	34,000	1,800,000	1,050,000	750,000

Source: National Safety Council, as reported by David M. Baldwin, "Accident Trends in Cities and City Traffic Engineering Staff, Budgets and Responsibilities," *Improved Street Utilization Through Traffic Engineering*, Highway Research Board Special Report No. 93 (Washington, D.C., 1967), p. 179.

urban noncollision accidents exceed those in rural noncollision accidents although the reverse is true for deaths.

Data are not available to permit examination of the kind of accident causing deaths and nonfatal injuries by size of city. It may be that the differences in trends in deaths and death rates are related to changes in patterns of accidents.

It is also possible that accident prevention efforts have not been undertaken in all sizes of cities on an equally energetic or effective basis. It must be emphasized that there have not been any successful attempts to correlate accident experience with the commonly available measures of accident prevention. Thus, there is no opportunity to test an assumption that the largest cities may have carried on better safety programs. Least of all can we say that they have had better traffic engineering work, valuable as such a conclusion would be to those devoted to the discussion and encouragement of traffic engineering techniques.

In a very valuable multiple regression study, a correlation between death rates and certain isolated measures of safety work has been demonstrated. The one inescapable conclusion arising from this study is that many of the most logical measures do not show any correlation. Two possibilities exist: either we are doing the wrong things or we have not developed good measuring devices for what we are doing.

To look more closely at the actual situation in cities, ten municipalities were queried on a number of pertinent points. Three cities were selected in the 50,000 to 100,000 population range, three in the 100,000 to 200,000 range, two in the 500,000 to 700,000 range, and two in the 700,000 to 1,000,000 range. In each case, cities that reported having full-time traffic engineers were selected. Some cities with high-activity scores in the National Safety Council's annual inventory program as well as some with low scores were also selected. Responses are given in Table 22.12.

TABLE 22.12
Sample City Traffic Data

Item	City A		City B	City C	City D	City E
Population, 1966	53,000		72,000	72,500	172,000	142,130
MV registration, 1966	17,858		39,754	28,891	72,335	80,000
Street mileage:						
Freeways	—		7.7 (3%)	—	21.1 (5%)	28* (7%)
Arterials	?		73.4 (25%)	29.2 (16%)	103.2 (23%)	55.3 (15%)
Others	?		213.4 (72%)	157.3 (84%)	330.7 (78%)	294.7 (78%)
Total miles	140		294.5	186.5	455.0	378

	1962	1966	1961 1966	1961 1966	1961 1966	1960 1966
Accident experience:						
Fatal	2	2	4 9	1 4	6 14	10 21
Non-fatal injury	376†	696‡	260 395	? ?	825 1,207	771 715
Property damage	?	?	1,371 1,779	? ?	3,212 4,382	? ?
Total	1,263	1,365	1,635 2,183	2,135 2,540	4,043 5,603	4,362 5,031
Population death rate	4.0	3.7	9.7 17.9	1.5 5.9	4.7 8.9	6.7 14.8‡
Registration death rate	0.9	0.8	2.0 3.1	0.3 1.0	1.0 1.8	1.7 2.6‡
Traffic engineer reports to	City Manager		City Manager	City Engineer	Director of Public Works	Director of Public Services
Professional TE staff	1		2	1	6	2
TE operating budget	$41,258		$233,870	$30,600	$885,351	$259,152
Capital improvement budget	$850		$1,140,000	$30,000	$119,000	$49,865

*Under construction.
†Non-fatal injuries.
‡Estimated.

Greatest needs in the field of traffic engineering:

City A—Getting parking off main streets to improve flow. Wider streets (arterial and main) and traffic signals at several locations to permit arterial street vehicles to cross main streets. Adequate street lighting system on main and arterial streets.

City B—Construction of additional major street systems together with traffic control to improve the flow of traffic. Also, several street intersections need improving by channelizing for more capacity.

City C—Creation of a separate traffic engineering department with a staff; separate budget; full responsibility and authority to initiate and put into effect regulations improving traffic safety. Funds be made available for inter-connection and coordination of traffic signals, all of which are isolated at present. That the City Council and Mayor be made aware of the importance of applying traffic engineering techniques, and give the traffic engineer their support.

City D—More rigid enforcement of traffic regulations. Functioning safety council. Street lighting on arterial streets in outlying and intermediate areas.

City E—One of the problems encountered is the availability of qualified personnel (especially on the sub-professional level). Feel that more undergraduate courses should be offered in the field of traffic engineering.

Source: David M. Baldwin, "Accident Trends in Cities and City Traffic Engineering Staff, Budgets and Responsibilities," p. 180.

TABLE 22.12 (Cont.)
Sample City Traffic Data

Item	City F	City G	City H	City I	City J
Population, 1966	131,000	500,000	670,000	713,214	940,000
MV registration, 1966	37,000	230,000	207,000	377,498	238,000
Street mileage:					
Freeways	15.2 (2%)	15 (1%)	9 (1%)	98.5 (4%)	17 (1%)
Arterials	109.3 (16%)	200 (20%)	265 (19%)	223 (10%)	450 (32%)
Others	566.6 (82%)	800 (79%)	1,126 (80%)	1,963 (86%)	933 (67%)
Total miles	691.1	1,015	1,400	2,285	1,400

	1961	1966	1961	1966	1961	1966	1961	1966	1961	1966
Accident experience:										
Fatal	15	31	42	73	58	89	50	89	84	118
Non-fatal injury	653	1,233	2,114	2,634	4,458	8,121	2,910	4,537	5,952	8,708
Property damage	2,475	3,140	19,796	23,029	15,174	23,215	11,602	16,422	11,504	15,357
Total	3,143	4,404	21,952	25,736	19,690	31,425	14,562	21,048	17,540	24,183
Population death rate	11.5	25.4	8.8	15.7	9.6	13.3*	9.5	14.1	9.7	13.3
Registration death rate	2.5	5.3	2.0	3.4	3.2	4.3*	2.4	3.4	2.7	3.4
Traffic engineer reports to	Mayor		Director of Public Utilities		Director of Streets		City Manager		Mayor and City Council	
Professional TE staff	4		12		4		6		17	
TE operating budget	$275,000		$320,000		$436,284		$478,890		$3,000,000	
Capital improvement budget	$30,000		$300,000		$110,000		$51,105		$160,000	

*Estimated.

Greatest needs in the field of traffic engineering:

City F—The greatest need in the field of traffic engineering is the acquisition and the retention of professional traffic engineers.

City G—More adequate financial and personnel resources for expanded program of upgrading, standardization, and improved maintenance of traffic control devices.

City H—To expand the scope and depth of present operations which would require an increase of staff and additional funding. Although the division is engaged in all aspects of traffic engineering, much more could be accomplished if more time and talent were available.

City I—Probably the greatest weakness is the inadequacy of the arterial street signal system. Much work needs to be done in providing progressive signal systems along the major arteries. We have surpassed or are rapidly approaching the absolute capacity of most major arteries. Widening and dividing these arteries is necessary to add more capacity and widening cannot be accomplished without acquiring more right-of-way. The Urban Transportation Study recommends a construction program of $145 million for freeways and $205 million for major arteries to provide a system that would be adequate until 1985. All we really need is money.

City J—Our greatest need would be the improvement of inter-departmental relationships. At times, it seems like we go around and around the same circle and never arrive at a decision. Another very important need would be a streamlined method of communicating with the Federal agencies, particularly in those areas which result in financial support. Lastly, we need some more vocal citizens speaking for the motorists. The anti-highway people manage to assemble very vocal groups, whereas with the exception of the motor club, there are very few groups who will stand up to be counted when highway plans are being fought over.

If any single point stands out, it is that urban accident experience increased substantially in the past five-year period. This would, of course, have been expected on the basis of the earlier report of national trends during the past decade. In only one city, the smallest one examined, did the registration rate fail to rise 25 percent. In several cities rates of increases were 75 percent or more.

Increases generally occurred in fatal accidents, nonfatal injury accidents, and in property damage cases, which suggests that the changes were actually in accident frequency rather than in accident severity. The data reinforce the conclusion that the urban accident problem is becoming more serious and rapidly changing in this direction.

Professional staff to cope with these problems is distressingly meager. On the average, the ten cities reported 1.6 professional traffic engineers per 100,000 population, with the highest rate being 3.5 per 100,000. Two registered rates were below 1.0 per 100,000. Traffic engineering budgets are similarly weak. On a per capita basis, the average for operating funds was $1.72. Table 22.13 indicates these relationships for each of the ten cities reporting.

TABLE 22.13
Accident Rates, Staff, and Operating Budgets in Selected Cities

Item	City									
	A	B	C	D	E	F	G	H	I	J
Population death rate										
1961	4.0*	9.7	1.5	4.7	6.7†	11.5	8.8	9.6	9.5	9.7
1966	3.7	17.9	5.9	8.9	14.8‡	25.4	15.7	13.3‡	14.1	13.3
Change, %	−8	+85	+294	+89	+121	+121	+78	+39	+49	+37
Registration death rate										
1961	0.9*	2.0	0.3	1.0	1.7†	2.5	2.0	3.2	2.4	2.7
1966	0.8	3.1	1.0	1.8	2.6‡	5.3	3.4	4.3‡	3.4	3.4
Change, %	−11	+55	+233	+80	+53	+112	+70	+34	+42	+26
Traffic engineering professional staff, 1966, No.	1	2	1	6	2	4	12	4	6	17
Per 100,000 population	1.9	2.8	1.4	3.5	1.4	3.1	2.4	0.6	0.8	1.8
TE operating budget, 1966 ($1,000's)	41	234	31	885	259	275	320	436	479	3,000
Per capita, $	0.77	3.25	0.43	5.15	1.83	2.10	0.64	0.65	0.67	3.19

*1962.
†1960.
‡Estimated.

Source: David M. Baldwin, "Accident Trends in Cities and City Traffic Engineering Staff, Budgets and Responsibilities," p. 181.

The significant conclusion in this study appears to be the relatively low levels of staff and budget, a conclusion that is further brought out by the comments of responsible traffic engineering people in each city in answer to the question: "What do you consider the greatest needs in your city in the field of traffic engineering?" The answers could probably best be summarized as money, men, and the authority to use them.

The application of safety benefit-cost ratios is discussed in Chapter 9, Traffic Accident Analysis.

TRAFFIC ENGINEERING EDUCATION

Although public officials have always been under pressure to reduce traffic accidents, it was the Highway Safety Act of 1966 that established standards for highway safety in the fifty states and their political subdivisions. Several of these standards relate to traffic engineering activities. The pressure to meet these standards will undoubtedly increase and communities that do not conform may lose available federal funds as well as render themselves vulnerable to damage suits. Technically qualified personnel are required to interpret these standards and put them into effect.

There is a shortage of both engineers and technicians in the traffic field. According to the Institute of Traffic Engineers, in addition to the 7,500 currently employed, there is a shortage of 1,900 traffic engineers in the United States. The shortage of technicians is even more critical because there is little opportunity for adequate training.

Engineers often attempt to bridge this gap with in-service training. The results are usually less than satisfactory because traffic technicians do not receive adequate training and, therefore, cannot perform as well as they should. Therefore, traffic engineers are often diverted from more useful tasks. Well-trained technicians, properly assigned, provide the traffic engineer more time for creative work. When trained technicians are used to their fullest potential, engineers are able to apply their own talents to a higher level of engineering work.

The traffic engineer is a technically trained person who works under his own direction. We can better understand the work of the traffic engineer if we first understand his task. The following is a paraphrase of the definition of civil engineer as given in the Report on Civil Engineering Technology Consultant's Workshop conducted by the American Association of Junior Colleges, May 17–20, 1967:

> The traffic engineer is a professional man who is responsible for safe, efficient, and convenient flow of traffic on our streets and highways. He must have a thorough grounding in fundamental mathematics, physical sciences, engineering design. To prepare for this responsibility, he must have at least a Bachelor of Science degree in engineering. More and more those involved in critical activities require Master's degrees in engineering. In addition to this formal education, he must, to hold many positions, meet certain state registration requirements as a registered professional engineer.[13]

The traffic engineer's work can best be understood in terms of the function of traffic engineering and its place in the highway transportation picture. The traffic engineer is responsible for developing a complete traffic system in a community, for planning and implementing long-range programs, for matters of public policy and public relations, and for the administration of the traffic engineering function.

Many people, both in and out of the profession, believe that traffic engineering is a form of human engineering instead of a part of the classical civil engineering branch of the profession. The traffic engineer has continuing contacts with people through his various manipulations of traffic movements through the application of his planning, geometric design, and traffic operations functions.

The young traffic engineer and the traffic engineering technician are particularly concerned with more repetitive tasks that involve collecting and analyzing data and preparing tentative recommendations for correcting problem locations in the roadway system. In order to accomplish these and related tasks, the following concepts and skills must be learned:

1. Communication skills (oral and written) since the traffic engineer will be dealing with people.
2. A knowledge of the driver, roadway, and vehicle characteristics and an understanding of physical laws as they relate to them.

---

[13] ADRIAN H. KOERT, *Traffic Engineering Technician Programs in the Community College* (Washington, D.C.: American Association of Junior Colleges, 1969), p. 2.

3. The ability to extract design information from manuals and apply it to specific problems.
4. A knowledge of data collection methods, tabulation, and analysis.
5. A knowledge of the operation and maintenance of traffic control devices and equipment.
6. The ability to prepare sketches and engineering drawings and to use graphics for illustrative purposes.
7. A knowledge of highway capacity analysis.
8. The basic principles of traffic and highway engineering.
9. An appreciation of the general concepts and principles of related fields—particularly urban planning and police traffic supervision.[14]

Because many traffic engineers are thrust into management activities at relatively early stages in their careers, formal and continuing education programs should provide a basis for performing the following activities:

1. General business administration (including general supervision, office procedures, policy guidance, public relations, and planning coordination with other agencies).
2. Personnel management (including direct supervision, employee rating, hiring-firing, and individual training).
3. Business finance (including contracts, specifications, purchasing, and budgeting).
4. Operations (including scheduling, program development, and maintenance).[15]

Although most traffic engineers work for governmental agencies, a large potential market for employment of traffic engineers exists with manufacturers of traffic engineering equipment and with consulting organizations. Traffic engineers are also employed in private industry. Several supermarket chains, for example, employ traffic engineers who assist with the location of new stores and with the planning and operation of traffic facilities in existing establishments.

One potential field of professional employment that has hardly been tapped is the growing number of suburban areas around central cities. These communities, whose traffic and transportation problems equal in complexity, if not in numbers, those of the central city, often operate without qualified, competent people who could solve their problems. In view of a growing shortage of traffic engineers, these communities could benefit greatly by employing traffic engineering technicians who might work under the direction of a city manager or a director of public works. Other agencies that have potential uses for traffic engineering technicians include: regional planning commissions, traffic sections of police departments, metropolitan transit departments, consulting firms, real estate and marketing firms, and public school systems.

A suggested two-year curriculum has been designed to provide a sound academic sequence of courses to prepare traffic engineering technicians. It is used only as a guide to suggest a specific program that might serve the needs of the local community. The courses listed in Tables 22.14, 22.15, and 22.16 are representative of some options for

---

[14] *Ibid.*, p. 3.
[15] "Management Training for Traffic Engineers," Institute of Traffic Engineers Informational Report, Washington, D.C., 1973 [summarized in *Traffic Engineering*, XLIII No. 4 (1973), 16–22].

TABLE 22.14
Traffic Engineering Technology Curriculum, First Year

Course Title	Class Hours	Lab Hours	Semester Credit
First Semester			
Introduction to traffic engineering	1	3	2
Engineering drawing	1	6	3
Technical mathematics I	4	0	4
Technical physics I	3	3	4
Communication skills	3	0	3
Physical education	0	2	1
	12	14	17
Second Semester			
Principles of traffic administration and safety	2	0	2
Graphics	1	6	3
Technical mathematics II	4	0	4
Technical physics II	3	3	4
Communication skills	3	0	3
Physical education	0	2	1
	13	11	17

Source: Adrian H. Koert, *Traffic Engineering Technician Programs in the Community College* (Washington, D.C.: American Association of Junior Colleges, 1969), p. 20.

TABLE 22.15
Traffic Engineering Technology Curriculum, Second Year

Course Title	Class Hours	Lab Hours	Semester Credit
Third Semester			
Field traffic surveys	3	3	4
Control devices	3	0	3
Geometric design	3	3	4
Statistics	3	0	3
Social science (government, sociology) elective	3	0	3
	15	6	17
Fourth Semester			
Traffic studies	3	3	4
Traffic laws and regulations	3	0	3
Urban transportation planning	3	3	4
Data processing	2	3	3
Social science (govt., soc.) elective	3	0	3
	14	9	17
Totals for four semesters	54	40	68

Source: Adrian H. Koert, *Traffic Engineering Technician Programs in the Community College*, p. 21.

a traffic engineering technology curriculum and should not be interpreted as a specific educational program guide. The eight courses described are only representative of some of the educational requirements that might be included in a traffic engineering technician curriculum.

TABLE 22.16
Curriculum Summary in Semester Credits

Technical Courses	Credits
Technical Skills	
*Introduction to traffic engineering	2
*Principles of traffic administration and safety	2
Engineering drawing	3
Graphics	3
Data processing	3
	13
Technical Specialties	
Field surveys	4
Control devices	3
Geometric design	4
Traffic studies	4
Traffic laws and regulations	3
Urban transportation planning	4
	22

*Might be considered technical specialty—depending on course emphasis.

Basic Science Courses	
Physical science	
Technical physics I, II	8
Mathematics	
Technical mathematics, I, II	8
Statistics	3
	19

Nontechnical Courses	
Communication skills	6
Social sciences	6
Health and physical education	2
	14
Totals	68

Source: Adrian H. Koert: *Traffic Engineering Technician Programs in the Community College*, p. 22.

COORDINATION OF TRAFFIC ENGINEERING ACTIVITIES

Because of the close relationship among the numerous traffic engineering activities involved, it is obvious that there is a real need for a high degree of horizontal and vertical coordination among all agencies concerned with traffic in a given area or region. When there is a single unit of government, the degree to which these activities can be combined into one major department is important in achieving this objective.

COORDINATION IN STATES

At the state or provincial level it is generally found that all of the engineering aspects are located in the highway department or division. Therefore, coordination is realized through the head of the department. There still remains a great need for coordination among the highway department and such agencies as the state enforcement and driver licensing departments. In some states coordination currently is being achieved through the formation of State Departments of Transportation. In many states coordination is achieved through an official state safety commission or coordinating committee. In other states it is on a personal and voluntary basis.

COORDINATION IN CITIES

Efforts to provide coordination in cities have taken on a number of forms. Basically, they are the following:

1. Officials of the government who individually have responsibility for some aspect of the overall program.
2. Limited to citizen and business leaders.
3. A combination of officials and citizen leaders.

Usually, such committees or commissions have advisory powers only. Basically, they serve to coordinate traffic engineering activities and to assist in safety education and law enforcement.

MODEL TRAFFIC ORDINANCE RECOMMENDATION

The *Model Traffic Ordinance*[16] recommends a commission for the purposes of coordination as follows:

§ 2-12—**Traffic commission—powers and duties**

(a) There is hereby established a traffic commission to serve without compensation, consisting of the city traffic engineer, the chief of police or in his discretion as his representative the chief of the traffic division, the chairman of the city council traffic committee, and one representative each from the city engineer's office, and the city attorney's office and such number of other city officers and representatives of unofficial bodies as may be determined and appointed by the mayor. The chairman of the commission shall be appointed by the mayor and may be removed by him.

(b) It shall be the duty of the traffic commission, and to this end it shall have the authority within the limits of the funds at its disposal, to coordinate traffic activities (to carry on educational activities in traffic matters), to supervise the preparation and publication of traffic reports, to receive complaints having to do with traffic matters, and to recommend to the legislative body of this city and to the city traffic engineer, the chief of the traffic division, and other city officials ways and means for improving traffic conditions and the administration and enforcement of traffic regulations.[5]

---

[5] There are two types of official traffic commissions, each of which has been found effective under certain conditions. The first type consists of a small number of city officials directly concerned with traffic administration, serving ex officio, with perhaps the addition of one or two citizen members. The principal function of this commission is to coordinate official traffic activities of the several departments of the city administration. The safety educational activities in the community are then conducted or coordinated by an unofficial organization such as a safety council, or a safety committee of the chamber of commerce, motor club, or similar organization.

The other type of traffic commission (sometimes called the safety commission or the traffic safety commission) is considerably larger in size, including other public officials and a number of citizen members in addition to the officials mentioned above. Such a commission and its subcommittees not only perform the functions mentioned above, but also carry on a comprehensive program of public safety education. If this type of commission is desired then there should be retained in § 2-12(b) the part reading, "(to carry on educational activities in traffic matters,)" but if the first type of commission is desired the statement with respect to educational activities should be omitted.

---

[16] *Model Traffic Ordinance* (1968), pp. 5–6.

The type of traffic commission most effective in any particular community will depend on the local conditions. Before organizing any such commission, advice should be had from one of the national organizations in this field.

In the event the second type of commission is desired it is suggested that it might properly include the following personnel in addition to those official representatives mentioned in § 2-12:

(1) The judicial official who handles most of the traffic cases.
(2) A representative of the board of education.
(3) A representative of the city planning commission.
(4) A representative of the fire department.
(5) A representative of the public utilities regulatory body, if any.
(6) A number of citizens vitally interested, including the following:
   (a) Representatives of the mass-transportation companies.
   (b) One or more representatives of business organizations.
   (c) Representatives of civic and professional groups such as the automobile club, engineers club, local safety council, chamber of commerce and junior chamber, and the parent-teachers association.
   (d) A representative of trucking interests.
   (e) A representative of taxicab companies.
   (f) A representative of automobile insurance companies.
   (g) One or two newspaper editors.

URBAN PLANNER-TRAFFIC ENGINEER COOPERATION

The close cooperation of professional groups is indispensable in any urban transportation program. In fact, it is required in all metropolitan areas of over 50,000 population in the United States as part of the continuing comprehensive and cooperative urban transportation planning process which is discussed in Chapter 12, Urban Transportation Planning. Recognizing that the individuals may have legal responsibilities to carry out certain specific assignments, there is a real opportunity for cooperative action in four major areas: data collection, data analysis, program development, and program implementation.

THE ENFORCEMENT FUNCTION[17]

Administrators of both urban and rural transportation programs seek to accomplish three things: (1) the creation of the best possible system of physical facilities within available resources, (2) the attainment of the highest and most efficient use of the facilities, and (3) the achievement of maximum public compliance with the regulations imposed. The first objective is attained through the functions of planning, design, construction, and maintenance, the second objective is achieved through street operations, and the third objective is realized through the enforcement work of the police and the courts.

Traffic regulations are designed to equitably distribute the time or space use of transportation facilities among the largest possible numbers of users. Unless these regulations are complied with, they are of no value at all. The enforcement function is therefore a vital part of urban and rural transportation programs.

Enforcement effort in the field of traffic regulations is customarily concentrated at selected locations and times and against particular violations. This policy is calculated to produce the most effective results for the enforcement effort available. Some, however, claim that the policy constitutes an abdication of responsibility with respect to the unemphasized places, times, and violations.

Because police officers must exercise much discretion in the enforcement of traffic regulations, they must be broadly trained. Individual judgment and intelligence are

---

[17] SEBURN AND MARSH, "Urban Transportation Administration," pp. 74–92.

in high demand. Therefore, governments must continually seek more highly qualified recruits for police work. As a result, higher salaries must be offered and more training provided, which, in turn, will increase the overall cost of enforcing traffic regulations.

The overall quality of police administration at the local level is steadily improving. There is an increasing tendency for cities to make use of trained police administrators instead of elevating members of the force with the longest service records to positions of command. Merit systems, competitive promotion procedures, and equitable pension plans are taking the place of the old system in most police departments.

Separate traffic divisions are commonplace in most police departments. This kind of organization is recognized as good practice, since the nature of the function differs markedly from other police activities. There also seems to be a move toward the development of judicious police specialization among the staff and service units. Women are being used increasingly for parking enforcement, crossing guard duty, communications, and record keeping.

The activities carried out by the traffic divisions within the police department differ from city to city, but they generally include three basic duties: (1) traffic direction to aid traffic flow, (2) police traffic accident investigation, and (3) apprehension and warning of violators. These duties do not lend themselves to division by functional unit. It is not feasible to have a formal unit for accident investigation, another for traffic direction, and a third for law enforcement. Since accidents do not occur at predictable times and places, the activities of direction and enforcement cannot be separated. Any officer might be called upon at almost any time to perform any one of the three functions. At an accident site he might investigate, direct traffic, and determine violations almost simultaneously.

It is important that the police agency, the public prosecutor, and the traffic court judge have common objectives in the enforcement process. Each plays an important part in this program. Good police work is too often neutralized by weak prosecution or lack of appropriate follow-through by the courts. Administrative coordination is extremely important.

The work of the traffic engineer can obviously affect the manpower distribution within the police department. Sound traffic control installations may reduce the need for intersection officers and make manpower available for patrol or other duty. When there are fewer citizen complaints and when there are fewer accidents, demands on police time are also reduced.

There appears to be a high degree of voluntary cooperation between traffic engineering and police traffic enforcement units. These men generally seem to work well together. Obviously, it is in the best interest of each unit to cooperate because there is little basis for competition or professional jealousy. Police personnel realize more than do most people that streets have physical limitations beyond which enforcement pressure cannot increase the overall traffic carrying capacity. On one hand, the traffic engineer can be the policeman's greatest ally in efforts to make traffic flow more easily. On the other hand, the traffic engineer is aided in his objectives by good police work. Without good enforcement traffic controls are ineffective. The traffic engineer is dependent on police work for traffic accident data upon which he may base corrective action. Faster flow, more capacity, and fewer accidents are results that reflect credit upon both agencies. There are fewer opportunities for professional disagreement or difference in objectives, as may be the case between traffic engineering and highway design or between traffic engineering and street maintenance.

Coordination between the traffic engineering and police departments can be enhanced when top supervisory staff members periodically meet to discuss mutual problems and jointly observe problem locations.

There is a serious need for police assistance in the street maintenance and construction forces. Traffic supervision is needed in the vicinity of work projects. Parking elimination is essential to the street cleaning or snow removal operations. Good police work is also needed in setting up and supervising detours.

The best possible system of transportation facilities can only be obtained through planning, design, construction, and effective use of the street operations system. Nevertheless, the transportation program may still be ineffective if public compliance with traffic regulations is not secured. The traffic enforcement function is partially dependent on the success of several groups outside government. These include safety councils, school organizations, and other agencies and volunteer groups concerned with driver education and the courts.

Since failure in any one function may react against the total enforcement program, there is a need for close cooperation between the police and these groups. The court system sometimes proves to be the weakest link in the traffic enforcement program through its mixing of traffic and criminal cases in the same court. The same is true of the practice of several courts having jurisdiction over traffic violation cases in the same area. It is highly desirable for the violations bureau to be administered by the court and for steps to be provided to keep records of habitual offenders.

RESEARCH ACTIVITIES

A vital element in the program of the traffic engineering administration is in the field of research involving traffic characteristics and behavior. This can be divided into two general categories, namely applied research and fundamental research.

APPLIED RESEARCH

Applied research is the application of proven techniques to the daily problems in the transportation field. It is probably best represented by the "before and after" studies. These studies are useful in determining the effects of various traffic control devices and measures upon the capacity and safety of the street system. Typical studied include:

1. Accident records and analysis
2. Driver and pedestrian observance
3. Cordon counts
4. Overall route speeds
5. Parking studies

A more complete listing of individual studies and methods of conducting these studies can be found in Chapter 10, Traffic Studies.

FUNDAMENTAL RESEARCH

Fundamental research is the basic research of the various characteristics or elements of the traffic stream. Typical of these characteristics are the following:

1. Traffic patterns (hourly, daily, weekly, etc.)
2. Arrivals and departures at terminals
3. Vehicle operating characteristics
4. Pedestrian travel characteristics
5. Parking characteristics
6. Driver psychology

Typical national research agencies are the Highway Research Board (U.S.) and the Transport and Road Research Laboratory (U.K.). The general purpose and scope of these organizations and their committees are to suggest, encourage, correlate, and evaluate research within the special fields of interest.

Applied research is mostly conducted in cities and states under the direction of the responsible public officials. However, the daily pressures of public duties usually prevent the traffic engineer from participating to any considerable extent in fundamental research. This work is normally carried on in universities, colleges, and by special research agencies and consulting firms. A great deal of this work is financed by grants made available from the federal government. In addition, much valuable research has also been conducted by the U.S. DOT Federal Highway Administration and National Highway Traffic Safety Administration.

REQUIRED LEGISLATIVE ACTION

Before taking action on the recommended traffic engineering improvement program, it is customary for the appropriate legislative bodies to hold public hearings. The frequency, length, and scheduling of these hearings will be dictated by legal requirements, the character and scope of the project, and any subsequent recommendations (also see Chapter 14, Geometric Design).

Obviously, there is a need for giving everyone an opportunity to present his views. Not all information presented will be clearly understood by everyone. Therefore, as much pertinent data as possible on the traffic engineering improvement program should be made available at the public hearing.

After the public hearings are completed, the legislative body or a committee of that group may desire to meet with members of the technical staff in order to review the proceedings. These meetings may result in modifications of the recommended program as indicated by the public opinion expressed at the hearings. It is absolutely essential that when the appropriate legislative bodies approve and adopt the specific recommendations they must also appropriate sufficient funds to implement the traffic engineering improvement program.

If special legal procedures are necessary to provide the necessary funds, these procedures should be initiated as soon as possible. After the report is adopted, any legal action necessary should be taken as soon as practical in order to implement the recommendations. The program should be started while the facts are still fresh in the public's mind and completed at the earliest possible date.

PUBLIC RELATIONS

Traffic engineering public relations takes many forms. Various media contacts, complaint handling, and published reports are only three of the means available to the traffic engineer to accomplish this communication with the public.

PUBLIC INFORMATION MEDIA[18]

The general public is intimately acquainted with the various traffic and environmental problems associated with the motor vehicle. Consequently, most citizens are vitally interested in any programs or activities that are intended to expedite traffic movement. This interest is shared by various public information media including newspapers, magazines, radio, and television. In many instances the press can be the backbone of a traffic engineer's public information efforts, for it is through the press that the most lengthy and detailed coverage of proposed traffic improvement programs can be presented.

The big challenge lies in satisfying the specialized needs of each of the different media. Advance news releases and appropriate illustrative materials are essential for the press. Appropriately animated, photographic information and supplementary descriptive material are needed for television. Radio scripts must be written that will present verbal pictures.

In each case, it is essential that the information be developed with the public in mind. The information must be brief and interesting as well as accurate and capable of being substantiated, if necessary. Representatives of the different information media should be part of the development of a traffic improvement program from its very inception. In fact, they may even help set the stage for the development of a long-range traffic plan.

Professional personnel, governmental officials, and citizens associated in some way with the various aspects of the program development should be willing to participate in interviews, panel discussions, etc. If a person is contacted and he is unable to participate, he should make every effort to secure a responsible substitute.

In dealing with different representatives of the various media, those who are responsible for the development and implementation of the traffic engineering program must be careful not to show favoritism toward one person or group. For example, if there are both morning and afternoon papers, every effort should be made to equally distribute the various news breaks between the two publication deadlines. Whenever possible, the media representatives should be alerted to potential stories, for example, the operation of a new one-way street system or special bus lane.

The development of a new, intriguing technique is always newsworthy. Human interest stories that are always occurring provide a local interest approach. If there is any possibility of being misquoted or misunderstood, it is highly desirable to have prepared releases ready for distribution. If illustrative material is distributed, it should be of such quality that it can readily be reproduced.

Personnel responsible for the development of traffic improvement programs must feel a personal responsibility for keeping the public informed on the progress of their efforts. Therefore, they should make personal contacts with the media representatives and be willing to spend as much time as possible answering what may seem to be repetitious or inconsequential questions. The better the reporter or the announcer understands what the traffic improvement program is, the better he can describe it to the general public.

It should be recognized that sometimes not all the technical information can be released without prior approval of the various cooperating agencies. This kind of in-

---

[18] JOHN E. BAERWALD, unpublished lecture notes, University of Illinois (Urbana, 1971).

formation should be kept to a minimum and be released as quickly as possible. In the meantime, all other information of interest to the public should be made available. Every effort should be made to present the traffic program in nontechnical language. The various benefits that the public will enjoy when the program is implemented should be described, for example, decreased travel time, reductions in deaths and injuries, increased economic benefits, reductions in operating costs, and increased comfort and convenience. Viable alternative solutions should be thoroughly described. After the final report on a major program or project has been presented, some newspapers may wish to prepare feature articles on certain phases of the program. Radio and television stations may want to present panel discussions or special programs dealing with the traffic engineering projects.

It is improbable that all of the various public information media will be equally receptive to all of the conclusions and recommendations. Similarly, their coverage of those who oppose the improvement will vary. In either case, a clear, concise, accurate, and comprehensive report will minimize the opportunities for unjustified criticism.

Another valuable vehicle used to inform the public about the development of traffic improvement programs is the presentation to civic service organizations. Most of these organizations are always eager to have qualified speakers to discuss local problems and plans for the future. Appropriate use of visual-aid material will be very helpful in these presentations as well as in answering questions.

The more the public knows and understands the purposes for a proposed traffic engineering program, the more willing it will be to accept these recommendations.

HANDLING COMPLAINTS

In a left-handed way, a complainer really is offering a compliment. He assumes that it is within your power to correct or improve things. If he thought that it was beyond your capacity to correct the problem, he would not waste his or your time by bringing it to your attention.

Effective means for handling complaints include:

1. Do not react personally, even if the complaint takes the form of a personal attack.
2. Whether the matter seems trivial or ridiculous, remember that it is important to the complainer.
3. It is not only courteous but also good business to be genuinely interested in the problem.
4. Give the complainer an opportunity to blow off steam.
5. Show that you understand the problem.
6. Do not make excuses and do not argue.
7. Do not be drawn into a discussion of past history.
8. Deal with the future, not the past.
9. If the complaint is valid, see that the problem is corrected as quickly as possible.
10. If time is needed, explain that the problem is too important to be handled by a snap judgment. Hopefully, this will satisfy the complainer's ego and extend his patience.

THE PUBLISHED REPORT[19]

The initiation, planning, study, and analysis phases of any traffic study are directed toward a single goal. The objective is the development of an orderly, factual, and realistic plan for any transportation improvement. The purpose of the report is to furnish either a direct or indirect opportunity for administrative action. Often, this will include official approval and the provision of funds necessary to implement the improvement.

If the report is made part of an informal presentation, it may be presented either in whole or in part by using a summary format. Periodic reports may be made from time to time as the work progresses. These reports need not be published, but they can be presented orally, often with the aid of illustrations or other visual aids. Periodic reports may be summarized in printed form in order to indicate the progress of the work.

The final report should obviously be cleared with all cooperating agencies. This may be handled by having a special preview for key officials. Similarly, a confidential orientation of public information agencies may also be an important part of the acceptability of the document. Obviously, a qualified representative fully familiar with the contents of the report should make the presentation. In addition to having the authority to represent the organization, the representative must *show authority* in the presentation. The four following steps, if used, will give him confidence:

1. Determine the specific audience and the real purpose of the meeting.
2. Analyze the conferees, their business or profession, their attitude toward the conferrer, and their current or potential interest in the subject of the meeting.
3. Prepare the material well.
4. Prepare the techniques or strategies for presenting the material and managing the conference.

Speaking ability is essential. Glibness and rhetoric do not constitute effective presentation, for audiences are quick to sense the intrinsic values in personality and character. The speaker must have something to say and discussion of previously expressed thoughts of others do not carry the conviction and sincerity required for acceptance. He must know how to present the report, he must know basic speech construction, and he must have the vocal skills necessary to express his ideas effectively. He should have personal knowledge of the problem and should be thoroughly familiar with the study. He should also have complete knowledge of the recommendations and how they relate to local conditions.

The following considerations are most important in the preparation of any report:

1. Statement of the specific purpose of the report
2. Gathering of the appropriate material
3. Outlining the report

Important considerations in the actual presentation of the report include the following:

---

[19] *Ibid.*

1. Use the elements of effective delivery necessary in any public speech.
2. Recognize that the report is being presented to an audience that has more than average interest in the contents. The speaker need not sacrifice clarity for the element of interest.
3. Adapt his vocabulary and style to that of the technical-layman capacity of his audience.
4. Valuable audio and visual aids include the following:
   a. illustrations
   b. examples
   c. charts
   d. demonstrations
   e. special devices

A speaker's manual or a kit of basic script and visual-aid materials should be prepared if numerous presentations are to be made by one or more individuals. This will not only save time, but it will also ensure compatibility of the individual presentations.

Because there usually is a large "anti" group and because it is usually the most vocal segment of the audience, the speaker should be prepared to intelligently answer the following questions:

1. When will the proposed action start?
2. How will right of way be selected, acquired, and paid for?
3. What will happen to displaced persons and businesses?
4. When will my street be improved?
5. Will taxes be increased? If so, how much and when?
6. Why does all of the traffic have to pass in front of my property?
7. Can you reroute the traffic so it will pass my business?
8. How will the proposed location affect proposed schools, parks, etc.?
9. Why can't we just ban all cars and walk like we did in the good old days?

Members of appropriate legislative bodies will find it worthwhile to attend public meetings because they provide an opportunity to hear firsthand the questions raised. Although officials may note the amount of public interest and the nature of the opposition, they must remember, however, that those who oppose a program are more inclined to attend these meetings and are also more inclined to speak up. Therefore, a show of hands or the trend of the discussion may not fully represent public interest or support.

GENERAL COMMENTS

An intensive study to determine how traffic decisions are made in cities that employ full-time traffic engineers was conducted in 17 U.S. cities ranging in population from 50,000 to 750,000. The following inferences were drawn from the study:

1. In cities where the elected representatives attempted to conduct independent evaluations of the technical aspects of traffic problems, the performance of the traffic engineering function was seriously impaired.

2. Successful traffic engineers were aware of the political complexities of their programs and adjusted their recommendations to account for these factors without demeaning their professional integrity.
3. In cities where the traffic engineer did not have direct access to the city's policy-making body, the success of the traffic engineering program was limited.
4. Traffic engineers who recognized the importance of cooperating with business groups, traffic or safety commissions, community leaders, and other city officials to develop their programs had a better record of implementing improvements than those engineers who attempted to have their programs evaluated solely on the basis of their technical merits.
5. Traffic engineers who utilized modern management and administrative techniques operated their departments in an efficient manner and were generally able to cope with the variety of decision-making situations encountered.
6. In communities where the traffic engineer recognized the importance of establishing good public relations and maintaining a positive and helpful image with the public, the information media, and the elected officials, the traffic engineers were generally successful in implementing desired projects.[20]

Successful administration of a traffic engineering program is a full-time responsibility. It can only be accomplished by hard hitting continuous attention to all elements of the program. Nothing should be taken for granted and no stone left unturned in order to maintain a strong traffic engineering administration within the limits of manpower, money, and other resources.

REFERENCES FOR FURTHER READING

BALDWIN, DAVID M., "Accident Trends in Cities and City Traffic Engineering Staff, Budgets and Responsibilities," *Improved Street Utilization Through Traffic Engineering*, Highway Research Board Special Report No. 93. Washington, D.C.: Highway Research Board, 1967.

KOERT, ADRIAN H., *Traffic Engineering Technician Programs in the Community College*. Washington, D.C.: American Association of Junior Colleges, 1969.

LIPINSKI, MARTIN E., *A Model of the Traffic Engineering Decision-Making Process*, Highway Traffic Safety Center Research Report No. 6. Urbana, Ill.: University of Illinois, 1973.

Model Traffic Ordinance, National Committee on Uniform Traffic Laws and Ordinances. Washington, D.C.: The Michie Company, Charlottesville, Va., 1968; Supplement I, 1972.

SEBURN, THOMAS J. and BERNARD L. MARSH, *Urban Transportation Administration*. New Haven: Yale University, Bureau of Highway Traffic, 1959.

"Traffic Engineering Functions in Cities of 50,000 to 200,000 Population," *Traffic Engineering*, XXXIII, No. 10 (1963).

Uniform Vehicle Code, National Committee on Uniform Traffic Laws and Ordinances. Charlottesville, Va.: The Michie Company, 1968; Supplement I, 1972.

WEBB, GEORGE W., "Organization and Administration of State Traffic Department," *Traffic Engineering*, Vol. 35, No. 5 (1965).

---

[20] MARTIN E. LIPINSKI, *A Model of the Traffic Engineering Decision-Making Process*, Highway Traffic Safety Center Research Report No. 6 (Urbana: University of Illinois, 1973), pp. 160–61.

Chapter 23

APPLICATION OF SYSTEM CONCEPTS

ROBERT H. WORTMAN. Associate Professor of Civil Engineering, Civil Engineering Department, University of Connecticut, Storrs, Connecticut.

The purpose of this chapter is to provide the traffic engineer with an introduction to the philosophies, concepts, and analyses that are pertinent to dealing with a systems problem. It is a general overview of the systems approach and an outline of the systems problem. Sections in the chapter are devoted to discussions of:

1. The definition of goals, objectives, and criteria
2. Modeling concepts
3. The application of economic analysis

Finally, several examples are presented to illustrate the application of systems concepts to traffic and transportation problems.

Traffic engineers must cope with a variety of problems that vary in both scope and complexity. Some of these problems may be simple, and the solutions are readily apparent. For other problems, which are more formidable and complex, the best solution is not readily apparent; and questions may be raised on how to deal with these problems. Engineers also are becoming more aware of the complexity of traffic problems and the limitations of the traditional approaches and solutions to these problems.

The application of systems concepts can yield a better understanding of traffic and transportation problems. This understanding will give direction to rational solutions to these problems and provide the traffic engineer with insights in dealing with the complex problems that he must face.

There is a vast inventory of numerical methods and techniques that may be applied in the course of analyzing a system. These include mathematical methods and models such as statistical analysis, optimization, network analysis, queuing models, mathematical programming, and other operations research techniques. It is not possible to include all of this information here, but a list of references for further reading is included at the end of this chapter. This list contains publications that provide a comprehensive documentation of these mathematical methods and techniques. It shall be recognized that these quantitative methods represent potential tools that may be utilized in the analysis of a system. The key to the use of these tools is predicated on an adequate understanding and formulation of the systems problem. The emphasis in this chapter, therefore, is on the actual application of the systems approach.

THE NATURE OF TRAFFIC PROBLEMS

The traditional approach to a traffic and transportation problem has been to view it as a problem in itself, and alternate solutions have been considered in this context. Certainly, this approach has been quite successful as is evident from the increases in capability to move persons and goods. It has been realized, however, that increased mobility has not necessarily eliminated all of the problems associated with traffic and transportation. This same analogy can be applied to all traffic problems with the resulting conclusion that there are many internal and external influences that must be considered.

Transportation is generally considered a service because it provides the mobility that is required for society. In this respect, transportation problems are created by the need to move persons and goods in response to these demands. This service is provided by a number of different modes of transport which serve as integral parts of an overall transportation system. Although each mode of transport serves as a component of the transportation system, the modes also represent individual operating systems that contain a number of interacting components. For example, the highway mode can be described as having vehicle, roadway, and terminal components. Furthermore, each of these components consists of a number of subcomponents. This examination reveals that transportation is composed of and defined as a hierarchy of systems.

It may be recognized that there are interactions occurring between the components at each level which can affect the operation of other components. For example, the improvement or modification of one form of transportation may affect the use of and demand for another mode; and the change in the design of a vehicle may affect the roadway requirements needed. Also, policies or modifications that are made at any system level may have effects on other levels in the hierarchy.

Transportation, therefore, can be viewed as an overall operating system that contains a number of subsystems, or smaller systems which all work ultimately to provide for the movement of persons and goods. A system is described as a group of interacting components or elements that function to accomplish some task or fulfill some specific purpose. In dealing with traffic problems, the traffic engineer must determine the level at which a problem occurs in the hierarchy of systems. Having made this determination, he must ascertain the specific function of the system involved and the requirements placed on the components of that system. In viewing the problem in this manner, the traffic engineer is able to appraise the situation as it actually exists in the physical and organizational framework.

Traffic and transportation problems are complex because of the interactive nature of the systems and the components involved. When presented with a problem, the engineer must identify the environment in which it occurs and the elements that are pertinent in seeking and evaluating a solution. An approach is required that will enable the engineer to address the problem in a comprehensive manner. This approach should result in a comprehensive understanding of the problem whereby the underlying cause is ascertained, alternative solutions are identified, and the consequences of the various solutions are determined.

A comprehensive understanding of a problem enhances the position of the engineer when he considers the decisions that must be made in selecting a best solution. Furthermore, the engineer is frequently in a position to make recommendations to decision-making bodies or a team of professionals who are concentrating on a common problem. In either case, the comprehensive view of the problem enables the development of a better appreciation of the problems of the decision makers and the concerns of other disciplines as well as permitting an examination of the full range of alternate solutions that may be considered.

THE SYSTEMS APPROACH

The systems approach represents a broad-based, systematic approach to problems that involve a system or systems. In essence, it is a problem-solving philosophy that may be used to formulate or structure, analyze, and solve systems oriented problems. The application of such an approach is particularly useful when the problem is complex and involves many aspects or elements that must be considered. Obviously, the usefulness of the approach increases with the complexity of the problem because the rational solution or solutions become less apparent.

The initial emphasis in the systems approach is focused on the definition of the problem. This definition should include the establishment of the system or systems that are involved, the purpose or function of the systems, the components or elements in the systems, and the interactions that occur among the components or elements. In this respect, the systems approach differs from other approaches to problem solving because the problem definition includes the formulation of the entire structure of the problem. In addition, the system that is involved is examined with respect to higher- or lower-order systems. An attempt is made to view the problem in its total environment rather than merely to consider it in an isolated or fragmented context. In this way, a better understanding results because the broad scope and nature of the problem are exposed and give greater assurance that the analysis will consider these comprehensive aspects.

The use of the systems approach certainly does not represent a panacea for solving traffic and transportation problems. The use of the approach in itself will not ensure that reasonable, rational, or even better solutions will emerge. In fact, there are a number of pitfalls to be avoided. Quade[1] lists the following as common sources of error:

1. Underemphasis on problem formulation
2. Inflexibility in the face of evidence
3. Adherence to cherished beliefs
4. Parochialism
5. Communication failure
6. Overconcentration on the model
7. Excessive attention to detail
8. Neglect of the question
9. Incorrect use of the model
10. Disregard of the limitations
11. Concentration on statistical uncertainty
12. Inattention to uncertainties
13. Use of side issues as criteria
14. Substitution of a model for the decision maker
15. Neglect of the subjective elements
16. Failure to reappraise the work

---

[1] E. S. QUADE, "Pitfalls and Limitations," *Systems Analysis and Policy Planning* (New York: American Elsevier Publishing Company, Inc., 1968) pp. 345–63.

In addition, there are several major disadvantages or limitations to the approach that must be recognized. First, it is unlikely that the total problem can be investigated because of resource constraints on the study; thus, a true comprehensive solution can never be achieved. Second, many relationships will often be revealed which may not be understood. The lack of knowledge about the behavior of these interactions severely restricts their consideration and hampers the engineer in predicting the consequences of various solutions that may be applied. In this respect, it is difficult to measure the overall effectiveness of the system. Finally, the problem may involve aspects or elements of a system that lie outside the control of the engineer. For example, there may be interactions with organizations or other disciplines over which the engineer has no decision-making authority. This may pose further restrictions on the possible solutions that may be considered.

In spite of these pitfalls and disadvantages, there are many positive views that encourage the use of the more comprehensive approach. The major advantage is that the full problem may be examined and the total spectrum of variables considered instead of its being examined in an isolated context. More specifically, de Neufville and Stafford[2] list the following functional characteristics of systems analysis that emphasize its advantages:

1. It sharpens the designer's awareness of his objectives by forcing him to make explicit statements about what they are and how they are to be measured.
2. It seeks mechanisms for predicting the future demands on a system, which often are not observable in advance but must be determined from an interaction of social and economic factors.
3. It establishes procedures for generating a large number of possible solutions and for determining efficient methods to search through them.
4. It assembles optimization techniques that can pick out favorable alternatives.
5. It suggests strategies of decision making that can be used to select among possible alternatives.

SYSTEMS ANALYSIS

Systems analysis is the basic problem-solving framework utilized in achieving a solution to a systems problem. It includes the entire process by which:

1. The systems problem is defined and formulated
2. The alternate solutions are posed and evaluated
3. A choice of the best solution is made

A superficial perusal of systems analysis would indicate that it probably does not differ greatly from other problem-solving techniques, but a closer examination would reveal that major differences do exist. These differences are primarily associated with the comprehensive nature of the approach and the emphasis on the system and the consequences of various alternatives on the system and its environment.

There may be some misconception that systems analysis is restricted to problems that can be solved in a strict numerical sense. Certainly, the quantification of relation-

---

[2] RICHARD DE NEUFVILLE AND JOSEPH N. STAFFORD, *Systems Analysis for Engineers and Managers* (New York: McGraw-Hill Book Company, 1971).

ships permits the use of a variety of numerical analysis techniques as aids in the solution of a problem. It is likely, however, that the engineer will encounter relationships between components in a system in which the behavior is unknown or cannot be quantified at the present time. The fact that such situations arise is no basis for discarding the use of systems concepts. Here the engineer must decide how the relationship will be treated in the analysis of the problem. Either he must disregard the interaction or consider it in subjective terms if quantification cannot be achieved.

Systems analysis can be segmented into a number of specific tasks or steps which must be undertaken in the analysis and solution of a problem. Thomas and Schofer[3] indicated these steps to be as follows:

1. Develop clear problem statement
2. Model the basic system
3. Examine and allocate study resources
4. Mathematize and calibrate the model
5. Establish criteria for design and evaluation
6. Test and evaluate the given system
7. Design alternate solutions
8. Test and evaluate the system and alternatives

Figure 23.1 depicts the systems analysis process and schematically presents these steps and their place in the process. It should be noted that throughout the process there is a feedback of information to earlier steps that provides for necessary revisions

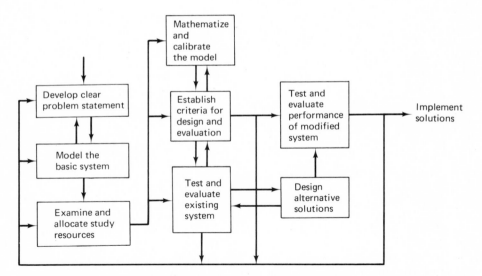

Figure 23.1. The systems analysis process. [Source: Edwin H. Thomas and Joseph L. Schofer, *Strategies for the Evaluation of Alternative Transportation Plans*, NCHRP Report 96 (Washington, D.C.: Highway Research Board, 1970), p. 7.]

---

[3] EDWIN H. THOMAS AND JOSEPH L. SCHOFER, *Strategies for the Evaluation of Alternative Transportation Plans*, National Cooperative Highway Research Program Report 96 (Washington, D.C.: Highway Research Board, 1970).

in the analysis as new information or phenomena concerning the problem are revealed or determined. In this respect, the process is iterative because the engineer may wish to reconsider his conception of the description of the problem and the model of the system as the analysis progresses.

PROBLEM DEFINITION

As previously indicated, a major emphasis in the systems approach and the first step in the analysis process involve a comprehensive definition and the development of a clear problem statement. Although it is a quite obvious initial step in addressing any problem, the attainment of a clear definition has been a frequently neglected and overlooked task. In dealing with any complex problem, such as those associated with traffic and transportation, there is a tendency to immediately pose alternate solutions without first establishing the exact nature and scope of the problem. Careful thought and an adequate understanding of a problem in this early stage of the analysis are mandatory because the manner in which a problem is formulated gives guidance and direction to its solution and dictates the nature of the results that are achieved.

The engineer should distinguish between the cause of a problem and the symptoms or phenomena that are the results of an underlying cause. Frequently, solutions are directed to apparent symptoms of conflict or difficulties and the basic cause is overlooked. For example, traffic congestion at an intersection might be considered a symptom of a problem instead of a problem. In this case, congestion is a phenomenon that is seen and can be measured. The cause of the congestion might be associated with inadequate zoning controls, economic activity in the area, lack of alternate forms of transportation, inadequate road network design, etc. The traffic engineer should definitely recognize that the problem can be much broader than the traffic demand being greater than the intersection capacity. Of course, one solution is to increase the capacity of the intersection, but there may be other solutions that should be given consideration. Although the cause of the problem may be associated with an area or activity outside the control of the engineer, he has a professional responsibility to recognize and consider the best possible solution. In defining the problem, he should not be constrained or limited by tradition.

SYSTEM MODEL

Once the problem has been initially defined and a statement of the problem developed, the next step in the analysis process is to model, or depict, the system that is involved. At this point in the analysis, the model is a conceptual depiction of the system and the environment in which it exists, the components in the system, and the interactions between the components. In this description of the system, the emphasis is placed on the recognition and relation of the system and its components; at this time a quantification of the components' interactions need not be quantified. The system model should be as close a representation of the actual problem situation as is possible.

It is common practice to simplify a problem in order to obtain a solution; however, if some pertinent aspects of the system are ignored, the validity of the solution is questionable. A simplification of the system may be necessary in later stages in the analysis because it may not be possible to cope with complexities of the full system. The engineer should not undertake this simplification without full knowledge of how it affects the reality of his statement of the problem.

Further information on the development of the problem statement and model may be found in the discussion of the subject by Meredith, *et al.*[4]

Actually, the steps involving the development of the problem statement and modeling the basic system are closely related. As the system is conceptualized, a reconsideration of the problem may be necessary. This will most likely result in a clearer statement of the problem.

Figures 23.2 and 23.3 show conceptual models for two different traffic problems. The first is a representation of the traffic components involved in a highway safety problem and the second models the terminal as a component of the transportation system. A review of each of these figures indicates which factors are included in each problem and component interactions that occur. Further illustrations of system models are presented with the examples at the end of this chapter.

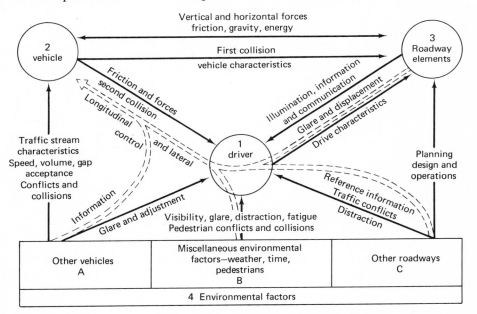

Figure 23.2. A conceptual model of traffic system components. [Source: Charles J. Keese and Louis J. Horn, "Efficiency Produces Safety for the Nation's Highways," *Texas Transportation Researcher*, VI, No. 2 (1970), p. 4.]

STUDY RESOURCE ALLOCATION

One of the distinct hazards and frustrations of undertaking a comprehensive examination of a problem is that a large number of aspects or facets are revealed that require investigation and analysis if a totally rational solution is to be achieved. Obviously, in utilizing a comprehensive approach, the time required for the complete study increases greatly with an increase in problem complexity. Nevertheless, the engineer must obtain a solution usually within a given time or under financial constraint. An examination of the available resources must be made and allocated to analysis

---

[4] D. D. MEREDITH, K. W. WONG, R. W. WOODHEAD, AND R. H. WORTMAN, *Design and Planning of Engineering Systems* (Englewood Cliffs, N.J.: Prentice-Hall, Inc., 1973).

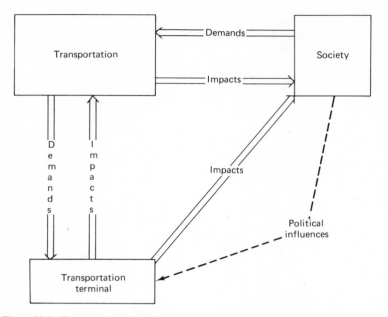

Figure 23.3. Conceptual model of the relation of the transportation terminal to the transportation system and society. (Source: Robert H. Wortman, "A Study of Terminals as a Component of the Transportation System," unpublished Ph.D. dissertation, University of Illinois, Urbana, 1970, p. 91.)

effort. At this step in the process, the engineer must decide on the scope and detail of the system that is to be studied, the degree that effort can be expended on further modeling of the system, and the number of alternatives that will be considered. A commitment must be made on the scope and depth of the study. This allocation of resources is particularly important because it guides the level of effort put into the study of the system. An attempt is made to balance the problem studied and the available resources.

ANALYSIS AND EVALUATION

Several interrelated steps may now be undertaken in the systems analysis process. These steps are generally associated with the development of specific analytical methods, techniques, or models that are to be used in the analysis and evaluation of the system. The following steps (shown in Figure 23.1) are included:

1. Mathematize and calibrate the model
2. Establish criteria for design and evaluation
3. Test and evaluate the existing system

Whenever possible, the interactions and relationships depicted in the conceptual model must now be quantified and mathematical models must be developed to represent these interactions. As indicated previously, the quantification of the relationships may not always be possible in purely objective terms. In such cases, subjective methods will have to be considered. The behavior of the interaction must be described for each interaction between components. In the course of the analysis it is likely that some relationships between components will be encountered for which there is insufficient knowledge of the interaction behavior. This places a constraint on the analysis, and the engineer must decide how situations are to be treated. For example, traffic engineers frequently must deal with problems that involve drivers and their response to roadway and traffic conditions. In many cases, the modeling of these situations is either limited or impossible because current knowledge does not permit a full understanding of the behavior at this time. The analysis of the problem must recognize this deficiency; however, this does not necessarily mean that this aspect or element of the problem should be ignored. A similar example could be cited in the area of planning in which the effect of a new transportation facility on a neighborhood is not clearly understood. It is known that there are certain impacts, but in many cases technology cannot reliably model the behavior of these impacts.

Another example might be the use of the gravity model for estimating zonal trip interchanges. The gravity model has generally been accepted in transportation planning and it is utilized in model trip behavior. Yet, the model must be calibrated for each area in which it is used. The reason for the calibration is related to the inability to explain the differences in travel behavior between different urban areas. The calibration is required to ensure that the gravity model provides a reasonable representation of travel for the particular study.

Concurrent with the development of the analysis methods and models, criteria should be established for measuring the performance of the system and any alternatives that may be proposed. In essence, a basis must be determined for judging how well a system functions according to a measurable scale. This scale may vary from objective to subjective, depending on the ability to develop quantifiable relationships of the component interactions. The criteria selected must represent a reliable basis for judging system performance, even though this judgment may be constrained by a lack of knowledge of system behavior. The measures of system performance should reflect the goals and objectives that may be established for the system. Further discussion of goals, objectives, and criteria is presented in a later section of this chapter.

The existing system should be evaluated in terms of the criteria that have been selected and in utilizing the models that have been developed. This test provides a basis for judging the reality of the models and the ability of the criteria to measure the functioning of the system. The test of the existing system should also reveal which components in the system can be revised to improve overall performance. This enables the engineer to systematically examine each component in the system and view the total spectrum of alternatives that may be considered. The list of alternative solutions can be generated from this evaluation. Referring to the problem of intersection congestion, we see that the list of alternatives would include consideration of solutions that are controlled by the traffic engineer as well as those controlled by other disciplines, for example, the urban planner.

SELECTION OF AN ALTERNATIVE

The final step in the analysis process is to test and evaluate the performance of the modified system. The modifications would reflect the alternatives that have been posed as possible solutions to the problem. Again, the criteria would be used as the basis for evaluating and selecting the alternative for implementation. It should be recognized that the best alternative is only in terms of the scope of the system that was actually considered.

GOALS, OBJECTIVES, AND CRITERIA

The definition of goals, objectives, and criteria represents an important aspect in the formulation and solution of the problem. The goals and objectives are statements reflecting an understanding of the problem, and the criteria are measures that may be applied to evaluate the achievement of goals and objectives. Engineers and other disciplines have historically been plagued with the difficulty of defining meaningful goals and objectives that will guide planning and design efforts. This is particularly true when dealing with complex problems such as transportation which involve sociopolitical considerations and interface with other professional disciplines. In such cases, the stated goals and objectives tend to be abstract and lack definitive meaning. Furthermore, the importance of goals and objectives in problem solving has frequently been neglected. In planning studies, for example, the development of goals and objectives is frequently undertaken as part of the evaluation of alternate solutions. At this point in the studies, data has already been collected and a commitment has been made on the nature of the analysis. This commitment constrains the scope and nature of the investigation, the alternatives that are considered, and the evaluation and selection of the best alternative.

A distinct advantage of utilizing the systems approach and systems analysis is that the development of goals, objectives, and criteria is embedded in the analysis process. In this way they are contained as an integral part of the analysis and guide the process of selecting the best alternative. An examination of goals, objectives, and criteria reveals that a hierarchical relationship exists among the three, and this relationship is also reflected by their use in the analysis process.

GOALS

A goal is a general statement of the end state that is to be achieved; thus, a goal should reflect the defined purpose or function that a system is to fulfill. If the system has a multiple purpose, more than one goal will exist. Transportation, for example, is a system that provides a service to society; the goal could be stated in terms of the necessary mobility for some kind of social, political, or economic function.

For a hierarchical system, goals can be defined for each system level. Further discussion of the definition of goals, objectives, and criteria in relation to hierarchical systems is presented in one of the examples in a later section.

The goal or goals can be developed at the time a problem is defined and the basic system is modeled. The definition of the goals at this early stage in the analysis process provides a statement of the general terms for evaluating the success of alternate solutions to a problem.

OBJECTIVES

An objective is a statement of the manner or ways in which a goal may be fulfilled or the system purpose achieved. For any given system, the components contained in that system represent the parts that can be modified or utilized in order to fulfill the goal. The objectives, therefore, would be related to the components and would be an indication of the alternate means of satisfying the goal. Again using the transportation system example, we see that the components of the system would be the different modes of transport that could be utilized to provide the required mobility. The objectives would be related to the provision of rail, highway, air, or water transport systems.

The conceptual model of the basic system should reveal the components involved, and these components could be utilized to develop the objectives for that particular problem. It is likely that some objectives may be in conflict, and it will be necessary to establish a priority or preference for objective fulfillment.

CRITERIA

Criteria are an indication of how the objective fulfillment is actually to be measured, and they generally represent the terms in which the objective can be quantified. For example, mobility can be measured in terms such as travel time, transport system capacity potential, and travel cost; safety can be measured by accidents or potential conflicts.

The definition of criteria is particularly important to data collection efforts in a study because criteria should provide an indication of the data that is required for the study. The use of systems analysis permits an appraisal of the information or data that must be assimilated for the study. Criteria are also the basis for analytical models of the system which are utilized in the evaluation of the system and of possible alternatives that may be considered.

Since some objectives cannot be quantified, it will be impossible to establish purely objective criteria. For example, in planning studies, it has been impossible to establish true measures of the social impact of transportation facilities. Driver preferences for different designs have been difficult to measure; therefore, a definite criterion has not been established to measure differences in design solutions. In such cases, subjective measures must be applied when the engineer seeks the opinions of others or simply uses his professional judgment.

MODELS

Throughout the analysis of a system there is considerable emphasis on the use of models; and in the course of the systems analysis process, a variety of different models may be applied or utilized. A model can be defined as a representation of some object, process, or situation. The engineer should be familiar with the use of models because they are widely applied in engineering practice. A major difference occurs in systems analysis, however, in that the engineer may be required to develop his own model rather than use an existing model. Certainly, this is not always the case, but it is particularly true in depicting a system and its structure.

Referring to the description of the systems analysis process, we see that one of the first tasks is to develop a model of the basic system because it is unlikely that such a model would exist for that particular system. As the analysis progresses, it may be possible to apply models that explain component behavior or interaction. If such models do not exist, the engineer will again be required to develop the required models.

A review of literature reveals that there are a number of different ways to classify models. Almost all authors tend to develop a classification that best suits their particular needs. For the purpose of systems analysis, perhaps a classification according to the model is appropriate because different models may be applied at different points in the analysis process. The following is a classification by model type:

1. Iconic
2. Analog
3. Symbolic

ICONIC MODELS

An iconic model is an actual physical representation of an object. An example of such a model would be a scale model of a facility or device. Although the application of iconic models is limited, traffic engineers have found them useful in specific situations. For example, a scale model of a highway interchange can be useful in the geometric characteristics of ramps or signing with respect to the design of the interchange.

ANALOG MODELS

The analog model is a representation of a dynamic situation or process. The analog model is useful in systems analysis because of its application in the representation of the system and the behavior of the system. The conceptual models of the systems shown in Figures 23.2 and 23.3 are examples of analog models. Figure 23.4 is another example of the analog model in which an analysis process is depicted. In addition, other examples would include supply and demand curves and statistical distribution curves.

SYMBOLIC MODELS

The symbolic model is a representation that utilizes symbols or mathematics. Thus, the vast spectrum of mathematical models falls under this heading. In traffic engineering, there may be found widespread use of this model. For example, the following expression is a symbolic representation of the minimum stopping sight distance for a vehicle as described in Chapter 14, Geometric Design:

$$SSD = 1.47\ PV + \frac{V^2}{30(f \pm g)}$$

Also, the equation

$$T_{ij} = \frac{P_i A_j F_{ij} K_{ij}}{\sum_{j=1}^{n} (A_j F_{ij} K_{ij})}, \quad i = 1, 2, \ldots, n$$

is used to depict the gravity model applied to travel forecasting problems (see Chapter 12, Urban Transportation Planning).

Further subdivisions of mathematical models can be achieved by grouping the models by such criteria as purpose or characteristics. Some of the models may fall into more than one group. Some of the common classifications are:

1. *Stochastic vs. deterministic:* The stochastic model involves variables that are subject to uncertainty and are expressed in probabilistic terms; the variables in a deterministic model are stated in terms of specific values.
2. *Descriptive vs. optimization:* The descriptive model simply describes a situation or a behavior; the optimization model is used to select between alternatives that may be applied.
3. *Analytical vs. design:* The analytical model is used in the analysis of a system or problem and is applied to analyze the state of a system at some given time; the design model is intended for use as an aid in design or to achieve some design solution.

Figure 23.4. Analog model representing trip end modal split forecasting process. [Source: Martin J. Fertal, Edward Weiner, Arthur J. Balek, and Ali F. Sevin, *Modal Split* (Washington, D.C.: U.S. Department of Transportation, Federal Highway Administration, Office of Highway Planning, (1970), p. 6.]

THE USE OF ECONOMIC ANALYSIS

Traffic and transportation engineers are involved in the development of facilities or programs that require the investment or allocation of funds. Economic analysis is an analytical tool that may be used to aid making decisions on economic feasibility or funding.

An extensive treatment of economic analysis as it relates specifically to highways has been presented by Winfrey and Zellner.[5] Their publication is a basic reference on economic analysis and contains a thorough explanation of its application and a number of examples.

Winfrey and Zellner indicate that economic analysis contains the following two broad areas:

1. Cost-benefit analysis
2. Effectiveness analysis

The main difference between the two analyses is the degree to which the outputs can be described in monetary terms. In the first case, the costs and the benefits are expressed in some monetary value, and the analysis is related to the analysis of the benefits that result from various expenditures or costs.

Effectiveness analysis can be used in the analysis of projects that contain consequences difficult to quantify. The concept of effectiveness analysis is based on an evaluation of the effectiveness of a project in terms of the overall goals that are established. It is important to recognize that effectiveness analysis provides an extremely useful tool for evaluating a comprehensive problem that contains aspects which normally are incorporated into the use of engineering economics.

In conjunction with these analyses, there are a number of analytical methods that may be utilized. They include:

1. Benefit cost ratio
2. Rate of return
3. Current worth of costs
4. Annual costs
5. Net current value
6. Annual net return

These methods represent techniques that can be utilized to evaluate and present economic information. A documentation of these methods can be found in numerous texts and references.

Economic analysis is not only useful in evaluating alternate improvements or modifications to a system, but the application of systems concepts in formulating a problem can also guide the economic analysis. The conceptual model of the system indicates the comprehensive scope of the elements that must be considered in the course of the economic analysis. The investigation of the full problem enables the engineer also to trace the impact of various strategies and alternatives in order to determine the consequences that result.

The engineer is also concerned with the functioning of the system over time. Frequently, decisions are made on the least expensive initial cost. The use of a systems model reveals the consequences of such decisions because of the influence on the total effectiveness of the system. For example, an engineer can purchase and install traffic signs that may have a variation in initial cost. Each sign will require maintenance that can also be expressed as a cost. Also, the deterioration of the sign with age has

---

[5] Robley Winfrey and Carl Zellner, "Summary and Evaluation of Economic Consequences of Highway Improvements," National Cooperative Highway Research Program Report 122 (Washington, D.C.: Highway Research Board, 1971).

potential consequences to the roadway users. An analysis of all of these factors yields a cost profile associated with each sign. The use of systems concepts enables the engineer to undertake a rational evaluation of the problem by considering the full nature of the total costs that are involved.

EXAMPLES OF THE APPLICATION OF SYSTEMS CONCEPTS

The purpose of this section is to provide several examples in which systems concepts have been applied to traffic and transportation problems. These examples are intended to demonstrate the merits of using the comprehensive systematic approach in dealing with such problems. Also, the examples illustrate that the approach can be applied to a wide spectrum of problems that vary in type and scope. The first example presents an overview of the use of a hierarchical system in structuring a framework for the analysis of transportation planning and design problems. The second example discusses the consideration of parking as a systems problem and emphasizes the need for its inclusion in planning analyses. The third example focuses on roadway design and operations and recognizes the manner in which systems concepts can be utilized in viewing the problem.

EXAMPLE 1: THE PLANNING OF TRANSPORTATION FACILITIES

One of the functions of the traffic engineer is to participate in the short- and long-range planning of transportation facilities. Because of the factors involved, the foresights required in accomplishing these tasks are extremely complex. Traditionally, the approach to planning has generally been to extend historical demand trends into the future. Facilities are then planned to meet this forecast demand. Yet, there has been criticism because the transportation problem continues to persist and some of the consequences of the existing trends have been considered undesirable. The engineer is obviously confronted with an extremely complex problem for which the attainment of a solution appears to be an overwhelming task. In dealing with such problems, the engineer requires an approach that will permit him to query higher level policies on the situation and to examine the comprehensive spectrum of alternatives and their consequences.

Systems concepts can be utilized to formulate and structure such a problem whereby each component part is viewed in relation to the overall system. The approach guides the development of rational goals and objectives for each system level and assists in the definition of data that will be required in the course of the study.

Having been given the planning problem, one of the first steps that needs to be accomplished is the structuring of the system and the components that are involved. Earlier in the chapter, it was indicated that transportation was a service function for society and provided the mobility that is required for societal activities. Furthermore, transportation is composed of a number of different forms of transport that all operate to provide this service. The engineer in dealing with the planning of transportation facilities can structure the problem as one in which there is a hierarchy of systems. This hierarchy of systems is depicted in Figure 23.5.

At the top level of the hierarchy is the entire society system of which transportation is a component or a subsystem. The problems at this level are associated with social, political, and economic activities of society, and the goals at this level would address the solution of these problems. The objectives for this level would relate to the alter-

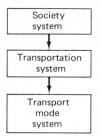

Figure 23.5. Hierarchy of systems associated with transportation.

natives available for the solution of societal problems. Such alternatives might be associated with education, housing, welfare, transportation, etc. In the case of transportation, the objective would be to improve mobility. At this level the question that must be answered concerns whether or not increased mobility, which is provided by transportation, is the best solution to the societal problem.

At the second level in the hierarchy of systems is the transportation system, which has as its components the different forms of transportation. At this level the question that must be addressed is related to the form or forms of transport required to fulfill the demands for mobility. The goals at this level, therefore, are focused on the provision of the required mobility, and the objectives indicate the characteristics of the mobility requirements. These characteristics are concerned with such aspects as safety, economics, convenience, and comfort. In essence, these characteristics should give use to measure the mobility capabilities of the different forms of transport. The criteria or actual measures in this would reflect the objectives that had been stated.

The hierarchy levels as shown in Figure 23.5 could be even further expanded with a component breakdown for each transport mode. For example, each mode would have vehicle, terminal, and roadway components which form the mobility capabilities for that specific mode.

For a given problem, there will also be interactions between components at that level. For example, at the societal level there would be an interaction between housing and transportation. The development of housing would have potential requirements for accessibility or mobility. The engineer would want to further develop Figure 23.5 by depicting these interactions for the specific problem involved.

The entire structure and framework in which the problem exists can be defined, and the engineer can develop and determine the questions that must be addressed at each system level. This structure or framework provides a basis for analyzing the consequences of various planning strategies or alternatives that are outlined because it depicts the potential impacts on other components and system levels. Furthermore, it permits an evaluation of transportation oriented policies at the various levels. By using such an approach, the engineer has a grasp of the comprehensive nature of the problem and a framework for analysis at whatever level the problem occurs.

EXAMPLE 2: PARKING

Historically, parking problems have been considered in an isolated context in that they have generally been studied independently at the roadway network. A review of parking studies reveals that emphasis has been placed on the adequacy of parking without considering the remaining system components and their requirements and capabilities. A rational solution to parking problems must recognize the systems nature of the problem.

As mentioned in Example 1, a specific mode of transportation contains vehicle, roadway, and terminal components. In the case of highway transportation, parking represents the terminal component. A balance in the system would require that the capabilities or supply of facilities for both the terminal and the roadway are sufficient to fulfill the demands imposed. In this case, the vehicle component is ignored in the magnitude of facilities required because the traffic engineer is responsible for the provision of the roadway and terminal components. The lack of adequate parking facilities in relation to the capability of the roadway results in constraints on the operation of the system; the impact may be changes in the use of the automobile, reduction in the demand for movement, or possible neglect of parking regulations.

In his analysis of parking, Schulman[6] depicts the problem in the more comprehensive framework. For example, Figure 23.6 depicts the analysis of parking with respect to the demands on the total transport system as well as the individual modes. In this figure the analysis is based on a trip interchange modal split model. Thus, a deficiency in parking is shown to have a potential impact on the actual number of trips generated, the choice of destination of a trip maker, and even the mode by which the trip is made.

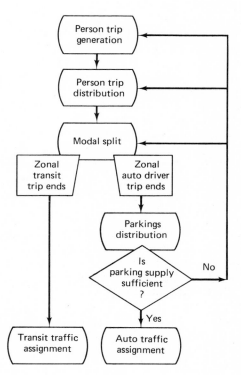

Figure 23.6. Parking as a component in the planning process. [Source: Lawrence L. Schulman, "Parking as an Element Within the Comprehensive Transportation Planning Process," *Public Roads*, XXXV, No. 1 (1968), p. 21.]

The actual determination of parking deficiency is shown in the analysis presented in Figure 23.7. Each trip is evaluated with respect to the supply of parking at the destination or within acceptable walking distances of the destination. If the demand for

---

[6] LAWRENCE L. SCHULMAN, "Parking as an Element Within the Comprehensive Transportation Planning Process," *Public Roads*, XXXV, No. 1 (1968), pp. 18–26.

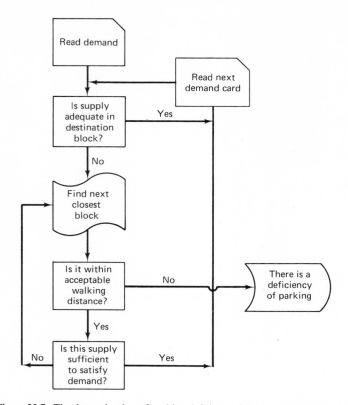

Figure 23.7. The determination of parking deficiency. (Source: Schulman, p. 23.)

space exceeds the supply, a parking deficiency occurs with the consequences being determined by the process shown in Figure 23.6. Certainly, such an analysis not only recognizes the systems aspects of the problem but also emphasizes the decisions that must be made and the consequences that result.

EXAMPLE 3: ROADWAY OPERATION AND DESIGN

In this particular example, the system components directly involved are the roadway, the vehicle, and the driver. The driver is actually a component of the vehicle and provides the control functions for the operation of the vehicle. There are interactions between the driver and the roadway; thus, a system results which is depicted by the conceptual model in Figure 23.2. This system is generally representative of traffic operations and roadway design problems. It is likely that the model of the system may require further detail for use with a specific problem.

An examination of the system reveals that the traffic engineer is confronted with a number of constraints when seeking solutions to problems involving this system. Since it is likely that he will have little control over the driver or the vehicle components, he may be forced ultimately to focus on roadway oriented solutions. It is his professional responsibility, however, to consider the other components in seeking the best solution to a problem.

As a specific example of the use of this system, assume that a traffic engineer has been asked to investigate the use of roadway lighting as a means of reducing accidents at intersections. Obviously, the basic roadway-vehicle-driver system would be involved. The lighting would be a component of the roadway system along with other components such as geometric features, pavements, and traffic control. The purpose of lighting is to provide a visual environment for the roadway for the driver.

Although the engineer has been asked to investigate lighting, he must recognize that the ultimate goal is to reduce accidents in the best and most feasible way. Lighting represents one way of potentially reducing accidents, and it should be considered in the context of the overall system. Certainly, other possible ways of reducing accidents include changes or improvements in the vehicle and its operation or other roadway components. For example, improved vehicle lighting or intersection geometric changes might be considered. By investigating the full problem instead of viewing lighting in an isolated context, the engineer may select a course of action that yields the greatest potential for reducing accidents.

REFERENCES FOR FURTHER READING

Au, T., and T. E. Stelson, *Introduction to Systems Engineering, Deterministic Models*. Reading, Mass.: Addison-Wesley Publishing Company, 1969.

Churchman, C. W., *The Systems Approach*. New York: Dell Publishing Company, 1968.

de Neufville, R., and J. H. Stafford, *Systems Analysis for Engineers and Managers*. New York: McGraw-Hill Book Company, Inc., 1972.

Meredith, D. D., K. W. Wong, R. W. Woodhead, and R. H. Wortman, *Design and Planning of Engineering Systems*. Englewood Cliffs, N.J.: Prentice-Hall, Inc., 1973.

Thomas, Edwin H., and Joseph L. Schofer, "Strategies for the Evaluation of Alternative Transportation Plans," NCHRP Report 96. Washington, D.C.: Highway Research Board, 1970.

Winfrey, Robley, and Carl Zellner, "Summary and Evaluation of Economic Consequences of Highway Improvements," NCHRP Report 122. Washington, D.C.: Highway Research Board, 1971.

Wohl, Martin, and Brian Martin, *Traffic System Analysis for Engineers and Planners*. New York: McGraw-Hill Book Company, 1967.

INDEX